# Springer Monographs in Mathematics

T0180557

For other titles published in this series, go to
http://www.springer.com/series/3733

John Rhodes · Benjamin Steinberg

# The q-theory of Finite Semigroups

 Springer

John Rhodes
University of California
Department of Mathematics
1000 Centennial Dr.
Berkeley CA 94720-3840
USA
blvdbastille@aol.com

Benjamin Steinberg
Cereleton University
School of Mathematics and Statistics
1125 Colonel by Drive
Ottawa ON K1S 5B6
Canada
bsteinbg@math.carleton.ca

ISSN 1439-7382
ISBN 978-1-4419-3536-6        ISBN 978-0-387-09781-7 (eBook)
DOI 10.1007/978-0-387-09781-7

Mathematics Subject Classification (2000): 20M07, 20M35, 20F20, 20F32, 20A15, 20A26, 16Y60, 06B35, 06E15, 06A15

springer.com

In memory of Bret Tilson

# Preface

When people are trying to learn mathematics for the purpose of research, they usually start at the forefront and then go backwards as needed in order to understand the results more fully. Yet with a mathematics book, it is not uncommon for people to start on page one and read onwards. This book is a research manuscript, and we heartily encourage the reader to delve in, read what is of interest, and go back as necessary. We hope that the material at the end of the book — the charts, tables and indices — will make this easier going.

Not many books have appeared in recent years dedicated to state-of-the-art Finite Semigroup Theory. There was the famous "Arbib" book in the 1960s [171] containing the lectures of Kenneth Krohn, John Rhodes and Bret Tilson. Samuel Eilenberg's treatise [85], with two chapters by Tilson [362,363], appeared more than 30 years ago. It revolutionized semigroup theory with the introduction of pseudovarieties of semigroups and varieties of languages. However, the most recent book on the subject is that of Jorge Almeida [7], which was originally published in 1992! Almeida's book made profinite methods in semigroup theory accessible. Howard Straubing's 1994 book [346] does touch on semigroup theory, but it is more concerned with applications to Computer Science than with semigroups themselves.

This volume is intended both to introduce a quantized version of Eilenberg's theory, by going to operators on pseudovarieties and relational morphisms, and to fill in some of the vacuum that has accrued from years that have passed with no new books on finite semigroups. Also, we have tried to include a number of classical results that profit from a recasting in a modern language to increase accessibility.

The philosophy of the book and an explanation of the individual chapters is covered in the Introduction, which we strongly recommend reading before entering into the body of the text. Here we try to provide a brief guide to the structure of the current volume.

Let us begin with what this book does not do. We do not intend in any way to give a basic introduction to Semigroup Theory. We have included for the reader's convenience an appendix on the basic structure theory, such as Green's relations and Rees's Theorem, but it is by no means complete. The classic treatise of A. H. Clifford and G. B. Preston [68] still serves admirably for learning this material, as does the "Arbib" book [171]. One can also turn to the books of J. M. Howie [139] and P. M. Higgins [133]. For the current text, it is handy to be familiar with a little bit of category theory [185], basic topology [158], and perhaps some combinatorial group theory [184].

This is a book about *finite* semigroups. Except for free semigroups, free groups and profinite semigroups, the reader should not expect to encounter infinite semigroups. Readers interested in the Prime Decomposition Theorem for infinite semigroups should consult the relevant papers on the topic [3, 50–52, 88, 125, 214, 215, 276, 277, 279]. Outside of a brief foray to prove Schützenberger's Theorem, we do not enter into Formal Language Theory and Automata Theory. Formal Language Theory is an important aspect of Finite Semigroup Theory, and much of the motivation for problems in the field derive from it. But there is already a sizable literature of excellent books devoted to this facet (for instance, [85, 177, 224, 229, 346]). Nonetheless, it would be an important task in the future to reinterpret the results of this book from the language theoretic point of view. Another important aspect of Finite Semigroup Theory that we touch upon only slightly is the finite basis problem. Mark Sapir, Marcel Jackson and Mikhail Volkov, among others, have done important work on this subject [77, 142–144, 146, 159, 246, 306–309, 372, 374].

The book is divided into four parts. The first part, entitled *The q-operator and Pseudovarieties of Relational Morphisms*, is the heart of q-theory. Chapter 1 is foundational material, and much of it may be familiar to the reader. However, at the end of the chapter the important notions of division of relational morphisms and relational morphism between relational morphisms are introduced. Chapter 2 introduces the classes of relational morphisms that are of central importance in this volume: continuously closed classes and pseudovarieties of relational morphisms. We introduce the operator q and its image is characterized. Maximal and minimal models, meaning maximal and minimum classes defining a given operator, are studied, and we give examples showing that all the commonly occurring operators in Finite Semigroup Theory fit into our framework. The Derived and Kernel Semigroupoids are presented and their basic properties explored in order to introduce the pseudovarieties of relational morphisms $\mathbf{V}_D$ and $\mathbf{V}_K$, whose images under q are the operators $\mathbf{V} * (-)$ and $\mathbf{V} ** (-)$, thanks to the Derived Semigroupoid and Kernel Semigroupoid Theorems.

The last chapter of Part I, Chapter 3, develops the equational theory for pseudovarieties of relational morphisms. Because the development parallels Reiterman's Theorem and uses heavily profinite semigroups, we have provided a brief introduction to the classical theory of pseudoidentities and profinite and pro-**V** semigroups. Then we establish the generalization of Reiterman's Theorem to pseudovarieties of relational morphisms and give examples of pseudoidentity bases for the pseudovarieties of relational morphisms associated to the most commonly studied operators: semidirect products, Mal'cev products and joins. Inevitable substitutions are introduced and studied, leading to a basis theorem for the composition of two pseudovarieties of relational morphisms. Classical basis theorems from the literature are deduced as consequences. We discuss when the basis theorem for the semidirect product [27] applies and provide a new basis theorem covering the general case.

The second part of the book is called *Complexity in Finite Semigroup Theory*. Chapter 4 provides a comprehensive look at group complexity starting from the very beginning with a proof of the Prime Decomposition Theorem and ending with such advanced topics as Ash's Theorem [33], the Ribes and Zalesskii Theorem [300], Rhodes's Presentation Lemma [43,334], Tilson's 2 $\mathscr{J}$-class Theorem [360] and Henckell's Aperiodic Pointlikes Theorem [121,129]. Also, some important and hard-to-find results from the literature are featured, including Graham's Theorem [107] on the idempotent-generated subsemigroup of a Rees matrix semigroup. Using the language and viewpoint of q-theory, a simplified presentation of the computability of complexity for semigroups in **DS** is presented [170,269,368]. A large part of this chapter is dedicated to an updated presentation of the results in the "Arbib" book [171] concerning Mal'cev products and subdirectly indecomposable semigroups. We do not make any attempt to duplicate the material covered in Tilson's chapters in Eilenberg's book [362,363], and so as a consequence the Fundamental Lemma of Complexity is not proved here. Few of the results in Chapter 4 are new, although some of the proofs are novel.

Chapter 5 introduces our general scheme for defining the complexity hierarchy associated to a single operator or a pair of operators on the lattice of pseudovarieties. We discuss many examples from the literature and present some new ones as well. The focus of the chapter is on two-sided complexity. The decomposition theory for maximal proper surmorphisms [293,296] is included for the convenience of the reader and because the language of pseudovarieties of relational morphisms sheds some new light on these results. The remainder of the chapter takes the first steps in generalizing results from group complexity to two-sided complexity. The Ideal Theorem is proved and the two-sided complexity of the full transformation monoid is computed.

Part III of the book, *The Algebraic Lattice of Semigroup Pseudovarieties*, provides a study of lattice theoretic aspects of the algebraic lattice of pseudovarieties of semigroups. Chapter 6 is essentially a condensed survey of algebraic and continuous lattices [97]. It is intended to establish notation and to introduce ideas that may not be familiar to a semigroup audience. Some examples related to q-theory are provided. For instance, it is shown that any continuous lattice of countable weight is the fixed-point lattice of an idempotent continuous operator on the lattice of pseudovarieties of semigroups. Chapter 7 is dedicated to the abstract spectral theory of the lattice of

pseudovarieties of semigroups (and to a much lesser degree the lattices of pseudovarieties of relational morphisms and continuous operators). In particular,
a multitude of new results about join irreducibility are obtained. Semigroups
$S$ with the property that $S \in \mathbf{V} \vee \mathbf{W}$ implies $S \in \mathbf{V}$ or $S \in \mathbf{W}$ are studied,
as well as their exclusion pseudovarieties. We introduce several novel families
of such semigroups, leading to the following application: if $\mathbf{H}$ is a pseudovariety of groups containing a non-nilpotent group, then the pseudovariety $\overline{\mathbf{H}}$
of semigroups whose subgroups belong to $\mathbf{H}$ is finite join irreducible. This
generalizes results of Margolis, Sapir and Weil [195]. Also, the classification
of Kovács-Newman semigroups that we began in an earlier paper [287] is
completed.

The final part of the book, entitled **Quantales, Idempotent Semirings,
Matrix Algebras and the Triangular Product**, consists of two chapters.
Chapter 8 introduces our slight weakening of the notion of a quantale and
develops the general algebraic theory of these objects. We also study the
equivalence between Boolean bialgebras and profinite semigroups, leading to
a bialgebra structure on the Boolean algebra of regular languages over a finite alphabet. Chapter 9 contains almost entirely new material. Viewing finite
quantales as idempotent semirings, we develop a decomposition theory for finite semirings in general via the triangular product of Plotkin [240], following
the model of group complexity. The centerpiece of the chapter is formed by a
pair of results we refer to as the Triangular Decomposition Theorem and the
Ideal Decomposition Theorem. They basically give semiring analogues of the
results of Munn and Ponizovskiĭ on semigroup algebras over fields [210,245] by
embedding the semigroup algebra (over a semiring) of a finite semigroup inside
an iterated triangular product of matrix algebras over the group algebras of its
Schützenberger subgroups. Applying these theorems in the context of idempotent semirings yields the decomposition half of the Prime Decomposition
Theorem for idempotent semirings. A large segment of the chapter is devoted
to proving that matrix algebras over the power semigroup of a group are irreducible with respect to the triangular product. The moral of the story is that
much more of ring theory works for semirings than one might expect. Finally,
Chapter 9 ends with applications of the decomposition theory of idempotent
semirings to computing the group complexity of power semigroups.

There is not a strict logical sequence for reading the various parts of the
book (hopefully, the dependency graph is at least acyclic!). All readers should
be familiar with the material in Chapter 1 to proceed. The basic definitions
and terminology introduced in Chapter 2, at least as far as to the end of
Section 2.3, are needed for Chapters 3–8, although specific results are rarely
required. The new material in Chapter 3 is highly technical and is not required
for most of the rest of the book, except for a brief reappearance in Chapter 7;
the reader already familiar with pseudoidentities and profinite semigroups
may skip it entirely on a first reading. Much of Part II requires only Chapter 1 and the language of Chapter 2, not the results. The key exceptions are
the Derived and Kernel Semigroupoid Theorems, which are used repeatedly.

Part III requires Part I, but Chapter 6 can be read entirely independently of Part II. Chapter 7 occasionally appeals to results from Chapters 4 and 5. Chapter 8 depends on Part I and Chapter 6, whereas Chapter 9 depends only on a basic knowledge of definitions from Chapters 1 and 4, familiarity with the Fundamental Lemma of Complexity and on a small part of Chapter 8, namely quantic nuclei and some examples.

We have interspersed throughout the text a large number of exercises, some routine and others more difficult. The exercises form an integral part of the subject matter, and the reader is encouraged to solve as many of them as possible.

The current volume has greatly benefitted from the comments and criticisms of our colleagues and students. The following mathematicians deserve special thanks for their hard work and thoughtfulness: Karl Auinger, Bridget Brimacombe, Attila Egri-Nagy, Karl Hofmann, Gabor Horvath, Mark Kambites, Jimmie Lawson, Stuart Margolis, Chrystopher Nehaniv, Boris and Eugene Plotkin, Luis Ribes, Kimmo Rosenthal and Mikhail Volkov. The anonymous reviewers should also be acknowledged for their careful reading of the text. All errors and inaccuracies that remain are the sole responsibility of the authors.

The first author would like to thank U.C. Berkeley's generous retirement funds. The second author has been funded by an NSERC Discovery Grant during the course of this research. Both authors gratefully acknowledge support from FCT through the *Centro de Matemática da Universidade do Porto* and both have received partial support from the FCT and POCTI approved project POCTI/32817/MAT/2000 in participation with the European Community Fund FEDER. Some of this work was done while the second author was at the University of Porto, Portugal. The second author would also like to thank his colleagues at the University of Porto, in particular Jorge Almeida, Manuel Delgado and Pedro Silva, for their friendship and collaboration throughout the years and for making Porto an enjoyable environment to work in.

The first author would like to thank his wonderful wife, Laura Morland, for everything, especially her editorial skills. The second author is grateful to Ariane Masuda for her love and support through the long process of bringing this book to fruition, as well as for her careful reading of the appendix. *Obrigado por tudo.*

The inspirational spark for q-theory itself might never have been ignited, were it not for the superb mathematics library at "Chevaleret" (Paris VI–CNRS–Paris VII), and we thank its librarians for their *disponibilité*. However, none of this work would have been possible if not for the many excellent cafés that graciously allowed us to occupy their tables for overly excessive time periods. Our special appreciation goes to the Paris cafés: Les Monts d'Auvergne, Café Lowrider, Au Père Tranquille, Aux Cadrans; the Porto cafés: Café Aviz, Café Majestic; the Berkeley café: Au Coquelet; and the Ottawa café: The Wild Oat.

This book is dedicated to Bret Tilson in gratitude to his contributions to Semigroup Theory, as well as his early participation in the research that led to this book. He was its first reader. Bret has had a great impact on all our lives and is sorely missed.

Paris, August 2008                                                    *John Rhodes*
Ottawa, August 2008                                          *Benjamin Steinberg*

# Contents

## Part II Complexity in Finite Semigroup Theory

# List of Tables

# List of Figures

# Introduction

What is q-theory? To explain the title of the current volume, we first need to put the whole field of Finite Semigroup Theory into a historical context. So let us begin with a theorem-based history of *finite* semigroups (biased, of course, by the authors' own viewpoint, and certainly by no means complete). Early theorems in the subject followed in the footsteps of other branches of algebra by aiming to classify semigroups up to isomorphism. An early success in this direction, and arguably the first theorem about finite semigroups, was the Rees-Suschkewitsch Theorem characterizing, up to isomorphism, simple [350] and 0-simple semigroups [258] as certain matrix semigroups over groups, a Wedderburn Theorem for semigroups, if you like. The sequel to this work was J. A. Green's fundamental paper [108] where the relations that now bear his name were introduced. This set up the framework to understand finite semigroups in terms of local coordinates at each (regular) $\mathscr{J}$-class: locally speaking, finite semigroups are Rees matrices over groups. Out of Green's paper came the famous eggbox pictures of Alfred Clifford and Gordon Preston [68]. However, a crucial missing ingredient was to understand in coordinates how the elements above a regular $\mathscr{J}$-class $J$ act on it. This gap in our knowledge was filled in by Marcel-Paul Schützenberger via his famous representation by monomial matrices, now called the Schützenberger representation [311, 312]. From a more global viewpoint, his representation gives wreath product coordinates to the action of a semigroup on the left or right of a $\mathscr{J}$-class, and thus the wreath product entered into the subject quite early on.

Another way in which early semigroup theory trod the well-beaten paths blazed by other areas of algebra was via the representation theory of finite semigroups. The pioneering work of Clifford [66,67], W. D. Munn [210,211] and I. S. Ponizovskiĭ [245] completely determined the irreducible representations of a finite semigroup, modulo group theory, as well as characterized when the semigroup algebra is semisimple. A more global viewpoint of these results that takes advantage of the Schützenberger representation can be found in John Rhodes and Yechezkel Zalcstein's paper in *Monoids and semigroups with applications* [297]. Donald McAlister, in his 1971 survey paper [198], pointed out that at the time, the sole significant application of representation theory to finite semigroups was Rhodes's [270], where the congruence on a finite semigroup induced by the Jacobson radical of its semigroup algebra was shown to coincide with (in modern terminology) the largest $\mathbb{L}\mathbf{1}$-congruence, and a character theoretic interpretation was given to the computation of the complexity of completely regular semigroups [171, Chapter 9]. (This theme has been reprised only quite recently in the paper [18], where representation theory was used to simplify some key results in Formal Language Theory.) In recent years, Mohan Putcha [251–255] has modernized semigroup representation theory, exploring relations with algebraic groups, algebraic monoids [250,262], quasi-hereditary algebras [70] and P. Gabriel's theory of quivers [92–94]. (See also [1,57,58,335,336] for recent connections of finite semigroup representation theory with Probability Theory, random walks, hyperplane arrangements and Coxeter groups.) This brings to a close the first chapter of our history, which takes us up to the early 1960s. Most of the results that we have discussed thus far can be found in the treatise of Clifford and Preston [68].

The direction initiated by the Rees-Suschkewitsch Theorem eventually had to be abandoned, for the program of classifying finite semigroups up to isomorphism is a hopeless one: there are simply too many of them. From an asymptotic viewpoint, the class of 3-nilpotent semigroups, i.e., semigroups satisfying the identity $xyz = 0$, covers almost all finite semigroups up to isomorphism [162]. Clearly it serves no purpose to try to classify such semigroups. More precisely, if the semigroups of order $n$ are distributed uniformly, then the probability that a labeled random semigroup of order $n$ is 3-nilpotent goes to 1 as $n$ goes to infinity. Intuitively, this happens because any multiplication table where the product of three elements is zero is automatically associative, and therefore 3-nilpotent semigroups are easy to construct. In fact, there are purported to be $1,843,120,128$ semigroups of order 8 up to isomorphism and anti-isomorphism, and approximately 99% are nilpotent according to S. Satoh, K. Yama, and M. Tokizawa [310]. So, whereas groups are gems, all of them precious, the garden of semigroups is filled with weeds. One needs to yank out these weeds to find the interesting semigroups. (On the other hand, semigroups have wider applicability than do groups, especially in Computer Science.)

Thus by the mid-1960s, it was time for a revolution in thinking about finite semigroups, and it would have to start from outside the world of algebra. Automata and sequential machines [161,209] had already made their appearance on the stage in the late 1950s, and the idea that there should be an algebraic theory of automata and machines was very much in the air. In particular, the work of Stephen Kleene [161], reinterpreted via the syntactic monoid, stated that regular languages (studied in Logic and Computer Science) are precisely subsets of the free monoid saturated by a finite index congruence. But it was not until Kenneth Krohn and Rhodes introduced the fundamental notion of division that a successful algebraic decomposition theory of machines and semigroups could be achieved [169]. Other attempts [118,119] at an algebraic decomposition theorem for sequential machines failed precisely because they did not use division: the full transformation monoid $T_n$ cannot embed in a semidirect product without embedding in one of the factors.

The statement of the Krohn-Rhodes Prime Decomposition Theorem can be formulated to any student who has taken a basic course in Group Theory: *every* finite semigroup divides (is a quotient of a subsemigroup of) an iterated wreath product of its simple group divisors and the three element monoid of transformations of the set $\{0, 1\}$ consisting of the constant maps and the identity map. This monoid is the semigroup analogue to the *flip-flop* sequential machine from Computer Science and Electrical Engineering [171]. The Prime Decomposition Theorem is an example of a "mature" theorem. Many early theorems in semigroup theory (and too many recent ones!) involved inventing a class of semigroups and then studying it. Not so for the Prime Decomposition Theorem: all the notions in the statement already existed, at least implicitly: wreath/semidirect products, simple groups (we like the terminology SNAGS for simple non-abelian groups) and division (the composition of the $\mathbb{H}$ and $\mathbb{S}$ operators from Garrett Birkhoff's Universal Algebra).

Let us briefly remind the reader of how the Jordan-Hölder program for Finite Group Theory goes. If $G$ is a finite group, there is a composition series

$$\{1\} = N_m < N_{m-1} < \cdots < N_1 = G$$

for $G$ where $N_{i+1} \lhd N_i$ and $N_i/N_{i+1}$ is simple, all $i$. These simple groups, called the composition factors of $G$, are unique, although the order in which they appear is not. The "monomial map" [113, 141, 381], going back to the work of Frobenius on induced representations, then places $G$ inside the iterated wreath product of these simple group divisors. Therefore, if you understand simple groups (and according to the Classification of Finite Simple Groups [101–106], we supposedly do) and you understand wreath products and sequential coordinates, then you understand all finite groups.

The Prime Decomposition Theorem transported the whole Jordan-Hölder program to the realm of semigroups [283]! This led naturally to the notion of group complexity of a finite semigroup: when decomposing a finite semigroup $S$ into a wreath product of groups and aperiodic (i.e., group-free) semigroups, how many groups do you need? The minimal number of groups is deemed the complexity of the semigroup. This notion was formalized by Krohn and Rhodes in their 1968 Annals paper [170], where they computed the complexity of union of groups semigroups (also called completely regular semigroups). Since that time, the major driving open problem in Finite Semigroup Theory has been to find an algorithm to compute group complexity. The problem has been open for more than 40 years! The literature on the subject — which comprises decomposition theorems, partial results, lower bounds and upper bounds — is much too vast to do it justice in this introduction. Chapter 4 of the current volume, together with Bret Tilson's chapters [362, 363] in Volume B of Samuel Eilenberg's *Automata, languages, and machines* [85], provides the most complete treatment of the subject to date.

Contemporary with the Prime Decomposition Theorem was Schützenberger's celebrated 1965 theorem on star-free languages [313]. It characterized star-free languages as the languages recognized by aperiodic semigroups (semigroups of complexity 0). In fact, the difficult direction of this theorem, decomposing the language accepted by an aperiodic semigroup, can be achieved quite easily as a consequence of the Prime Decomposition Theorem [207] and the relationship between sequential machines, wreath products and languages developed in [169], [171, Chapter 5] and summarized in [85, Chapter VI]. Moreover, the Krohn-Rhodes prime decomposition/sequential machine approach has been used successfully to characterize many other classes of languages, in particular by Howard Straubing [343, 344, 346]. Nowadays, the name "the wreath product principle" is often attached to this approach. This should not diminish the importance of Schützenberger's Theorem, which indicated that there is a relationship between classes of languages and classes of finite semigroups. Other theorems appeared soon thereafter that expressed that naturally occurring classes of regular languages corresponded via their syntactic monoids to naturally occurring classes of semigroups. These include the characterization of locally testable languages as being recognized by local semilattices [60, 204, 383, 384] (all of which appeared in the early 1970s) and Simon's characterization of piecewise testable languages as those recognized by $\mathscr{J}$-trivial monoids (1975) [318].

Once isomorphism was off the table, what was sorely lacking was a framework for classification in finite semigroup theory. The Prime Decomposition Theorem and the language results would seem to suggest that Universal Algebra could provide a possible framework. But classical Universal Algebra, with its reliance on free objects and equations and the requirement for closure under infinite products, wasn't quite the answer.

A fruitful tendency in modern mathematics is to turn theorems into definitions — after which the original theorem becomes a verification that some object satisfies the definition. For instance, the Heine-Borel Theorem originally stated that any covering of a closed interval by open intervals has a finite subcover. This led to the modern definition of a compact space, which eventuated in the Heine-Borel Theorem becoming the statement that closed intervals are compact. Another example is Rees's Theorem, which originally characterized 0-simple semigroups as Rees matrix semigroups over groups. This then led to the definition of Rees matrix semigroups over arbitrary semigroups, which in turn led to infinite iterated Rees matrix semigroups and the Synthesis Theorem [3, 50–52, 276, 277].

Eilenberg turned the language characterization theorems into a definition [85]. Together with Schützenberger [85, 86] he defined a pseudovariety of finite semigroups to be a class of finite semigroups closed under forming *finite* products and taking divisors. He defined the companion notion of a variety of formal languages and provided a correspondence between semigroup pseudovarieties and varieties of languages. From the point of view of this book, Eilenberg proved that the algebraic lattice **PV** of pseudovarieties of finite semigroups is isomorphic to the algebraic lattice of varieties of languages. Schützenberger's Theorem then became the theorem that the variety of star-free languages corresponds under the Eilenberg Correspondence to the pseudovariety of aperiodic semigroups. The Prime Decomposition Theorem turned into the statement that a finite semigroup belongs to the smallest pseudovariety of semigroups closed under semidirect product containing its simple group divisors and the flip-flop.

At this point (around 1976), Finite Semigroup Theory essentially became the study and classification of pseudovarieties of finite semigroups. Notice that the pseudovariety notion nearly wipes out the 3-nilpotent weeds: whereas almost all semigroups are, probabilistically speaking, 3-nilpotent, there are only a small finite number of pseudovarieties of 3-nilpotent semigroups. The starring role in the new theory, created in large part under the impetus of attacking the problem of group complexity, was then taken by the semidirect product of pseudovarieties of semigroups, an associative multiplication on the lattice **PV**. However, other non-associative products, such as the Mal'cev product [59], the two-sided semidirect product [293] and the Schützenberger product [85], have also garnered quite a bit of attention in the literature, not to mention the join operation on **PV**, which corresponds to direct products of semigroups. This was the state of play at the end of the 1970s.

In the 1980s, it became evident that there were two deficiencies in Eilenberg's book [85]. The resolution of these issues led to Finite Semigroup Theory as we know it today. The first deficiency was foundational, or one of scope. In Eilenberg [85] there is a result called the Tilson Trace-Delay Theorem. This result was used to prove $\mathbf{V} * \mathbf{D} = \mathbb{L}\mathbf{V}$ for various pseudovarieties **V** generated by monoids. Key tools used in these results were the so-called "graph congruences" and the derived transformation semigroup. Also in his Chapter XII of Eilenberg's book [362], Tilson introduced the derived semigroup, obtaining a one-way connection between the semidirect product and the derived semigroup. The derived semigroup was a salient feature of Tilson's simplified proof of Rhodes's Fundamental Lemma of Complexity [268], although it was only in combination with the Rhodes expansion [54] that the full power of the technique was revealed. However, graph congruences and semigroups were not the proper setting for these results and did not allow for a complete understanding of what was really behind them.

In a seminal paper in 1987 [364], Tilson, influenced by personal conversations with Stuart Margolis, as well as by the work of Margolis and Jean-Eric Pin [192–194], Denis Thérien and A. Weiss [358], Straubing [345] and Robert Knast [163], realized that categories would place the aforementioned results in their proper context. He replaced the derived semigroup with the derived category and proved the all-important Derived Category Theorem. If **V** and **W** are pseudovarieties of monoids and $\mathbf{V} * \mathbf{W}$ is their semidirect product, then Tilson established $S \in \mathbf{V} * \mathbf{W}$ if and only if there is a relational morphism $\varphi \colon S \to T$ with $T \in \mathbf{W}$ such that the derived category $D_\varphi$ of $\varphi$ divides a monoid in **V** [364]. This led to a theory of pseudovarieties of categories and the fundamental global versus local problem: when is category membership in the pseudovariety generated by a collection of monoids determined by looking at the local monoids of a category?

The Derived Category Theorem created a paradigm that was followed by much of later work, and which in the current text we turn into a definition. Tilson's 1987 "Categories as algebra" paper [364] was followed two years later by its two-sided analogue in his joint paper with Rhodes [293], which introduced the kernel category of a relational morphism and showed that a monoid $S$ belongs to the two-sided semidirect product $\mathbf{V} ** \mathbf{W}$ of pseudovarieties $\mathbf{V}$ and $\mathbf{W}$ if and only if there is a relational morphism $\varphi \colon S \to T$ with $T \in \mathbf{W}$ and the kernel category $K_\varphi$ of $\varphi$ dividing a monoid in $\mathbf{V}$. Combined with Rhodes's classification of maximal proper surmorphisms [267], this led to a decomposition theory [293, 296] that has had sweeping applications to Formal Language Theory [55, 234, 346, 349, 378, 379]. The advent of categories has removed much of the necessity for ad hoc wreath product decompositions and machine equations [169] from the theory.

Whereas the work of Tilson, Margolis and Pin brought the derived category to the attention of finite semigroup theorists, it should be mentioned that the derived category was already a well-known construct to category theorists; for instance, Daniel Quillen [257] used it to formulate a criterion for when the geometric realization of a functor between categories induces a homotopy equivalence between their nerves; the derived category of a functor $F \colon C \to D$ (pre-identifications) is precisely the category of elements [186] of the composition of $F$ with the Yoneda embedding $Y \colon D \to \mathbf{Set}$ (the category of elements was also put to good effect by P. J. Higgins [132] to construct groupoid coverings); and it is well-known to stand in an adjoint relationship with the Grothendieck construction [186], called by some the semidirect product of categories. William Nico also studied the relationship between the derived category and wreath products [219] before Tilson did, in fact at the categorical level, but his theory lacked the crucial ingredient of division. M. Loganathan's little known paper [183] on the cohomology of inverse semigroups gave a proof of McAlister's $P$-theorem [199] — and its generalization by L. O'Carroll [223] — using the derived category of a functor between categories even earlier than the paper of Margolis and Pin on the same subject [193].

The second deficiency in the theory of pseudovarieties espoused by Eilenberg and Schützenberger was the equational theory. They established [86] that pseudovarieties have ultimate equational descriptions, but in practice this approach is useless. An ultimate equational description of a pseudovariety $\mathbf{V}$ is a sequence of identities such that a semigroup belongs to $\mathbf{V}$ if and only if it satisfies all but finitely many of the identities in the sequence. One would like to be able to say what it means for a pseudovariety of semigroups to be defined by a finite number of "identities," that is, to be finitely based; the ultimate equational descriptions do not allow for this. Jan Reiterman in 1982 came up with the correct solution to the problem: pseudoidentities [261]. A usual identity is a formal equality between elements of a free semigroup. Reiterman's idea was that the role of a free semigroup in the finite world can be taken by a free profinite semigroup. A pseudoidentity is then a formal equality between elements of a free profinite semigroup.

Reiterman's Theorem shows that pseudovarieties are exactly the classes defined by pseudoidentities. However, it was only under the impetus of Jorge Almeida that profinite semigroups and the syntactic approach became a fundamental tool in Semigroup Theory, particularly in the 1990s. Much of Almeida's early work on the subject is encapsulated in his volume *Finite semigroups and universal algebra* [7]; further references can be found in the bibliography of the current text. The profinite approach was generalized to pseudovarieties of categories [27, 149], although there are a number of subtleties in this context. One goal then became to try and find a basis of pseudoidentities for $\mathbf{V} * \mathbf{W}$ given a basis of pseudoidentities for the categories dividing elements of $\mathbf{V}$ and knowledge of $\mathbf{W}$ [27] and similarly for other products [235]. This led to the notions of hyperdecidability [9] and tameness [19,20]. Nowadays, profinite semigroups are an indispensable tool in semigroup theory, in particular for studying pointlikes [9, 235, 322, 327, 330], stabilizer pairs (here one should compare the profinite argument in [130] with the argument in [124]) and related notions [9, 322, 330]. Many of these latter notions had their roots in the early work of Rhodes and Tilson on Type I/Type II semigroups [295] and Rhodes and Karsten Henckell on pointlike sets [121], but it was mostly Almeida who pushed these notions, especially with regard to the profinite approach [9, 19, 20].

The explosion of techniques in the 1980s resulted in profound work such as Ash's Theorem [33], solving the Rhodes Type II conjecture. The conjecture was proved independently by Luis Ribes and Pavel Zalesskii [300], using the method of profinite groups acting on profinite trees, via a translation of the problem into the profinite topology on a free group by Pin and Christophe Reutenauer [232]. The Type II Theorem describes which elements of a finite semigroup always relate to 1 under a relational morphism to a group. A review of the innumerable consequences of the resolution of the Rhodes conjecture appears in the *IJAC* survey paper of Henckell, Margolis, Pin and Rhodes [126]. Some highlights include characterizations of the pseudovarieties of semigroups generated by inverse semigroups, orthodox semigroups and power semigroups of groups. Actually, the first two cases were handled by special cases of the conjecture established earlier by Christopher Ash [32] and by Jean-Camille Birget, Margolis and Rhodes [53].

Ash, most likely due to his background as a logician, injected an important new technique into Finite Semigroup Theory: Ramsey Theory. The basic idea is that if one takes a long product of generators of a finite semigroup, then there must be a lot more repetition of idempotents than you might expect. Nowadays, there is a non-profinite proof of Ash's Theorem that does not rely on Ramsey Theory [34], but the technique remains invaluable. Karl Auinger and Benjamin Steinberg managed to synthesize the techniques of Ash and Ribes and Zalesskii to study related problems over other pseudovarieties of groups [37,39–41,328,329,331], in particular establishing intimate connections between semidirect product decompositions of semigroups and the geometry of profinite groups. This is based on earlier work of Margolis, Mark Sapir and Pascal Weil [196] amalgamating combinatorial group theory, via Stallings Folding [321] and M. Hall's Theorem [112], with inverse semigroup theory.

We are now close to being able to state the thesis of this book. Tilson's Derived Category Theorem [364] showed how to define the semidirect product operator $\mathbf{V} * (-)$ in terms of relational morphisms. Similarly, its sequel, the Kernel Category Theorem [293], showed how to define the two-sided semidirect product operator $\mathbf{V} ** (-)$ in terms of relational morphisms. Various authors [27, 293, 296] used this idea to define operators $\mathbf{V} * (-)$ where $\mathbf{V}$ is a pseudovariety of categories and even to define a semidirect product of pseudovarieties of categories [150]. Steinberg, in his Ph.D. thesis [322, 330], provided a necessary and sufficient condition on a relational morphism so that it could be factored as a division followed by the projection from a direct product. This resulted in a relational morphism description of the operator $\mathbf{V} \vee (-)$, as well as a number of new decidability results for joins of pseudovarieties. The Mal'cev product $\mathbf{V} \textcircled{m} \mathbf{W}$ is defined by declaring that $S \in \mathbf{V} \textcircled{m} \mathbf{W}$ if and only if there is a relational morphism $\varphi \colon S \to T$ with $T \in \mathbf{W}$ and so that, for each idempotent $e \in T$, the semigroup $e\varphi^{-1}$ belongs to $\mathbf{V}$. This product is then, by virtue of its construction, defined in terms of a class of relational morphisms.

In "Categories as algebra. II," Steinberg and Tilson "beefed up" the entire theory of the derived category of a monoid relational morphism to the level of categories [339]. One of the principal results of that paper states the following: Let $\varphi \colon S \to T$ be a relational morphism of monoids. Then the derived category $D_\varphi$ of $\varphi$ divides a monoid in $\mathbf{V} * \mathbf{W}$ if and only if $\varphi = \varphi_1 \varphi_2$ where $D_{\varphi_1}$ divides a monoid in $\mathbf{V}$ and $D_{\varphi_2}$ divides a monoid in $\mathbf{W}$. This result was dubbed the Composition Theorem. To express it in a more compact and elegant manner, they defined the class $\mathbf{V}_D$ of all relational morphisms whose derived category belongs to $\mathbf{V}$. One can compose classes of relational morphisms in an obvious way, and the Composition Theorem then admits the following reformulation:

$$(\mathbf{V} * \mathbf{W})_D = \mathbf{V}_D \mathbf{W}_D.$$

In the course of their research, Steinberg and Tilson tossed around the idea of defining pseudovarieties of relational morphisms. In particular, Steinberg had a notion of division of relational morphisms and pseudoidentities for relational morphisms. Pseudovarieties of relational morphisms could then be used to define operators on the lattice of semigroup pseudovarieties. But Tilson argued that without further evidence, it was better to delay bestowing the name pseudovariety on a class of relational morphisms that later on might not prove to be worthy of the name. In fact, Tilson had obviously flirted with the idea of defining pseudovarieties of relational morphisms in the past, as evidenced by his notion of a weakly closed class and the complexity of a relational morphism in [362]. Tilson also was attached to the idea that a semigroup

$S$ should be identified with a unique relational morphism: its collapsing map $S \to 1$. He identified a pseudovariety of semigroups $\mathbf{V}$ with its set of collapsing morphisms, and his viewpoint was that the action of a set R of relational morphisms on $\mathbf{V}$ should be the composition $R\mathbf{V}$, which is a collection of collapsing morphisms corresponding to a pseudovariety of semigroups. The disadvantage to this approach is that the axioms that were being considered for classes of relational morphisms were *not* satisfied by the set of collapsing morphisms of a pseudovariety and so one could not identify a pseudovariety of semigroups with a pseudovariety of relational morphisms in this way. In the end Tilson and Steinberg abandoned this line of research.

On June 9, 2000, Rhodes and Steinberg met to discuss semigroups at *Les Monts d'Auvergne*, a Parisian café in the neighborhood of Chevaleret (at the rue Maurice et Louis de Broglie). Rhodes had been reading "Categories as algebra. II" [339] and under its influence had also arrived at the idea of defining operators via relational morphisms. However, he was more interested in the operators themselves than in classes of relational morphisms. Inspired by the Composition Theorem, he believed that the *whole* of semigroup theory should be viewed as studying the composition of continuous operators on the lattice $\mathbf{PV}$ of pseudovarieties. The reasoning is as follows: the two-sided semidirect product and the Mal'cev product are non-associative, but the composition of operators is associative. How one chooses to bracket, say, iterated two-sided semidirect products is actually forced by associativity once one decides which factor is the operator and which factor is the variable. For instance, if your operators are $\alpha = \mathbf{A} ** (-)$ and $\beta = \mathbf{G} ** (-)$, then one can perfectly well write down nice associative expressions like $\alpha\beta\alpha$, or even $(\alpha\beta)^{\omega}\alpha$. Only after choosing a pseudovariety $\mathbf{V}$ on which to operate (think of this as making an observation in quantum physics) does a choice of bracketing and non-associativity appear. Thus taking $\mathbf{V}$ to be the trivial pseudovariety $\mathbf{1}$ yields

$$\alpha\beta\alpha(\mathbf{1}) = \mathbf{A} ** (\mathbf{G} ** (\mathbf{A} ** \mathbf{1})).$$

Suppose now that $\alpha' = (-) ** \mathbf{A}$ and $\beta' = (-) ** \mathbf{G}$. Then we can again form pleasant associative expressions like $\alpha'\beta'\alpha'$, only this time performing the experiment of evaluating at $\mathbf{1}$ results in

$$\alpha'\beta'\alpha'(\mathbf{1}) = ((\mathbf{1} ** \mathbf{A}) ** \mathbf{G}) ** \mathbf{A}.$$

One could also "mix" variables and consider $\alpha'' = \mathbf{A} \textcircled{m}(-)$ and $\beta'' = (-)\textcircled{m}\mathbf{G}$.

Getting ahead of ourselves for the moment, if we stick with the operators $\alpha = \mathbf{A} ** (-)$ and $\beta = \mathbf{G} ** (-)$, then the two-sided complexity hierarchy is obtained by taking the operator $(\alpha^{\omega}\beta^{\omega})^{n}\alpha^{\omega}$ and sampling it at the trivial pseudovariety $\mathbf{1}$. After the quantum effect of applying the operator to the trivial pseudovariety, we arrive at the level $n$ two-sided complexity pseudovariety given by

$$\mathbf{K}_n = (\alpha^{\omega}\beta^{\omega})^{n}\alpha^{\omega}(\mathbf{1}) = \mathbf{A} **^{\omega} (\mathbf{G} **^{\omega} (\cdots(\mathbf{A} **^{\omega} (\mathbf{G} **^{\omega} \mathbf{A}))\cdots))$$

where **G** appears $n$ times. This quantum idea of replacing points by operators and then only getting the points back by performing an observation (or an experiment) by evaluating at a point is why this book is called q-theory: q as in quantum! In fact, this book is very much a *quantized* version of Eilenberg [85]; pseudovarieties of semigroups are replaced by operators on the lattice of pseudovarieties; relational morphism and division of semigroups is replaced by relational morphism and division of relational morphisms. Tilson does identify (in his Chapter XII of Eilenberg [362]) a semigroup with its collapsing morphism, but this is not a true quantization; in this book, we quantize a semigroup to all relational morphisms having it as the domain.

Returning to our conversation at *Les Monts d'Auvergne*, Rhodes told Steinberg that he wanted to define a homomorphism from classes of relational morphisms to operators and characterize which operators were in the image. The viewpoint was that both **PV** and the lattice of relational morphism pseudovarieties were complete lattices, so they have the least upper bound property like the real numbers, and that one could define natural topologies for which our homomorphism was continuous. This begins to functor us over to soft analysis and the abstract spectral theory of continuous lattices [97]. Steinberg, who had already been influenced away from the idea by Tilson, spoke about the issues involved in deciding between the various candidates for the title of pseudovariety of relational morphisms and why he and Tilson had dropped the idea. It was fair to say he was at that moment against developing the notion. The following day the authors again met at *Les Monts d'Auvergne*, only this time their positions had switched under the influence of each other's arguments of the previous day: Steinberg thought the idea was great and wanted to develop it; Rhodes believed it should be dropped. Nonetheless, from this meeting the current volume was born. Shortly thereafter Tilson became a coauthor, but several months later creative differences arose, and he divorced himself from the project. However, Tilson was certainly a major influence on the early development of this book, and he followed its progress with interest up until his death.

Our original program went something like this: First we defined division of relational morphisms and pseudovarieties of relational morphisms. Pseudovarieties could be composed and a homomorphism q taking pseudovarieties to operators was defined. Then we characterized which operators on the lattice **PV** of pseudovarieties arose from pseudovarieties of relational morphisms. These turned out to be Scott continuous functions (in the sense of [97]) satisfying an additional condition that we termed the global Mal'cev condition. For instance, the Schützenberger product satisfies the global Mal'cev condition and hence can be defined by a pseudovariety of relational morphisms. Next our goal was to take Steinberg's notion of pseudoidentities for relational morphisms and prove a Reiterman's Theorem in this context. This was all completed by Fall 2000. This notion of a pseudoidentity led to a general notion of inevitable substitutions that encompassed the ideas used previously to study pointlikes [121, 322, 330], idempotent-pointlikes [235] and inevitable graphs [9, 33]. Since application of operators is a special instance of composition (where a pseudovariety is identified with a constant map), the basis theorems of Almeida and Weil [27] and Pin and Weil [235] became instances of a basis theorem for composition of pseudovarieties of relational morphisms. Complexity hierarchies associated to iteration of operators were to be defined, with group complexity as the model, and encompassing group complexity [170], dot-depth [71], $p$-length [368] and two-sided complexity. The goal was then to generalize Almeida and Steinberg's notion of tameness [19, 20] to obtain decidability results for arbitrary hierarchies arising from iteration of operators. For instance, if we could find a good basis of pseudoidentities for the pseudovariety of relational morphisms defining the Schützenberger product operator, this would provide a program to attack the dot-depth problem. Unfortunately, we realized in early 2001 that some additional hypotheses were needed on the factors for the basis theorem to work and that, in particular, there were missing hypotheses in [27], thereby invalidating many of the results of [19, 20].

To understand the issues underlying the instances when the basis theorem holds and when it doesn't, we were led to begin a systematic study of lattice theoretic considerations with regard to $\mathbf{PV}$ and continuous operators on $\mathbf{PV}$. In particular, we introduced a new class of relational morphisms, called a *continuously closed class*, and showed that all continuous operators are the q-image of a continuously closed class. We were also led to study order-theoretic properties of the map q itself. As a map of partially ordered sets it turns out to have both a left and right adjoint, and hence, given any continuous operator satisfying the global Mal'cev property, there is a unique maximal and a unique minimal pseudovariety of relational morphism defining it: each such operator is the q-image of a closed interval in the lattice of pseudovarieties of relational morphisms. This leads to many interesting questions, such as whether $\mathbf{V}_D$ is the minimal pseudovariety of relational morphisms defining $\mathbf{V} * (-)$ (we term this the "Tilson Problem"). Trying to resolve the case of the trivial pseudovariety $\mathbf{1}$, we were led to the fascinating question of whether every finite semigroup embeds in a relatively free finite semigroup. This was answered positively by George Bergman, who established that $\mathbf{1}_D$, the pseudovariety of divisions, is indeed the unique minimal pseudovariety of relational morphisms defining the identity operator [47]. Imagine what techniques will be needed to address the general case!

We then entered into the so-called "abstract spectral theory" of lattices [97], a sweeping generalization of Stone's duality between Boolean algebras and profinite spaces [61, 97, 117, 147, 341, 342]. This in turn brought us to quantales [303], the so-called quantum locales. These are complete lattices with a semigroup multiplication preserving all sups. They generalize the well-known locales of pointless topology [97, 147, 186] and play a role in the search for a non-commutative Gelfand space for $C^*$-algebras [172, 173, 303]. The monoid of pseudovarieties of relational morphisms and the monoid of continuous operators on $\mathbf{PV}$ are quantales in a slightly weakened sense. Finite quantales (in the classical sense) are nothing more than finite idempotent semirings, which have already been considered by Libor Polák in the context of Formal Language Theory [241–244]. It seemed natural to start applying the complexity of operators program in this context. To achieve this, a product was needed. Neither the usual wreath product nor the wreath product of ordered semigroups [237] works in this context. It turns out that it is the triangular product of Boris Plotkin [239,240,376] that does the job. Plotkin's triangular product is an axiomatization of the block triangular form obtained for a matrix representation by taking a Jordan-Hölder composition series. Our decision to use this product was influenced very much by the viewpoint of Almeida, Margolis, Steinberg, and Mikhail Volkov [18].

The establishment of a Prime Decomposition Theorem for Idempotent Semirings leads to a large number of open questions, including computing the complexity of a finite idempotent semiring and completing the classification of irreducible idempotent semirings. Also, this theorem opens up a new avenue of attack on the dot-depth problem as the Schützenberger product is a special case of the triangular product of semigroups. Our decomposition theorem works in this context as well and can be used to obtain lower triangular and block lower triangular Boolean matrix representations. The problem of dot-depth two is equivalent to deciding whether a finite semigroup divides a semigroup of lower triangular Boolean matrices [233]. The power set of a finite semigroup is an idempotent semiring; the Prime Decomposition Theorem for Idempotent Semirings works as a powerful tool for studying the complexity of power semigroups. In particular, it leads to an improved version of results of Cary Fox and Rhodes [91], allowing us to compute the exact group complexity of the power semigroup of a finite inverse semigroup, as well as an asymptotically tight bound on the complexity of the power semigroup of the full transformation of degree $n$.

The contents of the current volume are summarized in the Preface. Let us add that at the end of the text we compile a list of 74 problems generated by the results of this book. We invite our readers to solve them all!

## Part I

The q-operator and Pseudovarieties of
Relational Morphisms

# 1

# Foundations for Finite Semigroup Theory

This chapter sets up the foundations for Finite Semigroup Theory. Semidirect products, wreath products and two-sided semidirect products are introduced. The fundamental notion for comparing semigroups, division [169], is presented. This leads quickly to the notion of a pseudovariety [86], i.e., a collection of finite semigroups closed under taking finite direct products and divisors (that is, subsemigroups and quotients). The language of pseudovarieties serves as the unifying organizational principle in Finite Semigroup Theory. But equally important is the fact that the collection of all pseudovarieties is a complete, in fact, algebraic lattice with various associative and non-associative multiplications. Finite semigroup theory is as much about studying the lattice of all pseudovarieties, equipped with all this extra structure, as it is about studying individual semigroups. After all, probabilistically speaking, most semigroups have a zero element and satisfy $xyz = 0$ [162] (i.e., are 3-nilpotent). But there are only finitely many pseudovarieties of 3-nilpotent semigroups (as there are only finitely many identities one can write down with both sides having length at most 3, and every subpseudovariety of the locally finite pseudovariety of 3-nilpotent semigroups is defined by a set of such identities).

The notion of division also leads to the more general notion of relational morphism. Relational morphisms were introduced more than 35 years ago by the first author and B. Tilson. Tilson was very much responsible for pushing the viewpoint that relational morphisms are not merely diagrams, but arrows that can be composed thereby giving birth to the category of semigroups with arrows relational morphisms. Today more often than not, it is relational morphisms that play the key role in Finite Semigroup Theory, not homomorphisms. In this chapter, we explore some categorical facets of the category of semigroups with relational morphisms as arrows. In this setup, there are only weak products and pullbacks. We also introduce a tool for comparing relational morphisms by defining divisions of relational morphisms. With a notion of products and division for relational morphisms, you can guess that

J. Rhodes, B. Steinberg, *The q-theory of Finite Semigroups*,
Springer Monographs in Mathematics, DOI 10.1007/978-0-387-09781-7_1,
© Springer Science+Business Media, LLC 2009

we are heading toward defining pseudovarieties of relational morphisms, but that is the topic of the next chapter!

## 1.1 Basic Notation

### 1.1.1 General and philosophical remarks

We follow here the principle that Category Theory, especially via adjunctions, provides the "guiding light" for correct foundations. In particular, we make functorial choices in adjoining identities to semigroups and going from monoid constructs to their semigroup analogues.

For these reasons we establish in detail the categorical and Galois connection terminology of [185] and also the theory of continuous lattices, as expounded in [97] (see also [147, Chapter VII]), in a compatible way. The theory of quantales [303] will enter the picture, as well, creating a fusion of semigroups, lattices, categories and semirings.

We therefore provide a basic discussion of Galois connections and adjunctions. More can be found in [185] and [97]. A more detailed treatment of the subject is presented in Section 6.3.

### Standing notational conventions

Let us introduce some of our standing notational conventions. We usually will write morphisms between algebraic objects $f\colon S \to T$ with the variable on the left: $sf = t$. However, for functors $F\colon C \to D$ between categories viewed as classifying objects and for morphisms of partially ordered sets, especially Galois connections and pseudovariety operators, we shall write the variable on the right of the function: $F(x) = y$. Composition is written accordingly: we write $sfg$ or $FG(x)$ as appropriate.

The apparent exception to these conventions is the operator $\mathsf{q}$, to be defined below. This operator is both a morphism of algebraic lattices and a monoid homomorphism. We write it on the right of its argument, reflecting that it is a monoid homomorphism. This notation is convenient as $\mathsf{q}$ can, in fact, be viewed as a function of two variables, one written on the left and the other on the right. Hence when composing $\mathsf{q}$ with other maps between partially ordered sets, we follow the convention for composition of functions written on the right of their arguments.

The identity morphism of an object $S$ will be denoted $1_S$ or $1$ if $S$ is clear from context. We use **Set** to denote the category of all (small) sets and **Sgp** to denote the category of (small) semigroups. We shall denote by **FSet** the full subcategory of **Set** whose objects are the finite sets. In general, given a concrete category, we shall affix **F** to the name of the category to denote the full subcategory of its finite members. So, for instance, **FSgp** will denote the category of finite semigroups.

Following [132, 364], we often consider finite categories and finite semi-groupoids as algebraic structures, generalizing monoids and semigroups. When viewing categories in this way, functors shall be composed according to our conventions for morphisms of algebraic objects. We also shall use the convention that if $f\colon c \to d$ and $g\colon d \to e$ are arrows of a finite category or semi-groupoid, then $fg\colon c \to e$ is the composition. For categories, semigroupoids, posets, monoids, semigroups, etc., we use $S^{op}$ to denote the dual (or reverse) object to $S$.

If $S$ is an object of some category, then we shall use $T \leq S$ to indicate that $T$ is a subobject of $S$. For example if $S$ is a semigroup, then $T \leq S$ indicates that $T$ is a subsemigroup whereas $X \subseteq S$ merely asserts that $X$ is a subset. If $C$ is a category, the endomorphism or local monoid $C(c,c)$ at an object $c$ will be denoted simply $C(c)$.

If $f\colon S \to T$ is a function, then the kernel of $f$ is the equivalence relation

$$\ker f = \{(s, s') \in S^2 \mid sf = s'f\}.$$

If $f$ is a homomorphism of algebraic structures, then $\ker f$ is a congruence and $Sf \cong S/\ker f$. However, for a group homomorphism $f$, the notation $\ker f$ will retain its traditional meaning as the preimage of the identity.

Let us turn to two further conventions to which the authors will adhere in order to avoid adding trivial hypotheses and cluttering theorem statements.

*Convention 1.1.1 (Empty semigroup).* Let us make the following convention concerning the empty semigroup. Universal Algebra and Category Theory tell us we must admit the empty semigroup, and so we do. If not, the category of semigroups is not cocomplete and the subsemigroups of a semigroup do not form a complete (algebraic) lattice. However, to avoid making special statements or entering into special cases just to deal with the empty semigroup, we often omit the hypothesis non-empty when no confusion should arise.

*Convention 1.1.2 (Finiteness).* In this book, nearly all semigroups are finite or profinite, with the exception of free monoids and free groups. In many cases we shall omit the adjective *finite* if it is clear that we are in a context where only finite semigroups are being discussed. In other words, a theorem may use properties of finite semigroups in the proof without having the word *finite* in the statement. In such situations the word *finite* should be considered *implicit*. We have tried our best to inform the reader at the beginning of relevant sections or chapters whether we are assuming finiteness.

Hopefully, the reader will come to appreciate that these conventions ease the exposition by keeping theorem statements more concise.

### 1.1.2 Galois connections and adjunctions

Let $S$ and $T$ be partially ordered sets (posets), i.e., sets with a transitive, reflexive, anti-symmetric relation. Then functions $g\colon S \to T$ and $d\colon T \to S$ form a *Galois connection* if they are order preserving and

$$d(t) \leq s \iff t \leq g(s) \tag{1.1}$$

for all $s \in S$, $t \in T$, or equivalently

$$dg \leq 1_S \text{ and } 1_T \leq gd. \tag{1.2}$$

One calls $d$ the *left adjoint* and $g$ the *right adjoint*. We remark that (1.2) says that $dg$ is a kernel operator and $gd$ is a closure operator. Recall that a *closure* (*kernel*) operator on a poset is an order preserving, non-decreasing (non-increasing) idempotent map. Closure operators and kernel operators form part of the subject of Chapter 6. Diagrammatically, we draw Galois connections as follows:

$$T \xleftarrow[\text{$d$ (left adjoint)}]{\text{$g$ (right adjoint)}} S \tag{1.3}$$

or

$$d \text{ (left adjoint)} \left\downarrow \right\uparrow g \text{ (right adjoint)}.$$

with $T$ on top and $S$ on bottom.

A partially ordered set $(S, \leq)$ can be considered a category with objects $S$ and arrows $s_1 \to s_2$ if and only if $s_1 \leq s_2$ with composition of abutting arrows the only one possible. Note that in the constructed category, $\mathsf{Hom}(s_1, s_2)$ has one or no arrows, depending on whether $s_1 \leq s_2$, or not. Order preserving maps between partially ordered sets become functors. Now (1.3) becomes, when replacing the partially ordered sets by their associated categories, an adjunction of categories. Let us review the notion.

Our notation for D. Kan's concept of an adjunction will be

$$C \xleftarrow[\text{$F$ (left adjoint)}]{\text{$G$ (right adjoint)}} D. \tag{1.4}$$

We remark that in [185] the left adjoint is drawn on top, whereas we place it on the bottom. Recall that functors $F, G$ as per (1.4) form an *adjoint pair* if there is a natural isomorphism between the functors $D(F(-), -)$ and $C(-, G(-))$ (whence the terminology). One says that $F, G$ give an *adjunction* between $C$ and $D$. This is one of the central notions of category theory; see [185]. It is often more natural to think of adjunctions in terms of the unit and the counit. The *unit* is a natural transformation $\eta \colon 1_C \to GF$ with each component $\eta_c \colon c \to GF(c)$ universal among arrows from $c$ to objects of the form $G(d)$, and dually the *counit* $\varepsilon \colon FG \to 1_D$ is given by components $\varepsilon_d \colon FG(d) \to d$ universal among arrows from an object in the image of $F$ to $d$. We remark that $F$ uniquely determines $G$, if $G$ exists, and conversely (see [185]). The counit is analogous to the notion of a semigroup expansion [54] in the case where each component is surjective.

The example to keep in mind is the case where $C = \mathbf{Set}$ is the category of sets and $D = \mathbf{Sgp}$ is the category of semigroups. One takes $F$ to be the functor $F(X) = X^{+}$, the free semigroup generated by $X$. The functor $G$ is the underlying set functor. The component of the unit at $X$ is the canonical embedding of $X$ into the underlying set of $X^{+}$. The component of the counit at a semigroup $S$ is the canonical projection $S^{+} \twoheadrightarrow S$. In general the unit $\eta_c$ is the image of $1_{F(c)} \colon F(c) \to F(c)$ under the isomorphism

$$D(F(c), F(c)) \cong C(c, GF(c))$$

and the counit $\varepsilon_d$ is the image of $1_{G(d)} \colon G(d) \to G(d)$ under the isomorphism $C(G(d), G(d)) \cong D(FG(d), d)$.

A key point is that the left adjoint of an adjunction preserves all colimits and the right adjoint preserves all limits, see [185] for the definitions and details. Freyd's Adjoint Functor Theorem [185] says that if $C$ has all colimits and $F$ preserves all colimits, then $F$ has a right adjoint $G$ and dually.

If $S$ and $T$ are partially ordered sets viewed as categories in the manner discussed above, then condition (1.1) for a Galois connection just says that the functor $S(d(-), -)$ is naturally isomorphic to $T(-, g(-))$; that is, $d, g$ form an adjoint pair. Condition (1.2) describes the unit and counit of the adjunction.

*Remark 1.1.3.* Our choice of the notation $g$ and $d$ for Galois connections strictly adheres to [97]. Apparently $g$ originally stood for *gauche* and $d$ for *droite* because at one time some of the authors of [97] turned posets into categories by having an arrow $s \to s'$ if $s \geq s'$ and so $g$ was the left adjoint and $d$ the right adjoint. The authors of [97] suggest remembering that $g$ is *greater* and so $g(s) \geq t$ whereas $d$ is *downward* so $d(t) \leq s$.

For a lattice, that is a poset in which each pair of elements has a join and a meet, (finite) meets correspond to (finite) products in the associated category, whereas (finite) joins correspond to (finite) coproducts. A top element $\mathsf{T}$ corresponds to a terminal object, whereas a bottom $\mathsf{B}$ corresponds to an initial object. Thus lattices with a bottom $\mathsf{B}$ and a top $\mathsf{T}$ give rise (via the constructed category) to categories with all finite products and coproducts. A complete lattice determines a category closed under all products and co-products (i.e., complete and cocomplete). The condition that left adjoints and right adjoints preserve, respectively, colimits and limits then translates into saying that the left adjoint of a Galois connections preserves all sups and the right adjoint preserves all infs.

Several kinds of order preserving maps between lattices shall be of importance in this text. We shall impose on maps conditions of the form

$$f\left( \bigvee X \right) = \bigvee f(X) \tag{1.5}$$

(or dually replacing joins with meets).

We say that the map $f \colon S \to T$ is **sup** if, for all subsets $X \subseteq S$, (1.5) holds. If $f$ just satisfies (1.5) for all non-empty subsets $X$ of $S$, then we say that $f$ is $\mathbf{sup_B}$. The notation is to indicate that the bottom $\mathsf{B}$ need not be preserved.

A subset $D \subseteq S$ is said to be *directed* if every finite subset of $D$ has an upper bound. The fact that the empty subset has an upper bound implies that $D$ is non-empty. Thus $D \subseteq S$ is directed if and only if it is non-empty and, for all $d_1, d_2 \in D$, there exists $d_3 \in D$ with $d_1, d_2 \leq d_3$. We shall use the terms *downwards directed* or *inversely directed* for the dual concept.

The following definition, going back to D. Scott [97], is key for this book.

**Definition 1.1.4 (Continuous map of posets).** *A map* $f: S \to T$ *between posets is continuous if, for all directed subsets* $D \subseteq S$,

$$f\left(\bigvee D\right) = \bigvee f(D) \tag{1.6}$$

*whenever* $\bigvee D$ *exists in* $S$.

The motivation for the terminology will become apparent in Section 6.4 when we consider topologies on complete lattices. In any event, directed sets $D$ are analogous to nets, and (1.6) says essentially that $f$ preserves nets. A central theme of this book is that Finite Semigroup Theory is very much the study of continuous operators on the lattice of semigroup pseudovarieties.

**Proposition 1.1.5.** *A continuous map of posets is order preserving.*

*Proof.* Let $f: S \to T$ be a continuous map between posets and suppose $s \leq s'$. Then $\{s, s'\}$ is directed, so

$$f(s') = f(s \vee s') = f(s) \vee f(s') \geq f(s)$$

as required.                                                                  □

A map $f: S \to T$ of join semilattices with identity is called a $\vee$-map if (1.5) holds for all finite sets $X$. Equivalently, $f$ is a $\vee$-map if $f(\mathsf{B}) = \mathsf{B}$ and $f(x \vee y) = f(x) \vee f(y)$, all $x, y \in S$. It is easy to verify that $f: S \to T$ is $\mathbf{sup}$ if and only if it is continuous and a $\vee$-map. The dual concepts are denoted $\mathbf{inf}$, $\mathbf{inf_T}$, dual continuous and $\wedge$-map, respectively.

We use $\mathbf{Sup}$ for the category of complete lattices with morphisms the $\mathbf{sup}$ maps. The category of complete lattices with continuous maps as morphisms is denoted $\mathbf{Cnt}$. The category of partially ordered sets with order preserving maps is denoted $\mathbf{OP}$.

An element of a lattice is called *compact* if whenever it is below the join of a collection of elements of the lattice, it is actually below the join of a finite subcollection. Clearly a finite join of compact elements is again compact. A complete lattice is called *algebraic* if each element is a join of compact elements. Because the set of compact elements is closed under finite joins, every element of an algebraic lattice is a directed supremum of compact elements. For terminology on algebraic and continuous lattices, the reader is referred to Chapter 6 where these notions are treated in detail. See also [97, 203].

**Exercise 1.1.6.** Verify that a finite join of compact elements is compact.

In the context of complete lattices, Freyd's Adjoint Functor Theorem [185] says that if $d\colon S \to T$ is a **sup** map, then it has a right adjoint $g\colon T \to S$. Of course the dual result holds. Let us formulate this more precisely. The details can be found in Section 6.3.

**Proposition 1.1.7.** *Let $S$ and $T$ be complete lattices and $g, d$ be order preserving maps. Then*

$$T \xleftarrow{\;g\;} \xrightarrow[d]{} S \qquad (1.7)$$

*is an adjunction or Galois connection if and only if $g$ is* **inf** *(respectively, $d$ is* **sup***) and one determines the other via the formulas:*

$$g(s) = \bigvee \{t \mid d(t) \leq s\}$$
$$d(t) = \bigwedge \{s \mid t \leq g(s)\} \qquad (1.8)$$

*for $s \in S$, $t \in T$.*

*Furthermore, if $g, d$ form a Galois connection, then the following hold:*

1. *$dg$ is a kernel operator and $gd$ is a closure operator;*
2. *$gdg = g$ and $dgd = d$;*
3. *$g$ is injective if and only if $d$ is surjective, if and only if $dg = 1$; moreover, this occurs precisely when $g(s) = \max d^{-1}(s)$;*
4. *$d$ is injective if and only if $g$ is surjective, if and only if $gd = 1$; moreover, this occurs precisely when $d(t) = \min g^{-1}(t)$.*

*Finally, if $S$ and $T$ are partially ordered semigroups, then*

$$g(s_1)g(s_2) \leq g(s_1 s_2) \text{ if and only if } d(t_1 t_2) \leq d(t_1)d(t_2). \qquad (1.9)$$

**Exercise 1.1.8.** Prove (1.9).

Establishing the existence of a Galois connection (1.7) is a convenient way of proving that $g$ is **inf** or $d$ is **sup**, as we shall see shortly.

A map $d$ (respectively $g$) of partially ordered semigroups satisfying (1.9) is called a *prehomomorphism* (respectively *dual prehomomorphism*) [176]. Dual prehomomorphisms come up in the study of $F$-inverse monoids and their generalization, $F$-morphisms [178]. Recall that a semigroup $S$ is called *inverse* if, for all $s \in S$, there exists a unique element $s^* \in S$ with $ss^*s = s$, $s^*ss^* = s^*$. Inverse semigroups have a natural partial order compatible with multiplication defined by setting $s \leq t$ if $s = et$ with $e$ an idempotent of $S$ [68, 176]. An inverse semigroup homomorphism is an $F$-morphism if it has a right adjoint when viewed as a map of partially ordered sets.

## 1.2 Foundations

### 1.2.1 Some adjunctions

Recall that we denoted by **Sgp** the category with objects (small) semigroups and arrows semigroup homomorphisms. The initial object is the empty semigroup and the terminal object is the trivial semigroup $\{1\}$, sometimes denoted simply by 1. Following our earlier convention, **FSgp** is the full subcategory obtained by restricting the objects to be finite semigroups. Similarly one defines **Mon** and **FMon**, the categories of all monoids and finite monoids; of course, arrows in **Mon** send identities to identities, so **Mon** is a subcategory of **Sgp**, but *not* a full subcategory. In these latter categories $\{1\}$ is both initial and terminal.

Given a semigroup $S$ there are two popular ways to make $S$ into a monoid: $S^I$ and $S^\bullet$ (this latter is often written $S^1$). The construction $S^I$ adjoins a new identity $I$ to $S$, even if $S$ already has an identity, whereas $S^\bullet$ adds an identity only when $S$ has no identity. The first is the object part of a functor **Sgp** $\rightarrow$ **Mon**, the second is *not* a functor; even better, $S \mapsto S^I$ is the left adjoint of the forgetful functor from **Mon** to **Sgp**. So via the philosophy espoused in Section 1.1.1, the construction $S \mapsto S^I$ is the correct (or canonical) choice. This has the effect that for $G$ a group, $G^I$ is not a group. That's tough luck: from the categorical viewpoint, we cannot avoid this.

We proceed to consider a miscellany of adjunctions to give the reader a flavor for this abstract notion. First consider:

$$\mathbf{Sgp} \underset{S \longmapsto S^I}{\overset{\text{forgetful functor}}{\rightleftarrows}} \mathbf{Mon}. \tag{1.10}$$

The unit $\eta_S \colon S \hookrightarrow S^I$ is the inclusion, the counit $\varepsilon_M \colon M^I \twoheadrightarrow M$ extends the identity map by $I \mapsto 1_M$ and (1.10) restricts to **FSgp** and **FMon**.

By a *transformation semigroup* $(X, S)$ we mean a semigroup $S$ acting on the right of a set $X$ by (totally defined) functions, not necessarily faithfully (although faithfulness will be required in Chapter 4). If $M$ is a monoid, we call $(X, M)$ a *transformation monoid* if the identity of $M$ acts as the identity transformation. If $G$ is a group, then $(X, G)$ is called a *transformation group* or *permutation group* if $G$ acts on $X$ by permutations, or equivalently $(X, G)$ is a transformation monoid. Partial transformation semigroups and monoids are defined analogously but we allow the action to be via partial functions.

Another important adjunction is

$$\mathbf{Sgp} \underset{S \longmapsto (I, S^I, S)}{\overset{\text{forgetful functor}}{\rightleftarrows}} \mathbf{TSgp}^* \tag{1.11}$$

where **TSgp**$^*$ is the category of pointed transformation semigroups. Thus $(p, X, S)$ is an object of **TSgp**$^*$ if $X$ is a non-empty set, $p \in X$, and $(X, S)$

a right transformation semigroup. A morphism of $(p_1, X_1, S_1)$ to $(p_2, X_2, S_2)$ is a pair of maps $f \colon X_1 \to X_2$, $g \colon S_1 \to S_2$ with $p_1 f = p_2$ and $(x_1 \cdot s_1)f = x_1 f \cdot s_1 g$. This gives a category with initial object $(I, \{I\}, \emptyset)$ and terminal object $\{I, \{I\}, I\}$.

The counit component for $\mathcal{O} = (p, X, S)$ is $\varepsilon_{\mathcal{O}} \colon (I, S^I, S) \twoheadrightarrow (p, X, S)$ determined by $(f, g)$ where $If = p$, $sf = p \cdot s$ and $g$ is the identity map. The adjunction (1.11) restricts to finite and to faithful pointed transformation semigroups.

Next, we turn to the corresponding adjunction for monoids:

$$\mathbf{Mon} \xleftarrow{\text{forgetful functor}} \xrightarrow[M \longmapsto (1_M, M, M)]{} \mathbf{TMon}^* \tag{1.12}$$

with $\mathbf{TMon}^*$ the category of pointed transformation monoids, where now morphisms must respect the identity. The adjunction (1.12) also restricts to finite and to faithful transformation monoids. The bottom arrow is the right regular representation of $M$.

Now consider the adjunction

$$\mathbf{TSgp}^* \xleftarrow{\text{forgetful functor}} \xrightarrow[(p, X, S) \longmapsto (p, X, S^I)]{} \mathbf{TMon}^*, \tag{1.13}$$

which restricts to finite but *not* to faithful transformation semigroups.

Next we can compose abutting adjunctions yielding new adjunctions. The two ways from $\mathbf{Sgp}$ to $\mathbf{TMon}^*$ agree and equal the adjunction

$$\mathbf{Sgp} \xleftarrow{\text{forgetful functor}} \xrightarrow[S \longmapsto (I, S^I, S^I)]{} \mathbf{TMon}^*, \tag{1.14}$$

which restricts to finite and also to faithful transformation semigroups. This adjunction will be the basis for how we go from monoid constructions to semigroup constructions.

## 1.2.2 Wreath and semidirect products

In this section we define the semidirect product of monoids and of semigroups, as well as the wreath product of transformation semigroups. Wreath products of monoids also make an appearance here, but we delay the definition of wreath products of semigroups until the next section as this requires some care.

Suppose that $M$ and $N$ are monoids. Working in the category of monoids, we say that $N$ *acts* on the left of $M$ if there is a map $N \times M \to M$, written $(n, m) \mapsto n \cdot m$, satisfying:

- $n \cdot (m + m') = n \cdot m + n \cdot m'$;
- $n \cdot 0_M = 0_M$;
- $(nn') \cdot m = n \cdot (n' \cdot m)$;
- $1_N \cdot m = m$,

where we use additive notation for $M$, although we do not require commutativity. Equivalently, one has a *monoid* homomorphism from $N$ into the endomorphism monoid of $M$ (in the category of monoids). Sometimes we shall use multiplicative notation for $M$, in which case we write $^n m$ for the action of $n \in N$ on $m \in M$, i.e., we use exponential notation for the action.

The *semidirect product* $M \rtimes N$ with respect to such a left action is the set $M \times N$ equipped with the multiplication given by

$$(m, n)(m', n') = (m + n \cdot m', nn').  \tag{1.15}$$

It is sometimes fruitful to view $(m, n)$ as the matrix $\begin{pmatrix} 1 & 0 \\ m & n \end{pmatrix}$ and then (1.15) becomes usual matrix multiplication:

$$\begin{pmatrix} 1 & 0 \\ m & n \end{pmatrix} \begin{pmatrix} 1 & 0 \\ m' & n' \end{pmatrix} = \begin{pmatrix} 1 & 0 \\ m + n \cdot m' & nn' \end{pmatrix}.  \tag{1.16}$$

The resulting semigroup $M \rtimes N$ is in fact a monoid with identity $(0_M, 1_N)$. The projection $M \rtimes N \twoheadrightarrow N$ is a surjective homomorphism of monoids.

**Exercise 1.2.1.** Verify (1.15) is an associative product and $(0_M, 1_N)$ is the identity for $M \rtimes N$.

We are deliberately avoiding usage of the symbol $*$ for the semidirect product of two semigroups (as is frequently done in the literature [7, 27, 85, 364]) to avoid confusion with the free product of monoids and also because $*$ is too symmetric a symbol for a non-symmetric product. We shall continue to use $*$ on the pseudovariety level for the semidirect product, as the notation seems too well-entrenched to be changed at this point in time (for those not familiar with them, pseudovarieties will be defined in Definition 1.2.30). The symbol $\rtimes$ comes from group theory where $N \rtimes H$ indicates $H$ acts on the normal subgroup $N \lhd (N \rtimes H)$, whence the direction of $\rtimes$.

One can also define the *reverse semidirect product* in a dual fashion: If $M$ acts on the right of $N$, then $M \ltimes N$ is the set $(m, n)$ with multiplication given by $(m, n)(m', n') = (mm', nm' + n')$ where we write the product in $N$ additively. In matrix notation this boils down to

$$\begin{pmatrix} m & 0 \\ n & 1 \end{pmatrix} \begin{pmatrix} m' & 0 \\ n' & 1 \end{pmatrix} = \begin{pmatrix} mm' & 0 \\ nm' + n' & 1 \end{pmatrix}.$$

The projection map this time is onto $M$.

Similarly, if $S$ and $T$ are semigroups, then a left action of $T$ on $S$ is a map $T \times S \to S$, written $(t, s) \mapsto t \cdot s$, such that:

- $t \cdot (s + s') = t \cdot s + t \cdot s'$;
- $(tt') \cdot s = t \cdot (t' \cdot s)$,

where we use additive notation for $S$, although, again, we do not require commutativity. Equivalently, one has a homomorphism from $T$ into the endomorphism monoid of $S$ (in the category of semigroups).

The *semidirect product* $S \rtimes T$ with respect to this action is the set $S \times T$ with multiplication given by

$$(s, t)(s', t') = (s + t \cdot s', tt').$$

Again, the projection $S \rtimes T \twoheadrightarrow T$ is a surjective homomorphism.

The wreath product of transformation semigroups is an important example of a semidirect product.

**Definition 1.2.2 (Wreath product).** *Let $(X, S)$ and $(Y, T)$ be right transformation semigroups. Their wreath product is the transformation semigroup*

$$(X, S) \wr (Y, T) = (X \times Y, S^Y \rtimes T)$$

*defined as follows. The action of $T$ on $S^Y$ is given by $y^t f = ytf$ for $t \in T$, $y \in Y$ and $f \in S^Y$. For $(x, y) \in X \times Y$, define $(x, y)(f, t) = (x(yf), yt)$. The action semigroup is denoted $S \wr (Y, T)$ (as it does not depend on $X$).*

Contrary to Eilenberg [85], in this text, the notation $S \wr (Y, T)$ always denotes a semigroup, not a transformation semigroup; the corresponding transformation semigroup is denoted $(X, S) \wr (Y, T)$ where $X$ is a set acted on by $S$. We remark that many sources [85, 364] use $\circ$ for the wreath product. But again this is a symmetric symbol being used for an asymmetric operation, and also it is best to reserve $\circ$ for composition. In all other areas of mathematics $\wr$ is used for the wreath product, so we use this symbol, as well. The following series of exercises establishes some well-known properties of wreath products [85].

**Exercise 1.2.3.** Verify that $(X, S) \wr (Y, T)$ is a well-defined right transformation semigroup.

**Exercise 1.2.4.** Show that if $(X, S)$ and $(Y, T)$ are faithful, then so is $(X, S) \wr (Y, T)$.

**Exercise 1.2.5.** Show that if $(X, M)$ and $(Y, N)$ are right transformation monoids, then $(X, M) \wr (Y, N)$ is a transformation monoid.

**Exercise 1.2.6.** Show that if $(X, G)$ and $(Y, H)$ are right transformation groups, then $(X, G) \wr (Y, H)$ is a transformation group.

**Exercise 1.2.7.** Show that the wreath product of transformation semigroups is associative: that is, if $(X, S)$, $(Y, T)$ and $(Z, U)$ are transformation semigroups, then $[(X, S) \wr (Y, T)] \wr (Z, U) \cong (X, S) \wr [(Y, T) \wr (Z, U)]$.

If $M$ and $N$ are monoids, then we can view them as faithful transformation semigroups via their right regular representations $(M, M)$ and $(N, N)$. So it is natural to define the wreath product of two monoids to be the wreath product of their regular representations. To emphasize that we are dealing with monoids, we shall call this the *unitary wreath product* or the wreath product of monoids.

**Definition 1.2.8 (Unitary wreath product).** *If $M$ and $N$ are monoids, we define the unitary wreath product $M \wr N$ of $M$ with $N$ to be the monoid $M \wr (N, N) = M^N \rtimes N$.*

In particular, the unitary wreath product of two groups is again a group. The definition of the wreath product of semigroups will be deferred until the next section.

**Exercise 1.2.9.** Verify that the unitary wreath product of monoids is not associative.

### 1.2.3 Going from monoid constructions to semigroup constructions

Often we have a construction for monoids or categories, for instance the wreath product or the derived category and their complementary set of adjunctions (thanks to category theory and notably to the work of Grothendieck in the late 1950s [186], Quillen in the 1970s [257] and Tilson in the 1980s [364]). The question arises what is the *correct* corresponding semigroup and semigroupoid construction? By a semigroupoid, we mean a structure satisfying the axioms of a category except that we relax the condition demanding existence of identities at each object. So, for instance, a semigroup is a semigroupoid with a single object.

Many papers in the literature do not use the functorial companion constructions to the monoid/category constructions. The non-functoriality enters into the picture in a manner similar to the case of $S^I$ vs. $S^\bullet$. This has led to some flaws in several arguments, as we shall see later.

It is then natural to ask what is the correct way to obtain functorial constructions. Obviously, one cannot make exceptions depending on whether certain semigroups involved are monoids. The answer, in our opinion, to obtaining the correct, functorial companion constructions is to apply the philosophy of Section 1.1.1 via the adjunctions of Section 1.2.1. That is, given a category like **Sgp**, **TSgp**$^*$, etc., apply the relevant adjunctions of Section 1.2.1, do the relevant known functorial construction in **Mon**, **TMon**$^*$, etc., and then "scrape" the new identities $I$ off all the arrows (or semigroup elements), but leave the objects and sets alone, including any new identities introduced. Then check that, after performing this "scraping" process, the result yields an adjunction. We illustrate this philosophy with some examples. More important examples shall be considered later in the text.

For instance, one can ask: What is the right regular representation of a semigroup $S$? Our belief is that the answer is obtained by applying the adjunction (1.14), then "scraping" off $I$ as alluded to above, yielding the adjunction:

$$\mathbf{Sgp} \xrightarrow[\underset{S \longmapsto (I, S^I, S)}{}]{\overset{\text{forgetful functor}}{\longleftarrow}} \mathbf{TSgp}^*. \tag{1.17}$$

This adjunction restricts to finite and to faithful transformations semigroups. In particular, if $S$ is a group, the regular representation still has an adjoined identity on the set. Constructions, like the Cayley graph, that are based on the regular representation should use this definition.

Now that we have defined the right regular representation, we can define the wreath product of two semigroups $S$ and $T$ by taking the wreath product $S \wr (T^I, T)$ of their regular representations. This gives a different wreath product of semigroups than the "non-functorial" wreath product of semigroups defined, for instance, in [364] or [7]. Let us work this out in detail.

**Definition 1.2.10 (Wreath product of semigroups).** *If $S$ and $T$ are semigroups, then their wreath product $S \wr T$ is $S \wr (T^I, T) = S^{T^I} \rtimes T$.*

Note that if $S$ is non-trivial, then $S \wr \{1\} \ncong S$. This fact is important for this book, as we shall see later when the global Mal'cev condition (2.25) is introduced.

**Exercise 1.2.11.** Verify that $S \wr \{1\} \ncong S$ if $|S| > 1$.

**Exercise 1.2.12.** Show that the wreath product of semigroups is not associative.

**Exercise 1.2.13.** Show that $(S_1 \times S_2) \wr T$ embeds in $(S_1 \wr T) \times (S_2 \wr T)$.

Many papers and books do not do the foundations for the wreath product this functorial way, cf. [7, 27, 293, 364]. Notice that with our definition, the wreath product of two non-trivial groups (considered as semigroups) is never a group. This is necessary if one is to have wreath products and semidirect products (as defined by Eilenberg [85]) agree on the pseudovariety level. It is Eilenberg's definition that gives a well-behaved semidirect product operator; the competing definitions have problems and we shall point out later some of the errors that have arisen due to mistakenly translating monoid results to semigroup results without checking properly whether they work with non-functorial foundations. Universal Algebra tells us that when studying semigroups from a varietal viewpoint, we must put on blinders and not check whether our semigroup has extra structure, such as being a group or monoid. When studying a particular problem, one must decide whether one should be working in the category of semigroups or the category of monoids.

**Exercise 1.2.14.** Verify that the wreath product of two groups (considered as semigroups) is never a group unless the left-hand factor is trivial. In fact, the wreath product of two monoids (considered as semigroups) is never a monoid unless the left-hand factor is trivial.

Our next remark is intended for readers already familiar with pseudovarieties and their semidirect products.

*Remark 1.2.15.* As a pseudovariety of monoids $\mathbf{Sl} * \mathbf{G}$, the semidirect product of the pseudovariety of semilattices with the pseudovariety of groups, is the pseudovariety of monoids with commuting idempotents by a deep theorem of Ash [32, 126]. However, if one views $\mathbf{Sl}$ and $\mathbf{G}$ as semigroup pseudovarieties, then this is no longer the case. In this setting $\mathbf{Sl} * \mathbf{G}$ will contain left zero semigroups. The point is that this formulation of Ash's Theorem is really a theorem about monoids not semigroups. If you want a semigroup version of the theorem, you have to content yourself with stating that the Mal'cev product $\mathbf{Sl} \ⓜ\ \mathbf{G}$ is the pseudovariety of semigroups with commuting idempotents.

Let us see how our functorial method leads to the embedding theorem for semidirect products into wreath products. First we need an adjunction. Let $N$ be a fixed monoid and $\mathbf{Mon}^{N^{op}}$ be the category of monoids with left actions by $N$. If we view $N$ as a one-object category, such a left action is the same thing as a functor from $N^{op}$ to $\mathbf{Mon}$, whence the notation. There is then an adjunction

$$\mathbf{Mon}^{N^{op}} \xleftarrow{\ \ M^N \longleftarrow M\ \ }_{\text{forgetful functor}} \mathbf{Mon}. \qquad (1.18)$$

The counit $\eta\colon M^N \to M$ is given by evaluation at $1_N$. If $f\colon M' \to M$ is a morphism, with $M'$ in $\mathbf{Mon}^{N^{op}}$, the induced map $f_*\colon M' \to M^N$ is given by

$$n(m'f_*) = (nm')f. \qquad (1.19)$$

If $f$ is injective, then by evaluating at $1_N$, one verifies that $f_*$ is injective, as well.

**Exercise 1.2.16.** Show that if $M' \in \mathbf{Mon}^{N^{op}}$ and $f\colon M' \to M$ is an injective homomorphism of monoids, then $f_*\colon M' \to M^N$ given by (1.19) is an injective morphism in $\mathbf{Mon}^{N^{op}}$.

This semidirect product is a functor from $\mathbf{Mon}^{N^{op}}$ to the comma category (see [185, Chapter II, Section 6]) $\mathbf{Mon} \downarrow N$ of monoids with a distinguished homomorphism to $N$. That is, we associate $M \in \mathbf{Mon}^{N^{op}}$ to the semidirect product projection $M \rtimes N \twoheadrightarrow N$. Moreover, the semidirect product functor preserves injections meaning that if $\psi\colon M \to M'$ is an injective morphism in $\mathbf{Mon}^{N^{op}}$, there results a morphism

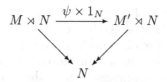

in **Mon** $\downarrow N$ with $\psi \times 1_N$ injective. We now turn to the embedding theorem.

**Theorem 1.2.17 (Embedding theorem).** *Let $M$ and $N$ be monoids and suppose that $M$ is equipped with a left action of $N$. Then there is an embedding*

$$M \rtimes N \hookrightarrow M \wr N$$
$$(m, n) \longmapsto (n_0 \mapsto n_0 \cdot m, n) \tag{1.20}$$

*where $M \wr N$ is the unitary wreath product.*

*Proof.* This result follows because if $M$ is in $\mathbf{Mon}^{N^{op}}$, then the identity map $1_M \colon M \to M$ induces an injective morphism $M \to M^N$ in $\mathbf{Mon}^{N^{op}}$ and hence an embedding of semidirect products $N \rtimes M \hookrightarrow N \wr M$, in fact, the one given above.                                                                         □

The above embedding is in fact a natural transformation from the functor $(-) \rtimes N$ on $\mathbf{Mon}^{N^{op}}$ to the composition of the forgetful functor from $\mathbf{Mon}^{N^{op}}$ to **Mon** and the wreath product functor $(-) \wr N$. We remark that the key ingredient in verifying that (1.20) is injective is to evaluate at $1_N$.

If $T$ is a semigroup, one can similarly define a category $\mathbf{Sgp}^{T^{op}}$ of semigroups with left actions by $T$. Note that even if $T$ is a monoid, $T$ is not required to act as a monoid. One then has a similar adjunction:

$$\mathbf{Sgp}^{T^{op}} \xleftarrow[\text{forgetful functor}]{S^{T^I} \longleftarrow S} \mathbf{Sgp}. \tag{1.21}$$

Again this functor preserves injective morphisms, as is seen by evaluating at $I$. Because the semidirect product functor also preserves injective maps, we obtain the semigroup version of the embedding theorem.

**Theorem 1.2.18 (Embedding theorem).** *Let $S$ and $T$ be semigroups and suppose that $T$ acts on the left of $S$. Then there is an embedding*

$$S \rtimes T \hookrightarrow S \wr T$$
$$(s, t) \longmapsto (t_0 \mapsto t_0 \cdot s, t) \tag{1.22}$$

*where $t_0$ takes values in $T^I$.*

The map here is injective as $I \mapsto I \cdot s = s$ (which is one reason why the "extra" $I$'s are left on the objects and sets). This embedding is again a natural transformation between correctly chosen functors.

Note that for monoids $N \rtimes \{1\} \cong N$, but for semigroups $T \rtimes \{1\}$ need not be isomorphic to $T$, as 1 need not act as the identity. Appendix A of [364] does not handle wreath products of semigroups in a functorial way. The book [7] does not treat the wreath product in this functorial way and, moreover, requires left actions of monoids on semigroups to have the identity act as an identity, apparently to be consistent with the wreath product defined in [364], again a non-functorial definition.

The problem with this latter definition is apparent when one considers a homomorphism $\varphi \colon T' \to T$. If $T$ acts on a semigroup $S$, then we can pull back the action to $T'$. That is $T'$ acts on $S$ via $t' \cdot s = t'\varphi \cdot s$. If $T'$ is a monoid but $T$ is not, then a left action of $T$ on a semigroup $S$ will not in general pull back to an action of $T'$ where the identity acts as an identity. The most important case for us is when $\varphi$ is an inclusion: if $T' \leq T$, it is only reasonable that an action of $T$ on a semigroup $S$ should restrict to an action of $T'$ on $S$. This problem propagates when one tries to rework [339] for semigroupoids using these non-functorial foundations.

### 1.2.4 Block and two-sided semidirect products

The semidirect product is not a left-right dual notion, which is unfortunate in many contexts, especially language theory [234, 378, 379]. Thus we turn to the two-sided semidirect product and the block product, introduced by the first author and Tilson [293].

**Definition 1.2.19 (Two-sided semidirect product).** *Suppose $S$ and $T$ are semigroups and $T$ has commuting left and right actions on $S$. For convenience we write $S$ additively. Then the two-sided semidirect product $S \bowtie T$ is the set $S \times T$ with multiplication*

$$(s,t)(s',t') = (st' + ts', tt').$$

*If $S$ and $T$ are monoids, the semidirect product is called unitary if both the left and right actions are monoid actions. In this case $S \bowtie T$ is a monoid with identity $(0_S, 1_T)$.*

Here by commuting actions we mean $(ts)t' = t(st')$ for all $s \in S$ and $t, t' \in T$.

Again, if $S$ is written multiplicatively, then we shall use exponential notation for the actions of $T$ on $S$. It is often convenient, and conceptually useful, to view the multiplication formula for the two-sided semidirect product as an instance of matrix multiplication. That is, if we identify $(s,t)$ with the matrix $\begin{pmatrix} t & 0 \\ s & t \end{pmatrix}$, then we have the formula:

$$\begin{pmatrix} t & 0 \\ s & t \end{pmatrix} \begin{pmatrix} t' & 0 \\ s' & t' \end{pmatrix} = \begin{pmatrix} tt' & 0 \\ st' + ts' & tt' \end{pmatrix}.$$

Notice that $(S \bowtie T)^{op} \cong S^{op} \bowtie T^{op}$ and so the two-sided semidirect product is self-dual. The projection $\pi \colon S \bowtie T \to T$ is a homomorphism.

The two-sided analogue to the wreath product of monoids is the block product of monoids. If $X, Y$ are sets and $f \colon Y \times Y \to X$, it is often convenient to write $y_1 f y_2$ instead of $(y_1, y_2)f$ for the image of $(y_1, y_2)$.

**Definition 1.2.20 (Unitary block product).** *Let $M$ and $N$ be monoids. Then the (unitary) block product $M \,\square\, N$ of $M$ and $N$ is $M^{N \times N} \bowtie N$ where $n_1{}^n f^{n'} n_2 = n_1 n f n' n_2$.*

If $S$ and $T$ are semigroups, then we follow our prescribed method to define their block product.

**Definition 1.2.21 (Block product).** *Let $S$ and $T$ be semigroups. Then their block product $S \,\square\, T$ is the subsemigroup $S^{T^I \times T^I} \bowtie T$ of the unitary block product $S^I \,\square\, T^I$.*

Again, there is a projection $\pi \colon S \,\square\, T \to T$. Similarly to the case of the wreath product, there is an embedding theorem for the block product, whose proof we leave to the reader.

**Theorem 1.2.22 (Embedding theorem).** *Any two-sided semidirect product $S \bowtie T$ embeds in $S \,\square\, T$.*

**Exercise 1.2.23.** Prove Theorem 1.2.22.

### 1.2.5 Limits in FSgp

The following adjunction is very well-known:

$$
\mathbf{Set} \xrightarrow[\;X \longmapsto X^{+}\;]{\overset{\text{forgetful functor}}{\longleftarrow}} \mathbf{Sgp} \tag{1.23}
$$

where $X^{+}$ is the free semigroup generated by $X$.

Recall that an empty product in a category $C$ is a terminal object. In this book $\Delta$ will always denote a diagonal mapping, which one should be clear from the context. One way to look at finite products for a category $C$ is via the adjunction

$$
C \xrightarrow[\;\Delta\;]{\overset{\times}{\longleftarrow}} C \times C. \tag{1.24}
$$

That is, when the right adjoint to $\Delta$ exists, and there is a terminal object, we say that $C$ has finite products; see [185].

When the category $C$ has finite products, then given arrows $f \colon c_1 \to d_1$ and $g \colon c_2 \to d_2$, one can define $f \times g \colon c_1 \times c_2 \to d_1 \times d_2$ by considering the arrow $(p_1 f, p_2 g) : (c_1 \times c_2)\Delta \to (d_1, d_2)$. Indeed, from the adjunction we infer the existence of a unique arrow $f \times g$ making the following diagram commute:

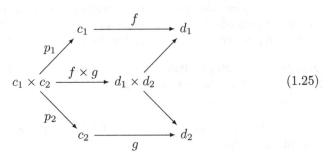

$$(1.25)$$

Of course, in the concrete categories of interest to us $(x, y)(f \times g) = (xf, yg)$, as usual.

It is easy to see (or could be deduced from (1.23) and abstract nonsense):

**Proposition 1.2.24.** *The category* **FSgp** *has all finite limits, and so a terminal object* $\{1\}$. *Moreover the underlying set (forgetful) functor preserves all limits.*

The product of $S_1$ and $S_2$ is $S_1 \times S_2$ endowed with pointwise multiplication. If $f : S_1 \to T_1$ and $g : S_2 \to T_2$, then $f \times g : S_1 \times S_2 \to T_1 \times T_2$ is given by $(s_1, s_2)(f \times g) = (s_1 f, s_2 g)$. We use $p_{S_i} : S_1 \times S_2 \to S_i$ for the projections, $i = 1, 2$. Note that **FSgp** also has an initial object $\emptyset$.

Thus the categorical product of arrows in **FSgp** agrees with the Universal Algebra notion of product of semigroups and morphisms. However, the category **FSgp** has a partial ordering stemming from the Prime Decomposition Theorem of Krohn and the first author [169], and also, of course, from Universal Algebra, via the work of Birkhoff.

### 1.2.6 Division and pseudovarieties of semigroups

Krohn and Rhodes introduced division into semigroup theory in order to obtain a wreath product decomposition theorem powerful enough to decompose the full transformation monoid [169].

**Definition 1.2.25 (Division of semigroups).** *Let* $S, T \in$ **Sgp**. *Then we say that* $S$ *divides* $T$, *written* $S \prec T$, *if there exists a subsemigroup* $T_1 \leq T$ *and a surjective homomorphism* $\varphi : T_1 \twoheadrightarrow S$.

The following proposition is easy to verify.

**Proposition 1.2.26.** *The pair* (**FSgp**, $\prec$) *is a partially ordered set (if semigroups are viewed up to isomorphism).*

In Universal Algebra $\mathbb{H}$, $\mathbb{S}$, $\mathbb{P}$, $\mathbb{P}_{\mathrm{fin}}$ denote the operators of closing a set of algebraic objects under, respectively: homomorphic images, subalgebras, arbitrary products and finite products. Notice then that closure under $\prec$ corresponds to the operator $\mathbb{HS}$.

**Exercise 1.2.27.** Prove Proposition 1.2.26.

**Exercise 1.2.28.** Prove that $(\mathbf{FSgp}, \prec)$ is neither a meet nor a join semilattice. Does it have a top or bottom?

**Exercise 1.2.29.** Show that if $S \prec S'$ and $T \prec T'$, then $S \wr T \prec S' \wr T'$.

The notion of a pseudovariety, due to Eilenberg and Schützenberger [85, 86], has become central to our subject. The language of pseudovarieties provides a framework into which most results in Finite Semigroup Theory fit. Of particular importance is Eilenberg's correspondence between pseudovarieties of semigroups and varieties of languages [85, 224], an aspect we barely touch upon in this text.

**Definition 1.2.30 (Pseudovariety of semigroups).** *A subset* $\mathbf{V} \subseteq \mathbf{FSgp}$ *is a pseudovariety of finite semigroups if* $\mathbf{V} = \mathbb{HSP}_{\mathrm{fin}}(\mathbf{V})$, *that is,* $\mathbf{V}$ *is closed under formation of finite products and taking divisors. Equivalently,* $\mathbf{V}$ *is a pseudovariety if and only if the following properties hold:*

- $\{1\} \in \mathbf{V}$
- $S_1, S_2 \in \mathbf{V}$ *implies* $S_1 \times S_2 \in \mathbf{V}$
- $T \in \mathbf{V}$ *and* $S \prec T$ *implies* $S \in \mathbf{V}$.

*We denote by* $\mathbf{PV}$ *the collection of all pseudovarieties of finite semigroups.*

*Remark 1.2.31.*

1. If $\mathbf{V}$ is a pseudovariety and $S \in \mathbf{V}$, then isomorphic copies of $S$ also belong to $\mathbf{V}$.
2. If $\mathbf{V}$ is a pseudovariety, then $\emptyset$ and $\{1\}$ belong to $\mathbf{V}$ as the empty product is the terminal object $\{1\}$ and $\emptyset \prec \{1\}$. In particular, there is no empty pseudovariety.
3. We shall see momentarily that $(\mathbf{PV}, \subseteq)$, with $\subseteq$ inclusion of subsets, is a complete algebraic lattice. We often write $\mathbf{V} \leq \mathbf{W}$ to mean $\mathbf{V} \subseteq \mathbf{W}$.

Pseudovarieties of monoids are defined similarly, only one considers divisions of monoids (so a monoid $M$ is a divisor of a monoid $N$ if it is a quotient of a submonoid of $N$). There is an important Galois connection relating pseudovarieties of monoids and pseudovarieties of semigroups. Let $\mathbf{MPV}$ be the complete lattice of monoid pseudovarieties. Then there is a Galois connection

$$\mathbf{PV} \xrightarrow[\mathbf{V} \longmapsto \mathbf{V} \cap \mathbf{FMon}]{\overset{\mathbb{L}\mathbf{V} \longleftarrow\!\!\!| \, \mathbf{V}}{\phantom{xxxxxxxxxxxxxx}}} \mathbf{MPV} \qquad (1.26)$$

where $\mathbb{L}\mathbf{V}$ is the pseudovariety of all semigroups $S$ so that the *local monoid* (or *localization*) $eSe$ of $S$ at $e$ belongs to $\mathbf{V}$ for each idempotent $e$ of $S$. Note that $(-) \cap \mathbf{FMon}$ also has a left adjoint sending a pseudovariety of monoids to the pseudovariety of semigroups it generates. The map $\mathbb{L}$ is continuous and hence is a CL-morphism in the sense of [97] (see Chapter 6).

If $\mathbf{V}$ and $\mathbf{W}$ are pseudovarieties of semigroups, then their *semidirect product* $\mathbf{V} * \mathbf{W}$ is the pseudovariety generated by all semidirect products of the form $V \rtimes W$ with $V \in \mathbf{V}$ and $W \in \mathbf{W}$.

**Exercise 1.2.32.** Show that $\mathbf{V} * \mathbf{W}$ consists of all divisors of semidirect products $V \rtimes W$ with $V \in \mathbf{V}$ and $W \in \mathbf{W}$.

**Exercise 1.2.33.** Show that $\mathbf{V} * \mathbf{W}$ consists of all divisors of wreath products $V \wr W$ with $V \in \mathbf{V}$ and $W \in \mathbf{W}$.

Similarly, if $\mathbf{V}$ and $\mathbf{W}$ are pseudovarieties of semigroups, their *two-sided semidirect product* $\mathbf{V} ** \mathbf{W}$ is the pseudovariety generated by all two-sided semidirect products of the form $V \bowtie W$ where $V \in \mathbf{V}$ and $W \in \mathbf{W}$.

**Exercise 1.2.34.** Show that $\mathbf{V} ** \mathbf{W}$ consists of all divisors of two-sided semidirect products $V \bowtie W$ where $V \in \mathbf{V}$ and $W \in \mathbf{W}$, or equivalently, of all divisors of block products $V \square W$ with $V \in \mathbf{V}$ and $W \in \mathbf{W}$.

A semigroup $S$ is said to be a *subdirect product* of $T_1$ and $T_2$, written $S \ll T_1 \times T_2$, if $S$ is a subsemigroup of $T_1 \times T_2$ mapping onto both $T_1$ and $T_2$ via the projections. For instance, $S\Delta \ll S \times S$. If $\equiv_i$, $i = 1, 2$, are congruences on a semigroup $S$ with trivial intersection, then $S \ll S/\equiv_1 \times S/\equiv_2$.

**Exercise 1.2.35.** Verify that if $S \ll T_1 \times T_2$, then $S$ belongs to a pseudovariety $\mathbf{V}$ if and only if $T_1, T_2 \in \mathbf{V}$.

### 1.2.7 Terminology from lattice theory

In Chapter 6 we discuss in detail the terminology concerning complete lattices, closure and kernel operators, algebraic and continuous lattices, etc., used here. We review some of the relevant definitions now. The reader can refer to Chapter 6, via the index, for more information or to [97]. For a complete lattice $L$ we use the notation $\vee_{\mathrm{det}}$ for the join operation determined in the natural way by the infinite meet. A dual notation is used for the determined meet from an infinite join. More precisely,

$$\bigvee_{\mathrm{det}} X = \bigwedge \{y \mid \forall x \in X,\ x \leq y\}.$$

If $L$ is a lattice, then $K(L)$ is used to denote the set of compact elements of $L$. The following results from [97] constitute Theorem 6.3.15 and Corollary 6.3.16. They shall be used frequently in the next chapter.

**Theorem 1.2.36 (Continuous closure operator theorem).** *Let $L$ be an algebraic lattice and suppose that $c\colon L \to L$ is a continuous closure operator. Then $c(L)$ is an algebraic lattice where the meet is induced from $L$ and the join is determined. Moreover, $K(c(L)) = c(K(L))$.*

**Corollary 1.2.37.** *Let $L_1$ be an algebraic lattice. If $d\colon L_1 \twoheadrightarrow L_2$ is a surjective* **sup** *morphism such that the right adjoint $g\colon L_2 \hookrightarrow L_1$ of $d$ is continuous, then $L_2$ is an algebraic lattice and $d(K(L_1)) = K(L_2)$.*

We shall adopt the notation $(S)$ for $\mathbb{HSP}_{\mathrm{fin}}(\{S\})$, when $S$ is a finite semigroup; so $(S)$ is the pseudovariety generated by $S$.

**Proposition 1.2.38.** *The 4-tuple $(2^{\mathbf{FSgp}}, \subseteq, \cup, \cap)$ is a complete algebraic lattice. Moreover, there is a Galois connection*

$$(2^{\mathbf{FSgp}} \subseteq, \cup, \cap) \underset{\mathbb{HSP}_{\mathrm{fin}}}{\overset{i}{\longleftrightarrow}} (\mathbf{PV}, \subseteq, \vee_{\mathrm{det}}, \cap) \tag{1.27}$$

*where $i$ is the inclusion. So $i$ is* **inf** *and $\mathbb{HSP}_{\mathrm{fin}}$ is* **sup.** *Moreover, $i$ is continuous. The determined join is given by:*

$$\bigvee_{\mathrm{det}} (\{\mathbf{V}_a \mid a \in A\}) = \mathbb{HSP}_{\mathrm{fin}}\left(\bigcup \mathbf{V}_a\right)$$

$$= \{T \mid \exists S_1, \ldots, S_n \in \bigcup \mathbf{V}_a, \ T \prec S_1 \times \cdots \times S_n\}.$$

*Hence $(\mathbf{PV}, \subseteq, \vee_{\mathrm{det}}, \cap)$ is an algebraic lattice with the compact elements the finitely generated, or equivalently one-generated, pseudovarieties.*

*Proof.* The first part is clear. The second part is a consequence of Corollary 1.2.37, the fact that the finite subsets are the compact elements of $2^{\mathbf{FSgp}}$ and

$$\mathbb{HSP}_{\mathrm{fin}}(\{S_1, \ldots, S_n\}) = \mathbb{HSP}_{\mathrm{fin}}(\{S_1 \times \cdots \times S_n\}) = (S_1 \times \cdots \times S_n).$$

This completes the proof of the proposition.   □

**Exercise 1.2.39.** Show that $S \in \mathbf{V} \vee \mathbf{W}$ if and only if $S$ is a quotient of a subdirect product $T \ll V \times W$ with $V \in \mathbf{V}$ and $W \in \mathbf{W}$.

**Exercise 1.2.40.** Use Exercise 1.2.13 to show that

$$(\mathbf{U} \vee \mathbf{V}) * \mathbf{W} = (\mathbf{U} * \mathbf{W}) \vee \mathbf{V} * \mathbf{W}.$$

Deduce that the operator $(-) * \mathbf{W}$ is $\mathbf{sup}_{\mathrm{B}}$, that is, it preserves all non-empty sups.

*Remark 1.2.41.* We will see that the set **PV** has lots and lots of structure. We just saw that **PV** is a complete algebraic lattice, a line of development that will be continued in Chapter 7. We can also turn **PV** into a topological space, an aspect that will be developed in Chapter 6. Various associative and non-associative multiplications have been introduced on **PV**, such as the semidirect, two-sided semidirect and Mal'cev products (see [85, 126, 293, 364]), and of course the lattice multiplications ∩ and $\vee_{\mathrm{det}}$. This line of development is continued in Chapter 2, and then again in Chapters 6–8.

Finally, via the q-theory, continuous self-maps of **PV** are introduced, leading to algebraic and continuous lattices (in the sense of [97]), topological semigroups (under composition) and various types of quantales (complete lattices with associative multiplications) [303]. "Coordinate systems" for continuous operators will be pseudovarieties of relational morphisms, created by B. Tilson and the second author (with some influence of the first author), which we shall proceed to describe shortly.

Also questions of decidability and undecidability for all the above can be considered; see [2, 36, 281, 285].

All this leads to a mélange of Logic, Category Theory, Universal Algebra, Algebraic and Continuous Lattices, Topology and Topological Algebra, Quantales, etc.

The continuous lattice and quantale structures lead to "abstract spectral theory," which is developed to some extent in Chapter 7 for important collections of operators on **PV**. This "abstract spectral theory," similar to that of the commutative rings of algebraic geometry (Zariski spectrum) or of $C^*$-algebras (actual spectral theory for operators on separable Hilbert spaces), leads us closer to quantum ideas, hence the q of q-theory!

The complexity of two continuous operators will be developed in Chapter 5, generalizing the group complexity of finite semigroups, defining the two-sided complexity of finite semigroups as well as generalizing other hierarchies such as dot-depth. In fact we generalize and place in context virtually all known definitions of complexity in the theories of finite automata, finite semigroups, and regular languages.

So let us turn to defining our basic notions. This is where the new material begins!

## 1.3 Relational Morphisms

### 1.3.1 The category FSgp with arrows relational morphisms

Let $S, T$ be finite semigroups. A *relational morphism* $\varphi \colon S \to T$ is a function $\varphi \colon S \to 2^T$ satisfying the following properties:

- $s\varphi \neq \emptyset$, for all $s \in S$ (i.e., $\varphi$ is fully defined);
- $s_1\varphi s_2\varphi \subseteq (s_1 s_2)\varphi$, for $s_1, s_2 \in S$.

Alternatively, a relational morphism $\varphi\colon S \to T$ is a relation from $S$ to $T$ such that the *graph*

$$\#\varphi = \{(s,t) \mid t \in s\varphi\} \leq S \times T$$

is a subsemigroup of $S \times T$ projecting *onto* $S$. The *image* of $\varphi$ is the semigroup

$$\#\varphi p_T = \{t \in T \mid \exists s \in S,\ t \in s\varphi\}$$

denoted Im $\varphi$. Recall that $p_S$ and $p_T$ are the projections from $S \times T$ to $S$ and $T$, respectively.

We can compose relational morphisms as binary relations. So if $\varphi\colon S \to T$ and $\psi\colon T \to U$, then $\varphi\psi$ is defined by

$$u \in s\varphi\psi \text{ if there exists } t \in s\varphi \text{ such that } u \in t\psi.$$

Composition of relations is well-known to be associative.

**Exercise 1.3.1.** Verify that the composition of two relational morphisms is again a relational morphism.

If $\varphi\colon S \to T$ is a relational morphism, then $p_S^{-1}p_T$

$$\begin{array}{ccc} \#\varphi & \xrightarrow{\ p_T\ } & T \\ {\scriptstyle p_S}\downarrow & \diagup\ {\scriptstyle \varphi} & \\ S & & \end{array} \tag{1.28}$$

is termed the *canonical factorization* of $\varphi$.

Suppose that one has a diagram

$$\begin{array}{ccc} U & \xrightarrow{\ \beta\ } & T \\ {\scriptstyle \alpha}\downarrow & \diagup\ {\scriptstyle \alpha^{-1}\beta} & \\ S & & \end{array} \tag{1.29}$$

where $\beta$ and $\alpha$ are homomorphisms, the latter onto. Then there is a natural map $\Delta(\alpha \times \beta)\colon U \to S \times T$. Let $R$ be the image, so we have

$$R = \{(s,t) \in S \times T \mid \exists u \in U \text{ so that } u\alpha = s, u\beta = t\}$$

Then $R$ is the graph of the relational morphism $\alpha^{-1}\beta$ and there is a diagram

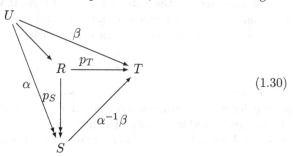

$$\tag{1.30}$$

with the bottom-right triangle the canonical factorization of the relational morphism $\alpha^{-1}\beta\colon S \to T$.

In this way, relational morphisms are "adding arrows" to the category **FSgp** by "turning around surmorphisms" (a homotopy-theoretic idea [257]).

The category (**FSgp,RM**) is then defined as the category with objects finite semigroups and with arrows from $S$ to $T$ consisting of relational morphisms, that is, $\mathsf{Hom}(S,T) = \{\varphi\colon S \to T \mid \varphi \text{ is a relational morphism}\}$. The initial object is still $\emptyset$ and the terminal object is still $\{1\}$. Note that $\emptyset \to T$ is the unique arrow from $\emptyset$ to $T$. The existence of an arrow $S \to \emptyset$ implies $S = \emptyset$.

*Convention 1.3.2 (Morphism vs. relational morphism).* Initially there was some resistance to the idea of relational morphism being the main type of arrow between finite semigroups. Now it has become an accepted tool and so is worthy of the abbreviation morphism. Unfortunately, the word *homomorphism* is a bit unwieldy and so it is convenient to abbreviate it to *morphism*. For this reason, in the current volume the word *morphism* unmodified, that is not preceded by the word *relational*, will mean a homomorphism.

### 1.3.2 Weak products and pullbacks

Unlike the category **FSgp** with arrows homomorphisms, the category (**FSgp, RM**) does not have (finite) products. It does have "weak products" (see [185, Chapter X.2]) in the sense that in (1.24), for $C = (\mathbf{FSgp}, \mathbf{RM})$, there is a functor $C \times C \to C$, which we also denote $\times$, that has a weak universal arrow with respect to $\Delta$ and, among all these weak universal arrows on $C$, there is a unique largest one (in the sense defined below).

On objects, the weak product on (**FSgp, RM**) agrees with the product of **FSgp**; that is, the weak product of $S$ and $T$ is $S \times T$. For relational morphisms $\varphi_1\colon S_1 \to T_1$ and $\varphi_2\colon S_2 \to T_2$, their weak product is the relational morphism $\varphi_1 \times \varphi_2\colon S_1 \times S_2 \to T_1 \times T_2$ given by

$$(s_1, s_2)(\varphi_1 \times \varphi_2) = s_1\varphi_1 \times s_2\varphi_2.$$

Notice that if $\varphi_1, \varphi_2$ are homomorphisms, then their weak product $\varphi_1 \times \varphi_2$ coincides with their product in **FSgp**.

The hom sets of the category $C = (\mathbf{FSgp}, \mathbf{RM})$ have the structure of a partially ordered set given by inclusion (of the graphs). More precisely if $f, g\colon S \to T$ are relational morphisms, we write

$$f \subseteq_s g \iff \#f \subseteq \#g. \tag{1.31}$$

This means that, for each $s \in S$, $sf \subseteq sg$. The hom set $C(S,T)$ is then a join semilattice with maximum. The join of $f$ and $g$ has graph the subsemigroup of $S \times T$ generated by $\#f, \#g$. The maximum is the "*universal relation*" given by $s\varphi = T$ for all $s \in S$. Moreover, the partial order $\subseteq_s$ is compatible with multiplication, that is, $f \subseteq_s f'$ and $g \subseteq_s g'$ implies $fg \subseteq_s f'g'$ when the composition is defined. It is with respect to this ordering that the weak universal arrow of the weak product is maximum. More precisely, if $\varphi_1 : S \to T_1$ and $\varphi_2 : S \to T_2$ are relational morphisms, then $\Delta(\varphi_1 \times \varphi_2) : S \to T_1 \times T_2$ is the largest relational morphism $\varphi : S \to T_1 \times T_2$, in this ordering, such that $\varphi p_{T_1} = \varphi_1$ and $\varphi p_{T_2} = \varphi_2$, i.e., such that the diagram

commutes, for $i = 1, 2$.

**Exercise 1.3.3.** Verify the above assertion.

Abusing language, this weak product will often simply be called the *product*.

We should mention that there is a more general ordering on relational morphisms. Namely if $f : S \to T$ and $g : S' \to T'$ are relational morphisms with $S' \leq S$, $T' \leq T$, then we write

$$g \subseteq f \iff \#g \subseteq \#f. \tag{1.32}$$

Note that the subscript $s$ in the notation $g \subseteq_s f$ indicates $f$ and $g$ are coterminal, and often we write it when we want to emphasize this fact.

Similarly, $C = (\mathbf{FSgp}, \mathbf{RM})$ has weak pullbacks. That is if $f_1 : S_1 \to T$ and $f_2 : S_2 \to T$ are relational morphisms, then the *weak pullback* (which we usually just call the *pullback*) is the semigroup:

$$S_1 \times_{f_1, f_2} S_2 = \{(s_1, s_2) \in S_1 \times S_2 \mid s_1 f_1 \cap s_2 f_2 \neq \emptyset\}. \tag{1.33}$$

There are natural projections $p_{S_i} : S_1 \times_{f_1, f_2} S_2 \to S_i$, $i = 1, 2$ such that, for all $x \in S_1 \times_{f_1, f_2} S_2$, $x p_{S_1} f_1 \cap x p_{S_2} f_2 \neq \emptyset$. There is the following not necessarily commutative diagram:

$$
\begin{array}{ccc}
S_1 \times_{f_1, f_2} S_2 & \xrightarrow{\ p_{S_2}\ } & S_2 \\
\Big\downarrow{\scriptstyle p_{S_1}} & & \Big\downarrow{\scriptstyle f_2} \\
S_1 & \xrightarrow[\ f_1\ ]{} & T.
\end{array}
\tag{1.34}
$$

Moreover, if $g_i : U \to S_i$, $i = 1, 2$, are such that $u g_1 f_1 \cap u g_2 f_2 \neq \emptyset$ for all $u \in U$, then there is a unique largest relational morphism $g : U \to S_1 \times_{f_1, f_2} S_2$ such that $g p_{S_i} \leq g_i$, $i = 1, 2$. This relational morphism is given by

$$u \longmapsto \{(s_1, s_2) \in S_1 \times_{f_1, f_2} S_2 \mid s_1 \in u g_1, s_2 \in u g_2\}. \tag{1.35}$$

*Remark 1.3.4.* The diagram (1.34) does commute for homomorphisms. The notion of a commuting diagram has several possible generalizations to the relational setting. One possibility is that going one way around the diagram is smaller than going the other way. The weakest possible generalization of commuting is that the two possible ways of tracing the diagram have a common lower bound (in the poset of coterminal relational morphisms). This is the situation here.

There is also a relational morphism $f_1 \times_T f_2 \colon S_1 \times_{f_1,f_2} S_2 \to T$ given by

$$(s_1, s_2)(f_1 \times_T f_2) = s_1 f_1 \cap s_2 f_2. \tag{1.36}$$

One sometimes calls $f_1 \times_T f_2$ the *pullback* of $f_1$ and $f_2$ (along $T$). For homomorphisms, things reduce to the usual notion of pullbacks.

Now we turn to some examples that will be important later on. Let $f \colon S \to T$ and $g \colon U \to V$ be relational morphisms. Consider $f' \colon S \to T \times V$ and $g' \colon U \to T \times V$ given by

$$f' = f p_T^{-1} \text{ and } g' = g p_V^{-1}.$$

Then one easily checks:

$$f \times g = f' \times_{(T \times V)} g'. \tag{1.37}$$

Let $f \colon S \to T$ be a relational morphism. Observe that

$$S \times_{f, 1_T} T = \{(s,t) \in S \times T \mid t \in sf\} = \#f \text{ and } f \times_T 1_T = p_T. \tag{1.38}$$

More generally, $f \times_T g$ is the graph of the relation $fg^{-1}$.

One can define the pullback of an arbitrary family of maps $\{f_i \colon S_i \to T\}$ in a similar fashion. We use the notation $\prod_T f_i$, or the notation $f_1 \times_T \cdots \times_T f_n$ if the family is finite. The pullback of an empty family is $1_{\{1\}}$. A class of relational morphisms closed under pullbacks of pairs and of the empty family can easily be shown to be closed under all finite pullbacks.

### 1.3.3 Divisions

An important subcategory of (**FSgp**, **RM**) has as object set the collection of all finite semigroups, but the only arrows allowed are divisions.

**Definition 1.3.5 (Division).** *A relational morphism $d \colon S \to T$ of semigroups is a division if and only if in the canonical factorization*

$$\#d \xrightarrow{\ p_T\ } T$$
$$p_S \downarrow \quad \nearrow d$$
$$S$$

$p_T$ *is injective, or equivalently, for all* $s_1, s_2 \in S$, *the implication*

$$s_1 d \cap s_2 d \neq \emptyset \implies s_1 = s_2$$

*holds.*

It is easy to verify that the composition of divisions results in a division. Of course the identity map of a finite semigroup is a division. The resulting subcategory is denoted $(\mathbf{FSgp}, \mathbf{Div})$.

Notice that if $d\colon S \to T$ is a division, then $S \prec T$; in fact $S \prec T$ if and only if there exists a division $d\colon S \to T$. The pre-ordered set $(\mathbf{FSgp}, \prec)$ (where isomorphic semigroups are not identified) is then the quotient of $(\mathbf{FSgp}, \mathbf{Div})$ obtained by identifying coterminal arrows.

*Remark 1.3.6.* Note that if $d\colon S \to T$ is a division, then $\#d$ is a subsemigroup of $T$ in a natural way and $p_S\colon \#d \twoheadrightarrow S$ is a surjective homomorphism. Observe that $p_S$ and $d$ "go in different directions," which can cause conceptual confusion if one is not careful. Compare with, for instance, the definitions of divisions and covers of transformations semigroups in [85].

**Exercise 1.3.7.** Show that if $d, d'$ are divisions, then $d \times d'$ is also a division.

### 1.3.4 Relational morphisms between relational morphisms

We next want to introduce arrows between the arrows of $(\mathbf{FSgp}, \mathbf{RM})$ leading to the category $\mathbf{RM}$. The following definitions, due to Tilson and the second author [339], need care when extended to categories and semigroupoids.

Here, by definition, if $f\colon S \to T$ and $g\colon U \to V$ are relational morphisms, then a *relational morphism* from $f$ to $g$ is a pair $(\alpha, m)$, where $\alpha\colon S \to U$ is a relational morphism and $m\colon T \to V$ is a homomorphism such that

$$
\begin{array}{ccc}
S & \xrightarrow{\;\alpha\;} & U \\
{\scriptstyle f}\big\downarrow & \subseteq & \big\downarrow{\scriptstyle g} \\
T & \xrightarrow[\;m\;]{} & V
\end{array}
\qquad (1.39)
$$

meaning $fm \subseteq \alpha g$. A diagram such as (1.39) with $fm \subseteq \alpha g$ will be called *subcommutative*.

By definition the pair $(\alpha, m)$ is an arrow of $\mathbf{RM}$ from $f$ to $g$ (sometimes written $(\alpha, m)\colon f \to g$). Composition of arrows $(\alpha, m)\colon f \to g$ and $(\alpha', m')\colon g \to h$ is given by $(\alpha\alpha', mm')$. The subcommutative diagram

$$
\begin{array}{ccccc}
S & \xrightarrow{\;\alpha\;} & U & \xrightarrow{\;\alpha'\;} & W \\
{\scriptstyle f}\big\downarrow & \subseteq & {\scriptstyle g}\big\downarrow & \subseteq & \big\downarrow{\scriptstyle h} \\
T & \xrightarrow[\;m\;]{} & V & \xrightarrow[\;m'\;]{} & Z
\end{array}
\qquad (1.40)
$$

shows that the composition of arrows is well defined. Clearly, composition is associative. The identity at $f\colon S \to T$ is the pair $(1_S, 1_T)\colon f \to f$. Thus $\mathbf{RM}$ is a category. A relational morphism (1.39) is called *strong* if $m$ is onto. The subcategory $\mathbf{SRM}$ of $\mathbf{RM}$ consisting of strong relational morphisms will also be of some interest.

Given arrows $(\alpha, m)\colon f \to g$ and $(\alpha', m')\colon f' \to g'$, one can form an arrow $(\alpha \times \alpha', m \times m')\colon f \times f' \to g \times g'$. This arrow is denoted $(\alpha, m) \times (\alpha', m')$ and is often called the *product* of $(\alpha, m)$ and $(\alpha', m')$. However, as was the case in $(\mathbf{FSgp}, \mathbf{RM})$, the assignment

$$
\begin{aligned}
(f, g) &\longmapsto f \times g \\
((\alpha, m), (\alpha', m')) &\longmapsto (\alpha, m) \times (\alpha', m')
\end{aligned}
\tag{1.41}
$$

constitutes the unique maximum weak product in $\mathbf{RM}$ (where we define $(\alpha, m) \leq (\alpha', m')\colon f \to g$ if $\alpha \leq \alpha'$ in the previous sense and $m = m'$, that is, we use the product ordering).

**Proposition 1.3.8.** *The category $\mathbf{RM}$ has initial object $1_\emptyset$, terminal object $1_{\{1\}}$ and weak product as in (1.41) (so the weak product agrees on objects with the weak product of Section 1.3.2). Thus the empty product of relational morphisms is $1_{\{1\}}$.*

*Proof.* Exercise. □

An important subcategory is the full subcategory of $\mathbf{RM}$ whose objects are homomorphisms. The further subcategory in which all the arrows of (1.39) are homomorphisms is $\mathbf{FSgp}^2$ as per [185].

### 1.3.5 Divisions between relational morphisms

We next introduce divisions between relational morphisms by declaring a relational morphism of relational morphisms $(\alpha, m)\colon f \to g$ to be a *division* if $\alpha$ is a division. If there exists a division from the relational morphism $f\colon S \to T$ to the relational morphism $g\colon U \to V$, we write $f \prec g$ and say that $f$ *divides* $g$. So $f \prec g$ if and only if there is a subcommutative diagram

$$
\begin{array}{ccc}
S & \xrightarrow{\ d\ } & U \\
{\scriptstyle f}\big\downarrow & \subseteq & \big\downarrow{\scriptstyle g} \\
T & \xrightarrow[\ m\ ]{} & V
\end{array}
\tag{1.42}
$$

with $d$ a division and $m$ a homomorphism. A frequently arising special case is when $d = 1_S$, in which case we have a subcommutative triangle

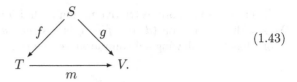

$$(1.43)$$

Another important instance of the notion of division is when $f \subseteq g$. Indeed, suppose $f \colon S \to T$ and $g \colon S' \to T'$ with $f \subseteq g$; hence $S \leq S'$ and $T \leq T'$. Then we have the subcommutative diagram

$$
\begin{array}{ccc}
S & \hookrightarrow & S' \\
f \downarrow & \subseteq & \downarrow g \\
T & \hookrightarrow & T'
\end{array}
$$

where the horizontal maps are the inclusions.

Clearly, the composition of divisions is a division.

**Proposition 1.3.9.** *Suppose* $(d, m) \colon f \to g$ *and* $(d', m') \colon g \to h$ *are divisions. Then* $(dd', mm')$ *is a division.*

*Proof.* We already know $(dd', mm')$ is a relational morphism. Because $dd'$ is a division, it follows $(dd', mm')$ is a division. $\qquad\square$

There is then a subcategory of **RM**, which we denote by $(\mathbf{RM}, \mathbf{Div})$ of **RM**, with the same object set, but now only divisions are permitted as arrows.

Let us remark one could define quotient and injective arrows in **RM** and show divisions are precisely the inverses of quotients followed by injective arrows, but we have no need to do this for our purposes.

**Exercise 1.3.10.** Show that the product $(d, m) \times (d', m')$ of divisions $(d, m)$ and $(d', m')$ of relational morphisms is again a division of relational morphisms.

The following fundamental fact about divisions of relational morphisms will be used throughout, often without comment.

**Fact 1.3.11.** *Let* $f \prec g$ *be a division as per* (1.42). *Then if* $W \leq T$, *we have* $W f^{-1} \prec W m g^{-1}$.

*Proof.* Define a relational morphism $d' \colon W f^{-1} \to W m g^{-1}$ by

$$\#d' = \#d \cap (W f^{-1} \times W m g^{-1}).$$

In other words, $sd' = sd \cap W m g^{-1}$ for $s \in W f^{-1}$. To show that $d'$ is a relational morphism, it suffices to show that it is fully defined. If $s \in W f^{-1}$, then there exists $w \in W$ such that $w \in sf$. Hence $wm \in sfm \subseteq sdg$. Thus there exists $u_s \in sd$ such that $wm \in u_s g$. Therefore $(s, u_s) \in \#d'$ and so $d'$ is fully defined. Because $d$ is a division, it follows that $d' \subseteq d$ is a division. $\qquad\square$

There is also a more restrictive notion of division that is frequently useful. We say that a division $(d, m)\colon f \to g$ is a *strong division* if $m$ is **onto**, that is, one has the following subcommutative diagram:

$$
\begin{array}{ccc}
S & \xrightarrow{\;d\;} & U \\
f \downarrow & \subseteq & \downarrow g \\
T & \xrightarrow[m]{} & V.
\end{array}
\tag{1.44}
$$

In this case we write $f \prec_s g$. Of course, $f \prec_s g$ implies $f \prec g$. Notice that a composition of strong divisions is again a strong division, so that one obtains a subcategory $(\mathbf{RM}, \mathbf{SDiv})$ of $(\mathbf{RM}, \mathbf{Div})$.

A commonly occurring special case is a commutative triangle

$$
\begin{array}{ccc}
S & \xrightarrow{\;d\;} & S' \\
 & f \searrow \quad \swarrow g & \\
 & T &
\end{array}
\tag{1.45}
$$

with $d$ a division. That is, $dg \prec_s g$ for any division $d$. The special case $f \subseteq_s g$ will be used frequently as well. Another case of particular interest is that of a subcommutative triangle

$$
\begin{array}{ccc}
 & S & \\
 f \swarrow & & \searrow g \\
 T & \xrightarrow[m]{} & V
\end{array}
\tag{1.46}
$$

with $m$ an onto morphism.

We now motivate the definition of $\prec$ for relational morphisms. The article [339] provides further motivation. A pseudovariety $\mathbf{V}$ is said to be *locally finite* if, for each finite set $A$, $\mathbf{V}$ contains a relatively free semigroup on $A$ denoted $F_{\mathbf{V}}(A)$. That is, there is a biggest element of $\mathbf{V}$ generated by any given finite set $A$. For example, compact pseudovarieties, being generated by a single finite semigroup, are locally finite by a well-known theorem of Birkhoff. Indeed, the free object on $A$ in $(S)$ is the subsemigroup of $S^{S^A}$ generated by the $S^A$-tuples $(a\sigma)_{\sigma \in S^A}$ with $a \in A$.

**Exercise 1.3.12.** Verify the free object on $A$ in $(S)$ is the subsemigroup of $S^{S^A}$ generated by $(a\sigma)_{\sigma \in S^A}$ with $a \in A$. In particular if $S$ and $A$ are finite, then this free object is finite.

This implies that given a finite semigroup $S$ and a finite semigroup $T$, one can decide whether $S$ belongs to the pseudovariety generated by $T$. One just needs to check whether $S$ divides $T^{T^S}$. This algorithm is doubly exponential. Jackson and McKenzie in fact showed that membership in the pseudovariety generated by a finite semigroup can be NP-hard [145].

*Remark 1.3.13.* The definition of division was designed for the following reason: suppose $g\colon S \to T$ is a relational morphism with $S$ generated by $A$ (which we view as a subset of $S$). Suppose $T$ belongs to a locally finite pseudovariety $\mathbf{W}$. For each $a \in A$, choose $t_a \in ag$. Then there is a homomorphism $m\colon F_{\mathbf{W}}(A) \to T$ induced by sending $a$ to $t_a$. Let $\rho_{\mathbf{W}}\colon S \to F_{\mathbf{W}}(A)$ be the relational morphism whose graph is the subsemigroup generated by the image of $\Delta\colon A \to S \times F_{\mathbf{W}}(A)$. Then there is a subcommutative triangle

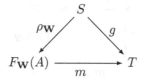

and so $\rho_{\mathbf{W}} \prec g$ (cf. (1.43)). *This argument and its profinite analogue are at the heart of the equational theory in Chapter 3.*

The division discussed in Remark 1.3.13 is so fundamental that we record it as a lemma. Let us first make a definition.

**Definition 1.3.14 (Canonical relational morphism).** *If $S$ and $T$ are $A$-generated semigroups, the canonical relational morphism $f\colon S \to T$ is the relational morphism whose graph is $\langle A\Delta \rangle \leq S \times T$; so the graph of $f$ is generated by all pairs $(a, a)$, with $a \in A$.*

**Lemma 1.3.15.** *Let $\mathbf{W}$ be a locally finite pseudovariety and let $g\colon S \to T$ be a relational morphism with $S$ an $A$-generated semigroup and $T \in \mathbf{W}$. Let $\rho_{\mathbf{W}}\colon S \to F_{\mathbf{W}}(A)$ be the canonical relational morphism. Then $\rho_{\mathbf{W}} \prec g$.*

The reason for the definition of $f \prec_s g$ is similar, but we want to guarantee that the codomain of $g$ divides the codomain of $f$.

**Proposition 1.3.16.** *The following results hold:*

*(a)* $\prec$, $\prec_s$ *are transitive and reflexive relations. Moreover,* $\prec_s$ *is anti-symmetric (up to isomorphism). However,* $\prec$ *is not anti-symmetric.*

*(b)* *Suppose $f$ and $g$ are relational morphisms with non-empty domains. Then $f, g \prec f \times g$. On the other hand, if $f$ and $g$ have a common domain, $\Delta(f \times g) \prec_s f, g$.*

*(c)* *Suppose that $f\colon S_1 \to T$ and $g\colon S_2 \to T$. Then $f \times_T g \prec f \times g$.*

*Proof.* We start with (a). Transitivity and reflexivity are evident. As to anti-symmetry of $\prec_s$, suppose that $f\colon S \to T$ and $f'\colon S' \to T'$ are such that $f \prec_s f'$ and $f' \prec_s f$. Then $S \prec S'$, $S' \prec S$, $T \prec T'$, and $T' \prec T$, whence

$S \cong S'$ and $T \cong T'$. Consider a strong division $(d, m): f \to g$. By cardinality considerations, $d$ and $m$ must be isomorphisms (the latter assertion uses that $m$ is onto).

We proceed to show that $\prec$ is not anti-symmetric. Indeed, suppose that $S, T, T'$ are finite semigroups with $T \not\cong T'$. Let $f: S \to T$ and $g: S \to T'$ be the universal relations (that is, $\#f = S \times T$ and $\#g = S \times T'$). Then it is easy to see that $f \prec g$ and $g \prec f$ although $f$ and $g$ are not isomorphic.

For (b), suppose $f: S_1 \to T_1$ and $g: S_2 \to T_2$. Let $e$ be an idempotent of $S_2 g$ and let $m: T_1 \to T_1 \times T_2$ be given by $t \mapsto (t, e)$. Then we have the following subcommutative diagram:

$$
\begin{array}{ccc}
S_1 & \xrightarrow{\; p_{S_1}^{-1} \;} & S_1 \times S_2 \\
{\scriptstyle f} \downarrow & & \downarrow {\scriptstyle f \times g} \\
T_1 & \xrightarrow[\; m \;]{} & T_1 \times T_2
\end{array}
$$

showing that $f \prec f \times g$. Indeed, if $e \in s_2 g$ and $t_1 \in s_1 f$, then one has $t_1 m = (t_1, e) \in (s_1, s_2)(f \times g)$, yielding the desired subcommutativity. The division $g \prec f \times g$ is dual.

Suppose now that $S_1 = S_2 = S$. Then the commutative triangle

$$
\begin{array}{ccc}
 & S & \\
{\scriptstyle \Delta(f \times g)} \swarrow & & \searrow {\scriptstyle f} \\
T_1 \times T_2 & \xrightarrow[\; p_{T_1} \;]{} & T_1
\end{array}
$$

establishes that $\Delta(f \times g) \prec_s f$. The proof that $\Delta(f \times g) \prec_s g$ is dual.

To prove (c), note that the following subcommutative diagram shows that $f \times_T g \prec f \times g$

$$
\begin{array}{ccc}
S_1 \times_{f,g} S_2 & \hookrightarrow & S_1 \times S_2 \\
{\scriptstyle f \times_T g} \downarrow & \subseteq & \downarrow {\scriptstyle f \times g} \\
T & \xrightarrow[\; \Delta \;]{} & T \times T
\end{array}
$$

where subcommutativity follows as $(s_1, s_2)(f \times_T g) = s_1 f \cap s_2 g$, and so if $t \in (s_1, s_2)(f \times_T g)$, then $(t, t) \in (s_1, s_2)(f \times g)$.                □

*Remark 1.3.17.* If $f: S \to T$ and $g: S' \to T'$ are relational morphisms, it is not in general true that $f \prec_s f \times g$. In fact, in order for this to occur, $T$ must map onto $T \times T'$ and so $T'$ must be trivial.

Proposition 1.3.16(a) shows $\prec$ is a preorder. This leads to the following notion of equivalence.

**Definition 1.3.18.** *If $f, g \in \mathbf{RM}$ are such that $f \prec g$ and $g \prec f$, then we write $f \sim g$ and say that $f$ and $g$ are divisionally equivalent.*

Hence we have the pre-ordered set $(\mathbf{RM}, \prec)$. This is the quotient of $(\mathbf{RM}, \mathbf{Div})$ obtained by identifying coterminal arrows.

**Exercise 1.3.19.** Show that if $f_1 \prec_s g_1$ and $f_2 \prec_s g_2$, then $f_1 \times f_2 \prec_s g_1 \times g_2$. Prove a similar result for $\prec$.

## Notes

Most of the material on Semigroup Theory in this chapter, e.g., the semidirect and wreath products, division, pseudovarieties, etc., is now standard and can be found in texts such as [7, 85, 224]. The category theoretic machinery discussed in this chapter is also classical and can be found in MacLane [185]. The theory of Galois connections is an even older subject. The preliminary chapter of [97] provides a good reference for this material, as well as the basic theory of algebraic lattices; see also [203] for the latter.

Division was first introduced into Finite Semigroup Theory by Krohn and Rhodes [169]. Relational morphisms of semigroups were invented by Rhodes and Tilson, although it seems they first appeared in print in Eilenberg [85]. However, a notion of relational morphism of groups already appears in work of Wedderburn [377]. Pseudovarieties were introduced by Eilenberg and Schützenberger [85, 86]. The two-sided semidirect product and the block product made its first appearance in Rhodes and Tilson [293], although it had roots in the triple product [85, 313] and the decomposition results of [168].

The definition of a relational morphism between relational morphisms was developed by Steinberg and Tilson in "Categories as algebra. II" [339], but in that paper a less restrictive notion than division, called a division diagram, was considered. The notions of division and strong division presented here are due to Steinberg. It was first observed by Steinberg and Tilson [339] that many universal arrows in the category of semigroups with arrows homomorphisms become weak universal arrows when the category is enriched to include relational morphisms.

# 2

# The q-operator

As mathematics evolves, theorems become definitions. The Heine-Borel theorem says that every open cover of a closed interval has a finite subcover. This is now the definition of compactness. Coxeter classified finite groups generated by reflections via presentations. These sorts of presentations then became Tits's definition of a Coxeter group. Tilson, in his seminal paper [364], showed that a semigroup $S$ belongs to $\mathbf{V} * \mathbf{W}$ if and only if there is a relational morphism $\varphi \colon S \to T$ with $T \in \mathbf{W}$ satisfying a certain condition. This result is called the Derived Semigroupoid Theorem. In this chapter, we essentially make this theorem into a definition.

More precisely, to a collection R of relational morphisms satisfying certain axioms, we associate a continuous operator $\alpha \colon \mathbf{PV} \to \mathbf{PV}$ so that $S \in \alpha(\mathbf{W})$ if and only if there is a relational morphism $\varphi \colon S \to T$ with $T \in \mathbf{W}$ and $\varphi \in \mathsf{R}$. Tilson's Derived Semigroupoid Theorem then translates into a result saying that the operator $\mathbf{V} * (-)$ arises in this manner from a class of relational morphisms, denoted $\mathbf{V}_D$, defined in terms of the derived semigroupoid [339]. The Mal'cev product [126] is intrinsically defined this way because $S \in \mathbf{V} \textcircled{m} \mathbf{W}$ means that there is a relational morphism $\varphi \colon S \to T$ with $T \in \mathbf{W}$ so that $e\varphi^{-1} \in \mathbf{V}$ for each idempotent $e \in T$. The Kernel Category Theorem [293] shows how to define the two-sided semidirect product of pseudovarieties in terms of relational morphisms. The Slice Theorem [322] describes the class of relational morphisms associated to the join operator.

Having defined in the previous chapter division and products for relational morphisms, a notion of pseudovariety of relational morphisms is crying out to be defined. However, we had both division and strong division and that means that there is more than one class of relational morphisms to be defined. One can then try to define an equational theory, à la Reiterman, for pseudovarieties of relational morphisms. This will be carried out in Chapter 3, where many pseudoidentity basis theorems in the literature [27, 235] will be derived as consequences of the equational theory of pseudovarieties of relational morphisms.

J. Rhodes, B. Steinberg, *The q-theory of Finite Semigroups*,
Springer Monographs in Mathematics, DOI 10.1007/978-0-387-09781-7_2,
© Springer Science+Business Media, LLC 2009

It turns out that more than one class of relational morphisms can define the same operator. One would then like to construct maximal and minimal classes that define a given operator. This leads us to define the map q from classes of relational morphisms to continuous operators, which will form part of several Galois connections and become a centerpiece of the theory.

Of course, continuous operators can be composed. Moreover, by identifying a pseudovariety with its constant map, we may fruitfully view the action of an operator on a pseudovariety as an instance of composition. This means it is worthwhile to consider the lattice structure on the set of all continuous operators. The associative operation of composition can then replace the various non-associative structures that exist on **PV**. And of course, we would like also to have a composition of classes of relational morphisms that corresponds with the composition of operators under the map q.

In this chapter, all semigroups should be assumed finite unless stated otherwise. Without further ado, let us turn to the axioms for classes of relational morphisms.

## 2.1 Axioms for Sets of Relational Morphisms

The following axioms for relational morphisms of finite semigroups have antecedents in Tilson's [362] and unpublished joint work of Tilson and the second author (with some input from the first author). They axiomatize the properties enjoyed by the collection $\mathbf{V}_D$ of relational morphisms with derived semigroupoid dividing a semigroup in **V**.

### 2.1.1 The axioms

**Axiom (Finite products ($\times$)).** *A subset $X \subseteq \mathbf{RM}$ satisfies Axiom ($\times$) if it is closed under finite (including empty) products, i.e., $f, g \in X$ implies $f \times g \in X$, and $1_{\{1\}} \in X$; see Section 1.3.2.*

**Axiom (Strong division ($\prec_s$)).** *A subset $X \subseteq \mathbf{RM}$ satisfies Axiom ($\prec_s$) if $g \in X$ and $f \prec_s g$ implies $f \in X$; see (1.44).*

**Axiom (Division ($\prec$)).** *A subset $X \subseteq \mathbf{RM}$ satisfies Axiom ($\prec$) if $g \in X$ and $f \prec g$ implies $f \in X$; see (1.42).*

**Axiom (Range extension (r-e)).** *A subset $X \subseteq \mathbf{RM}$ satisfies Axiom (r-e) if given $f\colon S \to T$ belonging to $X$ and $j\colon T \hookrightarrow \overline{T}$, where $j$ is an injective homomorphism, one has that $fj$ also lies in $X$.*

**Axiom (Identity maps (id)).** *A subset $X \subseteq \mathbf{RM}$ satisfies Axiom (id) if, for all finite semigroups $S$, the identity map $1_S\colon S \to S$ belongs to $X$.*

*Remark 2.1.1.* Notice that $f \prec g$ if and only if there is a homomorphism $m$ and a division $d$ such that $fm \subseteq_s dg$; an analogous remark holds for $\prec_s$, only $m$ must be onto in this case.

### 2.1.2 Continuously closed classes and pseudovarieties of relational morphisms

We are now ready to define the classes of relational morphisms that are important for this book. Let us begin by giving the formal definitions. The intuition behind the choices will be explained as the chapter progresses.

**Definition 2.1.2 (Continuously closed class).** *A collection $X \subseteq \mathbf{RM}$ of relational morphisms is said to be continuously closed if it satisfies Axiom* $(\times)$, *Axiom* $(\prec_s)$ *and Axiom* (r-e). *The collection of all continuously closed classes is denoted* $\mathbf{CC}$.

**Definition 2.1.3 ($\mathbf{CC}^+$).** *If $X$ in $\mathbf{CC}$ also satisfies Axiom* (id), *we say $X$ belongs to $\mathbf{CC}^+$ and call $X$ positive.*

The plus sign corresponds with the fact that the associated operator to an element of $\mathbf{CC}^+$ will be non-decreasing.

Roughly speaking, continuously closed classes are sufficiently rich to describe any continuous operator on the lattice $\mathbf{PV}$. There is a variant notion, which has the advantage of an equational theory (see Chapter 3) but some possible defects, which will be discussed later on.

**Definition 2.1.4 (Equational continuously closed class).** *A continuously closed class is said to be equational if it satisfies the additional axiom:*

**Axiom (Finite pullbacks (pb)).** *A subset $X \subseteq \mathbf{RM}$ satisfies Axiom* (pb) *if $f \colon S_1 \to T, g \colon S_2 \to T$ in $X$ implies $f \times_T g \in X$, and $1_{\{1\}} \in X$; see Section 1.3.2, specifically* (1.36).

Now we turn to the central notion of the chapter: pseudovarieties of relational morphisms. These have stronger closure properties than continuously closed classes and axiomatize those operators that are amenable to the methods of global semigroup theory as espoused in [278].

**Definition 2.1.5 (Pseudovariety of relational morphisms).** *A collection $X \subseteq \mathbf{RM}$ of relational morphisms is a pseudovariety of relational morphisms if it satisfies Axiom* $(\times)$, *Axiom* $(\prec)$ *and Axiom* (r-e). *Equivalently, $X$ is a pseudovariety of relational morphisms if it is closed under finite products, division and range extensions. The collection of all pseudovarieties of relational morphisms is denoted* $\mathbf{PVRM}$.

We also consider a positive version.

**Definition 2.1.6 ($\mathbf{PVRM}^+$).** *If $X \in \mathbf{PVRM}$ satisfies Axiom* (id), *we say $X$ belongs to $\mathbf{PVRM}^+$. In this case $X$ is said to be a positive pseudovariety of relational morphisms.*

It is not too difficult to verify that the members of **PVRM**$^+$ are exactly what Tilson calls weakly closed classes in [362], although he formulates the axioms in a slightly different way. However, we caution the reader that **PVRM** is *not* obtained from Tilson's axiom system by removing Axiom (id). His notion of restriction, in the case one does not have Axiom (id), is much weaker than our division.

**Exercise 2.1.7.** Prove that members of **PVRM**$^+$ are precisely the weakly closed classes of [362]. It may help to use Proposition 2.1.8 below.

In Chapter 3, an equational theory is developed for **PVRM**. A less rich equational theory is also developed in that chapter for equational continuously closed classes. Let us adjoin another axiom to our list, which will help us to understand just what is the difference between division and strong division.

**Axiom (Corestriction (co-re)).** *A subset* $X \subseteq \mathbf{RM}$ *satisfies Axiom (co-re) if* $f \colon S \to T$ *in* $X$ *implies* $f_{\mathrm{Im}} \colon S \to \mathrm{Im} f$ *belongs to* $X$ *where* $f_{\mathrm{Im}}$ *is the corestriction of* $f$ *to its image.*

Notice that if a class $X$ satisfies both Axioms (r-e) and (co-re), then a relational morphism belongs to $X$ if and only if its corestriction to its image does.

We now turn to establishing some basic properties of continuously closed classes and pseudovarieties of relational morphisms.

**Proposition 2.1.8.** *The following results hold:*

(a) *A collection* $X$ *of relational morphisms is a continuously closed class if and only if it satisfies Axiom* $(\times)$ *and whenever* $g \in X$ *and* $f \subseteq_s d_1 g d_2$ *with* $d_1, d_2$ *divisions, then also* $f \in X$.

(b) *Suppose* $X$ *is a collection of relational morphisms closed under Axiom (co-re) and Axiom (r-e). Let* $f \in X$ *with* $f \colon S \to T$ *and suppose* $\mathrm{Im} f \subseteq T' \subseteq T$. *Then the corestriction of* $f$ *to* $T'$ *belongs to* $X$.

(c) *A continuously closed class is a pseudovariety of relational morphisms if and only if it satisfies Axiom (co-re).*

(d) *If* $X$ *satisfies Axiom (pb) and Axiom* $(\prec_s)$, *then* $X$ *satisfies Axiom* $(\times)$.

(e) *Every pseudovariety of relational morphisms is an equational continuously closed class, that is, satisfies Axiom (pb).*

*Proof.* To prove (a) suppose first $X \in \mathbf{CC}$. Assume $g \in X$ and $f \subseteq_s d_1 g d_2$ with $d_1, d_2$ divisions. Because evidently $f \prec_s d_1 g d_2$, while $d_1 g d_2 \prec_s g d_2$ by (1.45), it suffices to show $g d_2 \in X$. Write $d_2 = s^{-1}i$ with $s$ a surjective homomorphism and $i$ an injective map. By closure of $X$ under Axiom (r-e), it suffices to show that $g s^{-1} \in X$. But $g s^{-1} s = g$ and $s$ is onto, so $g s^{-1} \prec_s g$ by (1.46). We conclude $g s^{-1} \in X$. This completes the proof that $f \in X$.

For the converse, we need to show that $X$ satisfies Axiom (r-e) and Axiom $(\prec_s)$. Suppose first $f \colon S \to T$ belongs to $X$ and $j \colon T \hookrightarrow T'$ is an injective homomorphism. Then $j$ is a division and so $fj \in X$ by assumption. Suppose $f \prec_s g$ with $g \in X$ via a strong division

$$
\begin{array}{ccc}
S & \xrightarrow{\ d\ } & U \\
f \downarrow & \subseteq & \downarrow g \\
T & \xrightarrow[\ m\ ]{} & V
\end{array}
$$

where $m$ is an onto homomorphism. Then $f \subseteq_s dgm^{-1}$ and so by assumption $f \in X$. This completes the proof that $X$ is a continuously closed class.

For (b), let $f' \colon S \to T'$ be the corestriction. Then clearly $f'$ is a range extension of the corestriction $f_{\text{Im}}$ of $f$ to its image.

To prove (c), first observe that clearly $\mathbf{PVRM} \subseteq \mathbf{CC}$. Let $f \colon S \to T$ be a relational morphism and let $f_{\text{Im}} \colon S \to \text{Im}\, f$ be the corestriction. Setting $j \colon \text{Im}\, f \hookrightarrow T$ to be the inclusion, we have the commutative triangle

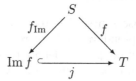

establishing that $f_{\text{Im}} \prec f$. Thus pseudovarieties of relational morphisms satisfy Axiom (co-re). Conversely, suppose that $X$ is a continuously closed class satisfying Axiom (co-re). We show that $X$ is a pseudovariety of relational morphisms. To do this it suffices to prove closure under division. So suppose $g \in X$ and we have a division

$$
\begin{array}{ccc}
S & \xrightarrow{\ d\ } & U \\
f \downarrow & \subseteq & \downarrow g \\
T & \xrightarrow[\ m\ ]{} & V.
\end{array}
$$

Then $fm \prec_s g$ and hence $fm \in X$. Clearly, $\text{Im}\, fm \subseteq \text{Im}\, m$ and so by (b) the corestriction $h \colon S \to \text{Im}\, m$ of $fm$ belongs to $X$. But $h = fm'$ where $m' \colon T \twoheadrightarrow \text{Im}\, m$ is the corestriction of $m$. Because $m'$ is onto, $f \prec_s fm' = h$ (cf. (1.46)) and so $f \in X$. This establishes that $X$ is a pseudovariety of relational morphisms.

To establish (d), suppose $f \colon S \to T$ and $g \colon U \to V$ are in $X$. Then (1.37) shows $f \times g = f' \times_{(T \times V)} g'$ where $f' = f p_T^{-1}$ and $g' = g p_U^{-1}$. But $f' \prec_s f$, $g' \prec_s g$, so $f \times g \in X$.

Finally (e) is an immediate consequence of Proposition 1.3.16(c). $\qquad\square$

The following useful lemma, proved by Tilson for $\mathbf{PVRM}^+$, will be used frequently throughout the book.

**Lemma 2.1.9 (Tilson's Lemma).** *Suppose $X$ is a positive equational continuously closed class and let $f\colon S \to T$ be a relational morphism with canonical factorization $f = p_S^{-1}p_T$ where $p_S\colon \#f \to S$ and $p_T\colon \#f \to T$ are the projections. Then $f \in X$ if and only if $p_T \in X$. This applies in particular if $X$ is a positive pseudovariety of relational morphisms.*

*Proof.* Because $f = p_S^{-1}p_T$, clearly $f \prec_s p_T$ and hence $p_T \in X$ implies $f \in X$. Conversely, it was shown in (1.38) that $p_T = f \times_T 1_T$, so if $f \in X$, then $p_T \in X$ using Axiom (id) and Axiom (pb). The final statement follows because every pseudovariety of relational morphisms is an equational continuously closed class by Proposition 2.1.8(e). □

*Remark 2.1.10.* If $X \in \mathbf{CC}$, then $1_{\{1\}} \in X$ by Axiom ($\times$). Because $\emptyset \prec \{1\}$, we may also conclude $\emptyset \to \{1\}$ belongs to $X$. By Proposition 2.1.8(a) and the fact that any relational morphism from $\{1\}$ is a division, we see that $X$ contains, for each finite semigroup $S$, the unique relation $\emptyset \to S$ and each relational morphism of the form $\{1\} \to S$.

We shall endeavor to use san serif letters, such as R, to denote continuously closed classes and pseudovarieties of relational morphisms, whereas boldface symbols like $\mathbf{V}$ will be used to denote pseudovarieties of semigroups and semigroupoids. Let us proceed to show that $\mathbf{CC}$ is an algebraic lattice.

**Proposition 2.1.11.** *One has that $(2^{\mathbf{RM}}, \subseteq, \cup, \cap)$ is a complete algebraic lattice. Moreover, there is a Galois connection*

$$(2^{\mathbf{RM}} \subseteq, \cup, \cap) \xleftarrow{\quad j \quad}_{\mathbb{CC}} (\mathbf{CC}, \subseteq, \vee_{\text{det}}, \cap) \tag{2.1}$$

*where $j$ is inclusion and*

$$\mathbb{CC}(X) = \bigcap \{\mathsf{R} \in \mathbf{CC} \mid \mathsf{R} \supseteq X\}.$$

*So $j$ is **inf**, and $\mathbb{CC}$ is **sup** and onto. Furthermore, $j$ is continuous. The determined join is given by*

$$\bigvee_{\text{det}} \{\mathsf{R}_a\}_{a \in A} = \mathbb{CC}\left(\bigcup_{a \in A} \mathsf{R}_a\right).$$

*Hence $(\mathbf{CC}, \subseteq, \vee_{\text{det}}, \cap)$ is a complete algebraic lattice with the compact elements the finitely generated elements, i.e., $\mathsf{R} \in \mathbf{CC}$ is compact if and only if $\mathsf{R} = \mathbb{CC}(F)$, where $F \subseteq \mathbf{RM}$ is finite.*

*Proof.* The proof is straightforward. The last part uses Corollary 1.2.37. □

**Exercise 2.1.12.** Prove Proposition 2.1.11.

*Remark 2.1.13.* It is not the case that the continuously closed class generated by a finite set of relational morphisms is in fact generated by a single relational morphism as, in general, $f \not\prec_s f \times g$; see Remark 1.3.17.

We often drop the notation $j$ and think of $\mathbb{CC}$ as a closure operator on $2^{\mathbf{RM}}$. Let us provide some alternative descriptions of the closure operator $\mathbb{CC}$. The first description shows that to obtain the continuously closed class generated by a set of relational morphisms, it suffices to first close under product, then under strong division and finally under range extension.

**Proposition 2.1.14.** *Let $X \subseteq \mathbb{CC}$. Then*

$$\mathbb{CC}(X) = \{fi \mid f \prec_s f_1 \times \cdots \times f_n,\ i\ an\ inclusion,\ f_i \in X\} \tag{2.2}$$
$$= \{f \mid f \subseteq_s d_1(f_1 \times \cdots \times f_n)d_2,\ d_1, d_2\ divisions,\ f_i \in X\}. \tag{2.3}$$

*Proof.* We prove (2.2), leaving (2.3) to the reader. Clearly, the right-hand side, call it R, of (2.2) is contained in the left, so it suffices to prove R is a continuously closed class. It clearly contains the empty product $1_{\{1\}}$. Suppose that $f \prec_s f_1 \times \cdots \times f_n$ and $g \prec_s g_1 \times \cdots \times g_m$ with $f_1, \ldots, f_n, g_1 \ldots, g_m \in X$. Let $i, j$ be inclusions such that $fi$ and $gj$ are defined. Then $i \times j$ is an inclusion, $fi \times gj = (f \times g)(i \times j)$ and

$$f \times g \prec_s f_1 \times \cdots \times f_n \times g_1 \times \cdots \times g_m$$

by Exercise 1.3.19. This shows that R is closed under products. Clearly, R is closed under range extension.

Suppose that $i$ is an inclusion, $g \prec_s fi$ and $f \prec_s f_1 \times \cdots \times f_n$ with the $f_k \in X$. So there is a division $d$ and an onto homomorphism $m$ such that $gm \subseteq_s dfi$. Let $f \colon S \to T$, $g \colon A \to B$ and $i \colon T \to T'$; so $m \colon B \twoheadrightarrow T'$. Set $C = Tm^{-1}$. Because $gm \subseteq_s dfi$, we must have that the image of $gm$ is contained in the image of $fi$, which in turn is contained in $T$. So the image of $g$ is contained in $C$. Let $g^0 \colon A \to C$ be the corestriction. Then there results a strong division

$$
\begin{array}{ccc}
A & \xrightarrow{\ d\ } & S \\
{\scriptstyle g^0}\downarrow & \subseteq & \downarrow{\scriptstyle f} \\
C & \xrightarrow[m|_C]{} & T.
\end{array}
$$

Hence, by transitivity of strong division, $g^0 \prec_s f_1 \times \cdots \times f_n$. Because $g$ is $g^0 j$ for a certain inclusion $j$, we see that $g \in$ R, as desired. This completes the proof that R is continuously closed, establishing the proposition. $\qquad\square$

**Exercise 2.1.15.** Prove (2.3).

There is one more variant on continuously closed classes that shall be considered in the sequel in order to provide minimal classes defining a given continuous operator, see Section 2.3.3.

**Definition 2.1.16 (Birkhoff continuously closed class).** *A continuously closed class is called a Birkhoff continuously closed class if it satisfies:*

**Axiom (Free).** *A subset* $X \subseteq \mathbf{RM}$ *satisfies Axiom* (free) *if whenever* $f \colon S \to T^n$ *belongs to* $X$, *for some* $n \geq 1$, *one has that the canonical relational morphism* $\rho_{(T)} \colon S \to F_{(T)}(S \times T)$ *(see Definition 1.3.14) is in* $X$ *where we view* $S$ *as generated by* $S \times T$ *via the projection* $p_S \colon S \times T \twoheadrightarrow S$.

*The collection of all Birkhoff continuously closed classes will be denoted* $\mathbf{BCC}$.

Of course, $\mathbf{BCC}^+$ is defined analogously, but with the addition of Axiom (id). Lemma 1.3.15 implies that every pseudovariety of relational morphisms is a Birkhoff continuously closed class, hence:

**Proposition 2.1.17.** *The inclusions* $\mathbf{PVRM} \leq \mathbf{BCC} \leq \mathbf{CC}$ *hold.*

### 2.1.3 First examples

Important examples of members of $\mathbf{CC}$ and $\mathbf{PVRM}$ can be found in Section 2.4. The reason for postponing the examples is to introduce the operator q and its basic properties first. However, the reader can skip ahead and read the examples without too much problem. We present here certain fundamental examples that will be needed shortly.

If $\mathbf{V}$ is a pseudovariety of semigroups, let us define:

$$(\mathbf{V}, \mathbf{1}) = \{f \colon S \to T \mid e \in E(T) \text{ implies } ef^{-1} \in \mathbf{V}\} \tag{2.4}$$

where $E(T)$ denotes the set of idempotents of $T$.

The class $(\mathbf{V}, \mathbf{1})$ is a pseudovariety of relational morphisms. More generally, if $\mathbf{V}$ and $\mathbf{W}$ are any pseudovarieties of semigroups, then we have the following pseudovariety of relational morphisms:

$$(\mathbf{V}, \mathbf{W}) = \{f \colon S \to T \mid T' \leq T, \text{ with } T' \in \mathbf{W}, \text{ implies } T'f^{-1} \in \mathbf{V}\}. \tag{2.5}$$

The previous example (2.4) is the case where $\mathbf{W} = \mathbf{1}$. Let us prove $(\mathbf{V}, \mathbf{W}) \in \mathbf{PVRM}$ as an illustration of working with the axioms.

**Proposition 2.1.18.** *Let* $\mathbf{V}$ *and* $\mathbf{W}$ *be pseudovarieties of semigroups. Then* $(\mathbf{V}, \mathbf{W}) \in \mathbf{PVRM}$. *Moreover,* $(\mathbf{V}, \mathbf{W}) \in \mathbf{PVRM}^+$ *if and only if* $\mathbf{V} \geq \mathbf{W}$.

*Proof.* Clearly $1_{\{1\}} \in (\mathbf{V}, \mathbf{W})$. Suppose $f \colon S \to T$, $g \colon S' \to T'$ belong to $(\mathbf{V}, \mathbf{W})$ and $T \times T' \geq W \in \mathbf{W}$. Let $W_1 = Wp_T$ and $W_2 = Wp_{T'}$. Then $W_1, W_2 \in \mathbf{W}$. Thus $W_1 f^{-1}, W_2 g^{-1} \in \mathbf{V}$. But then we have

$$W(f \times g)^{-1} \subseteq W_1 f^{-1} \times W_2 g^{-1} \in \mathbf{V}. \tag{2.6}$$

Indeed, if $(t, t') \in W$ satisfies $(t, t') \in (s, s')(f \times g)$, then $t \in sf$ and $t' \in s'g$. Since $t \in W_1$ and $t' \in W_2$, it follows $(s, s') \in W_1 f^{-1} \times W_2 g^{-1}$. From (2.6), we obtain $f \times g \in (\mathbf{V}, \mathbf{W})$ and so $(\mathbf{V}, \mathbf{W})$ is closed under finite products.

Suppose $f \prec g$ is given by

$$
\begin{array}{ccc}
S & \xrightarrow{\ d\ } & U \\
{\scriptstyle f}\downarrow & \subseteq & \downarrow{\scriptstyle g} \\
T & \xrightarrow[\ m\ ]{} & V
\end{array}
$$

where $d$ is a division, $m$ is a homomorphism and $g \in (\mathbf{V}, \mathbf{W})$. Let $W \leq T$ with $W \in \mathbf{W}$. Then $Wm \in \mathbf{W}$, so $Wmg^{-1} \in \mathbf{V}$. By Fact 1.3.11, it follows $Wf^{-1} \prec Wmg^{-1} \in \mathbf{V}$. We conclude $Wf^{-1} \in \mathbf{V}$ and so $(\mathbf{V}, \mathbf{W})$ is closed under division.

Finally, suppose $f \colon S \to T$ is in $(\mathbf{V}, \mathbf{W})$ and $i \colon T \to T'$ is injective. If $T' \geq W \in \mathbf{W}$, then $W(fi)^{-1} = (W \cap T)f^{-1} \in \mathbf{V}$ (as $W \cap T \in \mathbf{W}$) and so $fi \in (\mathbf{V}, \mathbf{W})$. Hence $(\mathbf{V}, \mathbf{W})$ is closed under range extension.

If $\mathbf{V} \geq \mathbf{W}$, then evidently $1_S \in (\mathbf{V}, \mathbf{W})$ for each semigroup $S$. Conversely, if $S \in \mathbf{W}$, but $S \notin \mathbf{V}$, then $1_S \notin (\mathbf{V}, \mathbf{W})$. This completes the proof.      $\square$

## 2.2 Continuous Operators

We now turn to the study of continuous operators on the lattice $\mathbf{PV}$. The connection with continuously closed classes will become apparent in the next section when we define the q-operator, which takes continuously closed classes to continuous operators. Recall from just after Definition 1.1.4 that $\mathbf{Cnt}$ denotes the category of lattices with continuous maps as morphisms and hence $\mathbf{Cnt}(\mathbf{PV})$ is the monoid of all continuous self-maps of $\mathbf{PV}$. The study of $\mathbf{Cnt}(\mathbf{PV})$ is an important special case of the main theme of [97] of studying all D. Scott continuous operators on a complete lattice.

Now it is easy to check $\mathbf{Cnt}(\mathbf{PV})$ is a complete lattice with pointwise ordering. That is, if $\alpha, \beta \in \mathbf{Cnt}(\mathbf{PV})$, then one has

$$
\alpha \leq_{\mathrm{pw}} \beta \iff \alpha(\mathbf{V}) \leq \beta(\mathbf{V})
$$

for all pseudovarieties $\mathbf{V}$ of finite semigroups. We often omit the subscript pw and write simply $\leq$. The join operation is just pointwise join (sometimes written $\vee_{\mathrm{pw}}$), but the determined meet $\wedge_{\mathrm{det}}$ is more mysterious and is the subject of Section 2.3.1. However, one can easily verify that the finite meets in $\mathbf{Cnt}(\mathbf{PV})$ are just pointwise meets, that is, a finite pointwise meet of continuous operators is continuous.

**Exercise 2.2.1.** Show that $\mathbf{Cnt}(\mathbf{PV})$ is a complete lattice with the pointwise ordering. Prove that joins are pointwise and that finite meets are pointwise.

The monoid of non-decreasing continuous operators on $\mathbf{PV}$, i.e., the continuous self-maps $\alpha$ with $\alpha \geq 1_{\mathbf{PV}}$, will be denoted $\mathbf{Cnt}(\mathbf{PV})^+$. Such operators will frequently be called *positive*.

Let $\mathbf{1} = \{\emptyset, \{1\}\}$ be the trivial pseudovariety. Recall that $\mathbf{OP}$ denotes the category of partially ordered sets with morphisms order preserving maps and that if $S$ is a finite semigroup, then $(S)$ denotes the pseudovariety generated by $S$. If $P$ is a poset and $L$ is a complete lattice, the collection $\mathbf{OP}(P, L)$ of order preserving maps from $P$ to $L$ is a complete lattice with pointwise operations.

**Proposition 2.2.2.** *The following hold:*

*(a) As a complete lattice* $\mathbf{Cnt(PV)}$ *is isomorphic to* $\mathbf{OP}(K(\mathbf{PV}), \mathbf{PV})$. *The isomorphism is given by sending* $\alpha \in \mathbf{Cnt(PV)}$ *to* $\alpha|_{K(\mathbf{PV})}$. *The inverse is defined by taking* $\beta \in \mathbf{OP}(K(\mathbf{PV}), \mathbf{PV})$ *to* $\overline{\beta}$ *where*

$$\overline{\beta}(\mathbf{V}) = \bigvee\{\beta((T)) \mid T \in \mathbf{V}\} = \bigvee\{\beta(\mathbf{W}) \mid \mathbf{V} \geq \mathbf{W} \in K(\mathbf{PV})\}. \quad (2.7)$$

*This "famous" formula is precisely what a function* $\overline{\beta} \colon \mathbf{PV} \to \mathbf{PV}$ *must satisfy to be continuous with* $\overline{\beta}|_{K(\mathbf{PV})} = \beta$.

*(b) Define, for* $S, T \in \mathbf{FSgp}$, $d(S, T) \colon K(\mathbf{PV}) \to \mathbf{PV}$ *by*

$$d(S, T)((T_1)) = \begin{cases} (S) & \text{if } T \in (T_1) \\ \mathbf{1} & \text{else.} \end{cases} \quad (2.8)$$

*Then* $\delta(S, T) = \overline{d}(S, T) \colon \mathbf{PV} \to \mathbf{PV}$ *is continuous and is given by*

$$\delta(S, T)(\mathbf{V}) = \begin{cases} (S) & \text{if } T \in \mathbf{V} \\ \mathbf{1} & \text{else.} \end{cases} \quad (2.9)$$

*The finite joins of elements of the form* $\delta(S, T)$ *are the compact elements of* $\mathbf{Cnt(PV)}$ *and* $\mathbf{Cnt(PV)}$ *is an algebraic lattice.*

*Proof.* To prove (a) observe that because $\mathbf{PV}$ is algebraic, each $\mathbf{V} \in \mathbf{PV}$ is a directed supremum of elements of $K(\mathbf{PV})$. Thus any $\alpha \in \mathbf{Cnt(PV)}$ is determined by its restriction to $K(\mathbf{PV})$ via the formula

$$\alpha(\mathbf{V}) = \alpha\left(\bigvee_{T \in \mathbf{V}} (T)\right) = \bigvee_{T \in \mathbf{V}} \alpha((T))$$

because $\{(T) \mid T \in \mathbf{V}\}$ is directed and $\alpha$ is continuous. The remaining details are straightforward and left to the reader.

Turning to (b), let $S$ and $T$ be finite semigroups. We first show $\delta(S, T)$ is compact. Assume we have $\delta(S, T) \leq \bigvee_{a \in A} \alpha_a$ with the $\alpha_a \in \mathbf{Cnt(PV)}$. Then

$$(S) = \delta(S, T)((T)) \leq \bigvee_{a \in A} \alpha_a((T)).$$

By compactness of $(S)$, there exists a finite subset $F \subseteq A$ such that $(S) \leq \bigvee_{f \in F} \alpha_f((T))$. We claim that $\delta(S, T) \leq \bigvee_{f \in F} \alpha_f$. Let $\mathbf{V} \in \mathbf{PV}$. When evaluating $\delta(S, T)(\mathbf{V})$, there are two cases: if $T \notin \mathbf{V}$, then

$$\delta(S,T)(\mathbf{V}) = \mathbf{1} \le \bigvee_{f \in F} \alpha_f(\mathbf{V});$$

if $T \in \mathbf{V}$ (and so $(T) \le \mathbf{V}$), then

$$\delta(S,T)(\mathbf{V}) = (S) \le \bigvee_{f \in F} \alpha_f((T)) \le \bigvee_{f \in F} \alpha_f(\mathbf{V}).$$

Therefore, $\delta(S,T) \le \bigvee_{f \in F} \alpha_f$, as required.

Because finite joins of compact elements are compact, to prove the remaining assertions, it suffices to show that each element $\alpha \in \mathbf{Cnt(PV)}$ is a supremum of elements of the form $\delta(S,T)$ with $S, T$ finite semigroups. To do this we establish the equality

$$\alpha = \bigvee \{ \delta(S,T) \mid S, T \in \mathbf{FSgp}, \ S \in \alpha((T)) \}. \tag{2.10}$$

Let $\beta$ be the right-hand side of (2.10). To prove (2.10) it suffices to show that both sides agree on compact pseudovarieties by (a). So let $T$ be a finite semigroup. Suppose that $S \in \alpha((T))$. Then $\delta(S,T)$ is in the join defining $\beta$ so $\delta(S,T) \le \beta$. But $S \in \delta(S,T)((T))$ so $S \in \beta((T))$. Thus $\alpha \le \beta$. For the reverse inclusion suppose now that $S \in \beta((T))$. If $S \in \mathbf{1}$, then clearly $S \in \alpha((T))$ so assume $S \notin \mathbf{1}$. Because $(S)$ is compact, there exist $S_1, \ldots, S_n$ and $T_1, \ldots, T_n$ such that

$$S \in \delta(S_1, T_1)((T)) \vee \cdots \vee \delta(S_n, T_n)((T))$$

where $S_i \in \alpha((T_i))$, for all $i = 1, \ldots, n$. Without loss of generality, we may assume $T_1, \ldots, T_k \in (T)$ and $T_{k+1}, \ldots, T_n \notin (T)$ where $k \ge 1$ as $S \notin \mathbf{1}$. Then $S \in (S_1) \vee \cdots \vee (S_k)$. Because $T_i \in (T)$, for $1 \le i \le k$, we have $S_i \in \alpha((T))$, all $1 \le i \le k$. Therefore, $S \in \alpha((T))$, establishing $\beta \le \alpha$. This completes the proof of (2.10) and hence of the proposition.  □

Proposition 2.2.2(a) can best be understood as follows: if $\alpha \colon \mathbf{PV} \to \mathbf{PV}$ is a continuous operator, then $S \in \alpha(\mathbf{V})$ if and only if there is a semigroup $T \in \mathbf{V}$ with $S \in \alpha((T))$.

There is a natural infinite power associated to an operator $\alpha \in \mathbf{Cnt(PV)}^+$. Namely define

$$\alpha^\omega = \bigvee_{n \in \mathbb{N}} \alpha^n. \tag{2.11}$$

Because $\alpha$ is positive, if $\mathbf{V} \in \mathbf{PV}$, then the $\alpha^n(\mathbf{V})$ form an ascending chain and so

$$\alpha^\omega(\mathbf{V}) = \bigcup_{n \in \mathbb{N}} \alpha^n(\mathbf{V}).$$

It follows that $(\alpha^\omega)^n = (\alpha^n)^\omega = (\alpha^\omega)^\omega = \alpha^\omega$, for all $n > 0$.

An important aspect of the theory is that there are order preserving operators on $\mathbf{PV}$ that are *not* continuous. If $n > 0$, then the *Brandt semigroup* $B_n$ is the semigroup of $n \times n$ matrix units (together with zero). So, for instance,

$$B_2 = \left\{ \begin{pmatrix} 1 & 0 \\ 0 & 0 \end{pmatrix}, \begin{pmatrix} 0 & 1 \\ 0 & 0 \end{pmatrix}, \begin{pmatrix} 0 & 0 \\ 1 & 0 \end{pmatrix}, \begin{pmatrix} 0 & 0 \\ 0 & 1 \end{pmatrix}, \begin{pmatrix} 0 & 0 \\ 0 & 0 \end{pmatrix} \right\}.$$

Let $\mathbf{DS} = \{S \mid$ the regular $\mathscr{J}$-classes of $S$ are subsemigroups$\}$. It is well-known and not difficult to verify that $\mathbf{DS}$ is a pseudovariety [7, Chapter 8]. Furthermore, $\mathbf{DS}$ is not compact as it contains the pseudovariety $\mathbf{G}$ of all finite groups and hence is not locally finite. The following lemma is well-known, so we only sketch the proof.

**Lemma 2.2.3.** *The equalities*

$$\mathbf{DS} = \bigvee \{\mathbf{V} \mid B_2 \notin \mathbf{V}\} = \{S \mid B_2 \nprec S^2\}$$

*hold.*

*Proof.* Clearly $B_2 \notin \mathbf{DS}$. Conversely, if $S \notin \mathbf{DS}$, then there must be $\mathscr{J}$-equivalent idempotents $e, f \in S$ with $fe \notin J$, where $J$ is the $\mathscr{J}$-class of $S$. Then we can choose a Rees matrix representation for $J^0$ (see Appendix A) so that the submatrix corresponding to the eggbox picture

| $H_e$ | $R_e \cap L_f$ |
|---|---|
| $R_f \cap L_e$ | $H_f$ |

is of the form $\begin{pmatrix} 1 & 0 \\ 0 & 1 \end{pmatrix}$ or $C = \begin{pmatrix} 1 & 1 \\ 0 & 1 \end{pmatrix}$. In the first case, we have $B_2$ is a divisor of $S$. In the second case, let $V$ be the Rees matrix semigroup $\mathscr{M}^0(\{1\}, 2, 2, C)$ (see Definition A.4.10 for notation and definitions concerning Rees matrix semigroups). Then $V^2 \prec S^2$ and $V^2/(V \times 0) \cup (0 \times V) \cong \mathscr{M}^0(\{1\}, 4, 4, C \otimes C)$ where $C \otimes C$ is the tensor product matrix:

$$\begin{pmatrix} 1 & 1 \\ 0 & 1 \end{pmatrix} \otimes \begin{pmatrix} 1 & 1 \\ 0 & 1 \end{pmatrix} = \begin{pmatrix} 1 & 1 & 1 & 1 \\ 0 & \overline{1} & 0 & 1 \\ 0 & \underline{0} & \underline{1} & 1 \\ 0 & 0 & 0 & 1 \end{pmatrix}.$$

We conclude that $B_2 \prec V^2 \prec S^2$, completing the proof. $\square$

**Exercise 2.2.4.** Complete the details of the proof of Lemma 2.2.3 by verifying that if $U = \mathscr{M}^0(G, A, B, C)$ and $V = \mathscr{M}^0(G', A', B', C')$, then

$$U \times V/[(U \times 0) \cup (0 \times V)] \cong \mathscr{M}^0(G \times G', A \times A', B \times B', C \otimes C').$$

Lemma 2.2.3 can be reinterpreted as saying that if $\mathbf{V}$ is a pseudovariety that is not contained in $\mathbf{DS}$, then $B_2 \in \mathbf{V}$.

**Proposition 2.2.5.** *There exists a non-decreasing, order preserving self-map $\alpha \colon \mathbf{PV} \to \mathbf{PV}$ that is not continuous.*

*Proof.* Let $\alpha \colon \mathbf{PV} \to \mathbf{PV}$ be defined by

$$\alpha(\mathbf{V}) = \begin{cases} \mathbf{DS} \vee (B_2) & \text{if } \mathbf{V} = \mathbf{DS} \\ \mathbf{V} & \text{otherwise.} \end{cases} \tag{2.12}$$

Then $\alpha$ is non-decreasing and order preserving, but not continuous. In fact, $\alpha(\mathbf{V}) = \mathbf{V}$ for all compact pseudovarieties $\mathbf{V}$, so if $\alpha$ were continuous, then it would have to be the identity operator, which is patently not the case.

It remains to prove that $\alpha$ is order preserving, that is, we must verify $\mathbf{V} \leq \mathbf{V}'$ implies $\alpha(\mathbf{V}) \leq \alpha(\mathbf{V}')$. If $\mathbf{V} = \mathbf{V}'$, there is nothing to prove so assume $\mathbf{V} < \mathbf{V}'$.

*Case 1.* $\mathbf{V} \neq \mathbf{DS}$, then $\alpha(\mathbf{V}) = \mathbf{V} \leq \mathbf{V}' \leq \alpha(\mathbf{V}')$.

*Case 2.* $\mathbf{V} = \mathbf{DS} < \mathbf{V}'$. Then, by Lemma 2.2.3, $B_2 \in \mathbf{V}'$ so

$$\mathbf{V} = \mathbf{DS} < \mathbf{DS} \vee (B_2) \leq \mathbf{V}'.$$

Thus $\alpha(\mathbf{V}) = \mathbf{DS} \vee (B_2) \leq \mathbf{V}' = \alpha(\mathbf{V}')$, completing the proof. □

## 2.3 Definition of the q-operator

Now things get a little more interesting. Roughly speaking, we want to axiomatize which operators on $\mathbf{PV}$ can be defined by classes of relational morphisms. We begin by associating a continuous operator to each continuously closed class. The following is the key new definition in this book.

**Definition 2.3.1 (q-operator).** *The map* $\mathsf{q} \colon \mathbf{CC} \to \mathbf{Cnt}(\mathbf{PV})$ *is defined by sending* $\mathsf{R} \in \mathbf{CC}$ *to the continuous operator* $\mathsf{Rq} \colon \mathbf{PV} \to \mathbf{PV}$ *given by*

$$\mathsf{Rq}(\mathbf{V}) = \{S \in \mathbf{FSgp} \mid \text{there exists } f \in \mathsf{R} \text{ and } T \in \mathbf{V} \text{ with } f \colon S \to T\} \tag{2.13}$$

*for* $\mathbf{V} \in \mathbf{PV}$.

Sometimes we also write $\mathsf{Rq}\mathbf{V}$ instead of $\mathsf{Rq}(\mathbf{V})$.

**Proposition 2.3.2.** *The map* $\mathsf{q}$ *is well defined and order preserving.*

*Proof.* First we must show, for $\mathsf{R} \in \mathbf{CC}$ and $\mathbf{V} \in \mathbf{PV}$, that $\mathsf{Rq}(\mathbf{V})$ is a pseudovariety: i.e., it contains $\{1\}$ and is closed under $\times$ and $\prec$.

Because $\{1\} \in \mathbf{V}$ and $1_{\{1\}} \in \mathsf{R}$, clearly $\{1\} \in \mathsf{Rq}(\mathbf{V})$. Suppose next that $S_1, S_2 \in \mathsf{Rq}(\mathbf{V})$. Then there exist $f_1 \colon S_1 \to T_1$ and $f_2 \colon S_2 \to T_2$ such that $T_1, T_2 \in \mathbf{V}$ and $f_1, f_2 \in \mathsf{R}$. By Axiom ($\times$), $f_1 \times f_2 \colon S_1 \times S_2 \to T_1 \times T_2$ belongs to $\mathsf{R}$. As $T_1 \times T_2 \in \mathbf{V}$, it follows from (2.13) that $S_1 \times S_2 \in \mathsf{Rq}(\mathbf{V})$. This establishes closure of $\mathsf{Rq}(\mathbf{V})$ under finite products.

Suppose $S_0 \in \mathsf{Rq}(\mathbf{V})$ and $d \colon S \to S_0$ is a division. Then there exists $f \colon S_0 \to T$ with $T \in \mathbf{V}$ and $f \in \mathsf{R}$. Composition then yields the relational morphism $df \colon S \to T$, which belongs to R because $df \prec_s f$ (cf. (1.45)). Thus $S \in \mathsf{Rq}(\mathbf{V})$, and so $\mathsf{Rq}(\mathbf{V})$ is closed under taking divisors.

It is immediate from (2.13) that $\mathsf{Rq}$ is order preserving. We proceed to show that $\mathsf{Rq}$ is continuous. Consider a directed set $\{\mathbf{V}_a\}_{a \in A}$ of pseudovarieties. Then we have the equality

$$\bigvee_{\det} \{\mathbf{V}_a \mid a \in A\} = \bigcup \{\mathbf{V}_a \mid a \in A\}$$

as $\mathbb{HSP}_{\text{fin}}$ is a continuous closure operator. The equality

$$\mathsf{Rq}\left(\bigcup\{\mathbf{V}_a \mid a \in A\}\right) = \bigcup\{\mathsf{Rq}(\mathbf{V}_a) \mid a \in A\} \tag{2.14}$$

follows directly from Definition 2.3.1. Now the right-hand side of (2.14) is directed, as $\mathsf{Rq}$ is order preserving, and thus equals $\bigvee\{\mathsf{Rq}(\mathbf{V}_a) \mid a \in A\}$ as directed sups in $\mathbf{PV}$ are just unions. Thus $\mathsf{Rq}$ is continuous and so $\mathsf{q}$ is well defined. It is clear that $\mathsf{q}$ is order preserving.                         □

*Remark 2.3.3.* If $\mathbf{V}$ is just closed under $\mathbb{P}_{\text{fin}}$, then the above proof shows $\mathsf{Rq}(\mathbf{V})$ is still a pseudovariety. Also the above proof only requires R to satisfy Axiom ($\times$) and to be closed under precomposition by divisions.

Note that $\mathsf{q}$ can be viewed as a function $\mathbf{CC} \times \mathbf{PV} \to \mathbf{PV}$ via the map $(\mathsf{R}, \mathbf{V}) \mapsto \mathsf{Rq}(\mathbf{V})$. If $\mathbf{V}$ is held fixed, then $(-)\mathsf{q}\mathbf{V} \colon \mathbf{CC} \to \mathbf{PV}$ turns out to be continuous.

**Exercise 2.3.4.** Verify that the map $\mathsf{q} \colon \mathbf{CC} \times \mathbf{PV} \to \mathbf{PV}$ is continuous where $\mathbf{CC} \times \mathbf{PV}$ is given the product ordering.

*Remark 2.3.5.* We make a brief historical digression here. Since the advent of Eilenberg's book [85], pseudovarieties have been defined, both implicitly and explicitly, by classes of relational morphisms. The Mal'cev product, for instance, is most commonly defined in terms of relational morphisms. In [224, 293, 362, 379] one defines a $\mathbf{V}$-relational morphism to be a relational morphism such that the inverse image of a semigroup in $\mathbf{V}$ still belongs to $\mathbf{V}$. In our notation, the collection of $\mathbf{V}$-relational morphisms is $(\mathbf{V}, \mathbf{V})$. The collection of semigroups with a $\mathbf{V}$-relational morphism to $\mathbf{W}$ forms a pseudovariety (which shall be denoted later $(\mathbf{V}, \mathbf{V}) \, \textcircled{m} \, \mathbf{W}$, although $\mathbf{V}^{-1}\mathbf{W}$ is sometimes used in the literature [224, 379]).

Tilson's seminal paper [364] allowed for a definition of the semidirect product of pseudovarieties in terms of relational morphisms. This idea was taken up by Jones and Pustejovsky [150] to define a semidirect product of pseudovarieties of categories via relational morphisms; they didn't see how to directly define the semidirect product of two categories (see [339] for the correct way to do this).

Similarly, one can define in terms of relational morphisms the semidirect product of a pseudovariety of semigroupoids with a pseudovariety of semigroups. Such definitions were given (or at least have forbearers) in [27, 234, 293, 357].

As another example, the second author showed in [322, 330] how to define the join operator in terms of relational morphisms.

In [339] Tilson and the second author allowed certain classes of relational morphisms to act on the collection of varieties (and pseudovarieties) of categories. The explicit definition of q was tossed around at that time, but not written down for two reasons: firstly, they could not decide what should be the domain of the q function; secondly, their "internal" picture or formulation was of "pseudovarieties" of relational morphisms as acting on the lattice of pseudovarieties and what we now call q as merely the transition morphism of the action.

In this book, much under the impetus of the first author, we place the main focus on the map q itself. Here q is considered simultaneously as a monoid homomorphism, as the left adjoint of a Galois connection between algebraic lattices (and hence a morphism of such; in fact a quantale morphism) and as a continuous map of topological spaces. The action point of view, only considers the monoid homomorphism and so loses important information. For instance, from the action point of view, one does not even have the notation to formulate the question as to whether or not q preserves infinite meets. The fact that q sends finite meets to pointwise meets, but not infinite meets, underlies an unfortunate error in the literature [27, Section 4.2, p. 51]. This book deals systematically with order and lattice theoretic aspects of the theory of finite semigroups to clarify such points.

From the Galois connection point of view, one easily characterizes which operators are in the image of q and shows that each such operator comes from a maximal continuously closed class. This is important when applying the theory, as it provides a criterion to determine whether the theory applies to a particular operator.

We have seen that each continuously closed class gives rise to a continuous operator via q. We now aim to show that the converse holds, hence the name continuously closed! This will also give us a new proof that $\mathbf{Cnt}(\mathbf{PV})$ is an algebraic lattice. Let us define a section $M \colon \mathbf{Cnt}(\mathbf{PV}) \to \mathbf{CC}$ (the capital $M$ is for max) to q by the formula:

$$M(\alpha) = \{f \colon S \to T \mid S \in \alpha((T))\}. \tag{2.15}$$

Notice that $M(\alpha)$ is entirely determined by the pairs $(S, T)$ with $S \in \alpha((T))$, i.e., the "graph" of $\alpha$ (flipped).

Remark 2.3.6. It should be remarked that $M(\alpha) = \{f \colon S \to T \mid \delta(S, T) \le \alpha\}$ where $\delta(S, T)$ is as per Proposition 2.2.2.

In the following theorems, we use our convention for composition for functions acting on the right. This is because q will turn out to be a monoid homomorphism (in fact a quantale morphism) and also because we write it on the right of its argument.

**Theorem 2.3.7.** *The map* $M\colon \mathbf{Cnt}(\mathbf{PV}) \to \mathbf{CC}$ *is a well-defined continuous map and* $M\mathfrak{q} = 1_{\mathbf{Cnt}(\mathbf{PV})}$. *In particular,* $\mathfrak{q}$ *is* **onto** *and* $M$ *is injective. Moreover,* $M(\alpha)$ *is an equational Birkhoff continuously closed class for* $\alpha \in \mathbf{Cnt}(\mathbf{PV})$.

*Proof.* Let $\alpha \in \mathbf{Cnt}(\mathbf{PV})$. Define $M$ by (2.15). We show $M(\alpha)$ is an equational Birkhoff continuously closed class. To verify Axiom $(\prec_s)$, suppose $(d,k)\colon f \to g$ is a strong division with $g \in M(\alpha)$. Say $f\colon S \to T$ and $g\colon S' \to T'$. Then $S \prec S' \in \alpha((T'))$ and $T' \prec T$ (as $k$ is onto). Thus $S \in \alpha((T))$, so $f \in M(\alpha)$.

We now turn to Axiom (pb) (note that this implies Axiom ($\times$) by Proposition 2.1.8(d)). Because $\{1\} \in \alpha(\mathbf{1})$, $1_{\{1\}} \in M(\alpha)$. Suppose that $f_1\colon S_1 \to T$, $f_2\colon S_2 \to T$ are in $M(\alpha)$. Then $S_1 \in \alpha((T))$, $S_2 \in \alpha((T))$, and so

$$S_1 \times_{f_1,f_2} S_2 \leq S_1 \times S_2 \in \alpha((T)).$$

Thus $f_1 \times_T f_2 \in M(\alpha)$.

For Axiom (r-e), suppose $f\colon S \to T$ belongs to $M(\alpha)$ and $i\colon T \to T'$ is an inclusion. Then $S \in \alpha((T)) \leq \alpha((T'))$, so $fi \in M(\alpha)$. We may conclude that $M(\alpha) \in \mathbf{CC}$, and is in fact equational. To see that it is Birkhoff, we establish that $M(\alpha)$ satisfies Axiom (free). Indeed, if $f\colon S \to T^n$ belongs to $M(\alpha)$, some $n \geq 1$, then $S \in \alpha((T))$. But $(T) = (F_{(T)}(S \times T))$ as the projection $p_T\colon S \times T \twoheadrightarrow T$ extends to a surjective homomorphism $F_{(T)}(S \times T) \twoheadrightarrow T$, showing $T \in (F_{(T)}(S \times T))$; the other inclusion is of course immediate. Therefore, $S \in \alpha((F_{(T)}(S \times T)))$ and so $\rho_{(T)}\colon S \to F_{(T)}(S \times T)$ belongs to $M(\alpha)$, as required.

It is immediate from the definition that $M$ is order preserving. We now verify $M$ continuous. Let $\{\alpha_d\}_{d\in D} \subseteq \mathbf{Cnt}(\mathbf{PV})$ be directed and set $\alpha = \bigvee_{d\in D} \alpha_d$; notice that $\{M(\alpha_d) \mid d \in D\}$ is directed because $M$ is order preserving. Then, for all $(T) \in K(\mathbf{PV})$, $\{\alpha_d((T)) \mid d \in D\}$ is directed and so, as directed suprema in $\mathbf{PV}$ are unions, we have

$$\alpha((T)) = \bigvee_{d\in D} \alpha_d((T)) = \bigcup_{d\in D} \alpha_d((T)).$$

So $f\colon S \to T$ is in $M(\alpha)$ if and only if $S \in \alpha((T))$, if and only if $S \in \alpha_d(T)$ some $d \in D$, if and only if $f \in M(\alpha_d)$ for some $d \in D$, if and only if $f$ belongs to $\bigvee_{d\in D} M(\alpha_d) = \bigcup_{d\in D} M(\alpha_d)$. Therefore, $M$ is continuous.

We now show that if $\alpha \in \mathbf{Cnt}(\mathbf{PV})$, then $M(\alpha)\mathfrak{q} = \alpha$. By continuity, it suffices to check that they agree on compact elements. Suppose $(T)$ is compact and $S \in \alpha((T))$. Let $f\colon S \to T$ be the universal relational morphism, that is,

$\#f = S \times T$. Then $f \in M(\alpha)$ and so $S \in M(\alpha)\mathsf{q}((T))$. Conversely, suppose $S \in M(\alpha)\mathsf{q}((T))$. Then there exists $f\colon S \to T'$ with $f \in M(\alpha)$ and $T' \in (T)$. But then, by construction, $S \in \alpha((T')) \leq \alpha((T))$, as desired. $\qquad\square$

The following lemma shows that $\mathsf{q}$ preserves finite meets of **CC**. In fact, because $\mathsf{q}$ is onto, it also provides another proof that finite meets in $\mathbf{Cnt}(\mathbf{PV})$ are pointwise meets.

**Lemma 2.3.8.** *Let* $\mathsf{R}_1, \dots, \mathsf{R}_n \in \mathbf{CC}$*. Then we have*

$$(\mathsf{R}_1 \cap \cdots \cap \mathsf{R}_n)\mathsf{q}(\mathbf{V}) = \mathsf{R}_1\mathsf{q}(\mathbf{V}) \cap \cdots \cap \mathsf{R}_n\mathsf{q}(\mathbf{V}).$$

*In particular, finite meets in* $\mathbf{Cnt}(\mathbf{PV})$ *are pointwise and* $\mathsf{q}$ *preserves finite meets.*

*Proof.* Because $\mathsf{q}$ preserves order, evidently we have

$$(\mathsf{R}_1 \cap \cdots \cap \mathsf{R}_n)\mathsf{q}(\mathbf{V}) \leq \mathsf{R}_1\mathsf{q}(\mathbf{V}) \cap \cdots \cap \mathsf{R}_n\mathsf{q}(\mathbf{V})$$

for every pseudovariety $\mathbf{V}$ of semigroups. For the other direction, suppose $S \in \mathsf{R}_i\mathsf{q}(\mathbf{V})$, all $i$. Hence there are relational morphisms $f_i\colon S \to T_i$ with $T_i \in \mathbf{V}$ and $f_i \in \mathsf{R}_i$, all $i$. Proposition 1.3.16(b) then shows that

$$g = \Delta(f_1 \times \cdots \times f_n) \prec_s f_1, \dots, f_n$$

so $g \in \bigcap \mathsf{R}_i$. Moreover, $T_1 \times \cdots \times T_n \in \mathbf{V}$, so $S \in (\mathsf{R}_1 \cap \cdots \cap \mathsf{R}_n)\mathsf{q}(\mathbf{V})$, as desired. In particular, it follows that the pointwise meet of the $\mathsf{R}_i\mathsf{q}$ is a continuous operator and hence must be the determined meet of the $\mathsf{R}_i\mathsf{q}$ in $\mathbf{Cnt}(\mathbf{PV})$. Therefore, $\mathsf{q}$ preserves finite meets. Because $\mathsf{q}$ is onto, it in fact follows that all meets in $\mathbf{Cnt}(\mathbf{PV})$ must be pointwise. $\qquad\square$

The stage is now set to establish that $M$ and $\mathsf{q}$ form a Galois connection.

**Theorem 2.3.9.** *There is a Galois connection:*

$$(\mathbf{CC}, \subseteq, \vee_{\det}, \cap) \xrightleftharpoons[\mathsf{q}]{M} (\mathbf{Cnt}(\mathbf{PV}), \leq_{\mathrm{pw}}, \vee_{\mathrm{pw}}, \wedge_{\det}). \qquad (2.16)$$

*Hence* $\mathsf{q}$ *is* **sup**, *$M$ is both* **inf** *and continuous, and* $M\mathsf{q} = 1_{\mathbf{Cnt}(\mathbf{PV})}$*. Thus* $\mathbf{Cnt}(\mathbf{PV})$ *is an algebraic lattice whose compact elements are the* $\mathsf{q}$*-images of the compact elements of* **CC**. *Moreover, the map* $\mathsf{q}$ *preserves* **finite** *meets. In particular,* $M(\alpha)$ *is the maximum continuously closed class with* $\mathsf{q}$*-image* $\alpha$.

*Proof.* Lemma 2.3.8 shows that $\mathsf{q}$ preserves finite meets. By Proposition 1.1.7, to establish that (2.16) is a Galois connection, it suffices to verify that, for $\alpha \in \mathbf{Cnt}(\mathbf{PV})$, one has that $M(\alpha)$ is the largest element of **CC** mapping to $\alpha$ under $\mathsf{q}$. So suppose $\mathsf{R}\mathsf{q} = \alpha$ and let $f\colon S \to T$ be in R. Then $S \in \mathsf{R}\mathsf{q}((T)) = \alpha((T))$ and so $f \in M(\alpha)$ by the definition of $M$. We conclude $\mathsf{R} \leq M(\mathsf{R}\mathsf{q})$. The remaining assertions follow from the general theory of Galois connections (Proposition 1.1.7), Theorem 2.3.7 and Corollary 1.2.37. $\qquad\square$

The above theorem can easily be adapted to the positive setting.

**Theorem 2.3.10.** *There is a Galois connection*

$$(\mathbf{CC}^+, \subseteq, \vee_{\det}, \cap) \xrightleftharpoons[\mathfrak{q}]{M} (\mathbf{Cnt}(\mathbf{PV})^+, \leq_{\mathrm{pw}}, \vee_{\mathrm{pw}}, \wedge_{\det}). \qquad (2.17)$$

*Here $\mathfrak{q}$ is onto, **sup** and preserves **finite** meets. The map $M$ is injective, **inf** and continuous. Moreover, $M(\alpha)$ is an equational Birkhoff continuously closed class, and so it satisfies the conclusion of Tilson's Lemma. Thus $\mathbf{Cnt}(\mathbf{PV})^+$ is an algebraic lattice with set of compact elements the $\mathfrak{q}$-image of the set of compact elements of $\mathbf{CC}^+$.*

*Proof.* The key point here is to verify that $\mathfrak{q}$ and $M$ restrict properly. Suppose that $\mathsf{R} \in \mathbf{CC}^+$. Let $\mathbf{V}$ be a pseudovariety and let $S \in \mathbf{V}$. Then $1_S \in \mathsf{R}$ and so we obtain that $S \in \mathsf{R}\mathfrak{q}(\mathbf{V})$. Thus $\mathbf{V} \leq \mathsf{R}\mathfrak{q}(\mathbf{V})$ and so $\mathsf{R}\mathfrak{q} \in \mathbf{Cnt}(\mathbf{PV})^+$.

If $\alpha \in \mathbf{Cnt}(\mathbf{PV})^+$, then for any finite semigroup $S$, we have $S \in \alpha((S))$, and thus $1_S \colon S \to S$ belongs to $M(\alpha)$. Hence $M(\alpha) \in \mathbf{CC}^+$. The remaining details of the theorem are straightforward and we omit them.    □

To show that Lemma 2.3.8 is sharp, we shall need the following lemma concerning one-generated continuously closed classes.

**Lemma 2.3.11.** *Let $f\colon S \to T$ be a relational morphism and let $g\colon S' \to T'$ belong to the continuously closed class generated by $f$. If $\emptyset \neq S' \neq \{1\}$, then $T \prec T'$.*

*Proof.* According to Proposition 2.1.14, we must have $g \subseteq_s d_1 f^n d_2$ for some $n \geq 0$ where $d_1, d_2$ are divisions. We cannot have $n = 0$ by our assumption on $S'$. If $n \geq 1$, then $d_2 \colon T^n \to T'$ and so $T \prec T^n \prec T'$, as required.    □

*Example 2.3.12 ($\mathfrak{q}$ is not **inf** on $\mathbf{CC}$).* This example shows that $\mathfrak{q}$ is not **inf** on $\mathbf{CC}$. Fix non-trivial semigroups $S$ and $T$. For $n \geq 1$, let $u_n \colon S \to T^n$ be the universal relation (so $\#u_n = S \times T^n$). Let $\mathsf{R}_n = (u_n)$. We claim $\bigcap \mathsf{R}_n$ consists of all relational morphisms with domain $\emptyset$ or $\{1\}$, the bottom of $\mathbf{CC}$. Indeed, all such relational morphisms belong to any continuously closed class. On the other hand, if $g\colon S' \to T'$ belongs to $\bigcap \mathsf{R}_n$ with $\emptyset \neq S' \neq \{1\}$, then Lemma 2.3.11 implies $T^n \prec T'$ for all $n \geq 0$, which is impossible.

Now $(\bigcap \mathsf{R}_n)\mathfrak{q}$ is easily seen to be the constant map $\mathfrak{q}_1$ to the trivial pseudovariety $\mathbf{1}$ (see Example 2.4.2). Recall $\delta(S,T)\colon \mathbf{PV} \to \mathbf{PV}$, defined by

$$\delta(S,T)(\mathbf{V}) = \begin{cases} (S) & T \in \mathbf{V} \\ \mathbf{1} & \text{otherwise} \end{cases}$$

is continuous (Proposition 2.2.2). Let us verify $\delta(S,T) \leq \mathsf{R}_n\mathfrak{q}$ for all $n \geq 1$. Indeed, if $T \in \mathbf{V}$, then because $T^n \in \mathbf{V}$ and $u_n \colon S \to T^n$ is in $\mathsf{R}_n$, we have $S \in \mathsf{R}_n\mathfrak{q}(\mathbf{V})$. This proves $\bigwedge \mathsf{R}_n\mathfrak{q} \geq \delta(S,T) > \mathfrak{q}_1$. Thus $\mathfrak{q}$ is not **inf** on $\mathbf{CC}$.

To give an analogous example for $\mathbf{CC}^+$, we have the following version of Lemma 2.3.11.

**Lemma 2.3.13.** *Let $f\colon S \to T$ be a relational morphism and let $g\colon S' \to T'$ belong to the positive continuously closed class generated by $f$. If $g$ is not a division, then $T \prec T'$.*

*Proof.* It is not hard to see there are divisions $d_1$, $d_2$ so that $g \subseteq_s d_1(1_R \times f^n)d_2$ for some $n \geq 0$ and some semigroup $R$. We cannot have $n = 0$, because in this case $g$ would be a division. If $n \geq 1$, then $d_2\colon R \times T^n \to T'$ and so $T \prec R \times T^n \prec T'$, as required. $\qquad\square$

*Example 2.3.14 ($\mathsf{q}$ is not **inf** on $\mathbf{CC}^+$).* This example shows that $\mathsf{q}$ is not **inf** on $\mathbf{CC}^+$. Fix non-trivial semigroups $S$ and $T$ with $S \notin (T)$. For $n \geq 1$, let $u_n\colon S \to T^n$ be the universal relation (so $\#u_n = S \times T^n$). Let $\mathsf{R}_n = (u_n)$, where this time we are considering the positive continuously closed class generated by $u_n$. We claim $\bigcap \mathsf{R}_n$ consists of all divisions and hence is the bottom of $\mathbf{CC}^+$. For if $g\colon S' \to T'$ in $\bigcap \mathsf{R}_n$ is not a division, then Lemma 2.3.13 implies $T^n \prec T'$ for all $n \geq 0$, which is impossible.

Now $(\bigcap \mathsf{R}_n)\mathsf{q}$ is easily seen to be the identity map on $\mathbf{PV}$; see Example 2.4.1. On the other hand $\mathsf{R}_n \geq \delta(S,T) \vee 1_{\mathbf{PV}}$ because if $T \in \mathbf{V}$, then $S \in \mathsf{R}_n\mathsf{q}(\mathbf{V})$ for all $n$. But because joins in $\mathbf{Cnt}(\mathbf{PV})^+$ are pointwise,

$$(\delta(S,T) \vee 1_{\mathbf{PV}})((T)) = (S) \vee (T) > (T).$$

Thus $\bigwedge \mathsf{R}_n\mathsf{q} > 1_{\mathbf{PV}}$ and so $\mathsf{q}$ is not **inf** on $\mathbf{CC}^+$.

### 2.3.1 Remarks on the determined meet of Cnt(PV) versus the pointwise meet

Proposition 2.3.9 tells us that $\mathbf{Cnt}(\mathbf{PV})$ is an algebraic lattice with the join taken pointwise over arbitrary subsets. But what is the determined meet $\wedge_{\mathrm{det}}$? As mentioned earlier, the pointwise meet of a finite number of continuous operators is continuous (and hence is the determined meet of this finite collection). But for infinite collections, the pointwise meet need *not* be continuous (see Proposition 2.3.15 below).

If meets in $\mathbf{Cnt}(\mathbf{PV})$ were determined pointwise, it would be quite easy to find a basis of pseudoidentities for $\mathsf{Rq}(\mathbf{V})$, given a basis of pseudoidentities (in the sense to be defined in the next chapter) for $\mathsf{R}$ in the case $\mathsf{R}$ is a pseudovariety of relational morphisms. The same would be true for composition of classes of relational morphisms (defined later in this chapter) and so there would be an avenue to understand iteration of operators via their equational theories. Consequently, Finite Semigroup Theory is a more difficult subject because the meet in $\mathbf{Cnt}(\mathbf{PV})$ is not pointwise.

The assumption that the meet in $\mathbf{Cnt}(\mathbf{PV})$ is pointwise was implicitly made in [27, Section 4.2, p. 51] and underlies the proof of the Basis Theorem

for the Semidirect Product [27, Thm. 5.2]. This unfortunate error invalidates
the main result of the joint work of Almeida and the second author [19], and
hence the first author's proof of the decidability of complexity via the notion
of "tameness" [282].

Let us now turn to an example showing that the meet in $\mathbf{Cnt(PV)}$ is not
pointwise.

**Proposition 2.3.15.** *In* $\mathbf{Cnt(PV)}^{+}$ *there exists a decreasing sequence of ele-*
*ments* $\alpha_1 \geq \alpha_2 \geq \cdots$ *so that* $\bigwedge_{pw} \alpha_i$ *is* **not** *continuous. Hence in the algebraic*
*lattice* $(\mathbf{Cnt(PV)}^{+}, \leq_{pw}, \vee_{pw}, \wedge_{det})$ *the determined meet is not the pointwise*
*meet.*

*Proof.* Consider the order preserving, discontinuous operator $\alpha$ from the proof
of Proposition 2.2.5 defined by (2.12). We obtain $\alpha$ as the pointwise meet of
a decreasing sequence of operators from $\mathbf{Cnt(PV)}^{+}$.

To begin with, we enumerate the pseudovariety $\mathbf{DS}$. Let us suppose that
$\mathbf{DS} = \{T_1, T_2, T_3, \ldots\}$. Next, define $g_j \colon K(\mathbf{PV}) \to \mathbf{PV}$, $j \geq 1$, by

$$g_j((T)) = \begin{cases} (T) \vee (B_2) & \text{if } T_1, \ldots, T_j \in (T) \\ (T) & \text{otherwise.} \end{cases} \tag{2.18}$$

Then by Proposition 2.2.2(a), $\bar{g}_j \in \mathbf{Cnt(PV)}$. To decongest notation let us
set $\alpha_j = \bar{g}_j$. It is easy to check that

$$\alpha_j(\mathbf{V}) = \begin{cases} \mathbf{V} \vee (B_2) & \text{if } T_1, \ldots, T_j \in \mathbf{V} \\ \mathbf{V} & \text{otherwise.} \end{cases} \tag{2.19}$$

From (2.19) one readily verifies $\alpha_1 \geq \alpha_2 \geq \cdots$.

We claim that the equality

$$\bigwedge_{pw} \alpha_j = \alpha \tag{2.20}$$

holds. Clearly, $\alpha_j \geq \alpha$ for all $j \geq 1$. If $\mathbf{V} \not\leq \mathbf{DS}$, then

$$\alpha(\mathbf{V}) = \alpha_j(\mathbf{V}) = \begin{cases} \mathbf{DS} \vee (B_2) & \text{for } \mathbf{V} = \mathbf{DS} \\ \mathbf{V} & \text{otherwise} \end{cases}$$

as every pseudovariety non-contained in $\mathbf{DS}$ contains $B_2$ by Lemma 2.2.3.
Thus we need only prove $\bigcap \alpha_i(\mathbf{V}) = \alpha(\mathbf{V}) = \mathbf{V}$ for $\mathbf{V} < \mathbf{DS}$. Because $\mathbf{V} < \mathbf{DS}$
there exists $j$ so that $T_j \notin \mathbf{V}$ and thus

$$\alpha_{j+n}(\mathbf{V}) = \mathbf{V} = \alpha(\mathbf{V}), \ n \geq 0.$$

This proves (2.20) and hence Proposition 2.3.15. $\qquad\qquad\square$

Notice that the determined meet of the $\alpha_j$ is in fact the identity opera-
tor. Indeed, $1_{\mathbf{V}} \leq \alpha$ and is the largest continuous operator below $\alpha$ because
$\mathbf{DS} \vee (B_2)$ covers $\mathbf{DS}$.

We now give a formula for the determined meet in $\mathbf{Cnt}(\mathbf{PV})$ (and
$\mathbf{Cnt}(\mathbf{PV})^+$). In the process, we shall see that the "problem" arising in Proposition 2.3.15 does not occur for compact elements.

**Proposition 2.3.16.** *Suppose* $\{\alpha_i \mid i \in I\} \subseteq \mathbf{Cnt}(\mathbf{PV})$. *Then*

$$\bigwedge_{\det} \alpha_i = \left( \bigcap_I M(\alpha_i) \right) \mathfrak{q}. \tag{2.21}$$

*Moreover, if* $\mathbf{V} \in K(\mathbf{PV})$, *there results an equality*

$$\bigwedge_{\det} \alpha_i(\mathbf{V}) = \bigcap_I \alpha_i(\mathbf{V}). \tag{2.22}$$

*That is,* $\wedge_{\det}$ *is pointwise on* **compact** *elements and determined by Proposition 2.2.2(a) elsewhere.*

*Proof.* Clearly, the right-hand side of (2.21) is below the left-hand side as
$M(\alpha_i)\mathfrak{q} = \alpha_i$. On the other hand, the left-hand side is plainly below the
pointwise meet of the $\alpha_i$. Thus, by Proposition 2.2.2, to prove (2.21) and
(2.22), it suffices to prove

$$\bigcap_I \alpha_i((T)) \subseteq \left( \bigcap_I M(\alpha_i) \right) \mathfrak{q}((T))$$

for any finite semigroup $T$. Suppose $S \in \bigcap_I \alpha_i((T))$ for all $i$ and let $f \colon S \to T$
be any relational morphism (say the universal relation). By formula (2.15),
$f \in M(\alpha_i)$ all $i$. Thus $S \in (\bigcap_I M(\alpha_i)) \mathfrak{q}((T))$, as desired. $\qquad \square$

### 2.3.2 The generalized Mal'cev product and global Mal'cev condition

We shall state and prove the analogues of Proposition 2.1.11, Theorem 2.3.9
and Proposition 2.3.16 for **PVRM** after a few important definitions. In particular, we must define the type of continuous operators one obtains from
pseudovarieties of relational morphisms.

**Definition 2.3.17 (Generalized Mal'cev product).** *Let* $\mathbf{U}, \mathbf{V}, \mathbf{W} \in \mathbf{PV}$.
*Then by definition their generalized Mal'cev product is* $(\mathbf{U}, \mathbf{V})\mathfrak{q}\mathbf{W}$; *see* (2.5).
*More precisely, we have*

$$(\mathbf{U}, \mathbf{V}) \textcircled{m} \mathbf{W} = \{S \mid \exists f \colon S \to W \text{ a relational morphism with}$$
$$W \in \mathbf{W} \text{ so that } W \geq V \in \mathbf{V} \text{ implies } Vf^{-1} \in \mathbf{U}\}. \tag{2.23}$$

Note that

$$(\mathbf{V}, 1)\mathsf{q}(\mathbf{W}) = (\mathbf{V}, 1) \ \textcircled{m} \ \mathbf{W} = \mathbf{V} \ \textcircled{m} \ \mathbf{W}$$

is the usual Mal'cev product [126] of $\mathbf{V}$ and $\mathbf{W}$ (whence the name). We could vary any of the three variables $\mathbf{U}, \mathbf{V}, \mathbf{W}$ in $(\mathbf{U}, \mathbf{V}) \ \textcircled{m} \ \mathbf{W}$ to obtain various continuous operators on $\mathbf{PV}$.

*Remark 2.3.18.* In general, one cannot replace relational morphism by homomorphism in the definition of $\mathsf{q}$, even for pseudovarieties of relational morphisms of the form $(\mathbf{V}, \mathbf{V})$. Indeed, if $\mathbf{V} = \mathbf{Sl}$ (the pseudovariety of semilattices), then every inverse semigroup belongs to $(\mathbf{Sl}, \mathbf{Sl}) \ \textcircled{m} \ \mathbf{G}$, but an inverse semigroup $S$ has a morphism $f \colon S \to G$ with $G$ a group and $f \in (\mathbf{Sl}, \mathbf{Sl})$ if and only if $S$ is $E$-unitary [176]. In Section 4.6 we consider certain pseudovarieties $\mathbf{V}$ where one only needs to consider homomorphisms.

In Chapter 3, we shall give a basis of pseudoidentities for $(\mathbf{U}, \mathbf{V}) \ \textcircled{m} \ \mathbf{W}$, as well as several generalizations of results of [235] for the usual Mal'cev product. The following easy formula will be used without further comment throughout this text.

**Proposition 2.3.19.** *Let $\mathbf{U}, \mathbf{V}, \mathbf{W}, \mathbf{Z}$ be pseudovarieties. Suppose $f \in (\mathbf{U}, \mathbf{V})$ and $g \in (\mathbf{V}, \mathbf{W})$ are composable relational morphisms. Then $fg \in (\mathbf{U}, \mathbf{W})$. As a consequence, the following inequality holds*

$$(\mathbf{U}, \mathbf{V}) \ \textcircled{m} \ ((\mathbf{V}, \mathbf{W}) \ \textcircled{m} \ \mathbf{Z}) \subseteq (\mathbf{U}, \mathbf{W}) \ \textcircled{m} \ \mathbf{Z}. \qquad (2.24)$$

*Proof.* Let $f \colon S \to T$ and $g \colon T \to Z$ be relational morphisms with $f \in (\mathbf{U}, \mathbf{V})$ and $g \in (\mathbf{V}, \mathbf{W})$. If $Z \geq W \in \mathbf{W}$, then, because $Wg^{-1} \in \mathbf{V}$, it follows that $W(fg)^{-1} = Wg^{-1}f^{-1} \in \mathbf{U}$. Thus $fg \in (\mathbf{U}, \mathbf{W})$, as required.

To prove (2.24), suppose $S \in (\mathbf{U}, \mathbf{V}) \ \textcircled{m} \ ((\mathbf{V}, \mathbf{W}) \ \textcircled{m} \ \mathbf{Z})$. Let $f \colon S \to T$ with $T \in (\mathbf{V}, \mathbf{W}) \textcircled{m} \mathbf{Z}$ and $f \in (\mathbf{U}, \mathbf{V})$. Then there exists $g \colon T \to Z$ with $Z \in \mathbf{Z}$ and with $g \in (\mathbf{V}, \mathbf{W})$. By the above, $fg \colon S \to Z$ belongs to $(\mathbf{U}, \mathbf{W})$, witnessing $S \in (\mathbf{U}, \mathbf{W}) \ \textcircled{m} \ \mathbf{Z}$. $\qquad \square$

Proposition 2.3.19 admits a useful generalization, which we leave as an exercise.

**Exercise 2.3.20.** Show that if $\mathbf{U}, \mathbf{V}, \mathbf{W}, \mathbf{X}, \mathbf{Y} \in \mathbf{PV}$, then

$$(\mathbf{U}, \mathbf{V}) \ \textcircled{m} \ ((\mathbf{W}, \mathbf{X}) \ \textcircled{m} \ \mathbf{Y}) \subseteq ((\mathbf{U}, \mathbf{V}) \ \textcircled{m} \ \mathbf{W}, \mathbf{X}) \ \textcircled{m} \ \mathbf{Y}.$$

Deduce $\mathbf{U} \ \textcircled{m} \ (\mathbf{V} \ \textcircled{m} \ \mathbf{W}) \subseteq (\mathbf{U} \ \textcircled{m} \ \mathbf{V}) \ \textcircled{m} \ \mathbf{W}$ and (2.24).

Note that Definition 2.3.17 can be used in the context of monoids. However, one must then take $V$ to be a submonoid of $W$ and $\mathbf{U}$, $\mathbf{V}$, and $\mathbf{W}$ to be pseudovarieties of monoids. For example, in the monoidal setting $(\mathbf{V}, 1) \ \textcircled{m} \ \mathbf{W}$ consists of all monoids $M$ with a relational morphism to $\mathbf{W}$ such that the inverse image of 1 belongs to $\mathbf{V}$. This is, in general, different than $\mathbf{V} \ \textcircled{m} \ \mathbf{W}$

(which is defined in terms of the inverse images of idempotents, as usual, and here $\mathbf{V}$ is a pseudovariety of semigroups). For this reason, q-theory works best as a theory of semigroups, rather than as a theory of monoids. We shall see this difference arise again in Section 3.5.2.

**Definition 2.3.21 (Global Mal'cev condition).** *A continuous operator $\alpha$ on* $\mathbf{PV}$ *satisfies the global Mal'cev condition (GMC) if, for all* $\mathbf{V}, \mathbf{W} \in \mathbf{PV}$,

$$\alpha(\mathbf{W}) \leq (\alpha(\mathbf{V}), \mathbf{V}) \,\textcircled{m}\, \mathbf{W}. \tag{2.25}$$

An important special case is GMC at $\mathbf{1}$, that is, the case where $\mathbf{V}$ is the trivial pseudovariety $\mathbf{1}$. In this case, (2.25) becomes

$$\alpha(\mathbf{W}) \leq \alpha(\mathbf{1}) \,\textcircled{m}\, \mathbf{W} \tag{2.26}$$

for all pseudovarieties $\mathbf{W}$. Example 2.4.28 will demonstrate that a continuous operator can satisfy (2.26) but fail to satisfy (2.25). We remark that continuity constitutes part of the definition of GMC.

It turns out the global Mal'cev condition is precisely the extra condition needed for a continuous operator to come from a pseudovariety of relational morphisms via q. First we consider some basic properties of GMC.

**Proposition 2.3.22.** *Suppose* $\alpha \colon \mathbf{PV} \to \mathbf{PV}$ *satisfies GMC and* $\mathbf{W}$ *is a pseudovariety of semigroups. Then*

$$\alpha(\mathbf{W}) = \bigcap_{\mathbf{V} \in \mathbf{PV}} ((\alpha(\mathbf{V}), \mathbf{V}) \,\textcircled{m}\, \mathbf{W}). \tag{2.27}$$

*Proof.* By the global Mal'cev condition, the left-hand side of (2.27) is contained in the right-hand side. Suppose $S$ belongs to the right-hand side. Taking $\mathbf{V} = \mathbf{W}$ in (2.25) yields $S \in (\alpha(\mathbf{W}), \mathbf{W}) \,\textcircled{m}\, \mathbf{W}$. Then there is a relational morphism $f \colon S \to T$ with $T \in \mathbf{W}$ and such that the inverse image of any subsemigroup of $T$ belonging to $\mathbf{W}$ is in $\alpha(\mathbf{W})$. In particular, it follows that $S = Tf^{-1} \in \alpha(\mathbf{W})$. □

For operators satisfying GMC, there is a simple criterion for preserving local finiteness.

**Proposition 2.3.23.** *Suppose that* $\alpha \colon \mathbf{V} \to \mathbf{V}$ *satisfies GMC. Then* $\alpha$ *preserves the set of locally finite pseudovarieties if and only if* $\alpha(\mathbf{1})$ *is locally finite.*

*Proof.* Because $\mathbf{1}$ is locally finite, necessity is clear. For sufficiency, suppose $\mathbf{V}$ is locally finite. Then by GMC, we have

$$\alpha(\mathbf{V}) \leq \alpha(\mathbf{1}) \,\textcircled{m}\, \mathbf{V}. \tag{2.28}$$

Brown's Theorem [59] implies that the Mal'cev product of locally finite pseudovarieties is locally finite, so the right-hand side of (2.28) is locally finite, and hence the left-hand side is, as well. □

*Remark 2.3.24.* A proof of Brown's Theorem will be given in Section 4.2.

We mention that $\alpha(\mathbf{1})$ compact does not imply that $\alpha$ preserves compact pseudovarieties. Recall that a semigroup is called a *band* if all its elements are idempotent. A band is called a *rectangular band* if it satisfies the identity $xyx = x$. The pseudovariety $\mathbf{RB}$ of rectangular bands is compact. If we set $\alpha = \mathbf{RB} \circledm (-)$, then $\alpha(\mathbf{Sl}) = \mathbf{B}$ where $\mathbf{Sl}$ is the pseudovariety of semilattices (a compact pseudovariety) and $\mathbf{B}$ is the pseudovariety of bands. However, $\mathbf{B}$ is well-known not to be compact [7, Section 5.5]. Note, however, that $\mathbf{B}$ is locally finite, so the above proposition is verified in this case.

The notation for the set of all $\alpha \in \mathbf{Cnt}(\mathbf{PV})$ satisfying GMC will be $\mathbf{GMC}(\mathbf{PV})$. Following our previous conventions, we set

$$\mathbf{GMC}(\mathbf{PV})^+ = \mathbf{GMC}(\mathbf{PV}) \cap \mathbf{Cnt}(\mathbf{PV})^+.$$

Also let us define $\mathbf{Cnt}(\mathbf{PV})^-$ to consist of those $\alpha \in \mathbf{Cnt}(\mathbf{PV})$ such that $\alpha \leq 1_{\mathbf{PV}}$. Of course, $\mathbf{Cnt}(\mathbf{PV})^- \cap \mathbf{GMC}(\mathbf{PV})$ is denoted $\mathbf{GMC}(\mathbf{PV})^-$. For example $\mathbf{V} \cap (-)$ belongs to $\mathbf{GMC}(\mathbf{PV})^-$. Now we can explain our choices of axioms systems for $\mathbf{CC}$ and $\mathbf{PVRM}$.

*Remark 2.3.25.* Let us attempt to motivate the definitions of $\mathbf{CC}$ and $\mathbf{PVRM}$ (and their positive analogues). The class $\mathbf{Cnt}(\mathbf{PV})$ contains basically all pseudovariety operators ever considered in the literature. We defined $\mathbf{CC}$ so that $\mathbf{CCq} = \mathbf{Cnt}(\mathbf{PV})$. One could very well have obtained this result by merely demanding that a class belong to $\mathbf{CC}$ if and only if it satisfies Axiom ($\times$) and is closed under precomposition by divisions, see Remark 2.3.3. However, one can add Axiom (pb), Axiom ($\prec_s$) and Axiom (r-e) without changing the fact that $\mathbf{CCq} = \mathbf{Cnt}(\mathbf{PV})$. Adding in some of these axioms improves the behavior of $\mathbf{CC}$. For example, Axiom ($\prec_s$) ensures that $\mathsf{q}$ preserves finite meets ($\mathsf{q}$ is **sup** no matter which variation on the definition of $\mathbf{CC}$ is chosen). Later work may reveal other axioms should be added to the definition as well; perhaps Axiom (free) should be added. This too would not change the image of $\mathsf{q}$, although there may be complications when dealing with composition.

However, adding any of the other axioms, such as Axiom ($\prec$), Axiom (id) or Axiom (co-re), changes the image of the $\mathsf{q}$-operator. For instance, adding in Axiom (id) changes the image under $\mathsf{q}$ to $\mathbf{Cnt}(\mathbf{PV})^+$, the non-decreasing operators. If one adds in all the axioms, one obtains the class $\mathbf{PVRM}^+$ whose image under $\mathsf{q}$ turns out to be $\mathbf{GMC}(\mathbf{PV})^+$.

The set $\mathbf{GMC}(\mathbf{PV})$ consists of those operators that are amenable to the techniques of global semigroup theory [278] and to profinite methods [9, 19, 20, 380]. Such operators can be written as an infinite intersection of generalized Mal'cev products (2.27). The axioms for $\mathbf{PVRM}$ were defined so that

$$\mathbf{PVRMq} = \mathbf{GMC}(\mathbf{PV})$$

and in order that the canonical theology of global semigroup theory [278], plus the modern notion of inevitability [9, 33, 295], could be applied. Elements of **PVRM** satisfy all the axioms considered earlier except Axiom (id). One obtains **PVRM** from **CC** by allowing Axiom (co-re).

Another consideration in choosing the definitions we have given is the equational theory for **CC** and **PVRM**. The equational theory for **PVRM** is completely successful; see Chapter 3. The equational theory for **CC** is possible only by adding Axiom (pb). This was not done in the formal definition because Axiom (pb) does not appear to behave well with respect to composition, another consideration in choosing the correct axioms; see Section 2.8. Remark 2.3.38 offers a continuation of this discussion.

We would like to show that **GMC(PV)** is closed under all infs and sups of **Cnt(PV)** and is an algebraic lattice, as well as proving analogues of Proposition 2.1.11 and Theorem 2.3.9 for **PVRM**. First we need some preliminary results that are of interest in their own right. These results indicate the strength of Axiom ($\prec$). We begin with the following two results, which motivated in part the definition of division. They will be generalized in Chapter 3. We say that a set is decidable if it has decidable membership.

**Proposition 2.3.26.** *Let* R *be a pseudovariety of relational morphisms and let* **W** *be a locally finite pseudovariety of semigroups. Let* $S$ *be a finite $A$-generated semigroup and let* $\rho_{\mathbf{W}} \colon S \to F_{\mathbf{W}}(A)$ *be the canonical relational morphism from* $S$ *to the free $A$-generated semigroup in* **W**. *Then* $S \in \mathsf{Rq}(\mathbf{W})$ *if and only if* $\rho_{\mathbf{W}} \in \mathsf{R}$. *In particular, if* $F_{\mathbf{W}}(A)$ *is computable as a function of* $A$, *and* R *is decidable, then* $\mathsf{Rq}(\mathbf{W})$ *is decidable.*

*Proof.* Clearly if $\rho_{\mathbf{W}} \in \mathsf{R}$, then $S \in \mathsf{Rq}(\mathbf{W})$. Conversely if $S \in \mathsf{Rq}(\mathbf{W})$, then there is a relational morphism $f \colon S \to T$ with $f \in \mathsf{R}$ and $T \in \mathbf{W}$. Lemma 1.3.15 tells us $\rho_{\mathbf{W}} \prec f$ and so $\rho_{\mathbf{W}} \in \mathsf{R}$, as required.    □

**Corollary 2.3.27.** *Let* $\{\mathsf{R}_a\}$ *be a collection of pseudovarieties of relational morphisms and let* **W** *be a **locally finite** pseudovariety of semigroups. Then*

$$\bigcap (\mathsf{R}_a \mathsf{q}(\mathbf{W})) = \left( \bigcap \mathsf{R}_a \right) \mathsf{q}(\mathbf{W}). \qquad (2.29)$$

*Proof.* The right-hand side of (2.29) is clearly contained in the left-hand side. Let $S$ be a finite $A$-generated semigroup in the left-hand side and let $\rho_{\mathbf{W}} \colon S \to F_{\mathbf{W}}(A)$ be the canonical relational morphism. Then $\rho_{\mathbf{W}} \in \bigcap \mathsf{R}_a$ by Proposition 2.3.26 and so $S$ belongs to the right-hand side.    □

The above result can fail if **W** is not locally finite; see Example 3.6.38. Now we prove that **GMC(PV)** is a complete lattice.

**Proposition 2.3.28.** *The subset* **GMC(PV)** *of* **Cnt(PV)** *is closed under arbitrary sups. Therefore,* **GMC(PV)** *is a complete lattice and the inclusion* $k \colon \mathbf{GMC(PV)} \hookrightarrow \mathbf{Cnt(PV)}$ *is* **sup**, *and so admits a right adjoint.*

*Proof.* Let $\{\alpha_a\}_{a \in A} \subseteq \mathbf{GMC(PV)}$. Then, for all $\mathbf{V}, \mathbf{W} \in \mathbf{PV}$, we have $\alpha_a(\mathbf{W}) \leq (\alpha_a(\mathbf{V}), \mathbf{V}) \, \circledm \, \mathbf{W}$. Hence

$$\bigvee_{a \in A} \alpha_a(\mathbf{W}) \leq \bigvee_{a \in A} ((\alpha_a(\mathbf{V}), \mathbf{V}) \, \circledm \, \mathbf{W}) \leq ( \bigvee_{a \in A} \alpha_a(\mathbf{V}), \mathbf{V}) \, \circledm \, \mathbf{W}$$

as desired. The last inequality holds because the generalized Mal'cev product preserves order in the first and last coordinates. This completes the proof. $\square$

We shall see in Theorem 2.3.36 that $\mathbf{GMC(PV)}$ is also closed under all infs in $\mathbf{Cnt(PV)}$ (note these are not pointwise infs, as an example in Chapter 3 will show). Let us commence with the first in a series of results that will culminate in the main theorem of this chapter.

**Proposition 2.3.29.** *Let* $\mathsf{R} \in \mathbf{PVRM}$. *Then*

$$\mathsf{R} \leq \bigcap_{\mathbf{V} \in \mathbf{PV}} (\mathsf{Rq}(\mathbf{V}), \mathbf{V}). \tag{2.30}$$

*Proof.* Let $f \colon S \to T$ belong to $\mathsf{R}$ and let $\mathbf{V}$ be a pseudovariety of semigroups. Suppose $T \geq V \in \mathbf{V}$. Consider the relational morphism $f' \colon V f^{-1} \to V$ whose graph is given by

$$\#f' = \{(s, t) \in \#f \mid t \in V\}.$$

Then $f' \prec f$ and so $f' \in \mathsf{R}$, whence $V f^{-1} \in \mathsf{Rq}(\mathbf{V})$. Thus $f \in (\mathsf{Rq}(\mathbf{V}), \mathbf{V})$, establishing (2.30), as $\mathbf{V}$ was arbitrary. $\square$

**Corollary 2.3.30.** *The map* $\mathsf{q}$ *restricts to a mapping* $\mathsf{q} \colon \mathbf{PVRM} \to \mathbf{GMC}$.

*Proof.* Let $\mathsf{R} \in \mathbf{PVRM}$ and let $\mathbf{V}$ be a pseudovariety of semigroups. Then, by Proposition 2.3.29, $\mathsf{R} \leq (\mathsf{Rq}(\mathbf{V}), \mathbf{V})$. Because $\mathsf{q}$ is order preserving (by Theorem 2.3.9), for each $\mathbf{W} \in \mathbf{PV}$, we have the inequality

$$\mathsf{Rq}(\mathbf{W}) \leq (\mathsf{Rq}(\mathbf{V}), \mathbf{V})\mathsf{q}(\mathbf{W}) = (\mathsf{Rq}(\mathbf{V}), \mathbf{V}) \, \circledm \, \mathbf{W}$$

establishing that GMC holds. $\square$

We now want to define a section $\max \colon \mathbf{GMC(PV)} \to \mathbf{PVRM}$. Let $\alpha \in \mathbf{GMC(PV)}$. Define a pseudovariety of relational morphisms

$$\max(\alpha) = \bigcap_{\mathbf{V} \in \mathbf{PV}} (\alpha(\mathbf{V}), \mathbf{V}). \tag{2.31}$$

In what follows, we again use our convention for composition of functions acting on the right of their arguments for $\mathsf{q}$.

**Theorem 2.3.31.** *The map* $\max \colon \mathbf{GMC(PV)} \to \mathbf{PVRM}$ *is well-defined and* $\max \mathsf{q} = 1_{\mathbf{GMC(PV)}}$. *In particular,*

$$\mathsf{q} \colon \mathbf{PVRM} \to \mathbf{GMC(PV)}$$

*is* **onto** *and* $\max$ *is injective.*

*Proof.* Let $\alpha \in \mathbf{GMC}(\mathbf{PV})$. Set $\mathsf{R}_\alpha = \max(\alpha)$. We prove that, for any pseudovariety $\mathbf{W}$ of semigroups, $\alpha(\mathbf{W}) = \mathsf{R}_\alpha\mathsf{q}(\mathbf{W})$. Suppose $S \in \mathsf{R}_\alpha\mathsf{q}(\mathbf{W})$. Then there is a relational morphism $f\colon S \to T$ with $T \in \mathbf{W}$ and $f \in \bigcap_\mathbf{U}(\alpha(\mathbf{U}), \mathbf{U})$. In particular, $f \in (\alpha(\mathbf{W}), \mathbf{W})$. Thus $S = Tf^{-1} \in \alpha(\mathbf{W})$, as $T \in \mathbf{W}$.

Conversely, suppose $S \in \alpha(\mathbf{W})$. Then, by continuity, there exists $T \in \mathbf{W}$ such that $S \in \alpha((T))$ (cf. (2.7)). By Proposition 2.3.22 it then follows

$$S \in \bigcap_{\mathbf{V} \in \mathbf{PV}} ((\alpha(\mathbf{V}), \mathbf{V}) \,\textcircled{m}\, (T)) = \bigcap_{\mathbf{V} \in \mathbf{PV}} ((\alpha(\mathbf{V}), \mathbf{V})\mathsf{q}((T))).$$

By Corollary 2.3.27, because $(T)$ is locally finite, we have

$$\bigcap_{\mathbf{V} \in \mathbf{PV}} ((\alpha(\mathbf{V}), \mathbf{V})\mathsf{q}((T))) = \left( \bigcap_{\mathbf{V} \in \mathbf{PV}} (\alpha(\mathbf{V}), \mathbf{V}) \right) \mathsf{q}((T))$$

$$= \mathsf{R}_\alpha\mathsf{q}((T)) \subseteq \mathsf{R}_\alpha\mathsf{q}(\mathbf{W}).$$

We conclude that $S \in \mathsf{R}_\alpha\mathsf{q}(\mathbf{W})$, completing the proof that $\mathsf{R}_\alpha\mathsf{q} = \alpha$.    $\square$

The following intrinsic characterization of $\max(\alpha)$ will be needed later on.

**Proposition 2.3.32.** *Suppose $\alpha \in \mathbf{GMC}(\mathbf{PV})$ and $f\colon S \to T$ is a relational morphism. Then $f \in \max(\alpha)$ if and only if, for each subsemigroup $W$ of $T$, one has $Wf^{-1} \in \alpha((W))$.*

*Proof.* If $f \in \max(\alpha)$ and $W \leq T$, then $f \in (\alpha((W)), (W))$ and so $Wf^{-1} \in \alpha((W))$. For the converse, suppose $\mathbf{W}$ is a pseudovariety and $T \geq W \in \mathbf{W}$. Then, by assumption, $Wf^{-1} \in \alpha((W)) \leq \alpha(\mathbf{W})$, showing that $f \in (\alpha(\mathbf{W}), \mathbf{W})$. Thus $f \in \max(\alpha)$, as $\mathbf{W}$ was arbitrary.    $\square$

As a corollary of Proposition 2.3.32 and Theorem 2.3.31 we obtain the following membership criterion.

**Corollary 2.3.33.** *Suppose $\alpha \in \mathbf{GMC}(\mathbf{PV})$ and $\mathbf{V}$ is a pseudovariety of semigroups. Then $S \in \alpha(\mathbf{V})$ if and only if there is a relational morphism $f\colon S \to T$ with $T \in \mathbf{V}$ such that, for each subsemigroup $W$ of $T$, one has $Wf^{-1} \in \alpha((W))$.*

For a set $X$ of relational morphisms of finite semigroups, let $\mathbb{PVRM}(X)$ be the pseudovariety of relational morphisms generated by $X$. If $f$ is a relational morphism, then we shall just write $(f)$ instead of $\mathbb{PVRM}(\{f\})$. The proof of the following proposition is almost identical to the proof of Proposition 2.1.14 and so will be left to the reader as an exercise.

**Proposition 2.3.34.** *For $X \subseteq \mathbf{RM}$,*

$$\mathbb{PVRM}(X) = \{fi \mid f \prec f_1 \times \cdots \times f_n, \ f_i \in X \text{ and } i \text{ an inclusion}\}$$
$$= \{f \mid f \subseteq d_1(f_1 \times \cdots \times f_n)d_2, \ d_1, d_2 \text{ divisions and } f_i \in X\}.$$

**Exercise 2.3.35.** Prove Proposition 2.3.34.

We are now ready to prove the main theorem of this chapter.

**Theorem 2.3.36.** *The following hold:*

*(a) The set* $\mathbf{Cnt(PV)}$ *is a monoid with respect to composition and* $\mathbf{GMC(PV)}$ *is a submonoid of* $\mathbf{Cnt(PV)}$.

*(b) The following are Galois connections:*

$$(2^{\mathbf{RM}}\, \subseteq, \cup, \cap) \;\underset{\mathbb{CC}}{\overset{j}{\xrightarrow{\hspace{1cm}}}}\; (\mathbf{CC}, \subseteq, \vee_{\det}, \cap) \;\underset{\mathbb{D}}{\overset{i}{\xrightarrow{\hspace{1cm}}}}\; (\mathbf{PVRM}, \subseteq, \vee_{\det}, \cap) \quad (2.32)$$

*where* $i, j$ *are the inclusions and where, for* $R \in \mathbf{CC}$,

$$\mathbb{D}(R) = \bigcap \{S \in \mathbf{PVRM} \mid S \supseteq R\}.$$

*Moreover,* $\mathbb{PVRM} = \mathbb{D} \circ \mathbb{CC}$. *So* $\mathbb{CC}$ *and* $\mathbb{D}$ *are* **sup** *and* $i$ *and* $j$ *are* **inf***. Furthermore,* $i$ *and* $j$ *are continuous, whence* $\mathbf{CC}$ *and* $\mathbf{PVRM}$ *are algebraic lattices and* $\mathbb{CC}(K(2^{\mathbf{RM}})) = K(\mathbf{CC})$ *and*

$$\mathbb{PVRM}(K(2^{\mathbf{RM}})) = K(\mathbf{PVRM}) = \mathbb{D}(K(\mathbf{CC})).$$

*In addition, the compact elements of* $\mathbf{PVRM}$ *are the one-generated pseudovarieties.*

*(c) We have the following four Galois connections in which all the right adjoints are continuous inclusions and the left adjoints are onto:*

$$
\begin{array}{ccc}
 & M & \\
\mathbf{CC} & \xleftarrow{\hspace{1.2cm}} & \mathbf{Cnt(PV)} \\
\mathbb{D} \downarrow \uparrow i & \quad \mathsf{q} \quad & \widehat{k} \downarrow \uparrow k \\
 & \overrightarrow{\hspace{1.2cm}} & \\
 & \max & \\
\mathbf{PVRM} & \xleftarrow{\hspace{1.2cm}} & \mathbf{GMC(PV)}. \\
 & \mathsf{q} &
\end{array}
\qquad (2.33)
$$

*Here* $\widehat{k}$ *is the determined left adjoint of* $k$. *In particular,* $\mathsf{q}$ *is* **sup***. Moreover, the inclusion map* $k\colon \mathbf{GMC(PV)} \hookrightarrow \mathbf{Cnt(PV)}$ *is* **sup** *(and so* $k$ *also admits a right adjoint). Finally,* $\mathsf{q}\colon \mathbf{CC} \to \mathbf{Cnt(PV)}$ *preserves finite meets while the restriction* $\mathsf{q}\colon \mathbf{PVRM} \to \mathbf{GMC(PV)}$ *is* **inf***. Therefore, there is a Galois connection*

$$\mathbf{GMC(PV)} \;\underset{\min}{\overset{\mathsf{q}}{\xrightleftharpoons{\hspace{1cm}}}}\; \mathbf{PVRM} \qquad (2.34)$$

*where* $\min(\alpha)$ *is the minimum element of* $\mathbf{PVRM}$ *mapping to* $\alpha$ *under* $\mathsf{q}$. *Thus every element of* $\mathbf{GMC(PV)}$ *is the* $\mathsf{q}$-*image of a closed interval in* $\mathbf{PVRM}$.

*(d) All four complete lattices in (2.33) are algebraic and the left adjoint maps of the respective Galois connections carry compact elements onto compact elements.*

*Proof.* To prove (a), first we show that the composition of continuous operators is continuous. Let $\alpha, \beta \in \mathbf{Cnt}(\mathbf{PV})$ and $D \subseteq \mathbf{PV}$ be directed. Then $\beta(D)$ is directed. Thus

$$\alpha\beta \left( \bigvee D \right) = \alpha \left( \bigvee \beta(D) \right) = \bigvee \alpha\beta(D)$$

as required. Clearly the identity map is continuous so $\mathbf{Cnt}(\mathbf{PV})$ is a monoid under composition. The identity map obviously satisfies GMC. To verify the set of operators satisfying GMC (2.25) is closed under composition, note that if $\alpha, \beta \in \mathbf{GMC}(\mathbf{PV})$, then, for $\mathbf{V}, \mathbf{W} \in \mathbf{PV}$, we have

$$
\begin{aligned}
\beta(\alpha(\mathbf{W})) \;&\leq\; (\beta(\alpha(\mathbf{V})), \alpha(\mathbf{V})) \; \textcircled{m} \; \alpha(\mathbf{W}) && \text{by GMC} \\
&\leq\; (\beta\alpha(\mathbf{V}), \alpha(\mathbf{V})) \; \textcircled{m} \; ((\alpha(\mathbf{V}), \mathbf{V}) \; \textcircled{m} \; \mathbf{W}) && \text{by GMC} \\
&\leq\; (\beta\alpha(\mathbf{V}), \mathbf{V}) \; \textcircled{m} \; \mathbf{W} && \text{by (2.24).}
\end{aligned}
$$

This proves (a).

The proof of (b) is mostly obvious from the definitions and Corollary 1.2.37. The fact that the compact elements of $\mathbf{PVRM}$ are generated by a single relational morphism follows from Proposition 1.3.16(b), which implies $\mathbb{PVRM}(f_1, \ldots, f_n) = (f_1 \times \cdots \times f_n)$.

For (c) we already know that $\mathsf{q} \colon \mathbf{PVRM} \to \mathbf{GMC}(\mathbf{PV})$ is onto by Theorem 2.3.31. Let us begin by showing that max is right adjoint to $\mathsf{q}$. It suffices by Proposition 1.1.7 to show, for $\alpha \in \mathbf{GMC}(\mathbf{PV})$, that $\max(\alpha)$ is the maximum element of $\mathbf{PVRM}$ mapping to $\alpha$ under $\mathsf{q}$. But if $\mathsf{R} \in \mathbf{PVRM}$ with $\mathsf{R}\mathsf{q} = \alpha$, then $\mathsf{R} \leq \max(\alpha)$ by Proposition 2.3.29 and so max is indeed a right adjoint to $\mathsf{q}$.

Next we show that max is continuous. It is clearly order preserving, so suppose $\alpha = \bigvee_{d \in D} \alpha_d$ with $\{\alpha_d\}_{d \in D}$ directed. We again use the notation $\mathsf{R}_\beta$ for $\max(\beta)$, $\beta \in \mathbf{GMC}(\mathbf{PV})$. Clearly $\bigvee \mathsf{R}_{\alpha_d} \leq \mathsf{R}_\alpha$. Suppose that $f \in \mathsf{R}_\alpha$ with $f \colon S \to T$. For any subsemigroup $W$ of $T$, $Wf^{-1} \in \alpha((W))$ by Proposition 2.3.32. Also the collection $\{\alpha_d((W))\}_{d \in D}$ is directed with union $\alpha((W))$. Thus $Wf^{-1} \in \alpha_d((W))$ some $d \in D$. Because $T$ has only finitely many subsemigroups $W$, there exists $d_0 \in D$ such that $Wf^{-1} \in \alpha_{d_0}((W))$ for all subsemigroups $W$ of $T$. But then $f \in \mathsf{R}_{\alpha_{d_0}}$ by another application of Proposition 2.3.32. Thus $f \in \bigvee_{d \in D} \mathsf{R}_{\alpha_d}$ showing that

$$\max(\alpha) = \bigvee_{d \in D} \max(\alpha_d)$$

as required.

We have now proved the assertions of Theorem 2.3.36(c) about $(\mathbb{D}, i)$ (from part (b)), $(M, \mathsf{q})$ (in Theorem 2.3.9) and $(\max, \mathsf{q})$, namely that they are Galois

connections with the right adjoints continuous. We also know already that
$q\colon \mathbf{CC} \to \mathbf{Cnt}(\mathbf{PV})$ preserves finite meets. It remains to consider the Galois
connection $(\widehat{k}, k)$ and to show that $q\colon \mathbf{PVRM} \to \mathbf{GMC}(\mathbf{PV})$ is $\mathbf{inf}$. It is the
content of Proposition 2.3.28 that $k$ is $\mathbf{sup}$.

We show that both $q\colon \mathbf{PVRM} \to \mathbf{GMC}(\mathbf{PV})$ and $k$ are $\mathbf{inf}$ in one fell
swoop. We do this by first showing that if $\{R_a \mid a \in A\} \subseteq \mathbf{PVRM}$, then
$(\bigcap_{a \in A} R_a)\, q$ is the determined meet in $\mathbf{Cnt}(\mathbf{PV})$ of the operators $R_a q$. It will
then follow that $(\bigcap_{a \in A} R_a)\, q$ is the determined meet of the $R_a q$ in $\mathbf{GMC}(\mathbf{PV})$,
and hence $q$ is $\mathbf{inf}$. Because $q\colon \mathbf{PVRM} \to \mathbf{GMC}(\mathbf{PV})$ is onto, that is every
GMC operator $\alpha$ is of the form $Rq$ for some $R \in \mathbf{PVRM}$, we shall also be
able to conclude that $k$ is $\mathbf{inf}$.

By Proposition 2.2.2 we just need to compare the two operators in question on compact pseudovarieties of semigroups. But by Proposition 2.3.16
the determined meet in $\mathbf{Cnt}(\mathbf{PV})$ is pointwise on compact pseudovarieties of
semigroups. So to prove the equality

$$\left( \bigcap_{a \in A} R_a \right) q = \bigwedge_{a \in A} R_a q \quad \text{(the meet taken in } \mathbf{Cnt}(\mathbf{PV})) \tag{2.35}$$

it suffices to show, for $\mathbf{V} \in K(\mathbf{PV})$, that

$$\left( \bigcap_{a \in A} R_a \right) q(\mathbf{V}) = \bigcap_{a \in A} (R_a q(\mathbf{V})).$$

But as compact pseudovarieties of semigroups are locally finite, this follows
from Corollary 2.3.27. This completes the proof of (c).

Part (d) is an immediate consequence of Corollary 1.2.37.     □

*Remark 2.3.37.* One can deduce from the proofs of Corollary 2.3.27 and Theorem 2.3.36 that $\min(\alpha)$ is generated by all canonical relational morphisms
$\rho_{\mathbf{V}}\colon S \to F_{\mathbf{V}}(A)$ where $S$ is $A$-generated, $\mathbf{V}$ is compact and $S \in \alpha(\mathbf{V})$. In
fact, one may assume that $S$ is freely generated by $A$ in the pseudovariety
generated by $S$ because pseudovarieties of relational morphisms are closed
under precomposition by divisions, thanks to Proposition 2.1.8(a). More on
min and max will appear in Section 3.5.2. In particular, we show they do not
always coincide.

Analogous results to Theorem 2.3.36 hold for the positive versions; we
leave their formulations and proofs to the reader.

*Remark 2.3.38.* Theorems 2.3.7 and 2.3.9 are not too difficult; the important
fact that $q\colon \mathbf{CC} \to \mathbf{Cnt}(\mathbf{PV})$ is onto is easy because *which* relational morphisms from $S$ to $T$ for $S \in \alpha((T))$ we consider does not matter: just the
domain and codomain, which is very crude. In the proof we took all relational
morphisms with correct domain and codomain. But every "reasonable opera-
tor," that is every continuous operator on $\mathbf{PV}$, is of the form $Rq$ with $R \in \mathbf{CC}$.
The *generality* is important because "everything should fit into the theory!"

Theorems 2.3.31 and 2.3.36, especially (2.33), are important, telling us that $\alpha$ satisfies GMC if and only if there exists $R \in \mathbf{PVRM}$ such that $Rq = \alpha$. Now many important operators fit into this framework. For example,

$$\mathbf{W} \longmapsto \mathbf{V} \curlywedge \mathbf{W}$$

where $\mathbf{V} \in \mathbf{PV}$ is fixed, $\mathbf{W}$ varies and $\curlywedge$ represents any of $*$ (the semidirect product), $**$ (the two-sided semidirect product), ⓜ (the Mal'cev product), $\vee$ or $\wedge$, satisfies GMC, as we shall see later in the chapter. The axioms for $\mathbf{PVRM}$ are stronger, so the *onto* part of Theorem 2.3.31 is deeper and in fact an *equational theory* for $R \in \mathbf{PVRM}$ exists (see Chapter 3). However, for example, if we consider the operator

$$\mathbf{V} \longmapsto \mathbf{V} \curlywedge \mathbf{W}$$

with $\curlywedge$ as above, but this time with $\mathbf{W}$ fixed and $\mathbf{V}$ varying, we obtain a continuous operator that does not necessarily satisfy GMC so the $\mathbf{CC}$ approach applies, but in general not the $\mathbf{PVRM}$ approach. See Section 2.9 on what to do for this type of operator. Whereas there is an equational theory for those members of $\mathbf{CC}$ that satisfy Axiom (pb), it is not as closely tied to the operator as in the $\mathbf{PVRM}$ case.

### 2.3.3 Minimal models for Cnt(PV)

We now wish to show that if one restricts to Birkhoff continuously closed classes, then q becomes both **sup** and **inf** and yet remains onto. This means that each continuous operator can be defined via q by a unique *minimal* continuously closed class satisfying Axiom (free). This makes a good case for adding Axiom (free) to the definition of a continuously closed class. On the other hand, this axiom does not appear very natural at first sight. Also it does not seem to be implied by the equational theory for equational continuously closed classes. Moreover, it is not clear how this axiom behaves with respect to the composition of continuously closed classes that will be defined later. For this reason, we did not make it part of the official definition of a continuously closed class, but instead introduced Birkhoff continuously closed classes.

We now proceed to establish analogues of Proposition 2.3.26 and Corollary 2.3.27 for Birkhoff continuously closed classes.

**Proposition 2.3.39.** *Let* R *be a Birkhoff continuously closed class and let* $S, T$ *be finite semigroups. Then* $S \in \mathsf{Rq}((T))$ *if and only if the canonical relational morphism* $\rho_{(T)} \colon S \to F_{(T)}(S \times T)$ *belongs to* R, *where we view* $S$ *as generated by* $S \times T$ *via the projection.*

*Proof.* If $\rho_{(T)} \in R$, then plainly $S \in \mathsf{Rq}((T))$. Conversely, suppose $S \in \mathsf{Rq}((T))$. Then $f \colon S \to T'$ with $f \in R$ and $T' \in (T)$. Let $d \colon T' \to T^n$ be a division, with $n \geq 1$. Then by Proposition 2.1.8(a), $fd \colon S \to T^n$ belongs to R. Axiom (free) now guarantees that $\rho_{(T)} \in R$.    $\square$

**Corollary 2.3.40.** *Let* $\{R_a\}$ *be a collection of Birkhoff continuously closed classes and let* $T$ *be a finite semigroup. Then*

$$\bigcap(R_a q((T))) = \left(\bigcap R_a\right) q((T)).$$

*Proof.* Trivially, the right-hand side of (2.29) is contained in the left-hand side. Let $S$ be a semigroup in the left-hand side and suppose $\rho_{(T)}\colon S \to F_{(T)}(S \times T)$ is the canonical relational morphism, where as before we view $S$ as $S \times T$-generated via the projection. Then $\rho_{(T)} \in \bigcap R_a$ by Proposition 2.3.26 and so $S$ belongs to the right-hand side.    $\square$

We may now state the analogue of Theorem 2.3.36 for **BCC**.

**Theorem 2.3.41.** *The map* $q\colon \mathbf{BCC} \to \mathbf{Cnt(PV)}$ *is* ***onto***, **inf** *and* **sup**. *Thus there are Galois connections*

$$\mathbf{BCC} \xleftarrow[\quad q \quad]{\quad M \quad} \mathbf{Cnt(PV)} \tag{2.36}$$

*and*

$$\mathbf{Cnt(PV)} \xleftarrow[\quad m \quad]{\quad q \quad} \mathbf{BCC} \tag{2.37}$$

*where* $m(\alpha)$ *is the minimal Birkhoff continuously closed class with* $m(\alpha)q = \alpha$.

*Proof.* Let $M$ be defined as in (2.15). Then Theorem 2.3.7 shows that $M(\alpha)$ is a Birkhoff continuously closed class. Therefore, (2.36) is a consequence of Theorem 2.3.9. By Proposition 2.3.16, to establish that $q$ is **inf** it suffices to show that, for each compact pseudovariety $(T)$, we have

$$\left(\bigcap R_a\right) q((T)) = \bigcap(R_a q((T))).$$

But this is the content of Corollary 2.3.40. The existence of the Galois connection (2.37) now follows from the general theory.    $\square$

One can deduce from the proofs that $m(\alpha)$ is generated by relational morphisms of the form $\rho_{(T)}\colon S \to F_{(T)}(S \times T)$ where $S \in \alpha((T))$.

Of course, analogous results hold for $\mathbf{BCC}^+$. We omit the details.

## 2.4 Key Examples

We now proceed to the motivating examples for the theory. In the examples below, we take $\mathbf{V}, \mathbf{W} \in \mathbf{PV}$.

### 2.4.1 Important examples of members of PVRM and GMC(PV)

We begin with examples satisfying the generalized Mal'cev condition. Recall from Theorem 2.3.36 that if $\alpha \in$ **GMC(PV)**, then there is a unique minimum pseudovariety of relational morphisms $\min(\alpha)$ and a unique maximum pseudovariety of relational morphisms $\max(\alpha)$ mapping to $\alpha$ under $\mathsf{q}$. That is, $\alpha \mathsf{q}^{-1}$ is the closed interval $[\min(\alpha), \max(\alpha)]$ in **PVRM**. Finding min is in practice very difficult. We shall see in Chapter 3 that even $\min(1_{\mathbf{PV}})$ is highly non-trivial to compute.

Our first example is the simplest one illustrating the theory.

*Example 2.4.1 (Divisions & Identity Map).* The collection of all divisions D is a pseudovariety of relational morphisms. One easily checks that $\mathsf{Dq} = 1_{\mathbf{PV}}$. Indeed, if $S$ is a finite semigroup, then $S \in \mathbf{W}$ if and only if there is a division $d: S \to T$ with $T \in \mathbf{W}$. Clearly, the operator $1_{\mathbf{PV}} \in \mathbf{GMC(PV)}^+$.

However, $\mathsf{D} \neq \max(1_{\mathbf{PV}})$ as the following example shows. Let $N_3 = \{a, b, 0\}$ be the three-element null semigroup (so the product of any two elements is zero) and let $N_2 = \{x, 0\}$ be the two-element null semigroup. The reader should verify that $(N_2) = (N_3)$. Let $f: N_3 \to N_2$ be the homomorphism given by $af = x = bf$ and $0f = 0$; of course, $f$ is not a division. The only non-empty subsemigroups of $N_2$ are $\{0\}$ and $N_2$. Clearly, $\{0\} = \{0\}f^{-1}$ and $N_3 = N_2 f^{-1}$. Because $N_3 \in (N_2)$, it follows $f \in \max(1_{\mathbf{PV}})$ by Proposition 2.3.32. This shows that in general $\max \neq \min$. We shall see later that $\min(1_{\mathbf{PV}}) = \mathsf{D}$.

The following example is quite useful because it will allow us to identify pseudovarieties of semigroups with certain pseudovarieties of relational morphisms. It is one of the principal reasons we work with **PVRM** rather than **PVRM**$^+$.

*Example 2.4.2 (Constants).* If $\mathbf{V} \in \mathbf{PV}$, then set

$$\widetilde{\mathbf{V}} = \{f: S \to T \mid S \in \mathbf{V}\}.$$

It is straightforward to verify that $\widetilde{\mathbf{V}} \in \mathbf{PVRM}$. However, if $\mathbf{V}$ is a proper pseudovariety it does not belong to **PVRM**$^+$. Set $q_{\mathbf{V}} = \widetilde{\mathbf{V}}\mathsf{q}$. Then $q_{\mathbf{V}}: \mathbf{PV} \to \mathbf{PV}$ is given by $q_{\mathbf{V}}(\mathbf{W}) = \widetilde{\mathbf{V}}\mathsf{q}(\mathbf{W}) = \mathbf{V}$. That is, $q_{\mathbf{V}}$ is the constant function sending all pseudovarieties to $\mathbf{V}$, a member of **GMC(PV)** but not **GMC(PV)**$^+$. Frequently $\widetilde{\mathbf{V}}$ is just denoted abusively by $\mathbf{V}$ for reasons to become clear in Chapter 3 (see also Section 2.8).

**Exercise 2.4.3.** Show that **GMC(PV)**$^-$ = $[\widetilde{\mathbf{1}}, \mathsf{D}]\mathsf{q}$ (where we consider the interval in question as a subinterval of **PVRM**). Hint: Use that $\mathsf{q}$ preserves finite meets.

Our next example is one of the motivating examples of the theory, being essentially the first operator to be defined by relational morphisms.

*Example 2.4.4 (Mal'cev product).* The classical Mal'cev product operator associated to $\mathbf{V}$, that is $\mathbf{V} \textcircled{m} (-)$, belongs to $\mathbf{GMC}(\mathbf{PV})^+$. Recall that

$$(\mathbf{V}, \mathbf{1}) = \{f \colon S \to T \mid \text{for } e \in E(T), \ ef^{-1} \in \mathbf{V}\}$$

belongs to $\mathbf{PVRM}^+$ and $(\mathbf{V}, \mathbf{1})\mathsf{q} = \mathbf{V} \textcircled{m} (-)$.

Of course the generalized Mal'cev products fit into this framework as well.

*Example 2.4.5 (Generalized Mal'cev product).* Recall that

$$(\mathbf{V}, \mathbf{W}) = \{f \colon S \to T \mid \text{for } T \geq W \in \mathbf{W}, \ Wf^{-1} \in \mathbf{V}\}$$

belongs to $\mathbf{PVRM}$ (Proposition 2.1.18). It belongs to $\mathbf{PVRM}^+$ if and only if $\mathbf{V} \geq \mathbf{W}$. Also $(\mathbf{V}, \mathbf{W})\mathsf{q} = (\mathbf{V}, \mathbf{W}) \textcircled{m} (-)$. Let us remark that GMC at $\mathbf{1}$ is just the statement that

$$(\mathbf{V}, \mathbf{W}) \textcircled{m} \mathbf{U} \leq ((\mathbf{V}, \mathbf{W}) \textcircled{m} \mathbf{1}) \textcircled{m} \mathbf{U} = \mathbf{V} \textcircled{m} \mathbf{U}$$

as $(\mathbf{V}, \mathbf{W}) \textcircled{m} \mathbf{1} = \mathbf{V}$.

We now turn to the join operation on $\mathbf{PV}$. First we need a definition. Let $\mathbf{V}$ be a pseudovariety of semigroups and $S$ a finite semigroup. A subset $X \subseteq S$ is said to be $\mathbf{V}$-*pointlike* if, for all relational morphisms $f \colon S \to V$ with $V \in \mathbf{V}$, there exists $v \in V$ such that $X \subseteq vf^{-1}$. In other words, $X$ "behaves like a point" with respect to $\mathbf{V}$. For instance, if $\mathbf{G}$ is the pseudovariety of groups, then the set $E(S)$ of idempotents of $S$ is $\mathbf{G}$-pointlike, as is the subsemigroup $\langle E(S) \rangle$ generated by the idempotents. The collection of $\mathbf{V}$-pointlike subsets of $S$ is denoted $\mathsf{PL}_{\mathbf{V}}(S)$.

**Exercise 2.4.6.** In this exercise, we investigate various properties and examples of pointlike sets. Throughout $S$ is a finite semigroup.

1. Show that the collection of $\mathbf{V}$-pointlike subsets $\mathsf{PL}_{\mathbf{V}}(S)$ of $S$ is a subsemigroup of the power set $P(S)$ of $S$ containing the singletons.
2. Show that $S \in \mathbf{V}$ if and only if the only $\mathbf{V}$-pointlike subsets of $S$ are the singletons.
3. Prove that if $S$ is a finite semigroup, then $\langle E(S) \rangle$ is $\mathbf{G}$-pointlike.
4. Prove that if $\mathbf{N}$ is the pseudovariety of nilpotent semigroups, then a non-singleton subset of $S$ is $\mathbf{N}$-pointlike if and only if it is contained in $SE(S)S$.

*Example 2.4.7 (Slice & Joins).* For $\mathbf{V} \in \mathbf{PV}$, we define $\mathbf{V}_\vee$ to consists of those relational morphisms $f \colon S \to T$ satisfying the "Slice Condition" of [322, 330]. That is, $\mathbf{V}_\vee$ consists of all relational morphisms $f \colon S \to T$ such that there are no $\mathbf{V}$-pointlike subsets of $\#f$ of the form $\{(s_1, t), (s_2, t)\}$ with $s_1 \neq s_2$. We ask the reader in the next exercise to verify that $\mathbf{V}_\vee$ belongs to $\mathbf{PVRM}^+$. An alternative proof appears in Chapter 3 where a basis of relational pseudoidentities is given for this class.

The results of [322, 330] show that the equality

$$\mathbf{V}_\vee \mathsf{q} = \mathbf{V} \vee (-)$$

holds, that is, $\mathbf{V}_\vee \mathsf{q}(\mathbf{W}) = \mathbf{V} \vee \mathbf{W}$. Details can be found in Section 2.7. We remark that $\mathbf{1}_\vee = \mathsf{D}$.

The generalized Mal'cev condition for the join is trivial to verify and so the join operator $\mathbf{V} \vee (-)$ belongs to $\mathbf{GMC}(\mathbf{PV})^+$ for any pseudovariety $\mathbf{V}$.

**Exercise 2.4.8.** Prove that $\mathbf{V}_\vee \in \mathbf{PVRM}^+$.

The semidirect product and its companion collection of relational morphisms, defined via the derived semigroupoid [339, 364], is another one of the motivating examples for this work. However, the definition of the derived semigroupoid in [364] does not correspond to our functorial version of the wreath product. In Section 2.5 we provide the correct version.

*Example 2.4.9 (Semidirect Product).* For a pseudovariety of semigroups $\mathbf{V}$, denote by $\mathbf{V}_D$ the collection of all relational morphisms $f \colon S \to T$ with derived semigroupoid [339, 364] dividing a semigroup in $\mathbf{V}$. The precise definition appears in Section 2.5 as well as a proof that this class belongs to $\mathbf{PVRM}^+$. The principal result of [364], adapted to semigroups (see Section 2.5.2), then states that $\mathbf{V}_D \mathsf{q} = \mathbf{V} * (-)$. In particular, the operator $\mathbf{V} * (-)$ belongs to $\mathbf{GMC}(\mathbf{PV})^+$. So, for instance, one has

$$\mathbf{V} * \mathbf{W} \leq (\mathbf{V} * \mathbf{1}) \, \textcircled{m} \, \mathbf{W}.$$

Note that in general $\mathbf{V} * \mathbf{1} \neq \mathbf{V}$.

We verify directly that $\mathbf{V} * (-)$ satisfies GMC. It suffices to show that if $V \rtimes W$ is a semidirect product with $V \in \mathbf{V}$ and $W \in \mathbf{W}$, then

$$V \rtimes W \in (\mathbf{V} * \mathbf{U}, \mathbf{U}) \, \textcircled{m} \, \mathbf{W}$$

for all pseudovarieties $\mathbf{U}$. Consider the projection $\pi \colon V \rtimes W \to W$. We claim that $\pi \in (\mathbf{V} * \mathbf{U}, \mathbf{U})$ for all pseudovarieties $\mathbf{U}$. Let $U \leq W$ with $U \in \mathbf{U}$. Then $U$ acts on the left of $V$ by restricting the action of $W$ and, in fact, $U\pi^{-1} = V \rtimes U$ and so belongs to $\mathbf{V} * \mathbf{U}$, as required. This proves that $V \rtimes W \in (\mathbf{V} * \mathbf{U}, \mathbf{U}) \textcircled{m} \mathbf{W}$, establishing that $\mathbf{V} * (-)$ satisfies GMC.

*Question 2.4.10.* What is $\mathbf{V} * \mathbf{1}$? It is easy to see that if $S$ is a semigroup in $\mathbf{V} * \mathbf{1}$, then the quotient of $S$ by the kernel of its action on the right of itself belongs to $\mathbf{V}$. The converse holds if $\mathbf{gV}$ is definable by pseudoidentities over strongly connected graphs. (See Section 2.5 for definitions.) Is the converse always valid?

**Exercise 2.4.11.** Verify that if $S$ is a semigroup in $\mathbf{V} * \mathbf{1}$, then the quotient of $S$ by the kernel of its action on the right of itself belongs to $\mathbf{V}$. In particular, if $S$ has a left identity, then $S \in \mathbf{V} * \mathbf{1}$ if and only if $S \in \mathbf{V}$.

Let us establish a key property of $\mathbf{V} * \mathbf{1}$. If $E$ is a set of pseudoidentities, then $\llbracket E \rrbracket$ denotes the pseudovariety of semigroups satisfying $E$.

**Proposition 2.4.12.** *Let $\mathbf{V}$ be a pseudovariety of semigroups. Then the inequality $\mathbf{V} * \mathbf{1} \leq \llbracket xy = xz \rrbracket \textcircled{m} \mathbf{V}$ holds.*

*Proof.* Let $S \in \mathbf{V}$ and suppose that $S \rtimes 1$ is a semidirect product. The binary operation of $S$ will be written additively, although we do not assume commutativity. Define $\psi \colon S \rtimes 1 \to S$ by $(s, 1)\psi = 1 \cdot s$. Then we compute

$$((s, 1)(s', 1))\, \psi = (s + 1 \cdot s', 1)\psi = 1 \cdot (s + 1 \cdot s') = 1 \cdot s + 1 \cdot s' = (s, 1)\psi + (s', 1)\psi$$

so $\psi$ is a homomorphism. Let us verify $\psi \in (\llbracket xy = xz \rrbracket, \mathbf{1})$. We establish more generally that if $(t, 1)\psi = (t', 1)\psi$, then $(s, 1)(t, 1) = (s, 1)(t', 1)$ for any $s \in S$. Indeed, one computes

$$(s, 1)(t, 1) = (s + 1 \cdot t, 1) = (s + 1 \cdot t', 1) = (s, 1)(t', 1)$$

as $1 \cdot t = (t, 1)\psi = (t', 1)\psi = 1 \cdot t'$. This completes the proof.   □

We may now deduce a well-known upper bound for the semidirect product.

**Corollary 2.4.13.** *Let $\mathbf{V}$ and $\mathbf{W}$ be pseudovarieties of semigroups. Then*

$$\mathbf{V} * \mathbf{W} \leq (\llbracket xy = xz \rrbracket \textcircled{m} \mathbf{V}) \textcircled{m} \mathbf{W}$$

*holds.*

*Proof.* The global Mal'cev condition provides $\mathbf{V} * \mathbf{W} \leq (\mathbf{V} * \mathbf{1}) \textcircled{m} \mathbf{W}$. Combining this with the previous proposition yields the desired result.   □

Analogously, one can consider the two-sided semidirect product and its companion collection of relational morphisms, defined in terms of the kernel semigroupoid [293].

*Example 2.4.14 (Two-sided Semidirect Product).* Given a pseudovariety of semigroups $\mathbf{V}$, denote by $\mathbf{V}_K$ the collection of all relational morphisms $f \colon S \to T$ whose kernel semigroupoid [293] divides an element of $\mathbf{V}$. Again the definition must be adapted to semigroups; more details can be found in Section 2.6. Suffice it to say that $\mathbf{V}_K \in \mathbf{PVRM}^+$ and $\mathbf{V}_K \mathsf{q} = \mathbf{V} ** (-)$. GMC at $\mathbf{1}$ then provides $\mathbf{V} ** \mathbf{W} \leq (\mathbf{V} ** \mathbf{1}) \textcircled{m} \mathbf{W}$. Again in general $\mathbf{V} ** \mathbf{1} \neq \mathbf{V}$.

One can verify directly that the operator $\mathbf{V} ** (-)$ satisfies GMC. Indeed, if $\pi \colon S \rtimes T \to T$ is a two-sided semidirect product projection with $S \in \mathbf{V}$ and if $U \leq T$ belongs to $\mathbf{U}$, then $U\pi^{-1} = S \rtimes U$ belongs to $\mathbf{V} ** \mathbf{U}$, establishing $\pi \in (\mathbf{V} ** \mathbf{U}, \mathbf{U})$ for all pseudovarieties $\mathbf{U}$. Let us turn to the two-sided analogue of Proposition 2.4.12.

**Proposition 2.4.15.** *The inequality $\mathbf{V} ** \mathbf{1} \leq \llbracket xyw = xzw \rrbracket \textcircled{m} \mathbf{V}$ holds for any pseudovariety of semigroups $\mathbf{V}$.*

*Proof.* Let $S \in \mathbf{V}$ and suppose that $S \bowtie 1$ is a two-sided semidirect product. We use additive notation for $S$, although we do not assume commutativity. Define $\psi \colon S \bowtie 1 \to S$ by $(s, 1)\psi = 1 \cdot s \cdot 1$. Then we compute

$$((s, 1)(s', 1)) \psi = (s \cdot 1 + 1 \cdot s', 1)\psi = 1 \cdot (s \cdot 1 + 1 \cdot s') \cdot 1$$
$$= 1 \cdot s \cdot 1 + 1 \cdot s' \cdot 1 = (s, 1)\psi + (s', 1)\psi$$

and so $\psi$ is a homomorphism. To prove that $\psi \in (\llbracket xyw = xzw \rrbracket, 1)$, we establish the more general fact that $(t, 1)\psi = (t', 1)\psi$ implies the equality $(s, 1)(t, 1)(s', 1) = (s, 1)(t', 1)(s', 1)$ for any $s, s' \in S$. Indeed, routine computation yields

$$(s, 1)(t, 1)(s', 1) = (s \cdot 1 + 1 \cdot t \cdot 1 + 1 \cdot s', 1)$$
$$= (s \cdot 1 + 1 \cdot t' \cdot 1 + 1 \cdot s', 1)$$
$$= (s, 1)(t', 1)(s', 1)$$

as $1 \cdot t \cdot 1 = (t, 1)\psi = (t', 1)\psi = 1 \cdot t' \cdot 1$. This completes the proof. $\qquad\square$

Similarly to the case of the semidirect product, we may now deduce:

**Corollary 2.4.16.** *Let* $\mathbf{V}$ *and* $\mathbf{W}$ *be pseudovarieties of semigroups, then*

$$\mathbf{V} *\!* \mathbf{W} \leq (\llbracket xyw = xzw \rrbracket \; \textcircled{m} \; \mathbf{V}) \; \textcircled{m} \; \mathbf{W}$$

*holds.*

We now turn to the meet operation.

*Example 2.4.17 (Meet).* The operator $\mathbf{V} \wedge (-)$ belongs to $\mathbf{GMC(PV)}$ but not to $\mathbf{GMC(PV)}^+$ (if $\mathbf{V}$ is proper). The associated pseudovariety of relational morphisms is $\widetilde{\mathbf{V}} \cap \mathsf{D}$, that is the collection of all divisions $f \colon S \to T$ with $S \in \mathbf{V}$. Indeed, $(\widetilde{\mathbf{V}} \cap \mathsf{D})\mathsf{q} = \widetilde{\mathbf{V}}\mathsf{q} \wedge \mathsf{Dq} = \mathsf{q}_{\mathbf{V}} \wedge 1_{\mathbf{PV}} = \mathbf{V} \wedge (-)$ because finite meets in $\mathbf{GMC(PV)}$ are taken pointwise.

Our next example is as much a new result as an example. First we need a definition. If $M_1, \ldots, M_n$ are monoids, then their *Schützenberger product* $\Diamond(M_1, \ldots, M_n)$ is the set of all $n \times n$ upper triangular matrices over the semiring $P(M_1 \times \cdots \times M_n)$ such that the $i, j$-entry belongs to

$$P(\{1\} \times \cdots \times \{1\} \times M_i \times \cdots \times M_j \times \{1\} \times \cdots \times \{1\})$$

where $P(X)$ denotes the power set of $X$, and if $i = j$, then the entry is a singleton. For semigroups $S_1, \ldots, S_n$, we define $\Diamond(S_1, \ldots, S_n)$ via our usual "scraping" philosophy: one takes the subsemigroup of $\Diamond(S_1^I, \ldots, S_n^I)$ consisting of those matrices whose $i, j$-entry belongs to

$$P(\{I\} \times \cdots \times \{I\} \times S_i \times \cdots \times S_j \times \{I\} \times \cdots \times \{I\})$$

and again if $i = j$, then the entry is a singleton.

For $\mathbf{V} \in \mathbf{PV}$, we denote by $\Diamond\mathbf{V}$ the pseudovariety of all semigroups generated by Schützenberger products of elements of $\mathbf{V}$. The operator $\mathbf{V} \mapsto \Diamond\mathbf{V}$ corresponds via the Eilenberg correspondence [85] to the operation of taking the (Boolean) polynomial closure of a variety of languages [229,230,236,325,379]. In the next example, we need the pseudovariety $\mathbf{J}$ of $\mathscr{J}$-trivial semigroups (i.e., semigroups whose $\mathscr{J}$-classes are singletons).

*Example 2.4.18 (Schützenberger product).* We begin by showing that the operator $\mathbf{V} \mapsto \Diamond\mathbf{V}$ belongs to $\mathbf{GMC(PV)}^{+}$. This is a new result and so we state it as a proposition. Because $\Diamond_1(S) = S$, clearly the operator $\Diamond$ is non-decreasing.

**Proposition 2.4.19.** *The Schützenberger product operator $\Diamond$ satisfies GMC.*

*Proof.* Let $\mathbf{V}$ be a pseudovariety of semigroups. Pin and Weil proved in [230, 236] that $\Diamond\mathbf{V}$ is the pseudovariety of semigroups generated by the pseudovariety of ordered semigroups $\mathbb{LJ}^{+} \ⓜ\ \mathbf{V}$, where $\mathbb{LJ}^{+} = [\![x^{\omega}yx^{\omega} \le x^{\omega}]\!]$; see [229,230,236,325] for more on pseudovarieties of ordered semigroups and their relationship with $\Diamond$. But then, as the Mal'cev product clearly satisfies GMC in the ordered context, we have the inequality

$$\mathbb{LJ}^{+} \ⓜ\ \mathbf{V} \subseteq (\mathbb{LJ}^{+} \ⓜ\ \mathbf{U}, \mathbf{U}) \ⓜ\ \mathbf{V} \subseteq (\Diamond\mathbf{U}, \mathbf{U}) \ⓜ\ \mathbf{V}$$

for any pseudovariety $\mathbf{U}$. Hence $\Diamond\mathbf{V} \subseteq (\Diamond\mathbf{U}, \mathbf{U}) \ⓜ\ \mathbf{V}$, as desired.    $\square$

Note that the case of GMC at $\mathbf{1}$, namely

$$\Diamond\mathbf{V} \le \Diamond\mathbf{1} \ⓜ\ \mathbf{V} = (\mathbf{J} * \mathbb{L}\mathbf{1}) \ⓜ\ \mathbf{V},$$

is a well-known result, see [229,230,325,345]. It now follows from the results of the previous section that $\Diamond$ can be defined by an element of $\mathbf{PVRM}^{+}$ (in fact there is a whole spectrum, including a maximum and a minimum, of such pseudovarieties of relational morphisms).

**Exercise 2.4.20.** Show directly that the projection

$$\pi \colon \Diamond_n(S_1, \ldots, S_n) \to S_1 \times \cdots \times S_n$$

to the diagonal satisfies $e\pi^{-1} \in \mathbb{LJ}^{+}$ where $\Diamond_n(S_1, \ldots, S_n)$ is ordered coordinate-wise by reverse inclusion.

*Question 2.4.21.* Give an element of $\mathbf{PVRM}^{+}$ analogous to $\mathbf{V}_D$ or $\mathbf{V}_K$ yielding $\Diamond(-)$ under $\mathfrak{q}$.

Now compositions of the above examples provide an infinite number of examples!

## 2.4.2 Compact elements of GMC(PV)

Theorem 2.3.36 implies that the compact elements of **PVRM** are the finitely generated members. But as $f_1, \ldots, f_n \prec f_1 \times \cdots \times f_n$ by Proposition 1.3.16, the compact elements of **PVRM** are the one-generated elements. Recall that if $f \colon S \to T$ is a relational morphism, $(f)$ denotes the pseudovariety of relational morphisms generated by $f$. Theorem 2.3.36 also implies that the operators of the form $(f)\mathsf{q}$ are precisely the compact elements of **GMC(PV)**. We now give a description of $(f)\mathsf{q}$.

**Proposition 2.4.22 (Formula for $(f)\mathsf{q}(\mathbf{V})$).** *Let $f \colon S \to T$ be a relational morphism. Then we have*

$$(f)\mathsf{q}(\mathbf{V}) = \mathbb{HSP}_{\mathrm{fin}}\{V f^{-1} \mid V \leq T, \ V \in \mathbf{V}\}. \tag{2.38}$$

*Proof.* The right-hand side of (2.38) is clearly contained in the left. Indeed, if $V \leq T$, with $V \in \mathbf{V}$, then define $f^0 \colon V f^{-1} \to V$ to be the restriction. Because $f^0 \prec f$, we have $f^0 \in (f)$ and therefore $V f^{-1} \in (f)\mathsf{q}(\mathbf{V})$.

Conversely, suppose that $M \in (f)\mathsf{q}(\mathbf{V})$. Then there is a relational morphism $g \colon M \to N$ with $N \in \mathbf{V}$ and $g \in (f)$. By Proposition 2.3.34, $g = hi$ where $i$ is an inclusion and $h \prec f^n$, some $n$. Because pseudovarieties of relational morphisms are closed under corestriction, without loss of generality, we may assume that $g = h$, that is, $g \prec f^n$ for some $n$. So there is a division $d \colon M \to S^n$ and a homomorphism $m \colon N \to T^n$ so that $gm \subseteq df^n$.

Let $V = Nm \leq T^n$. Then $V \in \mathbf{V}$. Let $U = V(f \times \cdots \times f)^{-1}$. Then as $gm \subseteq df^n$, we must have that $Ud^{-1} = M$. Thus $M \prec U$. So it suffices to show that $U$ is in the right-hand side of (2.38). Let $p_i \colon T^n \to T$ be the projection to the $i^{th}$ factor. Then $V p_i \in \mathbf{V}$ and $V p_i \leq T$. But

$$U = V(f \times \cdots \times f)^{-1} \leq V p_1 f^{-1} \times \cdots \times V p_n f^{-1}$$

(cf. the argument for (2.6)). This shows $U \in \mathbb{HSP}_{\mathrm{fin}}\{V p_i f^{-1} \mid i = 1, \ldots, n\}$ and hence belongs to the right-hand side of (2.38), as required.    □

The following corollary is quite useful.

**Corollary 2.4.23.** *Let $f \colon S \to T$ be a relational morphism. Then $(f)\mathsf{q}(\mathbf{V})$ is a compact subpseudovariety of $(S)$ for any pseudovariety $\mathbf{V}$ of semigroups. Moreover, if $\mathbf{V}$ is a fixed decidable pseudovariety, then there is an algorithm that given $f$ and a semigroup $U$ determines whether $U \in (f)\mathsf{q}(\mathbf{V})$.*

*Proof.* According to Proposition 2.4.22, $(f)\mathsf{q}(\mathbf{V})$ is generated by all semigroups $V f^{-1}$ with $V \leq T$, $V \in \mathbf{V}$. Because $T$ has only finitely many subsemigroups, it follows that $(f)\mathsf{q}(\mathbf{V})$ is a compact subpseudovariety of $(S)$. If $\mathbf{V}$ is decidable, then the set of such $V f^{-1}$ is effectively computable from $f$. As compact pseudovarieties have decidable membership, we are done.    □

It will be proved that $(f)$ has decidable membership in Theorem 3.5.16 using the equational theory developed in the next chapter.

### 2.4.3 Examples of continuous operators and continuously closed classes

We now provide some examples of continuous operators that do *not* satisfy GMC and describe some of their associated continuously closed classes. Recall from Theorem 2.3.36 that each continuous operator $\alpha$ has a unique maximum continuously closed class $M(\alpha)$ mapping to it under q. There is no minimum continuously closed class mapping to $\alpha$ (as q is not **inf** in this context), but there is a minimum Birkhoff continuously closed class $m(\alpha)$ (Theorem 2.3.41).

To get started let us fix some notation. If $S$ is a semigroup, then:

- $E(S)$ denotes the set of idempotents of $S$;
- $\mathrm{Reg}(S)$ denotes the set of regular elements;
- $\mathsf{K_G}(S)$ denotes the set of Type II elements, also known as group kernel (whence the notation) of $S$ (see [126, 295] and Section 4.12.1);
- $s^\omega$ is the idempotent power of $s \in S$.

A deep theorem of Ash [33], confirming a conjecture of the first author and proved independently by Ribes and Zalesskiĭ [300], shows that $\mathsf{K_G}(S)$ is the smallest subsemigroup of $S$ containing $E(S)$ and closed under the operation $x \mapsto axb$ where $a, b \in S$ are such that $aba = a$ or $bab = b$.

Our first example is the non-functorial wreath product operator of [364], called the semidirect product operator in [7, 27].

*Example 2.4.24 (Old Wreath Product).* For semigroups $S, T$, let

$$S \circ T = S^{T^\bullet} \rtimes T = S \wr (T^\bullet, T)$$

be the semidirect product with respect to the natural left action. This is the wreath product defined in [364]. Then $\mathbf{V} \circ \mathbf{W}$ is the pseudovariety generated by all semigroups of the form $S \circ T$ with $S \in \mathbf{V}$ and $T \in \mathbf{W}$ (equivalently, it is generated by all semidirect products $S \rtimes T$ of semigroups with $S \in \mathbf{V}$ and $T \in \mathbf{W}$ with the extra stipulation that the identity of $T$ must act as an identity whenever $T$ is a monoid). Note that $\mathbf{V} \circ \mathbf{W} \subseteq \mathbf{V} * \mathbf{W}$. Equality holds if and only if $\mathbf{W}$ is not a pseudovariety of groups [364].

Clearly, $\mathbf{V} \circ (-)$ is a continuous operator. It does not, however, satisfy GMC as Example 2.5.22 below shows. Let us describe briefly the associated continuously closed class. Let $\mathbf{V} \in \mathbf{PV}$. Then $\mathbf{V}_W$ consists of all relational morphisms $f \colon S \to T$ such that the derived semigroupoid of $f$ as defined in [364] divides a semigroup in $\mathbf{V}$. Then one can verify that $\mathbf{V}_W \in \mathbf{CC}^+$ (cf. Section 2.5.2). The results of [364] imply that

$$\mathbf{V}_W \mathsf{q} = \mathbf{V} \circ (-).$$

Example 2.5.21 will show that $\mathbf{V}_W \notin \mathbf{PVRM}$.

Another naturally occurring continuous operator failing to satisfy GMC is the power operator.

*Example 2.4.25 (Power operator).* Define the operator **P** by setting **P(V)** equal to the pseudovariety generated by power semigroups $P(S)$ of elements $S$ of **V**. Then **P** belongs to $\mathbf{Cnt(PV)}^+$ but not to **GMC(PV)**. Indeed, **P(1) = Sl**, so if **P** satisfied GMC, it would follow

$$\mathbf{P(G)} \leq \mathbf{P(1)} \,\textcircled{m}\, \mathbf{G} = \mathbf{Sl} \,\textcircled{m}\, \mathbf{G}.$$

But every semigroup in **Sl** ⓜ **G** has commuting idempotents, however the idempotents $\langle (01) \rangle$ and $\langle (12) \rangle$ of $P(S_3)$ do not commute (where $S_3$ is the symmetric group of degree 3), and so $P(S_3) \notin$ **Sl** ⓜ **G**. Therefore, **P** fails GMC at **1**. In fact, it is a very deep result that **P(G) = J** ⓜ **G** [126]. It would be nice to find a "natural" continuously closed class mapping to **P** under q.

*Example 2.4.26 (Regular $\mathscr{D}$-classes).* Recall that in a finite semigroup, Green's relations $\mathscr{J}$ and $\mathscr{D}$ coincide. If **V** is a pseudovariety of semigroups, **DV** denotes the pseudovariety of all semigroups whose regular $\mathscr{D}$-classes form a subsemigroup in **V**. The operator **D** is clearly continuous, so **D** $\in$ **Cnt(PV)**. It does not belong to $\mathbf{Cnt(PV)}^+$. Again we would like a "natural" continuously closed class that maps to **D** under q.

If **CS** denotes the pseudovariety of (completely) simple semigroups, then there is a Galois connection

$$[\mathbf{1}, \mathbf{DS}] \xrightleftharpoons[\mathbf{V} \mapsto \mathbf{V} \cap \mathbf{CS}]{\mathbf{D}} [\mathbf{1}, \mathbf{CS}]. \tag{2.39}$$

where we follow traditional usage and write **DS** for **DFSgp**. We should mention that the bottom map in (2.39) also has a left adjoint: the inclusion map.

Next we show that **D** does not satisfy **GMC(PV)**. The proof idea is based on an observation of K. Auinger. First note that **D1** is precisely the pseudovariety **J** of $\mathscr{J}$-trivial semigroups. Let **A** denote the pseudovariety of *aperiodic* semigroups; these are semigroups whose subgroups are all trivial. Now **DA = DRB** where **RB** is the pseudovariety of rectangular bands. So if **D** satisfied GMC, then

$$\mathbf{DA} = \mathbf{DRB} \leq \mathbf{D1} \,\textcircled{m}\, \mathbf{RB} = \mathbf{J} \,\textcircled{m}\, \mathbf{RB}.$$

We prove this is not the case. Let $U_2 = \{I, a, b\}$ be the two-element right zero semigroup with adjoined identity. Then $U_2$ is a band and so in **DA**. Let $\varphi \colon U_2 \to R$ be a relational morphism with $R$ a rectangular band. Suppose $a' \in a\varphi$, $b' \in b\varphi$ and $c' \in I\varphi$. Then $a'c' = a'b'c'$ and $a'c' \in a\varphi I\varphi \subseteq a\varphi$ while $a'b'c' \in a\varphi b\varphi I\varphi \subseteq b\varphi$. In other words, we have $a, b \in a'b'c'\varphi^{-1}$. Because the subsemigroup $\{a, b\}$ is not $\mathscr{J}$-trivial, this argument shows that $U_2 \notin \mathbf{J} \textcircled{m} \mathbf{RB}$.

*Example 2.4.27 (Local operator).* Recall there is a Galois connection (1.26)

$$\mathbf{PV} \xrightleftharpoons[\mathbf{V} \longmapsto \mathbf{V} \cap \mathbf{FMon}]{\mathbb{L}\mathbf{V} \longleftarrow \mathbf{V}} \mathbf{MPV}$$

between the lattices of semigroup and monoid pseudovarieties. Moreover, $\mathbb{L}$ is continuous and hence there is a continuous closure operator on $\mathbf{PV}$ given by $\mathbf{V} \mapsto \mathbb{L}(\mathbf{V} \cap \mathbf{FMon})$. It will be convenient to abuse notation and simply write $\mathbb{L}\mathbf{V}$ for $\mathbb{L}(\mathbf{V} \cap \mathbf{FMon})$. So if $\mathbf{V}$ is a pseudovariety of semigroups, then

$$\mathbb{L}\mathbf{V} = \{S \in \mathbf{FSgp} \mid eSe \in \mathbf{V}, \forall e \in E(S)\}.$$

A pseudovariety of semigroups $\mathbf{V}$ is said to be *local* in the sense of Eilenberg [85] if the equality $\mathbb{L}\mathbf{V} = \mathbf{V}$ is verified. The operator $\mathbb{L}$ belongs to $\mathbf{Cnt}(\mathbf{PV})^+$. Let us show that $\mathbb{L}$ fails GMC at $\mathbf{1}$. Namely, we establish $\mathbb{L}\mathbf{Sl} \not\leq \mathbb{L}\mathbf{1} \textcircled{m} \mathbf{Sl}$. Indeed, $\mathbb{L}\mathbf{1} \textcircled{m} \mathbf{Sl} = \mathbf{DA}$ (see for instance Corollary 4.6.51); on the other hand, $B_2 \in \mathbb{L}\mathbf{Sl}$.

*Example 2.4.28 (Idempotents).* Let $\mathbf{V}$ be a pseudovariety of semigroups. Then

$$\mathbf{EV} = \{S \mid \langle E(S) \rangle \in \mathbf{V}\}.$$

The operator $\mathbf{E}$ clearly belongs to $\mathbf{Cnt}(\mathbf{PV})^+$. The operator $\mathbf{E} \colon \mathbf{PV} \to \mathbf{PV}$ is in fact a continuous closure operator preserving all infs so it is what we shall call in Chapter 6 a CL-morphism. It is also part of a Galois connection, as we shall see momentarily. We would like a "naturally defined" member of $\mathbf{CC}^+$ mapping to $\mathbf{E}$.

Let us verify that $\mathbf{E} \notin \mathbf{GMC}(\mathbf{PV})$. Clearly, $\mathbf{E1}$ consists of all semigroups with a unique idempotent. That is $\mathbf{E1} = \mathbf{G} \textcircled{m} \mathbf{N}$. Because $\mathbf{E1} \leq \mathbb{L}\mathbf{G}$, we have $\mathbf{E1} \textcircled{m} \mathbf{V} \leq \mathbb{L}\mathbf{G} \textcircled{m} \mathbf{V}$. In particular,

$$\mathbf{E1} \textcircled{m} \mathbf{Sl} \leq \mathbb{L}\mathbf{G} \textcircled{m} \mathbf{Sl} = \mathbf{DS}.$$

On the other hand, $\mathbf{ESl}$ contains $B_2$ and hence is not contained in $\mathbf{DS}$. Thus $\mathbf{ESl} \not\leq \mathbf{E1} \textcircled{m} \mathbf{Sl}$ and hence $\mathbf{E}$ fails GMC at $\mathbf{1}$.

For $\mathbf{V} \in \mathbf{PV}$, define the pseudovariety $\mathbf{eV}$ by

$$\mathbf{eV} = \mathbb{HSP}_{\mathrm{fin}}(\{\langle E(S) \rangle \mid S \in \mathbf{V}\}).$$

Then $\mathbf{e} \in \mathbf{Cnt}^-$, that is $\mathbf{e}$ is continuous and non-increasing. In fact $\mathbf{e}$ is a $\mathbf{sup}$ kernel operator. The operators $\mathbf{E}$ and $\mathbf{e}$ are related by the following Galois connection:

$$\mathbf{PV} \underset{\mathbf{e}}{\overset{\mathbf{E}}{\rightleftarrows}} \mathbf{PV}.$$

Notice that $\mathbf{e1} = \mathbf{1}$ and that

$$\mathbf{eV} \leq \mathbf{V} \leq \mathbf{e1} \textcircled{m} \mathbf{V},$$

so $\mathbf{e}$ satisfies GMC at $\mathbf{1}$. But we claim it does not satisfy GMC, thus giving an example where GMC is not implied by GMC at $\mathbf{1}$. In fact we show that

$$\mathbf{eFSgp} \not\leq (\mathbf{e}(\mathbf{E1}), \mathbf{E1}) \textcircled{m} \mathbf{FSgp}. \tag{2.40}$$

Let $T_n$ be the full transformation monoid of degree $n$. Then it is well-known [137] (cf. Lemma 4.12.26) that $\langle E(T_n) \rangle = \{1\} \cup T_n \setminus S_n$ where $S_n$ denotes the symmetric group of degree $n$. In particular, $T_{n-1}$ is isomorphic to a subsemigroup of $\langle E(T_n) \rangle$. Thus $\mathbf{eFsgp} = \mathbf{Fsgp}$. Now clearly $e(\mathbf{E1}) = \mathbf{1}$, so to establish (2.40), it suffices to show that there is no relational morphism $f \colon S_n \to T$ with $T$ a finite semigroup and $f \in (\mathbf{1}, \mathbf{E1})$. Suppose such a relational morphism $f$ exists. It is well-known that if $U$ is a minimal subsemigroup of $T$ such that $S_n \leq U f^{-1}$, then $U$ is a group and hence belongs to $\mathbf{E1}$. Indeed, $S_n \leq (S_n f \cap U) f^{-1}$, so $U = S_n f \cap U$ by minimality. If $u \in U$ with $u \in \sigma f$, then $S_n = \sigma S_n \leq (uU) f^{-1}$ and $S_n = S_n \sigma \leq (Uu) f^{-1}$, so $uU = U = Uu$ by minimality. But a non-empty semigroup that is both left and right simple is a group (see Lemma A.3.3). Because $U \in \mathbf{E1}$ and $S_n \leq U f^{-1}$, $f$ cannot be in $(\mathbf{1}, \mathbf{E1})$ if $n \geq 2$. This proves (2.40) and hence shows that $\mathbf{e}$ fails GMC. Thus $\mathbf{e} \in \mathbf{Cnt}(\mathbf{PV})^- \setminus \mathbf{GMC}(\mathbf{PV})^-$.

We now turn to two similar examples.

*Example 2.4.29 (Regular elements).* Define, for a pseudovariety $\mathbf{V}$,

$$\mathbf{R}(\mathbf{V}) = \{S \mid \langle \mathrm{Reg}(S) \rangle \in \mathbf{V}\}.$$

The operator $\mathbf{R}$ clearly belongs to $\mathbf{Cnt}(\mathbf{PV})^+$ and is a continuous meet-preserving closure operator on $\mathbf{V}$ (and hence a CL-morphism as per Chapter 6). There is an associated Galois connection, defined as follows. For $\mathbf{V} \in \mathbf{PV}$, let

$$\mathbf{rV} = \mathbb{HSP}_{\mathrm{fin}}(\{\langle \mathrm{Reg}(S) \rangle \mid S \in \mathbf{V}\}).$$

Then $\mathbf{r} \in \mathbf{Cnt}(\mathbf{PV})^-$, in fact $\mathbf{r}$ is a continuous, $\mathbf{sup}$ kernel operator and we have the Galois connection

$$\mathbf{PV} \xleftarrow{\mathbf{R}} \mathbf{PV}.$$
$$\xrightarrow[\mathbf{r}]{}$$

Let us show that the operator $\mathbf{R}$ fails GMC at $\mathbf{1}$. The idea arose via discussions with M. Volkov. First note $\mathbf{R}(\mathbf{1}) = \mathbf{N}$ because a finite semigroup has a unique regular element if and only if it is nilpotent. Next observe $\mathbf{R}(\mathbf{CS}) = \mathbb{LG}$. Indeed, if $S \in \mathbf{R}(\mathbf{CS})$, then the regular elements must form a simple semigroup, so the only regular elements in $S$ are the elements of the minimal ideal; hence $eSe = H_e$ is a group for each idempotent $e \in S$. Conversely, if $S \in \mathbb{LG}$, then the minimal ideal of $S$ is exactly the set $\mathrm{Reg}(S)$ of regular elements (because if $e \in E(S)$ does not belong to the minimal ideal $I$ of $S$ and $x \in I$, then $(exe)^\omega$ is a second idempotent in $eSe$) and so $S \in \mathbf{R}(\mathbf{CS})$. Thus if GMC were to hold, then

$$\mathbb{LG} = \mathbf{R}(\mathbf{CS}) \leq \mathbf{R}(\mathbf{1}) \ \textcircled{m}\ \mathbf{CS} = \mathbf{N} \ \textcircled{m}\ \mathbf{CS}.$$

But this inequality is false. Let $S$ be the semigroup $\{a, b, aa, ab, ba, bb\}$ where $\{aa, ab, ba, bb\}$ is a set of right zeroes. In other words, $S$ is the quotient of the free semigroup $\{a, b\}^+$ by the congruence identifying two words if they have the same suffix of length 2. Then $\text{Reg}(S) = \{aa, ab, ba, bb\}$ is completely simple, so $S \in \mathbf{R(CS)}$. Let us show $S \notin \mathbf{N} \text{ⓜ} \mathbf{CS}$. Let $\varphi \colon S \to T$ be a relational morphism with $T$ completely simple. Let $a' \in a\varphi$, $b' \in b\varphi$. Then $(a')^\omega = (a'b'a')^\omega$ and $(a')^\omega \in a^2\varphi$ while $(a'b'a')^\omega \in ba\varphi$. Because $\{a^2, ba\}$ is a right zero semigroup, it follows that $\varphi \notin (\mathbf{N}, \mathbf{1})$. We conclude $S \notin \mathbf{N} \text{ⓜ} \mathbf{CS}$ and hence $\mathbf{R}$ fails GMC.

*Example 2.4.30 (Group kernel).* The operator $(-) \text{ⓜ} \mathbf{G}$ is of course a member of $\mathbf{Cnt(PV)}^+$. A semigroup $S$ belongs to $\mathbf{V} \text{ⓜ} \mathbf{G}$ if and only if $\mathsf{K_G}(S) \in \mathbf{V}$. Also $(-) \text{ⓜ} \mathbf{G} \le \mathbf{E}$. To see that this operator does not satisfy GMC, we recall that $\mathbf{ESl} = \mathbf{Sl} \text{ⓜ} \mathbf{G}$. Also $\mathbf{1} \text{ⓜ} \mathbf{G} = \mathbf{E1}$. So the same proof that we used in Example 2.4.28 for the operator $\mathbf{E}$ shows that $(-) \text{ⓜ} \mathbf{G} \notin \mathbf{GMC(PV)}$. We remark that $(-) \text{ⓜ} \mathbf{G}$ is not an idempotent operator [80, 295].

Now define, for $\mathbf{V} \in \mathbf{PV}$,

$$\mathsf{K_G}(\mathbf{V}) = \mathbb{HSP}_{\text{fin}}(\{\mathsf{K_G}(S) \mid S \in \mathbf{V}\}).$$

This operator belongs to $\mathbf{Cnt(PV)}^-$, but not $\mathbf{GMC(PV)}^-$ (one can use the same argument as in Example 2.4.28 for the operator $\mathbf{e}$). The operators $(-) \text{ⓜ} \mathbf{G}$ and $\mathsf{K_G}$ form a Galois connection

$$\mathbf{PV} \underset{\mathsf{K_G}}{\overset{(-) \text{ⓜ} \mathbf{G}}{\rightleftarrows}} \mathbf{PV}.$$

*Example 2.4.31 (Semidirect closure).* For a pseudovariety of semigroups $\mathbf{V}$, we define $\mathbf{V}^\omega$ to be the semidirect product closure of $\mathbf{V}$. That is,

$$\mathbf{V}^\omega = \bigcap \{\mathbf{W} \ge \mathbf{V} \mid \mathbf{W} * \mathbf{W} = \mathbf{W}\}.$$

Alternatively, for $\mathbf{V} \in \mathbf{PV}$, define inductively $\mathbf{V}^0 = \mathbf{1}$, $\mathbf{V}^n = \mathbf{V}^{n-1} * \mathbf{V}$. Then $\mathbf{V}^\omega = \bigcup \mathbf{V}^n$. The operator $\mathbf{V} \mapsto \mathbf{V}^\omega$ is clearly in $\mathbf{Cnt(PV)}^+$. It does not belong to $\mathbf{GMC(PV)}$. Indeed, $\mathbf{1}^\omega = \mathbf{1}$. So GMC at $\mathbf{1}$ would say that $\mathbf{V}^\omega \le \mathbf{1} \text{ⓜ} \mathbf{V}$. But this is clearly false. For instance, if $\mathbf{V} = \mathbf{Sl}$, then Stiffler's theorem [340] (our Theorem 4.5.2) shows that $\mathbf{Sl}^\omega = \mathbf{R}$, the pseudovariety of $\mathscr{R}$-trivial semigroups. On the other hand, $\mathbf{1} \text{ⓜ} \mathbf{Sl} = \mathbf{Sl}$ so GMC at $\mathbf{1}$ fails.

If we let $\lambda$ be any of $*$, $**$ or $\text{ⓜ}$, then we may consider the operator arising from the "other" variable: $\mathbf{V} \mapsto \mathbf{V} \lambda \mathbf{W}$. In all the above cases, the operator $(-) \lambda \mathbf{W}$ belongs to $\mathbf{Cnt(PV)}^+$ and so comes from a positive continuously closed class, which is usually not a pseudovariety of relational morphisms. Compositions of the various examples considered give an infinite number of further examples.

We mention that one can consider the operators $\mathbf{V} \lambda^\omega (-)$ and $(-)^\omega \lambda \mathbf{W}$. Unwinding definitions, one has

$$\mathbf{V} \curlywedge^{\omega} \mathbf{W} = \bigcup \mathbf{V} \curlywedge (\cdots \mathbf{V} \curlywedge (\mathbf{V} \curlywedge \mathbf{W}) \cdots)$$

$$\mathbf{V}\,^{\omega}\!\curlywedge \mathbf{W} = \bigcup (\cdots (\mathbf{V} \curlywedge \mathbf{W}) \curlywedge \mathbf{W} \cdots) \curlywedge \mathbf{W}. \tag{2.41}$$

For instance, the two-sided Prime Decomposition Theorem [293, 296] gives the equality $\mathbf{Sl} **^{\omega} \mathbf{Sl} = \mathbf{A}$ (see Corollary 5.3.22). On the other hand, a result of Thérien and Straubing [349] shows that $\mathbf{Sl}\,^{\omega}** \mathbf{Sl} = \mathbf{DA}$. Further discussion on which side to make "the variable" will appear in Section 2.9.

## 2.5 The Derived Semigroupoid Theorem

For $\mathbf{V} \in \mathbf{PV}$, we want to construct $\mathbf{V}_D \in \mathbf{PVRM}^+$ such that $\mathbf{V}_D\mathfrak{q} = \mathbf{V} * (-)$ as advertised in Example 2.4.9. Here we are using the semidirect product operator $*$ on $\mathbf{PV}$ as defined in [85] or Section 1.2.3. So $\mathbf{V} * \mathbf{W}$ is

$$\mathbb{HSP}_{\mathrm{fin}}(\{V \rtimes W \mid V \in \mathbf{V}, \ W \in \mathbf{W}\})$$

where $W$ acts on the left of $V$. We reiterate: if $W$ is a monoid, the identity of $W$ need *not* act as an identity. The semidirect product of pseudovarieties is associative [85] but $(\mathbf{PV}, *)$ is not a monoid. One does have $\mathbf{1} * \mathbf{V} = \mathbf{V}$, but in general $\mathbf{V} * \mathbf{1} \neq \mathbf{V}$. However, note $\mathbf{V} * \mathbf{W} \leq (\mathbf{V} * \mathbf{1}) \ \textcircled{m}\ \mathbf{W}$ by GMC at $\mathbf{1}$.

### 2.5.1 The derived semigroupoid

The Derived Category Theorem of [364] (respectively [339]) sets up the proper relationship between the semidirect product of monoid (respectively category) pseudovarieties and relational morphisms. Earlier works suggesting such a relationship should exist include [85, 345, 358, 361, 362]. Now we may apply our philosophy from Section 1.2.3 to obtain the Derived Semigroupoid Theorem (this will be different than the theorem of the same name in [364]).

Let us recall the monoid construction as presented in [339]. If $f \colon M \to N$ is a relational morphism of monoids, then we define a category $\mathbf{Der}(f)$ by:

$$\mathrm{Obj}(\mathbf{Der}(f)) = N;$$

$$\mathrm{Arr}(\mathbf{Der}(f)) = \{n_1 \xrightarrow{(m,n)} n_1 n \mid n_1 \in N, (m,n) \in \#f\}.$$

The pair $(m, n)$ is termed the *label* of the arrow $n_1 \xrightarrow{(m,n)} n_1 n$. It is convenient to write this arrow using the more compact notation $(n_1, (m, n))$. Composition is given by the rule:

$$n_1 \xrightarrow{(m,n)} n_1 n \xrightarrow{(m',n')} n_1 n n' = n_1 \xrightarrow{(mm',nn')} n_1 n n'.$$

The identity at the object $n$ is the arrow $(n, (1,1))$. For the cognoscenti, $\mathbf{Der}(f)$ is the category of elements [186] of the composition of the projection

$p_N\colon \#f \to N$ with the regular representation $Y\colon N \to \mathbf{Set}$; see Exercise 2.5.1 below. It first appeared in the work of Quillen [257] on algebraic $K$-theory and subsequently resurfaced in work of Tilson [361, 362] (in the guise of the derived semigroup), Loganathan [183] and Nico [219] before making its definitive entrance into semigroup theory in [364] (see also [192–194]).

The *derived category* of $f$, denoted $D_f$, is then the quotient of $\mathbf{Der}(f)$ obtained by identifying coterminal arrows $(n_1, (m, n))$ and $(n_1, (m', n'))$ (so $n_1 n = n_1 n'$) if the two maps $n_1 f^{-1} \to (n_1 n) f^{-1}$ given by

$$m_0 \longmapsto m_0 m \text{ and } m_0 \longmapsto m_0 m' \tag{2.42}$$

coincide. We use $\sigma_f\colon \mathbf{Der}(f) \twoheadrightarrow D_f$ for the canonical quotient map. If $f$ is clear from context, we just write $\sigma$. From a more global viewpoint, there is a natural representation of $\mathbf{Der}(f)$ as a concrete category via

$$n \longmapsto nf^{-1}, \quad (n_1, (m, n)) \longmapsto \cdot m\colon n_1 f^{-1} \to n_1 n f^{-1}$$

and simply put, $D_f$ is the quotient by the kernel of this representation.

**Exercise 2.5.1.** Let $C$ be a small category. The *category of elements* [186] $\int F$ of a functor $F\colon C \to \mathbf{Set}$ has object set consisting of all pairs $(x, c)$ with $c \in \mathrm{Obj}(C)$ and $x \in cF$. Arrows are of the form $(x, c) \xrightarrow{m} (x(mF), c')$ where $m\colon c \to c'$ is an arrow of $C$; this arrow can be more compactly represented by the pair $(x, m)$. The composition of arrows $(x, m)$ and $(x(mF), m')$ is the arrow $(x, mm')$. The identity at $(x, c)$ is $(x, 1_c)$.

1. Verify that $\mathbf{Der}(f)$ is the category of elements $\int p_N Y$.
2. A functor $P\colon D \to C$ is called a *discrete fibration* if, for each object $d \in D$ and each arrow $n$ with domain $dP$ in $C$, there is a unique arrow $m$ of $D$ with domain $d$ such that $mP = n$. Prove that the map $P\colon \int F \to C$ defined by projecting to the second coordinate on objects and arrows is a discrete fibration.
3. Let $P\colon D \to C$ be a discrete fibration. Define a functor $M_P\colon C \to \mathbf{Set}$ as follows: $cM_P = cP^{-1}$ on an object $c$; on an arrow $n\colon c \to c'$, the function $nM_P\colon cP^{-1} \to c'P^{-1}$ is given by sending $d \in cP^{-1}$ to the codomain of the unique arrow $m$ of $D$ with domain $d$ such that $mP = n$. Verify that $M_P$ is a functor.
4. Let $C$ be a category. Define the category $\mathscr{B}C$ of discrete fibrations over $C$ (known as called the *classifying topos* of $C$) and prove that $\int$ and $M$ provide an equivalence between the categories $\mathbf{Set}^C$ and $\mathscr{B}C$.

Now our philosophy says: given $f\colon S \to T$, a relational morphism of semigroups, construct $f^I\colon S^I \to T^I$ with $f^I$ given by extending $f$ to $S^I$ by $I \mapsto I$. One then defines $\mathbf{Der}(f)$ to be the result of deleting from $\mathbf{Der}(f^I)$ all arrows with labels $(I, I)$. It is absolutely crucial that $\mathbf{Der}$ is a functor in the appropriate context (cf. [339]): if one were to add new identities only when the semigroups in question are not monoids (as in [27, 364]), then $\mathbf{Der}$ would *not* be a functor. The result of removing the arrows with label $(I, I)$ yields a semigroupoid. A *semigroupoid* is defined precisely like a category [185], but one relaxes the axiom demanding identities at each object. We can view a semigroup as a semigroupoid with a unique object. The *derived semigroupoid* of $f$, denoted $D_f$, is defined to be the image of $\mathbf{Der}(f)$ in $D_{f^I}$.

We now show that this derived semigroupoid relates to the semidirect product of semigroups [85] as the derived category of [364] (modified as per [339]) relates to the semidirect product of monoids. But note that the different derived semigroupoid of [364] relates to what Tilson calls the wreath product of pseudovarieties (and [7, 27] call the semidirect product), the non-functorial formulation of Example 2.4.24.

First we establish the main properties of the derived semigroupoid. We follow the scheme of [339]. By a *relational morphism* $f \colon C \to D$ *of semigroupoids*, we mean a function $f \colon \mathrm{Obj}(C) \to \mathrm{Obj}(D)$ and hom set relations $f_{c,c'} \colon C(c, c') \to D(cf, c'f)$ such that:

- For $s \colon c \to c'$, $s f_{c,c'} \neq \emptyset$;
- For $s \colon c \to c'$ and $t \colon c' \to c''$, one has

$$s f_{c,c'} t f_{c',c''} \subseteq (st) f_{c,c''}.$$

We often omit the subscript $f_{c,c'}$ and just write $f$. A relational morphism of semigroupoids is called a *division* if $f$ is injective on coterminal arrows, that is, $s, t \colon c \to c'$ and $sf \cap tf \neq \emptyset$ implies $s = t$. In this case one says $C$ *divides* $D$ and writes $C \prec D$. We remark that $\prec$ is only a preorder on the class of finite semigroupoids. Of course, analogous notions can be defined for categories, but here the identities must be respected. Let us explore some basic properties of $\mathbf{Der}(f)$ and $D_f$.

**Lemma 2.5.2.** *Let* $f \colon S \to T$ *be a relational morphism of semigroups. Then* $(I, (s, t))\sigma = (I, (s', t'))\sigma$ *if and only if* $s = s'$, $t = t'$.

*Proof.* The if direction is trivial. Let us turn to the only if direction. First of all, because $\sigma$ is a quotient map, in order to identify the two arrows above they must be coterminal, that is, we need $t = It = It' = t'$. Also, because $I \in I(f^I)^{-1}$, the equality of the functions $\cdot s$ and $\cdot s'$ on $I(f^I)^{-1}$ implies that $s = Is = Is' = s'$, thereby completing the proof.                          $\square$

Arrows of the form $(I, (s, t))$ are quite special and are called *core arrows* of $\mathbf{Der}(f)$. They play a key role in the theory developed in [339].

**Definition 2.5.3 (Core injective).** *Let* $f \colon S \to T$ *be a relational morphism of semigroups and* $g \colon \mathbf{Der}(f) \to C$ *be a relational morphism of semigroupoids. Then* $g$ *is called* core injective *if*

$$(I, (s, t))g \cap (I, (s', t))g \neq \emptyset \implies (I, (s, t)) = (I, (s', t)). \qquad (2.43)$$

*Notice that the right-hand side of* (2.43) *is equivalent to* $s = s'$.

The content of Lemma 2.5.2 is that $\sigma$ is core injective. In particular, we can identify a core arrow of $\mathbf{Der}(f)$ with its image in $D_f$. The following is the key lemma relating $\mathbf{Der}(f)$ and $D_f$. It shows that $\sigma$ is in some sense the universal core injective quotient of $\mathbf{Der}(f)$. See [339] for a more precise formulation (namely that $\sigma$ induces the largest congruence that is injective on core arrows).

**Lemma 2.5.4.** *Let* $f: S \to T$ *be a relational morphism of semigroups and let* $g: \mathbf{Der}(f) \to C$ *be a relational morphism of semigroupoids. Consider the diagram*

$$
\begin{array}{ccc}
\mathbf{Der}(f) & \xrightarrow{\;g\;} & C \\[4pt]
{\scriptstyle\sigma}\big\downarrow & \nearrow_{\sigma^{-1}g} & \\[4pt]
D_f & &
\end{array}
$$

*Then* $g$ *is core injective if and only if* $\sigma^{-1}g: D_f \to C$ *is a division.*

*Proof.* Suppose that $\sigma^{-1}g$ is a division. Lemma 2.5.2 shows that $\sigma$ is core injective. Therefore, the restriction of $g$ to the core arrows of $\mathbf{Der}(f)$ is $\sigma(\sigma^{-1}g)$. As $\sigma$ is core injective and $\sigma^{-1}g$ is a division, we see that $g$ is core injective.

For the converse, we need to show that if $(t_1, (s, t))$, $(t_1, (s', t'))$ are coterminal arrows (and hence $t_1 t = t_1 t'$) with $(t_1, (s, t))g \cap (t_1, (s', t'))g \neq \emptyset$, then $(t_1, (s, t))\sigma = (t_1, (s', t'))\sigma$. If $t_1 = I$, then as $g$ is core injective, we must have the two arrows are equal, so there is nothing to prove. If $t_1 \neq I$, then $t_1(f^I)^{-1} = t_1 f^{-1}$. Suppose $s_1 \in t_1 f^{-1}$. Then $(I, (s_1, t_1))$ is an arrow of $\mathbf{Der}(f)$. Let $m \in (I, (s_1, t_1))g$ and $n \in (t_1, (s, t))g \cap (t_1, (s', t'))g$. Then

$$
mn \in (I, (s_1 s, t_1 t))g \cap (I, (s_1 s', t_1 t'))g. \tag{2.44}
$$

Because the two arrows appearing in (2.44) are coterminal (as $t_1 t = t_1 t'$), we must have by core injectivity of $g$ that $s_1 s = s_1 s'$. This yields the desired equality $(t_1, (s, t))\sigma = (t_1, (s', t'))\sigma$.  $\square$

The derived semigroupoid construction respects divisions of relational morphisms.

**Proposition 2.5.5.** *Suppose* $f: S \to T$ *and* $g: S' \to T'$ *are relational morphisms such that* $f \prec g$. *Then* $D_f \prec D_g$.

*Proof.* Suppose that we have a division

$$
\begin{array}{ccc}
S & \xrightarrow{\;d\;} & S' \\[4pt]
{\scriptstyle f}\big\downarrow & \subseteq & \big\downarrow{\scriptstyle g} \\[4pt]
T & \xrightarrow{\;m\;} & T'.
\end{array}
\tag{2.45}
$$

Define a relational morphism $\overline{d}\colon \mathbf{Der}(f) \to \mathbf{Der}(g)$ by $m^I$ on objects and by

$$(t_1, (s, t))\overline{d} = \{(t_1 m^I, (s', tm)) \mid s' \in sd\}$$

on arrows. Notice that $\overline{d}$ takes core arrows of $\mathbf{Der}(f)$ to core arrows of $\mathbf{Der}(g)$. Hence if we can prove that $\overline{d}$ is core injective, then $\overline{d}\sigma_g\colon \mathbf{Der}(f) \to D_g$ will be core injective (using Lemma 2.5.2) and hence, by Lemma 2.5.4, will induce a division $\sigma_f^{-1}\overline{d}\sigma_g\colon D_f \to D_g$, as required.

The verification that $\overline{d}$ is a well-defined relational morphism is straightforward (cf. [339]) and we leave it to the reader. To see that it is core injective, suppose that $(I, (s, t))\overline{d} \cap (I, (s', t))\overline{d} \neq \emptyset$. Then a common element is of the form $(I, (s_0, tm))$ with $s_0 \in sd \cap s'd$. Because $d$ is a division, we conclude that $s = s'$. Therefore, $\overline{d}$ is core injective, completing the proof.   □

**Exercise 2.5.6.** Check that $\overline{d}$, as defined in the above proof, is a relational morphism.

*Remark 2.5.7.* It is enough for (2.45) to be a division diagram in the sense of [339] in order to ensure $D_f \prec D_g$.

Our next goal: to determine how the derived semigroupoid behaves with respect to products.

**Proposition 2.5.8.** *Let $f\colon S \to T$ and $f'\colon S' \to T'$ be relational morphisms. Then $D_{f \times f'} \prec D_f \times D_{f'}$.*

*Proof.* Define a morphism $g\colon \mathbf{Der}(f \times f') \to D_f \times D_{f'}$ on objects by $I \mapsto (I, I)$ and $(t, t') \mapsto (t, t')$ for $(t, t') \in T \times T'$. Define $g$ on arrows by

$$(I, ((s, s'), (t, t')))g = ((I, (s, t))\sigma_f, (I, (s', t'))\sigma_{f'})$$
$$((t_1, t_1'), ((s, s'), (t, t')))g = ((t_1, (s, t))\sigma_f, (t_1', (s', t'))\sigma_{f'}). \qquad (2.46)$$

It is straightforward to check that $g$ is a functor. To establish the proposition, it therefore suffices, by Lemma 2.5.4, to show that $g$ is core injective. This follows easily from (2.46) and the fact that $\sigma_f$ and $\sigma_{f'}$ are core injective.   □

**Exercise 2.5.9.** Show that $g$ in the above proof is a core injective functor.

To prove the Derived Semigroupoid Theorem, we shall need to calculate the derived semigroupoid of a semidirect product projection.

**Lemma 2.5.10.** *Let $S$, $T$ be semigroups and suppose that $T$ has a left action on $S$. Let $\pi\colon S \rtimes T \to T$ be the semidirect product projection. Then $D_\pi \prec S$.*

*Proof.* We write the binary operation of $S$ additively (although, as usual, commutativity is not assumed) and the action of $T$ on $S$ multiplicatively. The action of $T$ on $S$ extends to a monoid action of $T^I$ on $S$ via $Is = s$ for all $s \in S$. Define a functor $g\colon \mathbf{Der}(\pi) \to S$ by the unique map on objects. On arrows, set $(t_1, ((s, t), t))g = t_1 s$. To check that $g$ is a functor, we verify

$$(t_1, ((s,t),t))g + (t_1 t, ((s',t'),t'))g = t_1 s + t_1 t s'$$
$$= t_1(s + t s')$$
$$= (t_1, ((s + t s', t t'), t t'))g$$
$$= ((t_1, ((s,t),t))(t_1 t, ((s',t'),t')))g.$$

Now we check $g$ is core injective. Suppose $(I, ((s,t),t))g = (I, ((s',t),t))g$. Then $s = Is = Is' = s'$. Thus $g$ is core injective and so $D_\pi \prec S$ by Lemma 2.5.4. □

We may now state and prove the Derived Semigroupoid Theorem, the correct semigroupoid analogue of Tilson's Derived Category Theorem [339, 364]. Recall from Section 1.2.3 that if $S$ and $T$ are semigroups, then $S \wr T = S^{T^I} \rtimes T$ with action $t_0{}^t f = t_0 t f$ for $t_0 \in T^I$.

### Theorem 2.5.11 (Derived Semigroupoid Theorem).

(a) Let $f: S \to T$ be a relational morphism and suppose that $f$ factors as $d\pi$ with $d: S \to U \rtimes T$ a division and $\pi: U \rtimes T \to T$ the semidirect product projection, as per the diagram

Then $D_f \prec U$.

(b) Let $f: S \to T$ be a relational morphism and $U$ a semigroup such that $D_f \prec U$. Then there is a division $d: S \to U \wr T$ such that $f = d\pi$ with $\pi: U \wr T \to T$ the wreath product projection, diagrammed as follows:

*Proof.* We begin with (a). If $f = d\pi$ with $d$ a division, then $f \prec \pi$ and so $D_f \prec D_\pi \prec U$ by Proposition 2.5.5 and Lemma 2.5.10.

To prove (b), let $g: D_f \to U$ be a division. We write $U$ additively (again not assuming commutativity). Define a division $d: S \to U \wr T$ by

$$sd = \{(h,t) \mid t \in sf, \ t_0 h \in (t_0, (s,t))\sigma g, \ \text{for } t_0 \in T^I\}.$$

The verification that $d$ is a relational morphism is straightforward (cf. [339, 364]). To see that $d$ is a division, suppose that $(h,t) \in sd \cap s'd$. Then $t \in sf \cap s'f$ and $Ih \in (I, (s,t))\sigma g \cap (I, (s',t))\sigma g$. As $g$ is a division, and the two arrows in question are coterminal, we obtain $(I, (s,t))\sigma = (I, (s',t))\sigma$ and hence $s = s'$ by Lemma 2.5.2. Thus $d$ is a division, as desired. Clearly, $d\pi = f$. □

**Exercise 2.5.12.** Show that $d$ in the above proof is a well-defined relational morphism.

**Corollary 2.5.13.** Let $f\colon S \to T$ be a relational morphism of semigroups. Then $f$ is a division if and only if $D_f$ divides the trivial semigroup.

*Proof.* If $D_f \prec \{1\}$, then by (b) of the Derived Semigroupoid Theorem, $f = d\pi$ where $d\colon S \to \{1\} \wr T$ is a division and $\pi\colon \{1\} \wr T \to T$ is the projection. But $\pi$ is an isomorphism, so $f$ is a division.

   Conversely, if $f$ is a division, then $d\colon S \to \{1\} \wr T$ given $sd = \{\overline{1}\} \times sf$ is a division (where $\overline{1}\colon T^I \to \{1\}$ is the unique function) and $f = d\pi$ with $\pi$ the projection. Thus, by (a) of the Derived Semigroupoid Theorem, we obtain $D_f \prec \{1\}$.     □

### 2.5.2 The construction of the pseudovariety $\mathbf{V}_D$

A *pseudovariety of semigroupoids* is a class of semigroupoids closed under taking finite (including empty) products, coproducts (equal disjoint unions) and divisors [364]. For $\mathbf{V} \in \mathbf{PV}$, denote by $\mathbf{gV}$ the pseudovariety of semigroupoids generated by elements of $\mathbf{V}$, viewed as one object semigroupoids, and by $\ell\mathbf{V}$ the pseudovariety of semigroupoids $C$ whose local semigroups $C(c)$ belong to $\mathbf{V}$. Then $\ell$ and $\mathbf{g}$ are the right adjoints and left adjoints, respectively, of the map $\mathbf{V} \mapsto \mathbf{V} \cap \mathbf{FSgp}$ from semigroupoid pseudovarieties to semigroup pseudovarieties.

   Let $\mathbf{V}$ be a pseudovariety of semigroups. We define (analogously to [339]) $\mathbf{V}_D$ to consist of all those relational morphisms $f\colon S \to T$ of finite semigroups such that the $D_f \in \mathbf{gV}$. That is,

$$\mathbf{V}_D = \{f \in \mathbf{RM} \mid D_f \in \mathbf{gV}\}. \tag{2.47}$$

It will also be useful to define, for a pseudovariety of *monoids* $\mathbf{V}$,

$$\ell\mathbf{V}_D = \{f \in \mathbf{RM} \mid D_f \text{ divides a category in } \ell\mathbf{V}\}. \tag{2.48}$$

We remark that if $\mathbf{V}$ is a pseudovariety of monoids, then it is possible that a semigroupoid $S$ has local semigroups that are monoids in $\mathbf{V}$ without $S$ dividing a category in $\ell\mathbf{V}$. In Chapter 5, it will be critical that we use the formulation of $\ell\mathbf{V}_D$ in (2.48) because of this issue. The class $\ell\mathbf{1}_D$ will play a particularly important role. For instance, a semigroupoid can be locally trivial (meaning the local semigroups are trivial semigroups) without it dividing a locally trivial category. The problem is that the idempotent in the local semigroup at an object need not act as an identity on arrows coming in or out of the object. It is a deep theorem of Tilson that every locally trivial category belongs to any non-trivial pseudovariety of categories [364].

   If we were dealing with the monoid/category context, it would follow immediately from [339] that $\mathbf{V}_D \in \mathbf{PVRM}^+$. In fact, everything carries over in a straightforward fashion from the monoidal setting except for Axiom (r-e). This is because the bonded component theorem is used in the Image Factorization Theorem of [339]. We remark that Axiom (co-re) fails if one uses the derived semigroupoid of [27,364]; see Example 2.5.21 below.

So let us prove $\mathbf{V}_D \in \mathbf{PVRM}^+$. Proposition 2.5.5 shows that $\mathbf{V}_D$ satisfies Axiom ($\prec$), whereas Proposition 2.5.8 and Corollary 2.5.13 establish Axiom ($\times$) and Axiom (id). Verifying Axiom (r-e) requires some familiarity with the notion of pseudoidentities over graphs.

In Chapters 2 and 3, by a *graph* we mean the following.

**Definition 2.5.14 (Graph).** *A graph $\Gamma$ consists of a vertex set $V$, an edge set $E$ and two functions $\alpha, \omega \colon E \to V$ selecting the initial and terminal vertices, respectively, of the edge.*

Later, in Chapter 4, we shall consider graphs in the sense of Serre [317]. There is a slightly different formalism in this case and so we shall use $\iota$ and $\tau$ for the functions selecting the initial and terminal vertices for graphs in the sense of Serre to help the reader distinguish which sort of graph we are considering.

The papers [27,149] should be consulted for details in what follows. The reader is advised to skip over the technical details on a first reading.

If $\mathbf{V}$ is a pseudovariety of semigroupoids, then Reiterman's Theorem, properly generalized [27,149], shows that $\mathbf{V}$ is defined by pseudoidentities over graphs, that is, by formal equalities between elements of finitely generated, free profinite semigroupoids. If $G$ is a graph, $\widehat{G^+}$ will denote the free profinite semigroupoid generated by $G$ [27,149]. An element $u \in \widehat{G^+}$ has a well-defined notion of a content (or support), denoted $c(u)$. These are precisely the edges of $G$ "used" in $u$. For instance, if $u$ belongs to the free semigroupoid $G^+$ on $G$, i.e., $u$ is a path, then $c(u)$ is the set of edges traversed by $u$. In general one can show, by considering relatively free semigroupoids in $\mathbf{gSl}$, that any sequence of paths converging to $u$ uses the same set of edges from some point onwards. So the content (or set of edges) used by $u$ is well defined.

If $u = v$ is a pseudoidentity over a finite graph $G$ (a formal equality of elements of $\widehat{G^+}$), then one says that $u = v$ is a path pseudoidentity if $c(u) \cup c(v) = G$. An important result of [364] shows that pseudovarieties of *categories* are defined by path pseudoidentities (actually, Tilson [364] uses the language of eventual equational descriptions; the pseudoidentity version can be found in [27,149]). Tilson's argument applies to any pseudovariety of semigroupoids that is generated by categories. However, in general, pseudovarieties of semigroupoids need not be definable by path pseudoidentities.

Consider, for instance, the identity:

$$(\circ \xrightarrow[y]{x} \circ \xrightarrow{z} \circ, \; x = y). \tag{2.49}$$

Note that $z$ is not in the content of either side of the equation. Now consider the semigroupoid:

$$S = \circ \quad \xrightarrow[\;b\;]{\;a\;} \quad \circ.$$

It satisfies (2.49) vacuously because there is no way to map the graph in (2.49) to $S$: $z$ has "no place to go." It follows that (2.49) is not equivalent to

$$(\circ \quad \xrightarrow[\;y\;]{\;x\;} \quad \circ, \ x = y).$$

In fact, one can verify that the pseudovariety of semigroupoids satisfying (2.49) cannot be defined by path pseudoidentities. Consider the graph in (2.49) as a semigroupoid $T$ with $xz = yz$. One can easily verify that $T$ does not satisfy (2.49), but satisfies exactly the same path pseudoidentities as $S$.

However, we now prove the following new (as far as we know) result.

**Theorem 2.5.15.** *Let $\mathbf{V} \in \mathbf{PV}$. Then $\mathbf{gV}$ is definable by path pseudoidentities.*

*Proof.* Let $(G, u = v)$ be a pseudoidentity satisfied by $\mathbf{gV}$ over a finite graph $G$. We show that $(c(u) \cup c(v), u = v)$ is satisfied by $\mathbf{gV}$. Because $(G, u = v)$ is clearly a consequence of $(c(u) \cup c(v), u = v)$, it will then follow that $\mathbf{gV}$ is definable by path pseudoidentities. Less formally, we can always remove all edges of the graph not belonging to the content of either side of the pseudoidentity.

The proof relies on the following simple observation: a pseudovariety $\mathbf{W}$ of semigroupoids given by generators satisfies a pseudoidentity if and only if each of the generators satisfy it. Thus, we just need to show that every semigroup $S \in \mathbf{V}$ satisfies $(c(u) \cup c(v), u = v)$.

But given a map of $c(u) \cup c(v)$ to $S$, we can extend it to $G$ in any old way, as a map from a graph into a semigroup (equal one object semigroupoid) is the same thing as a map from its edge set into the semigroup, that is, all edges have "a place to go." Because $S \models (G, u = v)$, it follows that the induced map $\psi \colon (\widehat{c(u) \cup c(v)})^{+} \to S$ satisfies $u\psi = v\psi$. So $S \models (c(u) \cup c(v), u = v)$ and the theorem follows. □

As a corollary, we show that $\mathbf{V}_D$ satisfies Axiom (r-e).

**Corollary 2.5.16.** *The class $\mathbf{V}_D$ satisfies Axiom (r-e).*

*Proof.* Suppose $f \colon S \to T$ is in $\mathbf{V}_D$ and $i \colon T \to T'$ is an inclusion. Then $D_f \in \mathbf{gV}$. To show that $D_{fi} \in \mathbf{gV}$ it suffices, by Theorem 2.5.15, to show that if $(G, u = v)$ is a path pseudoidentity satisfied by $\mathbf{gV}$, then $D_{fi} \models (G, u = v)$. Observe that $D_f$ is a subsemigroupoid of $D_{fi}$. Moreover, it is easily verified to have the property that any arrow of $D_{fi}$ with initial vertex in $D_f$ belongs to $D_f$. In particular, any path in $D_{fi}$ starting at a vertex of $D_f$ stays in $D_f$.

Let $q$ be the initial vertex of $u$ (equal the initial vertex of $v$). Suppose $\varphi \colon G \to D_{fi}$ is a morphism. If $q\varphi \in T^I$ (so $q\varphi \in D_f$), then by the above paragraph and because $(G, u = v)$ is a path pseudoidentity, $G\varphi \subseteq D_f$ and hence the image of $u$ and $v$ coincide because $D_f \in \mathbf{gV}$. If $q\varphi \in (T')^I \setminus T^I$, then $q\varphi(fi)^{-1} = \emptyset$ and so, by construction of $D_{fi}$, any hom set $D_{fi}(q\varphi, t')$ has at most one element. Thus the images of $u$ and $v$ must coincide (being coterminal and starting at $q\varphi$). □

In summary, we have the following result.

**Theorem 2.5.17 (Derived Semigroupoid Theorem: Pseudovarieties).**
*Let* $\mathbf{V}$ *be a pseudovariety of semigroups. Then the class* $\mathbf{V}_D$ *is a positive pseudovariety of relational morphisms. Moreover, for any pseudovariety* $\mathbf{W}$ *of semigroups, the equality*

$$\mathbf{V}_D\mathsf{q}(\mathbf{W}) = \mathbf{V} * \mathbf{W}$$

*holds. That is,* $\mathbf{V}_D\mathsf{q} = \mathbf{V} * (-)$.

*Proof.* We have already verified that $\mathbf{V}_D$ satisfies the axioms to belong to $\mathbf{PVRM}^+$. Suppose $S \in \mathbf{V} * \mathbf{W}$. Then there is a division $d\colon S \to V \rtimes W$ with $V \in \mathbf{V}$ and $W \in \mathbf{W}$. Let $f = d\pi\colon S \to W$ be the composition of $d$ with the projection. Then $D_f \prec V$ by the Derived Semigroupoid Theorem and so $f \in \mathbf{V}_D$, establishing $S \in \mathbf{V}_D\mathsf{q}(\mathbf{W})$. Conversely, if $S \in \mathbf{V}_D\mathsf{q}(\mathbf{W})$, then there is a relational morphism $f\colon S \to W$ with $W \in \mathbf{W}$ and $f \in \mathbf{V}_D$. Thus $D_f \in \mathbf{gV}$ and hence $D_f \prec V$ for some $V \in \mathbf{V}$. Another application of the Derived Semigroupoid Theorem shows that $S \prec V \wr W$. Because $V \wr W$ is the semidirect product $V^{W^I} \rtimes W$ and $V^{W^I} \in \mathbf{V}$, we see that $V \wr W \in \mathbf{V} * \mathbf{W}$ and hence $S \in \mathbf{V} * \mathbf{W}$. This completes the proof of the theorem. □

*Remark 2.5.18.* Corollary 2.5.13 shows that $\mathbf{1}_D = \mathsf{D}$, the pseudovariety of divisions.

An important open question is the following.

*Question 2.5.19 (Tilson Question).* Is $\mathbf{V}_D = \min(\mathbf{V} * (-))$? We shall explain in the next chapter how a non-trivial result of Bergman [47] implies that $\mathbf{1}_D = \min(\mathbf{1} * (-))$.

It follows from a result of Tilson that if $B_2$ belongs to the pseudovariety $\mathbf{V}$, then $\mathbf{gV}$ is decidable if and only if $\mathbf{V}$ is decidable [364]. It is also known that if $\mathbf{V}$ has decidable pointlikes, then $\mathbf{gV}$ is decidable [9,327] although the converse is not true [36]. Of course if $\mathbf{gV}$ is decidable, then so is $\mathbf{V}_D$. It remains an open question whether $\mathbf{V}$ decidable always implies $\mathbf{gV}$ is decidable.

*Question 2.5.20.* Is it true that a pseudovariety of semigroups $\mathbf{V}$ is decidable if and only if $\mathbf{gV}$ is decidable?

We now show $\mathbf{V}_W \notin \mathbf{PVRM}$ where $\mathbf{V}_W$ is defined as per Example 2.4.24. This shows that [27, Prop. 3.1] (which is stated without proof as being similar to the monoid case) is not correct with the definition of the derived semigroupoid given therein.

*Example 2.5.21.* Let $S = \{a, b\}^{\ell}$ be a left zero semigroup and $U_1 = \{0, 1\}$ with multiplication. Consider $f \colon S \to U_1$ mapping all elements to 0. Let $c_S \colon S \to \{0\}$ be the collapsing homomorphism. Then $c_S$ is the corestriction of $f$ to its image (and hence $c_S \prec f$). The derived semigroupoid from [27, 364] of $c_S$ is isomorphic to $S$ (being a collapsing morphism) whereas that of $f$ divides a locally trivial category (and hence any sufficiently large monoid, in particular any big enough group [364]). Hence $f \in \mathbf{G}_W$, whereas $c_S \notin \mathbf{G}_W$.

We also show that the wreath product operator of [364] (that is, the semidirect product operator of [7, 27]) does not satisfy GMC.

*Example 2.5.22.* The operator $\mathbf{G} \circ (-)$ does not satisfy GMC at $\mathbf{1}$, as the following example shows. Because $\mathbf{G} \circ \mathbf{1} = \mathbf{G}$, GMC demands $\mathbf{G} \circ \mathbf{Sl} \subseteq \mathbf{G} \textcircled{m} \mathbf{Sl}$. The two-element left zero semigroup $\{a, b\}^{\ell}$ does not belong to $\mathbf{G} \textcircled{m} \mathbf{Sl}$. Indeed, if $\varphi \colon \{a, b\}^{\ell} \to E$ is a relational morphism with $E$ a semilattice and $a' \in a\varphi$, $b' \in b\varphi$, then $a'b' \in a\varphi b\varphi \subseteq (ab)\varphi = a\varphi$ and similarly $b'a' \in b\varphi$. But $a'b' = b'a'$ is an idempotent of $E$, so $\varphi \notin (\mathbf{G}, \mathbf{1})$.

On the other hand, $\{a, b\}^{\ell}$ belongs to $\mathbf{G} \circ \mathbf{Sl}$. This follows from the calculation of the [27, 364]-version of the derived semigroupoid of the relational morphism $f \colon \{a, b\}^{\ell} \to U_1$ considered in Example 2.5.21. Alternatively, one can allow $U_1$ to act on $\mathbb{Z}_2$ by letting 1 act as the identity and 0 as the trivial endomorphism. Then the elements $(0, 0)$ and $(1, 0)$ in $\mathbb{Z}_2 \rtimes U_1$ form a left zero semigroup. However, as the action of $U_1$ on $\mathbb{Z}_2$ is a monoid action, it follows $\mathbb{Z}_2 \rtimes U_1 \in \mathbf{G} \circ \mathbf{Sl}$ (see [7, 27]).

### 2.5.3 Digression on pseudovarieties of semigroupoids

We observe that if $\mathbf{V}$ is a pseudovariety of semigroupoids, one could try to define an element $\mathbf{V}_D$ of $\mathbf{PVRM}^+$ by considering all relational morphisms $f \colon S \to T$ such that $D_f \in \mathbf{V}$. Such an approach is essentially considered in [27] (but with the derived semigroupoid of [364]). This approach works quite nicely in the monoidal context but may fail to verify Axiom (r-e) if $\mathbf{V}$ is not defined by path pseudoidentities.

There are two ways to surmount this difficulty: one can either define a pseudovariety of semigroupoids to be a collection definable by path pseudoidentities (Theorem 2.5.15 shows that $\mathbf{gV}$ is still a pseudovariety under this definition); or one can define $\mathbf{V}_D$ to consist of all relational morphisms $f \colon S \to T$ such that $D_{f_{\mathrm{Im}}} \in \mathbf{V}$ where $f_{\mathrm{Im}} \colon S \to \mathrm{Im}\, f$ is the corestriction.

The first choice suffers from the fact that we do not have an axiomatization of such classes. Nonetheless, this approach seems, in some sense, the "correct" approach based on the following thesis of this book: "the appropriateness of an axiom system is determined in part by whether it has a reasonable equational theory." That is, people were interested first in classes of algebras defined by identities, not in $\mathbb{HSP}$. We chose our axiom system for $\mathbf{PVRM}$ because it gives the best equational theory. An important research task would be to find an axiom to add to the definition of a pseudovariety of semigroupoids to ensure definability by path pseudoidentities. It should be something like $S \in \mathbf{V}$ if and only if any subsemigroupoid that is the support of some path belongs to $\mathbf{V}$ (see [286] for a description of such subsemigroupoids).

The second proposal is functional but suffers from a lack of elegance. It basically amounts to closing under Axiom (r-e).

In any event, if $\mathbf{V}$ is a pseudovariety of semigroupoids definable by path identities, then we shall use $\mathbf{V}_D$ to denote the pseudovariety of relational morphisms $f \colon S \to T$ with $D_f \in \mathbf{V}$. If $\mathbf{V}$ is a pseudovariety of semigroups, then $\ell\mathbf{V}$ is defined by the same pseudoidentities defining $\mathbf{V}$, viewed as pseudoidentities over one object graphs. Hence $\ell\mathbf{V}$ is definable by path identities and so $\ell\mathbf{V}_D$ is a positive pseudovariety of relational morphisms. There is some ambiguity here in the notation as we said earlier that if $\mathbf{V}$ is a pseudovariety of monoids, then $\ell\mathbf{V}$ denotes the pseudovariety of semigroupoids generated by categories whose local monoids belong to $\mathbf{V}$. If $\mathbf{V}$ is a pseudovariety of groups (in particular, if $\mathbf{V} = \mathbf{1}$), then we shall always mean $\ell\mathbf{V}$ in this monoidal sense unless we say otherwise.

A pseudovariety of semigroups $\mathbf{V}$ is said to be *local* in the sense of Tilson [364] if $\ell\mathbf{V} = \mathbf{g}\mathbf{V}$. Locality reduces membership in $\mathbf{V}_D$ to membership in $\mathbf{V}$. Many pseudovarieties are known to be local [8, 60, 151, 152, 338, 364]. However, the reader should be cautioned that locality as a pseudovariety of monoids is not the same thing as locality as a pseudovariety of semigroups. For instance, every non-trivial pseudovariety of groups is local as a pseudovariety of monoids, but this is not true if they are viewed as pseudovarieties of semigroups. The problem arises because the local semigroup can be a group without the idempotent of the group being an identity for incoming or outgoing arrows. There are also many examples of non-local pseudovarieties. See [364] for more information.

## 2.6 The Kernel Semigroupoid Theorem

In this section we present the kernel category from [293] and its semigroupoid companion construct. It provides the analogue of the derived category in the context of the two-sided semidirect product and the block product. One can then define, for each pseudovariety of semigroups $\mathbf{V}$, the class $\mathbf{V}_K$ of all relational morphisms $f \colon S \to T$ of finite semigroups with kernel semigroupoid $K_f$ belonging to $\mathbf{g}\mathbf{V}$. The class $\mathbf{V}_K$ will belong to $\mathbf{PVRM}^+$ and $\mathbf{V}_K\mathsf{q} = \mathbf{V} ** (-)$.

### 2.6.1 The kernel semigroupoid

The kernel category is the two-sided analogue of the derived category. Its properties were first studied in [293]. The corresponding semigroup construct is then obtained via our usual procedure for turning monoid constructions into semigroup constructions. Some details will be left to the reader as they can be found clearly exposited in [293].

Let $f: M \to N$ be a relational morphism of monoids. Define a category $\mathbf{Ker}(f)$ as follows:

$$\mathrm{Obj}(\mathbf{Ker}(f)) = N \times N;$$
$$\mathrm{Arr}(\mathbf{Ker}(f)) = N \times \#f \times N.$$

The domain and range of an arrow are determined by

$$(n_L, (m, n), n_R) \colon (n_L, nn_R) \to (n_L n, n_R).$$

Sometimes we write

$$(n_L, nn_R) \xrightarrow{(m,n)} (n_L n, n_R) \tag{2.50}$$

and the pair $(m, n)$ is called the *label* of the arrow (2.50). Composition is given by multiplying the labels:

$$(n_L, nn'n_R) \xrightarrow{(m,n)} (n_L n, n'n_R) \xrightarrow{(m',n')} (n_L nn', n_R)$$
$$= (n_L, nn'n_R) \xrightarrow{(mm',nn')} (n_L nn', n_R).$$

The identity at an object $(n_L, n_R)$ is $(n_L, (1, 1), n_R)$.

If $f: S \to T$ is a relational morphism of semigroups, then one defines $\mathbf{Ker}(f)$ to be the result of deleting from $\mathbf{Ker}(f^I)$ all arrows with labels $(I, I)$. This process yields a semigroupoid.

The *kernel category* of a relational morphism $f: M \to N$, denoted $K_f$, is the quotient of $\mathbf{Ker}(f)$ obtained by identifying coterminal arrows $(n_L, (m, n), n_R)$ and $(n_L, (m', n'), n_R)$ if, for all $m_L \in n_L f^{-1}$, $m_R \in n_R f^{-1}$, the equality

$$m_L m m_R = m_L m' m_R \tag{2.51}$$

holds. We use $\sigma_f \colon \mathbf{Ker}(f) \twoheadrightarrow K_f$ for the canonical quotient map. If $f$ is clear from context, we just write $\sigma$.

**Exercise 2.6.1.** Verify that (2.51) yields a congruence on $\mathbf{Ker}(f)$.

If $f: S \to T$ is a relational morphism of semigroups, the *kernel semigroupoid* $K_f$ is the image of $\mathbf{Ker}(f)$ in $K_{f^I}$. We now investigate some basic properties of $\mathbf{Ker}(f)$ and $K_f$. From now on we work with the kernel semigroupoid; the analogous results for the kernel category are obtained by simple modifications of the proofs and in any event can be found in [293]. However, in Chapter 5 we shall use the monoid/category versions without hesitation.

Observe that the kernel construction is left-right dual. That is, if $f: S \to T$ is a relational morphism, then $K_f^{op} = K_{f^{op}}$ where $f^{op}: S^{op} \to T^{op}$ is the dual relational morphism.

**Exercise 2.6.2.** Verify that $K_f^{op} = K_{f^{op}}$.

**Lemma 2.6.3.** *Let $f\colon S \to T$ be a relational morphism of semigroups. Then $(I,(s,t),I)\sigma = (I,(s',t'),I)\sigma$ if and only if $s = s'$, $t = t'$.*

*Proof.* The if direction is trivial. For the only if direction, as $\sigma$ is a quotient map, in order to identify the two arrows above, they must be coterminal; in particular, $t = It = It' = t'$. Hence, from $I \in I(f^I)^{-1}$, it follows that $s = IsI = Is'I = s'$. This completes the proof.    □

Arrows of the form $(I,(s,t),I)$ play a distinguished role and are called *core arrows* of $\mathbf{Ker}(f)$.

**Definition 2.6.4 (Core injective).** *Let $f\colon S \to T$ be a relational morphism of semigroups and $g\colon \mathbf{Ker}(f) \to C$ be a relational morphism of semigroupoids. Then $g$ is said to be core injective if*

$$(I,(s,t),I)g \cap (I,(s',t),I)g \neq \emptyset \implies (I,(s,t),I) = (I,(s',t),I). \quad (2.52)$$

*Notice that the right-hand side of (2.52) is equivalent to $s = s'$.*

Lemma 2.6.3 says that $\sigma$ is core injective. In particular, we can identify a core arrow of $\mathbf{Ker}(f)$ with its image in $K_f$. The following is the key lemma relating $\mathbf{Ker}(f)$ and $K_f$. It shows that $\sigma$ is in some sense the universal core injective quotient of $\mathbf{Ker}(f)$. A more precise result says $\sigma$ induces the largest congruence on $\mathbf{Ker}(f)$, which is injective on core arrows.

**Lemma 2.6.5.** *Let $f\colon S \to T$ be a relational morphism of semigroups and let $g\colon \mathbf{Ker}(f) \to C$ be a relational morphism of semigroupoids. Consider the diagram*

$$\mathbf{Ker}(f) \xrightarrow{g} C$$

$$\sigma \downarrow \quad \nearrow \sigma^{-1}g$$

$$K_f$$

*Then $g$ is core injective if and only if $\sigma^{-1}g\colon K_f \to C$ is a division.*

*Proof.* Suppose that $\sigma^{-1}g$ is a division. Lemma 2.6.3 shows that $\sigma$ is core injective. Therefore, the restriction of $g$ to the core arrows of $\mathbf{Ker}(f)$ is $\sigma(\sigma^{-1}g)$. As $\sigma$ is core injective and $\sigma^{-1}g$ is a division, we see that $g$ is core injective.

For the converse, we need to show that if $y = (t_L,(s,t),t_R)$ and $y' = (t_L,(s',t'),t_R)$ are coterminal arrows with $yg \cap y'g \neq \emptyset$, then $y\sigma = y'\sigma$. Let $n \in yg \cap y'g$. There are four cases.

If $t_L = I = t_R$, then these are core arrows and there is nothing to prove. Assume next that $t_R = I$, but $t_L \neq I$. Let $s_L \in t_L f^{-1}$. We must show $s_L s = s_L s'$. First note that by coterminality $t_L t = t_L t'$ and $y, y' \in \mathbf{Ker}(f)((t_L,t),(t_L t,I))$. Consider the arrow

$$x = (I,(s_L,t_L),t)\colon (I,t_L t) \to (t_L,t)$$

and let $m \in xg$. Then computing $xy$, $xy'$ yields

$$mn \in (I, (s_L s, t_L t), I)g \cap (I, (s_L s', t_L t), I)g.$$

By core injectivity we have

$$(I, (s_L s, t_L t), I) = (I, (s_L s', t_L t), I)$$

whence $s_L s = s_L s'$, as required. The case $t_L = I$, $t_R \neq I$ is dual, and we leave it to the reader.

Finally, assume $t_L \neq I \neq t_R$. Suppose $s_L \in t_L f^{-1}$ and $s_R \in t_R f^{-1}$. We must show $s_L s s_R = s_L s' s_R$. By coterminality $t_L t = t_L t'$, $t t_R = t' t_R$ and $y, y' \in \mathbf{Ker}(f)((t_L, t t_R), (t_L t, t_R))$. Then $x_L = (I, (s_L, t_L), t t_R)$ and $x_R = (t_L t, (s_R, t_R), I)$ are arrows of $\mathbf{Ker}(f)$. Let $m_L \in x_L g$ and $m_R \in x_R g$. Then $x_L y x_R$ and $x_L y' x_R$ are defined and routine computations show

$$m_L n m_R \in (I, (s_L s s_R, t_L t t_R), I)g \cap (I, (s_L s' s_R, t_L t' t_R), I)g. \qquad (2.53)$$

As the two core arrows in (2.53) are coterminal, we must have by core injectivity of $g$ the sought after equality $s_L s s_R = s_L s' s_R$. Thus $y\sigma = y'\sigma$, completing the proof. $\qquad \square$

**Exercise 2.6.6.** Show that $\sigma$ induces the largest core injective congruence on $\mathbf{Ker}(f)$.

It is useful to compare the kernel semigroupoid with the derived semigroupoid. The content of the next proposition is that the kernel semigroupoid is smaller in the division order than the derived semigroupoid.

**Proposition 2.6.7.** *Let $f: S \to T$ be a relational morphism of semigroups. Then $K_f \prec D_f$.*

*Proof.* Define a functor $p: \mathbf{Ker}(f) \to \mathbf{Der}(f)$ by $(t_L, t_R) \mapsto t_L$ on objects and $(t_L, (s, t), t_R) \mapsto (t_L, (s, t))$ on arrows. Notice $p$ takes core arrows to core arrows and is core injective. Hence the composite $\mathbf{Ker}(f) \to \mathbf{Der}(f) \twoheadrightarrow D_f$ is core injective by Lemma 2.5.2. We conclude $K_f \prec D_f$ by Lemma 2.6.5. $\qquad \square$

**Exercise 2.6.8.** Let $f: G \to H$ be a homomorphism of groups. Show $D_f$ and $K_f$ are divisionally equivalent to $\ker f$.

Next we prove the analogue of Proposition 2.5.5 for the kernel category.

**Proposition 2.6.9.** *Suppose $f: S \to T$ and $g: S' \to T'$ are relational morphisms of semigroups so that $f \prec g$. Then $K_f \prec K_g$.*

*Proof.* Suppose that we have a division

$$S \xrightarrow{\;d\;} S'$$

$$f \downarrow \quad \subseteq \quad \downarrow g$$

$$T \xrightarrow[m]{} T'.$$

Define a relational morphism $\overline{d}\colon \mathbf{Ker}(f) \to \mathbf{Ker}(g)$ on objects by $m^I \times m^I$ and on arrows by

$$(t_L, (s,t), t_R)\overline{d} = \{(t_L m^I, (s', tm), t_R m^I) \mid s' \in sd\}.$$

Because $\overline{d}$ takes core arrows of $\mathbf{Ker}(f)$ to core arrows of $\mathbf{Ker}(g)$, if we can establish that $\overline{d}$ is core injective, then $\overline{d}\sigma_g\colon \mathbf{Ker}(f) \to K_g$ will be core injective (using Lemma 2.6.3) and so, by Lemma 2.6.5, will induce the required division $\sigma_f^{-1}\overline{d}\sigma_g\colon K_f \to K_g$.

The reader will perform the routine verification that $\overline{d}$ is a well-defined relational morphism. As to core injectivity, assume $(I, (s,t), I)\overline{d} \cap (I, (s',t), I)\overline{d} \neq \emptyset$. Then a common element is of the form $(I, (s_0, tm), I)$ with $s_0 \in sd \cap s'd$. As $d$ is a division, we conclude that $s = s'$. Therefore, $\overline{d}$ is core injective, completing the proof. □

**Exercise 2.6.10.** Check that $\overline{d}$ as defined in the above proof is a relational morphism.

The next step toward verifying $\mathbf{V}_K$ is a pseudovariety of relational morphisms is to consider the behavior with respect to products.

**Proposition 2.6.11.** *Let* $f\colon S \to T$ *and* $f'\colon S' \to T'$ *be relational morphisms. Then* $K_{f \times f'} \prec K_f \times K_{f'}$.

*Proof.* Define a morphism $g\colon \mathbf{Ker}(f \times f') \to K_f \times K_{f'}$ on objects by

$$(I, I) \longmapsto ((I,I), (I,I))$$
$$(I, (t_R, t'_R)) \longmapsto ((I, t_R), (I, t'_R))$$
$$((t_L, t'_L), I) \longmapsto ((t_L, I), (t'_L, I))$$
$$((t_L, t'_L), (t_R, t'_R)) \longmapsto ((t_L, t_R), (t'_L, t'_R))$$

for $t_L, t_R \in T$, $t'_L, t'_R \in T'$. Define $g$ on arrows by

$$((I, ((s,s'), (t,t')), I))g = ((I, (s,t), I)\sigma_f, (I, (s',t'), I)\sigma_{f'})$$
$$(((t_L, t'_L), ((s,s'), (t,t')), I))g = ((t_L, (s,t), I)\sigma_f, (t'_L, (s',t'), I)\sigma_{f'})$$
$$((I, ((s,s'), (t,t')), (t_R, t'_R)))g = ((I, (s,t), t_R)\sigma_f, (I, (s',t'), t'_R)\sigma_{f'})$$
$$(((t_L, t'_L), ((s,s'), (t,t')), (t_R, t'_R)))g = ((t_L, (s,t), t_R)\sigma_f, (t'_L, (s',t'), t'_R)\sigma_{f'}).$$

It is straightforward to check that $g$ is a functor. To establish the proposition, it therefore suffices, by Lemma 2.6.5, to show that $g$ is core injective. This follows easily from the fact that $\sigma_f$ and $\sigma_{f'}$ are core injective. □

**Exercise 2.6.12.** Show that $g$ in the above proof is a core injective functor.

Let us now compute the kernel category of the projection from a two-sided semidirect product.

**Lemma 2.6.13.** *Let $S$, $T$ be semigroups and suppose that $T$ has commuting left and right actions on $S$. Let $\pi\colon S \bowtie T \to T$ be the two-sided semidirect product projection. Then $K_\pi \prec S$.*

*Proof.* We write the binary operation of $S$ additively (although, as usual, commutativity is not assumed) and the actions of $T$ on $S$ multiplicatively. The actions of $T$ on $S$ extend to monoid actions of $T^I$ on $S$ via $Is = s = sI$ for all $s \in S$. Define a functor $g\colon \mathbf{Ker}(\pi) \to S$ by the unique map on objects. We define $g$ on arrows by $(t_L, ((s,t),t), t_R)g = t_L s t_R$. To check that $g$ is a functor, we verify on the one hand

$$(t_L, ((s,t),t), t't_R)g + (t_Lt, ((s',t'),t'), t_R)g = t_L s t' t_R + t_L t s' t_R$$
$$= t_L(st' + ts')t_R$$

and on the other

$$\big((t_L, ((s,t),t), t't_R) \cdot (t_Lt, ((s',t'),t'), t_R)\big)g = (t_L, ((st'+ts', tt'), tt'), t_R)g$$
$$= t_L(st' + ts')t_R,$$

establishing that $g$ is a functor. Next we demonstrate that $g$ is core injective. Suppose $(I, ((s,t),t), I)g = (I, ((s',t),t), I)g$. Then $s = IsI = Is'I = s'$. Thus $g$ is core injective and so $K_\pi \prec S$ by Lemma 2.6.5.    $\square$

We may now state and prove the Kernel Semigroupoid Theorem, the semigroupoid analogue of the Kernel Category Theorem [293].

**Theorem 2.6.14 (Kernel Semigroupoid Theorem).**

*(a) Let $f\colon S \to T$ be a relational morphism and suppose that $f$ factors as $d\pi$ with $d\colon S \to U \bowtie T$ a division and $\pi\colon U \bowtie T \to T$ the two-sided semidirect product projection as per*

*Then $K_f \prec U$.*

*(b) Let $f\colon S \to T$ be a relational morphism and $U$ a semigroup such that $K_f \prec U$. Then there is a division $d\colon S \to U \square T$ such that $f = d\pi$ with $\pi\colon U \square T \to T$ the block product projection, as per*

$$U \square T \xrightarrow{\pi} T$$
$$d\uparrow \quad \nearrow f$$
$$S$$

*Proof.* For (a), if $f = d\pi$ with $d$ a division, then $f \prec \pi$ and so $K_f \prec K_\pi \prec U$ by Proposition 2.6.9 and Lemma 2.6.13.

Turning to (b), let $g\colon K_f \to U$ be a division. We write $U$ additively (again not assuming commutativity). Define a division $d\colon S \to U \square T$ by:

$$sd = \{(h,t) \mid t \in sf,\ t_L h t_R \in (t_L, (s,t), t_R)\sigma g,\ \text{for } t_L, t_R \in T^I\}.$$

The verification that $d$ is a relational morphism is straightforward (cf. [293]). To see that $d$ is a division, suppose that $(h,t) \in sd \cap s'd$. Then $t \in sf \cap s'f$ and $IhI \in (I,(s,t),I)\sigma g \cap (I,(s',t),I)\sigma g$. As $g$ is a division, and the two arrows in question are coterminal, we obtain $(I,(s,t),I)\sigma = (I,(s',t),I)\sigma$ and so $s = s'$ by Lemma 2.6.3. Therefore, $d$ is a division, as required. Evidently, $d\pi = f$. □

**Exercise 2.6.15.** Show that $d$ in the above proof is a well-defined relational morphism.

**Corollary 2.6.16.** *Let $f\colon S \to T$ be a relational morphism of semigroups. Then $f$ is a division if and only if $K_f$ divides the trivial semigroup.*

*Proof.* If $K_f \prec \{1\}$, then by (b) of the Kernel Semigroupoid Theorem, $f = d\pi$ where $d\colon S \to \{1\} \square T$ is a division and $\pi\colon \{1\} \square T \to T$ is the projection. But $\pi$ is an isomorphism, so $f$ is a division.

Conversely, if $f$ is a division, then $D_f \prec \{1\}$ by Corollary 2.5.13 and so $K_f \prec \{1\}$ by Proposition 2.6.7. □

Suppose $\varphi\colon S \to U$ is a semigroup homomorphism. Then it is easily verified that $\varphi \square T\colon S \square T \to U \square T$ defined by $(f,t)(\varphi \square T) = (f\varphi, t)$ is a homomorphism.

**Exercise 2.6.17.** Verify $\varphi \square T$ is a homomorphism.

The next lemma is rather tedious, so we omit the proof. The reader may either prove it as a non-trivial exercise or consult [293, Prop. 7.2].

**Lemma 2.6.18.** *Let $\varphi\colon S \to U$ be a homomorphism. Then $K_{\varphi \square T} \prec K_\varphi^{T^I \times T^I}$.*

**Exercise 2.6.19.** Prove Lemma 2.6.18.

We remark that an analogous result holds for the derived semigroupoid and the semidirect product.

### 2.6.2 The construction of the pseudovariety $\mathbf{V}_K$

Let $\mathbf{V}$ be a pseudovariety of semigroups. We now are prepared to define the analogue of $\mathbf{V}_D$ in the context of the two-sided semidirect product. Namely, define $\mathbf{V}_K$ to be the collection of all relational morphisms $f\colon S \to T$ of finite semigroups such that the $K_f \in \mathbf{gV}$. That is,

$$\mathbf{V}_K = \{f \in \mathbf{RM} \mid K_f \in \mathbf{gV}\}. \tag{2.54}$$

It is also convenient to define, for a pseudovariety of monoids $\mathbf{V}$,

$$\ell\mathbf{V}_K = \{f \in \mathbf{RM} \mid K_f \text{ divides a category in } \ell\mathbf{V}\}. \tag{2.55}$$

Our aim is to prove $\mathbf{V}_K \in \mathbf{PVRM}^+$. Axiom ($\prec$) follows from Proposition 2.6.9. Proposition 2.6.11 and Corollary 2.6.16 yield, respectively, Axiom ($\times$) and Axiom (id). Verifying Axiom (r-e), as was the case for $\mathbf{V}_D$, requires using pseudoidentities over graphs.

**Corollary 2.6.20.** *The class* $\mathbf{V}_K$ *satisfies* Axiom (r-e).

*Proof.* Suppose $f \colon S \to T$ is in $\mathbf{V}_K$ and $i \colon T \to T'$ is an inclusion. Then $K_f \in \mathbf{gV}$. To show that $K_{fi} \in \mathbf{gV}$ it suffices, by Theorem 2.5.15, to show that if $(G, u = v)$ is a path pseudoidentity satisfied by $\mathbf{gV}$, then $K_{fi} \models (G, u = v)$. Observe that $K_f$ is a subsemigroupoid of $K_{fi}$. Moreover, it is easily verified to have the property that any arrow of $K_{fi}$ of the form $(t_L, (s, t), t_R)$ with $t_L, t_R \in T^I$ belongs to $K_f$. In particular, any path in $K_{fi}$ starting at a vertex of the form $(t_L, t_0)$ and ending at a vertex of the form $(t_1, t_R)$ with $t_L, t_R \in T^I$ is contained in $K_f$.

Let $q$ be the initial vertex of $u$ (equal the initial vertex of $v$) and $r$ be the terminal vertex of $u$ (equal the terminal vertex of $v$). Suppose $\varphi \colon G \to K_{fi}$ is a morphism. If $u\varphi = (t_L, t_0)$, $r\varphi = (t_1, t_R)$ and $t_L$ or $t_R$ is not in $T^I$, then either $t_L f i^{-1}$ or $t_R f i^{-1}$ is empty and so by definition of $K_{fi}$ (cf. (2.51)) $u\varphi = v\varphi$. If both $t_L, t_R \in T^I$, then as $(G, u = v)$ is a path pseudoidentity, by the discussion in the previous paragraph $G\varphi \subseteq K_f$ and hence $u\varphi = v\varphi$, as $K_f \in \mathbf{gV}$. $\qquad\qquad\square$

As $\ell\mathbf{V}$ is clearly defined by path identities where the underlying graph consists of loops, a similar but easier proof shows that $\ell\mathbf{V}_K$ is a pseudovariety of relational morphisms.

**Exercise 2.6.21.** Let $\mathbf{V}$ be a pseudovariety of monoids. Verify that $\ell\mathbf{V}_K$ is a pseudovariety of relational morphisms.

In summary, we have the following result.

**Theorem 2.6.22 (Kernel Semigroupoid Theorem: Pseudovarieties).**
*Let* $\mathbf{V}$ *be a pseudovariety of semigroups. Then* $\mathbf{V}_K$ *is a positive pseudovariety of relational morphisms. More to the point, for any pseudovariety* $\mathbf{W}$ *of semigroups, the equality*

$$\mathbf{V}_K\mathsf{q}(\mathbf{W}) = \mathbf{V} ** \mathbf{W}$$

*holds. That is,* $\mathbf{V}_K\mathsf{q} = \mathbf{V} ** (-)$.

*Proof.* We have already verified that $\mathbf{V}_K$ satisfies the axioms to belong to
$\mathbf{PVRM}^+$. Suppose $S \in \mathbf{V} ** \mathbf{W}$. Then there is a division $d: S \to V \bowtie W$
with $V \in \mathbf{V}$ and $W \in \mathbf{W}$. Let $f = d\pi: S \to W$ be the composition of $d$ with
the projection. Then $K_f \prec V$ by the Kernel Semigroupoid Theorem, whence
$f \in \mathbf{V}_K$. This shows that $S \in \mathbf{V}_K\mathsf{q}(\mathbf{W})$. Conversely, if $S \in \mathbf{V}_K\mathsf{q}(\mathbf{W})$, then
there is a relational morphism $f: S \to W$ with $W \in \mathbf{W}$ and $f \in \mathbf{V}_K$. Thus
$K_f \in \mathbf{gV}$ and hence $K_f \prec V$ for some $V \in \mathbf{V}$. Applying again the Kernel
Semigroupoid Theorem provides a division $S \prec V \square W$. Because $V \square W$ is
the two-sided semidirect product $V^{W^I \times W^I} \bowtie W$ and $V^{W^I \times W^I} \in \mathbf{V}$, it follows
$V \square W \in \mathbf{V} ** \mathbf{W}$ and hence $S \in \mathbf{V} ** \mathbf{W}$. This completes the proof of the
theorem.                                                                          □

We remark that Corollary 2.6.16 says $\mathbf{1}_K = \mathsf{D} = \mathbf{1}_D$, the pseudovariety of
divisions. Observe that $\mathbf{V}_D \le \mathbf{V}_K$ and $\ell\mathbf{V}_D \le \ell\mathbf{V}_K$ for any pseudovariety $\mathbf{V}$
by Proposition 2.6.7.

As we did for $\mathbf{V}_D$, we ask whether $\mathbf{V}_K$ is minimal.

*Question 2.6.23 (Tilson Question Version 2).* Is $\mathbf{V}_K = \min(\mathbf{V} ** (-))$?

### 2.6.3 Non-associativity of the two-sided semidirect product

The reader should be warned that the two-sided semidirect product $**$ of
pseudovarieties is not associative. First we establish the containment that
does hold. Then we provide a counterexample to associativity.

**Proposition 2.6.24.** *Let $\varphi \in (\mathbf{V} ** \mathbf{W})_K$. Then $\varphi$ decomposes as $\varphi_1\varphi_2$
where $\varphi_1 \in \mathbf{V}_K$, $\varphi_2 \in \mathbf{W}_K$.*

*Proof.* Suppose $\varphi: S \to T$ and $K_\varphi \prec V \square W$ with $V \in \mathbf{V}$ and $W \in \mathbf{W}$. By
the Kernel Semigroupoid Theorem, we can factor $\varphi = d\psi$ with $d$ a division
and $\psi: (V \square W) \square T \to T$ the projection. Let $\pi_W: V \square W \to W$ and
$\pi_T: W \square T \to T$ be the projections. Setting $\alpha = \pi_W \square T$, it is routine
to verify $\psi = \alpha\pi_T$. By the Kernel Semigroupoid Theorem, $K_{\pi_W} \in \mathbf{gV}$ and
$K_{\pi_T} \in \mathbf{gW}$. Lemma 2.6.18 provides a division $K_\alpha \prec K_{\pi_W}^{T^I \times T^I} \in \mathbf{gV}$. So
choosing $\varphi_1 = d\alpha$ and $\varphi_2 = \pi_T$ does the job.                          □

**Exercise 2.6.25.** Prove the analogue of Proposition 2.6.24 for the semidirect
product.

**Corollary 2.6.26.** *Let $\mathbf{U}$, $\mathbf{V}$ and $\mathbf{W}$ be pseudovarieties of semigroups. Then
the inequality*

$$(\mathbf{U} ** \mathbf{V}) ** \mathbf{W} \le \mathbf{U} ** (\mathbf{V} ** \mathbf{W}) \tag{2.56}$$

*holds.*

*Proof.* Suppose $S$ belongs to the left-hand side of (2.56). Then there is a re-
lational morphism $\varphi: S \to T$ with $T \in \mathbf{W}$ and $\varphi \in (\mathbf{U} ** \mathbf{W})_K$. Proposition 2.6.24 then provides a decomposition $\varphi = \varphi_1\varphi_2$ with $\varphi_1 \in \mathbf{U}_K$,
$\varphi_2 \in \mathbf{V}_K$. If $\varphi_1: S \to R$ and $\varphi_2: R \to T$, then $R \in \mathbf{V}_K\mathsf{q}(\mathbf{W})$ and hence
$S \in \mathbf{U}_K\mathsf{q}(\mathbf{V}_K\mathsf{q}\mathbf{W}) = \mathbf{U} ** (\mathbf{V} ** \mathbf{W})$, as required.                        □

As advertised, $**$ is not associative. In fact, the following is true:

$$(\mathbf{Sl} ** \mathbf{Sl}) ** \mathbf{Sl} < \mathbf{Sl} ** (\mathbf{Sl} ** \mathbf{Sl}). \tag{2.57}$$

To prove this we first establish, using Corollary 2.4.16, the well-known fact that $\mathbf{DA} ** \mathbf{J} = \mathbf{DA}$ [27, 349].

**Lemma 2.6.27.** *The equalities* $\mathbf{DA} = \mathbb{L}\mathbf{1} \,\circledm\, \mathbf{DA}$ *and* $\mathbf{DA} = \mathbf{DA} \,\circledm\, \mathbf{J}$ *hold.*

*Proof.* For the non-trivial inclusion of the first equality, let $\varphi \colon S \to T$ be a homomorphism with $\varphi \in (\mathbb{L}\mathbf{1}, \mathbf{1})$ and $T \in \mathbf{DA}$. Then clearly $S$ is aperiodic, so to establish $S \in \mathbf{DA}$ it suffices to show $S \in \mathbf{DS}$. Let $J$ be a regular $\mathscr{J}$-class of $S$ containing an idempotent $e$ and suppose $x, y \in J$. Then $(exye)\varphi = e\varphi x\varphi y\varphi e\varphi = e\varphi$ because $T$ is in $\mathbf{DA}$ and $e\varphi \,\mathscr{J}\, x\varphi \,\mathscr{J}\, y\varphi$. Thus $exye, e$ belong to the locally trivial semigroup $e\varphi\varphi^{-1}$, and so $exye = e(exye)e = e$. It follows $e \leq_{\mathscr{J}} xy$ and so $xy \in J$. The first equality is now proved. Let us turn to the second.

Again we prove only the non-trivial inclusion. Suppose $\varphi \colon S \to T$ is a homomorphism where $T \in \mathbf{J}$ and $\varphi \in (\mathbf{DA}, \mathbf{1})$. Let $J$ be a regular $\mathscr{J}$-class of $S$. As $J$ is regular and $T$ is $\mathscr{J}$-trivial, $J\varphi = \{f\}$ for some idempotent $f$ of $T$. Thus $J \subseteq f\varphi^{-1}$ and $f\varphi^{-1} \in \mathbf{DA}$. Let $x, y \in J$. Because $J$ is regular, there exist $u, v, w, z \in J$ with $uxv = y$ and $wyz = x$. Thus $J$ is a $\mathscr{J}$-class of $f\varphi^{-1}$ and hence an aperiodic semigroup. This concludes the proof that $S \in \mathbf{DA}$. $\square$

**Theorem 2.6.28.** $\mathbf{DA} ** \mathbf{J} = \mathbf{DA}$.

*Proof.* According to Corollary 2.4.16, the inequality

$$\mathbf{DA} ** \mathbf{J} \leq (\llbracket xyw = xzw \rrbracket \,\circledm\, \mathbf{DA}) \,\circledm\, \mathbf{J}$$

holds. Clearly, $\llbracket xyw = xzw \rrbracket \leq \mathbb{L}\mathbf{1}$ and so two applications of Lemma 2.6.27 yield $\mathbf{DA} ** \mathbf{J} \leq \mathbf{DA}$. The reverse inclusion is trivial. $\square$

**Exercise 2.6.29.** Let $\mathbf{DO}$ be the pseudovariety of all finite semigroups whose regular $\mathscr{J}$-classes are orthodox semigroups (i.e., whose regular $\mathscr{J}$-classes are semigroups in $\mathbf{EB}$). Prove that $\mathbf{DO} = \mathbf{DO} ** \mathbf{J}$.

In fact, Straubing and Thérien proved [349] that $\mathbf{DA}$ is the smallest pseudovariety $\mathbf{V}$ satisfying $\mathbf{V} = \mathbf{V} ** \mathbf{Sl}$. As a consequence of Theorem 2.6.28, $(\mathbf{Sl} ** \mathbf{Sl}) ** \mathbf{Sl} \leq \mathbf{DA}$, and so to establish (2.57) it suffices to show $B_2 \in \mathbf{Sl} ** (\mathbf{Sl} ** \mathbf{Sl})$.

Let $U_1 = (\{0, 1\}, \cdot)$ be the two-element semilattice. Let $(\{1, 2\}, R)$ be the transformation semigroup of all constant maps on $\{1, 2\}$. Let us establish a division $B_2 \prec U_1 \wr (\{1, 2\}, R)$. Define $f_i \colon \{1, 2\} \to U_1$, for $i = 1, 2$, by

$$jf_i = \begin{cases} 1 & i = j \\ 0 & i \neq j \end{cases}$$

and set $e_{ij} = (f_i, \bar{j})$, for $i, j = 1, 2$, where $\bar{j}$ is the constant map to $j$. Let $z \colon \{1, 2\} \to U_1$ be the constant map to 0 and $u \colon \{1, 2\} \to U_1$ be the constant map to 1. Then there results a division $d \colon B_2 \to U_1 \wr (\{1, 2\}, R)$ given by $E_{ij}d = e_{ij}$, for $1 \leq i, j \leq 2$ and $0d = \{(z, \bar{0}), (z, \bar{1})\}$ where $E_{ij}$ is the standard $2 \times 2$ matrix unit. This follows directly from the computation

$$e_{ij}e_{k\ell} = \begin{cases} e_{i\ell} & j = k \\ (z, \bar{\ell}) & j \neq k. \end{cases} \tag{2.58}$$

To verify (2.58), we compute $(f_i, \bar{j})(f_k, \bar{\ell}) = (f_i \cdot {}^{\bar{j}}f_k, \bar{\ell})$, and if $j = k$, then ${}^{\bar{j}}f_k = u$, whereas if $j \neq k$, then ${}^{\bar{j}}f_k = z$. It is now immediate to deduce (2.58). Thus $B_2 \in \mathbf{Sl} * (R) \leq \mathbf{Sl} ** (R)$.

Next we show that $R^{op} \prec U_1 \rtimes 1$ where 1 acts by the endomorphism of $U_1$ sending all elements to 0. Indeed, $(0, 1)(1, 1) = (0 + 1 \cdot 1, 1) = (0, 1)$ and $(1, 1)(0, 1) = (1 + 1 \cdot 0, 1) = (1, 1)$. It follows $R^{op} \in \mathbf{Sl} * 1 \leq \mathbf{Sl} ** 1$ and hence $R \in \mathbf{Sl} ** 1$. We conclude that

$$B_2 \in \mathbf{Sl} ** (\mathbf{Sl} ** 1) \leq \mathbf{Sl} ** (\mathbf{Sl} ** \mathbf{Sl})$$

completing the proof of (2.57).

In Chapter 5, we show that the smallest pseudovariety $\mathbf{V}$ satisfying $\mathbf{Sl} ** \mathbf{V} = \mathbf{V}$ is $\mathbf{A}$. This means that $\mathbf{Sl} **^{\omega} \mathbf{Sl} = \mathbf{A}$, whereas $\mathbf{Sl} {}^{\omega}** \mathbf{Sl} = \mathbf{DA}$, so the two-sided semidirect is very far indeed from being associative!

## 2.7 The Slice Theorem and Joins

In this section, we present the Slice Theorem [322, 330] from the second author's thesis. It is the analogue of the Derived Semigroupoid and Kernel Semigroupoid Theorems for joins.

**Theorem 2.7.1 (Slice Theorem).** *Let $f \colon S \to T$ be a relational morphism. Then $f$ factors as $d\pi$ with $d \colon S \to U \times T$ a division and $\pi \colon U \times T \to T$ the projection as per*

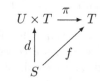

*if and only if there is a relational morphism $g \colon \#f \to U$ satisfying the Slice Condition: $(s, t)g \cap (s', t)g \neq \emptyset$ implies $s = s'$.*

*Proof.* Suppose first that one has a division $d \colon S \to U \times T$ with $d\pi = f$. Define $g \colon \#f \to U$ by $(s, t)g = \{u \mid (u, t) \in sd\}$. The reader verifies directly that $g$ is a relational morphism. Suppose $u \in (s, t)g \cap (s', t)g$. Then $(u, t) \in sd \cap s'd$ and so $s = s'$.

Conversely, suppose $g\colon \#f \to U$ satisfies the Slice Condition. Define $d\colon S \to U \times T$ by

$$sd = \{(u,t) \mid t \in sf \text{ and } u \in (s,t)g\}.$$

The reader easily checks that $d$ is a relational morphism with $d\pi = f$. Suppose $(u,t) \in sd \cap s'd$. Then $t \in sf \cap s'f$ and $u \in (s,t)g \cap (s',t)g$. Thus $s = s'$ (by the Slice Condition) and $d$ is a division, as required. $\qquad\square$

We can now prove that $\mathbf{V}_\vee$ defines the operator $\mathbf{V} \vee (-)$. See Example 2.4.7 for notation.

**Corollary 2.7.2.** *The equality $\mathbf{V}_\vee \mathsf{q} = \mathbf{V} \vee (-)$ holds.*

*Proof.* Suppose $S \in \mathbf{V} \vee \mathbf{W}$. Then there is a division $d\colon S \to V \times W$ with $V \in \mathbf{V}$ and $W \in \mathbf{W}$. Let $\pi\colon V \times W \to W$ be the projection and $f = d\pi$. Then the Slice Theorem provides a relational morphism $g\colon \#f \to V$ such that the Slice Condition holds. It follows that no pair $\{(s,t),(s',t)\}$ of $\#f$ with $s \neq s'$ is $\mathbf{V}$-pointlike, so $f \in \mathbf{V}_\vee$.

Conversely, if $S \in \mathbf{V}_\vee \mathsf{q}(\mathbf{W})$, then there is a relational morphism $f\colon S \to W$ with $W \in \mathbf{W}$ and $f \in \mathbf{V}_\vee$. Hence no pair $p = \{(s,t),(s',t)\}$ in $\#f$ with $s \neq s'$ is $\mathbf{V}$-pointlike. For each such pair $p$, there is a relational morphism $g_p\colon \#f \to V_p$ with $V_p \in \mathbf{V}$ such that $(s,t)g_p \cap (s',t)g_p = \emptyset$. Taking $\Delta \prod g_p$, where the product ranges over all such pairs, gives a relational morphism $g\colon \#f \to V$ with $V \in \mathbf{V}$ satisfying the Slice Condition and so $S \prec V \times W$ by the Slice Theorem. $\qquad\square$

It was observed by the second author [329] (and independently by Delgado [79]) that if $\mathbf{H}$ is a decidable pseudovariety of groups and $G$ is a group, then the $\mathbf{H}$-pointlike subsets of $G$ are computable; actually this follows directly from a result of Wedderburn [377]. It is a consequence of Zel'manov's solution to the restricted Burnside problem [371, 388, 389] that one can compute the free object on a finite generating set for any decidable locally finite pseudovariety of groups. It then follows from the Slice Theorem that $\mathbf{H} \vee \mathbf{K}$ is decidable whenever $\mathbf{H}$ is a decidable pseudovariety of groups and $\mathbf{K}$ is a decidable locally finite pseudovariety of groups [329]. It is also well-known that if $\mathbf{H}, \mathbf{K}$ are decidable pseudovarieties of groups, then $\mathbf{H} * \mathbf{K}$ is decidable as it consists of all groups that are extensions of a group in $\mathbf{H}$ by a group in $\mathbf{K}$ [85]. It is therefore natural to ask the following question.

*Question 2.7.3.* Is it true that if $\mathbf{H}, \mathbf{K}$ are decidable pseudovarieties of groups, then $\mathbf{H} \vee \mathbf{K}$ is decidable?

## 2.8 Composition

We next place associative multiplications on **CC** and **PVRM** in order to turn $\mathsf{q}$ into a monoid homomorphism.

### 2.8.1 Composition of sets of relational morphisms

Other than $\mathbf{PVRM}^+$, the classes that we have been considering are not closed under the obvious notion of composition of sets of relational morphisms considered in [339, 362]. A different tactic must hence be taken to define composition of these classes. First we introduce one last axiom, which is obviously implied by closure under strong division. Recall that $f \subseteq_s g$ implies the domains and codomains of $f$ and $g$ coincide.

**Axiom (Strong containment ($\subseteq_s$)).** *A subset $X \subseteq \mathbf{RM}$ satisfies Axiom ($\subseteq_s$) if given $g \in X$, one has that $f \subseteq_s g$ implies $f \in X$.*

Now we are prepared to define composition of continuously closed classes of relational morphisms.

**Definition 2.8.1 (Composition).**

*(a) Let $\mathsf{K}, \mathsf{L}$ be collections of relational morphisms. Their composition is defined by the rule*

$$\mathsf{KL} = \{h\colon S \to U \mid h = fg \text{ with } f\colon S \to T \text{ in } \mathsf{K}, \ g\colon T \to U \text{ in } \mathsf{L}\}. \quad (2.59)$$

*(b) Let $\mathsf{K}, \mathsf{L} \in \mathbf{X}$, where $\mathbf{X}$ is any of $\mathbf{CC}$, $\mathbf{CC}^+$, $\mathbf{PVRM}$, $\mathbf{PVRM}^+$. Then by definition $\mathsf{K} \odot \mathsf{L}$ is the smallest member of $\mathbf{X}$ containing $\mathsf{KL}$. Equivalently,*

$$\mathsf{K} \odot \mathsf{L} = c(\mathsf{KL}) \quad (2.60)$$

*where $c\colon 2^{\mathbf{RM}} \to \mathbf{X}$ is the associated closure operator (i.e., $\mathbb{CC}$, $\mathbb{PVRM}$, etc.).*

At first sight (2.60) seems to depend on the class $\mathbf{X}$ being considered, but this turns out not to be the case.

**Proposition 2.8.2.** *The following hold.*

*(a) $(\mathsf{K}, \mathsf{L}) \mapsto \mathsf{KL}$ is an associative multiplication on $2^{\mathbf{RM}}$.*
*(b) For $\mathsf{K}, \mathsf{L}$ both in $\mathbf{CC}$, $\mathbf{CC}^+$ or $\mathbf{PVRM}$*

$$\mathsf{K} \odot \mathsf{L} = \{h \in \mathbf{RM} \mid h \subseteq_s fg \text{ with } f \in \mathsf{K}, \ g \in \mathsf{L}\}, \quad (2.61)$$

*that is, $\mathsf{K} \odot \mathsf{L}$ is the closure of $\mathsf{KL}$ under Axiom ($\subseteq_s$).*
*(c) For $\mathsf{K}, \mathsf{L} \in \mathbf{PVRM}^+$, $\mathsf{K} \odot \mathsf{L} = \mathsf{KL}$.*
*(d) The definition of $\mathsf{K} \odot \mathsf{L}$ is independent of the class of which we view $\mathsf{K}$ and $\mathsf{L}$ as members. Moreover, $\odot$ is associative. For $\mathsf{K}, \mathsf{L}, \mathsf{M}$ in $\mathbf{CC}$, $\mathbf{CC}^+$, or $\mathbf{PVRM}$, their threefold product*

$$(\mathsf{K} \odot \mathsf{L}) \odot \mathsf{M} = \mathsf{K} \odot (\mathsf{L} \odot \mathsf{M})$$

*is the closure of $\mathsf{KLM}$ under Axiom ($\subseteq_s$).*

*(e) The collection of all divisions* D *is in* **PVRM**$^+$ *and, for any* K $\in$ **CC**,

$$DK = K = KD.$$

*(f) The set* **CC** *is a monoid under* $\odot$ *with identity* D, *and* **CC**$^+$, **PVRM**, **PVRM**$^+$ *are submonoids. Moreover, multiplication is continuous in each variable.*

*Proof.* Part (a) follows immediately from the associativity of composition of relations.

To prove (b) suppose first that K, L $\in$ **CC**. Clearly, the right-hand side of (2.61) is contained in the left-hand side as continuously closed classes satisfy Axiom ($\subseteq_s$). We show that the class R on the right-hand side of (2.61) is a continuously closed class. The result will then follow for the case of **CC**.

To verify that R satisfies Axiom ($\times$), first observe that $1_{\{1\}}$ is clearly in KL. Suppose $h \subseteq_s fg$ and $h' \subseteq_s f'g'$ with $f, f' \in$ K and $g, g' \in$ L. Then $h \times h' \subseteq_s fg \times f'g' = (f \times f')(g \times g')$ and $f \times f' \in$ K, $g \times g' \in$ L. Thus $h \times h' \in$ R.

Next we show that R satisfies Axiom (r-e). So let $h \colon S \to T$ belong to R and assume $j \colon T \hookrightarrow T'$ is an injective homomorphism. Then there exist $f \in$ K and $g \in$ L with $h \subseteq_s fg$. But $g$ has codomain $T$ and so $gj$ is defined and belongs to L by Axiom (r-e). Thus $hj \subseteq_s f(gj)$ and hence $hj$ belongs to R.

Finally, we establish that R satisfies Axiom ($\prec_s$). Suppose that $h \colon S \to T$ is in R and

$$\begin{array}{ccc} S' & \xrightarrow{d} & S \\ {\scriptstyle k}\downarrow & \subseteq & \downarrow{\scriptstyle h} \\ T' & \xrightarrow[m]{} & T \end{array}$$

is a strong division. By definition of R, we have $h \subseteq_s fg$ with $f \in$ K and $g \in$ L. Then $df \prec_s f$ and $gm^{-1} \prec_s g$ (the latter as $gm^{-1}m = g$ and $m$ is onto) and so $df \in$ K, $gm^{-1} \in$ L. Hence

$$k \subseteq_s dhm^{-1} \subseteq_s (df)(gm^{-1})$$

establishing $k \in$ R. Thus R is a continuously closed class. Moreover, if K, L belong to **CC**$^+$, then KL contains all identities, so R $\in$ **CC**$^+$.

Suppose now that K, L $\in$ **PVRM**. By Proposition 2.1.8(c), it remains only to verify Axiom (co-re) for R. Suppose $h \subseteq_s fg$ with $f \colon S \to T$ in K and $g \colon T \to U$ in L. Let $U' = \text{Im } h$. Let $g' \colon U'g^{-1} \to U'$ be the restriction of $g$ defined by $\#g' = \{(t, u) \in \#g \mid u \in U'\}$. Then $g' \prec g$. Let $f' \colon S \to U'g^{-1}$ be the relation with graph $\{(s, t) \in \#f \mid t \in U'g^{-1}\}$. This is a relational morphism because $U'g^{-1}f^{-1} = S$ (as $h \subseteq_s fg$ and $U' = \text{Im } h$). Also $f' \prec f$. Now we show that the corestriction $h_{\text{Im}} \colon S \to U'$ satisfies $h_{\text{Im}} \subseteq_s f'g'$. Indeed, suppose $u \in sh_{\text{Im}}$. Then there exists $t \in sf$ such that $u \in tg$. But then $u \in U'$, whence $t \in U'g^{-1}$, and so $u \in tg'$ and $t \in sf'$. Therefore, $h_{\mathrm{Im}} \subseteq_s f'g'$. Because $f' \in \mathsf{K}$ and $g' \in \mathsf{L}$, it follows $h_{\mathrm{Im}} \in \mathsf{R}$.

Part (c) is due to Tilson [362]. Clearly, $\mathsf{KL}$ contains all the identity maps. Also $\mathsf{KL}$ is closed under range extension because if $f \in \mathsf{K}$, $g \in \mathsf{L}$ and $j$ is an injective homomorphism with $fgj$ defined, then $gj \in \mathsf{L}$ and so $fgj = f(gj) \in \mathsf{KL}$. Because $\mathsf{K} \odot \mathsf{L}$ is closed under corestriction, it therefore suffices to show that if $h \in \mathsf{K} \odot \mathsf{L}$ is onto, then $h \in \mathsf{KL}$. From (b) it follows $h \subseteq_s fg$ with $f \colon S \to T$ in $\mathsf{K}$ and $g \colon T \to U$ in $\mathsf{L}$. Because $h$ is onto, so is $g$. First, consider the canonical factorization $g = p_T^{-1} p_U$. Then, because $fp_T^{-1} \prec_s f$ (as $fp_T^{-1}p_T = f$ and $p_T$ is onto, cf. (1.46)), it follows $fp_T^{-1} \in \mathsf{K}$ by Axiom $(\prec_s)$. On the other hand, an application of Tilson's Lemma shows that $p_U \in \mathsf{L}$. As $fg = (fp_T^{-1})p_U$, it follows we may assume without loss of generality that $g$ is a surjective homomorphism. Let $f' \colon S \to T$ be given by $sf' = \{t \in sf \mid tg \in sh\}$. It is easy to verify that $f'$ is a relational morphism. Clearly, $f' \prec f$ (as $f' \subseteq f$), so $f' \in \mathsf{K}$. We show that $f'g = h$. It is immediate $h \subseteq_s f'g$ because $h \subseteq_s fg$ implies that if $u \in sh$, then there exists $t \in sf$ with $tg = u$ and so $t \in sf'$ by definition. Assume conversely $u \in sf'g$. Then there exists $t \in sf'$ with $tg = u$. But, by definition of $f'$, $tg \in sh$, so $u \in sh$.

To establish (d), observe that it follows readily from (b) and (c) that the definition of $\odot$ is independent of context as all the classes in question satisfy Axiom $(\subseteq_s)$. We prove $\mathsf{K} \odot (\mathsf{L} \odot \mathsf{M}) = (\mathsf{K} \odot \mathsf{L}) \odot \mathsf{M}$ by showing that both sides consist of all relational morphisms $\varphi \colon S \to T$ such that $\varphi \subseteq_s \alpha\beta\gamma$ with $\alpha \colon S \to R$ in $\mathsf{K}$, $\beta \colon R \to L$ in $\mathsf{L}$, and $\gamma \colon L \to T$ in $\mathsf{M}$. Indeed, any such relational morphism $\varphi$ clearly belongs to both sides.

By symmetry, we only need to consider $(\mathsf{K} \odot \mathsf{L}) \odot \mathsf{M}$. Suppose $\varphi \colon S \to T$ belongs to $(\mathsf{K} \odot \mathsf{L}) \odot \mathsf{M}$. Then there exists by (b) $\rho \colon S \to L$ in $\mathsf{K} \odot \mathsf{L}$ and $\gamma \colon L \to T$ in $\mathsf{M}$ such that $\varphi \subseteq_s \rho\gamma$. But then, again by (b), we have that there exist $\alpha \colon S \to R$ in $\mathsf{K}$ and $\beta \colon R \to L$ in $\mathsf{L}$ such that $\rho \subseteq_s \alpha\beta$. But then $\varphi \subseteq_s \alpha\beta\gamma$ as desired.

Part (e) follows from Proposition 2.1.8(a). The first part of (f) is clear. For the second, suppose $\{\mathsf{R}_i\}$ is a directed set of continuously closed classes and $\mathsf{R}$ is a continuously closed class. Clearly, $\bigvee (\mathsf{R} \odot \mathsf{R}_i) \leq \mathsf{R} \odot \bigvee \mathsf{R}_i$. Conversely, if $f \in \mathsf{R} \odot \bigvee \mathsf{R}_i$, then there exists $g \in \mathsf{R}$ and $h \in \bigvee \mathsf{R}_i$ with $f \subseteq_s gh$. Because the $\mathsf{R}_i$ are directed, there exists $i$ with $h \in \mathsf{R}_i$. Thus $f \in \mathsf{R} \odot \mathsf{R}_i \leq \bigvee(\mathsf{R} \odot \mathsf{R}_i)$. We conclude $\bigvee(\mathsf{R} \odot \mathsf{R}_i) = \mathsf{R} \odot \bigvee \mathsf{R}_i$, as required. Continuity in the other variable is proved identically. □

It should be mentioned that $\mathsf{KL}$ is not in general $\mathsf{K} \odot \mathsf{L}$ (that this might be the case was pointed out to us by B. Tilson; the example is ours). Consider $\psi_1 \colon \{0,1\} \to \mathbb{Z}_2$ and $\psi_2 \colon \mathbb{Z}_2 \to \{0,1\}$ the universal relational morphisms (where $\{0,1\}$ is considered with multiplication). Then take $\mathsf{K}$ and $\mathsf{L}$ to be the continuous closed classes generated by $\psi_1$ and $\psi_2$, respectively (the reader should calculate the members of $\mathsf{K}$, $\mathsf{L}$). One then verifies $1_{\{0,1\}} \notin \mathsf{KL}$, but is in $\mathsf{K} \odot \mathsf{L} = \mathbb{CC}(\mathsf{KL})$.

Let us highlight a key idea in the proof of (c).

**Lemma 2.8.3.** *Let* $R_1, \ldots, R_n \in \mathbf{PVRM}^+$. *Then a relational morphism* $\varphi$ *belongs to* $R_1 \odot \cdots \odot R_n = R_1 \cdots R_n$ *if and only if* $\varphi = d\theta_1 \cdots \theta_n$ *with* $d$ *a division and* $\theta_i \in R_i$ *a homomorphism for all* $i$.

*Proof.* Sufficiency is clear. For necessity, Proposition 2.8.2(c) allows us to factor $\varphi = \psi_1 \cdots \psi_n$ with the $\psi_i \in R_i$, each $i$. By considering the canonical factorization and using Tilson's Lemma, we have $\psi_n = d\theta_n$ with $d$ a division and $\theta_n$ a homomorphism in $R_n$. Then $\psi_{n-1}d \in R_{n-1}$ by Proposition 2.1.8(a), so replace $\psi_n$ by $\theta_n$ and $\psi_{n-1}$ by $\psi_{n-1}d$. Suppose inductively $\varphi = \psi_1 \cdots \psi_i \theta_{i+1} \cdots \theta_n$ with the $\theta_j \in R_j$ homomorphisms. Then using the canonical factorization and Tilson's Lemma, we may write $\psi_i = d\theta_i$ with $d$ a division and $\theta_i \in R_i$. If $i = 1$, we are done. Else, $\psi_{i-1}d \in R_{i-1}$ and so replacing $\psi_i$ by $\theta_i$ and $\psi_{i-1}$ by $\psi_{i-1}d$ we may continue the process.    □

As a simple example of composition, suppose that $\mathbf{U}, \mathbf{V}, \mathbf{W}$ are pseudovarieties of semigroups. Then

$$(\mathbf{U}, \mathbf{V}) \odot (\mathbf{V}, \mathbf{W}) \subseteq (\mathbf{U}, \mathbf{W}). \tag{2.62}$$

Indeed, Proposition 2.3.19 shows that

$$(\mathbf{U}, \mathbf{V})(\mathbf{V}, \mathbf{W}) \subseteq (\mathbf{U}, \mathbf{W})$$

from which (2.62) immediately follows. In particular, for any pseudovariety of semigroups $\mathbf{V}$,

$$(\mathbf{V}, \mathbf{V}) \odot (\mathbf{V}, \mathbf{V}) = (\mathbf{V}, \mathbf{V})(\mathbf{V}, \mathbf{V}) = (\mathbf{V}, \mathbf{V})$$

and so the operator $(\mathbf{V}, \mathbf{V})\textcircled{m}(-)$ is idempotent. In the literature, this operator is sometimes denoted $\mathbf{V}^{-1}(-)$ [224, 379]. The case when $(\mathbf{V}, \mathbf{V}) = (\mathbf{V}, \mathbf{1})$ is of particular importance and has received quite a bit of attention [224, 293, 379]; see also Section 4.6. Sometimes a homomorphism in $(\mathbf{V}, \mathbf{V})$ is called a $\mathbf{V}$-*morphism*.

There is an infinite power defined on $\mathbf{CC}^+$ and $\mathbf{PVRM}^+$. Namely, if $R \in \mathbf{CC}^+$, define $R^0 = D$ and $R^{n+1} = R^n \odot R$. If $R \in \mathbf{PVRM}^+$, then $R^n$ consists of $n$-fold products of elements of $R$ by Proposition 2.8.2(c). In any event, as $R \in \mathbf{CC}^+$, we immediately have $R^{n+1} \geq R^n$ and hence the sequence $R^n$ is directed. Define

$$R^\omega = \bigvee_{n \in \mathbb{N}} R^n = \bigcup_{n \in \mathbb{N}} R^n. \tag{2.63}$$

Because the join is directed and directed joins in $\mathbf{CC}^+$ and $\mathbf{PVRM}^+$ coincide, (2.63) restricts to an operation on $\mathbf{PVRM}^+$.

The continuity of multiplication, afforded by Proposition 2.8.2(f), implies

$$R^\omega R = RR^\omega = R^\omega$$

and

$$(R^\omega)^n = (R^n)^\omega = (R^\omega)^\omega = R^\omega$$

for all $n > 0$.

**Exercise 2.8.4.** Verify the asserted properties of the map $R \mapsto R^\omega$.

## 2.8.2 The map q is a homomorphism

We may now complete our description of the fundamental properties of q, namely we show that it is a monoid homomorphism.

**Proposition 2.8.5 (q is a homomorphism).** *The map* $\mathsf{q}\colon \mathbf{CC} \twoheadrightarrow \mathbf{Cnt(PV)}$ *is an onto monoid homomorphism. Hence we have the following onto monoid homomorphisms via restriction:*

- $\mathsf{q}\colon \mathbf{CC} \twoheadrightarrow \mathbf{Cnt(PV)}$;
- $\mathsf{q}\colon \mathbf{CC}^+ \twoheadrightarrow \mathbf{Cnt(PV)}^+$;
- $\mathsf{q}\colon \mathbf{PVRM} \twoheadrightarrow \mathbf{GMC(PV)}$;
- $\mathsf{q}\colon \mathbf{PVRM}^+ \twoheadrightarrow \mathbf{GMC(PV)}^+$.

*Moreover, in the positive setting* $\mathsf{q}$ *preserves the* $\omega$*-power.*

*Proof.* We already know that $\mathsf{q}$ is onto in all the above settings. Suppose that $\mathbf{V}$ is a pseudovariety of semigroups and $S \in \mathsf{Kq}(\mathsf{Lq}(\mathbf{V}))$. Then there exists $f\colon S \to T$ with $f \in \mathsf{K}$ and $T \in \mathsf{Lq}(\mathbf{V})$. Hence (by Definition 2.3.1) there exists $g\colon T \to U$ with $g \in \mathsf{L}$ and $U \in \mathbf{V}$. But then $fg \in \mathsf{KL} \subseteq \mathsf{K} \odot \mathsf{L}$ and so $S \in (\mathsf{K} \odot \mathsf{L})\mathsf{q}(\mathbf{V})$. It follows $\mathsf{Kq} \cdot \mathsf{Lq} \leq (\mathsf{K} \odot \mathsf{L})\mathsf{q}$.

For the other direction, suppose $S \in (\mathsf{K} \odot \mathsf{L})\mathsf{q}(\mathbf{V})$ and $h\colon S \to U$ with $h \in \mathsf{K} \odot \mathsf{L}$, $U \in \mathbf{V}$. Then $h \subseteq_s fg$ with $f\colon S \to T$ in $\mathsf{K}$ and $g\colon T \to U$ in $\mathsf{L}$ by Proposition 2.8.2(b). So $T \in \mathsf{Lq}(\mathbf{V})$ and $S \in \mathsf{Kq}(\mathsf{Lq}(\mathbf{V}))$, establishing that $(\mathsf{K} \odot \mathsf{L})\mathsf{q} \leq \mathsf{Kq} \cdot \mathsf{Lq}$. The final remark follows as $\mathsf{q}$ is **sup**. □

We may deduce that if $f$ is one of the right adjoints $M$ or max of the Galois connections (2.33), then

$$f(\alpha_1) \odot f(\alpha_2) \leq f(\alpha_1 \cdot \alpha_2)$$

by Proposition 1.1.7.

Applying Proposition 2.8.5 and (2.62) provides another proof of (2.24).

## 2.8.3 Modeling q by composition

The reader is referred back to Section 2.4 for notation. One of the important contributions of q-theory is to replace iteration of non-associative operators, such as the Mal'cev and two-sided semidirect products, with the associative operation of composition. This tells us, for instance, the "right" way to bracket iterated Mal'cev and two-sided semidirect products. For instance,

$$(\mathbf{U}_K \odot \mathbf{V}_K)\mathsf{q}(\mathbf{W}) = \mathbf{U} \mathbin{**} (\mathbf{V} \mathbin{**} \mathbf{W}).$$

Hence this is the "natural" way to bracket iterated two-sided semidirect products [293].

In fact, any operator on the lattice of pseudovarieties of semigroups can be fruitfully modeled by composition of operators in the following obvious way:

if **V** is a pseudovariety of semigroups, we can consider the constant operator $q_{\mathbf{V}}$ defined by $q_{\mathbf{V}}(\mathbf{W}) = \mathbf{V}$ for all pseudovarieties **W** of finite semigroups. For any operator $\alpha$ on the lattice of semigroup pseudovarieties, $\alpha q_{\mathbf{V}} = q_{\alpha(\mathbf{V})}$. This gives an immediate way to translate results about composition of operators into results about operators, themselves, by identifying **V** with $q_{\mathbf{V}}$.

Now we want to lift this to **PVRM**. If **V** is a pseudovariety of semigroups, it is often convenient (cf. Example 2.4.2) to identify **V** with the pseudovariety of relational morphisms

$$\widetilde{\mathbf{V}} = \{\varphi\colon S \to T \mid S \in \mathbf{V}\} =: \mathbf{V}. \tag{2.64}$$

The abuse of notation (2.64) will allow us to state results about composition and operators simultaneously. One of the main reasons we do not restrict ourselves to positive pseudovarieties is in order to allow consideration of the pseudovariety $\widetilde{\mathbf{V}}$. We shall verify shortly that $\widetilde{\mathbf{V}}$ is the unique member of **CC** whose image under $\mathsf{q}$ is $q_{\mathbf{V}}$. First we observe that $\widetilde{\mathbf{V}}$ is maximal with this property (i.e., $\widetilde{\mathbf{V}} = M(q_{\mathbf{V}})$), because if $\mathsf{Rq} = q_{\mathbf{V}}$ and $f\colon S \to T$ is in R, then $S \in \mathsf{Rq}((T)) = \mathbf{V}$, whence $f \in \widetilde{\mathbf{V}}$. This shows $\mathsf{R} \subseteq \widetilde{\mathbf{V}}$.

We remark that, in [362] and [339], a pseudovariety of semigroups was identified systematically with its collection of collapsing morphisms

$$C_{\mathbf{V}} = \{c_S\colon S \to \{1\} \mid S \in \mathbf{V}\}.$$

However, this latter class forms neither a pseudovariety of relational morphisms nor a continuously closed class. Hence the approach of the aforementioned works to modeling operators via composition does not give the "right" answers in the following sense: theorems about composition do not immediately turn into theorems about operators.

Observe, however, that if $\mathsf{R} \in \mathbf{CC}$ and $\mathsf{Rq} = q_{\mathbf{V}}$, then by considering the equation $\mathsf{Rq}(\mathbf{1}) = \mathbf{V}$, we see $C_{\mathbf{V}} \subseteq \mathsf{R}$. Moreover, $C_{\mathbf{V}}$ generates (2.64) as a continuously closed class. To see this, suppose that $f\colon S \to T$ is any relational morphism. Then the commutative triangle

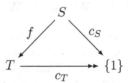

shows that $f \prec_s c_S$. It follows that if $\mathsf{R} \in \mathbf{CC}$ is such that $\mathsf{Rq} = q_{\mathbf{V}}$, then $\mathsf{R} \supseteq \widetilde{\mathbf{V}}$. We conclude that $\widetilde{\mathbf{V}}$ is the unique **CC** mapping to $q_{\mathbf{V}}$.

Next we show how to model pseudovariety operators via composition.

**Proposition 2.8.6.** *Suppose* $\mathsf{R} \in \mathbf{CC}$ *and* $\mathbf{V} \in \mathbf{PV}$. *Then, under identification* (2.64), *we have that*

$$\mathsf{R} \odot \mathbf{V} = \mathsf{Rq}(\mathbf{V}).$$

*More precisely, the equality*

$$R \odot \tilde{\mathbf{V}} = \widetilde{Rq(\mathbf{V})}$$

*holds.*

*Proof.* Because q is a homomorphism, we have

$$(R \odot \tilde{\mathbf{V}})q = Rq\tilde{\mathbf{V}}q = Rqq_{\mathbf{V}} = q_{Rq(\mathbf{V})}.$$

Hence $\widetilde{Rq(\mathbf{V})} = R \odot \tilde{\mathbf{V}}$ as constant maps have a unique preimage under q. $\square$

With Proposition 2.8.6 in hand, we now have a dictionary for translating theorems about composition into theorems about operators. So whenever we prove results about composition of continuously closed classes or pseudovarieties of relational morphisms, we can then specialize them to obtain results about continuous operators acting on **PV**.

### 2.8.4 The composition theorem for the semidirect product

The main result of Tilson and Steinberg [339] states that in the monoidal setting

$$\mathbf{V}_D \odot \mathbf{W}_D = \mathbf{V}_D \mathbf{W}_D = (\mathbf{V} * \mathbf{W})_D.$$

We now sketch a proof showing that this result holds in the semigroup setting (using the semidirect product as we have defined it — Axiom (co-re) plays an important role in the proof). The proof assumes familiarity with [339] and may be omitted on a first reading.

**Theorem 2.8.7.** *Suppose* **V** *and* **W** *are pseudovarieties of semigroups. Then*

$$\mathbf{V}_D \odot \mathbf{W}_D = \mathbf{V}_D \mathbf{W}_D = (\mathbf{V} * \mathbf{W})_D. \tag{2.65}$$

*Proof.* We first mention that the result [324, 339]

$$\mathbf{gV} * \mathbf{gW} = \mathbf{g}(\mathbf{V} * \mathbf{W}) \tag{2.66}$$

holds in the semigroup context. Here we define the semidirect product of semigroupoids and semigroupoid varieties as per [339] with modifications to the semigroup case as per Section 2.5. That is, one adds new identities to all semigroupoids, performs the [339] category construction and then removes the new identities from the arrows (but not the objects). One can also redo [324] this way for semigroups. Unlike what is proved in [339] for the category situation, it is not clear that the two different derived semigroupoid constructions so obtained generate the same pseudovariety of semigroupoids. But on the pseudovariety level, both papers yield the same definition of $\mathbf{U} * \mathbf{Z}$ for pseudovarieties of semigroupoids **U** and **Z**. This is because both definitions of $\mathbf{U} * (-)$ come from classes of relational morphisms closed under division; hence membership reduces to the case of canonical (generator-to-generator) relational morphisms; in this case, both derived constructions agree. The proof that (2.66) holds given in [324] then goes through without change (note: in the definition of a Cayley graph in [324, 339] one must always adjoin new identities).

Let us return to our context where $\mathbf{V}$ and $\mathbf{W}$ are pseudovarieties of semigroups. Now because $\mathbf{V}_D$ and $\mathbf{W}_D$ are positive pseudovarieties, it suffices (by Proposition 2.8.2) to show $\mathbf{V}_D \mathbf{W}_D = (\mathbf{V} * \mathbf{W})_D$. The direction $(\mathbf{V} * \mathbf{W})_D \subseteq \mathbf{V}_D \mathbf{W}_D$ is as per [339] for the monoid case (using the wreath product of Section 1.2.3).

For the other direction, suppose $f: S \to T$, $g: T \to U$ are such that $f \in \mathbf{V}_D$ and $g \in \mathbf{W}_D$. Then, as we proved $\mathbf{V}_D$ and $\mathbf{W}_D$ are in $\mathbf{PVRM}^+$, the reductions of [339] allow us to assume $\operatorname{Im} f = T$ and $\operatorname{Im} g = U$. The proof then proceeds as in [339], using the equality $\mathbf{g}(\mathbf{V} * \mathbf{W}) = \mathbf{gV} * \mathbf{gW}$.      □

This result implies that one has the equalities:

$$\mathbf{V}_D = \mathbf{V}_D \mathsf{D} = \mathbf{V}_D \mathbf{1}_D = (\mathbf{V} * \mathbf{1})_D.$$

In [339], one deduces from this that $\mathbf{V} = \mathbf{V} * \mathbf{1}$. Why can this deduction be made for monoids and not for semigroups? Because in the monoidal context the derived category $D_{c_M}$ of the collapsing morphism $c_M: M \to 1$ is isomorphic to $M$. Thus $\mathbf{V}$ (in the monoid setting) can be recovered by considering the collapsing morphisms in $\mathbf{V}_D$. This is not the case for the derived semigroupoid. In fact, if $c_S: S \to 1$ is the collapsing morphism, then $D_{c_S}$ has an empty local semigroup at $I$, whereas the local semigroup at $1$ is isomorphic to the transition semigroup of the right representation $(S, S)$ (not the right regular representation). In particular, if $S$ is a left zero semigroup, then $D_{c_S}$ is easily verified to divide a locally trivial category and hence belongs to $\mathbf{V}_D$ for any non-trivial pseudovariety of $\mathbf{V}$ generated by monoids [364].

On the other hand, we have

$$\mathbf{V}_K \odot \mathbf{W}_K \neq (\mathbf{V} ** \mathbf{W})_K$$

as this would imply associativity of $**$ (by modeling $\mathbf{V}_K \mathsf{q}$ by composition). One direction of the composition theorem is true for the two-sided semidirect product. Namely, Proposition 2.6.24 shows the following.

**Theorem 2.8.8.** *Let* $\mathbf{V}$ *and* $\mathbf{W}$ *be pseudovarieties of semigroups. Then*

$$(\mathbf{V} ** \mathbf{W})_K \leq \mathbf{V}_K \odot \mathbf{W}_K = \mathbf{V}_K \mathbf{W}_K.$$

## 2.9 Reverse Global Mal'cev Condition

By some coincidence of fate, most of the energy spent in studying semidirect, double semidirect and Mal'cev products in the literature has been focused on operators of the form $\mathbf{V} * (-)$, $\mathbf{V} ** (-)$ and $\mathbf{V} \, \text{ⓜ} \, (-)$ and not operators of the form $(-) * \mathbf{W}$, $(-) ** \mathbf{W}$ and $(-) \, \text{ⓜ} \, \mathbf{W}$. Notable exceptions are the operator $(-) * \mathbf{D}$ [12,345,364], the operators $(-) * \mathbf{G}$ [126] and $(-) \, \text{ⓜ} \, \mathbf{G}$ [80,126, 295] and the work of Straubing and Thérien on "weakly bracketed" two-sided semidirect products [349]. The operators of the first sort are GMC, whereas those of the latter sort are not. Nonetheless, these latter operators do satisfy a sort of dual global Mal'cev condition, which we term the *reverse global Mal'cev condition*. At the current time we only have the definition and the motivating examples. We do not yet have corresponding classes of relational morphisms.

**Definition 2.9.1 (Reverse GMC).** *Let* $\mathbf{U}$ *be a pseudovariety of semigroups with* $\mathbf{U} \, \text{ⓜ} \, \mathbf{U} = \mathbf{U}$. *Then a positive continuous operator* $\alpha \in \mathbf{Cnt}(\mathbf{PV})^+$ *is said to satisfy the reverse global Mal'cev condition (reverse GMC) relative to* $\mathbf{U}$ *if*

$$\alpha(\mathbf{V}) \le (\mathbf{U} \, \text{ⓜ} \, \mathbf{V}) \,^{\omega}\text{ⓜ}\, \alpha(\mathbf{1}) \tag{2.67}$$

*all* $\mathbf{V} \in \mathbf{PV}$. *The collection of all such operators is denoted* $\mathbf{GMC}^{\varrho}_{\mathbf{U}}(\mathbf{PV})$ *(the $\varrho$ stands for reverse).*

The reader should consult (2.41) for the meaning of the $\omega$-power. Of course satisfying reverse GMC relative to $\mathbf{FSgp}$ is no restriction, so one wants in general for $\mathbf{U}$ to be as small as possible.

**Theorem 2.9.2.** *Let* $\mathbf{U}$ *be a pseudovariety of semigroups closed under Mal'cev product. Then* $\mathbf{GMC}^{\varrho}_{\mathbf{U}}(\mathbf{PV})$ *is a submonoid of* $\mathbf{Cnt}(\mathbf{PV})^+$ *closed under all sups (and so is in particular a complete lattice).*

*Proof.* Recalling that joins in $\mathbf{Cnt}(\mathbf{PV})^+$ are pointwise, suppose $\{\alpha_j\}_{j \in J}$ is a collection of elements of $\mathbf{GMC}^{\varrho}_{\mathbf{U}}(\mathbf{PV})$. Then we have

$$\bigvee_{j \in J} \alpha_j(\mathbf{V}) \le \bigvee_{j \in J} [(\mathbf{U} \, \text{ⓜ} \, \mathbf{V}) \,^{\omega}\text{ⓜ}\, \alpha_j(\mathbf{1})]$$

$$\le (\mathbf{U} \, \text{ⓜ} \, \mathbf{V}) \,^{\omega}\text{ⓜ}\, \bigvee_{j \in J} \alpha_j(\mathbf{1}),$$

establishing $\bigvee_{j \in J} \alpha_j \in \mathbf{GMC}^{\varrho}_{\mathbf{U}}(\mathbf{PV})$.

Clearly, the identity map satisfies (2.67). For closure under composition, suppose $\alpha, \beta \in \mathbf{GMC}^{\varrho}_{\mathbf{U}}(\mathbf{PV})$. From positivity we may deduce $\alpha\beta \ge \alpha, \beta$. A routine computation using continuity and Exercise 2.3.20 shows

$$\begin{aligned}
\alpha\beta(\mathbf{V}) &\le \alpha\big((\mathbf{U} \, \text{ⓜ} \, \mathbf{V}) \,^{\omega}\text{ⓜ}\, \beta(\mathbf{1})\big) \\
&\le (\mathbf{U} \, \text{ⓜ} \, [(\mathbf{U} \, \text{ⓜ} \, \mathbf{V}) \,^{\omega}\text{ⓜ}\, \beta(\mathbf{1})]) \,^{\omega}\text{ⓜ}\, \alpha(\mathbf{1}) \\
&\le ([\mathbf{U} \, \text{ⓜ} \, (\mathbf{U} \, \text{ⓜ} \, \mathbf{V})] \,^{\omega}\text{ⓜ}\, \beta(\mathbf{1})) \,^{\omega}\text{ⓜ}\, \alpha(\mathbf{1}) \\
&\le ([(\mathbf{U} \, \text{ⓜ} \, \mathbf{U}) \, \text{ⓜ} \, \mathbf{V}] \,^{\omega}\text{ⓜ}\, \beta(\mathbf{1})) \,^{\omega}\text{ⓜ}\, \alpha(\mathbf{1}) \\
&= ([\mathbf{U} \, \text{ⓜ} \, \mathbf{V}] \,^{\omega}\text{ⓜ}\, \beta(\mathbf{1})) \,^{\omega}\text{ⓜ}\, \alpha(\mathbf{1}) \\
&\le ([\mathbf{U} \, \text{ⓜ} \, \mathbf{V}] \,^{\omega}\text{ⓜ}\, \alpha\beta(\mathbf{1})) \,^{\omega}\text{ⓜ}\, \alpha\beta(\mathbf{1}) \\
&= (\mathbf{U} \, \text{ⓜ} \, \mathbf{V}) \,^{\omega}\text{ⓜ}\, \alpha\beta(\mathbf{1})
\end{aligned}$$

where the first equality uses $\mathbf{U} \circledm \mathbf{U} = \mathbf{U}$. We conclude $\alpha\beta \in \mathbf{GMC}_{\mathbf{U}}^{\varrho}(\mathbf{PV})$, completing the proof of the theorem. □

It remains to show that our motivating examples satisfy a reverse global Mal'cev condition with respect to a relatively small pseudovariety. In fact, they all satisfy reverse GMC relative to the pseudovariety $\mathbb{L}1$ of locally trivial semigroups; it is straightforward to verify $\mathbb{L}1 \circledm \mathbb{L}1 = \mathbb{L}1$.

**Exercise 2.9.3.** Check $\mathbb{L}1 \circledm \mathbb{L}1 = \mathbb{L}1$.

*Example 2.9.4 (Mal'cev product).* Because $\mathbf{U} \leq \mathbf{1} \circledm \mathbf{U}$ for any pseudovariety $\mathbf{U}$, clearly

$$\mathbf{V} \circledm \mathbf{W} \leq (\mathbf{1} \circledm \mathbf{V}) \,{}^{\omega}\!\!\circledm\, (\mathbf{1} \circledm \mathbf{W})$$

showing that $(-) \circledm \mathbf{W}$ satisfies reverse GMC relative to the trivial pseudovariety.

*Example 2.9.5 (Semidirect product).* We claim $(-) * \mathbf{W}$ satisfies reverse GMC relative to $\mathbb{L}1$. As $\mathbf{1} * \mathbf{W} = \mathbf{W}$ and $[\![xy = xz]\!]$ is a locally trivial pseudovariety, this follows from the inequality

$$\mathbf{V} * \mathbf{W} \leq ([\![xy = xz]\!] \circledm \mathbf{V}) \circledm \mathbf{W}$$

of Corollary 2.4.13.

*Example 2.9.6 (Two-sided semidirect product).* The situation for $(-) ** \mathbf{W}$ is very similar. Corollary 2.4.16 yields

$$\mathbf{V} ** \mathbf{W} \leq ([\![xyz = xwz]\!] \circledm \mathbf{V}) \circledm \mathbf{W} \tag{2.68}$$

from which it follows that $(-) ** \mathbf{W}$ satisfies reverse GMC relative to $\mathbb{L}1$.

**Exercise 2.9.7.** Show that $(-) \vee \mathbf{W}$ satisfies reverse GMC relative to the trivial pseudovariety.

*Question 2.9.8.* Find a natural lattice in $\mathbf{CC}^+$ that maps to $\mathbf{GMC}_{\mathbb{L}1}^{\varrho}(\mathbf{PV})$ under q.

*Remark 2.9.9.* Suppose $\alpha \in \mathbf{GMC}(\mathbf{PV})^+$ satisfies $\alpha(\mathbf{1}) \leq \mathbf{U}$ where $\mathbf{U}$ is closed under Mal'cev product. Then $\alpha \in \mathbf{GMC}_{\mathbf{U}}^{\varrho}(\mathbf{PV})$. Indeed,

$$\alpha(\mathbf{V}) \leq \alpha(\mathbf{1}) \circledm \mathbf{V} \leq (\mathbf{U} \circledm \mathbf{V}) \,{}^{\omega}\!\!\circledm\, \alpha(\mathbf{1})$$

with the first inequality following from GMC at $\mathbf{1}$.

This ends our discussion of the "other" variable and brings to a close this chapter.

# Notes

This chapter has its roots in Tilson's Chapter XII [362] of Eilenberg [85], where weakly closed classes were introduced. They turn out to be precisely what we have called positive pseudovarieties of relational morphisms. However, there is no notion of defining an operator via a class of relational morphisms in [362]. Steinberg and Tilson defined in [339] the action of $\mathbf{V}_D$ on $\mathbf{PV}$ and discussed the idea at the time of allowing certain classes of relational morphisms to act on $\mathbf{PV}$. The idea was abandoned but then brought to back to life under the impetus of the first author as described in the introduction. For a time Tilson was a coauthor of the current volume, and the first part of this chapter, up until Proposition 2.3.2, should be considered joint work with Tilson. The definition of the global Mal'cev condition is due to Rhodes.

A major theme in Finite Semigroup Theory is to determine membership in $\alpha(\mathbf{V})$ for a continuous operator $\alpha$ and a pseudovariety $\mathbf{V}$ given some additional information on $\mathbf{V}$. It was once hoped that if $\alpha$ were any of the commonly studied operators and $\mathbf{V}$ a decidable pseudovariety, then $\alpha(\mathbf{V})$ would have decidable membership. It was first shown in Albert, Baldinger and Rhodes [2] that the join of decidable pseudovarieties need not be decidable. Rhodes afterwards established that the semidirect product, the two-sided semidirect product, the Schützenberger product and the Mal'cev product do not preserve decidability [281]. A much simpler proof of all these undecidability results was later supplied by Auinger and Steinberg, who also proved that the power operator $\mathbf{P}$ does not preserve decidability [36]. The key idea of q-theory is to use pseudovarieties of relational morphisms and their associated equational theory to try and determine when pseudovarieties of the form $\alpha(\mathbf{V})$ are decidable.

Most of the material in this chapter is new with the exception of the sections on the derived and kernel semigroupoids and the Slice Theorem. Sections 2.5–2.6 are adapted from [339, 364] and [293], respectively. Theorem 2.5.15 seems to be novel. The lack of associativity of the two-sided semidirect product is well-known. Theorem 2.6.28 is a piece of semigroup folklore [27, 348], although the proof given here using Mal'cev products appears to be new. The history of the derived category and the kernel category can be found in the introduction. The Slice Theorem is from Steinberg's doctoral thesis [322, 330].

# 3

# The Equational Theory

There is now a body of literature establishing pseudoidentity basis theorems for operators [19,20,27,235]. All of these results can be seen as instances of a more general result giving a basis for the composition of two pseudovarieties of relational morphisms. The previous results then fall out via our method of modeling operators via composition. In particular, in this chapter we provide a new basis theorem for the semidirect product.

To do this, we must explore the equational theory of pseudovarieties of relational morphisms and equational continuously closed classes. In particular, the notion of a relational pseudoidentity is introduced and we prove Theorem 3.5.14, an analogue of Reiterman's Theorem [261] for pseudovarieties of relational morphisms.

For example, if $\mathbf{V}$ is a pseudovariety of semigroups, then a basis of path pseudoidentities for $\mathbf{gV}$ immediately translates to bases of relational pseudoidentities for $\mathbf{V}_D$ and $\mathbf{V}_K$. Methods to construct bases for $(\mathbf{V}, \mathbf{W})$ and $\mathbf{V}_\vee$ are also given.

To effect this goal, we must parallel the development of the syntactic approach to finite semigroups [7,25]. However, where profinite semigroups appear in that theory, we need relational morphisms of profinite semigroups. Thus the level of abstraction increases. To make the book as self-contained as possible we give a brief introduction to compact and profinite semigroups and prove Reiterman's Theorem.

Throughout this chapter, we shall tend to focus on pseudovarieties of relational morphisms and continue to use san serif letters such as V and W to denote them. Results about pseudovarieties of semigroups will then be obtained by specialization via identification (2.64).

The reader should be advised that the results in this chapter are quite technical and will only reappear in Chapter 7. On a first reading, the reader may therefore skip this chapter, or perhaps just read the definitions and theorem statements, omitting the proofs.

J. Rhodes, B. Steinberg, *The q-theory of Finite Semigroups*,
Springer Monographs in Mathematics, DOI 10.1007/978-0-387-09781-7_3,
© Springer Science+Business Media, LLC 2009

## 3.1 Profinite Semigroups

Because this chapter will make heavy usage of profinite semigroups, we include here an introduction to the subject. Perforce we shall be brief and many of the details will be left to the reader. More details can be found in [7,25] or in the first chapter of [298].

### 3.1.1 Compact semigroups

A *topological semigroup* for us is a Hausdorff topological space $S$ equipped with a continuous associative multiplication $S \times S \to S$. We are mostly interested in compact semigroups. The theory of compact semigroups was developed by A. D. Wallace and his school, most notably in the 1950s. A relatively complete accounting of this work can be found in [63,64,135]. Let us briefly review here some basic results concerning compact semigroups. The first proposition shows that groups lift under continuous surjective homomorphisms, generalizing a well-known result for finite semigroups. If $S$ is a topological semigroup, we always view $S^I$ as a topological semigroup with $I$ as an isolated point.

**Proposition 3.1.1.** *Let $S$ be a compact semigroup, $G$ a topological group and $\varphi\colon S \twoheadrightarrow G$ a continuous surjective homomorphism. Then there is a closed subgroup $H \leq S$ such that $H\varphi = G$.*

*Proof.* Let $\mathscr{C}$ be the set of closed subsemigroups $T$ of $S$ with $T\varphi = G$. Then $S \in \mathscr{C}$, so $\mathscr{C}$ is non-empty. Suppose that $\{T_\alpha\}$ is a descending chain from $\mathscr{C}$. Then $T = \bigcap T_\alpha$ is a closed subsemigroup. We show that it belongs to $\mathscr{C}$. Indeed, if $g \in G$, then $C_\alpha = g\varphi^{-1} \cap T_\alpha$ is a non-empty closed subset of $S$ and the $C_\alpha$ form a descending chain. By compactness of $S$, it follows $\emptyset \neq \bigcap C_\alpha = g\varphi^{-1} \cap T$. We conclude that $T\varphi = G$ and so $T \in \mathscr{C}$. By Zorn's lemma, $\mathscr{C}$ has a minimal element $T$. We show that $T$ is left and right simple, and hence a group (Lemma A.3.3). Indeed, if $t \in T$, then $tT, Tt \leq T$ are closed subsemigroups of $S$ and

$$(tT)\varphi = t\varphi T\varphi = t\varphi G = G = Gt\varphi = T\varphi t\varphi = (Tt)\varphi.$$

We conclude by minimality that $tT = T = Tt$, as required. □

As a consequence, we deduce the existence of idempotents in compact semigroups.

**Corollary 3.1.2.** *Every non-empty compact semigroup contains an idempotent.*

*Proof.* Let $S$ be a non-empty compact semigroup and consider the collapsing homomorphism $c_S\colon S \twoheadrightarrow \{1\}$. It is trivially a continuous surjective homomorphism to a group. Proposition 3.1.1 then asserts the existence of a closed subgroup of $S$, the identity of which is our desired idempotent. □

One consequence of the existence of idempotents in non-empty compact semigroups is that one can lift idempotents under surjective homomorphisms.

**Corollary 3.1.3.** *Let $\varphi\colon S \twoheadrightarrow T$ be a continuous surjective homomorphism of compact semigroups. Let $e \in T$ be an idempotent. There there is an idempotent $f \in S$ with $f\varphi = e$.*

*Proof.* As $e\varphi^{-1}$ is a closed non-empty subsemigroup of $S$, it contains an idempotent by Corollary 3.1.2. $\qquad\square$

Another consequence is the fact that compact subsemigroups of topological groups must be subgroups.

**Corollary 3.1.4.** *Let $G$ be a topological group and let $S \leq G$ be a non-empty compact subsemigroup. Then $S$ is a subgroup.*

*Proof.* Let $s \in S$. Because $sS$ and $Ss$ are closed subsemigroups of $S$, they are compact. Therefore, they have an idempotent by Corollary 3.1.2. But the only idempotent in $G$ is the identity so $1 \in sS \cap Ss$. Thus $1 \in S$ and $s$ has a left and right inverse. We conclude that $S$ is a subgroup. $\qquad\square$

**Exercise 3.1.5.** Show that a cancellative compact semigroup is a group.

When studying equivalence relations and congruences on compact semigroups, one must demand that they respect the topology.

**Definition 3.1.6 (Closed equivalence relation).** *An equivalence relation $R$ on a topological space $X$ is said to be closed if $R$ is a closed subspace of $X \times X$.*

**Exercise 3.1.7.** Show that if $X$ is a topological space and $R \subseteq X \times X$ is a closed equivalence relation, then each equivalence class of $R$ is closed.

If $R$ is an equivalence relation on a topological space, then $X/R$ is a topological space with the quotient topology.

**Proposition 3.1.8.** *Let $X$ be a compact Hausdorff space. Then an equivalence relation $R$ on $X$ is closed if and only if $X/R$ is compact Hausdorff.*

*Proof.* Let $p\colon X \to X/R$ be the projection. If $X/R$ is compact Hausdorff, then the diagonal $(X/R)\Delta$ is a closed subspace of $X/R \times X/R$. But $R = (X/R)\Delta(p \times p)^{-1}$ and hence is closed. Now assume that $R$ is a closed equivalence relation. Let $\pi_i\colon R \to X$ be the projection to the $i^{th}$ factor, $i = 1, 2$. First we claim that $p$ is a closed mapping. Indeed, let $Y$ be a closed subspace of $X$. Then $Ypp^{-1} = Y\pi_2^{-1}\pi_1$ and hence is closed as $R$ and $X$ are compact Hausdorff. Thus $Yp$ is closed, establishing that $p$ is a closed map. Suppose now that $[x] \neq [y]$ are equivalence classes in $X/R$. By Exercise 3.1.7 $[x]$ and $[y]$ are closed as subsets of $X$. Because $X$ is compact Hausdorff, and hence normal, there are disjoint neighborhoods $U$ and $V$ in $X$ of the closed subsets

$[x], [y]$. Set $U' = X/R \setminus (X \setminus U)p$ and $V' = X/R \setminus (X \setminus V)p$. Because $p$ is a closed mapping, it follows that $U', V'$ are open subsets of $X/R$. By choice of $U$ and $V$, we have $[x] \in U'$ and $[y] \in V'$. Moreover, $U'$ and $V'$ are disjoint as $U'p^{-1} \subseteq U$ and $V'p^{-1} \subseteq V$. Thus $X/R$ is Hausdorff; of course it is compact being the continuous image of a compact space.    $\square$

Our next proposition shows that Green's relations are closed in a compact semigroup.

**Proposition 3.1.9.** *Let $S$ be a compact semigroup. Then each of Green's relations is closed.*

*Proof.* We just handle $\mathscr{R}$, as the other arguments are similar. Because $S^I$ is also a compact semigroup, we may assume that $S$ is a monoid. Suppose that we have a convergent net $(x_\alpha, y_\alpha)$ in $S \times S$ with limit $(x, y)$ such that $x_\alpha \mathrel{\mathscr{R}} y_\alpha$ for all $\alpha$. Then we can find $u_\alpha, v_\alpha \in S$ with $x_\alpha u_\alpha = y_\alpha$ and $y_\alpha v_\alpha = x_\alpha$. By passing to a subnet, we may assume that $u_\alpha \to u$ and $v_\alpha \to v$. Then $xu = y$ and $yv = x$, yielding $x \mathrel{\mathscr{R}} y$ as required.    $\square$

As a consequence, the equivalences classes of Green's relations are closed in a compact semigroup. Next we prove compact semigroups are stable. A semigroup $S$ is said to be *stable* if

$$sx \mathrel{\mathscr{J}} s \iff sx \mathrel{\mathscr{R}} s \text{ and also } xs \mathrel{\mathscr{J}} s \iff xs \mathrel{\mathscr{L}} s.$$

In a stable semigroup, Green's relations $\mathscr{J}$ and $\mathscr{D}$ coincide (Corollary A.2.5).

**Proposition 3.1.10.** *Compact semigroups are stable and so Green's relations $\mathscr{J}$ and $\mathscr{D}$ coincide in a compact semigroup.*

*Proof.* Let $S$ be a compact semigroup. Without loss of generality, we may assume that $S$ is a monoid. We show $sx \mathrel{\mathscr{J}} s \iff sx \mathrel{\mathscr{R}} s$; the proof of the other equivalence is dual. As $\mathscr{R} \subseteq \mathscr{J}$, it suffices to consider the case $sx \mathrel{\mathscr{J}} s$. Because $sx \leq_{\mathscr{R}} s$, we just need the reverse inequality. We can $\underline{\text{find}}$ $u, v \in S$ such that $usxv = s$. Then $s = u^n s(xv)^n$ for all $n$. Let $T = \overline{\langle xv \rangle}$. As $T$ is a closed subsemigroup of $S$, it contains an idempotent $e$ by Corollary 3.1.2. Therefore, there is a net $\{(xv)^{n_\alpha}\}$ with the $n_\alpha > 0$ so that $e = \lim(xv)^{n_\alpha}$. By passing to a subnet, we may assume that the nets $\{(xv)^{n_\alpha-1}\}$ and $\{u^{n_\alpha}\}$ converge to elements $t$ and $u_0$, respectively, of the monoid $S$. Then $e = xvt$ and $s = u_0 se$. Therefore, $(sx)vt = se = u_0 see = u_0 se = s$ and so $s \leq_{\mathscr{R}} sx$, as required.    $\square$

A consequence of Proposition 3.1.8 is that if $S$ is a compact semigroup and $R$ is a congruence on $S$, then $S/R$ is a compact semigroup if and only if $R$ is a closed equivalence relation. This leads to the following definition.

**Definition 3.1.11 (Closed congruence).** *A closed congruence on a compact semigroup is a congruence that is a closed equivalence relation.*

**Exercise 3.1.12.** Let $S$ be a compact semigroup and $I$ a closed ideal. Verify that the Rees quotient $S/I$ is a compact semigroup.

**Exercise 3.1.13.** Let $G$ be a compact group and let $N$ be a normal subgroup. Show that $G/N$ is a compact group if and only if $N$ is closed.

We end this section with a lemma about lifting regular $\mathscr{J}$-classes under continuous surjective homomorphisms. In Chapter 4, we shall reprove this lemma for finite semigroups in greater detail and consider all of $\mathscr{R}$, $\mathscr{L}$ and $\mathscr{H}$ (Lemma 4.6.10), but we record it here with this level of generality as it may be useful in future work.

**Lemma 3.1.14.** *Let $S$, $T$ be compact semigroups and let $\varphi\colon S \twoheadrightarrow T$ be a continuous surjective homomorphism. Let $J$ be a $\mathscr{J}$-class of $T$. Then there are $\leq_{\mathscr{J}}$-minimal $\mathscr{J}$-classes $J'$ of $S$ with $J'\varphi\cap J \neq \emptyset$. If $J'$ is such a minimal $\mathscr{J}$-class, then $J'\varphi = J$. Moreover, if $J$ is regular, then $J'$ is regular and unique.*

*Proof.* Let $\mathscr{C}$ be the set of closed ideals $K$ of $S$ such that $K\varphi \cap J \neq \emptyset$. As $\varphi$ is onto, $S \in \mathscr{C}$. Let $\{K_\alpha\}$ be a descending chain from $\mathscr{C}$ and set $K = \bigcap K_\alpha$. Evidently $K$ is a closed ideal. Then because $J$ is closed by Proposition 3.1.9, $C_\alpha = J\varphi^{-1} \cap K_\alpha$ is a non-empty closed subset of $S$ and the $C_\alpha$ form a descending chain. By compactness $\emptyset \neq \bigcap C_\alpha = J\varphi^{-1} \cap K$. Therefore, $K \in \mathscr{C}$. Zorn's lemma now implies that $\mathscr{C}$ has minimal elements. Let $K$ be such a minimal element and let $s \in K \cap J\varphi^{-1}$. Then $S^I s S^I$ is a closed ideal contained in $K$, which intersects $J\varphi^{-1}$ and so $K = S^I s S^I$ by minimality. If $t <_{\mathscr{J}} s$, then $S^I t S^I$ is a closed ideal properly contained in $K$ and so it cannot intersect $J\varphi^{-1}$ by minimality of $K$. Therefore, $J_s$ is a minimal $\mathscr{J}$-class of $S$ with $J_s \cap J\varphi^{-1} \neq \emptyset$.

Suppose now that $J'$ is a minimal $\mathscr{J}$-class of $S$ with $J' \cap J\varphi^{-1} \neq \emptyset$. Let $s \in J' \cap J\varphi^{-1}$ and set $t = s\varphi$. Suppose $u \mathscr{J} t$. Then there are $x, y \in T^I$ with $xty = u$. Because $\varphi$ is onto, we can find $x', y' \in S^I$ with $x'\varphi = x$, $y'\varphi = y$ (extending $\varphi$ to $S^I$ in the obvious way). Then $(x'sy')\varphi = u$. As $x'sy' \leq_{\mathscr{J}} s$, we conclude that $x'sy' \in J'$ by minimality and so $J'\varphi = J$. Finally, suppose that $J$ is regular and let $J'$, $J''$ be two minimal $\mathscr{J}$-classes intersecting $J\varphi^{-1}$ (possibly $J' = J''$). As $J$ is regular, it contains an idempotent $e$. Choose $e' \in J'$ and $e'' \in J''$ with $e'\varphi = e = e''\varphi$. Then $(e'e'')\varphi = e \in J$ and $e'e'' \leq_{\mathscr{J}} J', J''$. We conclude by minimality that $J' = J''$ and that $J'$ is regular. This completes the proof. $\quad\square$

As a corollary, we deduce the well-known fact that a compact semigroup has a minimal ideal, which is necessarily closed.

**Corollary 3.1.15 (Existence of minimal ideals).** *Let $S$ be a non-empty compact semigroup. Then it has a minimal ideal, which is necessarily closed.*

*Proof.* Consider the collapsing map $c_S\colon S \to \{1\}$. Then Lemma 3.1.14 implies that $S$ has a $\leq_{\mathscr{J}}$-minimal $\mathscr{J}$-class $J$, which is necessarily the minimal ideal of $S$. Moreover, $J$ is closed by Exercise 3.1.7. $\quad\square$

### 3.1.2 Inverse limits, profinite spaces and profinite semigroups

Let $D$ be a directed set. Then an *inverse system* indexed by $D$ in a category $C$ is a collection of objects $\{c_\alpha\}_{\alpha \in D}$ of $C$ and morphisms $\{\pi_{\beta,\alpha}\colon c_\beta \to c_\alpha, \alpha \leq \beta\}$ such that:

$$\text{if } \alpha \leq \beta \leq \gamma, \text{ then } \pi_{\gamma,\alpha} = \pi_{\gamma,\beta}\pi_{\beta,\alpha} \text{ and } \pi_{\alpha,\alpha} = 1_{c_\alpha}.$$

A *cone* of morphisms from an object $c$ to the inverse system $\{c_\alpha\}_{\alpha \in D}$ is a family of morphisms $\{\varphi_\alpha\colon c \to c_\alpha\}_{\alpha \in D}$ such that, for $\alpha \leq \beta$, the diagram

commutes.

*Remark 3.1.16.* There is a more categorical viewpoint on inverse systems. If the directed set $D$ is viewed as a category (as was done in Chapter 1), then an inverse system indexed by $D$ in a category $C$ is nothing more than a functor $F\colon D^{op} \to C$. A cone from $c$ to the system is a natural transformation $\varphi = \{\varphi_\alpha\}_{\alpha \in D}$ from the diagonal functor $c\Delta$ (which takes on the value $c$ at all objects of $D$ and the value $1_c$ on all arrows of $D$) to $F$.

The *inverse limit* (or *projective limit*) $\varprojlim_D c_\alpha$ is a universal object in $C$ equipped with a cone of morphisms $\{\pi_\alpha\colon \varprojlim_D c_\gamma \to c_\alpha\}_{\alpha \in D}$, in particular, the diagram

$$\varprojlim c_\gamma \xrightarrow{\ \pi_\beta\ } c_\beta$$

$$\pi_\alpha \searrow \quad \downarrow \pi_{\beta,\alpha}$$

$$c_\alpha$$

commutes for $\alpha \leq \beta$. The universal property states that if $c$ is any object of $C$ with a cone of morphisms $\{\varphi_\alpha\colon c \to c_\alpha\}_{\alpha \in D}$, then there is a unique morphism $\varphi\colon c \to \varprojlim_D c_\gamma$ (sometimes denoted $\varprojlim \varphi_\alpha$) such that

$$c \xrightarrow{\ \varphi\ } \varprojlim c_\gamma$$

$$\varphi_\alpha \searrow \quad \downarrow \pi_\alpha$$

$$c_\alpha$$

commutes for all $\alpha \in D$.

*Remark 3.1.17.* Categorically speaking, the category $C$ admits inverse limits indexed by $D$ if and only if there is an adjunction

$$C \xleftarrow{\varprojlim_D} \xrightarrow{\Delta} C^{D^{op}}.$$

See [185, Chapter III] for more details.

In categories like sets, semigroups, topological spaces and topological semi-groups, the inverse limit admits the following concrete description:

$$\varprojlim_D X_\alpha = \{(x_\alpha) \in \prod_D X_\alpha \mid x_\beta \pi_{\beta,\alpha} = x_\alpha, \ \alpha \le \beta\}.$$

In particular, if the $X_\alpha$ are compact Hausdorff spaces and the maps of the system are continuous, then $\varprojlim_D X_\alpha$ is compact Hausdorff by Tychonoff's Theorem and the next exercise.

**Exercise 3.1.18.** Show that if the $X_\alpha$, $\alpha \in D$, are Hausdorff spaces, then $\varprojlim_D X_\alpha$ is a closed subspace of $\prod_D X_\alpha$.

The next exercise provides a basis for the topology on an inverse limit.

**Exercise 3.1.19.** Let $X = \varprojlim_D X_\alpha$ be an inverse limit of topological spaces. Show that the sets of the form $V\pi_\alpha^{-1}$, with $\alpha \in D$ and $V \subseteq X_\alpha$ open, form a basis for the topology of $S$, where $\pi_\alpha \colon X \to X_\alpha$ is the projection.

Recall that a compact Hausdorff space is *totally disconnected* if and only if it has a basis of simultaneously open and closed, that is *clopen*, subsets for its topology [117, 147, 158]. Spaces with a basis of clopen subsets are often called *zero-dimensional*. Compact zero-dimensional spaces are known variously as Boolean spaces and Stone spaces. If each space $X_\alpha$ is zero-dimensional, then so is $\varprojlim_D X_\alpha$, as the next exercise indicates.

**Exercise 3.1.20.** Prove that the class of zero-dimensional Hausdorff spaces is closed under forming direct products and taking closed subspaces. Hence an inverse limit of zero-dimensional Hausdorff spaces is zero-dimensional.

**Exercise 3.1.21.** Verify that if $\{X_\alpha\}_{\alpha \in D}$ is an inverse system of subspaces of a space $X$ with the maps of the system the inclusions, then $\varprojlim X_\alpha = \bigcap X_\alpha$.

A fundamental fact about inverse limits of non-empty compact Hausdorff spaces is that they are non-empty. This is essentially König's lemma.

**Lemma 3.1.22.** *Let $\varprojlim_D X_\alpha$ be an inverse limit of non-empty compact Hausdorff spaces. Then $\varprojlim_D X_\alpha \neq \emptyset$.*

*Proof.* Let $\{\pi_{\beta,\alpha} \colon X_\beta \to X_\alpha\}$ be the maps of the inverse system. For each $\delta \in D$, let $C_\delta = \{(x_\alpha) \in \prod_D X_\alpha \mid x_\beta \pi_{\beta,\alpha} = x_\alpha, \forall \alpha, \beta \le \delta\}$. Then $C_\delta$ is non-empty as each $X_\alpha$ is non-empty. Also $\varprojlim_D X_\alpha = \bigcap_D C_\alpha$. But the $C_\alpha$ are closed because the $X_\alpha$ are Hausdorff and the $\pi_{\beta,\alpha}$ are continuous. Also if $\alpha_1, \ldots, \alpha_n \le \beta$, then $C_\beta \subseteq C_{\alpha_1} \cap \cdots \cap C_{\alpha_n}$, so the family $\{C_\alpha\}_{\alpha \in D}$ satisfies the finite intersection condition. Compactness of $\prod_D X_\alpha$, yields $\varprojlim_D X_\alpha = \bigcap_D C_\alpha \neq \emptyset$.    □

**Exercise 3.1.23.** Let $S = \varprojlim_D S_\alpha$ where the $S_\alpha$ are compact semigroups. Let $\mathscr{K}$ be one of Green's relations. Prove that $s \mathscr{K} t$ in $S$ if and only if, for all $\alpha \in D$, one has $s\pi_\alpha \mathscr{K} t\pi_\alpha$. Hint: For instance to handle $\mathscr{R}$, set $C_\alpha = \{u \in S_\alpha^I \mid s\pi_\alpha u = t\pi_\alpha\}$. Show that the $C_\alpha$ with $\alpha \in D$ form an inverse system of compact Hausdorff spaces. Now apply Lemma 3.1.22.

We will use freely the well-known fact that if $\{S_\alpha\}_{\alpha \in A}$ is an inverse system and $B \subseteq A$ is *cofinal*, that is, every element of $A$ is smaller than some element of $B$, then $\varprojlim_A S_\alpha = \varprojlim_B S_\alpha$. The reader is referred to [298, Lemma 1.1.9] for details.

**Exercise 3.1.24.** Prove the above fact.

In the theory of profinite semigroups, inverse systems in which each of the maps is a quotient map are particularly important.

**Definition 3.1.25 (Inverse quotient system).** *An inverse system of topological spaces, semigroups, etc., is called an* inverse quotient system *if each map of the system is a quotient map.*

**Lemma 3.1.26.** *Suppose that $\{\pi_{\beta,\alpha}\colon X_\beta \twoheadrightarrow X_\alpha, \alpha \leq \beta\}$ is an inverse quotient system of compact Hausdorff spaces indexed by a directed set $D$. Then the projections $\pi_\beta\colon \varprojlim_D X_\alpha \to X_\beta$ are quotient maps for all $\beta \in D$.*

*Proof.* By passing to a cofinal subset, we may assume without loss of generality that $\beta$ is the smallest element of $D$. Let $x \in X_\beta$. Define an inverse system indexed by $D$ by setting $Y_\gamma = x\pi_{\gamma,\beta}^{-1}$ and restricting the maps of the original system. Then each $Y_\gamma$ is a compact Hausdorff space and is non-empty because the $\pi_{\gamma,\beta}$ are surjective. So $\varprojlim_D Y_\alpha \neq \emptyset$ by Lemma 3.1.22. Clearly, one has $\varprojlim_D Y_\alpha = x\pi_\beta^{-1}$. Thus $\pi_\beta$ is surjective and hence a quotient map as all spaces involved are compact Hausdorff. $\qquad\square$

The following lemma is of a similar nature.

**Lemma 3.1.27.** *Let $X$ be a compact Hausdorff space and $\{X_\alpha\}_{\alpha \in D}$ an inverse system of Hausdorff spaces. Suppose $\{\varphi_\alpha\colon X \to X_\alpha\}_{\alpha \in D}$ is a cone of continuous surjective maps to $\{X_\alpha\}_{\alpha \in D}$. Let $\varphi\colon X \to \varprojlim X_\alpha$ be the induced map. Then $\varphi$ is surjective.*

*Proof.* Let $\xi = (x_\alpha) \in \varprojlim X_\alpha$. For each $\alpha \in D$, let $C_\alpha = \{x \in X \mid x\varphi_\alpha = x_\alpha\}$. Then $C_\alpha$ is a closed subspace of $X$ and hence compact Hausdorff. It is non-empty by assumption. The $C_\alpha$ form an inverse system with respect to the inclusion maps. Hence $\bigcap_D C_\alpha = \varprojlim_D C_\alpha \neq \emptyset$. But $\bigcap_D C_\alpha = \xi\varphi^{-1}$, so $\varphi$ is surjective. $\qquad\square$

A *profinite space* is an inverse limit of finite (discrete) sets. It is well-known that profinite spaces are precisely the compact Hausdorff totally disconnected spaces or equivalently the zero-dimensional compact Hausdorff spaces [298, Thm. 1.1.12].

**Exercise 3.1.28.** Let $L$ be a lattice with top and bottom and let $\operatorname{Spec} L$ be the set of all maps $\chi \colon L \to \{0,1\}$ preserving finite (including empty) sups and infs. Show that $\operatorname{Spec} L$ is a closed subspace of $\{0,1\}^L$ and hence profinite. Show that if $L$ is a Boolean algebra, then the lattice of clopen subsets of $\operatorname{Spec} L$ is isomorphic to $L$. This is the heart of the famous Stone duality [117, 147, 341, 342] between Boolean algebras and profinite spaces.

**Definition 3.1.29 (Profinite semigroup).** *A profinite semigroup is an inverse limit of finite semigroups.*

Notice that if $S$ is a profinite semigroup, then so is the monoid $S^I$ obtained by adjoining an identity $I$ to $S$ (if $S = \varprojlim S_\alpha$, then $S^I = \varprojlim S_\alpha^I$). The category of profinite semigroups with continuous homomorphisms will be denoted **PSgp**.

**Exercise 3.1.30.** Show that if a group $G$ is a profinite semigroup, then it is a profinite group (i.e., a projective limit of finite groups).

If $S$ is a profinite semigroup, then $S$ is said to be *topologically generated* by a topological space $A$ if there is a continuous map $\varphi \colon A \to S$ such that the subsemigroup generated by $A\varphi$ is dense.

**Exercise 3.1.31.** Show that a profinite group is topologically generated as a group by a topological space $X$ if and only if it is topologically generated by $X$ as a semigroup.

If $A$ is a set, then $A^+$ denotes the free semigroup on $A$ and $A^*$ denotes the free monoid. A profinite semigroup $T$ is said to a *free profinite semigroup* on a topological space $A$ if there is a continuous map $\iota \colon A \to T$ such that given any continuous map $\varphi \colon A \to S$ with $S$ a profinite semigroup, there is a unique continuous homomorphism $\psi \colon T \to S$ such that

commutes. The free profinite semigroup $\widehat{A^+}$ generated by a topological space $A$ exists and is unique up to isomorphism [7]. Roughly speaking, it is the inverse limit of all topologically $A$-generated profinite semigroups with respect to the canonical morphisms. Usually one is only interested in the case where $A$ is profinite. The most important case is when the space is a finite set $A$. Then $\widehat{A^+}$ be can constructed by taking the inverse limit of all finite $A$-generated semigroups with the natural maps between them. That is, if $\mathscr{F}$ is the directed set of all finite index congruences on the free semigroup $A^+$, ordered by reverse inclusion, then $\widehat{A^+} = \varprojlim_{\mathscr{F}} A^+/\!\equiv$ [7] (a more general result will be proved in Theorem 3.2.7 below). Elements of a free profinite semigroup $\widehat{A^+}$, with $A$

finite, are often termed *implicit operations* of arity $|A|$ [7, 25, 261]; from this viewpoint, elements of $A^+$ are called *explicit operations*. The next exercise justifies this terminology. The approach sketched here is inspired by [46].

**Exercise 3.1.32.** Let $U \colon \mathbf{FSgp} \to \mathbf{Set}$ and $U' \colon \mathbf{PSgp} \to \mathbf{Set}$ be the underlying set functors.

1. Show that if $A$ is a finite set and $w \in \widehat{A^+}$, then one can define a natural transformation $w \colon U^A \to U$ as follows. For each finite semigroup $S$, the component $w_S \colon S^A \to S$ is given by setting $w_S((s_a)_{a \in A})$ to be the image of $w$ under the unique continuous extension to $\widehat{A^+}$ of the map $a \mapsto s_a$.
2. Verify that each natural transformation $U^A \to U$ extends uniquely to a natural transformation $(U')^A \to U'$.
3. Show that, for any set $A$, the functor $(U')^A$ is representable with representing object $\widehat{A^+}$ (see [185, Chapter III, Section 2] for definitions).
4. Use Yoneda's Lemma [185, Chapter III, Section 2] to establish a bijection between elements of $\widehat{A^+}$ and natural transformations $U^A \to U$.

An alternative construction of $\widehat{A^+}$, when $A$ is finite, is to complete $A^+$ with respect to the profinite metric. If $u, v \in A^+$, define $\sigma(u, v)$ to be the minimum index of a congruence $\equiv$ such that $u \not\equiv v$. We take $\sigma(u, u) = \infty$. Then the *profinite metric* on $A^+$ is given by $d(u, v) = e^{-\sigma(u,v)}$. See [7, 25] for details.

**Exercise 3.1.33.** Verify that $d$ is an ultrametric. That is, $d$ satisfies the strong form of the triangle inequality $d(u, v) \leq \max\{d(u, w), d(w, v)\}$.

The following exercise is the content of [298, Cor. 1.1.8].

**Exercise 3.1.34.** Let $X = \varprojlim_D X_\alpha$ be a projective limit of finite discrete spaces $X_\alpha$, $\alpha \in D$. Let $\pi_\alpha \colon X \to X_\alpha$ be the projection. Show that $Y \subseteq X$ is closed if and only if $Y = \varprojlim_D Y\pi_\alpha$. Show more generally that if $Y \subseteq X$, then $\overline{Y} = \varprojlim_D Y\pi_\alpha$.

Let $A$ be a topological space. Define $\mathbf{PSgp}_A$ to be the category of topologically $A$-generated profinite semigroups. To define this formally, we take the objects of $\mathbf{PSgp}_A$ to be pairs $(S, \varphi)$ with $S$ a profinite semigroup and $\varphi \colon \widehat{A^+} \to S$ a continuous surjective homomorphism. An arrow $\psi \colon (S, \varphi) \to (T, \rho)$ of $\mathbf{PSgp}_A$ is a continuous homomorphism $\psi \colon S \to T$ respecting the generating maps, that is, so that

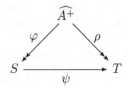

commutes. Because $\varphi$ and $\rho$ are surjective, $\psi$ must be surjective as well: that is, all maps of $\mathbf{PSgp}_A$ are surjective. Moreover, there is at most one arrow between $(S, \varphi)$ and $(T, \rho)$: either $\varphi^{-1}\rho$ is a homomorphism, in which case it is the unique arrow, or there is none.

*Remark 3.1.35.* One can view $\mathbf{PSgp}_A$ as the full subcategory of the comma category $(\widehat{A^+} \downarrow \mathbf{PSgp})$ with objects the surjective morphisms from $\widehat{A^+}$ [185, Chapter 2, Section 6].

The following corollary says that if $A$ is a topological space, then the inverse limit of topologically $A$-generated profinite semigroups is again topologically $A$-generated.

**Corollary 3.1.36.** *Let $A$ be a topological space. Then the category $\mathbf{PSgp}_A$ has inverse limits and the forgetful functor to $\mathbf{PSgp}$ preserves inverse limits. More precisely, if $\{(S_\alpha, \varphi_\alpha)\}_{\alpha \in D}$ is an inverse system in $\mathbf{PSgp}_A$, then $(\varprojlim_D S_\alpha, \varprojlim \varphi_\alpha) \in \mathbf{PSgp}_A$ and $\varprojlim(S_\alpha, \varphi_\alpha) = (\varprojlim_D S_\alpha, \varprojlim \varphi_\alpha)$.*

*Proof.* Let $\{(S_\alpha, \varphi_\alpha)\}_{\alpha \in D}$ be an inverse system in $\mathbf{PSgp}_A$. Note that because there is at most one map in $\mathbf{PSgp}_A$ between objects, the maps of the inverse system are understood. The maps $\varphi_\alpha$ form a cone of continuous surjective homomorphisms from $\widehat{A^+}$ to the inverse system $\{S_\alpha\}_{\alpha \in D}$ in $\mathbf{PSgp}$ and so, by Lemma 3.1.27, the induced map $\varphi \colon \widehat{A^+} \to \varprojlim S_\alpha$ is surjective. This establishes $(\varprojlim S_\alpha, \varphi) \in \mathbf{PSgp}_A$. If $(T, \rho) \in \mathbf{PSgp}_A$ has a cone $\{\lambda_\alpha\}$ of morphisms in $\mathbf{PSgp}_A$ to the inverse system, then it is easily verified that the induced map $\varprojlim \lambda_\alpha \colon T \to \varprojlim S_\alpha$ constitutes an arrow of $\mathbf{PSgp}_A$. The corollary follows.   $\square$

The following lemma is one of the fundamental compactness results that makes profinite semigroups so useful. An analogous result holds of course for profinite spaces.

**Lemma 3.1.37.** *Let $S = \varprojlim_D S_\alpha$ be a projective limit of an inverse quotient system of compact semigroups and let $\pi_\alpha \colon S \to S_\alpha$, $\alpha \in D$, be the projection. Let $T$ be a semigroup with the discrete topology and suppose $\varphi \colon S \to T$ is a continuous morphism. Then there exists an index $\alpha$ and a continuous morphism $\varphi_\alpha \colon S_\alpha \to T$ such that $\varphi = \pi_\alpha \varphi_\alpha$.*

*Proof.* The sets of the form $V\pi_\alpha^{-1}$, with $\alpha \in D$ and $V \subseteq S_\alpha$ open, form a basis for the topology of $S$ by Exercise 3.1.19. As $\varphi$ is continuous and $T$ is discrete, it follows that, for each $s \in S$, there is an index $\alpha_s$ and an open neighborhood $V_s$ of $s\pi_{\alpha_s}$ in $S_{\alpha_s}$ such that $V_s\pi_{\alpha_s}^{-1} \subseteq s\varphi\varphi^{-1}$. Because $S$ is compact and $S = \bigcup V_s\pi_{\alpha_s}^{-1}$, there exist elements $s_1, \ldots, s_n \in S$, $\alpha_1, \ldots, \alpha_n \in D$, and open sets $V_1 \ldots, V_n$ of $S_{\alpha_1}, \ldots, S_{\alpha_n}$, respectively, with $s_i \in V_i\pi_{\alpha_i}^{-1} \subseteq s_i\varphi\varphi^{-1}$ for $i = 1, \ldots, n$ and
$$S = V_1\pi_{\alpha_1}^{-1} \cup \cdots \cup V_n\pi_{\alpha_n}^{-1}.$$
As the $S_\alpha$ form an inverse system, there exists $\beta \in D$ with $\beta \geq \alpha_i$ for all $i = 1, \ldots, n$. Now we have that, for each $s \in S$, there exists $j$ so that $s \in V_j\pi_{\alpha_j}^{-1}$.

Thus $s \in s\pi_\beta\pi_\beta^{-1} \subseteq s\pi_{\alpha_j}\pi_{\alpha_j}^{-1} \subseteq V_j\pi_{\alpha_j}^{-1} \subseteq s_j\varphi\varphi^{-1}$. Hence $s_j\varphi = s\varphi$ and so $s\pi_\beta\pi_\beta^{-1} \subseteq s\varphi\varphi^{-1}$, all $s \in S$. In other words, $\ker\pi_\beta \subseteq \ker\varphi$. As $\pi_\beta$ is onto by Lemma 3.1.26, it follows $\varphi$ factors as $\varphi = \pi_\beta\varphi_\beta$ where $\varphi_\beta \colon S_\beta \to T$ is a homomorphism.

It remains to verify that $\varphi_\beta$ is continuous. To do this we just need to verify that $t\varphi_\beta^{-1}$ is closed in $S_\beta$ for each $t \in T$. But, by hypothesis, $t\varphi^{-1}$ is closed in $S$ and hence compact. Because $\pi_\beta$ is continuous, it follows $t\varphi_\beta^{-1} = t\varphi^{-1}\pi_\beta$ is compact and hence closed.                                                                                 □

*Remark 3.1.38.* It was pointed out to us by L. Ribes that Lemma 3.1.37 holds without assuming that the system is an inverse quotient system if each $S_\alpha$ is assumed to be profinite.

Suppose $S$ is a topological semigroup and $\equiv$ is a finite index congruence on $S$. The quotient topology on $S/\equiv$ will be the discrete topology if and only if each congruence class is open. In this case, as the complement of a congruence class is a finite union of congruence classes, each class is also closed and hence clopen. Conversely, if $S$ is a compact semigroup, any congruence whose congruence classes are open must have finite index because the congruence classes form an open covering. In other words, continuous finite quotients (with the discrete topology) of a compact semigroup are in bijection with congruences whose classes are both open and closed. This leads to the following definition.

**Definition 3.1.39 (Clopen congruence).** *A congruence on a compact semigroup $S$ is said to be clopen if each of its congruence classes are open.*

Clopen equivalence relations on compact spaces are defined similarly and analogous remarks apply.

**Exercise 3.1.40.** Verify that a clopen equivalence relation on a compact space is closed.

**Exercise 3.1.41.** Let $X$ be a profinite space. Show that the set of clopen equivalence relations on $X$, ordered by reverse inclusion, forms a directed set $\mathscr{F}$ with trivial intersection and that $X = \varprojlim_\mathscr{F} X/\equiv$. State and prove the analogous result for profinite semigroups and clopen congruences.

**Exercise 3.1.42.** Show that a compact semigroup $S$ is profinite if and only if the intersection of all the clopen congruences on $S$ is trivial.

It is well-known that the quotient of a profinite group by a closed normal subgroup is always profinite [298, Prop. 2.2.1]. Unfortunately, there is no analogue of this result for profinite semigroups: there are quotients of a profinite semigroup (i.e., quotients by closed congruences) that are not profinite. Recall that the closed interval $[0, 1]$ is a continuous image of the Cantor set $\{0, 1\}^\omega$ via the map sending a sequence to the element of $[0, 1]$ with the corresponding binary expansion. Make both of these spaces into compact semigroups using the right zero semigroup structure. Now $\{0,1\}^\omega = \varprojlim_{n\in\mathbb{N}}\{0,1\}^n$ where $\{0,1\}^n$ is given the right zero multiplication and the maps $\pi_{n,n-1}\colon \{0,1\}^n \to \{0,1\}^{n-1}$ project to the leftmost $n-1$ positions (the remaining maps of the system are determined by composition) and hence is profinite. This shows that a continuous image of a profinite semigroup need not be profinite.

*Remark 3.1.43.* In fact it is well-known that *every* compact metric space is a continuous quotient of a Cantor set [158], so the above example is just the tip of the iceberg.

In this book, by a *profinite congruence* or a congruence on a profinite semigroup, we mean a closed congruence whose quotient (equipped with the quotient topology) is profinite. It turns out a congruence is profinite if and only if it is an intersection of clopen congruences [147, Prop. 2.6].

**Proposition 3.1.44.** *Let $P$ be a profinite semigroup and let $\equiv$ be a congruence on $P$. Then $\equiv$ is a profinite congruence if and only if it is the intersection of the family $\mathscr{F}$ of clopen congruences containing it. In this case $P/\equiv$ is topologically isomorphic to $\varprojlim_{\mathscr{F}} P/\sim$.*

*Proof.* Suppose first $\equiv$ is profinite. Then the intersection of all clopen congruences on $P/\equiv$ is trivial by Exercise 3.1.41. As surjective maps between compact Hausdorff spaces are always quotient maps, a clopen congruence on $P/\equiv$ is induced by a clopen congruence $\sim$ on $P$ containing $\equiv$ (and conversely). Thus the intersection of $\mathscr{F}$ is $\equiv$. Conversely, suppose the intersection of $\mathscr{F}$ is $\equiv$. Clearly, $\equiv$ is then a closed congruence. Next observe that $\mathscr{F}$ ordered by reverse inclusion is directed. The natural map $\varphi\colon P \to \varprojlim_{\mathscr{F}} P/\sim$ is surjective by Lemma 3.1.27 and has kernel $\bigcap\mathscr{F}$, which is $\equiv$. Because a continuous surjection between compact Hausdorff spaces is automatically a quotient map, it follows $P/\equiv$ is isomorphic to $\varprojlim_{\mathscr{F}} P/\sim$ as a topological semigroup and hence is profinite. $\qquad\square$

From the above proposition, it follows that the collection of profinite congruences on a profinite semigroup $P$ is closed under arbitrary intersection. Hence given a relation $R$ on $P$, there is a smallest profinite congruence on $P$ containing $R$, called the congruence *generated* by $R$ and denoted $\langle R\rangle$. This means one can define a presentation of a profinite semigroup.

**Definition 3.1.45 (Profinite semigroup presentation).** *A profinite semigroup presentation $\langle A \mid R\rangle$ consists of a topological space $A$ and a relation $R$ on $\widehat{A^+}$. The profinite semigroup presented by $\langle A \mid R\rangle$ is the quotient $\widehat{A^+}/\langle R\rangle$ of $\widehat{A^+}$ by the profinite congruence generated by $R$.*

**Exercise 3.1.46.** Let $G$ be a profinite group and $N$ a closed normal subgroup. Show that $N$ is the intersection of the set $\mathscr{N}$ of all open subgroups containing it. Show that $G/N \cong \varprojlim_{\mathscr{N}} G/U$ and hence is profinite.

**Exercise 3.1.47.** Let $\varphi\colon P \to Q$ be a continuous homomorphism of profinite semigroups. Show that $\ker \varphi$ is a profinite congruence and $P/\ker \varphi$, equipped with the quotient topology, is topologically isomorphic to $\varphi(P)$.

**Exercise 3.1.48 (Eilenberg-Schützenberger [86]).** Let $A$ be a finite set and $\equiv$ a finite index congruence on $A^+$. Show that $\equiv$ is finitely generated (that is, there is a finite subset $R \subseteq A^+ \times A^+$ generating $\equiv$ as a congruence). Conclude that any clopen congruence on $\widehat{A^+}$ is finitely generated for $A$ a finite set.

Being profinite turns out to be a purely topological property for compact semigroups. The proof relies on the following classical lemma [135].

**Lemma 3.1.49.** *Let $S$ be a compact semigroup and let $\equiv$ be a clopen equivalence relation on $S$. Then there is a clopen congruence $\sim$ contained in $\equiv$.*

*Proof.* Consider the "syntactic congruence" $\sim$ given by $s \sim t$ if and only if, for all $u, v \in S^I$, $usv \equiv utv$. We claim that $\sim$ is the desired clopen congruence. It is straightforward to verify that $\sim$ is a congruence contained in $\equiv$. To see that it is clopen, let $s \in S$. We verify the $\sim$-equivalence class $[s]$ of $s$ is open. Suppose that $\{t_\alpha\}$ is a net of elements not equivalent to $s$ with $t_\alpha \to t$. For each $\alpha$, there exist $u_\alpha, v_\alpha \in S^I$ such that $u_\alpha t_\alpha v_\alpha \not\equiv u_\alpha s v_\alpha$. By passing to a subnet, we may assume that $u_\alpha \to u$ and $v_\alpha \to v$. Then $u_\alpha s v_\alpha \to usv$ and $u_\alpha t_\alpha v_\alpha \to utv$. Now the $\equiv$-class of $usv$ is a neighborhood $U$ of $usv$, so there exists $\alpha_0$ so that $u_\alpha s v_\alpha \equiv usv$ for $\alpha \geq \alpha_0$. However, $U$ does not contain $u_\alpha t_\alpha v_\alpha$ for $\alpha \geq \alpha_0$ and hence does not contain $utv$. Thus $utv \not\equiv usv$ and so $t \not\sim s$. This shows that $[s]$ is open, completing the proof. $\qquad\square$

**Exercise 3.1.50.** Verify that if $\equiv$ is an equivalence relation on a semigroup $S$, then the largest congruence $\sim$ on $S$ contained in $\equiv$ is given by $s \sim t$ if and only if, for all $u, v \in S^I$, $usv \equiv utv$.

**Theorem 3.1.51 (Numakura [222]).** *A topological semigroup $S$ is profinite if and only its underlying space is profinite.*

*Proof.* Clearly the underlying space of a profinite semigroup is profinite. For the converse, it suffices by Exercise 3.1.41 to show that the clopen congruences are cofinal among the clopen equivalence relations. But this is an immediate consequence of the previous lemma.

**Exercise 3.1.52.** Let $S$ be a compact semigroup and suppose that $x, y \in S$ belong to different connected components of $S$. Prove that there is a finite semigroup $T$ and a continuous homomorphism $\varphi\colon S \to T$ such that $x\varphi \neq y\varphi$.

*Remark 3.1.53.* It is a deep theorem of Nikolov and Segal that every homomorphism between finitely generated profinite groups is automatically continuous [220, 221]. In particular, every homomorphism from a finitely generated profinite group to a finite group is continuous. The same cannot be said for profinite semigroups. Consider $\mathbb{N}^\infty = \mathbb{N} \cup \{\infty\}$ with addition (so $n + \infty = \infty = \infty + n$ all $n \in \mathbb{N}^\infty$) where $\mathbb{N}^\infty$ is given the topology of the one-point compactification of $\mathbb{N}$. It is straightforward to verify that $\mathbb{N}^\infty$ is a topological semigroup. As its underlying space is profinite, it is in fact a profinite semigroup. It is topologically generated by 1 because $\mathbb{N}$ is dense. We can define a homomorphism $\mathbb{N}^\infty \to (\{0,1\}, \cdot)$ by sending all natural numbers to 1 and $\infty$ to 0. This homomorphism is discontinuous because the preimage of 1 is $\mathbb{N}$, which is not closed in $\mathbb{N}^\infty$. Also notice that $\mathbb{N}^\infty$ is a countably infinite profinite semigroup. This should be contrasted with the fact that an infinite profinite group must be uncountable [298].

The next lemma gives a sufficient condition for a profinite space (semigroup) to be an inverse limit of a sequence of spaces.

**Lemma 3.1.54.** *Let $A$ be a profinite space (semigroup) such that $A$ has only finitely many clopen equivalence relations (congruences) of any given index. Then $A$ is the inverse limit of an inverse quotient system $\{A_n\}_{n \in \mathbb{N}}$ of finite spaces (semigroups) indexed by the natural numbers.*

*Proof.* We just have to show that $A$ admits a cofinal sequence of clopen equivalence relations (congruences). Define $\equiv_n$ to be the intersection of all clopen equivalence relations (congruences) of index at most $n$. Because there are only finitely many such equivalence relations (congruences), it follows that $\equiv_n$ is clopen. Clearly, the $\equiv_n$ are cofinal. □

In fact, [298, Cor 1.1.13] shows that an infinite profinite space $X$ is second countable if and only if it is an inverse limit of a sequence of finite discrete spaces. This is equivalent to metrizability for a profinite space [117].

**Corollary 3.1.55.** *Let $P$ be a finitely generated profinite semigroup. Then $P = \varprojlim_{n \in \mathbb{N}} P_n$ where the $P_n$ are certain finite continuous images of $P$.*

*Proof.* Because every semigroup of order $n$ embeds into the full transformation semigroup $T_{n+1}$ of degree $n + 1$, the number of clopen congruences of index $n$ on $P$ is bounded by the number of homomorphisms from $P$ to $T_{n+1}$. If $P$ is generated by an $m$-element set, then $P$ has at most $|T_{n+1}|^m = (n+1)^{(n+1)m}$ continuous homomorphisms into $T_{n+1}$ and hence at most $(n+1)^{(n+1)m}$ clopen congruences of index $n$. Lemma 3.1.54 provides the desired conclusion. □

Because inverse limits commute with products in any category, we have:

**Proposition 3.1.56.** *Let $S = \prod_A S_\alpha$ be a direct product of profinite semigroups and let $\equiv$ be a clopen congruence on $S$. Then there are indices $\alpha_1, \ldots, \alpha_n$ and clopen congruences $\equiv_i$ on $S_{\alpha_i}$, for $i = 1, \ldots, n$, such that the projection $S \to S/\equiv$ factors through the composite of projections*

$$S \to S_{\alpha_1} \times \cdots \times S_{\alpha_n} \to S_{\alpha_1}/\equiv_1 \times \cdots \times S_{\alpha_n}/\equiv_n.$$

*Proof.* Let $\mathscr{F}$ be the set of finite subsets of $A$ ordered by inclusion and set $T_F = \prod_F S_\alpha$, for $F \in \mathscr{F}$. Then the $T_F$ form an inverse quotient system and $S = \varprojlim_{\mathscr{F}} T_F$. Therefore, Lemma 3.1.37 allows us to assume without loss of generality that $A$ is finite. So assume that $S = S_1 \times \cdots \times S_n$. Let us write $S_i = \varprojlim_{D_i} S_{i,\alpha}$ as an inverse quotient system of finite semigroups, $i = 1, \ldots, n$. Let $D = D_1 \times \cdots \times D_n$ with the product order; note that $D$ is directed. For $\alpha = (\alpha_1, \ldots, \alpha_n) \in D$, set $T_\alpha = S_{1,\alpha_1} \times \cdots \times S_{n,\alpha_n}$. The $T_\alpha$ form an inverse quotient system. We claim that $S \cong \varprojlim_D T_\alpha$. There is a natural cone of continuous surjective maps from $S_1 \times \cdots \times S_n$ to the inverse system $T_\alpha$ by taking products of the appropriate projections. Hence there is a continuous surjective homomorphism $\varphi \colon S \to \varprojlim_D T_\alpha$ by Lemma 3.1.27. For injectivity, if $(s_1, \ldots, s_n)$ and $(s'_1, \ldots, s'_n)$ belong to $S_1 \times \cdots \times S_n$, then we can find clopen congruences $\equiv_i$ on $S_i$, for $i = 1, \ldots, n$, such that $s_i \not\equiv_i s'_i$ whenever $s_i \neq s'_i$ and $\equiv_i$ is arbitrary otherwise. The projection to $S_1/\equiv_1 \times \cdots \times S_n/\equiv_n$ then separates these points. We conclude $S \cong \varprojlim_D T_\alpha$. An application of Lemma 3.1.37 to $S = \varprojlim_D T_\alpha$ completes the proof.    $\square$

## 3.2 Reiterman's Theorem

In this section, we prove Reiterman's Theorem [261], which states that pseudovarieties of semigroups can be defined by pseudoidentities, that is, by formal equalities between elements of a free profinite semigroup. It was via pseudoidentities that profinite semigroups first became of interest to researchers in Finite Semigroup Theory.

### 3.2.1 pro-V semigroups

The all important notion of a pro-**V** semigroup bridges the theory of profinite semigroups and the theory of pseudovarieties.

**Definition 3.2.1 (pro-V semigroup).** *Let* **V** *be a pseudovariety of semigroups. Then a profinite semigroup $S$ is called pro-***V*** if $S = \varprojlim S_\alpha$ with the $S_\alpha \in$ **V***.

The next lemma shows that there is no dependence on the inverse system.

**Lemma 3.2.2.** *Let $S$ be a profinite semigroup and* **V** *a pseudovariety. Then the following are equivalent:*

1. *$S$ is pro-***V***;*
2. *$S$ is an inverse limit of an inverse quotient system of elements of* **V***;*
3. *All finite continuous homomorphic images of $S$ belong to* **V***.*

*Proof.* For 1 implies 2, let $S = \varprojlim S_\alpha$ with $S_\alpha \in \mathbf{V}$. Let $\pi_\alpha \colon S \to S_\alpha$ be the projection. One checks directly from the universal property that $S = \varprojlim S\pi_\alpha$. Because $\mathbf{V}$ is closed under taking subsemigroups, each $S\pi_\alpha \in \mathbf{V}$. Clearly, $\{S\pi_\alpha\}$ — with the restricted mappings — is an inverse quotient system.

The implication 2 implies 3 follows directly from Lemma 3.1.37 and closure of $\mathbf{V}$ under homomorphic images. Clearly, 3 implies 1 as $S$ is an inverse limit of its finite continuous homomorphic images (Exercise 3.1.41). □

Next we show that the class of pro-$\mathbf{V}$ semigroups enjoys analogous closure properties to those of $\mathbf{V}$.

**Proposition 3.2.3.** *Let $\mathbf{V}$ be a pseudovariety of semigroups. Then the class of pro-$\mathbf{V}$ semigroups is closed under forming arbitrary products, taking closed subsemigroups and continuous profinite images.*

*Proof.* Closure under products is immediate from Proposition 3.1.56. If $S$ is a pro-$\mathbf{V}$ semigroup then $S = \varprojlim_D S_\alpha$ with the $S_\alpha \in \mathbf{V}$. Let $\pi_\alpha \colon S \to S_\alpha$ be the projection. If $T$ is a closed subsemigroup of $S$, then $T = \varprojlim_D T\pi_\alpha$ by Exercise 3.1.34 and so $T$ is pro-$\mathbf{V}$, as $T\pi_\alpha \leq S_\alpha \in \mathbf{V}$. Finally, suppose that $\varphi \colon S \twoheadrightarrow T$ is a continuous surjective morphism of profinite semigroups and that $S$ is pro-$\mathbf{V}$. By Lemma 3.2.2, each finite continuous image of $S$ belongs to $\mathbf{V}$. As each finite continuous image of $T$ is also a finite continuous image of $S$, we conclude that $T$ is pro-$\mathbf{V}$ by another application of Lemma 3.2.2. □

The following lemma collects a few more equivalent conditions to being pro-$\mathbf{V}$.

**Lemma 3.2.4.** *Let $S$ be a profinite semigroup and $\mathbf{V}$ a pseudovariety. Then the following are equivalent:*

1. *$S$ is pro-$\mathbf{V}$;*
2. *Points of $S$ can be separated by continuous homomorphisms into members of $\mathbf{V}$;*
3. *$S$ is topologically isomorphic to a closed subsemigroup of a direct product of semigroups from $\mathbf{V}$;*
4. *$S$ is a continuous quotient of a closed subsemigroup of a direct product of semigroups from $\mathbf{V}$.*

*Proof.* One sees that 1 implies 2 by considering the projections from $S$ to the inverse system. The implication 2 implies 3 follows by considering the product of all continuous homomorphisms of $S$ into members of $\mathbf{V}$ and from the compactness of $S$. The implication 3 implies 4 is trivial; the implication 4 implies 1 follows directly from Proposition 3.2.3. □

### 3.2.2 Pseudoidentities and Reiterman's Theorem

We now review the notion of a pseudoidentity. The reader is referred to the book [7] for further details and applications.

**Definition 3.2.5 (Pseudoidentity).** *Let $A$ be a finite set (called an alphabet in this context). A pseudoidentity $u = v$ in variables $A$ is a formal equality of two elements $u$ and $v$ of $\widehat{A^+}$.*

Sometimes we write $(A, u = v)$ to emphasize the alphabet $A$.

We say that a profinite semigroup $S$ *satisfies* $u = v$, written $S \models u = v$, if, for every map $\varphi \colon A \to S$, the induced continuous homomorphism $\varphi \colon \widehat{A^+} \to S$ satisfies $u\varphi = v\varphi$. Let $\mathscr{E}$ be a set of pseudoidentities. We say a profinite semigroup $S$ *satisfies* $\mathscr{E}$, written $S \models \mathscr{E}$, if $S$ satisfies every member of $\mathscr{E}$. Set

$$\llbracket \mathscr{E} \rrbracket = \{ S \in \mathbf{FSgp} \mid S \models \mathscr{E} \}.$$

The following is the easy half of Reiterman's Theorem.

**Proposition 3.2.6.** *If $\mathscr{E}$ is a set of pseudoidentities, then $\llbracket \mathscr{E} \rrbracket$ is a pseudovariety.*

*Proof.* Clearly, $\llbracket \mathscr{E} \rrbracket$ is closed under products and subsemigroups. It remains to prove a quotient of a semigroup satisfying $\mathscr{E}$ also satisfies $\mathscr{E}$. So suppose $u = v$ is a pseudoidentity of $\mathscr{E}$ in variables $A$ and that we have a diagram

where $R \models \mathscr{E}$. Then by choosing a section of $\tau$, we can lift $\varphi$ to a map $\gamma \colon A \to T$ making the diagram

commute. But as $R \models u = v$, it follows that $u\gamma = v\gamma$, and thus we have $u\varphi = u\gamma\tau = v\gamma\tau = v\varphi$. This establishes the proposition. □

The converse is the difficult half of Reiterman's Theorem [261], which is our next goal. If $\mathbf{V}$ is a pseudovariety and $A$ is a topological space, then $\widehat{F_{\mathbf{V}}}(A)$ denotes the free pro-$\mathbf{V}$ semigroup generated by $A$. It is a pro-$\mathbf{V}$ semigroup defined by the universal property that there is a continuous map $\iota \colon A \to \widehat{F_{\mathbf{V}}}(A)$ such that any continuous map $\sigma \colon A \to S$ into a pro-$\mathbf{V}$ semigroup $S$ extends uniquely to a continuous homomorphism $\widehat{\sigma} \colon \widehat{F_{\mathbf{V}}}(A) \to S$ such that

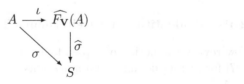

commutes (actually it is enough to consider maps to finite semigroups $S$ in $\mathbf{V}$). So for $\mathbf{V}$ the pseudovariety of all finite semigroups, one has $\widehat{F_\mathbf{V}}(A) = \widehat{A^+}$. Let us construct the free pro-$\mathbf{V}$ semigroup for the case of a finite set $A$.

**Theorem 3.2.7.** *Let $\mathbf{V}$ be a pseudovariety of semigroups and let $A$ be a finite set. Let $\mathscr{F}_\mathbf{V}$ be the collection of all congruences $\equiv$ on $A^+$ such that $A^+/\equiv \, \in \mathbf{V}$, ordered by reverse inclusion. Then $\widehat{F_\mathbf{V}}(A) = \varprojlim_{\mathscr{F}_\mathbf{V}} A^+/\equiv$, where we take $\iota\colon A \to \varprojlim_{\mathscr{F}_\mathbf{V}} A^+/\equiv$ to be the natural map.*

*Proof.* First note that $T = \varprojlim_{\mathscr{F}_\mathbf{V}} A^+/\equiv$ is topologically $A$-generated by Corollary 3.1.36. So uniqueness of the extension is automatic. Let $S = \varprojlim S_\alpha$ be a pro-$\mathbf{V}$ semigroup, with the $S_\alpha \in \mathbf{V}$, and let $\pi_\alpha\colon S \to S_\alpha$ be the projection. Let $\varphi\colon A \to S$ be a map. Suppose for the moment that each map $\varphi\pi_\alpha\colon A \to S_\alpha$ induces a map $\varphi_\alpha\colon T \to S_\alpha$ such that $\iota\varphi_\alpha = \varphi\pi_\alpha$. By uniqueness of the extensions, one easily verifies that the $\varphi_\alpha$ form a cone of maps to the inverse system and so there is an induced map $\widehat{\varphi}\colon T \to S$ with $\iota\widehat{\varphi} = \varphi$. Therefore, without loss of generality, we may assume that $S$ is a finite semigroup in $\mathbf{V}$. Let $\psi\colon A^+ \to S$ be the induced map. Then as $A^+\psi \in \mathbf{V}$, the kernel of $\psi$ belongs to $\mathscr{F}_\mathbf{V}$. Thus the composite $\varprojlim_{\mathscr{F}_\mathbf{V}} A^+/\equiv \,\twoheadrightarrow A^+/\ker\psi \hookrightarrow S$, where the first map is the projection and the second is the natural map induced by $\psi$, extends $\varphi$. $\qquad\square$

The following exercise is for those who did Exercise 3.1.28 and are familiar with $\mathbf{V}$-recognizable sets [7,85,224]. This will be expanded upon in Section 8.4.

**Exercise 3.2.8 (Almeida [7]).** Let $A$ be a finite set and let $\mathbf{V}(A^+)$ be the Boolean algebra of $\mathbf{V}$-recognizable subsets of $A$; so $L \in \mathbf{V}(A^+)$ if and only if there is a homomorphism $\varphi\colon A^+ \to S$ with $S \in \mathbf{V}$ and $L = L\varphi\varphi^{-1}$. Prove that $\mathrm{Spec}\,\mathbf{V}(A^+) \cong \widehat{F_\mathbf{V}}(A)$. Hint: Associate to each element $x \in \widehat{F_\mathbf{V}}(A)$ the map $\chi_x\colon \mathbf{V}(A^+) \to \{0,1\}$ in $\mathrm{Spec}\,\mathbf{V}(A^+)$ given by

$$\chi_x(L) = \begin{cases} 1 & x \in \overline{L} \\ 0 & x \notin \overline{L}. \end{cases}$$

Conversely, given a map $\chi\colon \mathbf{V}(A^+) \to \{0,1\}$ in $\mathrm{Spec}\,\mathbf{V}(A^+)$, show that the collection $\mathscr{F} = \{\overline{L} \mid L \in \mathbf{V}(A^+),\ \chi(L) = 1\}$ satisfies the finite intersection condition, and $|\bigcap\mathscr{F}| = 1$. It may be useful to use that exactly one of $\chi(L)$ and $\chi(A^+ \setminus L)$ is 1 for any $L \in \mathbf{V}(A^+)$.

Explicit descriptions of free pro-$\mathbf{V}$ semigroups have been obtained for a number of pseudovarieties of semigroups, mostly by the Portuguese School. The interested reader should consult the following references [4, 7, 24–26, 30, 73–75]. Let us just state one result to give the flavor.

**Theorem 3.2.9 (Almeida [4, 7]).** *Let $A$ be a finite set with a fixed order. Then every element $u \in \widehat{F_\mathbf{J}}(A)$ has a unique factorization of the form $u = u_0 v_1^\omega u_1 v_2^\omega \cdots u_{n-1} v_n^\omega u_n$ where:*

1. $u_0, \ldots, u_n \in A^*$;
2. $v_1, \ldots, v_n \in A^+$ are linear words (words having no repeated letters) whose letters appear in increasing order;
3. If $u_i = 1$, then $v_i$ and $v_{i+1}$ have incomparable support;
4. The last letter of $u_{i-1}$ and the first letter of $u_i$ do not appear in $v_i$.

**Exercise 3.2.10.** Let $\mathbf{N}$ be the pseudovariety of nilpotent semigroups. Show that $\widehat{F_{\mathbf{N}}}(A) = A^+ \cup \{0\}$ endowed with the topology of the one-point compactification of $A^+$.

It is now time to prove that pseudovarieties are precisely classes of finite semigroups defined by pseudoidentities.

**Theorem 3.2.11 (Reiterman [261]).** *Pseudovarieties of finite semigroups are precisely the classes of the form $[\![\mathscr{E}]\!]$ where $\mathscr{E}$ is a set of pseudoidentities.*

*Proof.* Proposition 3.2.6 shows that classes of the form $[\![\mathscr{E}]\!]$ are pseudovarieties. Conversely, let $\mathbf{V}$ be a pseudovariety. Choose a fixed countable set $X$ and let $\mathscr{E}$ be the set of all pseudoidentities in a finite subset of $X$ satisfied by every member of $\mathbf{V}$. Set $\mathbf{W} = [\![\mathscr{E}]\!]$. Clearly $\mathbf{V} \leq \mathbf{W}$. We prove the other inclusion. Let $S \in \mathbf{W}$. Then $S$ can be generated by a finite subset $A$ of $X$. As $S$ is a continuous image of $\widehat{F_{\mathbf{W}}}(A)$, we just need to show, by Lemma 3.2.2, that $\widehat{F_{\mathbf{W}}}(A)$ is pro-$\mathbf{V}$. To do this, it suffices by Lemma 3.2.4 to show that distinct points of $\widehat{F_{\mathbf{W}}}(A)$ can be separated by continuous homomorphisms to elements of $\mathbf{V}$.

Let $\eta\colon \widehat{A^+} \to \widehat{F_{\mathbf{W}}}(A)$ be the canonical projection and suppose $u\eta \neq v\eta$. We want to separate $u\eta$ and $v\eta$ by a continuous homomorphism to an element of $\mathbf{V}$. Because $u\eta$ and $v\eta$ can be separated in a finite continuous image of $\widehat{F_{\mathbf{W}}}(A)$, which necessarily lies in $\mathbf{W}$ (by Lemma 3.2.2) and hence satisfies $\mathscr{E}$ by definition of $\mathbf{W}$, it follows $u = v \notin \mathscr{E}$. By definition of $\mathscr{E}$, there is a semigroup $S \in \mathbf{V}$ and a map $\sigma\colon A \to S$ so that the induced homomorphism $\psi\colon \widehat{A^+} \to S$ does not satisfy $u\psi = v\psi$. Because $S \in \mathbf{W}$, the map $\psi$ factors through $\eta\colon \widehat{A^+} \to \widehat{F_{\mathbf{W}}}(A)$ and so $u\eta, v\eta$ can be separated by a continuous homomorphism to an element of $\mathbf{V}$. We conclude that $\widehat{F_{\mathbf{W}}}(A)$ has enough continuous homomorphisms to members of $\mathbf{V}$ to separate points, completing the proof. ☐

If $\mathscr{E}$ is a set of pseudoidentities such that $\mathbf{V} = [\![\mathscr{E}]\!]$, then $\mathscr{E}$ is called a *basis of pseudoidentities* for $\mathbf{V}$.

For instance, the sequence $\{x^{n!}\}$ from $\widehat{\{x\}^+}$, when evaluated in a finite semigroup, converges to the unique idempotent power of the image of $x$. Its limit $x^{\omega} = \lim_{n \to \infty} x^{n!}$ is the unique idempotent in $\widehat{\{x\}^+}$. In general, one writes $x^{\omega+k}$ for $x^{\omega} x^k$ and $x^{\omega-1}$ for the inverse of $x^{\omega+1}$ in the maximal subgroup of $\overline{\langle x \rangle}$. One can verify that $x^{\omega-1} = \lim_{n \to \infty} x^{n!-1}$.

It is easy to see that $\mathbf{A} = [\![x^{\omega+1} = x^{\omega}]\!]$ and $\mathbf{G} = [\![x^{\omega}y = y = yx^{\omega}]\!]$. The pseudovariety $\mathbf{N}$ of nilpotent semigroups is given by $[\![x^{\omega}y = x^{\omega} = yx^{\omega}]\!]$.

Although, strictly speaking, we cannot use constants like 0 and 1 in semigroup pseudoidentities, it is often convenient to do so as an abbreviation. For instance, the pseudoidentities defining groups can be abbreviated to the single pseudoidentity $x^\omega = 1$, and the pseudoidentities defining nilpotent semigroups can be abbreviated to $x^\omega = 0$. It turns out that the sequence $\{x^{p^{n!}}\}$ converges in every finite semigroup to the generator of the complement of the $p$-primary component of the maximal subgroup of $\langle x \rangle$ [10, 25] (see also Proposition 7.1.16). Setting $x^{p^\omega} = \lim x^{p^{n!}}$, one has that the pseudovariety of $p$-groups is defined by the pseudoidentity $x^{p^\omega} = 1$.

**Exercise 3.2.12.** Verify that $\{x^{p^{n!}}\}$ converges in every finite semigroup.

The reader should perform the straightforward task of verifying that the pseudovariety $\mathbf{R}$ of $\mathscr{R}$-trivial semigroups is $[\![(xy)^\omega x = (xy)^\omega]\!]$. The pseudovariety $\mathbf{D}$ of semigroups whose idempotents are right zeroes is defined by the pseudoidentity $xy^\omega = y^\omega$. As another example, $\mathbf{ESl} = [\![x^\omega y^\omega = y^\omega x^\omega]\!]$.

Although pseudoidentities were introduced by Reiterman [261], it was principally J. Almeida who actively pushed forward the idea that pseudoidentities and profinite semigroups were an essential ingredient in Finite Semigroup Theory [7].

We remark that Reiterman's Theorem applies equally well to pro-$\mathbf{V}$ semigroups. This will not be the case for pseudovarieties of relational morphisms.

**Proposition 3.2.13.** *Let $\mathbf{V} = [\![\mathscr{E}]\!]$ be a pseudovariety of semigroups. Then a profinite semigroup $S$ is pro-$\mathbf{V}$ if and only if $S \models \mathscr{E}$.*

*Proof.* Clearly, if $S \models \mathscr{E}$, then so do all its finite images and so $S$ is pro-$\mathbf{V}$ by Lemma 3.2.2. Conversely, if $S$ is pro-$\mathbf{V}$, $(A, u = v) \in \mathscr{E}$ and $\sigma\colon A \to S$ is a substitution, then for any continuous homomorphism $\varphi\colon S \to T$ with $T \in \mathbf{V}$, one has $u\sigma\varphi = v\sigma\varphi$ As such homomorphisms separate points of $S$ by Lemma 3.2.4, we conclude $u\sigma = v\sigma$ and so $S \models \mathscr{E}$.    □

## 3.3 On pro-V Relational Morphisms

Having developed the framework of pseudovarieties of relational morphisms, the natural next step is to try and define pseudoidentities in this context and to prove an analogue of Reiterman's Theorem [261]. To do this, we are forced to consider relational morphisms of profinite semigroups and to define what is meant by a pro-V relational morphism of profinite semigroups when V is a pseudovariety of relational morphisms. We also shall need to define generators for a relational morphism.

It is easy to see that a homomorphism $\varphi\colon S \to T$ of profinite semigroups is continuous if and only if its graph $\#\varphi$ is closed in $S \times T$. This leads to the following definition.

**Definition 3.3.1 (Relational morphism of profinite semigroups).** *If S and T are profinite semigroups, then by a relational morphism $\varphi\colon S \to T$ of profinite semigroups, we mean a relational morphism $\varphi\colon S \to T$ of semigroups such that the graph $\#\varphi$ is a closed subsemigroup of the profinite semigroup $S \times T$. A division of profinite semigroups is a relational morphism of profinite semigroups that is a division in the usual sense.*

In this book, all relational morphisms between profinite semigroups shall be assumed to be as per Definition 3.3.1 and all homomorphisms shall be assumed continuous.

**Exercise 3.3.2.** Verify that a homomorphism $\varphi\colon S \to T$ between profinite semigroups is continuous if and only if its graph $\#\varphi$ is closed in $S \times T$.

**Exercise 3.3.3.** Show that the composition of relational morphisms of profinite semigroups is again a relational morphism of profinite semigroups.

Let us discuss the canonical factorization in this context.

**Proposition 3.3.4.** *A relation $\varphi\colon S \to T$ between profinite semigroups is a relational morphism if and only if there exists a profinite semigroup $R$ and a factorization $\varphi = \eta^{-1}\tau$ where $\eta\colon R \twoheadrightarrow S$, $\tau\colon R \to T$ are continuous homomorphisms, with $\eta$ onto, as per*

$$
\begin{array}{ccc}
R & \xrightarrow{\ \tau\ } & T \\
{\scriptstyle\eta}\Big\downarrow & \nearrow{\scriptstyle\varphi} & \\
S & &
\end{array}
\tag{3.1}
$$

*Proof.* The only if is clear: just choose the canonical factorization (1.28). Suppose such a factorization exists. Then because $R$ and $S$ are Hausdorff, the graph $\#\eta^{-1}$ is a closed subspace of $S \times R$. Then one easily verifies $\#\eta^{-1}(1_S \times \tau) = \#\varphi$. Because $1_S \times \tau$ is continuous, $S \times R$ is compact and $S \times T$ is Hausdorff, it follows that $\#\varphi$ is closed.    □

We then have the following generalization of the fact that continuous maps send compact sets to compact sets.

**Corollary 3.3.5.** *Let $\varphi\colon S \to T$ be a relational morphism of profinite semigroups. Then for $A \subseteq S$ closed, $A\varphi$ is closed and for $B \subseteq T$ closed, $B\varphi^{-1}$ is closed.*

*Proof.* Consider a factorization of $\varphi$ as per (3.1). Then if $A \subseteq S$ is closed, so is $A\eta^{-1}$ by continuity and hence $A\varphi = A\eta^{-1}\tau$ by compactness and continuity of $\tau$. Similarly, $B\varphi^{-1}$ is closed when $B$ is closed.    □

Suppose that $\varphi\colon S \to T$ is a relational morphism of semigroups, $\equiv$ is a congruence on $S$ and $\sim$ is a congruence on $T$. Let $\eta\colon S \twoheadrightarrow S/\equiv$, $\rho\colon T \twoheadrightarrow T/\sim$ be the projections. Then we call $\eta^{-1}\varphi\rho\colon S/\equiv \to T/\sim$ the relational morphism induced by $\equiv$ and $\sim$. If $\sim$ is the trivial congruence, then we just say that $\eta^{-1}\varphi$ is induced by $\equiv$.

**Definition 3.3.6 (pro-V relational morphism).** *If* V *is a pseudovariety of relational morphisms, a relational morphism* $\varphi\colon P \to Q$ *is called pro-*V *if*

$$P = \varprojlim P/\!\equiv_i \tag{3.2}$$

*where the* $\equiv_i$ *are clopen congruences so that, for each* $i$, *there is a clopen congruence* $\sim_i$ *on* $Q$ *such that the induced relational morphism* $\varphi_i\colon P/\!\equiv_i \to Q/\!\sim_i$ *diagrammed as per*

$$
\begin{array}{ccc}
P & \xrightarrow{\ \varphi\ } & Q \\
{\scriptstyle p_i}\downarrow & & \downarrow{\scriptstyle q_i} \\
P/\!\equiv_i & \xrightarrow[\varphi_i]{} & Q/\!\sim_i
\end{array}
$$

*is in* V. *In more detail, if* $p_i\colon P \to P/\!\equiv_i$, $q_i\colon Q \to Q/\!\sim_i$ *are the quotient morphisms, then* $\varphi_i = p_i^{-1}\varphi q_i \in$ V.

The next proposition implies that the choice of representation of $P$ as an inverse limit in (3.2) is not essential and so the notion of a pro-V relational morphism is well defined.

**Proposition 3.3.7.** *A relational morphism* $\varphi\colon P \to Q$ *of profinite semigroups is pro-*V *if and only if, for every clopen congruence* $\equiv$ *on* $P$, *there exists a clopen congruence* $\sim$ *on* $Q$ *such that the induced relational morphism* $\varphi'\colon P/\!\equiv\ \to Q/\!\sim$ *is in* V.

*Proof.* Sufficiency is clear as $P$ is the inverse limit of its finite images. We prove necessity. Let the $\equiv_i$ be as in (3.2) and let $\equiv$ be a clopen congruence on $P$. Lemma 3.1.37 guarantees that $\equiv_i\ \subseteq\ \equiv$ for some $i$. Then there exists a clopen congruence $\sim_i$ on $Q$ such that the induced relational morphism $\varphi_i\colon P/\!\equiv_i \to Q/\!\sim_i$ is in V. Let $\pi\colon P/\!\equiv_i \twoheadrightarrow P/\!\equiv$ be the projection. Then from the diagram

$$
\begin{array}{ccc}
P & \xrightarrow{\ \varphi\ } & Q \\
{\scriptstyle p_i}\downarrow & & \downarrow{\scriptstyle q_i} \\
P/\!\equiv_i & \xrightarrow{\ \varphi_i\ } & Q/\!\sim_i \\
{\scriptstyle \pi}\downarrow & \nearrow{\scriptstyle \psi} & \\
P/\!\equiv & &
\end{array}
$$

we see that the induced relational morphism $\psi\colon P/\!\equiv\ \to Q/\!\sim_i$ is $\pi^{-1}\varphi_i$, which belongs to V as $\pi^{-1}\varphi_i \prec_s \varphi_i$.  $\square$

**Exercise 3.3.8.** Prove that if V is a positive pseudovariety of relational morphisms, then the identity map $1_P$ is pro-V for any profinite semigroup $P$.

We now verify that, for a relational morphism $\varphi\colon S \to T$ of finite semi-groups, being pro-V is equivalent to belonging to V.

**Proposition 3.3.9.** *Let* $\varphi\colon S \to T$ *be a relational morphism of finite semigroups and* V *a pseudovariety of relational morphisms. Then* $\varphi$ *is pro-V if and only if* $\varphi \in$ V.

*Proof.* If $\varphi$ is pro-V, then by considering the trivial congruence on $S$ we see there is a surjective homomorphism $f\colon T \to T'$ such that $\varphi f$ is in V. But $\varphi \prec_s \varphi f$, so $\varphi \in$ V.

Conversely, if $\varphi$ in V and $f\colon S \to S'$ is a surjective homomorphism, then $f^{-1}\varphi \prec_s \varphi$, whence $f^{-1}\varphi \in$ V. We may deduce $\varphi$ is pro-V.    $\square$

Of a similar vein we have:

**Proposition 3.3.10.** *Suppose* $\varphi\colon P \to T$ *is a relational morphism of profinite semigroups with* $T$ *finite. Then* $\varphi$ *is pro-V if and only if* $P = \varprojlim P/{\equiv_i}$ *with the* $\equiv_i$ *clopen and with the induced relational morphisms* $\varphi_i\colon P/{\equiv_i} \to T$ *in* V, *all* $i$, *if and only if, for each clopen congruence* $\equiv$ *on* $P$, *the induced relational morphism* $\varphi'\colon P/{\equiv} \to T$ *is in* V.

*Proof.* We handle only the first equivalence, the second being similar. Sufficiency follows from the definition of a pro-V relational morphism: just choose the trivial congruence on $T$ each time.

For necessity, assume $P = \varprojlim P_i$ such that, for each $i$, there is a congruence $\sim_i$ on $T$ such that the induced relational morphism $\psi_i\colon P/{\equiv_i} \to T/{\sim_i}$ belongs to V. Then we have the diagram

$$
\begin{array}{ccc}
P & \xrightarrow{\ \varphi\ } & T \\
\Big\downarrow{\scriptstyle\varphi_i\nearrow} & & \Big\downarrow{\scriptstyle\pi_i} \\
P/{\equiv_i} & \xrightarrow[\ \psi_i\ ]{} & T/{\sim_i}
\end{array}
$$

where $\varphi_i\pi_i = \psi_i$. Thus $\varphi_i \prec_s \psi_i \in$ V and so $\varphi_i \in$ V, as required.    $\square$

**Proposition 3.3.11.** *Let* $\varphi\colon S \to Q$ *be a relational morphism of profinite semigroups with* $S$ *finite and let* V *be a pseudovariety of relational morphisms. Then* $\varphi$ *is pro-V if there exists a clopen congruence* $\sim$ *on* $Q$ *such that the induced relational morphism* $\psi\colon S \to Q/{\sim}$ *is in* V.

*Proof.* Necessity is clear from consideration of the trivial congruence on $S$. For sufficiency let $\equiv$ be a congruence on $S$ and let $\pi\colon S \to S/{\equiv}$ be the projection. Then we have

$$
\begin{array}{ccc}
S & \xrightarrow{\ \varphi\ } & Q \\
\Big\downarrow{\scriptstyle\pi\ \ \searrow\psi} & & \Big\downarrow{} \\
S/{\equiv} & \xrightarrow[\ \tau\ ]{} & Q/{\sim}
\end{array}
$$

where $\tau$ is the induced relational morphism. Because $\tau = \pi^{-1}\psi$, it follows $\tau \prec_s \psi$ and hence $\tau \in V$. We conclude $\varphi$ is pro-V. $\qquad\square$

The following corollary furnishes a generalization of the fact that a semigroup $S$ is pro-**V** if and only if all its finite images are in **V** (cf. Lemma 3.2.2).

**Corollary 3.3.12.** *Suppose* $\varphi\colon P \to Q$ *is a relational morphism of profinite semigroups and* V *is a pseudovariety of relational morphisms. Then* $\varphi$ *is pro-*V *if and only if for all clopen congruences* $\equiv$ *on* $P$, *the induced relational morphism* $\varphi_\equiv\colon P/\equiv\ \to Q$ *diagrammed*

$$
\begin{array}{ccc}
P & \xrightarrow{\ \varphi\ } & Q \\
\Big\downarrow & \nearrow{\scriptstyle\varphi_\equiv} & \\
P/\equiv & &
\end{array}
$$

*is pro-*V. *In particular, if* $Q$ *is finite, then* $\varphi$ *is pro-*V *if and only if each such* $\varphi_\equiv$ *is in* V.

*Proof.* The key point here is that if $\sim$ is a clopen congruence on $Q$, then the relational morphisms $P/\equiv\ \to Q/\sim$ induced by $\varphi$ and $\varphi_\equiv$ are one and the same. So suppose $\varphi$ is pro-V and let $\equiv$ be a clopen congruence on $P$. Then, by Proposition 3.3.7, there is a clopen congruence $\sim$ on $Q$ such that the induced relational morphism $\psi\colon P/\equiv\ \to Q/\sim$ is in V. Thus $\varphi_\equiv$ satisfies the hypothesis of Proposition 3.3.11 and hence is pro-V. Conversely, if each $\varphi_\equiv$ is pro-V, then given a clopen congruence $\equiv$ on $P$, we can find a clopen congruence $\sim$ on $Q$ such that the induced relational morphism $\psi\colon P/\equiv\ \to Q/\sim$ is in V. But this is the very definition that $\varphi$ is pro-V. The final statement is immediate from Proposition 3.3.9. $\qquad\square$

Our next goal is to show that the collection of pro-V relational morphisms enjoys many of the same properties as V itself. We begin with products.

**Proposition 3.3.13.** *Suppose* $\{\varphi_\alpha\colon P_\alpha \to Q_\alpha\}$ *is a collection of pro-*V *relational morphisms. Then* $\prod \varphi_\alpha$ *is pro-*V.

*Proof.* It follows from Proposition 3.1.56 that $\prod P_\alpha$ is the inverse limit of products of the form $S_1 \times \cdots \times S_n$ where $S_i$ is a finite continuous image of some $P_{\alpha_i}$ ($\alpha_i \neq \alpha_j$, when $i \neq j$). Consider the clopen congruence $\equiv$ associated to the projection from $\prod P_\alpha$ onto such a product $S_1 \times \cdots \times S_n$. By Proposition 3.3.7 (as each $\varphi_\alpha$ is pro-V), for each $i$, there is a clopen congruence $\sim_i$ on $Q_{\alpha_i}$ such that the induced relational morphism $\varphi_i\colon S_i \to Q_{\alpha_i}/\sim_i$ is in V. Let $\sim$ be the clopen congruence on $\prod Q_\alpha$ associated to the projection to $\prod_{i=1}^{n} Q_{\alpha_i}/\sim_i$. Then $\varphi_1 \times \cdots \times \varphi_n \in V$ and is easily checked to be the induced relational morphism $\left(\prod P_\alpha\right)/\equiv\ \to \left(\prod Q_\alpha\right)/\sim$. Indeed, if

$$
\psi\colon S_1 \times \cdots \times S_n \longrightarrow Q_{\alpha_1}/\sim_1 \times \cdots \times Q_{\alpha_n}/\sim_n
$$

is the induced relational morphism $(\prod P_\alpha)/\equiv \to (\prod Q_\alpha)/\sim$, then

$$(\bar{q}_1, \ldots, \bar{q}_n) \in (s_1, \cdots, s_n)\psi$$

if and only if there exists $(p_\alpha) \in \prod P_\alpha$ such that $p_{\alpha_i}$ projects to $s_i$ in $S_i$, for $i = 1, \ldots, n$, and there exists $(q_\alpha) \in (p_\alpha)(\prod \varphi_\alpha)$ such that $q_{\alpha_i}$ projects to $\bar{q}_i$ in $Q_{\alpha_i}/\sim_i$, for $i = 1, \ldots, n$. But this happens if and only if $\bar{q}_i \in s_i\varphi_i$, for $i = 1, \ldots, n$, that is, if and only if

$$(\bar{q}_1, \ldots, \bar{q}_n) \in (s_1, \ldots, s_n)(\varphi_1 \times \cdots \times \varphi_n)$$

as was required

This completes the proof that $\prod \varphi_\alpha$ is pro-V.    $\square$

We break division up into two cases: inclusion and inverse surmorphism.

**Proposition 3.3.14.** *Suppose* $\varphi f \subseteq_s \iota\psi$ *where* $\varphi\colon P \to Q$ *is a relational morphism,* $f\colon Q \to Q'$ *is a homomorphism,* $\iota\colon P \to P'$ *is an inclusion, and* $\psi\colon P' \to Q'$ *is pro-*V. *Then* $\varphi$ *is pro-*V.

*Proof.* Suppose $P' = \varprojlim P'/\equiv_i$ where the $\equiv_i$ are clopen. Then, abusing notation, $P = \varprojlim P/\equiv_i$. Fix $i$. Let $\sim_i$ be a clopen congruence on $Q'$ such that the induced relational morphism $\psi_i\colon P'/\equiv_i \to Q'/\sim_i$ is in V. Let $\sim_i'$ be the clopen congruence associated to the composite homomorphism $Q \xrightarrow{f} Q' \to Q'/\sim_i$ and let $\varphi_i\colon P/\equiv_i \to Q/\sim_i'$ be the induced relational morphism. Then we have the following subcommutative diagram.

$$\begin{array}{ccc} P/\equiv_i & \hookrightarrow & P'/\equiv_i \\ \varphi_i \downarrow & \subseteq & \downarrow \psi_i \\ Q/\sim_i' & \hookrightarrow & Q'/\sim_i \end{array} \qquad (3.3)$$

where the horizontal arrows are the inclusions. Therefore, $\varphi_i \prec \psi_i$ and hence $\varphi_i \in$ V. As $i$ was arbitrary, we may conclude that $\varphi$ is pro-V. Note if $f$ is surjective, then the bottom map in (3.3) is surjective, as well.    $\square$

**Proposition 3.3.15.** *Suppose* $\varphi\colon P \to Q$ *is pro-*V *and* $\psi\colon P \to P'$ *is a continuous surjective homomorphism. Then* $\psi^{-1}\varphi$ *is pro-*V.

*Proof.* Let $\varphi' = \psi^{-1}\varphi$. Suppose $\equiv'$ is a clopen congruence on $P'$. Let $\equiv$ be the clopen congruence on $P$ associated to the continuous surjective homomorphism

$$P \xrightarrow{\psi} P' \to P'/\equiv'.$$

Then, by Proposition 3.3.7, there is a clopen congruence $\sim$ on $Q$ such that the induced relational morphism $\psi'\colon P/\equiv \to Q/\sim$ is in V. The situation is diagrammed as follows:

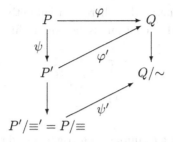

But $\psi'$ is then the induced relational morphism $P'/\equiv' \to Q/\sim$. Hence $\varphi'$ is pro-V, as required.                                                                  □

Putting together Propositions 3.3.13, 3.3.14 and 3.3.15, we obtain the following.

**Corollary 3.3.16.** *The class of pro-V relational morphisms is closed under arbitrary products and (profinite) divisions. If V is a positive pseudovariety, then the class of pro-V relational morphisms includes all identity maps of profinite semigroups.*

*Proof.* We prove the non-obvious statements. Suppose $(d, f)\colon \varphi \to \psi$ is a division with $\psi$ pro-V; so $\varphi f \subseteq_s d\psi$. Let $d = g^{-1}\iota$ with $g$ a continuous surjection and $\iota$ a continuous inclusion. Then $\iota\psi$ is pro-V by Proposition 3.3.14, whence $d\psi = g^{-1}\iota\psi$ is pro-V by Proposition 3.3.15. With $d\psi$ in the role of $\psi$ and the identity in the role of $\iota$, another application of Proposition 3.3.14 shows that $\varphi$ is pro-V. The last statement follows from Exercise 3.3.8.    □

The alert reader may have noticed that all the definitions and results given so far work equally well for continuously closed classes. However, those of the next section only go over, to some extent, to equational continuously closed classes; see Section 2.1.2 for the relevant definitions. This will be discussed in detail in Section 3.9.

# 3.4 Generators and Free pro-V Relational Morphisms

To discuss free objects and prove a Birkhoff/Reiterman type theorem, we need to define a notion of a generating set for a relational morphism. A useful expedient seems to be considering a generating set for the graph of the relational morphism.

## 3.4.1 Generators for relational morphisms

If $A$ is a profinite set, we say that a relational morphism $\varphi\colon P \to Q$ of profinite semigroups is *A-generated* if its graph $\#\varphi$ is (topologically) *A*-generated.

Equivalently, $\varphi$ is $A$-generated if there is a surjective continuous homomorphism $\alpha\colon \widehat{A^+} \to P$ and a continuous homomorphism $\beta\colon \widehat{A^+} \to Q$ so that $\varphi = \alpha^{-1}\beta$ as per:

A frequently occurring example of an $A$-generated relational morphism is a canonical relational morphism of $A$-generated profinite semigroups. Suppose that $P$ and $Q$ are $A$-generated profinite semigroups; then by the *canonical relational morphism* $\varphi\colon P \to Q$, we mean the relational morphism whose graph $\#\varphi$ is the closed subsemigroup of $P \times Q$ generated by all pairs $(a, a)$ (abusing notation) with $a \in A$. Equivalently, if $\alpha\colon \widehat{A^+} \twoheadrightarrow P$ and $\beta\colon \widehat{A^+} \twoheadrightarrow Q$ are the natural projections, then $\varphi = \alpha^{-1}\beta$. Notice $\varphi$ is $A$-generated as a relational morphism.

The following words of caution seem to be in order. Suppose that $S$, $R$, and $T$ are $A$-generated profinite semigroups and that $\alpha\colon S \to R$, $\beta\colon R \to T$ and $\gamma\colon S \to T$ are the canonical $A$-generated relational morphisms between them. It is not necessarily true that $\gamma = \alpha\beta$, although one does have $\gamma \subseteq_s \alpha\beta$. However, if $\alpha^{-1}$ is a surjective homomorphism or $\beta$ is a homomorphism, it is straightforward to verify that $\gamma = \alpha\beta$.

**Exercise 3.4.1.** Verify this last assertion.

Let $\psi\colon \widehat{A^+} \to Q$ be a continuous homomorphism of profinite semigroups. An $A$-generated relational morphism $\varphi\colon P \to Q$ is said to *respect* $\psi$ if

commutes. This is equivalent to $\varphi$ admitting a factorization $\varphi = \alpha^{-1}\psi$ with $\alpha\colon \widehat{A^+} \twoheadrightarrow P$ a surjective continuous homomorphism. Thus $A$-generated relational morphisms to $Q$, respecting $\psi$, are in bijection with (profinite) congruences on $\widehat{A^+}$. If $\equiv$ is such a congruence, we use $\varphi_\equiv\colon \widehat{A^+}/\equiv \to Q$ for the induced relational morphism. In detail, if $p_\equiv\colon \widehat{A^+} \to \widehat{A^+}/\equiv$ is the quotient morphism, then $\varphi_\equiv = p_\equiv^{-1}\psi$ as per

$$\widehat{A^+} \xrightarrow{\psi} Q$$

with $p_\equiv$ and $\varphi_\equiv$ to $\widehat{A^+}/\equiv$

**Proposition 3.4.2.** *Let* V *be a pseudovariety of relational morphisms and* $\psi\colon \widehat{A^+} \to Q$ *a continuous homomorphism of profinite semigroups. Then the collection* $\mathscr{C}_{\mathsf{V},Q}$ *of congruences* $\equiv$ *on* $\widehat{A^+}$ *such that* $\varphi_\equiv\colon \widehat{A^+}/\!\equiv\ \to Q$ *is pro-*V *is closed under arbitrary intersections.*

*Proof.* Let $C \subseteq \mathscr{C}_{\mathsf{V},Q}$ and set $\sim\ =\ \bigcap C$. Let $\iota\colon \widehat{A^+}/\!\sim\ \to\ \prod \widehat{A^+}/\!\equiv$ be the natural inclusion. There results a division $(\iota, \Delta)\colon \varphi_\sim \to \prod \varphi_\equiv$ diagrammed as follows:

$$
\begin{array}{ccc}
\widehat{A^+}/\!\sim & \overset{\iota}{\lhook\joinrel\longrightarrow} & \prod\limits_{\mathscr{C}_{\mathsf{V},Q}} \widehat{A^+}/\!\equiv \\[2ex]
\varphi_\sim \downarrow & \subseteq & \downarrow \prod \varphi_\equiv \\[2ex]
Q & \underset{\Delta}{\lhook\joinrel\longrightarrow} & \prod Q.
\end{array}
$$

Indeed, if $s \in \widehat{A^+}/\!\sim$, then $q \in s\varphi_\sim$ means there exists $w \in \widehat{A^+}$ projecting to $s$ with $w\psi = q$. Then $q \in ([w]_\equiv)\varphi_\equiv$ all $\equiv\ \in C$, where $[w]_\equiv$ is the $\equiv$-class of $w$. Hence $q\Delta \in ([w]_\equiv)\prod\varphi_\equiv = s\iota\prod\varphi_\equiv$, establishing $\varphi_\sim\Delta \subseteq_s \iota\prod\varphi_\equiv$. Because each $\varphi_\equiv$ is pro-V, so is $\varphi_\sim$ by Corollary 3.3.16. □

For equational continuously closed classes, we will make use of closure under pullbacks to prove the analogue of this proposition.

**Definition 3.4.3 (Free pro-V relational morphism).** *Let* $\psi\colon \widehat{A^+} \to Q$ *be a fixed continuous homomorphism of profinite semigroups and* V *a pseudovariety of relational morphisms. Define*

$$\equiv_{\mathsf{V},Q}\ =\ \bigcap \mathscr{C}_{\mathsf{V},Q}\ \text{ and } Q^{\mathsf{V}} = \widehat{A^+}/\!\equiv_{\mathsf{V},Q}. \tag{3.4}$$

*Setting* $\varphi_{\equiv_{\mathsf{V},Q}} = \varphi_{\mathsf{V},Q}$, *we call* $\varphi_{\mathsf{V},Q}\colon Q^{\mathsf{V}} \to Q$ *the free A-generated pro-*V *relational morphism to* $Q$.

Of course, this notation is abusive because the dependence on $\psi$ is ignored.

*Convention 3.4.4.* In the context of a fixed $A$-generated relational morphism of profinite semigroups $\varphi\colon P \to Q$, we make the standing assumption that in this context, $\varphi_{\mathsf{V},Q}$ and $Q^{\mathsf{V}}$ are defined relative to the map $\psi$ obtained via the composition $\widehat{A^+} \longrightarrow \#\varphi \overset{p_Q}{\longrightarrow} Q$ where the first map is the generating map.

If $1_Q$ is pro-V, in particular if $\mathsf{V} \in \mathbf{PVRM^+}$, then $\ker\psi \in \mathscr{C}_{\mathsf{V},Q}$ (recall that $\psi\colon \widehat{A^+} \to Q$ is our fixed map). Therefore, $\equiv_{\mathsf{V},Q} \subseteq \ker\psi$, and so $\varphi_{\mathsf{V},Q}$ is actually a continuous surjective homomorphism. Thus if V is a positive pseudovariety of relational morphisms and $\psi$ is onto, then the assignment $Q \mapsto Q^{\mathsf{V}}$ is an expansion of profinite semigroups cut-to-generators in the sense of [286] (in fact, virtually all the expansions considered in [286] are of this form; see also [87]). We now justify the terminology *free* used in Definition 3.4.3. Recall that we only consider profinite congruences in this text.

**Theorem 3.4.5.** *Let* $\psi\colon \widehat{A^+} \to Q$ *be a continuous homomorphism and* $\equiv$ *be a congruence on* $\widehat{A^+}$. *Then* $\varphi_{\equiv}\colon \widehat{A^+}/{\equiv} \to Q$ *is pro-*V *if and only if* ${\equiv_{V,Q}} \subseteq {\equiv}$.

*Proof.* Necessity is clear from (3.4) and the definition of $\mathscr{C}_{V,Q}$. For sufficiency, suppose ${\equiv_{V,Q}} \subseteq {\equiv}$. Then $\varphi_{\equiv} \prec_s \varphi_{V,Q}$, in fact we have the following commutative triangle

$$
\begin{array}{ccc}
\widehat{A^+}/{\equiv} & \xrightarrow{\;\;d\;\;} & \widehat{A^+}/{\equiv_{V,Q}} = Q^V \\
& \searrow{\scriptstyle \varphi_{\equiv}} \quad \swarrow{\scriptstyle \varphi_{\equiv_{V,Q}}} & \\
& Q &
\end{array}
$$

where $d^{-1}$ is the projection. Thus, $\varphi_{\equiv}$ is pro-V by Corollary 3.3.16.    □

Because the collection of pro-V relational morphisms of finite semigroups consists precisely of the members of V, we have the following corollary.

**Corollary 3.4.6.** *Let* $A$ *be a finite set,* $\varphi\colon S \to T$ *an* $A$-*generated relational morphism of finite semigroups and* $\pi\colon \widehat{A^+} \to S$ *the canonical surjection. Then* $\varphi \in V$ *if and only if* ${\equiv_{V,T}} \subseteq \ker\pi$.

If we consider a pseudovariety **V** of semigroups, viewed as a pseudovariety of relational morphisms via (2.64), then a relational morphism $\varphi\colon P \to Q$ is pro-**V** if and only if $P$ is pro-**V** in the usual sense. In this case, for any profinite set $A$ and continuous morphism $\psi\colon \widehat{A^+} \to Q$, one has that $Q^V$ is the free pro-**V** semigroup on $A$.

**Definition 3.4.7 (Locally finite pseudovariety).** *A pseudovariety of relational morphisms* V *is said to be locally finite if, for every finite set* $A$ *and continuous homomorphism* $\psi\colon \widehat{A^+} \to T$ *with* $T$ *finite,* $T^V$ *is finite.*

It is clear that if $W \subseteq V$ and $\psi\colon \widehat{A^+} \to Q$ with $Q$ profinite, then $Q^W$ is a quotient of $Q^V$. Hence any subpseudovariety of a locally finite pseudovariety is locally finite. Observe that if **V** is a pseudovariety of semigroups, then **V** is locally finite in this sense if and only if it is in the usual sense, as $T^V$ is a free pro-**V** semigroup on $A$.

**Proposition 3.4.8.** *Suppose that* V *is generated by a single relational morphism* $\varphi\colon S \to T$. *Then* V *is locally finite.*

*Proof.* The domain of any relational morphism in V must be in $(S)$. In fact, $V \subseteq (S)$ under identification (2.64). But $(S)$ is locally finite.    □

**Corollary 3.4.9.** *Every pseudovariety of relational morphisms is a directed join of locally finite pseudovarieties.*

# 3.5 Relational Pseudoidentities

Recall that a *pseudoidentity* is a formal equality $u = v$ between elements $u$ and $v$ of a free profinite semigroup $\widehat{A^+}$ on a finite set $A$. If $u = v$ is a pseudoidentity over $A$ and $\sigma\colon A \to S$ is a substitution ($S$ a profinite semigroup), we write $\sigma \vdash u = v$ if the induced morphism $\sigma\colon \widehat{A^+} \to S$ satisfies $u\sigma = v\sigma$. For a set $E$ of pseudoidentities, $\sigma \vdash E$ means $\sigma \vdash u = v$ for all $u = v \in E$. If $E$ and $E'$ are sets of pseudoidentities over the same alphabet $A$, we write $E \vdash E'$ if, for any substitution $\sigma\colon A \to S$ with $S$ a finite semigroup, one has $\sigma \vdash E$ implies $\sigma \vdash E'$. For example

$$\{x^2 = x, y^2 = y, xy = yx\} \vdash \{(xy)^2 = xy\}.$$

However,

$$\{x^2 = x, xy = yx\} \not\vdash \{(xy)^2 = xy\}$$

as we no longer have that the left-hand side implies $y^2 = y$. That is, $E \vdash E'$ is implication of *relations* not identities.

If $S$ is a profinite semigroup, then recall that $S$ is said to *satisfy* $u = v$ (a pseudoidentity over an alphabet $A$), written $S \models u = v$, if, for every substitution $\sigma\colon A \to S$, one has $\sigma \vdash u = v$. One defines $S \models E$ similarly for a set $E$ of pseudoidentities. By $E \models E'$, one means that $S \models E$ implies $S \models E'$ for any finite semigroup $S$. In other words, any finite semigroup that satisfies the set of pseudoidentities $E$ also satisfies the pseudoidentities $E'$. If $\mathbf{V}$ is a pseudovariety of semigroups, then we write $\mathbf{V} \models E$ if all semigroups in $\mathbf{V}$ satisfy $E$. If $E$ is a set of pseudoidentities, then recall that the collection of all finite semigroups satisfying $E$ is a pseudovariety, denoted $[\![E]\!]$. Reiterman's Theorem (Theorem 3.2.11) says that the converse holds: all pseudovarieties are defined by some set of pseudoidentities. An interesting phenomenon is that if $S$ is a profinite semigroup and $E$ is a basis of pseudoidentities for the pseudovariety $\mathbf{V}$, then $S$ is pro-$\mathbf{V}$ if and only if it satisfies $E$ — this unfortunately turns out *not* be the case for pseudovarieties of relational morphisms.

In summary, we use $\vdash$ for implication of relations and $\models$ for implication of (pseudo)identities.

## 3.5.1 Relational pseudoidentities

A *relational pseudoidentity* consists of a triple $e = (A, u = v, E)$, where $A$ is a finite set, $u = v$ is a pseudoidentity over $A$ and $E$ is a set of pseudoidentities over $A$. If $E'$ is a set of pseudoidentities over $A$, we use $(A, E', E)$ to denote the set of all relational pseudoidentities of the form $(A, u = v, E)$ with $u = v \in E'$.

Given a relational morphism of finite semigroups $\varphi\colon S \to T$ and a relational pseudoidentity $e = (A, u = v, E)$, we write $\varphi \models e$ (read $\varphi$ *satisfies* $e$) if, for every substitution $\psi\colon A \to \#\varphi$, one has $\psi p_T \vdash E$ implies $\psi p_S \vdash u = v$. Equivalently, given substitutions $\sigma, \tau$ from $A$ to $S$, $T$, respectively, which are

$\varphi$-*related* (meaning $a\tau \in a\sigma\varphi$ for all $a \in A$), $\tau \vdash E$ implies $\sigma \vdash u = v$. Observe that an identity morphism $1_S$ satisfies $e$ if and only if whenever $\sigma \colon A \to S$ is a substitution such that $\sigma \vdash E$, then also $\sigma \vdash u = v$; in other words "$E \vdash u = v$" for $S$.

*Remark 3.5.1.* Suppose $\varphi \colon S \to T$ is a relational morphism. Then a substitution $\sigma \colon A \to S$ can be viewed as an element of $S^A$. It is then straightforward to verify that $\sigma$ is $\varphi$-related to $\tau \in T^A$ if and only if $\tau \in \sigma\varphi^A$ where, for a relational morphism $\alpha$, we denote by $\alpha^A$ the $A$-fold product of $\alpha$ with itself.

Intuitively speaking, a relational morphism $\varphi \colon S \to T$ satisfies a relational pseudoidentity $(A, u = v, E)$ if whenever an $A$-tuple of elements of $T$ satisfies $E$, then any $A$-tuple of preimages in $S$ satisfies $u = v$. For example, $\varphi \colon S \to T$ satisfies the relational pseudoidentity $(\{x\}, x = x^2, \{x = x^2\})$ if each preimage of an idempotent of $T$ is an idempotent, i.e., $\varphi \in (\mathbf{B}, \mathbf{1})$. The choice for our ordering of the triple $(A, u = v, E)$ is because if $\varphi \colon S \to T$ is a relational morphism, then we always are evaluating $u = v$ in the domain $S$ and $E$ in the codomain $T$, so the ordering reflects where we evaluate the pseudoidentities. Notice that the implication however goes the other way: if $\tau \colon A \to T$ satisfies $E$ and $\sigma \colon A \to S$ is $\varphi$-related to $\tau$, then $\sigma$ satisfies $u = v$.

One can then define $\varphi \models \mathcal{E}$ for a set $\mathcal{E}$ of relational pseudoidentities to mean that $\varphi \models e$ for all $e \in \mathcal{E}$. If $\mathsf{V}$ is a pseudovariety of relational morphisms and $\mathcal{E}$ is a set of pseudoidentities, we write $\mathsf{V} \models \mathcal{E}$ to mean that all members of $\mathsf{V}$ satisfy $\mathcal{E}$.

We proceed to verify that if $\mathcal{E}$ is a set of relational pseudoidentities, then

$$[\![\mathcal{E}]\!] = \{\varphi \colon S \to T \mid \varphi \models \mathcal{E}\} \tag{3.5}$$

is a pseudovariety of relational morphisms.

**Proposition 3.5.2.** *Let $\mathcal{E}$ be a set of relational pseudoidentities. Then $[\![\mathcal{E}]\!]$ is a pseudovariety of relational morphisms.*

*Proof.* Let $\mathsf{V} = [\![\mathcal{E}]\!]$ and suppose $\varphi \colon S \to T, \psi \colon S' \to T'$ are relational morphisms in $\mathsf{V}$. Let $(A, u = v, E) \in \mathcal{E}$ and $\sigma \colon A \to \#(\varphi \times \psi)$ be a substitution such that $\sigma p_{T \times T'} \vdash E$. Let $p_S, p_{S'}, p_T, p_{T'}$ be the respective projections from $S \times S'$ and $T \times T'$. Then $\sigma p_{T \times T'} p_T, \sigma p_{T \times T'} p_{T'} \vdash E$ and are $\varphi$-, respectively, $\psi$-related to $\sigma p_{S \times S'} p_S$, $\sigma p_{S \times S'} p_{S'}$, respectively. It follows

$$\sigma p_{S \times S'} p_S \vdash u = v, \quad \sigma p_{S \times S'} p_{S'} \vdash u = v$$

and so $\sigma p_{S \times S'} \vdash u = v$. We conclude $\varphi \times \psi \in \mathsf{V}$. Clearly, $1_{\{1\}}$ is in $\mathsf{V}$. This establishes closure of $\mathsf{V}$ under Axiom $(\times)$.

Suppose $\varphi \colon S \to T$ is in $\mathsf{V}$ and $\iota \colon T \to T'$ is an inclusion. Consider a relational pseudoidentity $(A, u = v, E) \in \mathcal{E}$ and let $\sigma \colon A \to \#(\varphi\iota)$ be a substitution such that $\sigma p_{T'} \vdash E$. Observe that $\#\varphi = \#(\varphi\iota)$ as sets, so viewing $\sigma$ as a map to $\#\varphi$, we see $\sigma p_T \vdash E$, and thus $\sigma p_S \vdash u = v$. Therefore, $\varphi\iota \models \mathcal{E}$ and hence $\mathsf{V}$ satisfies Axiom (r-e).

Assume $\varphi\colon S \to T$ is a relational morphism, $\psi\colon S' \to T'$ is in V, $\varphi \prec \psi$, and $(A, u = v, E) \in \mathscr{E}$. Then $\varphi f \subseteq_s d\psi$ for some division $d\colon S \to S'$ and homomorphism $f\colon T \to T'$. Let $\sigma\colon A \to \#\varphi$ be a substitution such that $\sigma p_T \vdash E$. Then $\sigma p_T f \vdash E$. Because $\varphi f \subseteq_s d\psi$, for each $a \in A$, there exists $s'_a \in S'$ such that $s'_a \in a\sigma p_S d$ and $a\sigma p_T f \in s'_a \psi$. Define $\sigma'\colon A \to S'$ by $a\sigma' = s'_a$; then $\sigma'$ $\psi$-relates to $\sigma p_T f$. We have the diagram

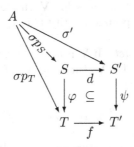

where $\sigma' \subseteq \sigma p_S d$. Because $\psi \models (A, u = v, E)$ and $\sigma p_T f \vdash E$, it follows that $\sigma' \vdash u = v$. Because $d$ is a division and $u\sigma' = v\sigma' \in u\sigma p_S d \cap v\sigma p_S d$, we have $u\sigma p_S = v\sigma p_S$, i.e., $\sigma p_S \vdash u = v$. Thus we have shown $\varphi \models \mathscr{E}$ and hence belongs to V. This completes the proof that V is a pseudovariety of relational morphisms. $\qquad\square$

If V is a pseudovariety of relational morphisms, then $\mathscr{E}$ is said to be a *basis* of relational pseudoidentities for V if $V = [\![\mathscr{E}]\!]$.

We now give some examples. Many of the details shall be left to the reader. The reader is referred back to Section 2.4.1 for notation.

*Example 3.5.3 (Divisions).* The pseudovariety D of all divisions is defined by the relational pseudoidentity

$$e = (\{x, y\}, x = y, \{x = y\}).$$

Indeed, $\varphi\colon S \to T$ satisfies $e$ if and only if given $s_1, s_2 \in S$ and $t_i \in s_i\varphi$, $i = 1, 2$, the implication $t_1 = t_2$ implies $s_1 = s_2$ holds. But this is exactly what it means for $\varphi$ to be a division.

*Example 3.5.4 (Slice).* More generally, for each pseudovariety V of semigroups, one can check using the well-known result appearing in Corollary 3.7.2 below that

$$V_\vee = [\![\{(A, u = v, \{u = v\})\} \mid V \models u = v\}]\!].$$

The previous example is where $V = 1$.

*Example 3.5.5 (Semigroup pseudovarieties).* Any pseudoidentity $u = v$ over $A$ can be identified with the relational pseudoidentity $(A, u = v, \emptyset)$. Clearly, if $\varphi\colon S \to T$, then $\varphi \models u = v$ if and only if $S \models u = v$. Hence, if V is a pseudovariety of semigroups and $\mathscr{E}$ is a basis of pseudoidentities for V, then

$\mathscr{E}$ is also a basis of relational pseudoidentities for $\widetilde{\mathbf{V}}$ (under this identification). Moreover, for any relational pseudoidentity $e = (A, u = v, E)$, we have $\widetilde{\mathbf{V}} \models e$ if and only if $\mathbf{V} \models u = v$ (consider collapsing morphisms). This gives further justification for the identification of $\widetilde{\mathbf{V}}$ and $\mathbf{V}$ as per (2.64).

*Example 3.5.6 (Intersection).* If $\mathbf{V}$ is a pseudovariety of semigroups, then a basis for $\mathbf{V} \cap \mathsf{D}$ is a union of a basis for $\mathbf{V}$ (under identification in Example 3.5.5) and the relational pseudoidentity $(\{x, y\}, x = y, \{x = y\})$.

*Example 3.5.7 (Mal'cev product).* If $\mathbf{V}$ is a pseudovariety of semigroups and $\mathscr{E}$ is a basis of pseudoidentities for $\mathbf{V}$, then the set of all relational pseudoidentities of the form

$$\{(\{a_1, \ldots, a_n\}, u = v, \{a_1^2 = a_1 = a_2 = \ldots = a_n\}) \mid u = v \in \mathscr{E}\}$$

is a basis of relational pseudoidentities for $(\mathbf{V}, \mathbf{1})$. Indeed, $\varphi \colon S \to T$ satisfies these pseudoidentities if and only if given an idempotent $e \in T$, elements $s_1, \ldots, s_n \in e\varphi^{-1}$ and a pseudoidentity $u = v \in \mathscr{E}$ in $n$ variables, one has $s_1, \ldots, s_n$ satisfy $u = v$. This is clearly equivalent to asking $e\varphi^{-1} \in \mathbf{V}$, all $e \in E(T)$.

*Example 3.5.8 (Generalized Mal'cev product).* If $\mathbf{V}$ and $\mathbf{W}$ are pseudovarieties of semigroups and $\mathscr{E}$ is a basis for $\mathbf{V}$, then a basis of relational pseudoidentities for $(\mathbf{V}, \mathbf{W})$ consists of all relational pseudoidentities of the form $(A, u = v, E_{\mathbf{W}, |A|})$, where $u = v \in \mathscr{E}$ and $\langle A \mid E_{\mathbf{W}, |A|} \rangle$ is a presentation of a free pro-$\mathbf{W}$ semigroup on $A$ (see Definition 3.1.45). Alternatively, one could use all relational pseudoidentities $(A, u = v, E)$ where $u = v \in \mathscr{E}$ and $\langle A \mid E \rangle$ is a finite presentation of a semigroup in $\mathbf{W}$.

*Example 3.5.9 (Semidirect product).* We provide here the construction of a basis for $\mathbf{V}_D$ for a general pseudovariety of semigroups $\mathbf{V}$. Specific examples, worked out in detail, will follow shortly. In Theorem 3.5.11 we provide a proof that the construction works. If $\mathscr{E}$ is a basis for $\mathbf{gV}$ consisting of path pseudoidentities, then a basis of relational pseudoidentities for $\mathbf{V}_D$ is given by all relational pseudoidentities of the form $(V \cup E, u = v, C_\Gamma)$ obtained as follows. Start with a pseudoidentity $(\Gamma, x = y) \in \mathscr{E}$. Set $V = V(\Gamma)$, $E = E(\Gamma)$ and let $C_\Gamma$ consist of all identities of the form $(e\alpha)e = e\omega$ with $e\alpha \xrightarrow{e} e\omega$ an edge of $\Gamma$; such equations are called the *consistency equations* for $\Gamma$. Then put $u = (x\alpha)x$, $v = (y\alpha)y$. We also include all relational pseudoidentities of the form $(V \setminus \{x\alpha\} \cup E, x = y, C'_\Gamma)$ where $(\Gamma, x = y) \in \mathscr{E}$ does not have any edges coming into $x\alpha$ and $C'_\Gamma$ consists of all equations of the form $(e\alpha)e = e\omega$ such that $e\alpha \neq x\alpha$ and of the form $e = e\omega$ where $e\alpha = x\alpha$. These relational pseudoidentities correspond to substitutions of $\Gamma$ into the derived semigroupoid sending the initial vertex of $x$ to the adjoined identity $I$. We also consider $C'_\Gamma$ to be consistency equations for $\Gamma$.

*Example 3.5.10 (Two-sided semidirect product).* If $\mathbf{V}$ is a pseudovariety of semigroups, a basis for $\mathbf{V}_K$ can be described analogously to the case of $\mathbf{V}_D$. If $\mathscr{E}$ is a basis for $\mathbf{gV}$ consisting of path pseudoidentities, then a basis of relational pseudoidentities for $\mathbf{V}_D$ is given by all relational pseudoidentities of the form $(V_L \cup V_R \cup E, u = w, K_\Gamma)$ obtained as follows. Start with a pseudoidentity $(\Gamma, x = y) \in \mathscr{E}$ where $\Gamma$ has vertex set $V$ and edge set $E$. Let $V_L$ and $V_R$ be sets in bijection with $V$ via maps $v \mapsto v_L$ and $v \mapsto v_R$, respectively. Let $K_\Gamma$ consist of all identities of the form $(e\alpha)_L e = (e\omega)_L$, $(e\alpha)_R = e(e\omega)_R$ with $e\alpha \xrightarrow{e} e\omega$ an edge of $\Gamma$; such equations are called the *bilateral consistency equations* for $\Gamma$. Then put $u = (x\alpha)_L x(x\omega)_R$, $w = (y\alpha)_L y(y\omega)_R$. If $\Gamma$ has no edges coming into $x\alpha$, one must add in the relational pseudoidentities corresponding to substituting $(x\alpha)_L$ by $I$; if $\Gamma$ has no edges coming out of $x\omega$, one must add in the relational pseudoidentities corresponding to substituting $(x\omega)_R$ by $I$; if $\Gamma$ has neither edges coming into $x\alpha$ nor edges coming out of $x\omega$, then one must add in the relational pseudoidentities corresponding to substituting both $(x\alpha)_L$ and $(x\omega)_R$ by $I$.

Let us consider concrete instances of Examples 3.5.7–3.5.10. Let $\mathbf{A}$ be the pseudovariety of aperiodic semigroups. Then

$$(\mathbf{A}, \mathbf{1}) = [\![(\{x\}, x^\omega = x^{\omega+1}, \{x^2 = x\})]\!].$$

Indeed, if $\varphi \colon S \to T$ is a relational morphism, then a map $\sigma \colon \{x\} \to \#\varphi$ is a choice of elements $s \in S$, $t \in T$ with $s \in t\varphi^{-1}$. The condition $\sigma p_T \vdash x^2 = x$ is equivalent to $t$ being an idempotent. So $\varphi$ satisfies this identity if and only if whenever $t \in T$ is an idempotent and $s \in t\varphi^{-1}$, then $s^\omega = s^{\omega+1}$. But this is equivalent to $\varphi \in (\mathbf{A}, \mathbf{1})$.

For $\mathbf{Com}$, the pseudovariety of finite commutative semigroups, we have:

$$(\mathbf{Com}, \mathbf{1}) = [\![(\{x, y\}, xy = yx, \{x^2 = x = y\})]\!].$$

Next we consider the pseudovariety of relational morphisms corresponding to the generalized Mal'cev product $(\mathbf{Com}, \mathbf{Sl})$. In this case

$$(\mathbf{Com}, \mathbf{Sl}) = [\![(\{x, y\}, xy = yx, \{x^2 = x, y^2 = y, xy = yx\})]\!].$$

Notice that, in the middle coordinate, one puts a basis of pseudoidentities for $\mathbf{Com}$, whereas in the third coordinate one puts a presentation of a free semilattice. This is because if one has a relational morphism $\varphi \colon S \to T$ and a substitution $\tau \colon \{x, y\} \to T$, one needs to know from $\tau$ alone that $x\tau$, $y\tau$ generate a semilattice.

Next we turn to some examples of the form $\mathbf{V}_D$. Because $\mathbf{Sl}$ is local by Simon's Theorem [85, Chapter VIII, Thm 7.1] [60,364], it follows $\mathbf{gSl}$ is defined by $[\![x^2 = x, xy = yx]\!]$ where the underlying graphs have one vertex $a$ and $x, y$ are "loop" edges. Then $\mathbf{Sl}_D$ is defined by the relational pseudoidentities

$$(\{a, x\}, ax^2 = ax, \{ax = a\}), \ (\{a, x, y\}, axy = ayx, \{ax = a = ay\}).$$

It was shown by Thérien and Weiss [358, 364] that **gCom** is defined by the identity

$$\left( \underset{y}{\overset{x,\,z}{ⓐ \underset{\longleftarrow}{\longrightarrow} ⓑ}},\ xyz = zyx \right) \tag{3.6}$$

and the local identity $xy = yx$ (defined over a one-vertex graph). A basis for $\mathbf{Com}_D$ then consists of the relational pseudoidentities

$$(\{a, b, x, y, z\}, axyz = azyx, \{ax = b = az, by = a\})$$
$$(\{a, x, y\}, axy = ayx, \{ax = a = ay\}).$$

Next we construct a basis for $\mathbf{J}_D$, where $\mathbf{J}$ is the pseudovariety of $\mathscr{J}$-trivial semigroups. Recall that $\mathbf{gJ}$ is defined by Knast's pseudoidentity [163, 364]

$$\left( \underset{y,\,z}{\overset{w,\,x}{ⓐ \underset{\longleftarrow}{\longrightarrow} ⓑ}},\ (wy)^\omega wz(xz)^\omega = (wy)^\omega (xz)^\omega \right)$$

together with the local pseudoidentities $x^\omega = x^{\omega+1}$, $(xy)^\omega = (yx)^\omega$ (over one-vertex graphs). A basis for $\mathbf{J}_D$ then consists of the relational pseudoidentity

$$(\{a, b, w, x, y, z\}, a(wy)^\omega wz(xz)^\omega = a(wy)^\omega (xz)^\omega,$$
$$\{aw = b, ax = b, by = a, bz = a\})$$

together with the relational pseudoidentities

$$(\{a, x\}, ax^\omega = ax^{\omega+1}, \{ax = a\}), (\{a, x, y\}, a(xy)^\omega = a(yx)^\omega, \{ax = a = ay\}).$$

As an instance of Example 3.5.10, we consider $\mathbf{Com}_K$. The reader is referred back to (3.6) for a basis for **gCom**. A basis for $\mathbf{Com}_K$ consists of the relational pseudoidentities

$$(\{a_L, a_R, b_L, b_R, x, y, z\}, a_L xyz b_R = a_L zyx b_R,$$
$$\{a_L x = b_L = a_L z, b_L y = a_L, x b_R = a_R = z b_R, y a_R = b_R\})$$

and

$$(\{a_L, a_R, x, y\}, a_L xy a_R = a_L yx a_R, \{a_L x = a_L = a_L y, x a_R = a_R = y a_R\}).$$

**Theorem 3.5.11 (Basis for $\mathbf{V}_D$).** *Let* $\mathbf{V}$ *be a pseudovariety of semigroups and let $\mathscr{E}$ be a basis of path pseudoidentities for* $\mathbf{gV}$. *Then the set of relational pseudoidentities presented in Example 3.5.9 is a basis for* $\mathbf{V}_D$.

*Proof.* Let $\mathbf{V}$ be a pseudovariety of semigroups and suppose $\mathscr{E}$ is a basis for $\mathbf{gV}$ consisting of path identities. Assume first that $\varphi\colon S \to T$ satisfies the relational pseudoidentities described in Example 3.5.9; we prove $D_\varphi \models \mathscr{E}$ and hence $D_\varphi \in \mathbf{gV}$. Let $(\Gamma, x = y) \in \mathscr{E}$ and let $m\colon \Gamma \to \mathbf{Der}(\varphi)$ be a graph morphism. Let $\sigma\colon \mathbf{Der}(\varphi) \to D_\varphi$ be the projection. We must show $xm\sigma = ym\sigma$. Suppose $\Gamma$ has vertex set $V$ and edge set $E$. If $x\alpha m$ is not in the image of $\varphi^I$, then automatically $xm\sigma = ym\sigma$, so assume that $x\alpha m$ is in the image of $\varphi^I$. Then because $(\Gamma, x = y)$ is a path identity, $vm$ is in the image of $\varphi^I$ for each vertex $v \in V$ (cf. the proof of Corollary 2.5.16). Set $vm = t_v$ and $em = (t_{e\alpha}, (s_e, t_e))$ for $v \in V$ and $e \in E$. Then $t_{e\alpha}t_e = t_{e\omega}$ for each edge $e$ as $m$ is a graph morphism. Thus the substitution $\tau\colon V \cup E \to T$ given by $z\tau = t_z$, for $z \in V \cup E$, satisfies the consistency equations $C_\Gamma$ for $\Gamma$ (where if $t_{x\alpha} = I$, then we delete $x\alpha$ from the domain of $\tau$ and work with $C'_\Gamma$). To ease notation, suppose $xm = (t_0, (s_x, t_x))$ and $ym = (t_0, (s_y, t_y))$. We need to prove that if $s_0 \in t_0\varphi^{-1}$, then $s_0 s_x = s_0 s_y$. Choose, for each vertex $v \in V \setminus x\alpha$, an element $s_v \in t_v\varphi^{-1}$; set $s_{x\alpha} = s_0$. (If $t_{x\alpha} = I$, then $s_0 = I$.) Define $\beta\colon V \cup E \to S$ by $z\beta = s_z$ for $z \in V \cup E$ (or if $t_{x\alpha} = I$, then $\beta$ is defined on $V \setminus \{x\alpha\} \cup E$). Then $\tau$ is $\varphi$-related to $\beta$ and $\tau \vdash C_\Gamma$ (or if $t_{x\alpha} = I$, then $\tau \vdash C'_\Gamma$). Notice that $x\beta = s_x$ and $y\beta = s_y$ by definition of multiplication in $\mathbf{Der}(\varphi)$. Because $\varphi$ satisfies the relational pseudoidentity $(V \cup E, (x\alpha)x = (y\alpha)y, C_\Gamma)$, it follows that $\beta \vdash (x\alpha)x = (y\alpha)y$, that is, $s_0 s_x = s_0 s_y$, as required. In the situation that $t_v = I$, then because $\varphi$ satisfies the relational pseudoidentity $(V \setminus \{x\alpha\} \cup E, x = y, C'_\Gamma)$, it follows $\beta \vdash x = y$ and hence $s_x = s_y$. Thus $s_0 s_x = I s_x = I s_y = s_0 s_y$, as desired.

Conversely, suppose that $\varphi \in \mathbf{V}_D$, where $\varphi\colon S \to T$. Let $(\Gamma, x = y)$ be an element of $\mathscr{E}$. We must show $\varphi$ satisfies $(V \cup E, (x\alpha)x = (y\alpha)y, C_\Gamma)$ where $V$ is the vertex set and $E$ is the edge set of $\Gamma$. The reader is left to handle the case of pseudoidentities of the form $(V \setminus \{x\alpha\} \cup E, x = y, C'_\Gamma)$. Let $\tau\colon V \cup E \to \#\varphi$ be a substitution such that $\tau p_T \vdash C_\Gamma$; write $z\tau = (s_z, t_z)$ for $z \in V \cup E$. Then $\tau p_T \vdash C_\Gamma$ means precisely that

$$t_{e\alpha}t_e = t_{e\omega} \tag{3.7}$$

for all edges $e \in E$. Define a morphism $m\colon \Gamma \to \mathbf{Der}(\varphi)$ by $vm = t_v$ on vertices and $em = (t_{e\alpha}, (s_e, t_e))$ on edges. Because $(s_e, t_e) \in \#\varphi$ and (3.7) holds, $m$ is indeed a morphism. Writing $x\tau = (s_x, t_x)$, $y\tau = (s_y, t_y)$ and $x\alpha\tau = (s_0, t_0)$ we need to show $s_0 s_x = s_0 s_y$. Now $xm\sigma = ym\sigma$ because $D_\varphi \in \mathbf{gV}$ and hence $D_\varphi \models (\Gamma, x = y)$. But $xm = (t_0, (s_x, t_x))$, $ym = (t_0, (s_y, t_y))$ and $s_0 \in t_0\varphi^{-1}$, so by definition of $\sigma$ it follows $s_0 s_x = s_0 s_y$, as required.    $\square$

*Remark 3.5.12.* Notice that the second part of the above proof did not use that $\mathscr{E}$ consists of path identities.

**Exercise 3.5.13.** Prove that the relational pseudoidentities in Example 3.5.10 define $\mathbf{V}_K$.

It is easy to check that $[\![\mathscr{E}]\!]$ is a positive pseudovariety if and only if, for all $(A, u = v, E) \in \mathscr{E}$, one has $E \vdash u = v$. This occurs in all the above examples except Examples 3.5.5, 3.5.6 and 3.5.8; in Example 3.5.8, one must impose the condition $\mathbf{V} \geq \mathbf{W}$ to obtain a positive pseudovariety. However, we note that one cannot, in general, determine whether a pseudoidentity defines a positive pseudovariety. Indeed, by the undecidability of the uniform word problem for finite semigroups, there is a finite set of $E \subseteq A^+ \times A^+$ such that it is undecidable, given $u, v \in A^+$, whether, for all substitutions $\sigma \colon A \to S$ with $S$ finite, $\sigma \vdash E$ implies $\sigma \vdash u = v$ (that is, $E \vdash u = v$); see [2] for more on undecidability problems and for precise references.

We now prove the analogue in our context to Reiterman's Theorem [261]; namely, we show that all pseudovarieties of relational morphisms are defined by relational pseudoidentities. The reader should refer back to Definition 3.4.3 for the definition of $\equiv_{\mathsf{V},T}$.

**Theorem 3.5.14.** *Let* $\mathsf{V}$ *be a pseudovariety of relational morphisms and let* $\mathscr{E}$ *consist of all relational pseudoidentities of the form* $(A, u = v, E)$ *where* $\langle A \mid E \rangle$ *is a finite presentation of a finite semigroup* $T$ *and* $(u, v) \in \equiv_{\mathsf{V},T}$ *(defined relative to the canonical projection* $\widehat{A^+} \twoheadrightarrow T$*). Then* $\mathsf{V} = [\![\mathscr{E}]\!]$.

*Proof.* First we show $\mathsf{V} \subseteq [\![\mathscr{E}]\!]$. Let $e = (A, u = v, E) \in \mathscr{E}$ and set $T = \langle A \mid E \rangle$. Suppose $\varphi \colon S \to U$ is in $\mathsf{V}$ and $\psi \colon A \to \#\varphi$ is a substitution such that $\psi p_U \vdash E$. Then we have a subcommutative diagram

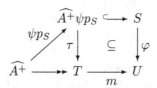

where $\tau$ is the canonical $A$-generated relational morphism and $m$ is the homomorphism obtained by factoring $\psi p_U$ through $T$ (as $\psi p_U \vdash E$). Thus $\tau \prec \varphi$ and so $\tau \in \mathsf{V}$. Corollary 3.4.6 then implies $\equiv_{\mathsf{V},T} \subseteq \ker \psi p_S$ and so $\psi p_S \vdash u = v$. We conclude $\varphi \models e$. This completes the proof that $\mathsf{V} \leq [\![\mathscr{E}]\!]$.

Suppose now that $\varphi \colon S \to T \models \mathscr{E}$. Our goal is to show $\varphi \in \mathsf{V}$. Let $\varphi_{\mathrm{Im}} \colon S \to T_{\mathrm{Im}}$ be the corestriction of $\varphi$ to its image. Then $\varphi_{\mathrm{Im}} \prec \varphi$ and $\varphi$ is a range extension of $\varphi_{\mathrm{Im}}$. Hence $\varphi$ and $\varphi_{\mathrm{Im}}$ belong to exactly the same pseudovarieties of relational morphisms. Thus without loss of generality, we may assume $\varphi$ is surjective. Let $A$ be a finite set of generators of $\#\varphi$ and $\pi \colon \widehat{A^+} \to \#\varphi$ the canonical surjection. Let $\psi \colon \widehat{A^+} \to T$ be given by $\psi = \pi p_T$. By Corollary 3.4.6, to show $\varphi \in \mathsf{V}$, it suffices to establish $\equiv_{\mathsf{V},T} \subseteq \ker \pi p_S$, or equivalently, $\pi p_S \vdash \equiv_{\mathsf{V},T}$, where we identify a pair in $\equiv_{\mathsf{V},T}$ with the corresponding pseudoidentity. Let $E$ be a finite set of generators of the kernel of $\psi \colon A^+ \to T$ (recall that every finite index congruence on a finitely generated semigroup is finitely generated [85, Chapter V, Prop. 2.2]). If $(u, v) \in \equiv_{\mathsf{V},T}$, then $(A, u = v, E) \in \mathscr{E}$ and $\pi p_T = \psi \vdash E$ by definition of $E$. Thus, by assumption that $\varphi \models \mathscr{E}$, it follows $\pi p_S \vdash u = v$. Hence $\pi p_S \vdash \equiv_{\mathsf{V},T}$ yielding $\varphi \in \mathsf{V}$. This shows $[\![\mathscr{E}]\!] \leq \mathsf{V}$, completing the proof.     □

**Exercise 3.5.15.** In [339], a relational morphism $(\alpha, m) \colon f \to g$ diagrammed as per

$$
\begin{array}{ccc}
S & \xrightarrow{\ \alpha\ } & U \\
{\scriptstyle f}\big\downarrow & \subseteq & \big\downarrow{\scriptstyle g} \\
T & \xrightarrow[\ m\ ]{} & V
\end{array}
\qquad\qquad (3.8)
$$

is called a *division diagram* if the implication

$$ t \in s_1 f \cap s_2 f, \; tm \in (s_1\alpha \cap s_2\alpha)g \implies s_1 = s_2 $$

holds. Notice that division is a special case of a division diagram. Given a division diagram as per (3.8), show using the equational theory that if $\mathsf{V}$ is a positive pseudovariety of relational morphisms and $g \in \mathsf{V}$, then $f \in \mathsf{V}$.

The fact that classes of positive pseudovarieties defined by relational pseudoidentities admit division diagrams was noted by B. Tilson. The second author then proved directly (unpublished) that positive pseudovarieties admit division diagrams.

### An application: Membership in compact pseudovarieties of relational morphisms

As an application of the equational theory, we prove that compact elements of **PVRM** have decidable membership. These are precisely the pseudovarieties generated by a single relational morphism.

Let $f\colon S \to T$ be a relational morphism. Recall that we can view the pseudovariety $(S)$ generated by $S$ as a pseudovariety of relational morphisms by identifying it with the collection of all relational morphisms $g\colon S' \to T'$ with $S' \in (S)$. Then the pseudovariety $(f)$ generated by $f$ is a subpseudovariety of $(S)$. If $A$ is a finite set and $\psi\colon \widehat{A^+} \to Q$ is any homomorphism with $Q$ a finite semigroup, then we can actually calculate the free pro-$(S)$ relational morphism generated by $A$ to $Q$. Namely, if $F_{(S)}(A)$ is the free $A$-generated semigroup in $(S)$ and $\varphi\colon \widehat{A^+} \to F_{(S)}(A)$ is the canonical projection, then $\varphi^{-1}\psi$ is the free pro-$(S)$ relational morphism generated by $A$ to $Q$. This is a relational morphism of finite semigroups and can be explicitly calculated: its graph is the subsemigroup of $F_{(S)}(A) \times Q$ generated by all pairs $(a\varphi, a\psi)$ with $a \in A$. It follows that the $A$-generated free pro-$(f)$ relational morphism to $Q$ will be a canonical relational morphism $h\colon S_Q \to Q$ of $A$-generated finite semigroups with $S_Q$ a finite quotient of $F_{(S)}(A)$. So to obtain the decidability of $(f)$, it suffices to be able to calculate $S_Q$ given $A$. Choose, for each $x \in F_{(S)}(A)$, a word $w_x \in A^+$ with $w_x\varphi = x$. Then $S_Q$ is the quotient of $F_{(S)}(A)$ by the congruence consisting of all pairs $(x, y)$ such that $(A, (w_x, w_y), E)$ is a relational pseudoidentity satisfied by $f$, where $E$ is a finite set of generators for the congruence associated to the map $\psi\colon A^+ \to Q$ (just do the relative version of Theorem 3.5.14 with the same proof). Because this is a finite set of relational pseudoidentities, we can compute this congruence and hence $S_Q$. This establishes:

**Theorem 3.5.16.** *Let* $f\colon S \to T$ *be a relational morphism. Then the pseudo-variety of relational morphisms generated by* $f$ *has decidable membership.*

**Exercise 3.5.17.** Write a detailed proof of the argument sketched above for Theorem 3.5.16.

We do not know whether the corresponding result is true for $\mathbf{PVRM}^+$.

### 3.5.2 Digression on quasivarieties and min vs. max

Recall that a *quasivariety* of finite semigroups is a class of finite semigroups closed under taking finite direct products and subsemigroups. The algebraic lattice of all quasivarieties is denoted $\mathbf{QV}$. As an example, notice that if $\mathsf{R} \in \mathbf{PVRM}$, then

$$\mathsf{qv}(\mathsf{R}) = \{S \mid 1_S \in \mathsf{R}\} \tag{3.9}$$

is a quasivariety.

**Exercise 3.5.18.** Verify that if $\mathsf{R} \in \mathbf{PVRM}$, then $\mathsf{qv}(\mathsf{R})$ is a quasivariety.

It is well-known that quasivarieties are precisely classes of finite semigroups defined by implicational pseudoidentities (see [61, Chapter V, Thm. 2.25] for the analogous result for implicational identities). Recall that an implicational pseudoidentity $E \vdash u = v$ consists of a set of pseudoidentities $E$ over $A$ and a pseudoidentity $u = v$ over $A$. A semigroup $S$ satisfies $E \vdash u = v$, written $S \models (E \vdash u = v)$, if, for all substitutions $\sigma\colon A \to S$ such that $\sigma \vdash E$, also $\sigma \vdash u = v$.

Notice that implicational pseudoidentities are in bijection with relational pseudoidentities by corresponding $E \vdash u = v$ over the alphabet $A$ with the relational pseudoidentity $(A, u = v, E)$. Moreover, $1_S \models (A, u = v, E)$ if and only if $S \models (E \vdash u = v)$. Hence if $\mathscr{E}$ is a set of relational pseudoidentities defining $\mathsf{R} \in \mathbf{PVRM}$, then $\mathscr{E}$ (viewed as a set of implicational pseudoidentities) defines $\mathsf{qv}(\mathsf{R})$.

Conversely, given a quasivariety $\mathbf{V}$, one can define $\mathsf{pvrm}(\mathbf{V})$ to be the pseudovariety of relational morphisms generated by the identity morphisms $1_S$, $S \in \mathbf{V}$.

**Proposition 3.5.19.** *Let* $\mathbf{V} \in \mathbf{QV}$ *and let* $\mathscr{E}$ *be the set of all implicational pseudoidentities satisfied by* $\mathbf{V}$. *Then* $\mathscr{E}$ *defines* $\mathsf{pvrm}(\mathbf{V})$ *as a set of relational pseudoidentities. Moreover,*

$$\mathsf{qv}(\mathsf{pvrm}(\mathbf{V})) = \mathbf{V}.$$

*Proof.* Clearly, $\mathsf{pvrm}(\mathbf{V}) \leq [\![\mathscr{E}]\!]$ as its generators satisfy $\mathscr{E}$. Conversely, if $\mathsf{pvrm}(\mathbf{V}) \models (A, u = v, E)$, then it is clear that every identity morphism of $\mathsf{pvrm}(\mathbf{V})$ satisfies $(A, u = v, E)$. Therefore, $\mathbf{V} \models (E \vdash u = v)$ and so $(E \vdash u = v) \in \mathscr{E}$. It follows from Theorem 3.5.14 that $\mathsf{pvrm}(\mathbf{V}) = [\![\mathscr{E}]\!]$. We immediately ascertain that $1_S \in \mathsf{pvrm}(\mathbf{V})$ if and only if $S \in \mathbf{V}$ and so $\mathsf{qv}(\mathsf{pvrm}(\mathbf{V})) = \mathbf{V}$.    $\square$

Notice that if $\mathbf{V}$ is a proper quasivariety, pvrm($\mathbf{V}$) will be proper, as well.

**Theorem 3.5.20.** *There is a Galois connection*

$$\mathbf{QV} \xleftrightarrow[\text{pvrm (sup)}]{\text{qv (inf)}} \mathbf{PVRM}. \qquad (3.10)$$

*Moreover,* qv $\circ$ pvrm $= 1_{\mathbf{QV}}$.

*Proof.* Clearly, qv is **inf** because the meet in both lattices is simply intersection. It is also clear from Proposition 3.5.19 that pvrm($\mathbf{V}$) is the smallest pseudovariety of relational morphisms R with qv(R) $= \mathbf{V}$. The result follows.
□

Set $\mathsf{D}' = \min(1_{\mathbf{PV}})$. By (2.34), it is the smallest element of $\mathbf{PVRM}$ that induces the identity operator under q. Note that D is generated by the collection of all identity maps, that is, $\mathsf{D} = $ pvrm($\mathbf{FSgp}$). On the other hand, we know from Remark 2.3.37 that $\mathsf{D}'$ is generated by all divisions $d\colon S \to T$ where $S$ and $T$ are relatively free $A$-generated semigroups in compact pseudovarieties and $d$ respects generators. Clearly, $d \prec 1_T$, so $\mathsf{D}'$ is in fact generated by identity morphisms $1_T$ with $T$ relatively free in a compact pseudovariety. Hence, if $\mathbf{F}$ is the quasivariety generated by all relatively free semigroups in compact pseudovarieties, then $\mathsf{D}' = $ pvrm($\mathbf{F}$). Because pvrm is injective, the question as to whether $\mathsf{D} = \mathsf{D}'$ is equivalent to whether $\mathbf{F} = \mathbf{FSgp}$. We posed this question to G. Bergman, who proved the following theorem [47]:

**Theorem 3.5.21 (Bergman).** *Every finite semigroup embeds in a relatively free finite semigroup.*

**Corollary 3.5.22.** $\mathsf{D} = \min(1_{\mathbf{PV}})$.

As Bergman points out [47], an analogous result is not possible in the monoidal context because if a relatively free monoid has a non-trivial group of units, then it must be a group, and hence a non-group monoid with a non-trivial group of units cannot be embedded in a relatively free monoid via a monoid embedding. In particular, the monoid $A_5 \cup 0$ is subdirectly indecomposable (see Definition 4.7.1) and cannot be embedded in a relatively free finite monoid.

More generally, if $\mathbf{V}$ is any pseudovariety of semigroups, then recall from Example 2.4.17 that $(\mathbf{V} \cap \mathsf{D})\mathsf{q} = \mathbf{V} \cap (-)$. It is then natural to ask whether

$$\mathbf{V} \cap \mathsf{D} = \min(\mathbf{V} \cap (-)). \qquad (3.11)$$

An argument analogous to the one provided above for $\mathbf{V} = \mathbf{FSgp}$ shows that $\min(\mathbf{V} \cap (-))$ is the pseudovariety of relational morphisms generated by all identity maps of relatively free finite semigroups in $\mathbf{V}$ and so (3.11) holds if and only if the quasivariety generated by such semigroups is all of $\mathbf{V}$; in other words if and only if each subdirectly indecomposable semigroup in **V** embeds in a relatively free semigroup from **V**. We show that this is not always possible; the example is due to the first author.

Let $S$ be the syntactic semigroup of the language $\{ab\}$. That is, $S = \{a, b, ab, 0\}$ with all multiplications other than $ab$ equal to 0; for this reason $S$ is sometimes denoted $S_{ab \neq 0}$. Notice that $S$ is subdirectly indecomposable: the unique minimal congruence on $S$ identifies $ab$ with 0. We consider $\mathbf{V} = (S)$. Suppose $F$ is a relatively free semigroup in **V** and $f \colon S \hookrightarrow F$ is an embedding with $\alpha = af, \beta = bf, z = 0f$. Because $S$ is 3-nilpotent, so is $F$. As $z = z^3$, it follows that $z$ is the 0 of $F$. Because $\alpha\beta \neq z$ and $F$ is 3-nilpotent, it follows that $\alpha$ and $\beta$ cannot be factored in $F$; hence they are part of any generating set of $F$. Thus there is an automorphism $g$ of $F$ switching $\alpha$ and $\beta$ and fixing all other generators. Then, as automorphisms preserve zeroes, $zg = z$. But

$$z = zg = (\beta\alpha)g = \alpha\beta,$$

a contradiction. We have thus established that, in general, $\min \neq \max$ as maps from **GMC(PV)** to **PVRM**.

**Theorem 3.5.23.** *The maps* $\min, \max \colon \mathbf{GMC(PV)} \to \mathbf{PVRM}$ *are different.*

*Question 3.5.24.* Compute $\min(\alpha)$ for any operator that is not a constant operator or of the form $\mathbf{V} \cap (-)$.

## 3.6 Composition and Inevitable Substitutions

In this section, we study composition of pseudovarieties of relational morphisms from the equational point of view. This material is more advanced than that of the previous sections of this chapter and generalizes ideas from [9, 19, 20, 24, 27, 33, 126, 235].

### 3.6.1 Free relational morphisms over compositions

Let us turn to the study of free relational morphisms over compositions. First we prove two lemmas relating composition to the correspondence $\mathbf{V} \mapsto \text{pro-}\mathbf{V}$.

**Lemma 3.6.1.** *Suppose that* $\alpha \colon P \to Q$ *is* pro-V *and* $\beta \colon Q \to R$ *is* pro-W. *Then* $\alpha\beta$ *is* pro-V $\odot$ W.

*Proof.* Let $\equiv$ be a clopen congruence on $P$. Then there is a clopen congruence $\sim$ on $Q$ such that the induced relational morphism $\alpha' \colon P/\equiv \to Q/\sim$ is in V. But then there must be a clopen congruence $\sim'$ on $R$ such that the induced relational morphism $\beta' \colon Q/\sim \to R/\sim'$ is in W. We then have the diagram

$$P \xrightarrow{\alpha} Q \xrightarrow{\beta} R$$

$$\pi_P \downarrow \qquad \pi_Q \downarrow \qquad \qquad \downarrow \pi_R$$

$$P/\equiv \xrightarrow[\alpha']{} Q/\sim \xrightarrow[\beta']{} R/\sim'$$

where $\alpha' = \pi_P^{-1}\alpha\pi_Q$ and $\beta' = \pi_Q^{-1}\beta\pi_R$. Let $\gamma\colon P/\equiv \to R/\sim'$ be the relational morphism induced by $\alpha\beta$ and the congruences $\equiv$ and $\sim'$, so $\gamma = \pi_P^{-1}\alpha\beta\pi_R$. We claim $\gamma \subseteq_s \alpha'\beta'$. Indeed,

$$\gamma = \pi_P^{-1}\alpha\beta\pi_R \subseteq_s \pi_P^{-1}\alpha\pi_Q\pi_Q^{-1}\beta\pi_R = \alpha'\beta'.$$

Hence $\gamma$ is in $\mathsf{V} \odot \mathsf{W}$ and the lemma follows.    □

**Lemma 3.6.2.** *Suppose $A$ is a finite set and $\varphi\colon S \to T$ in $\mathsf{V} \odot \mathsf{W}$ is $A$-generated. Then there exist $A$-generated relational morphisms $\alpha\colon S \to R$ in $\mathsf{V}$ and $\beta\colon R \to T$ in $\mathsf{W}$ such that $\varphi \subseteq_s \alpha\beta$.*

*Proof.* By Proposition 2.8.2, we know $\varphi \subseteq_s \alpha'\beta'$ with $\alpha'\colon S \to R'$ in $\mathsf{V}$ and $\beta'\colon R' \to T$ in $\mathsf{W}$. Let $\psi\colon A \to \#\varphi$ be the generating map. As $\varphi \subseteq_s \alpha'\beta'$, for each $a \in A$, there exists $r_a \in R'$ such that $a\psi p_S \; \alpha' \; r_a \; \beta' \; a\psi p_T$. Let $R$ be the subsemigroup of $R'$ generated by the $r_a$ and let $\alpha\colon S \to R$, $\beta\colon R \to T$ be the relational morphisms generated by the pairs $(a\psi p_S, r_a)$, $(r_a, a\psi p_T)$, $a \in A$, respectively. Then $\alpha \in \mathsf{V}$, $\beta \in \mathsf{W}$ are $A$-generated and $\varphi \subseteq_s \alpha\beta$.    □

We now prove the main result of this subsection.

**Theorem 3.6.3.** *Let $\psi\colon \widehat{A^+} \to Q$ be a continuous homomorphism with $A$ finite and let $\mathsf{V}$, $\mathsf{W}$ be pseudovarieties of relational morphisms. Then the free pro-$\mathsf{V} \odot \mathsf{W}$ relational morphism $\varphi_{\mathsf{V}\odot\mathsf{W},Q}\colon Q^{\mathsf{V}\odot\mathsf{W}} \to Q$ is the canonical relational morphism $\pi\colon (Q^{\mathsf{W}})^{\mathsf{V}} \to Q$, that is, the relational morphism to $Q$ corresponding to the congruence $\equiv_{\mathsf{V},Q^{\mathsf{W}}}$ on $\widehat{A^+}$, where this latter congruence is defined relative to the canonical projection $\widehat{A^+} \to Q^{\mathsf{W}}$.*

*Proof.* First we show $\pi$ is pro-$\mathsf{V} \odot \mathsf{W}$. Indeed, we have the diagram

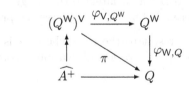

where the upper right triangle is subcommutative, that is, $\pi \subseteq_s \varphi_{\mathsf{V},Q^{\mathsf{W}}}\varphi_{\mathsf{W},Q}$. By definition, $\varphi_{\mathsf{V},Q^{\mathsf{W}}}$ is pro-$\mathsf{V}$ and $\varphi_{\mathsf{W},Q}$ is pro-$\mathsf{W}$. Therefore, $\varphi_{\mathsf{V},Q^{\mathsf{W}}}\varphi_{\mathsf{W},Q}$ is pro-$\mathsf{V} \odot \mathsf{W}$ by Lemma 3.6.1, and so $\pi$ is pro-$\mathsf{V} \odot \mathsf{W}$ by Corollary 3.3.16.

Next let $\equiv$ be a congruence on $\widehat{A^+}$ such that $\varphi_{\equiv}\colon \widehat{A^+}/\equiv \to Q$ is pro-$\mathsf{V} \odot \mathsf{W}$. We need to show that $\equiv_{\mathsf{V},Q^{\mathsf{W}}}\subseteq \equiv$. As $\equiv$ is the intersection of the clopen congruences containing it, it suffices to show that if $\equiv'$ is a clopen congruence on $\widehat{A^+}$ containing $\equiv$, then $\equiv_{\mathsf{V},Q^{\mathsf{W}}} \subseteq \equiv'$. That is, without loss of generality we may assume that $\equiv$ is clopen.

Then there must be a clopen congruence $\sim$ on $Q$ such that the induced $A$-generated relational morphism $\rho\colon \widehat{A^+}/\equiv \to Q/\sim$ is in $\mathsf{V}\odot\mathsf{W}$. By Lemma 3.6.2, there are $A$-generated relational morphisms $\alpha\colon \widehat{A^+}/\equiv \to \widehat{A^+}/\sim'$ in $\mathsf{V}$ and $\beta\colon \widehat{A^+}/\sim' \to Q/\sim$ in $\mathsf{W}$ such that $\rho \subseteq_s \alpha\beta$. Let $\gamma\colon Q \to Q/\sim$ be the natural surjection. Diagrammatically, we have

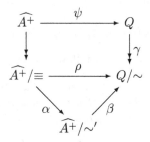

with $\rho \subseteq_s \alpha\beta$. So $\beta\gamma^{-1} \prec \beta$ (as $\beta\gamma^{-1}\gamma = \beta$) and therefore, by Corollary 3.3.16, $\beta\gamma^{-1}$ is pro-$\mathsf{W}$. As the canonical $A$-generated relational morphism $\widehat{A^+}/\sim' \to Q$ is contained in $\beta\gamma^{-1}$, it, too, is pro-$\mathsf{W}$ by the selfsame corollary. Thus $\equiv_{\mathsf{W},Q} \subseteq \sim'$.

We now have the diagram

$$\begin{array}{ccc} \widehat{A^+} & \xrightarrow{\mu} & Q^{\mathsf{W}} \\ {\scriptstyle\tau}\downarrow & & \downarrow{\scriptstyle\kappa} \\ \widehat{A^+}/\equiv & \xrightarrow{\alpha} & \widehat{A^+}/\sim' \end{array}$$

where $\kappa\colon Q^{\mathsf{W}} \to \widehat{A^+}/\sim'$ is the natural projection coming from $\equiv_{\mathsf{W},Q} \subseteq \sim'$, $\mu$ is the quotient map and $\tau^{-1}\mu\kappa = \alpha$. Now $\alpha\kappa^{-1} \prec \alpha$, and the canonical relational morphism of $A$-generated semigroups $\lambda\colon \widehat{A^+}/\equiv \to Q^{\mathsf{W}}$ is contained in $\alpha\kappa^{-1}$. Because $\alpha$ is pro-$\mathsf{V}$, Corollary 3.3.16 shows that $\lambda$ is pro-$\mathsf{V}$. As $\lambda$ is the relational morphism to $Q^{\mathsf{W}}$ associated to the congruence $\equiv$, we conclude that $\equiv_{\mathsf{V},Q^{\mathsf{W}}} \subseteq \equiv$ by Theorem 3.4.5. This completes the proof.    □

Specializing the above result to the case where $\mathsf{W}$ is a pseudovariety of semigroups $\mathbf{W}$ (viewed as a pseudovariety of relational morphisms via ((2.64))), we obtain the following.

**Theorem 3.6.4.** *Let* $\mathsf{V}$ *be a pseudovariety of relational morphisms and* $\mathbf{W}$ *a pseudovariety of semigroups. Then, for any finite set* $A$, *the equality*

$$\widehat{F_{\mathsf{Vq}(\mathbf{W})}}(A) = (\widehat{F_{\mathbf{W}}}(A))^{\mathsf{V}}$$

*holds.*

### 3.6.2 Inevitable substitutions

We remind the reader of some definitions. If $S$ is a finite semigroup and $s \in S$, then $s$ was defined to be Type II by Rhodes and Tilson [126, 295] if it relates to 1 under all relational morphisms from $S$ to a finite group. Ash [33] (and independently Ribes and Zalesskii [300]) proved a conjecture of the first author describing the Type II semigroup. The Type II Theorem yields decidability results for Mal'cev products of the form $\mathbf{V} \textcircled{m} \mathbf{G}$ where $\mathbf{G}$ is the pseudovariety of finite groups. A proof is given in Section 4.17.

If $\mathbf{V}$ is a pseudovariety of semigroups, then Henckell and Rhodes [121] defined a subset $X$ of a semigroup $S$ to be $\mathbf{V}$-*pointlike* if, for all relational morphisms $\varphi\colon S \to V$ with $V \in \mathbf{V}$, there exists $v \in V$ such that $X \subseteq v\varphi^{-1}$. These notions were generalized by Pin and Weil [235, 330] to the notion of $\mathbf{V}$-*idempotent pointlikes*. The definition is as per $\mathbf{V}$-pointlikes, only one requires $v \in V$ to be an idempotent. Hence the $\mathbf{G}$-idempotent pointlikes are just the Type II elements. Idempotent pointlikes can be used to determine membership in Mal'cev products.

Ash [33] gave a simultaneous generalization of pointlike sets with respect to $\mathbf{G}$ and Type II elements called *inevitable graphs*. Almeida [9] defined a notion of inevitable graphs that works for any pseudovariety $\mathbf{V}$ (and is equivalent with Ash's notion for pseudovarieties of groups). The reader is referred to Definition 2.5.14 for what we mean by a graph in this chapter. A *labeling* $\ell$ of a finite graph $\Gamma$ with vertex set $V(\Gamma)$ and edge set $E(\Gamma)$ over a semigroup $S$ consists of two maps, both called $\ell$ by abuse of notation, $\ell\colon V(\Gamma) \to S^I$ and $\ell\colon E(\Gamma) \to S$. The labeling $\ell$ is said to *commute* if $e\alpha\ell \cdot e\ell = e\omega\ell$ for each edge $e \in E(\Gamma)$, or equivalently $\ell \vdash C_\Gamma$ where $C_\Gamma$ is the set of consistency equations for $\Gamma$ from Example 3.5.9 (putting aside the issue of elements mapping to $I$). If $\varphi\colon S \to T$ is a relational morphism and $\ell, \ell'$ are labelings of $\Gamma$ over $S$ and $T$, respectively, then $\ell'$ is said to be $\varphi$-*related* to $\ell$ if $x\ell' \in x\ell\varphi$ for all vertices and edges $x$ of $\Gamma$. A labeling $\ell$ of a finite graph $\Gamma$ by a finite semigroup $S$ is $\mathbf{V}$-*inevitable* in the sense of Almeida [9] if, for any relational morphism $\varphi\colon S \to V$ with $V \in \mathbf{V}$, there exists a commuting labeling $\ell'$ of $\Gamma$ over $V$, which is $\varphi$-related to $\ell$. A pseudovariety $\mathbf{V}$ is said to be *hyperdecidable* [9] if $\mathbf{V}$-inevitability of graphs labeled over finite semigroups is decidable. The motivation was to obtain membership results and basis theorems for semidirect products of pseudovarieties [9, 19, 20, 27].

As an example, if $\Gamma$ is a graph consisting of a single vertex $v$ and edge set $E$ (necessarily consisting of loop edges), then a labeling $\ell\colon \Gamma \to S^I$ is $\mathbf{V}$-inevitable if and only if, for every relational morphism $\varphi\colon S \to T$ with $T \in \mathbf{V}$, there exists $t \in T^I$ such that $t \in v\ell\varphi^I$ and $E\ell \subseteq \mathrm{Stab}(t)\varphi^{-1}$ where $\mathrm{Stab}(t)$ is the right stabilizer of $t$ in $T$. A non-empty subset $X \subseteq S$ is $\mathbf{V}$-pointlike if and only if the labeling of the graph with two vertices $v_1, v_2$ and $|X|$ edges from $v_1$ to $v_2$ given by sending $v_1$ to $I$, $v_2$ to any element of $X$ and the $|X|$ edges to the elements of $X$ is $\mathbf{V}$-inevitable.

**Exercise 3.6.5.** Verify these last two assertions.

Our work on pseudovarieties of relational morphisms leads us in a natural way to consider substitutions that are inevitable with respect to a system of equations. All the above notions will then be special cases.

We shall often need to discuss collections of sets of pseudoidentities in this chapter. To avoid ambiguity, we introduce a special terminology for sets of pseudoidentities and collections of such sets.

**Definition 3.6.6 (System of pseudoidentities).** *By a system of pseudoidentities we mean formally a pair $(A, E)$ where $A$ is a finite alphabet and $E$ is a set of pseudoidentities defined over $A$. Normally we drop $A$ from the notation.*

The next definition is crucial for the remainder of this chapter.

**Definition 3.6.7 (Inevitable substitution).** *Let $E$ be a system of pseudoidentities over a finite alphabet $A$, $\mathsf{W}$ a pseudovariety of relational morphisms and $\varphi\colon S \to T$ a relational morphism of finite semigroups. A substitution $\sigma\colon A \to S$ is said to be $(\mathsf{W}, E)$-inevitable for $\varphi$ if, for every relational morphism $(\rho, f)\colon \varphi \to \psi$ with $\psi\colon S' \to T'$ in $\mathsf{W}$, there exists a substitution $\tau\colon A \to S'$ such that $\tau$ $\rho$-relates to $\sigma$ and $\tau \vdash E$.*

Note that the inevitability of a substitution depends *not* only on $S$, but on $\varphi$. However, if $\mathsf{W}$ is a pseudovariety of semigroups (viewed as a pseudovariety of relational morphisms via (2.64)), then the notion depends only on $S$. More formally, for a pseudovariety of semigroups $\mathsf{W}$, we define a substitution $\sigma\colon A \to S$ to be $(\mathsf{W}, E)$-*inevitable* if, for all relational morphisms $\rho\colon S \to S'$ with $S' \in \mathsf{W}$, there exists a substitution $\tau\colon A \to S'$, which is $\varphi$-related to $\sigma$ and such that $\tau \vdash E$.

The intuition is that if $\varphi\colon S \to T$ is a fixed relational morphism, then an $A$-tuple of elements of $S$ is $(\mathsf{W}, E)$-inevitable for $\varphi$ if, for all relational morphisms from $(\rho, f)\colon \varphi \to \psi$ with $\psi\colon S' \to T'$ in $\mathsf{W}$, we can always find an $A$-tuple of elements of $S'$ that satisfies $E$ (as relations) and that pulls back to our original $A$-tuple in $S$. Let us proceed with some examples of the notion.

*Example 3.6.8 (Pointlikes).* Fix a pseudovariety of semigroups $\mathsf{W}$ and a finite semigroup $S$. If $A_n = \{x_1, \ldots, x_n\}$ and $E_n = \{x_1 = \cdots = x_n\}$, then a substitution $\sigma\colon A \to S$ is $(\mathsf{W}, E)$ inevitable if and only if $A\sigma$ is $\mathsf{W}$-pointlike. Indeed, inevitability of the substitution is equivalent to saying that, for any relational morphism $\varphi\colon S \to T$ with $T \in \mathsf{W}$, there is a substitution $\tau\colon A \to T$ so that $x_1\tau = x_2\tau = \cdots = x_n\tau$ and $x_i\tau \in x_i\sigma\varphi$, all $i = 1, \ldots, n$. But this is exactly the same as saying $A\sigma$ is $\mathsf{W}$-pointlike because all the $x_i\tau$ are one and the same element!

*Example 3.6.9 (Idempotent pointlikes).* Again, let $\mathsf{W}$ be a pseudovariety of semigroups, $S$ a finite semigroup, $A = \{x_1, \ldots, x_n\}$ and $\sigma\colon A \to S$ a substitution. If $E = \{x_1 = \cdots = x_n = x_1^2\}$, then $\sigma$ is $(\mathsf{W}, E)$-inevitable if and only if $A\sigma$ is $\mathsf{W}$-idempotent pointlike.

*Example 3.6.10 (Inevitable graphs).* Let $\Gamma$ be a graph with vertex set $V$ and edge set $E$. A labeling $\ell$ of $\Gamma$ over $S$ yields, abusing notation, a substitution $\ell\colon V \cup E \to S$ (where if a vertex is mapped to $I$, the corresponding vertex is deleted from the variable set). Such a labeling is **W**-inevitable in the sense of Almeida [9] if and only if $\ell$ is a $(\mathbf{W}, C_\Gamma)$-inevitable substitution where $C_\Gamma$ is the set of consistency equations of the graph (where labeling a vertex by $I$ entails replacing the corresponding vertex in $V$ by the empty string in $C_\Gamma$).

The following definition is crucial.

**Definition 3.6.11 (Witness).** *Let $\varphi\colon S \to T$ be a relational morphism of finite semigroups and $E$ a system of pseudoidentities over a finite alphabet $A$. A relational morphism $(\rho, f)\colon \varphi \to \psi$ with $\psi\colon S' \to T'$ in* W *is said to witness* $(\mathbf{W}, E)$-*inevitability for $\varphi$ if, for every substitution $\sigma\colon A \to S$, we have that $\sigma$ is $\rho$-related to a substitution $\tau\colon A \to S'$ such that $\tau \vdash E$ if and only if $\sigma$ is* $(\mathbf{W}, E)$-*inevitable for $\varphi$.*

That is, intuitively speaking, a substitution is $(\mathbf{W}, E)$-inevitable for $\varphi$ if and only if the relational morphism $(\rho, f)$ says it is so; said differently, we only need to consider $(\rho, f)$ when checking $(\mathbf{W}, E)$-inevitability for $\varphi$ of a substitution $\sigma\colon A \to S$. See [9, 33, 121, 126, 322, 330] for more on this idea. In the next few results, we keep the above notation. We begin our study with a compactness result.

**Proposition 3.6.12.** *Let $E$ be a system of pseudoidentities over a finite alphabet $A$ and let $\varphi\colon S \to T$ be a relational morphism of finite semigroups. Then there exists a relational morphism $(\rho, f)\colon \varphi \to \psi$ with $\psi \in$ W witnessing* $(\mathbf{W}, E)$-*inevitability for $\varphi$.*

*Proof.* As $A$ and $S$ are finite, there are only finitely many substitutions from $A$ to $S$. For each substitution $\sigma\colon A \to S$ that is not $(\mathbf{W}, E)$-inevitable for $\varphi$, choose a relational morphism $(\rho_\sigma, f_\sigma)\colon \varphi \to \psi_\sigma$ with $\psi_\sigma\colon S_\sigma \to T_\sigma$ in W such that $\sigma$ is not $\rho_\sigma$-related to any substitution $\tau\colon A \to S_\sigma$ such that $\tau \vdash E$. It is then easily verified that $(\Delta \prod \rho_\sigma, \Delta \prod f_\sigma)\colon \varphi \to \prod \psi_\sigma$ witnesses $(\mathbf{W}, E)$-inevitability for $\varphi$.   $\square$

The next couple of results allow us to find relational morphisms witnessing $(\mathbf{W}, E)$-inevitability of a nicer form, in terms of generators.

**Lemma 3.6.13.** *The conclusion of Proposition 3.6.12 can be strengthened to demand $f$ is surjective.*

*Proof.* Let $(\rho, f)\colon \varphi \to \psi$ with $\psi \in$ W witness $(\mathbf{W}, E)$-inevitability for $\varphi$, as per Proposition 3.6.12. Say $\psi\colon S' \to T'$. For each $s \in S$ and $t \in s\varphi$, choose $s'_{s,t} \in s\rho$ such that $tf \in s'_{s,t}\psi$; such an element exists because $\varphi f \subseteq_s \rho\psi$. Let $S'' = \langle s'_{s,t} \mid (s,t) \in \#\varphi \rangle$ and set $T'' = \text{Im } f$. Denote by $\psi'\colon S'' \to T''$ the restriction of $\psi$. This is a relational morphism because $tf \in s'_{s,t}\psi'$ and the $s'_{s,t}$ generate $S''$. Now define a relational morphism $(\rho', f^0)\colon \varphi \to \psi'$ by setting the graph of $\rho'\colon S \to S''$ equal to $\#\rho \cap (S \times S'')$ and letting $f^0\colon T \twoheadrightarrow T''$ be the corestriction. By construction, $\rho'$ is a relational morphism and $\varphi f^0 \subseteq_s \rho'\psi'$. Notice $\rho' \subseteq \rho$.

Suppose $\sigma\colon A \to S$ is a substitution that $\rho'$-relates to $\tau\colon A \to S''$. Moreover, assume $\tau \vdash E$. Let $\iota\colon S'' \hookrightarrow S'$ be the inclusion. Then $\sigma$ $\rho$-relates to $\tau\iota$ and $\tau\iota \vdash E$. Hence $\sigma$ is $(\mathsf{W}, E)$-inevitable for $\varphi$, as required. We conclude that $(\rho', f^0)$ witnesses $(\mathsf{W}, E)$-inevitability for $\varphi$. $\qquad\square$

**Lemma 3.6.14.** *Let $(\rho, f)\colon \varphi \to \psi$ and $(\gamma, f')\colon \psi \to \psi'$ be relational morphisms with $\psi, \psi' \in \mathsf{W}$ and $\gamma$ a homomorphism. Suppose, furthermore, that $(\rho\gamma, ff')$ witnesses $(\mathsf{W}, E)$-inevitability for $\varphi$. Then $(\rho, f)$ witnesses $(\mathsf{W}, E)$-inevitability for $\varphi$.*

*Proof.* Suppose $\varphi\colon S \to T$ and $\psi\colon S' \to T'$. Assume $\sigma\colon A \to S$ is a substitution that is $\rho$-related to a substitution $\tau\colon A \to S'$ such that $\tau \vdash E$. Then $\tau\gamma \vdash E$ and $\sigma$ $\rho\gamma$-relates to $\tau\gamma$, so $\sigma$ is $(\mathsf{W}, E)$-inevitable for $\varphi$. $\qquad\square$

We remark that if $\gamma$ is injective, then $\psi \prec \psi'$ and so we need only assume that $\psi' \in \mathsf{W}$. An immediate corollary of the previous results is

**Corollary 3.6.15.** *Let $\varphi\colon S \to T$ be a relational morphism and $\mathsf{W}$ a pseudovariety of relational morphisms. Then there is a relational morphism*

$$(\rho, 1_T)\colon \varphi \to \psi$$

*with $\psi \in \mathsf{W}$ witnessing $(\mathsf{W}, E)$-inevitability for $\varphi$.*

*Proof.* Let $(\rho, f)\colon \varphi \to \psi'$ with $\psi' \in \mathsf{W}$ and $f$ surjective witness $(\mathsf{W}, E)$-inevitability for $\varphi$ as per Proposition 3.6.12 and Lemma 3.6.13. Let us say $\psi'\colon S' \to T'$. Define $\psi\colon S' \to T$ by $\psi = \psi'f^{-1}$. Then because $\psi \prec \psi'$ (as $\psi f = \psi'f^{-1}f = \psi'$), it follows $\psi \in \mathsf{W}$. Also $(\rho, 1_T)\colon \varphi \to \psi$ is a relational morphism, as

$$\varphi \subseteq \varphi ff^{-1} \subseteq_s \rho\psi'f^{-1} = \rho\psi.$$

Because $(\rho, f) = (\rho, 1_T)(1_{S'}, f)$, we see that the pair $(\rho, 1_T)$ witnesses $(\mathsf{W}, E)$-inevitability for $\varphi$ by Lemma 3.6.14. $\qquad\square$

**Lemma 3.6.16.** *Suppose $(\rho, f)\colon \varphi \to \psi$ is a relational morphism with $\psi \in \mathsf{W}$ and that $\rho \subseteq_s \rho'$ (note: this implies $(\rho', f)\colon \varphi \to \psi$ is a relational morphism). Then if $(\rho', f)$ witnesses $(\mathsf{W}, E)$-inevitability for $\varphi$, so does $(\rho, f)$.*

*Proof.* Assume that $\psi\colon S' \to T'$ and so $\rho\colon S \to S'$. Suppose that $\sigma\colon A \to S$ is a substitution that $\rho$-relates to $\tau\colon A \to S'$. Moreover, assume $\tau \vdash E$. Then $\sigma$ $\rho'$-relates to $\tau$ and hence $\sigma$ is $(\mathsf{W}, E)$-inevitable for $\varphi$, as desired. $\qquad\square$

**Corollary 3.6.17.** *Let $\varphi\colon S \to T$ be a B-generated relational morphism. Then there exist B-generated relational morphisms $\rho\colon S \to R$ and $\psi\colon R \to T$ such that $\varphi \subseteq_s \rho\psi$, $\psi \in \mathsf{W}$, and $(\rho, 1_T)$ witnesses $(\mathsf{W}, E)$-inevitability for $\varphi$.*

*Proof.* Let $(\beta, 1_T)\colon \varphi \to \psi'$ be a relational morphism witnessing $(\mathsf{W}, E)$-inevitability for $\varphi$ with $\psi'\colon R' \to T \in \mathsf{W}$ as per Corollary 3.6.15. Let $\alpha\colon B \to \#\varphi$ be the generating map. Because $\varphi \subseteq_s \beta\psi'$, for each $b \in B$, there exists $r_b \in R'$ such that $b\alpha p_S \; \beta \; r_b \; \psi' \; b\alpha p_T$. Let $R$ be the subsemigroup of $R'$ generated by the $r_b$, $b \in B$, let $\rho\colon S \to R$ be the relational morphism whose graph is generated by the pairs $(b\alpha p_S, r_b)$, $b \in B$, and let $\psi\colon R \to T$ be the relational morphism whose graph is generated by the pairs $(r_b, b\alpha p_T)$, $b \in B$. So we have the following subcommutative diagram

$$
\begin{array}{ccccc}
S & \xrightarrow{\;\rho\;} & R & \overset{\iota}{\hookrightarrow} & R' \\
{\scriptstyle\varphi}\downarrow & {\subseteq} & \downarrow{\scriptstyle\psi} & {\subseteq} & \downarrow{\scriptstyle\psi'} \\
T & \xrightarrow[1_T]{} & T & \xrightarrow[1_T]{} & T
\end{array}
$$

and, moreover, $\rho\iota \subseteq_s \beta$. Because $\psi \prec \psi'$, we have $\psi \in \mathsf{W}$. By Lemma 3.6.16, $(\rho\iota, 1_T)$ witnesses $(\mathsf{W}, E)$-inevitability for $\varphi$. Thus $(\rho, 1_T)$ witnesses $(\mathsf{W}, E)$-inevitability for $\varphi$ by Lemma 3.6.14, completing the proof of the corollary.  □

For locally finite pseudovarieties of relational morphisms, finding a witness reduces to looking at canonical relational morphisms to free objects, as the next corollary shows.

**Corollary 3.6.18.** *Suppose that* $\mathsf{W}$ *is a locally finite pseudovariety and let* $\varphi\colon S \to T$ *be a $B$-generated relational morphism of semigroups ($B$ finite). Let* $\rho\colon S \to T^{\mathsf{W}}$ *be the canonical $B$-generated relational morphism. Then* $(\rho, 1_T)\colon \varphi \to \varphi_{\mathsf{W},T}$ *witnesses $(\mathsf{W}, E)$-inevitability for* $\varphi$.

*Proof.* By Corollary 3.6.17, there exists a $B$-generated relational morphism $\rho'\colon S \to R$ so that the canonical $B$-generated relational morphism $\psi'\colon R \to T$ belongs to $\mathsf{W}$ and $(\rho', 1_T)$ witnesses $(\mathsf{W}, E)$-inevitability for $\varphi$. By Corollary 3.4.6, the relational morphism $\rho'$ factors as $\rho\gamma$ with $\gamma\colon T^{\mathsf{W}} \to R$ the canonical surjective homomorphism as per

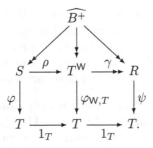

The result then follows from Lemma 3.6.14.  □

Recall from Definition 3.6.6 that by a system of pseudoidentities, we mean formally a pair $(A, E)$ where $A$ is a finite alphabet and $E$ is a set of pseudoidentities in variables $A$.

**Definition 3.6.19 (Collection of systems of pseudoidentities).** *By a collection of systems of pseudoidentities, we mean formally a set $\mathscr{C}$ whose elements are systems of pseudoidentities in the sense of Definition 3.6.6.*

We do not require in this definition that each system of pseudoidentities has the same alphabet, nor that there is any bound on the sizes of the alphabets of elements of $\mathscr{C}$. For example, if $\Gamma$ is a finite graph with vertex set $V$ and edge set $E$, then the set of consistency equations $C_\Gamma$ for $\Gamma$ is a system of pseudoidentities because the elements of $C_\Gamma$ are all defined over the finite set $V \cup E$; formally speaking $(V \cup E, C_\Gamma)$ is the system of pseudoidentities but we omit $V \cup E$ from the notation. On the other hand if $\{\Gamma_i \mid i \in J\}$ is a set of finite graphs, then $\mathscr{C} = \{C_{\Gamma_i} \mid i \in J\}$ is a collection of systems of pseudoidentities. As another example, $E_n = \{x_1 = x_2, x_1 = x_3, \dots, x_1 = x_n\}$ is a system of pseudoidentities over the alphabet $\{x_1, \dots, x_n\}$; again formally the pair $(\{x_1, \dots, x_n\}, E_n)$ is a system of pseudoidentities, but again we drop the alphabet from the notation. The collection $\mathscr{C}' = \{E_n \mid n \in \mathbb{N}\}$ is a collection of systems of pseudoidentities.

Let $\mathscr{C}$ be a collection of systems of pseudoidentities, $\varphi\colon S \to T$ a relational morphism and $\mathsf{W} \in \mathbf{PVRM}$. We say a relational morphism $(\rho, f)\colon \varphi \to \psi$ with $\psi\colon S' \to T'$ in $\mathsf{W}$, *witnesses* $(\mathsf{W}, \mathscr{C})$-*inevitability for $\varphi$* if it witnesses $(\mathsf{W}, E)$-inevitability for $\varphi$ for each system of pseudoidentities $E$ in the collection $\mathscr{C}$. The following is an immediate consequence of Corollary 3.6.18.

**Theorem 3.6.20.** *Let $\mathscr{C}$ be a collection of systems of pseudoidentities and $\mathsf{W}$ a locally finite pseudovariety of relational morphisms. Suppose $\varphi\colon S \to T$ is a $B$-generated relational morphism. Let $(\rho, 1_T)\colon \varphi \to \varphi_{\mathsf{W},T}$ be the canonical relational morphism (considered above). Then $(\rho, 1_T)$ witnesses $(\mathsf{W}, \mathscr{C})$-inevitability for $\varphi$.*

In general, one cannot always find a relational morphism that witnesses $(\mathsf{W}, \mathscr{C})$-inevitability as we shall see in Example 3.6.38. However, one does have the following result.

**Proposition 3.6.21.** *Suppose $\mathscr{C}$ is a finite collection of systems of pseudoidentities and $\varphi\colon S \to T$ is a relational morphism. Then, for any pseudovariety $\mathsf{W}$ of relational morphisms, there exists a relational morphism $(\rho, 1_T)\colon \varphi \to \psi$ with $\psi \in \mathsf{W}$ witnessing $(\mathsf{W}, \mathscr{C})$-inevitability for $\varphi$. Moreover, if $\varphi$ is $B$-generated, then $\rho$ and $\psi$ may be assumed to be $B$-generated.*

*Proof.* Choose, for each $E \in \mathscr{C}$, a relational morphism $(\rho_E, f_E)\colon \varphi \to \psi_E$ with $\psi_E \in \mathsf{W}$ witnessing $(\mathsf{W}, E)$-inevitability for $\varphi$. Then $(\Delta_S \prod \rho_E, \Delta_T \prod f_E)$ witnesses $(\mathsf{W}, \mathscr{C})$-inevitability for $\varphi$ by Lemma 3.6.14 applied to each $(A, E) \in \mathscr{C}$. The remaining assertions are then proved as in the case of a single system of pseudoidentities. $\square$

In fact, one can often do better. But to do so requires some rather technical definitions, which are introduced in the next subsection.

### 3.6.3 Finitely equivalent collections

An extended example may help in understanding the somewhat technical, but important, notion that we are about to define. This example is based on Example 3.6.8. Let $A_n$ be the alphabet $\{x_1, \ldots, x_n\}$ and let $E_n$ be the system of pseudoidentities $x_1 = \cdots = x_n$ over $A_n$. We saw in Example 3.6.8 that a substitution $\sigma \colon A_n \to S$ is $(\mathbf{W}, E_n)$-inevitable for a pseudovariety $\mathbf{W}$ if and only if $A_n\sigma$ is a $\mathbf{W}$-pointlike set. Let us set $\mathscr{C} = \{E_n \mid n \in \mathbb{N}\}$. What does it mean for a relational morphism $\varphi \colon S \to T$ with $T \in \mathbf{W}$ to witness $(\mathbf{W}, \mathscr{C})$-inevitability for $S$? It means that a substitution $\sigma \colon A_n \to S$ with $n \geq 1$ is $(\mathbf{W}, E_n)$-inevitable if and only if there is a substitution $\tau \colon A_n \to T$ so that $x_1\tau = x_2\tau = \cdots = x_n\tau = t$ (so $\tau$ satisfies $E_n$ as relations) with $x_i\tau = t \in x_i\sigma\varphi$, each $i = 1, \ldots, n$. Equivalently, $\varphi$ is a relational morphism witnessing $\mathbf{W}$-pointlikes in the sense that $X \subseteq S$ is $\mathbf{W}$-pointlike if and only if $X \subseteq t\varphi^{-1}$ some $t \in T$.

The collection $\mathscr{C}$ is an infinite set so Proposition 3.6.21 does not apply to find a relational morphism $\varphi \colon S \to T$ with $T \in \mathbf{W}$ witnessing $(\mathbf{W}, \mathscr{C})$-inevitability. Nonetheless, it is well-known that there always exists a relational morphism witnessing $\mathbf{W}$-pointlikes [121]. Namely, for each subset $X \subseteq S$ that is not $\mathbf{W}$-pointlike, choose a relational morphism $\varphi_X \colon S \to W_X$ with $W_X \in \mathbf{W}$ and so that $X \not\subseteq w\varphi_X^{-1}$ for any $w \in W_X$. Then $\Delta \prod_{X \notin \mathrm{PL}_{\mathbf{W}}(S)} \varphi_X$ is easily checked to be the desired witness, meaning that a subset $X \subseteq S$ is $\mathbf{W}$-pointlike if and only if $X \subseteq w\varphi^{-1}$ some $w \in \prod W_X$. What makes this argument work is that there is some finiteness going on here despite $\mathscr{C}$ being infinite, namely $S$ has only finitely many subsets. Substitutions defined on alphabets that are too big do not yield any new information.

Continuing with our example, if $S$ is a finite semigroup of order $n$ and $m > n$, then any substitution $\sigma \colon A_m \to S$ is not injective. We should therefore be able to reduce our considerations to an alphabet of size at most $n$. Let us formalize this. Suppose $S = \{s_1, \ldots, s_n\}$. Define $\beta \colon A_m \to A_n$ by $a_i\beta = a_j$ if $a_i\sigma = s_j$. For each $a_i \in A_n$, choose a preimage $a_i\gamma$ of $a_i$ under $\beta$, if there is one, and otherwise choose $\gamma$ arbitrarily. So $\gamma \colon A_n \to A_m$ is a substitution. Notice that $\beta\gamma\sigma = \sigma$ by construction. Then it is not hard to see that $\sigma$ is $(\mathbf{W}, E_m)$-inevitable if and only if $\gamma\sigma$ is $(\mathbf{W}, E_n)$-inevitable. Indeed, $A_m\sigma = A_n\gamma\sigma$ and so inevitability of both subsets corresponds with this one set being $\mathbf{W}$-pointlike. So to witness $(\mathbf{W}, \mathscr{C})$-inevitability we need only consider the finite subcollection $\{E_1, \ldots, E_n\}$ of $\mathscr{C}$. Slightly hidden under the rug in this argument is that the sets of equations $E_n$ and $E_m$ are not entirely unrelated.

Our next definition attempts to abstract this argument and several others that have appeared in the literature (cf. [9, 126, 235]). Let us introduce the following notation: if $\sigma \colon A \to \widehat{B^+}$ is a substitution and $E$ is a system of pseudoidentities over $A$, we set

$$E\sigma = \{u\sigma = v\sigma \mid u = v \in E\}.$$

Note that if $\tau\colon B \to S$, then $\tau \vdash E\sigma$ if and only if $\sigma\tau \vdash E$.

**Definition 3.6.22 (Finitely equivalent collection).** *A collection $\mathscr{C}$ of systems of pseudoidentities is said to be finitely equivalent if, for each $n$, there are only finitely many systems of pseudoidentities $(A, E) \in \mathscr{C}$ with $|A| \leq n$ and there is a number $f(n)$ such that the following holds. For each system of pseudoidentities $(A, E) \in \mathscr{C}$ with $|A| > f(n)$ and each substitution $\sigma\colon A \to S$ with $S$ a finite semigroup of order $n$, there exists $(C, E') \in \mathscr{C}$ with $|C| \leq f(n)$ and substitutions $\beta\colon A \to C$ and $\gamma\colon C \to A$ such that: $\beta\gamma\sigma = \sigma$, $E \vdash E'\gamma$, and $E' \vdash E\beta$.*

If $f$ is a computable function, we say $\mathscr{C}$ is *computably finitely equivalent*. Note that any finite collection $\mathscr{C}$ is computably finitely equivalent; just set $f(n)$ to be greater than the cardinality of the largest set of variables appearing in any system of pseudoidentities $E \in \mathscr{C}$.

*Example 3.6.23 (Pointlikes).* Our motivating example is a computably finitely equivalent collection. Recall that $A_n = \{x_1, \ldots, x_n\}$ and $E_n$ consists of the pseudoidentities $x_1 = \cdots = x_n$. The collection in question is $\mathscr{C} = \{E_n \mid n \geq 1\}$. There is only one system of pseudoidentities in this collection in $n$ variables. The argument before Definition 3.6.22 shows precisely that $\mathscr{C}$ is finitely equivalent and that we may take $f(n) = n$. The reader should verify $E \vdash E'\gamma$, and $E' \vdash E\beta$.

We now give some further examples of finitely equivalent collections $\mathscr{C}$.

*Example 3.6.24 (V-like sets).* Let $\mathbf{V}$ be a pseudovariety of semigroups and $\mathscr{C}_{\mathbf{V}}$ consist of all systems of pseudoidentities of the form $(A_n, E_{\mathbf{V},n})$ where $\langle A_n \mid E_{\mathbf{V},n} \rangle$ is a presentation of a free pro-$\mathbf{V}$ semigroup over the $n$ element alphabet $A_n = \{x_1, \ldots, x_n\}$; see Definition 3.1.45. We are abusing here the distinction between the system of pseudoidentities $E_{\mathbf{V},n}$ over $A_n$ and the subset $E_{\mathbf{V},n} \subseteq \widehat{A^+} \times \widehat{A^+}$. We claim that $\mathscr{C}_{\mathbf{V}}$ is computably finitely equivalent; indeed one can take $f(n) = n$.

Suppose $(A_k, E_{\mathbf{V},k}) \in \mathscr{C}_{\mathbf{V}}$ with $k > n$ and let $\sigma\colon A_k \to S$ be a substitution where $|S| = n$. Set $C = A_n$ and choose a bijection $f\colon S \to C$. Let $E' = E_{\mathbf{V},n}$, $\beta = \sigma f$ and let $\gamma\colon C \to A_k$ be any map such that $c\gamma\beta = c$ for all $c$ in the image of $\beta$. Then, for all $a \in A_k$, we have

$$a\beta\gamma\sigma = a\beta\gamma\beta f^{-1} = a\beta f^{-1} = a\sigma$$

so $\beta\gamma\sigma = \sigma$. Standard Universal Algebra shows $\mathbf{V} \models E_{\mathbf{V},k}\beta$ and $\mathbf{V} \models E_{\mathbf{V},n}\gamma$, whence $E_{\mathbf{V},n} \vdash E_{\mathbf{V},k}\beta$ and $E_{\mathbf{V},k} \vdash E_{\mathbf{V},n}\gamma$. This establishes that $\mathscr{C}_{\mathbf{V}}$ is finitely equivalent.

The collection of systems of pseudoidentities defining idempotent pointlikes in Example 3.6.9 is a special case of the above situation, where $\mathbf{V} = \mathbf{1}$. More generally, we have the following definition.

**Definition 3.6.25 (V-like).** *A subset $X$ of a finite semigroup $S$ is said to be* **V**-*like with respect to* **W** *if, for all relational morphisms $\varphi\colon S \to W$ with $W \in \mathbf{W}$, there exists $W \geq T \in \mathbf{V}$ such that $X \subseteq T\varphi^{-1}$.*

A subset is **1**-like with respect to **W** if and only if it is **W**-idempotent pointlike. One says **W** has *decidable* **V**-like sets, if one can decide whether a subset of a finite semigroup is **V**-like with respect to **W**.

We claim a substitution $\sigma\colon A_k \to S$ is $(\mathbf{W}, E_{\mathbf{V},k})$-inevitable if and only if $A_k\sigma$ is **V**-like with respect to **W**. Indeed, if $A_k\sigma$ is **V**-like with respect to **W** and $\varphi\colon S \to W$ is a relational morphism with $W \in \mathbf{W}$, then there exists $W \geq T \in \mathbf{V}$ such that $A_k\sigma \subseteq T\varphi^{-1}$. Choose, for each $a \in A_k$, $t_a \in a\sigma\varphi \cap T$. Then $\{t_a \mid a \in A_k\}$ generates a semigroup in **V**. Hence, defining $\tau\colon A_k \to W$ by $a\tau = t_a$, we see $\tau \vdash E_{\mathbf{V},k}$ as $\langle A_k \mid E_{\mathbf{V},k}\rangle$ presents a *free* pro-**V** semigroup on $A_k$.

Conversely, if $\sigma$ is $(\mathbf{W}, E_{\mathbf{V},k})$-inevitable and $\varphi\colon S \to W$ with $W \in \mathbf{W}$ is a relational morphism, then there is a substitution $\tau\colon A_k \to W$ such that $\tau$ $\varphi$-relates to $\sigma$ and $\tau \vdash E_{\mathbf{V},k}$. But then $T = \widehat{A_k^+}\tau$ is in **V** (as $E_{\mathbf{V},k}$ presents a free pro-**V** semigroup) and $A_k\sigma \subseteq T\varphi^{-1}$. We conclude $A_k\sigma$ is **V**-like with respect to **W**.

We now prove that there is a relational morphism witnessing $(\mathbf{W}, \mathscr{C}_{\mathbf{V}})$-inevitable substitutions for a finite semigroup $S$. This will be generalized in Proposition 3.6.29 for arbitrary finitely equivalent $\mathscr{C}$. The argument here imitates the case of pointlike sets considered above, Type I semigroups [295] and Type II semigroups [126, 295]. Because $S$ has only finitely many subsets $A$ that are not **V**-like with respect to **W** ($S$ being finite), we can find, for each such subset $X$, a relational morphism $\varphi_X\colon S \to W_X$ with $W \in \mathbf{W}$ such that there is no subsemigroup $T$ of $W_X$ with $T \in \mathbf{V}$ and $X \subseteq T\varphi_X^{-1}$. Then $\rho = \Delta \prod \varphi_X$ clearly witnesses **V**-like subsets with respect to **W**.

If the **V**-like subsets with respect to **W** are decidable, **W** is recursively enumerable, and **V** has decidable membership, then one can effectively find such a relational morphism. Indeed, one can enumerate all relational morphisms from a finite semigroup $S$ to an element of **W**. For each such relational morphism, calculate the inverse images of semigroups in **V** under such relational morphisms and see if the relational morphism witnesses **V**-like subsets. If it does, stop; otherwise, continue. This algorithm must eventually stop. This argument will be generalized later in Theorem 3.6.46.

The following proposition motivates Theorem 3.6.30; compare with [126, 235].

**Proposition 3.6.26.** *Suppose* $\mathbf{U}, \mathbf{V}, \mathbf{W}$ *are pseudovarieties of semigroups and assume $\rho\colon S \to W$, with $W \in \mathbf{W}$, witnesses **V**-like subsets of $S$ with respect to **W**. Then $S \in (\mathbf{U}, \mathbf{V}) \, \textcircled{m} \, \mathbf{W}$ if and only if $\rho \in (\mathbf{U}, \mathbf{V})$, if and only if, for each **V**-like **subsemigroup** $X$ of $S$, one has $X \in \mathbf{U}$. Hence if **V**-like subsets with respect to **W** are decidable and **U** is decidable, then $(\mathbf{U}, \mathbf{V}) \, \textcircled{m} \, \mathbf{W}$ is decidable.*

*Proof.* If $\rho \in (\mathbf{U}, \mathbf{V})$, then clearly $S \in (\mathbf{U}, \mathbf{V}) \, \textcircled{m} \, \mathbf{W}$. Conversely, suppose there exists a relational morphism $\psi \colon S \to W'$ with $\psi \in (\mathbf{U}, \mathbf{V})$ and $W' \in \mathbf{W}$. Let $W \geq T \in \mathbf{V}$. Then $X = T\rho^{-1}$ is $\mathbf{V}$-like with respect to $\mathbf{W}$. Hence there exists $W' \geq T' \in \mathbf{V}$ such that $X \subseteq T'\psi^{-1}$. Because $T'\psi^{-1} \in \mathbf{U}$, we see $X \in \mathbf{U}$. Thus $\rho \in (\mathbf{U}, \mathbf{W})$, as required.

It is clear $\rho \in (\mathbf{U}, \mathbf{W})$ if and only if each $\mathbf{V}$-like subsemigroup $X$ of $S$ (with respect to $\mathbf{W}$) belongs to $\mathbf{U}$. The last statement then follows. $\qquad \square$

*Example 3.6.27 (Finite vertex rank collections of graphs).* As another example, let $\mathscr{G}_k$ be the collection of all finite graphs with at most $k$ vertices (up to isomorphism) such that there are two coterminal paths $p$ and $q$ in $\varGamma$ with $(\varGamma, p = q)$ a path identity. Let $\mathscr{C}_{\mathscr{G}_k}$ be the collection of all the systems of pseudoidentities expressing the consistency equations $C_\varGamma, C'_\varGamma$ of graphs in $\mathscr{G}_k$; see Example 3.5.9. We claim $\mathscr{C}_{\mathscr{G}_k}$ is computably finitely equivalent. Let $f(n)$ be the sum of the number of vertices and edges in a graph with $k$ vertices and $n$ edges connecting each pair of vertices. Suppose $F \in \mathscr{C}_{\mathscr{G}_k}$ is the set of equations expressing the consistency equations of a graph $\varGamma$ with vertex set $V$ and edge set $E$ where $|V| + |E| > f(n)$. Let $A$ be the union of $V$ and $E$, less the missing vertex if $F$ is a set of consistency equations $C'_\varGamma$. Suppose $\sigma \colon A \to S$ is a substitution where $|S| = n$. We think of $\sigma$ as labeling of $A$ over $S$. By choice of $f(n)$, there are coterminal edges with the same label. Let $C = V \cup E'$ be the subgraph obtained from $A$ by leaving only one of any collection of coterminal edges with the same label and let $F'$ be the consistency equations for $C$; so $F' \in \mathscr{C}_{\mathscr{G}_k}$. Then, by choice of $f(n)$, $|C| \leq f(n)$. Define $\gamma \colon C \to A$ to be the inclusion and $\beta \colon A \to C$ the map that identifies coterminal edges with the same label. Then $\beta\gamma\sigma = \sigma$. It is clear that $F \vdash F'\gamma$ and $F' \vdash F\beta$.

*Remark 3.6.28.* Analogous results hold for the bilateral consistency equations.

The following technical proposition, along with the above examples, justifies the interest in finitely equivalent collections.

**Proposition 3.6.29.** *Suppose $\mathscr{C}$ is a finitely equivalent collection of systems of pseudoidentities and let $\mathscr{C}' \subseteq \mathscr{C}$. Let $\varphi \colon S \to T$ be a relational morphism of finite semigroups. Then there exists a relational morphism $(\rho, 1_T) \colon \varphi \to \psi$ with $\psi \in \mathsf{W}$ witnessing $(\mathsf{W}, \mathscr{C}')$-inevitability for $\varphi$. Moreover, if $\varphi$ is $B$-generated, then $\rho$ and $\psi$ may be assumed to be $B$-generated.*

*Proof.* Clearly any relational morphism witnessing $(\mathsf{W}, \mathscr{C})$-inevitability for $\varphi$ a *fortiori* witnesses $(\mathsf{W}, \mathscr{C}')$-inevitability for $\varphi$, so we just work with $\mathscr{C}$. Let $n = |S|$ and let $\mathscr{C}_n$ be the set of all $E \in \mathscr{C}$ over at most $f(n)$ variables; this is a finite set by assumption. By Proposition 3.6.21, there exists a relational morphism $(\rho, 1_T) \colon \varphi \to \psi$, with $\psi \colon S' \to T$ in $\mathsf{W}$, witnessing $(\mathsf{W}, \mathscr{C}_n)$-inevitability for $\varphi$. We show that, in fact, $(\rho, 1_T)$ witnesses $(\mathsf{W}, \mathscr{C})$-inevitability for $\varphi$. In the $B$-generated case, we may assume that $\rho$ and $\psi$ are $B$-generated.

Suppose $E \in \mathscr{C}$ is over $A$, $\sigma \colon A \to S$ is a substitution and that, moreover, there is a substitution $\tau \colon A \to S'$ such that $\tau \vdash E$ and $\sigma$ is $\rho$-related to $\tau$. If

$|A| \leq f(n)$, then, by choice of $(\rho, 1_T)$, $\sigma$ is $(W, E)$-inevitable for $\varphi$. So assume $|A| > f(n)$.

By the definition of a finitely equivalent collection, there exist $E' \in \mathscr{C}_n$ (over an alphabet $C$) and substitutions $\beta \colon A \to C$, $\gamma \colon C \to A$ with $\beta\gamma\sigma = \sigma$ such that $E \vdash E'\gamma$ and $E' \vdash E\beta$. As $\tau \vdash E \vdash E'\gamma$, we may deduce $\gamma\tau \vdash E'$. Also, because $\tau$ is $\rho$-related to $\sigma$, $\gamma\tau$ is $\rho$-related to $\gamma\sigma$. By choice of $(\rho, 1_T)$, we have that $\gamma\sigma$ is $(W, E')$-inevitable for $\varphi$. We use this to show that $\sigma$ is $(W, E)$-inevitable for $\varphi$.

Let $(\delta, h) \colon \varphi \to \psi'$ be a relational morphism with $\psi' \colon R \to U$ in W. Then there exists a substitution $\tau' \colon C \to R$ such that $\tau'$ $\delta$-relates to $\gamma\sigma$ and $\tau' \vdash E'$. But then $\tau' \vdash E\beta$, so $\beta\tau' \vdash E$. Now $\beta\tau'$ $\delta$-relates to $\beta\gamma\sigma = \sigma$. Because $(\delta, h)$ was arbitrary, we see that $\sigma$ is $(W, E)$-inevitable for $\varphi$.    □

### 3.6.4 Membership in V ⊙ W

Our interest in the case where one can witness $(W, \mathscr{C})$-inevitability is validated by the following result, generalizing results of [9, 19, 20, 126, 235].

**Theorem 3.6.30.** *Let* V *and* W *be pseudovarieties of relational morphisms and* $\mathscr{E}$ *be a basis of pseudoidentities for* V*. Define a collection of systems of pseudoidentities* $\mathscr{C}_{\mathscr{E}}$ *by*

$$\mathscr{C}_{\mathscr{E}} = \{E \mid (A, u = v, E) \in \mathscr{E}\}. \tag{3.12}$$

*Let* $\varphi \colon S \to T$ *be a fixed relational morphism of finite semigroups and let* $(\rho, f) \colon \varphi \to \psi$ *be a relational morphism, with* $\psi \colon S' \to T'$ *in* W*, witnessing* $(W, \mathscr{C}_{\mathscr{E}})$*-inevitability for* $\varphi$*. Then the following are equivalent:*

*(1)* $\varphi \in$ V ⊙ W*;*
*(2)* $\rho \in$ V*;*
*(3) for all* $(A, u = v, E) \in \mathscr{E}$ *and, for each* $(W, E)$*-inevitable substitution* $\sigma \colon A \to S$ *for* $\varphi$*, one has* $\sigma \vdash u = v$*.*

*Proof.* Suppose $\rho \in$ V. Then $\varphi f \subseteq_s \rho\psi$, and so $\varphi \prec \rho\psi$, yielding $\varphi \in$ V ⊙ W. Hence (2) implies (1).

Suppose now that $\varphi$ is in V ⊙ W. Then $\varphi \subseteq_s \alpha\beta$, where $\alpha \colon S \to R$ is in V and $\beta \colon R \to T$ is in W. Suppose $(A, u = v, E) \in \mathscr{E}$ and $\sigma \colon A \to S$ is $(W, E)$-inevitable for $\varphi$. Viewing $(\alpha, 1_T) \colon \varphi \to \beta$ as a relational morphism

$$\begin{array}{ccc} S & \xrightarrow{\alpha} & R \\ \varphi \downarrow & \subseteq & \downarrow \beta \\ T & \xrightarrow[1_T]{} & T \end{array}$$

we see that there exists $\tau \colon A \to R$ such that $\tau \vdash E$ and $\sigma$ $\alpha$-relates to $\tau$. Because $\alpha \in$ V, and so $\alpha \models \mathscr{E}$, it follows $\sigma \vdash u = v$, as desired. Therefore, (1) implies (3).

Suppose that (3) holds. Let $(A, u = v, E) \in \mathscr{E}$ and let $\sigma \colon A \to \#\rho$ be a substitution such that $\sigma p_{S'} \vdash E$. Then, by choice of $(\rho, f)$, $\sigma p_S$ is $(\mathsf{W}, E)$-inevitable for $\varphi$ so, by assumption, $\sigma p_S \vdash u = v$. We conclude that $\rho \models (A, u = v, E)$ and hence $\rho \in \mathsf{V}$, establishing that (3) implies (2).          $\square$

We may now draw several corollaries.

**Corollary 3.6.31.** *Suppose that* $\mathsf{V}$, $\mathsf{W}$, $\mathscr{E}$, *and* $\mathscr{C}_{\mathscr{E}}$ *are as in Theorem 3.6.30 above. Suppose, moreover, that* $\mathsf{V}$ *has decidable membership and that, for any relational morphism* $\varphi \colon S \to T$ *of finite semigroups, one can effectively compute a relational morphism* $(\rho, f) \colon \varphi \to \psi$ *with* $\psi \in \mathsf{W}$ *that witnesses* $(\mathsf{W}, \mathscr{C}_{\mathscr{E}})$*-inevitability for* $\varphi$. *Then* $\mathsf{V} \odot \mathsf{W}$ *has decidable membership.*

The following corollary generalizes Proposition 2.3.26 and could be proved directly, without using the equational theory.

**Corollary 3.6.32.** *Let* $\mathsf{V}$ *and* $\mathsf{W}$ *be pseudovarieties of relational morphisms with* $\mathsf{W}$ *locally finite. Let* $\varphi \colon S \to T$ *be a B-generated relational morphism and* $\rho_{\mathsf{W}} \colon S \to T^{\mathsf{W}}$ *be the canonical B-generated relational morphism. Then* $\varphi \in \mathsf{V} \odot \mathsf{W}$ *if and only if* $\rho_{\mathsf{W}} \in \mathsf{V}$. *In particular, if the finitely generated free pro-W relational morphisms to finite semigroups are computable (given the generating set and finite semigroup as input) and* $\mathsf{V}$ *has decidable membership, then* $\mathsf{V} \odot \mathsf{W}$ *has decidable membership.*

*Proof.* Theorem 3.6.20 says that $\rho_{\mathsf{W}}$ witnesses $(\mathsf{W}, \mathscr{C}_{\mathscr{E}})$ and so $\varphi \in \mathsf{V} \odot \mathsf{W}$ if and only if $\rho_{\mathsf{W}} \in \mathbf{V}$ by Theorem 3.6.30.          $\square$

Specializing to the case where $\mathbf{W}$ is a locally finite pseudovariety of semigroups recovers Proposition 2.3.26. Next, we generalize Corollary 2.3.27 to pseudovarieties of relational morphisms.

**Corollary 3.6.33.** *Let* $\{\mathsf{V}_\alpha\}$ *be a collection of pseudovarieties of relational morphisms and* $\mathsf{W}$ *a **locally finite** pseudovariety of relational morphisms. Then*

$$\bigcap(\mathsf{V}_\alpha \odot \mathsf{W}) = \left(\bigcap \mathsf{V}_\alpha\right) \odot \mathsf{W}. \qquad (3.13)$$

*Proof.* Clearly, the right-hand side is contained in the left-hand side. For the other direction, let $\varphi \colon S \to T$ be in the left-hand side of (3.13). Let $B$ be a generating set for $\varphi$ and $\rho_{\mathsf{W}} \colon S \to T^{\mathsf{W}}$ be as in Corollary 3.6.32. Then $\rho_{\mathsf{W}} \in \mathsf{V}_\alpha$, all $\alpha$, by the selfsame corollary; hence $\varphi$ belongs to the right-hand side of (3.13) by another application of Corollary 3.6.32.          $\square$

Specializing to when $\mathbf{W}$ is a locally finite pseudovariety of semigroups recovers Corollary 2.3.27. Example 3.6.38 below shows that (3.13) can fail if $\mathsf{W}$ is not locally finite, in fact Corollary 2.3.27 can fail for a non-locally finite pseudovariety of semigroups $\mathbf{W}$. Corollary 3.6.32 does admit the following generalization to the non-locally finite case.

**Theorem 3.6.34.** *Let* $V$ *and* $W$ *be pseudovarieties of relational morphisms. Let* $\varphi: S \to T$ *be a fixed* $B$-*generated relational morphism of profinite semigroups* ($B$ *finite) and* $\rho_W: S \to T^W$ *be the canonical* $B$-*generated relational morphism. Then* $\varphi$ *is pro-*$V \odot W$ *if and only if* $\rho_W$ *is pro-*$V$*. In particular, if* $S$ *and* $T$ *are finite,* $\varphi \in V \odot W$ *if and only if* $\rho_W$ *is pro-*$V$*.*

*Proof.* Assume $\rho_W$ is pro-$V$. Because $\varphi \subseteq_s \rho_W \varphi_{W,T}$, it is pro-$V \odot W$ by Lemma 3.6.1 and Corollary 3.3.16.

Conversely, suppose $\varphi$ is pro-$V \odot W$. Let $\pi: (T^W)^V \to T$ be the canonical relational morphism By Theorem 3.6.3, $\pi$ is the free $B$-generated pro-$V \odot W$ relational morphism to $T$. Hence, if $\psi: \widehat{B^+} \to S$ is the canonical surjective morphism, then $\equiv_{V,T^W} \subseteq \ker \psi$. But then, by Corollary 3.4.6, $\rho_W$ is pro-$V$ (as $\rho_W$ is the canonical relational morphism $S \to T^W$). The last statement follows from Proposition 3.3.9. $\qquad\square$

This gives a new proof of Corollary 3.6.32 by considering the case where $W$ is locally finite. Specializing to the case where $\mathbf{W}$ is a pseudovariety of semigroups we obtain:

**Corollary 3.6.35.** *Let* $V$ *be a pseudovariety of relational morphisms and* $\mathbf{W}$ *a pseudovariety of semigroups. Let* $S$ *be a (pro)finite* $A$-*generated semigroup and* $\rho: S \to \widehat{F_{\mathbf{W}}}(A)$ *be the canonical relational morphism. Then* $S \in V\mathsf{q}(\mathbf{W})$ ($S$ *is pro-*$V\mathsf{q}(\mathbf{W})$*) if and only if* $\rho$ *is pro-*$V$*.*

The latter two results do not immediately lead to decidability results because $\rho$ is not normally computable in any reasonable sense (exceptions are the cases considered in [322, 330]). One might be tempted to try and use these results to prove an analogue of Corollary 3.6.33. However, if one tries to imitate the proof, one ends up wanting to show that if $\{V_\alpha\}$ is an arbitrary collection of pseudovarieties, then a relational morphism of profinite semigroups is pro-$(\bigcap_\alpha V_\alpha)$ if and only if it is pro-$V_\alpha$ for all $\alpha$; this latter statement need *not* be true if the index set is infinite; see Example 3.6.38 below. However, we do have the following positive result.

**Corollary 3.6.36.** *Let* $\{V_\alpha\}$ *be a collection of pseudovarieties of relational morphisms with bases of relational pseudoidentities* $\mathscr{E}_\alpha$ *such that* $\mathscr{C}_{\mathscr{E}_\alpha}$, *see* (3.12), *is the same for all* $\alpha$; *call this common collection* $\mathscr{C}$*. Suppose* $W$ *is a pseudovariety of relational morphisms such that, for every relational morphism* $\varphi: S \to T$ *of finite semigroups, there is a relational morphism from* $\varphi$ *to a member of* $W$ *witnessing* $(W, \mathscr{C})$*-inevitability for* $\varphi$*. Then*

$$\bigcap (V_\alpha \odot W) = \left(\bigcap V_\alpha\right) \odot W. \qquad (3.14)$$

*Proof.* Again, the right-hand side of (3.14) is clearly contained in the left-hand side. For the other direction, let $\varphi: S \to T$ be in the left-hand side of (3.14) and let $(\rho, f): \varphi \to \psi$ witness $(W, \mathscr{C})$-inevitability for $\varphi$ with $\psi \in W$. Then, by Theorem 3.6.30, $\rho \in \bigcap V_\alpha$ and so $\varphi$ belongs to the right-hand side of (3.14), again by Theorem 3.6.30 (as $\bigcup \mathscr{E}_\alpha$ defines $\bigcap V_\alpha$). $\qquad\square$

Let $\mathbf{V}$ be a fixed pseudovariety of semigroups and let $\{\mathbf{U}_\alpha\}_{\alpha\in A}$ be a collection of pseudovarieties of semigroups. Then the above corollary applies to $\mathbf{V}_\alpha = (\mathbf{U}_\alpha, \mathbf{V})$. Indeed, $\mathscr{C}$ can be taken to consist of all systems of pseudoidentities of the form $E_{\mathbf{V},n}$ where $E_{\mathbf{V},n}$ *presents a free pro-$\mathbf{V}$ semigroup on $n$ generators*; this set was shown to be computably finitely equivalent in Example 3.6.24, so Proposition 3.6.29 applies. In particular, we obtain the following generalization of [235].

**Corollary 3.6.37.** *Let $\{\mathbf{U}_\alpha\}_{\alpha\in A}$ be a set of semigroup pseudovarieties and let $\mathbf{V}$ and $\mathbf{W}$ be pseudovarieties of semigroups. Then*

$$\left(\bigcap_{\alpha\in A}\mathbf{U}_\alpha, \mathbf{V}\right) \textcircled{m}\, \mathbf{W} = \bigcap_{\alpha\in A}\left((\mathbf{U}_\alpha, \mathbf{V}) \textcircled{m}\, \mathbf{W}\right). \tag{3.15}$$

It is also worth considering the following more illustrative proof of this result. Proposition 3.6.26 shows that $S$ belongs to $(\mathbf{U}, \mathbf{V}) \textcircled{m}\, \mathbf{W}$ if and only if the $\mathbf{V}$-like subsemigroups of $S$ with respect to $\mathbf{W}$ belong to $\mathbf{U}$. Hence $S$ belongs to either side of (3.15) if and only if its $\mathbf{V}$-like subsemigroups with respect to $\mathbf{W}$ belong to $\mathbf{U}_\alpha$, all $\alpha$ (see Definition 3.6.25).

We now give an example to show that Corollary 3.6.33 may fail if W is not locally finite. As a consequence, we prove meets are *not* pointwise in $\mathbf{GMC(PV)}$. We also show that intersection does not commute with the operator pro-$(-)$.

*Example 3.6.38 (Meets in $\mathbf{GMC(PV)}$ are not pointwise).* Theorem 2.3.31 tells us that $\mathfrak{q}\colon \mathbf{PVRM} \to \mathbf{GMC(PV)}$ is onto and **inf**. Hence to show that meets are not pointwise in $\mathbf{GMC(PV)}$, it suffices to find $\mathsf{V}_\alpha \in \mathbf{PVRM}$ and $\mathsf{V} \in \mathbf{PV}$ such that $(\bigcap \mathsf{V}_\alpha)\mathfrak{q}(\mathbf{V}) \neq \bigcap(\mathsf{V}_\alpha\mathfrak{q}(\mathbf{V}))$.

Let $\mathsf{V}_n$ be the pseudovariety of relational morphisms defined by the relational pseudoidentity

$$e_n = (A_n, x_1 = x_1^2, E_n)$$
$$= (\{x_1,\ldots,x_n\}, x_1 = x_1^2, \{x_1\cdots x_n = (x_1\cdots x_n)^2, x_1 = \cdots = x_n\}).$$

Let $\mathsf{V} = \bigcap \mathsf{V}_n$. First observe that $\varphi\colon S \to T$ is in $\mathsf{V}$ if and only if $S$ is a band. Indeed, if $S$ is a band, clearly $\varphi \in \mathsf{V}$. For the converse, let $k$ be an integer such that $t^k = t^{2k}$ for all $t \in T$. Let $s \in S$ and $t \in s\varphi$. Consider the substitutions $\sigma\colon A_k \to S$ and $\tau\colon A_k \to T$ given by $x_j\sigma = s$, $x_j\tau = t$, all $j$. Then $\tau \vdash E_k$ by choice of $k$. Thus $\sigma \vdash x_1 = x_1^2$ so $s = s^2$. It follows that $S$ is a band.

Let $\mathbf{G}$ be the pseudovariety of all finite groups. Using identification (2.64)

$$\left(\bigcap \mathsf{V}_n\right) \odot \mathbf{G} = \mathsf{V} \odot \mathbf{G} = \mathsf{V}\mathfrak{q}(\mathbf{G}) = \left(\bigcap \mathsf{V}_n\right)\mathfrak{q}(\mathbf{G})$$

is the pseudovariety of all bands. We show $\mathbb{Z}_2 \in \bigcap(\mathsf{V}_n \odot \mathbf{G}) = \bigcap(\mathsf{V}_n\mathfrak{q}(\mathbf{G}))$. This will prove that the meet is not pointwise in $\mathbf{GMC(PV)}$ and also establish that $(\bigcap \mathsf{V}_n) \odot \mathsf{W} \neq \bigcap(\mathsf{V}_n \odot \mathsf{W})$ for general W.

Indeed, consider the relational morphism $\rho_n \colon \mathbb{Z}_2 \to \mathbb{Z}_{2^n}$, which is the inverse of the canonical surjection. We view $\mathbb{Z}_2$ and $\mathbb{Z}_{2^n}$ as additive groups. Suppose $\sigma \colon A_n \to \mathbb{Z}_2$ is $\rho_n$-related to $\tau \colon A_n \to \mathbb{Z}_{2^n}$ where $\tau \vdash E_n$. Because $0\rho_n \cap 1\rho_n = \emptyset$ and $\tau \vdash x_1 = \cdots = x_n$, we see that either $A_n\sigma = 0$ or $A_n\sigma = 1$. In the former case, $\sigma \vdash x_1 = x_1^2$ because $0$ is an idempotent for addition; in the latter case, $A_n\tau = \{m\}$ for some odd number $m$ and $nm = 0 \bmod 2^n$ because $x_1\tau + \cdots + x_n\tau$ is the idempotent $0$ of $\mathbb{Z}_{2^n}$, as $\tau \vdash x_1 \cdots x_n = (x_1 \cdots x_n)^2$. Because $m$ is odd, we must have $2^n \mid n$, which is impossible; thus the latter case does not occur. We conclude $\rho_n \in \mathsf{V}_n$ and so $\mathbb{Z}_2 \in \mathsf{V}_n\mathsf{q}(\mathbf{G})$ all $n$, as desired.

Observe that the proof shows that if $\mathscr{E} = \{e_n \mid n \geq 1\}$, then there is no relational morphism of $\mathbb{Z}_2$ to an element of $\mathbf{G}$ witnessing $(\mathbf{G}, \mathscr{C}_{\mathscr{E}})$-inevitability.

Next we show that there is a relational morphism that is not pro-$\bigcap \mathsf{V}_n$, but which is pro-$\mathsf{V}_n$ for all $n$. Let $\widehat{\mathbb{Z}_2}$ be the additive group of 2-adic integers and $\psi \colon \widehat{\mathbb{Z}_2} \to \mathbb{Z}_2$ the canonical surjection. We show $\psi^{-1} \colon \mathbb{Z}_2 \to \widehat{\mathbb{Z}_2}$ is pro-$\mathsf{V}_n$ for all $n$, but is not pro-$\mathsf{V}$. Indeed, because $\mathsf{V}$ is the pseudovariety of bands (under identification (2.64)) and $\mathbb{Z}_2$ is not a band, $\psi^{-1}$ is not pro-$\mathsf{V}$. To see that $\psi^{-1}$ is pro-$\mathsf{V}_n$ all $n$, consider the projection $\varphi_n \colon \widehat{\mathbb{Z}_2} \to \mathbb{Z}_{2^n}$. Then we saw above $\rho_n = \psi^{-1}\varphi_n$ is in $\mathsf{V}_n$. Hence, as $\psi^{-1} \prec \psi^{-1}\varphi_n = \rho_n$, it follows $\psi^{-1}$ is pro-$\mathsf{V}_n$ by Proposition 3.3.11.

Let us state explicitly some corollaries of this example.

**Corollary 3.6.39.** *Let $\varphi \colon S \to T$ be a relational morphism of finite semigroups. If $\mathscr{C}$ is a collection of systems of pseudoidentities and $\mathsf{W}$ is a pseudovariety of relational morphisms, there need not be a relational morphism $(\rho, f) \colon \varphi \to \psi$ with $\psi \in \mathsf{W}$ witnessing $(\mathsf{W}, \mathscr{C})$-inevitability for $\varphi$.*

**Corollary 3.6.40.** *There exists a collection $\{\mathsf{V}_n\}_{n \geq 1}$ of pseudovarieties of relational morphisms and a pseudovariety $\mathbf{W}$ of semigroups so that*

$$\left(\bigcap \mathsf{V}_n\right) \mathsf{q}(\mathbf{W}) = \left(\bigcap \mathsf{V}_n\right) \odot \mathbf{W} \neq \bigcap (\mathsf{V}_n \odot \mathbf{W}) = \bigcap (\mathsf{V}_n\mathsf{q}(\mathbf{W})).$$

**Corollary 3.6.41.** *There exists a collection $\{\mathsf{V}_\alpha\}_{\alpha \in A}$ of pseudovarieties of relational morphisms such that there is a relational morphism $\varphi$ of profinite semigroups that is pro-$\mathsf{V}_\alpha$ for all $\alpha$ but that is not pro-$\bigcap \mathsf{V}_\alpha$.*

However, one does have the following positive result.

**Theorem 3.6.42.** *Suppose that $\mathsf{V}_1, \ldots, \mathsf{V}_n, \mathsf{W}$ are pseudovarieties of relational morphisms. Then*

$$\bigcap_{i=1}^{n} (\mathsf{V}_i \odot \mathsf{W}) = \left(\bigcap_{i=1}^{n} \mathsf{V}_i\right) \odot \mathsf{W}.$$

*Proof.* Clearly, the right-hand side is contained in the left-hand side. Suppose $\varphi\colon S \to T$ is in the left-hand side. Then there are relational morphisms $\alpha_i\colon S \to R_i$ in $\mathsf{V}_i$ and $\beta_i\colon R_i \to T$ in $\mathsf{W}$ such that $\varphi \subseteq_s \alpha_i\beta_i$ for all $i$. Let $\alpha = \Delta\prod\alpha_i$ and $\beta = \prod\beta_i$. Then $\alpha \prec \alpha_i$ for all $i$ by Proposition 1.3.16 and so $\alpha \in \bigcap_{i=1}^{n}\mathsf{V}_i$. Of course, $\beta \in \mathsf{W}$. As $\varphi\Delta \subseteq_s \alpha\beta$, we have $\varphi \prec \alpha\beta$, whence $\varphi \in (\bigcap_{i=1}^{n}\mathsf{V}_n)\odot\mathsf{W}$, as desired.    $\square$

By considering the case $\mathsf{W}$ is a pseudovariety of semigroups, we obtain the old result that $\mathsf{q}$ sends finite meets to pointwise meets (see Chapter 2).

*Question 3.6.43.* Give an example showing that meets are not pointwise in $\mathbf{GMC(PV)}^{+}$.

### 3.6.5 Some decidability results

If $E$ is a system of pseudoidentities over a finite alphabet $A$ and $\mathsf{W}$ is a pseudovariety of relational morphisms, then we say that $\mathsf{W}$ has *decidable $E$-substitutions* if, given any relational morphism $\varphi\colon S \to T$ and substitution $\sigma\colon A \to S$, it is decidable whether $\sigma$ is $(\mathsf{W}, E)$-inevitable for $\varphi$.

For instance if $E = \{x = y\}$ and $\mathbf{W}$ is a pseudovariety of semigroups, then $\mathbf{W}$ has decidable $E$-substitutions if and only if it has decidable pointlike pairs (in which case it has decidable membership). If $\mathscr{C}$ is a collection of systems of pseudoidentities, we say $\mathsf{W}$ has decidable $\mathscr{C}$-substitutions if it has decidable $E$-substitutions for all $E \in \mathscr{C}$. For instance, if $\mathscr{C}$ consists of the systems of pseudoidentities $(\{x_1,\ldots,x_n\}, \{x_1 = \cdots = x_n\})$ for all $n$, then $\mathbf{W}$ has decidable $\mathscr{C}$-substitutions if and only if $\mathbf{W}$ has decidable pointlike sets. As another example, a pseudovariety $\mathbf{W}$ has decidable $\mathscr{C}_{\mathbf{V}}$-substitutions if and only if the $\mathbf{V}$-like sets with respect to $\mathbf{W}$ are decidable; see Example 3.6.24.

A system of pseudoidentities $E$ over a finite set $A$ is said to be *computable* if given any substitution $\sigma\colon A \to S$ with $S$ a finite semigroup, it is decidable whether $\sigma \vdash E$. A relational pseudoidentity $(A, u = v, E)$ is said to be *computable* if $\{u = v\}$ and $E$ are computable. Theorem 3.5.14 says that pseudovarieties of relational morphisms can be defined by pseudoidentities of the form $(A, u = v, E)$ where $E$ is a finite set of formal equalities between words in $A^{+}$. Such $E$ are clearly computable.

As an example, we describe the $(\mathsf{D}, E)$-inevitable substitutions.

**Lemma 3.6.44.** *Suppose $E$ is a system of pseudoidentities over a set $A$ and let $\varphi\colon S \to T$ be a relational morphism. Then a substitution $\sigma\colon A \to S$ is $(\mathsf{D}, E)$-inevitable for $\varphi$ if and only if there is a substitution $\tau\colon A \to T$ such that $\tau \vdash E$ and $\tau$ $\varphi$-relates to $\sigma$.*

*Proof.* Necessity is clear by considering the relational morphism

$$(\varphi, 1_T)\colon \varphi \to 1_T.$$

For sufficiency, suppose such $\tau$ exists and that $(\rho, 1_T) \colon \varphi \to d$ is a relational morphism with $d \colon R \to T \in \mathsf{D}$ (the proof of Corollary 3.6.15 shows we need only consider such). For each $a \in A$, there exists $r_a \in R$ such that $a\sigma \ \rho \ r_a \ d \ a\tau$. Let $\tau' \colon A \to R$ be given by $a \mapsto r_a$. Clearly, $\tau'$ $\rho$-relates to $\sigma$. If $u = v \in E$, then we have $u\tau \in u\tau' d$, $v\tau \in v\tau' d$ and $u\tau = v\tau$, whence $u\tau' = v\tau'$ because $d$ is a division. Thus $\tau' \vdash E$, as desired.                                                □

**Corollary 3.6.45.** *Suppose $\mathscr{C}$ is a collection of computable systems of pseudoidentities. Then $\mathsf{D}$ has decidable $\mathscr{C}$-substitutions.*

*Proof.* Lemma 3.6.44 shows that, for $\varphi \colon S \to T$, $(\varphi, 1) \colon \varphi \to 1_T$ witnesses $(\mathsf{D}, \mathscr{C})$-inevitability for $\varphi$.                                                □

It follows easily from the results of [330] and Theorem 3.7.1, below, that if $\mathbf{W}$ is a pseudovariety of $\mathscr{J}$-trivial semigroups, then $\mathbf{W}$ has decidable $E$-substitutions for any computable system of pseudoidentities $E$ provided the word problem is decidable for the free pro-$\mathbf{W}$ semigroup as a $\sigma$-algebra where $\sigma$ is the implicit signature consisting of all implicit operations appearing in a pseudoidentity of $E$ (see [19, 20] for undefined terminology). This occurs, for instance, for the pseudovariety $\mathbf{J}$ of $\mathscr{J}$-trivial semigroups if $E$ contains only pseudoidentities involving words and the $\omega - 1$ unary operation [330]. See [19, 20] for more on $\sigma$-algebras and implicit signatures. It is shown in [82] that if $\mathbf{H}$ is a decidable pseudovariety of abelian groups, then $\mathbf{H}$ has decidable $E$-substitutions for any system of pseudoidentities $E$ whose elements can be built up from words and the implicit operation $x^{\omega - 1}$ via composition and multiplication. This generalizes an earlier result of Almeida and Delgado [15].

We now state the principal result of this section, generalizing old ideas of the first author, dating back to the development of Type I and Type II elements and pointlike sets [126, 156, 295]. The current formulation is based on some recent generalizations of these notions by Almeida, Pin, Weil and the second author [9, 19, 20, 33, 235, 330].

A collection $\mathscr{C}$ of systems of pseudoidentities is said to be *completely computable* if it is computably finitely equivalent, all members of $\mathscr{C}$ are computable, and there is an algorithm that, on input $n$, produces the finitely many systems of pseudoidentities of $\mathscr{C}$ over $n$ variables.

**Theorem 3.6.46.** *Let $\mathsf{V}$ and $\mathsf{W}$ be pseudovarieties of relational morphisms and suppose that $\mathscr{E}$ is a basis of relational pseudoidentities for $\mathsf{V}$ such that $\mathsf{W}$ has decidable $\mathscr{C}_{\mathscr{E}}$-substitutions and $\mathscr{C}_{\mathscr{E}}$ is completely computable (see (3.12)). Suppose, moreover, that $\mathsf{V}$ has decidable membership and that $\mathsf{W}$ is recursively enumerable. Then $\mathsf{V} \odot \mathsf{W}$ has decidable membership problem.*

*Proof.* By Corollary 3.6.31 (as $\mathsf{V}$ has decidable membership), it suffices to show that, given a relational morphism of finite semigroups $\varphi \colon S \to T$, we can effectively compute a relational morphism $(\rho, m) \colon \varphi \to \psi$ with $\psi \colon R \to T$ in $\mathsf{W}$ witnessing $(\mathsf{W}, \mathscr{C}_{\mathscr{E}})$-inevitability for $\varphi$. Let $n = |S|$ and $(\mathscr{C}_{\mathscr{E}})_n$ consist of all $E \in \mathscr{C}_{\mathscr{E}}$ over at most $f(n)$ variables. The proof of Proposition 3.6.29 then shows that it suffices to find a relational morphism that witnesses $(\mathsf{W}, (\mathscr{C}_{\mathscr{E}})_n)$-inevitability for $\varphi$, and that such a relational morphism must exist. Moreover, the assumptions on $\mathscr{C}_{\mathscr{E}}$ show that the finite set $(\mathscr{C}_{\mathscr{E}})_n$ can be effectively computed.

Because $\mathsf{W}$ is recursively enumerable, we can enumerate all relational morphisms from $\varphi$ to an element of $\mathsf{W}$. Because $(\mathscr{C}_{\mathscr{E}})_n$ is finite and consists of computable systems of pseudoidentities, for each such relational morphism $(\rho, m)$, where $\rho \colon S \to R$, and for each $E \in (\mathscr{C}_{\mathscr{E}})_n$ over $A$, we can actually compute whether there exist any $\rho$-related substitutions $\sigma \colon A \to S$, $\tau \colon A \to R$ such that $\tau \vdash E$. Because $\mathsf{W}$ has decidable $\mathscr{C}_{\mathscr{E}}$-substitutions, we can decide which such $\sigma$ are $(\mathsf{W}, E)$-inevitable for $\varphi$. It follows we can determine whether $(\rho, m)$ witnesses $(\mathsf{W}, (\mathscr{C}_{\mathscr{E}})_n)$-inevitability for $\varphi$. The algorithm stops when we finally find such a relational morphism witnessing $(\mathsf{W}, (\mathscr{C}_{\mathscr{E}})_n)$-inevitability for $\varphi$ (such exists *a priori*). $\qquad\square$

The above result applies, in particular, if $\mathbf{W}$ is a hyperdecidable pseudovariety of semigroups and $\mathsf{V}$ is of the form $\mathbf{V}_D$ where $\mathbf{gV}$ has finite vertex rank, and so generalizes the main result of [9]. A pseudovariety of semigroupoids is said to have *finite vertex rank* if it has a basis $\mathscr{E}$ of pseudoidentities over graphs with a bounded number $k$ of vertices. Without loss of generality, we may add to $\mathscr{E}$ all path pseudoidentities satisfied by $\mathbf{V}$ over graphs with at most $k$ vertices so that the finitely equivalent collection $\mathscr{C}_{\mathscr{G}_k}$ expresses the consistency equations of all graphs in $\mathscr{E}$. It also applies if $\mathbf{W}$ is a pseudovariety of semigroups with decidable idempotent pointlikes and $\mathsf{V}$ is of the form $(\mathbf{V}, \mathbf{1})$ [235, 330]. More generally, it can be applied to pseudovarieties of the form $(\mathbf{U}, \mathbf{V})$ provided $\mathbf{U}$ and $\mathbf{V}$ have decidable membership and $\mathbf{V}$-like subsemigroups with respect to $\mathbf{W}$ are computable. The argument above, like the argument of [9], is a generalization of the argument in [126, 156, 295] showing that the computability of the Type II elements of a semigroup implies the decidability membership in $\mathbf{V} \circledm \mathbf{G}$ for any decidable pseudovariety $\mathbf{V}$ of semigroups.

## 3.7 Basis Theorems

Our next goal is to provide a basis of relational pseudoidentities for $\mathsf{V} \odot \mathsf{W}$, given a basis of relational pseudoidentities $\mathscr{E}$ for $\mathsf{V}$ under the following assumption on $\mathsf{W}$: for each relational morphism $\varphi \colon S \to T$ of finite semigroups, there is a relational morphism $(\rho, f) \colon \varphi \to \psi$, with $\psi \in \mathsf{W}$, witnessing $(\mathsf{W}, \mathscr{C}_{\mathscr{E}})$-inevitability for $\varphi$. Specializing to the case where $\mathbf{W}$ is a pseudovariety of semigroups yields a basis for $\mathsf{Vq}(\mathbf{W})$. Afterwards, we give a basis theorem in general, but we will need to ask much more of $\mathscr{E}$.

### 3.7.1 A compactness theorem

First we need a compactness result, which simultaneously generalizes various results in the literature [9, 27, 235, 322, 330]. The relatively simple proof below follows along the lines of [130].

**Theorem 3.7.1.** *Let* $\mathsf{W}$ *be a pseudovariety of relational morphisms, $E$ a system of pseudoidentities over a finite alphabet $A$ and $\varphi\colon S \to T$ a $B$-generated relational morphism of finite semigroups. Then a substitution $\sigma\colon A \to S$ is $(\mathsf{W}, E)$-inevitable for $\varphi$ if and only if there is a substitution $\tau\colon A \to T^{\mathsf{W}}$ such that $\tau \vdash E$ and $\sigma$ is $\rho_{\mathsf{W}}$-related to $\tau$ where $\rho_{\mathsf{W}}\colon S \to T^{\mathsf{W}}$ is the canonical relational morphism of $B$-generated relational morphisms.*

*Proof.* As per Corollary 3.6.17, when testing for $(\mathsf{W}, E)$-inevitability for $\varphi$ we just need to consider relational morphisms $(\gamma, 1_T)\colon \varphi \to \psi$ with $\psi \in \mathsf{W}$ a $B$-generated relational morphism and with $\gamma\colon S \to R$ a canonical relational morphism of $B$-generated semigroups.

In this proof, we view substitutions of $A$ into a semigroup $U$ as elements of $U^A$. In particular, $\sigma \in S^A$. Consider $(T^{\mathsf{W}})^A$. As a direct product of profinite semigroups, $(T^{\mathsf{W}})^A$ is a profinite semigroup. In fact, if $T^{\mathsf{W}} = \varprojlim R_i$, where the $R_i$ are the finite quotients of $T^{\mathsf{W}}$, then $(T^{\mathsf{W}})^A = \varprojlim R_i^A$. Let us denote by $\pi_i\colon T^{\mathsf{W}} \to R_i$ the canonical projection and let $\beta_i = \pi_i^{-1}\varphi_{\mathsf{W},T}$, $\rho_i = \rho_{\mathsf{W}}\pi_i$.

$$
\begin{array}{ccc}
 & T^{\mathsf{W}} \xrightarrow{\varphi_{\mathsf{W},T}} T & \\
\rho_{\mathsf{W}} \nearrow \quad \pi_i \downarrow \quad \nearrow \beta_i & & \\
S \xrightarrow[\rho_i]{} R_i & &
\end{array}
$$

Then $\beta_i$ are in $\mathsf{W}$ (say by Corollary 3.3.16 and Proposition 3.3.9). Denote by $C_i$ the set of substitutions in $R_i^A$ satisfying $E$ that are $\rho_i$-related to $\sigma$; so

$$
C_i = \{\tau' \in R_i^A \mid \tau' \vdash E, \tau' \in \sigma\rho_i^A\}.
$$

Because the $\beta_i \in \mathsf{W}$ and the $R_i$ run over all finite images of $T^{\mathsf{W}}$, we see that $\sigma$ is $(\mathsf{W}, E)$-inevitable for $\varphi$ if and only if each $C_i$ is non-empty. Moreover, the $C_i$ form an inverse system. Observe that, for $\tau \in (T^{\mathsf{W}})^A$, we have that $\tau \vdash E$ if and only if $\tau\pi_i \vdash E$ for all $i$ (cf. [7, Thm 5.6.1]). Also, because $\sigma\rho_{\mathsf{W}}^A$ is closed (Corollary 3.3.5), $\sigma\rho_{\mathsf{W}}^A = \varprojlim \sigma\rho_i^A$ (Exercise 3.1.34). From this we conclude

$$
\varprojlim C_i = \{\tau \in (T^{\mathsf{W}})^A \mid \tau \vdash E, \tau \in \sigma\rho^A\}.
$$

So we can find the required $\tau \in (T^{\mathsf{W}})^A$ if and only if $\varprojlim C_i \neq \emptyset$. But Lemma 3.1.22 implies that $\varprojlim C_i \neq \emptyset$ if and only if $C_i$ is non-empty for each $i$, which we saw earlier was equivalent to $\sigma$ being $(\mathsf{W}, E)$-inevitable for $\varphi$. This completes the proof of the theorem. $\qquad\square$

First we state some known results as corollaries. The first one can be found in any of [9, 235, 322, 330] whereas the second can be found in [235, 330] and the third in [9]. All of these come from Theorem 3.7.1 by choosing W to be a pseudovariety **W** of semigroups and $E$ according to the motivating examples we have been considering.

**Corollary 3.7.2.** *Let $S$ be a $B$-generated semigroup and **W** a pseudovariety of semigroups. Let $\rho_{\mathbf{W}} \colon S \to \widehat{F_{\mathbf{W}}}(B)$ be the canonical relational morphism. Then $X \subseteq S$ is **W**-pointlike if and only if there exists $w \in \widehat{F_{\mathbf{W}}}(B)$ with $X \subseteq w\rho_{\mathbf{W}}^{-1}$.*

**Corollary 3.7.3.** *Suppose $S$ is a $B$-generated semigroup and **W** is a pseudovariety of semigroups. Let $\rho_{\mathbf{W}} \colon S \to \widehat{F_{\mathbf{W}}}(B)$ be the canonical relational morphism. Then $X \subseteq S$ is **W**-idempotent pointlike if and only if there exists an idempotent $e \in \widehat{F_{\mathbf{W}}}(B)$ with $X \subseteq e\rho_{\mathbf{W}}^{-1}$.*

**Corollary 3.7.4.** *Let $S$ be a $B$-generated semigroup and **W** a pseudovariety of semigroups. Let $\rho_{\mathbf{W}} \colon S \to \widehat{F_{\mathbf{W}}}(B)$ be the canonical relational morphism. A labeling $\ell$ of a graph $\Gamma$ by $S$ is **W**-inevitable if and only if there is a labeling $\tau \colon \Gamma \to \widehat{F_{\mathbf{W}}}(B)$ that is $\rho_{\mathbf{W}}$-related to $\ell$ and commutes (i.e., $e\alpha\tau \cdot e\tau = e\omega\tau$ for all edges $e$ of $\Gamma$).*

We may also deduce the following new result, generalizing the case of idempotent pointlikes [235].

**Corollary 3.7.5.** *Let $\mathbf{V}, \mathbf{W} \in \mathbf{PV}$. Suppose $S$ is an $A$-generated finite semigroup (with $A$ finite) and let $\rho_{\mathbf{W}} \colon S \to \widehat{F_{\mathbf{W}}}(A)$ be the canonical $A$-generated relational morphism. Then $X \subseteq S$ is **V**-like with respect to **W** if and only if there exists a pro-**V** (closed) subsemigroup $T \leq \widehat{F_{\mathbf{W}}}(A)$ such that $X \subseteq T\rho_{\mathbf{W}}^{-1}$.*

*Proof.* Let $\sigma \colon X \to S$ be the inclusion substitution. Then $\sigma$ is $(\mathscr{C}_{\mathbf{V}}, \mathbf{W})$-inevitable if and only if $X$ is **V**-like. By Theorem 3.7.1, $\sigma$ is inevitable if and only if there is a substitution $\tau \colon X \to \widehat{F_{\mathbf{W}}}(A)$ that is $\rho_{\mathbf{W}}$-related to $\sigma$ and $\tau \vdash E_{\mathbf{V}, |X|}$.

Suppose first $\sigma$ is inevitable. Then $\tau \vdash E_{\mathbf{V}, |X|}$ implies that $T = \overline{\langle X\tau \rangle}$, the closed subsemigroup generated by $X\tau$, is pro-**V** (as $E_{\mathbf{V}, |X|}$ presents a free pro-**V** semigroup and the class of pro-**V** semigroups is closed under continuous profinite images by Proposition 3.2.3). But then $X \subseteq T\rho_{\mathbf{W}}^{-1}$.

Conversely, suppose $X \subseteq T\rho_{\mathbf{W}}^{-1}$ with $\widehat{F_{\mathbf{W}}}(A) \geq T$ and $T$ pro-**V**. For each $x \in X$, choose $\pi_x \in x\rho_{\mathbf{W}} \cap T$. Then the substitution $\tau \colon X \to T \leq \widehat{F_{\mathbf{W}}}(A)$ given by $x \mapsto \pi_x$ induces a homomorphism $\widehat{\tau} \colon \widehat{F_{\mathbf{V}}}(X) \to \widehat{F_{\mathbf{W}}}(A)$. Thus $\tau \vdash E_{\mathbf{V}, |X|}$ and hence $\sigma$ is inevitable. $\square$

As a consequence, we obtain the following criterion for membership in $(\mathbf{U}, \mathbf{V}) \, \textcircled{m} \, \mathbf{W}$.

**Theorem 3.7.6.** *Let* $\mathbf{U}, \mathbf{V}, \mathbf{W} \in \mathbf{PV}$. *Suppose that* $S$ *is an* $A$-*generated finite semigroup* ($A$ *finite*) *and let* $\rho_{\mathbf{W}} \colon S \to \widehat{F_{\mathbf{W}}}(A)$ *be the canonical* $A$-*generated relational morphism. Then the following are equivalent:*

*(a)* $\rho_{\mathbf{W}}$ *is pro-*$(\mathbf{U}, \mathbf{V})$;
*(b) If* $\widehat{F_{\mathbf{W}}}(A) \geq T$ *with* $T$ *a pro-*$\mathbf{V}$ *closed subsemigroup, then* $T\rho_{\mathbf{W}}^{-1} \in \mathbf{U}$;
*(c)* $S \in (\mathbf{U}, \mathbf{V}) \,\textcircled{m}\, \mathbf{W}$.

*Proof.* Theorem 3.6.34 shows (a) and (c) are equivalent. Proposition 3.6.26 shows that $S \in (\mathbf{U}, \mathbf{V}) \,\textcircled{m}\, \mathbf{W}$ if and only if the $\mathbf{V}$-like subsemigroups of $S$ with respect to $\mathbf{W}$ belong to $\mathbf{U}$. But Corollary 3.7.5 says that such semigroups are exactly the semigroups $X \leq S$ such that $X \subseteq T\rho_{\mathbf{W}}^{-1}$ with $T$ a pro-$\mathbf{V}$ closed subsemigroup of $\widehat{F_{\mathbf{W}}}(A)$. This gives the equivalence of (b) and (c). $\qquad\square$

### 3.7.2 The basis theorems

Let $\mathsf{V}$ and $\mathsf{W}$ be pseudovarieties of relational morphisms. Our goal is to describe a relational pseudoidentity basis for $\mathsf{V} \odot \mathsf{W}$ in terms of a basis for $\mathsf{V}$ and some information about $\mathsf{W}$. When $\mathsf{W}$ is a pseudovariety of semigroups $\mathbf{W}$ and $\mathsf{V}$ is either $\mathsf{V}_D$ or $\mathsf{V}_K$, we recover the basis theorems of [27] for $\mathbf{V} * \mathbf{W}$ and $\mathbf{V} ** \mathbf{W}$ (when valid), whereas if $\mathsf{V} = (\mathbf{V}, \mathbf{1})$, then we recover the basis theorem of [235] for $\mathbf{V} \,\textcircled{m}\, \mathbf{W}$.

**Definition 3.7.7** ($\mathscr{E}_{\mathsf{W}}$). *Suppose* $\mathsf{V}$ *and* $\mathsf{W}$ *are pseudovarieties of relational morphisms and* $\mathscr{E}$ *is a basis of relational pseudoidentities for* $\mathsf{V}$. *Let* $\mathscr{E}_{\mathsf{W}}$ *be the collection of all relational pseudoidentities obtained in the following way: start with* $(A, u = v, E) \in \mathscr{E}$ *and a substitution* $\sigma \colon A \to \widehat{B^+}$ (*with* $A$ *and* $B$ *finite sets*) *such that* $\mathsf{W} \models (B, E\sigma, E')$ *where* $E'$ *is a set of pseudoidentities over* $B$; *then place* $(B, u\sigma = v\sigma, E')$ *in* $\mathscr{E}_{\mathsf{W}}$.

We begin our work by establishing that $\mathsf{V} \odot \mathsf{W} \models \mathscr{E}_{\mathsf{W}}$. Afterwards, we try to determine when the converse holds.

**Proposition 3.7.8.** *Let* $\mathsf{V}$, $\mathsf{W}$, $\mathscr{E}$, *and* $\mathscr{E}_{\mathsf{W}}$ *be as in Definition 3.7.7. Then* $\mathsf{V} \odot \mathsf{W} \models \mathscr{E}_{\mathsf{W}}$.

*Proof.* It suffices to show that if $\alpha \colon S \to R$ is in $\mathsf{V}$ and $\beta \colon R \to T$ is in $\mathsf{W}$, then $\alpha\beta \models \mathscr{E}_{\mathsf{W}}$. Let $(A, u = v, E) \in \mathscr{E}$, $\sigma \colon A \to \widehat{B^+}$ and $E'$ be as in Definition 3.7.7. Let $\psi \colon B \to S$ be a substitution and let $\tau \colon B \to T$ be a substitution $\alpha\beta$-related to $\psi$ such that $\tau \vdash E'$. Then, for each $b \in B$, there exists $r_b \in R$ such that $b\psi \; \alpha \; r_b \; \beta \; b\tau$. Let $\tau' \colon B \to R$ be given by $b\tau' = r_b$. Then $\tau'$ is $\alpha$-related to $\psi$ and $\beta$-related to $\tau$. As $\mathsf{W} \models (B, E\sigma, E')$ and $\tau'$ is $\beta$-related to $\tau$, we see that $\tau' \vdash E\sigma$, whence $\sigma\tau' \vdash E$. As $\sigma\tau'$ is $\alpha$-related to $\sigma\psi$ and $\mathsf{V} \vdash (A, u = v, E)$, we see that $\sigma\psi \vdash u = v$, and thus $\psi \vdash u\sigma = v\sigma$, as required. $\qquad\square$

Our next step is to show that if there is a relational morphism witnessing $(\mathsf{W}, \mathscr{C}_{\mathscr{E}})$-inevitability for $\varphi$, then $\varphi \in \mathsf{V} \odot \mathsf{W}$ if and only if $\varphi \models \mathscr{E}_{\mathsf{W}}$.

**Theorem 3.7.9.** *Let $\mathsf{V}$ be a pseudovariety of relational morphisms with basis of relational pseudoidentities $\mathscr{E}$ and let $\mathsf{W}$ be a pseudovariety of relational morphisms. Suppose $\varphi\colon S \to T$ is a relational morphism of finite semigroups such that there is a relational morphism from $\varphi$ to a member of $\mathsf{W}$, witnessing $(\mathsf{W}, \mathscr{C}_{\mathscr{E}})$-inevitability for $\varphi$. Then $\varphi \in \mathsf{V} \odot \mathsf{W}$ if and only if $\varphi \models \mathscr{E}_{\mathsf{W}}$.*

*Proof.* We have seen in Proposition 3.7.8, that if $\varphi \in \mathsf{V} \odot \mathsf{W}$, then $\varphi \models \mathscr{E}_{\mathsf{W}}$. For the other direction, assume $\varphi$ is $B$-generated and let $(\rho_{\mathsf{W}}, 1_T)\colon \varphi \to \varphi_{\mathsf{W},T}$ be the canonical $B$-generated relational morphism. Recall $\varphi_{\mathsf{W},T}\colon T^{\mathsf{W}} \to T$ is the free pro-$\mathsf{W}$ relational morphism to $T$.

By Theorem 3.6.30, to show that $\varphi \in \mathsf{V} \odot \mathsf{W}$, it suffices to show that if $(A, u = v, E) \in \mathscr{E}$ and $\sigma\colon A \to S$ is $(\mathsf{W}, E)$-inevitable for $\varphi$, then $\sigma \vdash u = v$. So suppose $\sigma\colon A \to S$ is $(\mathsf{W}, E)$-inevitable for $\varphi$. By Theorem 3.7.1, there exists $\tau\colon A \to T^{\mathsf{W}}$ such that $\tau \vdash E$ and $\tau$ is $\rho_{\mathsf{W}}$-related to $\sigma$. Let $\gamma\colon \widehat{B^+} \to T^{\mathsf{W}}$ and $\beta\colon \widehat{B^+} \to S$ be the canonical surjections, and let $E'$ be a finite subset of $B^+ \times B^+$ generating the kernel of the canonical homomorphism $\delta\colon B^+ \to T$. Observe $\rho_{\mathsf{W}} = \beta^{-1}\gamma$. Hence, for each $a \in A$, choose $\pi_a \in \widehat{B^+}$ such that $\pi_a\gamma = a\tau$ and $\pi_a\beta = a\sigma$. In this manner, we have defined a substitution $\pi\colon A \to \widehat{B^+}$ such that $\pi\beta = \sigma$ and $\tau = \pi\gamma \vdash E$, whence $\gamma \vdash E\pi$. So $E\pi \subseteq \equiv_{\mathsf{W},T}$ and we may apply Theorem 3.5.14 to conclude $\mathsf{W} \models (B, E\pi, E')$. Therefore, $(B, u\pi = v\pi, E') \in \mathscr{E}_{\mathsf{W}}$ by construction. Now $\beta|_B$ $\varphi$-relates to $\delta|_B$ (as $\varphi = \beta|_{B^+}^{-1}\delta$) and $\delta \vdash E'$. Thus, because $\varphi \models \mathscr{E}_{\mathsf{W}}$, $\beta \vdash u\pi = v\pi$, and so $\pi\beta \vdash u = v$. But $\pi\beta = \sigma$, therefore $\sigma \vdash u = v$, as desired. $\qquad\square$

Note that the proof shows that we need only place in $\mathscr{E}_{\mathsf{W}}$ all pseudoidentities of the form $(B, u\sigma = v\sigma, E')$ where $\sigma\colon A \to \widehat{B^+}$ is a substitution, $(A, u = v, E) \in \mathscr{E}$ and $\mathsf{W} \models (B, E\sigma, E')$ with $\langle B \mid E' \rangle$ a finite presentation of a finite semigroup.

It now follows that if it is always possible to find a relational morphism witnessing $(\mathsf{W}, \mathscr{C}_{\mathscr{E}})$-inevitability for every relational morphism $\varphi$ of finite semigroups, then $\mathscr{E}_{\mathsf{W}}$ is, in fact, a basis of pseudoidentities for $\mathsf{V} \odot \mathsf{W}$. Thus Proposition 3.6.29, Corollary 3.6.32 and Theorem 3.7.9 yield the following theorem, generalizing several important results in the literature [27, 235], which we discuss after the theorem.

**Theorem 3.7.10 (Basis Theorem for Composition).** *Let $\mathsf{V}$ be a pseudovariety of relational morphisms with basis of relational pseudoidentities $\mathscr{E}$ and $\mathsf{W}$ a pseudovariety of relational morphisms. Suppose further that either $\mathscr{C}_{\mathscr{E}}$ from (3.12) is contained in a finitely equivalent collection or $\mathsf{W}$ is locally finite. Then the set of all relational pseudoidentities of the form $(B, u\sigma = v\sigma, E')$ where $\sigma\colon A \to \widehat{B^+}$ is a substitution with $B$ finite, $(A, u = v, E) \in \mathscr{E}$ and $\mathsf{W} \models (B, E\sigma, E')$ is basis for $\mathsf{V} \odot \mathsf{W}$.*

Specializing to the case $\mathbf{W}$ is a pseudovariety of semigroups $\mathbf{W}$ results in the following basis theorem for $\mathsf{Vq}(\mathbf{W})$.

**Theorem 3.7.11 (Basis Theorem for Operators).** *Suppose* $\mathsf{V}$ *is a pseudovariety of relational morphisms with basis of relational pseudoidentities $\mathscr{E}$ and that $\mathbf{W}$ is a pseudovariety of semigroups. Assume either $\mathscr{C}_{\mathscr{E}}$ as per (3.12) is contained in a finitely equivalent collection or $\mathbf{W}$ is locally finite. Then a basis of pseudoidentities for $\mathsf{Vq}(\mathbf{W})$ consists of all identities of the form $u\sigma = v\sigma$ where $\sigma\colon A \to \widehat{B^+}$ (A and B finite) is a substitution and $(A, u = v, E) \in \mathscr{E}$ is such that $\mathbf{W} \models E\sigma$.*

*Example 3.7.12 (Intersection).* The simplest case is when $\mathbf{V}$ and $\mathbf{W}$ are pseudovarieties of semigroups and $\mathsf{V} = \mathbf{V} \cap \mathbf{D}$. Then $\mathsf{V} \odot \mathbf{W} = \mathbf{V} \cap \mathbf{W}$. Let $\mathscr{E}'$ be a basis of pseudoidentities for $\mathbf{V}$. Let $\mathscr{E} = \mathscr{E}' \cup \{(\{x,y\}, x = y, x = y)\}$. Then $\mathscr{C}_{\mathscr{E}}$ is finite, so Theorem 3.7.11 applies. The reader can verify that $\mathscr{E}_{\mathbf{W}}$ contains all pseudoidentities of $\mathscr{E}'$ (by considering the identity substitution and pseudoidentities of the form $(A, u = v) \in \mathscr{E}'$ and using that $\mathbf{W} \models \emptyset$) and all pseudoidentities of the form $u = v$ satisfied by $\mathbf{W}$; these latter pseudoidentities are obtained by considering a substitution $\sigma\colon \{x,y\} \to \widehat{A^+}$ sending $x$ to $u$ and $y$ to $v$; if $\mathbf{W} \models u = v$, then $u = v \in \mathscr{E}_{\mathbf{W}}$. Clearly, such a collection is a basis for $\mathbf{V} \cap \mathbf{W}$.

By considering the case where $\mathbf{W}$ is a pseudovariety of semigroups and $\mathsf{V}$ is of the form $(\mathbf{V}, \mathbf{1})$, for a pseudovariety of semigroups $\mathbf{V}$, we obtain the main result of [235].

**Theorem 3.7.13 (Pin-Weil [235]).** *Let $\mathbf{V}, \mathbf{W}$ be pseudovarieties of semigroups and let $\mathscr{E}$ be a basis of pseudoidentities for $\mathbf{V}$. Then a basis for $\mathbf{V} \textcircled{m} \mathbf{W}$ consists of all pseudoidentities of the form $u\sigma = v\sigma$ with $(A, u = v) \in \mathscr{E}$ and $\sigma\colon A \to \widehat{B^+}$ a substitution such that $\mathbf{W} \models a\sigma = a^2\sigma = a'\sigma$ for all $a, a' \in A$.*

*Proof.* First note that $(\mathbf{V}, \mathbf{1}) \odot \mathbf{W} = (\mathbf{V}, \mathbf{1})\mathsf{q}(\mathbf{W}) = \mathbf{V} \textcircled{m} \mathbf{W}$. Let $\mathscr{E}$ be a basis of pseudoidentities for $\mathbf{V}$; we assume that each of the pseudoidentities is defined over a finite subset of a fixed countable alphabet. According to Example 3.5.7, a basis for $(\mathbf{V}, \mathbf{1})$ consists of all relational pseudoidentities of the form $(A_n, u = v, E_n)$ where $u = v$ is a pseudoidentity over $A_n = \{x_1, \ldots, x_n\}$ in $\mathscr{E}$ and $E_n$ consists of the pseudoidentities $x_1 = x_1^2 = x_2 = \cdots = x_n$. The collection $\{E_n \mid n \in \mathbb{N}\}$ is finitely equivalent by Example 3.6.24 and so Theorem 3.7.11 says that a basis for $\mathbf{V} \textcircled{m} \mathbf{W}$ consists of all pseudoidentities of the form $u\sigma = v\sigma$ with $(A_n, u = v) \in \mathscr{E}$ and $\sigma\colon A_n \to \widehat{B^+}$ a substitution such that $\mathbf{W} \models E_n\sigma$. This establishes the theorem.  $\square$

More generally, we have the following new result.

**Theorem 3.7.14 (Basis Theorem for Generalized Mal'cev Product).** *Suppose that $\mathbf{U}, \mathbf{V}, \mathbf{W} \in \mathbf{PV}$ and let $\mathscr{E}$ be a basis of pseudoidentities for $\mathbf{U}$. Then a basis for $(\mathbf{U}, \mathbf{V}) \textcircled{m} \mathbf{W}$ is given by all pseudoidentities of the form*

$(B, u\sigma = v\sigma)$ where $(A, u = v) \in \mathscr{E}$, $\sigma \colon A \to \widehat{B^+}$ is a substitution with $B$ finite, and $\mathbf{W} \models E_{\mathbf{V},|A|}\sigma$. (This latter condition says that if $\pi \colon \widehat{B^+} \to \widehat{F_{\mathbf{W}}}(B)$ is the canonical surjection, then $A\sigma\pi$ generates a closed pro-$\mathbf{V}$ subsemigroup.)

*Proof.* Here, $(\mathbf{U}, \mathbf{V}) \circledm \mathbf{W} = (\mathbf{U}, \mathbf{V})\mathsf{q}(\mathbf{W}) = (\mathbf{U}, \mathbf{V}) \odot \mathbf{W}$. Let $\mathscr{E}$ be a basis of pseudoidentities for $\mathbf{U}$; again we assume that each of the pseudoidentities is defined over a finite subset of a fixed countable alphabet. Example 3.5.8 says that a basis for $(\mathbf{U}, \mathbf{V})$ consists of all relational pseudoidentities of the form $(A_n, u = v, E_{\mathbf{V},n})$ where $u = v$ is a pseudoidentity over $A_n = \{x_1, \dots, x_n\}$ in $\mathscr{E}$ and $\langle A_n \mid E_{\mathbf{V},n} \rangle$ presents a free pro-$\mathbf{V}$ semigroup on $A_n$. The collection $\{E_{\mathbf{V},n} \mid n \in \mathbb{N}\}$ is finitely equivalent by Example 3.6.24, and hence Theorem 3.7.11 provides a basis for $(\mathbf{U}, \mathbf{V}) \circledm \mathbf{W}$ consisting of all pseudoidentities of the form $u\sigma = v\sigma$ with $(A_n, u = v) \in \mathscr{E}$ and $\sigma \colon A_n \to \widehat{B^+}$ a substitution such that $\mathbf{W} \models E_{\mathbf{V},n}\sigma$. This establishes the theorem. $\qquad\square$

One could also prove Theorem 3.7.14 directly from Theorem 3.7.6. The case of Theorem 3.7.11 where $\mathbf{W}$ is a pseudovariety of semigroups and $\mathbf{V}$ is of the form $\mathbf{V}_D$ for a pseudovariety $\mathbf{V}$ of semigroups (or semigroupoids) of finite vertex rank yields the valid version of [27, Thms. 5.2 and 5.3].

**Theorem 3.7.15 (Almeida-Weil [27]).** *Let $\mathbf{V}$ and $\mathbf{W}$ be pseudovariety of semigroups. Let $\mathscr{E}$ be a basis of path pseudoidentities for $g\mathbf{V}$. Suppose that either $\mathbf{W}$ is locally finite or there is a bound on the number of vertices in any graph appearing in an element of $\mathscr{E}$. Then $\mathbf{V} * \mathbf{W}$ has a basis of pseudoidentities consisting of all pseudoidentities of the form $(p\alpha\ell)p\ell = (q\alpha\ell)q\ell$ such that there exist $(\Gamma, p = q) \in \mathscr{E}$ and a labeling $\ell \colon \Gamma \to \widehat{B^+}$ so that $\mathbf{W} \models (e\alpha\ell)e\ell = e\omega\ell$ for each edge $e$ of $\Gamma$.*

*Proof.* We just deal with the case that there exists a bound on the number of vertices in any graph from a pseudoidentity in $\mathscr{E}$. The case $\mathbf{W}$ is locally finite is left to the reader. This time, we use $\mathbf{V} * \mathbf{W} = \mathbf{V}_D \odot \mathbf{W}$. Without loss of generality we may assume that all graphs from $\mathscr{E}$ belong to a fixed set containing an isomorphic copy of each finite graph. Notice that if $(\Gamma, p = q) \in \mathscr{E}$ and $\ell \colon \Gamma \to \widehat{B^+}$ labels some vertex $v \neq p\alpha$ by $I$, then because $(\Gamma, p = q)$ is a path pseudoidentity, there is an edge $e$ with $e\omega = v$ and hence $\mathbf{W} \models (e\alpha\ell)e\ell = e\omega\ell$ is impossible. Thus in all labelings considered, only $p\alpha$ can be labeled by $I$ (and this only if there is no edge going into $p\alpha$ by the same reasoning). Example 3.5.9 shows that $\mathbf{V}_D$ has a basis consisting of all pairs $(V \cup E, (p\alpha)p = (q\alpha)q, C_\Gamma)$ where $(\Gamma, p = q) \in \mathscr{E}$, $\Gamma$ has vertex set $V$ and edge set $E$ and $C_\Gamma$ is the set of consistency equations of $\Gamma$; also one has the relational pseudoidentities corresponding to substituting $p\alpha$ by $I$ and using $C'_\Gamma$. Suppose that no graph appearing in $\mathscr{E}$ has more than $k$ vertices. Then the collection of consistency equations for the graphs appearing in $\mathscr{E}$ is contained in the finitely equivalent collection $\mathscr{C}_{\mathscr{G}_k}$ from Example 3.6.27. According to Theorem 3.7.11, a basis for $\mathbf{V} * \mathbf{W}$ consists of pseudoidentities of the form $((p\alpha)p)\sigma = ((q\alpha)q)\sigma$ where $\sigma \colon V \cup E \to \widehat{B^+}$ is a substitution with $B$ a finite alphabet such that $\mathbf{W} \models C_\Gamma \sigma$ (and the analogous pseudoidentities obtained by replacing $p\alpha$ by $I$ and using $C_\Gamma'$). But such a substitution is precisely the same thing as a labeling $\ell \colon \Gamma \to \widehat{B^+}$ (where only $p\alpha$ is permitted to be labeled by $I$) and the condition $\mathbf{W} \models C_\Gamma \sigma$ (or $\mathbf{W} \models C_\Gamma' \sigma$ if $p\alpha\ell = I$) is equivalent to $\mathbf{W} \models (e\alpha\ell)e\ell = e\omega\ell$ for each edge $e$ of $\Gamma$. This completes the proof. $\qquad\square$

The analogous results for the two-sided semidirect product (cf. [27, Thms. 6.2 and 6.3] with the missing finite vertex rank hypotheses) are obtained by taking $\mathbf{V} = \mathbf{V}_K$. We just state the result, leaving the proof to the reader.

**Theorem 3.7.16 (Almeida-Weil [27]).** *Let $\mathbf{V}$ and $\mathbf{W}$ be pseudovariety of semigroups. Let $\mathcal{E}$ be a basis of path pseudoidentities for $\mathbf{gV}$. Suppose that either $\mathbf{W}$ is locally finite or there is a bound on the number of vertices in any graph appearing in an element of $\mathcal{E}$. Then $\mathbf{V} \ast\ast \mathbf{W}$ has a basis of pseudoidentities consisting of all pseudoidentities of the form*

$$p\alpha\ell_L p\ell p\omega\ell_R = q\alpha\ell_L q\ell q\omega\ell_R$$

*such that there exist $(\Gamma, p = q) \in \mathcal{E}$ and functions $\ell_L, \ell_R \colon V(\Gamma) \to \widehat{B^*}$, $\ell \colon E(\Gamma) \to \widehat{B^+}$ so that $\mathbf{W} \models (e\alpha\ell_L)e\ell = e\omega\ell_L, e\alpha\ell_R = e\ell e\omega\ell_R$ for each edge $e$ of $\Gamma$.*

The general case of [27, Thm. 5.2] relies on a flawed argument [27, Section 4.2, p. 51] based on the idea that the pointwise meet is the determined meet in **GMC(PV)**. Given that its obvious generalization to the setting of pseudovarieties of relational morphisms fails, it seems unlikely to be valid. However, as far as we know, no explicit counterexample has been found. We provide in Section 3.8 a basis for the semidirect product $\mathbf{V} \ast \mathbf{W}$ in the general case. The missing pseudoidentities are due to the fact that a certain configuration can appear in every derived semigroupoid of a relational morphism from a finite semigroup $S$ to $\mathbf{W}$, but it could come from different graphs each time. Because the sizes of the graphs can be unbounded, there is no compactness argument to obtain a labeling of a single graph over $\widehat{F_{\mathbf{W}}}(A)$.

**Exercise 3.7.17.** Prove Theorem 3.7.16.

We give here a simple argument showing that Theorem 3.7.10 and (therefore its corollaries) can fail without the assumption of $\mathscr{C}_{\mathcal{E}}$ being contained in a finitely equivalent collection. To do this, we claim that if Theorem 3.7.10 were to hold unconditionally, then

$$\bigcap (\mathsf{V}_\alpha \odot \mathsf{W}) = \left( \bigcap \mathsf{V}_\alpha \right) \odot \mathsf{W} \tag{3.16}$$

would hold for any collection $\{\mathsf{V}_\alpha\}$ of pseudovarieties of relational morphisms. However, Corollary 3.6.40 shows that (3.16) fails in general and hence Theorem 3.7.10 does not hold without some assumptions on $\mathscr{C}_{\mathcal{E}}$. To prove the claim, observe that if $\mathcal{E}_\alpha$ is a basis of pseudoidentities for $\mathsf{V}_\alpha$, then $\bigcup \mathcal{E}_\alpha$ is a basis for $\bigcap \mathsf{V}_\alpha$. If Theorem 3.7.10 always held, then $(\bigcup \mathcal{E}_\alpha)_{\mathsf{W}}$ would be a basis for the right-hand side of (3.16) and $\bigcup (\mathcal{E}_\alpha)_{\mathsf{W}}$ would be a basis for the left-hand side. But these two sets of relational pseudoidentities are clearly equal.

### 3.7.3 A discussion of tameness

In [19,20], a notion called tameness was introduced with respect to inevitability of graphs, or in our terminology: inevitability with respect to the consistency equations of graphs. These notions can easily be generalized to our context, and we let the interested reader do so on his own. However, we now argue that perhaps it is not worth the effort in doing so.

The point of this notion of tameness was that, given a recursively enumerable basis $\mathscr{E}$ of computable relational pseudoidentities for V and a tameness assumption on W, one could cut $\mathscr{E}_W$ down to a recursively enumerable basis of computable pseudoidentities and then iterate the process. Our original hope was to generalize [19, 20] to the calculation of iterated compositions of pseudovarieties of relational morphisms, thereby allowing us to generalize the program espoused therein for computing the classical complexity of finite semigroups to a program for deciding the complexity theory determined by any two operators, as defined in Chapter 5. This led us to find the error in [27], invalidating the reduction theorem for complexity in [19,20]. Without Theorem 3.7.10 holding in general, one cannot use tameness to deal with iterated compositions. Thus tameness merely serves as a sufficient condition for decidability of $(W, E)$-inevitable substitutions and a way to cut the basis of Theorem 3.7.10 down to a more manageable (but still infinite) size provided the hypotheses of that theorem are satisfied.

Perhaps one could improve on tameness using the results that follow, but one will need to ask for a lot more.

### 3.7.4 A projective basis theorem

In this section, we show how to adapt Theorem 3.7.10 to obtain a basis theorem in general. One has to demand something stronger from the basis $\mathscr{E}$ chosen for V.

If **V** is a pseudovariety of finite semigroups and $\mathscr{E}$ is a basis of pseudoidentities for **V**, then, for a profinite semigroup $S$, the following are equivalent: $S \models \mathscr{E}$ and $S$ is pro-**V** (see Proposition 3.2.13). This is not the case for pseudovarieties of relational morphisms. Indeed, Example 3.6.38 provides pseudovarieties $V_\alpha$ and a relational morphism $\varphi$ of profinite semigroups that is pro-$V_\alpha$ for all $\alpha$, but which is not pro-$\bigcap V_\alpha$. If $\mathscr{E}_\alpha$ is a basis for $V_\alpha$, then $\bigcup \mathscr{E}_\alpha$ is a basis for $\bigcap V_\alpha$. As $\varphi$ is pro-$V_\alpha$ for all $\alpha$, Proposition 3.7.18 below yields $\varphi \models \mathscr{E}_\alpha$ for all $\alpha$. Hence $\varphi$ satisfies a basis for $\bigcap V_\alpha$ but is not pro-$\bigcap V_\alpha$.

**Proposition 3.7.18.** *Suppose* $V \models (A, u = v, E)$ *and* $\varphi \colon P \to Q$ *is a pro-*V *relational morphism. Then* $\varphi \models (A, u = v, E)$.

*Proof.* If $\sigma \colon A \to P$ is a substitution, then $\sigma \vdash u = v$ if and only if, for every continuous surjective homomorphism $\rho \colon P \twoheadrightarrow S$ with $S$ finite, $\sigma\rho \vdash u = v$. Suppose $\tau \colon A \to Q$ is $\varphi$-related to $\sigma$ and $\tau \vdash E$. Consider $\rho \colon P \twoheadrightarrow S$ a continuous onto homomorphism with $S$ finite. Then as $\varphi$ is pro-V, there is a continuous onto homomorphism $\pi\colon Q \twoheadrightarrow T$ with $T$ finite and $\psi = \rho^{-1}\varphi\pi \in \mathsf{V}$. Now $\sigma\rho$ is $\psi$-related to $\tau\pi$ and $\tau\pi \vdash E$. Hence $\sigma\rho \vdash u = v$, as required.     □

Let $\mathsf{V}$ be a pseudovariety of relational morphisms. We call a relational morphism $\varphi\colon S \to T$ *onto* if $\operatorname{Im} \varphi = T$.

**Definition 3.7.19 (Projective basis).** *A collection $\mathscr{E}$ of relational pseudoidentities is a projective basis for $\mathsf{V}$ if a finitely generated onto relational morphism of profinite semigroups satisfies $\mathscr{E}$ if and only if it is pro-$\mathsf{V}$.*

Because a pseudovariety of relational morphisms admits Axiom (co-re), it follows that any projective basis is a basis.

We first prove that any pseudovariety of relational morphisms is defined by a projective basis; this can be viewed as an improvement on Theorem 3.5.14. However, the proof is a mere adaptation that we leave to the reader. The reader should consult (3.4) for the notation.

**Theorem 3.7.20.** *Let $\mathsf{V}$ be a pseudovariety of relational morphisms and let $\mathscr{E}$ consist of all relational pseudoidentities of the form $(A, u = v, E)$ with $A$ finite and $(u, v) \in \equiv_{\mathsf{V},\widehat{A^+}/\langle E\rangle}$. Then $\mathscr{E}$ is a projective basis for $\mathsf{V}$.*

**Exercise 3.7.21.** Adapt the proof of Theorem 3.5.14 to prove Theorem 3.7.20.

*Remark 3.7.22.* If $\mathbf{V}$ is a pseudovariety of semigroups, it would seem a basis for $\mathbf{gV}$ does not in general yield a projective basis for $\mathbf{V}_D$. In Section 3.8 we try a different tactic.

To turn Theorem 3.7.10 into a general theorem, we make use of Theorem 3.6.34. The reader is referred back to Section 3.7.2 for notation.

**Theorem 3.7.23 (Projective Basis Theorem for Composition).** *Suppose that $\mathsf{V}$ is a pseudovariety of relational morphisms with projective basis of relational pseudoidentities $\mathscr{E}$ and that $\mathsf{W}$ is a pseudovariety of relational morphisms. Then the set of all relational pseudoidentities of the form $(B, u\sigma = v\sigma, E')$ where $\sigma\colon A \to \widehat{B^+}$ a substitution, $(A, u = v, E) \in \mathscr{E}$ and $\mathsf{W} \models (B, E\sigma, E')$ is a projective basis for $\mathsf{V} \odot \mathsf{W}$.*

*Proof.* Propositions 3.7.8 and 3.7.18 show any pro-$\mathsf{V} \odot \mathsf{W}$ relational morphism satisfies $\widehat{\mathscr{E}}$. For the other direction, assume $\varphi\colon S \to T$ is a $B$-generated onto relational morphism of profinite semigroups (with $B$ finite) satisfying $\widehat{\mathscr{E}}$. Let $(\rho_\mathsf{W}, 1_T)\colon \varphi \to \varphi_{\mathsf{W},T}$ be the canonical $B$-generated relational morphism. By Theorem 3.6.34, to show that $\varphi$ is pro-$\mathsf{V} \odot \mathsf{W}$, it suffices to show $\rho_\mathsf{W}$ is pro-$\mathsf{V}$. Because $\rho_\mathsf{W}$ is onto and finitely generated, it suffices to show $\rho_\mathsf{W} \models \mathscr{E}$.

Suppose $(A, u = v, E) \in \mathscr{E}$ and $\sigma\colon A \to S$, $\tau\colon A \to T^\mathsf{W}$ are $\rho_\mathsf{W}$-related to substitutions where $\tau \vdash E$. Let $\gamma\colon \widehat{B^+} \to T^\mathsf{W}$ and $\beta\colon \widehat{B^+} \to S$ be the canonical surjections, and let $E' \subseteq \widehat{B^+} \times \widehat{B^+}$ generate the kernel of the canonical surjection $\delta\colon \widehat{B^+} \to T$ (that is, $\langle B \mid E'\rangle$ is a presentation of $T$ as a $B$-generated profinite semigroup). Note $\rho_W = \beta^{-1}\gamma$ and so, for each $a \in A$, choose $\pi_a \in \widehat{B^+}$ such that $\pi_a\gamma = a\tau$ and $\pi_a\beta = a\sigma$. In this manner, we have defined a substitution $\pi\colon A \to \widehat{B^+}$ such that $\pi\beta = \sigma$ and $\tau = \pi\gamma \vdash E$, and so $\gamma \vdash E\pi$. Thus $E\pi \subseteq \equiv_{W,T}$, so Theorem 3.7.20 implies $W \models (B, E\pi, E')$. Therefore, by construction, $(B, u\pi = v\pi, E') \in \mathscr{E}_W$. Now $\beta|_B$ $\varphi$-relates to $\delta|_B$ (as $\varphi = \beta^{-1}\delta$) and $\delta|_B \vdash E'$. Thus, because $\varphi \models \mathscr{E}_W$, $\beta \vdash u\pi = v\pi$, and so $\pi\beta \vdash u = v$. But $\pi\beta = \sigma$, therefore $\sigma \vdash u = v$, as desired. We conclude that $\rho_W$ is pro-V and so $\varphi$ is pro-V $\odot$ W.                                                     $\square$

The proof shows more. Let $\varphi\colon S \to T$ be as in the theorem and $\langle B \mid E' \rangle$ be a presentation of $T$ as a $B$-generated profinite semigroup. Then $\varphi$ was shown to be in V $\odot$ W if and only if the canonical projection to $S$ satisfies all relations $(B, u\sigma = v\sigma)$ where $\sigma\colon A \to \widehat{B^+}$ is a substitution, $(A, u = v, E) \in \mathscr{E}$ and $W \models (B, E\sigma, E')$. Hence we obtain the following.

**Theorem 3.7.24.** *Suppose* V *is a pseudovariety of relational morphisms with projective basis of relational pseudoidentities $\mathscr{E}$ and that* W *is a pseudovariety of relational morphisms. Let $\varphi\colon S \to T$ be a $B$-generated onto relational morphism of finite semigroups ($B$ finite). Let $\rho_W\colon S \to T^W$ be the canonical relational morphism. Then the following are equivalent:*

*(1)* $\varphi \in$ V $\odot$ W;
*(2)* $\rho_W$ *is pro-V;*
*(3) for all $(A, u = v, E) \in \mathscr{E}$ and, for each $(W, E)$-inevitable substitution $\sigma\colon A \to S$ for $\varphi$, one has $\sigma \vdash u = v$.*

*Proof.* Theorem 3.6.34 shows that (1) and (2) are equivalent. We show (1) and (3) are equivalent. The proof of Theorem 3.6.30 shows (1) implies (3).

Suppose (3) holds. Let $\langle B \mid E' \rangle$ be a finite presentation of $T$ as a $B$-generated semigroup. As observed above, it suffices to show $S$ satisfies all relations $(B, u\gamma = v\gamma)$ where $\gamma\colon A \to \widehat{B^+}$ is a substitution, $(A, u = v, E) \in \mathscr{E}$ and $W \models (B, E\gamma, E')$. Let $(B, u\gamma = v\gamma)$ be such a relation. Let $\alpha\colon \widehat{B^+} \to S$, $\beta\colon \widehat{B^+} \to T^W$ be the canonical surjections; so $\rho_W = \alpha^{-1}\beta$. Then $\gamma\alpha$ $\rho_W$-relates to $\gamma\beta$. As $\langle B \mid E' \rangle$ is a finite presentation of $T$ and $W \models (B, E\gamma, E')$, it follows easily that $\beta \vdash E\gamma$ by Proposition 3.7.18. Therefore, $\gamma\beta \vdash E$ and so $\sigma = \gamma\alpha$ is $(W, E)$-inevitable for $\varphi$ by Theorem 3.7.1. By assumption we have $\gamma\alpha = \sigma \vdash u = v$, whence $\alpha \vdash u\gamma = v\gamma$, as desired.                    $\square$

Let us say that $(W, \mathscr{C})$-inevitability is *decidable* if given any relational morphism $\varphi\colon S \to T$ of finite semigroups and a system of pseudoidentities $(A, E) \in \mathscr{C}$, it is decidable whether a substitution $\sigma\colon A \to S$ is $(W, E)$-inevitable for $\varphi$. We then obtain the following corollary, fixing an incorrect result of [19, 20] (namely [19, Thm. 5.2]).

**Corollary 3.7.25.** *Suppose that* V *and* W *are recursively enumerable pseudovarieties of relational morphisms. Suppose further that* V *has a recursively enumerable projective basis $\mathscr{E}$ of computable relational pseudoidentities such that $(W, \mathscr{C}_\mathscr{E})$-inevitability is decidable. Then* V $\odot$ W *is decidable.*

*Proof.* Because V and W are recursively enumerable, so is V ⊙ W. We now show it is co-recursively enumerable.

Let $\mathscr{E}$ be a recursively enumerable projective basis for V of computable relational pseudoidentities such that $(W, \mathscr{C}_{\mathscr{E}})$-inevitability is decidable. Without loss of generality, we may assume $\varphi$ onto.

We give the following algorithm to detect whether $\varphi \notin V \odot W$ based on testing whether condition (3) of Theorem 3.7.24 fails to hold. For each $(A, u = v, E) \in \mathscr{E}$, we can determine all $(W, E)$-inevitable substitutions to $S$ for $\varphi$ and see which satisfy $u = v$. If $\varphi \notin V \odot W$, we shall eventually find such $u = v$ that is not satisfied.    □

The notion of tameness [19, 20] was in a large part inspired by this result and one would hope things could be generalized to this context.

Specializing Theorem 3.7.23 to the case where **W** is a pseudovariety of finite semigroups, we obtain the following theorem.

**Theorem 3.7.26 (Projective Basis Theorem for Operators).** *Suppose* V *is a pseudovariety of relational morphisms with projective basis of relational pseudoidentities* $\mathscr{E}$ *and that* **W** *is a pseudovariety of semigroups. Then a basis of pseudoidentities for* V**q**(**W**) *consists of all identities of the form* $u\sigma = v\sigma$ *where* $\sigma \colon A \to \widehat{B^+}$ *(A and B finite) is a substitution and* $(A, u = v, E) \in \mathscr{E}$ *is such that* **W** $\models E\sigma$.

**Corollary 3.7.27.** *Suppose* V *is a recursively enumerable pseudovariety of relational morphisms and* **W** *is a pseudovariety of semigroups. Suppose further that* V *has a recursively enumerable projective basis* $\mathscr{E}$ *of computable relational pseudoidentities such that* $(W, \mathscr{C}_{\mathscr{E}})$-*inevitability is decidable. Then* V**q**(**W**) *is decidable.*

*Question 3.7.28.* Find a projective basis for $V_D$.

## 3.8 Flows and the Basis Theorem for Semidirect Products

We present here a basis theorem for the semidirect product of pseudovarieties, which includes the pseudoidentities missing from the Almeida-Weil pseudoidentities [27] (our Theorem 3.7.15). A similar result holds for the two-sided semidirect product. The modifications for the monoidal setting are straightforward.

If $\varphi \colon S \to T$ is a relational morphism, we shall abuse notation and also use $\varphi$ for the functorial extension $\varphi \colon S^I \to T^I$ sending $I$ to $I$. Note that if $S$ is a semigroup, then the power set $P(S)$ is a semigroup under setwise multiplication. Observe $P(S)^I$ can be identified with the subset $P(S) \cup \{\{I\}\}$ of $P(S^I)$.

Fix a finite semigroup $S$ and a graph $\Gamma$. The following definition of a flow on $\Gamma$ is crucial [128, 280].

**Definition 3.8.1 (Flow).** *A (set) flow $f = (f_V, f_E)$ on $\Gamma$ consists of functions $f_V \colon V(\Gamma) \to P(S)^I$ and $f_E \colon E(\Gamma) \to S$ such that $e\alpha f_V \cdot e f_E \subseteq e\tau f_V$. Usually, we write $f \colon \Gamma \to S$, omitting the subscripts.*

Let us provide two examples of flows, followed immediately afterwards by some intuition.

*Example 3.8.2 (One vertex graph).* If $\Gamma$ is a graph with a single vertex $v$ and edge set $E$, then a flow on $\Gamma$ consists of a subset $X$ of $S^I$ (placed at the vertex) and an assignment of elements $s_e \in S$ for each edge $e \in E$ such that $X s_e \subseteq X$ for all $e \in E$.

*Example 3.8.3 (Relational morphism flow).* Let $\varphi \colon S \to T$ be a relational morphism. Let $\Gamma$ be the Cayley graph of $T$ with vertex set $T^I$ and edge set $T^I \times T$ where the edge $(t_L, t)$ is drawn: $t_L \xrightarrow{t} t_L t$. Define a flow $f = (f_V, f_E)$ on $\Gamma$ by $t f_V = t\varphi^{-1}$ and by choosing $(t_L, t) f_E$ to be any element $s \in t\varphi^{-1}$. Then $f$ is a flow by the definition of a relational morphism.

If $\varphi \colon S \to T$ is a relational morphism, then there is a representation $\mathbf{Der}(\varphi) \to \mathbf{Set}$ defined by

$$t \longmapsto t\varphi^{-1} \qquad \text{on objects } t \in T^I$$

$$t_L \xrightarrow{(s,t)} t_L t \longmapsto t_L\varphi^{-1} \xrightarrow{\cdot s} t_L t\varphi^{-1} \qquad \text{on arrows.}$$

The derived semigroupoid is just the quotient of $\mathbf{Der}(\varphi)$ by the kernel of this representation. Intuitively, a flow is a finite subgraph of the "image" of the representation. This image loses track of $T$, all you have are subsets of $S$ moving into each other via right multiplication. The construction $D_\varphi$ makes these sets formally disjoint by indexing them by elements of $T^I$. The next several definitions concern how to recover $T$.

Recall that a *labeling* $\ell \colon \Gamma \to T$ of a graph $\Gamma$ over a semigroup $T$ is a pair $\ell = (\ell_V, \ell_E)$ where $\ell_V \colon V(\Gamma) \to T^I$ and $\ell_E \colon E(\Gamma) \to T$. Again we drop the subscripts. Notice that a labeling is a (singleton) set flow if and only if it is consistent in the sense of [9]. We prefer to say that a labeling that is a singleton set flow *commutes* as a labeling commutes if and only if the diagram

$$
\begin{array}{ccc}
E(\Gamma) & \xrightarrow{\ \Delta(\alpha\ell \times \ell)\ } & T^I \times T \\
& \searrow{\scriptstyle \omega\ell} & \downarrow{\scriptstyle \mu} \\
& & T
\end{array}
$$

commutes, where $\mu$ is the semigroup multiplication.

**Definition 3.8.4 (Computing flows).** *Let $\varphi \colon S \to T$ be a relational morphism and $\ell \colon \Gamma \to T$ a labeling. We say a flow $f \colon \Gamma \to S$ is **computed** by the relational morphism $\varphi$ and the labeling $\ell$ if $vf \subseteq v\ell\varphi^{-1}$ all $v \in V(\Gamma)$ and*

$ef \in e\ell\varphi^{-1}$ for all $e \in E(\Gamma)$. A *relational morphism* $\varphi\colon S \to T$ *is said to* **compute** *a flow* $f$ *on* $\Gamma$ *if there exists a commuting labeling* $\ell$ *of* $\Gamma$ *over* $T$ *such that* $\varphi$ *and* $\ell$ *compute* $f$. *We shall sometimes say* $\ell$ **computes** $f$ *for* $\varphi$ *if* $\varphi$ *and* $\ell$ *compute* $f$.

Let us reprise our previous two examples.

*Example 3.8.5 (Computing a flow on a one vertex graph).* A relational morphism $\varphi\colon S \to T$ computes the flow from Example 3.8.2 if and only if there exists $t \in T$ so that $X \subseteq t\varphi^{-1}$ and $s_e \in \mathrm{Stab}(t)\varphi^{-1}$ for each edge $e \in E$, where $\mathrm{Stab}(t)$ is the right stabilizer of $t$.

*Example 3.8.6 (Relational morphism flow revisited).* The flow in Example 3.8.3 is computed by $\varphi$ via the commuting labeling of $\Gamma$ over $T$ obtained by sending a vertex $t_L$ to itself and sending the edge $t_L \xrightarrow{t} t_L t$ to $t$.

If $f\colon E(\Gamma) \to T$ is a function, we also use $f$ to denote the unique extension $f\colon \widehat{\Gamma^+} \to T$ (where $\widehat{\Gamma^+}$ is the free profinite semigroupoid generated by $\Gamma$).

The following definition is key to dealing with the semidirect product of pseudovarieties. It is based on the quantum mechanical idea of sampling: we care about the behavior of a certain system and not the exact manner in which the behavior came about. The Almeida-Weil approach [27] requires one to understand exactly how the behavior came about.

**Definition 3.8.7 (Flow configuration).** *A* **flow configuration** *for a semigroup* $S$ *is an element of* $P(S)^I \times S \times S$. *Flow configurations are partially ordered by setting* $(X, s_1, s_2) \leq (X', s_1', s_2')$ *if* $X \subseteq X'$ *and* $s_i = s_i'$, $i = 1, 2$.

Flow configurations $(X, s_1, s_2)$ arise from a flow on a graph $\Gamma$ and two coterminal paths $p_1, p_2$ in $\Gamma$. Here $X$ is the set attached to the initial vertex of $p_1$, and $s_i$ is the product of the labels of the edges of $p_i$, $i = 1, 2$. Roughly speaking, a relational morphism computes a flow configuration if it computes the corresponding flow on $\Gamma$. Here is the formal definition.

**Definition 3.8.8 (Computing flow configurations).** *Let* $\mathscr{E}$ *be a set of graph pseudoidentities and* $\varphi\colon S \to T$ *a relational morphism. We say* $\varphi$ **computes** *the flow configuration* $(X, s_1, s_2)$ *with respect to* $\mathscr{E}$ *if there exists* $(\Gamma, p_1 = p_2) \in \mathscr{E}$ *and a flow* $f\colon \Gamma \to S$ *computed by* $\varphi$ *such that* $X \subseteq p_1 \alpha f$ *and* $s_i = p_i f$, $i = 1, 2$. *We also shall say* $f$ **computes** $(X, s_1, s_2)$ *for* $\varphi$. *If* $\ell$ *is a commuting labeling computing* $f$ *for* $\varphi$, *we also say that* $\ell$ *and* $f$ **compute** $(X, s_1, s_2)$.

Decongesting notation, to say that $\varphi$ computes $(X, s_1, s_2)$ is to assert there exist a flow $f\colon \Gamma \to S$ with $X \subseteq p_1 \alpha f$ and $s_i = p_i f$, $i = 1, 2$, and a commuting labeling $\ell\colon \Gamma \to T$ such that, for each vertex $v$ of $\Gamma$, we have $vf \subseteq v\ell\varphi^{-1}$, and for each edge $e$, we have $ef \in e\ell\varphi^{-1}$. Notice that if $\varphi$ computes a flow configuration, then it computes all smaller flow configurations.

**Definition 3.8.9 (Inevitable flow configuration).** *Let* **V** *be a pseudovariety of semigroups and* $\mathscr{E}$ *a set of graph pseudoidentities. A flow configuration* $(X, s_1, s_2)$ *is termed* $\mathscr{E}$*-inevitable with respect to* **V** *(or simply* $(\mathbf{V}, \mathscr{E})$*-inevitable) if every relational morphism* $\varphi\colon S \to T$ *with* $T \in \mathbf{V}$ *computes* $(X, s_1, s_2)$ *with respect to* $\mathscr{E}$.

If **V** and $\mathscr{E}$ are understood, then we shall just use the word *inevitable*; if just $\mathscr{E}$ is understood, we shall say **V***-inevitable*. Notice that if $(X, s_1, s_2)$ is $(\mathbf{V}, \mathscr{E})$-inevitable, then so is any smaller flow configuration. For this reason, usually we are only interested in the maximal ones.

*Example 3.8.10 (Flow configurations via graph labelings).* Let $(\Gamma, p_1 = p_2)$ belong to $\mathscr{E}$ and assume $\ell\colon \Gamma \to S$ is a labeling. The associated flow configuration is $(\{p_1 \alpha \ell\}, p_1 \ell, p_2 \ell)$. It is not hard to see that if $\ell$ is **V**-inevitable in the sense of Almeida [9], then $(\{p_1 \alpha \ell\}, p_1 \ell, p_2 \ell)$ is a **V**-inevitable flow configuration. If the converse were true, then Theorem 3.7.15 would be correct without the hypothesis of a bound on the number of vertices in the graphs (as is claimed in [27]). The key difference in our notion is that we do not need to know which graph gives rise to the flow configuration for a given semigroup in **V**.

**Exercise 3.8.11.** Fill in the details for Example 3.8.10.

As with all notions of inevitability, there is the companion idea of witnessing inevitability.

**Definition 3.8.12 (Witness).** *A relational morphism* $\varphi\colon S \to T$ *with* $T \in \mathbf{V}$ *witnesses* $(\mathbf{V}, \mathscr{E})$*-inevitable flow configurations if all flow configurations computed by* $\varphi$ *are* $(\mathbf{V}, \mathscr{E})$*-inevitable.*

The next few results establish that witnesses exist.

**Lemma 3.8.13.** *Suppose* $\varphi\colon S \to T$ *is a relational morphism computing* $(X, s_1, s_2)$.

*(a) If* $m\colon T \to T'$ *is a homomorphism, then* $\varphi m$ *computes* $(X, s_1, s_2)$.
*(b) If* $\varphi'\colon S \to T'$ *is such that* $\varphi \subseteq \varphi'$*, then* $\varphi'$ *computes* $(X, s_1, s_2)$.

*Proof.* For (a), suppose $\ell$ and $f$ compute $(X, s_1, s_2)$ for $\varphi$. Then $\ell m$ and $f$ compute $(X, s_1, s_2)$ for $\varphi m$. For (b), if $\ell$ and $f$ compute $(X, s_1, s_2)$ with respect to $\varphi$, then they also compute it with respect to $\varphi'$.    □

**Corollary 3.8.14.** *If* $S$ *is a semigroup,* $\mathscr{E}$ *is a collection of graph pseudoidentities and* **V** *is a pseudovariety of semigroups, then there exists a relational morphism* $\varphi\colon S \to T$ *with* $T \in \mathbf{V}$ *witnessing* $(\mathbf{V}, \mathscr{E})$*-inevitable flow configurations.*

*Proof.* For each flow configuration $x$ that is not inevitable, choose a relational morphism $\varphi_x\colon S \to T_x$ with $T_x \in \mathbf{V}$ that does not compute it. Let $\varphi = \Delta \prod \varphi_x$. Then any flow configuration computed by $\varphi$ is computed by each $\varphi_x$ by Lemma 3.8.13. It follows that any flow configuration computed by $\varphi$ is inevitable.    □

This next corollary shows that inevitable flow configurations lift and push. Early versions of this can be found in [275].

**Corollary 3.8.15.** *Suppose $m\colon R \twoheadrightarrow S$ is an onto homomorphism of finite semigroups. Then $(X, s_1, s_2)$ is a $(\mathbf{V}, \mathscr{E})$-inevitable flow configuration for $S$ if and only if there is an inevitable flow configuration $(Y, r_1, r_2)$ for $R$ with $X = Ym$ and $s_i = r_i m$, $i = 1, 2$.*

*Proof.* Suppose first $(Y, r_1, r_2)$ is a $\mathbf{V}$-inevitable flow configuration for $R$ with $X = Ym$ and $s_i = r_i m$, for $i = 1, 2$, and let $\varphi\colon S \to T$ with $T \in \mathbf{V}$ witness $\mathbf{V}$-inevitable flow configurations. Suppose $f$ is a flow computing $(Y, r_1, r_2)$ for $m\varphi$. Then $fm$ computes $(Ym, r_1 m, r_2 m)$ for $\varphi$. We conclude $(X, s_1, s_2)$ is a $(\mathbf{V}, \mathscr{E})$-inevitable flow configuration for $S$ by choice of $\varphi$

Conversely, suppose $(X, s_1, s_2)$ is $\mathbf{V}$-inevitable and let $\varphi\colon R \to T$ with $T \in \mathbf{V}$ witness $\mathbf{V}$-inevitable flow configurations. Then we can find a pseudoidentity $(\varGamma, p_1 = p_2) \in \mathscr{E}$, a flow $f\colon \varGamma \to S$ and a commuting labeling $\ell\colon \varGamma \to T$ computing $(X, s_1, s_2)$ for $m^{-1}\varphi$. So in particular, $X \subseteq p_1 \alpha \ell \varphi^{-1} m$. Define a flow $f'\colon \varGamma \to R$ on $R$ by setting $f' = \ell \varphi^{-1}$ on vertices and by defining $ef'$, for an edge $e$, to be an element $r$ of $R$ with $rm = ef$ and $r \in e\ell\varphi^{-1}$ (we can do this because $ef \in e\ell\varphi^{-1}m$). To see that $f'$ is a flow, observe

$$(e\alpha)f'ef' \subseteq (e\alpha\ell)\varphi^{-1}(e\ell)\varphi^{-1} \subseteq e\tau\ell\varphi^{-1}$$

because $\ell$ commutes. Moreover, $\ell$ computes $f'$ for $\varphi$ and hence the flow configuration $(p_1\alpha f', p_1 f', p_2 f')$ is inevitable. Note that $p_i f'm = s_i$, $i = 1, 2$, because $m$ is a homomorphism. Furthermore, $X \subseteq p_1\alpha f \subseteq p_1\alpha\ell\varphi^{-1}m = p_1\alpha f'm$. So, for each $s \in X$, there is an element $r_s \in R$ with $r_s m = s$ and $r_s \in p_1\alpha f'$. Hence if $Y = \{r_s \mid s \in X\}$, then $(Y, p_1 f', p_2 f')$ is $\mathbf{V}$-inevitable and is the desired lift.    $\square$

Let $\varphi\colon S \to T$ be a relational morphism. The reader is referred back to Section 2.5 for the definitions of $\mathbf{Der}(\varphi)$ and $D_\varphi$. Recall that $\mathbf{Der}(\varphi)$ has a natural representation $\sigma\colon \mathbf{Der}(\varphi) \to \mathbf{Set}$ (which motivates the definition of flows) given by $t_L\sigma = t_L\varphi^{-1}$ on objects and

$$(t_L, (s, t))\sigma = \cdot s\colon t_L\varphi^{-1} \to (t_L t)\varphi^{-1}$$

on arrows. Then $D_\varphi = \mathbf{Der}(\varphi)/\ker\sigma$. Actually, in Section 2.5 we used $\sigma$ for the projection $\mathbf{Der}(\varphi) \to D_\varphi$, but as they give rise to the same congruence no confusion should occur in what follows.

Notice there are functors $p_S\colon \mathbf{Der}(\varphi) \to S$ and $p_T\colon \mathbf{Der}(\varphi) \to T$ given by $(t_L, (s, t)) \mapsto s$ and $(t_L, (s, t)) \mapsto t$, respectively. The following lemma connects flows with the derived semigroupoid.

**Lemma 3.8.16.** *Fix a graph $\varGamma$ and a relational morphism $\varphi\colon S \to T$.*

*(a) Let $m\colon \varGamma \to \mathbf{Der}(\varphi)$ be a morphism. Then the labeling $\ell = (m, mp_T)$ of $\varGamma$ over $T$ commutes and $f = (m\sigma, mp_S)$ is a flow computed by $\varphi$ and $\ell$.*

*(b) Let $f\colon \Gamma \to S$ be a flow computed by $\varphi$ and $\ell$. Define $m\colon \Gamma \to \mathbf{Der}(\varphi)$ by $vm = v\ell$ on vertices and $em = (e\alpha\ell, (ef, e\ell))$ on edges. Then $m$ is a morphism and $vf \subseteq vm\sigma = vm\varphi^{-1}$ for all $v \in V(\Gamma)$.*

*Proof.* To establish (a), suppose $v_0 \xrightarrow{e} v_1$ is an edge of $\Gamma$ and $em = t_L \xrightarrow{(s,t)} t_L t$. Then we have $v_0 f = t_L \varphi^{-1} = v_0 \ell \varphi^{-1}$, $ef = s \in t\varphi^{-1} = e\ell\varphi^{-1}$ and $v_1 f = (t_L t)\varphi^{-1} = v_1 \ell \varphi^{-1}$. Also, $v_0 fef = t_L \varphi^{-1} s \subseteq t_L \varphi^{-1} t \varphi^{-1} \subseteq (t_L t)\varphi^{-1} = v_1 f$. Thus $f$ is a flow computed by $\varphi$ and $\ell$. Because $v_0 \ell e\ell = t_L t = v_1 \ell$, we see that $\ell$ commutes. This proves (a).

For (b), let $v_0 \xrightarrow{e} v_1$ be an edge of $\Gamma$. Because $f$ is computed by $\varphi$ and $\ell$, we have $ef \in e\ell\varphi^{-1}$. As $\ell$ commutes, $v_0 \ell e\ell = v_1 \ell$. Thus $(v_0 \ell, (ef, e\ell))$ is an arrow of $\mathbf{Der}(\varphi)$ from $v_0 \ell$ to $v_1 \ell$. This shows that $m$ is a morphism. Because $f$ is computed by $\varphi$ and $\ell$, for each vertex $v$, one has $vf \subseteq v\ell\varphi^{-1} = vm\sigma$. This completes the proof of (b).    □

We are now in a position to characterize membership in semidirect products of pseudovarieties in terms of inevitable flow configurations. This will allow us to obtain a basis of pseudoidentities for the semidirect product. In fact, we consider the following more general situation. Recall that if $\mathbf{V}$ is a pseudovariety of semigroupoids, then $\mathbf{V}_D$ is the pseudovariety of relational morphisms generated by relational morphisms with derived semigroupoid in $\mathbf{V}$. If $\mathbf{V}$ is of the form $\mathbf{gU}$ (or more generally defined by path pseudoidentities), then we saw in Section 2.5.2 that we do not need to add any new relational morphisms, otherwise we must close under range extension. In any event, $S \in \mathbf{V}_D \mathsf{q}(\mathbf{W})$ if and only if there is a relational morphism $\varphi\colon S \to T$ so that $T \in \mathbf{W}$ and $D_\varphi \in \mathbf{V}$, as range extension does not change the operator. Of course, $(\mathbf{gU})_D \mathsf{q}(\mathbf{W}) = \mathbf{U} * \mathbf{W}$ so all our results apply to the semidirect product.

**Theorem 3.8.17.** *Let $S$ be a finite semigroup and let $\mathbf{V}$ be a pseudovariety of semigroupoids with basis $\mathscr{E}$ of pseudoidentities. Then the following conditions are equivalent:*

*(a) $S \in \mathbf{V}_D \mathsf{q}(\mathbf{W})$;*
*(b) The equality $\cdot s_1 = \cdot s_2 \colon X \to S$ holds for all $(\mathbf{W}, \mathscr{E})$-inevitable flow configurations $(X, s_1, s_2)$;*
*(c) The equality $s s_1 = s s_2$ holds for all $(\mathbf{W}, \mathscr{E})$-inevitable flow configurations $(\{s\}, s_1, s_2)$;*
*(d) If $\varphi\colon S \to T$, with $T \in \mathbf{W}$, witnesses $(\mathbf{W}, \mathscr{E})$-inevitable flow configurations, then $D_\varphi \in \mathbf{V}$.*

*Proof.* We begin with (a) implies (b). Suppose $\varphi\colon S \to T \in \mathbf{W}$ with $D_\varphi \in \mathbf{V}$. Because $(X, s_1, s_2)$ is inevitable, there exist $(\Gamma, p_1 = p_2) \in \mathscr{E}$, a commuting labeling $\ell\colon \Gamma \to T$ and a flow $f\colon \Gamma \to S$ computed by $\varphi$ and $\ell$, which computes $(X, s_1, s_2)$. By Lemma 3.8.16(b), there is a morphism $m\colon \Gamma \to \mathbf{Der}(\varphi)$ defined by $\ell$ on vertices and $e \mapsto (e\alpha\ell, (ef, e\ell))$ on edges such that $vf \subseteq vm\sigma$ all

$v \in V(\Gamma)$. Hence $X \subseteq p_1 \alpha m \sigma$. Because $D_\varphi \in \mathbf{V}$, it follows $D_\varphi \models (\Gamma, p_1 = p_2)$ and so $p_1 m \sigma = p_2 m \sigma$. As $p_i m p_S = p_i f = s_i$, it then follows $\cdot s_1 = \cdot s_2$ as maps $p_1 \alpha m \sigma \rightarrow p_1 \tau m \sigma$. Thus $\cdot s_1$ and $\cdot s_2$ coincide on $X$, as desired. The implications (b) implies (c) and (d) implies (a) are obvious.

It remains to prove that (c) implies (d). Suppose that $\varphi \colon S \rightarrow T$ with $T \in \mathbf{W}$ witnesses $(\mathbf{W}, \mathscr{E})$-inevitable flow configurations. Let $(\Gamma, p_1 = p_2) \in \mathscr{E}$ and let $g \colon \Gamma \rightarrow D_\varphi$ be a morphism. Because $D_\varphi$ is a quotient of $\mathbf{Der}(\varphi)$, we can lift $g$ to a morphism $m \colon \Gamma \rightarrow \mathbf{Der}(\varphi)$. Let $s_i = p_i m p_S$, $i = 1, 2$. By Lemma 3.8.16(a), $(p_1 \alpha m \sigma, s_1, s_2)$ is a flow configuration computed by $\varphi$ and is hence $(\mathbf{W}, \mathscr{E})$-inevitable. To show that $p_1 g = p_2 g$ it suffices to show that, for all $s \in p_1 \alpha m \sigma$, one has $s s_1 = s s_2$. But because $(\{s\}, s_1, s_2) \leq (p_1 \alpha m \sigma, s_1, s_2)$, it follows $(\{s\}, s_1, s_2)$ is $(\mathbf{W}, \mathscr{E})$-inevitable and so $s s_1 = s s_2$ by (c). We conclude $D_\varphi \in \mathbf{V}$.                                                                        $\square$

This theorem shows that computing semidirect products $\mathbf{V} * \mathbf{W}$ amounts to computing inevitable flow configurations with respect to some basis for $\mathbf{gV}$ (canonically we can use the collection of all pseudoidentities for $\mathbf{gV}$). This is implicit in our approach to the complexity problem [128].

Let us state and prove the Basis Theorem for Semidirect Products. An analogous result holds for $\mathbf{V}_D \mathbf{q}(\mathbf{W})$ where $\mathbf{V}$ is a pseudovariety of semigroupoids.

**Theorem 3.8.18 (Basis Theorem for Semidirect Products).** *Let* $\mathbf{V}$ *and* $\mathbf{W}$ *be pseudovarieties of semigroups. Let* $\mathscr{E}$ *be a pseudoidentity basis for* $\mathbf{gV}$. *Then* $\mathbf{V} * \mathbf{W}$ *is defined by all pseudoidentities of the form* $(A, \pi \pi_1 = \pi \pi_2)$ *(where* $\pi = I$ *is allowed) such that, for each finite $A$-generated semigroup $S$, one has* $(\{\pi \rho_S\}, \pi_1 \rho_S, \pi_2 \rho_S)$ *is a* $(\mathbf{W}, \mathscr{E})$-*inevitable flow configuration, where* $\rho_S \colon \widehat{A^+} \rightarrow S$ *is the canonical projection.*

*Proof.* Suppose first $S \in \mathbf{V} * \mathbf{W}$ and $\pi, \pi_1, \pi_2$ are as in the theorem statement. We show that $S$ satisfies the above pseudoidentities. Let $m \colon A \rightarrow S$ be a substitution. To show that $\pi \pi_1 m = \pi \pi_2 m$, it suffices to consider the subsemigroup generated by the image of $m$. Thus we may assume that $m = \rho_S$. Then $(\{\pi \rho_S\}, \pi_1 \rho_S, \pi_2 \rho_S)$ is $(\mathbf{W}, \mathscr{E})$-inevitable and so $\pi \rho_S \pi_1 \rho_S = \pi \rho_S \pi_2 \rho_S$ by Theorem 3.8.17(c).

For the converse, suppose that an $A$-generated finite semigroup $S$ satisfies the above pseudoidentities. By Theorem 3.8.17(c), to show $S \in \mathbf{V} * \mathbf{W}$, it suffices to show $s s_1 = s s_2$, for each $(\mathbf{W}, \mathscr{E})$-inevitable flow configuration $(\{s\}, s_1, s_2)$. So suppose $s, s_1, s_2$ give rise to such an inevitable flow configuration. For each finite $A$-generated semigroup $T$, let $C_T$ be the set of triples $(t, t_1, t_2)$ such that $(\{t\}, t_1, t_2)$ is a $(\mathbf{W}, \mathscr{E})$-inevitable flow configuration. It follows from Corollary 3.8.15 that the $C_T$ form an inverse quotient system. Hence $C = \varprojlim C_T \subseteq \widehat{A^+} \times \widehat{A^+} \times \widehat{A^+}$ maps onto $C_S$ by Lemma 3.1.26. In particular, there is a triple $(\pi, \pi_1, \pi_2)$ so that $(\{\pi \rho_T\}, \pi_1 \rho_T, \pi_2 \rho_T)$ is $(\mathbf{W}, \mathscr{E})$-inevitable for all finite $A$-generated semigroups $T$ and $\pi \rho_S = s$, $\pi_i \rho_S = s_i$, $i = 1, 2$. Because $S$ satisfies $\pi \pi_1 = \pi \pi_2$, we have $s s_1 = s s_2$, as required.                $\square$

*Remark 3.8.19 (Comparison of basis theorems).* How does the basis in Theorem 3.8.18 compare with the basis proposed by Almeida and Weil [27] (our Theorem 3.7.15 but without assuming finite vertex rank)? Almeida and Weil consider all pseudoidentities of the form $\pi\pi_1 = \pi\pi_2$ such that there exist a pseudoidentity $(\Gamma, p_1 = p_2) \in \mathcal{E}$ and a labeling $\ell\colon \Gamma \to \widehat{A^+}$ such that $p_1\alpha\ell = \pi$, $p_i\ell = \pi_i$ for $i = 1, 2$ and $\ell\rho_{\mathbf{W}}$ commutes where $\rho_{\mathbf{W}}\colon \widehat{A^+} \to \widehat{F_{\mathbf{W}}}(A)$ is the canonical projection. But if $S$ is a finite semigroup and $\rho_S\colon \widehat{A^+} \to S$ is the projection, then Corollary 3.7.4 implies $\ell\rho_S$ is a $\mathbf{W}$-inevitable labeling of $\Gamma$ by $S$. The discussion in Example 3.8.10 then shows that $(\{\pi\rho_S\}, \pi_1\rho, \pi_2\rho)$ is a $(\mathbf{W}, \mathcal{E})$-inevitable flow configuration. Thus $\pi\pi_1 = \pi\pi_2$ belongs to the basis in Theorem 3.8.18.

*Question 3.8.20 (Almeida-Weil Basis Question).* Find an example of pseudovarieties $\mathbf{V}$ and $\mathbf{W}$ so that pseudoidentities in Theorem 3.7.15 (i.e., [27, Thms. 5.2 and 5.3]) do *not* define $\mathbf{V} * \mathbf{W}$. For such an example give an explicit example of a member of the basis from Theorem 3.8.18 that is not a consequence of the pseudoidentities in Theorem 3.7.15.

## 3.9 The Equational Theory for Continuously Closed Classes

We now want to generalize the previous results (to the extent that we can) to continuously closed classes. These results will not be used elsewhere in the text. If $\mathbf{V} \in \mathbf{CC}$, then the definition of a pro-$\mathbf{V}$ relational morphism is exactly as in the setting of pseudovarieties of relational morphisms. Propositions 3.3.7, 3.3.9, 3.3.10, 3.3.13 and 3.3.15 go through without change. Proposition 3.3.14 also goes through if we assume $f$ is onto. All in all, we have the following.

**Proposition 3.9.1.** *Let $\mathbf{V} \in \mathbf{CC}$. Then the class of pro-$\mathbf{V}$ relational morphisms is closed under arbitrary products and (profinite) strong divisions. If $\mathbf{V} \in \mathbf{CC}^+$, then all identity maps of profinite semigroups are pro-$\mathbf{V}$.*

From now on we also assume $\mathbf{V}$ is equational, that is, $\mathbf{V}$ is closed under pullbacks; see Section 1.3.2 for more on pullbacks. We need the following lemma.

**Lemma 3.9.2.** *Let $\varphi_i\colon P_i \to Q$, $i \in I$, be a collection of relational morphisms of profinite semigroups. Then the pullback $\prod_{\varphi_i} P_i$ is profinite and is, in fact, the inverse limit of the inverse system consisting of all pullbacks indexed over a finite subset $J$ of $I$ of the form $\prod_{\psi_j} S_j$ with $S_j$ a finite image of $P_j$ and $\psi_j\colon S_j \to Q$ the relational morphism induced by $\varphi_j$.*

*Proof.* For the first part of the lemma, it suffices to show that $\prod_{\varphi_i} P_i$ is a closed subset of $\prod_i P_i$. Indeed, suppose $\{(s_{i,\alpha})_\alpha\}$ is a net in $\prod_{\varphi_i} P_i$ that converges to $(s_i)$. Then there exists, for each $\alpha$, $q_\alpha \in Q$ that $\varphi_i$-relates to all $s_{i,\alpha}$. By going to a subnet and by compactness of $Q$, we may assume that $q_\alpha$ converges to $q \in Q$. Because each relational morphism has a closed graph, we see that $q$ $\varphi_i$-relates to each $s_i$. Hence $(s_i) \in \prod_{\varphi_i} P_i$.

For the second part, we know that $\prod_i P_i$ is the inverse limit of the system consisting of all products indexed by finite subsets $J$ of $I$ of the form $\prod_j S_j$ with $S_j$ a quotient of $P_j$. Moreover, the image of $\prod_{\varphi_i} P_i$ is contained in $\prod_{\psi_j} S_j$. Hence there is a natural inclusion from $\prod_{\varphi_i} P_i$ to the inverse limit of the system, call it $L$, in the statement of the lemma. Moreover, $L$ is a closed subsemigroup of $\prod_i P_i$. A straightforward compactness argument shows that if $(s_i) \in L$, then $\bigcap s_i \varphi_i \neq \emptyset$. Thus $L = \prod_{\varphi_i} P_i$.    □

We may now establish the following.

**Proposition 3.9.3.** *Let* V *be an equational continuously closed class and let* $\varphi_i \colon P_i \to T$ *be a collection of pro-*V *relational morphisms with* $T$ *finite. Then the pullback* $\prod_T \varphi_i \colon \prod_{\varphi_i} P_i \to T$ *is pro-*V.

*Proof.* Lemma 3.9.2 shows that $P = \prod_{\varphi_i} P_i$ is profinite and is the inverse limit of the system described in that lemma. Let $\equiv$ be the congruence associated with the projection to such a pullback $\prod_{\psi_j} S_j$ and $\psi \colon P/\equiv\ \to T$ be the induced relational morphism. By the analogue for continuously closed classes of Proposition 3.3.10, the relational morphisms $\psi_i$ are all in V. Hence, by assumption on V, the pullback $\prod_T \psi_j \in$ V. But $\psi \prec_s \prod_T \psi_j$, so $\psi \in$ V. Thus $\prod_T \varphi_i$ is pro-V.    □

Suppose $T$ is a finite semigroup and $\psi \colon \widehat{A^+} \to T$ is a continuous homomorphism with $A$ a profinite set (note that $\psi$ is *not* assumed surjective). Recall that the $A$-generated relational morphisms to $T$ (respecting $\psi$) are of the form $\varphi_\equiv$ where $\equiv$ is a profinite congruence on $\widehat{A^+}$ and $\varphi_\equiv$ is the induced relational morphism. Let $\mathscr{C}_{\mathsf{V},T}$ denote the set of all (profinite) congruences $\equiv$ on $\widehat{A^+}$ such that $\varphi_\equiv$ is pro-V. We may now prove the analogue in our context of Proposition 3.4.2.

**Proposition 3.9.4.** *Let* V *be an equational continuously closed class and let* $\psi \colon \widehat{A^+} \to T$ *be a continuous homomorphism with* $T$. *Then the collection* $\mathscr{C}_{\mathsf{V},T}$ *is closed under arbitrary intersections.*

*Proof.* Let $C \subseteq \mathscr{C}_{\mathsf{V},Q}$ and let $\sim\ =\ \bigcap_C \equiv$. One verifies that the diagram

$$
\begin{array}{ccc}
\widehat{A^+}/\!\sim & \lhook\joinrel\longrightarrow & \prod \widehat{A^+}/\!\equiv \\[4pt]
& {}^{\varphi_\equiv} & \\[-6pt]
{}_{\varphi_\sim}\searrow & & \swarrow {}_{\prod_T \varphi_\equiv} \\[4pt]
& T &
\end{array}
$$

is subcommutative and therefore $\varphi_\sim \prec_s \prod_T \varphi_\equiv$. Propositions 3.9.1 and 3.9.3 then yield $\varphi_\sim$ is pro-V.    □

One can then define

$$\equiv_{\mathsf{V},T} = \bigcap \mathscr{C}_{\mathsf{V},T} \text{ and } T^{\mathsf{V}} = \widehat{A^+}/\equiv_{\mathsf{V},T}. \tag{3.17}$$

We again call $\varphi_{\mathsf{V},T} = \varphi_{\equiv_{\mathsf{V},T}}$ the *free pro-*$\mathsf{V}$ *relational morphism* to $T$. If $\mathsf{V}$ is positive, $\varphi_{\mathsf{V},T}$ will be a continuous homomorphism. One then obtains, analogously to Section 3.4, the following result justifying this terminology.

**Theorem 3.9.5.** *Let* $\mathsf{V}$ *be an equational continuously closed class and let* $\psi\colon \widehat{A^+} \to T$ *be a continuous homomorphism with* $T$ *finite.*

1. *Let* $\equiv$ *be a congruence on* $\widehat{A^+}$. *Then* $\varphi_{\equiv}$ *is pro-*$\mathsf{V}$ *if and only if* $\equiv_{\mathsf{V},T} \subseteq \equiv$.
2. *Suppose* $A$ *is finite and* $\varphi\colon S \to T$ *is an* $A$*-generated relational morphism of finite semigroups. Let* $\rho\colon \widehat{A^+} \to S$ *be the canonical surjection. Then* $\varphi \in \mathsf{V}$ *if and only if* $\equiv_{\mathsf{V},T} \subseteq \ker\varphi$.

The definitions and results concerning locally finite continuously closed equational classes then apply in this context.

### 3.9.1 Pseudoidentities for continuously closed classes

We now define a suitable notion of pseudoidentities in the context of equational continuously closed classes. A *strong relational pseudoidentity* over a finite alphabet $A$ consists of a triple $(A, u = v, \psi)$ where $u = v$ is a pseudoidentity over $A$ and $\psi\colon \widehat{A^+} \to Q$ is a continuous homomorphism of profinite semigroups. A *relational pseudoidentity* $(A, u = v, E)$ can be viewed as the strong relational pseudoidentity $(A, u = v, \psi\colon \widehat{A^+} \to \widehat{A^+}/\langle E \rangle)$. However, the definition of satisfaction in this context will be slightly different.

Let $e = (A, u = v, \psi\colon \widehat{A^+} \to Q)$ be a strong relational pseudoidentity. A relational morphism $\varphi\colon S \to T$ of finite semigroups satisfies $e$, written $\varphi \models e$, if, for all subcommutative diagrams,

$$\begin{array}{ccc} \widehat{A^+} & \xrightarrow{\sigma} & S \\ {\scriptstyle\psi}\big\downarrow & \subseteq & \big\downarrow{\scriptstyle\varphi} \\ Q & \xrightarrow[f]{} & T \end{array} \tag{3.18}$$

one has $u\sigma = v\sigma$ where $\sigma$ is a continuous homomorphism and $f$ a surjective continuous homomorphism. That is, $\varphi \models e$ if and only if given a continuous homomorphism $\sigma\colon \widehat{A^+} \to S$ and a continuous surjective morphism $f\colon Q \to T$ such that $\psi f \subseteq_s \sigma\varphi$, one has $u\sigma = v\sigma$. If we view a relational pseudoidentity as a strong relational pseudoidentity, the difference in the definition of satisfaction comes down to the requirement that $f$ be *onto* (i.e., when we substitute $A$ into $T$ we require that the image of $A$ generates $T$).

For a set of strong relational pseudoidentities $\mathscr{E}$ and a relational morphism $\varphi$ of finite semigroups, we write $\varphi \models \mathscr{E}$, if $\varphi \models e$ for all $e \in \mathscr{E}$. We set

$$[\![\mathscr{E}]\!] = \{\varphi \colon S \to T \mid \varphi \models \mathscr{E}\}.$$

If $\mathscr{E}_1$ and $\mathscr{E}_2$ are sets of relational pseudoidentities, we write $\mathscr{E}_1 \models \mathscr{E}_2$ if, for all relational morphisms $\varphi$ of finite semigroups, $\varphi \models \mathscr{E}_1$ implies $\varphi \models \mathscr{E}_2$.

If $f \colon P \to Q$ is a homomorphism and $Q'$ is a closed subsemigroup of $Q$, then we use $_{Q'|}f$ for the corestriction of $f$ to $Q'$.

A collection $\mathscr{E}$ of strong relational pseudoidentities is said to be *closed* if $(A, u = v, \psi \colon \widehat{A^+} \to Q) \in \mathscr{E}$ and $\operatorname{Im} \psi \subseteq Q' \subseteq Q$ with $Q'$ a clopen subsemigroup of $Q$ implies that $\mathscr{E} \models (A, u = v, _{Q'|}\psi \colon \widehat{A^+} \to Q')$. This condition ensures that if a relational morphism satisfies $\mathscr{E}$, then all its range extensions do, as well.

**Proposition 3.9.6.** *Let $\mathscr{E}$ be a closed set of strong relational pseudoidentities. Then $[\![\mathscr{E}]\!]$ is an equational continuously closed class.*

*Proof.* Clearly, $1_{\{1\}}$ satisfies all pseudoidentities of $\mathscr{E}$. Suppose that $\varphi \colon S \to T$ and $\varphi' \colon S' \to T$ belong to $[\![\mathscr{E}]\!]$. We show that the pullback $\varphi \times_T \varphi'$ is in $\mathscr{E}$. Let $e = (A, u = v, \psi \colon \widehat{A^+} \to Q) \in \mathscr{E}$ and suppose we have a diagram

$$
\begin{array}{ccc}
\widehat{A^+} & \xrightarrow{\ \sigma\ } & S \times_{\varphi,\varphi'} S' \\[2pt]
\psi \downarrow & \subseteq & \downarrow \varphi \times_T \varphi' \\[2pt]
Q & \xrightarrow[\ f\ ]{} & T
\end{array}
$$

(with $\sigma, f$ homomorphisms). Let $p_S, p_{S'}$ be the projections. Then we have $\psi f \subseteq_s \sigma p_S \varphi, \sigma p_{S'} \varphi'$. Because $\varphi, \varphi' \models e$, it follows $u \sigma p_S = v \sigma p_S$ and $u \sigma p_{S'} = v \sigma p_{S'}$, whence $u \sigma = v \sigma$.

Suppose $\varphi \colon S \to T$ belongs to $[\![\mathscr{E}]\!]$ and $\varphi' m \subseteq_s d\varphi$ with $d$ a division and $m$ surjective, i.e., $\varphi' \prec_s \varphi$. Suppose $\varphi' \colon S' \to T'$. Let $e = (A, u = v, \psi \colon \widehat{A^+} \to Q)$ belong to $\mathscr{E}$ and suppose $\sigma \colon \widehat{A^+} \to S'$ is a continuous homomorphism and $f \colon Q \twoheadrightarrow T'$ is a continuous onto homomorphism with $\psi f \subseteq_s \sigma \varphi'$. For each $a \in A$, choose $s_a \in a \sigma d$ such that $s_a$ is $\varphi$-related to $a \psi f m$. This can be done because $\psi f m \subseteq \sigma \varphi' m \subseteq \sigma d \varphi$. Let $\sigma' \colon \widehat{A^+} \to S$ be the map induced by $a \mapsto s_a$. Then because $\varphi \models e$ and $\psi f m \subseteq_s \sigma' \varphi$, it follows that $u \sigma' = v \sigma'$ and hence, because $d$ is a division, $u \sigma = v \sigma$.

Suppose $\varphi \colon S \to T$ belongs to $[\![\mathscr{E}]\!]$ and $\iota \colon T \to T'$ is injective. Let $e = (A, u = v, \psi \colon \widehat{A^+} \to Q) \in \mathscr{E}$ and suppose $\sigma \colon \widehat{A^+} \to S$, $f \colon Q \twoheadrightarrow T'$ are homomorphisms such that $\psi f \subseteq_s \sigma \varphi \iota$. Let $Q' = Tf^{-1}$. Then $Q'$ is a clopen subsemigroup of $Q$. We show $\operatorname{Im} \psi \subseteq Q'$. Indeed, let $x \in \widehat{A^+}$. Then $x \psi f \in x \sigma \varphi \iota \subseteq T$. So $x \psi \in Tf^{-1}$. Because $\mathscr{E}$ is closed, it follows $\mathscr{E} \models (A, u = v, _{Q'|}\psi)$. By considering $\sigma$ and $_{T|}f$, we see $u \sigma = v \sigma$. Thus V satisfies Axiom (r-e). This completes the proof that V is an equational continuously closed class. □

Note that $[\![\mathscr{E}]\!]$ is positive if and only if, for all $(A, u = v, \psi) \in \mathscr{E}$, $u\psi = v\psi$. If $\mathscr{E}$ is a closed set of strong relational pseudoidentities, we say $\mathscr{E}$ is a basis for a continuously closed class V if $V = [\![\mathscr{E}]\!]$.

We now provide some examples. Suppose V is a pseudovariety of relational morphisms defined by relational pseudoidentities $\mathscr{E}$. Let $\widetilde{\mathscr{E}}$ consist of all strong relational pseudoidentities of the form $(A, u = v, \psi \colon \widehat{A^+} \to Q)$ such that there exists $(A, u = v, E) \in \mathscr{E}$ with $E \subseteq \ker \psi$. It is easy to verify that this is a closed collection of strong relational pseudoidentities.

**Proposition 3.9.7.** $V = [\![\widetilde{\mathscr{E}}]\!]$.

*Proof.* Suppose first $\varphi \colon S \to T$ is in V and let $(A, u = v, \psi \colon \widehat{A^+} \to Q) \in \widetilde{\mathscr{E}}$. Then there exists $(A, u = v, E) \in \mathscr{E}$ with $E \subseteq \ker \psi$. Suppose one has a diagram as per (3.18). We view $\sigma$ and $\psi f$ as $\varphi$-related substitutions. Because $E \subseteq \ker \psi$, it follows that $\psi f \vdash E$, whence $\sigma \vdash u = v$, as desired.

Conversely, suppose $\varphi \in [\![\widetilde{\mathscr{E}}]\!]$ and $(A, u = v, E) \in \mathscr{E}$. Suppose $\sigma \colon A \to S$ and $\psi \colon A \to T$ are $\varphi$-related substitutions such that $\psi \vdash E$. Then $E \subseteq \ker \psi$, so $(A, u = v, \psi) \in \widetilde{\mathscr{E}}$. Because $\psi 1_T \subseteq_s \sigma\varphi$ and $\varphi \models (A, u = v, \psi)$, $u\sigma = v\sigma$, as desired. □

As another example, recall from (2.15) that, for $\alpha \colon \mathbf{PV} \to \mathbf{PV}$ continuous,

$$M(\alpha) = \{\varphi \colon S \to T \mid S \in \alpha(T)\}$$

is an equational continuously closed class and is the largest continuously closed class V with $Vq = \alpha$. We claim that a basis of strong relational pseudoidentities for $M(\alpha)$ is given as follows. For each finite semigroup $T$, choose a basis $E_T$ of pseudoidentities for $\alpha((T))$. If $u = v \in E_T$ is over a finite alphabet $A$, then we include all triples $(A, u = v, \psi)$ where $\psi \colon \widehat{A^+} \to T$ is a continuous homomorphism.

Our next result proves the converse of Proposition 3.9.6 and can be viewed as Reiterman's Theorem [261] for equational continuously closed classes.

**Theorem 3.9.8.** *Let V be an equational continuously closed class and let $\mathscr{E}$ consist of all relational pseudoidentities of the form $(A, u = v, \psi \colon \widehat{A^+} \to T)$ where $T$ is a finite semigroup and $(u, v) \in \equiv_{V,T}$. Then $\mathscr{E}$ is closed and the equality $V = [\![\mathscr{E}]\!]$ holds.*

*Proof.* First we show that $\mathscr{E}$ is closed. To do this, it suffices to show that if $\iota \colon T \to T'$ is injective and $\psi \colon \widehat{A^+} \to T$ is a homomorphism, then $\equiv_{V,T'} \subseteq \equiv_{V,T}$. To prove this latter statement, we need only show (by Proposition 3.3.10) that if $\equiv$ is a clopen congruence on $\widehat{A^+}$ such that $\varphi_{\equiv} \colon \widehat{A^+}/\equiv \to T$ is in V, then $\varphi_{\equiv}\iota$ is in V. But this follows from the fact that V is closed under range extension (Axiom (r-e)).

Next we establish $V \subseteq [\![\mathscr{E}]\!]$. Let $e = (A, u = v, \psi \colon \widehat{A^+} \to T) \in \mathscr{E}$. Suppose $\varphi \colon S \to U$ is in $V$ and we have a subcommutative diagram

$$
\begin{array}{ccc}
\widehat{A^+} & \xrightarrow{\ \sigma\ } & S \\
{\scriptstyle \psi}\Big\downarrow & \subseteq & \Big\downarrow{\scriptstyle \varphi} \\
T & \xrightarrow[\ f\ ]{} & U
\end{array}
$$

(with $\sigma, f$ homomorphisms). Denote $\ker \sigma$ by $\equiv$. Then for $\varphi_\equiv \colon \widehat{A^+}\sigma \to T$, we have $\varphi_\equiv \prec_s \varphi$. Indeed, if $\iota \colon \widehat{A^+}\sigma \to S$ is the inclusion, then $\varphi_\equiv f \subseteq_s \iota\varphi$ and $f$ is onto. It follows $\varphi_\equiv \in V$. Hence, by Theorem 3.9.5, $u \equiv v$ and so $u\sigma = v\sigma$.

Suppose now that $\varphi \colon S \to T \models \mathscr{E}$; we show $\varphi \in V$. Let $A$ be a finite set of generators of $\#\varphi$ and $\psi \colon \widehat{A^+} \to \#\varphi$ be the canonical surjection. Consider $\equiv_{V,T}$ with respect to $\psi p_T \colon \widehat{A^+} \to T$. By Theorem 3.9.5, it suffices to show $\equiv_{V,T} \subseteq \ker \psi p_S$. If $(u,v) \in \equiv_{V,T}$, then $(A, u = v, \psi p_T) \in \mathscr{E}$ by construction. We also have the subcommutative diagram

$$
\begin{array}{ccc}
\widehat{A^+} & \xrightarrow{\ \psi p_S\ } & S \\
{\scriptstyle \psi p_T}\Big\downarrow & \subseteq & \Big\downarrow{\scriptstyle \varphi} \\
T & =\!=\!= & T.
\end{array}
$$

Thus, by the assumption that $\varphi \models \mathscr{E}$, we have $u\psi p_S = v\psi p_S$. It follows $\equiv_{V,T} \subseteq \ker \psi p_S$, establishing Theorem 3.9.8.   $\square$

**Exercise 3.9.9.** A relational morphism $(\alpha, m) \colon f \to g$ diagrammed as per

$$
\begin{array}{ccc}
S & \xrightarrow{\ \alpha\ } & U \\
{\scriptstyle f}\Big\downarrow & \subseteq & \Big\downarrow{\scriptstyle g} \\
T & \xrightarrow[\ m\ ]{} & V
\end{array}
\qquad\qquad (3.19)
$$

with $m$ onto is called a *strong division diagram* if the implication

$$
t \in s_1 f \cap s_2 f, \ tm \in (s_1\alpha \cap s_2\alpha)g \implies s_1 = s_2
$$

holds. Notice that strong division is a special case of a strong division diagram. Given a strong division diagram as per (3.8), show using the equational theory that if $V$ is a positive equational continuously closed class and $g \in V$, then $f \in V$.

We leave it to interested parties to develop the theory of pseudoidentities for compositions in this context.

# Notes

The material on compact semigroups in Section 3.1.1 goes back to A. D. Wallace and his school and can be found in such sources as [63,64,135]. However, until fairly recently, profinite semigroups received relatively little attention except for the result of Numakura [222] that a compact semigroup is profinite if and only if it is totally disconnected and for some work of Hunter [140]. The foundational material on inverse limits and profinite spaces/semigroups in Section 3.1.2 can be found in [7,298] among other places. The categorical viewpoint is from [185]. Stone duality between Boolean algebras and profinite spaces is the subject of [117,147,341,342] and in fact profinite spaces are alternatively known as Boolean spaces and Stone spaces. Reiterman's Theorem [261] was originally stated in the language of implicit operations; the profinite viewpoint was first emphasized by Banaschewski [44]. However, it was really J. Almeida who pushed the pseudoidentity/profinite approach in Finite Semigroup Theory [7,25]. Recent results concerning the structure of free profinite semigroups can be found in [11,21–23,130,286,290,337].

All the material concerning pseudovarieties of relational morphisms and continuously closed classes is new. The notion of a relational pseudoidentity was invented by Steinberg when he was working with Tilson on [339]. It should be noted that free pro-$\mathbf{V}$ relational morphisms generalize the profinite expansions considered by the authors in [286], based on earlier work of Elston [87]. Theorem 3.6.4 can be viewed as a generalization of the results of [24] on free objects over semidirect products. Inevitable substitutions arose naturally in the context of relational pseudoidentities. It has its antecedents in the work of Rhodes and Tilson [295] on Type I–Type II semigroups, of Rhodes and Henckell [121] on pointlike sets, of Ash on inevitable graphs [33], of Pin and Weil on idempotent-pointlikes [235], of Almeida on hyperdecidability [9,14,28,29] (as well as the second author [322,327,330]) and of Almeida and Steinberg on tameness [19,20]. The authors have shown that decidability of a pseudovariety is not enough to guarantee even decidability of pointlikes [285]; Auinger and Steinberg constructed a decidable pseudovariety of metabelian groups with undecidable pointlikes [36]. Theorem 3.7.1 generalizes simultaneously several compactness results in the literature [9,27,130,235,322,330]. Theorem 3.7.10, the Basis Theorem for Composition, seems to include as special cases all the basis theorems in the literature including the basis theorems of Almeida and Weil [27] for semidirect and two-sided semidirect products (with appropriate finite vertex rank assumptions) and the basis theorem of Pin and Weil [235] for Mal'cev products. The Basis Theorem for Semidirect Products (Theorem 3.8.18) is new. The idea of a flow is due to Rhodes [280]. It is based on the viewpoint that relational morphisms should be oriented the other way: instead of considering $\varphi \colon S \to T$, one should consider $\varphi^{-1} \colon T \to S$.

## Part II

Complexity in Finite Semigroup Theory

# 4

# The Complexity of Finite Semigroups

Our eventual goal is to generalize Krohn-Rhodes complexity of finite semi-groups to arbitrary continuous operators. As the notion of complexity in Finite Semigroup Theory begins with the Prime Decomposition Theorem [169], we first present a proof. Schützenberger's Theorem on star-free languages is then deduced as a consequence of the Prime Decomposition Theorem [313]. Simon's proof [319, 320] of Brown's Theorem [59] using the techniques of the Prime Decomposition Theorem is given as well. The rest of the chapter surveys Krohn-Rhodes complexity theory, providing proofs whenever possible. In particular, we give a modern proof of the decidability of complexity for semigroups in $\mathbf{LG} \text{ⓜ} \mathbf{A}$, first proved in [170] for union of groups semigroups (i.e., completely regular semigroups), and then more generally in [269, 368]. In the process, we redevelop the semilocal theory from [171] with an updated presentation and provide an improved version of the classification of subdirectly indecomposable finite semigroups from [171]. Various upper bounds and lower bounds in the complexity literature are discussed with many illustrative examples. In particular, we discuss in detail the Type I–Type II lower bound [126, 295]. The results of Stiffler's Advances in Mathematics paper [340] are also treated.

More advanced topics include: Graham's description of the idempotent-generated subsemigroup of a 0-simple semigroup [107], the Presentation Lemma [43, 334], Ash's Type II Theorem [33], the Ribes and Zalesskii Theorem [300] and Henckell's Theorem on aperiodic pointlikes [121, 129].

Many of the older results in this chapter are surveyed in Tilson [359]. The next chapter will consider two-sided complexity in the broader context of the complexity of a pair of operators.

The first section of this chapter, on the Prime Decomposition Theorem, can be read immediately after Chapter 1. The remainder of the chapter will make use of the language of Chapter 2, but not the results with the exception of the Derived Semigroupoid Theorem and Tilson's Lemma. It should be possible to read this chapter without first reading Chapter 2 by referring back, via the index, as needed.

J. Rhodes, B. Steinberg, *The q-theory of Finite Semigroups*,
Springer Monographs in Mathematics, DOI 10.1007/978-0-387-09781-7_4,
© Springer Science+Business Media, LLC 2009

Throughout this chapter we shall make use without comment of the fact that finite semigroups are *stable*. This means

$$sx \ \mathscr{J} \ s \iff sx \ \mathscr{R} \ s \text{ and also } xs \ \mathscr{J} \ s \iff xs \ \mathscr{L} \ s.$$

Stability yields $\mathscr{J} = \mathscr{D}$ and underlies the Green-Rees structure theory (cf. Appendix A). In this chapter, semigroups should be assumed finite with the exception of free semigroups, free groups and when we explicitly say otherwise.

## 4.1 The Prime Decomposition Theorem

The Prime Decomposition Theorem [169] states that every finite semigroup divides an iterated wreath product of its finite simple group divisors and copies of the three element aperiodic monoid consisting of two right zeroes and an identity (recall that a finite semigroup is *aperiodic* if all of its subgroups are trivial). In other words, the basic building blocks of finite semigroups are the finite simple groups and semigroups of constant maps with an adjoined identity.

The earliest proof of the Prime Decomposition Theorem in book form appears in [171]. A variant of the original proof appears in Eilenberg [85], but from a transformation semigroup point of view. Eilenberg also gives a tighter proof, which he calls the Holonomy Theorem [85], that is based on Zeiger's proof of the Prime Decomposition Theorem via weakly preserved covers [386, 387].

Many recent books follow Lallement's proof [174] to avoid using transformation semigroups. However, certain statements that are true at the transformation semigroup level do not transfer so easily to abstract semigroups and we should warn the reader that the proof given in [174] is slightly flawed: it claims incorrectly that the augmentation of the wreath product of abstract monoids embeds in the wreath product of augmented monoids. A correction was published shortly afterwards [175], but unfortunately the incorrect version made its way into some more recent books [7, 110]. Another problem is that the arguments in [7, 110, 174] claim that if $N$ is a submonoid of $M$, then the augmented monoid of $N$ is a submonoid of the augmented monoid of $M$. This is not quite accurate, one only has division; we shall provide a counterexample below. Probably the approach via transformation semigroups taken in Eilenberg [85] is the most natural as the wreath product is associative in this context and the decompositions are smaller. Here we use a hybrid approach to keep the proofs as transparent as possible. Our proof more or less follows the lines of [171].

Let us recall that, for us, the wreath product of semigroups $S$ and $T$ was defined as $S \wr T = S^{T^I} \rtimes T$ where $T^I$ is $T$ with an adjoined identity. However, the Prime Decomposition Theorem is essentially a theorem about monoids in the sense that the decomposition result for semigroups follows from the result for monoids and the components of a prime decomposition are monoids. Therefore, it will be most convenient to work in the category of monoids. If $M$ and $N$ are monoids, we shall continue to refer to the wreath product $M \wr N = M^N \rtimes N$ in the category of monoids as the *unitary wreath product*. Recall that any unitary semidirect product $M \rtimes N$ of monoids embeds in the unitary wreath product $M \wr N$; see Theorem 1.2.17.

In this chapter, a transformation semigroup should be assumed faithful unless stated otherwise. Also if we do not use the modifier partial, then the transformation semigroup should be understood as being total. If $(Q, T)$ is a transformation semigroup and $S$ is a semigroup, then we write $S \wr (Q, T)$ for $S^Q \rtimes T$, the action semigroup for $(S^I, S) \wr (Q, T)$. We remark that unlike Eilenberg [85], we do not systematically identify a semigroup with its regular representation and so for us $S \wr (Q, T)$ is an abstract semigroup and not a transformation semigroup, as is the case in [85]. If we want to consider the transformation semigroup, we shall explicitly write $(S^I, S) \wr (Q, T)$.

Later, when dealing with the Schützenberger representation and the Presentation Lemma, we shall need wreath products of partial transformation semigroups, so we recall the notion [85]. If $(P, S)$ is a partial right transformation semigroup, then $(P, S)^0$ denotes the *completion* of $(P, S)$. This is the transformation semigroup $(P^0, S)$, where $P^0 = P \cup \{0\}$, with 0 an element not belonging to $P$, and where $qs = 0$ if $qs$ was undefined and, of course, $0s = 0$ for all $s \in S$. The element 0 is sometimes referred to as a *sink*.

**Definition 4.1.1 (Partial transformation wreath product).** *The wreath product of partial transformation semigroups $(P, S) \wr (Q, T)$ is the partial transformation semigroup $(P \times Q, W)$ where $W$ is the quotient of the wreath product $S \wr (Q, T)^0$ that identifies $(f, t)$ and $(g, t)$ if $f$ and $g$ coincide on the domain of $t$; equivalently the value of the function $f$ need only be specified on the domain of $t$. The action is given by declaring that $(p, q)(f, t)$ is defined if and only if $p(qf)$ and $qt$ are both defined, in which case $(p, q)(f, t) = (p(qf), qt)$.*

Notice that if $(P, S)$ and $(Q, T)$ are complete, then the resulting wreath product is isomorphic to the usual one. Again, if $S$ is a semigroup and $(Q, T)$ is a partial transformation semigroup, then $S \wr (Q, T)$ denotes the action semigroup of the partial transformation semigroup $(S^I, S) \wr (Q, T)$. By definition $S \wr (Q, T)$ is a quotient of $S \wr (Q, T)^0 = S^{Q^0} \rtimes T$ and so if $S \in \mathbf{V}$ and $T \in \mathbf{W}$, then $S \wr (Q, T)$ belongs to $\mathbf{V} * \mathbf{W}$. The reader is referred to [85] for details. There is an alternative description of the wreath product of partial transformation semigroups in terms of row monomial matrices. This will make an appearance in Section 5.5 and the reader is welcome to skip ahead if he or she so desires for more details.

### 4.1.1 Some wreath product decompositions

Our first decomposition is a straightforward, but important, result that we shall use without comment throughout the book.

**Proposition 4.1.2.** *Given divisions of semigroups $S_1 \prec T_1$ and $S_2 \prec T_2$, there results a division $S_1 \wr S_2 \prec T_1 \wr T_2$. An analogous result holds in the category of monoids.*

*Proof.* Let $d_i \colon S_i \to T_i$ be a division, for $i = 1, 2$. Let $d_2^I \colon S_2^I \to T_2^I$ be the extension. Then $(d_2^I)^{-1} \colon T_2^I \to S_2^I$ is a partial surjective function. Define a division $d \colon S_1 \wr S_2 \to T_1 \wr T_2$ by

$$(f, s)d = \{(g, t) \mid t \in sd_2, \ t'g \in t'(d_2^I)^{-1} f d_1 \text{ whenever } t'(d_2^I)^{-1} \text{ is defined}\}.$$

First we verify $d$ is a relational morphism. To see that $d$ is fully defined, observe that if $(f, s) \in S_1 \wr S_2$, then there exists $t \in sd_2$. Define

$$t'g = \begin{cases} \text{an element of } t'(d_2^I)^{-1} f d_1 & \text{if } t'(d_2^I)^{-1} \text{ is defined} \\ \text{arbitrary} & \text{else.} \end{cases}$$

Then $(g, t) \in (f, s)d$.

Suppose $(g_i, t_i) \in (f_i, s_i)d$, $i = 1, 2$. Then, writing $T_1^{T_2^I}$ additively,

$$(g_1, t_1)(g_2, t_2) = (g_1 + {}^{t_1}g_2, t_1 t_2).$$

Evidently $t_1 t_2 \in (s_1 s_2)d_2$. Suppose $t' \in T$. Then $t'(g_1 + {}^{t_1}g_2) = t'g_1 + t't_1 g_2$. If $t'(d_2^I)^{-1}$ is undefined, there is nothing to check. If $t'(d_2^I)^{-1} = s'$, that is to say $t' \in s'd_2^I$, then $t't_1 \in (s's_1)d_2$. Hence $s's_1 = (t't_1)d_2^{-1}$ and so $t'g_1 \in s'f_1 d_1$ and $t't_1 g_2 \in s's_1 f_2 d_1$. Therefore, $t'g_1 + t't_1 g_2 \in (s'f_1 + s's_1 f_2)d_1$. From

$$(f_1, s_1)(f_2, s_2) = (f_1 + {}^{s_1}f_2, s_1 s_2)$$

it now follows $(g_1, t_1)(g_2, t_2) \in ((f_1, s_1)(f_2, s_2)) \, d$, as required. Thus $d$ is a relational morphism.

Finally, we verify $d$ is a division. Suppose $(g, t) \in (f, s)d \cap (f', s')d$. Then $t \in sd_2 \cap s'd_2$ and so $s = s'$, as $d_2$ is a division. Let $s_0 \in S_2^I$ and choose $t_0 \in s_0 d_2^I$. Then $t_0 g \in s_0 f d_1 \cap s_0 f' d_1$, whence $s_0 f = s_0 f'$ as $d_1$ is a division. Because $s_0$ was arbitrary, $f = f'$ and thus $(f, s) = (f', s')$. This completes the proof that $d$ is a division.

The case of monoids and unitary wreath products is left to the reader.   $\square$

*Remark 4.1.3.* Of course an analogous result holds true for the block product.

Next we consider a standard decomposition result (cf. [69]) involving automorphism groups of partial transformation semigroups. If $(Q, S)$ is a right partial transformation semigroup, then an automorphism of $(Q, S)$ is a bijection $\varphi \colon Q \to Q$ (acting on the left of $Q$) such that $\varphi(qs) = \varphi(q)s$, for all $q \in Q$ and $s \in S$, where the equality means that either both sides are defined and agree, or neither side is defined. Recall that a group $G$ acts *freely* on the left of a set $Q$ if, for $g \in G$, $q \in Q$, the equality $gq = q$ implies $g = 1$. The set of orbits will be denoted $G\backslash Q$. Notice that this notation is similar to the notation for complementing a set, but no confusion should arise. If $(Q, S)$ is a partial right transformation semigroup and $G$ is a group of automorphisms of $(Q, S)$ acting on the left, then $S$ acts naturally on the right of $G \backslash Q$. The resulting faithful partial transformation semigroup will be denoted $(G \backslash Q, \widetilde{S})$. The image of an element $s \in S$ in $\widetilde{S}$ will be denoted $\widetilde{s}$. Notice that $Gq\widetilde{s}$ is defined if and only if $qs$ is defined, in which case $(Gq)\widetilde{s} = G(qs)$.

**Proposition 4.1.4.** *Let $(Q, S)$ be a faithful partial right transformation semigroup and let $G$ be a group of automorphisms of $(Q, S)$ acting freely on the left. Then $S$ embeds in $G \wr (G \backslash Q, \widetilde{S})$.*

*Proof.* Fix a transversal $T$ for $G \backslash Q$ in $Q$. Denote by $[Gq]$ the representative from $T$ of the orbit $Gq$. Define $\psi \colon S \to G^{(G \backslash Q)^0} \rtimes \widetilde{S}$ by $s\psi = (f_s, \widetilde{s})$ where, for $q$ in the domain of $s$, $(Gq)f_s$ is the unique (by freeness) element $g \in G$ such that $[Gq]s = g[Gqs]$; the definition of $f_s$ outside the domain of $s$ is irrelevant. To see that $\psi$ is injective, suppose $s\psi = t\psi$. Because $\widetilde{s} = \widetilde{t}$, it follows that $s$ and $t$ have the same domain by the comment just before the proposition. Also we have $Gqs = Gqt$ for all $q$ in their common domain. Let $q$ be in the common domain of $s$ and $t$. Write $q = g[Gq]$ with $g \in G$. Then

$$qs = g[Gq]s = g(Gq)f_s[Gqs] = g(Gq)f_t[Gqt] = g[Gq]t = qt.$$

Because $(Q, S)$ is faithful, we have $s = t$. Thus $\psi$ is injective.

To see that $\psi$ is a homomorphism, suppose $s, t \in S$. Then $(f_s, \widetilde{s})(f_t, \widetilde{t}) = (f_s\,{}^{\widetilde{s}}\!f_t, \widetilde{s}\widetilde{t})$. Now $Gq$ is in the domain of $st$ if and only if $Gqs$ and $(Gqs)t$ are defined, which occurs if and only if $[Gq]s$ and $[Gqs]t$ are defined. In this case,

$$[Gq]st = (Gq)f_s[Gqs]t = (Gq)f_s(Gqs)f_t[Gqst]$$

and so $(Gq)f_{st} = (Gq)f_s(Gqs)f_t = (Gq)(f_s\,{}^{\widetilde{s}}\!f_t)$, establishing that $\psi$ is a homomorphism. $\qquad\square$

If $G$ is a group and $N \lhd G$ a normal subgroup, then $N$ acts freely on the left of $G$ as a group of automorphisms of $(G, G)$. Thus we obtain the following well-known theorem, called alternatively the monomial map [113, 381] or the Krasner-Kaloujnine embedding [167].

**Corollary 4.1.5.** *Let $G$ be a group and $N \lhd G$ a normal subgroup. Then $G$ embeds in the unitary wreath product $N \wr G/N$.*

A simple induction along a Jordan-Hölder composition series then yields:

**Corollary 4.1.6.** *Let $G$ be a group, then $G$ embeds in an iterated unitary wreath product of its simple group divisors.*

*Proof.* Consider a composition series

$$\{1\} = N_m < N_{m-1} < \cdots < N_1 = G$$

for $G$. So $N_{i+1} \lhd N_i$ and $N_i/N_{i+1}$ is simple, all $i$. Then Corollary 4.1.5 provides an embedding of $G$ into the unitary wreath product $N_2 \wr G/N_2$. The group $G/N_2$ is a simple divisor of $G$ so the result follows by applying induction to $N_2$. Alternatively, by the associativity of the wreath product of transformation groups, we have that $G$ embeds in the action group of the wreath product

$$(N_{m-1}, N_{m-1}) \wr (N_{m-2}/N_{m-1}, N_{m-2}/N_{m-1}) \wr \cdots \wr (G/N_2, G/N_2)$$

of its simple group divisors.    □

Our next result decomposes a monoid along a "normal" submonoid and the group of units.

**Proposition 4.1.7.** *Let $M$ be a monoid with group of units $G$ and suppose $M = NG$ with $N$ a submonoid of $M$, which is invariant under conjugation by $G$. Then there is a quotient map $\varphi \colon N \rtimes G \twoheadrightarrow M$, where $G$ acts on $N$ via conjugation. Moreover, $\varphi$ is idempotent-separating (i.e., is injective when restricted to the set of idempotents).*

*Proof.* Define $\varphi \colon N \rtimes G \to M$ by $(n, g)\varphi = ng$. Because $NG = M$, $\varphi$ is onto. To see that $\varphi$ is a homomorphism, notice that

$$(n, g)\varphi(n', g')\varphi = ngn'g' = n(gn'g^{-1})gg' = (n^g n', gg')\varphi = [(n, g)(n', g')]\varphi$$

showing that $\varphi$ is a homomorphism. Clearly, $E(N \rtimes G) = E(N) \times \{1\}$ and so $\varphi$ is idempotent-separating.    □

The above proposition has numerous applications. Recall that an *inverse semigroup* is a semigroup $S$ such that, for each $s \in S$, there is a unique $s'$ (called the *inverse* of $s$) such that $ss's = s$ and $s'ss' = s'$ [68, 176]. A typical example is the *symmetric inverse monoid* $I_n$ of degree $n$. This is the monoid of all partial permutations of an $n$ element set. It is well-known that if $S$ is an inverse semigroup of order $n$, then $S$ is an inverse subsemigroup of $I_n$ [68, 176]. An inverse monoid $I$ with group of units $G$ is called *factorizable* if $I = E(I)G$ [176]; for example, the symmetric inverse monoid $I_n$ is factorizable as every partial permutation of a finite set is a restriction of a permutation. Because $E(I)$ is closed under conjugation, the proposition applies in this situation. Let us agree more generally to call a monoid $M$ with group of units $G$ *factorizable* if $M = \langle E(M) \rangle G$. The monoid $M_n(K)$ of $n \times n$-matrices over a field is factorizable (in fact, any reductive algebraic monoid is factorizable in this sense [250, 262]). The full transformation monoid $T_n$ is also factorizable. Because $\langle E(M) \rangle$ is closed under conjugation, the proposition applies in this case, as well. The final statement of the next corollary was first proved by Tilson in his thesis (see also [294, 295]); it follows from the fact that each finite inverse semigroup embeds in $I_n$ [68, 176], for some $n$, and that $I_n$ is factorizable.

**Corollary 4.1.8.** *Let $M$ be a factorizable monoid with group of units $G$. Then $M \prec \langle E(M) \rangle \rtimes G$. Consequently, every inverse semigroup divides a unitary semidirect product of a semilattice and a group.*

Another application, due to Pin and Margolis [191, 227], concerns the monoid $P(G)$ of subsets of a group $G$. Let $P_1(G)$ be the set of subsets of $G$ containing 1 and let $P_1^0(G) = P_1(G) \cup \{\emptyset\}$. Then $P(G) = P_1^0(G) \cdot G$ (identifying $G$ with the singleton subsets) and $P_1^0(G)$ is closed under conjugation. Thus $P(G) \prec P_1^0(G) \rtimes G$. Because $(P_1(G), \supseteq)$ is an ordered monoid with the identity the biggest element, it is $\mathscr{J}$-trivial (see, for instance, Proposition 8.2.1) and hence so is $P_1^0(G)$. Thus we have the following result of Pin and Margolis [191].

**Corollary 4.1.9.** *Let $\mathbf{H}$ be a pseudovariety of groups. Then $\mathbf{P(H)} \leq \mathbf{J} * \mathbf{H}$.*

Let $\mathbf{LRB}$ be the pseudovariety of *left regular bands*. These are precisely the $\mathscr{R}$-trivial bands. Left regular bands are defined by the identities: $x^2 = x$ and $xyx = xy$. Recently, left regular bands have played an important role in the theory of random walks on chambers of hyperplane arrangements [49, 57, 58] and in the representation theory of finite Coxeter groups [1]. The primary example is the Rhodes expansion [85] of a semilattice, which can be viewed as a semigroup structure on the flag complex (also known as the order complex) of the semilattice. The following result is one of Stiffler's switching rules [340].

**Proposition 4.1.10.** *Let $G \rtimes E$ be a semidirect product of a group and a semilattice. Then $G \rtimes E \prec B \rtimes G$ for a left regular band $B$. Hence if $\mathbf{H}$ is a pseudovariety of groups, then $\mathbf{H} * \mathbf{Sl} \leq \mathbf{LRB} * \mathbf{H}$.*

*Proof.* By adjoining an identity to $E$ that acts trivially on $G$, we may assume without loss of generality that $E$ has an identity $I$ and the action is unitary. For $e \in E$, set $K_e = \{g \in G \mid {}^e g = 1\}$. Let $B = \{(g, e) \in G \rtimes E \mid g \in K_e\}$. We claim that $B$ is a submonoid of $G \rtimes E$. Clearly, $(1, I) \in B$. If $(g, e), (h, f) \in B$, then $(g, e)(h, f) = (g {}^e h, ef)$ and ${}^{ef}(g {}^e h) = {}^f({}^e g) {}^e({}^f h) = 1$, establishing that $B$ is closed under products. Because $B$ is a monoid, to show that it is a left regular band we just need to show it satisfies $xyx = xy$. So suppose $(g, e)$ and $(h, f)$ are in $B$. Then we have

$$(g, e)(h, f)(g, e) = (g {}^e h {}^{ef} g, efe) = (g {}^e h, ef) = (g, e)(h, f)$$

as ${}^e g = 1$. So $B$ is a left regular band.

Clearly, $G \cong G \rtimes \{I\}$ is the group of units of $G \rtimes E$. So to prove the result, it suffices, by Proposition 4.1.7, to show that $B$ is closed under conjugation and that $BG = G \rtimes E$. For the latter, observe that $(g, e) = (g {}^e(g^{-1}), e)(g, I)$ and $(g {}^e(g^{-1}), e) \in B$. For the former, suppose that $(g, e) \in B$. Then $(h, I)(g, e)(h^{-1}, I) = (hg {}^e(h^{-1}), e)$. But ${}^e(hg {}^e(h^{-1})) = {}^e h {}^e g {}^e(h^{-1}) = 1$, as ${}^e g = 1$. This shows $B$ is closed under conjugation, completing the proof. □

A final corollary to Proposition 4.1.7, which will be used in the proof of the Prime Decomposition Theorem, is

**Corollary 4.1.11.** *Let $M$ be a monoid with group of units $G$ and $J = M \setminus G$. Then $J$ is an ideal and $M \prec J^I \rtimes G$; moreover the semidirect product is unitary.*

*Proof.* To see that $J$ is an ideal, we observe that the $\mathscr{J}$-class of 1 is $G$ because $m \mathrel{\mathscr{J}} 1$ implies $m \mathrel{\mathscr{H}} 1$ by stability. Therefore, $M \setminus G$ is an ideal. It follows that the submonoid $J \cup 1$ (which we identify with $J^I$) is closed under conjugation. Clearly, $M = J^I G$. Proposition 4.1.7 then yields the desired division.    □

### 4.1.2 Augmented monoids

If $Q$ is a set, then $\overline{Q}$ will denote the set of constant maps on $Q$; for $q \in Q$, we use $\overline{q}$ for the constant map with image $q$. If $(Q, S)$ is a **total** transformation semigroup, the *augmented transformation semigroup* is $\overline{(Q, S)} = (Q, S \cup \overline{Q})$. If $M$ is a monoid, then the action monoid $M^\sharp$ of $\overline{(M, M)}$ is called the *augmented monoid* of $M$. Notice that if $m \in M$ is a right zero (that is, $nm = m$ for all $n \in M$), then $\overline{m} = m$. Again, contrary to Eilenberg [85], for us $M^\sharp$ is a monoid, not a transformation monoid. If $Q$ is a set, then $\overline{Q}$ denotes the abstract semigroup consisting of the constant maps on $Q$, i.e., a $|Q|$ element right zero semigroup. This should not be confused with the transformation semigroup $\overline{(Q, \emptyset)} = (Q, \overline{Q})$.

We now want to show augmentation preserves division. First we give an example to show that if $N$ is a submonoid of $M$, then it is not necessarily the case that $N^\sharp$ is a submonoid of $M^\sharp$, as claimed in [7, 110, 174].

*Example 4.1.12.* Let $N = \{1, x, a, b\}$ where $a, b$ are right zeroes, $x^2 = ax = bx = a$ and 1 is an identity. Let $M = N \cup \{0\}$ where 0 is a zero. Then $N^\sharp$ does not embed in $M^\sharp$. Indeed, suppose $\psi \colon N^\sharp \to M^\sharp$ is an embedding. Then $\{a, b, \overline{1}\}\psi$ must be a three element right zero subsemigroup of $M^\sharp$ and hence must be contained in the set $\overline{M}$ of constant maps of $M^\sharp$. In particular, $(x\psi)^2$ must then be in $\overline{M}$. Observe that $x\psi$ is not a constant map and $M^\sharp$ contains no non-constant map whose square is a constant map.

In light of the above example, we must have a more complicated proof that augmentation preserves division; essentially the proof is a transformation semigroup proof, couched in an abstract semigroup theoretic language.

**Proposition 4.1.13.** *Let $S$ and $T$ be monoids and suppose that $S \prec T$. Then $S^\sharp \prec T^\sharp$.*

*Proof.* Let $d \colon S \to T$ be a division of monoids. Let us use the notation $x \cdot y$ to denote the action of $y$ on $x$ in either of the faithful transformation monoids $\overline{(S, S)}$ or $\overline{(T, T)}$. Define a division $\overline{d} \colon S^\sharp \to T^\sharp$ by setting, for $s \in S^\sharp$,

$$s\overline{d} = \{t \in T^{\sharp} \mid \forall s' \in S, \ t' \in s'd \implies t' \cdot t \in (s' \cdot s)d\}.$$

To see that $\overline{d}$ is fully defined, first note, for $s \in S$, $\emptyset \neq sd \subseteq s\overline{d}$. In particular, $1 \in 1\overline{d}$. If $s \in S$ and $t \in sd$, then we claim that $\overline{t} \in \overline{s}\overline{d}$. Indeed, if $s' \in S$ and $t' \in s'd$, then $t' \cdot \overline{t} = t \in sd = (s' \cdot \overline{s})d$.

Now suppose that $t_1 \in s_1\overline{d}$ and $t_2 \in s_2\overline{d}$ with $s_1, s_2 \in S^{\sharp}$. Let $s' \in S$ and $t' \in s'd$. Then $t' \cdot t_1 \in (s' \cdot s_1)d$ and so

$$t' \cdot (t_1 t_2) = (t' \cdot t_1) \cdot t_2 \in ((s' \cdot s_1) \cdot s_2)\, d = (s' \cdot (s_1 s_2))d$$

showing $t_1 t_2 \in (s_1 s_2)\overline{d}$. Thus $\overline{d}$ is a relational morphism. To see that it is a division, suppose that $t \in s_1\overline{d} \cap s_2\overline{d}$. Let $s' \in S$ and choose $t' \in s'd$. Then $t' \cdot t \in (s' \cdot s_1)d \cap (s' \cdot s_2)d$ and so $s' \cdot s_1 = s' \cdot s_2$. Because $s'$ was arbitrary and $\overline{(S, S)}$ is faithful, we conclude that $s_1 = s_2$ and hence $\overline{d}$ is a division.    □

If $G$ is a group, then notice that $G^{\sharp} \setminus G$ is the set $\overline{G}$ of constant maps on $G$. Consequently, we obtain from Corollary 4.1.11 our next proposition.

**Proposition 4.1.14.** *Let $G$ be a group. Then $G^{\sharp} \prec \overline{G}^I \rtimes G$.*

Let us make a definition concerning transformation semigroups.

**Definition 4.1.15 (Embedding of transformation semigroups).** *We say that a faithful transformation semigroup $(X, S)$ embeds in a faithful transformation semigroup $(Y, T)$, written $(X, S) \leq (Y, T)$, if there is a bijection $f \colon X \to Y$ and a map $\psi \colon S \to T$ such that $xsf = xfs\psi$ for all $x \in X$, $s \in S$. An embedding of transformation monoids requires that $1\psi = 1$.*

One easily verifies that $\psi$ must be an injective homomorphism.

**Exercise 4.1.16.** Verify $\psi$ is an injective homomorphism.

In order to use the augmentation construction effectively, it is necessary to establish a decomposition result for augmented monoids of semidirect products. Let us proceed in three steps. The first step concerns the wreath product embedding theorem at the level of transformation monoids.

**Lemma 4.1.17.** *Let $S \rtimes T$ be a unitary semidirect product of monoids. There is an embedding of transformation monoids $(S \times T, S \rtimes T) \hookrightarrow (S, S) \wr (T, T)$.*

*Proof.* The embedding $\psi \colon S \rtimes T \to S \wr T$ of Theorem 1.2.17 sends $(s, t)$ to $(f_s, t)$ where $t_0 f_s = {}^{t_0}s$ for $t_0 \in T$. The computation

$$(s_0, t_0)(s, t)\psi = (s_0, t_0)(f_s, t) = (s_0(t_0 f_s), t_0 t) = (s_0 {}^{t_0}s, t_0 t) = (s_0, t_0)(s, t)$$

shows that $\psi$ gives rise to an embedding of faithful transformation monoids (where we take the identity map as our bijection).    □

Next we wish to show that augmentation of transformation semigroups distributes over wreath products.

**Lemma 4.1.18.** *Let $(P, S)$ and $(Q, T)$ be faithful transformation semigroups. Then there results an embedding of faithful transformation semigroups*

$$\overline{(P,S) \wr (Q,T))} \leq \overline{(P,S)} \wr \overline{(Q,T)}.$$

*Proof.* For $p \in P$, define $f_p \colon Q \to S \cup \overline{P}$ by $q f_p = \overline{p}$ for all $q \in Q$. Then

$$(p_0, q_0)(f_p, \overline{q}) = (p_0(q_0 f_p), q_0 \overline{q}) = (p_0 \overline{p}, q) = (p, q)$$

establishing that $(f_p, q) = \overline{(p, q)}$. Thus the right-hand side contains all the constant maps; it clearly contains all other members of the left-hand side.  □

Now we can achieve our desired decomposition.

**Proposition 4.1.19.** *Suppose that $S \rtimes T$ is a unitary semidirect product of monoids. Then $(S \rtimes T)^{\sharp}$ embeds in $S^{\sharp} \wr \overline{(T, T)} = (S^{\sharp})^T \rtimes T^{\sharp}$.*

*Proof.* By Lemmas 4.1.17 and 4.1.18 we have:

$$\overline{(S \times T, S \rtimes T)} \leq \overline{(S, S) \wr (T, T)} \leq \overline{(S, S)} \wr \overline{(T, T)}.$$

The action monoid of the left-hand side is $(S \rtimes T)^{\sharp}$ whereas the action monoid of the right-hand side is $S^{\sharp} \wr \overline{(T, T)} = (S^{\sharp})^T \rtimes T^{\sharp}$.  □

Recall that a *left zero* semigroup is a semigroup satisfying the identity $xy = x$. If $A$ is a set, then $A^{\ell}$ denotes the unique left zero semigroup structure on $A$. Right zero semigroups are defined dually and are of the form $\overline{A}$ where $A$ is a set. Sometimes we use $A^r$ to denote the right zero semigroup $\overline{A}$.

If $n$ is an integer, then we set $\mathbf{n} = \{0, \ldots, n-1\}$. The semigroup $\overline{\mathbf{n}}$ is then an $n$-element right zero semigroup. Following Eilenberg [85], we set $U_n = \overline{\mathbf{n}}^I$; so $U_n$ is the monoid obtained by adjoining an identity to the $n$ element right zero semigroup. Notice that $n$ counts the number of right zeroes, not the order of the semigroup. In particular, $U_1$ is the two-element semilattice and $U_2 = \overline{\mathbf{2}}^I$. The semigroup $U_2$ is often called the *flip-flop* because the associated automaton consisting of an identity and two resets models the flip-flop circuit from Electrical Engineering.

**Lemma 4.1.20.** *For $n \geq 1$, $\overline{\mathbf{n}} \leq \overline{\mathbf{2}}^n$. Hence $U_n \leq U_2^n$.*

*Proof.* The first embedding sends $\overline{k}$ to the function $\chi_k \colon \mathbf{n} \to \overline{\mathbf{2}}$ given by

$$j \chi_k = \begin{cases} \overline{1} & j = k \\ \overline{0} & j \neq k. \end{cases}$$

The second embedding is immediate from the first.  □

We are now in a position to show that passing to augmented monoids presents no difficulties with respect to proving the Prime Decomposition Theorem.

**Proposition 4.1.21.** *Let $\mathscr{C}$ be a collection of groups and let $S$ be a monoid dividing an iterated unitary wreath product of groups from $\mathscr{C}$ and copies of $\overline{2}^I$. Then the same is true for $S^\sharp$.*

*Proof.* This follows by iterated application of Proposition 4.1.19, along with Propositions 4.1.13 and 4.1.14, Lemma 4.1.20, Proposition 4.1.2 and the observation that $U_2{}^\sharp = U_3$. □

Next we decompose left zero semigroups.

**Lemma 4.1.22.** *Let $A$ be a finite set and $M$ be any monoid with cardinality greater than that of $A$. Then $A^\ell$ embeds in $M \wr \{0\}$ and $(A^\ell)^I$ embeds in the unitary wreath product $M \wr U_1$. In particular, $(A^\ell)^I$ embeds in the unitary wreath product $U_n \wr U_1$ where $n = |A|$.*

*Proof.* First observe that $M \wr \{0\}$ can be identified with the subsemigroup $M^{U_1} \times \{0\}$ of the unitary wreath product $M \wr U_1$ by viewing 1 as the adjoined identity to $\{0\}$. By choice of $M$, we can define an injective map $a \mapsto m_a$ from $A$ to $M \setminus 1$. For $a \in A$, define $f_a \colon \{0,1\} \to M$ by $0f_a = 1$, $1f_a = m_a$. We claim that $\{(f_a, 0) \mid a \in A\}$ is a subsemigroup isomorphic to $A^\ell$. As $1f_a = m_a$, these elements are all distinct. So we just need to check the left zero multiplication: $(f_a, 0)(f_b, 0) = (f_a{}^0 f_b, 0) = (f_a, 0)$ because $0f_a(0 \cdot 0)f_b = 1$ and $1f_a(1 \cdot 0)f_b = m_a$. Finally, we can add in the identity of the unitary wreath product $M \wr U_1$ to obtain a copy of $(A^\ell)^I$. □

We shall apply several times the following well-known lemma of Krohn and Rhodes [85, 171] concerning total transformation semigroups to simplify the construction of divisions.

**Lemma 4.1.23.** *Let $(P, S)$ and $(Q, T)$ be faithful transformation semigroups (monoids). Suppose that $f \colon Q \twoheadrightarrow P$ is a partial surjective map so that, for each $s \in S$, there exists $\widehat{s} \in T$ such that*

$$q\widehat{s}f = qfs, \quad \text{for all } q \in Q \text{ with } qf \text{ defined}. \tag{4.1}$$

*Then $S \prec T$ (as monoids).*

*Proof.* First we remark that it is implicit in (4.1) that if $qf$ is defined, then $q\widehat{s}f$ must also be defined. Define a relational morphism $\varphi \colon S \to T$ by setting $s\varphi$ to be the set of all elements $\widehat{s} \in T$ satisfying (4.1). Clearly, $\varphi$ is fully defined and $1 \in 1\varphi$ in the monoidal context. Suppose $\widehat{s}_0 \in s_0\varphi$, $\widehat{s}_1 \in s_1\varphi$ and $q$ belongs to the domain of $f$. Then $q\widehat{s}_0$ is in the domain of $f$ and hence so is $(q\widehat{s}_0)\widehat{s}_1$. We then compute $q\widehat{s}_0\widehat{s}_1 f = q\widehat{s}_0 f s_1 = qf s_0 s_1$ and so $\widehat{s}_0\widehat{s}_1 \in (s_0 s_1)\varphi$, showing that $\varphi$ is indeed a relational morphism. To see that it is a division, suppose $t \in s\varphi \cap s'\varphi$. Let $p \in P$ and choose $q \in pf^{-1}$. Then

$$ps = qfs = qtf = qfs' = ps'$$

and so $s = s'$ by faithfulness. □

We remark that the proof of the previous lemma shows that if $A$ is a generating set for $S$, it suffices to specify $\widehat{s}$ satisfying (4.1) for each $s \in A$, because we can then take $\widehat{s_1 \cdots s_n} = \widehat{s}_1 \cdots \widehat{s}_n$, for $s_1, \ldots, s_n \in A$. We shall most often use Lemma 4.1.23 in the case where $f$ is a total function.

The following fundamental decomposition result is known as the "$V \cup T$ Lemma" and is key to this proof of the Prime Decomposition Theorem.

**Proposition 4.1.24 ($V \cup T$ Lemma).** *Let $S$ be a semigroup and suppose that $S = V \cup T$ where $V$ is a left ideal and $T$ is a subsemigroup of $S$. Then there is a division of monoids $S^I \prec V^I \wr \overline{(T^I, T^I)} = (V^I)^{T^I} \rtimes (T^I)^\sharp$.*

*Proof.* In order to apply Lemma 4.1.23 to the faithful transformation monoids $(S^I, S^I)$ and $(V^I, V^I) \wr \overline{(T^I, T^I)}$, we define $f \colon V^I \times T^I \to S^I$ by $(v, t)f = vt$. As $S^I = V \cup T^I$, $f$ is clearly surjective. Let $i \colon T^I \to V^I$ be the constant map taking on the value $I$. For $t \in T^I$, we set $\widehat{t} = (i, t)$, whereas, for $v \in V$, we set $\widehat{v} = (f_v, \overline{I})$, where $f_v \colon T^I \to V^I$ is given by $t f_v = tv$ for $t \in T^I$. Then, for $t_0 \in T^I$, $v_0 \in V$ and $(v, t) \in V^I \times T^I$, we have

$$(v, t)\widehat{t_0}f = (v, t)(i, t_0)f = (v, tt_0)f = vtt_0 = (v, t)ft_0$$

$$(v, t)\widehat{v_0}f = (v, t)(f_{v_0}, \overline{I})f = (v(tf_{v_0}), I)f = (vtv_0, I)f = vtv_0 = (v, t)fv_0.$$

As $S^I = V \cup T^I$, Lemma 4.1.23 now applies to yield the desired division.    $\square$

### 4.1.3 Proof of the Prime Decomposition Theorem

We begin with a lemma of Krohn and Rhodes [171]; the proof presented here is due to Clifford. Recall that a semigroup $S$ is said to be *left simple* if it has no proper left ideals.

**Lemma 4.1.25.** *Let $S$ be a finite semigroup. Then either:*

1. *$S$ is left simple;*
2. *$S$ is cyclic;*
3. *There exists a proper left ideal $V < S$ and a proper subsemigroup $T < S$ such that $S = V \cup T$.*

*Proof.* If $S$ is left simple, we are in the first case so assume that it is not. By finiteness, $S$ contains a maximal proper left ideal $L$. Let $a \in S \setminus L$. Then $S = L \cup S^I a$. If $S^I a \neq S$, then we may take $V = L$ and $T = S^I a$ and we are in the third case. So assume that $S^I a = S$. If $a \notin Sa$, then $S = Sa \cup \langle a \rangle$, where $\langle a \rangle$ is the subsemigroup generated by $a$. If $\langle a \rangle = S$, we are in the second case; else $V = Sa$ is a proper left ideal and $T = \langle a \rangle$ is a proper subsemigroup and we are in the third case again. Therefore, we may assume that $a \in Sa$ and so $S = S^I a = Sa$. Let us write $a = s_0 a$ with $s_0 \in S$.

Let $La^{-1} = \{s \in S \mid sa \in L\}$. Because $L \subseteq S = Sa$, we must have that $La^{-1} \neq \emptyset$. Clearly, $La^{-1}$ is a left ideal; moreover, it is proper because

$s_0 \notin La^{-1}$. By maximality of $L$ either $L \cup La^{-1} = L$ or $L \cup La^{-1} = S$. In the latter case, we may take $V = L$ and $T = La^{-1}$, and so again we are in the third case. So we may assume that $La^{-1} \subseteq L$ for all $a \in S \setminus L$, because otherwise we are done by the cases already considered. We claim that $S \setminus L$ is a subsemigroup. Let $a, b \in S \setminus L$. Then $ba \in L$ implies $b \in La^{-1} \subseteq L$, a contradiction. Thus taking $V = L$ and $T = S \setminus L$ finishes the proof.  $\square$

We begin with the first two cases of the lemma as a basis for induction.

**Lemma 4.1.26.** *Let $S$ be left simple. Then $S^I$ divides an iterated unitary wreath product of a subgroup and copies of $\overline{\mathbf{2}}^I$.*

*Proof.* By Rees's Theorem, $S$ is a direct product $L \times G$ of a left zero semigroup $L$ and a maximal subgroup $G$ (Corollary A.4.17). So $S^I \leq L^I \times G$ and the result then follows from Lemmas 4.1.20 and 4.1.22.  $\square$

Let us denote by $C_{i,n}$ the cyclic monoid of index $i$ and period $n$, that is, $C_{i,n}$ is the monoid with presentation $\langle a \mid a^i = a^{i+n} \rangle$. It is well-known and easy to see that $C_{i,n} \leq C_{i,1} \times C_{0,n}$ and $C_{0,n}$ is a cyclic subgroup of $C_{i,n}$ of order $n$. Thus we only need decompose $C_{i,1}$. Of course, $C_{1,1} = U_1 \prec \overline{\mathbf{2}}^I$.

**Lemma 4.1.27.** *The cyclic monoid $C_{i,1}$ divides the unitary wreath product $C_{i-1,1} \wr U_1$, for $i > 1$.*

*Proof.* In order to apply Lemma 4.1.23 to the faithful transformation monoids $(C_{i,1}, C_{i,1})$ and $(C_{i-1,1}, C_{i-1,1}) \wr (\mathbf{2}, U_1)$, define a surjective partial function $f \colon C_{i-1,1} \times \mathbf{2} \to C_{i,1}$ by $(a^j, 0)f = a^{j+1}$ and $(a^0, 1)f = a^0$. Notice if $j \geq i-1$, then $a^{j+1} = a^i$ in $C_{i,1}$, so $f$ is well defined. By the remark after Lemma 4.1.23, it suffices to define $\widehat{a}$; we do this by setting $\widehat{a} = (g, 0)$ where $0g = a$, $1g = 1$. Then we compute

$$(a^j, 0)\widehat{a}f = (a^j(0g), 0)f = (a^j a, 0)f = (a^{j+1}, 0)f = a^{j+2} = (a^j, 0)fa \quad (4.2)$$
$$(a^0, 1)\widehat{a}f = (a^0(1g), 0)f = (a^0, 0)f = a = (a^0, 1)fa. \quad (4.3)$$

This completes the proof.  $\square$

**Corollary 4.1.28.** *A finite cyclic monoid $C$ divides an iterated unitary wreath product of a subgroup and copies of $U_1$.*

*Proof.* This follows immediately from the decomposition $C_{i,n} \leq C_{i,1} \times C_{0,n}$ and an easy induction argument using Lemma 4.1.27.  $\square$

**Exercise 4.1.29.** Let $d \colon S \to M$ be a division with $M$ a monoid and suppose $1 \in sd$. Show that $S$ is a monoid with identity $s$. Conclude that if $S \prec M$ with $M$ a monoid and $S$ not a monoid, then $S^I \prec M$.

We are now ready to prove the Prime Decomposition Theorem of Krohn and Rhodes [169]. Throughout we use implicitly that semidirect products embed in wreath products.

**Theorem 4.1.30 (Prime Decomposition Theorem [169]).** *Let $S$ be a finite semigroup. Then $S$ divides an iterated unitary wreath product of its simple group divisors and copies of $U_2 = \overline{\mathbf{2}}^I$ (if $S$ is a monoid, the division can be taken to be monoidal).*

*Proof.* It suffices to prove the theorem in the category of finite monoids because $S \prec S^I$ and they both have the same simple group divisors; in particular, in this proof we consider only unitary wreath products. The proof proceeds by induction on the size of the monoid $M$, the case $|M| = 1$ being trivial. Let $G$ be the group of units of $M$ and set $S = M \setminus G$. If $G = M$, then the result follows from Corollary 4.1.6. So assume $G \neq M$, that is, $S \neq \emptyset$. Suppose first that $G \neq \{1\}$. Then $M \prec S^I \rtimes G$ by Corollary 4.1.11. As $|S^I| < |M|$, the result follows by induction and Corollary 4.1.6. Next suppose $G = \{1\}$; so $M = S^I$. By Lemma 4.1.25, $S$ is either left simple, cyclic or there exists a proper left ideal $V$ and a proper subsemigroup $T$ such that $S = V \cup T$. In the first case the result follows from Lemma 4.1.26 and Corollary 4.1.6; the second case is handled by Corollaries 4.1.28 and 4.1.6. So assume that the third case applies. Proposition 4.1.24 then shows $M \prec (V^I)^{T^I} \rtimes (T^I)^{\sharp}$. Because we have $|V^I|, |T^I| < |S^I| = |M|$, the result holds for $V^I, T^I$ by induction. Proposition 4.1.21 then provides a decomposition for $(T^I)^{\sharp}$ of the required form. Proposition 4.1.2 then yields the sought after decomposition for $M$.    □

The minimal length of a decomposition of a semigroup into a wreath product of finite simple groups and copies of $\overline{\mathbf{2}}^I$ turns out to be a useful induction parameter, and we shall apply it several times, in particular to aperiodic semigroups. We shall formalize this parameter in Section 4.3.

One might ask whether the Prime Decomposition Theorem really is a decomposition into "prime" components. This is indeed the case, as we proceed to demonstrate. A semigroup $S$ is said to be $\rtimes$-*prime* if whenever $S \prec T \rtimes U$, then $S \prec T$ or $S \prec U$. We aim to prove that the $\rtimes$-prime semigroups are the finite simple groups and the divisors of $U_2$. The proof uses the following technical lemma, which measures to what extent $S$ behaves like the kernel of the projection $\pi \colon S \rtimes T \to T$, and which will be used throughout this chapter. We recall that the idempotents of a semigroup $S$ form a partially ordered set with respect to the order given by $e \leq f$ if and only if $ef = e = fe$.

**Lemma 4.1.31.** *Let $S \rtimes T$ be a semidirect product with associated projection $\pi \colon S \rtimes T \to T$. Let $e, f \in T$ be idempotents with $f \leq e$. Define $\psi \colon \{e, f\}\pi^{-1} \to S$ by $(x, g)\psi = {}^f x$. Then $\psi$ is a homomorphism. Moreover, $\psi$ is injective on: subsemigroups of $f\pi^{-1}$ containing a left identity (for instance right zero semigroups) and copies of $U_n$, with $n \geq 2$, intersecting $f\pi^{-1}$ non-trivially.*

*Proof.* Suppose $(x, g), (y, h) \in \{e, f\}\pi^{-1}$. Then, because $fe = f = f^2$,

$$[(x, g)(y, h)]\psi = (x\,{}^g y, gh)\psi = {}^f(x\,{}^g y) = {}^f x\,{}^f y = (x, g)\psi(y, h)\psi$$

and so $\psi$ is a homomorphism.

We are left with checking injectivity. Suppose first that $N \leq f\pi^{-1}$ has a left identity $(y, f)$ and $(s, f) \in N$. Then we compute

$$(s, f) = (y, f)(s, f) = (y^f s, f) = (y(s, f)\psi, f)$$

and so $s$ is determined by $(s, f)\psi$, whence $\psi$ is injective. Finally, suppose that $N \leq \{f\}\varphi^{-1}$ is isomorphic to $\overline{\mathbf{n}}$ with $n \geq 2$ and $N \cup \{(x, e)\} \cong U_n$. We know from the previous case that $\psi$ is injective on $N$. Suppose by way of contradiction that $(x, e)\psi = (s, f)\psi$. Choose $(s_0, f) \in N \setminus \{(s, f)\}$. Then we compute $(s, f) = (s_0, f)(s, f) = (s_0{}^f s, f) = (s_0(s, f)\psi, f)$ and

$$(s_0, f) = (s_0, f)(x, e) = (s_0{}^f x, f) = (s_0(x, e)\psi, f) = (s_0(s, f)\psi, f) = (s, f)$$

a contradiction. We deduce that $\psi$ is injective on $N \cup \{(x, e)\}$.    □

**Corollary 4.1.32.** *If* $\mathbf{V}$ *and* $\mathbf{W}$ *are pseudovarieties, then* $\mathbf{V} * \mathbf{W} \leq \mathbb{L}\mathbf{V} \textcircled{m} \mathbf{W}$. *In particular,* $\mathbf{A} * \mathbf{V} \subseteq \mathbf{A} \textcircled{m} \mathbf{V}$.

We also need a lemma on projective semigroups. Let us first give the definition.

**Definition 4.1.33 (Projective semigroup).** *A (pro)finite semigroup* $T$ *is said to be projective if, for any (continuous) onto homomorphism* $\varphi \colon S \twoheadrightarrow T$ *of (pro)finite semigroups, there exists a splitting homomorphism* $\psi \colon T \to S$ *such that* $\psi\varphi = 1_T$.

*Remark 4.1.34 (Discussion of projective semigroups).* A simple compactness argument shows that a finite semigroup is projective as a finite semigroup if and only if it is projective as a profinite semigroup. It may be an interesting question to determine the projective finite semigroups. Such semigroups can embed into a free profinite semigroup (in fact they are precisely the finite retracts of free profinite semigroups) and so understanding the projective finite semigroups is tantamount to understanding finite subsemigroups of free profinite semigroups. For free profinite groups, all closed subgroups are projective. As a consequence, free profinite groups are torsion-free. One can easily verify that not all closed subsemigroups of a free profinite semigroup are projective [21]. However, the authors have shown that all closed subgroups of a free profinite semigroup are projective and that all finite subsemigroups are bands [290]. In particular, every projective finite semigroup is a band.

It is shown in [130, 286] that left and right stabilizers in a free profinite semigroup are $\mathscr{R}$-chains, respectively $\mathscr{L}$-chains of idempotents. This leads to the following situation. If $S$ is a finite subsemigroup with a zero of a free profinite semigroup, then $S$ is a chain of idempotents (in the natural partial order on idempotents) and hence a semilattice isomorphic to $(\{1, \ldots, n\}, \min)$ for some $n$. If the minimal ideal of $S$ is a left zero semigroup, then $S$ must be an $\mathscr{L}$-chain of idempotents and in particular $\mathscr{R}$-trivial; if the minimal ideal of $S$

is a right zero semigroup, then $S$ must be an $\mathscr{R}$-chain of idempotents and in particular $\mathscr{L}$-trivial. The authors can classify up to isomorphism all projective finite semigroups whose minimal ideal does not contain a $2 \times 2$-rectangular band. For a long time the authors believed that a $2 \times 2$ rectangular band could not embed in a projective finite semigroup, but recently we have found an example. Namely, the singular square semigroup of Nambooripad [212], also considered by Leech in his work on cohomology [181, 182], is projective. This semigroup has underlying set $S = (\{a, b\} \times \{a, b\}) \cup \{e\}$ with multiplication given by $(x, y)(z, w) = (x, w)$, $e(x, y) = (x, y)$, $(x, y)e = (x, a)$ and $e^2 = e$; in particular the minimal ideal of $S$ is a $2 \times 2$ rectangular band. On the other hand, the $2 \times 2$ rectangular band $\{a, b\} \times \{a, b\}$ is not projective because the natural quotient map from $\mathscr{M}^0 \left( \{\pm 1\}, 2, 2, \begin{pmatrix} 1 & 1 \\ 1 & -1 \end{pmatrix} \right)$ to it does not split. In particular, subsemigroups of projective finite semigroups need not be projective.

A possible program to try and classify projective semigroups is as follows. According to a result of Rhodes [267] (see our Lemma 5.2.10), each surjective homomorphism $\varphi$ factors as a composition $\theta_1 \cdots \theta_n$ of surjective morphisms such that either $\theta_i$ separates $\mathscr{H}$-classes (i.e., is an $\mathscr{H}$-morphism) or is injective when restricted to $\mathscr{H}$-classes. Therefore, a semigroup $P$ is projective if and only if every morphism of these two types onto $P$ splits. Leech's cohomology theory [181,182] should afford a possible approach to deal with the case of $\mathscr{H}$-morphisms; perhaps expansions will help with morphisms that are injective on $\mathscr{H}$-classes.

*Question 4.1.35.* Is it decidable whether a finite semigroup is projective?

**Exercise 4.1.36.** Show that a finite semigroup $S$ is projective if and only if it is a (continuous) retract of a free profinite semigroup.

**Exercise 4.1.37.** Show that a finite semigroup $S$ is projective if and only if whenever there is a diagram of homomorphisms

where $A$ and $B$ are finite semigroups, there is a lift $\widetilde{\varphi} \colon S \to A$ of $\varphi$ (i.e., $\widetilde{\varphi} \psi = \varphi$). Hint: Either use the result of the previous exercise or consider the pullback of $\varphi$ and $\psi$.

**Exercise 4.1.38.** Verify directly that a non-trivial cyclic semigroup is not projective.

**Lemma 4.1.39.** *The semigroup $\overline{n}$, $n \geq 1$, and its dual are projective. Also if $P$ is projective, then so is $P^I$. Hence $U_n$ and its divisors are projective, all $n \geq 1$.*

*Proof.* Let $\varphi\colon S \twoheadrightarrow \overline{\mathbf{n}}$ be an onto homomorphism. Let $S'$ be a minimal sub-semigroup with $S'\varphi = \overline{\mathbf{n}}$. First observe that $S'$ is right simple because if $s \in S'$, then $(sS')\varphi = \overline{\mathbf{n}}$ and so $sS' = S'$ by minimality. Hence $S' = R \times G$ with $R$ a right zero semigroup and $G$ a group by Rees's Theorem. One can then define $\psi$ by choosing an idempotent preimage of $\overline{i}$ in $S'$, for each $0 \leq i \leq n-1$. This proves the first statement.

For the second statement, suppose $\varphi\colon S \twoheadrightarrow P^I$ is an onto homomorphism. Let $e \in E(S)$ be a preimage of $I$. Then $eSe$ is a monoid with identity $e$ and $(eSe)\varphi = P^I$. So there is a splitting $\psi\colon P \to eSe$ of $\varphi|_{eSe \cap P\varphi^{-1}}$. As $e\varphi = I$, we must have that $e \notin P\psi$. Thus we can extend $\psi$ to $P^I$ by defining $I\psi = e$ to obtain the desired splitting. The final statement is now immediate. $\square$

We now show that the singular square semigroup $S$, described in Remark 4.1.34, is projective. Leech proved [182] that all $\mathscr{H}$-morphisms onto $S$ split using his cohomology theory. We handle arbitrary morphisms in a completely elementary fashion.

**Theorem 4.1.40.** *The singular square semigroup is projective.*

*Proof.* Denote by $S$ the singular square semigroup from above. Setting $x = (a,a), y = (b,a), z = (a,b), w = (b,b)$ yields $xe = x, ye = y, ze = x, we = y$. Let $\varphi\colon T \twoheadrightarrow S$ be a surjective homomorphism. Choose an idempotent $e' \in T$ with $e'\varphi = e$. Then $e'T\varphi = S$ and so it suffices to show that the restriction of $\varphi$ to $e'T$ splits. Thus without loss of generality we may assume that $e'$ is a left identity for $T$. By Lemma 4.1.39 we can find a left zero subsemigroup $\{z', w'\}$ of $T$ with $z'\varphi = z, w'\varphi = w$. Define $x' = z'e'$ and $y' = w'e'$. Direct computation yields $x'\varphi = x, y'\varphi = y$. We claim that $\{e', x', y', z', w'\}$ is a subsemigroup mapped by $\varphi$ isomorphically to $S$. First note that $x'x' = z'e'z'e = z'e' = x'$ because $e'$ is a left identity and similarly $y'y' = w'e'w'e = w'e = y'$. From $z' \mathscr{L} w'$ it follows $x' \mathscr{L} y'$ and hence $\{x', y'\}$ is a left zero semigroup. Next we compute $x'z' = z'e'z' = z'z' = z'$, $z'x' = z'z'e' = x'$ and likewise $y'w' = w', w'y' = y'$. Finally, we check $x'w' = z'e'w' = z'w' = z'$, $w'x' = w'z'e' = w'e' = y'$ and similarly $z'y' = z'w'e' = z'e' = x'$, $y'z' = w'e'z' = w'z' = w'$. Thus $\{e', x', y', z', w'\}$ is isomorphic to $S$ via $\varphi$, as required. $\square$

Because, as was observed earlier, a $2 \times 2$ rectangular band is not projective, we obtain our first corollary.

**Corollary 4.1.41.** *Subsemigroups of projective finite semigroups need not be projective.*

To the best of our knowledge the following consequence seems to be new.

**Corollary 4.1.42.** *A $2 \times 2$ rectangular band embeds in a free profinite semigroup on two or more generators.*

One can generalize this to obtain arbitrarily large rectangular bands as subsemigroups of projective (and hence free profinite) semigroups, as the following exercise shows.

**Exercise 4.1.43.** Let $B = \{a_1, \ldots, a_m\} \times \{b_1, \ldots, b_n\}$ be an $m \times n$ rectangular band and let $\{e_1, \ldots, e_{n-1}\}$ be a right zero semigroup such that $e_i$ acts on the left of $B$ as the identity and on the right of $B$ by $(x, y)e_i = (x, b_i)$. Show that $S = \{e_1, \ldots, e_{n-1}\} \cup B$ is a projective semigroup.

The following proposition already appeared in Chapter 3 for compact semigroups, but we repeat it here for finite semigroups for the reader's convenience.

**Proposition 4.1.44.** *Let $S$ be a finite semigroup and $\varphi \colon S \twoheadrightarrow G$ a surjective homomorphism with $G$ a group. Then there is a subgroup $H \le S$ such that $H\varphi = G$.*

*Proof.* Let $T$ be a minimal subsemigroup of $S$ with $T\varphi = G$. We show that $T$ is left and right simple, and hence a group by Lemma A.3.3. Indeed, if $t \in T$, then $tT, Tt \le T$ are subsemigroups of $S$ and

$$(tT)\varphi = t\varphi T\varphi = t\varphi G = G = Gt\varphi = T\varphi t\varphi = (Tt)\varphi.$$

By minimality of $T$, we obtain $tT = T = Tt$, as desired. ☐

**Theorem 4.1.45 (Krohn-Rhodes [169]).** *The $\rtimes$-prime semigroups are the simple groups and the divisors for $\overline{\mathbf{2}}^I$.*

*Proof.* The Prime Decomposition Theorem implies that the aforementioned semigroups are the only candidates to be $\rtimes$-prime. Let us begin with a simple group $G$. Suppose $G \prec S \rtimes T$. Proposition 4.1.44 assures us there is a subgroup $H \le S \rtimes T$ and a normal subgroup $N$ of $H$ such that $G = H/N$. Let $\pi \colon S \rtimes T \to T$ be the projection and let $K = \ker \pi|_H$. As $G$ is simple, either $KN = N$ or $KN = H$. In the first case, $K \le N$ and so $G = H/N \prec H/K \le T$. In the second case, $G = H/N = KN/N \cong K/K \cap N \prec K \prec S$, the last division being a consequence of Lemma 4.1.31. Thus $G$ is $\rtimes$-prime.

We next show $\overline{\mathbf{2}}^I$ is $\rtimes$-prime. If $\overline{\mathbf{2}}^I \prec S \rtimes T$, then, by Lemma 4.1.39, $S \rtimes T$ contains an isomorphic copy $\{e, f, x\}$ of $\overline{\mathbf{2}}^I$, where $x$ is the identity. Suppose $x = (s_x, t_x)$, $e = (s_e, t_e)$ and $f = (s_f, t_f)$. If $t_e \ne t_f$, then $\{t_x, t_e, t_f\}$ are all distinct and so $\overline{\mathbf{2}}^I \prec T$. Indeed, if $t_e = t_x$, then from $fx = f$, we have $(s_f, t_f)(s_x, t_e) = (s_f, t_f)$ and so $t_f = t_f t_e = t_e$. Similarly, if $t_f = t_x$, then $t_e = t_f$. So suppose that $t_e = t_f$. Then $\{x, e, f\}$ embeds in $S$ by Lemma 4.1.31.

Similarly, if $\overline{\mathbf{2}} \prec S \rtimes T$, we may find, by Lemma 4.1.39, idempotents $e = (s_e, t_e)$ and $f = (s_f, t_f)$ of $S \rtimes T$ with $ef = f$, $fe = e$. Then if $t_e \ne t_f$, $\overline{\mathbf{2}} \prec T$, else $\overline{\mathbf{2}} \prec S$ by Lemma 4.1.31. Finally, if $U_1 \prec S \rtimes T$, we can find idempotents $x = (s_x, t_x)$, $e = (s_e, t_e)$ of $S \rtimes T$ such that $xe = ex = e$ by Lemma 4.1.39. If $t_e \ne t_x$, then $U_1 \prec T$, otherwise $U_1 \prec S$ by Lemma 4.1.31. This completes the proof. ☐

To state some consequences of the Prime Decomposition Theorem, we need the following definition from group theory.

**Definition 4.1.46 (Extension-closed).** *A pseudovariety* **H** *of groups is said to be closed under extension (or extension-closed) if whenever*

$$1 \to N \to G \to H \to 1$$

*is an exact sequence of groups with* $N, H \in \mathbf{H}$, *necessarily* $G \in \mathbf{H}$.

It is a consequence of Corollary 4.1.5 that **H** is extension-closed if and only if it is closed under wreath product (or equivalently semidirect product) in the category of groups. Such pseudovarieties are determined by their simple members (by Corollary 4.1.6). The next corollary of the Prime Decomposition Theorem extends this to the realm of semigroups. If **H** is a pseudovariety of groups, then $\overline{\mathbf{H}}$ is the pseudovariety of semigroups whose subgroups belong to **H** (the fact that $\overline{\mathbf{H}}$ is a pseudovariety is a consequence of Proposition 4.1.44).

**Corollary 4.1.47.** *Let* **H** *be a pseudovariety of groups closed under extension. Then* $\overline{\mathbf{H}}$ *is the smallest pseudovariety closed under semidirect product containing the simple groups from* **H** *and the monoid* $\overline{\mathbf{2}}^I$. *In particular, the pseudovariety* **A** *of aperiodic semigroups is the smallest pseudovariety containing* $\overline{\mathbf{2}}^I$ *and closed under semidirect product.*

*Proof.* Let **V** be the smallest pseudovariety closed under semidirect product and containing the simple groups from **H** and the monoid $\overline{\mathbf{2}}^I$. The Prime Decomposition Theorem immediately implies $\overline{\mathbf{H}} \leq \mathbf{V}$. For the converse, it suffices to show that $\overline{\mathbf{H}}$ is closed under semidirect product. By Corollary 4.1.6, and because **H** is closed under extension, a group $G \in \mathbf{H}$ if and only if its simple group divisors belong to **H**. Hence a semigroup belongs to $\overline{\mathbf{H}}$ if and only if its simple group divisors belong to **H**. So suppose $S, T \in \overline{\mathbf{H}}$ and $G$ is a simple group dividing $S \rtimes T$. Because simple groups are $\rtimes$-prime by Theorem 4.1.45, $G$ divides $S$ or $T$ and hence belongs to **H**. We conclude $S \rtimes T \in \overline{\mathbf{H}}$. This establishes $\mathbf{V} \leq \overline{\mathbf{H}}$. $\qquad\square$

Let us give, by way of example, a well-known decomposition of the full transformation monoid $T_n$ of degree $n$ into a wreath product of augmented symmetric groups of rank at most $n$.

**Proposition 4.1.48.** *For* $n \geq 2$, *there is a division* $T_n \prec T_{n-1} \wr \overline{(\mathbf{n}, S_n)}$. *Hence* $T_n$ *divides a unitary iterated wreath product of augmented symmetric groups of degree at most* $n$.

*Proof.* We apply Lemma 4.1.23 to the faithful transformation semigroups $(\mathbf{n}, T_n)$ and $(\mathbf{n} - \mathbf{1}, T_{n-1}) \wr \overline{(\mathbf{n}, S_n)}$. The key idea is to encode subsets of size $n - 1$ by the missing element (this is essentially a primitive form of so-called Zeiger coding). Define $f \colon \mathbf{n} - \mathbf{1} \times \mathbf{n} \to \mathbf{n}$ by

$$(i,j)f = \begin{cases} i & i < j \\ i+1 & i \geq j. \end{cases}$$

Here the coordinate $j$ encodes the $n-1$-element subset $\mathbf{n} \setminus \{j\}$; the coordinate $i$ then encodes the $i^{th}$ element of this set (in the usual ordering). Clearly, $f$ is onto. Let $s \in T_n$. If $s \in S_n$, define $\widehat{s} = (g_s, s)$ where $g_s \colon \mathbf{n} \to T_{n-1}$ is given by

$$i(jg_s) = \begin{cases} is & i < j, \ is < js \\ is - 1 & i < j, \ is > js \\ (i+1)s & i \geq j, \ (i+1)s < js \\ (i+1)s - 1 & i \geq j, \ (i+1)s > js. \end{cases}$$

Notice that if $i < j$, then because $s$ is a permutation, $is \neq js$, whereas if $i \geq j$, then $(i+1)s \neq js$. If $s \in T_n \setminus S_n$, fix $j_s \notin \mathbf{n}s$. Then define $f_s \colon \mathbf{n} \to T_{n-1}$ by

$$i(jf_s) = \begin{cases} is & i < j, \ is < j_s \\ is - 1 & i < j, \ is > j_s \\ (i+1)s & i \geq j, \ (i+1)s < j_s \\ (i+1)s - 1 & i \geq j, \ (i+1)s > j_s. \end{cases}$$

This definition works because $j_s \notin \mathbf{n}s$. In this case, set $\widehat{s} = (f_s, \overline{j_s})$. Let $s \in S_n$ and $(i,j) \in \mathbf{n-1} \times \mathbf{n}$. There are four cases to check, we verify only one of them. Suppose $i < j$ and $is > js$. Then $(i,j)fs = is$ and

$$(i,j)\widehat{s}f = (i(jg_s), js)f = (is - 1, js)f = is.$$

Similarly, if $s \in T_n \setminus S_n$, there are four cases. We check only $i \geq j$ and $(i+1)s < j_s$. Then $(i,j)fs = (i+1)s$ and

$$(i,j)\widehat{s}f = (i(jf_s), j_s)f = ((i+1)s, j_s)f = (i+1)s.$$

The remaining details are left to the reader.                                    □

**Exercise 4.1.49.** Verify the remaining cases in the above proposition.

## 4.2 Brown's Theorem

As a further application of the techniques underlying the Prime Decomposition Theorem, we provide a proof of Brown's Theorem [59]. Our proof is based on an elegant argument due to Simon [319, 320]; we also use a lemma from [179] where another proof of Brown's Theorem is given. An application of the proof scheme of the Prime Decomposition Theorem to zeta functions appears in [265]. The results in this section will appeal to the Derived Category Theorem [364] and an application will touch upon profinite semigroups.

The reader who skipped earlier chapters should then consult the necessary definitions, or may entirely skip this section as it will not be used further in this chapter. However, it was used in Proposition 2.3.23. For this section, we relax the implicit assumption that all semigroups are finite.

**Definition 4.2.1 (Locally finite).** *A semigroup is called locally finite if all of its finitely generated subsemigroups are finite.*

We use completely analogous terminology for monoids, categories and semigroupoids. Note that a profinite semigroup $S$ is locally finite if and only if every topologically finitely generated closed subsemigroup of $S$ is finite because such subsemigroups are precisely the closures of (abstractly) finitely generated subsemigroups. Our first lemma, from [179], states that a category whose local monoids are locally finite is a locally finite category (so being locally finite is a local property). The idea is based on the McNaughton-Yamada proof [206] of Kleene's Theorem [161]. We use here the convention that if $X$ is a subset of a monoid $M$, then $X^*$ denotes the submonoid generated by $X$.

**Lemma 4.2.2 (Le Saëc, Pin, Weil).** *Let $C$ be a category, all of whose local monoids are locally finite. Then $C$ is locally finite.*

*Proof.* Let $\Gamma$ be a finite graph with vertex set $V$ and let $\varphi\colon \Gamma^* \to C$ be a morphism, which is injective on vertices. For each $X \subseteq V$ and $v, v' \in V$, define $\Gamma^X_{v,v'}$ to be the set of all paths (including the empty path if $v = v'$) in $\Gamma$ from $v$ to $v'$ visiting only vertices from $X$ outside of its initial and terminal vertices. Then $\Gamma^*\varphi = \bigcup_{v,v'} \Gamma^V_{v,v'}\varphi$ and so it suffices to show each $\Gamma^V_{v,v'}\varphi$ is finite. We proceed by establishing each $\Gamma^X_{v,v'}\varphi$ is finite by induction on $|X|$. Because $\Gamma^\emptyset_{v,v'}$ is just the set of edges from $v$ to $v'$ and $\Gamma$ is finite, the case $|X| = 0$ is handled. Suppose $X \neq \emptyset$ and choose $x \in X$. Plainly,

$$\Gamma^X_{v,v'}\varphi = (\Gamma^{X\setminus\{x\}}_{v,v'})\varphi \cup (\Gamma^{X\setminus\{x\}}_{v,x})\varphi \left(\Gamma^{X\setminus\{x\}}_{x,x}\varphi\right)^* (\Gamma^{X\setminus\{x\}}_{x,v'})\varphi \qquad (4.4)$$

and by induction all sets $\Gamma^{X\setminus\{x\}}_{q,q'}\varphi$ are finite. Because the local monoid $C(x\varphi)$ is locally finite, we conclude $\left(\Gamma^{X\setminus\{x\}}_{x,x}\varphi\right)^*$ is also finite. Finiteness of $\Gamma^X_{v,v'}\varphi$ now follows from (4.4). $\qquad\square$

Another notion that we shall need is the consolidation of a semigroupoid.

**Definition 4.2.3 (Consolidation).** *Let $S$ be a semigroupoid. Then the consolidation of $S$ is the semigroup with underlying set consisting of the arrows of $S$ and $0$. Of course, $0$ is a multiplicative zero; the product of arrows from $S$ is as in $S$ when defined and is otherwise $0$.*

For example, the semigroup $B_2$ is the consolidation of the category with two objects, all of whose hom sets have cardinality 1. Notice that there is a faithful functor from any semigroupoid to its consolidation given by the unique map on objects and by the inclusion on arrows. Hence a semigroupoid divides its consolidation.

We now turn to Brown's Theorem; the original proof is combinatorial in nature [59, 110]. Our proof follows [319, 320].

**Theorem 4.2.4 (Brown).** *Let $S'$ be a locally finite semigroup and $\varphi \colon S \to S'$ a homomorphism such that $e\varphi^{-1}$ is locally finite for each idempotent $e \in S'$. Then $S$ is locally finite.*

*Proof.* Without loss of generality, we may assume that $S$ is finitely generated, say by a finite set $X$, and $\varphi$ is onto. In this case $S'$ is finite and our goal is to prove $S$ is finite. We proceed by induction on $|S'|$. If $S'$ is trivial, there is nothing to do, so assume this is not the case. By the dual of Lemma 4.1.25 there are three cases: $S'$ is right simple; $S'$ is cyclic; or $S' = V' \cup T'$ where $V'$ is a proper right ideal and $T'$ is a proper subsemigroup.

*Case 1 ($S'$ is right simple).* Then $S' = G \times E$ where $G$ is a finite group and $E$ is a finite right zero semigroup (see Corollary A.4.17). Let us verify that the local monoids of $\mathbf{Der}(\varphi^I)$ are locally finite. Indeed, $\mathbf{Der}(\varphi^I)(I)$ is trivial. On the other hand a routine computation establishes

$$\mathbf{Der}(\varphi^I)(g, e) \cong \{s \in S \mid (g, e)s\varphi = (g, e)\}^I.$$

Let $s\varphi = (h, f)$. Then $(g, e)s\varphi = (g, e)(h, f) = (gh, f)$, from which we conclude $(g, e)s\varphi = (g, e)$ if and only if $s\varphi = (1, e)$. As a consequence, we obtain $\mathbf{Der}(\varphi^I)(g, e) \cong [(1, e)\varphi^{-1}]^I$, which is locally finite as $(1, e) \in E(S')$. It is a straightforward exercise to verify that $\mathbf{Der}(\varphi^I)$ is generated as a category by the finite subgraph consisting of all the objects and the arrows of the form $(s', (x, x\varphi))$ with $s' \in (S')^I$ and $x \in X$ (i.e., the Cayley graph of $S'$ with respect to $X$). Hence an application of Lemma 4.2.2 yields $\mathbf{Der}(\varphi^I)$ is a finite category. Because $D_\varphi$ is a quotient of a subsemigroupoid of $\mathbf{Der}(\varphi^I)$, we conclude $D_\varphi$ is finite. Therefore, the consolidation $D$ of $D_\varphi$ is a finite semigroup divided by $D_\varphi$. The Derived Semigroupoid Theorem then provides a decomposition $S \prec D \wr S'$. As $D \wr S'$ is evidently finite, we obtain $S$ is finite, as required.

*Case 2 ($S'$ is cyclic).* Suppose $S' = \langle a \rangle$. Let $m$ be the least positive integer such that $a^m = a^{m+n}$ for some integer $n$. Let $N = m^2 + m - 1$. We claim that all integers $n \geq N$ belong to the (additive) subsemigroup $\langle m, m + 1 \rangle$ of $\mathbb{N}$ generated by $m$ and $m + 1$. Indeed, write $n = qm + r$ with $0 \leq r < m$. By choice of $N$, we must have $q \geq m$. Therefore, $q - r > 0$ and so

$$n = (q - r)m + r(m + 1) \in \langle m, m + 1 \rangle.$$

It follows that $S_0 = \langle X^m \cup X^{m+1} \rangle = \bigcup_{c, d > 0} X^{cm + d(m+1)}$ contains $X^n$ for all $n \geq N$, and hence all but finitely many elements of $S$. It therefore suffices to show that $S_0$ is finite. Now $X^m \cup X^{m+1}$ is finite and $S_0\varphi \subseteq \{a^i \mid i \geq m\}$, which is a finite group. The argument from Case 1 then shows that $S_0$ is finite.

*Case 3 ($S' = V' \cup T'$).* Let $V = V'\varphi^{-1}$ and $T = T'\varphi^{-1}$. Then $V$ is a right ideal, $T$ is a subsemigroup and $S = V \cup T$. Let $Y = X \cap V$ and $Z = X \cap T$. Then $\langle Z \rangle \varphi \leq T'$ and so induction yields $\widetilde{T} = \langle Z \rangle$ is finite. Clearly, $S = \widetilde{T}^I \langle Y \widetilde{T}^I \rangle \cup \widetilde{T}$ so it suffices to show $V_0 = \langle Y \widetilde{T}^I \rangle$ is finite. But $V_0 \subseteq V$, as $V$ is a right ideal, and so $V_0 \varphi \leq V'$. Because $Y \widetilde{T}^I$ is finite, induction yields $V_0$ is finite, as required.

This completes the proof of the theorem. $\qquad\qquad\qquad\qquad\qquad\qquad$ □

**Exercise 4.2.5.** Verify in Case 1 that $\mathbf{Der}(\varphi^I)$ is generated as a category by the subgraph consisting of all the objects and the arrows of the form $(s', (x, x\varphi))$ with $s' \in (S')^I$ and $x \in X$.

*Remark 4.2.6.* Notice that a pseudovariety $\mathbf{V}$ is locally finite if and only if every pro-$\mathbf{V}$ semigroup is locally finite.

The following consequence of Brown's Theorem was used in Proposition 2.3.23.

**Corollary 4.2.7.** *Let $\mathbf{V}$ and $\mathbf{W}$ be locally finite pseudovarieties. Then the Mal'cev product $\mathbf{V} \; \circledm \; \mathbf{W}$ is locally finite.*

*Proof.* Let $X$ be a finite set and $\widehat{F}$ be the free pro-$\mathbf{V} \; \circledm \; \mathbf{W}$ semigroup on $X$. Recalling that $\widehat{F_{\mathbf{W}}}(X) = F_{\mathbf{W}}(X)$, there is a natural continuous projection $\pi \colon \widehat{F} \to F_{\mathbf{W}}(X)$. We claim that $e\pi^{-1}$ is a pro-$\mathbf{V}$ semigroup, for all idempotents $e \in F_{\mathbf{W}}(X)$, and hence locally finite. Brown's Theorem will then imply that $\widehat{F}$ is finite, as desired.

To prove the claim, we use that $\widehat{F} = \varprojlim F_i$ where the $F_i \in \mathbf{V} \; \circledm \; \mathbf{W}$ are finite quotients of $\widehat{F}$. By passing to a cofinal system, we may assume without loss of generality the projection $\pi$ factors through each of the projections to the $F_i$, as $F_{\mathbf{W}}(X)$ is finite (Lemma 3.1.37). The natural surjection $\rho_i \colon F_i \to F_{\mathbf{W}}(X)$ is the canonical relational morphism from $F_i$ to $F_{\mathbf{W}}(X)$ and so Proposition 2.3.26 implies $\rho_i \in (\mathbf{V}, \mathbf{1})$. Therefore, $e\rho_i^{-1} \in \mathbf{V}$ for each idempotent $e \in F_{\mathbf{W}}(X)$. Now if $e \in E(F_{\mathbf{W}}(X))$, then $e\pi^{-1}$ is closed and so $e\pi^{-1} = \varprojlim e\rho_i^{-1}$ is pro-$\mathbf{V}$, as required. This completes the proof. $\qquad$ □

# 4.3 The Definition of Complexity

Let us present first a reasonably general definition of a complexity hierarchy and a hierarchical complexity function. Because we will be defining such things in several contexts (e.g., $\mathbf{PV}$, $\mathbf{CC}^+$ and $\mathbf{PVRM}^+$), we phrase the definition in terms of lattices in general. First a lemma about complete lattices.

**Lemma 4.3.1.** *Let $L$ be a complete lattice and set*

$$\mathscr{H}(L) = \{(\ell_i) \in L^{\mathbb{N}} \mid \ell_i \leq \ell_{i+1}, \; i \geq 0\}.$$

*Then $\mathscr{H}(L)$ is a complete lattice with pointwise ordering and pointwise join and meet.*

*Proof.* The constant sequences belong to $\mathscr{H}(L)$ and hence we have closure under empty joins and meets. Let $(\ell_i^{(\alpha)})$, $\alpha \in A$, be elements of $\mathscr{H}(L)$. Then, for all $\alpha_0 \in A$,

$$\bigwedge_{\alpha \in A} \ell_i^{(\alpha)} \le \ell_i^{(\alpha_0)} \le \ell_{i+1}^{(\alpha_0)} \le \bigvee_{\alpha \in A} \ell_{i+1}^{(\alpha)}.$$

Thus $\bigwedge_{\alpha \in A} \ell_i^{(\alpha)} \le \bigwedge_{\alpha \in A} \ell_{i+1}^{(\alpha)}$ and $\bigvee_{\alpha \in A} \ell_i^{(\alpha)} \le \bigvee_{\alpha \in A} \ell_{i+1}^{(\alpha)}$, showing that $\mathscr{H}(L)$ is closed under all joins and meets. $\qquad\square$

**Definition 4.3.2 (Complexity hierarchy).** *A complexity hierarchy for a lattice $L$ is a non-decreasing sequence $\ell_0 \le \ell_1 \le \cdots$ of elements of $L$, that is, a member of $\mathscr{H}(L)$.*

The letter $\mathscr{H}$ is used to stand for the word "hierarchies." We are of course interested in the case $L = \mathbf{PV}$, and later $L = \mathbf{PVRM}$. A complexity hierarchy for $\mathbf{PV}$ will simply be called a complexity hierarchy.

We want to give an equivalent description of complexity hierarchies in terms of hierarchical complexity functions. Set $\mathbb{N}^\infty = \mathbb{N} \cup \{\infty\}$ where $\infty$ is a top element for the order.

**Definition 4.3.3 (Hierarchical complexity function).** *We term a function $f \colon \mathbf{FSgp} \to \mathbb{N}^\infty$ a hierarchical complexity function if it satisfies the following three axioms:*

*(C1) $f(\{1\}) = 0$;*
*(C2) $S \prec T \implies f(S) \le f(T)$;*
*(C3) $f(S \times T) = \max\{f(S), f(T)\}$.*

*The associated $f$-complexity pseudovarieties are defined by*

$$\mathbf{V}_n = \{S \in \mathbf{FSgp} \mid f(S) \le n\}$$

*for $0 \le n < \infty$. We sometimes write*

$$\mathbf{V}_\infty = \bigcup_{n \ge 0} \mathbf{V}_n = \{S \in \mathbf{FSgp} \mid f(S) < \infty\}.$$

We establish a bijection between complexity hierarchies and hierarchical complexity functions. Recall that $\mathbb{N}^\infty$ is a complete lattice with max as the join. The empty meet is $\infty$.

**Proposition 4.3.4.** *Let $f \colon \mathbf{FSgp} \to \mathbb{N}^\infty$ be a hierarchical complexity function. Then the $f$-complexity pseudovarieties form a complexity hierarchy. Conversely, suppose $\mathbf{V}_0 \subseteq \mathbf{V}_1 \subseteq \cdots$ is a complexity hierarchy. Then the mapping $f \colon \mathbf{FSgp} \to \mathbb{N}^\infty$ given by*

$$f(S) = \min\{n \in \mathbb{N} \mid S \in \mathbf{V}_n\}$$

*is the unique hierarchical complexity function whose $f$-complexity pseudova-rieties are the $\mathbf{V}_n$.*

*Proof.* Suppose first that $f$ is a hierarchical complexity function. Then the axioms immediately imply that $\mathbf{V}_n$ is a pseudovariety for all $n \geq 0$. Indeed, (C1) and (C3) imply closure under taking finite products, whereas (C2) implies closure under taking divisors. Clearly, $\mathbf{V}_0 \subseteq \mathbf{V}_1 \subseteq \cdots$ is a complexity hierarchy.

Conversely, suppose $(\mathbf{V}_n)$, $n \geq 0$, is a complexity hierarchy. Then $\{1\} \in \mathbf{V}_0$ yields axiom (C1). If $S \prec T$ and $T \in \mathbf{V}_n$, then $S \in \mathbf{V}_n$. Thus we have $f(S) \leq f(T)$, establishing (C2). Finally, $S \times T \in \mathbf{V}_n$ if and only if $S, T \in \mathbf{V}_n$, from which it is immediate that $f(S \times T) = \max\{f(S), f(T)\}$. Uniqueness of $f$ is clear.    □

**Definition 4.3.5 (Complete lattice of hierarchical complexity functions).** *By the complete lattice of hierarchical complexity functions, we mean the set $\mathscr{C}$ of all hierarchical complexity functions, which we order via the pointwise ordering. Then $\mathscr{C}$ is a complete lattice with pointwise joins and the determined meet.*

We remark that the determined meet is not the pointwise meet. Let us prove that $\mathscr{C}$ is indeed a complete lattice.

**Proposition 4.3.6.** *The partially ordered set $\mathscr{C}$ is a complete lattice and the correspondence between $\mathscr{C}$ and $\mathscr{H}(\mathbf{PV})$ given in Proposition 4.3.4 is an isomorphism between $\mathscr{C}$ and $\mathscr{H}(\mathbf{PV})^{op}$.*

*Proof.* It suffices to prove that the correspondence in Proposition 4.3.4 is an order isomorphism for the dual ordering on $\mathscr{H}(\mathbf{PV})$ and to verify that the pointwise join in $\mathscr{C}$ corresponds to the pointwise meet in $\mathscr{H}(\mathbf{PV})$.

Indeed, if $f, g \in \mathscr{C}$ have corresponding complexity hierarchies $\mathbf{V}_n$ and $\mathbf{W}_n$ respectively, then $f(S) \leq g(S)$ if and only if $S \in \mathbf{W}_n$ implies $S \in \mathbf{V}_n$, for all $n \geq 0$. Hence $f \leq g$ if and only if $\mathbf{W}_n \leq \mathbf{V}_n$, all $n \geq 0$. Let $f_\alpha$, $\alpha \in A$, be a family from $\mathscr{C}$ with associated complexity hierarchies $\mathbf{V}_{\alpha,n}$. Then $\bigvee_{\alpha \in A} f_\alpha(S) \leq n$ if and only if $f_\alpha(S) \leq n$ for all $\alpha \in A$, if and only if $S \in \bigcap_{\alpha \in A} \mathbf{V}_{\alpha,n}$. This completes the proof.    □

We remark that the problem of computability for a hierarchical complexity function $f$ is equivalent to a uniform algorithm for deciding membership in the $f$-complexity pseudovarieties and an algorithm for membership in $\mathbf{V}_\infty$. By a uniform algorithm, we mean an algorithm, which given $n \geq 0$ and a semigroup $S$ determines whether $S \in \mathbf{V}_n$. The next set of exercises shows how to extend these ideas to arbitrary algebraic lattices.

**Exercise 4.3.7.** Show that there is a lattice isomorphism between hierarchical complexity functions and the lattice of maps $f \colon K(\mathbf{PV}) \to \mathbb{N}^\infty$ preserving finite joins (with pointwise ordering).

**Exercise 4.3.8.** Show that there is a lattice isomorphism between the lattice $\mathscr{C}$ of hierarchical complexity functions and $\mathbf{Sup}(\mathbf{PV}, \mathbb{N}^\infty)$ given by extending a hierarchical complexity function $f$ to $\mathbf{PV}$ via the formula $f(\mathbf{V}) = \max\{f(S) \mid S \in \mathbf{V}\}$. Let $(\mathbf{V}_n)$ be complexity hierarchy associated to $f$. Show that the map $n \mapsto \mathbf{V}_n$ is the right adjoint $\mathbb{N}^\infty \to \mathbf{PV}$ to the extended map $f$.

**Exercise 4.3.9.** Let $L$ be an algebraic lattice with set of compact elements $K$. Define a hierarchical complexity function on $L$ to be a map $f\colon K \to \mathbb{N}^\infty$ preserving all finite joins. Define the associated complexity hierarchy by $\ell_n = \bigvee\{k \in K \mid f(k) \leq n\}$, for $n \geq 0$. Show that the lattice of hierarchical complexity functions with pointwise ordering and join is isomorphic to $\mathscr{H}(L)^{op}$.

To define group complexity, we define a complexity hierarchy.

**Definition 4.3.10 (Complexity pseudovarieties).** *The (group) complexity pseudovarieties are defined inductively according the following scheme:*

$$\mathbf{C}_0 = \mathbf{A}$$
$$\mathbf{C}_{n+1} = \mathbf{A} * \mathbf{G} * \mathbf{C}_n$$
$$= \mathbf{C}_n * \mathbf{G} * \mathbf{A}, \ n \geq 0$$

*where $\mathbf{A}$ is the pseudovariety of aperiodic semigroups and $\mathbf{G}$ is the pseudovariety of groups.*

The Prime Decomposition Theorem implies $\mathbf{FSgp} = \bigcup_{n \geq 0} \mathbf{C}_n$ and so the associated hierarchical complexity function takes on only finite values.

**Definition 4.3.11 (Group complexity function).** *The group complexity function $c\colon \mathbf{FSgp} \to \mathbb{N}$ is the hierarchical complexity function associated to the complexity hierarchy $(\mathbf{C}_n)$, $n \geq 0$. The number $c(S)$ is called the group complexity of $S$ or just the complexity of $S$, for short. The pseudovarieties $\mathbf{C}_n$ are simply called the complexity pseudovarieties.*

It is a major open question to determine whether the group complexity function is computable. Certainly each complexity pseudovariety is recursively enumerable. Let us now turn to characterizing group complexity as the largest hierarchical complexity function satisfying some additional axioms [272].

**Theorem 4.3.12 (Rhodes).** *The group complexity $c\colon \mathbf{FSgp} \to \mathbb{N}$ is the greatest hierarchical complexity function $f$ satisfying:*

*1. $f(S \rtimes T) \leq f(S) + f(T)$;*
*2. $f(U_2) = 0$;*
*3. $f(G) \leq 1$ for every group $G$.*

*Proof.* Suppose $f$ satisfies 1–3 and let $(\mathbf{V}_n)$, $n \geq 0$, be the associated $f$-complexity pseudovarieties. By Proposition 4.3.6, it suffices to show $\mathbf{C}_n \leq \mathbf{V}_n$. We proceed by induction on $n$. First note that by the Prime Decomposition Theorem and the assumptions on $f$, we have that $f(S) = 0$ for each aperiodic semigroup $S$. Thus $\mathbf{C}_0 = \mathbf{A} \leq \mathbf{V}_0$. Also $\mathbf{G} \leq \mathbf{V}_1$ by assumption 3. Note that the first hypothesis implies that $\mathbf{V}_i * \mathbf{V}_j \leq \mathbf{V}_{i+j}$. Hence if $\mathbf{C}_n \leq \mathbf{V}_n$, then $\mathbf{C}_{n+1} = \mathbf{A} * \mathbf{G} * \mathbf{C}_n \leq \mathbf{V}_0 * \mathbf{V}_1 * \mathbf{V}_n \leq \mathbf{V}_{n+1}$. □

*Remark 4.3.13 (Heuristics behind the complexity axioms).* The idea behind the complexity axioms is that if $S \prec T$, then $T$ can simulate $S$ and so $S$ should have no more complexity than $T$. Multiplying in the direct product $S \times T$ is a parallel computation in $S$ and $T$, so the complexity of $S \times T$ should be the maximum of the complexities of $S$ and $T$. Multiplication in a semidirect product $S \rtimes T$ is a serial computation: first you must compute in $T$ and then pass the answer to $S$. So the complexity of $S \rtimes T$ should be at most the sum of the complexity of the parts. The semigroup $U_2$ is essentially "junk": it can just reset bits or leave its input alone. So a complexity function should give it a value of 0. Finally, groups are gems. Group complexity is counting how many groups (=reversible computations) you need to build up your semigroup in series-parallel together with flip-flops. The group complexity function is the largest function that satisfies these axioms. See [283] for a further discussion.

A nice property of the axiomatic approach is that in order to show that a function $f\colon \mathbf{FSgp} \to \mathbb{N}$ is a lower bound for complexity, one just needs to show that it satisfies the axioms. To illustrate the technique we show that adjoining identities and augmentation do not change complexity.

**Proposition 4.3.14.** *If $S$ is a semigroup, then $c(S) = c(S^I) = c((S^I)^{\sharp})$.*

*Proof.* Because $S \leq S^I \leq (S^I)^{\sharp}$, clearly $c(S) \leq c(S^I) \leq c((S^I)^{\sharp})$. To show that $c((S^I)^{\sharp}) \leq c(S)$, we define a function $f\colon \mathbf{FSgp} \to \mathbb{N}^{\infty}$ by $f(S) = c((S^I)^{\sharp})$. We show that $f$ is a hierarchical complexity function satisfying the axioms of Theorem 4.3.12. It will then follow that $f \leq c$, giving the reverse inequality. Clearly $f(\{1\}) = c(U_1^{\sharp}) = 0$. If $S \prec T$, then Proposition 4.1.13 implies $(S^I)^{\sharp} \prec (T^I)^{\sharp}$ and hence $c((S^I)^{\sharp}) \leq c((T^I)^{\sharp})$. So $f(S) \leq f(T)$. If $M$ and $N$ are monoids, then $(M \times N, M \times N) \leq (M, M) \times (N, N)$ because $(\overline{m}, \overline{n})$ is a constant map to $(m, n)$. Thus $(M \times N)^{\sharp} \prec M^{\sharp} \times N^{\sharp}$ and so $c((M \times N)^{\sharp}) \leq \max\{c(M^{\sharp}), c(N^{\sharp})\}$. But $M, N \prec M \times N$, which gives the reverse inequality in light of Proposition 4.1.13. It is then immediate that $f(S \times T) = \max\{f(S), f(T)\}$. We have therefore proved that $f$ is a hierarchical complexity function.

As $(U_2^I)^{\sharp}$ is aperiodic, $f(U_2) = 0$. If $M$ is a monoid, then $M^I \leq M \times U_1$ via $m \mapsto (m, 0)$ and $I \mapsto (1, 1)$. So, by the above,

$$c(M^{\sharp}) \leq c((M^I)^{\sharp}) \leq c((M \times U_1)^{\sharp}) = \max\{c(M^{\sharp}), c(U_1^{\sharp})\} = c(M^{\sharp}).$$

It follows for a group $G$, $c((G^I)^\sharp) = c(G^\sharp) \leq 1$, as $G^\sharp \prec \overline{G}^I \rtimes G$ by Proposition 4.1.14.

Finally, $(S \rtimes T)^I \leq S^I \rtimes T^I$ and $(S^I \rtimes T^I)^\sharp \prec [(S^I)^\sharp]^{T^I} \rtimes (T^I)^\sharp$ by Proposition 4.1.19. It follows that $f(S \rtimes T) \leq f(S) + f(T)$. This completes the proof that $f$ satisfies the axioms and so is smaller than $c$.                    □

The main open question concerning the group complexity function is whether it is computable.

*Question 4.3.15 (Complexity Problem).* Is the group complexity function $c$ computable? More precisely, is there a Turing machine that given a finite semigroup $S$ by its multiplication table as input can output $c(S)$?

We believe that complexity is decidable and are currently working on a proof of it [128].

Another important hierarchical complexity function is the $U_2$-length of an aperiodic semigroup.

**Definition 4.3.16 ($U_2$-length).** *The $U_2$-length complexity hierarchy is defined by $\mathbf{U}_0 = \mathbf{1}$ and $\mathbf{U}_n = (U_2) * \mathbf{U}_{n-1}$. Let $c_\mathbf{A}$ be the associated hierarchical complexity function; it is called the $U_2$-length.*

The Prime Decomposition Theorem immediately implies that $\mathbf{U}_\infty = \mathbf{A}$, that is, $c_\mathbf{A}(S) < \infty$ if and only if $S$ is aperiodic. The following characterization of $U_2$-length is proved in a similar way to Theorem 4.3.12. We leave the proof to the reader.

**Theorem 4.3.17.** *The hierarchical complexity function $c_\mathbf{A}$ is the largest hierarchical complexity function $f \colon \mathbf{FSgp} \to \mathbb{N}^\infty$ such that:*

1. *$f(U_2) = 1$;*
2. *$f(S \rtimes T) \leq f(S) + f(T)$.*

**Exercise 4.3.18.** Prove Theorem 4.3.17.

**Exercise 4.3.19.** Verify that $c_\mathbf{A}(S) = c_\mathbf{A}(S^I)$ if $S$ is non-trivial.

We remark that $U_2$-length is computable because each $\mathbf{U}_n$ is locally finite with computable bounds on the size of the free objects as a function of the number of generators.

## 4.4 Aperiodics and Schützenberger's Theorem

Schützenberger characterized the star-free languages as the languages recognized by aperiodic monoids [313]. This is often considered to be one of the most significant results in the algebraic theory of languages. We give a variant on one of the proofs of this theorem via the Prime Decomposition Theorem; other variants can be found in [85, 207, 346]. Recall that $A^*$ denotes the free monoid on a set $A$.

**Definition 4.4.1 (Recognizable sets).** *A subset $L \subseteq A^*$ is recognizable if there is a finite monoid $M$ and a homomorphism $\eta \colon A^* \to M$ such that $L\eta\eta^{-1} = L$. One says that $\eta$ recognizes $L$, or sometimes that $M$ recognizes $L$.*

It is a well-known theorem of Kleene [85, 161] that the recognizable subsets are those that can be obtained from finite subsets by taking unions, products and generating submonoids; such subsets are alternatively known as the *regular sets* or *rational sets*. Also it is well-known that the collection of recognizable subsets of $A^*$ is closed under Boolean operations [85]. If $\mathbf{V}$ is a pseudovariety of monoids, then a subset $L \subseteq A^*$ is said to be $\mathbf{V}$-*recognizable* if it can be recognized by a monoid from $\mathbf{V}$. The $\mathbf{V}$-recognizable subsets of $A^*$ form a Boolean algebra [85, 224]. A subset of $A^*$ is called *star-free* if it can be built up from finite subsets via Boolean operations and product [205]. For instance, $A^*$ is star-free because it is the complement of $\emptyset$. Schützenberger's Theorem says that the $\mathbf{A}$-recognizable subsets are exactly the star-free subsets [313]. More about the connections between Finite Semigroup Theory and the theory of recognizable sets can be found in [85, 177, 224, 229, 346]. Let us agree that a set $L$ is a Boolean combination of elements of a collection $\mathscr{C}$ of sets if it can be obtained from sets in $\mathscr{C}$ by using finitely many times union, intersection and complement.

The first step in proving Schützenberger's Theorem is to establish the closure of the set of $\mathbf{A}$-recognizable subsets under product. The original approach was via Schützenberger products [85, 313]. We use the more modern approach via aperiodic relational morphisms. For this purpose, we need some tools.

**Definition 4.4.2 (Syntactic monoid).** *If $L \subseteq A^*$, then the syntactic monoid $M_L$ of $L$ is $A^*/{\equiv_L}$ where $x \equiv_L y$ if, for all $u, v \in A^*$,*

$$uxv \in L \iff uyv \in L.$$

*The quotient map $\eta_L \colon A^* \to M_L$ is called the syntactic morphism.*

It is easy to verify $\equiv_L$ is a congruence. It is well-known that $M_L$ is finite if and only if $L$ is recognizable and, more generally, $M_L \in \mathbf{V}$ if and only if $L$ is $\mathbf{V}$-recognizable. The reader is referred to [85, 177, 224] for details.

Recall that if $\mathbf{V}$ and $\mathbf{W}$ are pseudovarieties of semigroups, then $(\mathbf{V}, \mathbf{W})$ denotes the collection of all relational morphisms $\varphi \colon S \to T$ such that if $W \leq T$ with $W \in \mathbf{W}$, then $W\varphi^{-1} \in \mathbf{V}$.

**Definition 4.4.3 (Aperiodic relational morphism).** *A relational morphism is called aperiodic if it belongs to $(\mathbf{A}, \mathbf{A})$. A congruence is called aperiodic if the associated quotient morphism is aperiodic.*

Aperiodic morphisms are termed $\gamma$-maps in [171]. Let us give a characterization of aperiodic relational morphisms from [224]; see also [334], whose approach we follow. The version of this result for morphisms can be found in [171].

Let $\varphi\colon S \to T$ be a relational morphism and let $X \subseteq S$. Then $\varphi$ is said to be *injective* on $X$ if, for all $x_1, x_2 \in X$,

$$x_1\varphi \cap x_2\varphi \neq \emptyset \implies x_1 = x_2. \tag{4.5}$$

Notice that a relational morphism is injective on $S$ if and only if it is a division. We denote by $s^{\omega-1}$ the inverse of $s^{\omega+1} = s^\omega s$ in the group $\langle s^{\omega+1}\rangle$. Recall $s^\omega$ is the unique idempotent positive power of $s$.

**Lemma 4.4.4 (Aperiodicity Lemma).** *Let $\varphi\colon S \to T$ be a relational morphism. Then the following are equivalent:*

1. *$\varphi$ is aperiodic, i.e., $\varphi \in (\mathbf{A}, \mathbf{A})$;*
2. *$\varphi \in (\mathbf{A}, \mathbf{1})$;*
3. *$\varphi$ is injective on subgroups;*
4. *$\varphi$ is injective on regular $\mathscr{H}$-classes.*

*Proof.* The implications 4 implies 3 implies 2 are clear. To show that 2 implies 1, suppose $T' \leq T$ is aperiodic. Set $S' = T'\varphi^{-1}$ and let $\varphi'\colon S' \to T'$ be the restriction. Then $\varphi' = \alpha^{-1}\beta$ where $\alpha\colon R \twoheadrightarrow S'$ is a surjective homomorphism, $\beta\colon R \to T'$ is a homomorphism and $R$ is finite (by taking for instance the canonical factorization (1.28)). Let $G \leq S'$ be a subgroup. Then there is a subgroup $G'$ of $R$ with $G'\alpha = G$ (Proposition 4.1.44). Because $T'$ is aperiodic, $G'\beta$ is an idempotent $e$ and hence $G \leq e\varphi^{-1}$. It follows that $G$ is trivial by 2. So $S' = T'\varphi^{-1}$ is aperiodic.

Now we turn to 1 implies 4. Suppose $x, y \in S$ are regular, $x \mathscr{H} y$ and $t \in x\varphi \cap y\varphi$. Let $x'$ be an inverse of $x$ and choose $t' \in x'\varphi$. Then $xx'$, $x'x$ are idempotents. Because $xx' \mathscr{R} x \mathscr{R} y$ and $x'x \mathscr{L} x \mathscr{L} y$, we have $xx'y = y$ and $yx'x = y$. Using relational notation, we compute:

$$x = (xx')^\omega x \; \varphi \; (tt')^\omega t$$
$$y = (xx')^\omega y \; \varphi \; (tt')^\omega t$$
$$x' = x'(xx')^{\omega-1} \; \varphi \; t'(tt')^{\omega-1}.$$

Thus $xx'$ and $yx'$ $\varphi$-relate to the idempotent $(tt')^\omega$. Moreover, $yx' \mathscr{H} xx'$ (they both belong to $R_x \cap L_{x'}$). Because $(tt')^\omega\varphi^{-1}$ is aperiodic and $xx'$ is an idempotent (and so $\langle yx'\rangle$ is a subgroup of $(tt')^\omega\varphi^{-1}$ with identity $xx'$), we conclude that $yx' = xx'$. Hence $y = yx'x = xx'x = x$, as desired. □

For example, if $S$ is a finite semigroup and $I$ is an aperiodic ideal, then $S \to S/I$ is an aperiodic morphism.

**Exercise 4.4.5.** Verify that if $S \rtimes T$ is a semidirect product with projection $\pi\colon S \rtimes T \to T$ and $S$ is aperiodic, then $\pi$ is aperiodic.

An important application of aperiodic relational morphisms to language theory is the following simple lemma. It can also be deduced from standard properties of Schützenberger products; the direct proof can be found in [343].

**Lemma 4.4.6.** *Let $L_1, L_2 \subseteq A^*$ be recognizable subsets and put $L = L_1 L_2$. Then there is an aperiodic relational morphism $\varphi \colon M_L \to M_{L_1} \times M_{L_2}$. In particular, if $M_{L_1}$ and $M_{L_2}$ are aperiodic, then so is $M_L$.*

*Proof.* The second statement is an immediate consequence of the first. Let $\varphi = \eta_L^{-1} \Delta(\eta_{L_1} \times \eta_{L_2})$. We show that $\varphi$ is aperiodic. Let $n \geq 3$ be such that $m^n = m^\omega$ for each $m \in M_L$. It suffices, by the Aperiodicity Lemma, to show that if $w \in A^*$ such that $w \equiv_{L_i} w^2$, $i = 1, 2$, then $w^n \equiv_L w^{n+1}$.

Suppose first that $uw^n v \in L$. Then $uw^n v = xy$ with $x \in L_1$ and $y \in L_2$. Because $n \geq 3$, we must have either the first $w$ is a factor of $x$ or the last $w$ is a factor of $y$; say that the first is a factor of $x$, the other case being dual. Then $x = uws$ with $sy = w^{n-1}v$. Therefore, $uw^2 s \in L_1$ because $w \equiv_{L_1} w^2$ and $uws \in L_1$. Thus $uw^{n+1}v = uw^2 sy \in L_1 L_2 = L$, as required.

Conversely, suppose that $uw^{n+1}v \in L$. Then $uw^{n+1}v = xy$ with $x \in L_1$ and $y \in L_2$. Because $n \geq 3$, either the first $w^2$ is a factor of $x$ or the last $w^2$ is a factor of $y$; we handle the second case, as the first is dual. Then $y = sw^2v$ with $xs = uw^{n-1}$. From $sw^2v \in L_2$, it follows $swv \in L_2$, as $w \equiv_{L_2} w^2$. Therefore, $uw^n v = xswv \in L_1 L_2 = L$, as desired. □

The proof that each **A**-recognizable set is star-free goes by induction along the length of a prime decomposition of the aperiodic semigroup. Straubing [346] follows a similar approach, using the block product and the two-sided Prime Decomposition Theorem. Eilenberg [85] (based on [207]) also inducts on the length of a prime decomposition, but he iterates the wreath product in the other variable. We begin with a lemma on languages recognized by wreath products with $U_2$.

**Lemma 4.4.7.** *Let $M$ be a monoid. Then any subset of $A^*$ recognized by the unitary wreath product $\overline{\mathbf{2}}^I \wr M$ is a Boolean combination of languages of the form $L'$, $L'aA^*$ and $L'a\left(A^* \setminus \bigcup_{i \in J} L_i b_i A^*\right)$, where $a, b_i \in A$, $L'$ and the $L_i$ are subsets of $A^*$ recognized by $M$ and the index set $J$ is finite.*

*Proof.* Let $\varphi \colon A^* \to \overline{\mathbf{2}}^I \wr M$ be a homomorphism and let $\pi \colon \overline{\mathbf{2}}^I \wr M \to M$ be the projection. Then $\psi = \varphi\pi \colon A^* \to M$ is a homomorphism. For $m \in M$, set $L_m = \{w \in A^* \mid w\psi = m\}$; clearly, $L_m$ is recognized by $M$.

Any subset recognized by $\varphi$ is a finite union of sets of the form $(f, m)\varphi^{-1}$ and so we need only consider sets of this form. Let us set $w\varphi = (f_w, m_w)$. Define, for $m \in M$ and $x \in \overline{\mathbf{2}}^I$, $L(m, x) = \{w \in A^* \mid mf_w = x\}$. Then $(f, m)\varphi^{-1} = L_m \cap \bigcap_{t \in M} L(t, tf)$. Thus we just need to show that each subset $L(m, x)$, $m \in M$, is a Boolean combination of languages of the required form. Direct computation yields, for $w = a_1 \cdots a_n$ with $a_i \in A$, $i = 1, \ldots, n$, that

$$mf_w = mf_{a_1}(ma_1)f_{a_2} \cdots (ma_1 \cdots a_{n-1})f_{a_n} \qquad (4.6)$$

where we abuse notation by identifying $a_i$ with its image in $M$ under $\psi$.

First we consider $L(m, I)$. Actually, it is enough to prove $A^* \setminus L(m, I)$ is a Boolean combination of languages of the required form. From (4.6), we see that $mf_w \neq I$ if and only if there is a proper (possibly empty) prefix $p = a_1 \cdots a_i$ of $w = a_1 \cdots a_n$ such that $(mp\psi)f_{a_{i+1}} \neq I$. Thus

$$A^* \setminus L(m, I) = \bigcup_{\{t \in M, a \in A | (mt)f_a \neq I\}} L_t a A^*.$$

Because $L(m, \overline{1}) = A^* \setminus \left( L(m, I) \cup L(m, \overline{0}) \right)$, we are left with the case $L(m, \overline{0})$. From (4.6), we see that $mf_w = \overline{0}$ if and only if there is a proper (possibly empty) prefix $p = a_1 \cdots a_i$ of $w = a_1 \cdots a_n$ such that $(mp\psi)f_{a_{i+1}} = \overline{0}$ and, for $i + 1 \leq k < n$, $(mpa_{i+1} \cdots a_k \psi)f_{a_{k+1}} \neq \overline{1}$. Therefore, we conclude

$$L(m, \overline{0}) = \bigcup_{\{t \in M, a \in A | (mt)f_a = \overline{0}\}} L_t a \left[ A^* \setminus \left( \bigcup_{\{u \in M, b \in A | (mtau)f_b = \overline{1}\}} L_u b A^* \right) \right],$$

which is a Boolean combination of the desired form, completing the proof.  □

Let us now turn to a proof of Schützenberger's Theorem [313] via the Prime Decomposition Theorem.

**Theorem 4.4.8 (Schützenberger).** *A subset $L \subseteq A^*$ is star-free if and only if it is **A**-recognizable.*

*Proof.* Clearly, any finite subset $L \subseteq A^*$ is **A**-recognizable. Indeed, if $n$ is larger than the length of any word in $L$, then $L$ can be recognized by the quotient morphism $A^* \to A^*/I_n$, where $I_n$ is the ideal of words of length at least $n$. But $A^*/I_n$ is evidently a finite nilpotent semigroup with an adjoined identity (and hence aperiodic). Lemma 4.4.6 implies the closure of the set of **A**-recognizable subsets under product. Because the **A**-recognizable sets are closed under Boolean operations, we conclude that each star-free subset of $A^*$ is **A**-recognizable.

For the converse, assume $L \subseteq A^*$ is **A**-recognizable and $\eta \colon A^* \to N$, with $N \in \mathbf{A}$, recognizes $L$. We proceed by induction on the $U_2$-length $c_{\mathbf{A}}(N)$ of $N$; however, we use the monoidal version of $c_{\mathbf{A}}$ defined in terms of the semidirect product of monoid pseudovarieties. If $c_{\mathbf{A}}(N) = 0$, then $N$ is trivial. Therefore, $L = \emptyset$ or $L = A^* = A^* \setminus \{\emptyset\}$ and so $L$ is star-free. If $N \in \mathbf{A}$ with $c_{\mathbf{A}}(N) > 0$, then there is a division $d \colon N \to U_2^m \wr M$ into a unitary wreath product where $c_{\mathbf{A}}(M) < c_{\mathbf{A}}(N)$. We assume then inductively that each subset recognized by $M$ is star-free. For each $a \in A$, choose $(f_a, m_a) \in a\eta d$ and consider the morphism $\varphi \colon A^* \to U_2^m \wr M$ given by $a \mapsto (f_a, m_a)$. We claim $\varphi$ recognizes $L$. Notice if $w \in A^*$, then $w\varphi \in w\eta d$. So if $u \in L\varphi\varphi^{-1}$, then there exists $v \in L$ such that $u\varphi = v\varphi$. Hence $u\varphi \in u\eta d \cap v\eta d$, and so $u\eta = v\eta$, from which we obtain $u \in L\eta\eta^{-1} = L$. Thus $\varphi$ recognizes $L$. Next observe $U_2^m \wr M \leq (U_2 \wr M)^m$ by Exercise 1.2.13 and any language recognized by a direct power of a monoid is a Boolean combination of languages recognized by the monoid in question [85, Chapter VII, Prop. 5.2]. Lemma 4.4.7 then decomposes $L$ as a Boolean combination of languages of the form $L'$, $L'aA^*$ and $L'a\left(A^* \setminus \bigcup_{i \in J} L_i b_i A^*\right)$, where $a, b_i \in A$, $L'$ and the $L_i$ are subsets of $A^*$ recognized by $M$ and $J$ is finite. By induction, $L'$ and the $L_i$ are star-free, from which we easily deduce $L$ is star-free. This completes the induction.     □

**Exercise 4.4.9.** Verify that if $L \subseteq A^*$ is recognized by $\prod M_i$, then it is a Boolean combination of languages recognized by the $M_i$.

## 4.5 Stiffler's Theorem

In this section, we prove Stiffler's Theorem characterizing the smallest pseudovariety closed under semidirect product that contains $U_1$. It can be viewed as a refinement of the decomposition of aperiodic semigroups into primes. Our modern proof of Stiffler's Theorem relies on Simon's Theorem [60, 85], which says that **Sl** is local and the Derived Semigroupoid Theorem. An alternate proof is given in Chapter 5. Let **R** denote the pseudovariety of $\mathscr{R}$-trivial semigroups. Of course, $U_1 \in \mathbf{R}$.

**Lemma 4.5.1.** *The pseudovariety of $\mathscr{R}$-trivial semigroups is closed under semidirect product.*

*Proof.* Suppose $S \rtimes T$ is a semidirect product with $S, T \in \mathbf{R}$ and assume $(s, t) \mathscr{R} (s', t')$ with $(s, t) \neq (s', t')$. Then $t \mathscr{R} t'$ and so $t = t'$. Suppose $(s, t)(u, v) = (s', t')$. Then $s' = s^t u$ and so $s' \leq_{\mathscr{R}} s$. Dually, $s \leq_{\mathscr{R}} s'$ and so $s = s'$. This contradiction shows that $S \rtimes T \in \mathbf{R}$.     □

**Theorem 4.5.2 (Stiffler).** *A semigroup $S$ is $\mathscr{R}$-trivial if and only if it divides an iterated wreath product of two-element semilattices. More precisely, if we set $\mathbf{R}_1 = \mathbf{Sl}$ and $\mathbf{R}_n = \mathbf{Sl} * \mathbf{R}_{n-1}$, for $n \geq 2$, then $\mathbf{R} = \bigcup_{n \geq 1} \mathbf{R}_n$.*

*Proof.* Lemma 4.5.1 implies that $\mathbf{R}_n \leq \mathbf{R}$ for all $n \geq 1$. For the converse, we prove by induction on the order of $S$ that if $S \in \mathbf{R}$, then $S \in \mathbf{R}_n$ for some $n \geq 1$. The case $|S| = 1$ is trivial. Let $J$ be a 0-minimal ideal of $S$ (see Definition A.1.4) and consider the quotient morphism $\varphi \colon S \to S/J$. By induction, $S/J \in \mathbf{R}_n$ for some $n \geq 1$ ($S/J$ could be trivial). We show that $D_\varphi \in \ell\mathbf{Sl}$. Because $\ell\mathbf{Sl} = \mathbf{gSl}$ by Simon's Theorem [85, Chapter VIII, Thm 7.1] [60, 364], it will then follow from the Derived Semigroupoid Theorem (Theorem 2.5.11) that $S \in \mathbf{Sl} * \mathbf{R}_n = \mathbf{R}_{n+1}$.

Suppose first that $0 \neq t \in (S/J)^I$. Then $|t\varphi^{-1}| \leq 1$, and so the local semigroup $D_\varphi(t)$ is trivial or empty. So the only interesting local semigroup is $D_\varphi(0)$. Now $0\varphi^{-1} = J$ and $0s\varphi = 0$, for all $s \in S$. So $D_\varphi(0)$ can be identified with the quotient $S'$ of $S$ by the kernel of its action on the right of $J$. We show that $S' \leq U_1$. Indeed, if $x \in J \setminus \{0\}$ and $s \in S$, then either: $xs = 0$; or by stability $xs \mathscr{R} x$, and hence $xs = x$ as $S$ is $\mathscr{R}$-trivial. Suppose that $xs = x$. Because $S$ is $\mathscr{R}$-trivial, $J \setminus \{0\}$ is in fact a single $\mathscr{L}$-class and hence if $y \in J$, then $y = tx$ for some $t \in S^I$. But then $ys = txs = tx = y$. It follows that each element of $S$ acts on $J$ as either the zero map or the identity, and so $S' \leq U_1$. This concludes the proof that $D_\varphi(0) \in \ell\mathbf{Sl}$ and hence the proof of the theorem.    □

Stiffler's Theorem admits the following useful switching rule as a corollary.

**Corollary 4.5.3 (Stiffler).** *Let $\mathbf{H}$ be a pseudovariety of groups. Then the inequality $\mathbf{H} * \mathbf{R} \leq \mathbf{R} * \mathbf{H}$ holds.*

*Proof.* Because $\mathbf{R} = \bigcup \mathbf{R}_n$ by Theorem 4.5.2, $\mathbf{H} * \mathbf{R} = \bigcup (\mathbf{H} * \mathbf{R}_n)$ and so it suffices to show that $\mathbf{H} * \mathbf{R}_n \subseteq \mathbf{R} * \mathbf{H}$ for each $n \geq 1$. For $n = 1$, we have

$$\mathbf{H} * \mathbf{R}_1 = \mathbf{H} * \mathbf{Sl} \leq \mathbf{LRB} * \mathbf{H} \leq \mathbf{R} * \mathbf{H}$$

using Proposition 4.1.10 and the fact that left regular bands are precisely the $\mathscr{R}$-trivial bands. Using that $\mathbf{R}_n = \mathbf{R}_{n-1} * \mathbf{Sl}$, the induction proceeds via

$$\mathbf{H} * \mathbf{R}_n = \mathbf{H} * \mathbf{R}_{n-1} * \mathbf{Sl} \leq \mathbf{R} * \mathbf{H} * \mathbf{Sl} \leq \mathbf{R} * \mathbf{LRB} * \mathbf{H} \leq \mathbf{R} * \mathbf{H}$$

where the last inequality uses that $\mathbf{R}$ is closed under semidirect product.    □

**Exercise 4.5.4.** Show that if $\mathbf{H}$ is a non-trivial pseudovariety of groups closed under extension, then $\mathbf{H} * \mathbf{H} = \mathbf{LZ} \vee \mathbf{H}$.

**Exercise 4.5.5.** Show that if $\mathbf{H}$ is an extension-closed pseudovariety of groups then $\mathbf{R} * \mathbf{H}$ is closed under semidirect product.

We end this section with another result of Stiffler [340]. Let $\mathbf{D}$ be the pseudovariety of semigroups whose idempotents are right zeroes. Stiffler showed that $\mathbf{D}$ is the smallest pseudovariety of semigroups containing $\overline{\mathbf{2}}$, which is closed under semidirect product.

**Lemma 4.5.6 (Pumping Lemma).** *Let $S$ be a semigroup of order $n$. Then $S^n = SE(S)S$.*

*Proof.* Clearly $SE(S)S \subseteq S^n$. Conversely, consider a product $s = s_1 \cdots s_n$ and let $t_i = s_1 \cdots s_i$. If the $t_i$ are all distinct, then because $|S| = n$, it follows $S = \{t_1, \ldots, t_n\}$ and so $t_i$ is an idempotent for some $i$. But then $s = t_i s_{i+1} \cdots s_n$ belongs to $SE(S)S$. If the $t_i$ are not all distinct, then there exist $i < j$ such that $t_i = t_j$. Then $t_i(s_{i+1} \cdots s_j)^\omega = t_i$ and so $s = s_1 \cdots s_i(s_{i+1} \cdots s_j)^\omega s_{j+1} \cdots s_n$ and hence belongs to $SE(S)S$. This completes the proof.    □

Set $\mathbf{D}_n = [\![ x_1 \cdots x_n = yx_1 \cdots x_n ]\!]$. Clearly, $\mathbf{D}_n \subseteq \mathbf{D}$. On the other hand, the Pumping Lemma implies that if $S \in \mathbf{D}$, then $S^n = SE(S)S = E(S)$ and hence consists of right zeroes. Therefore, $S \in \mathbf{D}_n$. Thus $\mathbf{D} = \bigcup_{n \geq 1} \mathbf{D}_n$.

**Theorem 4.5.7 (Stiffler).** *The smallest pseudovariety containing $\overline{\mathbf{2}}$ that is closed under semidirect product is $\mathbf{D}$.*

*Proof.* First we show that $\mathbf{D}$ is closed under semidirect product. Let $S, T \in \mathbf{D}$ and suppose $S \rtimes T$ is a semidirect product. Let $(s, t) \in S \rtimes T$ and suppose $(u, e) \in E(S \rtimes T)$. In particular, $e \in E(T)$. Then $(u, e)(u, e) = (u^e u, e)$ and so $u = u^e u$ and hence $^e u = (^e u)^2$. But then because $S \in \mathbf{D}$, it follows $u = u^e u = {}^e u$. Thus we obtain, using that $^e u$ and $e$ are right zeroes,

$$(s, t)(u, e) = (s, t)(u, e)(u, e) = (s^t u^{te} u, tee) = (s^t u^e u, e) = (^e u, e) = (u, e)$$

and so $S \rtimes T \in \mathbf{D}$. Therefore, $\mathbf{D}$ is closed under semidirect product.

For the converse, we claim that $\mathbf{D}_{n+1} \leq \mathbf{RZ} * \mathbf{D}_n$. As $\mathbf{D}_1 = \mathbf{RZ} = (\overline{\mathbf{2}})$, this will establish the theorem. For convenience, set $\mathbf{D}_0 = \mathbf{1}$. Then the claim holds for $n = 0$ trivially. Assume the claim is true for $n$. First observe that $\mathbf{D}_k$ is locally finite. If $A$ is a finite set, then $F_{\mathbf{D}_k}(A)$ is the quotient of $A^+$ by the congruence identifying two words if they have length at least $k$ and the same suffix of length $k$. Let $\tau_k \colon A^+ \to A^+$ be defined by

$$w\tau_k = \begin{cases} w & |w| \leq k \\ u & w = xu \text{ with } |u| = k. \end{cases}$$

Then we may view $F_{\mathbf{D}_k}(A)$ as the semigroup $D_k$ whose underlying set is $A \cup A^2 \cup \cdots \cup A^k$ equipped with the product $u \cdot v = (uv)\tau_k$. Also we may view $\tau_k$ as a homomorphism $A^+ \to D_k$. Notice that $D_k$ acts faithfully on the right of its minimal ideal $A^k$. For $w \in A^+$, let $w\alpha$ be the first letter of $w$.

We now apply Lemma 4.1.23 to the faithful transformation semigroups $(A^{n+1}, D_{n+1})$ and $(A, \overline{A}) \wr (A^n, D_n)$. Define $f \colon A \times A^n \to A^{n+1}$ by $(a, u)f = au$. Clearly, $f$ is a surjective function. Let $w \in D_{n+1}$. Set $\widehat{w} = (f_w, w\tau_n)$ where $uf_w = (uw)\tau_{n+1}\alpha$ for $u \in A^n$. Let us verify $\widehat{w}$ covers $w$. Indeed, we compute

$$(a, u)\widehat{w}f = (auf_w, (uw)\tau_n)f = (a\overline{(uw)\tau_{n+1}\alpha}, (uw)\tau_n)f$$
$$= ((uw)\tau_{n+1}\alpha, (uw)\tau_n)f = (uw)\tau_{n+1} = (a, u)fw$$

as required. It now follows $D_{n+1} \prec \overline{A} \wr (A^n, D_n)$ thereby completing the inductive verification that $\mathbf{D}_{n+1} \leq \mathbf{RZ} * \mathbf{D}_n$ and establishing the theorem. $\square$

## 4.6 The Semilocal Theory

The material from this section is an updated presentation of the results of [171, Chapter 8] and [170, 269, 368]. This collection of results is sometimes referred to informally as the "toolbox." We assume finiteness throughout this section.

By the local theory, we mean the Green-Rees structure theory of a regular $\mathscr{J}$-class, which gives a local coordinates picture of the $\mathscr{J}$-class via a Rees matrix representation. The semilocal theory is concerned with how the different regular $\mathscr{J}$-classes are pieced together.

### 4.6.1 $\mathscr{K}'$-morphisms

Given that Green's relations play such a critical role in semigroup theory, it should come as no surprise that homomorphisms respecting these relations are pivotal. As the role of Green's relations is most crucial for regular elements, it turns out that respecting these relations on regular elements is the key idea.

**Definition 4.6.1 ($\mathscr{K}'$-morphism).** *Let $S$ be a semigroup and let $\mathscr{K}$ be any of Green's equivalence relations $\mathscr{J}$, $\mathscr{R}$, $\mathscr{L}$ or $\mathscr{H}$. A homomorphism $\varphi$ defined on $S$ is called a $\mathscr{K}'$-morphism if whenever $s, t \in S$ are regular and $s\varphi = t\varphi$, then $s \mathscr{K} t$. A congruence on $S$ is called a $\mathscr{K}'$-congruence if the associated quotient map is a $\mathscr{K}'$-morphism.*

Analogously, $\varphi$ is called a $\mathscr{K}$-*morphism* if $s\varphi = t\varphi$ implies $s \mathscr{K} t$, any $s, t \in S$. Because $\mathscr{K}$-morphisms are not closed under restriction, whereas $\mathscr{K}'$-morphisms are, they do not play such an important role in the theory.

**Exercise 4.6.2.** Verify that a congruence on $S$ is idempotent-separating (i.e., injective on $E(S)$) if and only if it is an $\mathscr{H}'$-congruence.

The notion of a $\mathscr{K}'$-morphism is closely tied to Mal'cev products and pseudovarieties of relational morphisms. If $\mathbf{V}$ is a pseudovariety, we set $\mathbf{V}^\mathbf{N} = \mathbf{V} \text{ⓜ} \mathbf{N}$; it consists of all semigroups $S$ such that $SE(S)S \in \mathbf{V}$. Alternatively, the members of $\mathbf{V}^\mathbf{N}$ can be thought of as nilpotent extensions of semigroups in $\mathbf{V}$.

**Exercise 4.6.3.** Show that $S \in \mathbf{V}^\mathbf{N}$ if and only if $SE(S)S \in \mathbf{V}$.

Denote by $\mathbf{CS}$, $\mathbf{LS}$, $\mathbf{RS}$ and $\mathbf{G}$ the pseudovarieties of completely simple semigroups, left simple semigroups, right simple semigroups and groups, respectively. Note that $\mathbf{CS}^\mathbf{N} = \mathbb{L}\mathbf{G}$, the pseudovariety of local groups.

**Exercise 4.6.4.** Show $\mathbf{CS}^\mathbf{N} = \mathbb{L}\mathbf{G}$.

**Proposition 4.6.5.** *A semigroup $S$ belongs to $\mathbb{L}\mathbf{G}$ if and only if $S$ does not contain a subsemigroup isomorphic to $U_1$.*

*Proof.* Clearly, $U_1 \notin \mathbb{L}\mathbf{G}$. For sufficiency, suppose that $S$ does not contain a copy of $U_1$. Let $f$ be any idempotent of $S$. Then $fSf$ is a monoid with identity $f$. If $e \in fSf$ is any idempotent different from $f$, then $\{e, f\} \cong U_1$, which is impossible. Thus $fSf$ is a monoid with a unique idempotent, i.e., a group. □

**Exercise 4.6.6.** Prove a finite monoid with a unique idempotent is a group.

The following exercise provides several analogues of Proposition 4.6.5.

**Exercise 4.6.7.**

1. Show that $S \in \mathbf{LS}^\mathbf{N}$ if and only if $S$ contains neither an isomorphic copy of $U_1$ nor of $\overline{\mathbf{2}}$.

2. Show that $S \in \mathbf{RS^N}$ if and only if $S$ contains neither an isomorphic copy of $U_1$ nor of $\{a,b\}^\ell$.
3. Show that $S \in \mathbf{G^N}$ if and only if $S$ contains a unique idempotent, or equivalently $S$ does not contain an isomorphic copy of $U_1$, $\overline{\mathbf{2}}$, or $\{a,b\}^\ell$.

**Exercise 4.6.8.**

1. Show that $S$ belongs to $\mathbf{LG}$ if and only if it has a unique regular $\mathscr{J}$-class.
2. Show that $S$ belongs to $\mathbf{LS^N}$ if and only if it has unique regular $\mathscr{L}$-class.
3. Show that $S$ belongs to $\mathbf{RS^N}$ if and only if it has unique regular $\mathscr{R}$-class.
4. Show that $S$ belongs to $\mathbf{G^N}$ if and only if it has unique regular $\mathscr{H}$-class.

Observe that $\mathbf{LG} \cap \mathbf{A} = \mathbf{L1}$, $\mathbf{RS^N} \cap \mathbf{A} = \mathbf{D}$, $\mathbf{LS^N} \cap \mathbf{A} = \mathbf{K}$ and $\mathbf{G^N} \cap \mathbf{A} = \mathbf{N}$ where $\mathbf{K}$ is the opposite pseudovariety of $\mathbf{D}$, i.e., $\mathbf{K} = \mathbf{D}^{op}$.

**Exercise 4.6.9.** Prove that $\mathbf{LS^N} = \mathbf{K} \vee \mathbf{G}$, $\mathbf{RS^N} = \mathbf{D} \vee \mathbf{G}$ and $\mathbf{G^N} = \mathbf{N} \vee \mathbf{G}$.

Before characterizing $\mathscr{K}'$-morphisms, we first prove a crucial lemma from [171] on lifting regular $\mathscr{K}$-classes. It is fundamental to many diverse aspects of Finite Semigroup Theory, from complexity to representation theory. Lemma 3.1.14 already considered the case of $\mathscr{J}$-classes in the context of compact semigroups.

**Lemma 4.6.10 (Lifting regular $\mathscr{K}$-classes).** *Let $\varphi \colon S \twoheadrightarrow T$ be an onto semigroup homomorphism and let $\mathscr{K}$ be one of Green's relations $\mathscr{J}$, $\mathscr{R}$ or $\mathscr{L}$. Let $K$ be a $\mathscr{K}$-class of $T$. Then:*

1. *$K\varphi^{-1}$ is a union of $\mathscr{K}$-classes of $S$;*
2. *if $K'$ is a $\leq_{\mathscr{K}}$-minimal $\mathscr{K}$-class of $K\varphi^{-1}$, then $K'\varphi = K$;*
3. *if $K$ is regular, then any $\leq_{\mathscr{K}}$-minimal $\mathscr{K}$-class of $K\varphi^{-1}$ is regular;*
4. *if $\mathscr{K} = \mathscr{J}$ and $K$ is regular, then there is a unique $\leq_{\mathscr{J}}$-minimal $\mathscr{J}$-class in $K\varphi^{-1}$.*

*If $H$ is a regular $\mathscr{H}$-class of $T$, then there is a regular $\mathscr{H}$-class $H'$ of $S$ with $H'\varphi = H$. In fact, if $J$ is a regular $\mathscr{J}$-class of $T$ and $J'$ is the unique $\leq_{\mathscr{J}}$-minimal $\mathscr{J}$-class of $S$ mapping into $J$, then **every** $\mathscr{H}$-class of $J'$ maps **onto** the $\mathscr{H}$-class of $J$ containing its image.*

*Proof.* We just prove 1–4 for $\mathscr{K} = \mathscr{J}$, as the other cases are entirely analogous. Clearly, $s \mathrel{\mathscr{J}} t$ implies $s\varphi \mathrel{\mathscr{J}} t\varphi$ so $J\varphi^{-1}$ is a union of $\mathscr{J}$-classes of $S$. Suppose that $J'$ is $\leq_{\mathscr{J}}$-minimal among these $\mathscr{J}$-classes. Let $s \in J'$ and let $t \in J$. Then $s\varphi \mathrel{\mathscr{J}} t$, so there exists $u,v \in T^I$ with $us\varphi v = t$. Choose $u',v' \in S^I$ with $u'\varphi = u$, $v'\varphi = v$. Then $(u'sv')\varphi = t$ and $u'sv' \leq_{\mathscr{J}} s$. By minimality we must have $u'sv' \in J'$ and so $J'\varphi = J$. Suppose now that $J$ is regular and let $J'$ and $J''$ be $\leq_{\mathscr{J}}$-minimal in $J\varphi^{-1}$. Let $e \in J$ be an idempotent and let $s_1 \in e\varphi^{-1} \cap J'$ and $s_2 \in e\varphi^{-1} \cap J''$. Then $(s_1 s_2)\varphi = e$ and $s_1 s_2 \leq_{\mathscr{J}} J', J''$. Thus $J' = J''$ by minimality and $J'$ is regular (as $s_1 s_2 \in J'J'$ if we take $J' = J''$).

The statement about regular $\mathscr{H}$-classes follows immediately from the final statement, which we now prove. First let $G'$ be a maximal subgroup of $J'$ with idempotent $e$. Let $G$ be the $\mathscr{H}$-class of $J$ containing the image of $G'$. If $g \in G$ and $s \in S$ is any preimage of $g$, then $(ese)\varphi = g$ and so, by minimality, $ese \in J'$. Stability shows in fact $ese \in G'$. Thus $G'\varphi = G$. Now let $H'$ be an $\mathscr{H}$-class $\mathscr{L}$-equivalent to $G'$ and let $H$ be the $\mathscr{H}$-class of $J$ into which $H'$ maps. Fix $h \in H'$; so $h\varphi \in H$. Green's Lemma implies that

$$H = h\varphi G = h\varphi G'\varphi = (hG')\varphi = H'\varphi.$$

Because every $\mathscr{H}$-class in a regular $\mathscr{J}$-class is $\mathscr{L}$-equivalent to a maximal subgroup, this completes the proof.                                    $\square$

There is an example in [171, Chapter 7, Remark 2.11] to show that when considering $\mathscr{H}$, one really needs to restrict to regular $\mathscr{H}$-classes.

**Proposition 4.6.11.** *Let $\varphi \colon S \twoheadrightarrow T$ be an onto homomorphism. Then the following are equivalent:*

1. *$\varphi \in (\mathbb{LG}, \mathbb{LG})$;*
2. *$\varphi \in (\mathbb{LG}, \mathbf{1})$;*
3. *$\varphi$ is injective on copies of $U_1$ (or equivalently on idempotents $e \leq f$);*
4. *If $J$ is a regular $\mathscr{J}$-class of $T$, then there is a unique regular $\mathscr{J}$-class $J'$ of $S$ with $J'\varphi \subseteq J$. Moreover, $J'\varphi = J$ and $J'' <_{\mathscr{J}} J'$ implies $J''\varphi \cap J = \emptyset$;*
5. *$\varphi$ is a $\mathscr{J}'$-morphism.*

*Moreover, if $\varphi$ is a $\mathscr{J}'$-morphism and $J$ is a regular $\mathscr{J}$-class of $S$, then $J\varphi$ is a regular $\mathscr{J}$-class of $T$.*

*Proof.* Clearly, 1 implies 2. Suppose 2 holds and that $e \leq f$ with $e\varphi = f\varphi$. As $e\varphi$ is an idempotent, we obtain $\{e, f\} \in \mathbb{LG}$ and so $e = f$. For 3 implies 4, let $J'$ be the unique $\leq_{\mathscr{J}}$-minimal $\mathscr{J}$-class in $J\varphi^{-1}$. It is regular and $J'\varphi = J$ by Lemma 4.6.10. Suppose $J'' >_{\mathscr{J}} J'$ is another regular $\mathscr{J}$-class in $J\varphi^{-1}$. Let $e \in J''$ be an idempotent and let $s \in J'$ be an element such that $e\varphi = s\varphi$. Then $(ese)^{\omega}\varphi = e\varphi \in J$ and so, by minimality, $f = (ese)^{\omega} \in J'$, and hence is not $e$. Clearly, $f = f^2$, $ef = fe = f$ and $e\varphi = f\varphi$, contradicting 3. Suppose now that 4 holds and that $s, t$ are regular with $s\varphi = t\varphi$. Then $s\varphi = t\varphi$ belongs to a regular $\mathscr{J}$-class and hence $s \mathrel{\mathscr{J}} t$ by 4, establishing that $\varphi$ is a $\mathscr{J}'$-morphism. Finally, to see that 5 implies 1, suppose that $U \leq T$ with $U \in \mathbb{LG}$. By Proposition 4.6.5, it suffices to show that $U\varphi^{-1}$ contains no copy of $U_1$. So suppose $e, f \in E(U\varphi^{-1})$ with $e \leq f$. Then $e\varphi \leq f\varphi$ and so $e\varphi = f\varphi$, as $U \in \mathbb{LG}$. Hence $e \mathrel{\mathscr{J}} f$ in $S$ by 5. But then $e = fef \in H_f$, which implies $e = f$.

Clearly, if $J$ is a regular $\mathscr{J}$-class of $S$, then $J\varphi$ is contained in a regular $\mathscr{J}$-class $J'$ of $T$. So it must be that $J$ is the unique regular $\mathscr{J}$-class in $J'\varphi^{-1}$ and so maps onto $J'$ by 4. This completes the proof.              $\square$

**Exercise 4.6.12.** Show that an onto homomorphism $\varphi\colon S \twoheadrightarrow T$ is injective on copies of $U_1$ and $\overline{\mathbf{2}}$ if and only if it is injective on idempotents $e \leq_{\mathscr{R}} f$.

We record the analogous facts for $\mathscr{R}'$- and $\mathscr{L}'$-morphisms.

**Proposition 4.6.13.** *Let* $\varphi\colon S \twoheadrightarrow T$ *be an onto homomorphism. Then the following are equivalent:*

1. $\varphi \in (\mathbf{LS^N}, \mathbf{LS^N})$;
2. $\varphi \in (\mathbf{LS^N}, \mathbf{1})$;
3. $\varphi$ *is injective on copies of* $U_1$ *and* $\overline{\mathbf{2}}$ *(or equivalently on idempotents* $e \leq_{\mathscr{R}} f$*);*
4. *If* $L$ *is a regular* $\mathscr{L}$*-class of* $T$, *then there is a unique regular* $\mathscr{L}$*-class* $L'$ *of* $S$ *with* $L'\varphi \subseteq L$. *Moreover,* $L'\varphi = L$ *and* $L'' <_{\mathscr{L}} L'$ *implies* $L''\varphi \cap L = \emptyset$;
5. $\varphi$ *is an* $\mathscr{L}'$*-morphism.*

*Moreover, if* $\varphi$ *is an* $\mathscr{L}'$*-morphism and* $L$ *is a regular* $\mathscr{L}$*-class of* $S$, *then* $L\varphi$ *is a regular* $\mathscr{L}$*-class of* $T$. *A dual result holds for* $\mathscr{R}$.

*Proof.* Clearly 1 implies 2 implies 3, because $\mathbf{LS^N}$ does not contain $U_1$ or $\overline{\mathbf{2}}$ (and using Exercise 4.6.12). For 3 implies 4, let $L'$ be an $\leq_{\mathscr{L}}$-minimal $\mathscr{L}$-class in $L\varphi^{-1}$. It is regular and $L'\varphi = L$ by Lemma 4.6.10. Suppose $L''$ is another regular $\mathscr{L}$-class in $L\varphi^{-1}$. By Proposition 4.6.11(3), we must have that $L'' \not{\mathscr{J}} L'$. Let $e \in L''$ be an idempotent and let $s \in L'$ be an element such that $e\varphi = s\varphi$. Then $(es)^{\omega}\varphi = e\varphi \in L$ and so, by minimality, $f = (es)^{\omega} \in L'$, and in particular is not $e$. Also $ef = f$ and hence, because $e \not{\mathscr{J}} f$, stability yields $e \mathscr{R} f$. Thus $fe = e$ and so $\{e, f\} \cong \overline{\mathbf{2}}$. We conclude that no such $L''$ exists. Suppose now that 4 holds and that $s, t$ are regular with $s\varphi = t\varphi$. Then $s\varphi = t\varphi$ belongs to a regular $\mathscr{L}$-class and hence $s \mathscr{L} t$ by 4, so $\varphi$ is an $\mathscr{L}'$-morphism. Finally, to see that 5 implies 1, suppose that $U \leq T$ with $U \in \mathbf{LS^N}$. By Exercise 4.6.7, it suffices to show that $U\varphi^{-1}$ does not contain a copy of $U_1$ or $\overline{\mathbf{2}}$. Because $\mathscr{L}'$-morphisms are $\mathscr{J}'$-morphisms, it suffices to consider $\overline{\mathbf{2}}$ by Proposition 4.6.11. So suppose $e, f \in E(U\varphi^{-1})$ with $ef = f$, $fe = e$. Then $\{e\varphi, f\varphi\}$ is a right zero semigroup, whence $e\varphi = f\varphi$, as $U \in \mathbf{LS^N}$. Therefore $e \mathscr{L} f$ in $S$ by 5. But then $e = fe = f$. The last assertion of the proposition is proved in the same way as the analogous statement in Proposition 4.6.11. $\quad\square$

The following corollary is important for the complexity theory of finite semigroups.

**Corollary 4.6.14.** *Let* $S \rtimes T$ *be a semidirect product with* $S \in \mathbf{LS^N}$. *Then the projection* $\pi\colon S \rtimes T \to T$ *is an* $\mathscr{L}'$*-morphism. In particular,* $\mathbf{G} * \mathbf{V} \leq \mathbf{LS^N} \circledm \mathbf{V}$ *for any pseudovariety* $\mathbf{V}$.

*Proof.* By Proposition 4.6.13, we just need to check that, for each idempotent $f \in T$, $f\pi^{-1}$ does not contain a copy of $U_1$ or $\overline{\mathbf{2}}$. But this is immediate from Lemma 4.1.31 and the fact that $S$ does not contain copies of these semigroups. Because $\mathbf{G} \leq \mathbf{LS^N}$, the second statement is immediate. $\quad\square$

We shall prove in Section 4.11 a result of Karnofsky and the first author [156] giving sufficient conditions for the reverse inequality to hold in the above corollary.

Finally, we consider the case of $\mathscr{H}'$-morphisms. Notice that $\varphi$ is an $\mathscr{H}'$-morphism if and only if it is both an $\mathscr{L}'$-morphism and an $\mathscr{R}'$-morphism. Hence most of the next proposition is a straightforward consequence of the previous one and so we omit the proof.

**Proposition 4.6.15.** *Let* $\varphi \colon S \twoheadrightarrow T$ *be an onto homomorphism. Then the following are equivalent:*

1. $\varphi \in (\mathbf{G^N}, \mathbf{G^N})$;
2. $\varphi \in (\mathbf{G^N}, \mathbf{1})$;
3. $\varphi$ *is injective on copies of* $U_1$, $\overline{\mathbf{2}}$ *and* $\{a, b\}^{\ell}$ *(equivalently on idempotents* $e, f$*);*
4. *If* $H$ *is a regular* $\mathscr{H}$*-class of* $T$, *then there is a unique regular* $\mathscr{H}$*-class* $H'$ *of* $S$ *with* $H'\varphi \subseteq H$. *Moreover,* $H'\varphi = H$ *and* $H'' <_{\mathscr{H}} H'$ *implies* $H''\varphi \cap H = \emptyset$;
5. $\varphi$ *is an* $\mathscr{H}'$*-morphism.*

*Moreover, if* $\varphi$ *is an* $\mathscr{H}'$*-morphism, then for a regular* $\mathscr{H}$*-class* $H$ *of* $S$, $H\varphi$ *is a regular* $\mathscr{H}$*-class of* $T$.

**Exercise 4.6.16.** Prove Proposition 4.6.15.

**Exercise 4.6.17.** Use Tilson's Lemma from Chapter 2 (or direct calculation) and Propositions 4.6.11–4.6.15 to deduce:

1. $(\mathbb{L}\mathbf{G}, \mathbb{L}\mathbf{G}) = (\mathbb{L}\mathbf{G}, \mathbf{1})$;
2. $(\mathbf{LS^N}, \mathbf{LS^N}) = (\mathbf{LS^N}, \mathbf{1})$;
3. $(\mathbf{RS^N}, \mathbf{RS^N}) = (\mathbf{RS^N}, \mathbf{1})$;
4. $(\mathbf{G^N}, \mathbf{G^N}) = (\mathbf{G^N}, \mathbf{1})$.

We next wish to characterize aperiodic $\mathscr{K}'$-morphisms. Let us begin with a straightforward lemma.

**Lemma 4.6.18.** *Let* $\mathbf{V}_1, \mathbf{V}_2$ *be pseudovarieties such that* $(\mathbf{V}_i, \mathbf{V}_i) = (\mathbf{V}_i, \mathbf{1})$. *Then* $(\mathbf{V}_1 \cap \mathbf{V}_2, \mathbf{V}_1 \cap \mathbf{V}_2) = (\mathbf{V}_1 \cap \mathbf{V}_2, \mathbf{1}) = (\mathbf{V}_1, \mathbf{V}_1) \cap (\mathbf{V}_2, \mathbf{V}_2)$.

*Proof.* A routine calculation yields

$$
\begin{aligned}
(\mathbf{V}_1 \cap \mathbf{V}_2, \mathbf{1}) &= (\mathbf{V}_1, \mathbf{1}) \cap (\mathbf{V}_2, \mathbf{1}) \\
&= (\mathbf{V}_1, \mathbf{V}_1) \cap (\mathbf{V}_2, \mathbf{V}_2) \\
&\subseteq (\mathbf{V}_1 \cap \mathbf{V}_2, \mathbf{V}_1 \cap \mathbf{V}_2) \\
&\subseteq (\mathbf{V}_1 \cap \mathbf{V}_2, \mathbf{1}),
\end{aligned}
$$

establishing the desired equalities. $\qquad\square$

As a consequence of the Aperiodicity Lemma (Lemma 4.4.4), Lemma 4.6.18 and Exercise 4.6.17, we obtain:

**Proposition 4.6.19.** *Let* $\varphi\colon S \to T$ *be a homomorphism.*

1. $\varphi$ *is an aperiodic* $\mathscr{J}'$*-morphism if and only if* $\varphi \in (\mathbb{L}1, \mathbb{L}1)$, *if and only if* $\varphi \in (\mathbb{L}1, \mathbf{1})$.
2. $\varphi$ *is an aperiodic* $\mathscr{L}'$*-morphism if and only if* $\varphi \in (\mathbf{K}, \mathbf{K})$, *if and only if* $\varphi \in (\mathbf{K}, \mathbf{1})$.
3. $\varphi$ *is an aperiodic* $\mathscr{R}'$*-morphism if and only if* $\varphi \in (\mathbf{D}, \mathbf{D})$, *if and only if* $\varphi \in (\mathbf{D}, \mathbf{1})$.
4. $\varphi$ *is an aperiodic* $\mathscr{H}'$*-morphism if and only if* $\varphi \in (\mathbf{N}, \mathbf{N})$, *if and only if* $\varphi \in (\mathbf{N}, \mathbf{1})$.

**Exercise 4.6.20.** Show that $\varphi\colon S \to T$ is an aperiodic $\mathscr{H}'$-morphism if and only if it is injective on the subset $\mathrm{Reg}(S)$ of regular elements of $S$.

### 4.6.2 Maximal $\mathscr{H}'$-congruences

In this subsection we shall construct the maximal $\mathscr{H}'$ and aperiodic $\mathscr{H}'$-congruences. First we define right/left mapping, right/left letter mapping and generalized group mapping semigroups [171]. Recall that an ideal of a semigroup is called 0-*minimal* if it is a minimal non-zero ideal; in particular, in a semigroup without zero we count the minimal ideal as a 0-minimal ideal. In the sequel we shall often treat things as if the semigroup in question had a zero. The obvious modifications for when this is not the case are left to the reader.

**Definition 4.6.21 (Mapping semigroups).** *A semigroup* $S$ *is called right (left) mapping if it contains a 0-minimal ideal* $I$ *such that* $S$ *acts faithfully on the right (left) of* $I$. *If* $S$ *is both right and left mapping, then it is called generalized group mapping.*

It is a result of Rhodes [270, 297] that every finite irreducible matrix semigroup is generalized group mapping and so the notion is important for semigroup representation theory [18].

**Proposition 4.6.22.** *If* $S$ *is a right or left mapping semigroup, then it has a unique 0-minimal ideal, which is necessarily regular.*

*Proof.* We just handle the case of right mapping. Let $I$ be the distinguished 0-minimal ideal from the definition of right mapping and $J$ any non-zero ideal. Then $IJ \neq 0$ by faithfulness of the action. Thus $IJ = I$ and so $I \leq J$, yielding both uniqueness and $I^2 = I$ (taking $I = J$), from which regularity follows.    $\square$

If $S$ is a right mapping semigroup with 0-minimal ideal $I$, then the regular $\mathscr{J}$-class $J = I \setminus \{0\}$ is called the *distinguished* $\mathscr{J}$-class of $S$. Notice that $S$ acts faithfully by partial transformations on the right of $J$. If $R$ is an $\mathscr{R}$-class of $J$, then $(R, S)$ is a faithful right partial transformation semigroup that, by Green's Lemma (Lemma A.3.1), depends only on $J$ up to isomorphism. Dual remarks hold for left mapping semigroups. Generalized group mapping semigroups are more subtle to understand, as is often the case for left-right dual notions. The following lemma should provide some insight.

**Lemma 4.6.23.** *Let $S$ be a generalized group mapping semigroup with distinguished $\mathscr{J}$-class $J$. Then, for $s, t \in S$, one has that $s = t$ if and only if $xsy = xty$ for all $x, y \in J$.*

*Proof.* Because $S$ is right mapping, $s = t$ if and only if $xs = xt$ for all $x \in J$. Because $S$ is left mapping, $xs = xt$ if and only if $xsy = xty$ all $y \in J$.    □

**Exercise 4.6.24.** Prove the converse of Lemma 4.6.23, that is, suppose $S$ is semigroup with a 0-minimal ideal $I$ such that, for $s, t \in S$, the equality $s = t$ holds if and only if $xsy = xty$, all $x, y \in I$. Show that $S$ is generalized group mapping with distinguished $\mathscr{J}$-class $I \setminus 0$.

Lemma 4.6.10 leads us to the following important notion, which also plays a crucial role in semigroup representation theory [18, 68, 270, 297].

**Definition 4.6.25 (Apex).** *Let $\varphi\colon S \twoheadrightarrow T$ be an onto homomorphism and suppose that $T$ is a right or left mapping semigroup with distinguished $\mathscr{J}$-class $J$. Then the unique $\leq_{\mathscr{J}}$-minimal $\mathscr{J}$-class of $S$ contained in $J\varphi^{-1}$ is termed the apex of $\varphi$, denoted $\mathrm{Apx}(\varphi)$. The $\mathscr{J}$-class $\mathrm{Apx}(\varphi)$ is regular and maps onto $J$.*

Following [171], we classify mapping semigroups according to whether or not their 0-minimal ideal is aperiodic.

**Definition 4.6.26 (Letter mapping and group mapping semigroups).** *A semigroup is called a right letter mapping semigroup if it is right mapping with an aperiodic 0-minimal ideal. Left letter mapping is defined dually. A generalized group mapping semigroup whose 0-minimal ideal is not aperiodic is called group mapping.*

*Remark 4.6.27.* Our terminology differs slightly from that of [171] in that we do not consider the trivial semigroup to be group mapping.

We now construct canonical right/left mapping semigroups associated to a semigroup $S$ and one of its regular $\mathscr{J}$-classes $J$. They are universal with respect to having apex $J$.

**Definition 4.6.28 (Schützenberger representation).** *Let $S$ be a semi-group and $J$ a $\mathscr{J}$-class. Then $S$ acts by partial transformations on the right of $J$. The resulting faithful right partial transformation semigroup is denoted $(J, \mathrm{RM}_J(S))$. We use $\rho_J \colon S \to \mathrm{RM}_J(S)$ for the associated quotient homomorphism. Sometimes $\rho_J$ is called the right Schützenberger representation of $S$ on $J$. We use $\mathrm{LM}_J(S)$ and $\lambda_J$ for the dual notions.*

Mostly we will be interested in the case where $J$ is a regular $\mathscr{J}$-class, but we shall need the more general case in Chapter 9. The associated congruence to $\rho_J$ is given by $s \equiv t$ if and only if, for all $x \in J$, $xs \in J \iff xt \in J$, in which case $xs = xt$. Equivalently, thanks to Green's Lemma and stability, it can be defined by choosing an $\mathscr{R}$-class $R$ of $J$ and then setting $s \equiv t$ if, for all $r \in R$, $rs \in R \iff rt \in R$, in which case $rs = rt$. Our next proposition establishes the universal property of $\mathrm{RM}_J(S)$.

**Proposition 4.6.29.** *Let $S$ be a semigroup and $J$ a regular $\mathscr{J}$-class. Then $\mathrm{RM}_J(S)$ is a right mapping semigroup with distinguished $\mathscr{J}$-class $J\rho_J$. Moreover, $\mathrm{Apx}(\rho_J) = J$ and if $\varphi \colon S \twoheadrightarrow T$ is any onto homomorphism with $T$ right mapping and with apex $J$, then $\varphi$ factors uniquely through $\rho_J$. A dual result holds for $\mathrm{LM}_J(S)$.*

*Proof.* First of all, notice that if $s \not\geq_{\mathscr{J}} J$, then $s\rho_J = 0$ and thus $J\rho_J \cup \{0\}$ is a 0-minimal ideal of $\mathrm{RM}_J(S)$ and $\mathrm{Apx}(\rho_J) = J$ (once we prove that $\mathrm{RM}_J(S)$ is right mapping). Suppose that, for all $j \in J$, the equality $j\rho_J s\rho_J = j\rho_J t\rho_J$ holds. Choose an idempotent $e$ in the $\mathscr{R}$-class of $j$. Then, as $ej = j$, we obtain $j(s\rho_J) = e(j\rho_J)(s\rho_J) = e(j\rho_J)(t\rho_J) = j(t\rho_J)$. Thus $s\rho_J = t\rho_J$. We conclude $\mathrm{RM}_J(S)$ is right mapping.

Suppose now that $\varphi \colon S \twoheadrightarrow T$ is an onto homomorphism with $T$ right mapping and with apex $J$. We must show that if $s\rho_J = t\rho_J$, then $s\varphi = t\varphi$. Because $T$ is right mapping and $J$ is the apex, it suffices to show that, for all $j \in J$, $j\varphi s\varphi = j\varphi t\varphi$ or equivalently that $(js)\varphi = (jt)\varphi$. But as $s\rho_J = t\rho_J$ either both $js, jt \notin J$, which implies $(js)\varphi = 0 = (jt)\varphi$, or $js = jt$ and so $(js)\varphi = (jt)\varphi$. We conclude that $\ker \rho_J \subseteq \ker \varphi$, as required.  $\square$

Our next construction provides the universal generalized group mapping image with a given apex.

**Definition 4.6.30.** *Let $S$ be a semigroup and $J$ a regular $\mathscr{J}$-class. Define a congruence $\equiv_J$ by $s \equiv_J t$ if, for all $x, y \in J$, $xsy \in J \iff xty \in J$, in which case $xsy = xty$. The quotient is denoted $\mathrm{GGM}_J(S)$ and the quotient map is denoted $\gamma_J \colon S \to \mathrm{GGM}_J(S)$.*

We now turn to the universal property of $\mathrm{GGM}_J(S)$.

**Proposition 4.6.31.** *Let $S$ be a semigroup and $J$ a regular $\mathscr{J}$-class. Then $\mathrm{GGM}_J(S)$ is generalized group mapping with distinguished $\mathscr{J}$-class $J\gamma_J$. Moreover, $\mathrm{Apx}(\gamma_J) = J$ and every onto homomorphism $\varphi \colon S \twoheadrightarrow T$ with $T$ generalized group mapping and $\mathrm{Apx}(\varphi) = J$ factors uniquely through $\gamma_J$.*

*Proof.* To see that $\equiv_J$ is a congruence, suppose that $s \equiv_J t$. Let $z \in S$ and $x, y \in J$. Then either $xz <_{\mathscr{J}} J$, in which case $xzsy, xzty \notin J$ or $xz \in J$ and so $(xz)sy \in J \iff (xz)ty \in J$, in which case $(xz)sy = (xz)ty$. Therefore, $zs \equiv_J zt$. The proof that $\equiv_J$ is a right congruence is dual.

Again, it is easy to see that if $s \not\geq_{\mathscr{J}} J$, then $s\gamma_J = 0$, so $J\gamma_J \cup \{0\}$ is a 0-minimal ideal of $\mathsf{GGM}_J(S)$ (and so $J$ will be the apex). We verify that $\mathsf{GGM}_J(S)$ is right mapping, the case of left mapping being dual. Suppose that, for all $j \in J$, $j\gamma_J s\gamma_J = j\gamma_J t\gamma_J$ and let $x, y \in J$. Choose an idempotent $e \in J$ with $ex = x$. Then $xsy = e(xs)y$ and $xty = e(xt)y$ and so using $(xs)\gamma_J = (xt)\gamma_J$, we conclude $xsy \in J$ if and only if $xty \in J$, in which case $xsy = xty$. Thus $s\gamma_J = t\gamma_J$ and $\mathsf{GGM}_J(S)$ is right mapping.

For the final statement, we show that $\ker \gamma_J \subseteq \ker \varphi$. Suppose $s\gamma_J = t\gamma_J$. Because $T$ is generalized group mapping, by Lemma 4.6.23, we have $s\varphi = t\varphi$ if and only if $x\varphi s\varphi y\varphi = x\varphi t\varphi y\varphi$ for all $x, y \in J$; equivalently, this occurs if and only if $(xsy)\varphi = (xty)\varphi$ all $x, y \in J$. Now if $xsy \notin J$, then $xty \notin J$ and $(xsy)\varphi = 0 = (xty)\varphi$, as $J$ is the apex of $\varphi$. Otherwise, $xsy = xty$ and so clearly $(xsy)\varphi = (xty)\varphi$. We conclude that $\varphi$ factors through $\gamma_J$. ☐

**Exercise 4.6.32.** Prove $\mathsf{GGM}_J(S) = \mathsf{LM}_{J\rho_J}(\mathsf{RM}_J(S)) = \mathsf{RM}_{J\lambda_J}(\mathsf{LM}_J(S))$.

**Exercise 4.6.33.** Let $J$ be a $\mathscr{J}$-class of a finite semigroup $S$ and suppose $x \in J$ and $s \in S$. Show that if $xs \in J$, then $L_x s = L_{xs}$.

Next we consider the letter mapping variants of these constructions.

**Definition 4.6.34.** *Let $S$ be a semigroup and $J$ a $\mathscr{J}$-class. Denote by $L(J)$ the set of $\mathscr{L}$-classes of $S$ in $J$. Exercise 4.6.33 implies that if $L \in L(J)$, then for $s \in S$, either $Ls \cap J = \emptyset$ or $Ls \in L(J)$. More precisely, if $x \in J$, then we can define an action by partial transformations of $S$ on $L(J)$ by*

$$L_x s = \begin{cases} L_{xs} & xs \in J \\ \text{undefined} & \text{else.} \end{cases} \tag{4.7}$$

*The resulting faithful right partial transformation semigroup is denoted by $(L(J), \mathsf{RLM}_J(S))$ and the quotient map by $\mu_J^R \colon S \to \mathsf{RLM}_J(S)$. The morphism $\mu_J^L$ and the left partial transformation semigroup $(\mathsf{LLM}_J(S), R(J))$ are defined dually.*

Again, we are mostly concerned with the case when $J$ is regular. The origin of the terminology left/right letter mapping and group mapping is that if one represents $J^0$ as a Rees matrix semigroup $\mathscr{M}^0(G, A, B, C)$ and elements of $J$ as triples $(a, g, b)$ with $a \in A$, $g \in G$ and $b \in B$, then $A$ is in bijection with the set of $\mathscr{R}$-classes and $B$ is in bijection with the set of $\mathscr{L}$-classes of $J$. One calls $A$ the "left letters" and $B$ the "right letters" of $J$, whereas $G$ is called the "group coordinate." It is shown in [171] that the actions of $S$ on the left and right of $J^0$ are determined by the natural map of $S$ into the direct product

$\mathsf{LLM}_J(S) \times \mathsf{GGM}_J(S) \times \mathsf{RLM}_J(S)$. Here $\mathsf{LLM}_J(S)$ and $\mathsf{RLM}_J(S)$ keep track of the actions of $S$ on the left and right letters, respectively, whereas $\mathsf{GGM}_J(S)$, intuitively speaking, keeps track of the action on the group coordinate. Unfortunately, $\mathsf{GGM}_J(S)$ also keeps track of whether an element falls out of the $\mathscr{J}$-class. If this were not the case, complexity would be easy to decide.

**Proposition 4.6.35.** *Let $S$ be a semigroup with $J$ a regular $\mathscr{J}$-class. Then $\mathsf{RLM}_J(S)$ is a right letter mapping semigroup with distinguished $\mathscr{J}$-class $J\mu_J^R$. The apex of $\mu_J^R$ is $J$ and if $\varphi\colon S \to T$ is any onto morphism with $T$ right letter mapping and with apex $J$, then $\varphi$ factors uniquely through $\mu_J^R$. The dual result holds for $\mathsf{LLM}_J(S)$.*

*Proof.* Yet again, if $s \not\geq_{\mathscr{J}} J$, then $s\mu_J^R = 0$ and thus $J\mu_J^R \cup \{0\}$ is a 0-minimal ideal of $\mathsf{RLM}_J(S)$ and $\mathrm{Apx}(\mu_J^R) = J$ (once we prove that $\mathsf{RLM}_J(S)$ is right letter mapping). To see that $\mathsf{RLM}_J(S)$ is right mapping, suppose that for all $j \in J$, $j\mu_J^R s\mu_J^R = j\mu_J^R t\mu_J^R$. Choose an idempotent $e$ in the $\mathscr{R}$-class of $j$. Then $L_j s\mu_J^R = L_{ej} s\mu_J^R = L_e j\mu_J^R s\mu_J^R = L_e j\mu_J^R t\mu_J^R = L_j t\mu_J^R$, establishing $s\mu_J^R = t\mu_J^R$. We conclude $\mathsf{RLM}_J(S)$ is right mapping. To see $J\mu_J^R$ is aperiodic, suppose $x \in J$. If $x^2 \notin J$, then $(x\mu_J^R)^2 = 0$. So assume $x^2 \in J$ and hence $x^2 \mathscr{H} x$. Then, for $j \in J$, we have $jx^2 \mathscr{R} jx$ and so $jx^2 \in J$ if and only if $jx \in J$. In the case $jx, jx^2 \in J$, it follows from stability $jx \mathscr{L} x \mathscr{L} x^2 \mathscr{L} jx^2$. Thus $L_j x\mu_J^R = L_j (x\mu_J^R)^2$ and so $J\mu_J^R$ is aperiodic.

Suppose now that $\varphi\colon S \to T$ is an onto homomorphism with $T$ right letter mapping and with apex $J$. We must show that if $s\mu_J^R = t\mu_J^R$, then $s\varphi = t\varphi$. Because $T$ is right mapping and $J$ is the apex, it suffices to show that, for all $j \in J$, $j\varphi s\varphi = j\varphi t\varphi$ or equivalently that $(js)\varphi = (jt)\varphi$. Because $s\mu_J^R = t\mu_J^R$, either both $js \notin J$, $jt \notin J$, in which case $(js)\varphi = 0 = (jt)\varphi$, or $js, jt \in J$. In the latter case $js \mathscr{R} j \mathscr{R} jt$ and hence $(js)\varphi \mathscr{R} (jt)\varphi$. Because $J\varphi$ is aperiodic, it suffices to show that $(js)\varphi \mathscr{L} (jt)\varphi$. But as $s\mu_J^R = t\mu_J^R$, we have $L_{js} = L_{jt}$ and so $(js)\varphi \mathscr{L} (jt)\varphi$. We conclude that $\ker \mu_J^R \subseteq \ker \varphi$, as required.    □

Finally, we consider generalized group mapping semigroups with aperiodic 0-minimal ideal, i.e., that are not group mapping. This construction does not appear in [171], but it can be found in [368].

**Definition 4.6.36.** *Let $S$ be a semigroup and $J$ a regular $\mathscr{J}$-class. Define a congruence $\sim_J$ by $s \sim_J t$ if $xsy \in J \iff xty \in J$, all $x, y \in J$. The quotient is denoted $\mathsf{AGGM}_J(S)$ and the quotient map by $\Gamma_J\colon S \to \mathsf{AGGM}_J(S)$.*

**Proposition 4.6.37.** *Let $S$ be a semigroup and $J$ a regular $\mathscr{J}$-class. Then $\mathsf{AGGM}_J(S)$ is generalized group mapping with aperiodic distinguished $\mathscr{J}$-class $J\Gamma_J$. Moreover, $\mathrm{Apx}(\Gamma_J) = J$ and every onto homomorphism $\varphi\colon S \to T$ with $T$ generalized group mapping, but not group mapping, and with $\mathrm{Apx}(\varphi) = J$, factors uniquely as $\Gamma_J \tau$ where $\tau$ is an isomorphism.*

*Proof.* To see that $\sim_J$ is a congruence, suppose that $s \sim_J t$. Let $z \in S$ and $x, y \in J$. Then either $xz <_{\mathscr{J}} J$, in which case $xzsy, xzty \notin J$ or $xz \in J$ and so $(xz)sy \in J \iff (xz)ty \in J$. Thus $zs \sim_J zt$. The proof that $\sim_J$ is a right congruence is dual.

Clearly, if $s \not\geq_{\mathscr{J}} J$, then $s\Gamma_J = 0$, so $J\Gamma_J \cup \{0\}$ is the unique 0-minimal ideal of $\mathsf{AGGM}_J(S)$ (and so $J$ will be the apex). We verify that $\mathsf{AGGM}_J(S)$ is right mapping, left mapping being dual. Suppose that, for all $j \in J$, $j\Gamma_J s\Gamma_J = j\Gamma_J t\Gamma_J$ and let $x, y \in J$. Choose an idempotent $e \in J$ with $ex = x$. Then $xsy = e(xs)y$ and $xty = e(xt)y$ and so using $(xs)\Gamma_J = (xt)\Gamma_J$ all $x \in J$, we conclude $xsy \in J \iff xty \in J$ and thus $s\Gamma_J = t\Gamma_J$. Therefore, $\mathsf{AGGM}_J(S)$ is right mapping. To see that $J\Gamma_J$ is a aperiodic, suppose $j \in J$. If $j^2 \notin J$, then $(j\Gamma_J)^2 = 0$. If $j^2 \in J$ and $x, y \in J$, then standard results from the theory of $\mathscr{J}$-classes in a finite semigroup imply $xjy \in J$ if and only if $L_x \cap R_j$ and $L_j \cap R_y$ contain idempotents; see for instance Theorem A.3.4(7)(8). Because $j \mathscr{H} j^2$ by stability, we see that this occurs if and only if $L_x \cap R_{j^2}$ and $L_{j^2} \cap R_y$ contain idempotents, if and only if $xj^2y \in J$. Thus $j\Gamma_J = (j\Gamma_J)^2$, establishing that $J\Gamma_J$ is aperiodic.

For the final statement, we show that $\ker \Gamma_J = \ker \varphi$. Suppose first $s\varphi = t\varphi$ and let $x, y \in J$. Then $(xsy)\varphi = (xty)\varphi$. Because $J$ is the apex of $\varphi$, clearly

$$xsy \in J \iff (xsy)\varphi \neq 0 \iff (xty)\varphi \neq 0 \iff xty \in J$$

yielding $\ker \varphi \subseteq \ker \Gamma_J$. Conversely, if $s\Gamma_J = t\Gamma_J$, then because $T$ is generalized group mapping, we have by Lemma 4.6.23 that $s\varphi = t\varphi$ if and only if $(xsy)\varphi = (xty)\varphi$ for all $x, y \in J$. By assumption, either $xsy, xty$ are both not in $J$, in which case $(xsy)\varphi = 0 = (xty)\varphi$, or $xsy, xty \in J$, in which case one has $xsy, xty \in R_x \cap L_y$. Thus $(xsy)\varphi \mathscr{H} (xty)\varphi$ and so, by aperiodicity of the 0-minimal ideal of $T$, we have $(xsy)\varphi = (xty)\varphi$, as required.    □

The construction $\mathsf{AGGM}_J(S)$ can be used to detect whether a regular $\mathscr{J}$-class is a subsemigroup.

**Lemma 4.6.38.** *Let $S$ be a semigroup and $J$ a regular $\mathscr{J}$-class. Then $J$ is a subsemigroup if and only if $\mathsf{AGGM}_J(S)$ is a semilattice, in which case it is either trivial or isomorphic to $U_1$.*

*Proof.* Suppose first that $\mathsf{AGGM}_J(S)$ is a semilattice. Then the distinguished $\mathscr{J}$-class $J\Gamma_J$ is an idempotent $e$. Assume $x, y \in J$. Then $(xy)\Gamma_J = x\Gamma_J y\Gamma_J = e^2 = e$. Because $J = \mathrm{Apx}(\Gamma_J)$ and $xy \leq_{\mathscr{J}} J$, it follows $xy \in J$.

For the converse, suppose $J$ is a subsemigroup and set

$$T(J) = \{s \in S \mid s \geq_{\mathscr{J}} J\}.$$

It suffices to show $T(J)$ is a subsemigroup. Then we shall have $T(J)$ maps to an idempotent of $\mathsf{AGGM}_J(S)$ and $S \setminus T(J)$, if non-empty, maps to zero, showing $\mathsf{AGGM}_J(S) \cong U_1$ or is trivial. So suppose $s, t \in T(J)$ and let $e \in J$ be an idempotent. Then $e = usv = xty$ with $u, v, x, y \in S^I$. Then $veus \mathscr{J} e$ because $us(veus)v = (usv)e(usv) = e$. Similarly, $tyex \mathscr{J} e$. Thus, as $J$ is a subsemigroup, $veustyex \in J$ and so $st \geq_{\mathscr{J}} J$, as required.    □

Notice that the above propositions imply

$$\ker \rho_J \subseteq \ker \mu_J^R \cap \ker \gamma_J, \quad \ker \gamma_J \subseteq \ker \Gamma_J \qquad (4.8)$$

and dually for $\lambda_J$ and $\mu_J^L$. In fact, the first containment is an equality, as the next exercise establishes.

**Exercise 4.6.39.** Show that $\ker \rho_J = \ker \mu_J^R \cap \ker \gamma_J$. Conclude there is a subdirect embedding $\mathsf{RM}_J(S) \ll \mathsf{GGM}_J(S) \times \mathsf{RLM}_J(S)$.

The semigroups $\mathsf{RM}_J(S)$ and $\mathsf{RLM}_J(S)$, for a regular $\mathscr{J}$-class $J$, turn out to be intimately related. The reader may profit by referring back to Proposition 4.1.4 and the discussion immediately preceding it for definitions and notation concerning free actions and wreath products of partial transformation semigroups. Let $R$ be an $\mathscr{R}$-class of $J$. Then $(R, \mathsf{RM}_J(S))$ is a faithful partial right transformation semigroup, as mentioned earlier. It turns out the maximal subgroup of $R$ acts freely on the left of $(R, \mathsf{RM}_J(S))$ by automorphisms. Even more is true, as the following proposition, which is at the heart of the classical Schützenberger representation by monomial matrices [68], says.

**Proposition 4.6.40.** *Let $S$ be a semigroup, $J$ a regular $\mathscr{J}$-class of $S$, $R$ an $\mathscr{R}$-class of $J$ and $e \in R$ an idempotent. Then the maximal subgroup $H_e$ at $e$ acts freely by automorphisms on the left of $(R, \mathsf{RM}_J(S))$ and $(H_e \backslash R, \widetilde{\mathsf{RM}_J(S)}) \cong (L(J), \mathsf{RLM}_J(S))$.*

*Proof.* First we check that $H_e$ acts freely on the left of $(R, \mathsf{RM}_J(S))$ by automorphisms. Let $h \in H_e$ and $r \in R$. Choose $y \in S^I$ such that $ry = e$ and let $h'$ be the inverse of $h$ in $H_e$. Then one checks $ehr = hr$ and $(hr)yh' = e$, showing that $hr \in R$. If $s \in S$, then $hrs \mathscr{L} ers = rs$. As $hr, r \in R$, stability implies that $hrs \in R$ if and only if $rs \in R$, in which case $h \cdot (rs\rho_J) = (hr)s\rho_J$, as both sides are $hrs$. Thus $h$ acts on the left of $(R, \mathsf{RM}_J(S))$ by automorphisms. Suppose that $hr = r$ with $h \in H_e$ and $r \in R$. Choose $y \in S^I$ such that $ry = e$. Then $h = he = hry = ry = e$, establishing freeness of the action.

Clearly, for $r \in R$, $h \in H_e$, we have $hr \mathscr{L} er = r$. Conversely, if $r, s \in R$ and $r \mathscr{L} s$, then $yr = s$ for some $y \in S^I$. So $s = es = eyr = eyer$ and $eye \in J \cap eSe = H_e$. Thus $H_e r = L_r \cap R = H_r$. Because each $\mathscr{L}$-class of $J$ intersects $R$, we see that $H_e \backslash R$ is in bijection with $L(J)$ via $H_e r \mapsto L_r$ and, moreover, a glance at (4.7) shows that this bijection is an isomorphism of partial transformation semigroups. This completes the proof. □

Proposition 4.1.4 then provides a wreath product decomposition for right mapping semigroups.

**Corollary 4.6.41.** *Let $S$ be a semigroup and $J$ a regular $\mathscr{J}$-class. Then $\mathsf{RM}_J(S) \leq H \wr (L(J), \mathsf{RLM}_J(S))$ where $H$ is a maximal subgroup of $J$.*

In fact, if $H$ has identity $e$ and one chooses $e$ as the representative of the $\mathscr{L}$-class $L_e$ of $e$, then the embedding in Corollary 4.6.41 maps $h \in H$ to $(f_h, h\mu_J^R)$ where $L_e f_h = h$. In fact, one has even more precise information about the embedding, as the following proposition provides.

**Proposition 4.6.42.** *Let $S$ be a semigroup, $J$ a regular $\mathscr{J}$-class of $S$ and $J^0 \cong \mathscr{M}^0(G, A, B, C)$ a Rees matrix coordinatization. Suppose $b_0 C a_0 = 1$ and let $H$ be the $\mathscr{H}$-class $a_0 \times G \times b_0$. Choose $(a_0, 1, b')$ as the representative of the $\mathscr{L}$-class $b'$ of $J$. Identifying $H$ with $G$, the composite*

$$ S \xrightarrow{\rho_J} \mathsf{RM}_J(S) \hookrightarrow G \wr (B, \mathsf{RLM}_J(S)) $$

*takes $(a, g, b) \in J^0$ to $(f, \widetilde{b})$ where if $b'Ca \neq 0$, then $b' \cdot \widetilde{b} = b$ and $b'f = (b'Ca)g$, and otherwise $b' \cdot \widetilde{b}$ is undefined and $b'f$ is arbitrary.*

*Proof.* This is a simple computation; if $b'Ca \neq 0$:

$$ (a_0, 1, b')(a, g, b) = (a_0, (b'Ca)g, b) = (a_0, (b'Ca)g, b_0)(a_0, 1, b). $$

Otherwise, the above product is undefined in $(a_0 \times G \times B, \mathsf{RM}_J(S))$. The result then follows from the construction in Proposition 4.1.4.    □

Of course dual results hold for left mapping and left letter mapping.

**Exercise 4.6.43.** Prove an analogue of Corollary 4.6.41 for a null $\mathscr{J}$-class $J$ by replacing $H$ with the Schützenberger group of $J$; see Section A.3.1.

**Definition 4.6.44.** *Let $S$ be a semigroup. Define a congruence $\mathsf{RM}$ by setting $s \mathsf{RM} t$ if and only if $s\rho_J = t\rho_J$ for all regular $\mathscr{J}$-classes $J$ of $S$. The congruences $\mathsf{LM}$, $\mathsf{GGM}$, $\mathsf{RLM}$, $\mathsf{LLM}$ and $\mathsf{AGGM}$ are defined analogously.*

Notice that $S/\mathsf{RM} \ll \prod \mathsf{RM}_J(S)$, where the product ranges over all regular $\mathscr{J}$-classes of $S$. Thus $S/\mathsf{RM}$ belongs to a pseudovariety $\mathbf{V}$ if and only if, for every regular $\mathscr{J}$-class $J$ of $S$, $\mathsf{RM}_J(S) \in \mathbf{V}$. Similar remarks apply to the other congruences from Definition 4.6.44. Our goal now is to show that these congruences give the maximal $\mathscr{K}'$ and aperiodic $\mathscr{K}'$-congruences. This will in turn lead to a simple description for certain important Mal'cev products.

**Proposition 4.6.45.** *The congruences $\mathsf{RM}$, $\mathsf{LM}$ and $\mathsf{GGM}$ are aperiodic $\mathscr{L}'$, $\mathscr{R}'$ and $\mathscr{J}'$-congruences, respectively. The congruences $\mathsf{RLM}$, $\mathsf{LLM}$ and $\mathsf{AGGM}$ are $\mathscr{L}'$, $\mathscr{R}'$ and $\mathscr{J}'$-congruences, respectively. Consequently, $\mathsf{RM} \cap \mathsf{LM}$ is an aperiodic $\mathscr{H}'$-congruence, whereas $\mathsf{RLM} \cap \mathsf{LLM}$ is an $\mathscr{H}'$-congruence.*

*Proof.* We begin by showing that $\mathsf{GGM}$ is aperiodic. Let $G$ be a subgroup of $S$ with identity $e$, belonging to a $\mathscr{J}$-class $J$. Suppose that $g \neq e$ is an element of $G$. Then $ege = g \neq e = eee$ so $g\gamma_J \neq e\gamma_J$. Thus $S \to S/\mathsf{GGM}$ is aperiodic by the Aperiodicity Lemma. Because (4.8) and its dual imply that

RM, LM $\subseteq$ GGM, the aperiodicity of GGM immediately implies the aperiodicity of RM and LM.

Next we show that AGGM is a $\mathscr{J}'$-congruence. Because GGM $\subseteq$ AGGM, by (4.8), it will then follow that GGM is a $\mathscr{J}'$-congruence. Assume $s$ and $t$ are regular with $s$ AGGM $t$. Let $e, f$ be idempotents with $e \mathscr{R} s$, $f \mathscr{L} s$. Then $esf = s \in J_s$ implies $etf \in J_s$, as $s\Gamma_{J_s} = t\Gamma_{J_s}$. Thus $s \leq_{\mathscr{J}} t$. A symmetric argument shows that $t \leq_{\mathscr{J}} s$ and so AGGM is a $\mathscr{J}'$-congruence.

Now we show that RLM is an $\mathscr{L}'$-congruence. Because (4.8) implies that RM $\subseteq$ RLM, this will imply that RM is also an $\mathscr{L}'$-congruence. Suppose $s, t$ are regular and $s$ RLM $t$. Choose an idempotent $e \in R_s$. Then $es = s$, so $L_e s\mu_J^R = L_{es} = L_s$. Therefore, as $s\mu_J^R = t\mu_J^R$, we obtain $L_s = L_e t\mu_J^R = L_{et}$, and so $s \leq_{\mathscr{L}} t$. By symmetry $t \leq_{\mathscr{L}} s$ and so $s \mathscr{L} t$, establishing RLM is an $\mathscr{L}'$-congruence. The results for LM and LLM are dual. $\qquad\square$

We are now ready for the main theorem of this section. These results are due to Krohn and Rhodes [170, 171], except for the case of AGGM, which is due to Rhodes and Tilson [269, 368].

**Theorem 4.6.46 (Krohn, Rhodes, Tilson).** *Let $S$ be a finite semigroup. Then* RM, LM *and* GGM *are the largest aperiodic $\mathscr{L}'$, $\mathscr{R}'$ and $\mathscr{J}'$-congruences on $S$, respectively, whereas* RLM, LLM *and* AGGM *are the largest $\mathscr{L}'$, $\mathscr{R}'$ and $\mathscr{J}'$-congruences on $S$, respectively. Hence* RM $\cap$ LM *is the largest aperiodic $\mathscr{H}'$-congruence on $S$ and* RLM $\cap$ LLM *is the largest $\mathscr{H}'$-congruence (i.e., idempotent-separating congruence) on $S$.*

*Proof.* Let us begin with RM (respectively RLM). Suppose $\varphi\colon S \twoheadrightarrow T$ is an onto aperiodic $\mathscr{L}'$-morphism (respectively $\mathscr{L}'$-morphism). It suffices to show that $\ker\varphi \subseteq \ker\rho_J$ (respectively $\ker\varphi \subseteq \ker\mu_J^R$) for each regular $\mathscr{J}$-class $J$ of $S$. So fix a regular $\mathscr{J}$-class $J$ of $S$ and suppose $s\varphi = t\varphi$. Proposition 4.6.11 then tells us $J\varphi$ is a regular $\mathscr{J}$-class of $T$ and $J$ is the unique $\leq_{\mathscr{J}}$-minimal $\mathscr{J}$-class mapping into $J\varphi$. Let $x \in J$. Then, as $(xs)\varphi = (xt)\varphi$ and $xs, xt \leq_{\mathscr{J}} J$, we have by minimality of $J$ that

$$ xs \in J \iff (xs)\varphi \in J\varphi \iff (xt)\varphi \in J\varphi \iff xt \in J. $$

Assume now we are in the case $xs, xt \in J$; so $xs$ and $xt$ are regular. Because $\varphi$ is an $\mathscr{L}'$-morphism, we conclude $xs \mathscr{L} xt$. Therefore, in the RLM case we have $L_x s\mu_J^R = L_{xs} = L_{xt} = L_x t\mu_J^R$ and so $\ker\varphi \subseteq \ker\mu_J^R$, as required. In the RM case, observe that $xs \mathscr{R} x \mathscr{R} xt$ by stability and so $xs \mathscr{H} xt$. Because $\varphi$ is aperiodic, we deduce, using Lemma 4.4.4, that $xs = xt$. Thus $s\rho_J = t\rho_J$, showing that $\ker\varphi \subseteq \ker\rho_J$. The result for LM and LLM is dual.

We turn now to GGM (respectively AGGM). So suppose that $\varphi\colon S \twoheadrightarrow T$ is an onto aperiodic $\mathscr{J}'$-morphism (respectively $\mathscr{J}'$-morphism). It suffices to show that $\ker\varphi \subseteq \ker\gamma_J$ (respectively $\ker\varphi \subseteq \ker\Gamma_J$) for all regular $\mathscr{J}$-classes $J$ of $S$. Choose a regular $\mathscr{J}$-class $J$ of $S$. Because $\varphi$ is a $\mathscr{J}'$-morphism, Proposition 4.6.11 shows that $J\varphi$ is a regular $\mathscr{J}$-class of $T$ and $J$ is the unique

$\leq_J$-minimal $\mathscr{J}$-class mapping into $J\varphi$. So suppose that $s\varphi = t\varphi$. Let $x, y \in J$. As $(xsy)\varphi = (xty)\varphi$ and $xsy, xty \leq_{\mathscr{J}} J$, the minimality of $J$ yields

$$xsy \in J \iff (xsy)\varphi \in J\varphi \iff (xty)\varphi \in J\varphi \iff xty \in J.$$

This shows in the **AGGM** case that $\ker \varphi \subseteq \ker \Gamma_J$. For the **GGM** case, assume $xsy, xty \in J$. Then $xsy, xty \in R_x \cap L_y$ by stability, i.e., $xsy \, \mathscr{H} \, xty$. Because $\varphi$ is aperiodic, Lemma 4.4.4 now implies that $xsy = xty$. This establishes that $\ker \varphi \subseteq \ker \gamma_J$, completing the proof of the theorem. $\qquad \square$

Let us introduce some alternative notation for these congruences.

**Definition 4.6.47.** *Let $S$ be a finite semigroup and let $\mathscr{K}$ be one of Green's relations. Then we shall often use, respectively, $\mathscr{K}'$ and $\mathscr{K}' \cap \mathbf{A}$ to denote the respective largest $\mathscr{K}'$ and aperiodic $\mathscr{K}'$-congruences on $S$. The quotients $S/\mathscr{K}'$ and $S/(\mathscr{K}' \cap \mathbf{A})$ shall be denoted $S^{\mathscr{K}'}$ and $S^{\mathscr{K}' \cap \mathbf{A}}$, respectively.*

The constructions $S \mapsto S^{\mathscr{K}'}$ and $S \mapsto S^{\mathscr{K}' \cap \mathbf{A}}$ turn out to be functorial in the category of semigroups with surjective morphisms.

**Lemma 4.6.48.** *Let $\mathscr{K}$ be one of Green's relations. Suppose that $\varphi \colon S \twoheadrightarrow T$ is an onto morphism. Then there are induced morphisms $\varphi^{\mathscr{K}'} \colon S^{\mathscr{K}'} \twoheadrightarrow T^{\mathscr{K}'}$ and $\varphi^{\mathscr{K}' \cap \mathbf{A}} \colon S^{\mathscr{K}' \cap \mathbf{A}} \twoheadrightarrow T^{\mathscr{K}' \cap \mathbf{A}}$ such that the diagrams*

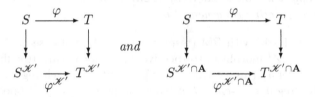

*commute.*

*Proof.* We just handle the case $\mathscr{J}' \cap \mathbf{A}$, as the other cases are treated in exactly the same fashion. Suppose that $s$ GGM $s'$. We need to show that $s\varphi$ GGM $s'\varphi$. To do this, it suffices to show that if $J$ is a regular $\mathscr{J}$-class of $T$, then $s\varphi\gamma_J = s'\varphi\gamma_J$. Now $\varphi\gamma_J \colon S \twoheadrightarrow \mathrm{GGM}_J(T)$ is an onto morphism with $\mathrm{GGM}_J(T)$ a generalized group mapping semigroup. So, setting $J' = \mathrm{Apx}(\varphi\gamma_J)$, we have that $\varphi\gamma_J$ factors through $\gamma_{J'}$ by Proposition 4.6.31. But $s\gamma_{J'} = s'\gamma_{J'}$, yielding $s\varphi\gamma_J = s'\varphi\gamma_J$, as required. $\qquad \square$

**Exercise 4.6.49.** Prove the remaining cases of Lemma 4.6.48.

Let us call $\mathbb{LG}$, $\mathbf{LS^N}$, $\mathbf{RS^N}$ and $\mathbf{G^N}$ the *pseudovarieties* associated to $\mathscr{J}$, $\mathscr{L}$, $\mathscr{R}$ and $\mathscr{H}$, respectively. The following theorem is the main reason that these maximal congruences are important. These results are due to Rhodes and Tilson [171, 269, 368], using a different language. They also considered a variation where $\mathbf{G}$ is replaced by an extension-closed pseudovariety of groups.

**Theorem 4.6.50.** *Let* **W** *be a pseudovariety of semigroups and* $S$ *a finite semigroup. Then:*

1. $S \in \mathbf{LG} \, \text{\textcircled{m}} \, \mathbf{W}$ *if and only if* $S^{\mathscr{J}'} \in \mathbf{W}$, *if and only if* $S/\mathrm{AGGM} \in \mathbf{W}$;
2. $S \in \mathbb{L}1 \, \text{\textcircled{m}} \, \mathbf{W}$ *if and only if* $S^{\mathscr{J}' \cap \mathbf{A}} \in \mathbf{W}$, *if and only if* $S/\mathrm{GGM} \in \mathbf{W}$;
3. $S \in \mathbf{LS}^{\mathbf{N}} \, \text{\textcircled{m}} \, \mathbf{W}$ *if and only if* $S^{\mathscr{L}'} \in \mathbf{W}$, *if and only if* $S/\mathrm{RLM} \in \mathbf{W}$;
4. $S \in \mathbf{K} \, \text{\textcircled{m}} \, \mathbf{W}$ *if and only if* $S^{\mathscr{L}' \cap \mathbf{A}} \in \mathbf{W}$, *if and only if* $S/\mathrm{RM} \in \mathbf{W}$;
5. $S \in \mathbf{RS}^{\mathbf{N}} \, \text{\textcircled{m}} \, \mathbf{W}$ *if and only if* $S^{\mathscr{R}'} \in \mathbf{W}$, *if and only if* $S/\mathrm{LLM} \in \mathbf{W}$;
6. $S \in \mathbf{D} \, \text{\textcircled{m}} \, \mathbf{W}$ *if and only if* $S^{\mathscr{R}' \cap \mathbf{A}} \in \mathbf{W}$, *if and only if* $S/\mathrm{LM} \in \mathbf{W}$;
7. $S \in \mathbf{G}^{\mathbf{N}} \, \text{\textcircled{m}} \, \mathbf{W}$ *if and only if* $S^{\mathscr{H}'} \in \mathbf{W}$, *if and only if* $S/(\mathrm{RLM} \cap \mathrm{LLM}) \in \mathbf{W}$;
8. $S \in \mathbf{N} \, \text{\textcircled{m}} \, \mathbf{W}$ *if and only if* $S^{\mathscr{H}' \cap \mathbf{A}} \in \mathbf{W}$, *if and only if* $S/(\mathrm{RM} \cap \mathrm{LM}) \in \mathbf{W}$.

*In particular, if* **W** *has decidable membership, then so does each of the above Mal'cev products.*

*Proof.* Let $\mathscr{K}$ be one of Green's relations and let **V** be the pseudovariety associated to $\mathscr{K}$. We handle just the case of $\mathscr{K}'$, as $\mathscr{K}' \cap \mathbf{A}$ is entirely analogous. Because the natural map $S \twoheadrightarrow S^{\mathscr{K}'}$ is a $\mathscr{K}'$-morphism, Propositions 4.6.11–4.6.15 show that if $S^{\mathscr{K}'} \in \mathbf{W}$, then $S \in \mathbf{V} \, \text{\textcircled{m}} \, \mathbf{W}$.

Conversely, suppose that $S \in \mathbf{V} \, \text{\textcircled{m}} \, \mathbf{W}$. Then there is a relational morphism $\varphi \colon S \to T$ with $T \in \mathbf{W}$ and $\varphi \in (\mathbf{V}, \mathbf{1})$. By Tilson's Lemma or direct calculation, if $\varphi = \alpha^{-1}\beta$ is the canonical factorization (1.28), then $\beta \in (\mathbf{V}, \mathbf{1})$. Suppose $\alpha \colon R \twoheadrightarrow S$ and $\beta \colon R \to T$. By restricting to the image of $\beta$, we may assume without loss of generality that $\beta$ is onto. Then $\beta$ is a $\mathscr{K}'$-morphism by Propositions 4.6.11–4.6.15 and so $\ker \beta \subseteq \mathscr{K}'$. Thus $R^{\mathscr{K}'} \prec T$ and hence $R^{\mathscr{K}'} \in \mathbf{W}$. But Lemma 4.6.48 implies that $S^{\mathscr{K}'} \prec R^{\mathscr{K}'}$, leading to the desired conclusion: $S^{\mathscr{K}'} \in \mathbf{W}$. $\qquad\square$

In [259], the congruences we have been studying were used to determine a basis of pseudoidentities for each of the pseudovarieties appearing in Theorem 4.6.50 in terms of a basis for **W**. In particular, finiteness and computability of the bases are preserved. In [259], many of the results on these congruences are attributed to [35], apparently unaware that these results already appeared in [171, 269, 368], perhaps due to the lack of a dictionary between the language of [171] and the language of pseudovarieties. The paper [35] does credit [171], but not [269, 368].

As a corollary, we obtain the well-known descriptions of **DS** and **DA**, due independently to Putcha [248] and Schützenberger [314].

**Corollary 4.6.51.** *The equalities* $\mathbf{DS} = \mathbf{LG} \, \text{\textcircled{m}} \, \mathbf{Sl}$ *and* $\mathbf{DA} = \mathbb{L}1 \, \text{\textcircled{m}} \, \mathbf{Sl}$ *hold.*

*Proof.* By Lemma 4.6.38, $S \in \mathbf{DS}$ if and only if $S^{\mathscr{J}'}$ is a semilattice and so Theorem 4.6.50 provides the equality $\mathbf{DS} = \mathbf{LG} \, \text{\textcircled{m}} \, \mathbf{Sl}$. For an aperiodic semigroup, $S^{\mathscr{J}'} = S^{\mathscr{J}' \cap \mathbf{A}}$. Conversely, if $S^{\mathscr{J}' \cap \mathbf{A}}$ is aperiodic, then $S$ is aperiodic and so again $S^{\mathscr{J}'} = S^{\mathscr{J}' \cap \mathbf{A}}$. Therefore, $S^{\mathscr{J}' \cap \mathbf{A}} \in \mathbf{Sl}$ if and only if $S$ is aperiodic and $S^{\mathscr{J}'} \in \mathbf{Sl}$, if and only if $S \in \mathbf{DS} \cap \mathbf{A} = \mathbf{DA}$. Theorem 4.6.50 then yields $\mathbf{DA} = \mathbb{L}1 \, \text{\textcircled{m}} \, \mathbf{Sl}$. $\qquad\square$

**Exercise 4.6.52.** Show that a semigroup $S$ is $\mathscr{R}$-trivial if and only if $\mathrm{RM}_J(S)$ is a semilattice for each regular $\mathscr{J}$-class $J$. Deduce that $\mathbf{R} = \mathbf{K} \textcircled{m} \mathbf{Sl}$, $\mathbf{L} = \mathbf{D} \textcircled{m} \mathbf{Sl}$ and $\mathbf{J} = \mathbf{N} \textcircled{m} \mathbf{Sl}$.

**Corollary 4.6.53.** *A semigroup $S$ belongs to* $\mathbb{LG}\textcircled{m}\mathbf{A}$ *if and only if* $S^{\mathscr{J}'} \in \mathbf{A}$. *In particular,* $\mathbb{LG} \textcircled{m} \mathbf{A}$ *is decidable and* $\mathbf{DS} \leq \mathbb{LG} \textcircled{m} \mathbf{A}$.

The following lemma shall be used later on.

**Lemma 4.6.54.** *Let $S$ be a right mapping semigroup with distinguished $\mathscr{J}$-class $J$. Then $S^{\mathscr{L}'} = \mathrm{RLM}_J(S)$. As a consequence, if $\mathbf{V}$ is a pseudovariety, then $S \in \mathbf{LS^N} \textcircled{m} \mathbf{V}$ if and only if $S \in \mathbf{G} * \mathbf{V}$.*    □

*Proof.* Because $\mathrm{RLM} \subseteq \ker \mu_J^R$ by definition, we just need to prove the reverse inclusion. As $\mathrm{RLM}$ is the largest $\mathscr{L}'$-congruence on $S$, it suffices to show $\mu_J^R$ is an $\mathscr{L}'$-morphism. By Proposition 4.6.13, $\mu_J^R$ is an $\mathscr{L}'$-morphism if and only if it is injective on idempotents $e \leq_{\mathscr{R}} f$. So suppose $e, f$ are idempotents with $e \leq_{\mathscr{R}} f$ and $e\mu_J^R = f\mu_J^R$. Then $e$ and $f$ have the same domains for their action on the right of $J$ by partial transformations. Now if $x \in J$ is in their common domain, then $L_{xe} = L_x e\mu_J^R = L_x f\mu_J^R = L_{xf}$. So $xf = yxe$ from some $y \in S^I$. Thus $xe = xfe = yxee = yxe = xf$. Because $S$ is right mapping, it follows $e = f$. This completes the proof of the first statement.

By Corollary 4.6.14, the inequality $\mathbf{G} * \mathbf{V} \leq \mathbf{LS^N} \textcircled{m} \mathbf{V}$ is always verified. For the reverse inclusion, if $S \in \mathbf{LS^N} \textcircled{m} \mathbf{V}$, then $\mathrm{RLM}_J(S) = S^{\mathscr{L}'} \in \mathbf{V}$ by Theorem 4.6.50. The desired result then follows from Corollary 4.6.41.    □

An immediate corollary is

**Corollary 4.6.55.** *Let $S$ be a subdirect product of right mapping semigroups and $\mathbf{V}$ a pseudovariety. Then $S \in \mathbf{LS^N} \textcircled{m} \mathbf{V}$ if and only if $S \in \mathbf{G} * \mathbf{V}$.*

*Proof.* If $S$ is a subdirect product of $S_1, \ldots, S_m$, then $S$ belongs to a pseudo-variety $\mathbf{W}$ if and only if $S_i \in \mathbf{W}$, all $i = 1, \ldots, m$. The corollary then follows directly from Lemma 4.6.54.    □

Let us end this section with an observation that will be used in Chapter 7. Namely we show that each local monoid of a generalized group mapping semigroup is generalized group mapping.

**Proposition 4.6.56.** *Let $S$ be a generalized group mapping semigroup with $0$-minimal ideal $I$ and let $e \in E(S)$ be a non-zero idempotent. Then the $eSe$ is generalized group mapping with $0$-minimal ideal $eIe$.*

*Proof.* Let $M = eSe$. First we show that $eIe$ is a $0$-minimal ideal of $M$. Clearly, it is an ideal as $(eSe)(eIe)(eSe) \subseteq eSISe \subseteq eIe$. Also if $x, y \in eIe \setminus 0$, then $x = uyv$ for some $u, v \in I$. Then $(eue)y(eve) = euyve = exe = x$. It follows that if $eIe \neq 0$, then $eIe$ is $0$-minimal. We proceed by establishing that $M$ acts faithfully on the left and right of $eIe$. Because $e \neq 0$, this will imply $eIe \neq 0$ and that $M$ is generalized group mapping with 0-minimal ideal $eIe$. We just handle faithfulness on the right, the other side being dual.

Set $J = I \setminus 0$. Because $e \neq 0$ and $S$ is generalized group mapping, we have $eJ \neq 0$. Fix $s \in eJ$. Let $x, y \in M$ with $x \neq y$. Then there exists $j \in J$ with $jx \neq jy$ because $S$ is generalized group mapping. Without loss of generality, assume $jx \neq 0$. Then $jx = jex$ and so $je \in J$. Also $jy = jey$. So replacing $j$ by $je$, we may assume $j \in Je$. By stability $jx \mathrel{\mathscr{R}} j$. Choose an idempotent $f \in R_j$ and choose $z \in R_s \cap L_f$. Observe that $z \in sJ \subseteq eJ$. Let $z'$ be an inverse of $z$ in $R_f = R_j$; so we have the following eggbox picture:

| $f = z'z$ | $j$ | $jx$ | $z'$ |
|---|---|---|---|
| $z$ | | | |

Then $zjx \neq 0$, as $L_z \cap R_{jx}$ contains the idempotent $f$. Moreover, $zjx \neq zjy$. Indeed, if $zjx = zjy$, then $jy \neq 0$ and so $jy \mathrel{\mathscr{R}} j \mathrel{\mathscr{R}} jx$. Therefore,

$$jx = fjx = z'zjx = z'zjy = fjy = jy$$

a contradiction. Now $zj \in eJe$. This shows $M$ acts faithfully on the right of $eIe$ and $eIe \neq 0$.                                                      □

Similar results hold for the other types of mapping semigroups.

**Exercise 4.6.57.** Let $S$ be a right letter mapping semigroup with 0-minimal ideal $I$. Prove that $eSe$ is right letter mapping with 0-minimal ideal $eIe$ for each non-zero idempotent $e \in S$.

**Exercise 4.6.58.** Show that $\mathbb{L}(\mathbb{L}1 \textcircled{m} \mathbf{V}) = \mathbb{L}1 \textcircled{m} \mathbf{LV}$ for any pseudovariety of monoids $\mathbf{V}$.

# 4.7 The Classification of Subdirectly Indecomposable Semigroups

An important application of the semilocal theory is to provide a classification scheme for subdirectly indecomposable semigroups. Because every finite semigroup is a subdirect product of subdirectly indecomposable semigroups, many an induction argument involves a reduction to subdirectly indecomposable semigroups (e.g., [156, 348]). The results of this section are mostly from [171, Chapter 8], although Theorems 4.7.10 and 4.7.20 seem to be new. Again all semigroups are assumed to be finite throughout this section.

**Definition 4.7.1 (Subdirectly indecomposable).** *A semigroup $T$ is called subdirectly indecomposable if $T \ll T_1 \times T_2$ implies that at least one of the projections $\pi_i \colon T \twoheadrightarrow T_i$ is an isomorphism.*

Let us record here a standard fact about subdirectly indecomposable semigroups that shall be used without hesitation in the sequel. First a word on terminology: by the *trivial congruence* we mean the one for which each equivalence class is a singleton; a congruence will be called *proper* if there is more than one equivalence class.

**Fact 4.7.2.** *A non-trivial finite semigroup $S$ is subdirectly indecomposable if and only if it has a unique minimal (non-trivial) congruence.*

*Proof.* Suppose first $S$ has a unique minimal congruence $\equiv$. Let $S \ll S_1 \times S_2$ and let $\pi_i \colon S \to S_i$ be the projection, $i = 1, 2$. Then $\ker \pi_1 \cap \ker \pi_2$ is the trivial congruence. By the uniqueness of $\equiv$, one of $\ker \pi_1$ or $\ker \pi_2$ must be trivial (else the intersection would contain $\equiv$), and hence one of the projections is an isomorphism.

Conversely, suppose $S$ has distinct minimal congruences $\equiv_1 \neq \equiv_2$. Then by minimality $\equiv_1 \cap \equiv_2$ is trivial and so the natural map $S \to S/\equiv_1 \times S/\equiv_2$ yields a subdirect embedding.     $\square$

**Exercise 4.7.3.** Show that each semigroup is a subdirect product of subdirectly indecomposable semigroups. Conclude that the membership problem for a pseudovariety **V** reduces to the case of subdirectly indecomposable semigroups.

Let us begin the classification by reminding the reader about the situation for groups. From the above fact, a group $G$ is subdirectly indecomposable if and only if it has a unique minimal (non-trivial) normal subgroup $M$, called its *monolith* [217] (or socle). Frequently the term *monolithic* is used as a synonym for subdirectly indecomposable in the context of groups [217]. It is well-known that a minimal normal subgroup of a finite group is a direct power of a simple group. Because we shall use the result later, we include a proof for the convenience of the reader. Recall that a subgroup $H$ of a group $G$ is called *characteristic* if $H\varphi = H$ for each automorphism $\varphi$ of $G$.

**Lemma 4.7.4.** *Let $G$ be a non-trivial finite group and $N$ a minimal normal subgroup of $G$. Then $N \cong Q^k$ for some finite simple group $Q$.*

*Proof.* Any characteristic subgroup of $N$ would have to be a normal subgroup of $G$. Hence $N$ contains no non-trivial characteristic subgroups, i.e., is characteristically simple. So it suffices to prove a characteristically simple finite group $N$ is a direct power of a simple group. Let $M$ be a minimal normal subgroup of $N$ and let $H = M\alpha_1 \cdots M\alpha_n$, where $1_N = \alpha_1, \ldots, \alpha_n$ range over the automorphism group of $N$. Then $H$ is characteristic, so $H = N$. Set $M_i = M\alpha_i$; each $M_i$ is a minimal normal subgroup of $N$. Set $j_1 = 1$ and suppose inductively we have found $j_1 < \cdots < j_k$ so that

$$M_1 M_2 \cdots M_{j_k} = M_{j_1} M_{j_2} \cdots M_{j_k} \cong M_{j_1} \times \cdots \times M_{j_k}.$$

Suppose $M_{j_1} M_{j_2} \cdots M_{j_k} \neq N$. Then there is a least index $j_{k+1}$ such that $M_{j_{k+1}} \nleq M_{j_1} M_{j_2} \cdots M_{j_k}$. Then $(M_{j_1} M_{j_2} \cdots M_{j_k}) \cap M_{j_{k+1}} = \{1\}$ by minimality of $M_{j_{k+1}}$. Therefore, we have by choice of $j_{k+1}$

$$M_1 M_2 \cdots M_{j_{k+1}} = M_{j_1} M_{j_2} \cdots M_{j_{k+1}} \cong M_{j_1} \times \cdots \times M_{j_{k+1}}.$$

So eventually we reach the case $M_{j_1} M_{j_2} \cdots M_{j_k} = N$. Because $M_{j_1} = M_1 = M$ is a direct factor of $N$, any normal subgroup of $M$ is a normal subgroup of $N$. By minimality of $M$, we conclude $M$ is a simple group $Q$. Because each $M_{j_i} \cong M = Q$, we obtain $N \cong Q^k$, as required.    □

We are now ready to begin the classification. The first ingredient is that subdirectly indecomposable semigroups have unique 0-minimal ideals.

**Lemma 4.7.5.** *A subdirectly indecomposable semigroup has a unique 0-minimal ideal.*

*Proof.* Let $S$ be a semigroup with two distinct 0-minimal ideals $I, J$. Then $S \ll S/I \times S/J$ and so $S$ is not subdirectly indecomposable.    □

The classification scheme divides subdirectly indecomposable semigroups into two classes according to the nature of the 0-minimal ideal $I$ of $S$. If $I^2 = 0$, then $S$ is said to be of *null type*. If $I$ is regular, then $S$ is said to be of *semisimple type*. Clearly, any regular semigroup is a subdirect product of subdirectly indecomposable semigroups of semisimple type. Quite often theorems about finite regular semigroups really only require this property [334].

At this moment in time, not much is known about subdirectly indecomposable semigroups of null type. It turns out that subdirectly indecomposable semigroups of semisimple type must be either right letter mapping, left letter mapping or group mapping [171], as the next theorem shows.

**Theorem 4.7.6.** *Let $S$ be a subdirectly indecomposable semigroup of semisimple type (i.e., with regular 0-minimal ideal). Then one of the following holds:*

1. *$S$ is right letter mapping;*
2. *$S$ is left letter mapping;*
3. *$S$ is group mapping.*

*Proof.* Let $I$ be the unique 0-minimal ideal of $S$. Consider the natural map $\varphi\colon S \to S/I \times S/\text{LLM} \times S/\text{GGM} \times S/\text{RLM}$ induced by the canonical projections. We claim $\varphi$ is injective and so there results a subdirect decomposition $S \ll S/I \times S/\text{LLM} \times S/\text{GGM} \times S/\text{RLM}$. Indeed, clearly $\varphi$ is injective on $S \setminus I$ and separates $I$ from $S \setminus I$ thanks to the first component of the map. As $I$ is regular, it suffices to show $\varphi$ is injective on regular elements. Suppose that $s, t \in S$ are regular elements of $S$. Then $s\varphi = t\varphi$ implies $s \mathscr{L} t$ and $s \mathscr{R} t$ by Theorem 4.6.46 because $S \to S/\text{LLM}$ is $\mathscr{R}'$ and $S \to S/\text{RLM}$ is $\mathscr{L}'$. Therefore,

$s \, \mathscr{H} \, t$. An application of the Aperiodicity Lemma (Lemma 4.4.4) and the fact that $S \to S/\mathsf{GGM}$ is aperiodic (Theorem 4.6.46) implies $s = t$.

Because $S$ is subdirectly indecomposable and $S \to S/I$ is not injective, we conclude that at least one of $S \to S/\mathsf{LLM}$, $S \to S/\mathsf{GGM}$ or $S \to S/\mathsf{RLM}$ is an isomorphism. By construction $S/\mathsf{LLM}$, $S/\mathsf{GGM}$ and $S/\mathsf{RLM}$ are subdirect products of respectively left letter mapping, generalized group mapping and right letter mapping semigroups. Therefore, $S$ must be one of these types. Because a generalized group mapping semigroup with aperiodic 0-minimal ideal is both right and left letter mapping, this completes the proof.    □

A subdirectly indecomposable semigroup without zero necessarily is of semisimple type and so Theorem 4.7.6 admits the following corollary.

**Corollary 4.7.7.** *Let $S$ be a non-trivial subdirectly indecomposable semigroup without zero and with minimal ideal $K(S)$. Then exactly one of the following occurs:*

1. *$S$ is group mapping with distinguished $\mathscr{J}$-class $K(S)$;*
2. *$K(S)$ is a right zero semigroup and $S$ acts faithfully on the right of $K(S)$;*
3. *$K(S)$ is a left zero semigroup and $S$ acts faithfully on the left of $K(S)$.*

Notice that if $K(S)$ is aperiodic, then $S$ cannot be generalized group mapping because a rectangular band does not act faithfully on both the left and right of itself.

Little is known about subdirectly indecomposable semigroups with an aperiodic 0-minimal ideal except for the case when $S$ is generalized group mapping (although some things can be inferred from the classification of maximal proper surmorphisms [296]). Our first goal is to give a complete characterization of group mapping subdirectly indecomposable semigroups. We need some preliminary lemmas, the first of which is a special case of results of [171, Chapter 8].

**Lemma 4.7.8.** *Let $S$ be a group mapping semigroup with distinguished $\mathscr{J}$-class $J$ and let $G$ be the maximal subgroup of $J$. Then a homomorphism $\varphi\colon S \to T$ is injective if and only if its restriction to $G$ is injective.*

*Proof.* Clearly, $\varphi$ injective implies $\varphi|_G$ is injective. For the converse, first observe $0\varphi \notin J\varphi$ because if $0\varphi = s\varphi$ for some $s \in J$, then $G\varphi \subseteq SsS\varphi = 0\varphi$, contradicting that $\varphi|_G$ is injective (as $G$ is non-trivial). Next observe that if $H$ is an $\mathscr{H}$-class of $J$, then $\varphi|_H$ is injective. Indeed, by Green-Rees structure theory, there exist $x, y \in J$ so that $s \mapsto xsy$ is a bijection from $H$ to $G$. Hence if $a, b \in H$ with $a\varphi = b\varphi$, then $(xay)\varphi = (xby)\varphi$ and so $xay = xby$ by assumption on $\varphi$. But this implies $a = b$ as $s \mapsto xsy$ is a bijection.

Suppose now $s\varphi = s'\varphi$ and let $x, y \in J$. Then $(xsy)\varphi = (xs'y)\varphi$ and, because $0\varphi \notin J\varphi$, either both $xsy \in J$ and $xs'y \in J$, or neither belongs to $J$. If they are both in $J$, then $xsy \, \mathscr{H} \, xs'y$ by stability (as they both belong to $R_x \cap L_y$). But we already showed that $\varphi$ is injective on $\mathscr{H}$-classes of $J$, so

$xsy = xs'y$. As $x, y \in J$ were arbitrary and $S$ is group mapping, Lemma 4.6.23 implies $s = s'$. This completes the proof of the lemma. □

Our second lemma constructs from a maximal subgroup $G$ of a semigroup $S$ and a normal subgroup $N \triangleleft G$ a homomorphism defined on $S$ whose kernel, when restricted to $G$, is exactly $N$.

**Lemma 4.7.9.** *Let $S$ be a semigroup, $G$ a maximal subgroup of $S$ and $N$ a normal subgroup of $G$. Then there is a semigroup $T$ and a homomorphism $\varphi\colon S \to T$ with $\ker \varphi|_G = N$.*

*Proof.* Let $J$ be the $\mathscr{J}$-class of $G$ in $S$ and write $J^0 \cong \mathscr{M}^0(G, A, B, C)$ with $b_0 C a_0 = 1$ where $G$ is the $\mathscr{H}$-class $a_0 \times G \times b_0$. Let $\varphi$ be the composite

$$S \xrightarrow{\rho_J} \mathsf{RM}_J(S) \hookrightarrow G \wr (B, \mathsf{RLM}_J(S)) \to G/N \wr (B, \mathsf{RLM}_J(S))$$

where the embedding is from Proposition 4.6.42 and the last map is from the functoriality of the wreath product in the left variable. According to Proposition 4.6.42, this composite takes $(a_0, g, b_0)$ to $(f, \widetilde{b}_0)$ where $\widetilde{b}_0$ is as in that proposition and $b'f = (b'Ca_0)gN$ if $b'Ca_0 \neq 0$ and is otherwise arbitrary. It follows $\ker \varphi|_G = N$. □

We are now ready to prove our result, half of which can also be deduced from the classification of maximal proper surmorphisms [293, 296].

**Theorem 4.7.10.** *Let $S$ be a group mapping semigroup with distinguished $\mathscr{J}$-class $J$ and let $G$ be the maximal subgroup of $J$. Then $S$ is subdirectly indecomposable if and only if $G$ is monolithic (i.e., subdirectly indecomposable).*

*Proof.* First suppose $G$ is not monolithic. Then $G$ has distinct non-trivial normal subgroups $N_1, N_2$ with $N_1 \cap N_2 = 1$. Lemma 4.7.9 provides onto homomorphisms $\varphi_i\colon S \to T_i$ with $\ker \varphi_i|_G = N_i$, $i = 1, 2$. In particular, $\Delta(\varphi_1 \times \varphi_2)$ is injective on $G$ and so by Lemma 4.7.8 induces a subdirect embedding $S \ll T_1 \times T_2$. Because neither $\varphi_1$, nor $\varphi_2$ is injective, this establishes $S$ is not subdirectly indecomposable.

Suppose now $G$ is monolithic. Let $S \ll S_1 \times S_2$ be a non-trivial subdirect product decomposition and let $\pi_i\colon S \to S_i$ be the projections, $i = 1, 2$. Then $G \ll G\pi_1 \times G\pi_2$ is a subdirect product decomposition of $G$. Because $G$ is subdirectly indecomposable, one of the projections, say $\pi_i$, is injective on $G$. Then $\pi_i$ is an isomorphism by Lemma 4.7.8. We conclude $S$ is subdirectly indecomposable. □

Next we wish to show that if $S$ is a generalized group mapping semigroup with aperiodic 0-minimal ideal, then $S$ is subdirectly indecomposable. This was pointed out to us by S. W. Margolis. To prove this, we first establish the well-known classification of congruence-free finite semigroups.

### 4.7.1 Congruence-free finite semigroups

Let us formally define the notion of a congruence-free semigroup.

**Definition 4.7.11 (Congruence-free).** *A semigroup is termed congruence-free if it admits no non-trivial proper congruences.*

Let us commence with a lemma concerning homomorphisms from 0-simple semigroups (see Definition A.4.1).

**Lemma 4.7.12.** *Let $S$ be a 0-simple semigroup and let $\varphi \colon S \twoheadrightarrow T$ be a non-trivial surjective morphism. Then $T$ is 0-simple and $\varphi$ is a $\mathscr{J}'$-morphism.*

*Proof.* First note $T^2 = S\varphi S\varphi = S^2\varphi = S\varphi = T$, so if $T \neq 0$, then $T^2 \neq 0$. Suppose $0 \neq s \in S$. Then $Ts\varphi T = (SsS)\varphi = S\varphi = T$. Thus $T$ is 0-simple and $\varphi$ separates the non-trivial $\mathscr{J}$-class of $S$ from 0, i.e., is a $\mathscr{J}'$-morphism.     □

**Corollary 4.7.13.** *Let $S$ be a 0-simple semigroup and $J = S \setminus 0$. Then the unique largest proper congruence on $S$ is the congruence $\ker \Gamma_J = \mathsf{AGGM}$. In particular, if $S$ is an aperiodic generalized group mapping semigroup, then $S$ is congruence-free.*

*Proof.* Lemma 4.7.12 implies each non-zero surjective homomorphism is a $\mathscr{J}'$-morphism. Because the zero homomorphism is clearly not a $\mathscr{J}'$-morphism, it follows from Theorem 4.6.46 that $\mathsf{AGGM}$ is the largest non-trivial congruence on $S$. Because the $\mathscr{J}$-class of 0 obviously does not contribute to $\mathsf{AGGM}$, clearly $\mathsf{AGGM} = \ker \Gamma_J$. The final statement is immediate from the previous ones as in this case $\ker \Gamma_J = \mathsf{AGGM}$ is the trivial congruence.     □

It now seems the opportune moment to characterize aperiodic generalized group mapping 0-simple semigroups in terms of their structure matrices.

**Proposition 4.7.14.** *Let $S$ be a 0-simple aperiodic semigroup. Then $S$ is right letter mapping if and only if $S \cong \mathcal{M}^0(\{1\}, A, B, C)$ where $C$ has no identical columns, or equivalently, for any two distinct $\mathscr{R}$-classes $R_1, R_2$ of $S$, there is an $\mathscr{L}$-class $L$ so that exactly one of $R_1 \cap L$ and $R_2 \cap L$ contains an idempotent. A dual result holds for the left letter mapping case. In particular, $S$ is generalized group mapping if and only if $S \cong \mathcal{M}^0(\{1\}, A, B, C)$ where $C$ has no identical rows or columns.*

*Proof.* The statement on Rees matrix semigroups is easily seen to be equivalent to the statement on Green's relations, so we just prove the abstract statement. Suppose $S$ is right letter mapping and $R_1, R_2$ are $\mathscr{R}$-classes such that $R_1 \cap L$ contains an idempotent if and only if $R_2 \cap L$ contains an idempotent, for each $\mathscr{L}$-class $L$. We must show $R_1 = R_2$. Fix an $\mathscr{L}$-class $L_0$ of $S$ and choose $x \in R_1 \cap L_0$ and $y \in R_2 \cap L_0$. Then if $j \in J$, we have

$$jx \neq 0 \iff L_j \cap R_x \cap E(S) \neq \emptyset$$
$$\iff L_j \cap R_y \cap E(S) \neq \emptyset \qquad \text{(by assumption)}$$
$$\iff jy \neq 0.$$

In the case that both products are non-zero, we have $jx \in R_j \cap L_0$ and $jy \in R_j \cap L_0$ and so $jx = jy$, as $S$ is aperiodic. But then $x = y$ because $S$ is right letter mapping. We conclude $R_1 = R_2$, as required.

Conversely, suppose $S$ satisfies the condition above on $\mathscr{R}$-classes and let $x, y \in S$. Choose an idempotent $e \in R_x$. Then $ex = x \in L_x$, whereas $ey \in L_y$ or $ey = 0$. So if $L_x \neq L_y$, then $x$ and $y$ act differently on the right of $S$. If $L_x = L_y$, but $x \neq y$, then $R_x \neq R_y$ because $S$ is aperiodic. Choose an $\mathscr{L}$-class $L$ so that exactly one of $R_x \cap L$ and $R_y \cap L$ contains an idempotent; without loss of generality assume $R_x \cap L$ has an idempotent. Then if $s \in L$, one has $sx \neq 0$ and $sy = 0$. Thus $S$ acts faithfully on the right of itself. This completes the proof.     $\square$

The above proposition shows, in fact, that a non-trivial aperiodic simple semigroup (i.e., a rectangular band) cannot be generalized group mapping.

**Exercise 4.7.15.** Show that a non-trivial rectangular band cannot be generalized group mapping.

The following lemma is well-known [68, 171].

**Lemma 4.7.16.** *Let $S$ be a non-trivial semigroup with $0$ containing no proper non-zero ideals. Then $S$ is $0$-simple or $S$ is a two-element null semigroup.*

*Proof.* If $S^2 \neq 0$, then $S$ is $0$-simple. Suppose $S^2 = 0$. Let $x \in S \setminus 0$. Then $\{x, 0\}$ is an ideal, so $S = \{x, 0\}$. As $x^2 \in S^2 = 0$, the lemma is proved.     $\square$

Now we may state and prove the classification theorem for congruence-free semigroups.

**Theorem 4.7.17 (Classification of congruence-free semigroups).** *The finite congruence-free semigroups are the simple groups, the two-element left zero semigroup $\mathbf{2}^{\ell}$, the two-element right zero semigroup $\overline{\mathbf{2}}$, the two-element null semigroup $N_2$ and the $0$-simple aperiodic generalized group mapping semigroups.*

*Proof.* Clearly, simple groups and semigroups of order $2$ are congruence-free. Corollary 4.7.13 shows that $0$-simple aperiodic generalized group mapping semigroups are congruence-free. Thus we are left with proving the converse.

Suppose $S$ is a congruence-free finite semigroup. First suppose that $S$ has a non-zero ideal $I$. Then the Rees congruence associated to $S \to S/I$ is a non-trivial congruence and hence $S = I$. From Lemma 4.7.16, we conclude that $S$ is either simple, $0$-simple or $S \cong N_2$.

Suppose first that $S$ is simple; so $S$ is isomorphic to a Rees matrix semigroup $\mathscr{M}(G, A, B, C)$. If $|B| \geq 2$, then the projection sending $(a, g, b) \to b$ is a homomorphism onto $\overline{B}$. We conclude $S \cong \overline{B}$. But if $|B| \geq 2$, then $\overline{B}$ admits a homomorphism onto $\overline{2}$ by identifying all but one element of $\overline{B}$. Thus $S \cong \overline{2}$. Dually, if $|A| \geq 2$ then $S$ is isomorphic to the two-element left zero semigroup $2^{\ell}$. If $|A| = 1 = |B|$, then $S$ is a group and so $S$ must be a simple group.

Finally, suppose $S$ is 0-simple and let $J = S \setminus 0$. Corollary 4.7.13 implies $\ker \Gamma_J = \mathsf{AGGM}$ and so $\Gamma_J \colon S \to \mathsf{AGGM}_J(S) = S/\ker \Gamma_J$ is a $\mathscr{J}'$-morphism and hence non-trivial. Consequently, $\Gamma_J$ is an isomorphism and $S$ is aperiodic generalized group mapping. This completes the proof of the theorem.    $\square$

**Exercise 4.7.18.** Define two matrices $C$ and $C'$ with entries in $\{0, 1\}$ to be equivalent if they have the same size and there are permutation matrices $P, Q$ so that $PCQ = C'$. Show that two congruence-free aperiodic 0-simple semigroups are isomorphic if and only if their structure matrices are equivalent.

We now return to the study of subdirectly indecomposable semigroups. Lemma 4.7.8 has the following analogue for right/left mapping semigroups.

**Lemma 4.7.19.** *Let $S$ be a right/left mapping semigroup with 0-minimal ideal $I$. Then $\varphi \colon S \to T$ is injective if and only if $\varphi|_I$ is injective.*

*Proof.* We prove the non-trivial direction for a right mapping semigroup. Assume $\varphi|_I$ is injective and $s\varphi = s'\varphi$. Let $x \in I$. Then $xs, xs' \in I$ and $xs\varphi = xs'\varphi$, and so $xs = xs'$ by injectivity of $\varphi$ on $I$. As $x \in I$ was arbitrary and $S$ is right mapping, we deduce $s = s'$ and so $\varphi$ is injective.    $\square$

Our next theorem can be viewed as a generalization of Corollary 4.7.13 to generalized group mapping semigroups with aperiodic 0-minimal ideal.

**Theorem 4.7.20.** *Let $S$ be a generalized group mapping semigroup with aperiodic 0-minimal ideal. Then $S$ is subdirectly indecomposable.*

*Proof.* If $S$ is trivial, we are done. Otherwise, the distinguished aperiodic 0-minimal ideal $I$ is non-trivial. Notice that $I$ acts faithfully on the left and right of itself, and so is in fact generalized group mapping in its own right. It is also simple or 0-simple aperiodic, being a regular 0-minimal ideal of $S$. Because $S$ is non-trivial, this means that $I$ contains a 0 and, moreover, $I$ is congruence-free by Theorem 4.7.17.

We claim that the Rees congruence associated to the quotient $S \to S/I$ is the unique minimal proper congruence on $S$. From this it follows $S$ is subdirectly indecomposable. Indeed, if $\varphi \colon S \to T$ is an onto homomorphism, then $\varphi|_I$ is either the identity or the trivial homomorphism. In the latter case, $\ker \varphi$ contains the Rees congruence modulo $I$; in the former $\varphi$ is an isomorphism by Lemma 4.7.19.    $\square$

As a corollary, we can completely describe finite inverse semigroups that are subdirectly indecomposable.

**Corollary 4.7.21.** *A finite inverse semigroup $S$ is subdirectly indecomposable if and only if it is generalized group mapping with monolithic maximal subgroup (perhaps trivial) in its distinguished $\mathscr{J}$-class.*

*Proof.* Sufficiency is clear from the above results. Necessity follows because inverse semigroups are regular and if $I$ is an ideal of $S$, then $I$ is closed under inverses (as $x^{-1} = x^{-1}xx^{-1}$). Therefore, $S$ acts faithfully on the right of $I$ if and only if it acts faithfully on the left of $I$. Thus $S$ must be generalized group mapping by Theorem 4.7.6. The corollary then follows from the above results. $\qquad\square$

In summary, subdirectly indecomposable semigroups can be divided into two types: null type and semisimple type. The semisimple type can be further subdivided into three subclasses: right letter mapping, left letter mapping and group mapping. The generalized group mapping (and hence group mapping) cases are fully understood, whereas the remaining cases deserve further investigation. To illustrate the combinatorial nature of this problem, we show that if $(X, S)$ is a 2-homogeneous transformation semigroup (meaning $S$ acts transitively on two-element subsets of $X$) containing all the constant maps, then $S$ is a subdirectly indecomposable, right letter mapping semigroup.

**Proposition 4.7.22.** *Let $(X, S)$ be a 2-homogeneous finite transformation semigroup containing all the constant maps. Then $S$ is a subdirectly indecomposable, right letter mapping semigroup. In particular, $T_n$ is subdirectly indecomposable.*

*Proof.* Clearly, the minimal ideal $I$ of $S$ consists of the constant maps on $X$ and $S$ acts faithfully on the right of this ideal, so $S$ is right letter mapping. We claim that the Rees congruence modulo $I$ is the unique minimal congruence on $S$. Indeed, suppose $\varphi\colon S \to T$ is a homomorphism. If $\varphi|_I$ is injective, the $\varphi$ is an isomorphism by Lemma 4.7.19. It follows that any proper homomorphism identifies a pair $c, c'$ of distinct elements of $I$. Let $k \in I$. Then $\{c, c'\}s = \{c, k\}$ for some $s \in S$ by 2-homogeneity. Thus $c\varphi = k\varphi$. It follows that $\varphi$ collapses $I$ to a point and so $\ker \varphi$ contains the Rees congruence modulo $I$. We conclude $S$ is subdirectly indecomposable.

Clearly, $T_1 = \{1\}$ is subdirectly indecomposable. For $n \geq 2$, it is trivial to see $T_n$ is 2-homogeneous and contains all the constant maps. $\qquad\square$

*Remark 4.7.23.* The above proof shows more generally that if $(X, S)$ is a primitive transformation semigroup containing all the constant maps, then $S$ is subdirectly indecomposable. Here primitive means there is no equivalence relation $\equiv$ on $X$ such that $x \equiv y$ implies $xs \equiv ys$, all $s \in S$.

**Exercise 4.7.24.** Prove the claim in the above remark.

The next exercise provides one of the key steps in the proof of Eilenberg's correspondence between pseudovarieties of monoids and varieties of languages.

**Exercise 4.7.25.** Let $A$ be a finite set and let $M$ be an $A$-generated monoid. For $m \in M$, let $L_m = \{w \in A^* \mid [w]_M = m\}$. Show that $M \ll \prod_{m \in M} M_{L_m}$. Deduce every subdirectly indecomposable monoid is a syntactic monoid.

## 4.8 The Exclusion Classes of $U_2$ and $\overline{2}$

If $\mathbf{V}$ is a pseudovariety of semigroups closed under semidirect product containing $U_2$, then $\mathbf{V} = \overline{\mathbf{H}}$ where $\mathbf{H}$ is the unitary wreath product closure of the simple groups in $\mathbf{V}$. So it is natural to consider the pseudovariety of semigroups that do not admit $U_2$ as a divisor. This pseudovariety of semigroups is closed under semidirect product and was first characterized by Stiffler [340]. We begin with the pseudovariety of semigroups excluding $\overline{2}$ as a divisor.

Recall that if $\mathbf{V}$ is a pseudovariety, then $\mathbf{EV}$ is the pseudovariety of all semigroups $S$ such that $\langle E(S) \rangle \in \mathbf{V}$.

**Exercise 4.8.1.** Show that $\mathbf{EV}$ is a pseudovariety.

The pseudovariety $\mathbf{EV}$ is the "obvious" upper bound for $\mathbf{V} \text{ⓜ} \mathbf{G}$.

**Lemma 4.8.2.** *Let $\mathbf{V}$ be a pseudovariety. Then $\mathbf{V} \text{ⓜ} \mathbf{G} \subseteq \mathbf{EV}$.*

*Proof.* Suppose $\varphi \colon S \to G$ is a relational morphism with $\varphi \in (\mathbf{V}, \mathbf{1})$ and $G \in \mathbf{G}$. Let $\varphi = \alpha^{-1}\beta$ be the canonical factorization with $\beta \colon R \to G$. Then $\beta \in (\mathbf{V}, \mathbf{1})$ by Tilson's Lemma and $S \prec R$. So it suffices to show $R \in \mathbf{EV}$. But if $e \in E(R)$, then $e\beta = 1$ as $1$ is the only idempotent of a group. Hence $\langle E(R) \rangle \leq 1\beta^{-1} \in \mathbf{V}$, completing the proof.    $\square$

It is a quite interesting question as to when the equality $\mathbf{EV} = \mathbf{V} \text{ⓜ} \mathbf{G}$ holds. It was shown by Rhodes and Tilson [295] that $\mathbf{A} \text{ⓜ} \mathbf{G} \neq \mathbf{EA}$ (see also [80]). Rhodes provided an example of a semigroup of complexity 2 that belongs to $(\mathbf{A} \text{ⓜ} \mathbf{G}) \text{ⓜ} \mathbf{G} \subset \mathbf{EA}$ [275] and examples of such semigroups of arbitrary complexity were provided by the authors [289] (see Corollary 4.16.2 of this book). But there are many interesting examples where equality does hold [16, 32, 53, 126]. One such example is the subject of our next theorem, which combines results from [294, 295, 340]. Our proof is modeled upon [328].

**Theorem 4.8.3.** *There is an equality of pseudovarieties:*

$$\mathbf{ER} = \mathbf{R} \text{ⓜ} \mathbf{G} = \mathbf{K} \text{ⓜ} (\mathbf{Sl} \text{ⓜ} \mathbf{G}). \tag{4.9}$$

*Moreover, the following are equivalent for a semigroup $S$:*

1. *$S \in \mathbf{ER}$;*
2. *$(J, \mathsf{RM}_J(S))$ is a semigroup of partial permutations for each regular $\mathscr{J}$-class $J$ of $S$;*
3. *$S$ does not admit $\overline{2}$ as a divisor;*
4. *$S$ does not contain a subsemigroup isomorphic to $\overline{2}$;*

5. *Each regular $\mathcal{R}$-class of $S$ has a unique idempotent.*

*In particular,* **ER** *is closed under semidirect product.*

*Proof.* By general properties of Mal'cev products (see Proposition 2.3.19),

$$\mathbf{K} \text{ⓜ} (\mathbf{Sl} \text{ⓜ} \mathbf{G}) \leq (\mathbf{K} \text{ⓜ} \mathbf{Sl}) \text{ⓜ} \mathbf{G} = \mathbf{R} \text{ⓜ} \mathbf{G}$$

where the last equality is a consequence of Exercise 4.6.52. Lemma 4.8.2 yields $\mathbf{R} \text{ⓜ} \mathbf{G} \leq \mathbf{ER}$. We are left with showing that if $S \in \mathbf{ER}$, then $S \in \mathbf{K} \text{ⓜ}$ $(\mathbf{Sl} \text{ⓜ} \mathbf{G})$. Clearly, $\overline{\mathbf{2}}$ divides no semigroup from $\mathbf{ER}$. So we in fact show that if $S$ does not have $\overline{\mathbf{2}}$ as a divisor, then $S \in \mathbf{K} \text{ⓜ} (\mathbf{Sl} \text{ⓜ} \mathbf{G})$. As $\overline{\mathbf{2}}$ is projective (Lemma 4.1.39), this is equivalent to assuming that $S$ does not have a subsemigroup isomorphic to $\overline{\mathbf{2}}$. Having a subsemigroup isomorphic to $\overline{\mathbf{2}}$ is clearly equivalent to having two distinct $\mathcal{R}$-related idempotents.

So let us assume $S$ is a semigroup such that each regular $\mathcal{R}$-class of $S$ has a unique idempotent. By Theorem 4.6.50, we just need to show $S/\mathrm{RM} = S^{\mathscr{L}' \cap \mathbf{A}} \in \mathbf{Sl} \text{ⓜ} \mathbf{G}$. Because $S/\mathrm{RM}$ is a subdirect product of the $\mathrm{RM}_J(S)$ as $J$ varies over the regular $\mathscr{J}$-classes, it suffices to prove $(J, \mathrm{RM}_J(S))$ is a semigroup of partial permutations. Indeed, in this case $\mathrm{RM}_J(S)$ embeds in an inverse monoid and hence divides a unitary semidirect product $E \rtimes G$ of a semilattice $E$ and group $G$ by Corollary 4.1.8. Such a semidirect product is clearly in $\mathbf{Sl} \text{ⓜ} \mathbf{G}$ (consider the projection $\pi \colon E \rtimes G \to G$).

So suppose that $r, s \in J$ and $rx = sx \in J$. Then $r \mathcal{R} rx = sx \mathcal{R} s$ by stability. So there exist $y, z \in S^I$ such that $rxy = r$ and $sxz = s$. Let $e = (xy)^\omega$ and $f = (xz)^\omega$. Then $r = re = se$ and $s = sf = rf$ using $rx = sx$. Thus $s = r(ef)^\omega$ and $r = r(ef)^\omega e$. Because $e$ is an idempotent, $[(ef)^\omega e]^2 = (ef)^\omega e$, from which we easily deduce that $(ef)^\omega, (ef)^\omega e$ are $\mathcal{R}$-equivalent idempotents of $S$. By assumption on $S$, we conclude that $(ef)^\omega e = (ef)^\omega$ and so $r = s$. Therefore, $\mathrm{RM}_J(S)$ is a semigroup of partial permutations of $J$. This completes the proof of (4.9) and 1–5. Because $\overline{\mathbf{2}}$ is $\rtimes$-prime, it now follows that $\mathbf{ER}$ is closed under semidirect product.                                        $\square$

Clearly, $\mathbf{R}, \mathbf{G} \leq \mathbf{ER}$, so it follows $\mathbf{R} * \mathbf{G} \leq \mathbf{ER}$. The converse is true as well and was first proved by Stiffler [85, 340]. It follows easily from (4.9) and the locality of $\mathbf{R}$ [8, 85, 332]. For the applications we have in mind, we just need the following simpler result [156, 295].

**Theorem 4.8.4 (Rhodes-Tilson).** *There is an equality* $\mathbf{A} * \mathbf{G} = \mathbf{A} \text{ⓜ} \mathbf{G}$.

*Proof.* The inclusion $\mathbf{A} * \mathbf{G} \leq \mathbf{A} \text{ⓜ} \mathbf{G}$ holds by Corollary 4.1.32. For the converse, suppose $\varphi \colon S \to G$ is an aperiodic relational morphism. Then the local semigroups of $\mathbf{Der}(\varphi)$ are all isomorphic to $1\varphi^{-1} \in \mathbf{A}$, excepting the empty semigroup at $I$. Hence $D_\varphi \in \ell\mathbf{A}$. Now if $D_\varphi^{cd} = \mathrm{Arr}(D_\varphi) \cup 0$ is the consolidation of $D_\varphi$ (see Definition 4.2.3), then the inclusion of $\mathrm{Arr}(D_\varphi)$ gives rise to a faithful semigroupoid morphism from $D_\varphi$ to $D_\varphi^{cd}$. Clearly, the square of any element of $D_\varphi^{cd}$ not belonging to a local semigroup is zero. We conclude

$D_\varphi^{cd}$ is aperiodic, from which it follows $D_\varphi \in \mathbf{gA}$. Thus $S \in \mathbf{A} * \mathbf{G}$ by the Derived Semigroupoid Theorem.                                                                   □

**Corollary 4.8.5 (Stiffler).** *Every semigroup in* **ER** *has complexity at most one.*

*Proof.* Because $\mathbf{ER} = \mathbf{R} \textcircled{m} \mathbf{G} \leq \mathbf{A} \textcircled{m} \mathbf{G}$, the result is immediate from Theorem 4.8.4.                                                                                      □

Theorem 4.8.3 leads one naturally to ask how complexity relates to the maximum number of $\mathscr{R}$-equivalent idempotents.

**Definition 4.8.6 (Tilson number).** *Let $S$ be a semigroup. Then the Tilson number $\tau(S)$ of $S$ is defined to be the maximum number of $\mathscr{R}$-equivalent idempotents of $S$, i.e., the largest cardinality $n$ so that $\overline{\mathbf{n}}$ is a subsemigroup of $S$.*

In particular, $\tau(S) = 1$ if and only if $S \in \mathbf{ER}$. Tilson conjectured in [362] that the Tilson number is an upper bound for complexity. This conjecture was confirmed by Margolis [187].

**Theorem 4.8.7 (Margolis).** *The Tilson number is an upper bound for complexity. That is, one has $c(S) \leq \tau(S)$. Equivalently, if $\overline{\mathbf{n}}$ does not divide $S$, then $c(S) < n$.*

Of course, the Tilson number is by no means a sharp bound: $\tau(\overline{\mathbf{n}}) = n$ whereas $c(\overline{\mathbf{n}}) = 0$. We next turn to the local version of Theorem 4.8.3 characterizing the semigroups not admitting $U_2$ as a divisor.

**Theorem 4.8.8.** *The following are equivalent for a semigroup $S$:*

1. *$S \in \mathbb{LER}$;*
2. *$S$ does not admit $U_2$ as a divisor;*
3. *$S$ does not contain a subsemigroup isomorphic to $U_2$.*

*In particular, $\mathbb{LER}$ is closed under semidirect product.*

*Proof.* The equivalence of items 2 and 3 are immediate because $U_2$ is projective. Clearly, $U_2$ does not belong to $\mathbb{LER}$. Conversely, suppose that $S$ is a semigroup that does not contain a subsemigroup isomorphic to $U_2$. Let $e \in S$ be an idempotent and suppose $eSe \notin \mathbf{ER}$. Theorem 4.8.3 then provides a two-element right zero subsemigroup $\{x, y\}$ of $eSe$. Because $\{e, x, y\} \cong U_2$, we have arrived at a contradiction. This completes the proof.                         □

Stiffler showed that every semigroup not containing $U_2$ as a divisor has complexity at most one [340]. Margolis and Tilson showed more generally [197] that if $S$ does not contain $U_n$, where $n \geq 2$, then $c(S) < n$. We prove these results using the Delay Theorem [364]. The original version of the Delay Theorem is Tilson's Trace-Delay Theorem [85, Chapter III, Thm 9.5]. In modern terminology, this result essentially shows that if $\mathbf{V}$ is a local pseudovariety of monoids, then $\mathbf{V} * \mathbf{D} = \mathbb{L}\mathbf{V}$; namely Tilson showed the local monoids of the derived category of the delay relational morphism divide the local monoids of the original semigroup. Straubing generalized this result [345] to describe membership in $\mathbf{V} * \mathbf{D}$ in terms of graph congruences. The modern formulation in terms of categories is due to Tilson [364]. Here if $\mathbf{V}$ is a pseudovariety of monoids and $\mathbf{W}$ is a pseudovariety of semigroups, then $\mathbf{V} * \mathbf{W}$ is the pseudovariety generated by all semidirect products of monoids in $\mathbf{V}$ with semigroups in $\mathbf{W}$, or equivalently it is the usual semidirect product of the pseudovariety of semigroups generated by $\mathbf{V}$ with $\mathbf{W}$.

Let $S$ be a semigroup. Following category theorists, we define the *idempotent splitting* of $S$ to be the category $S_E$ with object set $E(S)$ and with

$$S_E(e, f) = \{(e, s, f) \mid s \in eSf\}.$$

Multiplication is given by $(e, s, f)(f, s', g) = (e, ss', g)$ and the identity at $e$ is the arrow $(e, e, e)$. In particular, the local monoid of $S_E$ at an idempotent $e$ is isomorphic to the localization $eSe$. The idempotent splitting is also known as the *Karoubi envelope* or *Cauchy completion*.

Let $S_E^{cd}$ be the consolidation of $S_E$ and $B_E$ be an $E \times E$ Brandt semigroup. Then it is not hard to see that

$$T = \{((e, f), s) \in B_E \times S \mid s \in eSf\} \cup (0 \times S)$$

is a subsemigroup of $B_E \times S$ and $S_E^{cd} \cong T/(0 \times S)$. Therefore, $c(S_E^{cd}) \leq c(S)$. The converse is a consequence of the Delay Theorem as $S_E \prec (S_E^{cd})^I$.

**Theorem 4.8.9 (Delay Theorem [364]).** *Let $S$ be a semigroup and $\mathbf{V}$ a pseudovariety of monoids. Then $S \in \mathbf{V} * \mathbf{D}$ if and only if $S_E \in \mathbf{g}\mathbf{V}$. In particular, $S \in ((S_E^{cd})^I) * \mathbf{D}$ and so $c(S) = c(S_E^{cd})$.*

The following lemma relates the Tilson numbers $\tau(S)$ and $\tau(S_E^{cd})$.

**Lemma 4.8.10.** *Let $S$ be a semigroup. Then the equality*

$$\tau(S_E^{cd}) = \max\{\tau(eSe) \mid e \in E(S)\}$$

*holds.*

*Proof.* Because each non-zero idempotent of $S_E^{cd}$ is of the form $(e, f, e)$ with $f \in E(eSe)$, any copy of $\overline{\mathbf{n}}$ in $S_E^{cd}$ with $n \geq 2$ must be a subsemigroup of $eSe$ for some idempotent $e$. The result follows immediately from this. $\quad\square$

As a first corollary, we prove Stiffler's Theorem [340] on excluding $U_2$.

**Theorem 4.8.11 (Stiffler).** *The equality $\mathbf{LER} = \mathbf{ER} * \mathbf{D}$ holds. Consequently, each semigroup in $\mathbb{L}\mathbf{ER}$ has complexity at most one.*

*Proof.* The inequality $\mathbf{V} * \mathbf{D} \leq \mathbb{L}\mathbf{V}$ is verified for any pseudovariety of monoids [85, 364]. Recalling that a semigroup belongs to $\mathbf{ER}$ if and only if it has Tilson number one, Lemma 4.8.10 immediately implies that $S \in \mathbf{LER}$ if and only if $S_E^{cd} \in \mathbf{ER}$. The Delay Theorem then yields $\mathbf{LER} = \mathbf{ER} * \mathbf{D}$. Because $\mathbf{D} \leq \mathbf{A}$ and $\mathbf{ER} \leq \mathbf{C_1}$ (Corollary 4.8.5), it follows that members of $\mathbf{LER}$ have complexity at most one. $\qquad\square$

**Exercise 4.8.12.** Show that $\mathbf{V} * \mathbf{D} \leq \mathbb{L}\mathbf{V}$ by using that $\mathbf{V} * \mathbf{D} \leq \mathbb{L}\mathbf{V} \, \text{ⓜ} \, \mathbf{D}$.

Next we turn to the local version of the Tilson number upper bound. Analogously to Theorem 4.8.8, it is easy to see $\tau(eSe) \leq n$ for all idempotents $e \in S$ if and only if $S$ does not contain $U_{n+1}$ as a subsemigroup.

**Theorem 4.8.13 (Margolis-Tilson [197]).** *Let $S$ be a semigroup. Then*

$$c(S) \leq \max\{\tau(eSe) \mid e \in E(S)\}.$$

*Equivalently, if $U_n$ does not divide $S$, with $n \geq 2$, then $c(S) < n$.*

*Proof.* By the Delay Theorem, Lemma 4.8.10 and Theorem 4.8.7, we conclude $c(S) = c(S_E^{cd}) \leq \tau(S_E^{cd}) = \max\{\tau(eSe) \mid e \in E(S)\}$. The final statement follows from the remark immediately preceding the theorem. $\qquad\square$

A function $f\colon \mathbf{FSgp} \to \mathbb{N}$ is said to be a *local upper bound for complexity* if $S \prec T$ implies $f(S) \leq f(T)$ and $c(S) \leq \max\{f(eSe) \mid e \in E(S)\}$ for every finite semigroup $S$. The complexity function $c$ is not itself local, that is, in general $c(S) \neq \max\{c(eSe) \mid e \in E(S)\}$. The first author gave a complexity two counterexample, called the Tall Fork, in [275]; this example is discussed in Section 4.14. Examples of arbitrary complexity were provided by the authors in [289]; details can be found in Section 4.16.

We end this section with some exercises on the idempotent splitting, which give some indication as to its significance.

**Exercise 4.8.14.** Let $S$ be a semigroup and let $S_E$ be the idempotent splitting. Recall that two objects $c, c'$ of a category are isomorphic, written $c \cong c'$, if there are morphisms $g\colon c \to c'$, $g'\colon c' \to c$ with $gg' = 1_c$, $g'g = 1_{c'}$. In this case $g$ and $g'$ are said to be inverse morphisms. Let $e, f \in E(S)$.

1. Show that $e \cong f$ in $S_E$ if and only if $e \, \mathscr{D} \, f$.
2. Deduce $e \, \mathscr{D} \, f$ implies $eSe \cong fSf$ and hence $H_e \cong H_f$.
3. Show that if $e \, \mathscr{D} \, f$, then $R_e \cap L_f$ consists precisely of the invertible elements of $S_E(e, f)$ (identifying $S_E(e, f)$ with $eSf$).
4. Conclude that if $S$ is completely simple, then $S_E$ is a groupoid.

**Exercise 4.8.15.** A morphism of left $S$-sets $X$ and $Y$ is a map $f\colon X \to Y$ such that $(sx)f = s(xf)$ for all $x \in X$, $s \in S$. Let $\mathbf{Set}^{S^{op}}$ be the category of left $S$-sets. Show that the idempotent splitting $S_E$ is isomorphic to the full subcategory of $\mathbf{Set}^{S^{op}}$ whose objects are principal left ideals $Se$ with $e \in E(S)$. In particular, $eSe$ is the endomorphism monoid of the left action of $S$ on $Se$.

# 4.9 The Fundamental Lemma of Complexity

This section collects three of the main results in complexity theory: the Fundamental Lemma of Complexity, the Reduction Theorem and the Ideal Theorem. These comprise Tilson's chapters [362, 363] in Eilenberg [85]. We do not furnish proofs, as we have nothing to add to what is already done there.

The first of these results, and one of the most difficult results in the subject, is the Fundamental Lemma of Complexity, due to Rhodes [268, 271, 274]. Accessible proofs appear in Tilson's [361, 362]. Roughly speaking, the Fundamental Lemma allows one to replace semidirect products with Mal'cev products when working with complexity. Here is the precise statement:

**Theorem 4.9.1 (Fundamental Lemma of Complexity, Rhodes [268]).**
*Let $\varphi \colon S \twoheadrightarrow T$ be an onto aperiodic homomorphism. Then $c(S) = c(T)$.*

**Exercise 4.9.2.** Let $S$ be a semigroup and $I$ be an aperiodic ideal of $S$. Show using the Fundamental Lemma of Complexity that $c(S) = c(S/I)$.

Historically, Zeiger believed that the result of the previous exercise would be true, but not the Fundamental Lemma. The first author in fact proved that the Fundamental Lemma is equivalent to the special case in the exercise. The proof of the equivalence, which we leave to the reader as a challenging exercise, uses the MPS theory developed in the next chapter (see also [296]).

**Exercise 4.9.3.** Use the previous exercise, Exercise 4.7.3 and the classification of subdirectly indecomposable semigroups to reduce the problem of computing complexity to the case of a group mapping semigroup whose distinguished $\mathscr{J}$-class has a monolithic maximal subgroup.

In the language of pseudovarieties, the Fundamental Lemma has the following restatement.

**Corollary 4.9.4.** *For each $n \geq 1$, $\mathbf{C}_n = \mathbf{A} \textcircled{m} (\mathbf{G} * \mathbf{C}_{n-1})$. As a consequence $\mathbf{C}_n = \mathbf{A} \textcircled{m} \mathbf{C}_n$, $n \geq 0$.*

*Proof.* Corollary 4.1.32 shows that $\mathbf{C}_n = \mathbf{A} * \mathbf{G} * \mathbf{C}_{n-1} \leq \mathbf{A} \textcircled{m} (\mathbf{G} * \mathbf{C}_{n-1})$. Conversely, suppose that $S \in \mathbf{A} \textcircled{m} (\mathbf{G} * \mathbf{C}_{n-1})$. Then there is a relational morphism $\varphi \colon S \to T$ with $\varphi \in (\mathbf{A}, \mathbf{1})$ and $T \in \mathbf{G} * \mathbf{C}_{n-1}$. Let $\alpha^{-1}\beta$ be the canonical factorization of $\varphi$. By Tilson's Lemma or direct computation one verifies $\beta$ is aperiodic. Suppose $\alpha \colon R \twoheadrightarrow S$ and $\beta \colon R \to T$. Without loss of generality we may assume $\beta$ is onto. Then $c(S) \leq c(R) = c(T)$ where the equality results from the Fundamental Lemma of Complexity. We conclude $S \in \mathbf{C}_n$. The second statement follows from the fact $(\mathbf{A}, \mathbf{1}) = (\mathbf{A}, \mathbf{A})$, and the latter is closed under composition; thus, $\mathbf{A} \textcircled{m} (\mathbf{A} \textcircled{m} \mathbf{V}) = \mathbf{A} \textcircled{m} \mathbf{V}$ for any pseudovariety $\mathbf{V}$. $\qquad\square$

We now give a Mal'cev product characterization of the complexity hierarchy, due to Rhodes.

**Theorem 4.9.5 (Rhodes).** *Define a complexity hierarchy by* $\mathbf{C}'_0 = \mathbf{A}$ *and* $\mathbf{C}'_n = \mathbf{A} \, \textcircled{m} \, (\mathbf{LS}^{\mathbf{N}} \, \textcircled{m} \, \mathbf{C}'_{n-1})$ *for* $n \geq 1$. *Then* $\mathbf{C}_n = \mathbf{C}'_n$ *for all* $n \geq 0$. *Hence the group complexity function is the hierarchical complexity function associated to the hierarchy* $\mathbf{C}'_n$, $n \geq 0$.

*Proof.* The proof proceeds by induction on $n$, the case $n = 0$ being trivial. First we establish $\mathbf{C}_n \subseteq \mathbf{C}'_n$. Corollaries 4.6.14 and 4.9.4 imply

$$\mathbf{C}_n \leq \mathbf{A} \, \textcircled{m} \, (\mathbf{G} * \mathbf{C}_{n-1}) \leq \mathbf{A} \, \textcircled{m} \, (\mathbf{LS}^{\mathbf{N}} \, \textcircled{m} \, \mathbf{C}_{n-1}).$$

The inductive hypothesis then yields $\mathbf{C}_n \leq \mathbf{C}'_n$.

We claim that $\mathbf{LS}^{\mathbf{N}} \, \textcircled{m} \, \mathbf{C}'_{n-1} \leq \mathbf{A} \, \textcircled{m} \, (\mathbf{G} * \mathbf{C}'_{n-1})$. Suppose the truth of this claim for the moment. Then, by induction and Corollary 4.9.4, we obtain that

$$\mathbf{C}'_n = \mathbf{A} \, \textcircled{m} \, (\mathbf{LS}^{\mathbf{N}} \, \textcircled{m} \, \mathbf{C}'_{n-1}) \leq \mathbf{A} \, \textcircled{m} \, (\mathbf{A} \, \textcircled{m} \, (\mathbf{G} * \mathbf{C}_{n-1})) = \mathbf{C}_n$$

completing the proof modulo the claim. To prove the claim, suppose that $S \in \mathbf{LS}^{\mathbf{N}} \, \textcircled{m} \, \mathbf{C}'_{n-1}$. It suffices to show that $S/\mathrm{RM} = S^{\mathscr{L}' \cap \mathbf{A}} \in \mathbf{G} * \mathbf{C}'_{n-1}$, as RM is an aperiodic congruence. But $S/\mathrm{RM}$ is a subdirect product of right mapping semigroups, so Corollary 4.6.55 yields the claim.    □

This theorem is one of the main triumphs of the semilocal theory. The next section gives another. Theorem 4.9.5 admits the following reformulation in terms of the language of q-theory.

**Corollary 4.9.6.** *The complexity hierarchy is given by*

$$\mathbf{C}_n = \left( (\mathbf{A}, \mathbf{A})(\mathbf{LS}^{\mathbf{N}}, \mathbf{LS}^{\mathbf{N}}) \right)^n \mathsf{q}\mathbf{A}.$$

Notice how the use of pseudovarieties of relational morphisms allows one to avoid issues of bracketing that arise from non-associative products. Let us restate Corollary 4.9.6 in purely semigroup theoretic terms using Lemma 2.8.3.

**Corollary 4.9.7.** *Let $S$ be a finite semigroup. Then $c(S)$ is the minimum $n \in \mathbb{N}$ such that there are a division $d$, aperiodic morphisms $\gamma_1, \ldots, \gamma_n$ and $\mathscr{L}'$-morphisms $\lambda_1, \ldots, \lambda_n$ so that $d\gamma_1 \lambda_1 \cdots \gamma_n \lambda_n \colon S \to T$ with $T$ aperiodic.*

**Corollary 4.9.8.** *Let $S$ be a semigroup. Then $c(S^{\mathscr{L}'}) \leq c(S) \leq c(S^{\mathscr{L}'}) + 1$.*

*Proof.* The first inequality is clear. For the second, suppose that $S^{\mathscr{L}'} \in \mathbf{C}_n$. Then, by Theorem 4.6.50, $S \in \mathbf{LS}^{\mathbf{N}} \, \textcircled{m} \, \mathbf{C}_n \leq \mathbf{C}_{n+1}$, where the last inequality follows from Theorem 4.9.5. Therefore, $c(S) \leq c(S^{\mathscr{L}'}) + 1$, as required.    □

An alternative approach to complexity, via generalized Mal'cev products, is taken in [362]. Recall that $\mathbf{LER}$ is the pseudovariety of semigroups not admitting $U_2$ as a divisor. A relational morphism is said to be $U_2$-*free* if it belongs to the class $(\mathbf{LER}, \mathbf{LER})$. One of the principal results in [362] is

**Theorem 4.9.9.** *Let* $\varphi\colon S \twoheadrightarrow T$ *be a surjective* $U_2$-*free homomorphism. Then* $c(S) \leq c(T) + 1$.

The machinery underlying this theorem is the same as that underlying the Fundamental Lemma of Complexity. To draw some consequences, we need a lemma characterizing $U_2$-free morphisms.

**Lemma 4.9.10.** *Let* $\varphi\colon S \to T$ *be a homomorphism. Then* $\varphi$ *is* $U_2$-*free if and only if* $\varphi$ *is injective on copies of* $U_2$.

*Proof.* Suppose first $\varphi$ is $U_2$-free and $U \leq S$ is isomorphic to $U_2$ with $\varphi|_U$ not injective. Then $U\varphi$ does not admit $U_2$ as a divisor and hence belongs to $\mathbf{LER}$. Because $\varphi$ is $U_2$-free, we conclude $U\varphi\varphi^{-1} \in \mathbf{LER}$, a contradiction as $U \leq U\varphi\varphi^{-1}$. Conversely, suppose $\varphi$ is injective on copies of $U_2$. Let $T' \leq T$ belong to $\mathbf{LER}$. Then $T'$ contains no copy of $U_2$ and hence $T'\varphi^{-1}$ does not contain a copy of $U_2$ by our hypothesis on $\varphi$. But this is equivalent to $T'\varphi^{-1} \in \mathbf{LER}$ by Theorem 4.8.8. □

Notice that the natural projection $U_2 \to U_1$ identifying the two right zeroes is not $U_2$-free, but the inverse image of each idempotent belongs to $\mathbf{LER}$. Hence $(\mathbf{LER}, \mathbf{LER}) < (\mathbf{LER}, \mathbf{1})$ and so one must use generalized Mal'cev products and not just Mal'cev products.

**Corollary 4.9.11.** *The inclusion* $(\mathbf{LS^N}, \mathbf{1}) \leq (\mathbf{LER}, \mathbf{LER})$ *holds. Hence*

$$\mathbf{LS^N} \,\textcircled{m}\, \mathbf{V} \leq (\mathbf{LER}, \mathbf{LER}) \,\textcircled{m}\, \mathbf{V}$$

*for every pseudovariety* $\mathbf{V}$.

*Proof.* Because both pseudovarieties of relational morphisms in question are positive, by Tilson's Lemma it suffices to show that an $\mathscr{L}'$-morphism is $U_2$-free. By the Lemma 4.9.10, it suffices to show that $\mathscr{L}'$-morphisms are injective on copies of $U_2$. Because pseudovarieties of relational morphisms are closed under restriction, we may assume that $U_2$ is the domain of our $\mathscr{L}'$-morphism. But $U_2$ is $\mathscr{L}$-trivial and regular, hence any $\mathscr{L}'$-morphism, by its very definition, must be injective on $U_2$. The final statement is immediate because $\mathbf{q}$ is order preserving. □

The class of $U_2$-free morphisms is much bigger than the class of $\mathscr{L}'$-morphisms. For instance, the collapsing map $U_1 \to \{1\}$ is $U_2$-free but not an $\mathscr{L}'$-morphism.

As a consequence of Corollary 4.9.11 and Theorem 4.9.9, we obtain the following alternative characterization of the complexity hierarchy [362].

**Theorem 4.9.12.** *The complexity hierarchy is given by*

$$\mathbf{C}_0 = \mathbf{A}$$
$$\mathbf{C}_{n+1} = \mathbf{A} \,\textcircled{m}\, ((\mathbf{LER}, \mathbf{LER}) \,\textcircled{m}\, \mathbf{C}_n), \quad n \geq 0.$$

*Equivalently,* $\mathbf{C}_n = ((\mathbf{A}, \mathbf{A})(\mathbf{LER}, \mathbf{LER}))^n \,\mathbf{q}\mathbf{A}$.

*Proof.* Let $\mathbf{V}_0 = \mathbf{A}$ and $\mathbf{V}_{n+1} = \mathbf{A} \textcircled{m} ((\mathbf{LER}, \mathbf{LER}) \textcircled{m} \mathbf{V}_n)$, for $n \geq 0$. We must show $\mathbf{V}_n = \mathbf{C}_n$. The proof goes by induction on $n$ with the case $n = 0$ being trivial. First we have, by Theorem 4.9.5,

$$\mathbf{C}_{n+1} = \mathbf{A} \textcircled{m} (\mathbf{LS}^{\mathbf{N}} \textcircled{m} \mathbf{C}_n) \leq \mathbf{A} \textcircled{m} ((\mathbf{LER}, \mathbf{LER}) \textcircled{m} \mathbf{V}_n) = \mathbf{V}_{n+1}$$

where the last inequality uses Corollary 4.9.11 and induction. Conversely, suppose $S \in \mathbf{A} \textcircled{m} ((\mathbf{LER}, \mathbf{LER}) \textcircled{m} \mathbf{V}_n)$. Then $S \in ((\mathbf{A}, \mathbf{A})(\mathbf{LER}, \mathbf{ER})) \mathsf{q} \mathbf{C}_n$ (using induction) and so, by Lemma 2.8.3, there exist a division $d$, an aperiodic morphism $\varphi_1$ and a $U_2$-free morphism $\varphi_2$ so that $d\varphi_1\varphi_2 \colon S \to T$ with $T \in \mathbf{C}_n$. It follows by the Fundamental Lemma of Complexity and Theorem 4.9.9 that $c(S) \leq c(T) + 1$ and so $S \in \mathbf{C}_{n+1}$. This completes the proof. $\qquad\square$

Lemma 2.8.3 allows the following reformulation of Theorem 4.9.12.

**Corollary 4.9.13.** *Let $S$ be a finite semigroup. Then $c(S)$ is the minimum $n \in \mathbb{N}$ such that there exist a division $d$, aperiodic morphisms $\gamma_1, \ldots, \gamma_n$ and $U_2$-free morphisms $\upsilon_1, \ldots, \upsilon_n$ so that $d\gamma_1\upsilon_1 \cdots \gamma_n\upsilon_n \colon S \to T$ with $T$ aperiodic.*

Because the simple groups and $U_2$ (and its divisors) are precisely the $\rtimes$-prime semigroups (and aperiodic morphisms can be viewed as simple group-free), this corollary is a morphic version of the Prime Decomposition Theorem.

Let us now establish an upper bound for complexity. An alternative, more constructive approach to this result is via the Depth Decomposition Theorem [363]. A $\mathscr{J}$-class of a semigroup is called *essential* if it contains a nontrivial subgroup. The *depth* of a semigroup is the length of the longest chain of essential $\mathscr{J}$-classes; we use $\delta(S)$ to denote the depth of a semigroup $S$.

**Lemma 4.9.14.** *Let $\varphi \colon S \twoheadrightarrow T$ be an onto homomorphism. Then the depth satisfies $\delta(T) \leq \delta(S)$.*

*Proof.* For a semigroup $R$, denote by $\mathscr{U}(R)$ the partially ordered set of regular $\mathscr{J}$-classes of $R$. Then we can define, by Lemma 4.6.10, a map $\mathrm{Apx} \colon \mathscr{U}(T) \to \mathscr{U}(S)$ that maps $J \in \mathscr{U}(T)$ to the unique $\leq_{\mathscr{J}}$-minimal $\mathscr{J}$-class $\mathrm{Apx}(J)$ of $S$ mapping into $J$. We show that Apx is an order-isomorphism with its image. Clearly, if $\mathrm{Apx}(J) \leq_{\mathscr{J}} \mathrm{Apx}(J')$, then

$$J = \mathrm{Apx}(J)\varphi \leq_{\mathscr{J}} \mathrm{Apx}(J')\varphi = J'.$$

Conversely, to see that the map Apx is order preserving, suppose $J \leq_{\mathscr{J}} J'$. Let $e \in \mathrm{Apx}(J)$ be an idempotent and suppose $e\varphi = u\varphi s\varphi v\varphi$ with $u, v \in S^I$ and $s \in \mathrm{Apx}(J')$. Then $(eusv)\varphi = e\varphi$ and $eusv \leq_{\mathscr{J}} \mathrm{Apx}(J)$. Thus, by definition of $\mathrm{Apx}(J)$, we conclude $eusv \in \mathrm{Apx}(J)$ and so $\mathrm{Apx}(J) \leq_{\mathscr{J}} \mathrm{Apx}(J')$. Because $J$ essential implies $\mathrm{Apx}(J)$ is essential (by Lemma 4.6.10), we deduce that $\delta(T) \leq \delta(S)$. $\qquad\square$

We now establish that depth is an upper bound on the complexity of a semigroup.

**Theorem 4.9.15.** *Let $S$ be a semigroup. Then $c(S) \leq \delta(S)$.*

*Proof.* We proceed by induction on $|S|$, the case $|S| = 1$ being clear. Let $I$ be a 0-minimal ideal of $S$. If $I$ is aperiodic, then by the Fundamental Lemma of Complexity $c(S) = c(S/I) \leq \delta(S/I) \leq \delta(S)$ by induction and Lemma 4.9.14. So we may assume that all the minimal non-zero $\mathscr{J}$-classes of $S$ are essential. If $J$ is such a minimal non-zero $\mathscr{J}$-class, then the elements of $J$ act as partial constant maps on the right of the set of $\mathscr{L}$-classes of $J$ (by stability of finite semigroups) and are undefined in right letter mapping with respect to any other non-zero regular $\mathscr{J}$-class of $S$. Thus the image of $J$ in $S^{\mathscr{L}'}$ is inessential (and in particular $S^{\mathscr{L}'} \neq S$). It follows that $\delta(S^{\mathscr{L}'}) \leq \delta(S) - 1$ (use the map Apx from Lemma 4.9.14 and that all the minimal non-zero $\mathscr{J}$-classes of $S$ are essential, but none of them are in the image of an essential $\mathscr{J}$-class under Apx). Because $c(S) \leq c(S^{\mathscr{L}'}) + 1$ by Corollary 4.9.8, induction yields

$$c(S) \leq c(S^{\mathscr{L}'}) + 1 \leq \delta(S^{\mathscr{L}'}) + 1 \leq \delta(S)$$

completing the proof.                                                          □

Although the depth of a semigroup is an easy to compute upper bound for its complexity, it is far from a sharp upper bound. Corollary 4.1.8 implies that each inverse semigroup has complexity at most 1; in particular, the symmetric inverse monoid $I_n$ on $n$ letters has $c(I_n) = 1$, all $n \geq 2$. On the other hand, $\delta(I_n) = n - 1$. That the complexity of an inverse semigroup is at most one was first proved by Tilson in his thesis. Margolis later generalized this to show that a faithful partial transformation semigroup of $k$ to 1 maps has complexity at most $k$ [187].

The next major result we wish to state in this section is the Reduction Theorem [291, 363] of Rhodes and Tilson. The reader should consult [363] for a detailed proof.

**Theorem 4.9.16 (Reduction Theorem).** *Let $S$ be a finite semigroup, let $J_1, \ldots, J_m$ be the maximal essential $\mathscr{J}$-classes of $S$ and let $E$ be a set consisting of one idempotent from each $J_i$. Then $S \prec ESE \wr A$ where $A$ is an aperiodic (in fact, locally trivial) semigroup. In particular, $c(S) = c(ESE)$.*

An important consequence of the Reduction Theorem is that if $S$ is a semigroup with an ideal $I$ such that $S/I$ is aperiodic, then $c(S) = c(I)$. Indeed, one must have in this case that $ESE \leq I \leq S$ and so the Reduction Theorem implies $c(I) = c(S)$. This is a special case of the Ideal Theorem, a result closely related to the Fundamental Lemma of Complexity. The Ideal Theorem is joint work of the first author and Tilson, but first appeared in Tilson's [361]. In [362], a stronger result goes by the name of the Ideal Theorem. We state here the original version.

**Theorem 4.9.17 (Ideal Theorem of Rhodes and Tilson).** *Let $S$ be a finite semigroup and $I$ an ideal of $S$. Then $c(S) \leq c(I) + c(S/I)$.*

The proof is more complicated than one would guess: the derived category of $S \to S/I$ does not divide $I$ — consider $T_n/J_1$ where $J_1$ is the ideal of constant maps. For more on this, see [362, Prop. 6.3].

*Remark 4.9.18 (Discussion of Tilson's Ideal Theorem).* What Tilson calls the Ideal Theorem in [362], which we call here Tilson's Ideal Theorem, is a stronger statement, very much appropriate to the language of q-theory, and the reason Tilson invented weakly closed classes (equivalent to our positive pseudovarieties of relational morphisms). Namely, one defines a complexity hierarchy in $\mathbf{PVRM}^+$ by setting

$$\mathsf{C}_0 = (\mathbf{A}, \mathbf{A})$$
$$\mathsf{C}_{n+1} = (\mathbf{A}, \mathbf{A})(\mathbf{LER}, \mathbf{LER})\mathsf{C}_n$$
$$= \mathsf{C}_n(\mathbf{LER}, \mathbf{LER})(\mathbf{A}, \mathbf{A}), \ n \geq 0.$$

One can then define a hierarchical complexity function $c \colon \mathbf{PVRM}^+ \to \mathbb{N}$ by

$$c(\varphi) = \min\{n \mid \varphi \in \mathsf{C}_n\}.$$

It is fairly easy to show $c(\varphi) \leq c(D_\varphi^{cd})$ (the consolidation of $D_\varphi$) [362, Prop. 8.2], and hence is finite. Also it is immediate that $c(\varphi\psi) \leq c(\varphi) + c(\psi)$. Tilson's Ideal Theorem is then:

**Theorem 4.9.19 (Tilson's Ideal Theorem).** *Let $I$ be an ideal of a semigroup and $\varphi \colon S \to S/I$ the projection. Then $c(I) = c(\varphi)$.*

For instance, the collapsing morphism $c_S \colon S \to \{1\} = S/S$ satisfies the equality $c(c_S) = c(S)$. Therefore, if $\varphi \colon S \to T$ is a relational morphism, then

$$c(S) = c(c_S) = c(\varphi c_T) \leq c(\varphi) + c(c_T) = c(\varphi) + c(T). \qquad (4.10)$$

In particular, if $\varphi$ is aperiodic, then $c(\varphi) = 0$ and so (4.10) says $c(S) \leq c(T)$, which is the Fundamental Lemma of Complexity. If $\varphi$ is $U_2$-free, then $c(\varphi) \leq 1$ and so (4.10) says $c(S) \leq 1 + c(T)$, which is Theorem 4.9.9. Taking $\varphi$ the projection $S \to S/I$ yields $c(S) \leq c(I) + c(S/I)$, the original Ideal Theorem, via Tilson's Ideal Theorem.

Notice that Theorem 4.9.12 can be reformulated as stating $\mathbf{C}_n = \mathbf{C}_n\mathbf{q}(\mathbf{1})$.

We end this section with another upper bound for complexity. This bound turns out to be sharp for semigroups in $\mathbf{DS}$.

**Definition 4.9.20 (Group mapping–right letter mapping chain).** *A group mapping–right letter mapping chain for a semigroup $S$ is a surjective homomorphism $\varphi \colon S \to A$ with $A$ aperiodic, together with a factorization $\varphi = \gamma_1\mu_1\gamma_1 \cdots \gamma_n\mu_n$ of $\varphi$ into surjective morphisms where the codomain $S_i$ of each $\gamma_i$ is group mapping and $\mu_i \colon S_i \to T_i$ is the projection $S_i \to \mathsf{RLM}_{J_i}(S_i)$ where $J_i$ is the distinguished $\mathscr{J}$-class of $S_i$. The number $n$ is called the length of the chain. If $\varphi$ is the identity map on an aperiodic semigroup, then we consider it to be a group mapping–right letter mapping chain of length 0.*

**Theorem 4.9.21 (Rhodes).** *The complexity of a semigroup $S$ is bounded above by the maximum length $\chi(S)$ of a group mapping–right letter mapping chain for $S$.*

*Proof.* We proceed by induction on the order of $S$. If $S$ is aperiodic, there is nothing to prove. Else, let $X$ be the set of essential $\mathscr{J}$-classes of $S$ and consider the map $\alpha\colon S \to \prod_{J\in X} \mathsf{GGM}_J(S)$. This map is easily verified to be aperiodic. Indeed, if $G$ is a non-trivial maximal subgroup with identity $e$, then the projection $\gamma_{J_e}\colon S \to \mathsf{GGM}_{J_e}(S)$ is injective on $G$ (as $g\gamma_{J_e} = g'\gamma_{J_e}$ implies $ege = eg'e$ for $g, g' \in G$). By the Fundamental Lemma of Complexity, $S\alpha$ has the same complexity as $S$. Because $S\alpha \ll \prod_{J\in X} \mathsf{GGM}_J(S)$, there is an essential $\mathscr{J}$-class $J$ with $c(S) = c(\mathsf{GGM}_J(S))$. Because $\gamma_J$ is injective on the maximal subgroup $G$ of $J$, it follows $\mathsf{GGM}_J(S)$ is group mapping. Now, according to Corollary 4.6.41, $\mathsf{GGM}_J(S) \leq G \wr (B, \mathsf{RLM}_{J_0}(\mathsf{GGM}_J(S)))$, where $J_0 = J\gamma_J$ is the distinguished $\mathscr{J}$-class of $\mathsf{GGM}_J(S)$ and $B$ is the set of $\mathscr{L}$-classes of $J_0$. Thus $c(S) \leq 1 + c(\mathsf{RLM}_{J_0}(\mathsf{GGM}_J(S)))$. By stability, each element of the maximal subgroup of $J_0$ acts the same on $B$, hence $|\mathsf{RLM}_{J_0}(\mathsf{GGM}_J(S))| < |\mathsf{GGM}_J(S)| \leq |S|$. Induction then provides

$$\chi(S) \geq 1 + \chi(\mathsf{RLM}_{J_0}(\mathsf{GGM}_J(S))) \geq 1 + c(\mathsf{RLM}_{J_0}(\mathsf{GGM}_J(S))) \geq c(S)$$

as required.                                                                           $\square$

One can check that for the symmetric inverse monoid $I_n$, with $n \geq 2$, the group mapping–right letter mapping chain upper bound is $n - 1$.

**Exercise 4.9.22.** Show that the maximum length of a group mapping–right letter mapping chain for $I_n$ with $n \geq 2$ is $n - 1$.

One can also show that the group mapping–right letter mapping chain upper bound is a local upper bound [171, Chapter 9].

**Exercise 4.9.23.** Prove that the group mapping–right letter mapping chain upper bound is a local upper bound. You may find Proposition 4.6.56 and Exercise 4.6.57 helpful.

Notice that the pointwise meet of all local upper bounds to complexity is again an upper bound for complexity. It seems unlikely to be a local upper bound, but what is it?

*Question 4.9.24.* Determine the pointwise meet of all local upper bounds to complexity. Find more local upper bounds.

In the context of local upper bounds, it is natural to make the following definition.

**Definition 4.9.25 (Local complexity [292]).** *The local complexity function $c_\ell\colon \mathbf{FSgp} \to \mathbb{N}$ is the greatest hierarchical complexity function $f$ satisfying:*

1. $f(S \rtimes T) \leq f(S) + f(T)$;
2. $f(U_2) = 0$;
3. $f(G) \leq 1$ for every group $G$;
4. $f(S) = \max\{f(eSe) \mid e \in E(S)\}$.

Because $c_\ell$ satisfies all the axioms of the complexity function, it must be a lower bound to complexity, that is, $c_\ell \leq c$. A characterization of the local complexity function was obtained in [292]; it was shown to be computable in [126]. We will discuss this further in Section 4.12.

## 4.10 The Decidability of Complexity for DS and Related Pseudovarieties

In this section, we prove a result of Rhodes [170, 269, 368] that complexity is decidable for semigroups in $\mathbf{LG} \,\text{ⓜ}\, \mathbf{A}$; notice that this pseudovariety is decidable by Corollary 4.6.53. Our proof takes advantage of the language of pseudovarieties and Mal'cev products to greatly simplify things. One should compare with the original proof in the case of completely regular semigroups, from [171, Chapter 9] or [170], to see just how much things have been simplified. Recall that a semigroup is *completely regular* if each element belongs to a subgroup, or equivalently it is regular and belongs to **DS**, or it satisfies the pseudoidentity $x^{\omega+1} = x$. The pseudovariety of completely regular semigroups is denoted **CR**. We remark that $\mathbf{DS} = \mathbf{LG} \,\text{ⓜ}\, \mathbf{Sl} \leq \mathbf{LG} \,\text{ⓜ}\, \mathbf{A}$ and so in particular complexity is decidable for semigroups in **DS**.

**Definition 4.10.1.** *Define a complexity hierarchy by setting*

$$\widetilde{\mathbf{C}}_0 = \mathbf{A}$$
$$\widetilde{\mathbf{C}}_{n+1} = \mathbb{L}\mathbf{1} \,\text{ⓜ}\, (\mathbf{LS}^\mathbf{N} \,\text{ⓜ}\, \widetilde{\mathbf{C}}_n), \ n \geq 0.$$

*In other words, we have* $\widetilde{\mathbf{C}}_n = ((\mathbb{L}\mathbf{1}, \mathbb{L}\mathbf{1})(\mathbf{LS}^\mathbf{N}, \mathbf{LS}^\mathbf{N}))^n \mathbf{q}(\mathbf{A})$. *Denote by* $\widetilde{c} \colon \mathbf{FSgp} \to \mathbb{N}^\infty$ *the associated hierarchical complexity function.*

Our goal is to prove that $\widetilde{c}(S) < \infty$ if and only if $S \in \mathbf{LG} \,\text{ⓜ}\, \mathbf{A}$, in which case $\widetilde{c}(S) = c(S)$. First we show that $\widetilde{c}$ is computable.

**Definition 4.10.2.** *Define a function* $C \colon \mathbf{FSgp} \to \mathbf{FSgp}$ *by setting*

$$C(S) = (S^{\mathscr{J}' \cap \mathbf{A}})^{\mathscr{L}'}$$

*for a semigroup $S$.*

Lemma 4.6.48 implies that $C$ is a functor if we take as arrows the onto homomorphisms. We remark that $C$ is polynomial time computable.

**Exercise 4.10.3.** Show that GGM is computable in time $O(n^5)$ and RLM is computable in time $O(n^6)$, where $n$ is the order of the semigroup. Deduce that $C$ is computable in time $O(n^6)$.

As an immediate consequence of Theorem 4.6.50, we have:

**Proposition 4.10.4.** *Let* **V** *be a pseudovariety and* $S$ *a semigroup. Then* $S$ *belongs to* $\mathbb{L}\mathbf{1} \, \textcircled{m} \, (\mathbf{LS}^{\mathbf{N}} \, \textcircled{m} \, \mathbf{V})$ *if and only if* $C(S) \in \mathbf{V}$.

We may now compute $\widetilde{c}$ using the construction $C$.

**Proposition 4.10.5.** *The hierarchical complexity function* $\widetilde{c} \colon \mathbf{FSgp} \to \mathbb{N}^\infty$ *associated to the complexity hierarchy* $\widetilde{\mathbf{C}}_n$ *is given by the formula:*

$$\widetilde{c}(S) = \min\{n \in \mathbb{N} \mid C^n(S) \in \mathbf{A}\}. \tag{4.11}$$

*In particular,* $\widetilde{c}$ *is computable in polynomial time.*

*Proof.* We prove by induction on $n$ that $S \in \widetilde{\mathbf{C}}_n$ if and only if $C^n(S) \in \mathbf{A}$. Clearly, $S = C^0(S) \in \mathbf{A}$ if and only if $S \in \widetilde{\mathbf{C}}_0 = \mathbf{A}$. Assume the statement is true for $n - 1 \geq 0$. Proposition 4.10.4 and the induction hypothesis then yield:

$$S \in \widetilde{\mathbf{C}}_n = \mathbb{L}\mathbf{1} \, \textcircled{m} \, (\mathbf{LS}^{\mathbf{N}} \, \textcircled{m} \, \widetilde{\mathbf{C}}_{n-1}) \iff C(S) \in \widetilde{\mathbf{C}}_{n-1}$$
$$\iff C^n(S) = C^{n-1}(C(S)) \in \mathbf{A}.$$

This establishes (4.11). For the computability of $\widetilde{c}$, observe that $|C(S)| \leq |S|$ and that if $|C(S)| = |S|$, then $C^n(S) = S$ for all $n \geq 0$. So if $|S| = n$, then $C^{n-1}(S) = C^n(S)$. It follows that $\widetilde{c}(S) = \infty$ if and only if $C^{n-1}(S) \notin \mathbf{A}$ and otherwise, $C^k(S) \in \mathbf{A}$ for some $0 \leq k \leq n - 1$. Because $C$ is polynomial time computable, it follows that $\widetilde{c}$ is, as well. $\qquad\square$

**Exercise 4.10.6.** Show that $\widetilde{c}$ is computable in $O(n^8)$.

We are almost prepared to prove the main result of this section. First we need a basic property of Mal'cev products.

**Lemma 4.10.7.** *Let* **U**, **V** *and* **W** *be pseudovarieties. Then*

$$(\mathbf{U} \, \textcircled{m} \, \mathbf{W}) \cap (\mathbf{V} \, \textcircled{m} \, \mathbf{W}) = (\mathbf{U} \cap \mathbf{V}) \, \textcircled{m} \, \mathbf{W}.$$

*Proof.* Suppose that $f_i \colon S \to T_i$, $i = 1, 2$ with $f_1 \in (\mathbf{U}, \mathbf{1})$ and $f_2 \in (\mathbf{V}, \mathbf{1})$, respectively, and $T_1, T_2 \in \mathbf{W}$. Then $\Delta(f_1 \times f_2) \colon S \to T_1 \times T_2$ belongs to $(\mathbf{U} \cap \mathbf{V}, \mathbf{1})$ and $T_1 \times T_2 \in \mathbf{W}$. Indeed, if $(e_1, e_2) \in T_1 \times T_2$ is an idempotent, then $(e_1, e_2)(f_1 \times f_2)^{-1} \Delta^{-1} = e_1 f_1^{-1} \cap e_2 f_2^{-1}$. $\qquad\square$

**Theorem 4.10.8 (Rhodes).** *For all $n \geq 0$, $\widetilde{\mathbf{C}}_n = \mathbf{C}_n \cap (\mathbf{LG} \textcircled{m} \mathbf{A})$. Hence,*

$$\bigcup_{n \geq 0} \widetilde{\mathbf{C}}_n = \mathbf{LG} \textcircled{m} \mathbf{A} \ \text{and} \ c(S) = \widetilde{c}(S), \ \text{for} \ S \in \mathbf{LG} \textcircled{m} \mathbf{A}.$$

*Therefore, group complexity is polynomial time computable for semigroups in $\mathbf{LG} \textcircled{m} \mathbf{A}$, and in particular for semigroups in $\mathbf{DS}$.*

*Proof.* The proof of the first statement proceeds by induction on $n$, the case $n = 0$ being clear because $\mathbf{C}_0 = \mathbf{A} = \widetilde{\mathbf{C}}_0$. Assume the assertion holds for $n - 1 \geq 0$. First observe that as $\mathbf{L1}, \mathbf{LS}^{\mathbf{N}} \leq \mathbf{LG}$, $(\mathbf{LG}, \mathbf{1}) = (\mathbf{LG}, \mathbf{LG})$ and the latter class is closed under composition, $\widetilde{\mathbf{C}}_n \leq \mathbf{LG} \textcircled{m} \mathbf{A}$. Also,

$$\begin{aligned}
\widetilde{\mathbf{C}}_n &= \mathbf{L1} \textcircled{m} (\mathbf{LS}^{\mathbf{N}} \textcircled{m} \widetilde{\mathbf{C}}_{n-1}) \\
&\leq \mathbf{L1} \textcircled{m} (\mathbf{LS}^{\mathbf{N}} \textcircled{m} \mathbf{C}_{n-1}) \quad \text{[by induction]} \\
&\leq \mathbf{A} \textcircled{m} (\mathbf{LS}^{\mathbf{N}} \textcircled{m} \mathbf{C}_{n-1}) \\
&= \mathbf{C}_n \quad\quad\quad\quad\quad\quad\quad \text{[by Theorem 4.9.5]}.
\end{aligned}$$

For the opposite inequality, we have by Theorem 4.9.5, for $n \geq 1$,

$$\begin{aligned}
\mathbf{C}_n \cap (\mathbf{LG} \textcircled{m} \mathbf{A}) &= (\mathbf{A} \textcircled{m} (\mathbf{LS}^{\mathbf{N}} \textcircled{m} \mathbf{C}_{n-1})) \cap (\mathbf{LG} \textcircled{m} \mathbf{A}) \\
&\leq (\mathbf{A} \textcircled{m} (\mathbf{LS}^{\mathbf{N}} \textcircled{m} \mathbf{C}_{n-1})) \cap (\mathbf{LG} \textcircled{m} (\mathbf{LS}^{\mathbf{N}} \textcircled{m} \mathbf{C}_{n-1})) \\
&= (\mathbf{A} \cap \mathbf{LG}) \textcircled{m} (\mathbf{LS}^{\mathbf{N}} \textcircled{m} \mathbf{C}_{n-1})) \quad \text{[by Lemma 4.10.7]} \\
&= \mathbf{L1} \textcircled{m} (\mathbf{LS}^{\mathbf{N}} \textcircled{m} \mathbf{C}_{n-1}).
\end{aligned}$$

Now if $S$ belongs to $(\mathbf{L1} \textcircled{m} (\mathbf{LS}^{\mathbf{N}} \textcircled{m} \mathbf{C}_{n-1})) \cap (\mathbf{LG} \textcircled{m} \mathbf{A})$, then

$$C(S) \in \mathbf{C}_{n-1} \cap (\mathbf{LG} \textcircled{m} \mathbf{A}) = \widetilde{\mathbf{C}}_{n-1}$$

by Proposition 4.10.4 and the induction hypothesis. Another application of Proposition 4.10.4 then shows that $S \in \widetilde{\mathbf{C}}_n$. This completes the proof that $\widetilde{\mathbf{C}}_n = \mathbf{C}_n \cap (\mathbf{LG} \textcircled{m} \mathbf{A})$. The remaining assertions are immediate.    □

Theorem 4.10.8 and Proposition 4.10.5 yield:

**Corollary 4.10.9.** *Let $S \in \mathbf{LG} \textcircled{m} \mathbf{A}$. If $S \notin \mathbf{A}$, then*

$$c(S) = c((S^{\mathscr{J}' \cap \mathbf{A}})^{\mathscr{L}'}) + 1$$

*holds.*

**Exercise 4.10.10.** Use the Fundamental Lemma of Complexity and Corollary 4.9.8 to show that in general $c((S^{\mathscr{J}' \cap \mathbf{A}})^{\mathscr{L}'}) \leq c(S) \leq c((S^{\mathscr{J}' \cap \mathbf{A}})^{\mathscr{L}'}) + 1$ for a semigroup $S$. Show that the second inequality is strict for the symmetric inverse monoid.

As an application of Corollary 4.10.9, we show that the group mapping–right letter mapping chain upper bound $\chi$ yields exactly the complexity for a semigroup in $\mathbb{LG} \circledm \mathbf{A}$. This generalizes a result of the first author and Krohn from the completely regular case [170, 171].

**Theorem 4.10.11.** *Let* $S \in \mathbb{LG} \circledm \mathbf{A}$ *and let* $\chi$ *be as in Theorem 4.9.21. Then* $c(S) = \chi(S)$. *In particular, if* $S \in \mathbf{DS}$, *then* $c(S) = \chi(S)$.

*Proof.* The proof begins with an observation: going from group mapping to right letter mapping always causes a drop in complexity within $\mathbb{LG} \circledm \mathbf{A}$. If $T \in \mathbb{LG} \circledm \mathbf{A}$ is a group mapping semigroup (and hence not aperiodic) with distinguished $\mathscr{J}$-class $J$, then $\gamma_J : T \to \mathrm{GGM}_J(T)$ is an isomorphism and so GGM is the trivial congruence on $T$, that is, $T^{\mathscr{J}' \cap \mathbf{A}} = T$. Moreover, $\mathrm{RLM}_J(T) = T^{\mathscr{L}'}$ by Lemma 4.6.54. Corollary 4.10.9 then yields the equality $c(T) = c(\mathrm{RLM}_J(T)) + 1$.

We already know from Theorem 4.9.21 that $c(S) \leq \chi(S)$. For the other direction, consider a group mapping-right letter mapping chain for $S$

$$S \xrightarrow{\gamma_1} S_1 \xrightarrow{\mu_1} T_1 \longrightarrow \cdots \longrightarrow S_n \xrightarrow{\mu_n} T_n$$

of length $n = \chi(S)$. The observation in the first paragraph yields

$$c(S) \geq c(S_1) > c(T_1) \geq c(S_2) > c(T_2) \geq \cdots > c(T_{n-1}) \geq c(S_n) > c(T_n)$$

and so $c(S) \geq n = \chi(S)$, as required. $\qquad\square$

We end this section with an example.

*Example 4.10.12.* Consider the matrix

$$P = \begin{pmatrix} 1 & 1 \\ 1 & -1 \end{pmatrix}$$

over the multiplicative group $\{\pm 1\}$. Let $J = \mathscr{M}^0(\{\pm 1\}, \{a_1, a_2\}, \{b_1, b_2\}, P)$ be the corresponding Rees matrix semigroup. Let $G = \{e, g\}$ be a cyclic group of order 2 with identity $e$. We define a completely regular monoid $S = G \cup J$ by extending the multiplication of $J$ and $G$ to $S$ by defining:

$$(a_i, \epsilon, b_1)g = (a_i, \epsilon, b_2), \ i = 1, 2$$
$$(a_i, \epsilon, b_2)g = (a_i, \epsilon, b_1), \ i = 1, 2$$
$$g(a_1, \epsilon, b_i) = (a_1, \epsilon, b_i), \ i = 1, 2$$
$$g(a_2, \epsilon, b_i) = (a_2, -\epsilon, b_i), \ i = 1, 2.$$

It is straightforward to verify this multiplication is associative. Also a routine calculation shows that $S$ is group mapping with distinguished $\mathscr{J}$-class $J$. Thus $C(S) = S^{\mathscr{L}'} = \mathrm{RLM}_J(S)$. One immediately sees that $g$ switches the two $\mathscr{L}$-classes, and the elements from $J$ provide the two constant maps, so

$\mathsf{RLM}_J(S) \cong G^\sharp \prec \overline{\mathbf{2}}^I \rtimes G$ (by Proposition 4.1.14). We conclude that $\mathsf{RLM}_J(S)$ has complexity 1, and so $S$ has complexity 2 by Corollary 4.10.9. Now consider the reverse monoid $S^{op}$. It also is a completely regular, group mapping monoid with distinguished $\mathscr{J}$-class $J$. So $C(S^{op}) = \mathsf{RLM}_J(S^{op})$. This time, however, $G$ acts trivially on the $\mathscr{L}$-classes, so $\mathsf{RLM}_J(S^{op}) \cong \overline{\mathbf{2}}^I$ is aperiodic. Thus $c(S^{op}) = 1$. In particular, complexity is not reversal invariant.

**Exercise 4.10.13.** Find a group mapping–right letter mapping chain of length 2 for $S$ and of length 1 for $S^{op}$.

Zalcstein [385] exploited this example to construct a sequence of completely regular monoids $\{S_n\}$, $n \geq 2$, with $c(S_n) = n$ and $c(S_n^{op}) = 1$. The idea is to set $S_2 = S$ and set $S_n = G \cup \mathscr{M}^0(S_{n-1}, \{a_1, a_2\}, \{b_1, b_2\}, P')$, for $n \geq 3$, where $P'$ is obtained from $P$ by replacing 1 with $e$ and $-1$ by $g$ and the action of $G$ is defined analogously to above.

**Exercise 4.10.14.** Complete the definition of $S_n$ and verify $c(S_n) = n$, whereas $c(S_n^{op}) = 1$.

M. Kambites [153] used Theorem 4.10.8 to show that if $k$ is a finite field and $n > 1$, then the semigroup of $n \times n$ upper triangular matrices over $k$ has complexity $n - 1$ (it is a result of Putcha that upper triangular matrix semigroups belong to **DS** [18, 250]).

## 4.11 The Karnofsky-Rhodes Decompositions and G * A

In this section, we prove a result of Karnofsky and the first author [156] showing that $\mathbf{G} * \mathbf{V} = \mathbf{LS}^{\mathbf{N}} \, \textcircled{m} \, \mathbf{V}$ if $\mathbf{V} * \mathbf{V} = \mathbf{V}$ and $\mathbf{V}$ is closed under adjunction of identities. In particular, $S \in \mathbf{G} * \mathbf{A}$ if and only if $S^{\mathscr{L}'} \in \mathbf{A}$, which is a decidable criterion. The proof uses a small part of the classification of subdirectly indecomposable semigroups and two decomposition results.

The first result concerns semigroups with an aperiodic regular 0-minimal ideal. The statement given here is weaker than [156, Prop 2.1], but the proof is shorter and avoids transformation semigroups.

**Proposition 4.11.1.** *Let $S$ be a semigroup with an aperiodic regular 0-minimal ideal $I$. Then the natural projection $\varphi \colon S \to S/I \times \mathsf{RLM}_J(S)$ belongs to $\ell 1_D$, where $J = I \setminus 0$.*

*Proof.* By adjoining an identity, may assume without loss of generality that $S$ is a monoid and work with the derived category construction instead of the derived semigroupoid. First note that if $s \notin I$, then $|(s, s\mu_J^R)\varphi^{-1}| = 1$ and so $D_\varphi((s, s\mu_J^R))$ is trivial. Next we consider $D = \mathbf{Der}(\varphi)((0, s\mu_J^R))$ for $s \in I$. If $s = 0$, then because $J$ is regular, $\mu_J^R$ separates $J$ from 0. Thus $|((0, s\mu_J^R))\varphi^{-1}| = 1$ and so again $D_\varphi((0, s\mu_J^R))$ is trivial. Finally if $s \in J$, then the local monoid $D$ consists of elements $(s', s'\varphi)$ with $(ss')\mu_J^R = s\mu_J^R$. Suppose $s_0 \in (0, s\mu_J^R)\varphi^{-1}$. Then $s_0 \in J$. Choose an idempotent $e$ in the $\mathscr{R}$-class of $s$. Then we have

$$L_s = L_e s\mu_J^R = L_e s_0 \mu_J^R$$

and so $es_0 \in J$ and hence $es_0 \mathscr{L} s_0$ by stability. Thus $L_s = L_{es_0} = L_{s_0}$. Now let $(s', s'\varphi) \in D$. Then

$$L_s s' \mu_J^R = L_e (ss')\mu_J^R = L_e s\mu_J^R = L_s$$

and so $L_{ss'} = L_s$. Because $s_0 s' \mathscr{L} ss'$ we obtain $L_{s_0 s'} = L_{ss'} = L_s = L_{s_0}$. By stability, $s_0 s' \mathscr{R} s_0$. So $s_0 s' \mathscr{H} s_0$ and hence $s_0 s' = s_0$ as $I$ is aperiodic. Therefore $D_\varphi((0, s\mu_J^R))$ is trivial, as required.     $\square$

Next we deal with the case of a nilpotent ideal. First we present a lemma, which should be viewed as motivation for our next decomposition.

**Lemma 4.11.2.** *Let $S$ be a finite semigroup with $0$ and let $\mu\colon S \to S^{\mathscr{L}'}$ be the canonical projection. Then $0\mu^{-1}$ is the (unique) maximal nilpotent ideal of $S$.*

*Proof.* Clearly, $0\mu^{-1}$ is an ideal. To see that it is nilpotent, it suffices to show that $0$ is its only idempotent. Indeed, if $0 \neq e \in E(S)$, then $L_e e\mu_{J_e}^R = L_e \neq L_e 0\mu_{J_e}^R$. Thus $e\mu \neq 0\mu$, establishing $0\mu^{-1}$ is a nilpotent ideal. Suppose now $N$ is a nilpotent ideal of $S$ and let $J$ be a non-zero regular $\mathscr{J}$-class of $S$. Then $N \cap J = \emptyset$. Because $JN \subseteq N$, plainly $JN \cap J = \emptyset$. From the definition of $\mu_J^R$, it follows $N\mu_J^R = 0\mu_J^R$. We deduce $N \subseteq 0\mu^{-1}$, thereby establishing the lemma.     $\square$

We now establish the decomposition result. It would be interesting to find a proof using the derived category (again with a perhaps weaker statement). Our proof is taken from [156] and relies on Lemma 4.1.23. Intuition for the proof can be found in [156].

**Proposition 4.11.3.** *Let $S$ be a monoid with maximal nilpotent ideal $N \neq 0$. Let $k \geq 2$ be such that $N^k = 0$ and $J = N^{k-1} \neq 0$. Then*

$$S \prec (M \wr S/N) \times (S/J \wr S^{\mathscr{L}'})$$

*where $M$ is **any** monoid with $|M| > |N|$ and the wreath products are unitary.*

*Proof.* Choose an injective function $n \mapsto m_n$ from $N$ to $M \setminus 1$. We write $[s]$ for the image of $s$ in $S^{\mathscr{L}'}$. Lemma 4.11.2 shows $N = \{s \in S \mid [s] = [0]\}$. For convenience, we denote the image of $s$ in $S/N$ by $[s]_N$. Analogous notation is used for $S/J$. Define a surjective partial function

$$\varphi\colon M \times S/N \times S/J \times S^{\mathscr{L}'} \to S$$

by setting

$$(1, [s]_N, [1]_J, [s])\varphi = s \text{ for } s \in S \setminus N$$
$$(m_n, [0]_N, [s]_J, [0])\varphi = ns \text{ for } n \in N, s \in S.$$

Notice that $\varphi$ is well defined because if $n \in N$ and $s \in J$, then $ns \in NJ = NN^{k-1} = N^k = 0$, and so does not depend on the choice of $s \in J$. Let us first verify $\varphi$ is surjective. If $s \in S \setminus N$, then $s = (1, [s]_N, [1]_J, [s])\varphi$. If $n \in N$, then $n = (m_n, [0]_N, [1]_J, [0])\varphi$. So $\varphi$ is indeed surjective.

Next, for $s \in S$, set

$$\widehat{s} = ((g_s, [s]_N), (f_s, [s])) \in (M \wr S/N) \times (S/J \wr S^{\mathcal{L}'})$$

where $f_s \colon S^{\mathcal{L}'} \to S/J$ and $g_s \colon S/N \to M$ are given by

$$[s']f_s = \begin{cases} [1]_J & [s'] \neq [0] \\ [s]_J & [s'] = [0] \end{cases}$$

$$([s']_N)g_s = \begin{cases} m_{s's} & [s']_N \neq 0, \ s's \in N \\ 1 & \text{else.} \end{cases}$$

We need to show $q\widehat{s}\varphi = q\varphi s$ for all $q$ in the domain of $\varphi$. There are two cases.

*Case 1.* Suppose $s' \in S \setminus N$. Then

$$(1, [s']_N, [1]_J, [s'])\widehat{s}\varphi = \begin{cases} (1, [s's]_N, [1]_J, [s's])\varphi = s's & s's \notin N \\ (m_{s's}, [0]_N, [1]_J, [0])\varphi = s's & s's \in N \end{cases}$$

whereas $(1, [s']_N, [1]_J, [s'])\varphi s = s's$.

*Case 2.* Suppose $n \in N$ and $s' \in S$. Then

$$(m_n, [0]_N, [s']_J, [0])\widehat{s}\varphi = (m_n, [0]_N, [s's]_J, [0])\varphi = ns's$$

whereas $(m_n, [0]_N, [s']_J, [0])\varphi s = ns's$.

An application of Lemma 4.1.23 provides the sought after division.    □

We now prove the main theorem of this section.

**Theorem 4.11.4 (Karnofsky-Rhodes [156]).** *Let* **V** *be a pseudovariety of semigroups closed under semidirect product and adjoining identities. Then* $\mathbf{G} * \mathbf{V} = \mathbf{LS^N} \text{ ⓜ } \mathbf{V}$. *Hence* $S \in \mathbf{G} * \mathbf{V}$ *if and only if* $S^{\mathcal{L}'} \in \mathbf{V}$. *In particular, if* **V** *is decidable, then so is* $\mathbf{G} * \mathbf{V}$.

*Proof.* We already know from Corollary 4.6.14 that $\mathbf{G} * \mathbf{V} \leq \mathbf{LS^N} \text{ ⓜ } \mathbf{V}$. As every finite semigroup is a subdirect product of subdirectly indecomposable semigroups, for the reverse inequality, it suffices to show that if $S \in \mathbf{LS^N} \text{ ⓜ } \mathbf{V}$ is subdirectly indecomposable, then $S \in \mathbf{G} * \mathbf{V}$. By Theorem 4.6.50, this means $S^{\mathscr{L}'} \in \mathbf{V}$, or equivalently $\mathsf{RLM}_J(S) \in \mathbf{V}$ for all regular $\mathscr{J}$-classes $J$ of $S$ (Theorem 4.6.46).

Because $\mathbf{V}$ is closed under adjoining identities, we may assume without loss of generality $S$ is a monoid and we may work the semidirect product of monoid pseudovarieties. By Lemma 4.7.5, $S$ has a unique 0-minimal ideal. Suppose first that the ideal is regular, i.e., $S$ is of semisimple type. Then Theorem 4.7.6 implies that either $S$ is group mapping or $S$ has an aperiodic 0-minimal ideal $I$. In the first case, Lemma 4.6.54 implies $S \in \mathbf{G} * \mathbf{V}$. In the second case, we may assume by induction that $S/I \in \mathbf{G} * \mathbf{V}$ and hence $S/I \times \mathsf{RLM}_J(S) \in \mathbf{G} * \mathbf{V}$ where $J = I \setminus 0$. Proposition 4.11.1 implies that the derived category of the natural projection $S \to S/I \times \mathsf{RLM}_J(S)$ is locally trivial. Because $\ell\mathbf{1} \leq \mathbf{gG}$ [364], it follows from the Derived Category Theorem $S \in \mathbf{G} * \mathbf{G} * \mathbf{V} = \mathbf{G} * \mathbf{V}$.

Finally, suppose $S$ is of null type, that is, it has a null 0-minimal ideal. Then $S$ has a non-trivial maximal nilpotent ideal $N$. Retaining the notation from Lemma 4.11.3, we have $S \prec (M \wr S/N) \times (S/J \wr S^{\mathscr{L}'})$ for any sufficiently large monoid $M$, in particular for any sufficiently large group $M$. Moreover, the wreath products here are unitary. By induction we may assume $S/N$ and $S/J$ are in $\mathbf{G} * \mathbf{V}$. By assumption $S^{\mathscr{L}'} \in \mathbf{V}$ and hence

$$(S/J)^I \wr S^{\mathscr{L}'} \in \mathbf{G} * \mathbf{V} * \mathbf{V} = \mathbf{G} * \mathbf{V}.$$

On the other hand, because $S/N \in \mathbf{G} * \mathbf{V}$, we conclude $M \wr S/N \in \mathbf{G} * \mathbf{G} * \mathbf{V} = \mathbf{G} * \mathbf{V}$. Therefore, $S \in \mathbf{G} * \mathbf{V}$, completing the proof of the first statement. The remaining statements follow directly from Theorem 4.6.50.    $\square$

**Corollary 4.11.5 (Karnofsky-Rhodes [156]).** *The pseudovariety $\mathbf{G} * \mathbf{A}$ is decidable.*

The description of $\mathbf{G} * \mathbf{A}$ furnished by Theorem 4.11.4, namely $S \in \mathbf{G} * \mathbf{A}$ if and only if $S^{\mathscr{L}'} \in \mathbf{A}$, was used by the second author to prove that $\mathbf{G} * \mathbf{A}$ is local [333].

*Remark 4.11.6.* Notice that the closure of $\mathbf{V}$ under semidirect product was only used in the case of a null ideal. So if $S$ is a regular monoid, then the proof shows $S \in \mathbf{G} * \mathbf{V}$ if and only if $S \in \mathbf{LS}^{\mathbf{N}} \textcircled{m} \mathbf{V}$ for any pseudovariety $\mathbf{V}$.

## 4.12 Lower Bounds for Complexity

So far, we only know how to compute complexity for semigroups in $\mathbb{L}\mathbf{G} \textcircled{m} \mathbf{A}$. In this section, we would like to compute the complexity of the full transformation monoid, which is easily seen not to belong to $\mathbb{L}\mathbf{G} \textcircled{m} \mathbf{A}$. An upper bound follows directly from Proposition 4.1.48. To obtain lower bounds, we need some techniques. This section presents the Type I–Type II lower bound method of Rhodes and Tilson [294, 295].

Upper bounds for complexity can often be quite easy to obtain: for instance having an explicit wreath product decomposition, or a division into a semigroup of known complexity yields an upper bound; the depth and the (local) Tilson number are other easy to compute upper bounds. A subtle point in complexity theory is constructing lower bounds. The notion of inevitability, discussed in Chapter 3, has it origins in the work of Rhodes and Tilson [295] on lower bounds for complexity, the idea being that you have to find what obstructions always come up under relational morphisms to aperiodic semigroups or groups in order to compute a lower bound.

### 4.12.1 Constructing lower bounds for complexity

If $S$ is a semigroup and $\mathbf{V}, \mathbf{W}$ are a pseudovarieties, then a subsemigroup $U \leq S$ is said to be $\mathbf{V}$-*like* with respect to $\mathbf{W}$ if, for all relational morphisms $\varphi \colon S \to T$ with $T \in \mathbf{W}$, there exists $V \leq T$ with $V \in \mathbf{V}$ and $U \leq V\varphi^{-1}$; see Definition 3.6.25. In other words, $U$ is always covered by a semigroup from $\mathbf{V}$ under relational morphisms to $\mathbf{W}$. Rhodes and Tilson defined [295] a subsemigroup $U \leq S$ to be of *Type I* if it is $\mathbf{ER}$-like with respect to $\mathbf{A}$. They defined an element $s \in S$ to be of *Type II* if $s \in 1\varphi^{-1}$, for all relational morphisms $\varphi \colon S \to G$ with $G$ a group.

**Exercise 4.12.1.** Prove that a subgroup is always a Type I subsemigroup and a product of idempotents is always of Type II.

**Definition 4.12.2 (Absolute Type I semigroup).** *A semigroup $S$ is said to be an absolute Type I semigroup if it is a Type I subsemigroup of itself.*

Let us say that an element $s \in S$ is a *weak conjugate* of $t \in S$ if there exist $x, y \in S$ such that $xyx = x$ and $s = xty$ or $s = ytx$. One says that $x$ is a *weak inverse* of $y$. Notice that $y$ need not be regular. The following key observation is from [295, 365].

**Lemma 4.12.3.** *Suppose $\varphi \colon S \to G$ is a relational morphism with $G$ a group. Let $x, y \in S$ and assume $xyx = x$. Let $g \in y\varphi$ be arbitrary. Then $g^{-1} \in x\varphi$.*

*Proof.* Let $h \in y\varphi$. Observe $(xy)^k x = x$ for all $k \geq 1$. Let $n \geq 2$ be an exponent of $G$. Then $g^{-1} = (hg)^{n-1}h \in ((xy)^{n-1}x)\varphi = x\varphi$, as required.    □

Now we can establish some basic facts about the set of Type II elements.

**Proposition 4.12.4.** *The set $\mathsf{K}_{\mathbf{G}}(S)$ of all Type II elements of $S$ is a subsemigroup containing the idempotents $E(S)$, which is closed under taking weak conjugates.*

*Proof.* Let $\varphi \colon S \to G$ be a relational morphism with $G$ a group. Clearly, $x, y \in 1\varphi^{-1}$, implies $xy \in 1\varphi^{-1}$. It follows $\mathsf{K}_{\mathbf{G}}(S)$ is a subsemigroup. If $e \in S$ is an idempotent and $g \in e\varphi$, then $1 = g^\omega \in e\varphi$. We conclude $E(S) \leq \mathsf{K}_{\mathbf{G}}(S)$. Finally, suppose $s \in 1\varphi^{-1}$ and $xyx = x$ with $x, y \in S$. Choose $g \in y\varphi^{-1}$. By Lemma 4.12.3, we have $g^{-1} \in x\varphi^{-1}$. It follows that $1 \in (xsy)\varphi \cap (ysx)\varphi$. We conclude $\mathsf{K}_{\mathbf{G}}(S)$ is closed under taking weak conjugates.    □

The famous Rhodes Type II Conjecture asserted that $\mathsf{K_G}(S)$ is the smallest subsemigroup containing $\langle E(S) \rangle$, which is closed under taking weak conjugates. The conjecture was resolved in the affirmative by Ash [33], and independently by Ribes and Zalesskii [300]. We shall provide a proof of this difficult result in Section 4.17.

In order to establish a certain functoriality property of $\mathsf{K_G}$, we need a compactness result similar to the sort considered in Chapter 3.

**Proposition 4.12.5.** *Let $S$ be a semigroup. Then there exists a relational morphism $\varphi\colon S \to G$ with $G$ a group such that $1\varphi^{-1} = \mathsf{K_G}(S)$.*

*Proof.* For each $s \notin \mathsf{K_G}(S)$, we can find a relational morphism $\varphi_s\colon S \to G_s$ with $1 \notin s\varphi_s$. Then $\varphi = \Delta \prod_{s \notin \mathsf{K_G}(S)} \varphi_s\colon S \to \prod_{s \notin \mathsf{K_G}(S)} G_s$ is easily verified to do the job.                                                                    □

Let us turn to the desired functoriality:

**Proposition 4.12.6.** *Let $\varphi\colon S \to T$ be a semigroup homomorphism. Then $\mathsf{K_G}(S)\varphi \leq \mathsf{K_G}(T)$. Moreover, if $\varphi$ is onto, then $\mathsf{K_G}(S)\varphi = \mathsf{K_G}(T)$.*

*Proof.* Suppose $\psi\colon T \to G$ is a relational morphism with $G$ a group and let $s \in \mathsf{K_G}(S)$. Then $\varphi\psi\colon S \to G$ is a relational morphism and so $1 \in s\varphi\psi$. It follows that $s\varphi \in \mathsf{K_G}(T)$.

Suppose now $\varphi\colon S \twoheadrightarrow T$ is onto. Let $\psi\colon S \to G$ be a relational morphism with $1\psi^{-1} = \mathsf{K_G}(S)$, as per Proposition 4.12.5. Then $\varphi^{-1}\psi\colon T \to G$ is a relational morphism, so if $t \in \mathsf{K_G}(T)$, then $1 \in t\varphi^{-1}\psi$. Thus, there exists $s \in S$ with $t = s\varphi$ and $1 \in s\psi$. By choice of $\psi$, $s \in \mathsf{K_G}(S)$ and so $t \in \mathsf{K_G}(S)\varphi$, as required.                                                                    □

Our next lemma shows that Type I subsemigroups behave well with respect to the operation of adjoining an identity.

**Lemma 4.12.7.** *Let $U$ be a Type I subsemigroup of a semigroup $S$. Then $U^I$ is a Type I subsemigroup of $S^I$.*

*Proof.* Let $\varphi\colon S^I \to T$ be a relational morphism with $T$ aperiodic. Consider the canonical factorization $\varphi = \alpha^{-1}\beta$ where $\alpha\colon R \twoheadrightarrow S^I$ and $\beta\colon R \to T$. Let $e \in I\alpha^{-1}$ be an idempotent. Then $(eRe)\alpha = S^I$ and so replacing $T$ by $(eRe)\beta$ and restricting $\varphi$, we may assume without loss of generality that $T$ is a monoid and $\varphi$ is a relational morphism of monoids. Because $U$ is a Type I subsemigroup of $S$, there is a subsemigroup $V \leq T$ with $V \in \mathbf{ER}$ and $U \leq V\varphi^{-1}$. Then $V \cup \{1\} \in \mathbf{ER}$ and $U^I \leq (V \cup \{1\})\varphi^{-1}$, as required.                                                                    □

Our next theorem underlies the Type I–Type II lower bound of Rhodes and Tilson [295]. The idea is that by passing to a Type I subsemigroup, one can remove an aperiodic from the right end of a decomposition into a wreath product of groups and aperiodic semigroups to obtain a decomposition ending in a group. By passing further to the Type II subsemigroup, one can remove the group from the end, thereby reducing the complexity.

**Theorem 4.12.8 (Rhodes-Tilson).** *Let $S$ be a semigroup that is not aperiodic and let $U \leq S$ be a Type I subsemigroup. Then $c(\mathsf{K_G}(U)) < c(S)$.*

*Proof.* Suppose that $c(S) = n > 0$. Then, by Proposition 4.3.14, $c(S^I) = n$, as well. Because $\mathsf{K_G}(U)^I \leq \mathsf{K_G}(U^I)$, we may assume without loss of generality (applying Lemma 4.12.7) $S$ is a monoid and $U$ is a submonoid. By the definition of complexity $S \in \mathbf{C}_{n-1} * \mathbf{G} * \mathbf{A}$. Hence there is a semidirect product $T \rtimes A$ with $T \in \mathbf{C}_{n-1} * \mathbf{G}$ and $A \in \mathbf{A}$, and a division $d \colon S \to T \rtimes A$. Let $\pi \colon T \rtimes A \to A$ be the projection and consider the relational morphism $\varphi = d\pi \colon S \to A$. As $U$ is Type I, there is a subsemigroup $R \leq A$ belonging to $\mathbf{ER}$ such that $U \leq R\varphi^{-1} = R\pi^{-1}d^{-1}$. So $U \prec R\pi^{-1} = T \rtimes R$. This yields

$$U \prec T \rtimes R \in \mathbf{C}_{n-1} * \mathbf{G} * \mathbf{ER} = \mathbf{C}_{n-1} * \mathbf{ER} \leq \mathbf{C}_{n-1} * \mathbf{A} * \mathbf{G} = \mathbf{C}_{n-1} * \mathbf{G}$$

where the first equality uses that $\mathbf{ER}$ is closed under semidirect product and where the inequality comes by way of Theorem 4.8.4.

Thus there is a semidirect product $C \rtimes G$ with $G$ a group and $C \in \mathbf{C}_{n-1}$ and a subsemigroup $V \leq C \rtimes G$ mapping onto $U$ via a homomorphism $\psi \colon V \twoheadrightarrow U$. By choosing $e \in 1\psi^{-1}$ and cutting to $eVe$, we may assume without loss of generality that $V$ is a monoid. Let $\pi \colon C \rtimes G \to G$ be the projection. Set $V' = \mathsf{K_G}(V)$. On the one hand $V'\psi = \mathsf{K_G}(U)$ by Proposition 4.12.6; on the other hand $V'\pi = 1$, and so Lemma 4.1.31 implies that $V' \prec C$ as $V'$ is a submonoid. We conclude $\mathsf{K_G}(U) \in \mathbf{C}_{n-1}$, as required. $\square$

An immediate consequence is the following theorem of Rhodes and Tilson.

**Theorem 4.12.9 (Rhodes-Tilson [294, 295]).** *Let $S$ be a semigroup. Define a **Type I–Type II chain** of length $n$ to be a chain of subsemigroups*

$$S = U_0 \geq T_1 \geq U_1 \geq T_2 \geq U_2 \geq \cdots \geq T_{n-1} \geq U_{n-1} \geq T_n \geq U_n$$

*where $T_i$ is a Type I subsemigroup of $U_{i-1}$ and $U_i$ is a Type II subsemigroup of $T_i$, for $i = 1, \ldots, n$, with $U_i$ non-aperiodic, for $i = 0, \ldots, n-1$. Let $\#(S)$ be the maximum length of a Type I–Type II chain for $S$. Then $\#(S) \leq c(S)$.*

**Exercise 4.12.10.** Prove Theorem 4.12.9.

We remark that it was shown in [170,171] that $\#(S) = c(S)$ for completely regular semigroups. To employ Theorem 4.12.8 (or Theorem 4.12.9), we need a ready supply of Type I subsemigroups. Rhodes and Tilson gave a nice sufficient condition for a semigroup to be an absolute Type I semigroup [294, 295]. This condition is satisfied for instance by full transformation monoids and by full linear monoids. Later it was shown [126] this sufficient condition is necessary.

**Definition 4.12.11 ($\mathscr{T}_1$-semigroup).** *A semigroup $S$ is called a $\mathscr{T}_1$-semigroup if there exists an $\mathscr{L}$-chain $s_1 \geq_{\mathscr{L}} s_2 \geq_{\mathscr{L}} \cdots \geq_{\mathscr{L}} s_n$ of elements of $S$ such that $S = \langle s_1, s_2, \ldots, s_n \rangle$.*

For instance, we shall see below that the full transformation monoid $T_n$ can be generated by the permutations and a single idempotent, and hence is a $\mathcal{T}_1$-semigroup. As you may have guessed, the $\mathcal{T}_1$ stands for Type I. Let us begin with some basic properties of $\mathcal{T}_1$-semigroups. The first proposition says that they "lift and push" under homomorphisms.

**Proposition 4.12.12.** *Let* $\alpha\colon R \twoheadrightarrow S$ *be an onto homomorphism. If* $R$ *is a* $\mathcal{T}_1$-*semigroup, then so is* $S$. *If* $S$ *is a* $\mathcal{T}_1$-*semigroup, then there is a* $\mathcal{T}_1$-*subsemigroup* $R_0 \leq R$ *such that* $R_0\alpha = S$.

*Proof.* The first assertion is trivial. For the second, by replacing $R$ with a minimal subsemigroup $R_0$ of $R$ such that $R_0\alpha = S$, we may assume no proper subsemigroup of $R$ maps onto $S$ under $\alpha$. We shall prove under these hypotheses $R$ is a $\mathcal{T}_1$-semigroup. Let $S = \langle s_1, s_2, \ldots, s_n \rangle$ with $s_1 \geq_{\mathscr{L}} s_2 \geq_{\mathscr{L}} \cdots \geq_{\mathscr{L}} s_n$. Choose any $r_1 \in R$ with $r_1\alpha = s_1$. Assume $1 < i \leq n$ and we have found $r_1, r_2, \ldots, r_{i-1} \in R$ with $r_j\alpha = s_j$, for $1 \leq j \leq i-1$, and with $r_1 \geq_{\mathscr{L}} r_2 \geq_{\mathscr{L}} \cdots \geq_{\mathscr{L}} r_{i-1}$. Then $s_i \in S^I s_{i-1} = (R^I r_{i-1})\alpha$ and so we may find $r_i \leq_{\mathscr{L}} r_{i-1}$ such that $r_i\alpha = s_i$. This completes the induction. Now clearly $\langle r_1, \ldots, r_n \rangle\alpha = S$, so $R = \langle r_1, \ldots, r_n \rangle$ and hence is a $\mathcal{T}_1$-semigroup. $\qquad\square$

Next we establish that an aperiodic $\mathcal{T}_1$-semigroup belongs to **ER**. The proof goes by induction on the minimal length of a decomposition into a wreath product of copies of $\overline{\mathbf{2}}^I$. It would be interesting to find a direct proof. Define recursively a sequence of aperiodic semigroups $U_2^{(n)}$ by $U_2^{(0)} = \{1\}$ and $U_2^{(n+1)} = U_2^{(n)} \wr U_2$, for $n \geq 0$, where we use the unitary wreath product.

**Proposition 4.12.13.** *If* $S$ *is an aperiodic* $\mathcal{T}_1$-*semigroup, then* $S \in$ **ER**.

*Proof.* We first prove by induction on $n$ that any $\mathcal{T}_1$-subsemigroup of $U_2^{(n)}$ belongs to **ER**. The case of $U_2^{(0)}$ being trivial, suppose by way of induction that any $\mathcal{T}_1$-subsemigroup of $U_2^{(n)}$ belongs to **ER** and let $R \leq U_2^{(n+1)}$ be a $\mathcal{T}_1$-semigroup. Say $R = \langle r_1, \ldots, r_n \rangle$ where $r_1 \geq_{\mathscr{L}} r_2 \geq_{\mathscr{L}} \cdots \geq_{\mathscr{L}} r_n$. We need to show, by Theorem 4.8.3, that $R$ contains no copy of $\overline{\mathbf{2}}$. So suppose $x, y \in R$ form a right zero subsemigroup. Let $\pi\colon U_2^{(n)} \wr U_2 \to U_2$ be the projection. Then $R\pi$ is a $\mathcal{T}_1$-semigroup by Proposition 4.12.12. But the only non-singleton $\mathscr{L}$-chains of $U_2$ are $I >_{\mathscr{L}} \overline{0}$ and $I >_{\mathscr{L}} \overline{1}$. So there are idempotents $f \leq e$ of $U_2$ (possibly $e = f$) such that $R \leq \{e, f\}\pi^{-1}$. It follows that $x, y \in f\pi^{-1}$ or $f < e = I$ and $x, y \in I\pi^{-1}$. We handle the former case first.

Let $\psi\colon \{e, f\}\pi^{-1} \to (U_2^{(n)})^{U_2}$ be the homomorphism from Lemma 4.1.31. Applying Proposition 4.12.12, it is straightforward to verify that $R\psi$ is a subdirect product of $\mathcal{T}_1$-subsemigroups of the factors $U_2^{(n)}$, and so belongs to **ER** by induction. In particular, $R\psi$ does not contain a copy of $\overline{\mathbf{2}}$, so $x\psi = y\psi$. But $\psi$ is injective on right zero subsemigroups of $f\pi^{-1}$ by Lemma 4.1.31, so $x = y$ in this case.

For the case that $f < e = I$ and $x, y \in I\pi^{-1}$, choose the largest index $i$ such that $r_i\pi = I$. Then if $j > i$, we must have $r_i >_{\mathscr{J}} r_j$ because $r_j\pi = f <_{\mathscr{J}} I$.

It follows $x, y \in R' = \langle r_1, \ldots, r_i \rangle$ and $r_1 \geq_{\mathscr{L}} \cdots \geq_{\mathscr{L}} r_i$ in $R'$, so $R'$ is a $\mathscr{T}_1$-semigroup. Then $R' \leq I\pi^{-1}$ and so Lemma 4.1.31 provides a homomorphism $\psi' \colon R' \to (U_2^{(n)})^{U_2}$, which is injective on right zero semigroups. Similarly to the previous case, induction yields $R'\psi' \in \mathbf{ER}$ and so $x\psi' = y\psi'$, yielding $x = y$. Therefore, $R$ does not contain a copy of $\overline{2}$ and so belongs to $\mathbf{ER}$.

Suppose now $S$ is an aperiodic $\mathscr{T}_1$-semigroup. By the Prime Decomposition Theorem $S \prec U_2^{(n)}$ for some $n$. A routine application of Proposition 4.12.12 shows that $S$ is a quotient of a $\mathscr{T}_1$-subsemigroup of $U_2^{(n)}$ and so $S \in \mathbf{ER}$ by what we have already proved. $\qquad\square$

The previous results allow us to establish easily that $\mathscr{T}_1$-semigroups are absolute Type I semigroups.

**Theorem 4.12.14 (Rhodes-Tilson).** *Let $S$ be a $\mathscr{T}_1$-semigroup. Then $S$ is an absolute Type I semigroup.*

*Proof.* Suppose that $\varphi \colon S \to T$ is a relational morphism with $T$ aperiodic. Let $\varphi = \alpha^{-1}\beta$ be the canonical factorization where $\alpha \colon R \twoheadrightarrow S$ and $\beta \colon R \to T$. By Proposition 4.12.12, there is a $\mathscr{T}_1$-subsemigroup $R_0 \leq R$ such that $R_0\alpha = S$. Then $U = R_0\beta$ is a $\mathscr{T}_1$-semigroup, again by Proposition 4.12.12. Because $U$ is aperiodic, it belongs to $\mathbf{ER}$ by Proposition 4.12.13. As $S \leq U\beta^{-1}\alpha = U\varphi^{-1}$, we conclude $S$ is indeed an absolute Type I semigroup. $\qquad\square$

**Exercise 4.12.15.** Let $S$ be a monoid generated by its group of units and an idempotent. Show that $S$ is a **Sl**-like subsemigroup of itself with respect to $\mathbf{A}$.

**Exercise 4.12.16.** Suppose that $S$ is a semigroup generated by an $\mathscr{L}$-chain $s_1 \geq_{\mathscr{L}} \cdots \geq_{\mathscr{L}} s_n$ of elements satisfying $s_i = s_i^{\omega+1}$. Without using the Prime Decomposition Theorem, show that $S$ is an **LRB**-like subsemigroup of itself with respect to $\mathbf{A}$.

We can now state without proof the main result of [292], describing the local complexity function $c_\ell$.

**Theorem 4.12.17 (Rhodes-Tilson [292]).** *Let $S$ be a semigroup. Then the local complexity $c_\ell$ is the maximum length of a chain of subsemigroups*

$$S = U_0 \geq T_1 \geq U_1 \geq T_2 \geq U_2 \geq \cdots \geq T_{n-1} \geq U_{n-1} \geq T_n \geq U_n$$

*where $T_i$ is an absolute Type I semigroup and $U_i$ is the Type II subsemigroup of $T_i$, for $i = 1, \ldots, n$, and $U_i$ is non-aperiodic, for $i = 0, \ldots, n-1$.*

By Ash's Theorem [33] the Type II subsemigroup of a semigroup is computable. On the other hand, in [126] it was shown that it is decidable whether a semigroup is absolute Type I: namely the converse of Theorem 4.12.14 was established. Hence the local complexity function is computable. With this serving as motivation, let us prove the converse of Theorem 4.12.14, beginning with the following simple lemma.

**Lemma 4.12.18.** *Let $\varphi\colon S \twoheadrightarrow T$ be a surjective homomorphism. If $S$ is an absolute Type I semigroup, then so is $T$.*

*Proof.* Let $\psi\colon T \to A$ be a relational morphism to an aperiodic semigroup $A$. The $\varphi\psi\colon S \to A$ is a relational morphism, so there is a subsemigroup $U \leq A$ with $U \in \mathbf{ER}$ and $S \leq U\psi^{-1}\varphi^{-1}$. So $T = S\varphi \leq U\psi^{-1}\varphi^{-1}\varphi = U\psi^{-1}$, because $\varphi^{-1}\varphi = 1_T$ as $\varphi$ is surjective. Thus $T$ is an absolute Type I semigroup.     □

**Theorem 4.12.19 (Henckell-Margolis-Pin-Rhodes [126]).** *A semigroup $S$ is an absolute Type I semigroup if and only if it is a $\mathscr{T}_1$-semigroup.*

*Proof.* Theorem 4.12.14 provides one direction of the theorem. For the converse, we proceed by induction on the size of $S$. There is nothing to prove if $|S| \leq 1$.

Assume every absolute Type I semigroup having at most $k \geq 1$ elements is a $\mathscr{T}_1$-semigroup and let $S$ be an absolute Type I semigroup having $k + 1$ elements. Let $V$ be a 0-minimal ideal of $S$ (so, by our convention, $V$ is the minimal ideal if $S$ has no zero). Denote by $J$ the non-zero $\mathscr{J}$-class of $V$. Let $T = S/V$ be the Rees quotient. By Lemma 4.12.18, $T$ is an absolute Type I semigroup. Because $T$ has at most $k$ elements, by induction $T$ is generated by an $\mathscr{L}$-chain $X$ of its elements.

Let $K$ be the subsemigroup of $S$ generated by the non-zero elements of $X$ (viewed as elements of $S$) and 0 (if $S$ has a zero). Notice that if $S$ is completely simple, then $K$ is empty whereas if $S$ is 0-simple or a 2-element null semigroup, then $K = \{0\}$. If $K = S$, then $S$ is a $\mathscr{T}_1$-semigroup, and we are done. So assume $K \subsetneq S$. In any event $S \setminus K \subseteq J$. So let $Y$ be the set of all $\mathscr{L}$-classes of $J$ not contained in $K$. By assumption, $Y \neq \emptyset$. Let us define a preorder on $Y$ by $L \geq L'$ if there exists $k \in K^I$ with $Lk = L'$. Denote by $[L]$ the equivalence class of $L$ with respect to this preorder; we also use $\geq$ for the induced partial order on the equivalence classes.

We claim that $Y$ has a unique $\geq$-maximal equivalence class. To prove this claim, first let $[L_0], \ldots, [L_{n-1}]$ be the $\geq$-maximal classes of $Y$. Set

$$Z_i = \left\{ s \in \bigcup Y \mid [L_s] = [L_i] \right\}$$
$$Z = S \setminus \bigcup_{i=0}^{n-1} Z_i$$

for $0 \leq i \leq n - 1$. Let $X$ be the subset of $S \times U_n$ consisting of all pairs $(s, \bar{\imath})$ with $s \in Z_i$ and all pairs $(s, I)$ with $s \in Z$. Let $\varphi\colon S \to U_n$ be the relational morphism whose graph is generated by $X$. We first establish two claims.

*Claim.* Let $s \in Z_i \setminus K$. Then $I \notin s\varphi$.

*Proof.* Because $s \notin K$, it follows $0 \neq s \in J$. Then as $I \in s\varphi$, we can factor $s = s_1 \cdots s_m$ with $s_k \in Z$ all $1 \leq k \leq m$. Because $s \notin K$, there is an index $k$

with $s_k \notin K$. Let $r$ be the largest index with $s_r \notin K$; in particular, $s_r \in J$. Let $x = s_1 \cdots s_{r-1}$ and $y = s_{r+1} \cdots s_m$. Then $y \in K^I$. As $s, s_r \in J$, stability yields $s = x s_r y \mathscr{L} s_r y$. So $L_s = L_{s_r y} = L_{s_r} y$. As $y \in K^I$, this means $[L_{s_r}] \geq [L_s]$ (notice $s_r \notin K$ implies $L_{s_r} \in Y$). Thus $[L_s] = [L_{s_r}]$ because $[L_s] = [L_i]$ is maximal. Therefore, $s_r \in Z_i$, contradicting $s_r \in Z$. This proves the claim. $\quad\square$

*Claim.* Let $s \in Z_i \setminus K$. Then $s\varphi = \{\bar{i}\}$.

*Proof.* By definition of $\varphi$ we have $\bar{i} \in s\varphi$. For the converse, the previous claim asserts $I \notin s\varphi$. Suppose $\bar{j} \in s\varphi$ with $j \neq i$. Then $s = s_1 \cdots s_m$ where, for some $r$, we have $s_r \in Z_j$ and $s_k \in Z$ for $k > r$. Let $x = s_1 \cdots s_{r-1}$ and $y = s_{r+1} \cdots s_m$. Because $s, s_r \in J$, it follows that $s = x s_r y \mathscr{L} s_r y$ by stability. If $y \notin K^I$, then $y \in J$ and so stability yields $s = x s_r y \mathscr{L} y$. But then $y \in Z_i \setminus K$ and so the previous claim implies $I \notin y\varphi$, a contradiction as $I \in s_k \varphi$ for $k > r$. So suppose $y \in K^I$. Then $L_s = L_{s_r} y$ implies $[L_{s_r}] \geq [L_s]$ and so by maximality of $[L_s] = [L_i]$, it follows $[L_{s_r}] = [L_i]$. But $s_r \in Z_j$ implies $[L_{s_r}] = [L_j]$ yielding $i = j$, a contradiction. This establishes the claim. $\quad\square$

Because $S$ is absolute Type I and the maximal subsemigroups of $U_n$ that belong to $\mathbf{ER}$ are of the form $\{I, \bar{j}\}$, we conclude $S \leq \{I, \bar{j}\}\varphi^{-1}$ for some $0 \leq j \leq n - 1$. It is then easy to see that $[L_j]$ is the unique maximal $\geq$-equivalence class. Indeed, if $i \neq j$, then because $L_i \not\subseteq K$ by definition of $Y$ we can find $s \in L_i \setminus K$. Then $s \in Z_i \setminus K$, so the second claim yields $s\varphi = \{\bar{i}\}$. This contradicts $S \leq \{I, \bar{j}\}\varphi^{-1}$ and establishes the required uniqueness.

Let $[L]$ be the unique maximal $\geq$-equivalence class of $Y$. Suppose that $L_1 >_{\mathscr{L}} \cdots >_{\mathscr{L}} L_r(>_{\mathscr{L}} 0)$ is an $\mathscr{L}$-chain generating $K$. We may assume $L_r >_{\mathscr{J}} J$ by definition of $X$ and $K$. Without loss of generality, we may assume $L_r \not\subseteq \langle L_1, \ldots, L_{r-1} \rangle$, as otherwise we could omit it. The definition of $\geq$ yields $S = \langle K, L \rangle$. It suffices therefore to show that we may choose $L' \in [L]$ with $L_r >_{\mathscr{L}} L'$. Let $\varphi \colon S \to U_2$ be the relational morphism whose graph is generated by the pairs:

- $(s, I)$ with $s \in \langle L_1, \ldots, L_{r-1} \rangle$;
- $(s, \bar{0})$ with $s \in L_r$;
- $(s, \bar{1})$ with $s \in L$.

Because $L_r >_{\mathscr{J}} L$, we conclude $\bar{1} \notin L_r \varphi$. Also, as $L_r \not\subseteq \langle L_1, \ldots, L_{r-1} \rangle$, it follows there exists $s \in L_r$ with $I \notin s\varphi$. Therefore $s\varphi = \{\bar{0}\}$. Because $S$ is absolute Type I and $\{I, \bar{0}\}$ is the unique maximal subsemigroup of $U_2$ belonging to $\mathbf{ER}$ and containing $\bar{0}$, it follows $S \subseteq \{I, \bar{0}\}\varphi^{-1}$. By construction $I\varphi^{-1} \subseteq K$. Because $L$ is not contained in $K$, we can find $x \in L \setminus K$. As $I \notin x\varphi$, we must have $\bar{0} \in x\varphi$, so $x = yts$ where $t \in L_r$ and $s \in I\varphi^{-1} \subseteq K$ or is an adjoined identity $I$. As $s \in K^I$ but $x \notin K$, we conclude that $yt \notin K$. Clearly $[L_{yt}] \geq [L_{yts}] = [L_x] = [L]$, by the definition of $\geq$. Because $[L]$ was maximal, it follows $[L_{yt}] = [L]$. But $yt <_{\mathscr{L}} t \in L_r$. So taking $L' = L_{yt}$ completes the proof. $\quad\square$

In [124, 130], a more general notion than a Type I subsemigroup is considered and characterized.

### 4.12.2 The complexity of full transformation and linear monoids

A standard question, given a complexity hierarchy, is to determine whether it is strict. Rhodes proved that the full transformation monoid $T_n$ has group complexity $n - 1$ for all $n \geq 1$ [266] and so the complexity hierarchy is indeed strict. The proof requires the tools of the previous subsection as well as some further decomposition results that are of interest in their own right. We also calculate the complexity of the monoid $M_n(\mathbb{F}_q)$ of $n \times n$ matrices over the finite field $\mathbb{F}_q$ of order $q$.

#### Some further wreath product decompositions

To compute the complexity of $T_n$ and $M_n(\mathbb{F}_q)$, we need to show if $e \in E(S)$, then $c(SeS) = c(eSe)$. This is a primitive form, due to Rhodes and Allen [3], of the Reduction Theorem of Rhodes and Tilson [291, 363]. The idea is that multiplication $Se \times eSe \times eS \rightarrow SeS$ serves as a coordinatization.

**Proposition 4.12.20.** *Let $S$ be a semigroup and $e \in E(S)$. Then*

$$SeS \prec (Se)^\ell \times (eSe \wr (eS^I, \overline{eS})).$$

*Hence $c(SeS) = c(eSe)$.*

*Proof.* We apply Lemma 4.1.23 to the faithful transformation semigroups $((SeS)^I, SeS)$ and $((Se)^I, (Se)^\ell) \times (eSe, eSe) \wr ((eS)^I, \overline{eS})$. Define a partial surjective function $f \colon (Se)^I \times eSe \times (eS)^I \rightarrow (SeS)^I$ by $(I, e, I)f = I$, $(x, y, z)f = xyz$ if $x \neq I \neq z$ and otherwise $f$ is undefined. Clearly, $f$ is onto because $xey = f(x, e, y)$. For $s \in SeS$, write $s = uev$. Then set $\widehat{s} = (ue, (f_s, \overline{ev}))$, where $If_s = e$ and $zf_s = zue$ for $z \in eS$. The element $\widehat{s}$ covers $s$ because

$$(I, e, I)\widehat{s}f = (ue, e(If_s), ev)f = (ue, e, ev)f = uev = s = (I, e, I)fs$$
$$(x, y, z)\widehat{s}f = (x, y(zf_s), ev)f = (x, yzue, ev)f = xyzuev = xyzs = (x, y, z)fs$$

for $x \neq I \neq y$. This yields the desired division. The complexity result follows from $eSe \leq SeS$ and because $(Se)^\ell$ and $\overline{eS}$ are aperiodic. $\qquad\square$

The above result is often proved by first writing $SeS$ as a quotient of a Rees matrix semigroup over $eSe$ and then obtaining the wreath product decomposition [363]. This is related to Morita equivalence of semigroups: Talwar proved that $SeS$ and $eSe$ are Morita equivalent [352]. Further connections between Morita equivalence and the results of Rhodes and Allen [3] can be found in [352–354].

**Corollary 4.12.21.** *Let $S$ be a simple semigroup and $\mathbf{H}$ a non-trivial pseudovariety of groups containing the maximal subgroup of $S$. Then $S \in \mathbf{H} * \mathbf{RZ}$. If $S$ is 0-simple with maximal subgroup in a pseudovariety $\mathbf{H}$ (possibly trivial), then $S \in (\mathbf{H} \vee \mathbf{Sl}) * \mathbf{RZ}$. In particular, the complexity of a (0-)simple semigroup is at most one.*

*Proof.* If **V** is any pseudovariety of semigroups containing a non-trivial monoid, then $\mathbf{LZ} \leq \mathbf{V} * \mathbf{1}$ by Lemma 4.1.22. Now if $S$ is simple and $e \in E(S)$, then $S = SeS$ and $eSe$ is the maximal subgroup of $S$. By Proposition 4.12.20 $S \prec (Se)^{\ell} \times (eSe \wr (eS, \overline{eS}))$, so $S \in \mathbf{LZ} \vee (\mathbf{H} * \mathbf{RZ}) \leq \mathbf{H} * \mathbf{RZ}$, as **H** contains a non-trivial monoid.

Similarly, if $S$ is 0-simple with maximal subgroup $G \in \mathbf{H}$ and $e \in S$ is a non-zero idempotent, then $S = SeS$ and $eSe = G^0$. But

$$G^0 = (G \times U_1)/(G \times 0) \in \mathbf{H} \vee \mathbf{Sl}.$$

Proposition 4.12.20 yields $S \prec (Se)^{\ell} \times (eSe \wr (eS, \overline{eS}))$ and hence, as $\mathbf{H} \vee \mathbf{Sl}$ contains a non-trivial monoid,

$$S \in \mathbf{LZ} \vee ((\mathbf{H} \vee \mathbf{Sl}) * \mathbf{RZ}) \leq (\mathbf{H} \vee \mathbf{Sl}) * \mathbf{RZ}.$$

The statement about complexity is clear.    □

**Exercise 4.12.22.** Verify that if $G$ is a group and $R$ is a right zero semigroup, then $G \wr R$ is completely simple with maximal subgroup isomorphic to $G$. Deduce if **H** is a non-trivial pseudovariety of groups, then $\mathbf{CS} \cap \overline{\mathbf{H}} = \mathbf{H} * \mathbf{RZ}$. In particular, $\mathbf{CS} = \mathbf{G} * \mathbf{RZ}$.

Combining the various ideas of this section, we obtain the following result, which will be used to compute the complexity of full transformation and full linear monoids.

**Proposition 4.12.23.** *Suppose that $S$ is a monoid with non-trivial group of units $G$ such that $S = \langle G \cup e \rangle$ with $1 \neq e \in E(S)$ and $SeS \leq \langle E(S) \rangle$. Then one has $c(S) = c(eSe) + 1$.*

*Proof.* Certainly $c(eSe) \leq c(S)$. First observe that $SeS = S \setminus G$ because $S$ is generated by $G$ and $e$. Therefore $S \prec (SeS)^I \rtimes G$ by Corollary 4.1.11. Now $c((SeS)^I) = c(SeS) = c(eSe)$, where the first equality is from Proposition 4.3.14 and the second is from Proposition 4.12.20. We conclude that $c(S) \leq c(eSe) + 1$. So to establish the equality, it suffices to show that $c(eSe) < c(S)$. Because $G \geq_{\mathscr{L}} e$, by definition $S$ is a $\mathscr{T}_1$-semigroup. Theorem 4.12.14 then implies $S$ is an absolute Type I semigroup. Therefore, Theorem 4.12.8 tells us $c(\mathsf{K}_{\mathbf{G}}(S)) < c(S)$. Because $eSe \leq SeS \leq \langle E(S) \rangle \leq \mathsf{K}_{\mathbf{G}}(S)$, we obtain the desired conclusion: $c(eSe) < c(S)$.    □

We remark that one just needs the result of Exercise 4.12.15, and not the full strength of Theorem 4.12.14, to prove the proposition.

### Basic facts about full transformation and linear monoids

We review some standard facts about full transformation monoids that can be found, for instance, in [133]. For the sake of completeness we sketch the proofs.

An element of $T_n \setminus S_n$ is called *singular*. The *rank* of a transformation $f$ is the size of its image and is denoted by $\mathrm{rk}(f)$. Two elements are $\mathcal{J}$-equivalent if and only if they have the same rank. Also $f \mathrel{\mathcal{R}} g$ if and only if $\ker f = \ker g$ and $f \mathrel{\mathcal{L}} g$ if and only if $\mathbf{n}f = \mathbf{n}g$. Denote by $J_r$ the $\mathcal{J}$-class of rank $r$ transformations.

**Lemma 4.12.24.** *Let $e$ be a rank $n-1$ idempotent of $T_n$. Then $e T_n e \cong T_{n-1}$.*

*Proof.* The map $f \mapsto f|_{\mathbf{n}e}$ gives the desired isomorphism. □

**Lemma 4.12.25.** *$T_n \setminus S_n$ is generated by the rank $n-1$ transformations.*

*Proof.* One shows by induction on the parameter $d(f) = n - \mathrm{rk}(f)$ that if $f$ is singular, then $f$ is a product of rank $n-1$ transformations. If $d(f) = 1$, there is nothing to prove. Suppose the lemma holds for all $g \in T_n$ with $d(g) = r \geq 1$ and suppose $d(f) = r+1$. Let $x \notin \mathbf{n}f$ and let $i \neq j$ with $if = jf$. Define

$$ kf' = \begin{cases} x & k = i \\ kf & k \neq i \end{cases}, \qquad kg = \begin{cases} if & k = x \\ k & k \neq x. \end{cases} $$

Then $d(f') = d(f) - 1$ and $d(g) = 1$, so $f', g$ belong to $\langle J_{n-1} \rangle$ by induction. Now $if'g = xg = if$, whereas if $k \neq i$, then $kf'g = kfg = kf$, as $x \notin \mathbf{n}f$. Thus $f = f'g \in \langle J_{n-1} \rangle$, completing the induction. □

Next we prove that $J_{n-1}$ is generated by its set of idempotents. This observation is attributed to Howie [137]. For convenience, we shall denote by $[i, j)$ the unique rank $n-1$ idempotent that sends $j$ to $i$.

**Lemma 4.12.26.** *Every rank $n-1$ element of $T_n$ is a product of idempotents of rank $n-1$. Hence every singular transformation of degree $n$ is a product of idempotents of rank $n-1$. In particular, $\langle E(T_n) \rangle = (T_n \setminus S_n) \cup \{1\}$.*

*Proof.* We prove the first statement, as the remaining statements are consequences of it and Lemma 4.12.25. Fix the idempotent $e = [n-2, n-1)$ and denote by $H_e$ the $\mathcal{H}$-class of $e$. The kernel of a rank $n-1$ transformation is determined by the two elements it collapses, and so we can index $\mathcal{R}$-classes by two-element subsets of $\mathbf{n}$; so denote by $R_{i,j}$, for $i < j$, the $\mathcal{R}$-class of transformations $f \in T_n$ with $if = jf$. We let $L_i$ denote the $\mathcal{L}$-class of all transformations with range $\mathbf{n} \setminus \{i\}$. So $H_e = R_{n-2,n-1} \cap L_{n-1}$. If we choose $r_{i,j} \in R_{i,j} \cap L_{n-1}$, $i < j$, and $\ell_k \in L_k \cap R_{n-2,n-1}$, $k \in \mathbf{n}$, then we have

$$ J_{n-1} = \bigcup_{i < j, k \in \mathbf{n}} r_{i,j} H_e \ell_k \tag{4.12} $$

by standard Green-Rees theory [68, 171] (see Appendix A).

We begin by showing that $r_{i,j}$ and $\ell_k$ can be chosen in $\langle E(J_{n-1}) \rangle$. Indeed,

$$ r_{i,j} = [ij)[j, n-1) \in R_{i,j} \cap L_{n-1}, \quad \ell_k = [n-2, n-1)[n-1, k) \in R_{n-2,n-1} \cap L_k. $$

Now $H_e$ is isomorphic to $S_{n-1}$ via restriction to $\mathbf{n} - \mathbf{1}$. For $i \neq j \in \mathbf{n} - \mathbf{1}$,

$$[n-2, n-1)[n-1, j)[j, i)[i, n-1) \in \langle E(J_{n-1}) \rangle$$

has kernel $\{n-2, n-1\}$, range $\mathbf{n} \setminus \{n-1\}$ and restricts to $\mathbf{n} - \mathbf{1}$ as the transposition $(i\ j)$. Because $S_{n-1}$ is generated by transpositions, we see that $H_e \leq \langle E(J_{n-1}) \rangle$ and hence $J_{n-1} \leq \langle E(J_{n-1}) \rangle$ by (4.12) and our choices of the $r_{i,j}$ and $\ell_k$.    □

Our final lemma is classical as well.

**Lemma 4.12.27.** *The rank $n-1$ idempotents are all conjugate. Hence $T_n = \langle S_n \cup e \rangle$ where $e$ is any rank $n-1$ idempotent.*

*Proof.* It is easy to verify that $g^{-1}[i, j)g = [ig, jg)$ and hence the first part of the lemma follows from the fact that $S_n$ doubly transitive on $\mathbf{n}$. The second part of the lemma follows directly from the first and Lemma 4.12.26.    □

McAlister [201] showed that, for $n \geq 3$, if $G \leq S_n$ and $e \in E(J_{n-1})$, then $\langle G \cup e \rangle$ contains all singular transformations if and only if $G$ is 2-homogeneous (that is, transitive on two-element subsets).

**Exercise 4.12.28.** Prove McAlister's theorem. Hint: If $G$ is transitive, then $[i, j) \in \langle G \cup e \rangle$ if and only if $[j, i) \in \langle G \cup e \rangle$.

If $K$ is a field, denote by $M_n(K)$ the full linear monoid of degree $n$; its group of units is $GL_n(K)$. We state without proof the analogous results for $M_n(K)$ [89, 249, 256].

**Proposition 4.12.29.** *Let $K$ be a field and $e \in M_n(K)$ a rank $n-1$ idempotent.*

1. *Every singular matrix is a product of rank $n-1$ idempotents.*
2. *$\langle E(M_n(K)) \rangle = (M_n(K) \setminus GL_n(K)) \cup \{1\}$.*
3. *All the rank $n-1$ idempotents are conjugate.*
4. *$M_n(K) = \langle GL_n(K) \cup e \rangle$.*
5. *$eM_n(K)e \cong M_{n-1}(K)$.*

**Exercise 4.12.30.** Prove Proposition 4.12.29.

**Theorem 4.12.31 (Rhodes [266]).** *The complexities of $T_n$ and $M_n(\mathbb{F}_q)$ are given by $c(T_n) = n - 1 = c(M_n(\mathbb{F}_2))$ and $c(M_n(\mathbb{F}_q)) = n$ if $q > 2$. In particular, the complexity hierarchy is strict.*

*Proof.* Let $\{R_n\}$, $n \geq 1$, be one of the series $\{T_n\}$ or $\{M_n(\mathbb{F}_q)\}$. Let $G$ be the group of units of $R_n$ and let $e$ be a rank $n-1$ idempotent. Then we know that $R_n = \langle G \cup e \rangle$, $\langle E(R_n) \rangle \setminus \{1\} = R_n e R_n$ and $eR_n e \cong R_{n-1}$. Proposition 4.12.23 then shows that $c(R_n) = c(R_{n-1}) + 1$. Hence $c(R_n) = c(R_1) + n - 1$. But $c(T_1) = 0$, whereas $c(M_1(\mathbb{F}_q)) = 1$ unless $q = 2$, in which case $c(M_1(\mathbb{F}_2)) = 0$. The result follows.    □

Rhodes established that the semigroup of binary relations on an $n$ element set (equivalently the semigroup of $n \times n$ matrices over the two-element Boolean semiring) has complexity $n - 1$ [273]. An elegant proof is given in [362, Example 6.2]. A generalization, due to Fox and Rhodes [91], will appear in Theorem 9.5.4.

As a final illustration of the above techniques, let us compute the complexity of the wreath product $G \wr (\mathbf{n}, T_n)$ where $G$ is a group. This result is due to Fox and Rhodes [91] and will be used in Chapter 9.

**Theorem 4.12.32.** *Let $G$ be a group. Then $c(G \wr (\mathbf{n}, T_n)) = n - 1 + c(G)$.*

*Proof.* The proof goes by induction on $n$ following the scheme for $T_n$. Set $R_n = G \wr (\mathbf{n}, T_n)$. Because $R_1 = G$, we clearly have $c(R_1) = c(G)$. Assume $c(R_{n-1}) = n - 2 + c(G)$ for $n \geq 2$. First we observe that the group of units $H_n$ of $R_n$ is $G \wr (\mathbf{n}, S_n)$. Notice that, for $(f, \sigma) \in R_n$,

$$(f, \sigma) = (f, 1)(1, \sigma) \tag{4.13}$$

where 1 is the identity of $G^n$. It follows easily from (4.13) that if $\epsilon$ is a rank $n - 1$ idempotent of $T_n$, then $R_n = \langle H_n, (1, \epsilon) \rangle$, $R_n(1, \epsilon)R_n = R_n \setminus H_n$ and $(1, \epsilon)R_n(1, \epsilon) \cong R_{n-1}$. Our goal is to show that $R_n \setminus H_n \subseteq \langle E(R_n) \rangle$. Proposition 4.12.23 will then imply

$$c(R_n) = c(R_{n-1}) + 1 = n - 2 + c(G) + 1 = n - 1 + c(G).$$

Define the rank of an element $(f, \sigma) \in R_n$ to be the rank of $\sigma$. Notice $H_n$ is precisely the set of elements of rank $n$. If $(f, \sigma)$ has rank less than $n$ and $i\sigma = j\sigma$, then $(f, \sigma) = (f, [i, j))(1, \sigma)$, from which it follows that the elements of rank $n - 1$ generate $R_n \setminus H_n$ by the corresponding result for $T_n$. In fact, because $T_n \setminus S_n \subseteq \langle E(T_n) \rangle$, in order to prove $R_n \setminus H_n \subseteq \langle E(R_n) \rangle$, it suffices to show elements of the form $(f, [i, j))$ can be expressed as a product of idempotents. Because the idempotent-generated subsemigroup is closed under conjugation by $1 \times S_n$, it suffices to show that $(f, [n-2, n-1))$ is a product of idempotents. First suppose that $h \colon \mathbf{n} \to G$ is such that $ih = 1$ for $i < n - 1$. Then $(h, [n - 2, n - 1))$ is idempotent because

$$(n - 1)h \cdot (n - 1[n - 2, n - 1))h = (n - 1)h \cdot (n - 2)h = (n - 1)h.$$

To express $(f, [n-2, n-1))$ as a product of idempotents set, for $0 \leq j \leq n-2$,

$$if' = \begin{cases} (n - 1)f \cdot (n - 2)f^{-1} & i = n - 1 \\ 1 & \text{else,} \end{cases}$$

$$if_j = \begin{cases} if & i = j \\ 1 & \text{else.} \end{cases}$$

Then direct computation establishes

$$(f, [n-2, n-1)) = (f', [n-2, n-1))(f_0, [n-2, n-1)) \cdots (f_{n-2}, [n-2, n-1)).$$

Because $(f', [n-2, n-1))$ is idempotent by the above observation, it remains to establish each element $(f_j, [n-2, n-1))$ is a product of idempotents. Indeed,

$$(f_j, [n-2, n-1)) = (1, \sigma)(h, [n-2, n-1))(1, \tau)$$

where

$$i\sigma = \begin{cases} n-1 & i = j \\ j & i = n-2, n-1 \\ i & \text{else} \end{cases}$$

$$ih = \begin{cases} jf & i = n-1 \\ 1 & \text{else} \end{cases}$$

$$i\tau = \begin{cases} j & i = n-2, n-1 \\ n-2 & i = j \\ i & \text{else} \end{cases}$$

and $(h, [n-2, n-1))$ is idempotent. By our results for $T_n$, we can express $(1, \sigma)$ and $(1, \tau)$ as products of idempotents. This completes the proof. □

## 4.13 The Topology of Graphs and Graham's Theorem

In the previous section, we have seen that a detailed study of the idempotent-generated subsemigroup of a semigroup is important for understanding complexity. Such an analysis also plays a key role in Tilson's method for computing the complexity of a semigroup with at most two non-zero $\mathscr{J}$-classes [360]. In this section, we present Graham's Theorem [107], describing the idempotent-generated subsemigroup of a Rees matrix semigroup. These results were later rediscovered by Houghton [136], who gave a topological interpretation that has strongly influenced our approach. The groupoid interpretation is ours, although it clearly relates to the viewpoint of Nambooripad [212]. For applications, it is important that we consider arbitrary Rees matrix semigroups over groups and not just regular ones.

Let us begin with a well-known lemma, which apparently goes back to Fitz-Gerald, showing that understanding the regular elements of $\langle E(S) \rangle$ boils down to understanding, for each regular $\mathscr{J}$-class $J$, the set $\langle E(J) \rangle \cap J$.

**Lemma 4.13.1.** *Let $S$ be a semigroup and let $s \in \langle E(S) \rangle$ be an element of a regular $\mathscr{J}$-class $J$ of $S$. Then, there exist $e_1, e_2, \ldots, e_m \in E(J)$ such that $e_1 \mathscr{R} e_2 \mathscr{L} e_3 \mathscr{R} \cdots \mathscr{L} e_{m-1} \mathscr{R} e_m$ and $s = e_1 e_2 \cdots e_m$. Hence $\langle E(S) \rangle \cap J = \langle E(J) \rangle \cap J$.*

*Proof.* Let $e_1, \ldots, e_n \in E(S)$ be such that $s = e_1 \cdots e_n \in J$ and let $s'$ be an inverse of $s$. For $i = 1, \ldots, n$, set

$$f_i = e_i \cdots e_n s' e_1 \cdots e_{i-1}$$
$$g_i = f_i e_i = e_i \cdots e_n s' e_1 \cdots e_i = e_i f_{i+1}.$$

Then one can readily check that for $i = 1, \ldots, n$:

1. $f_i$ and $g_i$ are idempotents in $J$;
2. $f_i \mathscr{R} g_i$;
3. $g_i \mathscr{L} f_{i+1}$;
4. $s = f_1 g_1 \cdots f_n g_n$.

Indeed, $f_i^2 = e_i \cdots e_n s' e_1 \cdots e_n s' e_1 \cdots e_{i-1} = e_i \cdots e_n s' s s' e_1 \cdots e_{i-1} = f_i$, and $g_i^2 = f_i e_i f_i e_i = f_i^2 e_i = f_i e_i$, yielding 1. Assertions 2 and 3 follow from $g_i = f_i e_i = e_i f_{i+1}$ and the observations: $g_i f_i = f_i e_i f_i = f_i^2 = f_i$ and $f_{i+1} g_i = f_{i+1} e_i f_{i+1} = f_{i+1}^2 = f_{i+1}$. Finally, to see that 4 holds, note by 2:

$$f_1 g_1 \cdots f_n g_n = g_1 g_2 \cdots g_n = (e_1 \cdots e_n s' e_1)(e_2 \cdots e_n s' e_1 e_2) \cdots (e_n s' e_1 \cdots e_n)$$
$$= (ss')^n s = s.$$

This completes the proof. ☐

Fitz-Gerald [90] proved, using the same ideas as the proof of Lemma 4.13.1, that the idempotent-generated subsemigroup of a regular semigroup is again regular. Notice the lemma reduces it to the 0-simple case for finite semigroups, where it was first proved by Graham [107].

**Exercise 4.13.2 (Fitz-Gerald [90]).** Show that if $S$ is a regular semigroup, then $\langle E(S) \rangle$ is regular. Hint: Using the notation of the above proof, show that $f_n \cdots f_1 = e_n s'$ is an inverse to $s$.

**Exercise 4.13.3.** Show that if $s$ is a regular element of $S$ in $\langle E(S) \rangle$, then each weak inverse of $s$ belongs to $\langle E(S) \rangle$.

As a consequence of the above lemma, we obtain the following reduction for the membership problem in **EA**.

**Corollary 4.13.4.** *A semigroup $S$ belongs to* **EA** *if and only if $\langle E(J) \rangle \cap J$ contains only trivial subgroups for each regular $\mathscr{J}$-class $J$ of $S$.*

This leads us to seek a criterion to determine for a regular $\mathscr{J}$-class $J$ whether $\langle E(J) \rangle \cap J$ contains only trivial subgroups. Because $J^0$ is 0-simple, Rees's Theorem implies that $J^0 \cong \mathscr{M}^0(G, A, B, C)$ where $G$ is a maximal subgroup, $A$ and $B$ are sets in bijection with the set of $\mathscr{R}$-classes and the set of $\mathscr{L}$-classes of $J$, respectively, and $C \colon B \times A \to G^0$ is a matrix [68, 171] and so we are led to consider the topology of Rees matrix semigroups.

### 4.13.1 Topology of labeled graphs

In this subsection, we introduce some topological notions for graphs and labeled graphs. These notions will be used in this section, as well as in the sections on the Type II Theorem [33] and the Ribes and Zalesskii Theorem [300]. Good references are [184, 317, 321].

For the purposes of this chapter, by a *graph* we mean a graph in the sense of Serre [317]. So a graph $\Gamma$ consists of a set $V$ of vertices, a set $E$ of edges, functions $\iota, \tau \colon E \to V$ selecting, respectively, the initial and terminal vertices of an edge $e$ and a fixed-point free involution $e \mapsto \bar{e}$ such that $\bar{e}\iota = e\tau$, $\bar{e}\tau = e\iota$. An edge pair $\{e, \bar{e}\}$ is called a *geometric edge* and often graphs are described in terms of their geometric edges. An *orientation* of $\Gamma$ is a choice of a representative from each geometric edge. The chosen representative is said to be *positively oriented* and sometimes a graph is specified by its positively oriented edges; in particular, we usually only draw the positively oriented edges. Most of the graphs of interest to us in this section are bipartite. Recall that a graph $\Gamma$ is *bipartite* if its vertex set $V$ can be written as a disjoint union $V = X \uplus Y$ of non-empty subsets (called *parts*) so that each edge of $\Gamma$ has one of its vertices in $X$ and the other in $Y$. A bipartite graph admits two natural orientations: namely, orienting the geometric edges to point toward one part or the other. A graph is called *simplicial* if it contains no loop edges (i.e., edges $e$ with $e\iota = e\tau$) and no pairs of distinct, coterminal geometric edges.

A *path* in $\Gamma$ is a sequence $e_1 \cdots e_n$ of edges such that $e_i\tau = e_{i+1}\iota$. We admit an empty path $1_v$ at each vertex $v$. We extend the maps $\iota$ and $\tau$ to paths in the obvious way and we often write $p \colon v \to w$ if $p$ is a path from $v$ to $w$. We extend the involution to paths by setting, for a path $p = e_1 \cdots e_n$, the inverse path to be $\bar{p} = \bar{e}_n \cdots \bar{e}_1$. Paths $p, q$ can be concatenated to a path $pq$ if $p\tau = q\iota$. A graph is said to be *connected* if any two vertices in the graph can be connected by a path. The *connected components* of a graph are then the maximal connected subgraphs.

Two paths $p_1, p_2$ are said to be *elementarily homotopic* if $p_1$ can be obtained from $p_2$ by insertion or deletion of a subpath $e\bar{e}$; such a subpath is called a *backtrack*. Two paths $p, q$ are said to be *homotopic* if there exists a sequence $p = p_0, p_1, \ldots, p_n = q$ of paths such that $p_i$ is elementarily homotopic to $p_{i+1}$. Notice that $p$ and $q$ must necessarily be coterminal. We write $p \sim q$ and use $[p]$ to denote the homotopy class of $p_i$. Then one verifies directly that $p \sim q$, $p' \sim q'$ implies $\bar{p} \sim \bar{q}$ and $pp' \sim qq'$ if $pp'$ and $qq'$ are defined. Hence $[p][q] = [pq]$, when $p\tau = q\iota$, gives a well-defined concatenation of homotopy classes and $\overline{[p]} = [\bar{p}]$ gives a well-defined involution such that $[p]\overline{[p]} = [1_{p\iota}]$. Thus, if $v$ is a vertex of $\Gamma$, the collection of all homotopy classes of circuits at $v$ forms a group $\pi_1(\Gamma, v)$, called *fundamental group* of $\Gamma$ at $v$. More generally, one can consider the *fundamental groupoid* $\Pi_1(\Gamma)$, which has object set $V$ and arrows the homotopy classes of paths.

It is well-known that $\pi_1(\Gamma, v)$ is a free group [184, 321]. To explain this a little, recall that a path $p$ is called *reduced* if it contains no backtracks. It is well-known that every homotopy class contains a unique reduced path. A graph is called a *tree* if between any two vertices there is a unique reduced path. A *forest* is a graph whose connected components are trees. To calculate $\pi_1(\Gamma, v)$, we might as well assume that $\Gamma$ is connected. Then choose a spanning tree $T$ for $\Gamma$. For each vertex $w$ of $\Gamma$ one defines $p_w$ to be the unique reduced path in $T$ from the base point $v$ to $w$. If we orient $\Gamma$ and denote by $E^+$ the set of positively oriented edges, then $\pi_1(\Gamma, v)$ is freely generated by the homotopy classes of the paths $p_{e_\iota} e \overline{p}_{e_\tau}$, where $e \in E^+ \setminus T$ [184, 317, 321]. In particular, the rank of $\pi_1(\Gamma, v)$ is $|E^+| - |V| + 1$ as a spanning tree has $|V| - 1$ geometric edges. Notice that a connected graph $\Gamma$ is a tree if and only if $\pi_1(\Gamma, v)$ is trivial.

We need to consider a more general notion, where homotopy is determined according to some not necessarily free group.

**Definition 4.13.5 (G-labeled graph).** *Let $\Gamma$ be a graph (in the sense of Serre) and $G$ a group. A $G$-labeling of $\Gamma$ is a map $\ell\colon E \to G$ such that $\overline{e}\ell = (e\ell)^{-1}$. One calls $(\Gamma, \ell)$ a $G$-labeled graph. The set of all $G$-labelings of $\Gamma$ is denoted $Z(\Gamma, G)$.*

Clearly, one can extend $\ell$ to paths by $(e_1 \cdots e_n)\ell = e_1\ell \cdots e_n\ell$ and it is still true that $\overline{p}\ell = (p\ell)^{-1}$ for paths $p$. Moreover, $(pq)\ell = p\ell q\ell$ when $p\tau = q\iota$.

**Definition 4.13.6 (G-homotopy).** *Coterminal paths $p$ and $q$ in a $G$-labeled graph $(\Gamma, \ell)$ are called $G$-homotopic if $p\ell = q\ell$. We use $[p]_G$ to denote the $G$-homotopy class of $p$.*

Because, for a backtrack $e\overline{e}$, we have $(e\overline{e})\ell = e\ell(e\ell)^{-1} = 1$, it follows that homotopy (which might more aptly be called *free homotopy*) implies $G$-homotopy. In fact, a $G$-labeling of $\Gamma$ is precisely the same piece of data as a functor $\ell\colon \Pi_1(\Gamma) \to G$. In particular, if $v$ is a vertex, then there is a well-defined homomorphism $\ell\colon \pi_1(\Gamma, v) \to G$. We shall denote the subgroup $\pi_1(\Gamma, v)\ell$ of $G$ by $\pi_1(\Gamma, \ell, v)$ and call it the *fundamental group* of the $G$-labeled graph $(\Gamma, \ell)$ at $v$. Up to conjugacy $\pi_1(\Gamma, \ell, v)$ depends only on the connected component of $v$. Of course, if $x_1, \ldots, x_m$ generate $\pi_1(\Gamma, v)$, then $x_1\ell, \ldots, x_m\ell$ generate $\pi_1(\Gamma, \ell, v)$. In particular, if $\Gamma$ is a tree, then $\pi_1(\Gamma, \ell, v)$ is trivial. This leads to the following notion.

**Definition 4.13.7 (G-acyclic).** *Let $(\Gamma, \ell)$ be a $G$-labeled graph. We say that $\Gamma$ is $G$-acyclic if $\pi_1(\Gamma, \ell, v)$ is trivial for each vertex $v$ of $\Gamma$ (or equivalently, for some vertex $v$ of each connected component of $\Gamma$).*

Notice that if $\Gamma$ is a graph with set of positively oriented edges $E^+$, then it can be viewed as $FG(E^+)$-labeled, where $FG(E^+)$ is a free group on $E^+$. Simply label a positive edge $e$ by itself and a negative edge $\overline{e}$ by $e^{-1}$. It is straightforward to see that paths $p$ and $q$ are $FG(E^+)$-homotopic if and only if they are homotopic in the usual sense and so the $FG(E^+)$-fundamental group is the usual one, whereas being $FG(E^+)$-acyclic is equivalent to being a forest. We now introduce a notion of cohomology, due to Houghton [136], which boils down to comparing labelings up to natural equivalence.

**Definition 4.13.8 (*G*-cohomology).** *Let $\Gamma = V \cup E$ be a graph and define $B(\Gamma, G)$ to be the group of all functions $\delta \colon V \to G$ with pointwise multiplication. There is a left action of $B(\Gamma, G)$ on $Z(\Gamma, G)$ defined, for $\delta \in B(\Gamma, G)$ and $\ell \in Z(\Gamma, G)$, by $e^{\delta}\ell = e\iota\delta e\ell(e\tau\delta)^{-1}$. We say that two G-labelings of $\Gamma$ are cohomologous if they belong to the same orbit under the action of $B(\Gamma, G)$. The quotient $H(\Gamma, G) = B(\Gamma, G) \backslash Z(\Gamma, G)$ is called the G-cohomology of $\Gamma$. If $^{\delta}\ell = \ell'$, then we call $\delta$ a coboundary from $\ell$ to $\ell'$.*

*Remark 4.13.9.* It is not hard to see a coboundary $\delta$ from $\ell$ to $\ell'$ is nothing more than a natural transformation $\delta \colon \ell' \to \ell$ where labelings are viewed as functors $\Pi_1(\Gamma) \to G$. Indeed, the equality $e\ell' = e\iota\delta e\ell(e\tau\delta)^{-1}$ just says the diagram

$$
\begin{array}{ccc}
\bullet & \xrightarrow{\;e\ell'\;} & \bullet \\
{\scriptstyle e\iota\delta}\Big\downarrow & & \Big\downarrow{\scriptstyle e\tau\delta} \\
\bullet & \xrightarrow[\;e\ell\;]{} & \bullet
\end{array}
$$

commutes, where $\bullet$ is the unique object of $G$. Thus $\ell$ and $\ell'$ are homotopic from the viewpoint of [132].

We show now that, up to conjugacy, the $G$-homotopy groups depend only on the $G$-cohomology class of the labeling.

**Proposition 4.13.10.** *Suppose that $\ell$ and $\ell'$ are cohomologous labelings of a graph $\Gamma$ via a coboundary $\delta$ from $\ell$ to $\ell'$. Then*

$$\pi_1(\Gamma, \ell', v) = (v\delta)\pi_1(\Gamma, \ell, v)(v\delta)^{-1}.$$

*In particular, $(\Gamma, \ell)$ is G-acyclic if and only if $(\Gamma, \ell')$ is G-acyclic.*

*Proof.* Indeed, if $e_1 \cdots e_n$ is a circuit at $v$, then we have

$$
\begin{aligned}
e_1\ell' \cdots e_n\ell' &= (e_1\iota\delta)e_1\ell(e_1\tau\delta)^{-1}(e_2\iota\delta)e_2\ell \cdots (e_{n-1}\tau\delta)^{-1}(e_n\iota\delta)e_n\ell(e_n\tau\delta)^{-1} \\
&= (v\delta)e_1\ell e_2\ell \cdots e_n\ell(v\delta)^{-1}.
\end{aligned}
$$

This establishes the first assertion, from which the second follows.    □

The following result, due to Graham [107] (although he does not use this language), corresponds with the operation of contracting a spanning tree.

**Theorem 4.13.11 (Graham normalization).** *Let $\Gamma$ be a connected graph with a fixed vertex $v$ and a spanning tree $T$. Let $\ell \in Z(\Gamma, G)$ be a labeling. Then there is a labeling $\ell_0$ cohomologous to $\ell$ with $\pi_1(\Gamma, \ell, v) = \pi_1(\Gamma, \ell_0, v)$ and $e\ell_0 = 1$ for all $e \in T$. One calls $\ell_0$ a Graham normalization of $\ell$.*

*Proof.* For each vertex $w$, let $p_w$ be the unique reduced path in $T$ from $v$ to $w$. In particular, $p_v = 1_v$. Define $\delta\colon V \to G$ by $w\delta = p_w \ell$ and set $\ell_0 = {}^\delta\ell$. Then $\ell$ and $\ell_0$ are cohomologous. Moreover, because $v\delta = 1_v\ell = 1$, we have that $\pi_1(\Gamma, \ell, v) = \pi_1(\Gamma, \ell_0, v)$ by Proposition 4.13.10.

Suppose that $e$ is an edge of $T$ from $x$ to $y$. By switching to $\bar{e}$ if necessary, we may assume that $x$ is on the reduced path from $v$ to $y$. Then $p_y = p_x e$, and so $e\ell_0 = x\delta e\ell(y\delta)^{-1} = p_x \ell e\ell(p_y \ell)^{-1} = p_x \ell e\ell((p_x e)\ell)^{-1} = 1$, as required. $\quad\square$

*Remark 4.13.12.* What underlies Graham normalization is that $w \mapsto p_w^{-1}$ gives a natural equivalence from the identity functor on $\Pi_1(\Gamma)$ to the functor that collapses all vertices of $\Gamma$ to $v$, corresponding to contraction of the spanning tree $T$ [132]. Composition of this natural transformation with $\ell$ gives the Graham normalization.

The construction of a Graham normalization in the above proof can clearly be done in polynomial time in the size of the graph $\Gamma$, assuming that multiplication in $G$ costs one unit of time. We say that a labeling $\ell \in Z(\Gamma, G)$ is *Graham normalized* with respect to a spanning forest $F$ of $\Gamma$ if $e\ell = 1$ for all edges $e \in F$. We can always replace any labeling of $\Gamma$ by a cohomologous one that is Graham normalized with respect to $F$ by applying Theorem 4.13.11 to each of the connected components of $\Gamma$. By the *trivial labeling* of a graph over $G$, we mean the labeling that sends every edge to 1.

**Corollary 4.13.13.** *Suppose that $\Gamma$ is a connected graph with edge set $E$ and with spanning tree $T$, and suppose that $\ell \in Z(\Gamma, G)$ is Graham normalized with respect to $T$. Then $\pi_1(\Gamma, \ell, v) = \langle E\ell \rangle$ for all vertices $v$ of $\Gamma$. In particular, $(\Gamma, \ell)$ is $G$-acyclic if and only if $\ell$ is the trivial labeling.*

*Proof.* First of all, $E\ell$ consists of 1 and the labels of edges of $\Gamma$ that do not belong to $T$. We already know that $\pi_1(\Gamma, \ell, v)$ is generated by elements of the form $p\ell e\ell q\ell$ where $p$ and $q$ are certain paths in the spanning tree $T$, and $e$ is an edge not in $T$. Because $p\ell = 1 = q\ell$, by the assumption that $\ell$ is Graham normalized with respect to $T$, we conclude that $\pi_1(\Gamma, \ell, v) = \langle E\ell \rangle$. In particular, $\pi_1(\Gamma, \ell, v) = 1$ if and only if $E\ell = \{1\}$, that is, if and only if $\ell$ is the trivial labeling. $\quad\square$

The above corollary illustrates some of the advantages of working with Graham normalized labelings. It also leads to a cohomological characterization of $G$-acyclicity.

**Proposition 4.13.14.** *Let $(\Gamma, \ell)$ be a $G$-labeled graph. Then $\Gamma$ is $G$-acyclic if and only if $\ell$ is cohomologous to the trivial labeling.*

*Proof.* If $\ell$ is cohomologous to the trivial labeling, then Proposition 4.13.10 and the fact that the trivial labeling is always $G$-acyclic shows that $(\Gamma, \ell)$ is $G$-acyclic. Conversely, suppose that $(\Gamma, \ell)$ is $G$-acyclic and choose a spanning forest $F$ for $\Gamma$. Apply Theorem 4.13.11 to each connected component of $\Gamma$ in order to construct a labeling $\ell'$ cohomologous to $\ell$, which is Graham normalized with respect to $F$. Proposition 4.13.10 shows that $(\Gamma, \ell')$ is also $G$-acyclic and hence is the trivial labeling by Corollary 4.13.13 (applied to each connected component of $\Gamma$). $\qquad\qquad\qquad\qquad\qquad\qquad\qquad\qquad\qquad\qquad\square$

To show that our notion of $G$-cohomology of a graph is worthy of the name cohomology, let us show that trees are characterized by the triviality of $H(\Gamma, G)$ for some (equivalently any) non-trivial group $G$.

**Proposition 4.13.15.** *For a connected graph $\Gamma$, the following are equivalent:*

1. *$\Gamma$ is a tree;*
2. *$|H(\Gamma, G)| = 1$ for all groups $G$;*
3. *$|H(\Gamma, G)| = 1$ for some non-trivial group $G$.*

*Proof.* To see that 1 implies 2, note if $\Gamma$ is a tree, every $G$-labeling of $\Gamma$ is $G$-acyclic, and hence cohomologous to the trivial one by Proposition 4.13.14. Trivially, 2 implies 3. For 3 implies 1, choose a spanning tree $T$ for $\Gamma$. Label each edge of $T$ by 1 and each positively oriented edge of $\Gamma \setminus T$ by $1 \neq g \in G$ (and label its inverse edge accordingly). This labeling is Graham normalized with respect to $T$. By 3, it must be cohomologous to the trivial labeling. Corollary 4.13.13 then implies the labeling we constructed must be trivial. This can only happen if $\Gamma = T$, so $\Gamma$ is a tree. $\qquad\qquad\square$

### 4.13.2 The incidence graph of a Rees matrix semigroup

The reason for developing all these notions is to understand the idempotent-generated subsemigroup of a Rees matrix semigroup in terms of the topology of an associated labeled graph. This graph was first introduced by Graham [107], and then rediscovered several years later by Houghton [136].

**Definition 4.13.16 (Incidence graph).** *Let $S = \mathscr{M}^0(G, A, B, C)$ be a Rees matrix semigroup. We assume $A$ and $B$ are disjoint. The incidence graph of $S$, denoted $\Gamma(S)$, has vertex set $V = A \cup B$. The edge set is given by*

$$E = \{(a, b), (b, a) \in (A \times B) \cup (B \times A) \mid bCa \neq 0\}.$$

*The involution is given by transposing the coordinates: $\overline{(x, y)} = (y, x)$. One defines $(x, y)\iota = x$ and $(x, y)\tau = y$. The structure of a $G$-labeled graph is given to $\Gamma(S)$ by defining $\ell_C \colon E \to G$ by $(a, b)\ell_C = (bCa)^{-1}$, $(b, a)\ell_C = bCa$, for $a \in A$, $b \in B$. We orient $\Gamma(S)$ by taking $(B \times A) \cap E$ as the set of positively oriented edges.*

Observe that $\Gamma(S)$ is a simplicial bipartite graph with the two parts being $A$ and $B$. Our choice of orientation is the natural orientation under which positively oriented edges point toward $A$. Notice that $S^{op} \cong \mathscr{M}^0(G, B, A, \check{C}^T)$ where $\check{C}$ is the matrix obtained by replacing each entry of $C$ by its inverse.

Thus $\Gamma(S) = \Gamma(S^{op})$ as labeled graphs, but they have opposite orientations. This remark indicates that the orientation is an important component of the incidence graph.

*A priori* the graph $\Gamma(S)$ seems to depend on the Rees matrix representation of $S$. However, if $S$ is 0-simple, then $A$ is in bijection with the $\mathscr{R}$-classes of $S$, $B$ is in bijection with the $\mathscr{L}$-classes of $S$ and $bCa \neq 0$ if and only if the $\mathscr{H}$-class $a \cap b$ contains an idempotent. Thus the isomorphism type of the graph $\Gamma(S)$ depends only on the isomorphism type of $S$. The hypothesis that $S$ is 0-simple is equivalent to demanding that $\Gamma(S)$ has no isolated vertices. It turns out that after removing the isolated vertices, the isomorphism type of the graph again depends only on the isomorphism type of $S$.

**Exercise 4.13.17.** Show that $S$ is 0-simple if and only if $\Gamma(S)$ has no isolated vertices. Show that $S \setminus \{0\}$ is simple if and only if $\Gamma(S)$ is a complete bipartite graph. In this case, show that if $a_0 \in A$ and $b_0 \in B$ are fixed, then the set of all geometric edges with either $a_0$ or $b_0$ as an endpoint is a spanning tree.

**Exercise 4.13.18.** Let $C \colon B \times A \to G^0$ be a Rees matrix such that each row and column has a non-zero entry. Let $C'$ be obtained from $C$ by adding a row of zeroes and let $C''$ be obtained from $C$ by adding a column of zeroes. Show that the oriented graphs associated to $C'$ and $C''$ are isomorphic, but the associated Rees matrix semigroups are not.

Graham's fundamental observation, although he did not use the topological language, is that the oriented $G$-labeled graph $(\Gamma(S), \ell_C)$ encodes the idempotent-generated subsemigroup of $S = \mathscr{M}^0(G, A, B, C)$. In particular, we shall see that the subsemigroup $\langle E(S) \rangle$ is aperiodic if and only if $(\Gamma(S), \ell_C)$ is $G$-acyclic. First we show how the negatively oriented edges encode idempotents.

**Lemma 4.13.19.** *Let $S = \mathscr{M}^0(G, A, B, C)$ be a Rees matrix semigroup. Then there is a bijection between geometric edges of $\Gamma(S)$ and non-zero idempotents of $S$ defined by $(a, b)\psi = (a, (a, b)\ell_C, b) = (a, (bCa)^{-1}, b)$.*

*Proof.* Indeed, $(a, g, b)(a, g, b) = (a, g(bCa)g, b)$ and so $(a, g, b)$ is idempotent if and only if $bCa \neq 0$ and $g = (bCa)^{-1}$. The lemma is now immediate.   □

This correspondence extends to $G$-homotopy classes of paths.

**Theorem 4.13.20 (Graham).** *Let $S = \mathscr{M}^0(G, A, B, C)$ be a Rees matrix semigroup. Let $P$ be the set of $G$-homotopy classes of paths $p$ in $\Gamma(S)$ with $p\iota \in A$ and $p\tau \in B$. Then there is a bijection $\psi \colon P \to \langle E(S) \rangle \setminus 0$ given by $[p]_G \psi = (p\iota, p\ell_C, p\tau)$. Moreover, $[p]_G \psi [q]_G \psi = [p]_G \psi (p\tau C q\iota) [q]_G \psi$. This last product, if non-zero, is $[(p(p\tau, q\iota)q)]_G \psi$.*

*Proof.* The multiplication formula is a direct calculation, so we just prove the first assertion. Set $T = \langle E(S) \rangle \setminus 0$. Let us first prove that $\psi$ takes on values in $T$. Let $p$ be a representative of a $G$-homotopy class from $P$. We show by induction on the length of $p$ that $[p]_G\psi$ belongs to $T$. Because $p$ goes from $A$ to $B$, it must have odd length. The case where $p$ is an edge was already handled in Lemma 4.13.19. So suppose $\psi$ takes on values in $T$ for shorter paths than $p$. Because $p$ has odd length, $p = q(b,a)(a,b')$ where $[q]_G \in P$, $q\tau = b$, $p\tau = b'$ and $bCa \neq 0 \neq b'Ca$. Then $p\ell_C = q\ell_C(bCa)(b'Ca)^{-1}$. On the other hand, $e = (a, (b'Ca)^{-1}, b')$ is an idempotent by Lemma 4.13.19, and

$$[q]_G\psi e = (q\iota, q\ell_C, b)(a, (b'Ca)^{-1}, b') = (q\iota, q\ell_C(bCa)(b'Ca)^{-1}, b')$$
$$= (p\iota, p\ell_C, p\tau) = [p]_G\psi.$$

By induction, $(q\iota, q\ell_C, q\tau) \in T$ and so $[p]_G\psi \in T$.

Conversely, we prove by induction on $n$ that if $f_1, \ldots, f_n$ are idempotents and $f_1 \cdots f_n \neq 0$, then $f_1 \cdots f_n \in P\psi$. The case $n = 1$ was already dealt with in Lemma 4.13.19. Suppose that $f_1 \cdots f_{n-1} = [q]_G\psi$ and that $f_n = (a, (bCa)^{-1}, b)$. Then $f_1 \cdots f_{n-1} = (q\iota, q\ell_C, q\tau)$. Set $b' = q\tau$. Then, because $[q]_G\psi f_n \neq 0$, it follows $b'Ca \neq 0$. Putting $p = q(b',a)(a,b)$, we have $[q]_G\psi f_n = (q\iota, q\ell_C(b'Ca)(bCa)^{-1}, b) = (p\iota, p\ell_C, p\tau) = [p]_G\psi$. This completes the proof that $\psi$ is a surjective function.

Suppose now that $[p]_G \in P$. Then $[p]_G\psi = (p\iota, p\ell_C, p\tau)$ and hence $[p]_G\psi$ determines the initial and terminal vertices of $p$ as well $p\ell_C$. It is then immediate $[p]_G\psi$ determines $p$ up to $G$-homotopy, establishing injectivity of $\psi$.    □

Specializing the above bijection to fixed endpoints yields:

**Corollary 4.13.21.** *Let* $S = \mathcal{M}^0(G, A, B, C)$ *be a Rees matrix semigroup. If* $a \in A$ *and* $b \in B$, *then there is a bijection between* $\Pi_1(a,b)\ell_C$ *and elements* $(a, g, b) \in \langle E(S) \rangle$ *given by* $[p]_G \mapsto (a, p\ell_C, b)$.

Now we shall use $\Gamma(S)$ to understand Green's relations on $\langle E(S) \rangle$, in particular to understand the $\mathscr{J}$-relation and the maximal subgroups.

**Theorem 4.13.22 (Graham).** *Let* $S = \mathcal{M}^0(G, A, B, C)$ *be a Rees matrix semigroup and let* $\psi$ *be the bijection from Theorem 4.13.20. Let* $[p]_G, [q]_G \in P$. *Then, taking Green's relations in* $\langle E(S) \rangle$:

1. $[p]_G\psi \mathscr{R} [q]_G\psi$ *if and only if* $p\iota = q\iota$;
2. $[p]_G\psi \mathscr{L} [q]_G\psi$ *if and only if* $p\tau = q\tau$;
3. $[p]_G\psi \mathscr{H} [q]_G\psi$ *if and only if* $p$ *and* $q$ *are coterminal*;
4. $[p]_G\psi \mathscr{J} [q]_G\psi$ *if and only if* $p$ *and* $q$ *belong to the same connected component of* $\Gamma(S)$;
5. *The maximal subgroup of* $\langle E(S) \rangle$ *at an idempotent* $e = (a, (bCa)^{-1}, b)$ *is isomorphic to* $\pi_1(\Gamma(S), \ell_C, b)$ *via the map* $[p]_G \mapsto ([(a,b)]_G[p]_G)\psi$.

*As a consequence,* $\langle E(S) \rangle$ *is a regular semigroup.*

*Proof.* We start by tackling 1. Assume $p\iota = q\iota$. Suppose $q = re$ with $e$ the last edge of $q$. Because $e = (r\tau, q\tau)$, the multiplication formula from Theorem 4.13.20 implies

$$[q]_G\psi[\overline{r}p]_G\psi = [q\overline{q}p]_G\psi = [p]_G\psi$$

and so $[p]_G\psi \leq_{\mathscr{R}} [q]_G\psi$. Symmetry establishes $\mathscr{R}$-equivalence. Now suppose that $[p]_G\psi \mathrel{\mathscr{R}} [q]_G\psi$. Then $[p]_G\psi = [q]_G\psi[r]_G\psi$ some $r$ and the multiplication formula from Theorem 4.13.20 shows $[p]_G = [q(b,a)r]_G$ for an appropriate edge $(b,a)$. Thus $p\iota = q\iota$. A dual argument establishes 2, and 3 is immediate from 1 and 2. One can either deduce 4 from 1 and 2 and the fact that $\mathscr{J} = \mathscr{D} = \mathscr{R} \circ \mathscr{L} = \mathscr{L} \circ \mathscr{R}$, or prove it directly along the lines of the proof of 1. Details are left to the reader.

Turning to the proof of 5, denote by $H_e$ the maximal subgroup of $\langle E(S)\rangle$ at $e$ and define $\varphi\colon \pi_1(\Gamma(S), \ell_C, b) \to H_e$ by $[p]_G\varphi = ([(a,b)]_G[p]_G)\psi$. By 3 and Corollary 4.13.21, we see $\varphi$ indeed maps into $H_e$. To verify $\varphi$ is a homomorphism, observe that if $[p]_G, [q]_G \in \pi_1(\Gamma(S), \ell_C, b)$, then by Theorem 4.13.20

$$[p]_G\varphi[q]_G\varphi = [(a,b)p(b,a)(a,b)q]_G\psi = ([(a,b)]_G[pq]_G)\psi = ([p]_G[q]_G)\varphi$$

as $(b,a) = \overline{(a,b)}$. Because $\psi$ is injective, we have $[p]_G\varphi = [q]_G\varphi$ implies $[(a,b)]_G[p]_G = [(a,b)]_G[q]_G$ and so $[p]_G = [q]_G$. This establishes the injectivity of $\varphi$. For the surjectivity, let $h \in H_e$. Then, by Corollary 4.13.21, we have $h = [p]_G\psi$ where $p\colon a \to b$. Setting $q = (b,a)p$, we have $[q]_G \in \pi_1(\Gamma(S), \ell_C, b)$ and $[q]_G\varphi = ([(a,b)]_G[(b,a)]_G[p]_G)\psi = [p]_G\psi = h$. This completes the proof that $H_e \cong \pi_1(\Gamma(S), \ell_C, b)$.

Finally, we observe that $\langle E(S)\rangle$ is regular because if $[p]_G \in P$ and the first edge of $p$ is $(a,b)$, then $[p]_G\psi$ is $\mathscr{R}$-equivalent in $\langle E(S)\rangle$ to the idempotent $(a, (bCa)^{-1}, b) = [(a,b)]_G\psi$ by 1.    □

**Exercise 4.13.23.** Prove item 4 of Theorem 4.13.22.

**Corollary 4.13.24.** *Let* $S = \mathscr{M}^0(G, A, B, C)$ *be a Rees matrix semigroup. Then* $\langle E(S)\rangle$ *is aperiodic if and only if* $(\Gamma(S), \ell_C)$ *is G-acyclic.*

**Corollary 4.13.25 (Fitz-Gerald [90]).** *Let* $S$ *be a regular semigroup. Then* $\langle E(S)\rangle$ *is regular.*

*Proof.* By Lemma 4.13.1, $\langle E(S)\rangle$ is the union of the sets $\langle E(J)\rangle \cap J$ as $J$ runs over the $\mathscr{J}$-classes of $S$. An application of Theorem 4.13.22 then yields the desired result.    □

Let $S = \mathscr{M}^0(G, A, B, C)$ be a Rees matrix semigroup. Let $\Gamma'(S)$ be the graph obtained from $\Gamma(S)$ by removing all isolated vertices. Theorem 4.13.22 implies that the vertex set of $\Gamma'(S)$ is in bijection with the union of the set of $\mathscr{R}$-classes and the set of $\mathscr{L}$-classes of $\langle E(S)\rangle$, and that there is a geometric edge between a given pair consisting of an $\mathscr{L}$-class and an $\mathscr{R}$-class if and only if their intersection contains an idempotent. Hence the isomorphism type of $\Gamma'(S)$ as an oriented graph depends only on the isomorphism type of $S$ (and in fact on the isomorphism type of $\langle E(S)\rangle$). The isolated vertices of $\Gamma(S)$ correspond to zero rows and zero columns of $S$ and it is impossible to distinguish from the graph which of these possibilities corresponds to a given vertex.

Our next goal is to show that changing the labeling of $\Gamma(S)$ to a cohomologous labeling does not change the isomorphism type of $S$. In particular, we may always assume a Rees matrix semigroup is given with a Graham normalized labeling with respect to some spanning forest for $\Gamma(S)$. This will lead to important algebraic consequences.

Let $G$ be a group and $X$ a set. Define $\Delta(X, G)$ to be the group of diagonal matrices $\Omega\colon X \times X \to G^0$ with diagonal entries in $G$. The reader should note that $\Delta(X, G) \cong G^X$.

**Definition 4.13.26 (Cohomology of Rees matrices).** *Let $\Gamma$ be an oriented simplicial bipartite graph with parts $A$ and $B$ and with positive edges pointing to $A$. Then a Rees matrix $C\colon B \times A \to G^0$ is said to be compatible with $\Gamma$ if $\Gamma = \Gamma(S)$, where $S = \mathscr{M}^0(G, A, B, C)$. Clearly, there is a bijection between labelings of $\Gamma$ over $G$ and Rees matrices $C\colon B \times A \to G^0$ compatible with $\Gamma$ via $C \mapsto \ell_C$. There is a natural left action of $\Delta(B, G) \times \Delta(A, G)$ on the set of Rees matrices compatible with $\Gamma$ given by $C \mapsto \Omega C \Lambda^{-1}$ for $(\Omega, \Lambda) \in \Delta(B, G) \times \Delta(A, G)$. Two Rees matrices in the same orbit are said to be cohomologous.*

The importance of cohomology of Rees matrices is that cohomologous matrices give rise to isomorphic Rees matrix semigroups.

**Proposition 4.13.27.** *Let $C\colon B \times A \to G^0$ and $C'\colon B \times A \to G^0$ be cohomologous Rees matrices. Then $\mathscr{M}^0(G, A, B, C) \cong \mathscr{M}^0(G, A, B, C')$.*

*Proof.* Let $(\Omega, \Lambda) \in \Delta(G, B) \times \Delta(G, A)$ be such that $C' = \Omega C \Lambda^{-1}$. Define an isomorphism $\varphi\colon \mathscr{M}^0(G, A, B, C) \to \mathscr{M}^0(G, A, B, C')$ by $M\varphi = \Lambda M \Omega^{-1}$, where we view elements of $\mathscr{M}^0(G, A, B, C)$ and $\mathscr{M}^0(G, A, B, C')$ as $A \times B$ matrices with over $G^0$ at most one non-zero entry (see Section A.4).    □

The process of replacing a Rees matrix by a cohomologous Rees matrix is termed *renormalization*. Let us now show that the cohomology of labelings and Rees matrices coincide.

**Proposition 4.13.28.** *Let $\Gamma$ be an oriented simplicial bipartite graph with parts $A$ and $B$ such that the positive edges point to $A$ and let $G$ be a group. Then two Rees matrices $C, C'\colon B \times A \to G^0$ compatible with $\Gamma$ are cohomologous if and only if their respective labelings $\ell_C, \ell_{C'}$ are cohomologous. More precisely, $\Delta(B, G) \times \Delta(A, G) \cong B(\Gamma, G)$ and the action of $\Delta(B, G) \times \Delta(A, G)$ on Rees matrices compatible with $\Gamma$ is conjugate to the action of $B(\Gamma, G)$ on $Z(\Gamma, G)$.*

*Proof.* Define $\psi\colon B(\Gamma, G) \to \Delta(B, G) \times \Delta(A, G)$ by $\delta\psi = (\Omega_\delta, \Lambda_\delta)$ where $b\Omega_\delta b = b\delta$ and $a\Lambda_\delta a = a\delta$. It is evident that $\psi$ is a group isomorphism. If $\delta \in B(\Gamma, G)$ and $C\colon B \times A \to G^0$, then

$$(b, a)^\delta \ell_C = (b\delta)(b, a)\ell_C(a\delta)^{-1} = b(\Omega_\delta C \Lambda_\delta^{-1})a = (b, a)\ell_{\Omega_\delta C \Lambda_\delta^{-1}}.$$

This completes the proof of the proposition.                                $\square$

**Exercise 4.13.29.** Let $S = \mathscr{M}^0(G, A, B, C)$ and $T = \mathscr{M}^0(G', A', B', C')$ be 0-simple Rees matrix semigroups. Show that $S \cong T$ if and only if there is an orientation-preserving graph isomorphism $\psi\colon \Gamma(S) \to \Gamma(T)$ and a group isomorphism $\gamma\colon G \to G'$ such that $\psi\ell_{C'}$ is cohomologous to $\ell_C\gamma$. Reformulate this result in terms of matrices (hint: consider $B \times B$ and $A \times A$ monomial matrices with entries in $G$ acting on the left and right of Rees matrices).

Now we can finally give the long-awaited characterization of when $\langle E(S) \rangle$ is aperiodic.

**Theorem 4.13.30 (Graham).** *Let $S = \mathscr{M}^0(G, A, B, C)$ be a Rees matrix semigroup. Then $\langle E(S) \rangle$ is aperiodic if and only if $C$ is cohomologous to a zero-one matrix.*

*Proof.* Proposition 4.13.28 implies that $C$ being cohomologous to a zero-one matrix is equivalent to the labeling $\ell_C$ being cohomologous to the trivial labeling. This in turn is equivalent to $(\Gamma, \ell_C)$ being $G$-acyclic by Proposition 4.13.14. But, Lemma 4.13.24 asserts that $(\Gamma, \ell_C)$ is $G$-acyclic if and only if $\langle E(S) \rangle$ is aperiodic.                                $\square$

In light of Corollary 4.13.4, Theorem 4.13.30 can be used to describe all semigroups whose idempotent-generated subsemigroup is aperiodic. The following application was first noted by Rhodes and Tilson [295].

**Theorem 4.13.31 (Rhodes-Tilson).** *Let $\mathbf{H}$ be a pseudovariety of groups and suppose $S = \mathscr{M}^0(G, A, B, C)$ is a Rees matrix semigroup with $G \in \mathbf{H}$. Then $S \in \mathbf{EA}$ if and only if $S \in \mathbf{A} \vee \mathbf{H}$, if and only if $S \in (\mathbf{Sl} * \mathbf{RZ}) \vee \mathbf{H}$.*

*Proof.* Clearly $(\mathbf{Sl} * \mathbf{RZ}) \vee \mathbf{H} \leq \mathbf{A} \vee \mathbf{H} \leq \mathbf{EA}$. For the converse direction, suppose $S \in \mathbf{EA}$. By Theorem 4.13.30, we may assume without loss of generality that $C$ is a zero-one matrix. Then define a surjective map

$$\varphi\colon \mathscr{M}^0(\{1\}, A, B, C) \times G \twoheadrightarrow S$$

by $((a, 1, b), g) \mapsto (a, g, b)$ and $(0, g) \mapsto 0$. Because $C$ is a zero-one matrix, $\varphi$ is easily verified to be a homomorphism. A routine computation shows that if $e$ is a non-zero idempotent of $T = \mathscr{M}^0(\{1\}, A, B, C)$, then $eTe = \{0, e\}$ and $TeT = T$. Thus Proposition 4.12.20 implies $T \in \mathbf{Sl} * \mathbf{RZ}$ (as $\mathbf{LZ} \leq \mathbf{Sl} * \mathbf{RZ}$ by Lemma 4.1.22). This completes the proof.                                $\square$

The following corollary shall be used several times in the sequel.

**Corollary 4.13.32.** *Let $M$ be a monoid with group of units $G$ such that $M \backslash G$ is a Rees matrix semigroup. Then $M \in \mathbf{A} * \mathbf{G}$ if and only if $M \in \mathbf{EA}$.*

*Proof.* Because $\mathbf{A} * \mathbf{G} \leq \mathbf{EA}$, one direction is trivial. For the other direction, let $J = M \setminus G$. Then Corollary 4.1.11 shows that $M$ divides the unitary semidirect product $J^I \rtimes G$. Because $\langle E(J) \rangle \in \mathbf{A}$, we have $J$, and hence $J^I$, belongs to $\mathbf{A} \vee \mathbf{G}$ by Theorem 4.13.31. We conclude $S \in \mathbf{A} * \mathbf{G}$.     □

We use the theory just developed to give a topological proof of a variation on a result of Volkov [373] that the pseudovarieties $\mathbf{EA}$ and $\mathbf{A} \vee \mathbf{G}$ are not finitely based (see also [370] for related results). Recall that a pseudovariety is finitely based if it has a finite basis of pseudoidentities. A finitely based pseudovariety $\mathbf{V}$ can be defined by pseudoidentities using a bounded number of variables, say $n$ of them. It follows then that a semigroup belongs to $\mathbf{V}$ if and only if all its $n$-generated subsemigroups do. So to show that a pseudovariety $\mathbf{V}$ is not finitely based, the usual trick is to produce, for each $k \geq 1$, a semigroup $S_k \notin \mathbf{V}$, all of whose $k$-generated subsemigroups belong to $\mathbf{V}$.

**Theorem 4.13.33.** *Let $\mathbf{H}$ be a proper pseudovariety of groups and $G$ a group not belonging to $\mathbf{H}$. Then every pseudovariety belonging to the interval $[(\mathbf{Sl} * \mathbf{RZ}) \vee (G), \mathbf{E\overline{H}}]$ is not finitely based. In particular, $\mathbf{A} \vee \mathbf{G}$, $\mathbf{A} * \mathbf{G}$ and $\mathbf{EA}$ are not finitely based.*

*Proof.* Let $G = \langle g_1, \ldots, g_r \rangle$. For $k \geq 1$, consider the graph $\Gamma_k$, whose geometric edges form a wedge of $r$ cycles $C_1, \ldots, C_r$, each of length $2k + 2$. Then $\Gamma_k$ is a connected, simplicial bipartite graph. We orient $\Gamma_k$ by having the edges point away from the part containing the base point. We can choose a spanning tree $T$ for $\Gamma_k$ consisting of all the edges except one edge $e_i$ from each cycle $C_i$, $i = 1, \ldots, r$. Label the edge $e_i$ by $g_i$, $i = 1, \ldots, r$, and all other edges by 1; call this labeling $\ell_k$. By construction, $\ell_k$ is Graham normalized with respect to $T$. Let $S_k$ be the 0-simple Rees matrix semigroup associated to the oriented $G$-labeled graph $(\Gamma_k, \ell_k)$. Then the maximal subgroup of $\langle E(S_k) \rangle$ is $G$ by Theorem 4.13.22 and Corollary 4.13.13, so $S_k \notin \mathbf{E\overline{H}}$.

Suppose that $V \leq S_k$ is generated by a subset $X$ of $k$ or fewer elements. Then $X$ intersects at most $k$ of the $\mathscr{R}$-classes of $S_k$, and hence the same is true of $V$. In particular, $V$ must miss some $\mathscr{R}$-class from each of the $r$ cycles $C_1, \ldots, C_r$. Thus $V$ is contained in a Rees matrix semigroup $U$ corresponding to a $G$-labeled graph obtained from $\Gamma_k$ by removing one vertex (corresponding to an $\mathscr{R}$-class missing from $V$) and any incident edges from each cycle $C_i$, $i = 1, \ldots, r$. As removing a vertex from each cycle results in a forest, and hence a $G$-acyclic graph, $\langle E(U) \rangle$ is aperiodic by Corollary 4.13.24. Therefore, $U \in (\mathbf{Sl} * \mathbf{RZ}) \vee (G)$ by Theorem 4.13.31, yielding the desired result.

The final statement follows by taking $\mathbf{H}$ to be trivial.     □

We remark that the pseudovariety $\mathbf{Sl} * \mathbf{RZ}$ is well-known to be compact (see Theorem 7.3.7 or [116]).

Recall that a Rees matrix $C \colon B \times A \to G^0$ is said to be *regular* if it contains no zero rows or columns. A Rees matrix is said to be *connected* if its incidence graph is connected. Connected matrices are clearly regular.

**Theorem 4.13.34 (Graham).** *Let* $S = \mathcal{M}^0(G, A, B, C)$ *be a Rees matrix semigroup. Let* $A'$ *and* $B'$ *be the respective sets of isolated vertices of* $\Gamma(S)$ *from* $A$ *and* $B$. *Let* $(A_1 \cup B_1), \ldots, (A_n \cup B_n)$ *be the connected components of* $(A \setminus A') \cup (B \setminus B')$. *Then there is a* $B' \times A'$ *zero matrix* $C_N$ *and a regular Rees matrix* $C_R \colon (B \setminus B') \times (A \setminus A') \to G^0$ *such that:*

1. $S \cong \mathcal{M}^0(G, A, B, C_R \oplus C_N)$ *where* $C_R \oplus C_N = \begin{pmatrix} C_R & 0 \\ 0 & C_N \end{pmatrix}$;

2. *The matrix* $C_R$ *is block diagonal of the form:*

$$C_R = \begin{pmatrix} C_1 & 0 & \cdots & 0 \\ 0 & C_2 & \cdots & \vdots \\ \vdots & \vdots & \ddots & 0 \\ 0 & \cdots & 0 & C_n \end{pmatrix}$$

*where* $C_i \colon B_i \times A_i \to G^0$ *is a connected Rees matrix over* $G^0$, $i = 1, \ldots n$;

3. $\langle E(S) \rangle \cong \bigcup_{i=1}^n \mathcal{M}^0(G_i, A_i, B_i, C_i)$, *where* $G_i$ *is the subgroup generated by the entries of* $C_i$.

*Proof.* The isolated vertices of $\Gamma(S)$ correspond to zero rows and zero columns of $S$, explaining the construction of $C_N$. Choose a maximal forest $F$ for $\Gamma(S)$. We can then find a Graham normalization with respect to $F$, which is cohomologous to the original labeling $\ell_C$. Because replacing a Rees matrix with a cohomologous matrix does not change the isomorphism type (Proposition 4.13.27), we may as well assume that the original labeling already is Graham normalized with respect to $F$. The result is then an immediate consequence of Theorems 4.13.20 and 4.13.22 and Corollary 4.13.13.    $\square$

**Exercise 4.13.35.** Provide all the remaining details of the proof of Theorem 4.13.34.

A Rees matrix coordinatization of $S$ of the form guaranteed by Theorem 4.13.34 is termed a *Graham normalization* of $S$. One can be computed from the multiplication table of $S$ in polynomial time.

The regular Type II elements of any semigroup were described in coordinates in [295] making use of Graham normalizations. If $S$ is as in Theorem 4.13.34, then $\mathsf{K}_\mathbf{G}(S) \cong \bigcup_{i=1}^n \mathcal{M}^0(N, A_i, B_i, C_i)$, where $N$ is the normal closure of $\langle G_1, \ldots, G_n \rangle$ in $G$. In particular, if $S \in \mathbf{EA}$, then $\mathsf{K}_\mathbf{G}(S) = \langle E(S) \rangle$ (this also follows from Theorem 4.13.31). The following exercise was suggested by K. Auinger.

**Exercise 4.13.36.** Let $S$ be a 0-simple semigroup with $\langle E(S) \rangle$ aperiodic. Prove without using a Rees matrix representation of $S$ that the Type II sub-semigroup of $S$ coincides with $\langle E(S) \rangle$. Hint: let $a \in \langle E(S) \rangle$ and $x, x'$ be a pair of mutually inverse elements. Suppose $xax' \neq 0$; prove that $L_x \cap R_{x'}$ contains an inverse $a'$ of $a$, which by Lemma 4.13.1 and Exercise 4.13.2 belongs to $\langle E(S) \rangle$. Conclude that $a' = x'x$ and $xax' = (xax')^2$.

The next several exercises sketch a proof that **EDS** = **DS** ⑩ **G** and **EDA** = **DA** ⑩ **G**.

**Exercise 4.13.37.** Let $S = \mathscr{M}^0(G, A, B, C)$ be a 0-simple semigroup. Show that $\langle E(S) \rangle$ is completely regular if and only if each connected component of $\Gamma(S)$ is a complete bipartite graph. Show that $E(S)$ is a subsemigroup if and only if each connected component of $(\Gamma(S), \ell_C)$ is a complete bipartite $G$-acyclic graph.

**Exercise 4.13.38.** Show that a semigroup $S$ belongs to **EDS** if and only if, for each regular $\mathscr{J}$-class $J$, the connected components of the incidence graph $\Gamma(J^0)$ are complete bipartite graphs. Show that $S \in$ **EDA** if and only if, for each regular $\mathscr{J}$-class $J$ with maximal subgroup $G_J$, the connected components of the $G_J$-labeled incidence graph $\Gamma(J^0)$ are complete bipartite $G_J$-acyclic graphs.

**Exercise 4.13.39.** Let $S = \mathscr{M}^0(G, A, B, C)$ be a 0-simple semigroup and $J = S \setminus \{0\}$. Show that $\mathsf{AGGM}_J(S) = \mathscr{M}^0(\{1\}, A', B', C')$ where $C'$ is the matrix obtained from $C$ by replacing all non-zero entries by 1 and then identifying equal rows and equal columns. Conclude using Graham's Theorem that $\langle E(S) \rangle$ is completely regular if and only if $\mathsf{AGGM}_J(S)$ is an inverse semigroup.

**Exercise 4.13.40.** Let $S = \mathscr{M}^0(G, A, B, C)$ be a 0-simple semigroup and set $J = S \setminus \{0\}$. Say that two rows $b, b'$ of $C$ are proportional on the left, if there is an element $g \in G$ such that $bCa = gb'Ca$ for all $a \in A$. Two columns $a, a'$ of $C$ are proportional on the right if there is an element $g \in G$ such that $bCa = bCa'g$ for all $b \in B$. Show that $\mathsf{GGM}_J(S) = \mathscr{M}^0(G, A', B', C')$ where $C'$ is obtained from $C$ by identifying all rows that are proportional on the left and all columns that are proportional on the right. Deduce using Graham's Theorem that $E(S)$ is a subsemigroup if and only if $\mathsf{GGM}_J(S)$ is an inverse semigroup.

**Exercise 4.13.41.** Suppose that $S$ is a right mapping semigroup whose 0-minimal ideal $J^0$ is an inverse semigroup. Show that $(J, S)$ is a semigroup of partial permutations.

**Exercise 4.13.42.** Use the previous exercises to show $S \in$ **EDS** if and only if $S^{\mathscr{J}'} \in$ **Sl** ⑩ **G** and $S \in$ **EDA** if and only if $S^{\mathscr{J}' \cap \mathbf{A}} \in$ **Sl** ⑩ **G**. Deduce:

$$\mathbf{EDS} = \mathbf{LG} \text{ ⑩ } (\mathbf{Sl} \text{ ⑩ } \mathbf{G}) = \mathbf{DS} \text{ ⑩ } \mathbf{G}$$
$$\mathbf{EDA} = \mathbf{LI} \text{ ⑩ } (\mathbf{Sl} \text{ ⑩ } \mathbf{G}) = \mathbf{DA} \text{ ⑩ } \mathbf{G}.$$

**Exercise 4.13.43.** Suppose that $S = \mathcal{M}^0(G, A, B, C)$. Recall that $\Gamma(S)$ has a natural labeling $\ell$ over the free group $FG(E^+)$ on its set of positively oriented edges $E^+$. Let $\psi \colon FG(E^+) \to G$ be the map induced by $e \mapsto e\ell_C$. Let $T$ be the Rees matrix semigroup over $FG(E^+)$ associated to the labeling $\ell$. Prove there is a homomorphism $\alpha \colon T \to S$ given by $(a, w, b) \mapsto (a, w\psi, b)$. Show that $\alpha$ is onto if and only if $G = \langle E^+ \ell_C \rangle$.

The subsequent exercises consider again 0-simple semigroups.

**Exercise 4.13.44.** Suppose that $S = \mathcal{M}(G, A, B, C)$ is a simple semigroup. Let $a_0 \in A$ and $b_0 \in B$ be fixed. Choose a spanning tree $T$ for $\Gamma(S)$ as per Exercise 4.13.17. Let $C'$ be the Rees matrix associated to a Graham normalization of $\Gamma(S)$ with respect to $T$. Show that $C'$ has only ones in row $b_0$ and column $a_0$.

**Exercise 4.13.45.** The purpose of this exercise is to describe homomorphisms and partial homomorphisms between Rees matrix semigroups in our topological language. Let $S = \mathcal{M}^0(G, A, B, C)$ and $S' = \mathcal{M}^0(G', A', B', C')$ be 0-simple Rees matrix semigroups. A partial homomorphism from $S$ to $S'$ is a function $\varphi \colon S \setminus \{0\} \to S' \setminus \{0\}$ such that $s_1 s_2 \neq 0$ implies $s_1 \varphi s_2 \varphi = (s_1 s_2)\varphi$.

1. Give a bijection between the following data:
   (a) a partial homomorphism $\varphi \colon S \setminus \{0\} \to S' \setminus \{0\}$;
   (b) an orientation-preserving graph morphism $\psi \colon \Gamma(S) \to \Gamma(S')$ and a homomorphism $\gamma \colon G \to G'$ such that $\psi \ell_{C'}$ is cohomologous to $\ell_C \gamma$.
2. Show that with the above notation, $\varphi$ extends to a homomorphism $S \to S'$ if and only if $\Gamma(S)\psi$ is an induced subgraph of $\Gamma(S')$.
3. Suppose $\varphi$ extends to an onto homomorphism from $S$ to $S'$. Show that there is a labeling $\ell$ of $\Gamma(S)$ cohomologous to $\ell_C$ such that $\psi \ell_{C'} = \ell \gamma$.
4. Conclude that if $\varphi \colon S \twoheadrightarrow S'$ is an onto homomorphism, then there exist a Rees matrix $C_0 \colon B \times A \to G^0$, onto maps $\alpha \colon A \twoheadrightarrow A'$, $\beta \colon B \twoheadrightarrow B'$ and an onto homomorphism $\gamma \colon G \twoheadrightarrow G'$ such that $S \cong \mathcal{M}^0(G, A, B, C_0)$, $C' = C_0 \gamma$ and with respect to this Rees matrix coordinatization one has $(a, g, b)\varphi = (a\alpha, g\gamma, b\beta)$.

## 4.14 The Presentation Lemma

The techniques introduced so far are insufficient to deal with the complexity of arbitrary semigroups. Let us give an example. Consider the matrix

$$
P = \begin{array}{c|ccccccc}
 & a_0 & a_1 & a_2 & a_3 & a_4 & a_5 & a_6 \\
\hline
0 & 1 & 0 & 0 & 1 & 0 & 0 & 0 \\
1 & 1 & 1 & 0 & 0 & 0 & 0 & 0 \\
2 & 0 & 1 & 1 & 0 & 0 & 0 & 0 \\
3 & 0 & 0 & 1 & 1 & 0 & 0 & 0 \\
0' & 0 & 0 & 0 & 0 & 1 & 0 & 1 \\
2' & 0 & 0 & 0 & 0 & 0 & 1 & 1 \\
\end{array}
\tag{4.14}
$$

**Table 4.1.** Multiplication in the Tall Fork $F$

$$(a, \epsilon, i)h = (a, \epsilon, i+1 \bmod 4), \ i = 0, 1, 2, 3$$
$$(a, \epsilon, i')h = 0, \ i = 0, 2$$
$$h(a_i, \epsilon, b) = (a_{i-1 \bmod 4}, \epsilon, b), \ i = 0, 1, 2, 3$$
$$h(a_i, \epsilon, b) = 0, \ i \geq 4$$
$$(a, \epsilon, i')k = (a, \epsilon, (i+2 \bmod 4)'), \ i = 0, 2$$
$$(a, \epsilon, i)k = 0, \ i = 0, 1, 2, 3$$
$$k(a_i, \epsilon, b) = 0, \ i = 0, 1, 2, 3$$
$$k(a_4, \epsilon, b) = (a_5, \epsilon, b)$$
$$k(a_5, \epsilon, b) = (a_4, \epsilon, b)$$
$$k(a_6, \epsilon, b) = (a_6, \epsilon, b)$$
$$(a, \epsilon, i')t = 0, \ i = 0, 1, 2, 3$$
$$(a, \epsilon, 0')t = (a, -\epsilon, 0)$$
$$(a, \epsilon, 2')t = (a, \epsilon, 2)$$
$$t(a_i, \epsilon, b) = (a_4, -\epsilon, b), \ i = 0, 3$$
$$t(a_i, \epsilon, b) = (a_5, \epsilon, b), \ i = 1, 2$$
$$t(a_i, \epsilon, b) = 0, \ i = 4, 5, 6$$

and the Rees matrix semigroup $J^0 = \mathscr{M}^0(\{\pm 1\}, A, B, P)$ where we set $A = \{a_0, \ldots, a_6\}$ and $B = \{0, 1, 2, 3, 0', 2'\}$. The matrix $P$ underlies many interesting examples in complexity theory [43, 275, 288, 289]. The salient feature is that $\Gamma(J^0)$ contains two connected components. One component is a cycle of length 8 consisting of $\{0, a_0, 1, a_1, 2, a_2, 3, a_3\}$. Crucial here is that 0 and 2 have no common neighbors. The other connected component $\{0', 2', a_4, a_5, a_6\}$ is a tree, and $0'$ and $2'$ have the common neighbor $a_6$.

Let $H = \langle h \rangle$ and $K = \langle k \rangle$ be cyclic groups of respective orders 4 and 2. Let $N = \{k^i t h^j\}$ where $t$ is a new symbol and $0 \leq i \leq 1, 0 \leq j \leq 3$. Set $F = H \cup K \cup N \cup J^0$. We define $N^2 = HK = KH = HN = NK = 0$, whereas the actions of $K$ on the left of $N$ and $H$ on the right of $N$ are the obvious ones. We are left with defining the actions of $h$, $k$ and $t$ on $J^0$. The definitions are recorded in Table 4.1. The reader should verify that in order to have associativity for all products with $t$ in the middle, it is essential that 0 and 2 have no common neighbors in $\Gamma(J^0)$.

**Exercise 4.14.1.** Verify that the multiplication defined on $F$ in Table 4.1 is associative and that $F$ is a group mapping semigroup.

The semigroup $F$ is known as the *Tall Fork* because of the shape of the Hasse diagram of the poset of $\mathscr{J}$-classes of $F$; see Figure 4.1. Both the depth and Tilson number of $F$ are 2. On the other hand, define a relational morphism $\psi$ from $F$ to $\overline{\mathbf{2}}$ by

$$0\psi = \overline{\mathbf{2}}, \ K\psi = \overline{1}, \ H\psi = N\psi = \overline{0}, \ (a, \epsilon, b)\psi = \begin{cases} \overline{0} & b \in \{0, 1, 2, 3\} \\ \overline{1} & b \in \{0', 2'\}. \end{cases}$$

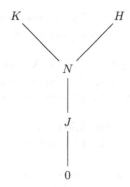

**Fig. 4.1.** Hasse diagram for the $\leq_{\mathscr{J}}$-order of the Tall Fork

One can verify easily that $\psi$ is a relational morphism. It follows that any Type I subsemigroup of $F$ is contained in $\overline{0}\psi^{-1}$ or $\overline{1}\psi^{-1}$. It is then straightforward to verify that the Type II subsemigroups of $\overline{0}\psi^{-1}$ and $\overline{1}\psi^{-1}$ are aperiodic.

**Exercise 4.14.2.** Show that the Type II subsemigroups of $\overline{0}\psi^{-1}$ and $\overline{1}\psi^{-1}$ are aperiodic.

So for the Tall Fork $F$, we see that the Type I–Type II lower bound $\#(F)$ is 1 (see Theorem 4.12.9). Therefore, we need some other technique to determine whether the actual complexity of $F$ is 1 or 2. This technique is the Presentation Lemma. In Section 4.14.2, we shall show that the complexity of the Tall Fork is 2. For historical comments on the Presentation Lemma, consult the notes at the end of the chapter. All semigroups in this section are again assumed to be finite.

### 4.14.1 Cross sections

We begin with the crucial notion of a cross section. We say that a relational morphism $\varphi \colon S \to T$ is *injective on $\mathscr{H}$-classes* of a subset $X \subseteq S$ if

$$s, s' \in X, \ s \mathrel{\mathscr{H}} s' \text{ and } s\varphi \cap s'\varphi \neq \emptyset \implies s = s'. \qquad (4.15)$$

Note that we only demand $s \mathrel{\mathscr{H}} s'$ in $S$, even if $X$ is a subsemigroup.

**Definition 4.14.3 (Cross section).** *Let $S$ be a semigroup and let $X \subseteq S$. Then a cross section for $X$ is a relational morphism $\varphi \colon S \to T$ that is injective on $\mathscr{H}$-classes of $X$. If there exists such a $\varphi$ with $T \in \mathbf{V}$, then we say $S$ has a cross section for $X$ over $\mathbf{V}$.*

To get a handle on this definition, we prove the following simple proposition.

**Proposition 4.14.4.** *Let* $\varphi\colon S \to T$ *be a cross section for* $X$ *and* $d\colon T \to U$ *a division. Then* $\varphi d$ *is also a cross section for* $X$.

*Proof.* Suppose $u \in s\varphi d \cap s'\varphi d$ with $s, s' \in X$ and $s \mathscr{H} s'$. Then we can find $t \in s\varphi$ and $t' \in s'\varphi$ such that $u \in td \cap t'd$. But $d$ is a division, so this forces $t = t'$. But then $s = s'$ because $\varphi$ is a cross section for $X$.    $\square$

The following elementary proposition says that when hunting a cross section for a $\mathscr{J}$-class, it suffices to find a cross section for one of its $\mathscr{R}$-classes.

**Proposition 4.14.5.** *Suppose that* $R$ *is an* $\mathscr{R}$-class *of a semigroup* $S$ *contained in a* $\mathscr{J}$-class $J$. *Let* $\varphi\colon S \to T$ *be a relational morphism. Then the following are equivalent:*

1. $\varphi$ *is a cross section for* $J$;
2. $\varphi$ *is a cross section for every* $\mathscr{R}$-class *of* $J$;
3. $\varphi$ *is a cross section for* $R$.

*Proof.* Clearly, 1 and 2 are equivalent from the definition of a cross section. Trivially 2 implies 3. To see that 3 implies 2, assume $\varphi$ is a cross section for $R$ and let $R'$ be an $\mathscr{R}$-class of $J$. Assume $s_1 \mathscr{H} s_2$ are elements of $R'$ with $t \in s_1\varphi \cap s_2\varphi$. Let $y \in L_{s_1} \cap R$. Then there exists $s' \in S^I$ with $s's_1 = y$. Green's Lemma (Lemma A.3.1) then says $s'\cdot\colon R' \to R$ is a bijection preserving $\mathscr{H}$. Hence $s's_1 \mathscr{H} s's_2$ and $y = s's_1, s's_2 \in R$. Moreover, if $t' \in s'\varphi$ (where we take $t' = I$ if $s' = I$), then $t't \in s's_i\varphi$, $i = 1, 2$. We conclude $s's_1 = s's_2$, as $\varphi$ is a cross section for $R$, whence $s_1 = s_2$ because $s'\cdot\colon R' \to R$ is a bijection.    $\square$

Of course, a dual result holds for $\mathscr{L}$-classes. Recall that $\mathrm{Reg}(S)$ denotes the set of regular elements of $S$. The Aperiodicity Lemma admits the following restatement.

**Proposition 4.14.6 (Aperiodicity Lemma: Cross section version).** *A relational morphism* $\varphi\colon S \to T$ *is a cross section for* $\mathrm{Reg}(S)$ *if and only if it is aperiodic.*

Cross sections lead to an alternative description of $\mathbf{A} \,\textcircled{m}\, \mathbf{V}$.

**Theorem 4.14.7 (Cross section Theorem).** *Let* $\mathbf{V}$ *be a pseudovariety. Then* $S \in \mathbf{A} \,\textcircled{m}\, \mathbf{V}$ *if and only if every regular* $\mathscr{R}$-class *admits a cross section over* $\mathbf{V}$.

*Proof.* If $\varphi\colon S \to T$ is an aperiodic relational morphism with $T \in \mathbf{V}$, then $\varphi$ is a cross section for $\mathrm{Reg}(S)$ and hence for every regular $\mathscr{R}$-class.

Conversely, if each regular $\mathscr{R}$-class $R$ admits a cross section $\varphi_R\colon S \to T_R$ with $T_R \in \mathbf{V}$, then $\Delta \prod_R \varphi_R\colon S \to \prod_R T_R$ is a cross section for $\mathrm{Reg}(S)$ and hence an aperiodic relational morphism.    $\square$

By a *module* $(X, S)$ we mean a semigroup $S$ acting by partial transformations on the right of a set $X$ (we do not assume faithfulness). Such objects were simply called partial transformation semigroups in Chapter 1, but in this chapter we have been reserving the latter term for the faithful case. Let us use the convention that if $xs$ is not defined, then we write $xs = \emptyset$. Our definition of a relational morphism follows the unpublished work of Tilson [367], and is essentially the notion of a relational covering from Eilenberg [85] with the arrows turned around (here we find once again the difference in viewpoint between coverings and relational morphisms in terms of direction).

**Definition 4.14.8 (Relational morphism of modules).** *A relational morphism of modules* $\varphi \colon (X, S) \to (Y, T)$ *is a fully defined relation* $\varphi \colon X \to Y$ *such that, for each* $s \in S$, *there exists* $\widehat{s} \in T$ *with the property:*

$$\forall y \in Y, \ y\varphi^{-1}s \subseteq y\widehat{s}\varphi^{-1}. \tag{4.16}$$

*Equivalently, for all* $x \in X$ *with* $xs$ *defined and all* $y \in x\varphi$, *the element* $y\widehat{s}$ *is defined and* $y\widehat{s} \in (xs)\varphi$. *One says in this situation that* $\widehat{s}$ *covers* $s$. *If, moreover,* $\varphi$ *is injective, then it is called a* division.

A relational morphism of modules between partial transformation semigroups will also be called a *relational morphism of partial transformation semigroups*.

If $\varphi \colon (R, S) \to (Q, T)$ is a relational morphism of modules, then the *companion relation* $f_\varphi \colon S \to T$ is defined by

$$sf_\varphi = \{t \in T \mid t \text{ covers } s\}. \tag{4.17}$$

**Exercise 4.14.9.** Prove that $f_\varphi$ is a relational morphism. Show that if $\varphi$ is a division and $(R, S)$ is faithful and total, then $f_\varphi$ is a division.

The reader should compare with Lemma 4.1.23.

The Fundamental Lemma of Complexity tells us that in order to compute complexity, we need to decide membership in pseudovarieties of the form $\mathbf{A} \, \textcircled{m} \, (\mathbf{G} * \mathbf{C}_{n-1})$; see Corollary 4.9.4. We would thus like to undertake a detailed study of what it means to have a cross section for a pseudovariety of the form $\mathbf{G} * \mathbf{V}$. This will lead us to formulate the Presentation Lemma.

*Remark 4.14.10.* Rhodes had originally defined a notion of sets, partitions and cross sections in Rees coordinates and applied it to complexity [43]. Tilson's notion of cross section [366] was an attempt to give a coordinate-free notion, whereas Steinberg's cross section idea [334] attempted to remove partial transformation semigroups from the picture as much as possible. See [334] for further discussion.

### 4.14.2 Presentations

To define presentations, we must first discuss automata. An *automaton* is a pair $(Q, X)$ with $Q$ a set and $X$ a set acting on $Q$ via a transition partial function $\delta\colon Q \times X \to Q$. That is, $(Q, X)$ is a deterministic partial automaton in the usual sense, but without initial or terminal states. As usual, $(q, x)\delta$ is written $qx$ and if $qx$ is undefined, then we write $qx = \emptyset$. The *transition semigroup* of an automaton $(Q, X)$ is the semigroup of partial transformations of $Q$ generated by the maps $q \mapsto qx$ with $x \in X$. An automaton is called *injective* if $X$ acts on $Q$ via partial permutations.

**Definition 4.14.11 (Automaton congruence).** *An automaton congruence on $(Q, X)$ is an equivalence relation $\sim$ on $Q$ such that $q \sim q'$ and $qx, q'x$ defined imply $qx \sim q'x$. An automaton congruence $\sim$ is called injective if $qx \sim q'x$ implies $q \sim q'$.*

The *quotient automaton* $(Q, X)/{\sim} = (Q/{\sim}, X)$ is defined by declaring $[q]x = [q'x]$ if $q'x$ is defined for some $q' \in [q]$; else $[q]x$ is undefined.

**Exercise 4.14.12.** Verify that if $\sim$ is an automaton congruence on $(Q, X)$, then $(Q, X)/{\sim}$ is a well-defined automaton. Show that $\sim$ is injective if and only if the quotient $(Q, X)/{\sim}$ is injective.

**Exercise 4.14.13.** Define notions of morphism and quotient morphism of automata. State and prove an equivalence between automaton congruences and quotient morphisms.

Let us recall the notion of a parameterized relational morphism of modules: a *parameterized relational morphism* of modules $\Phi\colon (Q, S) \to (P, T)$ consists of a relational morphism $\varphi_1\colon (Q, S) \to (P, T)$ and a relational morphism $\varphi_2\colon S \to T$ such that $\varphi_2 \subseteq f_{\varphi_1}$, where $f_{\varphi_1}$ is the companion relation (4.17) to $\varphi_1$.

**Definition 4.14.14 (Derived automaton).** *The derived automaton of a parameterized relational morphism of modules $\Phi\colon (R, S) \to (Q, T)$ is the automaton $\mathcal{D}_\Phi = (\#\varphi_1, D_\Phi)$ given by*

$$D_\Phi = \{(q, (s, t)) \in Q \times \#\varphi_2 \mid qt \neq \emptyset\}. \tag{4.18}$$

*The action is given by*

$$(r, q)(q_0, (s, t)) = \begin{cases} (rs, qt) & q_0 = q \text{ and } rs \text{ is defined} \\ \emptyset & \text{otherwise.} \end{cases} \tag{4.19}$$

*Remark 4.14.15.* The transition semigroup of the derived automaton is the derived transformation semigroup [85, 367].

We have now arrived at the all important definition of a presentation. The form given here is taken from the second author's work [334] and is a coordinate-free version of the first author's notion [43]. In what follows, we do not distinguish between a partition and its associated equivalence relation.

**Definition 4.14.16 (Presentation).** *Let $S$ be a semigroup and $R$ a regular $\mathscr{R}$-class. Then a presentation for $R$ is a pair $(\Phi, \mathscr{P})$ where:*

1. *$\Phi\colon (R, S) \to (Q, T)$ is a parameterized relational morphism;*
2. *$\mathscr{P}$ is an injective automaton congruence on $\mathcal{D}_\Phi$ such that*

$$(x, q) \; \mathscr{P} \; (x', q') \implies q = q'; \qquad (4.20)$$

3. *$(x, q) \; \mathscr{P} \; (x', q')$ and $x \; \mathscr{H} \; x'$ implies $(x, q) = (x', q')$.*

*If $\Phi = (\varphi_1, \varphi_2)$, then the sets $q\varphi_1^{-1}$, $q \in Q$, are called the **sets** of the presentation, and the partition $\mathscr{P}$ is called the **partition** of the presentation. Condition 3 is the **cross section** condition.*

*If $T$ belongs to the pseudovariety $\mathbf{V}$, then we say $S$ has a presentation for $R$ over $\mathbf{V}$.*

More generally, we call a partition $\mathscr{P}$ on $\mathcal{D}_\Phi$ admissible if it satisfies condition 2 of Definition 4.14.16, but not necessarily condition 3.

We remark that one can always take $(Q, T)$ to be faithful by composing $\Phi$ with the natural map from $(Q, T)$ to the associated faithful partial transformation semigroup. One can readily verify conditions 2 and 3 are still satisfied.

Our next goal is to show that $S$ has a presentation for a regular $\mathscr{R}$-class $R$ over $\mathbf{V}$ if and only if it has a cross section for $R$ over $\mathbf{G} * \mathbf{V}$. We actually prove a more general result concerning pointlikes. The proof has roots in [43, 322, 327, 365, 366] and is a major revision of the one presented in [334].

We break the proof up into two technical lemmas, which may be omitted on a first reading.

**Lemma 4.14.17.** *Let $R$ be a regular $\mathscr{R}$-class of a semigroup $S$ and suppose $\Phi\colon (R, S) \to (Q, T)$ is a parameterized relational morphism. Let $\mathscr{P}$ be an admissible partition on $\mathcal{D}_\Phi$ and suppose $\Phi = (\varphi_1, \varphi_2)$. Then there is a group $G$ and a relational morphism $\rho\colon S \to G \wr (Q, T)$ such that if $w \in G \wr (Q, T)$, then there exists $q \in Q$ such that: $w\rho^{-1} \cap R \subseteq q\varphi_1^{-1}$ and $(w\rho^{-1} \cap R) \times \{q\}$ is contained in a single block of $\mathscr{P}$.*

*Proof.* To establish notation, for $(r, q) \in \#\varphi_1$, let us denote by $[(r, q)]$ the partition block of $(r, q)$ in $\mathscr{P}$. Because any partial permutation of the finite set $P = \#\varphi_1/\mathscr{P}$ can be extended to a (total) permutation of $P$, for each $x \in \mathcal{D}_\Phi$, choose a permutation $\sigma_x$ extending the action of $x$ on $P$ in $\mathcal{D}_\Phi/\mathscr{P}$. Let $(P, G)$ be the permutation group obtained by setting

$$G = \langle \sigma_x \mid x \in \mathcal{D}_\Phi \rangle.$$

Define a relational morphism of modules

$$\alpha\colon (R, S) \to (P, G) \wr (Q, T) = (P \times Q, W)$$

by setting

$$r\alpha = \{(p,q) \mid q \in r\varphi_1,\ p = [(r,q)]\}. \tag{4.21}$$

To see that $\alpha$ is indeed a relational morphism, first observe if $r \in R$ and $q \in r\varphi_1$, then $([(r,q)],q) \in r\alpha$, and so $r\alpha \neq \emptyset$. Suppose $s \in S$ and choose $t \in s\varphi_2$. Define $f \in G^{Q^0}$ by

$$qf = \begin{cases} \sigma_{(q,(s,t))} & qt \neq 0 \\ \text{arbitrary} & \text{otherwise} \end{cases} \tag{4.22}$$

and let $\hat{s} \in G \wr (Q,T)$ be represented by $(f,t)$ (recall $G \wr (Q,T)$ is a quotient of $G \wr (Q^0,T)$). We must show that if $(p,q) \in P \times Q$, then $(p,q)\alpha^{-1} s \subseteq (p,q)\hat{s}\alpha^{-1}$. Let $r \in (p,q)\alpha^{-1}$. If $rs = \emptyset$, we are done. So assume $rs \in R$. Then, by definition of $\alpha$, we have that $q \in r\varphi_1$ and $p = [(r,q)]$. As $\varphi_2$ is contained in the companion relation to $\varphi_1$ and $t \in s\varphi_2$, it follows $rs \in q\varphi_1^{-1}s \subseteq qt\varphi_1^{-1}$ and consequently $qt$ is defined and belongs to $(rs)\varphi_1$. Now $(r,q)(q,(s,t)) = (rs,qt)$ and so by the definition of the quotient by an automaton congruence,

$$[(r,q)](q,(s,t)) = [(rs,qt)].$$

As $\sigma_{(q,(s,t))}$ agrees with $(q,(s,t))$ on its domain, we obtain

$$[(r,q)]\sigma_{(q,(s,t))} = [(rs,qt)].$$

Recalling $p = [(r,q)]$, (4.22) yields

$$(p,q)(f,t) = (p(qf),qt) = ([(r,q)]\sigma_{(q,(s,t))},qt) = ([(rs,qt)],qt)$$

with $qt \in (rs)\varphi_1$ and so $rs \in (p,q)(f,t)\alpha^{-1}$ (see (4.21)), establishing $\alpha$ is a relational morphism.

Let $f_\alpha \colon S \to W$ be the companion relation (4.17) to $\alpha$. We claim that $f_\alpha$ is the sought after relational morphism $\rho$. Let $w \in W$ and choose an idempotent $e \in R$. Fix $q \in e\varphi_1$ and set $p = [(e,q)]$. Then $(p,q) \in e\alpha$. Let $s \in wf_\alpha^{-1} \cap R$. By the definition of the companion relation, it follows

$$s = es \in (p,q)\alpha^{-1}s \subseteq (p,q)w\alpha^{-1}.$$

Consequently, $(p,q)w$ is defined and if $(p,q)w = (p',q')$, then the inclusion $wf_\alpha^{-1} \cap R \subseteq (p',q')\alpha^{-1}$ is verified as $s$ was arbitrary. By definition of $\alpha$, if $s \in wf_\alpha^{-1} \cap R$, then $q' \in s\varphi_1$ and $p' = [(s,q')]$. Thus $wf_\alpha^{-1} \cap R \subseteq q'\varphi_1^{-1}$ and $(wf_\alpha^{-1} \cap R) \times \{q'\}$ is contained in a single block of $\mathscr{P}$, as desired.     □

Our second technical lemma is essentially a converse to the previous one. All proofs of the Presentation Lemma rely on the fact that regular $\mathscr{J}$ or $\mathscr{R}$-classes can be lifted under surmorphisms to regular $\mathscr{J}$ or $\mathscr{R}$-classes.

**Lemma 4.14.18.** *Let $R$ be a regular $\mathscr{R}$-class of a semigroup $S$ and suppose $\rho \colon S \to G \rtimes T$ is a relational morphism with $G$ a group. Then there are a parameterized relational morphism $(\varphi_1,\varphi_2) = \Phi \colon (R,S) \to (Q,T)$ and an admissible partition $\mathscr{P}$ on $D_\Phi$ such that: if $X \subseteq q\varphi_1^{-1}$ and $X \times \{q\}$ is contained in a single block of $\mathscr{P}$, then there exists $(g,t) \in G \rtimes T$ with $X \subseteq (g,t)\rho^{-1}$.*

*Proof.* Let $\pi\colon G \rtimes T \twoheadrightarrow T$ denote the semidirect product projection and set $\psi = \rho\pi$. We use $\alpha\colon \#\psi \twoheadrightarrow S$ and $\beta\colon \#\psi \to T$ for the projections. Let $\overline{R}$ be an $\leq_{\mathscr{R}}$-minimal $\mathscr{R}$-class of $\#\psi$ mapping under $\alpha$ into $R$. Then, by Lemma 4.6.10, $\overline{R}\alpha = R$ and $\overline{R}$ is regular. Let $R'$ be the $\mathscr{R}$-class of $T$ defined by $\overline{R}\beta \subseteq R'$.

There results a parameterized relational morphism $\Phi\colon (R, S) \to (R', T)$ given by

$$r\varphi_1 = r\alpha|_{\overline{R}}^{-1}\beta = \{r' \in R' \mid (r, r') \in \overline{R}\} \text{ and } \varphi_2 = \psi.$$

Notice that $\#\varphi_1 = \overline{R}$; in particular, $\varphi_1$ is fully defined as $\overline{R}\alpha = R$. To verify $\Phi$ is a parameterized relational morphism, let $r' \in R'$ and $s \in S$. Then our goal is to show that for any $t \in s\psi$, we have

$$r'\varphi_1^{-1}s \subseteq r't\varphi_1^{-1}$$

as this will prove that $t$ covers $s$ and $\psi$ is contained in the companion relation to $\varphi_1$. So let $r \in r'\varphi_1^{-1}$. If $rs \notin R$, there is nothing to prove, so assume $rs \in R$. Then we have $(r, r') \in \overline{R}$, $(s, t) \in \#\psi$ and

$$(r, r')(s, t) = (rs, r't) \leq_{\mathscr{R}} (r, r').$$

From $(rs, r't)\alpha = rs \in R$ and minimality of $\overline{R}$, it follows $(rs, r't) \in \overline{R}$ and so $rs \in (r't)\varphi_1^{-1}$. This proves $\Phi$ is a parameterized relational morphism.

Because $\overline{R}$ is regular, we can find an idempotent $(e, f) \in \overline{R}$; in particular, $f \in e\varphi_1$. Let us fix this idempotent $(e, f)$ for the rest of the proof.

The remainder of the proof makes use of the semigroupoid $\mathbf{Der}(\psi)$ associated to a relational morphism of semigroups $\psi\colon S \to T$; see Section 2.5. If $\psi\colon S \to T$ is a relational morphism, observe there is a faithful functor $\mathbf{Der}(\psi) \to \#\psi$ given on arrows by

$$(t_L, (s, t)) \longmapsto (s, t). \tag{4.23}$$

Consider the $\mathscr{R}$-class $\widetilde{R}$ of the arrow $(f, (e, f))$ in $\mathbf{Der}(\psi)$. Notice that $(f, (e, f))\colon f \to f$ is an idempotent of $\widetilde{R}$, and so $\widetilde{R}$ is a regular $\mathscr{R}$-class.

We claim that the following equalities hold:

$$\widetilde{R} = \{(f, (s, t)) \mid t \in s\varphi_1\} = \{(f, (s, t)) \mid (s, t) \in \overline{R}\}. \tag{4.24}$$

To see this, suppose first that $(f, (e, f)) \mathscr{R} (f, (s, t))$. Then, by considering our faithful functor (4.23), we obtain $(e, f) \mathscr{R} (s, t)$ in $\#\psi$. It follows $(s, t) \in \overline{R}$, i.e., $t \in s\varphi_1$.

Conversely, if $t \in s\varphi_1$, then $(s, t) \in \overline{R} \subseteq \#\psi$ and so the arrow $(f, (s, t))$ of $\mathbf{Der}(\psi)$ is well defined. Because $(e, f)$ is an idempotent of the $\mathscr{R}$-class $\overline{R}$, it follows $(e, f)(s, t) = (s, t)$ and so

$$(f, (e, f))(f, (s, t)) = (f, (s, t))\colon f \to t.$$

Suppose $(s, t)(s', t') = (e, f)$ with $(s', t') \in \#\psi$. Then $(t, (s', t')) \in \mathbf{Der}(\psi)$ and

$$(f, (s, t))(t, (s', t')) = (f, (ss', tt')) = (f, (e, f))$$

proving $(f, (s, t)) \in \widetilde{R}$ and establishing (4.24). Notice that if $(f, (s, t)) \in \widetilde{R}$, then $ft = t$ as $(s, t) \in \overline{R}$ and $(e, f)$ is an idempotent of the $\mathscr{R}$-class $\overline{R}$. This observation will be used without comment throughout the rest of the proof.

Consider the relational morphism $\zeta \colon \mathbf{Der}(\psi) \to G$ defined by

$$(t_L, (s, t))\zeta = \{^{t_L}g \mid (g, t) \in s\rho\}. \tag{4.25}$$

It is easy to check that $\zeta$ is a relational morphism, (cf. [339, Section 1.4]).

Define a relation $\mathscr{P}$ on $\overline{R} = \#\varphi_1$ by setting

$$(s, t) \ \mathscr{P} \ (s', t') \iff t = t' \text{ and } (f, (s, t))\zeta \cap (f, (s', t'))\zeta \neq \emptyset. \tag{4.26}$$

By definition of $\mathscr{P}$ (4.20) holds. The fact that $\mathscr{P}$ is an equivalence relation is immediate from the following claim.

*Claim.* Let $(s, t), (s', t') \in \overline{R} = \#\varphi_1$. Then $(s, t) \ \mathscr{P} \ (s', t')$ if and only if $t = t'$ and $(f, (s, t))\zeta = (f, (s', t'))\zeta$.

*Proof.* It follows directly from the definition of $\mathscr{P}$ (4.26) that if $t = t'$ and $(f, (s, t))\zeta = (f, (s', t'))\zeta$, then $(s, t) \ \mathscr{P} \ (s', t')$. For the converse, suppose $(s, t) \ \mathscr{P} \ (s', t')$. Then $t = t'$ and, setting $a = (f, (s, t))$, $b = (f, (s', t'))$, one has $a\zeta \cap b\zeta \neq \emptyset$. Because $\widetilde{R}$ is regular, $a$ has an inverse $a' = (t, (s_0, t_0))$. Lemma 4.12.3 shows that if $h \in a\zeta$, then $h^{-1} \in a'\zeta$. Because $aa' \ \mathscr{R} \ a \ \mathscr{R} \ b$, one has $aa'b = b$. Suppose now that $g \in a\zeta \cap b\zeta$. Then from $g^{-1} \in a'\zeta$, we compute

$$a\zeta = a\zeta \cdot \{g^{-1}\} \cdot \{g\} \subseteq a\zeta \cdot a'\zeta \cdot b\zeta \subseteq (aa'b)\zeta = b\zeta.$$

The reverse inclusion $b\zeta \subseteq a\zeta$ is symmetric, establishing the claim.     □

To see that $\mathscr{P}$ is an automaton congruence, let

$$(s, t) \ \mathscr{P} \ (s', t') \text{ and } (t_L, (s_1, t_1)) \in D_\Phi.$$

Then $(s, t)(t_L, (s_1, t_1))$ and $(s', t')(t_L, (s_1, t_1))$ defined implies $t = t_L = t'$, $ss_1, s's_1 \in R$ and

$$(s, t)(t_L, (s_1, t_1)) = (ss_1, tt_1) \tag{4.27}$$
$$(s', t')(t_L, (s_1, t_1)) = (s's_1, tt_1). \tag{4.28}$$

By (4.26), there exists $g \in (f, (s, t))\zeta \cap (f, (s', t))\zeta$. Choose $g' \in (t, (s_1, t_1))\zeta$. Then clearly

$$gg' \in (f, (ss_1, tt_1))\zeta \cap (f, (s's_1, tt_1))\zeta$$

and so $(ss_1, tt_1) \ \mathscr{P} \ (s's_1, tt_1)$, as desired. We conclude $\mathscr{P}$ is an automaton congruence.

To see that $\mathscr{P}$ is injective, suppose

$$(s,t)(t_L,(s_1,t_1)) \text{ and } (s',t')(t_L,(s_1,t_1))$$

are defined and in the same $\mathscr{P}$-block. Then $t = t_L = t'$ and (4.27) and (4.28) hold. Because $(ss_1,tt_1) \,\mathscr{P}\, (s's_1,tt_1)$, there is an element

$$g \in (f,(ss_1,tt_1))\zeta \cap (f,(s's_1,tt_1))\zeta.$$

Choose $g_0 \in (t,(s_1,t_1))\zeta$. We aim to show $gg_0^{-1} \in (f,(s,t))\zeta \cap (f,(s',t'))\zeta$. This will establish $(s,t) \,\mathscr{P}\, (s',t')$, as $t = t'$, and hence the injectivity of $\mathscr{P}$.

The elements $(ss_1,tt_1)$ and $(s,t)$ belong to the regular $\mathscr{R}$-class $\overline{R}$ of $\#\psi$ and so there exists $(u,v) \in \#\psi$ such that $(ss_1,tt_1)(u,v) = (s,t)$. Let

$$(w,z) = (u,v)\big((s_1,t_1)(u,v)\big)^{\omega-1} \in \#\psi.$$

Then straightforward calculations show

$$(s,t)(s_1u,t_1v) = (s,t) \tag{4.29}$$

$$(ss_1,tt_1)(w,z) = (s,t) \tag{4.30}$$

$$(w,z)(s_1,t_1)(w,z) = (w,z). \tag{4.31}$$

Indeed, (4.29) is immediate from the choice of $(u,v)$, whereas (4.30) follows directly from (4.29) as $(s_1,t_1)(w,z) = (s_1u,t_1v)^\omega$. The final equation (4.31) is an instance of the general fact that

$$\big(x(yx)^{\omega-1}\big)\, y\, \big(x(yx)^{\omega-1}\big) = x(yx)^{\omega-1}(yx)^\omega = x(yx)^{\omega-1}$$

where for us $x = (u,v)$ and $y = (s_1,t_1)$.

From (4.30), it follows $tt_1 z = t$. Then (4.31) yields that in $\mathbf{Der}(\psi)$

$$(tt_1,(w,z))(t,(s_1,t_1))(tt_1,(w,z)) = (tt_1,(w,z))$$

and so Lemma 4.12.3 tells us $g_0^{-1} \in (tt_1,(w,z))\zeta$. Therefore,

$$gg_0^{-1} \in (f,(ss_1,tt_1))\zeta \cdot (tt_1,(w,z))\zeta \subseteq [(f,(ss_1,tt_1))(tt_1,(w,z))]\,\zeta = (f,(s,t))\zeta$$

where the equality uses (4.30). A symmetric argument applies to establish $gg_0^{-1} \in (f,(s',t'))\zeta$. Because $t = t'$ and $gg_0^{-1} \in (f,(s,t))\zeta \cap (f,(s',t'))\zeta$, we conclude $(s,t) \,\mathscr{P}\, (s',t')$ and hence $\mathscr{P}$ is an injective automaton congruence.

We have thus established that $\mathscr{P}$ is an admissible partition. Suppose now that $X \subseteq t\varphi_1^{-1}$ with and $X \times \{t\}$ contained in a single $\mathscr{P}$-block, some $t \in R'$. Fix $x_0 \in X$ and let $g \in (f,(x_0,t))\zeta$. Then, for all $x \in X$, we have by the claim that $(f,(x,t))\zeta = (f,(x_0,t))\zeta$ and therefore $g \in (f,(x,t))\zeta$. As $(e,f)$ is an idempotent of the $\mathscr{R}$-class $\overline{R}$, for all $(x,t) \in \overline{R}$, the equality $(e,f)(x,t) = (x,t)$ holds and hence $(1,(e,f))(f,(x,t)) = (1,(x,t))$. Choose $g' \in (1,(e,f))\zeta$. Then, for all $x \in X$, we have

$$g'g \in (1,(e,f))\zeta \cdot (f,(x,t))\zeta \subseteq (1,(x,t))\zeta.$$

Hence, by (4.25), $(g'g,t) \in x\rho$ for all $x \in X$, as required. This completes the proof of the lemma.     $\square$

We may now state and prove the Presentation Lemma.

**Theorem 4.14.19 (Presentation Lemma).** *Let $R$ be a regular $\mathscr{R}$-class of $S$ and $\mathbf{V}$ be a pseudovariety. Then $S$ has a cross section for $R$ over $\mathbf{G} * \mathbf{V}$ if and only if $S$ has a presentation for $R$ over $\mathbf{V}$. In particular, $S \in \mathbf{A} \textcircled{m}\,(\mathbf{G}*\mathbf{V})$ if and only if, for each regular $\mathscr{R}$-class $R$ of $S$, there is a presentation for $R$ over $\mathbf{G} * \mathbf{V}$.*

*Proof.* Suppose first $S$ has a presentation $(\Phi, \mathscr{P})$ over $\mathbf{V}$ for $R$; say $\Phi = (\varphi_1, \varphi_2)$. Then Lemma 4.14.17 provides a relational morphism $\rho \colon S \to W$ with $W \in \mathbf{G} * \mathbf{V}$ so that if $w \in W$, then there exists $q \in Q$ such that $w\rho^{-1} \cap R \subseteq q\varphi_1^{-1}$ and $(w\rho^{-1} \cap R) \times \{q\}$ is contained in a single block of $\mathscr{P}$. We claim $\rho$ is a cross section for $R$. Suppose $x, y \in R$ with $x \mathscr{H} y$ and $w \in x\rho \cap y\rho$. By choice of $\rho$ there exists $q \in Q$ such that $\{x, y\} \subseteq q\varphi_1^{-1}$ and $(x, q) \mathscr{P} (y, q)$. Because $(\Phi, \mathscr{P})$ is a presentation and $x \mathscr{H} y$, it follows $x = y$ and so $\rho$ is a cross section for $R$.

Conversely, suppose $\varphi \colon S \to W$ is a cross section for $R$ with $W \in \mathbf{G} * \mathbf{V}$. Then there is a division $d \colon W \to G \rtimes T$ with $T \in \mathbf{V}$ and $G$ a group. The composed relational morphism $\rho = \varphi d$ is still a cross section by Proposition 4.14.4. Lemma 4.14.18 provides a parameterized relational morphism $\Phi \colon (R, S) \to (Q, T)$ and an admissible partition $\mathscr{P}$ on $\mathcal{D}_\Phi$ such that: if $X \subseteq q\varphi_1^{-1}$ and $X \times \{q\}$ is contained in a single block of $\mathscr{P}$, then there exists $(g, t) \in G \rtimes T$ with $X \subseteq (g, t)\rho^{-1}$ where $\Phi = (\varphi_1, \varphi_2)$. We claim that $(\Phi, \mathscr{P})$ is a presentation for $R$ over $\mathbf{V}$. It suffices to verify the third condition in the definition of a presentation. So suppose $(x, q) \mathscr{P} (x', q')$ with $x \mathscr{H} x'$. Because $\mathscr{P}$ is admissible, $q = q'$ and so by construction of $\rho$ there exists $(g, t) \in G \rtimes T$ with $\{x, x'\} \subseteq (g, t)\rho^{-1}$. Because $\rho$ is a cross section for $R$, this yields the desired conclusion that $x = x'$. Thus $(\Phi, \mathscr{P})$ is a presentation for $R$ over $\mathbf{V}$.    $\square$

The Presentation Lemma can be "beefed up" to a statement about pointlikes for $\mathbf{G} * \mathbf{V}$ that is frequently useful in iterated matrix constructions of high complexity semigroups. The reader is referred to the discussion just before Example 2.4.7 or to Section 3.6.2 for the definition and properties of pointlike sets.

**Theorem 4.14.20 (Pointlikes via Presentations).** *Let $R$ be a regular $\mathscr{R}$-class of a semigroup $S$, $\mathbf{V}$ be a pseudovariety and $X \subseteq R$. Then $X$ is $\mathbf{G} * \mathbf{V}$-pointlike if and only if, for all parameterized relational morphisms $\Phi \colon (R, S) \to (Q, T)$ with $T \in \mathbf{V}$ and for all admissible partitions $\mathscr{P}$ on $\mathcal{D}_\Phi$, there exists $q \in Q$ such that $X \subseteq q\varphi_1^{-1}$ and $X \times \{q\}$ is contained in a single block of $\mathscr{P}$, where $\Phi = (\varphi_1, \varphi_2)$.*

*Proof.* Suppose first $X \subseteq R$ is $\mathbf{G} * \mathbf{V}$-pointlike and let $\Phi \colon (R, S) \to (Q, T)$ be a parameterized relational morphism with $T \in \mathbf{V}$ and let $\mathscr{P}$ be an admissible partition on $\mathcal{D}_\Phi$. Assume $\Phi = (\varphi_1, \varphi_2)$. Then Lemma 4.14.17 provides a relational morphism $\rho \colon S \to W$ with $W \in \mathbf{G} * \mathbf{V}$ so that if $w \in W$, then there exists $q \in Q$ such that $w\rho^{-1} \cap R \subseteq q\varphi_1^{-1}$ and $(w\rho^{-1} \cap R) \times \{q\}$ is contained in a single block of $\mathscr{P}$. Because $X$ is $\mathbf{G} * \mathbf{V}$-pointlike, there exists $w \in W$ with $X \subseteq w\rho^{-1}$. By construction of $\rho$, we can find $q \in Q$ such that $X \subseteq q\varphi_1^{-1}$ and $X \times \{q\}$ is contained in a single block of $\mathscr{P}$, as was required.

Conversely, assume $X$ satisfies the property asserted in the theorem. We must show that $X$ is $\mathbf{G}*\mathbf{V}$-pointlike. Let $\varphi \colon S \to W$ be a relational morphism with $W \in \mathbf{G} * \mathbf{V}$. Then there is a division $d \colon W \to G \rtimes T$ with $G$ a group and $T \in \mathbf{V}$. Let $\rho = \varphi d \colon S \to G \rtimes T$. According to Lemma 4.14.18, there exist a parameterized relational morphism $(\varphi_1, \varphi_2) = \Phi \colon (R, S) \to (Q, T)$ and an admissible partition $\mathscr{P}$ on $\mathcal{D}_\Phi$ such that: if $Y \subseteq q\varphi_1^{-1}$ and $Y \times \{q\}$ is contained in a single block of $\mathscr{P}$, then there exists $(g, t) \in G \rtimes T$ with $Y \subseteq (g, t)\rho^{-1}$. By our hypothesis on $X$, it follows $X \subseteq (g, t)\rho^{-1}$ for some $(g, t) \in G \rtimes T$. As $\rho = \varphi d$, for each $x \in X$, there exists $w_x \in x\varphi$ such that $(g, t) \in w_x d$. Because $d$ is a division, we conclude that $w_x = w_{x'}$, for all $x, x' \in X$. If we call this common element $w$, then $X \subseteq w\varphi^{-1}$. It follows $X$ is $\mathbf{G} * \mathbf{V}$-pointlike, completing the proof. ☐

**Exercise 4.14.21.** Derive the Presentation Lemma from Theorem 4.14.20.

We mention that there is a variant of the Presentation Lemma for group mapping semigroups [334]. Namely, if $S$ is a group mapping semigroup with distinguished $\mathscr{J}$-class $J$ and $\mathscr{R}$-class $R$, then $S \in \mathbf{C}_n$ ($n \geq 1$) if and only if $\mathsf{RLM}_J(S) \in \mathbf{C}_n$ and there is a presentation for $R$ over $\mathbf{C}_{n-1}$. Because $|\mathsf{RLM}_J(S)| < |S|$ for $S$ group mapping, by induction we see that the whole complexity problem reduces to determining whether there is presentation over $\mathbf{C}_n$ for an $\mathscr{R}$-class of the distinguished $\mathscr{J}$-class of a group mapping semigroup $S$. The following exercises sketch this variant.

**Exercise 4.14.22.** Let $\equiv$ be a congruence on a semigroup $S$ and let $\varphi \colon S \to T$ be a relational morphism such that $s \equiv t$ and $s\varphi \cap t\varphi \neq \emptyset$ implies $s = t$. Show that $S \prec T \times S/\equiv$.

**Exercise 4.14.23.** Use the previous exercise to show that if $S$ is right mapping semigroup with distinguished $\mathscr{J}$-class $J$ and $\varphi \colon S \to T$ is a cross section for $J$, then $S \prec T \times \mathsf{RLM}_J(S)$. Hint: Suppose $s\varphi \cap s'\varphi \neq \emptyset$ and $s\mu_J^R = s'\mu_J^R$ and let $x \in J$; show $xs \in J$ if and only if $xs' \in J$ and that if $xs, xs' \in J$, then $xs \mathscr{H} xs'$ and $xs\varphi \cap xs'\varphi \neq \emptyset$; conclude $xs = xs'$ and hence $s = s'$, as $S$ is right mapping.

**Exercise 4.14.24.** Use the Presentation Lemma and the previous exercise to show that if $S$ is a right mapping semigroup with distinguished $\mathscr{J}$-class $J$ that $S \in \mathbf{C}_n$ if and only if $\mathsf{RLM}_J(S) \in \mathbf{C}_n$ and an $\mathscr{R}$-class $R$ of $J$ has a presentation over $\mathbf{C}_{n-1}$.

The Presentation Lemma was originally proved by the first author using the description of homomorphisms and partial homomorphisms between Rees matrix semigroups sketched in Exercise 4.13.45. The partitions in this sort of proof come from the coboundarys involved in changing between cohomologous Rees matrix coordinatizations. See [43] for details.

## An example: The complexity of the Tall Fork

Although the proof of the Presentation Lemma is quite technical, it is not so difficult to use in practice. To illustrate this, we compute in detail the complexity of the Tall Fork semigroup $F$ from the beginning of Section 4.14.

It is immediate from Corollary 4.13.4 and Theorem 4.13.30 that $F \in \mathbf{EA}$ because $P$ is a zero-one matrix. Let us begin our analysis of $F$ by proving that every local monoid of $F$ belongs to $\mathbf{A} * \mathbf{G}$. If $x \in E(F) \cap J^0$, then $xFx$ is either zero or a group with adjoined zero. In either case, it divides a direct product of a group with $U_1$, and hence belongs to $\mathbf{A} * \mathbf{G}$. Let $e$ be the identity of $H$ and $f$ be the identity of $K$. Then $eFe = H \cup J_H^0$ and $fFf = K \cup J_K^0$ where $J_H^0 = \mathcal{M}^0(\{\pm 1\}, A_H, B_H, P_H)$ and $J_K^0 = \mathcal{M}^0(\{\pm 1\}, A_K, B_K, P_K)$ with $A_H = \{a_0, a_1, a_2, a_3\}$, $B_H = \{0, 1, 2, 3\}$, $P_H = P|_{B_H \times A_H}$, $A_K = \{a_4, a_5, a_6\}$, $B_K = \{0', 2'\}$ and $P_K = P|_{B_K \times A_K}$. As $eFe, fFf \in \mathbf{EA}$, Corollary 4.13.32 implies that $eFe, fFf \in \mathbf{A} * \mathbf{G}$. Thus the semigroup $F$ is locally complexity one, that is $F \in \mathbb{L}\mathbf{C}_1$ (in fact, $F \in \mathbb{L}(\mathbf{A} * \mathbf{G})$).

Recall that $\delta(F) = 2$ and so $c(F) \leq 2$. We now wish to show that, in fact, $c(F) = 2$. Let $R = a_0 \times \{\pm 1\} \times B$ and suppose that $\Phi \colon (R, F) \to (Q, T)$ is a parameterized relational morphism with $T$ aperiodic. Our job is to show every admissible congruence on $\mathcal{D}_\Phi$ necessarily has elements $(x, q) \mathrel{\mathscr{P}} (y, q)$ with $x \mathrel{\mathscr{H}} y$. The key point is that $t$ can cause a change in the group coordinate. We begin with a simple lemma.

**Lemma 4.14.25.** Let $\varphi \colon S \to T$ be a relational morphism with $T$ aperiodic. Let $G \leq S$ be a group. Then there exists an idempotent $e \in T$ with $G \leq e\varphi^{-1}$.

*Proof.* Let $\varphi = \alpha^{-1}\beta$ be the canonical factorization, say $\alpha \colon R \twoheadrightarrow S$. Then by Proposition 4.1.44 there is a subgroup $H \leq R$ with $H\alpha = G$. Because $T$ is aperiodic, $H\beta$ is an idempotent $e$. Then $e\varphi^{-1} = e\beta^{-1}\alpha \supseteq H\alpha = G$. $\qquad \square$

The reader may find it useful to draw a picture of $\mathcal{D}_\Phi$ as the proof progresses. Suppose $e', f' \in E(T)$ are such that $H \leq e'\varphi_2^{-1}$ and $K \leq f'\varphi_2^{-1}$ as per the lemma. Let $t' \in t\varphi_2$. Choose $q \in Q$ such that $q \in (a_0, 1, 0')\varphi_1$ and set $q' = qf'$. Then $\{(a_0, 1, 0'), (a_0, 1, 2')\} = (a_0, 1, 0')K \subseteq q'\varphi_1^{-1}$. Choose $x \in (a_6, 1, 0')\varphi_2$. Then $(a_0, 1, 0')(a_6, 1, 0') = (a_0, 1, 0') = (a_0, 1, 2')(a_6, 1, 0')$. Thus $q'x$ is defined and

$$((a_0, 1, 0'), q')(q', ((a_6, 1, 0'), x)) = ((a_0, 1, 0'), q'x)$$
$$= ((a_0, 1, 2'), q')(q', ((a_6, 1, 0'), x)).$$

Because $\mathscr{P}$ is an injective automaton congruence, we immediately obtain that $((a_0, 1, 0'), q') \mathrel{\mathscr{P}} ((a_0, 1, 2'), q')$. Let $q'' = q't'e'$. Then

$$a_0 \times \{-1\} \times \{0, 1, 2, 3\} = (a_0, 1, 0')tH \subseteq q''\varphi_1^{-1}$$
$$a_0 \times \{1\} \times \{0, 1, 2, 3\} = (a_0, 1, 2')tH \subseteq q''\varphi_1^{-1}.$$

This shows that $a_0 \times \{\pm 1\} \times \{0, 1, 2, 3\} \subseteq q''\varphi_1^{-1}$. Now

$$((a_0, 1, 0'), q')(q', (te, t'e')) = ((a_0, -1, 0), q'')$$
$$((a_0, 1, 2'), q')(q', (te, t'e')) = ((a_0, 1, 2), q'').$$

Because $\mathscr{P}$ is an automaton congruence, it follows

$$((a_0, -1, 0), q'') \ \mathscr{P} \ ((a_0, 1, 2), q''). \tag{4.32}$$

Now choose $y \in (a_2, 1, 0)\varphi_2$ and $z \in (a_3, 1, 0)\varphi_2$. Notice that

$$(a_0, 1, 2)(a_2, 1, 0) = (a_0, 1, 0) = (a_0, 1, 3)(a_2, 1, 0)$$
$$(a_0, 1, 3)(a_3, 1, 0) = (a_0, 1, 0) = (a_0, 1, 0)(a_3, 1, 0).$$

Because $\mathscr{P}$ is an injective automaton congruence, the computation

$$\begin{aligned}((a_0, 1, 2), q'')(q'', ((a_2, 1, 0), y)) &= ((a_0, 1, 0), q''y) \\ &= ((a_0, 1, 3), q'')(q'', ((a_2, 1, 0), y))\end{aligned}$$

shows that $((a_0, 1, 2), q'') \ \mathscr{P} \ ((a_0, 1, 3), q'')$, whereas

$$\begin{aligned}((a_0, 1, 3), q'')(q'', ((a_3, 1, 0), z)) &= ((a_0, 1, 0), q''z) \\ &= ((a_0, 1, 0), q'')(q'', ((a_3, 1, 0), z))\end{aligned}$$

yields $((a_0, 1, 3), q'') \ \mathscr{P} \ ((a_0, 1, 0), q'')$. Thus $((a_0, 1, 0), q'') \ \mathscr{P} \ ((a_0, 1, 2), q'')$. Combining this with (4.32), yields $((a_0, 1, 0), q'') \ \mathscr{P} \ ((a_0, -1, 0), q'')$. Because $(a_0, 1, 0) \ \mathscr{H} \ (a_0, -1, 0)$, this shows $(\Phi, \mathscr{P})$ is not a presentation. This completes the proof that $c(F) = 2$.

**Theorem 4.14.26 (Rhodes).** *The Tall Fork is a semigroup of complexity* 2 *in* **EA**, *which is locally complexity* 1 *and such that* $\#(S) = 1$.

**Exercise 4.14.27.** Show that no proper divisor of the Tall Fork has complexity two.

A semigroup $S$ is said to be *critical* if each proper divisor $T$ of $S$ has strictly smaller complexity. The previous exercise establishes that the Tall Fork is critical.

*Question 4.14.28.* What are the possible posets for the $\mathscr{J}$-order of a critical semigroup?

The next lemma is termed by Rhodes the "Tie-Your-Shoes" Lemma [43]. It abstracts the argument used several times in the computation of $c(F)$, where we acted by elements of the $\mathscr{J}$-class $J$ to force elements into a common partition block.

**Lemma 4.14.29 (Tie-Your-Shoes).** *Suppose* $R$ *is a regular* $\mathscr{R}$-*class of a semigroup* $S$, *belonging to a* $\mathscr{J}$-*class* $J$. *Assume* $J^0 \cong \mathscr{M}^0(G, A, B, C)$. *Let* $\Phi \colon (R, S) \to (Q, T)$ *a parameterized relational morphism and* $\mathscr{P}$ *an admissible* *partition on $\mathcal{D}_\Phi$. Suppose $R$ is the $\mathscr{R}$-class corresponding to $a \in A$ and suppose further: $b_1 C a_0 \neq 0 \neq b_2 C a_0$ and*

$$x = (a, g(b_1 C a_0)^{-1}, b_1), \ y = (a, g(b_2 C a_0)^{-1}, b_2) \in q\varphi_1^{-1}.$$

*Then $(x, q) \mathscr{P} (y, q)$.*

*Proof.* Set $z = (a_0, 1, b_1)$. Then $xz = (a, g, b_1) = yz \in R$, so if $t \in z\varphi_2$, then $qt$ is defined and $(x, q)(q, (z, t)) = (xz, qt) = (yz, qt) = (y, q)(q, (z, t))$. Because $\mathscr{P}$ is an injective automaton congruence on $\mathcal{D}_\Phi$, it follows $(x, q) \mathscr{P} (y, q)$.

The name Tie-Your-Shoes comes from consideration of the graph $\Gamma(J^0)$. The condition $b_1 C a_0 \neq 0 \neq b_2 C a_0$ says that there are edges $(b_1, a_0)$ and $(b_2, a_0)$ in $\Gamma(J^0)$, which are thought of as the shoelaces emanating from the vertex $a_0$ (the shoe). The lemma says that certain corresponding elements in the $\mathscr{L}$-classes of $b_1, b_2$ are then forced to be in the same partition block, which can be viewed as tying the shoelaces.

## 4.15 Tilson's Two $\mathscr{J}$-class Theorem

Tilson proved [360] that complexity is decidable for semigroups with at most two non-zero $\mathscr{J}$-classes; we call such a semigroup a $2 \mathscr{J}$-semigroup. Our aim is to provide a proof of this theorem using the machinery of presentations and cross sections that we have developed.

Because a $2 \mathscr{J}$-semigroup $S$ has, by virtue of the definition, depth of at most 2, we conclude from Theorem 4.9.15 that $c(S) \leq 2$. As complexity 0 is decidable, we just need to distinguish the case of complexity 1 from complexity 2. More careful depth considerations show that in order to have a chance of having complexity 2, $S$ must in fact have 2 essential $\mathscr{J}$-classes forming a chain; that is, the $\mathscr{J}$-structure of $S$ must be of the form: $J_1 >_{\mathscr{J}} J_2 >_{\mathscr{J}} 0$. We record this as a proposition.

**Proposition 4.15.1.** *Let $S$ be a $2 \mathscr{J}$-semigroup. Then $c(S) \leq 2$. A necessary condition for $c(S) = 2$ is that the two non-zero $\mathscr{J}$-classes of $S$ be essential and form a $\leq_{\mathscr{J}}$-chain.*

Let $e \in J_1$ (retaining the above notation) be an idempotent. Then $S = SeS$ and so Proposition 4.12.20 implies that $c(S) = c(eSe)$. We claim that $eSe$ is a $2 \mathscr{J}$-semigroup. Indeed, because $e \in J_1$, $eSe \cap J_1 = H_e$. Suppose $x, y \in eSe \cap J_2$. Then there exist $u, v, w, z \in S^I$ with $uxv = y$, $wyz = x$. Then $euexeve = y$ and $eweyeze = x$ establishing that $eSe \cap J_2$ is a $\mathscr{J}$-class of $eSe$. It follows that $eSe$ is a $2 \mathscr{J}$-semigroup, which is in fact a monoid.

We have thus reduced the problem to the case of a $2 \mathscr{J}$-semigroup consisting of a (non-trivial) group of units and a regular 0-minimal ideal. Such monoids are sometimes called *small monoids* [174]. Notice that by adjoining a 0, we may always assume that our small monoid has a 0. We are now in a position to state Tilson's theorem [360].

**Theorem 4.15.2 (Tilson).** *Let $M$ be a small monoid with non-trivial group of units $H$ and regular $0$-minimal ideal $I$. Then $M$ has complexity $1$ if and only if $LH \cup 0 \in \mathbf{EA}$, for each $\mathscr{L}$-class $L \subseteq I \backslash 0$. Otherwise, $M$ has complexity $2$.*

**Corollary 4.15.3 (Tilson).** *Complexity is decidable for semigroups with at most $2$ non-zero $\mathscr{J}$-classes.*

Note that $LH \cup 0$ is indeed a subsemigroup, in fact a left ideal, of $M$. This is because, by stability, $ML = L \cup 0$ as $x \in L$, $m \in M$ implies $mx = 0$ or $mx \mathscr{L} x$. Hence $LH \cup 0 = (L \cup 0)H$ is also a left ideal. Also observe that $LH \cup \{0\}$ is isomorphic to a Rees matrix semigroup, although we caution the reader that in general it will not be a *regular* Rees matrix semigroup. Fix an isomorphism $I \cong \mathscr{M}^0(G, A, B, C)$ where $B$ is the set of $\mathscr{L}$-classes. Then if we set $B' = \{b \in B \mid bH = LH\}$, it follows $LH = A \times G \times B'$ and so $LH \cup 0 \cong \mathscr{M}^0(G, A, B', C|_{B' \times A})$. The reason $C|_{B' \times A}$ might not be regular is that if $a \in A$, it could happen all the non-zero entries of column $a$ belong to rows in $B \backslash B'$. This is one reason we did not assume regularity in Graham's Theorem.

**Exercise 4.15.4.** Give an example where $C|_{B' \times A}$ is not regular.

*Remark 4.15.5.* Notice that if $M = H \cup I$ is a small monoid, then the maximal absolute Type I semigroups are exactly the submonoids $M_L = H \cup LH \cup 0$ with $L$ an $\mathscr{L}$-class of $I \backslash 0$. Moreover, Corollary 4.13.32 shows that $M_L \in \mathbf{EA}$ if and only if $M_L \in \mathbf{A} * \mathbf{G}$, if and only if $\mathsf{K}_{\mathbf{G}}(M_L) \in \mathbf{A}$. But it is easy to see that $\langle E(M_L) \rangle = 1 \cup \langle E(LH \cup 0) \rangle$. So the Type I–Type II lower bound yields 1 for the complexity of $M$ precisely when Tilson's Theorem says the complexity is 1 and yields 2 when Tilson's Theorem says the complexity is 2. Therefore, the Type I–Type II lower bound is sharp for small monoids.

Although the proof of Tilson's Theorem is somewhat technical, the result is extremely facile to use in practice. To illustrate this point, we revisit Example 4.10.12.

*Example 4.15.6.* Let $S$ be the semigroup from Example 4.10.12. Because $S$ is a small monoid, we can use Tilson's Theorem to give an alternate computation of its complexity. The semigroup $S$ does not belong to $\mathbf{EA}$. Moreover, the two $\mathscr{L}$-classes of the minimal ideal form a single orbit of the group of units $G$. Tilson's Theorem immediately implies that $c(S) = 2$. On the other hand, for the reverse monoid $S^{op}$, the two $\mathscr{L}$-classes are distinct, singleton orbits of $G$. Because each $\mathscr{L}$-class is left simple, and hence a member of $\mathbf{EA}$, we conclude that $c(S^{op}) = 1$.

One might next try to decide complexity for a semigroup with at most 3 non-zero $\mathscr{J}$-classes. Henckell and Rhodes had sketched a proof in the 1970s that the problem of deciding whether a semigroup has complexity one reduces to the case of a semigroup with at most 3 non-zero $\mathscr{J}$-classes. We state this as a conjecture.

*Question 4.15.7 (Henckell-Rhodes).* Is it true that the problem of deciding whether a semigroup has complexity one reduces to the case of a semigroup with at most 3 non-zero $\mathscr{J}$-classes? More generally, is it true that the problem of deciding whether a semigroup has complexity $n$ reduces to the case of a semigroup with at most $n+2$ non-zero $\mathscr{J}$-classes?

### The proof of Theorem 4.15.2

The proof presented here is a coordinate-free version of the proof from [334]. Let us first dispense with the necessity. Suppose that $LH \cup 0 \notin \mathbf{EA}$ and let $T = H \cup LH \cup 0$. Then $T$ is a submonoid of $M$ and is generated by its $\mathscr{L}$-chain $H >_{\mathscr{L}} L >_{\mathscr{L}} 0$. Therefore, $T$ is a $\mathscr{T}_1$-semigroup and hence absolute Type I by Theorem 4.12.14. Because $\langle E(LH \cup 0) \rangle \leq \langle E(T) \rangle \leq \mathsf{K}_{\mathbf{G}}(T)$, Theorem 4.12.8 shows $0 < c(\langle E(T) \rangle) < c(T) \leq c(M)$. As a consequence $c(M) \geq 2$. In light of Proposition 4.15.1, we see that $c(M) = 2$.

To prove sufficiency, let $M = H \cup I$ be a small monoid with non-trivial group of units $H$ such that $LH \cup 0 \in \mathbf{EA}$ for each $\mathscr{L}$-class $L$ of $J = I \setminus 0$. As a consequence of the Fundamental Lemma of Complexity, it suffices to prove $M \in \mathbf{A} \circledm (\mathbf{G} * \mathbf{A})$ (cf. Corollary 4.9.4). By Theorem 4.14.7, it suffices to show each regular $\mathscr{R}$-class has a cross section over $\mathbf{G} * \mathbf{A}$.

We define a cross section $\rho \colon M \to H$ for $H$ over $\mathbf{G} * \mathbf{A}$ by setting

$$m\rho = \begin{cases} m & m \in H \\ H & m \notin H. \end{cases}$$

This leaves us with finding a cross section for a regular $\mathscr{R}$-class $R$ of $J$. By the Presentation Lemma (Theorem 4.14.19), it suffices to find a presentation $(\Phi, \mathscr{P})$ for $R$ over $\mathbf{A}$.

Set $B = J/\mathscr{L}$. Notice that $(B, H\mu_J^R)$ is a right permutation group inside of $(B, \mathsf{RLM}_J(S))$. Let $B/H$ be the set of orbits under the $H$-action. Two $\mathscr{L}$-classes $L$ and $L'$ belong to the same orbit if and only if $LH = L'H$ as subsets of $J$, so no confusion should arise if we use the notation $LH$ for both the orbit of $L$ under the action of $H$ and the subset $LH$ of $J$.

We proceed to define a parameterized relational morphism

$$\Phi \colon (R, M) \to \overline{(B/H, \{1_{B/H}\})}.$$

Set $x\varphi_1 = L_x H$ for $x \in R$. To see that $\varphi_1 \colon (R, M) \to \overline{(B/H, \{1_{B/H}\})}$ is a relational morphism, suppose $x \in b\varphi_1^{-1}$ (with $x \in R$) and $m \in M$. We may take $m \neq 0$, as any element covers 0. To define a cover of $m$, set

$$\widehat{m} = \begin{cases} 1_{B/H} & m \in H \\ \overline{L_m H} & \text{otherwise.} \end{cases} \tag{4.33}$$

In the first case, $xm \in bH$ and so $xm \in bH 1_{B/H}\varphi_1^{-1} = bH\widehat{m}\varphi_1^{-1}$, as desired. For the second case, if $xm$ is undefined, we are done. If $xm$ is defined, then $L_{xm} = L_m$ by stability. But then

$$xm \in L_m H \varphi_1^{-1} = bH\overline{L_m H}\varphi_1^{-1} = bH\widehat{m}\varphi_1^{-1}.$$

So $\widehat{m}$ covers $m$ in all cases.

Define $\varphi_2 \colon M \to \overline{B/H} \cup \{1_{B/H}\}$ by

$$m\varphi_2 = \begin{cases} 1_{B/H} & m \in H \\ \overline{L_m H} & m \in J \\ \overline{B/H} & m = 0. \end{cases} \tag{4.34}$$

Using that $LH \cup \{0\}$ is a left ideal closed under right multiplication by $H$, for any $\mathscr{L}$-class $L$ of $J$, it is straightforward to verify that $\varphi_2$ is a relational morphism. The argument of the previous paragraph shows that $\Phi = (\varphi_1, \varphi_2)$ is a parameterized relational morphism.

Let us decongest notation for $\mathcal{D}_\Phi$. First observe that because $\varphi_1$ is a function, we can canonically identify $\#\varphi_1$ with $R$ via $x \leftrightarrow (x, L_x H)$, that is, we can take $\mathcal{D}_\Phi = (R, D_\Phi)$. Also $\varphi_2|_{M\setminus 0}$ is a function and so to ease notation, for $m \in M \setminus 0$, we write the arrow $(LH, (m, m\varphi_2))$ as simply $(LH, m)$.

We now define a partition $\mathscr{P}$ on $R$ by $x \mathrel{\mathscr{P}} y$ if and only if

$$y = xm \text{ where } m \in \langle E(L_x H \cup 0)\rangle. \tag{4.35}$$

Notice that (4.35) implies that $y \in L_x H$ and hence

$$x \mathrel{\mathscr{P}} y \implies L_x H = L_y H \tag{4.36}$$

or equivalently $x\varphi_1 = y\varphi_1$. So identifying $R$ with $\#\varphi_1$, it follows that if $(x, x\varphi_1) \mathrel{\mathscr{P}} (y, y\varphi_1)$, then $x\varphi_1 = y\varphi_1$, verifying (4.20). The following proposition shows that $\mathscr{P}$ is really a partition. The proof uses without comment the well-known fact that if $S'$ is a subsemigroup of a semigroup $S$ and $x, y \in S'$ are regular elements, then $x \mathrel{\mathscr{K}} y$ in $S'$ if and only if $x \mathrel{\mathscr{K}} y$ in $S$ for $\mathscr{K}$ any of Green's relations $\mathscr{R}$, $\mathscr{L}$ or $\mathscr{H}$ — this is not true for $\mathscr{J}$. (See Proposition A.1.16 for details.)

**Proposition 4.15.8.** *The relation $\mathscr{P}$ is an equivalence relation.*

*Proof.* For reflexivity, let $e \in L_x$ be an idempotent. Then $e \in \langle E(L_x H \cup 0)\rangle$ and $xe = x$. So $x \mathrel{\mathscr{P}} x$, as required. For symmetry, suppose $x \mathrel{\mathscr{P}} y$. Then $L_x H = L_y H$ and setting $N = L_x H \cup 0 = L_y H \cup 0$, there exists $m \in \langle E(N)\rangle$ with $y = xm$. Then $L_x \cap R_m$ contains an idempotent $e$ and $e \in \langle E(N)\rangle$. As $N$ is a Rees matrix semigroup, $\langle E(N)\rangle$ is regular by Theorem 4.13.22. Hence, because $m \mathrel{\mathscr{R}} e$, there is an inverse $m'$ to $m$ in $\langle E(N)\rangle$ with $m' \in L_e$ and $mm' = e$. Then $ym' = xmm' = xe = x$ (recall $e \in L_x$) and so $y \mathrel{\mathscr{P}} x$. Finally transitivity follows because $x \mathrel{\mathscr{P}} y \mathrel{\mathscr{P}} z$ implies $L_x H = L_y H$ by (4.36) and, moreover, if $N = L_x H \cup 0$, then there exist $m_1, m_2 \in \langle E(N)\rangle$ with $y = xm_1$, $z = ym_2$. But then $xm_1 m_2 = ym_2 = z$ and $m_1 m_2 \in \langle E(N)\rangle$. Thus $x \mathrel{\mathscr{P}} z$. This completes the proof $\mathscr{P}$ is an equivalence relation. $\qquad\square$

We next verify that item 3 of Definition 4.14.16 is satisfied.

**Lemma 4.15.9.** *Suppose $x, y \in R$ with $x \mathscr{H} y$ and $x \mathscr{P} y$. Then $x = y$.*

*Proof.* Let $L$ be the $\mathscr{L}$-class of $x, y$ and set $N = LH \cup 0$. Then it follows from $x \mathscr{P} y$ that $y = xm$ some $m \in \langle E(N) \rangle$. Choose an inverse $x'$ of $x$. Then $x'x \mathscr{R} x'y$ and $x'x \mathscr{L} x \mathscr{L} y \mathscr{L} x'y$ where the last $\mathscr{L}$-equivalence uses stability. Thus $x'x \mathscr{H} x'y = x'xm$. Recall that $N$ is a left ideal of $M$, so $x'x \in E(N)$ and hence $x'x$ and $x'xm$ are $\mathscr{H}$-equivalent elements of the regular semigroup $\langle E(N) \rangle$. But $N \in \mathbf{EA}$ by assumption on $M$ and so the $\mathscr{H}$-class of the idempotent $x'x$ in $\langle E(N) \rangle$ is trivial. It follows $x'x = x'xm$ and so $x = xx'x = xx'xm = xm = y$, establishing the lemma. $\qquad\square$

To complete the proof it suffices to establish $\mathscr{P}$ is an injective automaton congruence. We first show that $\mathscr{P}$ is an automaton congruence.

**Lemma 4.15.10.** *$\mathscr{P}$ is an automaton congruence.*

*Proof.* Suppose $x, y \in R$ with $x \mathscr{P} y$. Then, for some $b_0 \in B$, $L_x H = b_0 H = L_y H$ by (4.36). Set $N = b_0 H \cup 0$. Let $t \in D_\Phi$ with $xt, yt$ defined; we need to show $xt \mathscr{P} yt$. There are two cases.

First suppose $t = (b_1 H, h)$ with $h \in H$. The assumption that $xt, yt$ are defined in $\mathcal{D}_\Phi$ means that $b_0 H = y\varphi_1 = b_1 H = x\varphi_1$. In this case $xt = xh$, $yt = yh$. Note that $xh, yh \in b_0 H = L_x H = L_y H$. Because $x \mathscr{P} y$, we have $y = xm$ with $m \in \langle E(N) \rangle$. Then

$$yt = yh = xmh = xh(h^{-1}mh) = xt(h^{-1}mh).$$

So to show that $yt \mathscr{P} xt$, it suffices to prove $h^{-1}mh \in \langle E(N) \rangle$. Because $N$ is a left ideal of $M$ closed under right multiplication by $H$, it follows $N \cup H$ is a submonoid of $M$; moreover, $\langle E(N \cup H) \rangle = \langle E(N) \rangle \cup 1$. Because the idempotent-generated subsemigroup of any monoid is closed under conjugation by the group of units, this shows $h^{-1}mh \in \langle E(N) \rangle$.

Suppose now that $t = (b_1 H, s)$ with $s \in J$. Then $xt = xs \in R$ and $yt = ys \in R$. We claim $xt = yt$ and hence $xt \mathscr{P} yt$. Indeed, from $xs \in R$, there must be an idempotent $e \in L_x \cap R_s$. In particular, $e \in L_x \subseteq N$. Also, from $ys \in R$, there must be an idempotent in $L_y \cap R_s = L_y \cap R_e$. Hence $ye \in R$ and $ye \mathscr{P} y \mathscr{P} x$ as $e \in E(N)$. Now $ye \in R \cap L_e = R \cap L_x$ and so $ye \mathscr{H} x$. Lemma 4.15.9 then yields $ye = x$ and hence $xs = yes = ys$ as $e \in R_s$. Thus $xt = yt$, as was desired. This proves $\mathscr{P}$ is an automaton congruence. $\qquad\square$

We now complete the proof that $(\Phi, \mathscr{P})$ is a presentation by establishing that $\mathscr{P}$ is an injective congruence.

**Lemma 4.15.11.** *$\mathscr{P}$ is an injective congruence.*

*Proof.* Suppose $x, y \in R$ and $t \in D_\Phi$ with $xt \mathcal{P} yt$. We must show $x \mathcal{P} y$. Again we have two cases. If $t = (b_1 H, h)$, define $t' = (b_1 H, h^{-1})$. Then because $xt, yt$ are defined in $\mathcal{D}_\Phi$, it follows $x, y \in b_1 H$. Also $xh\varphi_1 = b_1 H = yh\varphi_1$, so $(xh)t', (yh)t'$ are defined in $\mathcal{D}_\Phi$ and

$$(xt)t' = xhh^{-1} = x \text{ and } (yt)t' = yhh^{-1} = y.$$

From the assumption $xt \mathcal{P} yt$, we conclude using the previous lemma that $x \mathcal{P} y$.

Thus we are left with the case $t = (b_1 H, s)$ with $s \in J$. Because $xt$ and $yt$ are defined in $\mathcal{D}_\Phi$, it follows $xs, ys \in R$. Stability then yields $xs \mathcal{L} s \mathcal{L} ys$ and hence $xs \mathcal{H} ys$. Moreover, we are assuming $xs \mathcal{P} ys$, whence $xs = ys$ by Lemma 4.15.9. Because $xt, yt$ are defined, $L_x H = b_1 H = L_y H$. From $xs \in R$, it follows there is an idempotent $e \in L_x \cap R_s$. Then $e \in L_x \subseteq L_x H = L_y H$ and $e = sz$ from some $z \in M$. Then $x = xe = xsz = ysz = ye$ and so $y \mathcal{P} x$ as $e \in \langle E(L_y H \cup 0) \rangle$. This establishes that $\mathcal{P}$ is an injective congruence. □

We have thus shown $(\Phi, \mathcal{P})$ is a presentation for $R$. As $\overline{(B/H, \{1_{B/H}\})}$ is plainly aperiodic, this completes the proof of Theorem 4.15.2. □

## 4.16 Complexity Pseudovarieties Are Not Local

The following examples are from [289] and further illustrate how to apply the Presentation Lemma. Similar ideas were used by the authors to prove that the complexity pseudovarieties $\mathbf{C}_n$ have no finite basis of pseudoidentities [288]. The results of this section are not used in the rest of the book and may be omitted.

**Theorem 4.16.1.** *For each $n > 0$, there exists a monoid $S_n$ of complexity $n + 1$ with $\mathsf{K}_{\mathbf{G}}(S_n) \in \mathbf{A} * \mathbf{G}$.*

An immediate corollary is

**Corollary 4.16.2.** *There exist monoids of arbitrary complexity in the pseudovariety $(\mathbf{A} \textcircled{m} \mathbf{G}) \textcircled{m} \mathbf{G}$.*

Because $\mathbf{A} \textcircled{m} (\mathbf{G} \textcircled{m} \mathbf{G}) = \mathbf{A} \textcircled{m} \mathbf{G} = \mathbf{A} * \mathbf{G}$ and every semigroup in $\mathbf{A} * \mathbf{G}$ has complexity 1, Theorem 4.16.1 shows that the Mal'cev product is non-associative in a very strong sense!

**Corollary 4.16.3.** *For each $n$, there exists a monoid $S_n$ of complexity $n + 1$ and an onto homomorphism $\varphi_n \colon S_n \twoheadrightarrow G \in \mathbf{G}$ such that $1\varphi^{-1} \in \mathbf{A} * \mathbf{G}$.*

The proof of Theorem 4.16.1 in fact shows that in the above corollary $G$ can be taken to be an elementary abelian 2-group. Our next corollary shows that $\mathbf{C}_n$ is not local in the sense of Tilson, that is, $\ell\mathbf{C}_n \neq \mathbf{g}\mathbf{C}_n$.

**Corollary 4.16.4.** *For $n > 0$, $\mathbf{C}_n$ is not local (in the sense of Tilson).*

*Proof.* Let $n > 0$ and let $S$ be a monoid of complexity $n+2$ in $(\mathbf{A} \textcircled{m} \mathbf{G}) \textcircled{m} \mathbf{G}$. Then there is a relational morphism $\varphi \colon S \to G \in \mathbf{G}$ with

$$1\varphi^{-1} \in \mathbf{A} \textcircled{m} \mathbf{G} = \mathbf{A} * \mathbf{G} \leq \mathbf{C}_1.$$

Hence the derived category $D_\varphi$ is locally in $\mathbf{C}_n$. If $\mathbf{C}_n$ were local, then the Derived Category Theorem [364] would imply $S \in \mathbf{C}_n * \mathbf{G} \leq \mathbf{C}_{n+1}$. But $S$ has complexity $n + 2$, so $\mathbf{C}_n$ cannot be local. $\qquad\square$

Recall that a pseudovariety of semigroups is said to be *local* in the sense of Eilenberg [85] if $\mathbb{L}\mathbf{V} = \mathbf{V}$. Notice that the complexity of a monoid viewed as a semigroup or as a monoid is the same by Proposition 4.3.14.

**Corollary 4.16.5.** *Let $n > 0$. Then $\mathbf{C}_n < \mathbb{L}\mathbf{C}_n$. That is, $\mathbf{C}_n$ is not local in the sense of Eilenberg.*

*Proof.* Because $\mathbf{D} \leq \mathbf{A}$, clearly $\mathbf{C}_n * \mathbf{D} = \mathbf{C}_n$ for all $n \geq 0$. By the Delay Theorem [364], if $\mathbf{V}$ is a non-trivial pseudovariety of monoids, then the equality $\mathbb{L}\mathbf{V} = \mathbf{V} * \mathbf{D}$ holds if and only if $\mathbf{V}$ is local in the sense of Tilson. Let $n > 0$. Then because $\mathbf{C}_n$ is not local in the sense of Tilson, $\mathbb{L}\mathbf{C}_n > \mathbf{C}_n * \mathbf{D} = \mathbf{C}_n$, as required. $\qquad\square$

### 4.16.1 Proof of Theorem 4.16.1

#### Construction of the $S_n$

The monoids $S_n$ will be constructed iteratively. They are based on the construction of the Tall Fork. For the moment, suppose $S$ is a monoid with zero and with non-trivial group of units $G$. Fix $1 \neq g \in G$. Define $F(S, g)$ as follows. Set $A = \{a_0, a_1, a_2, a_3, a_4, a_5, a_6\}$, $B = \{0, 1, 2, 3, 0', 2'\}$ and let $P$ be the Tall Fork matrix (4.14), viewed as a map $P \colon B \times A \to S$.

Let $S'$ be the quotient of the Rees matrix semigroup $\mathscr{M}(S, A, B, P)$ by the ideal $A \times 0 \times B$. Let $H = \langle h \rangle$ be a cyclic group of order 4 generated by $h$, written multiplicatively. Let $t$ be a new element whose action will be defined below and let $N = HtH = \{h^i t h^j \mid 0 \leq i, j \leq 3\}$. As a set, we define

$$F(S, g) = H \cup N \cup S'.$$

The group of units of $F(S, g)$ will be $H$. It is clear how $H$ multiplies against elements of $F(S, g) \setminus S'$. We now define how $H$ acts on $S'$; it suffices to consider $h$. Of course, $h0 = 0h = 0$. Define

$$
\begin{aligned}
(a, s, i)h &= (a, s, i + 1 \bmod 4), \quad i = 0, 1, 2, 3 \\
(a, s, i')h &= (a, s, (i + 2 \bmod 4)'), \quad i = 0, 2 \\
h(a_i, s, b) &= (a_{i-1 \bmod 4}, s, b), \quad i = 0, 1, 2, 3 \\
h(a_4, s, b) &= (a_5, s, b) \\
h(a_5, s, b) &= (a_4, s, b) \\
h(a_6, s, b) &= (a_6, s, b).
\end{aligned}
$$

It is clear how $N$ multiplies against $H$. Define $N^2 = 0$. It remains to declare how $N$ multiplies against $S'$. As we have already defined the action of $h$ on $S'$, it suffices to define how $t$ acts on $S'$. Define

$$(a, s, 0')t = (a, sg, 0) \tag{4.37}$$
$$(a, s, 2')t = (a, s, 2) \tag{4.38}$$
$$t(a_i, s, b) = (a_4, gs, b), \quad i = 0, 3 \tag{4.39}$$
$$t(a_i, s, b) = (a_5, s, b), \quad i = 1, 2 \tag{4.40}$$

and all other products involving $t$ and $S'$ to be 0. We remark that the multiplications by $g \neq 1$ in (4.37) and (4.39) (as opposed to no multiplications in the middle coordinates of (4.38) and (4.40)) are the key to making this construction work. It is straightforward to check associativity.

**Exercise 4.16.6.** Verify the associativity of $F(S, g)$.

We choose to identify $S$ with the subsemigroup $a_0 \times S \times 0$ of $F(S, g)$ (and we call this choice "canonical"). Notice that $F(G \cup 0, g)$ (recall that $G$ is the group of units of $S$) is a subsemigroup of $F(S, g)$ and the two "canonical" ways of viewing $G \cup 0$ as a subsemigroup of $F(S, g)$ (via $F(S, g)$ and via $F(G \cup 0, g)$) coincide.

Let $G_0 = \langle g_0 \rangle$ be a cyclic group of order 4. Let $S_1 = F(G_0 \cup 0, g_0)$ where we take $g = g_0$. Changing notation, we let $G_1 = H$, $g_1 = h$ and $N_1 = N$. Iteratively, we set $S_n = F(S_{n-1}, g_{n-1})$ where we set $H = G_n$ with $h = g_n$ and $N_n = N$. Following the conventions established above, we "canonically" identify $S_{n-1}$ with a certain subsemigroup of $S_n$. The reader is referred to [277, 288, 385] for further examples of such iterated matrix constructions.

## Complexity of $S_n$

We first ask the reader to verify inductively that the depth of $S_n$ is $n + 1$.

**Exercise 4.16.7.** Prove $\delta(S_n) = n + 1$.

As a consequence, we obtain, by Theorem 4.9.15, the following upper bound for the complexity of $S$.

**Proposition 4.16.8.** $c(S_n) \leq n + 1$.

Let $\mathbf{V}$ be a pseudovariety and $S$ a semigroup. We shall use frequently that if $X \subseteq S$ is $\mathbf{V}$-pointlike and $X = X^2$, then $X$ is $\mathbf{V}$-idempotent pointlike. Henckell has proved a sort of converse for certain pseudovarieties, including the complexity pseudovarieties [123, 130]. In particular, if $G \in \mathsf{PL}_{\mathbf{V}}(S)$ is a group, then $G$ is $\mathbf{V}$-idempotent pointlike.

**Exercise 4.16.9.** Verify that if $X \in \mathsf{PL}_{\mathbf{V}}(S)$ and $X^2 = X$, then $X$ is $\mathbf{V}$-idempotent pointlike.

The following proposition is essentially in [322] and is key to the proof we present here.

**Proposition 4.16.10.** *Suppose $S$ is a semigroup and $U \leq S$ is a subsemigroup. Suppose that $U$ is $\mathbf{W}$-idempotent pointlike in $S$ and $A \in \mathsf{PL}_{\mathbf{V}}(U)$. Then $A$ is $\mathbf{V} \circledm \mathbf{W}$-pointlike in $S$.*

*Proof.* Let $\varphi \colon S \to T$, with $T \in \mathbf{V} \circledm \mathbf{W}$, be a relational morphism. Suppose $\psi \colon T \to W \in \mathbf{W}$ is a relational morphism with $e\psi^{-1} \in \mathbf{V}$ for each idempotent $e \in W$. Because $U$ is $\mathbf{W}$-idempotent pointlike, there is an idempotent $e \in W$ such that $U \leq e\psi^{-1}\varphi^{-1}$. Let $V = e\psi^{-1}$. Then $V \in \mathbf{V}$ and $\varphi$ restricts to a relational morphism $\rho \colon U \to V$. Because $A \in \mathsf{PL}_{\mathbf{V}}(U)$, there exists $v \in V \leq T$ with $A \subseteq v\rho^{-1} \subseteq v\varphi^{-1}$. Thus $A$ is $\mathbf{V} \circledm \mathbf{W}$-pointlike in $S$, as desired. $\qquad \square$

**Corollary 4.16.11.** *Suppose $S$ is a semigroup and $G \leq S$ is a group. Suppose further $G \in \mathsf{PL}_{\mathbf{V}}(S)$. Then $G \in \mathsf{PL}_{\mathbf{A} \circledm \mathbf{V}}(S)$.*

*Proof.* As was observed earlier, $G$ must in fact be $\mathbf{V}$-idempotent pointlike in $S$. Because $G \in \mathsf{PL}_{\mathbf{A}}(G)$, Proposition 4.16.10 implies $G \in \mathsf{PL}_{\mathbf{A} \circledm \mathbf{V}}(S)$. $\qquad \square$

Now we turn to our main technical lemma, which will allow us to calculate the complexity of $S_n$ inductively. Essential use shall be made of Theorem 4.14.20. We retain the notation established earlier in this section, in particular the symbols $G$, $H$, $F(G \cup 0, g)$ and the matrix $P$ keep their previous meanings. The reader should compare the argument with our earlier computation of the complexity of the Tall Fork.

**Lemma 4.16.12.** *Suppose $G = \langle g \rangle$ is a cyclic group and $F(G \cup 0, g)$ is a subsemigroup of a semigroup $S$ such that $H \in \mathsf{PL}_{\mathbf{V}}(S)$. Then $G \in \mathsf{PL}_{\mathbf{A} \circledm (\mathbf{G} * \mathbf{V})}(S)$ (where we identify $G$ with a subsemigroup of $F(G \cup 0, g)$ in our "canonical" way).*

*Proof.* Let $R$ be the $\mathscr{R}$-class of $a_0 \times G \times B$ in $S$. By convention, we identify $G$ with $a_0 \times G \times 0$. By Corollary 4.16.11, it suffices to show $G \in \mathsf{PL}_{\mathbf{G} * \mathbf{V}}(S)$. Because $G = \{1, g\}^n$ for $n$ sufficiently large and $\mathbf{G} * \mathbf{V}$-pointlikes are closed under products, it suffices to show $Y = a_0 \times \{1, g\} \times 0 \in \mathsf{PL}_{\mathbf{G} * \mathbf{V}}(S)$. By Theorem 4.14.20, it suffices to show that, for all parameterized relational morphisms $\Phi = (\varphi_1, \varphi_2) \colon (R, S) \to (Q, T)$ with $T \in \mathbf{V}$ and for all admissible partitions $\mathscr{P}$ on $\mathcal{D}_{\Phi}$, there exists $q \in Q$ such that $Y \subseteq q\varphi_1^{-1}$ and $Y \times \{q\}$ is contained in a single block of $\mathscr{P}$.

So suppose $\Phi \colon (R, S) \to (Q, T)$ is a parameterized relational morphism with $T \in \mathbf{V}$ and let $\mathscr{P}$ be an admissible partition on $\mathcal{D}_{\Phi}$. Because $H$ is a group and is $\mathbf{V}$-pointlike, our above observations show that $H$ is in fact $\mathbf{V}$-idempotent pointlike. Therefore, there exists $e' \in E(T)$ such that $H \leq e'\varphi_2^{-1}$. Set $x = (a_0, 1, 0')$, $y = (a_0, 1, 2')$. Notice that $xH = yH = \{x, y\}$. Choose $q_0 \in x\varphi_1$. Then $xH \subseteq q_0\varphi_1^{-1}e'\varphi_2^{-1} \subseteq q_0e'\varphi_1^{-1}$. So setting $q = q_0e'$, we have

$x, y \in q\varphi_1^{-1}$. Because $0'Pa_6 = 1 = 2'Pa_6$, it follows from "Tie-Your-Shoes" (Lemma 4.14.29) that $(x, q) \mathscr{P} (y, q)$.

Choose $t' \in t\varphi_2$ and set $X = \{x, y\}tH$. Then because

$$\{x, y\}t = \{(a_0, g, 0), (a_0, 1, 2)\} \tag{4.41}$$

we see that $X = a_0 \times \{1, g\} \times \{0, 1, 2, 3\}$, that $q' = qt'e'$ is defined and that $X \subseteq q'\varphi_1^{-1}$. Let $e$ be the identity of $H$ (and hence of $F(G \cup 0, g)$). Consider

$$(x, q)(q, (te, t'e')) \text{ and } (y, q)(q, (te, t'e')).$$

Because $\mathscr{P}$ is an automaton congruence, $(x, q) \mathscr{P} (y, q)$ and $te = t$, it follows from (4.41) that

$$((a_0, g, 0), q') \mathscr{P} ((a_0, 1, 2), q').$$

Repeated application of "Tie-Your-Shoes" (Lemma 4.14.29) yields:

$$((a_0, g, 0), q') \mathscr{P} ((a_0, g, 1), q') \mathscr{P} ((a_0, g, 2), q') \mathscr{P} ((a_0, g, 3), q') \text{ and}$$
$$((a_0, 1, 2), q') \mathscr{P} ((a_0, 1, 3), q') \mathscr{P} ((a_0, 1, 0), q') \mathscr{P} ((a_0, 1, 1), q').$$

We conclude $(a_0 \times \{1, g\} \times 0) \times q'$ belongs to a single partition block of $\mathscr{P}$, as desired.  □

We may now compute the complexity of $S_n$.

**Theorem 4.16.13.** $c(S_n) = n + 1$.

*Proof.* By Proposition 4.16.8, $c(S_n) \leq n + 1$. We prove by downwards induction on $i$ that $G_i$ is $\mathbf{C}_{n-i}$-pointlike in $S_n$. Clearly $G_n$, being a group, is **A**-pointlike in $S_n$. Assume that $G_i$ is $\mathbf{C}_{n-i}$-pointlike in $S_n$. Then $G_i$ is the group of units of $S_i \leq S_n$. Moreover $F(G_{i-1} \cup 0, g_{i-1})$ is a subsemigroup of $S_i$ (and hence of $S_n$). Lemma 4.16.12 with $S = S_n$, $H = G_i$, $G = G_{i-1}$ and $g = g_{i-1}$ allows us to conclude $G_{i-1} \in \mathrm{PL}_{\mathbf{A}\textcircled{m}(\mathbf{G}*\mathbf{C}_{n-i})}(S_n)$, that is, $G_{i-1}$ is $\mathbf{C}_{n-(i-1)}$-pointlike in $S_n$, completing the induction. Consequently, $G_0$ is a $\mathbf{C}_n$-pointlike subset of $S_n$ and hence $c(S_n) > n$, establishing the theorem.  □

### 4.16.2 The Type II subsemigroup of $S_n$

Our next goal is to prove that $S_n \in (\mathbf{A} \textcircled{m} \mathbf{G}) \textcircled{m} \mathbf{G}$.

**Proposition 4.16.14.** *There is a relational morphism* $\varphi \colon S_1 \to G \in \mathbf{G}$ *with* $1\varphi^{-1} \in \mathbf{A} * \mathbf{G}$, $G_1\varphi = 1$ *and* $0\varphi = G$.

*Proof.* Let $G = \{1, -1\}$ be a cyclic group of order 2. Define a relational morphism $\varphi \colon S_1 \to G$ by

$$x\varphi = \begin{cases} 1 & x \in G_1 \\ -1 & x \in N_1 \\ G & \text{else.} \end{cases}$$

Then setting $T = 1\varphi^{-1}$, we have $T = S_1 \setminus N_1 = G_1 \cup \mathscr{M}^0(G_0, A, B, P)$. Because $P$ is a zero-one matrix, and hence $T \in \mathbf{EA}$, Corollary 4.13.32 yields $T \in \mathbf{A} * \mathbf{G}$, completing the proof.                                    □

We now prove that $S_n \in (\mathbf{A} \circledm \mathbf{G}) \circledm \mathbf{G}$ by induction.

**Theorem 4.16.15.** *For all $n \geq 1$, the monoid $S_n$ belongs to $(\mathbf{A} \circledm \mathbf{G}) \circledm \mathbf{G}$.*

*Proof.* We prove by induction on $n$ that there is a relational morphism $\varphi \colon S_n \to G \in \mathbf{G}$ such that $G_n\varphi = 1$, $0\varphi = G$ and $1\varphi^{-1} \in \mathbf{A} * \mathbf{G}$. The result will then follow. The case $n = 1$ is Proposition 4.16.14. Suppose the result holds for $n$. Let $\psi \colon S_n \to G \in \mathbf{G}$ be such that $G_n\psi = 1$, $0\psi = G$ and $1\psi^{-1} \in \mathbf{A} * \mathbf{G}$. Let $G' = \{1, -1\}$ be a cyclic group of order two. Define a relational morphism $\varphi \colon S_{n+1} \to G' \times G$ by

$$x\varphi = \begin{cases} (1,1) & x \in G_{n+1} \\ (-1,1) & x \in N_{n+1} \\ G' \times s\psi & x = (a,s,b), s \in S_n. \end{cases}$$

Note that 0 is included in the third case and so $0\varphi = G' \times G$, as $0\psi = G$. The only non-trivial verifications to show that $\varphi$ is a relational morphism are of the form $u\varphi x\varphi \subseteq (ux)\varphi$ for $u \in N_n$ or $u = (a', s', b')$ with $s' \in S_n$ and $x = (a, s, b)$ with $s \in S_n$ (or the dual situation). If $ux = 0$, things are trivial. If not, suppose first $u \in N_n$; then the middle coordinate of $x$ is either multiplied by 1 or by $g_n$. Because $G_n\psi = 1$, in either case we have $u\varphi x\varphi = G' \times s\psi \subseteq (ux)\varphi$. If $u = (a', s', b')$ with $s' \in S_n$ and $ux \neq 0$, then $ux = (a', s's, b)$ and so

$$u\varphi x\varphi = G' \times s'\psi s\psi \subseteq G' \times (s's)\psi = (ux)\varphi$$

as desired.

Now $G_{n+1}\varphi = (1,1)$ and $0\varphi = G' \times G$, so to finish the proof it suffices to show $(1,1)\varphi^{-1} \in \mathbf{A} * \mathbf{G}$. Let $K = 1\psi^{-1} \leq S_n$. By the induction hypothesis, $K \in \mathbf{A} * \mathbf{G}$. Notice that

$$(1,1)\varphi^{-1} = G_{n+1} \cup \mathscr{M}(K, A, B, P)/(A \times 0 \times B).$$

Corollary 4.1.11 implies that $(1,1)\varphi^{-1}$ divides a unitary semidirect product $(\mathscr{M}(K, A, B, P)/A \times 0 \times B)^I \rtimes G_{n+1}$ and so it suffices to show that $T = \mathscr{M}(K, A, B, P)/(A \times 0 \times B)$ belongs to $\mathbf{A} * \mathbf{G}$. The idea is that because $P$ is a zero-one matrix, we are able to split $K$ off from $T$ by a direct product, as was done in the proof of Theorem 4.13.31. Formally speaking, let $U = \mathscr{M}^0(\{1\}, A, B, P)$. Then the map $\alpha \colon U \times K \twoheadrightarrow T$ given by $((a,1,b), k)\alpha = (a, k, b)$ and $(0, k)\alpha = 0$ is an onto morphism, as $P$ is a zero-one matrix. Hence $T$ divides $U \times K \in \mathbf{A} * \mathbf{G}$.                                    □

Theorems 4.16.13 and 4.16.15 complete the proof of Theorem 4.16.1.

## 4.17 The Type II Theorem

The goal of this section is to prove the Rhodes Type II conjecture: $K_G(S)$ is the smallest subsemigroup of $S$ containing $E(S)$ and closed under taking weak conjugates. The conjecture was first proved by Ash [33], and independently by Ribes and Zalesskii [300], via a group theoretic reinterpretation due to Pin and Reutenauer [232]. A recent semigroup theoretic proof can be found in Auinger [34]; this proof has the advantage that it constructs an explicit, well-controlled group witnessing the Type II semigroup. Our approach here is to give the second author's geometric reduction of the Type II conjecture to the Ribes and Zalesskii Theorem about the profinite topology on a free group [328]. Then we give a proof of the Ribes and Zalesskii Theorem based on a proof by Auinger and the second author [39]. This proof also is constructive in that the groups involved are well-controlled.

It is clear that $K_G(S^I) = K_G(S)^I$. Hence it suffices to prove the Type II Theorem for monoids, and so in this section we restrict our attention to monoids. For a monoid $M$, if $e \in E(M)$, then $e$ is a weak inverse of itself and $e1e = e$. Hence, the result we want to prove in the monoidal context is that $K_G(M)$ is the smallest submonoid of $M$ closed under taking weak conjugates.

### 4.17.1 Stallings folding and inverse graphs: an excursion into combinatorial group theory

This section develops the topological and group theoretic techniques we shall need for our proof of the Type II Theorem. We begin with the tool of inverse graphs and Stallings folding [321]. The reader is referred back to Section 4.13.1 for the notion of a graph labeled over a group. Let $(Q, A)$ be a finite automaton. Then we can define a graph (in the sense of Serre) $\Gamma(Q, A)$ with vertex set $Q$ and with set of positively oriented edges $\{(q, a) \mid qa \neq \emptyset\}$. Set $(q, a)\iota = q$, $(q, a)\tau = qa$ and define a labeling $\ell \colon \Gamma(Q, A) \to FG(A)$ by $(q, a)\ell = a$, pictured $q \xrightarrow{a} qa$. In particular, if $M$ is an $A$-generated monoid, then the graph $\Gamma(M, A)$ associated to the automaton $(M, A)$ is called the *right Cayley graph* of $M$ with respect to $A$. The corresponding labeling is denoted $\ell_M$. This discussion leads us to the notion of an $A$-graph.

**Definition 4.17.1 ($A$-graph).** *Let $A$ be a set. Then an $A$-graph is a pair $(\Gamma, \ell)$ where $\Gamma$ is a graph (in the sense of Serre) and $\ell$ is an $FG(A)$-labeling of $\Gamma$ such that each edge is labeled by an element of $A \cup A^{-1}$ and no two coterminal edges have the same label. We always take the edges labeled by elements of $A$ as the positively oriented edges.*

Of special interest to us are so-called inverse $A$-graphs and covers.

**Definition 4.17.2 (Inverse graphs and covers).** *An $A$-graph is called an inverse $A$-graph if whenever $e, e'$ are edges with $e\iota = e'\iota$, then $e\ell = e'\ell$ implies $e = e'$. An $A$-graph is called a cover if, for each vertex $v$ and letter $a \in A \cup A^{-1}$, there is a unique edge $e$ with $e\iota = v$ and $e\ell = a$.*

Of course, covers are inverse $A$-graphs. Inverse $A$-graphs also go by the name *labeled graph immersions* [189, 321, 328]. They are precisely the graphs of the form $\Gamma(Q, A)$ coming from injective automata $(Q, A)$. The connection between inverse graphs, graph immersions and inverse semigroups was first drawn by Margolis and Meakin [189]; see [81, 196, 328] for further developments. Covers are sometimes called *permutation automata*, as they are precisely the graphs $\Gamma(Q, A)$ coming from actions of $A$ on $Q$ by permutations. By a morphism of $A$-graphs, we mean a graph morphism (in the obvious sense) that preserves the labelings. That is $\varphi \colon (\Gamma, \ell) \to (\Gamma', \ell')$ is a *morphism of $A$-graphs* if $\varphi$ sends vertices to vertices, edges to edges, preserves the involution and the incidence maps, and $\varphi\ell' = \ell$, i.e., the diagram

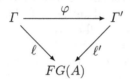

commutes.

**Proposition 4.17.3 (Stallings).** *Let $(\Gamma, \ell)$ be an inverse $A$-graph. Then for any reduced path $p = e_1 \cdots e_n$ in $\Gamma$, the word $e_1\ell \cdots e_n\ell$ is freely reduced. Hence $\ell \colon \pi_1(\Gamma, v) \to \pi_1(\Gamma, \ell, v)$ is an isomorphism.*

*Proof.* Suppose $p$ is reduced, but $e_1\ell \cdots e_n\ell$ is not freely reduced. Then there are consecutive edges $e_i e_{i+1}$ in $p$ with $e_i\ell = (e_{i+1}\ell)^{-1}$. Then $\bar{e}_i\ell = e_{i+1}\ell$ and $\bar{e}_i\ell = e_{i+1}\ell$. So, by the definition of an inverse $A$-graph, $\bar{e}_i = e_{i+1}$. Hence $p$ contains a backtrack, contradicting that $p$ is reduced.

For the second statement, suppose that $p = e_1 \cdots e_n$ is reduced and non-empty. Then because $e_1\ell \cdots e_n\ell$ is freely reduced, by the first part of the proof, and non-empty, $p\ell \neq 1$. Thus $\ell$ is injective on fundamental groups.    □

So from now on if $(\Gamma, \ell)$ is an inverse $A$-graph, we do not distinguish $\pi_1(\Gamma, v)$ and $\pi_1(\Gamma, \ell, v)$. As a consequence of Proposition 4.17.3, if $v$ is a vertex of an inverse $A$-graph and $g \in FG(A)$, then there is at most one reduced path starting at $v$ with label $g$. Indeed, if $p$ and $q$ are two such reduced paths, then $(p^{-1}q)\ell = 1$ and so $p^{-1}q$ is null homotopic by the proposition. Therefore, $[q] = [p][p^{-1}q] = [p]$ and so, by uniqueness of reduced paths in a homotopy class, $p = q$. If $(\Gamma, \ell)$ is a cover, it is easy to check that such a reduced path always exists.

## Basic properties of covers

We discuss some basic topological facts about graphs. The reader is referred to [184, 321] for more details.

**Proposition 4.17.4.** *If a finite $A$-graph $(\Gamma, \ell)$ is a cover, then $\pi_1(\Gamma, v)$ is a finite index subgroup of $FG(A)$.*

*Proof.* We may assume without loss of generality $\Gamma$ is connected. In this case, the right cosets of $\pi_1(\Gamma, v)$ are in bijection with the vertices of $\Gamma$ via the map $\pi_1(\Gamma, v)g \mapsto p_g\tau$ where $p_g$ is the reduced path labeled by $g$ starting at $v$.  □

**Exercise 4.17.5.** Verify the assertion in the proof of Proposition 4.17.4.

This leads to the following definition.

**Definition 4.17.6 (Monodromy group).** *Suppose an A-graph $(\Gamma, \ell)$ is a cover. Then $FG(A)$ acts on the vertex set $V$ of $\Gamma$ by defining, for $v \in V$ and $g \in FG(A)$, $vg$ to be the endpoint of the unique reduced path labeled by $g$ starting from $v$. In particular, $\pi_1(\Gamma, v)$ is the stabilizer of $v$. The resulting group $G$ of permutations of $V$ is called the monodromy group of the covering in topology and the transition group in automata theory. We shall prefer the former terminology due to historical precedence.*

A morphism $\varphi\colon (\Gamma, \ell) \to (\Gamma', \ell')$ of $A$-graphs is called a *covering* if it is onto on vertices and, for each vertex $v \in \Gamma$ and each edge $e \in \Gamma'$ with $e\iota = v\varphi$, there is a unique edge $\tilde{e} \in \Gamma$ such that $\tilde{e}\iota = v$ and $\tilde{e}\varphi = e$. In this case we say $\Gamma$ *covers* $\Gamma'$. Notice that if $\mathscr{B}_A$ is the graph with a single vertex and $|A|$ positively oriented edges labeled by letters of $A$ (called a *bouquet of circles*), then each $A$-graph admits a unique morphism of labeled graphs from $\Gamma$ to $\mathscr{B}_A$. The morphism associated to a labeling $\ell$ of $\Gamma$ is a covering if and only if $(\Gamma, \ell)$ is a cover in the sense of our earlier terminology. Notice that $\varphi\colon (\Gamma, \ell) \to (\Gamma', \ell')$ is a morphism of $A$-graphs if and only if the diagram

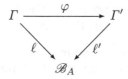

commutes. The following is a standard topological result [184, 321].

**Proposition 4.17.7.** *Suppose $(\Gamma, \ell)$ and $(\Gamma', \ell')$ are connected A-graphs that are covers. Let $v$, $v'$ be vertices of $\Gamma$, $\Gamma'$ respectively. Then there is a covering $\varphi\colon (\Gamma, \ell) \to (\Gamma', \ell')$ taking $v$ to $v'$ if and only if $\pi_1(\Gamma, v) \leq \pi_1(\Gamma', v')$ (viewed as subgroups of $FG(A)$). The covering $\varphi$ is unique when it exists.*

*Proof.* The existence of $\varphi$ easily implies $\pi_1(\Gamma, v) \leq \pi_1(\Gamma, v')$ because circuits at $v$ are mapped to circuits at $v'$. For the converse, suppose $\pi_1(\Gamma, v) \leq \pi_1(\Gamma', v')$. Define $\varphi$ on vertices as follows. If $w$ is a vertex of $\Gamma$, choose a path $p\colon v \to w$ and define $w\varphi$ to be the endpoint of the unique reduced path $p'$ in $\Gamma'$ starting from $v'$ with label $p\ell$. If $q\colon v \to w$ is another such path, with corresponding path $q'$ in $\Gamma'$ from $v'$, then $g = (pq^{-1})\ell \in \pi_1(\Gamma, v) \leq \pi_1(\Gamma', v')$. Hence $g$ labels a reduced circuit $r$ at $v'$ in $\Gamma'$ and $(p')^{-1}rq'$ is null homotopic. We conclude $p'\tau = q'\tau$ and so $\varphi$ is well defined on vertices. If $e$ is an edge, one defines $e\varphi$ to be the unique edge $e'$ emanating from $e\iota\varphi$ with label $e\ell$. If $p\colon v \to e\iota$ is a path, then $pe\colon v \mapsto e\tau$ is a path that can be used to define $e\tau\varphi$. Letting $p'$ be the reduced path in $\Gamma'$ associated to $p$ as above, we see that the reduced form of $p'e'$ is the path associated to $pe$. It follows that $\varphi$ is a graph morphism and it is straightforward to verify that $\varphi$ is a covering and is the unique such.    □

**Exercise 4.17.8.** Complete the remaining details of the proof of Proposition 4.17.7.

**Proposition 4.17.9.** *Let* $(\Gamma, \ell)$ *be a connected $A$-graph, which is a cover with monodromy group $G$. Suppose the projection $FG(A) \twoheadrightarrow G$ factors through an $A$-generated group $H$. Then, for any vertex $h$ of the Cayley graph $\Gamma(H, A)$ of $H$ and $v \in \Gamma$, there is a covering $\varphi\colon \Gamma(H, A) \to \Gamma$ taking $h$ to $v$.*

*Proof.* By Proposition 4.17.7, it suffices to show that if $g \in \pi_1(\Gamma(H, A), h)$, then $g \in \pi_1(\Gamma, v)$. But $g \in \pi_1(\Gamma(H, A), h)$ implies $g$ maps to the identity in $H$ and hence in $G$ as well. From the definition of the action of $G$ on $\Gamma$ it follows that the unique reduced path labeled by $g$ at $v$ is in fact a circuit, whence $g \in \pi_1(\Gamma, v)$, as required.    □

### The minimal injective congruence and Stallings folding

**Definition 4.17.10 ($A$-graph congruence).** *A congruence $\equiv$ on an $A$-graph $\Gamma$ is an equivalence relation (also denoted $\equiv$) on the vertex set $V$ of $\Gamma$. The quotient graph $(\Gamma/\equiv, \ell)$ has vertex set $V/\equiv$. For $a \in A \cup A^{-1}$, there is an edge from $[v_0]$ to $[v_1]$ labeled by $a$ if there exists an edge $e$ with $e\ell = a$, $e\iota \equiv v_0$ and $e\tau \equiv v_1$. A congruence $\equiv$ is called injective if whenever $e_1, e_2$ are edges with $e_1\ell = e_2\ell$ and $e_1\iota \equiv e_2\iota$, then $e_1\tau \equiv e_2\tau$.*

One can easily check that $\equiv$ is an injective $A$-graph congruence on $(\Gamma, \ell)$ if and only if $\Gamma/\equiv$ is an inverse $A$-graph. Also if $(Q, A)$ is an automaton, then an injective $A$-graph congruence on $\Gamma(Q, A)$ is the same thing as an injective automaton congruence on $(Q, A)$.

**Exercise 4.17.11.** Verify the previous two assertions.

The following construction of the minimal injective congruence can be found in [365] for automata and in [321] using the language of folding (see also [33, 328]).

**Theorem 4.17.12.** *Let $(\Gamma, \ell)$ be an $A$-graph. Define a congruence $\sim_{\mathscr{I}}$ on $(\Gamma, \ell)$ by $q \sim_{\mathscr{I}} q'$ if there exists a path $p\colon q \to q'$ in $\Gamma$ with $p\ell = 1$. Then $\sim_{\mathscr{I}}$ is an injective congruence and if $\equiv$ is any injective congruence on $(\Gamma, \ell)$, then $\sim_{\mathscr{I}}$ is contained in $\equiv$.*

*Proof.* Using that the set of paths with label 1 contains the empty paths and is closed under involution and product, it is easy to see $\sim_{\mathscr{I}}$ is an equivalence relation. Suppose $e_1, e_2$ are edges and $e_1\ell = e_2\ell$ with $e_1\iota \sim_{\mathscr{I}} e_2\iota$ and let

$p: e_1\iota \to e_2\iota$ be a path with $p\ell = 1$. Then $p' = e_1^{-1}pe_2$ is a path from $e_1\tau$ to $e_2\tau$ with $p'\ell = 1$, so $e_1\tau \sim_{\mathscr{I}} e_2\tau$, establishing $\sim_{\mathscr{I}}$ is an injective congruence.

To verify the minimality of $\sim_{\mathscr{I}}$, suppose $\equiv$ is an injective automaton congruence on $(\Gamma, \ell)$ and let $q \sim_{\mathscr{I}} q'$. We must verify $q \equiv q'$. Let $p: q \to q'$ be a path with $p\ell = 1$. The proof goes by induction on the length of $p$. If $p = 1_q$, then $q = q'$ and so trivially $q \equiv q'$. Suppose now that $p = e_1 \cdots e_n$ with $n \geq 1$. Because $p\ell = 1$, either $p = rs$ with $r\ell = 1$, $s\ell = 1$ and neither $r$ nor $s$ empty, or $p = e_1 r e_n$ with $e_1\ell = (e_n\ell)^{-1}$ and $r\ell = 1$, depending on whether $e_1\ell$ is cancelled in the middle or at the end. In the first case, $q \equiv r\tau = s\iota \equiv q'$ by induction and so $q \equiv q'$, as required. In the second case, $e_1\tau \equiv e_n\iota$ by induction. Then $\bar{e}_1\ell = e_n\ell$ and $\bar{e}_1\iota \equiv e_n\iota$, and so $q \equiv q'$ as $\equiv$ is an injective congruence. This completes the proof.    □

We denote the quotient inverse $A$-graph by $\mathscr{I}(\Gamma, \ell)$. Stallings [321] gave an iterative procedure for constructing $\mathscr{I}(\Gamma, \ell)$ when $\Gamma$ is finite.

**Definition 4.17.13 (Fold).** *A congruence on an $A$-graph $(\Gamma, \ell)$ is called a fold if there are edges $e_1, e_2$ such that: $e_1\iota = e_2\iota$, $e_1\ell = e_2\ell$ and the only non-trivial congruence class is $\{e_1\tau, e_2\tau\}$. See Figure 4.2*

It is immediate from the definitions that if $\equiv$ is a fold, then $\equiv$ is contained in $\sim_{\mathscr{I}}$. Moreover, it is clear that $(\Gamma, \ell)$ admits a fold if and only if it is not an inverse $A$-graph. Hence if $(\Gamma, \ell)$ is a finite $A$-graph, then $\mathscr{I}(\Gamma, \ell)$ can be obtained from $(\Gamma, \ell)$ by performing finitely many folds. This can be performed in quadratic time as each fold diminishes the number of vertices. Sometimes $\mathscr{I}(\Gamma, \ell)$ is called the result of performing *Stallings folding* on $(\Gamma, \ell)$. Notice that the order of folding the edges is irrelevant.

Stallings described the effect of folding on the fundamental group [321]. First we prove the following lemma.

**Lemma 4.17.14.** *Suppose $(\Gamma, \ell)$ is an $A$-graph and $q: [v] \to [w]$ is a path in $\mathscr{I}(\Gamma, \ell)$. Then there is a path $p: v \to w$ with $p\ell = q\ell$.*

*Proof.* Suppose that $q$ is given by $e_1 \cdots e_n$. By definition of $\sim_{\mathscr{I}}$ and of the quotient graph, there exist edges $f_1, \ldots, f_n$ in $\Gamma$ with $f_i\ell = e_i\ell$, for $i = 1, \ldots, n$, and paths $p_0, p_1, \ldots, p_n$ with $p_j\ell = 1$, $j = 0, \ldots, n$ such that $p = p_0 f_1 p_1 \cdots f_n p_n$ is a path from $v$ to $w$. Then $p\ell = q\ell$,    □

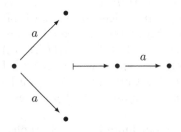

**Fig. 4.2.** A fold

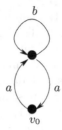

**Fig. 4.3.** The Stallings graph $\Gamma_H$ for $H = \langle aba^{-1}, a^2 \rangle$

**Corollary 4.17.15 (Stallings).** *Let $(\Gamma, \ell)$ be an A-graph. Then $\pi_1(\Gamma, \ell, v) = \pi_1(\mathscr{I}(\Gamma, \ell), [v])$.*

*Proof.* Recall we identify $\pi_1(\mathscr{I}(\Gamma, \ell), v)$ with its image in $FG(A)$. Because circuits at $v$ in $\Gamma$ map to circuits at $[v]$ in $\mathscr{I}(\Gamma, \ell)$, it is clearly the case

$$\pi_1(\Gamma, v)\ell \leq \pi_1(\mathscr{I}(\Gamma, \ell), [v]).$$

The reverse inclusion is immediate from Lemma 4.17.14. □

Stallings used Corollary 4.17.15 to give an algorithm to realize a finitely generated subgroup of $FG(A)$ as the fundamental group of a finite inverse A-graph.

**Definition 4.17.16 (Stallings graph).** *Let $H \leq FG(A)$ be a finitely generated subgroup and let $Y = \{w_1, \ldots, w_r\}$ be a finite set of reduced words such that $H = \langle Y \rangle$. Construct an A-graph $\Delta$ as follows. Take a graph whose geometric edges form a wedge of $r$ cycles $C_1, \ldots, C_r$ where $C_i$ is a cycle of length $|w_i|$; the base point of the wedge is denoted $v_0$. Now, for $i = 1, \ldots, r$, label the $j^{th}$ edge of the cycle $C_i$, in the counterclockwise direction, by the $j^{th}$ letter of $w_i$ (and label the inverse edges accordingly). If $\ell$ is the associated labeling, then $\pi_1(\Delta, \ell, v_0) = H$. Let $\Gamma_H = \mathscr{I}(\Delta, \ell)$. Then $\Gamma_H$ is an inverse A-graph and $\pi_1(\Gamma_H, [v_0]) = H$ by Corollary 4.17.15. We call $\Gamma_H$ the Stallings graph of $H$. Notice that $\Gamma_H$ is connected.*

An example of a Stallings graph is provided in Figure 4.3. There is a useful variation of the Stallings graph. Let $H \leq FG(A)$ be finitely generated and let $w \in FG(A)$ be of length $n$. Let $v_0$ be the base point of $\Gamma_H$. Attach to $v_0$ a graph whose geometric edges form a path of length $n$ and label it (starting from $v_0$) by $w$ (where the inverse edges are labeled accordingly). The resulting graph $(\Gamma', \ell')$ clearly still has $\pi_1(\Gamma', \ell', v_0) = H$ because the new path is contained entirely in a spanning tree for $\Gamma'$. Let $\Gamma_{H,w} = \mathscr{I}(\Gamma', \ell')$. Then $\pi_1(\Gamma_{H,w}, [v_0]) = H$ (by Corollary 4.17.15), $w$ labels a reduced path $p$ starting from $[v_0]$ and $w \in H$ if and only if $p$ is a circuit. See Figure 4.4 for an example.

One of the key facts about finite inverse A-graphs is that they extend to finite covers.

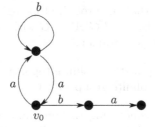

**Fig. 4.4.** The graph $\Gamma_{H,w}$ for $H = \langle aba^{-1}, a^2 \rangle$ and $w = a^2 ba$

**Lemma 4.17.17.** *Let $(\Gamma, \ell)$ be a finite inverse $A$-graph with vertex set $V$. Then there is a cover $(\widetilde{\Gamma}, \widetilde{\ell})$ with vertex set $V$ containing $\Gamma$ as a subgraph and with $\widetilde{\ell}|_{\Gamma} = \ell$.*

*Proof.* Because $\Gamma$ is inverse, it is of the form $\Gamma(V, A)$ for an injective automaton $(V, A)$. Extending the partial permutations of $(V, A)$ to permutations and interpreting the result as a cover $(\widetilde{\Gamma}, \widetilde{\ell})$ establishes the lemma.    □

If $(\Gamma, \ell)$ is an inverse $A$-graph and $\Gamma \subseteq \widetilde{\Gamma}$, where $(\widetilde{\Gamma}, \widetilde{\ell})$ is a cover such that $\widetilde{\ell}|_{\Gamma} = \ell$, then we say that $\widetilde{\Gamma}$ *extends* $\Gamma$. The following lemma will be used several times in the sequel; it also underlies the Margolis-Meakin expansion [188].

**Lemma 4.17.18.** *Let $(\Gamma, \ell)$ be an inverse $A$-graph and $(\widetilde{\Gamma}, \widetilde{\ell})$ be a cover extending $\Gamma$ with monodromy group $G$. Suppose the canonical morphism $FG(A) \twoheadrightarrow G$ factors through an $A$-generated group $H$. Let $p, q$ be reduced paths in the Cayley graph $\Gamma(H, A)$ such that $p\iota = q\iota$ and each geometric edge traversed by $q$ is also traversed by $p$. Then given a reduced path $p'$ in $\Gamma$ with $p'\ell = p\ell_H$, there is a reduced path $q'$ in $\Gamma$ with $q'\ell = q\ell_H$ such that $q'\iota = p'\iota$. Moreover, if $p\tau = q\tau$, then $q'\tau = p'\tau$.*

*Proof.* Let $h = p\iota = q\iota$ and let $\widetilde{\Gamma}_0$ be the connected component of $\widetilde{\Gamma}$ containing $p'$. By Proposition 4.17.9, there is a covering $\varphi \colon \Gamma(H, A) \to \widetilde{\Gamma}_0$ sending $h$ to $p'\iota$. As $p'\ell = p\ell_H$ and coverings send reduced paths to reduced paths, it follows $p\varphi = p'$. Define $q' = q\varphi$; it is a reduced path. Clearly, $q'\iota = p'\iota$ and $q'\ell = q\ell_H$. Because each geometric edge used by $q$ is also used by $p$, each geometric edge traversed by $q'$ is also used by $p'$, and hence belongs to $\Gamma$. If in addition, $p\tau = q\tau$, then $p'\tau = p\tau\varphi = q\tau\varphi = q'\tau$.    □

### 4.17.2 The profinite topology on a free group

The *profinite topology* on $FG(A)$ is the group topology obtained by taking as a fundamental system of neighborhoods of 1 all finite index subgroups. The profinite topology is the weakest topology so that any homomorphism $\varphi \colon FG(A) \to G$ with $G$ a finite (discrete) group is continuous.

**Exercise 4.17.19.** Verify that the profinite topology is the weakest topology such that any homomorphism $\varphi \colon FG(A) \to G$ with $G$ a finite group endowed with the discrete topology is continuous.

**Exercise 4.17.20.** Show that the profinite topology is the induced topology on $FG(A)$ from the free profinite group $\widehat{F}_{\mathbf{G}}(A)$.

A fundamental result concerning the profinite topology is Hall's Theorem [112]. The proof presented here is the well-known proof of Stallings [321].

**Theorem 4.17.21 (Hall).** *Finitely generated subgroups of $FG(A)$ are closed in the profinite topology.*

*Proof.* Suppose that $H$ is a finitely generated subgroup of $FG(A)$ and $w \notin H$. Consider the finite inverse $A$-graph $\Gamma_{H,w}$ with base point $v_0$. By construction, $w$ labels a unique reduced path $p$ starting at $v_0$ and $\pi_1(\Gamma_{H,w}, v_0) = H$. Consequently $p$ is not a circuit. Let $(\widetilde{\Gamma}_{H,w}, \widetilde{\ell})$ be a finite cover extending $\Gamma_{H,w}$ as in Lemma 4.17.17. Then $K = \pi_1(\widetilde{\Gamma}_{H,w}, v_0)$ is a finite index subgroup (Proposition 4.17.4), $H \leq K$ and $w \notin K$, as $p$ is still the unique reduced path in $\widetilde{\Gamma}_{H,w}$ starting at $v_0$ with label $w$. Because $Kw \cap H = \emptyset$ and $Kw$ is open, we conclude $H$ is closed.    $\square$

By taking $H$ to be the trivial group, we see that $FG(A)$ is residually finite and hence if $g \neq 1$, then there is a finite index subgroup $K \leq FG(A)$ such that $g \notin K$.

**Exercise 4.17.22.** The goal of this exercise is to show that the profinite topology on $FG(A)$ is metric when $A$ is a finite set. More specifically, define, a norm on $FG(A)$ by defining $\|1\| = 0$ and, for $1 \neq g \in FG(A)$,

$$\|g\| = 2^{-\min\{[FG(A):K] \mid g \notin K\}}.$$

Define a metric on $FG(A)$ by $d(g, h) = \|g^{-1}h\|$.

1. Prove that $\|g\| = \|g^{-1}\|$.
2. Prove $\|gh\| \leq \max\{\|g\|, \|h\|\}$.
3. Prove that $d$ is an ultrametric (meaning a metric satisfying the inequality $d(g, h) \leq \max\{d(g, k), d(k, h)\}$).
4. Prove $d(kg, kh) = d(g, h)$.
5. Prove that $d$ defines the profinite topology.

The following sweeping generalization of Hall's Theorem, due to Ribes and Zalesskii [300], was first conjectured by Pin and Reutenauer [232], who proved it implies the Rhodes Type II conjecture. This is a good example of how people from one area can come up with a great conjecture concerning another area. The special case of two subgroups was proved independently by Gitik and Rips [99] at approximately the same time (see [218]).

**Theorem 4.17.23 (Ribes and Zalesskii).** *Let $H_1, \ldots, H_n$ be finitely generated subgroups of $FG(A)$. Then $H_1 H_2 \cdots H_n$ is closed in the profinite topology.*

A proof of the theorem appears in the next section based on Auinger and Steinberg [39]. A model theoretic proof can be found in [131]. Generalizations and variations can be found in [37, 76, 98, 208, 299, 301, 331, 382]. In the sequel we shall mostly use the following variant of the theorem.

**Corollary 4.17.24.** *Let $H_0, \ldots, H_n$ be finitely generated subgroups of $FG(A)$, and $g_0, \ldots, g_{n+1} \in FG(A)$. Then $g_0 H_0 g_1 \cdots H_n g_{n+1}$ is closed.*

*Proof.* Let $N_i = g_{n+1}^{-1} \cdots g_{i+1}^{-1} H_i g_{i+1} \cdots g_{n+1}$. Routine computation yields

$$g_0 H_0 g_1 \cdots H_n g_{n+1} = g_0 \cdots g_{n+1} N_0 N_1 \cdots N_n.$$

Because each $N_i$ is finitely generated and left translation is a homeomorphism, Theorem 4.17.23 implies $g_0 \cdots g_{n+1} N_0 N_1 \cdots N_n$ is closed. $\square$

Let us consider two examples of convergent sequences in the profinite topology. They turn out to be the only ones we shall ever need.

**Proposition 4.17.25.** *Suppose $g \in FG(A)$. Then $g^{n!} \to 1$ and $g^{n!-1} \to g^{-1}$.*

*Proof.* Suppose that $[FG(A) : K] = m$. The action of $FG(A)$ on the right of $FG(A)/K$ provides a permutation representation $FG(A) \to S_m$. It then follows that, for all $n \geq m$, $Kg^{n!} = K$, i.e., $g^{n!} \in K$. As the finite index subgroups form a neighborhood basis of 1, we conclude $g^{n!} \to 1$. Then $g^{n!-1} \to g^{-1}$ because right translation by $g^{-1}$ is a homeomorphism. $\square$

Now we wish to relate the profinite topology to Type II elements. This connection is due to Pin [226].

**Proposition 4.17.26.** *Let $M$ be an $A$-generated monoid and $\alpha \colon A^* \twoheadrightarrow M$ the projection. Define $\varphi_{\mathbf{G}} \colon M \to FG(A)$ by $m\varphi_{\mathbf{G}} = \overline{m\alpha^{-1}}$ (viewing $A^*$ as a subset of $FG(A)$). Then $\varphi_{\mathbf{G}}$ is a relational morphism and $\mathsf{K}_{\mathbf{G}}(M) = 1\varphi_{\mathbf{G}}^{-1}$.*

*Proof.* By continuity of multiplication, if $X, Y \subseteq FG(A)$, then $\overline{X} \cdot \overline{Y} \subseteq \overline{XY}$. Combining this with the fact that $\alpha^{-1}$ is a relational morphism, we conclude that $\varphi_{\mathbf{G}}$ is a relational morphism. Suppose first $m \in 1\varphi_{\mathbf{G}}^{-1}$ and let $\psi \colon M \to G$ be a relational morphism with $G \in \mathbf{G}$. Choose, for each $a \in A$, $g_a \in a\alpha\psi$. Define a continuous homomorphism $\gamma \colon FG(A) \to G$ by $a\gamma = g_a$. By construction it is the case $\alpha^{-1}\gamma \subseteq \psi$. Because $K = \ker \gamma$ is of finite index in $FG(A)$ (and hence open), there exists $w \in m\alpha^{-1} \cap K$, as $1 \in \overline{m\alpha^{-1}}$. It then follows $1 = w\gamma \in m\alpha^{-1}\gamma \subseteq m\psi$, establishing $m \in \mathsf{K}_{\mathbf{G}}(M)$.

For the converse, suppose $m \in \mathsf{K}_{\mathbf{G}}(M)$. Because the finite index subgroups form a neighborhood basis of 1, we must show that if $K \leq FG(A)$ is a finite index subgroup, then $K \cap m\alpha^{-1} \neq \emptyset$. Because $K$ has finite index, it contains a finite index normal subgroup $N$ (namely the intersection of all its conjugates). Consider the quotient map $\gamma \colon FG(A) \to FG(A)/N$. There results a relational morphism $\alpha^{-1}\gamma \colon M \to FG(A)/N$, whence $1 \in m\alpha^{-1}\gamma$. Therefore, $K \cap m\alpha^{-1} \neq \emptyset$, establishing $1 \in m\varphi_{\mathbf{G}}$. This completes the proof that $\mathsf{K}_{\mathbf{G}}(M) = 1\varphi_{\mathbf{G}}^{-1}$. $\hfill\square$

**Exercise 4.17.27.** Show that $\{m_1, m_2\} \subseteq M$ is **G**-pointlike if and only if $1 \in m_1\varphi_{\mathbf{G}}(m_2\varphi_{\mathbf{G}})^{-1}$, if and only if $m_1\varphi_{\mathbf{G}} \cap m_2\varphi_{\mathbf{G}} \neq \emptyset$.

To prove the Type II conjecture, we need a good description of $1\varphi_{\mathbf{G}}^{-1}$. Notice that $m\varphi_{\mathbf{G}}$ is the closure of the rational subset $m\alpha^{-1}$ of $A^*$ inside of $FG(A)$. In [328], an efficient algorithm is given to compute the closure of an arbitrary rational subset of $A^*$ in the profinite topology on $FG(A)$ from a non-deterministic automaton (see also [232], where regular expressions are considered). Here we restrict to the minimum needed to deal with Type II elements.

**Definition 4.17.28 (Strongly connected components).** *If $(\Gamma, \ell)$ is an A-graph, a path $p$ is called directed if it only uses positively oriented edges. One says that $\Gamma$ is strongly connected if, for all vertices $v$, $v'$ of $\Gamma$, there is a directed path $p \colon v \to v'$. Maximal strongly connected subgraphs of an A-graph $\Gamma$ are termed strongly connected components. A positively oriented edge $e$ whose endpoints lie in different strongly connected components is called a transition edge.*

We observe that there is a natural partial order on the set of strongly connected components defined by $C \leq C'$ if there is a directed path from a vertex of $C'$ to a vertex of $C$. Our next theorem is a primitive version of the second author's work [328, Thm. 6.27]. We recommend drawing pictures!

**Theorem 4.17.29.** *Let $M$ be an A-generated monoid, $\alpha \colon A^* \to M$ be the canonical projection and $\varphi_{\mathbf{G}} \colon M \to FG(A)$ be the relational morphism given by $m\varphi_{\mathbf{G}} = \overline{m\alpha^{-1}}$. Then $g \in m\varphi_{\mathbf{G}}$ if and only if there is a path $p \colon 1 \to m$ in the Cayley graph $\Gamma(M, A)$ of $M$ with $p\ell_M = g$ such that $p$ only traverses transition edges of $\Gamma(M, A)$ in their positive direction. In particular, $m \in \mathsf{K}_{\mathbf{G}}(M)$ if and only if there is a path $p \colon 1 \to m$ with $p\ell_M = 1$ and which only traverses transition edges of $\Gamma(M, A)$ in their positive direction.*

*Proof.* Set $\ell = \ell_M$. Let $X$ be the set all paths $p \colon 1 \to m$ that only traverse transition edges in their positive direction. First we show $X\ell \subseteq m\varphi_{\mathbf{G}}$. Let $p \in X$. If $p$ is empty, then $m = 1$ and $1 \in 1\varphi_{\mathbf{G}}$ so there is nothing more to prove. So assume $p = e_1 \cdots e_n$ and suppose $e_i\ell = a^{-1}$, where $a \in A$. Then $\overline{e}_i$ is not a transition edge, so there is a directed path $r \colon e_i\iota \to e_i\tau$. Define $e_{i,n} = (r\overline{e}_i)^{n!-1}r$; it is a directed path from $e_i\iota$ to $e_i\tau$. Moreover,

$$\lim e_{i,n}\ell = \lim[(r\ell a)^{n!-1}r\ell] = (r\ell a)^{-1}r\ell = a^{-1} \qquad (4.42)$$

by Proposition 4.17.25. Now construct a sequence of directed paths $p_n \colon 1 \to m$, $n \geq 1$, by replacing each negatively oriented edge $e_i$ in $p$ with $e_{i,n}$. From (4.42) it follows $\lim p_n \ell = p\ell = g$. Because $p_n \ell \in m\alpha^{-1}$, we conclude $g \in m\varphi_{\mathbf{G}}$, establishing $X\ell \subseteq m\varphi_{\mathbf{G}}$.

For the reverse inclusion, we clearly have $\overline{m\alpha^{-1}} \subseteq X\ell$. So if we can show $X\ell$ is closed, then the inclusion $m\varphi_{\mathbf{G}} = \overline{m\alpha^{-1}} \subseteq X\ell$ follows. First some notation: for elements $m_1$ and $m_2$ in the same strongly connected component $\Gamma(m_1, m_2)$ of $\Gamma(M, A)$, define $L(m_1, m_2) \subseteq FG(A)$ to be the set of all labels of paths (not necessarily directed) from $m_1$ to $m_2$, which are entirely contained in $\Gamma(m_1, m_2)$. Then $X\ell$ is a finite union of sets of the form

$$L(1, e_1\iota)(e_1\ell)L(e_1\tau, e_2\iota)(e_2\ell)\cdots(e_n\ell)L(e_n\tau, m) \tag{4.43}$$

where the $e_i$ are transition edges (if $m$ is in the group of units of $M$, we admit in (4.43) the set $L(1, m)$). The finiteness comes from the fact that $n+1$ is bounded by the size of the longest chain in the partially ordered set of strongly connected components of $\Gamma(M, A)$. So to show that $X\ell$ is closed, it suffices to show that a set such as in (4.43) is closed. By Corollary 4.17.24, it suffices to show that if $m_1, m_2 \in M$ are in the same strongly connected component of $\Gamma(M, A)$, then $L(m_1, m_2) = Hg$ for some finitely generated subgroup $H$ of $FG(A)$ and some element $g \in FG(A)$. But if we fix a path $q\colon m_1 \to m_2$ in $\Gamma = \Gamma(m_1, m_2)$ with label $g$, then it is immediate that $L(m_1, m_2) = \pi_1(\Gamma, \ell, m_1)g$, because if $p\colon m_1 \to m_2$ is any path in $\Gamma(m_1, m_2)$, then $(pq^{-1})\ell \in \pi_1(\Gamma, \ell, m_1)$ and $p\ell = (pq^{-1})\ell q\ell$. Conversely, $\pi_1(\Gamma, \ell, m_1)g$ is clearly contained in $L(m_1, m_2)$. This completes the proof that $X\ell$ is closed and hence that $X\ell = m\varphi_{\mathbf{G}}$.

The final statement follows directly from Proposition 4.17.26.    $\square$

We are now ready to prove the Type II Theorem assuming the Ribes-Zalesskii Theorem. Again drawing pictures is strongly advised.

**Theorem 4.17.30 (Ash [33], Ribes-Zalesskii [300]).** *Let $M$ be a monoid. Then $\mathsf{K}_{\mathbf{G}}(M)$ is the least submonoid of $M$ closed under taking weak conjugates.*

*Proof.* We have already proved that $\mathsf{K}_{\mathbf{G}}(M)$ is a submonoid closed under taking weak conjugates in Proposition 4.12.4. Let $C(M)$ be the smallest submonoid of $M$ closed under taking weak conjugates. We must show $\mathsf{K}_{\mathbf{G}}(M) \leq C(M)$.

Choose a generating set $A$ for $M$ and let $\alpha\colon A^* \to M$ be the canonical projection. Setting $\ell = \ell_M$, we claim that if $p$ is a path in the Cayley graph $\Gamma(M, A)$ of $M$ that only traverses transition edges in their positive direction and $p\ell = 1$, then there is an element $s \in C(M)$ so that $p\iota s = p\tau$. The proof goes by induction on the length of $p$. If $p$ is empty, then $p\iota = p\tau$ and $1 \in C(M)$ does the job. Suppose that the claim holds for all shorter paths than $p$ and let $p = e_1 \cdots e_n$. There are two cases: either $p = rq$ where $r$ and $q$ are non-empty paths with $r\ell = 1$ and $q\ell = 1$; or $e_1\ell = (e_n\ell)^{-1}$ and $p = e_1re_n$ with $r\ell = 1$. In the first case, we have by induction elements $s, s' \in C(M)$ such that $r\iota s = r\tau$ and $q\iota s' = q\tau$. Then

$$p\iota ss' = r\iota ss' = r\tau s' = q\iota s' = q\tau = p\tau$$

and $ss' \in C(M)$. In the second case, we have by induction $s \in C(M)$ such that $r\iota s = r\tau$. Suppose that $e_1\ell = a \in A$. Then $e_n\ell = a^{-1}$ and $\bar{e}_n$ is not a transition edge. Assume $e_n\colon m_1 \to m_2$. Then there is a directed path $t\colon m_1 \to m_2$. Let $m = t\ell a$. The picture looks something like:

$$p\iota \xrightarrow{\ a\ } r\iota \xrightarrow{\ s\ } r\tau = m_1 \underset{a}{\overset{m}{\rightleftarrows}} m_2 = p\tau. \qquad (4.44)$$

It follows $m_2 a\alpha = m_1$ and $m_1 m = m_2$. So $m_1(ma\alpha)^{\omega-1}m = m_2$ and direct computation shows that $(ma\alpha)^{\omega-1}m$ is a weak inverse of $a\alpha$. Thus $(a\alpha)s[(ma\alpha)^{\omega-1}m] \in C(M)$ and $(p\iota)(a\alpha)s[(ma\alpha)^{\omega-1}m] = p\tau$, as a glance at (4.44) reveals. The case that $e_n\ell = a \in A$ is handled in the same way. This completes the induction.

In particular, if $m \in \mathsf{K_G}(M)$, then Theorem 4.17.29 provides a path $p\colon 1 \to m$ with $p\ell = 1$ and which only traverses transition edges in their positive direction. By the claim, there exists $s \in C(M)$ with $s = p\iota s = p\tau = m$. Thus $m \in C(M)$, as required. $\qquad \square$

The reader is referred to [126] for a survey of the numerous consequences of the Type II Theorem. We content ourselves with a small sampling. All semigroups here are taken to be finite.

**Corollary 4.17.31 (Ash [32]).** *A semigroup $S$ divides an inverse semigroup if and only if it has commuting idempotents.*

*Proof.* Because inverse semigroups have commuting idempotents and the collection of semigroups with commuting idempotents is the pseudovariety **ESl**, necessity is clear. For sufficiency, suppose that $S$ has commuting idempotents. Then $S^I$ also has commuting idempotents so we may assume without loss of generality that $S$ is a monoid. Because the idempotents of $S$ commute, $E(S)$ is a submonoid. We show that it is closed under taking weak conjugates. Let $sts = s$ with $s, t \in S$ and let $e$ be an idempotent. Then because $ts$ is an idempotent and idempotents commute, $setset = stset = set$. Similarly $tes$ is idempotent. We conclude $\mathsf{K_G}(S) = E(S) \in \mathbf{Sl}$. Let $\varphi\colon S \to G$ be a relational morphism with $G$ a group and $1\varphi^{-1} = \mathsf{K_G}(S) = E(S)$ (such exists by Proposition 4.12.5). Then every local monoid of the derived category $D_\varphi$ is a quotient of $1\varphi^{-1} = E(S)$. It follows that $D_\varphi \in \ell\mathbf{Sl} = \mathbf{gSl}$ (the equality coming by way of Simon's Theorem [60,85]). Thus $S$ divides a unitary wreath product $E \wr G$, with $E$ a semilattice, by the Derived Category Theorem [364]. But such a semidirect product is an inverse monoid. $\qquad \square$

**Exercise 4.17.32.** Show that a unitary semidirect product $E \rtimes G$ of a semilattice $E$ with a group $G$ is an inverse semigroup.

**Exercise 4.17.33.** Show that $\mathbf{ESl} = \mathbf{Sl} \textcircled{m} \mathbf{G} = \mathbf{Sl} \circ \mathbf{G}$.

A regular semigroup is called *orthodox* if its idempotents form a subsemigroup. An idempotent semigroup is called a *band*. The pseudovariety of bands is denoted by **B**.

**Exercise 4.17.34.** Show that if $E \rtimes G$ is a unitary semidirect product with $E$ a band and $G$ a group, then $E \rtimes G$ is orthodox.

**Corollary 4.17.35 (Birget-Margolis-Rhodes [53]).** *A semigroup $S$ divides an orthodox semigroup if and only if the idempotents of $S$ form a subsemigroup.*

*Proof.* Because the idempotents of an orthodox semigroup form a semigroup and the collection of semigroups whose idempotents form a subsemigroup is the pseudovariety **EB**, necessity is clear. For sufficiency, suppose that the idempotents of $S$ form a semigroup. Then the same is true for $S^I$, so we may assume without loss of generality that $S$ is a monoid. We show that the submonoid $E(S)$ is closed under taking weak conjugates. Let $sts = s$ with $s, t \in S$ and let $e$ be an idempotent. Then $ts$ is idempotent and so $tse$ is idempotent by hypothesis. Therefore,

$$set = s(tse)t = s(tsetse)t = setset$$

and similarly $(tes)^2 = tes$. We conclude $\mathsf{K}_{\mathbf{G}}(S) = E(S)$. Let $\varphi \colon S \to G$ be a relational morphism with $G$ a group and $1\varphi^{-1} = \mathsf{K}_{\mathbf{G}}(S) = E(S)$ (such exists by Proposition 4.12.5). Then every local monoid of the derived category $D_\varphi$ is a quotient of $1\varphi^{-1} = E(S)$. It follows that $D_\varphi \in \ell\mathbf{B} = \mathbf{gB}$ (equality holding by a result of Jones and Szendrei [151]). Therefore, $S$ divides a unitary wreath product $E \wr G$, with $E$ a band, by the Derived Category Theorem [364]. According to Exercise 4.17.34, such a semidirect product is orthodox. $\quad\Box$

**Exercise 4.17.36.** Show that $\mathbf{EB} = \mathbf{B} \, \textcircled{m} \, \mathbf{G} = \mathbf{B} \circ \mathbf{G}$.

The result of the following exercise was proved by Birget, Margolis and Rhodes [53].

**Exercise 4.17.37 (Birget-Margolis-Rhodes [53]).** A regular semigroup is called *E-solid* if the idempotents generate a completely regular semigroup. Show that a finite semigroup $S$ divides a finite $E$-solid semigroup if and only if it belongs to $\mathbf{CR} \, \textcircled{m} \, \mathbf{G}$. You may want to use that $\mathbf{CR}$ is local [148, 356] and that a unitary semidirect product of a completely regular semigroup and a group is $E$-solid.

More results about pseudovarieties generated by so-called *e*-varieties can be found in [42].

A further application of the solution to the Rhodes Type II conjecture is the celebrated $\mathbf{P}(\mathbf{G}) = \mathbf{BG}$ theorem [122, 126, 127, 191, 227]. Recall that a semigroup is called a *block group* if each element has at most one inverse. For instance, an inverse semigroup is the same thing as a regular block group. The collection of finite block groups is a pseudovariety denoted $\mathbf{BG}$.

**Theorem 4.17.38 (Henckell-Margolis-Pin-Rhodes [126]).** *The equality*

$$\mathbf{P}(\mathbf{G}) = \mathbf{J} \circ \mathbf{G} = \Diamond \mathbf{G} = \mathbf{J} \,\circledm\, \mathbf{G} = \mathbf{BG}$$

*holds.*

This theorem relies on: a description of the **G**-pointlike pairs, conjectured by Henckell and Rhodes [127] and coming out of Ash's Theorem [33] or the Ribes and Zalesskii Theorem [300]; Knast's Theorem [163]; Simon's Theorem on $\mathscr{J}$-trivial monoids [85,318]; and the Derived Category Theorem [364]. The simplest proof, which can be found in the second author's work [326], is based on Gitik and Rip's elegant proof that the product of two finitely generated subgroups of a free group is closed in the profinite topology [99].

If **H** is a pseudovariety of groups, then one always has

$$\mathbf{P}(\mathbf{H}) \leq \mathbf{J} \circ \mathbf{H} \leq \Diamond \mathbf{H} \leq \mathbf{J} \,\circledm\, \mathbf{H}. \tag{4.45}$$

The second author proved [325] that $\Diamond \mathbf{H} = \mathbf{J} \circ \mathbf{H}$ holds for all pseudovarieties of groups **H**. Pin asked to what extent the equality $\mathbf{PG} = \mathbf{BG}$ can be extended to other pseudovarieties of groups [227]. In a series of papers, Auinger and the second author completely characterized when the inequalities in (4.45) can be equalities [37,38,40,41,328,329,331]. A sample of their results is given in the following theorem.

**Theorem 4.17.39 (Auinger-Steinberg).** *Let **H** be a non-trivial pseudovariety of groups.*

1. *The equality $\mathbf{J} \circ \mathbf{H} = \mathbf{J} \,\circledm\, \mathbf{H}$ implies $\mathbf{P}(\mathbf{H}) = \mathbf{J} \circ \mathbf{H}$ and implies that the products of pro-**H** closed finitely generated subgroups of a free group are closed.*
2. *The equality $\mathbf{P}(\mathbf{H}) = \mathbf{J} \circ \mathbf{H}$ implies that **H** satisfies no non-trivial group identities and is finite join irreducible (see Definition 6.1.5).*
3. *If, for each $G \in \mathbf{H}$, there exists $p$ with $\mathbb{Z}_p \wr G \in \mathbf{H}$, then*

$$\mathbf{P}(\mathbf{H}) = \mathbf{J} \circ \mathbf{H} = \mathbf{J} \,\circledm\, \mathbf{H}.$$

*This applies in particular if **H** is extension-closed or if **H** is not locally finite and can be defined by a pseudoidentity in one variable.*

4. *If **H** is a pseudovariety of supersolvable groups, then the following are equivalent:*
   *(a) $\mathbf{P}(\mathbf{H}) = \mathbf{J} \circ \mathbf{H}$;*
   *(b) $\mathbf{J} \circ \mathbf{H} = \mathbf{J} \,\circledm\, \mathbf{H}$;*
   *(c) $\mathbf{H} = \mathbf{G}_p \circ \mathbf{V}$ where $p$ is a prime and $\mathbf{V}$ is a pseudovariety of abelian groups of exponent $p - 1$.*

In particular, all the inequalities in (4.45) are equalities for **H** the pseudovariety of solvable groups or the pseudovariety of $p$-groups, $p$ prime. On the other hand, all the inequalities in (4.45) are strict for the pseudovariety of nilpotent groups (as any nilpotent group is supersolvable). No examples are known for which $\mathbf{P}(\mathbf{H}) = \mathbf{J} \circ \mathbf{H}$ but $\mathbf{J} \circ \mathbf{H} \neq \mathbf{J} \,\circledm\, \mathbf{H}$.

# 4.18 The Ribes and Zalesskii Theorem

The aim of this section is to prove the Ribes and Zalesskii Theorem (Theorem 4.17.23). The proof given here is based on Auinger and the second author's paper [39], but we have formulated it in a more topological language. If $G$ is an $A$-generated group and $w \in FG(A)$, then $[w]_G$ will denote the image of $w$ in $G$ under the canonical projection $FG(A) \twoheadrightarrow G$.

## 4.18.1 Expansion by cyclic groups of prime order

A crucial ingredient in our proof is an expansion for groups of the sort studied in [87]. The reader is again referred to Section 4.13.1 for information on graphs labeled over groups. Fix a prime $p$. Let $\Gamma$ be a graph with set of positively oriented edges $E^+$. We can define a $\mathbb{Z}_p E^+$-labeling $\eta$ of $\Gamma$ by extending the inclusion map of $E^+$ to negatively oriented edges by defining $\bar{e}\eta = -e$, for $e \in E^+$. Hence we can define the label $q\eta$ of a path $q$ in $\Gamma$. We remark that if $\Gamma$ is connected, then $\pi_1(\Gamma, \eta, v)$ is independent of $v$ and is nothing more than $H_1(\Gamma, \mathbb{Z}_p)$, the first homology group of $\Gamma$ with $\mathbb{Z}_p$ coefficients. Notice that $q\eta = \sum_{e \in E^+} q(e)e$ where $q(e)$ counts the number of times, modulo $p$, that the geometric edge $\{e, \bar{e}\}$ is traversed in the path $q$, where $e$ is counted positively and $\bar{e}$ is counted negatively.

Let $G$ be an $A$-generated group (meaning generated by $A$ as a group, not necessarily as a semigroup). Let $E^+$ be the set of positively oriented edges of the Cayley graph $\Gamma(G, A)$, so $E^+ = G \times A$. Let $\eta_G$ be the associated labeling of $\Gamma(G, A)$ over $\mathbb{Z}_p E^+$. The natural left action of $G$ on $E^+$, given by $g(g_0, a) = (gg_0, a)$, extends uniquely to a left action of $G$ by automorphisms on $\mathbb{Z}_p E^+$. Let $G^{\mathbf{Ab}(p)}$ be the following $A$-generated group:

$$G^{\mathbf{Ab}(p)} = \langle (e\eta_G, [e\iota]_G) \mid e \in E^+, e\iota = 1 \rangle \leq \mathbb{Z}_p E^+ \rtimes G. \qquad (4.46)$$

It is an extension of an elementary abelian $p$-group by $G$. The reason for the notation is that $\mathbf{Ab}(p)$ denotes the pseudovariety of elementary abelian $p$-groups.

It will be notationally convenient to denote, for $g \in G$ and $a \in A$, the edge $\overline{(g[a^{-1}]_G, a)}$ by $(g, a^{-1})$. Moreover, if $g' \in G$, then $g'(g, a^{-1}) = (g'g, a^{-1})$. The next exercise explains why we have chosen this notation.

**Exercise 4.18.1.** Verify that if $a \in A$, then

$$((1, a)\eta_G, [a]_G)^{-1} = ((1, a^{-1})\eta_G, [a^{-1}]_G).$$

**Proposition 4.18.2.** Let $w \in FG(A)$ and let $p_w$ denote the unique reduced path in $\Gamma(G, A)$ from $1$ to $[w]_G$ labeled by $w$. Then $[w]_{G^{\mathbf{Ab}(p)}} = (p_w \eta_G, [w]_G)$.

*Proof.* Let $w = a_1 \cdots a_n$, with $a_1, \ldots, a_n \in A \cup A^{-1}$, be a reduced factorization. Then, taking into account Exercise 4.18.1,

$$[w]_{G^{\mathbf{Ab}(p)}} = ((1, a_1)\eta_G, [a_1]_G)((1, a_2)\eta_G, [a_2]_G) \cdots ((1, a_n)\eta_G, [a_n]_G)$$
$$= ((1, a_1)\eta_G + ([a_1]_G, a_2)\eta_G + \cdots + ([a_1 \cdots a_{n-1}]_G, a_n)\eta_G, [w]_G)$$
$$= (p_w\eta_G, [w]_G)$$

as required.                                                                 □

The assignment $G \mapsto G^{\mathbf{Ab}(p)}$ is an expansion of the sort considered in [87]. It has the nice property that it makes the Cayley graph become closer to a tree by separating paths. For instance, if $u, w$ label vertex simple paths in the Cayley graph of $G$ to the same vertex, then they label vertex simple paths in the Cayley graph of $G^{\mathbf{Ab}}(p)$ to distinct vertices. This expansion is hence a group theoretic analogue of the McCammond expansion from geometric semigroup theory [202].

It is well-known (c.f [87, Thm. 4.3], [286, Thm. 10.1], [38, Thm. 5.2]) that $G^{\mathbf{Ab}(p)}$ enjoys the following universal property: for any $A$-generated extension $H$ of an elementary abelian $p$-group by $G$, the canonical morphism $\alpha \colon G^{\mathbf{Ab}(p)} \to G$ factors through $H$. In the parlance of Chapter 3, $\alpha$ is the free $A$-generated pro-$\mathbf{Ab}(p)_D$ relational morphism to $G$. We shall not use the universal property in this book.

The power of the expansion $G \mapsto G^{\mathbf{Ab}(p)}$ is demonstrated in the following topological lemma extracted from [39], whose proof has origins in [33, 34, 37]. The Ribes and Zalesskii Theorem will be an easy consequence of the lemma. We encourage the reader to draw pictures.

**Lemma 4.18.3.** *Let $\Gamma$ be a finite inverse $A$-graph and $\widetilde{\Gamma}$ a finite cover extending $\Gamma$. Let $G$ be the monodromy group of $\widetilde{\Gamma}$ and let $m_1, \ldots, m_{n-1}$ be (not necessarily distinct) primes. Define inductively $G_0 = G$ and $G_k = G_{k-1}^{\mathbf{Ab}(m_k)}$, for $1 \le k \le n-1$. Suppose that $p_1, \ldots, p_n$ are reduced paths in $\Gamma$ such that*

$$[p_1\ell \cdots p_n\ell]_{G_{n-1}} = 1.$$

*Then there exist paths $q_1, \ldots, q_n$ in $\Gamma$ such that: $q_i$ is coterminal with $p_i$, for all $i = 1, \ldots, n$, and $q_1\ell \cdots q_n\ell = 1$.*

*Proof.* Throughout the proof we shall use $\ell$ for all labelings; no confusion should arise. The proof proceeds by induction on $n$. If $n = 1$, then $[p_1]_G = 1$ and so $p_1$ must be a circuit at a vertex $v$. Taking $q_1$ to be the empty path at $v$ then does the job.

Next we consider $n = 2$. Let $w_i = p_i\ell$, for $i = 1, 2$. So $[w_1w_2]_{G_1} = 1$ and therefore $[w_1w_2]_G = 1$. Let $\Delta_1$ and $\Delta_2$ be the subgraphs of the Cayley graph $\Gamma(G, A)$ of $G$ spanned by the reduced path $p_{w_1} \colon 1 \to [w_1]_G$ labeled by $w_1$ and by the reduced path $p_{w_2} \colon [w_1]_G \to 1$ labeled by $w_2$, respectively.

We claim the intersection $\Delta_1 \cap \Delta_2$ contains a reduced path $q \colon 1 \to [w_1]_G$. Suppose by way of contradiction that this is not the case. Let $\Upsilon$ be the connected component of $1$ in $\Delta_1 \cap \Delta_2$. Consider the "coboundary" of $\Upsilon$ in $\Delta_1$:

$$\Omega = \{e \in \Delta_1 \mid e\iota \in \Upsilon, \ e\tau \in \Delta_1 \setminus \Delta_2\}.$$

As usual, $\overline{\Omega}$ denotes the set of inverse edges to the edges of $\Omega$. Let $m$ be the number of edges from $\Omega$ and $\overline{m}$ be the number of edges from $\overline{\Omega}$ appearing in $p_{w_1}$. As $p_{w_1}: 1 \to [w_1]_G$ in $\Delta_1$ and $1 \in \Upsilon$, $[w_1]_G \notin \Upsilon$, we must have $m - \overline{m} = 1$ (that is, $p_{w_1}$ must leave $\Upsilon$ one more time than it enters). Let $\Omega^+ = \Omega \cap E^+$ and $\overline{\Omega}^+ = \overline{\Omega} \cap E^+$. Then

$$1 = m - \overline{m} \equiv \left[ \sum_{e \in \Omega^+} p_{w_1}(e) - \sum_{e \in \overline{\Omega}^+} p_{w_1}(e) \right] \mod m_1. \qquad (4.47)$$

It follows from (4.47) that, for some positively oriented edge $e \in \Omega \cup \overline{\Omega}$, $p_{w_1}(e) \neq 0$ (that is, the number of signed traversals of $e$ by $p_{w_1}$ is not divisible by the prime $m_1$). Because $e \notin \Delta_2$, by the definition of $\Omega$, the geometric edge $\{e, \overline{e}\}$ is not used by the path $p_{w_2}$. Thus the coefficient of $e$ in $(p_{w_1} p_{w_2}) \eta_G$ is non-zero (in fact equals $p_{w_1}(e)$) and hence, by Proposition 4.18.2, $[w_1 w_2]_{G_1} = ((p_{w_1} p_{w_2}) \eta_G, [w_1 w_2]_G) \neq 1$, a contradiction. It follows $\Delta_1 \cap \Delta_2$ contains a reduced path $q: 1 \to [w_1]_G$.

Because $p_1 \ell = w_1 = p_{w_1} \ell$ and $p_2 \ell = w_2 = p_{w_2} \ell$, Lemma 4.17.18 immediately implies that there are paths $q_1$ and $q_2$ in $\Gamma$ such that $q_i$ is coterminal with $p_i$, $i = 1, 2$, and $q_1 \ell = q\ell$, $q_2 \ell = \overline{q} \ell = (q\ell)^{-1}$. Obviously, $q_1 \ell q_2 \ell = 1$, completing the proof for $n = 2$.

Suppose now $n \geq 3$ and the lemma holds for all $i \in \{1, \ldots, n - 1\}$. Let $p_1, \ldots, p_n$ be reduced paths in $\Gamma$ with $[p_1 \ell \cdots p_n \ell]_{G_{n-1}} = 1$. For $i = 1, \ldots, n$, set $w_i = p_i \ell \in FG(A)$, $g_0 = 1$, $g_i = [w_1 \cdots w_i]_{G_{n-2}}$, and let $\Delta_i$ be the subgraph of the Cayley graph $\Gamma(G_{n-2}, A)$ spanned by the reduced path $p_{w_i}: g_{i-1} \to g_i$ labeled by $w_i$. Because $[w_1 \cdots w_n]_{G_{n-1}} = 1$, we have $g_n = 1$. So we can define $\Upsilon$ to be the connected component containing 1 of the graph $\Delta_1 \cap \Delta_n$. We claim there exists $i \in \{2, \ldots, n - 1\}$ such that $\Delta_i \cap \Upsilon \neq \emptyset$.

Indeed, suppose by way of contradiction that $\Delta_i \cap \Upsilon = \emptyset$ for all $i$ with $2 \leq i \leq n - 1$. Then, because $1 \in \Upsilon$ and $g_1 \in \Delta_1 \setminus \Upsilon$ (as $\Delta_2 \cap \Upsilon = \emptyset$), an argument analogous to the one used for the case $n = 2$ shows that there is a positively oriented edge $e \in \Delta_1$ such that: one endpoint of $e$ belongs to $\Upsilon$, the other endpoint belongs to $\Delta_1 \setminus \Delta_n$ and $p_{w_1}(e) \neq 0$ (that is, the number of signed traversals of $e$ by $p_{w_1}$ is not divisible by $m_{n-1}$). Because $e$ has a vertex in $\Upsilon$, our assumption implies $e \notin \Delta_i$ for all $2 \leq i \leq n - 1$. But also $e$ has a vertex not belonging to $\Delta_n$, so $e \notin \Delta_n$. Thus all occurrences of the geometric edge $\{e, \overline{e}\}$ in the path $p_{w_1} \cdots p_{w_n}$ come from $p_{w_1}$. It follows $(p_{w_1} \cdots p_{w_n}) \eta_{G_{n-2}}$ has a non-zero coefficient for $e$ (namely equal to $p_{w_1}(e)$) and so $[w_1 \cdots w_n]_{G_{n-1}} = ((p_{w_1} \cdots p_{w_n}) \eta_{G_{n-2}}, [w_1 \cdots w_n]_G) \neq 1$, a contradiction.

So, let $g$ be a vertex of $\Delta_i \cap \Upsilon$ for some $i \in \{2, \ldots, n - 1\}$. Let $u: g_{i-1} \to g$ be a reduced path contained in $\Delta_i$ and $x: g \to 1$ be a reduced path contained in $\Upsilon$. See Figure 4.5.

Because $\overline{x} \iota = 1 = p_{w_1} \iota$ and each edge of $\overline{x}$ is used by $p_{w_1}$, Lemma 4.17.18 provides a reduced path $x_0$ in $\Gamma$ with $x_0 \ell = \overline{x} \ell = (x\ell)^{-1}$ and $x_0 \iota = p_1 \iota$. Set $x' = \overline{x}_0$; then $x' \ell = x\ell$ and $x' \tau = p_1 \iota$. Similar applications of Lemma 4.17.18

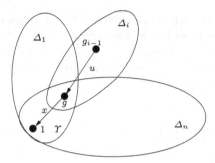

**Fig. 4.5.** Construction of $u$ and $x$

allow us to find reduced paths $u'$ and $x''$ in $\Gamma$ such that: $u'\ell = u\ell$, $u'\iota = p_i\iota$, $x''\ell = x\ell$ and $x''\tau = p_n\tau$. Let $p_1'$ be the reduced path homotopic to $x'p_1$, let $p_i'$ be the reduced path homotopic to $\overline{u'}p_i$ and let $p_n'$ be a reduced path homotopic to $p_n\overline{x''}$. Then $p_1'\ell = x\ell w_1$, $p_i'\ell = (u\ell)^{-1}w_i$ and $p_n'\ell = w_n(x\ell)^{-1}$. The reader may find the following pictures useful.

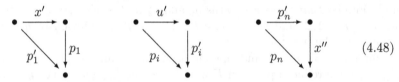

$$(4.48)$$

From Figure 4.5, it is evident that:

$$[(x\ell w_1) \cdot w_2 \cdots w_{i-1} \cdot u\ell)]_{G_{n-2}} = 1 = [((u\ell)^{-1}w_i) \cdot w_{i+1} \cdots (w_n(x\ell)^{-1})]_{G_{n-2}}.$$

Because $1 \le i-1, n-i \le n-2$, the groups $G_{i-1}$, $G_{n-i}$ are quotients of $G_{n-2}$. Therefore, we have

$$[(x\ell w_1) \cdot w_2 \cdots w_{i-1} \cdot u\ell]_{G_{i-1}} = 1$$
$$[((u\ell)^{-1}w_i) \cdot w_{i+1} \cdots (w_n(x\ell)^{-1})]_{G_{n-i}} = 1.$$

Applying the induction hypothesis to the sets of paths $p_1', p_2, \ldots, p_{i-1}, u'$ and to $p_i', p_{i+1}, \ldots p_{n-1}, p_n'$, respectively, yields paths $u_1, \ldots, u_i, u_i', u_{i+1}, \ldots, u_n$ in $\Gamma$ such that:

(1) $u_1$ is coterminal with $p_1'$, $u_i$ is coterminal with $u'$, $u_i'$ is coterminal with $p_i'$, and $u_n$ is coterminal with $p_n'$;
(2) $u_k$ is coterminal with $p_k$, for all $k \in \{2, \ldots, i-1, i+1, \ldots, n-1\}$;
(3) $u_1\ell \cdots u_i\ell = 1 = u_i'\ell u_{i+1}\ell \cdots u_n\ell$.

Let us set $q_1 = (x')^{-1}u_1$, $q_i = u_iu_i'$, $q_n = u_nx''$, and $q_k = u_k$ for $k \ne 1, i, n$. Then, by (1)–(3) and the choices of $x'$, $u'$, $x''$, $p_1'$, $p_i'$ and $p_n'$ (see (4.48)), one verifies, for all $j = 1, \ldots n$, that $q_j$ is a path in $\Gamma$ coterminal to $p_j$, and

$$q_1\ell \cdots q_n\ell = (x\ell)^{-1} \cdot u_1\ell \cdots u_i\ell \cdot u_i'\ell \cdot u_{i+1}\ell \cdots u_n\ell \cdot x\ell = 1.$$

This completes the proof of the lemma. $\qquad\qquad\qquad\qquad\qquad\qquad\square$

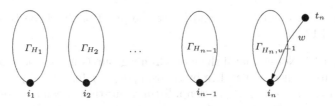

**Fig. 4.6.** The inverse $A$-graph $(\Gamma, \ell)$

## Proof of the Ribes and Zalesskii Theorem

Let $H_1, \ldots, H_n$ be finitely generated subgroups of $FG(A)$. To show that $H_1 \cdots H_n$ is closed, it suffices to find, for each $w \notin H_1 \cdots H_n$, a finite group $K$ and a homomorphism $\varphi \colon FG(A) \to K$, such that $w\varphi \notin (H_1 \cdots H_n)\varphi$. Then $N = \ker \varphi$ has finite index and $Nw$ is a neighborhood of $w$ disjoint from $H_1 \cdots H_n$.

Let $w \in FG(A)$, let $\Gamma_{H_1}, \ldots, \Gamma_{H_{n-1}}$ be the Stallings graphs associated to $H_1, \ldots, H_{n-1}$ and let $\Gamma_{H_n, w^{-1}}$ be the inverse $A$-graph associated to $H$ and $w^{-1}$ from Definition 4.17.16 and the discussion thereafter. Let $i_j$ be the base point of $\Gamma_{H_j}$, $j = 1, \ldots, n-1$, and let $i_n$ be the base point of $\Gamma_{H_n, w^{-1}}$. The endpoint of the unique reduced path in $\Gamma_{H_n, w^{-1}}$ starting at $i_n$ and labeled by $w^{-1}$ will be denoted by $t_n$. Because $\pi_1(\Gamma_{H_n, w^{-1}}, i_n) = H_n$, it is immediate that the coset $H_n w^{-1}$ consists of the set of all elements $h \in FG(A)$ labeling a path from $i_n$ to $t_n$ in $\Gamma_{H_n, w^{-1}}$. Note that $t_n \neq i_n$ if and only if $w \notin H_n$.

Next, let $(\Gamma, \ell)$ be the disjoint union of the inverse $A$-graphs $\Gamma_{H_1}, \ldots, \Gamma_{H_{n-1}}$ and $\Gamma_{H_n, w^{-1}}$ as per Figure 4.6. Let $(\widetilde{\Gamma}, \widetilde{\ell})$ be a finite covering extending $(\Gamma, \ell)$ as per Lemma 4.17.17 with monodromy group $G$. Let $m_1, \ldots, m_{n-1}$ be a sequence of not necessarily distinct primes and set $G_0 = G$, $G_k = G_{k-1}^{\mathbf{Ab}(m_k)}$, for $1 \leq k \leq n-1$. Denote by $\varphi \colon FG(A) \twoheadrightarrow G_{n-1}$ the canonical projection. Suppose that $w \notin H_1 \cdots H_n$. We claim that $w\varphi \notin (H_1 \cdots H_n)\varphi$. Indeed, suppose $w\varphi \in (H_1 \cdots H_n)\varphi$. Then $[w]_{G_{n-1}} = [h_1 \cdots h_n]_{G_{n-1}}$ where $h_i \in H_i$, $i = 1, \ldots, n$. As $\pi_1(\Gamma_{H_j}, i_j) = H_j$, there are reduced paths $p_1, \ldots, p_n$ in $\Gamma$ such that:

- $p_j \colon i_j \to i_j$, $j = 1, \ldots, n-1$;
- $p_j \ell = h_j$, $j = 1, \ldots, n-1$;
- $p_n \colon i_n \to t_n$ with $p_n \ell = h_n w^{-1}$.

As we have

$$[p_1 \ell \cdots p_n \ell]_{G_{n-1}} = [h_1 \cdots h_n w^{-1}]_{G_{n-1}} = 1,$$

Lemma 4.18.3 implies there are paths $q_1, \ldots, q_n$ in $\Gamma$ so that $q_i$ is coterminal with $p_i$, $i = 1, \cdots, n$, and $q_1 \ell \cdots q_n \ell = 1$. But then $q_j \ell \in \pi_1(\Gamma_j, i_j) = H_j$, for $j = 1, \ldots, n-1$ and $q_n \ell \in H_n w^{-1}$ as $q_n \colon i_n \to t_n$. We conclude that $1 \in H_1 \cdots H_n w^{-1}$ and so $w \in H_1 \cdots H_n$, a contradiction. This completes the proof of Theorem 4.17.23 and hence the proof of the Type II Theorem (Theorem 4.17.30). □

*Remark 4.18.4.* Notice that if the monodromy group $G$ in the above proof is a $p$-group, where $p$ is a prime, then by choosing $m_k = p$ for all $1 \leq j \leq n-1$, we can guarantee that $G_{n-1}$ is a $p$-group. It turns out that $G$ can be chosen to be a $p$-group if and only if $H_1, \ldots, H_n$ are closed in the pro-$p$ topology on $FG(A)$. In fact, the above proof can be adapted to show that if $\mathbf{H}$ is any pseudovariety of groups with the property that, for each $G \in \mathbf{H}$, there is a prime $p$ such that $G^{\mathbf{Ab}(p)} \in \mathbf{H}$, then products of pro-$\mathbf{H}$ closed subgroups of $FG(A)$ are again closed. This generalizes the result of Ribes and Zalesskii [301] for the extension-closed case. See [39] for details.

## 4.19 Henckell's Theorem

In [121], Henckell showed that aperiodic pointlikes are computable; the companion result for groups was proved by Ash [33]. The proof of Henckell's Theorem presented here is adapted from [129]. The reader is once again referred to the discussion just before Example 2.4.7 or to Section 3.6.2 for the definition and properties of pointlike sets.

Let $T$ be a semigroup. If $Z \in P(T)$, define $Z^{\omega+*} = Z^\omega \bigcup_{n \geq 1} Z^n$. Because products distribute over union in $P(T)$, it follows easily that

$$ZZ^{\omega+*} = Z^{\omega+*} = Z^{\omega+*}Z. \tag{4.49}$$

One deduces immediately from (4.49) that $Z^{\omega+*}$ is an idempotent. Let us prove a crucial observation of Henckell [121]

**Proposition 4.19.1.** *Let $\mathbf{V}$ be a pseudovariety of aperiodic semigroups. Then $\mathsf{PL}_{\mathbf{V}}(T)$ is closed under the operation $Z \mapsto Z^{\omega+*}$.*

*Proof.* Let $\varphi \colon T \to S$ be a relational morphism with $S \in \mathbf{V}$. Choose $s \in S$ with $Z \subseteq s\varphi^{-1}$. Then $Z^\omega Z^n \subseteq s^\omega s^n \varphi^{-1} = s^\omega \varphi^{-1}$, for all $n \geq 1$. Hence $Z^{\omega+*} \subseteq s^\omega \varphi^{-1}$. We conclude $Z^{\omega+*} \in \mathsf{PL}_{\mathbf{V}}(T)$.     □

Our goal is to prove Henckell's Theorem describing the aperiodic pointlike sets [121, 129].

**Theorem 4.19.2 (Henckell).** *Let $T$ be a finite semigroup. Denote by $CP(T)$ the smallest subsemigroup of $P(T)$ containing the singletons and closed under $Z \mapsto Z^{\omega+*}$. Then $\mathsf{PL}_{\mathbf{A}}(T)$ consists of all $X \in P(T)$ with $X \subseteq Y$ some $Y \in CP(T)$. In particular, pointlike sets are decidable for $\mathbf{A}$.*

Proposition 4.19.1 shows that $CP(T) \subseteq \mathsf{PL}_{\mathbf{A}}(T)$ and hence each of the subsets described in Theorem 4.19.2 is indeed $\mathbf{A}$-pointlike. The hard part of the theorem is proving the converse. A generalization of Henckell's Theorem to pseudovarieties of the form $\overline{\mathbf{G}}_\pi$, with $\pi$ a set of primes, was obtained by the authors with Henckell [129].

Notice the decidability of $\mathbf{A}$-pointlikes implies that the pseudovariety of relational morphisms $\mathbf{A}_\vee$ is decidable. This implies in turn the following result of the second author [322, 330].

**Corollary 4.19.3.** *Let* $\mathbf{V}$ *be a locally finite pseudovariety so that the order of* $F_\mathbf{V}(A)$ *is computable as a function of* $|A|$ *(for example, if* $\mathbf{V}$ *is compact). Then* $\mathbf{V} \vee \mathbf{A}$ *has decidable membership.*

*Proof.* According to Proposition 2.3.26, an $A$-generated semigroup $S$ belongs to $\mathbf{V} \vee \mathbf{A}$ if and only if the canonical relational morphism $\rho_\mathbf{V} \colon S \to F_\mathbf{V}(A)$ belongs to $\mathbf{A}_\vee$. Our hypotheses and Henckell's Theorem guarantee this is computable.                                            $\square$

For instance, $\mathbf{A} \vee (\mathbb{Z}_2)$ is decidable, a result of the second author answering a question first posed by Rhodes and M. Volkov.[1] More generally, it is a consequence of Zel'manov's solution to the restricted Burnside problem that the pseudovariety $[\![x^n = 1]\!]$ (and hence all its subpseudovarieties) are locally finite, and in fact computable bounds are known for the sizes of free objects in terms of the number of generators (assuming decidability for the case of a subpseudovariety) [111, 371, 388, 389]. Hence we have the following corollary.

**Corollary 4.19.4.** *Let* $\mathbf{H}$ *be a decidable pseudovariety of groups of bounded exponent. Then* $\mathbf{A} \vee \mathbf{H}$ *is decidable.*

In [36], Auinger and Steinberg provide an example of a pseudovariety $\mathbf{H}$ of metabelian groups with decidable membership such that $\mathbf{A} \vee \mathbf{H}$ is undecidable. An important open question, originally posed by Schützenberger, is the following.

*Question 4.19.5 (Schützenberger's Question).* Does the pseudovariety $\mathbf{A} \vee \mathbf{G}$ have decidable membership?

McAlister showed that membership in $\mathbf{A} \vee \mathbf{G}$ is decidable for regular semigroups [200]. It was later pointed out by the second author [334] that a proof of McAlister's result, in greater generality, could already be deduced from the results of [295]. Namely, the results there imply that, for a subdirect product $S$ of subdirectly indecomposable semigroups of semisimple type, one has $S \in \mathbf{A} \vee \mathbf{G}$ if and only if $S/(\mathsf{RLM} \cap \mathsf{LLM}) \in \mathbf{A}$ and $\mathsf{K}_\mathbf{G}(S) \in \mathbf{A}$. In light of the results of [156] (our Theorem 4.11.4), these latter conditions are equivalent to saying $S \in (\mathbf{G} * \mathbf{A}) \cap (\mathbf{G} * \mathbf{A})^{op} \cap \mathbf{A} * \mathbf{G}$. As any regular semigroup $S$ is a subdirect product of subdirectly indecomposable semigroups of semisimple type, this implies McAlister's result. On the other hand, Volkov proved $\mathbf{A} \vee \mathbf{G}$ is not finitely based [373]; see also Theorem 4.13.33.

We now turn to the proof of Henckell's Theorem. The proof is highly technical and can be skipped on a first reading.

---

[1] Private communication.

### 4.19.1 An aperiodic variant of the Rhodes expansion

Our goal in this section is to associate a finite aperiodic semigroup $S^{\mathscr{A}}$ to each finite semigroup $S$. The case $S = CP(T)$ will yield an aperiodic semigroup and a relational morphism that establishes Theorem 4.19.2.

Fix a finite semigroup $S$ for this section. Elements of the free monoid $S^*$ will be written as strings $\mathbf{x} = (x_n, x_{n-1}, \ldots, x_1)$. The empty string is denoted $\varepsilon$. We omit parentheses for strings of length 1. If $n \geq \ell$, define

$$(x_n, x_{n-1}, \ldots, x_1)\alpha_\ell = (x_\ell, x_{\ell-1}, \ldots, x_1)$$

where we identify $\mathbf{x}\alpha_1$ with the "first" letter of $\mathbf{x}$. By convention $\mathbf{x}\alpha_0 = \varepsilon$. Set $(x_n, \ldots, x_1)\omega = x_n$. We use $\mathbf{b} \cdot \mathbf{a}$ for the concatenation of $\mathbf{b}$ and $\mathbf{a}$. As the notation suggests, we read strings from right to left.

If $P$ is a pre-ordered set, then a *flag* of elements of $P$ is a strict chain $p_n < p_{n-1} < \cdots < p_1$. We also allow an empty flag. Denote by $\mathscr{F}(S)$ the set of flags for the $\mathscr{L}$-order on $S$. Of course, $\mathscr{F}(S)$ is a finite set. A typical flag $s_n <_{\mathscr{L}} s_{n-1} <_{\mathscr{L}} \cdots <_{\mathscr{L}} s_1$ shall be denoted $(s_n, s_{n-1}, \ldots, s_1)$. We shall also consider the set $\overline{\mathscr{F}}(S)$ of $\mathscr{L}$-chains, that is, all strings $(s_n, s_{n-1}, \ldots, s_1) \in S^*$ (including the empty string) such that $s_{i+1} \leq_{\mathscr{L}} s_i$ for all $i$. Of course, the inclusions $\mathscr{F}(S) \subseteq \overline{\mathscr{F}}(S) \subseteq S^*$ hold.

We call an element $s \in S$ *aperiodic* if its $\mathscr{H}$-class is a singleton. Likewise, a string $(s_n, \ldots, s_1)$ shall be called *aperiodic* if each $s_i$ is aperiodic. We use $\mathscr{F}_{\mathscr{A}}(S)$ and $\overline{\mathscr{F}}_{\mathscr{A}}(S)$ to denote the respective subsets of $\mathscr{F}(S)$ and $\overline{\mathscr{F}}(S)$ consisting of aperiodic strings.

There is a natural retraction from $\overline{\mathscr{F}}(S)$ to $\mathscr{F}(S)$ (mapping $\overline{\mathscr{F}}_{\mathscr{A}}(S)$ onto $\mathscr{F}_{\mathscr{A}}(S)$), which we proceed to define. An *elementary reduction* is a rule of the form $(s', s) \to s'$ where $s' \mathscr{L} s$. Elementary reductions are length-decreasing. It is well-known and easy to prove that the elementary reductions form a confluent rewriting system and so each element $\mathbf{x} \in \overline{\mathscr{F}}(S)$ can be reduced to a unique flag $\mathbf{x}\rho \in \mathscr{F}(S)$, called its *reduction* [362]. Clearly, the reduction map is a retract and takes $\overline{\mathscr{F}}_{\mathscr{A}}(S)$ to $\mathscr{F}_{\mathscr{A}}(S)$. The Rhodes expansion defines a multiplication on $\mathscr{F}(S)$ via the reduction map [110, 362]. Our constructions are motivated by properties of the Rhodes expansion, but we shall not need this expansion per se. Some key properties of the reduction map, which are immediate from the definition, are recorded in the following lemma.

**Lemma 4.19.6.** *The reduction map $\rho$ enjoys the following properties:*

1. *For $\mathbf{x} \in \overline{\mathscr{F}}(S)$, $\mathbf{x}\rho\omega = \mathbf{x}\omega$;*
2. *Let $\mathbf{b}, \mathbf{a}, \mathbf{b} \cdot \mathbf{a} \in \overline{\mathscr{F}}(S)$ and suppose $\mathbf{a}\rho = (a_\ell, a_{\ell-1}, \ldots, a_1)$. Then one has $(\mathbf{b} \cdot \mathbf{a})\rho\alpha_\ell = (x_\ell, a_{\ell-1}, \ldots, a_1)$ where $x_\ell \mathscr{L} a_\ell$.*

Let us now turn to defining an auxiliary semigroup that will play a role in the proof. Denote by $\check{S}$ the monoid of all functions $f\colon S \to S$ such that $sf \leq_{\mathscr{R}} s$ for all $s \in S$. Notice that the faithful quotient of the natural action of $S^I$ on the right of $S$ consists of elements of $\check{S}$.

Next we wish to discuss a formalism for infinite iterated wreath products due to Grigorchuk and Nekrashevych [109,216]. Let $(X,T)$ be a finite faithful right transformation semigroup and denote by $T^\infty$ the action monoid of the infinite wreath product $\wr^\infty(X,T)$ of right transformation monoids $(X,T)$. There is a natural action of $T^\infty$ on the free monoid $X^*$ by length-preserving sequential functions via the projections $\wr^\infty(X,T) \to \wr^n(X,T)$; to obtain the action on a word of length $n$, project first to $\wr^n(X,T)$ and then act. If $F \in T^\infty$ and $\mathbf{a} \in X^*$, then there is a unique element $_\mathbf{a}F \in T^\infty$ such that $(\mathbf{b}\cdot\mathbf{a})F = \mathbf{b}_\mathbf{a}F\cdot\mathbf{a}F$ for all $\mathbf{b} \in X^*$; for example, $_\varepsilon F = F$. Moreover, if $F \in T^\infty$, then there is a unique element $\sigma_F \in T$ such that, for $x \in X$, the equality $xF = x\sigma_F$ holds. In particular,

$$(x_n, \ldots, x_1)F = (x_n, \ldots, x_2)_{x_1}F \cdot x_1\sigma_F. \tag{4.50}$$

So $\sigma_F$ describes the action on the first letter, and must belong to $T$, whereas $_{x_1}F \in T^\infty$ is how $F$ acts on the rest of a string starting with $x_1$. In fact, (4.50) can serve as a recursive definition of what it means to belong to $T^\infty$ (cf. [109,216]).

Let us consider some examples to illustrate this formalism for infinite iterated wreath products, in particular the wreath recursion (4.50). For simplicity, we work with $\wr^\infty(\{0,1\}, T_2)$. Notice that the iterated wreath product $T_2^\infty$ is isomorphic to the semidirect product $(T_2^\infty)^2 \rtimes T_2$. As a first example, consider the 2-adic odometer, which adds one to the 2-adic expansion of an integer (where the least significant bit is the first one read from *right to left*) [109,216].

*Example 4.19.7 (Odometer).* Let $A$ be the 2-adic odometer considered above, acting on $\{0,1\}^*$, and let $I$ be the identity function on $\{0,1\}^*$. If a 2-adic integer has 0 as its least significant bit, we change the 0 to a 1 and then continue with the identity map the rest of the way; if the least significant bit is 1, we change it to 0 and we add 1 to what remains (i.e., perform a carry). So in terms of the wreath recursion (4.50), $\sigma_A = (01)$ and $_0A = I$, $_1A = A$. As a sample computation, consider

$$101A = 10A0 = 1I10 = 110.$$

If we identify $T_2^\infty$ with $(T_2^\infty)^2 \rtimes T_2$, then $A = ((I, A), (01))$.

Next we consider the two sections to the unilateral shift.

*Example 4.19.8 (Shift).* Consider the functions $F$, $G$ on $\{0,1\}^*$ that send $x_n x_{n-1} \cdots x_1$ to, respectively, $x_{n-1}x_{n-2}\cdots x_1 0$ and $x_{n-1}x_{n-2}\cdots x_1 1$. Both of these functions act by remembering the first letter, then resetting it to a predetermined symbol, and then resetting the second letter to the first and so on and so forth. Formally, the wreath recursion (4.50) is given by $\sigma_F = \bar{0}$, $\sigma_G = \bar{1}$ (where $\bar{x}$ is the constant map to $x$) and $_0F = F = {}_0G$, $_1F = G = {}_1G$. Identifying $T_2^\infty$ with $(T_2^\infty)^2 \rtimes T_2$, we have $F = ((F,G), \bar{0})$ and $G = ((F,G), \bar{1})$. So, for example, we compute

$$101F = 10G0 = 1F10 = 010.$$

Notice that on infinite bit strings, $F$ and $G$ are the two sections to the unilateral shift that erases the first letter of a left infinite string.

From these examples, the reader familiar with sequential functions should see the connection between iterated wreath products and sequential functions [84, 85, 109, 216].

**Exercise 4.19.9.** Show that $F, G$ from Example 4.19.8 freely generate a free semigroup of rank 2.

Again let $(X, T)$ be a right transformation semigroup. A subsemigroup $U$ of $T^\infty$ is called *self-similar* [216] if, for all $F \in U$ and $\mathbf{a} \in X^*$, one has $_\mathbf{a}F \in U$; so $T^\infty$ itself is self-similar. It is actually enough that, for each letter $x \in X$, one has $_xF \in U$. For instance, the monoid generated by the 2-adic odometer $A$ is self-similar in $\ell^\infty(\{0,1\}, T_2)$ because $_0A = I$, $_1A = A$ (in fact, the group generated by $A$ is self-similar). Similarly, the semigroup generated by the two sections $F, G$ to the shift is self-similar as $_0F = F = {}_0G$, $_1F = G = {}_1G$. We are interested in the case where $(X, T)$ is $(S, \check{S})$.

**Definition 4.19.10 ($\check{S}_0^\infty$).** *Denote by $\check{S}_0^\infty$ the collection of all transformations $F \in \check{S}^\infty$ such that whenever $(x_n, x_{n-1}, \ldots, x_1)F = (y_n, y_{n-1}, \ldots, y_1)$ with $n \geq 2$, $x_{n-1} \mathrel{\mathscr{R}} y_{n-1}$ and $x_n \mathrel{\mathscr{R}} y_n$, there exists $s \in S^I$ with $x_{n-1}s = y_{n-1}$ and $x_n s = y_n$.*

The element $s$ above can depend on the string $(x_n, \ldots, x_1)$.

**Proposition 4.19.11.** $\check{S}_0^\infty$ *is a self-similar submonoid of* $\check{S}^\infty$.

*Proof.* Clearly, it contains the identity. Suppose $F, G \in \check{S}_0^\infty$ and

$$(x_n, x_{n-1}, \ldots, x_1)FG = (z_n, z_{n-1}, \ldots, z_1)$$

with $x_{n-1} \mathrel{\mathscr{R}} z_{n-1}$ and $x_n \mathrel{\mathscr{R}} z_n$. Suppose that $(x_n, x_{n-1}, \ldots, x_1)F = (y_n, y_{n-1}, \ldots, y_1)$. Then, for $i = n-1, n$, we have $x_i \geq_{\mathscr{R}} y_i \geq_{\mathscr{R}} z_i \mathrel{\mathscr{R}} x_i$. Thus $x_i \mathrel{\mathscr{R}} y_i$ and $y_i \mathrel{\mathscr{R}} z_i$, $i = n-1, n$. By assumption, there exists $s, t \in S$ with $x_i s = y_i$ and $y_i t = z_i$, $i = n-1, n$. Then $x_i st = z_i$, for $i = n-1, n$. Hence $\check{S}_0^\infty$ is submonoid of $\check{S}^\infty$. Self-similarity is immediate from the equation $(x_n, \ldots, x_1)_\mathbf{a}F \cdot \mathbf{a}F = ((x_n, \ldots, x_1) \cdot \mathbf{a})F$ and the definition of $\check{S}_0^\infty$.     $\square$

If $s \in S^I$, define the diagonal operator $\Delta_s \colon S^* \to S^*$ by

$$(x_n, x_{n-1}, \ldots, x_1)\Delta_s = (x_n s, x_{n-1}s, \ldots, x_1 s). \qquad (4.51)$$

It is immediate that $\Delta_s$ belongs to $\check{S}_0^\infty$.

The following lemma shows that $\check{S}_0^\infty$ has the so-called Zeiger property.

**Lemma 4.19.12 (Zeiger property).** *Suppose* $\mathbf{x} = (x_n, \ldots, x_1) \in \overline{\mathscr{F}}(S)$ *and* $F \in \check{S}_0^\infty$ *are such that* $\mathbf{x}F = (y_n, y_{n-1}, \ldots, y_1)$ *with* $x_{n-1} = y_{n-1}$ *and* $x_n \mathscr{R} y_n$. *Then* $x_n = y_n$.

*Proof.* By definition of $\check{S}_0^\infty$, there exists $s \in S^I$ such that $x_{n-1}s = y_{n-1} = x_{n-1}$ and $x_n s = y_n$. Because $x_n \leq_\mathscr{L} x_{n-1}$, we can write $x_n = ux_{n-1}$ with $u \in S^I$. Then $y_n = x_n s = ux_{n-1}s = ux_{n-1} = x_n$, as required. $\qquad\square$

We are now ready to define an important transformation semigroup on $\overline{\mathscr{F}}(S)$.

**Definition 4.19.13 ($\mathscr{C}$).** *Let* $\mathscr{C}$ *consist of all transformations* $f \colon S^* \to S^*$ *such that* $\overline{\mathscr{F}}(S)f \subseteq \overline{\mathscr{F}}(S)$ *and there exists* $(\widehat{f}, \overline{f}) \in \check{S}_0^\infty \times (\overline{\mathscr{F}}(S) \setminus \{\varepsilon\})$ *with* $\mathbf{x}f = \mathbf{x}\widehat{f} \cdot \overline{f}$.

Observe that $\overline{f}$ is uniquely determined by the equation $\varepsilon f = \overline{f}$ and hence $\widehat{f}$ is determined by the equality $\mathbf{x}f = \mathbf{x}\widehat{f} \cdot \overline{f} = \mathbf{x}\widehat{f} \cdot \varepsilon f$. For instance, if $s \in S$, then one readily checks that $(\Delta_s, s)$ defines an element of $\mathscr{C}$ via the formula:

$$(x_n, x_{n-1}, \ldots, x_1)(\Delta_s, s) = (x_n s, x_{n-1}s, \ldots, x_1 s, s).$$

Such elements correspond to generators of the Rhodes expansion [362].

*Remark 4.19.14.* The reason we use the notation $\overline{f}$ here is that $\overline{f}$ is the "constant" part of $(\widehat{f}, \overline{f})$ in the sense that $\overline{f}$ is appended to the right-hand side of a string independently of the string in question. This is one of the key ideas behind the Rhodes expansion.

Notice that in order for $(\widehat{f}, \overline{f})$ to define an element of $\mathscr{C}$, one must have $x\widehat{f} \leq_\mathscr{L} \overline{f}\omega$ for every $x \in S$. It is essential that $\overline{f}$ is not empty in the definition of $\mathscr{C}$.

**Proposition 4.19.15.** *The set* $\mathscr{C}$ *is a semigroup.*

*Proof.* The first condition in the definition of $\mathscr{C}$ is evidently preserved by composition. Let $f, g \in \mathscr{C}$. We claim $\widehat{fg} = \widehat{f}_{\overline{f}}\widehat{g}$ and $\overline{fg} = (\overline{f})\widehat{g} \cdot \overline{g}$. Indeed,

$$\mathbf{x}fg = (\mathbf{x}\widehat{f} \cdot \overline{f})g = (\mathbf{x}\widehat{f} \cdot \overline{f})\widehat{g} \cdot \overline{g} = \mathbf{x}\widehat{f}_{\overline{f}}\widehat{g} \cdot (\overline{f})\widehat{g} \cdot \overline{g}.$$

As $\check{S}_0^\infty$ is self-similar, $_{\overline{f}}\widehat{g} \in \check{S}_0^\infty$ and so $fg \in \mathscr{C}$. $\qquad\square$

An immediate consequence of the definition is

**Lemma 4.19.16.** *If* $f \in \mathscr{C}$ *and* $\mathbf{a}, \mathbf{b}, \mathbf{b} \cdot \mathbf{a} \in \overline{\mathscr{F}}(S)$, *then* $(\mathbf{b} \cdot \mathbf{a})f = \mathbf{b}_\mathbf{a}\widehat{f} \cdot \mathbf{a}f$.

*Proof.* Indeed, a straightforward computation yields

$$(\mathbf{b} \cdot \mathbf{a})f = (\mathbf{b} \cdot \mathbf{a})\widehat{f} \cdot \overline{f} = \mathbf{b}_\mathbf{a}\widehat{f} \cdot (\mathbf{a}\widehat{f} \cdot \overline{f}) = \mathbf{b}_\mathbf{a}\widehat{f} \cdot \mathbf{a}f$$

proving the lemma. $\qquad\square$

Lemma 4.19.16 shows that $\mathscr{C}$ behaves very much like a wreath product, a property that we shall exploit repeatedly. In fact, an element $f$ of $\mathscr{C}$ is like an asynchronous transducer that outputs $\overline{f}$ with empty input and then computes synchronously. We are almost prepared to define our aperiodic semigroup. Recall that $\rho$ denotes the reduction map.

**Definition 4.19.17 ($\mathscr{C}^{\mathscr{A}}$).** *Let $\mathscr{C}^{\mathscr{A}}$ be the subset of $\mathscr{C}$ consisting of all transformations $f \in \mathscr{C}$ such that:*

1. $\overline{\mathscr{F}}_{\mathscr{A}}(S)f \subseteq \overline{\mathscr{F}}_{\mathscr{A}}(S)$;
2. $\mathbf{x}\rho f\rho = \mathbf{x}f\rho$ *for all* $\mathbf{x} \in \overline{\mathscr{F}}(S)$.

**Proposition 4.19.18.** *The set $\mathscr{C}^{\mathscr{A}}$ is a semigroup.*

*Proof.* Closure of the first item under composition is clear. The computation

$$\mathbf{x}\rho fg\rho = \mathbf{x}\rho f(\rho g\rho) = \mathbf{x}(\rho f\rho)g\rho = \mathbf{x}f\rho g\rho = \mathbf{x}fg\rho \qquad (4.52)$$

for $\mathbf{x} \in \overline{\mathscr{F}}(S)$ completes the proof that $\mathscr{C}^{\pi}$ is a semigroup.  □

Notice (4.52) allows us to define an action of $\mathscr{C}^{\mathscr{A}}$ on $\mathscr{F}_{\mathscr{A}}(S)$ by $\mathbf{x} \mapsto \mathbf{x}f\rho$ for $f \in \mathscr{C}^{\mathscr{A}}$. Let us denote the resulting faithful transformation semigroup by $(\mathscr{F}_{\mathscr{A}}(S), S^{\mathscr{A}})$. Observe that $\overline{\mathscr{F}}_{\mathscr{A}}(S)$ appears in Definition 4.19.17, whereas in the definition of $S^{\mathscr{A}}$ we use $\mathscr{F}_{\mathscr{A}}(S)$. Because $\mathscr{F}_{\mathscr{A}}(S)$ is finite, so is $S^{\mathscr{A}}$. Our goal is to prove $S^{\mathscr{A}}$ is aperiodic. First we make a simple observation.

**Lemma 4.19.19.** *Let $f \in \mathscr{C}$ and $\varepsilon \neq \mathbf{x} \in \overline{\mathscr{F}}(S)$. Then*

$$\mathbf{x}f\rho\omega = \mathbf{x}f\omega = \mathbf{x}\widehat{f}\omega \leq_{\mathscr{R}} \mathbf{x}\omega.$$

*Proof.* By definition of $\mathscr{C}$, we have $\mathbf{x}f\omega = (\mathbf{x}\widehat{f} \cdot \overline{f})\omega = \mathbf{x}\widehat{f}\omega$. Because we have $\widehat{f} \in \check{S}_0^{\infty} \subseteq \check{S}^{\infty}$, we conclude $\mathbf{x}\widehat{f}\omega \leq_{\mathscr{R}} \mathbf{x}\omega$. The lemma then follows from Lemma 4.19.6.  □

Notice that if $f \in \mathscr{C}^{\mathscr{A}}$, then $\mathbf{x}f\rho \neq \varepsilon$, for all $\mathbf{x} \in \overline{\mathscr{F}}(S)$. Indeed, $\overline{f} \neq \varepsilon$ implies $\mathbf{x}f\rho = (\mathbf{x}\widehat{f} \cdot \overline{f})\rho \neq \varepsilon$. Finally, we turn to the main result of this section.

**Theorem 4.19.20.** *Let $S$ be a finite semigroup. Then $S^{\mathscr{A}}$ is a finite aperiodic semigroup.*

*Proof.* To prove that $S^{\mathscr{A}}$ is aperiodic we must prove that, for all $f \in S^{\mathscr{A}}$, the equality $f^{n+1} = f$, for some $n \geq 2$, implies $f^2 = f$. In view of Definition 4.19.17, to do this it suffices to prove, for $f \in \mathscr{C}^{\mathscr{A}}$, the following:

(A) If $\varepsilon f^{n+1}\rho = \varepsilon f\rho$, then $\varepsilon f^2\rho = \varepsilon f\rho$;
(B) If $\mathbf{x}f^{n+1}\rho = \mathbf{x}f\rho$ for all $\mathbf{x} \in \mathscr{F}_{\mathscr{A}}(S)$, then $\mathbf{x}f^2\rho = \mathbf{x}f\rho$ all $\mathbf{x} \in \mathscr{F}_{\mathscr{A}}(S)$.

Let $f = (\widehat{f}, \overline{f}) \in \mathscr{C}^{\mathscr{A}}$ with $\widehat{f} \in \check{S}_0^{\infty}$ and $\overline{f} \in \mathscr{F}(S) \setminus \{\varepsilon\}$. As $\mathbf{x}f \neq \varepsilon$, if $\mathbf{x}f = \mathbf{x}f^{n+1}$ some $n \geq 1$, then Lemma 4.19.19 yields

$$\mathbf{x}f\omega \geq_{\mathscr{R}} \mathbf{x}f^2\omega \geq_{\mathscr{R}} \cdots \geq_{\mathscr{R}} \mathbf{x}f^n\omega \geq_{\mathscr{R}} \mathbf{x}f^{n+1}\omega = \mathbf{x}f\omega \qquad (4.53)$$

and so all the inequalities are in fact equivalences.

We now turn to the proof of (A). Suppose $\varepsilon f^{n+1}\rho = \varepsilon f\rho$ with $n \geq 2$. Then $\varepsilon f^k = \varepsilon f^{k-1} \widehat{f} \cdot \overline{f}$, for $k \geq 1$, and so $\varepsilon f^k \leq_{\mathscr{L}} \overline{f}\omega = \epsilon f\omega$ as $\varepsilon f^k \in \overline{\mathscr{F}}(S)$. Putting this together with (4.53) (where we take $\mathbf{x} = \varepsilon$) yields

$$\varepsilon f\omega \; \mathscr{H} \; \varepsilon f^2\omega \; \mathscr{H} \; \cdots \; \mathscr{H} \; \varepsilon f^n\omega. \qquad (4.54)$$

Because $f \in \mathscr{C}^{\mathscr{A}}$, it follows that $\varepsilon f \in \mathscr{F}_{\mathscr{A}}(S)$ and hence $\varepsilon f\omega$ is aperiodic. Consequently, we have $\varepsilon f\omega = \varepsilon f^2\omega$. Let us write $\overline{f} = \overline{f}\omega \cdot \mathbf{y}$ with $\mathbf{y} \in \mathscr{F}(S)$. Direct computation shows

$$\varepsilon f^2 = (\overline{f})\widehat{f} \cdot \overline{f} = (\overline{f}\omega \cdot \mathbf{y})\widehat{f} \cdot \overline{f} = \overline{f}\omega_{\mathbf{y}}\widehat{f} \cdot \mathbf{y}\widehat{f} \cdot \overline{f}\omega \cdot \mathbf{y}. \qquad (4.55)$$

Hence $\overline{f}\omega_{\mathbf{y}}\widehat{f} = \varepsilon f^2\omega = \varepsilon f\omega = \overline{f}\omega$ and so (4.55) implies that

$$\varepsilon f^2\rho = (\overline{f}\omega_{\mathbf{y}}\widehat{f} \cdot \mathbf{y}\widehat{f} \cdot \overline{f}\omega \cdot \mathbf{y})\rho = (\overline{f}\omega \cdot \mathbf{y})\rho = \overline{f}\rho = \varepsilon f\rho$$

as required.

Let us now establish (B). Assume that $\mathbf{x}f^{n+1}\rho = \mathbf{x}f\rho$ for all $\mathbf{x} \in \mathscr{F}_{\mathscr{A}}(S)$. We show that $\mathbf{x}f^2\rho = \mathbf{x}f\rho$ all $\mathbf{x} \in \mathscr{F}_{\mathscr{A}}(S)$ by induction on the length $|\mathbf{x}|$ of $\mathbf{x}$. By (A), $\varepsilon f^2\rho = \varepsilon f\rho$. Let $\varepsilon \neq \mathbf{x} \in \mathscr{F}_{\mathscr{A}}(S)$ and assume $\mathbf{z}f^2\rho = \mathbf{z}f\rho$ for all $\mathbf{z} \in \mathscr{F}_{\mathscr{A}}(S)$ with $|\mathbf{z}| = |\mathbf{x}| - 1$. We then have by Lemma 4.19.16

$$\mathbf{x}f = \mathbf{x}\widehat{f} \cdot \overline{f} \quad \text{and} \quad \mathbf{x}f^2 = (\mathbf{x}\widehat{f} \cdot \overline{f})f = \mathbf{x}\widehat{f}_{\overline{f}}\widehat{f} \cdot (\overline{f})f. \qquad (4.56)$$

In addition, $\mathbf{x}f^2 = (\mathbf{x}\widehat{f} \cdot \overline{f})\widehat{f} \cdot \overline{f}$ and

$$(\mathbf{x}\widehat{f} \cdot \overline{f})\widehat{f} = \mathbf{x}\widehat{f}_{\overline{f}}\widehat{f} \cdot (\overline{f})\widehat{f}. \qquad (4.57)$$

Let us first suppose that $|\mathbf{x}| = 1$. Then by (4.53) and (4.56), we have

$$\mathbf{x}\widehat{f} = \mathbf{x}f\omega \; \mathscr{R} \; \mathbf{x}f^2\omega = \mathbf{x}\widehat{f}_{\overline{f}}\widehat{f}. \qquad (4.58)$$

From the equality $\varepsilon f^2\rho = \varepsilon f\rho$, it follows $(\overline{f})f\rho = \overline{f}\rho$, i.e., $((\overline{f})\widehat{f} \cdot \overline{f})\rho = \overline{f}\rho$. This, together with Lemma 4.19.6, then implies

$$\overline{f}\omega = (\overline{f})\widehat{f}\omega \qquad (4.59)$$

whereas, together with (4.56), it yields $\mathbf{x}f^2\rho = (\mathbf{x}\widehat{f}_{\overline{f}}\widehat{f} \cdot (\overline{f})f\rho)\rho = (\mathbf{x}\widehat{f}_{\overline{f}}\widehat{f} \cdot \overline{f})\rho$. As $\widehat{f} \in \check{S}_0^{\infty}$ in light of (4.59), (4.57) and (4.58), Lemma 4.19.12 applies to the string $\mathbf{x}\widehat{f} \cdot \overline{f}$ to provide the equalities $\mathbf{x}\widehat{f} = (\mathbf{x}\widehat{f} \cdot \overline{f})\widehat{f}\omega = \mathbf{x}\widehat{f}_{\overline{f}}\widehat{f}$. It follows that

$$\mathbf{x}f^2\rho = (\mathbf{x}\widehat{f_{\overline{f}}}\widehat{f}\cdot\overline{f})\rho = (\mathbf{x}\widehat{f}\cdot\overline{f})\rho = \mathbf{x}f\rho$$

as required.

Finally, we consider the case $|\mathbf{x}| > 1$. We may write $\mathbf{x} = \mathbf{x}\omega\cdot\mathbf{z}$ for some $\mathbf{z} \in \mathscr{F}_{\mathscr{A}}(S)$. By the induction hypothesis, $\mathbf{z}f^2\rho = \mathbf{z}f\rho$ and consequently $\mathbf{z}f^2\omega = \mathbf{z}f\omega$. Analogously to (4.56), one has

$$\mathbf{z}f = \mathbf{z}\widehat{f}\cdot\overline{f} \quad\text{and}\quad \mathbf{z}f^2 = \mathbf{z}\widehat{f_{\overline{f}}}\widehat{f}\cdot(\overline{f})f, \tag{4.60}$$

whence, as $\mathbf{z} \neq \varepsilon$,

$$\mathbf{z}\widehat{f}\omega = \mathbf{z}f\omega = \mathbf{z}f^2\omega = \mathbf{z}\widehat{f_{\overline{f}}}\widehat{f}\omega. \tag{4.61}$$

Next we compute

$$\begin{aligned}\mathbf{x}\widehat{f_{\overline{f}}}\widehat{f} &= (\mathbf{x}\omega\cdot\mathbf{z})\widehat{f_{\overline{f}}}\widehat{f} = (\mathbf{x}\omega_{\mathbf{z}}\widehat{f}\cdot\mathbf{z}\widehat{f})_{\overline{f}}\widehat{f} = \mathbf{x}\omega_{\mathbf{z}}\widehat{f}_{\mathbf{z}\widehat{f}}(\overline{f}\widehat{f})\cdot\mathbf{z}\widehat{f_{\overline{f}}}\widehat{f}\\ &= \mathbf{x}\omega_{\mathbf{z}}\widehat{f}_{\mathbf{z}f}\widehat{f}\cdot\mathbf{z}\widehat{f_{\overline{f}}}\widehat{f}.\end{aligned} \tag{4.62}$$

As $\mathbf{x} = \mathbf{x}\omega\cdot\mathbf{z}$, Lemma 4.19.16 implies $\mathbf{x}f = \mathbf{x}\omega_{\mathbf{z}}\widehat{f}\cdot\mathbf{z}f = \mathbf{x}f\omega\cdot\mathbf{z}f$. Thus (4.53), (4.56) and (4.62) yield

$$\mathbf{x}\omega_{\mathbf{z}}\widehat{f} = \mathbf{x}f\omega \,\mathscr{R}\, \mathbf{x}f^2\omega = \mathbf{x}\omega_{\mathbf{z}}\widehat{f}_{\mathbf{z}f}\widehat{f}. \tag{4.63}$$

Because $_{\overline{f}}\widehat{f} \in \check{S}_0^{\infty}$ by Proposition 4.19.11 we have that Lemma 4.19.12 (applied to the string $\mathbf{x}\omega_{\mathbf{z}}\widehat{f}\cdot\mathbf{z}\widehat{f}$), together with (4.61), (4.62) and (4.63), implies

$$\mathbf{x}\omega_{\mathbf{z}}\widehat{f} = (\mathbf{x}\omega_{\mathbf{z}}\widehat{f}\cdot\mathbf{z}\widehat{f})_{\overline{f}}\widehat{f}\omega = \mathbf{x}\omega_{\mathbf{z}}\widehat{f}_{\mathbf{z}f}\widehat{f}$$

and so $\mathbf{x}f\omega = \mathbf{x}\omega_{\mathbf{z}}\widehat{f} = \mathbf{x}\omega_{\mathbf{z}}\widehat{f}_{\mathbf{z}f}\widehat{f} = \mathbf{x}f^2\omega$. Plugging this into (4.62) and using (4.56) and (4.60) we see that

$$\mathbf{x}f^2 = \mathbf{x}\widehat{f_{\overline{f}}}\widehat{f}\cdot(\overline{f})f = \mathbf{x}\omega_{\mathbf{z}}\widehat{f}_{\mathbf{z}f}\widehat{f}\cdot\mathbf{z}\widehat{f_{\overline{f}}}\widehat{f}\cdot(\overline{f})f = \mathbf{x}f^2\omega\cdot\mathbf{z}\widehat{f_{\overline{f}}}\widehat{f}\cdot(\overline{f})f = \mathbf{x}f^2\omega\cdot\mathbf{z}f^2.$$

Putting it all together, and using confluence of reduction, we obtain

$$\mathbf{x}f^2\rho = (\mathbf{x}f^2\omega\cdot\mathbf{z}f^2)\rho = (\mathbf{x}f\omega\cdot\mathbf{z}f^2\rho)\rho = (\mathbf{x}f\omega\cdot\mathbf{z}f\rho)\rho = (\mathbf{x}f\omega\cdot\mathbf{z}f)\rho = \mathbf{x}f\rho$$

where the middle equality uses the induction hypothesis.    □

### 4.19.2 Blowup operators

Fix a finite semigroup $T$ and let $S = CP(T)$. The key idea behind Henckell's construction is to find a retraction $\widehat{B}\colon \overline{\mathscr{F}}(S) \to \overline{\mathscr{F}}_{\mathscr{A}}(S)$ belonging to $\check{S}_0^{\infty}$. One then "conjugates" the action of the generators of the Rhodes expansion on $\mathscr{F}(S)$ by this retract to get an aperiodic action on $\mathscr{F}_{\mathscr{A}}(S)$.

**Definition 4.19.21 (Blowup operator).** *A preblowup operator on $S$ is a function $B\colon S \to S$ satisfying the following properties:*

1. $sB = s$ if $s$ is aperiodic;
2. $sB <_{\mathcal{H}} s$ if $s$ is not aperiodic;
3. $s \subseteq sB$ (the "blow up");
4. There exists a function $m\colon S \to S^I$, written $s \mapsto m_s$, such that $sB = sm_s$ and $m_s = m_{s'}$ whenever $s \mathscr{L} s'$.

An idempotent preblowup operator is called a blowup operator.

The element $m_s$ is called the *right multiplier* associated to $s$.

**Lemma 4.19.22.** *The collection of preblowup operators on $S$ is a finite semigroup. In particular, if there are any preblowup operators on $S$, then there is a blowup operator on $S$.*

*Proof.* The first three conditions are obviously preserved by composition. If $B$ and $B'$ are preblowup operators with respective right multipliers $s \mapsto m_s$ and $s \mapsto n_s$, then $sBB' = sm_s n_{sm_s}$. If $s \mathscr{L} s'$, then $m_s = m_{s'}$ and $sm_s \mathscr{L} s'm_s$. Therefore, $n_{sm_s} = n_{s'm_s} = n_{s'm_{s'}}$. This shows that $BB'$ is a preblowup operator with $m_s n_{sm_s}$ as the right multiplier associated to $s$. The final statement follows from the existence of idempotents in non-empty finite semigroups. $\square$

The next proposition collects some elementary properties of blowup operators.

**Proposition 4.19.23.** *Let $B\colon S \to S$ be a blowup operator. Then:*

1. *The image of $B$ is the set of aperiodic elements of $S$;*
2. *Suppose $y \leq_{\mathscr{L}} s$. Then $y \subseteq ym_s$;*
3. *If $s$ is aperiodic and $y \leq_{\mathscr{L}} s$, then $y = ym_s$.*

*Proof.* Observe that $B$ fixes an element $s$ if and only if $s$ is aperiodic. As $B$ is idempotent, its image is its fixed-point set. This establishes the first item. For the second, write $y = zs$ with $z \in S^I$. Then we have

$$y = zs \subseteq z(sB) = zsm_s = ym_s$$

as required. For the final item, we have $s = sB = sm_s$. Because $y = zs$, some $z \in S^I$, we have $ym_s = zsm_s = zs = y$. This completes the proof. $\square$

For the rest of this section, we assume the existence of a blowup operator $B$ on $S$; a construction appears in the next section. We proceed to define an "extension" $\widehat{B}$ of $B$ to $S^*$. Recall that $\Delta_s$ is the diagonal operator in $\check{S}^\infty$ corresponding to $s \in S^I$, see (4.51).

**Definition 4.19.24 ($\widehat{B}$).** *Define $\widehat{B}\colon S^* \to S^*$ recursively by*

- $\varepsilon\widehat{B} = \varepsilon$
- $(\mathbf{b} \cdot s)\widehat{B} = (\mathbf{b}\Delta_{m_s})\widehat{B} \cdot sB$ *for $\mathbf{b} \in S^*, s \in S$.*

This recursive definition is known as the Henckell formula.

Because $sB \leq_{\mathscr{L}} s$, we obtain the following lemma.

**Lemma 4.19.25.** *If* $\mathbf{x} \neq \varepsilon$, *then* $\mathbf{x}\widehat{B}\alpha_1 \leq_{\mathscr{L}} \mathbf{x}\alpha_1$.

We retain the notation from the previous section for the next proposition.

**Proposition 4.19.26.** *The map* $\widehat{B}$ *belongs to* $\check{S}_0^\infty$. *Moreover,* $\overline{\mathscr{F}}(S)\widehat{B} = \overline{\mathscr{F}}_{\mathscr{A}}(S)$ *and* $\widehat{B}|_{\overline{\mathscr{F}}(S)}$ *is idempotent.*

*Proof.* Because $sB \leq_{\mathscr{R}} s$ (and so $B \in \check{S}$) and $\Delta_{m_s} \in \check{S}^\infty$, it is immediate from the recursive definition that $\widehat{B} \in \check{S}^\infty$. Next we verify that $\overline{\mathscr{F}}(S)\widehat{B} \subseteq \overline{\mathscr{F}}_{\mathscr{A}}(S)$ by induction on length. The base case is trivial. In general, we have that

$$(\mathbf{b} \cdot x)\widehat{B} = (\mathbf{b}\Delta_{m_x})\widehat{B} \cdot xB.$$

Because $\Delta_{m_x}$ preserves $\overline{\mathscr{F}}(S)$, by induction $(\mathbf{b}\Delta_{m_x})\widehat{B} \in \overline{\mathscr{F}}_{\mathscr{A}}(S)$. As $xB$ is aperiodic (Proposition 4.19.23), if $\mathbf{b} = \varepsilon$ we are done. Otherwise, let $x_1$ be the first entry of $\mathbf{b}$. Then Lemma 4.19.25 shows that $(\mathbf{b}\Delta_{m_x})\widehat{B}\alpha_1 \leq_{\mathscr{L}} x_1 m_x$. As $xB = xm_x$ and $x_1 \leq_{\mathscr{L}} x$, we see that $x_1 m_x \leq_{\mathscr{L}} xm_x$ and so $(\mathbf{b}\Delta_{m_x})\widehat{B} \cdot xB$ belongs to $\overline{\mathscr{F}}_{\mathscr{A}}(S)$.

We show by induction on length that $\widehat{B}$ fixes $\overline{\mathscr{F}}_{\mathscr{A}}(S)$ element-wise, the case of length 0 being trivial. If $\mathbf{x} = \mathbf{b} \cdot x \in \overline{\mathscr{F}}_{\mathscr{A}}(S)$, then $\mathbf{x}\widehat{B} = (\mathbf{b}\Delta_{m_x})\widehat{B} \cdot xB$. But Proposition 4.19.23, together with the fact that $x$ is aperiodic and $\mathbf{x}$ is an $\mathscr{L}$-chain, implies $xB = x$ and $\mathbf{b}\Delta_{m_x} = \mathbf{b}$. So a simple induction yields $\mathbf{x}\widehat{B} = \mathbf{x}$. We conclude $\widehat{B}|_{\overline{\mathscr{F}}(S)}$ is idempotent.

Finally, we must verify $\widehat{B} \in \check{S}_0^\infty$. We proceed by induction on length. Suppose that we have $(x_2, x_1)\widehat{B} = (y_2, y_1)$ with $x_1 \mathscr{R} y_1$ and $x_2 \mathscr{R} y_2$. Then $y_1 = x_1 B = x_1 m_{x_1}$. Now

$$x_2 \mathscr{R} y_2 = (x_2 m_{x_1})B \leq_{\mathscr{R}} x_2 m_{x_1} \leq_{\mathscr{R}} x_2$$

and so $x_2 m_{x_1}$ must be aperiodic by the second condition in the definition of a blowup operator. Thus $y_1 = x_1 m_{x_1}$ and $y_2 = (x_2 m_{x_1})B = x_2 m_{x_1}$, showing that the condition in Definition 4.19.10 is satisfied. Suppose now $n > 2$ and $(x_n, x_{n-1}, \ldots, x_1)\widehat{B} = (y_n, y_{n-1}, \ldots, y_1)$ with $x_i \mathscr{R} y_i$ for $i = n - 1, n$. Then

$$(y_n, y_{n-1}, \ldots, y_1) = ((x_n, x_{n-1}, \ldots, x_2)\Delta_{m_{x_1}})\widehat{B} \cdot x_1 B.$$

Now $x_i \geq_{\mathscr{R}} x_i m_{x_1} \geq_{\mathscr{R}} y_i \mathscr{R} x_i$, for $i = n - 1, n$. Therefore, $x_i m_{x_1} \mathscr{R} y_i$, $i = n - 1, n$. Induction provides $s' \in S^I$ with $x_i m_{x_1} s' = y_i$, $i = n - 1, n$. Taking $s = m_{x_1} s'$ yields $x_i s = y_i$, for $i = n - 1, n$, completing the proof.   $\square$

Another crucial property of $\widehat{B}$ is that it "blows up $\mathscr{L}$-chains."

**Proposition 4.19.27.** *Let* $\mathbf{x} = (x_n, \ldots, x_1) \in \overline{\mathscr{F}}(S)$ *and let us set* $\mathbf{x}\widehat{B} = (y_n, \ldots, y_1)$. *Then* $x_i \subseteq y_i$ *for* $i = 1, \ldots, n$.

*Proof.* The proof is by induction on $n$. For $n = 0$, the statement is vacuously true. In general, $\mathbf{x}\widehat{B} = (x_n m_{x_1}, \ldots, x_2 m_{x_1})\widehat{B} \cdot x_1 B$. By the definition of a blowup operator $x_1 \subseteq x_1 B$. Because $\mathbf{x}$ is an $\mathscr{L}$-chain, Proposition 4.19.23 shows that $x_i \subseteq x_i m_{x_1}$ for $i = 2, \ldots, n$. Induction yields $x_i m_{x_1} \subseteq y_i$ for $i = 2, \ldots, n$, establishing that $x_i \subseteq y_i$. □

Recall that if $s \in S$, then $(\Delta_s, s)$ denotes the element of $\mathscr{C}$ that acts by $\mathbf{x}(\Delta_s, s) = \mathbf{x}\Delta_s \cdot s$.

**Proposition 4.19.28.** *The equalities* $\mathbf{x}\rho(\Delta_s, s)\rho = \mathbf{x}(\Delta_s, s)\rho$, *for* $s \in S$, *and* $\mathbf{x}\rho\widehat{B}\rho = \mathbf{x}\widehat{B}\rho$ *hold for all* $\mathbf{x} \in \overline{\mathscr{F}}(S)$.

*Proof.* First we claim that if $s \in S^I$ and $\mathbf{x} \in \overline{\mathscr{F}}(S)$, then

$$\mathbf{x}\Delta_s\rho = \mathbf{x}\rho\Delta_s\rho \tag{4.64}$$

holds. We proceed by induction on the length of $\mathbf{x}$, the cases $|\mathbf{x}| = 0, 1$ being trivial. Suppose that it is true for strings from $\overline{\mathscr{F}}(S)$ of length $n - 1$ with $n \geq 2$ and let $\mathbf{x} = (x_n, \ldots, x_1) \in \overline{\mathscr{F}}(S)$. Then $\mathbf{x}\Delta_s = (x_n s, \ldots, x_1 s)$. Choose $m$ maximum so that $x_m \mathscr{L} x_{m-1}$; if no such $m$ exists, then $\mathbf{x} = \mathbf{x}\rho$ and there is nothing to prove. Then $x_m s \mathscr{L} x_{m-1} s$, whence

$$\mathbf{x}\rho = (x_n, \ldots, x_m, x_{m-2}, \ldots, x_1)\rho$$
$$\mathbf{x}\Delta_s\rho = (x_n s, \ldots, x_m s, x_{m-2} s, \ldots, x_1 s)\rho$$
$$= (x_n, \ldots, x_m, x_{m-2}, \ldots, x_1)\Delta_s\rho$$

and so (4.64) follows by induction.

It is immediate from (4.64) and confluence of reduction that, for $\mathbf{x} \in \overline{\mathscr{F}}(S)$,

$$\mathbf{x}(\Delta_s, s)\rho = (\mathbf{x}\Delta_s \cdot s)\rho = (\mathbf{x}\Delta_s\rho \cdot s)\rho = (\mathbf{x}\rho\Delta_s\rho \cdot s)\rho = (\mathbf{x}\rho\Delta_s \cdot s)\rho = \mathbf{x}\rho(\Delta_s, s)\rho$$

as required.

Let us now turn to $\widehat{B}$. We prove by induction on length that if $\mathbf{x} \in \overline{\mathscr{F}}(S)$, then $\mathbf{x}\rho\widehat{B}\rho = \mathbf{x}\widehat{B}\rho$. If $|\mathbf{x}| = 0, 1$, then there is nothing to prove as $\mathbf{x} = \mathbf{x}\rho$. So assume the result is true for strings of length $n - 1$ in $\overline{\mathscr{F}}(S)$ with $n \geq 2$ and suppose $\mathbf{x} = (x_n, \ldots, x_1)$. Then we have

$$(x_n, \ldots, x_1)\widehat{B}\rho = (((x_n, \ldots, x_2)\Delta_{m_{x_1}})\widehat{B} \cdot x_1 B)\rho$$
$$= (((x_n, \ldots, x_2)\Delta_{m_{x_1}})\widehat{B}\rho \cdot x_1 B)\rho$$
$$= ((x_n, \ldots, x_2)\Delta_{m_{x_1}}\rho\widehat{B}\rho \cdot x_1 B)\rho$$
$$= ((x_n, \ldots, x_2)\rho\Delta_{m_{x_1}}\rho\widehat{B}\rho \cdot x_1 B)\rho$$
$$= ((x_n, \ldots, x_2)\rho\Delta_{m_{x_1}}\widehat{B}\rho \cdot x_1 B)\rho$$
$$= ((x_n, \ldots, x_2)\rho\Delta_{m_{x_1}}\widehat{B} \cdot x_1 B)\rho$$
$$= ((x_n, \ldots, x_2)\rho \cdot x_1)\widehat{B}\rho$$

where the third and fifth equalities come by way of the induction hypothesis
and the fourth one uses (4.64). Thus, resetting notation, we may assume that
$(x_n, \ldots x_2)$ is already reduced.

If $x_2$ is not $\mathscr{L}$-equivalent to $x_1$, then $\mathbf{x}$ is reduced and there is nothing to
prove. So assume $x_2 \mathscr{L} x_1$. Then, by the fourth property in the definition of
a blowup operator, $m_{x_1} = m_{x_2}$. Hence $x_2 B = x_2 m_2 = x_2 m_{x_1}$. Because $B$ is
idempotent, $x_2 m_{x_1} B = x_2 B^2 = x_2 B = x_2 m_{x_1}$ and hence $x_2 m_{x_1}$ is aperiodic
by item 2 of Definition 4.19.21. As, for $i \geq 3$, one has $x_i m_{x_1} \leq_{\mathscr{L}} x_2 m_{x_1}$, item 3
of Proposition 4.19.23 yields $x_i m_{x_1} m_{x_2} m_{x_1} = x_i m_{x_1}$, for $i \geq 3$. In addition,
$x_2 B = x_2 m_{x_1} \mathscr{L} x_1 m_{x_1} = x_1 B$. Therefore, we compute

$$(x_n, \ldots, x_1)\widehat{B}\rho = (((x_n, \ldots, x_3)\Delta_{m_{x_1} m_{x_2} m_{x_1}})\widehat{B} \cdot x_2 m_{x_1} B \cdot x_1 B)\rho$$

$$= (((x_n, \ldots, x_3)\Delta_{m_{x_1}})\widehat{B} \cdot x_2 B \cdot x_1 B)\rho$$

$$= (((x_n, \ldots, x_3)\Delta_{m_{x_1}})\widehat{B} \cdot x_2 B)\rho$$

$$= (x_n, \ldots, x_2)\widehat{B}\rho$$

$$= (x_n, \ldots, x_1)\rho\widehat{B}\rho$$

where the penultimate equality uses that $m_{x_1} = m_{x_2}$. □

For the next proposition, the reader is referred to Definition 4.19.17.

**Proposition 4.19.29.** *If $s \in S$, then $(\Delta, s)\widehat{B} \in \mathscr{C}^{\mathscr{A}}$.*

*Proof.* We saw in the proof of Proposition 4.19.18 that the set of transfor-
mations $f$ satisfying $\mathbf{x}\rho f \rho = \mathbf{x} f \rho$, for all $\mathbf{x} \in \overline{\mathscr{F}}(S)$, is a semigroup. As
$\widehat{B} \colon \overline{\mathscr{F}}(S) \to \overline{\mathscr{F}}_{\mathscr{A}}(S)$ (Proposition 4.19.26), it suffices by Proposition 4.19.28
to show that $(\Delta_s, s)\widehat{B} \in \mathscr{C}$. Both $(\Delta_s, s)$ and $\widehat{B}$ leave $\overline{\mathscr{F}}(S)$ invariant. Now

$$\mathbf{x}(\Delta_s, s)\widehat{B} = (\mathbf{x}\Delta_s \cdot s)\widehat{B} = (\mathbf{x}\Delta_s \Delta_{m_s})\widehat{B} \cdot sB = \mathbf{x}\Delta_{sm_s}\widehat{B} \cdot sB$$

and $\Delta_{sm_s}\widehat{B} \in \check{S}_0^\infty$ by Proposition 4.19.26. This establishes $(\Delta_s, s)\widehat{B} \in \mathscr{C}$,
completing the proof. □

Recall we are assuming that $S = CP(T)$ admits a blowup operator. We
define a relational morphism $\varphi \colon (T^I, T) \to (\mathscr{F}_{\mathscr{A}}(S), S^{\mathscr{A}})$ of faithful transformation semigroups as follows: set $I\varphi = \varepsilon$, and, for $t \in T$, define

$$t\varphi = \{\mathbf{x} \in \mathscr{F}_{\mathscr{A}}(S) \mid t \in \mathbf{x}\omega\}.$$

Notice that $\varphi^{-1}$ coincides with $\omega$ on non-empty strings.

**Lemma 4.19.30.** *The relation $\varphi \colon T^I \to \mathscr{F}_{\mathscr{A}}(S)$ gives rise to a relational
morphism $\varphi \colon (T^I, T) \to (\mathscr{F}_{\mathscr{A}}(S), S^{\mathscr{A}})$.*

*Proof.* Because $t \in \{t\} \subseteq \{t\}B$ and $\{t\}B \in \mathscr{F}_{\mathscr{A}}(S)$ for every $t \in T$, it follows that $\varphi$ is fully defined. Let $t \in T$. We set $\tilde{t} = (\Delta_{\{t\}}, \{t\})\widehat{B}\rho$. Proposition 4.19.29 shows that $\tilde{t} \in S^{\mathscr{A}}$. We need to show that $\mathbf{x}\varphi^{-1}t \subseteq \mathbf{x}\tilde{t}\varphi^{-1}$. If $\mathbf{x} = \varepsilon$, then $\mathbf{x}(\Delta_{\{t\}}, \{t\})\widehat{B}\rho = \{t\}B$. So $\varepsilon\varphi^{-1}t = \{I\}t = \{t\} \subseteq \{t\}B = \varepsilon\tilde{t}\varphi^{-1}$.

If $\mathbf{x} \neq \varepsilon$, then we need to show $\mathbf{x}\omega t \subseteq \mathbf{x}\tilde{t}\omega$. Abusing notation, we identify $\{t\}$ with $t$. Then we have $\mathbf{x}(\Delta_t, t)\widehat{B}\rho = \left( (\mathbf{x}\Delta_t \cdot t)\widehat{B} \right)\rho$. Lemma 4.19.6 tells us $\left( (\mathbf{x}\Delta_t \cdot t)\widehat{B} \right)\rho\omega = (\mathbf{x}\Delta_t \cdot t)\widehat{B}\omega$. An application of Proposition 4.19.27 yields

$$\mathbf{x}\tilde{t}\omega = (\mathbf{x}\Delta_t \cdot t)\widehat{B}\omega \supseteq (\mathbf{x}\Delta_t \cdot t)\omega = \mathbf{x}\omega t$$

completing the proof. $\qquad\square$

For the next proposition, the reader is referred to (4.17) for the definition of the companion relational morphism.

**Proposition 4.19.31.** *The companion relation $f_\varphi \colon T \to S^{\mathscr{A}}$ satisfies the inequality $gf_\varphi^{-1} \subseteq \varepsilon g\omega \in CP(T)$ for $g \in S^{\mathscr{A}}$.*

*Proof.* Let $t \in gf_\varphi^{-1}$. Then $t = It \in \varepsilon\varphi^{-1}t \subseteq \varepsilon g\varphi^{-1} = \varepsilon g\omega \in CP(T)$, as $\varepsilon g \neq \varepsilon$ and $\varphi^{-1}$ coincides with $\omega$ on non-empty strings. Hence $gf_\varphi^{-1} \subseteq \varepsilon g\omega$, as required. $\qquad\square$

**Corollary 4.19.32.** *If $CP(T)$ admits a blowup operator, then Theorem 4.19.2 holds. That is, $\mathsf{PL}_{\mathbf{A}}(T) = \{X \subseteq T \mid X \subseteq Y \text{ for some } Y \in CP(T)\}$.*

*Proof.* We already know $CP(T) \subseteq \mathsf{PL}_{\mathbf{A}}(T)$. As $S^{\mathscr{A}}$ is aperiodic by Theorem 4.19.20, we have that each $\mathbf{A}$-pointlike set is contained in $gf_\varphi^{-1}$ for some $g \in S^{\mathscr{A}}$. An application of Proposition 4.19.31 then completes the proof. $\qquad\square$

### 4.19.3 Construction of the blowup operator

We continue to work with our fixed finite semigroup $T$ and to denote $CP(T)$ by $S$. Our task now consists of constructing a blowup operator for $S$. By Lemma 4.19.22, it suffices to construct a preblowup operator. Our approach is a variation on Henckell's [121], which leads to a shorter proof. His approach may be more aesthetically appealing.

We shall make use of the Schützenberger group to do this; let us recall the notion, details can be found in [68, Section 2.4], [171, Chapter 7, Prop. 2.8] or Section A.3.1. Our notation, which we briefly recall, follows Section A.3.1 and the reader should consult that section before continuing. Let $H$ be an $\mathscr{H}$-class of $S$ and $\mathrm{Stab}(H) = \{s \in S^I \mid Hs \subseteq H\}$. Then $\mathrm{Stab}(H)$ acts on the right of $H$ and the associated faithful transformation monoid, denoted $(H, \Gamma_R(H))$, is in fact a transitive regular permutation group called the *Schützenberger group* of $H$. Recall a permutation group is regular if the group acts freely. If $H$ is a maximal subgroup, then $\Gamma_R(H) \cong H$. In general, one can always find a subgroup $\widetilde{\Gamma}_R(H) \subseteq \mathrm{Stab}(H)$ acting transitively on $H$ with faithful quotient $\Gamma_R(H)$. Indeed, apply Proposition 4.1.44 to the quotient $\mathrm{Stab}(H) \to \Gamma_R(H)$. The following proposition is immediate from Theorem A.3.11.

**Proposition 4.19.33.** *Let $H$, $H'$ be $\mathscr{L}$-equivalent $\mathscr{H}$-classes. Then one has $\mathrm{Stab}(H) = \mathrm{Stab}(H')$ and one can take $\widetilde{\Gamma}_R(H) = \widetilde{\Gamma}_R(H')$. Moreover, the kernel of the natural maps $\widetilde{\Gamma}_R(H) \to \Gamma_R(H)$ and $\widetilde{\Gamma}_R(H) \to \Gamma_R(H')$ coincide. In particular, $\Gamma_R(H) \cong \Gamma_R(H')$.*

Similarly, there is a left Schützenberger group $(\Gamma_L(H), H)$ and a subgroup $\widetilde{\Gamma}_L(H)$ of the left stabilizer of $H$ mapping onto $\Gamma_L(H)$. The groups $\Gamma_L(H)$ and $\Gamma_R(H)$ are isomorphic. In fact, if $h_0 \in H$ is a fixed base point and $g \in \Gamma_R(H)$, then the map $\gamma$ sending $g$ to the unique $g\gamma \in \Gamma_L(H)$ with $g\gamma h_0 = h_0 g$ is an anti-isomorphism. See [68, 171] or Section A.3.1 for details.

For each non-aperiodic $\mathscr{L}$-class $L$, choose an $\mathscr{H}$-class $H_L$ of $L$ and an element $g_L \in \widetilde{\Gamma}_R(H_L)$ representing a non-trivial element of $\Gamma_R(H_L)$. We are now prepared to define our preblowup operator $B$. If $s \in S$ is an aperiodic element, define $m_s = I$. If $s$ is not aperiodic, define $m_s = g_{L_s}^{\omega+*} \in S$. Notice that $g_{L_s}^{\omega+*} = \bigcup_{n \geq 1} g_{L_s}^n$, the subgroup generated by $g_{L_s}$, because $g_{L_s}$ is a group element and so $g_{L_s}^\omega g_{L_s} = g_{L_s}$. In particular, we see that $g_{L_s} \subseteq g_{L_s}^{\omega+*}$. Define an operator $B: S \to S$ by $sB = sm_s$.

**Proposition 4.19.34.** *The operator $B$ is a preblowup operator.*

*Proof.* If $s$ is aperiodic, then $m_s = I$ and $sB = sm_s = s$. The fourth item of Definition 4.19.21 is clearly satisfied by construction. We turn now to the third item. If $s$ is aperiodic, then trivially $s \subseteq sB$. Suppose $s$ is not aperiodic. Then by Proposition 4.19.33, $g_{L_s} \in \widetilde{\Gamma}_R(H_s)$ and hence $sg_{L_s}^\omega = s$. As $g_{L_s}^\omega \subseteq g_{L_s}^{\omega+*}$, it follows

$$s = sg_{L_s}^\omega \subseteq sg_{L_s}^{\omega+*} = sm_s = sB$$

as required. Finally, we turn to the second item of Definition 4.19.21. Suppose that $s \in S$ is not aperiodic. It is immediate from the definition that $sB = sm_s \leq_{\mathscr{R}} s$. Let $\gamma: \Gamma_R(H_s) \to \Gamma_L(H_s)$ be the anti-isomorphism given by $g\gamma s = sg$ for $g \in \Gamma_R(H_s)$. Choose $x \in \widetilde{\Gamma}_L(H_s)$ so that $x$ maps to $g_{L_s}\gamma$ in $\Gamma_L(H_s)$ (where we view $g_{L_s}$ as an element of $\Gamma_R(H_s)$ using Proposition 4.19.33 and the projection map). Then we have $\bigcup_{n\geq 1} x^n = x^{\omega+*} \in S$, as $x$ is a group element. We then calculate $sB$ as follows:

$$sB = sg_{L_s}^{\omega+*} = s\bigcup_{n\geq 1} g_{L_s}^n = \bigcup_{n\geq 1} sg_{L_s}^n = \bigcup_{n\geq 1} (g_{L_s}\gamma)^n s = \bigcup_{n\geq 1} x^n s = x^{\omega+*}s.$$

We conclude $sB \leq_{\mathscr{L}} s$. Therefore, we have $sB \leq_{\mathscr{H}} s$. To establish that $sB <_{\mathscr{H}} s$, we observe, using (4.49), that

$$sBg_{L_s} = sg_{L_s}^{\omega+*}g_{L_s} = sg_{L_s}^{\omega+*} = sB.$$

By construction $g_{L_s}$ represents a non-trivial element of $\Gamma_R(H_{L_s})$ and hence of $\Gamma_R(H_s)$ by Proposition 4.19.33. As $(H_s, \Gamma_R(H_s))$ is a regular permutation group, we deduce that $sB \notin H_s$. This concludes the proof that $sB <_{\mathscr{H}} s$ when $s$ is not aperiodic. Therefore, $B$ is a preblowup operator, as required.     □

In light of Corollary 4.19.32, we have now established Theorem 4.19.2.

## Notes

The proof of the Prime Decomposition Theorem presented here follows along the lines of the "Arbib" book [171, Chapter 6]. In substance, this is not much different than the original proof [169], except that [169] does everything in the language of sequential machines. The other approaches to the Prime Decomposition Theorem, such as the Holonomy Theorem [85, Chapter II, Thm. 7.1], have their origins in the work of Zeiger [386, 387] and the method of weakly preserved covers. Versions of the Prime Decomposition Theorem for infinite semigroups can be found in [3, 50–52, 88, 125, 214, 215, 276, 277, 279].

Brown's original proof of his theorem is entirely combinatorial in nature, as the title of his paper suggests [59]; see also the book of Grillet [110]. Our proof follows an idea of Simon [319, 320]. Another algebraic proof, using expansions, can be found in [179]. The axiomatic approach to complexity has roots in [171, Chapter 9] and [272]. The lattice theoretic viewpoint seems natural in the context of the rest of this book.

Schützenberger's Theorem [313] now has a variety of proofs. The two most common induction parameters are the $\mathscr{J}$-order and the length of a prime decomposition. The first published proof using the Prime Decomposition Theorem seems to be [207]. This same proof appears in [85]. The proof we have given inducts off the other end and seems to be easier in our opinion. The easiest prime decomposition style proof uses the two-sided semidirect product and can be found in Straubing's book [346].

In his far-reaching paper [340], Stiffler characterized the semidirect product closure of a number of classes of semigroups. Many of his results can be proved more simply with the help of the Derived Category Theorem and the locality of the pseudovariety of semilattices; see [332]. This is the approach we have adopted here. Eilenberg [85] proves these results using partial transformation semigroups. The semilocal theory is the subject of [171, Chapter 8]. The material here is essentially an updated presentation of the classical results, exploiting the language of Mal'cev products to provide a unifying framework. The supplemental results of [269, 368] on $\mathscr{J}'$-morphisms are also treated. Some of these results were rediscovered much later by other authors [35, 114], most likely due to the lack of a dictionary between modern semigroup theory and the language used in the book [171], a situation the current volume is attempting to remedy. In particular, it should be noted that the maximal idempotent-separating congruence on a finite semigroup was constructed by

Krohn and Rhodes in 1968 [170, 171]; essentially the same description and proof idea works for regular [114] and eventually regular semigroups [83, 133], and the other varied generalizations appearing in the literature.

A preliminary classification of subdirectly indecomposable semigroups forms a part of [171, Chapter 8]. We have added some novel improvements, notably Theorems 4.7.10 and 4.7.20 covering the generalized group mapping case. The latter of the two theorems was pointed out to us by S. W. Margolis. The classification of congruence-free semigroups is a part of semigroup folklore.

The exclusion classes of $U_2$ and $\overline{2}$ were characterized in Stiffler's paper [340]. Theorem 4.8.4 was originally proved by Rhodes and Tilson [295] via a different route, as the derived semigroup and the derived semigroupoid had not yet been introduced into semigroup theory at that time. The Tilson number was introduced in [362]. Theorem 4.8.7 is the main result of Margolis's thesis [187]. The generalization to the local setting by Margolis and Tilson used Tilson's Trace-Delay Theorem [197]. The approach via the Delay Theorem is new, but implicit in the original approach. The idempotent splitting seems to be a standard construction in category theory. Whereas all the components of the idempotent splitting have been around in semigroup theory for a long time (see, for instance, [363, Prop. 1.3]), it seems to only have been explicitly considered as a category in semigroup theory for the first time in [364]. The Delay Theorem has a long and sordid history, growing out of Tilson's Trace-Delay Theorem [85] and Straubing's work [345], before arriving at its final form [364]. Steinberg generalized the Delay Theorem to a statement about pointlikes with respect to $\mathbf{V} * \mathbf{D}$ [327].

The Fundamental Lemma of Complexity was first proved for completely regular semigroups in [170]. The general case was announced in [268] and appeared in [271, 274]. The original proofs were quite difficult. Rhodes later showed Tilson the Rhodes expansion, which combined with the derived semigroup, led to the improved exposition found in [361, 362]. Most of the results of Section 4.9 can be found in [362, 363] with the exception of Theorem 4.9.5 and its consequences, which are due to Rhodes.

Section 4.10 is a simplification by Steinberg of the original argument of Rhodes and Tilson [269, 368], which in turn was based on the arguments of Krohn and Rhodes [170], [171, Chapter 9] for the case of completely regular semigroups. The simplification comes about from the language of pseudovarieties and Mal'cev products and in particular takes advantage of the fact that the operator $(-) \ \textcircled{m} \ \mathbf{W}$ is $\mathbf{inf}$. Together with the results of Reilly and Zhang [259], it follows that the complexity pseudovarieties, when restricted to $\mathbf{DS}$, are finitely based. The full complexity pseudovarieties were shown by the authors not to be finitely based [288]. The Karnofsky-Rhodes decompositions are from [156].

The lower bounds results of Section 4.12 are mostly from Rhodes and Tilson [294, 295]. Theorem 4.12.19 appears in [126]. Recently, Henckell achieved an effective characterization of aperiodic stable pairs [124], a more general notion than a Type I subsemigroup. If $S$ is a semigroup, $T \leq S$ and $X \subseteq S$ is a subset, then $(X, T)$ is a **V**-stable pair if, for all relational morphisms $\varphi \colon S \to V$ with $V \in \mathbf{V}$, there exists $v \in V$ with $X \subseteq v\varphi^{-1}$ and $T \leq \mathrm{Stab}(v)\varphi^{-1}$. It is easy to see that if **U** is local and **V**-stable pairs are decidable, then $\mathbf{U} * \mathbf{V}$ is decidable. A short proof of the decidability of **A**-stable pairs via profinite methods is the content of [130]. The complexity of $T_n$ was first computed by Rhodes [266]. Theorem 4.12.32 is an unpublished result of Fox and Rhodes [91].

Graham's paper [107] seems to be a lost classic. He computes the idempotent-generated subsemigroup of a completely 0-simple semigroup via graph theoretic methods. These techniques were also used to compute other subsemigroups such as maximal nilpotent subsemigroups. The results of this paper have been republished at least twice [136, 138], both times occurring 10 years after the original publication! Houghton's paper [136] contributes a nice topological viewpoint that has influenced our presentation. The notion of a fundamental group of a graph labeled by a group seems novel. The fact that the intersection of the idempotent-generated subsemigroup with a regular $\mathscr{J}$-class $J$ is $\langle E(J) \rangle \cap J$ was certainly known to, and exploited by, Rhodes and Tilson in the late 1960s. Variations on Lemma 4.13.1 can be found in [114, 213].

The version of the Presentation Lemma presented here is adapted from Steinberg [334]. The original version is due to Rhodes [43]. The kernel of the idea of the Presentation Lemma appears in [295], where a condition for membership in $\mathbf{A} * \mathbf{G}$ is given in terms of finding, for each regular $\mathscr{R}$-class, a partition with certain properties. This idea further evolved in [360] where Tilson computed the complexity of a $2\,\mathscr{J}$-class semigroup, but apparently Tilson did not like very much this approach. Rhodes really took to it and developed it much further in [275], where many interesting counterexamples to conjectures in complexity theory were published, before arriving at a formulation of the Presentation Lemma that remained unpublished for a number of years until Nehaniv wrote it up [43]. This formulation is in terms of coordinates, and its relationship to our formulation is very similar to the relationship between the description of regular Type II elements in [295] and the description in [365]. Tilson also has an unpublished variation on the Presentation Lemma [366]. The first author has made it clear that the Presentation Lemma is at the core of any reasonable approach to complexity, and certainly it has been used for a number of results such as a proving complexity pseudovarieties are not local [289] and not finitely based [288]. A restatement in the language of flows on automata can be found in [128, 280]. See [43, 275, 334] for more history on this result. Tilson's $2\,\mathscr{J}$-class Theorem is from [360]. The proof there uses machine equations and coordinates. A proof of the decidability part of the result is given in [43], but not Tilson's elegant solution. Our treatment is a coordinate-free simplification of the argument in Steinberg [334].

A comprehensive survey of the consequences of Ash's Theorem can be found in [126]. The proof of the Type II Theorem [33] given here is based on the geometric approach of Auinger and Steinberg [34, 39, 328]. This approach requires a bit of combinatorial group theory, especially fundamental groups of graphs. It has the advantage of giving the tightest control over the groups involved. In particular, [34] gives the best choice to date of a relational morphism to a group witnessing the Type II subsemigroup. Ash's original approach made use of Ramsey Theory [32,33], and there was no real control over the groups involved. The topological conjecture of Pin and Reutenauer [232] followed on the heel of earlier topological versions of the Type II conjecture considered by Pin [225,226]. Ribes and Zalesskii's proof of their theorem [300] uses profinite groups and graphs and hence offers no effective control on the groups involved in witnessing the Type II semigroup. In [301], Ribes and Zalesskii generalized their approach to handle kernels relative to extension-closed pseudovarieties of groups. This was taken much further by Auinger and Steinberg in the series of papers [37,40,41,328], where the techniques of Ash and Ribes and Zalesskii are synthesized. Other generalizations of the Ribes-Zalesskii Theorem consider various classes of groups besides free groups and the closure of the "Ribes and Zalesskii" (or product separability) property under free constructions [76,98,208,299,382]. The graph immersion/inverse graph approach to free groups is due to Stallings [321]; a fairly comprehensive survey of its applications to algorithmic problems can be found in [155]. Its relevance to semigroup theory was first noted by Margolis and Meakin [189] (see also [196]).

Our treatment of Henckell's Theorem [121] on aperiodic pointlikes is a simplification of the proof from [129] obtained by restricting to the case of aperiodic semigroups. In [129], it is shown that if $\pi$ is a recursive set of primes, then the $\overline{\mathbf{G}}_\pi$-pointlike sets are decidable. The original proof of Henckell's Theorem relies on the complicated Zeiger coding of the Rhodes expansion into the wreath product [88,121,125,279]; the basic idea for the direct approach is an elaboration by Steinberg of an idea of Henckell, which was further simplified by the referee of [129]. The relevance of pointlike sets for deciding joins was discovered independently by Steinberg [322,330] and Almeida, Azevedo and Zeitoun [13]. A decidable pseudovariety of semigroups with undecidable pointlikes was first constructed in [285]. Examples of decidable pseudovarieties of groups with undecidable pointlikes appear in the work of Auinger and Steinberg [36].

# 5

# Two-Sided Complexity and the Complexity of Operators

The goal of this chapter is to discuss complexity hierarchies associated to operators, or more precisely iteration of operators. A general research program is: given two operators, try and prove analogues of the results of Chapter 4 for the complexity hierarchy associated to these operators. Tilson began this program for $p$-length [368]. We carry this program out to some extent in the current chapter for the two-sided complexity function. In the process many important techniques of Finite Semigroup Theory are presented including MPS theory [267, 293, 296] and translational hulls of 0-simple semigroups. The Berkeley School method for constructing examples is exposited. In this chapter all semigroups are assumed to be finite.

## 5.1 Complexity of Operators

### 5.1.1 Complexity of a single operator

Let $\alpha \in \mathbf{Cnt}(\mathbf{PV})^+$ and let $\mathbf{V}$ be a pseudovariety. Then we associate a complexity hierarchy to $\alpha$ with base pseudovariety $\mathbf{V}$ by setting

$$\mathbf{V}_0 = \mathbf{V},$$
$$\mathbf{V}_{n+1} = \alpha(\mathbf{V}_n) \text{ for } n \geq 0.$$

Notice that we need $\alpha \in \mathbf{Cnt}(\mathbf{PV})^+$ to guarantee that this is indeed a complexity hierarchy. Recall that $\mathbf{V}_\infty$ denotes the union $\bigcup \mathbf{V}_n$.

The dot-depth hierarchy [71] is an example of a complexity hierarchy arising from iteration of an operator: in this case the Schützenberger product operator $\Diamond(-)$. The base pseudovariety $\mathbf{V}$ for dot-depth is the trivial pseudovariety. See [85, 230] for more details. Pin also considers the analogue of dot-depth with $\mathbf{V} = \mathbf{G}$ as the base pseudovariety and $\alpha = \Diamond(-)$ [230]. In this case, one has $\mathbf{V}_\infty = \mathbf{A} * \mathbf{G}$.

Several authors have also studied iteration of the operator $\mathbf{P}$; a nearly complete treatment can be found in [7, Chapter 11]. In this case, it turns out

J. Rhodes, B. Steinberg, *The q-theory of Finite Semigroups*,
Springer Monographs in Mathematics, DOI 10.1007/978-0-387-09781-7_5,
© Springer Science+Business Media, LLC 2009

that if the base pseudovariety is not permutative, then within three iterations one ends up with the pseudovariety of all finite semigroups [7, Chapter 11].

Another example of a hierarchy arising from iteration of a single operator that has been considered in the literature comes from $\alpha = (-) \, \textcircled{m} \, \mathbf{G}$ with $\mathbf{A}$ as the base pseudovariety [80]. The associated complexity hierarchy is strict [295] and has union $\mathbf{EA}$ [80]. This hierarchy is computable because the Type II semigroup is computable.

If $\mathbf{V}$ is a pseudovariety of semigroups, then one can consider the operator $\alpha = \mathbf{V} * (-)$ and take the trivial pseudovariety as the starting pseudovariety. This leads to a natural hierarchy for the semidirect product closure of $\mathbf{V}$. This is of particular interest when $\mathbf{V} = (S)$ for a finite semigroup $S$. Applying this to $(U_2)$ gives the $U_2$-length hierarchical complexity function, whereas applying this to $\mathbf{Sl}$ gives a natural hierarchy for $\mathbf{R}$. Alternatively, one can consider the hierarchy associated to the operator $\alpha = \mathbf{V} ** (-)$ where the trivial pseudovariety is the starting pseudovariety. Taking $\mathbf{V} = \mathbf{Sl} = (U_1)$, one obtains a strict complexity hierarchy for $\mathbf{A}$, as the results of Section 5.3 will show. The following question was first asked by Tilson; see also [7].

*Question 5.1.1.* Given a finite semigroup $S$, is the semidirect product closure of $(S)$ decidable? How about the closure under the two-sided semidirect product?

From the Prime Decomposition Theorem, if $U_2$ divides $S$, then the semidirect product closure of $(S)$ is decidable: it consists of all semigroups whose simple group divisors divide $S$. From the two-sided Prime Decomposition Theorem (proved below), it follows that the two-sided semidirect product closure of $(S)$ is decidable if $U_1$ divides $S$ (and again consists of all semigroups whose simple group divisors divide $S$). So this question is a "low level" case analysis. Almeida's [7, Section 10.10] is dedicated to pseudovarieties closed under semidirect product. Teixeira characterized the semidirect product closures of the pseudovarieties $(B_2)$ and $(B_2^I)$, in particular establishing that they are decidable [355].

Sometimes it is convenient to consider the action of an operator $\beta$ on the complexity hierarchy associated to an operator $\alpha$. Namely, define the $\beta$-*shift* of the $\alpha$-complexity hierarchy $\{\mathbf{V}_n\}$ with base pseudovariety $\mathbf{V}$ to be the complexity hierarchy with $\mathbf{W}_n = \beta(\mathbf{V}_n)$. For instance, if $\alpha = \Diamond(-)$, $\mathbf{V} = \mathbf{1}$ and $\beta = (-) * \mathbf{G}$, then Pin showed [230] that in the monoidal context the $\beta$-shift of the $\alpha$-complexity hierarchy with base pseudovariety $\mathbf{1}$ is the $\alpha$-complexity hierarchy with base pseudovariety $\mathbf{G}$.

### 5.1.2 Complexity of two operators and the two-sided complexity function

Let $\alpha, \beta \in \mathbf{Cnt}(\mathbf{PV})^+$. Our goal is to associate a complexity hierarchy to these elements, or more precisely to the ordered pair $(\alpha, \beta)$. The approach is modeled on the group complexity hierarchy where $\alpha$ plays the role of $\mathbf{A} * (-)$ and $\beta$ plays the role of $\mathbf{G} * (-)$. Because $\mathbf{Cnt}(\mathbf{PV})^+$ is an ordered monoid whose identity is its smallest element, it is $\mathscr{J}$-trivial (cf. Proposition 8.2.1). From this, it immediately follows:

**Proposition 5.1.2.** *Let $\alpha, \beta \in \mathbf{Cnt}(\mathbf{PV})^+$. Then $(\alpha\beta)^\omega = (\beta\alpha)^\omega$.*

**Exercise 5.1.3.** Prove the above proposition.

Let $\mathbf{V}$ be a pseudovariety. Then there is a natural *complexity hierarchy* associated to the ordered pair $(\alpha, \beta)$ and $\mathbf{V}$ given by

$$\mathbf{V}_0 = \alpha(\mathbf{V}),$$
$$\mathbf{V}_{n+1} = \alpha\beta(\mathbf{V}_n) \text{ for } n \geq 0.$$

Here $\mathbf{V}$ is termed the *base pseudovariety*. Concretely speaking, we have $\mathbf{V}_{n+1} = (\alpha\beta)^n \alpha(\mathbf{V})$. In the case that $\mathbf{V}$ is the trivial pseudovariety, then the associated complexity hierarchy is called the *standard complexity hierarchy* associated to the ordered pair $(\alpha, \beta)$. Notice that $\mathbf{V}_\infty = (\alpha\beta)^\omega(\mathbf{V}) = (\beta\alpha)^\omega(\mathbf{V})$.

The classical complexity hierarchy is the standard complexity hierarchy associated to the pair $\alpha = \mathbf{A}*(-)$ and $\beta = \mathbf{G}*(-)$. The results of Section 4.9 give several other pairs of operators whose standard complexity hierarchy is the classical complexity hierarchy. For instance, Theorem 4.9.5 says the standard hierarchy associated to the operators $\alpha = \mathbf{A}\circledm(-)$ and $\beta = \mathbf{LS}^{\mathbf{N}}\circledm(-)$ is precisely the classical complexity hierarchy. On the other hand, Theorem 4.9.12 tells us that $\alpha = \mathbf{A}\circledm(-)$ and $\beta = (\mathbb{LER}, \mathbb{LER})\circledm(-)$ also give rise to the classical complexity hierarchy.

There is a generalization of the group complexity hierarchy, fitting into our scheme, that appeared in the paper [368]. Let $\pi$ be a collection of primes and let $\pi'$ be the complementary set of primes. A $\pi$-*group* (respectively $\pi'$-*group*) is a group whose order involves only primes from $\pi$ (respectively $\pi'$); the corresponding pseudovarieties are denoted $\mathbf{G}_\pi$ (respectively $\mathbf{G}_{\pi'}$). We extend this notion to semigroups by saying that $S$ is a $\pi$-semigroup (respectively a $\pi'$-semigroup) if it belongs to $\overline{\mathbf{G}}_\pi$ (respectively $\overline{\mathbf{G}}_{\pi'}$). A group $G$ is said to be $\pi$-*solvable* if it is in the unitary semidirect product closure of the pseudovarieties $\mathbf{G}_\pi$ and $\mathbf{G}_{\pi'}$ in the lattice of group pseudovarieties. A semigroup is said to be $\pi$-*solvable* if its maximal subgroups are $\pi$-solvable. As a consequence of the Prime Decomposition Theorem, the pseudovariety of $\pi$-solvable semigroups is the semidirect product closure of $\mathbf{G}_\pi$ and $\overline{\mathbf{G}}_{\pi'}$. The $\pi$-*length* of a $\pi$-solvable semigroup is the hierarchical complexity function $\ell_\pi$ associated to the pair of operators $\alpha = \overline{\mathbf{G}}_{\pi'} * (-)$ and $\beta = \mathbf{G}_\pi * (-)$. For instance, if $\pi$ is the set of all primes, then $\alpha = \mathbf{A} * (-)$ and $\beta = \mathbf{G} * (-)$ and one recovers the usual complexity hierarchy. Of special interest is when $\pi = \{p\}$. Then $\ell_p$ extends the usual definition of $p$-length in group theory to semigroups. A substantial part of the theory developed in Chapter 4 can be generalized to this context [368]. In particular, Tilson constructs the largest $\mathbb{LG}_\pi$-congruence on a finite semigroup and shows that $S \in \mathbb{L}\mathbf{G}_{\pi} \text{ⓜ} \mathbf{V}$ if and only if the quotient of $S$ by this congruence belongs to $\mathbf{V}$. For the case of a single prime, see the recent paper [18] for a representation theoretic treatment. See also [17] where pseudoidentities for $\mathbb{L}\mathbf{G}_p \text{ⓜ} \mathbf{V}$ are provided. The paper [129] computes the $\overline{\mathbf{G}}_{\pi}$-pointlike subsets whenever $\pi$ is a recursive set of primes.

A related hierarchy is for solvable semigroups. A semigroup is called *solvable* if its maximal subgroups are solvable. As a consequence of the Prime Decomposition Theorem, every solvable semigroup divides an iterated wreath product of solvable groups and aperiodic semigroups. So there is a natural complexity function for solvable semigroups by setting $\alpha = \mathbf{A} * (-)$ and $\beta = \mathbf{G}_{sol}*(-)$. Alternatively, one can let $\beta = \mathbf{Ab}*(-)$. This latter choice yields a complexity function generalizing the derived length of a solvable group.

The usual complexity of finite semigroups is not left-right dual. As was mentioned earlier, there are semigroups of arbitrary complexity whose opposite semigroups have complexity one [385]. This is essentially because the wreath product is not left-right dual. Two-sided complexity was developed to remove this asymmetry. One should compare this to the discussion of sequential machines versus bimachines in [84, 284].

A first guess at the definition of two-sided complexity would be to replace the operators $\mathbf{A} * (-)$ and $\mathbf{G} * (-)$ by $\mathbf{A} ** (-)$ and $\mathbf{G} ** (-)$, or from the more sophisticated point of view of q-theory, to replace $\mathbf{A}_D, \mathbf{G}_D$ by $\mathbf{A}_K, \mathbf{G}_K$. However, due to the non-associativity of the two-sided semidirect product of pseudovarieties, these latter operators are not idempotent (even in the monoidal context). One could take the idempotent infinite powers of these operators and use them to define a two-sided complexity hierarchy; this in fact will give the right answer but it will take further development to show this gives a satisfactory theory. Instead, inspired by Theorem 4.9.5, we shall use the expedient of Mal'cev products. The aforementioned theorem says that the standard complexity hierarchy is the hierarchy associated to the operators $\mathbf{A} \text{ⓜ} (-)$ and $\mathbf{LS}^{\mathbf{N}} \text{ⓜ} (-)$. The first operator is already self-dual. The second operator corresponds to $\mathscr{L}'$-maps and so to make it self-dual we should consider instead the operator corresponding to $\mathscr{J}'$-maps, $\mathbb{L}\mathbf{G} \text{ⓜ} (-)$. So two-sided complexity will be the complexity hierarchy associated to the ordered pair of operators $(\mathbf{A} \text{ⓜ} (-), \mathbb{L}\mathbf{G} \text{ⓜ} (-))$. Another justification for this choice is the following. The maximal aperiodic prime for the semidirect product is $U_2$ and $\mathbb{L}\mathbf{ER}$ is the pseudovariety of semigroups excluding $U_2$ as a divisor. On the other hand, we shall see that $U_1$ is the only non-trivial aperiodic prime for the two-sided semidirect product. Excluding $U_1$ leads to the pseudovariety $\mathbb{L}\mathbf{G}$. Thus $\mathbb{L}\mathbf{G} \text{ⓜ} (-)$ can be viewed as a two-sided analogue of $(\mathbb{L}\mathbf{ER}, \mathbb{L}\mathbf{ER}) \text{ⓜ} (-)$.

**Definition 5.1.4 (Two-sided complexity).** *Define the two-sided complexity hierarchy by*

$$\mathbf{K}_0 = \mathbf{A}$$
$$\mathbf{K}_{n+1} = \mathbf{A} \text{ⓜ} (\mathbb{L}\mathbf{G} \text{ⓜ} \mathbf{K}_n), \ n \geq 0.$$

*The corresponding hierarchical complexity function* $C\colon \mathbf{FSgp} \to \mathbb{N}$ *is called the two-sided complexity function.*

Notice that $\mathbf{C}_n \subseteq \mathbf{K}_n$ by Theorem 4.9.5 and so $C(S) \le c(S) < \infty$. It follows directly from the definition that $\mathbf{K}_n = \mathbf{K}_n^{op}$ and so $C(S) = C(S^{op})$ for any semigroup $S$. In the language of $\mathsf{q}$-theory, we can restate the definition this way:

$$\mathbf{K}_n = ((\mathbf{A}, \mathbf{A})(\mathbb{L}\mathbf{G}, \mathbb{L}\mathbf{G}))^n \, \mathsf{q}(\mathbf{A}).$$

In light of Lemma 2.8.3, the definition can also be reformulated in purely semigroup terms as follows.

**Proposition 5.1.5.** *Let $S$ be a finite semigroup. Then $C(S)$ is the minimum $n \in \mathbb{N}$ such that there exist a division d, aperiodic morphisms $\gamma_1, \ldots, \gamma_n$ and $\mathbb{L}\mathbf{G}$-morphisms $\theta_1, \ldots, \theta_n$ so that $d\gamma_1\theta_1 \cdots \gamma_n\theta_n \colon S \to T$ with $T$ aperiodic.*

By definition every semigroup in $\mathbb{L}\mathbf{G} \textcircled{m} \mathbf{A}$ has two-sided complexity at most 1. In particular, completely regular semigroups and semigroups in $\mathbf{DS}$ have two-sided complexity at most one. On the other hand, we saw in Chapter 4 that $\mathbb{L}\mathbf{G} \textcircled{m} \mathbf{A}$ contains semigroups of arbitrary complexity. In fact, if $S \in \mathbb{L}\mathbf{G} \textcircled{m} \mathbf{A}$ has complexity $n \ge 0$, then $T = S \times S^{op}$ and $T^{op}$ both have complexity at least $n$, but $C(T) = 1$. So two-sided complexity is not simply the meet of complexity and its dual.

In the sequel we shall recast the definition of two-sided complexity in terms of iterated two-sided semidirect products. To do this we need to consider maximal proper surmorphism (MPS) theory and the two-sided decomposition theory [293, 296]. In particular we will prove that standard complexity hierarchy associated to $\alpha = \mathbf{A} \ast\ast^\omega (-)$ and $\beta = \mathbf{G} \ast\ast^\omega (-)$ is the two-sided complexity hierarchy.

An alternative way to make the complexity function self-dual is to work with $\mathscr{H}'$-maps and the corresponding operator $\mathbf{G}^{\mathbf{N}} \textcircled{m} (-)$. That is, one can consider the complexity hierarchy associated to the pair $\alpha = \mathbf{A} \textcircled{m} (-)$ and $\beta = \mathbf{G}^{\mathbf{N}} \textcircled{m} (-)$. Let us denote the associated complexity function by $\kappa$. Then $c(S) \le \kappa(S)$ and at first sight it is not clear that $\kappa(S)$ need be finite. Rhodes proved in [267] that every homomorphism factors as an alternating product of morphisms that are respectively injective on $\mathscr{H}$-classes and which separate $\mathscr{H}$-classes. Applying this to the collapsing homomorphism $c_S \colon S \to 1$ shows $\kappa(S)$ is finite and bounded by $|S|$. This result will be proved in the next section. The two-sided complexity corresponds to two-sided semidirect products. No product on semigroups is known yielding $\kappa$.

*Question 5.1.6.* Study the complexity function $\kappa$ associated to the operators $\alpha = \mathbf{A} \textcircled{m} (-)$ and $\beta = \mathbf{G}^{\mathbf{N}} \textcircled{m} (-)$.

Another complexity hierarchy, implicitly considered in [296], comes from choosing $\alpha = \mathbf{A} \textcircled{m} (-)$ and $\beta = \mathbf{CS} \textcircled{m} (-)$. Let $\beta_1 = \mathbf{G} \ast (-)$ and let $\beta_2 = (-) \ast_r \mathbf{G}$ where $\ast_r$ denotes the reverse semidirect product of pseudovarieties.

The results of [296] (in particular, Proposition 3.8), show that $\beta = (\beta_1\beta_2)^\omega$; this requires a more refined version of the MPS classification than we present in this book. Let $\mu$ be the hierarchical complexity function associated to these operators. We claim $\mu = C$. This is an immediate consequence of the fact that

$$\mathbf{CS} \,\textcircled{m}\, (\mathbf{A} \,\textcircled{m}\, \mathbf{V}) = \mathbb{LG} \,\textcircled{m}\, (\mathbf{A} \,\textcircled{m}\, \mathbf{V})$$

for any pseudovariety $\mathbf{V}$. Because $\mathbf{CS} \leq \mathbb{LG}$, the inclusion from left to right is clear. For the reverse inclusion, suppose $S \in \mathbb{LG} \,\textcircled{m}\, (\mathbf{A} \,\textcircled{m}\, \mathbf{V})$. Then every $\mathbf{A} \,\textcircled{m}\, \mathbf{V}$-idempotent pointlike subsemigroup $X$ of $S$ belongs to $\mathbb{LG}$. By a result of Henckell [123, 130], each maximal $\mathbf{A} \,\textcircled{m}\, \mathbf{V}$-pointlike subsemigroup $X$ of $S$ satisfies $X^2 = X$. But a local group satisfying $X^2 = X$ is completely simple. Hence each maximal $\mathbf{A} \,\textcircled{m}\, \mathbf{V}$-idempotent pointlike subsemigroup $X$ of $S$ belongs to $\mathbf{CS}$ and so $S \in \mathbf{CS} \,\textcircled{m}\, (\mathbf{A} \,\textcircled{m}\, \mathbf{V})$. The same argument shows that $\mathbf{W}^\mathbf{N} \,\textcircled{m}\, (\mathbf{A} \,\textcircled{m}\, \mathbf{V}) = \mathbf{W} \,\textcircled{m}\, (\mathbf{A} \,\textcircled{m}\, \mathbf{V})$ for any pseudovariety $\mathbf{W}$. Consequently, $\alpha = \mathbf{A} \,\textcircled{m}\, (-)$ and $\beta = \mathbb{LS} \,\textcircled{m}\, (-)$ define the group complexity hierarchy.

One can also consider two-sided $\pi$-length. Let $\alpha = \overline{\mathbf{G}}_{\pi'} \,\textcircled{m}\, (-)$ and $\beta = \mathbb{LG}_\pi \,\textcircled{m}\, (-)$. The complexity hierarchy associated to $\alpha$ and $\beta$ has union all $\pi$-solvable semigroups. The results of Section 5.3 imply that this is the same as the hierarchy associated to $\overline{\mathbf{G}}_{\pi'} \,**^\omega\, (-)$ and $\mathbf{G}_\pi \,**^\omega\, (-)$. The associated complexity function is called the *two-sided $\pi$-length*.

Similarly there is a two-sided complexity function for solvable semigroups given by taking $\alpha = \mathbf{A} \,\textcircled{m}\, (-)$ and $\beta = \mathbb{LG}_{sol} \,\textcircled{m}\, (-)$.

Another interesting hierarchy for aperiodic semigroups, related to the dot-depth hierarchy, is obtained by choosing $\alpha = \mathbf{Sl} \,**\, (-)$ and $\beta = \mathbb{L1}\textcircled{m}(-)$. The results of Section 5.3.2 below imply that every aperiodic semigroup belongs to an element of this hierarchy. From a language theoretic point-of-view, the operator $\mathbf{Sl} \,**\, (-)$ corresponds to forming marked products $L_1aL_2$ (and closing under Boolean operations) and the operator $\mathbb{L1} \,\textcircled{m}\, (-)$ corresponds to forming arbitrary length unambiguous marked products $L_1a_1L_2 \cdots a_nL_{n+1}$ [229]. On the other hand, the Schützenberger product $\Diamond(-)$ corresponds to forming arbitrary marked products and closing under Boolean operations [229]. Because $\mathbb{L1} \,\textcircled{m}\, (-)$ preserves decidability and $\mathbf{Sl}$ is local, the associated hierarchy for $\mathbf{A}$ seems like a promising way to get a first approximation of dot-depth.

*Question 5.1.7.* Is the complexity function for $\mathbf{A}$ associated to the operators $\alpha = \mathbf{Sl} \,**\, (-)$ and $\beta = \mathbb{L1} \,\textcircled{m}\, (-)$ computable? How exactly does it compare to the dot-depth?

*Remark 5.1.8.* Notice that the complexity of two positive operators $\alpha$ and $\beta$ can be defined by taking a complexity hierarchy in $\mathbf{Cnt}(\mathbf{PV})^+$. That is, we can take $f_0 = \alpha$ and $f_{n+1} = \alpha\beta f_n = f_n\beta\alpha = (\alpha\beta)^n\alpha$, for $n \geq 0$. The sequence $\{f_n\}$ is increasing, because we are considering positive operators, and

$$\bigvee f_n = (\alpha\beta)^\omega = (\beta\alpha)^\omega.$$

The complexity hierarchy associated to $\alpha$ and $\beta$ with base pseudovariety $\mathbf{V}$ is given by $\mathbf{V}_n = f_n(\mathbf{V})$.

As a final remark, we should mention that the complexity of two opera-tors is very closely related to considering complexity hierarchies in $\mathbf{CC}^+$ or $\mathbf{PVRM}^+$. Namely, if A and B are members of $\mathbf{CC}^+$ (respectively $\mathbf{PVRM}^+$), one can define a complexity hierarchy in $\mathbf{CC}^+$ (respectively $\mathbf{PVRM}^+$) by

$$V_0 = A$$
$$V_{n+1} = A \odot B \odot V_n$$
$$= V_n \odot B \odot A$$
$$= (A \odot B)^n \odot A, \ n \geq 0.$$

If $\alpha = A\mathsf{q}$ and $\beta = B\mathsf{q}$, then the complexity hierarchy with base pseudovariety $\mathbf{V}$ is nothing more than the hierarchy $V_n\mathsf{q}(\mathbf{V})$, $n \geq 0$ (in fact, the $\mathsf{q}$-image of $\{V_n\}$ is the hierarchy considered in Remark 5.1.8). Notice computability of the standard hierarchy associated to $\alpha$ and $\beta$ (so taking $\mathbf{V} = \mathbf{1}$) is implied by computability of the hierarchy for A and B by our usual modeling process. Tilson considers hierarchies of this sort for $\mathbf{PVRM}^+$ in [362], although he only considers the case where A is idempotent. The reason he makes this assumption is that if A is idempotent, and $f$ is the associated hierarchical complexity function, then $f(\varphi\psi) \leq f(\varphi) + f(\psi)$.

In Remark 4.9.18 a one-sided complexity hierarchy was defined for rela-tional morphisms. The analogous two-sided complexity hierarchy would be given by

$$K_0 = (\mathbf{A}, \mathbf{A})$$
$$K_{n+1} = (\mathbf{A}, \mathbf{A})(\mathbb{LG}, \mathbb{LG})K_n$$
$$= K_n(\mathbb{LG}, \mathbb{LG})(\mathbf{A}, \mathbf{A}), \ n \geq 0,$$

that is, by the hierarchy associated to $(\mathbf{A}, \mathbf{A})$ and $(\mathbb{LG}, \mathbb{LG})$. The associated hierarchical complexity function for relational morphisms will be denoted by $C$. Notice $\mathbf{K}_n = K_n\mathsf{q}(\mathbf{1})$. With these definitions, if $\varphi\colon S \to T$, then

$$C(S) \leq C(\varphi) + C(T)$$

is a tautology as $S \in K_{C(\varphi)}\mathsf{q}(T)$. For one-sided complexity, this is a deep theorem.

## 5.2 Maximal Proper Surmorphisms

To relate Mal'cev products to two-sided semidirect products, we need to con-sider surjective morphisms ("surmorphisms") that do not factor. Let us agree to call a surjective homomorphism *proper* if it is not an isomorphism.

**Definition 5.2.1 (Maximal proper surmorphism (MPS)).** *A maximal proper surmorphism, or MPS, is a proper surjective homomorphism $\theta\colon S \twoheadrightarrow T$ of finite semigroups such that $\theta = \theta_1\theta_2$ implies $\theta_1$ or $\theta_2$ is an isomorphism.*

Notice that MPSs correspond with minimal congruences: that is, $\theta\colon S \twoheadrightarrow T$ is an MPS if and only if $\ker\theta$ is a minimal congruence on $S$. In particular, being subdirectly indecomposable is equivalent to having a unique MPS (up to isomorphism). For example, if $\theta\colon G \to H$ is an MPS of groups, then $\ker\theta$ is a direct power of a finite simple group, being a minimal normal subgroup (cf. Lemma 4.7.4).

**Exercise 5.2.2.** Show that $\theta\colon S \twoheadrightarrow T$ is an MPS if and only if $\ker\theta$ is a minimal non-trivial congruence on $S$.

**Definition 5.2.3 (MPS factorization).** *If $\theta\colon S \to T$ is a proper surjective morphism, a factorization $\theta = \theta_1\theta_2\cdots\theta_n$ with the $\theta_i$ MPSs is called an MPS factorization of $\theta$.*

**Lemma 5.2.4.** *A proper surjective morphism admits an MPS factorization.*

*Proof.* Let $\theta\colon S \twoheadrightarrow T$ be an onto homomorphism with non-trivial kernel. One works by induction on $|S| + |T|$. If $\theta$ is an MPS, there is nothing to prove. Otherwise, $\theta = \alpha\beta$ where $\alpha, \beta$ are proper surjective morphisms. One now factors $\alpha$ and $\beta$ by induction.                                              $\square$

MPSs were originally classified by the first author in [267] (see also [171]). In [296], a more detailed version of the classification is given. For the sake of brevity, we shall skip over the classification into types and prove only the results needed for the decomposition theorem. In this sense, our presentation follows [293] rather than [296].

If $\varphi\colon S \to T$ is a relational morphism and $X, Y \subseteq S$, then we say that $\varphi$ *separates* $X$ and $Y$ if $X\varphi \cap Y\varphi = \emptyset$. If $J$ is a $\mathscr{J}$-class of a semigroup $S$, define

$$A(J) = \{s \in S \mid s >_{\mathscr{J}} J\}$$
$$B(J) = S \setminus (A(J) \cup J).$$

Here $A$ is for above and $B$ for below. Notice that $S = A(J) \uplus J \uplus B(J)$ and that $B(J)$ and $J \cup B(J)$ are ideals; see Figure 5.1.

The key notion in MPS theory is that of a $\mathscr{J}$-singular class. Our terminology follows [293]; a weaker notion is considered in [296].

**Definition 5.2.5 ($\mathscr{J}$-singular).** *A relational morphism $\theta\colon S \to T$ of semigroups is said to be $\mathscr{J}$-singular if there is a $\mathscr{J}$-class $J$ of $S$ such that:*

1. *$\varphi$ is injective on $S \setminus J$;*
2. *$\varphi$ separates $J$ and $A(J)$.*

*The $\mathscr{J}$-class $J$ is said to be $\mathscr{J}$-singular for $\theta$.*

*Remark 5.2.6.* It follows from the definition that a $\mathscr{J}$-singular relational morphism that is not a division can have at most two $\mathscr{J}$-singular $\mathscr{J}$-classes and that these must be incomparable in the $\mathscr{J}$-order.

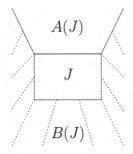

**Fig. 5.1.** $S = A(J) \uplus J \uplus B(J)$

**Exercise 5.2.7.** Show that if $\varphi \colon S \to T$ is $\mathscr{J}$-singular with $\mathscr{J}$-singular $\mathscr{J}$-class $J$, then $\varphi$ separates $A(J)$ from $J \cup B(J)$.

**Exercise 5.2.8.** Let $\varphi \colon S \to T$ be a $\mathscr{J}$-singular relational morphism. Show that $\varphi^I \colon S^I \to T^I$ is a $\mathscr{J}$-singular relational morphism (where $\varphi^I(I) = I$ and $\varphi^I|_S = \varphi$).

We now prove that MPSs are $\mathscr{J}$-singular.

**Lemma 5.2.9.** *Let $\theta \colon S \twoheadrightarrow T$ be an MPS. Let $I$ be a maximal proper ideal (perhaps empty) of $S$ on which $\theta$ is injective and let $J$ be a $\leq_{\mathscr{J}}$-minimal $\mathscr{J}$-class of $S$ not contained in $I$. Then $J$ is $\mathscr{J}$-singular for $\theta$.*

*Proof.* Because $\theta$ is not injective, such an ideal $I$ exists. Set $I' = J \cup I$; it is an ideal by minimality of $J$. Moreover, $A(J) \cap I' = \emptyset$. Because $I \subsetneq I'$, the restriction of $\theta$ to $I'$ is not injective. Hence the intersection of $\ker \theta$ and the Rees congruence $\sim_{I'}$ associated to $I'$ is non-trivial. By minimality of $\ker \theta$, it follows $\ker \theta \subseteq \sim_{I'}$. Hence $\theta$ separates $J$ and $A(J)$, $\theta|_{S \setminus I'}$ is injective and $\theta$ separates $I'$ from $S \setminus I'$. Because $\theta|_I$ is injective by choice of $I$, we conclude $\theta|_{S \setminus J}$ is injective. Therefore, $J$ is $\mathscr{J}$-singular for $\theta$, as required.     □

A homomorphism $\theta \colon S \to T$ is said to be an $\mathscr{H}$-*morphism* if $s\theta = s'\theta$ implies $s \mathrel{\mathscr{H}} s'$, i.e., $\ker \theta \subseteq \mathscr{H}$.

**Lemma 5.2.10.** *Let $\theta \colon S \twoheadrightarrow T$ be an MPS. Then either $\theta$ is an $\mathscr{H}$-morphism or $\theta$ is injective on $\mathscr{H}$-classes. In particular, an MPS is either a $\mathbf{G}^{\mathbf{N}}$-morphism (and hence an $\mathbb{L}\mathbf{G}$-morphism) or an aperiodic morphism.*

*Proof.* Because $\theta$ is proper, it cannot be both an $\mathscr{H}$-morphism and injective on $\mathscr{H}$-classes. Assume $\theta$ is not injective on $\mathscr{H}$-classes. Let $\sim$ be the intersection of $\ker \theta$ and $\mathscr{H}$. Then $\sim$ is a non-trivial equivalence relation contained in $\ker \theta$. If we can show that it is a congruence, then by minimality of $\ker \theta$ we shall obtain $\ker \theta \subseteq \mathscr{H}$, i.e., $\theta$ is an $\mathscr{H}$-morphism.

Let $J$ be a $\mathscr{J}$-singular class for $\theta$. If $s \sim s'$ and $s \neq s'$, then as $s \mathrel{\mathscr{H}} s'$, the definition of a $\mathscr{J}$-singular $\mathscr{J}$-class immediately implies $s, s' \in J$. We first verify $\sim$ is a left congruence. Let $u \in S$; automatically $us\theta = us'\theta$. We must show $us \; \mathscr{H} \; us'$. Clearly $us \; \mathscr{R} \; us'$, so either both belong to $J$ or neither. If neither $us$ nor $us'$ belongs to $J$, then by $\mathscr{J}$-singularity and from $us\theta = us'\theta$, we obtain $us = us'$ and in particular $us \; \mathscr{H} \; us'$. Next suppose $us, us' \in J$. Then by stability $us \; \mathscr{L} \; s \; \mathscr{L} \; s' \; \mathscr{L} \; us'$. Therefore, we have $us \; \mathscr{H} \; us'$, as required. The proof that $\sim$ is a right congruence is dual.     □

Notice this lemma gives another proof that the two-sided complexity of a semigroup is finite. Indeed, if $S$ is a finite semigroup and $\theta = \theta_1 \cdots \theta_n$ is an MPS factorization of the collapsing morphism $\theta \colon S \to \{1\}$, then Lemma 5.2.10 implies that $C(S) \leq n$ (in fact $\kappa(S) \leq n$).

The last fact that we shall need about MPSs is that if $\theta$ is an MPS, then the kernel of $\theta$ restricted to any maximal subgroup is either trivial or a minimal normal subgroup.

**Proposition 5.2.11.** *Let* $\theta \colon S \twoheadrightarrow T$ *be an MPS and let* $G$ *be a maximal subgroup of* $T$. *Then* $\theta|_G$ *is either injective or an MPS of groups.*

*Proof.* Assume $\theta|_G$ is not injective and suppose $N \leq \ker \theta|_G$ is a non-trivial normal subgroup of $G$. Lemma 4.7.9 provides a homomorphism $\varphi \colon S \to T$ with $\ker \varphi|_G = N$. Then $\ker \varphi \cap \ker \theta$ is non-trivial, and so $\ker \theta \subseteq \ker \varphi$ by minimality. Thus $\ker \theta|_G = N$, and so $\ker \theta|_G$ is a minimal normal subgroup, as was required.     □

*Remark 5.2.12 (Reduction to monoids).* Notice that $\varphi \colon S \to T$ is an MPS of semigroups if and only if $\varphi^I \colon S^I \to T^I$ is an MPS of monoids. Therefore, we may often reduce our study of MPSs to the case of monoids.

## 5.3 The Two-Sided Decomposition Theory

The goal of this section is to study membership of MPSs in pseudovarieties of the form $\mathbf{V}_K$ or $\ell\mathbf{1}_K$ [293, 296]. These results lead to an expression of certain Mal'cev products as infinite iterated two-sided semidirect products. There are innumerable applications to Formal Language Theory [234, 343, 378, 379]. The formalism of pseudovarieties of relational morphisms will allow us to state the results in a more elegant form than in the papers [293, 296].

### 5.3.1 The MPS Decomposition Theorem

It will be convenient to define the notion of the Mal'cev kernel of a relational morphism.

**Definition 5.3.1 (Mal'cev kernel).** *If* $\varphi \colon S \to T$ *is a relational morphism, then the Mal'cev kernel of* $\varphi$ *is the pseudovariety generated by the semigroups* $e\varphi^{-1}$ *with* $e \in T$ *an idempotent.*

Notice that two relational morphisms have the same Mal'cev kernel if and only if they belong to the same pseudovarieties of the form $(\mathbf{V}, 1)$.

By the results of Section 4.7, the congruence-free finite monoids are precisely the finite simple groups and $U_1$. Moreover, these monoids are self-dual. The two-sided Prime Decomposition Theorem for finite monoids will provide a decomposition into congruence-free divisors, affording a nicer parallel to the situation for groups.

The main result of this section is the MPS Decomposition Theorem [293, 296]. The reader is referred to (2.48) and (2.55) for the meaning of $\ell\mathbf{V}_D$ and $\ell\mathbf{V}_K$ when $\mathbf{V}$ is a pseudovariety of monoids.

**Theorem 5.3.2 (MPS Decomposition Theorem).** *Let $\theta\colon S \twoheadrightarrow T$ be an MPS. Then:*

1. *$\theta$ is an $\mathbb{LG}$-morphism, but not an aperiodic morphism, if and only if $\theta \in (Q)_K$ where $Q \neq 1$ is a simple group in the Mal'cev kernel of $\theta$;*
2. *$\theta$ is an $\mathbb{L1}$-morphism if and only if $\theta \in \ell 1_K$;*
3. *$\theta$ is an aperiodic morphism, but not an $\mathbb{L1}$-morphism, if and only if $\theta \in \mathbf{Sl}_K \ell 1_K$ and $U_1$ is in the Mal'cev kernel of $\theta$.*

*Moreover, exactly one of these three cases arises.*

In other words, the kernel semigroupoid of an MPS divides a category, which locally belongs to the pseudovariety generated by some congruence-free monoid in the Mal'cev kernel of the MPS. The proof of this theorem is built out of several decomposition results that are of interest in their own right, as they contain more explicit information than Theorem 5.3.2. The third case is by far the most difficult. First we handle the if directions.

**Lemma 5.3.3.** *Let $\varphi\colon S \to T$ be a relational morphism of semigroups and $e \in E(T)$. Let $S' = e\varphi^{-1}$ and $f \in E(S')$. Then $fS'f$ embeds in $K_\varphi((e,e))$.*

*Proof.* Define an injective homomorphism $\psi\colon S' \to \mathbf{Ker}(\varphi)((e,e))$ by sending $s$ to $(e, (s, e), e)$. We claim the projection $\sigma\colon \mathbf{Ker}(\varphi) \to K_\varphi$ is injective on the image of $fS'f$. Indeed, suppose $(e, (s, e), e)\sigma = (e, (s', e), e)\sigma$ with $s, s' \in fS'f$. Then as $f \in e\varphi^{-1}$, we deduce from (2.51) that $s = fsf = fs'f = s'$, as required. $\square$

The following immediate corollary establishes the if directions of Theorem 5.3.4.

**Corollary 5.3.4.** *Suppose $\mathbf{V}$ is a pseudovariety of semigroups or monoids and $\varphi \in \ell\mathbf{V}_K$. Then we have $\varphi \in (\mathbb{L}\mathbf{V}, 1)$.*

The next corollary will be used later to determine the primes with respect to the two-sided semidirect product.

**Corollary 5.3.5.** *Let $\pi\colon S \bowtie T \to T$ be the projection. Let $e \in E(T)$ and suppose $M \leq e\pi^{-1}$ is a monoid. Then $M \prec S$.*

*Proof.* We know $K_\pi \prec S$ by the Kernel Semigroupoid Theorem. Lemma 5.3.3 then yields the result. $\qquad\square$

For the remainder of this section, we shall use without comment the fact that pseudovarieties of groups are local in the monoidal context [364] and that the pseudovariety of semilattices is local [60,85,364]. The reader should recall that $\mathbf{V}_D \leq \mathbf{V}_K$ and $\ell\mathbf{V}_D \leq \ell\mathbf{V}_K$. Turning to the decomposition results, we begin with the case that $\theta$ is not aperiodic. Here we follow the coordinate-free proof given in [293]. This has the advantage of being shorter and using less machinery, but one does not really know how the result was thought up. A more conceptual proof using Rees coordinates can be found in [296].

**Proposition 5.3.6.** *Suppose $\theta\colon S \twoheadrightarrow T$ is an MPS that is not aperiodic. Then $\theta \in (Q)_D$ where $Q$ is a simple group in the Mal'cev kernel of $\theta$.*

*Proof.* By adjoining identities, we may assume without loss of generality $\theta$ is a monoid morphism. Because $\theta$ is not aperiodic, there is a maximal subgroup $G$ of $S$ such that $N = \ker\theta|_G$ is non-trivial. Then $N$ is a minimal normal subgroup of $G$ by Proposition 5.2.11, and hence $N = Q^m$ where $Q$ is a simple group (cf. Lemma 4.7.4). Notice $N$, and hence $Q$, belongs to the Mal'cev kernel of $\theta$. Our goal is to show that $D_\theta \in \ell(Q)$. It will then follow that $D_\theta \in \mathbf{g}(Q)$ by locality of group pseudovarieties in the monoidal context [364].

Let $J$ be the $\mathscr{J}$-class of $G$; then $J$ must be $\mathscr{J}$-singular. Because $\theta$ is not aperiodic, it must be an $\mathscr{H}$-morphism by Lemma 5.2.10. Consequently, if $t_0 \notin J\theta$, then $t_0\theta^{-1}$ is a singleton and hence the local monoid $D_\theta(t_0)$ is trivial.

Suppose $t_0 \in J\theta$. Set $M = \mathbf{Der}(\theta)(t_0)$ and let $\sigma\colon \mathbf{Der}(\theta) \to D_\theta$ be the quotient map. Because $\theta$ is an $\mathscr{H}$-morphism, $t_0\theta^{-1}$ is contained in an $\mathscr{H}$-class $H$ of $S$. Let $e \in G$ be the identity. Choose $r, \ell \in J$ such that $G\ \mathscr{L}\ r\ \mathscr{R}\ H$ and $G\ \mathscr{R}\ \ell\ \mathscr{L}\ H$. Then $rG\ell = H$. By Green's Lemma (or Rees's Theorem) we can find $r', \ell' \in J$ such that: $r'H\ell' = G$, $r'r = e = \ell\ell'$ and $rr'h = h$, $h\ell'\ell = h$, for all $h \in H$. (Choose $r'$ to be an inverse of $r$ in the $\mathscr{R}$-class of $G$ and $\ell'$ to be an inverse of $\ell$ in the $\mathscr{L}$-class of $G$ as per Proposition A.1.15.)

Define $\varphi\colon M \to N$ by $(t_0,(s,t))\varphi = \ell s\ell'$. We show that $\varphi$ induces an embedding of $D_\theta(t_0)$ into $N$. First of all we verify $\varphi$ takes values in $G$. Fix $s_0 \in t_0\varphi^{-1}$. Then $s_0 \in H$, and so $r's_0\ell' \in G$ and $s_0\ell'\ell = s_0$, yielding

$$r's_0s\ell' = r's_0\ell'\ell s\ell' = (r's_0\ell')(\ell s\ell'). \tag{5.1}$$

As $t_0t = t_0$, and so $(s_0s)\theta = t_0$, we also have $s_0s \in H$. Therefore, $r's_0s\ell' \in G$ and so $\ell s\ell' \in J$ by (5.1). Stability then yields $\ell s\ell' \in G$. Next we verify $\ell s\ell' \in N$. Using that $r's_0s\ell'$, $r's_0\ell'$ and $\ell s\ell'$ belong to the group $G$, we obtain

$$(r's_0\ell')\theta = (r'\theta)t_0(\ell'\theta) = (r'\theta)t_0t(\ell'\theta) = (r's_0s\ell')\theta = (r's_0\ell')\theta(\ell s\ell')\theta$$

from which we conclude $\ell s\ell' \in N$.

Now we check that $\varphi$ is a homomorphism. A routine computation shows this boils down to checking $\ell ss'\ell' = \ell s\ell'\ell s'\ell'$ assuming $t_0(s\theta) = t_0 = t_0(s'\theta)$.

Because $s_0 \in H$ and $\ell \mathscr{L} H$, it follows that $\ell = x s_0$ some $x \in S$. In addition, $s_0 s \in t_0 \theta^{-1} \subseteq H$, and so $s_0 s \ell' \ell = s_0 s$. Therefore, we have

$$\ell s \ell' \ell s' \ell = x s_0 s \ell' \ell s' \ell' = x s_0 s s' \ell' = \ell s s' \ell'$$

as required.

Finally, we must show that $(t_0, (s, t))\varphi = (t_0, (s', t'))\varphi$ implies $(t_0, (s, t))\sigma = (t_0, (s', t'))\sigma$. Suppose $h \in t_0 \varphi^{-1}$. Then, as $h \in H$, one verifies

$$r' h s \ell' = r' h \ell' \ell s \ell' = r' h \ell' (t_0, (s, t))\varphi$$
$$= r' h \ell' (t_0, (s', t'))\varphi = r' h \ell' \ell s' \ell' = r' h s' \ell'.$$

But from $t_0 t = t_0 = t_0 t'$ it follows $hs, hs' \in t_0 \varphi^{-1} \subseteq H$. Therefore,

$$hs = r r' h s \ell' \ell = r r' h s' \ell' \ell = h s'$$

whence $(t_0, (s, t))\sigma = (t_0, (s', t'))\sigma$, as required.     □

Next we turn to the case of an $\mathbb{L}1$-morphism. This is the only case where we need the kernel semigroupoid; all the other decompositions can be achieved via the derived semigroupoid or its dual.

**Proposition 5.3.7.** *Let $\theta \colon S \twoheadrightarrow T$ be a $\mathscr{J}$-singular surjective $\mathbb{L}1$-morphism. Then $\theta \in \ell 1_K$.*

*Proof.* Again, we may assume without loss of generality $\theta$ is a monoid morphism. Suppose $(t_L, (s, t), t_R) \in \mathbf{Ker}(\theta)((t_L, t_R))$; so $t_L t = t_L$, $t t_R = t_R$ and in particular $t_L t_R = t_L t t_R$. We need to show $(t_L, (s, t), t_R)\sigma = (t_L, (1, 1), t_R)\sigma$, or equivalently, for all $s_L \in t_L \theta^{-1}$ and $s_R \in t_R \theta^{-1}$,

$$s_L s s_R = s_L s_R. \tag{5.2}$$

Let $J$ be a $\mathscr{J}$-singular $\mathscr{J}$-class of $\theta$.

*Case 1.* Suppose $s_L \in A(J)$. Because $s_L s \theta = s_L \theta$ and $\theta$ separates $A(J)$ from $J \cup B(J)$ (cf. Exercise 5.2.7), clearly $s_L s \in A(J)$. As $\theta|_{A(J)}$ is injective, it follows $s_L s = s_L$, yielding (5.2). A dual argument holds if $s_R \in A(J)$.

*Case 2.* Suppose $s_L \in B(J)$. As $B(J)$ is an ideal, $s_L s \in B(J)$. Then, from $s_L s \theta = s_L \theta$ and the injectivity of $\theta$ on $B(J)$, we deduce $s_L s = s_L$, again establishing (5.2). The situation is dual if $s_R \in B(J)$.

*Case 3.* We are left with the case $s_L, s_R \in J$. First suppose $s_L s_R$ and $s_L s s_R$ both belong to $B(J)$. Then, as $\theta|_{B(J)}$ is injective and $s_L s_R \theta = s_L s s_R \theta$, we obtain (5.2). So we may assume at least one of $s_L s_R$ or $s_L s s_R$ belongs to $J$; in particular, $J$ must be a regular $\mathscr{J}$-class. But $\theta$ is a $\mathscr{J}'$-morphism and so must separate $J$ from $B(J)$. We conclude both $s_L s_R$ and $s_L s s_R \in J$. Stability then yields $s_L s_R \mathscr{H} s_L s s_R$ and so aperiodicity of $\theta$ establishes (5.2) (by the Aperiodicity Lemma).     □

Notice that if $\theta \colon S \to T$ is a $\mathscr{J}$-singular morphism with a null $\mathscr{J}$-singular $\mathscr{J}$-class, then $\theta$ is an $\mathbb{L}1$-morphism. This is because the inverse image of an idempotent in this case can only have at most one regular element. Now if $S$ is a semigroup with a null 0-minimal ideal $N$, then the quotient $S \to S/N$ is $\mathscr{J}$-singular with $\mathscr{J}$-singular $\mathscr{J}$-class $N$. Therefore, we have the following corollary of Proposition 5.3.7.

**Corollary 5.3.8.** *Suppose $S$ is a semigroup with a null 0-minimal ideal $N$ and let $\varphi \colon S \to S/N$ be the quotient. Then $\varphi \in \ell\mathbf{1}_K$.*

Before turning to the general case of an aperiodic MPS, we consider a special case that will afford us an easy proof of Stiffler's Theorem [340].

**Proposition 5.3.9.** *Let $\theta \colon S \to T$ be a $\mathscr{J}$-singular relational morphism which is injective on $\mathscr{R}$-classes. Then $\theta \in \mathbf{Sl}_D$. Consequently, a $\mathscr{J}$-singular relational morphism injective on $\mathscr{L}$-classes or $\mathscr{R}$-classes belongs to $\mathbf{Sl}_K$.*

*Proof.* As before, we may assume without loss of generality that $\theta$ is a relational morphism of monoids. By Simon's Theorem [60], it suffices to show $\theta \in \ell\mathbf{Sl}_D$. Let $J$ be a $\mathscr{J}$-singular class for $\theta$ and consider a local monoid $\mathbf{Der}(\theta)(t_0)$. If $t_0 \in A(J)\theta$, then $t_0\theta^{-1}$ is a singleton and so $D_\theta(t_0)$ is trivial. So we may assume $t_0 \in (J \cup B(J))\theta$.

Let $(t_0, (s, t)) \in \mathbf{Der}(\theta)(t_0)$; so $t_0 t = t_0$. Because $\theta|_{S \setminus J}$ is injective and $\theta$ separates $J$ from $A(J)$, there is at most one element $b$ in $t_0\theta^{-1} \setminus J$ and necessarily $b \in B(J)$. Because $bs \in B(J)$ and $t_0 = t_0 t \in b\theta s\theta \subseteq bs\theta$, we conclude $bs = b$ by injectivity of $\theta|_{B(J)}$. We claim that, for $s_0 \in t_0\theta^{-1} \cap J$, either $s_0 s = s_0$ or $s_0 s = b$. It then follows directly $D_\theta(t_0)$ is a semilattice (identify $D_\theta(t_0)$ with a subsemilattice of the semilattice of partial identities on $J \cap t_0\theta^{-1}$). To verify the claim, observe $s_0 s \in t_0\theta^{-1}$ and so either $s_0 s = b$ or $s_0 s \in J$. In the latter case $s_0 s \mathscr{R} s_0$ by stability, whence injectivity on $\mathscr{R}$-classes yields $s_0 s = s_0$.

The final statement of the proposition follows since membership in $\mathbf{Sl}_K$ is left-right dual.                                                                                   □

We now provide an alternate proof [296] of Stiffler's Theorem (Theorem 4.5.2). This proof technique was extended by the second author to prove the locality of $\mathbf{R}$ [332].

**Corollary 5.3.10.** *The pseudovariety of $\mathscr{R}$-trivial semigroups is the smallest pseudovariety of semigroups closed under semidirect product containing $\mathbf{Sl}$.*

*Proof.* We just prove the difficult direction: if $S \in \mathbf{R}$, then $S$ belongs to the semidirect product closure of $\mathbf{Sl}$. We proceed by induction on $|S|$. If $S$ is trivial, we are done. If not, $S$ admits a non-trivial congruence and hence there is an MPS $\theta \colon S \twoheadrightarrow T$. By induction, $T$ is in the semidirect product closure of $\mathbf{Sl}$. Because $S$ is $\mathscr{R}$-trivial, $\theta$ is injective on $\mathscr{R}$-classes. Proposition 5.3.9 then implies $\theta \in \mathbf{Sl}_D$, completing the proof.                              □

Our final decomposition result concerns the aperiodic case. Our proof again is modeled upon [293].

**Proposition 5.3.11.** *Let* $\theta \colon S \to T$ *be an MPS that is not an* $\mathbb{L}\mathbf{G}$*-morphism. Then* $\theta = \varphi\psi$ *with* $\varphi^{op} \in \mathbf{Sl}_D$ *a relational morphism and* $\psi \in \ell 1_D$ *a homomorphism.*

*Proof.* Without loss of generality, we may assume that $\theta$ is a monoid homomorphism. Let $J$ be a $\mathcal{J}$-singular $\mathcal{J}$-class of $\theta$. We use a construction termed the "Stiffler expansion" in [296]. Choose a monoid $M$ such that $M \setminus \{1\}$ is in bijection with the set of $\mathcal{R}$-classes of $J$. For $a \in M \setminus \{1\}$, we write $R_a$ for the corresponding $\mathcal{R}$-class of $J$. Because $\theta|_{A(J)}$ is injective, for each $t \in A(J)\theta$, there is a unique element in $A(J) \cap t\theta^{-1}$ — denote it by $\tilde{t}$. The decomposition $\theta = \varphi\psi$ will be achieved by means of the wreath product. Consider the unitary wreath product $M \wr T$. For each $s \in S$, define $f_s \colon T \to M$ by

$$tf_s = \begin{cases} a & t \in A(J)\theta, \ \tilde{t}s \in R_a \\ 1 & \text{else.} \end{cases} \tag{5.3}$$

Let $\varphi \colon S \to M \wr T$ be the relational morphism whose graph is generated by all pairs $(s, (f_s, s\theta))$ with $s \in S$. Let $\pi \colon M \wr T \to T$ be the projection and set $\psi = \pi|_{S\varphi}$. By construction, $\varphi\psi = \theta$.

First we show $\psi \in \ell 1_D$, i.e., $D_\psi \in \ell 1$. This amounts to showing that if $t_L \in T$ and $(f, t) \in S\varphi$ with $t_L t = t_L$, then for all $(f_L, t_L) \in t_L\psi^{-1}$ the equality $(f_L, t_L)(f, t) = (f_L, t_L)$ holds. A routine computation yields

$$(f_L, t_L)(f, t) = (f_L{}^{t_L}f, t_L t) = (f_L{}^{t_L}f, t_L)$$

and so it suffices to show ${}^{t_L}f$ is the identity of $M^T$, that is, $t_0{}^{t_L}f = 1$ for all $t_0 \in T$. Write $(f, t) = s\varphi$. Then by definition of $\varphi$, there exists a factorization $s = s_1 \cdots s_n$ such that

$$(f, t) = (f_{s_1}, s_1\theta) \cdots (f_{s_n}, s_n\theta). \tag{5.4}$$

Let $t_0 \in T$. A straightforward computation establishes

$$t_0{}^{t_L}f = (t_0 t_L) f_{s_1}(t_0 t_L s_1\theta) f_{s_2} \cdots (t_0 t_L s_1\theta \ldots s_{n-1}\theta) f_{s_n}. \tag{5.5}$$

We make two observations:

1. $T \setminus A(J)\theta = (J \cup B(J))\theta$ is an ideal of $T$;
2. $t_0 t_L s_1\theta \cdots s_i\theta \mathcal{R} t_0 t_L s_1\theta \cdots s_j\theta$ for all $0 \le i, j \le n$.

Indeed, 1 holds because $\theta$ separates $J \cup B(J)$ from $A(J)$, and the set $J \cup B(J)$ is an ideal. Hence $T \setminus A(J)\theta = (J \cup B(J))\theta$ is an ideal. The second observation follows as

$$t_0 t_L = t_0 t_L t = t_0 t_L s\theta = t_0 t_L s_1\theta \cdots s_n\theta.$$

Now in order for $t_0{}^{t_L}f \neq 1$ to occur, we must have, by (5.5),

$$(t_0 t_L s_1 \theta \cdots s_{k-1}\theta)f_{s_k} \neq 1$$

for some $1 \leq k \leq n$. But a glance at (5.3) shows that in this case $t_0 t_L s_1 \theta \cdots s_{k-1}\theta \in A(J)\theta$ and $t_0 t_L s_1 \theta \cdots s_k \theta \in J\theta$. But this is impossible in light of observations 1 and 2. We conclude $t_0{}^{t_L}f = 1$, as required. Thus $D_\psi \in \ell\mathbf{1}$.

Next we show that $\varphi$ is $\mathcal{J}$-singular and injective on $\mathcal{L}$-classes. An application of Proposition 5.3.9 will then yield $\varphi^{op} \in \mathbf{Sl}_D$. Because $\theta = \varphi\psi$, evidently $\varphi$ is $\mathcal{J}$-singular with $\mathcal{J}$-singular $\mathcal{J}$-class $J$. As $\varphi$ is injective on $S \setminus J$, it suffices to establish injectivity on $\mathcal{L}$-classes of $J$. As $\theta$ is not an $\mathbb{LG}$-morphism (and so, in particular, is not an $\mathcal{H}$-morphism), $\theta$ must be injective on $\mathcal{H}$-classes by Lemma 5.2.10. This latter property is inherited by $\varphi$. We show that if $s, s' \in J$ and $s\varphi \cap s'\varphi \neq \emptyset$, then $s \mathrel{\mathcal{R}} s'$. In particular, if $s \mathrel{\mathcal{L}} s'$ and $s\varphi \cap s'\varphi \neq \emptyset$, we will have $s \mathrel{\mathcal{H}} s'$ and hence $s = s'$ by injectivity of $\varphi$ on $\mathcal{H}$-classes.

To achieve our goal, we prove if $s \in J$ belongs to the $\mathcal{R}$-class $R_a$ and $(f, t) \in s\varphi$, then $1f = a$. Indeed, by definition of $\varphi$, there is a factorization $s = s_1 \cdots s_n$ so that (5.4) holds. As $s \in J$, there is a unique index $i$ so that $s_1 \cdots s_{i-1} \in A(J)$ and $s_1 \cdots s_i \in J$. Because $s_1 \cdots s_n = s \in J$, stability shows that $s_1 \cdots s_i \mathrel{\mathcal{R}} s$ and so $s_1 \cdots s_i \in R_a$. Now a routine computation yields

$$1f = 1f_{s_1}(s_1\theta)f_{s_2} \cdots (s_1\theta \cdots s_{n-1}\theta)f_{s_n}.$$

Then (5.3) implies that $(s_1\theta \cdots s_{i-1}\theta)f_{s_i} = a$ and $(s_1\theta \cdots s_{j-1}\theta)f_{s_j} = 1$ for $j \neq i$. This establishes $1f = a$ and completes the proof.     □

Propositions 5.3.6, 5.3.7 and 5.3.11 immediately yield Theorem 5.3.2

## 5.3.2 Consequences of the MPS Decomposition Theorem

The first consequence we shall draw of the MPS Decomposition Theorem is the two-sided Prime Decomposition Theorem [168]. It provides, in particular, an iterated block product decomposition of any finite monoid into congruence-free divisors.

**Theorem 5.3.12 (Two-sided Prime Decomposition Theorem).** *Let $S$ be a finite semigroup. Then $S$ divides an iterated block product of copies of the two-element semilattice $U_1$ and finite simple group divisors of $S$.*

*Proof.* We proceed by induction on the order of $S$. If $S$ is trivial, there is nothing to prove. Otherwise, take a minimal non-trivial congruence $\equiv$ on $S$ and consider the MPS $\varphi \colon S \to S/\equiv$. By the MPS Decomposition Theorem either $K_\varphi \in \mathbf{Sl}_K$ or $K_\varphi \in (Q)_K$ where $Q$ is a simple group dividing $S$. The result now follows by induction.     □

Notice that if $S$ is a group, then the above proof furnishes a decomposition without semilattices. A consequence of Theorem 5.3.12 is that the aperiodic semigroups are precisely the divisors of iterated block products of semilattices. This allows for an even easier proof of Schützenberger's Theorem; see [346]. It is interesting that the two-sided Prime Decomposition Theorem has a "modern," conceptual proof using category decomposition techniques, whereas the proof we gave for the Prime Decomposition Theorem is essentially the original proof, based on explicit wreath product decompositions. One would somehow have to obtain a decomposition of aperiodic MPSs into products of relational morphisms whose derived categories locally divide $U_2$. Factoring by a null ideal is already non-trivial and requires the Fundamental Lemma of Complexity.

*Question 5.3.13.* Find a proof of the classical Prime Decomposition Theorem using the Derived Category Theorem and some factorization theorem for relational morphisms.

Let us now prove that the congruence-free monoids are precisely the ⋈-prime semigroups.

**Definition 5.3.14 (⋈-prime).** *A finite semigroup $S$ is said to be ⋈-prime if $S \prec U \bowtie T$ implies $S \prec U$ or $S \prec T$.*

**Theorem 5.3.15.** *The ⋈-prime semigroups are precisely the finite simple groups and the two-element semilattice $U_1$, i.e., the congruence-free monoids.*

*Proof.* By the two-sided Prime Decomposition Theorem, these are the only candidates to be ⋈-prime. First we show $U_1$ is ⋈-prime. Because $U_1$ is projective, it suffices to show $U_1 \leq S \bowtie T$ implies $U_1 \prec S$ or $U_1 \prec T$. Let $\pi \colon S \bowtie T \to T$ be the projection. If $\pi|_{U_1}$ is injective, we are done. Else $U_1 \leq e\pi^{-1}$ some $e \in E(T)$. Corollary 5.3.5 then shows $U_1 \prec S$.

Next, suppose $Q$ is a finite simple group and $Q \prec S \bowtie T$. Then, there is a subgroup $G \leq S \bowtie T$ and a normal subgroup $N \lhd G$ such that $G/N \cong Q$. Let $\pi \colon S \bowtie T \to T$ be the projection and set $K = \ker \pi|_G$. Clearly, $G/K \prec T$. An application of Corollary 5.3.5 yields $K \prec S$. Because $G/N$ is simple, either $K \leq N$ or $NK = G$. In the first case $Q \cong G/N \prec G/K \prec T$; in the second $Q \cong NK/N \prec N \prec S$. This completes the proof.     □

The main consequence of the MPS Decomposition Theorem is that certain pseudovarieties of the form $(\mathbf{V}, \mathbf{V})$ can be written as infinite powers of pseudovarieties of the form $\mathbf{W}_K$. The reader is referred to Definition 4.1.46 for the notion of an extension-closed pseudovariety of groups.

**Lemma 5.3.16.** *Let $\mathbf{H}$ be an extension-closed pseudovariety of groups. Then $(\overline{\mathbf{H}}, \overline{\mathbf{H}}) = (\overline{\mathbf{H}}, \mathbf{1})$ and $(\mathbb{L}\mathbf{H}, \mathbb{L}\mathbf{H}) = (\mathbb{L}\mathbf{H}, \mathbf{1})$.*

*Proof.* Because $\mathbb{L}\mathbf{H} = \mathbb{L}\mathbf{G} \cap \overline{\mathbf{H}}$ and we already know $(\mathbb{L}\mathbf{G}, \mathbb{L}\mathbf{G}) = (\mathbb{L}\mathbf{G}, \mathbf{1})$ (Exercise 4.6.17), it suffices, by Lemma 4.6.18, to prove $(\overline{\mathbf{H}}, \overline{\mathbf{H}}) = (\overline{\mathbf{H}}, \mathbf{1})$. The inclusion from left to right is trivial, so we just handle the reverse inclusion. By Tilson's Lemma, we need only consider homomorphisms. So suppose that $\varphi\colon S \to T \in (\overline{\mathbf{H}}, \mathbf{1})$ is a homomorphism and $U \leq T$ is a subsemigroup with $U \in \overline{\mathbf{H}}$. We need to show that every subgroup of $U\varphi^{-1}$ belongs to $\mathbf{H}$. Let $G$ be a subgroup of $U\varphi^{-1}$ and $K = \ker\varphi|_G$. Because $\varphi \in (\overline{\mathbf{H}}, \mathbf{1})$, we must have $K \in \mathbf{H}$. But also $G\varphi \leq U$ implies $G\varphi \in \mathbf{H}$. As $\mathbf{H}$ is extension-closed, we conclude $G \in \mathbf{H}$, as required.                                     $\square$

We already know that pseudovarieties of the form $(\mathbf{V}, \mathbf{V})$ are closed under composition. For surjective homomorphisms, there is a dual fact.

**Lemma 5.3.17.** *Let* $\mathbf{V}$, $\mathbf{W}$ *be pseudovarieties of semigroups. Let* $\varphi_1\colon S \to T$ *and* $\varphi_2\colon T \to U$ *be onto homomorphisms such that* $\varphi_1\varphi_2 \in (\mathbf{V}, \mathbf{W})$. *Then* $\varphi_1, \varphi_2 \in (\mathbf{V}, \mathbf{W})$.

*Proof.* First we show $\varphi_2 \in (\mathbf{V}, \mathbf{W})$. Let $W \leq U$ be a subsemigroup with $W \in \mathbf{W}$. Then $W\varphi_2^{-1}\varphi_1^{-1} \in \mathbf{V}$ and so $W\varphi_2^{-1} = W\varphi_2^{-1}\varphi_1^{-1}\varphi_1 \in \mathbf{V}$. This establishes $\varphi_2 \in (\mathbf{V}, \mathbf{W})$. Next suppose that $X \leq T$ is a subsemigroup with $X \in \mathbf{W}$. Then $X\varphi_2 \in \mathbf{W}$ so $X\varphi_1^{-1} \leq X\varphi_2\varphi_2^{-1}\varphi_1^{-1} \in \mathbf{V}$. This establishes $\varphi_1 \in (\mathbf{V}, \mathbf{W})$, completing the proof of the lemma.             $\square$

**Corollary 5.3.18.** *Let* $\mathbf{V}$ *be a pseudovariety of semigroups and* $\theta\colon S \to T$ *be an onto morphism. If* $\theta = \theta_1 \cdots \theta_n$ *is an MPS factorization, then* $\theta \in (\mathbf{V}, \mathbf{V})$ *if and only if* $\theta_i \in (\mathbf{V}, \mathbf{V})$ *each* $i$.

We may now state the Prime Decomposition Theorem for relational morphisms.

**Theorem 5.3.19 (Prime Decomposition Theorem: Relational Morphisms).** *Let* $\varphi\colon S \to T$ *be a relational morphism. Then there is a factorization* $\varphi = d\sigma_1 \cdots \sigma_n$ *such that* $d$ *is a division and each* $\sigma_i$ *is a homomorphism either in* $\ell\mathbf{1}_K$ *or in* $(Q)_K$ *where* $Q$ *is a congruence-free monoid (that is, a two-element semilattice or a finite simple group) in the Mal'cev kernel of* $\sigma_i$ *(and hence of* $\varphi$*).*

*Proof.* Let $\varphi\colon S \to T$ be a relational morphism. Then $\varphi$ has canonical factorization $p_S^{-1}p_T$ and the Mal'cev kernel of $\varphi$ and $p_T$ coincide by Tilson's Lemma (as both $\varphi$ and $p_T$ belong to the same positive pseudovarieties of relational morphisms). So without loss of generality, we may assume $\varphi\colon S \to T$ is a homomorphism. In fact, using closure of pseudovarieties of relational morphisms under range extension, we may assume $\varphi$ is a surjective morphism. Hence it has an MPS factorization $\varphi = \theta_1 \cdots \theta_m$. Lemma 5.3.17 implies that the Mal'cev kernel of each $\theta_i$ is contained in the Mal'cev kernel of $\varphi$. The MPS Decomposition Theorem then shows that $\varphi = \tau_1 \cdots \tau_m$ where each $\tau_i$ is a relational morphism belonging either to $\ell\mathbf{1}_K$ or $(Q)_K$ where $Q$ is a congruence-free monoid in the Mal'cev kernel of both $\tau_i$ and $\varphi$. Lemma 2.8.3 allows us to replace the $\tau_i$ by morphisms at the price of putting a division out front.     $\square$

Notice that the two-sided Prime Decomposition Theorem can be obtained from the Prime Decomposition Theorem for relational morphisms by considering a "prime decomposition" for the collapsing morphism $S \to \{1\}$.

**Theorem 5.3.20.** *Let* $\mathbf{H}$ *be an extension-closed pseudovariety of groups.*

1. $(\overline{\mathbf{H}}, \overline{\mathbf{H}})$ *is the smallest pseudovariety of relational morphisms closed under composition and containing* $\mathbf{Sl}_K$ *and* $(Q)_K$*, where* $Q \in \mathbf{H}$ *is a simple group.*
2. *If* $\mathbf{H} \neq \mathbf{1}$*, then* $(\mathbb{L}\mathbf{H}, \mathbb{L}\mathbf{H})$ *is the smallest pseudovariety of relational morphisms closed under composition and containing* $(Q)_K$*, where* $Q \in \mathbf{H}$ *is a simple group.*
3. $(\mathbb{L}\mathbf{1}, \mathbb{L}\mathbf{1})$ *is the smallest pseudovariety of relational morphisms closed under composition and containing* $\ell\mathbf{1}_K$*.*

*Proof.* Using Lemma 5.3.16 and Corollary 5.3.18, the result is immediate from the Prime Decomposition Theorem for relational morphisms and the fact $\ell\mathbf{1}_K \leq \mathbf{gSl}_K \cap (Q)_K$ for any simple group $Q$ by the locality of $\mathbf{Sl}$ and $(Q)$, or the fact $\ell\mathbf{1}$ is the smallest non-trivial pseudovariety of categories [364]. $\square$

Let us spell out some particularly important special cases of Theorem 5.3.20. Recall that if $p$ is a prime, then $\mathbf{Ab}(p)$ is the pseudovariety of abelian groups of exponent $p$. Also recall if R is a positive pseudovariety of relational morphisms, then $\mathrm{R}^\omega = \bigcup_{n=0}^\infty \mathrm{R}^n$.

**Corollary 5.3.21.** *The following equalities hold.*

1. $(\mathbf{A}, \mathbf{A}) = \mathbf{Sl}_K^\omega = \mathbf{A}_K^\omega$.
2. $(\mathbb{L}\mathbf{1}, \mathbb{L}\mathbf{1}) = \ell\mathbf{1}_K^\omega$.
3. $(\mathbb{L}\mathbf{G}, \mathbb{L}\mathbf{G}) = \mathbf{G}_K^\omega$.
4. $(\mathbb{L}\mathbf{G}_p, \mathbb{L}\mathbf{G}_p) = \mathbf{Ab}(p)_K^\omega$.
5. $(\mathbb{L}\mathbf{G}_\pi, \mathbb{L}\mathbf{G}_\pi) = (\mathbf{G}_\pi)_K^\omega$
6. $(\mathbb{L}\mathbf{G}_{sol}, \mathbb{L}\mathbf{G}_{sol}) = \mathbf{Ab}_K^\omega$.
7. $(\overline{\mathbf{G}}_p, \overline{\mathbf{G}}_p) = (\mathbf{Sl}_K \mathbf{Ab}(p)_K)^\omega$.
8. $(\overline{\mathbf{G}}_{\pi'}, \overline{\mathbf{G}}_{\pi'}) = (\mathbf{Sl}_K (\mathbf{G}_{\pi'})_K)^\omega$.
9. $(\overline{\mathbf{G}}_{sol}, \overline{\mathbf{G}}_{sol}) = (\mathbf{Sl}_K \mathbf{Ab}_K)^\omega$.
10. $\mathbf{RM} = (\mathbf{Sl}_K \mathbf{G}_K)^\omega = (\mathbf{A}_K \mathbf{G}_K)^\omega$.

We remind the reader that if $\alpha \in \mathbf{Cnt}(\mathbf{PV})^+$, then $\alpha^\omega = \bigvee_{n \geq 0} \alpha^n$. Also, set $\ell\mathbf{1}_K \mathsf{q} = \ell\mathbf{1} ** (-)$. Then in terms of Mal'cev products and semidirect products, we can restate some of the results of the previous corollary in the following form.

**Corollary 5.3.22.** *Let* $\mathbf{V}$ *be a pseudovariety of semigroups. The following equalities hold.*

1. $\mathbb{L}\mathbf{1} \,\textcircled{m}\, \mathbf{V} = \ell\mathbf{1} **^\omega \mathbf{V}$.
2. $\mathbf{A} \,\textcircled{m}\, \mathbf{V} = \mathbf{A} **^\omega \mathbf{V} = \mathbf{Sl} **^\omega \mathbf{V}$.

3. $\mathbf{LG} \,\textcircled{m}\, \mathbf{V} = \mathbf{G} **^{\omega} \mathbf{V}$.
4. $\mathbf{LG}_p \,\textcircled{m}\, \mathbf{V} = \mathbf{Ab}(p) **^{\omega} \mathbf{V}$.
5. $\mathbf{LG}_{\pi} \,\textcircled{m}\, \mathbf{V} = \mathbf{G}_{\pi} **^{\omega} \mathbf{V}$.
6. $\mathbf{LG}_{sol} \,\textcircled{m}\, \mathbf{V} = \mathbf{Ab} **^{\omega} \mathbf{V}$.

*Remark 5.3.23.* The proof of the Fundamental Lemma of Complexity [362] shows that if $\varphi\colon S \to T$ is an aperiodic morphism and $\eta\colon \widehat{T}^{\mathscr{L}} \to T$ is the projection from the Rhodes expansion, then $\varphi\eta^{-1} \in \mathbf{A}_D$. On the other hand, it is well-known [87, 296] that $\eta^{op} \in \ell\mathbf{1}_D$. So

$$\varphi = \varphi\eta^{-1}\eta \in \mathbf{A}_D(\ell\mathbf{1}_D)^{op} \leq \mathbf{A}_K\mathbf{A}_K.$$

Thus $\mathbf{A}_K^{\omega} = (\mathbf{A}, \mathbf{A}) = \mathbf{A}_K\mathbf{A}_K$ and hence $\mathbf{A} \,\textcircled{m}\, \mathbf{V} = \mathbf{A} ** (\mathbf{A} ** \mathbf{V})$ for any pseudovariety $\mathbf{V}$ by applying the homomorphism $\mathfrak{q}$. Consequently the operator $\alpha = \mathbf{A} ** (-)$ satisfies $\alpha^2 = \alpha^3$.

*Question 5.3.24.* Is the operator $\mathbf{A} ** (-)$ idempotent? We suspect that it is not. This would be equivalent to $\mathbf{A}\textcircled{m}(-) = \mathbf{A} ** (-)$. If $PT_n$ is the semigroup of partial transformations of an $n$ element set and $I$ is the aperiodic ideal of maps of rank at most 1, then we suspect $PT_n \notin \mathbf{A} ** (PT_n/I)$ although the projection $PT_n \to PT_n/I$ is an aperiodic morphism.

From Corollary 5.3.22, we now see that two-sided complexity has an alternative formulation in terms of iterated two-sided semidirect products, parallel to the one-sided theory.

**Theorem 5.3.25.** *The two-sided complexity hierarchy is given by*

$$\mathbf{K}_0 = \mathbf{A}$$
$$\mathbf{K}_{n+1} = \mathbf{A} **^{\omega} (\mathbf{G} **^{\omega} \mathbf{K}_n), \qquad\qquad \text{for } n \geq 0$$

*or simply put:* $\mathbf{K}_n = (\mathbf{A}_K^{\omega}\mathbf{G}_K^{\omega})^n \mathfrak{q}(\mathbf{A})$.

As a corollary, we obtain the following property of the two-sided complexity function.

**Corollary 5.3.26.** *Let $m, n \geq 0$. Then $\mathbf{K}_m ** \mathbf{K}_n \leq \mathbf{K}_{m+n}$. Hence for semigroups $S$ and $T$, the inequality $C(S \bowtie T) \leq C(S) + C(T)$ holds.*

*Proof.* The second statement is an immediate consequence of the first. We prove the first by induction on $m$. If $m = 0$, then we are simply observing that $\mathbf{A} ** \mathbf{K}_n = \mathbf{K}_n$, which is immediate from Theorem 5.3.25. Suppose the corollary holds for $m$. Using repeatedly the inequality (2.56) and continuity of the operator $**$ in both variables, we compute

$$\begin{aligned}
\mathbf{K}_{m+1} ** \mathbf{K}_n &= (\mathbf{A} **^{\omega} (\mathbf{G} **^{\omega} \mathbf{K}_m)) ** \mathbf{K}_n \\
&\leq \mathbf{A} **^{\omega} ((\mathbf{G} **^{\omega} \mathbf{K}_m) ** \mathbf{K}_n) \\
&\leq \mathbf{A} **^{\omega} (\mathbf{G} **^{\omega} (\mathbf{K}_m ** \mathbf{K}_n)) \\
&\leq \mathbf{A} **^{\omega} (\mathbf{G} **^{\omega} \mathbf{K}_{m+n}) \qquad \text{by induction} \\
&= \mathbf{K}_{m+1+n}
\end{aligned}$$

as required.                                                          □

Further applications of the two-sided decomposition theory, in particular to Formal Language Theory, can be found in [152, 234, 332, 338, 378, 379].

## 5.4 The Ideal Theorem

The goal of this section is to prove the analogue of the Ideal Theorem (Theorem 4.9.17) for two-sided complexity. This is an unpublished result of the first author. The proof more or less follows that of the Ideal Theorem for usual complexity [362], but is actually easier because we have for free that aperiodic relational morphisms do not change two-sided complexity. The three tools used in the proof are the Kernel Semigroupoid Theorem, the Reduction Theorem and the Rhodes expansion. Let us begin by stating a corollary of the Reduction Theorem (Theorem 4.9.16).

**Corollary 5.4.1.** *Let $S$ be a finite semigroup, let $J_1, \ldots, J_m$ be the maximal essential $\mathscr{J}$-classes of $S$ and let $E$ be a set consisting of one idempotent from each $J_i$. Then $C(S) = C(ESE)$.*

*Proof.* The Reduction Theorem asserts $S \prec ESE \wr A$ with $A$ aperiodic. Corollary 5.3.26 then implies $C(S) \leq C(ESE)$; the reverse inequality is trivial.     □

In particular, if $I$ is an ideal of $S$ with $S/I$ is aperiodic, then $ESE \leq I \leq S$ and so we obtain the following corollary.

**Corollary 5.4.2.** *Let $S$ be a semigroup and $I$ an ideal of $S$ with $S/I$ aperiodic. Then $C(S) = C(I)$.*

### 5.4.1 Expansions

We shall need to exploit some properties of the Rhodes expansion in order to prove the ideal theorem. The notion of an expansion is due to the first author. See [54, 87, 286] for a plethora of expansions.

**Definition 5.4.3 (Expansion).** *An expansion consists of a pair $(F, \eta)$ where $F: \mathbf{FSgp} \to \mathbf{FSgp}$ is a functor and $\eta: F \twoheadrightarrow 1_{\mathbf{FSgp}}$ is a surjective natural transformation from $F$ to the identity functor. In other words, $F$ assigns to each finite semigroup $S$ a finite semigroup $F(S)$ and a surjective morphism $\eta_S: F(S) \twoheadrightarrow S$ so that, for every homomorphism $\varphi: S \to T$, the diagram*

$$
\begin{array}{ccc}
F(S) & \xrightarrow{\ F(\varphi)\ } & F(T) \\
{\scriptstyle \eta_S}\downarrow & & \downarrow{\scriptstyle \eta_T} \\
S & \xrightarrow[\ \varphi\ ]{} & T
\end{array}
$$

*commutes.*

If $S$ is a semigroup, then the *right stabilizer* of an element $s \in S$ consists of all elements $s' \in S$ with $ss' = s$. Understanding right stabilizers is important to understanding local monoids in the derived category of a relational morphism. The (left) Rhodes expansion is an expansion $S \mapsto \widehat{S}^{\mathscr{L}}$ enjoying the following properties:

(RE1) $\eta_S \colon \widehat{S}^{\mathscr{L}} \twoheadrightarrow S$ is an aperiodic morphism;
(RE2) Right stabilizers of $\widehat{S}^{\mathscr{L}}$ are aperiodic, in fact $\mathscr{R}$-trivial.

A construction of the Rhodes expansion, as well as proofs of the above properties, can be found in [110, 362]. The explicit construction shall not be used.

*Question 5.4.4.* Find an easier proof than the one in [362] that $c(\widehat{S}^{\mathscr{L}}) = c(S)$.

### 5.4.2 Proof of the Ideal Theorem

Let us now turn to the Ideal Theorem, a vast generalization of Corollary 5.4.2.

**Theorem 5.4.5 (Ideal Theorem).** *Let $S$ be a finite semigroup and $I$ an ideal of $S$. Then $C(S) \leq C(I) + C(S/I)$.*

*Proof.* Let $\theta \colon S \twoheadrightarrow S/I$ be the quotient map and set $\varphi = \theta \eta_{S/I}^{-1} \colon S \to \widehat{S/I}^{\mathscr{L}}$. By (RE1), $C(\widehat{S/I}^{\mathscr{L}}) = C(S/I)$. Let $K = \mathbf{Ker}(\varphi) \cup \{0\}$ be the consolidation of $\mathbf{Ker}(\varphi)$. There is an evident faithful functor from $\mathbf{Ker}(\varphi) \to K$ and so $K_\varphi \prec K$. The Kernel Semigroupoid Theorem provides a decomposition of the form $S \prec K \square \widehat{S/I}^{\mathscr{L}}$ and thus it suffices to show $C(K) \leq C(I)$ by Corollary 5.3.26.

Let $\widetilde{I} = \{(t_L, (i,t), t_R) \in \mathbf{Ker}(\varphi) \mid i \in I\} \cup \{0\}$. It is routine to verify $\widetilde{I}$ is an ideal of $K$. Let $A = I\varphi$; it is the inverse image of the zero of $S/I$ under $\eta_{S/I}$, and is hence aperiodic by (RE1). Define $\psi \colon \widetilde{I} \to I \times A$ by

$$(t_L, (i,t), t_R) \longmapsto (i, t), \quad 0 \longmapsto I \times A.$$

Clearly, $\psi$ is a relational morphism. In fact if $(i,t) \in I \times A$ is an idempotent, one easily verifies that the each of its inverse images is either an idempotent or squares to zero, so the Mal'cev kernel is contained in $[\![x^2 = x^3]\!]$. Consequently, $C(\widetilde{I}) \leq C(I)$.

Next we establish that $K/\widetilde{I}$ is aperiodic. Corollary 5.4.2 will then yield $C(K) = C(\widetilde{I}) \leq C(I)$, as required. Any non-trivial subgroup $G$ of $K/\widetilde{I}$ must be contained in the image of a local semigroup $\mathbf{Ker}(\varphi)((t_L, t_R))$. The map $(t_L, (s,t), t_R) \mapsto t$ is a homomorphism from $G$ into the right stabilizer of $t_L$, which is aperiodic by (RE2). Moreover, it is injective because $s \notin I$ implies that $s = t\eta_{S/I}$. It follows $K/\widetilde{I}$ is aperiodic, as required.    □

*Remark 5.4.6.* One can also establish the analogue of Tilson's Ideal Theorem for two-sided complexity: that is, if $\varphi \colon S \to S/I$ is the projection, then $C(\varphi) = C(I)$.

# 5.5 Translational Hulls and Ideal Extensions

Our next aim is to compute the two-sided complexity of a $2\mathscr{J}$-semigroup. This requires the notions of translational hulls and ideal extensions, so we now consider these in some detail. Most of these results can be found in [171, Chapter 7, Section 2] and [68, Section 4.4].

## 5.5.1 Translational hulls

If $S$ is a semigroup, then a *left translation* of $S$ is a map $\lambda\colon S \to S$, written on the left of its argument, such that $\lambda(st) = (\lambda s)t$. Left translations are composed via the convention of functions acting on the left. Right translations are defined dually and are composed with the convention for right actions. A left translation $\lambda$ and a right translation $\rho$ are said to be *linked* if $s(\lambda t) = (s\rho)t$, for all $s, t \in T$.

*Example 5.5.1 (Inner translations).* For each $s \in S$, we can define a left translation $\lambda_s$ and a right translation $\rho_s$ by $\lambda_s t = st$, $t\rho_s = ts$. The associativity of $S$ guarantees these translations are linked: $t(\lambda_s u) = t(su) = (ts)u = (t\rho_s)u$. Right/left translations of this type are said to be *inner*.

For a monoid, all translations are inner.

**Exercise 5.5.2.** Show that every left and right translation of a monoid is inner.

The collection of all pairs $(\lambda, \rho)$ of linked left and right translations is a semigroup $\Omega(S)$ called the *translational hull* of $S$. Multiplication is given by

$$(\lambda_1, \rho_1)(\lambda_2, \rho_2) = (\lambda_1\lambda_2, \rho_1\rho_2).$$

There is a natural map $S \to \Omega(S)$ given by $s \mapsto (\lambda_s, \rho_s)$, which in general is not injective.

**Exercise 5.5.3.** Verify the translational hull of $S$ is a semigroup.

The translational hull of a 0-simple semigroup is well understood. There are two formulations: one in terms of matrices and the other in terms of wreath products. First we provide the matrix formulation; afterwards we consider the wreath product formulation. Fix a 0-simple semigroup $S = \mathscr{M}^0(H, A, B, C)$. For the moment, we view $S$ as consisting of certain $A \times B$ matrices over $H^0$ as per Section A.4. A matrix over a semigroup with zero is said to be *row monomial* if each row has at most one non-zero entry. If each row has exactly one non-zero entry, we say the matrix is *total*. Column monomial matrices are defined dually. If $H$ is a group and $B$ is a set, then $RM_B(H)$ denotes the semigroup of $B \times B$ row monomial matrices over $H^0$. Dually, $CM_A(H)$ denotes the group of $A \times A$ column monomial matrices over $H^0$. Clearly, $RM_B(H)$ acts on the right of $S$ by right translations and $CM_A(H)$ acts on the left of $S$ by left translations via matrix multiplication. In fact, the following more detailed statement can be found in [171, Chapter 7, Facts 2.14–2.15].

**Theorem 5.5.4.** *Let $S = \mathcal{M}^0(H, A, B, C)$ be a 0-simple semigroup. Then:*

1. $RM_B(H)$ *is the monoid of right translations of $S$;*
2. $CM_A(H)$ *is the monoid of left translations of $S$;*
3. *Matrices $R \in RM_B(H)$ and $L \in CM_A(H)$ are linked if and only if the equality $RC = CL$ holds.*

The equation in the third item is called the *linked equations* (we say equations as a matrix equation entails a system of equations).

**Exercise 5.5.5.** Prove Theorem 5.5.4.

Because often we view $S$ as triples $(a, g, b)$, it is useful to describe left and right translations from this viewpoint. This leads us to the wreath product formulation of the translational hull and the linked equations. Indeed, it is not hard to show that $RM_B(H) \cong H \wr (B, PT_B)$, where $PT_B$ is the semigroup of partial transformations of $B$ acting on the right and we are considering the wreath product of partial transformation semigroups as defined at the beginning of Chapter 4 or in [85]. Under these identifications, the monoid of total row monomial matrices corresponds to the submonoid $H \wr (B, T_B)$. The details are nearly identical to Proposition 4.6.42. Let us describe the correspondence for $RM_B(H)$. If $(f, s) \in H \wr (B, PT_B)$, we associate the row monomial matrix

$$R(f, s) = \sum_{bs \neq \emptyset} (bf) E_{b, bs}$$

where $E_{ij}$ denotes the standard $B \times B$ matrix unit with 1 in position $i, j$ and zeroes elsewhere. The action of $(f, s)$ on $S$ is given by the formula

$$(a, g, b)(f, s) = \begin{cases} (a, gbf, bs) & bs \neq \emptyset \\ 0 & bs = \emptyset. \end{cases}$$

Conversely, suppose $R$ is a $B \times B$ row monomial matrix and define a partial function $s_R \colon B \to B$ by setting $bs_R$ to be the index of the unique column of $R$ with a non-zero entry in row $b$ if such exists, and undefined otherwise. Then

$$R = \sum_{bs_R \neq \emptyset} h_b E_{b, bs_R}$$

for certain $h_b \in H$. The corresponding element of $H \wr (B, PT_B)$ is $(f_R, s_R)$ where $f_R \colon B^0 \to H$ is given by

$$bf_R = \begin{cases} h_b & bs_R \neq \emptyset \\ \text{arbitrary} & \text{else} \end{cases}$$

where 0 is the sink. The reader is left to verify that we have just defined an isomorphism of $RM_B(H)$ and $H \wr (B, PT_B)$.

**Exercise 5.5.6.** Verify $RM_B(H) \cong H \wr (B, PT_B)$.

One can view $CM_A(H)$ in a dual manner using left transformation semigroups and the wreath product of such.

Let us describe the linked equations in the notation of wreath products. Suppose $(f_R, s_R) \in H \wr (B, PT_B)$ and $(s_L, f_L) \in (PT_A, A) \wr^{op} H$, where this latter semigroup is the wreath product of left partial transformation semigroups (defined dually to the usual wreath product). Then the linked equations are easily verified to take the form:

$$b f_R(b s_R Ca) = (bC s_L a) f_L a, \text{ all } a \in A, \ b \in B \tag{5.6}$$

where if $b s_R$, respectively $s_L a$, is undefined, then we interpret $b s_R Ca$, respectively, $bC s_L a$ as 0.

**Exercise 5.5.7.** Verify the linked equations are as in (5.6).

Let us describe inner translations of $S$ with these two formulations.

*Example 5.5.8 (Inner translations of 0-simple semigroups).* If $(a, g, b) \in S$, then

$$(a', g', b')(a, g, b) = \begin{cases} (a, g'(b'Ca)g, b) & b'Ca \neq 0 \\ 0 & \text{else.} \end{cases}$$

It follows that in wreath product coordinates, the right translation $\rho_{(a,g,b)}$ is represented by $(f, s)$ where $s$ is the partial constant map

$$b's = \begin{cases} b & b'Ca \neq 0 \\ \text{undefined} & \text{else} \end{cases}$$

and

$$b'f = \begin{cases} (b'Ca)g & b'Ca \neq 0 \\ \text{arbitrary} & \text{else.} \end{cases}$$

Let $C_a$ be column $a$ of $C$. Then the corresponding row monomial matrix has $C_a g$ as column $b$ and all other columns are 0. In particular, the action of $\mathcal{M}^0(G, A, B, C)$ on the right of itself is faithful if and only if $C$ has no two columns that are proportional on the right.

*Remark 5.5.9 (Linked equations and matrix representations).* There is an important connection between the linked equations and matrix representations of finite semigroups, emphasized by the first author and Zalcstein [297]. Let $J$ be a regular $\mathscr{J}$-class of a semigroup $S$ and suppose $J^0 \cong \mathcal{M}^0(G, A, B, C)$. Let $\rho: S \to RM_B(G)$ and $\lambda: S \to CM_A(G)$ be the right and left Schützenberger representations of $S$ on $J$, respectively. The linked equations say that

$$(s\rho)C = C(s\lambda) \tag{5.7}$$

for all $s \in S$. Let $k$ be a field such that $\operatorname{char} k \nmid |G|$ and set $V = kG^{|A|}$. Then $\lambda$ can be viewed as a representation of $S$ on $kG^{|A|}$ and so $V$ has the structure of a right $kS$-module and a left $kG$-module, in fact, $V$ is a $kG$-$kS$-bimodule. Let $W$ be the $kG$-span in $V$ of the rows of $C$. Then (5.7) implies that $W$ is a $kS$-submodule. Let $\varphi$ be an irreducible representation of $G$ and let $M$ be the corresponding simple right $kG$-module. Then the tensor product $M \otimes_{kG} W$ is a finite dimensional right $kS$-module, which according to [297] is simple. If $X$ is a matrix over $kG$, let $X \otimes \varphi$ be the matrix obtained by applying $\varphi$ to $X$ entrywise. Then the irreducible $k$-representation of $S$ associated to the simple module $M \otimes_{kG} W$ is obtained by restricting the representation $s \mapsto s\lambda \otimes \varphi$ to the span of the rows of $C \otimes \varphi$. Moreover, all irreducible representations of $S$ come about in this way. For a modern and coordinate-free approach to the irreducible representations of a finite semigroup over any commutative base ring, the reader is referred to [95].

Let $\Omega^{\mathrm{tot}}(S) \leq \Omega(S)$ be the semigroup consisting of linked total translations. It will be convenient also to consider the subsemigroup $\Omega_1^{\mathrm{tot}}(S)$ consisting of all linked total translations $(L, R)$ such that each non-zero entry of the matrices $L$ and $R$ is the identity 1 of $H$. Such translations can be identified with pairs of transformations $(\lambda, \rho) \in T_A \times T_B$ (where $\lambda$ acts on the left) satisfying the following simplified form of the linked equations:

$$b\rho Ca = bC\lambda a, \text{ all } a \in A, \ b \in B. \tag{5.8}$$

In what follows, we always view $\Omega_1^{\mathrm{tot}}(S) \leq T_A \times T_B$.

**Lemma 5.5.10.** *Let $S = \mathscr{M}^0(H, A, B, C)$ be a 0-simple semigroup and fix an element $h_0 \in H$. Let $C'$ be the matrix obtained by replacing each 0 of $C$ with $h_0$ and set $S' = \mathscr{M}(H, A, B, C')$. Then $\Omega_1^{\mathrm{tot}}(S) \leq \Omega_1^{\mathrm{tot}}(S')$.*

*Proof.* Suppose $(\lambda, \rho) \in \Omega_1^{\mathrm{tot}}(S)$. Let $a, b \in B$. Then $b\rho Ca = bC\lambda a$. If neither side is 0, then trivially $b\rho C'a = bC'\lambda a$. Otherwise, both sides are zero and so $b\rho C'a = h_0 = bC'\lambda a$. We conclude $\Omega_1^{\mathrm{tot}}(S) \leq \Omega_1^{\mathrm{tot}}(S')$.    $\square$

### 5.5.2 Ideal extensions

So why all the fuss about translational hulls? They are precisely what one needs to construct ideal extensions. We consider two types of ideal extensions. The first type will be called total and is the only type used in computing the two-sided complexity of $2\mathscr{J}$-semigroups. The second type is the general case and is important for constructing examples of semigroups in a "bottoms up" fashion. This was how the Tall Fork from Section 4.14 and the semigroups from Section 4.16 were constructed. These ideas will resurface in Section 7.4.

## Total ideal extensions

Let $S$ and $T$ be semigroups. Then a *total ideal extension* of $S$ by $T$ is a semigroup structure on the disjoint union $S \cup T$ containing $S$ and $T$ as subsemigroups with $S$ an ideal. For instance, if $M$ is a finite monoid with group of units $G$ and $S = M \setminus G$, then $M$ is a total ideal extension of $S$ by $G$. There is then a homomorphism $T \to \Omega(S)$ obtained by considering the actions of $T$ on the left and right of $S$ by inner translations in $S \cup T$. We call this the homomorphism *associated* to the total ideal extension.

Conversely, suppose $\varphi \colon T \to \Omega(S)$ is a homomorphism. Let us write $t\varphi = (\lambda_t, \rho_t)$. Then one can form the total ideal extension $S \cup T$ with multiplication given by extending the multiplications of $S$ and $T$ by defining, for $s \in S$ and $t \in T$, $st = s\rho_t$ and $ts = \lambda_t s$. The condition that $\lambda_t$ and $\rho_t$ are linked translations says exactly that $(st)s' = s(ts')$ for $s, s' \in S$ and $t \in T$. The remaining verifications for associativity are easy. Clearly, $S$ is an ideal of $S \cup T$ and $S$ and $T$ are subsemigroups.

**Exercise 5.5.11.** Verify $S \cup T$ is a semigroup.

## Ideal extensions: the general case

There is a more general notion of an ideal extension. Let $T$ be a semigroup with zero and let $S$ be a semigroup. Then a semigroup $U$ is said to be an *ideal extension* of $S$ by $T$ if $U$ contains $S$ as an ideal such that $U/S \cong T$. For instance, if $U = S \cup T$ is a total ideal extension of $S$ by $T$, then $U$ is an ideal extension of $S$ by $T^0$. However, the ideal extension concept is much more general because if $U$ is an ideal extension of $S$ by $T$, there is no reason that $T$ must be a subsemigroup of $U$.

**Definition 5.5.12 (Partial homomorphism).** *Let $T$ be a semigroup with zero and $S$ a semigroup. A map $\varphi \colon T \setminus 0 \to S$ is termed a* partial homomorphism *if $t\varphi t'\varphi = (tt')\varphi$ whenever $tt' \neq 0$.*

A fairly simple example of a partial homomorphism is as follows.

*Example 5.5.13 (Filling in some zeroes).* Let $S = \mathcal{M}^0(G, A, B, C)$ and suppose that $C' \colon B \times A \to G^0$ is obtained from $C$ by replacing some number of zero entries by non-zero entries. Set $S' = \mathcal{M}^0(G, A, B, C')$. Then the identity map on $A \times G \times B$ yields a partial homomorphism $S \setminus 0 \to S'$. Indeed, if $(a, g, b)(a', g', b') \neq 0$, then $bCa' \neq 0$ and so $bCa' = bC'a'$. We reprise this example in Example 5.5.17.

The following observation is trivial.

**Proposition 5.5.14.** *Let $U = S \cup (T \setminus 0)$ be an ideal extension of $S$ by $T$. Then the inclusion map is a partial homomorphism $T \setminus 0 \to U$.*

Let $\mu\colon T \times T \to T$ be the multiplication map. It turns out that an ideal extension of $S$ by $T$ is determined by a partial homomorphism $\varphi\colon T\backslash 0 \to \Omega(S)$ and a map $\xi\colon 0\mu^{-1} \to T$ satisfying some axioms. The following result is essentially [68, Thm. 4.21].

**Theorem 5.5.15.** *Let $T$ be a semigroup with zero and $S$ a semigroup. Let $\mu\colon T \times T \to T$ be the multiplication map.*

1. *Suppose that $U = S \cup (T \setminus 0)$ is an ideal extension of $S$ by $T$. Let $\varphi\colon T\backslash 0 \to \Omega(S)$ be given by $t\varphi = (\lambda_t, \rho_t)$ where $\lambda_t s = ts$, $s\rho_t = st$ and let $\xi\colon 0\mu^{-1} \to S$ be given by $(t, t')\xi = tt'$. Then $\varphi$ is a partial homomorphism and, for all $(t, t') \in 0\mu^{-1}$, the equality*

$$t\varphi t'\varphi = (\lambda_{(t,t')\xi}, \rho_{(t,t')\xi}) \qquad (5.9)$$

*holds.*

2. *Let $\varphi\colon T\backslash 0 \to \Omega(S)$ be a partial homomorphism given by $t \mapsto (\lambda_t, \rho_t)$ and let $\xi\colon 0\mu^{-1} \to S$ be a map such that (5.9) holds for all $(t, t') \in 0\mu^{-1}$. Then $U = S \cup (T \setminus 0)$ is an ideal extension of $S$ by $T$ with respect to the binary operation $\diamond$ defined as follows. For $s, s' \in S$, define $s \diamond s' = ss'$. For $t, t' \in T$, we set*

$$t \diamond t' = \begin{cases} tt' & tt' \neq 0 \\ (t, t')\xi & tt' = 0. \end{cases}$$

*Finally, for $s \in S$ and $t \in T$, we set:*

$$s \diamond t = s\rho_t \quad \text{and} \quad t \diamond s = \lambda_t s.$$

*Proof.* The first statement is an immediate consequence of associativity of the binary operation of $U$. We prove the second. The only thing that needs verification is the associativity of $\diamond$. From the fact that $\varphi$ is a partial homomorphism, the only case that is not completely trivial is $(t \diamond t') \diamond s = t \diamond (t' \diamond s)$ with $t, t' \in T$, $s \in S$ and $tt' = 0$ (and the dual equality). The left-hand side is $(t, t')\xi s$ and the right-hand side is $\lambda_t(\lambda_{t'} s)$. But (5.9) implies that $\lambda_t \lambda_{t'}$ is the inner translation $\lambda_{(t,t')\xi}$ and so $\lambda_t(\lambda_{t'} s) = (t, t')\xi s$. $\qquad\square$

One calls $\varphi$ the *associated partial homomorphism* of the ideal extension and $\xi$ the *ramification*.

**Exercise 5.5.16.** Supply the remaining details of the proof of Theorem 5.5.15.

An important special case is when $\psi\colon T\backslash 0 \to S$ is a partial homomorphism. Then $\varphi\colon T \setminus 0 \to \Omega(S)$ given by $t \mapsto (\lambda_{t\psi}, \rho_{t\psi})$ and $\xi\colon 0\mu^{-1} \to S$ given by $(t, t')\xi = t\psi t'\psi$ satisfy (5.9) and so $S \cup (T \setminus 0)$ is an ideal extension of $S$ by $T$.

*Example 5.5.17.* Let $V = \mathscr{M}^0(\{1\}, 2, 2, C)$ where

$$C = \begin{pmatrix} 1 & 1 \\ 0 & 1 \end{pmatrix}$$

and let $B_2$ be the 5-element Brandt semigroup. According to Example 5.5.13, the identity map is a partial homomorphism $B_2 \setminus 0 \to V$ and so we can build the ideal extension $V \cup (B_2 \setminus 0)$. Let us write the elements of $B_2$ as standard matrix units $E_{ij}$ and the elements of $V \setminus 0$ as pairs $(i,j)$. Then $(i,j)E_{k\ell} = (i,j)(k,\ell)$ (and dually), whereas

$$E_{ij}E_{k\ell} = \begin{cases} E_{i\ell} & j = k \\ (i,j)(k,\ell) & j \neq k. \end{cases}$$

For instance, $E_{21}E_{22} = (2,2)$ and $(1,1)E_{22} = (1,2)$.

### 5.5.3 Constructing Examples

Let us now describe the principal technique used by the Berkeley School in constructing examples of generalized group mapping semigroups using translational hulls. See Example 4.10.12 and Section 4.16 for some examples. We try to develop a general theory to this approach, but the reader should pay particular attention to the examples to see how things work.

The fundamental idea to take to heart is that one wants to construct sandwich matrices $C$ so that each right translation of $\mathcal{M}^0(G, A, B, C)$ is linked to a unique left translation. This makes it relatively easy to construct an ideal extension with prescribed Schützenberger representation. The safest way to do this is to put in all possible columns up to scalar multiplication.

Fix a group $G$ and a finite set $B$. Two (column) vectors $v, w$ in $(G^0)^{|B|}$ are said to be *proportional on the right* if there is an element $g \in G$ so that $vg = w$. By a *subspace* of $(G^0)^{|B|}$ we mean a subset $V$ closed under right scalar multiplication by $G^0$. If $T$ is a collection of $B \times B$ row monomial matrices, a subspace $V$ is said to be *$T$-invariant* if $RV \subseteq V$ for each $R \in T$. If $X \subseteq (G^0)^{|B|}$, then the *span* of $X$, denoted $\operatorname{Span} X$, is the smallest subspace containing $X$.

**Proposition 5.5.18 (Put in "all" columns).** *Let $B$ be a set and $G$ a group. Let $\mathbb{P}_B(G)$ be the projective space of all $|B| \times 1$ non-zero column vectors over $G^0$ modulo scalar multiplication on the right by elements of $G$. Let $C \colon B \times A \to G^0$ be a matrix whose columns form a transversal for $\mathbb{P}_B(G)$. Then each $B \times B$ row monomial matrix over $G$ is linked to a unique $A \times A$ column monomial matrix.*

*Proof.* Let $R \in RM_B(G)$. We need to show that the equation $RC = CL$ has a unique solution $L \in CM_A(G)$. Indeed, if $v_i$ is the $i^{th}$ column of $RC$, then either $v_i = 0$ or there is a unique column $c_{i\sigma}$ of $C$ and element $g_i$ of $G$ so that $v_i = c_{i\sigma}g_i$. Then $L = \sum_{\{i \in A | v_i \neq 0\}} g_i E_{i\sigma,i}$ is the desired column monomial matrix. $\square$

An analogous result holds for matrices $C \colon B \times A \to G$, only here one considers column vectors over $G$ rather than over $G^0$.

*Example 5.5.19.* The Rees matrix given in Example 4.10.12 contains all possible $2 \times 1$ column vectors over $\{\pm 1\}$ up to scalar multiplication. Therefore, the action of $g$ on the left of the minimal ideal is uniquely determined by its action on the right. However, notice that in this example the wreath product coordinates for the left action and the right action are quite different as there is only one orbit of $\mathscr{L}$-classes, but two orbits of $\mathscr{R}$-classes. This is why the complexity of $S$ and its reversal are different.

More generally, the proof of Proposition 5.5.18 shows that if $R$ is a row monomial matrix such that every non-zero column of $RC$ is proportional on the right to a unique column of $C$, then $R$ is linked to a unique column monomial matrix $L$. This is the case for the elements $h$, $k$ and $t$ of the Tall Fork (Table 4.1) and so again the action of these elements on the right of the 0-minimal ideal determines their action on the left. Let us state this as a proposition.

**Proposition 5.5.20.** *Let $B$ be a set and $G$ a group. Let $R$ be a $B \times B$ row monomial matrix over $G$ and let $C \colon B \times A \to G^0$ be a matrix such that each non-zero column of $RC$ is proportional on the right to exactly one column of $C$. Then $R$ is linked to a unique $A \times A$ column monomial matrix.*

Let us try to abstract this just a bit. Let us call a row monomial matrix $R \in RM_B(G)$ a *partial constant matrix* if it has at most one non-zero column. This means that the wreath product coordinates for $R$ are of the form $(f, s)$ where $s$ is a partial constant map and $f \colon B^0 \to G$. For instance, the inner translations of a 0-simple semigroup are represented by partial constant matrices according to Example 5.5.8. Let $T$ be a semigroup with 0. Then given a partial homomorphism $\rho \colon T \setminus 0 \to RM_B(G)$, or equivalently a partial homomorphism $\rho \colon T \setminus 0 \to G \wr (B, PT_B)$, we want to construct a 0-simple semigroup $S = \mathscr{M}^0(G, A, B, C)$ and an ideal extension $U$ of $S$ by $T$ so that $\rho$ is the restriction of the Schützenberger representation $\rho_J$ to $T \setminus 0$ where $J = S \setminus 0$ (see Definition 4.6.28).

So let $\rho \colon T \setminus 0 \to RM_B(G)$ be a partial homomorphism. We say that $\rho$ is *non-degenerate* if $t\rho$ is not a partial constant matrix for $t \in T \setminus 0$. Let $C \colon B \times A \to G^0$ be a matrix. We write $\mathrm{Span}\, C$ for the span of the columns of $C$. The condition that each non-zero column of $RC$ is proportional on the right to a column of $C$ is exactly that $\mathrm{Span}\, C$ is invariant under $R$. The condition that no two columns of $C$ are proportional is equivalent saying that $C$ is a transversal for the image of $\mathrm{Span}\, C \setminus 0$ in the projective space $\mathbb{P}_B(G)$.

**Theorem 5.5.21.** *Let $\rho \colon T \setminus 0 \to RM_B(G)$ be a partial homomorphism. Let $C \colon B \times A \to G^0$ be a regular Rees matrix such that:*

1. *No two columns of $C$ are proportional on the right;*
2. *Each non-zero column of $(t\rho)C$ is proportional on the right to a (necessarily unique) column of $C$ for each element $t \in T \setminus 0$;*

3. If $tt' = 0$, for $t, t' \in T$, then $t\rho t' \rho$ is either $0$ or a partial constant matrix whose non-zero column is proportional on the right to a (necessarily unique) column of $C$.

Let $S = \mathcal{M}^0(G, A, B, C)$. Then we can form an ideal extension $U = S \cup (T \setminus 0)$ such that $\rho_J|_{T \setminus 0} = \rho$ where $J = S \setminus 0$. If $\rho$ is injective and $t\rho$ is not an inner translation of $S$ for all $t \in T \setminus 0$ (for instance, if $\rho$ is non-degenerate), then $U$ is a right mapping semigroup.

*Proof.* Let us set $R_t = t\rho$ for $t \in T \setminus 0$. Then $R_t$ is linked to a unique column monomial matrix $L_t$ by Proposition 5.5.20 and if $tt' \neq 0$, then $L_t L'_t$ is linked to $R_t R_{t'} = R_{tt'}$ so $L_{tt'} = L_t L_{t'}$ by uniqueness. It follows $\varphi : T \setminus 0 \to \Omega(S)$ given by $t\varphi = (L_t, R_t)$ is a partial homomorphism. Now if $tt' = 0$, then there are two cases. If $t\rho t' \rho = 0$, set $(t, t')\xi = 0$. Else there exist $a \in A$, $g \in G$ and $b \in B$ so that the unique non-zero column of $t\rho t' \rho$ is in position $b$ and equals $C_a g$ where $C_a$ is column $a$ of $C$. In this case, set $(t, t')\xi = (a, g, b)$. Then $t\rho t' \rho$ is the row monomial matrix corresponding to right translation by $(t, t')\xi$. Because $\mathrm{Span}\, C$ is invariant under $T\rho$ and hence under $\langle T\rho \rangle$, Proposition 5.5.20 implies that $t\rho t' \rho$ is linked to a unique left translation, which must then be the inner left translation associated to $(t, t')\xi$. It follows that (5.9) is satisfied. Hence Theorem 5.5.15 implies that $U = S \cup (T \setminus 0)$ is an ideal extension of $S$ by $T$. By construction, $\rho_J|_{T \setminus 0} = \rho$. The final statement is immediate.     □

A quite general method for constructing a right mapping semigroup with specified structure is as follows. Let $T$ be a semigroup with $0$ and suppose that $\rho : T \to RM_B(G)$ is an injective non-degenerate partial homomorphism such that $t\varphi t' \varphi$ is a partial constant matrix (perhaps zero) whenever $tt' = 0$. Let $X$ be a set of non-zero $B \times 1$ column vectors over $G^0$, no two of which are proportional on the right. Let $Y$ be the set of non-zero columns of the partial constant matrices $t\varphi t' \varphi$ with $tt' = 0$. Let $V$ be the smallest $T\rho$-invariant subspace of $(G^0)^{|B|}$ containing $X \cup Y$. Choose a transversal $C$ containing $X$ for the image of $V$ in $\mathbb{P}_B(G)$, which we view as a matrix (say it is size $B \times A$) by choosing an ordering for the columns. Then, according to Theorem 5.5.21, we can form the ideal extension $U = \mathcal{M}^0(G, A, B, C) \cup (T \setminus 0)$ and it will be right mapping, with $T \setminus 0$ acting in the prescribed manner and with the vectors of $X$ among columns of $C$. Now one can obtain a generalized group mapping example by taking left letter mapping with respect to the distinguished $\mathscr{J}$-class. In practice, $X$ often consists of a single column vector.

Let us demonstrate this method with an example involving adjoining a group of units.

*Example 5.5.22 (Adjoining a group of units).* Let $B = \{1, 2, 3\}$ and $G = \{\pm 1\}$. We want to construct a group mapping small monoid with group of units a cyclic group $Z = \langle g \rangle$ of order 6 so that $g$ acts as the cyclic permutation $(1\ 2\ 3)$ on $B$ in right letter mapping and $v_1 = (1\ 1\ 0)^T$ is a column of the structure matrix of the 0-minimal ideal. Define a homomorphism $\varphi : Z \to RM_B(G)$ by

$$g\varphi = \begin{pmatrix} 0 & -1 & 0 \\ 0 & 0 & 1 \\ 1 & 0 & 0 \end{pmatrix}.$$

We take $X = \{v_1\}$ and $Y = \emptyset$ in the above construction. Now we need to compute a transversal for the $Z$-invariant subspace generated by $v_1$ modulo scalar multiplication. So we compute $v_2 = gv_1 = \begin{pmatrix} -1 & 0 & 1 \end{pmatrix}^T$, $v_3 = gv_2 = \begin{pmatrix} 0 & 1 & -1 \end{pmatrix}^T$ and $gv_3 = v_1(-1)$. So

$$C = \begin{pmatrix} 1 & -1 & 0 \\ 1 & 0 & 1 \\ 0 & 1 & -1 \end{pmatrix}$$

is the desired matrix. Notice that no two rows or columns of $C$ are proportional. Because $g$ generates a cyclic group of order 6, we have constructed a group mapping semigroup of the form $\mathbb{Z}_6 \cup \mathcal{M}^0(\{\pm 1\}, 3, 3, C)$ where the group of units acts in the prescribed way on the $\mathscr{L}$-classes. It is instructive for the reader to calculate the action of $g$ on the left of the 0-minimal ideal.

**Exercise 5.5.23.** Compute the column monomial matrix corresponding to the action of $g$ on the left of $\mathcal{M}^0(\{\pm 1\}, 3, 3, C)$.

This is only the first step in constructing interesting examples. The next step is to perform iterated Rees matrix constructions [3,276,277], often reusing the same structure matrix (as in Section 4.16).

**Exercise 5.5.24.** Construct an interesting small monoid $M$ using the ideal extension techniques of this section. Then construct a Rees matrix semigroup over $M$ and adjoin a new group of units.

## 5.6 Two-Sided Complexity of $2\mathscr{J}$-Semigroups

The main result of this section proves another unpublished result of the first author: the two-sided complexity of a $2\mathscr{J}$-semigroup is at most 1. Recall that by a $2\mathscr{J}$-semigroup, we mean a semigroup with at most two non-zero $\mathscr{J}$-classes.

### 5.6.1 Two-sided complexity for small monoids

The same arguments as in Section 4.15 reduce the problem of computing the two-sided complexity of a $2\mathscr{J}$-semigroup to the case of a small monoid $M = G \cup \mathcal{M}^0(H, A, B, C)$ with $J^0 = \mathcal{M}^0(H, A, B, C)$ a 0-simple semigroup and $G$ a group of units. Such a small monoid is a total ideal extension of $J^0$ by $G$. The associated homomorphism to this total ideal extension is then of the form $\varphi \colon G \to \Omega^{\text{tot}}(J^0)$. We call this the *associated homomorphism* of the small monoid. Our first lemma constructs a sort of universal covering of a small monoid. It has the effect of making the image of the associated homomorphism lie inside of $\Omega_1^{\text{tot}}$.

**Lemma 5.6.1.** *Let $M = G \cup J^0$ be a small monoid. Then there is a small monoid $\widetilde{M} = G \cup \widetilde{J}^0$, with associated homomorphism $G \to \Omega_1^{\text{tot}}(\widetilde{J}^0)$, and an onto aperiodic morphism $\psi\colon \widetilde{M} \twoheadrightarrow M$ that restricts to the identity on $G$.*

*Proof.* Suppose $J^0 = \mathscr{M}^0(H, A, B, C)$ and let $\varphi\colon G \to \Omega^{\text{tot}}(J^0)$ be the associated homomorphism. For $g \in G$, write

$$g\varphi = ((\lambda_g, L_g), (R_g, \rho_g))$$

with $(R_g, \rho_g) \in H \wr (B, T_B)$ and $(\lambda_g, L_g) \in (T_A, A) \wr^{op} H$. Notice we have total maps because $JG = J = GJ$. Set $\widetilde{J}^0 = \mathscr{M}^0(H, A \times H, H \times B, \widetilde{C})$ with $\widetilde{C}\colon (H \times B) \times (A \times H) \to H^0$ defined by

$$(h_L, b)\widetilde{C}(a, h_R) = h_L(bCa)h_R.$$

Notice that $\Omega^{\text{tot}}(J^0)$ can be viewed as a submonoid of $T_{A \times H} \times T_{H \times B}$ via the usual action of the wreath product. Under this identification, we claim $\Omega^{\text{tot}}(J^0) \leq \Omega_1^{\text{tot}}(\widetilde{J}^0)$. This amounts to verifying $g\varphi$ satisfies the simplified linked equations (5.8) for $\widetilde{C}$. But we have, using the linked equations for $g\varphi$,

$$
\begin{aligned}
(h_L, b)(R_g, \rho_g)\widetilde{C}(a, h_R) &= (h_L(bR_g), b\rho_g)\widetilde{C}(a, h_R) \\
&= h_L(bR_g)(b\rho_g Ca)h_R \\
&= h_L(bC\lambda_g a)(L_g a)h_R \qquad \text{by (5.6)} \\
&= (h_L, b)\widetilde{C}(\lambda_g a, (L_g a)h_R) \\
&= (h_L, b)\widetilde{C}(\lambda_g, L_g)(a, h_R)
\end{aligned}
$$

showing that $g\varphi$ is indeed linked with respect to $\widetilde{C}$ as an element of $\Omega_1^{\text{tot}}(\widetilde{J}^0)$. Thus we can view $\varphi$ as a map $\varphi\colon G \to \Omega_1^{\text{tot}}(\widetilde{J}^0)$ and form the total ideal extension $\widetilde{M} = G \cup \widetilde{J}^0$. To decongest notation, basically we have defined, for $g \in G$,

$$((a, h_L), h, (h_R, b))g = ((a, h_L), h, (h_R(bR_g), b\rho_g))$$

and dually for left multiplication. Define a homomorphism $\psi\colon \widetilde{M} \to M$ by the identity on $G$ and on $\widetilde{J}^0$ by $((a, h_L), h, (h_R, b)) \longmapsto (a, h_L h h_R, b)$. To check that $\psi$ is a homomorphism, we first consider two elements of $\widetilde{J}^0$:

$$
\begin{aligned}
((a, h_L), h, (h_R, b))\psi((a', h_L'), h', (h_R', b'))\psi & \\
&\hspace{-6cm}= (a, h_L h h_R, b)(a', h_L' h' h_R', b') \\
&\hspace{-6cm}= (a, h_L h h_R (bCa') h_L' h' h_R', b') \\
&\hspace{-6cm}= ((a, h_L), h h_R (bCa') h_L' h', (h_R', b'))\psi \\
&\hspace{-6cm}= ((a, h_L), h \left[ (h_R, b)\widetilde{C}(a', h_L') \right] h', (h_R', b'))\psi \\
&\hspace{-6cm}= (((a, h_L), h, (h_R, b))((a', h_L'), h', (h_R', b')))\psi.
\end{aligned}
$$

Next we must consider a product of an element of $G$ with an element of $\widetilde{J}^0$. Let $g \in G$. Then

$$
\begin{aligned}
((a, h_L), h, (h_R, b))\psi g\psi &= (a, h_L h h_R, b)(R_g, \rho_g) \\
&= (a, h_L h h_R(bR_g), b\rho_g) \\
&= ((a, h_L), h, (h_R(bR_g), b\rho_g))\psi \\
&= (((a, h_L), h, (h_R, b))g)\psi.
\end{aligned}
$$

The case of left multiplication by $g$ is dealt with in a similar fashion. Evidently $\psi$ is onto because $((a, 1), h, (1, b))\psi = (a, h, b)$. It remains to verify $\psi$ is an aperiodic morphism. Because $\psi$ is injective on $G$, it suffices to check injectivity on the $\mathscr{H}$-classes of $\widetilde{J}^0$. But $((a, h_L), h, (h_R, b))\psi = (a, g, b)$ implies $h = h_L^{-1} g h_R^{-1}$ and so $\psi$ is injective on $\mathscr{H}$-classes. This completes the proof.    $\square$

We are now ready for the main theorem of this section.

**Theorem 5.6.2 (Rhodes).** *Let $S$ be a semigroup with at most two non-zero $\mathscr{J}$-classes. Then $C(S) \leq 1$.*

*Proof.* Without loss of generality, we may reduce to the case of a small monoid $M = G \cup \mathscr{M}^0(H, A, B, C)$ by the same considerations used in the proof of Theorem 4.15.2. An application of Lemma 5.6.1 allows us to assume the associated homomorphism of the small monoid is of the form $\varphi \colon G \to \Omega_1^{\text{tot}}(J^0)$. Our goal is to construct an aperiodic relational morphism to a completely regular semigroup. Because completely regular semigroups have two-sided complexity at most one, this will serve to prove the theorem. Fix $h_0 \in H$ and let $C' \colon B \times A \to H$ be the matrix obtained by replacing each $0$ in $C$ by $h_0$. Set $J' = \mathscr{M}(H, A, B, C')$. Then Lemma 5.5.10 implies $\Omega_1^{\text{tot}}(J^0) \leq \Omega_1^{\text{tot}}(J')$. Thus we can form the total ideal extension $M' = G \cup J'$ using $\varphi$; of course, $M'$ is completely regular and so $C(M') \leq 1$. Define a relational morphism $\alpha \colon M \to M'$ by the identity on $G \cup (A \times H \times B)$ and by $0\alpha = J'$. Because $bCa \neq 0$ implies $bCa = bC'a$, it is immediate that $\alpha$ is a relational morphism. Notice that $\alpha$ is a bijection on the complement of $0$ and hence is clearly aperiodic. Hence $C(M) \leq C(M') \leq 1$, completing the proof of the theorem.    $\square$

As a consequence of Theorem 5.6.2 we obtain an upper bound on the two-sided complexity of a semigroup whose $\mathscr{J}$-classes form a chain.

**Corollary 5.6.3 (Rhodes).** *Let $S$ be a semigroup whose non-zero $\mathscr{J}$-classes form a chain $J_1 < J_2 < \cdots < J_n$. Then $C(S) \leq \left\lceil \frac{n}{2} \right\rceil$.*

*Proof.* The proof goes by induction on $n$, the cases $n = 1, 2$ being covered by Theorem 5.6.2. Assume by induction the corollary is true for $n \geq 2$ and $S$ is a semigroup whose nonzero $\mathscr{J}$-classes form a chain $J_1 < J_2 < \cdots < J_{n+1}$. Let $I$ be the ideal $\{0\} \cup J_1 \cup J_2$. The Ideal Theorem asserts $C(S) \leq C(I) + C(S/I)$. Induction gives $C(S/I) \leq \left\lceil \frac{n-1}{2} \right\rceil$, whereas Theorem 5.6.2 establishes $C(I) \leq 1$. We conclude $C(S) \leq \left\lceil \frac{n+1}{2} \right\rceil$, as required.    $\square$

*Question 5.6.4.* Suppose that $S$ is a semigroup such that the longest chain of non-zero $\mathcal{J}$-classes of $S$ has length at most 2. Must $C(S) \leq 1$?

## 5.7 Lower Bounds

In this section, we present an analogue of the Type I–Type II lower bound for two-sided complexity. Recall from Proposition 3.6.26 that $S \in \mathbf{V} \textcircled{m} \mathbf{W}$ if and only if each $\mathbf{W}$-idempotent pointlike subsemigroup of $S$ belongs to $\mathbf{V}$. So understanding iteration of idempotent pointlikes is tantamount to deciding two-sided complexity. The following simple observation about idempotent pointlikes for Mal'cev products is from [322, 330]. It is a slight variation on Proposition 4.16.10.

**Proposition 5.7.1.** *Let* $\mathbf{V}$ *and* $\mathbf{W}$ *be pseudovarieties of semigroups. Let $S$ be a semigroup and suppose $S' \leq S$ is $\mathbf{W}$-idempotent pointlike and $S'' \leq S'$ is $\mathbf{V}$-idempotent pointlike as a subsemigroup of $S'$. Then $S''$ is $\mathbf{V} \textcircled{m} \mathbf{W}$-idempotent pointlike as a subsemigroup of $S$.*

*Proof.* Let $\varphi \colon S \to T$, with $T \in \mathbf{V} \textcircled{m} \mathbf{W}$, be a relational morphism. Suppose $\psi \colon T \to W \in \mathbf{W}$ is a relational morphism in $(\mathbf{V}, \mathbf{1})$. Then because $S'$ is $\mathbf{W}$-idempotent pointlike, there is an idempotent $e \in W$ such that $S' \leq e\psi^{-1}\varphi^{-1}$. Let $V = e\psi^{-1}$. Then $V \in \mathbf{V}$ and $\varphi$ restricts to a relational morphism $\rho \colon S' \to V$. Because $S''$ is $\mathbf{V}$-idempotent pointlike as a subsemigroup of $S'$, we conclude there exists an idempotent $f \in V \leq T$ with $S'' \subseteq f\rho^{-1} \subseteq f\varphi^{-1}$. Thus $S''$ is $\mathbf{V} \textcircled{m} \mathbf{W}$-idempotent pointlike in $S$, as required. $\qquad\square$

As was mentioned in Section 4.16.1, if $X$ is an idempotent in $\mathsf{PL}_{\mathbf{V}}(S)$, then $X$ is $\mathbf{V}$-idempotent pointlike. Henckell showed the converse for $\mathbf{A}$ [123, 130]. As he also proved $\mathbf{A}$-pointlikes are decidable [121, 129] (our Theorem 4.19.2), we have:

**Theorem 5.7.2 (Henckell).** *Idempotent pointlike sets are decidable for $\mathbf{A}$.*

Whereas we do not know whether $\mathbb{L}\mathbf{G}$-idempotent pointlikes are decidable, the second author proved that $\mathbb{L}\mathbf{G}$-pointlike sets are decidable [327]. In particular, one can compute the idempotent elements of $\mathsf{PL}_{\mathbb{L}\mathbf{G}}(S)$. Now let us describe our lower bound.

**Theorem 5.7.3.** *Let $S$ be a semigroup that is not aperiodic and let $S' \leq S$ be an $\mathbf{A}$-idempotent pointlike subsemigroup such that $S'' \leq S'$ is an $\mathbb{L}\mathbf{G}$-idempotent pointlike subsemigroup of $S'$. Then $C(S'') < C(S)$.*

*Proof.* Suppose $C(S) = n > 0$. Then

$$S \in ((\mathbf{A}, \mathbf{A})(\mathbb{L}\mathbf{G}, \mathbb{L}\mathbf{G}))^{n} \mathsf{q}(\mathbf{A}) = \left[((\mathbf{A}, \mathbf{A})(\mathbb{L}\mathbf{G}, \mathbb{L}\mathbf{G}))^{n-1}(\mathbf{A}, \mathbf{A})\right] \mathsf{q}(\mathbb{L}\mathbf{G} \textcircled{m} \mathbf{A})$$

so there exist a semigroup $U \in \mathbb{LG} \textcircled{m} \mathbf{A}$ and relational morphisms $\varphi_1 \colon S \to T$ and $\varphi_2 \colon T \to U$ with $\varphi_1 \in ((\mathbf{A}, \mathbf{A})(\mathbb{LG}, \mathbb{LG}))^{n-1}$ and $\varphi_2$ aperiodic. Proposition 5.7.1 yields $S''$ is $\mathbb{LG} \textcircled{m} \mathbf{A}$-idempotent pointlike. Thus there exists $e \in E(U)$ with $S'' \leq e\varphi_2^{-1}\varphi_1^{-1}$. Setting $T' = e\varphi_2^{-1} \leq T$, we have $T' \in \mathbf{A}$, as $\varphi_2$ is aperiodic. Let $\alpha \colon S'' \to T'$ be the restriction of $\varphi_1$. Then $\alpha \in ((\mathbf{A}, \mathbf{A})(\mathbb{LG}, \mathbb{LG}))^{n-1}$ and so $C(S'') \leq n-1 < C(S)$. This proves the theorem. □

**Definition 5.7.4 (A–LG chain).** *Let $S$ be a semigroup. Define an* **A–LG** *chain of length $n$ to be a chain of subsemigroups*

$$S = U_0 \geq T_1 \geq U_1 \geq T_2 \geq U_2 \geq \cdots \geq T_{n-1} \geq U_{n-1} \geq T_n \geq U_n$$

*where $T_i$ is an* **A**-*idempotent pointlike subsemigroup of $U_{i-1}$ and $U_i$ is an* **LG**-*idempotent pointlike subsemigroup of $T_i$, for $i = 1, \ldots, n$, with $U_i$ non-aperiodic, for $i = 0, \ldots, n-1$. Let $\dagger(S)$ be the maximum length of an* **A–LG** *chain for $S$.*

**Corollary 5.7.5.** *One has $\dagger(S) \leq C(S)$.*

The next two lemmas provide some examples of **A** and **LG**-idempotent pointlikes.

**Lemma 5.7.6.** *Let $S$ be a semigroup. Suppose $G$ is a subgroup with identity $f$ and let $e \leq f$ be an idempotent. Then $\langle GeG \rangle$ is* **A**-*idempotent pointlike.*

*Proof.* Let $\varphi \colon S \to T$ be a relational morphism with $T \in \mathbf{A}$. We can find idempotents $e', f' \in T$ with $e \in e'\varphi^{-1}$ and $G \leq f'\varphi^{-1}$ by Lemma 4.14.25. Then $e \in (f'e'f')^\omega \varphi^{-1}$ and $(f'e'f')^\omega \leq f'$. So without loss of generality, we may assume $e' \leq f'$. In this case $GeG \leq (f'e'f')\varphi^{-1} = e'\varphi^{-1}$, whence $\langle GeG \rangle \leq e'\varphi^{-1}$, as desired. □

Our old friend, the Type II semigroup, allows us to construct some **LG**-idempotent pointlikes.

**Lemma 5.7.7.** *Let $S$ be a semigroup and $e \in S$ an idempotent. Then $\mathsf{K}_{\mathbf{G}}(eSe)$ is* **LG**-*idempotent pointlike.*

*Proof.* Let $\varphi \colon S \to T$ be a relational morphism with $T \in \mathbb{LG}$. Then there is an idempotent $e' \in e\varphi$. We can define a relational morphism $\psi \colon eSe \to e'Te'$ by restriction: $t \in s\psi$ if and only if $t \in s\varphi$, for $s \in eSe$, $t \in e'Te'$. Clearly, $\psi$ is fully defined because if $s = es'e$ and $t' \in s'\varphi$, then $e't'e' \in s\psi$. It now follows easily $\psi$ is a relational morphism. Because $e'Te'$ is a group with identity $e'$, we conclude $\mathsf{K}_{\mathbf{G}}(eSe) \subseteq e'\psi^{-1} \subseteq e'\varphi^{-1}$. □

All the preparatory work has now been completed to compute $C(T_n)$. Our computation establishes the strictness of the two-sided complexity hierarchy: there are semigroups of all two-sided complexities. This is another unpublished result of the first author.

**Theorem 5.7.8 (Rhodes).** *The two-sided complexities of $T_n$ and $M_n(\mathbb{F}_q)$ are given by $C(T_n) = \lceil \frac{n-1}{2} \rceil = C(M_n(\mathbb{F}_2))$ and $C(M_n(\mathbb{F}_q)) = \lceil \frac{n}{2} \rceil$ if $q > 2$. In particular, the two-sided complexity hierarchy is strict.*

*Proof.* Let $\{R_n\}$, $n \geq 1$, be one of the series $\{T_n\}$ or $\{M_n(\mathbb{F}_q)\}$. Then the non-zero $\mathscr{J}$-classes of $R_n$ form a chain of length $n$. In the case of $T_n$ and $M_n(\mathbb{F}_2)$, the 0-minimal ideal is aperiodic, and so we can factor it out to obtain a semigroup of the same two-sided complexity whose $\mathscr{J}$-classes form a chain of length $n - 1$. Corollary 5.6.3 then provides the upper bound. Now we turn to the lower bound.

We claim, for $n \geq 3$, that $C(R_n) > C(R_{n-2})$. Let $G$ be the group of units of $R_n$ and let $e$ be a rank $n - 1$ idempotent. Then we know $S = \langle GeG \rangle$ consists precisely of the singular elements of $R_n$ and moreover is **A**-idempotent pointlike by Lemma 5.7.6. Furthermore, $\mathsf{K}_{\mathbf{G}}(eSe)$ is $\mathbb{L}\mathbf{G}$-idempotent pointlike by Lemma 5.7.7. Clearly, $\langle E(eSe) \rangle \leq \mathsf{K}_{\mathbf{G}}(eSe)$. Because $eSe \cong R_{n-1}$ and $\langle E(R_{n-1}) \rangle$ contains all singular elements of $R_{n-1}$, and in particular contains a copy of $R_{n-2}$, an application of Theorem 5.7.3 establishes

$$C(R_{n-2}) \leq C(\mathsf{K}_{\mathbf{G}}(eSe)) < C(R_n).$$

To establish the lower bound it now suffices to compute $C(R_1)$ and $C(R_2)$. The two-sided complexities of these semigroups coincide with their usual complexities, computed in Theorem 4.12.31, with the exception of $M_2(\mathbb{F}_q)$ with $q > 2$. This semigroup is a $2\mathscr{J}$-semigroup and so has two-sided complexity 1 by Theorem 5.6.2. This completes the proof. $\qquad\qquad\square$

This brings to an end our brief introduction to two-sided complexity. Let us conclude with some questions, the most important of which is that of computability.

*Question 5.7.9.* Is the two-sided complexity function $C$ computable? Start with a semigroup with at most 3 non-zero $\mathscr{J}$-classes.

Because the description of two-sided complexity in terms of semidirect products involves iteration of the operator $\mathbf{G} ** (-)$, the analogue of the presentation lemma is not so clear in this context.

*Question 5.7.10.* Is there some sort of Presentation Lemma to describe the existence of a cross section with respect to $\mathbb{L}\mathbf{G} \ \text{ⓜ} \ \mathbf{V}$?

None of our examples give a semigroup whose two-sided complexity is at least two and agrees with its one-sided complexity.

*Question 5.7.11.* Are there arbitrarily large numbers $n$ such that there is a semigroup $S$ with $c(S) = n = C(S)$? Probably there are.

*Question 5.7.12.* What is the two-sided complexity of the semigroup of binary relations on an $n$ element set?

# Notes

Complexity hierarchies have long been a part of semigroup theory [71, 170, 171, 230, 269, 368]. The complexity hierarchy associated to a single operator is essentially a standard idea; the two operator notion is modeled on group complexity and is one of the reasons for writing this book. MPS theory was first introduced in [267] by Rhodes. A case-by-case classification appears in [296]. The MPS decomposition results are from [293, 296]. Our treatment follows very closely [293] in terms of proof, but [296] in terms of presentation of the consequences. The availability of the language of pseudovarieties of relational morphisms adds a new ingredient into the mix. The consequences of the two-sided decomposition theory are too numerous to mention them all, however all the following papers make use of it [55, 152, 234, 338, 378, 379], as does the book of Straubing [343].

Two-sided complexity was invented by Rhodes but never before published. In part it is motivated by the theory of bimachines [84, 168, 284]. The fact that the primes for the two-sided semidirect product are exactly the congruence-free monoids, as well as its inherent left-right duality, makes the theory more elegant than the classical complexity theory. The Ideal Theorem for two-sided complexity is due to Rhodes, although there is no essential difference in the proof from the one-sided case. Translational hulls of 0-simple semigroups are a classical aspect of semigroup theory [68, 171]. The trick for constructing ideal extensions by adding "all" possible columns is due to the father of the Berkeley School. The results on 2-sided complexity, as well as the definition, are due to Rhodes and have been Berkeley School folklore for years. The lower bound for two-sided complexity in terms of idempotent-pointlikes is an improvement by Steinberg of a previous lower bound (unpublished) of Rhodes. It would be quite interesting to try and compute the two-sided complexity of a $3 \mathscr{J}$-semigroup.

# The Algebraic Lattice of Semigroup Pseudovarieties

Part III

The Algebraic Lattice of Semigroup
Pseudovarieties

# 6

# Algebraic Lattices, Continuous Lattices and Closure Operators

Algebraic lattices and continuous operators on algebraic lattices played a salient role in Chapter 2. In order to enter into the study of pseudovarieties of relational morphisms as quickly as possible, we tried to work there with a bare minimum of the theory. In this chapter, we delve further into the theory of algebraic lattices, and their close cousins, continuous lattices. We shall see that the fixed-point set of an idempotent continuous operator on **PV** is a continuous lattice of countable weight. Conversely, every continuous lattice of countable weight is the fixed-point set of an idempotent of **Cnt(PV)**. This chapter also introduces the promised topology on an algebraic lattice for which our continuous operators become the continuous order preserving maps. It should be mentioned that it is a celebrated result [97, 134, 147] that the category of algebraic lattices with what we shall call CL-morphisms is equivalent to the category of profinite meet semilattices with identities (which is in turn equivalent to the opposite of the category of join semilattices with identity), which gives another motivation as to why the concept should be of interest to semigroup theorists.

Closure operators also play an important role in semigroup theory. Many natural pseudovariety operators are closure operators. More importantly, closure operators play a role in constructing the Type II subsemigroup [33, 295] and in constructing the aperiodic pointlikes; see Section 4.19. Closure operators seem likely to play an important foundational role in the proof of the decidability of Krohn-Rhodes complexity [128]. In particular, if $L$ is a complete lattice, then an interesting semigroup structure can be placed on the set of closure operators on $L \times L$, turning it into an ordered semigroup; see [128].

## 6.1 Complete and Algebraic Lattices

We now turn to the lattice theoretic notions that have been discussed in connection with our favorite examples in the previous chapters, but which have yet to be formally treated.

J. Rhodes, B. Steinberg, *The q-theory of Finite Semigroups*,
Springer Monographs in Mathematics, DOI 10.1007/978-0-387-09781-7_6,
© Springer Science+Business Media, LLC 2009

### 6.1.1 Algebraic lattices

A partially ordered set (poset) $L$ is a *lattice* if each pair of elements admits a least upper bound (also called a sup or join) and a greatest lower bound (also called an inf or meet). If $L$ just admits meets (respectively, joins), then $L$ is called a meet (respectively, join) semilattice. Notice the empty set is a lattice with our definition. A totally ordered set is called a *chain*.

A lattice $L$ is called *complete* if arbitrary (including empty) subsets of $L$ have both sups and infs. In particular, a complete lattice always has a top and a bottom and so is never empty. If we view $L$ as a category in the natural way [185], then sups correspond to coproducts and infs to products. In particular, the empty sup is the bottom of $L$ and the empty inf is the top. We shall use the convention that $\mathsf{T}$ and $\mathsf{B}$ denote the top and bottom, respectively, of a complete lattice. Sometimes we shall use 0 for the bottom when we are viewing the lattice as a monoid with the join operation. Notice that a non-empty finite lattice is automatically complete.

In a complete lattice, the meet and join determine each other. Namely, if $X \subseteq L$, then

$$\bigvee X = \bigwedge \{y \mid y \geq x, \forall x \in X\} \tag{6.1}$$

$$\bigwedge X = \bigvee \{y \mid y \leq x, \forall x \in X\}. \tag{6.2}$$

If $L$ is a partially ordered set admitting arbitrary infs, then $L$ must be a complete lattice and (6.1) can serve as the definition of the sup. When we use (6.1) to define the join, we call it the *determined join* and denote it $\vee_{\mathrm{det}}$. Dual remarks and notation apply to the case where $L$ admits arbitrary sups; in particular, (6.2) can be used to define the *determined meet* $\wedge_{\mathrm{det}}$.

Recall that an element $k$ of a lattice $L$ is called *compact* if $k \leq \bigvee I$, $I \subseteq L$, implies there is a *finite* subset $F \subseteq I$ such that $k \leq \bigvee F$. For instance, finitely generated (equals one-generated) pseudovarieties are compact in **PV**. If $X$ is a set, then the compact elements of $2^X$ are the finite subsets. Every element of a finite lattice is compact. If a lattice has a bottom element, then the bottom is always compact. The compact elements of a lattice form a join semilattice, as the following exercise shows.

**Exercise 6.1.1.** Let $L$ be a lattice and $K(L)$ be the set of compact elements of $L$. Show that $K(L)$ is closed under all finite sups (including the empty sup if $L$ has a bottom).

As mentioned in Chapter 1, a complete lattice $L$ is called an *algebraic lattice* if each element is a directed supremum of compact elements. More precisely, if $\ell \in L$, then there is a directed set $D$ of compact elements with $\ell = \bigvee D$. Virtually all of the lattices we have been considering in the current volume are algebraic lattices: **PV**, **PVRM**, **CC**, **Cnt(PV)**, **GMC(PV)**, etc. If $X$ is a set, $2^X$ is an algebraic lattice. To be algebraic, it suffices for each element to be a join of compact elements thanks to Exercise 6.1.1.

## Basic properties of algebraic lattices

We now turn to some basic properties of algebraic lattices. The following exercise is [203, Lemma 4.49ii].

**Exercise 6.1.2.** Let $L$ be a complete (algebraic) lattice. If $u \leq v$ in $L$, then $[u, v]$ is a complete (algebraic) lattice.

If $L$ is a lattice, for $a, b \in L$, we write $a \nearrow b$ if $a \leq b$ and there are no elements between $a$ and $b$. In this case, one says that $b$ *covers* $a$ or that the interval $[a, b]$ is a *gap*. The following result is standard [203, Lemma 4.49]

**Proposition 6.1.3 (Gaps are dense).** *Let $L$ be an algebraic lattice.*

1. *If $\ell < c$ and $c$ is compact, there exists $m \in L$ such that $\ell \leq m \nearrow c$.*
2. *If $\ell < n$ in $L$, there exists $a, b \in L$ such that $\ell \leq a$, $b \leq n$ and $a \nearrow b$.*

*Proof.* For 1, first observe that the half open interval $[\ell, c)$ is closed under directed sups because $c$ is compact. Hence, by Zorn's Lemma, it has a maximal element $m$. Clearly $\ell \leq m \nearrow c$.

We now turn to 2. By Exercise 6.1.2, $[\ell, n]$ is algebraic. Hence, because $n$ is a join of compact elements from $[\ell, n]$, we must have a compact element $b$ of $[\ell, n]$ with $\ell < b \leq n$. Then, by 1, there is an element $a \in [\ell, n]$ with $\ell \leq a \nearrow b \leq n$.    □

We have the following specializations for the lattices of interest to us.

**Corollary 6.1.4.**

1. *For **PV** gaps are dense and, for each non-trivial finite semigroup $S$, the pseudovariety $(S)$ generated by $S$ has a maximal proper subpseudovariety. Furthermore, $\mathbf{W} \leq \mathbf{V}$ implies $[\mathbf{W}, \mathbf{V}]$ is an algebraic lattice.*
2. *For **PVRM** gaps are dense and, for each relational morphism $f$ with non-trivial domain, the pseudovariety $(f)$ generated by $f$ has a maximal property subpseudovariety. Furthermore, $\mathsf{R}_1 \leq \mathsf{R}_2$ implies $[\mathsf{R}_1, \mathsf{R}_2]$ is an algebraic lattice.*
3. *(J. Goodwin[1]) Let $S$ be an arbitrary semigroup and let $\mathscr{I}$ be the collection of all ideals of $S$ including $\emptyset$. Then $\mathscr{I}$ is an algebraic lattice with compact elements the finitely generated ideals. Let $C$ be a maximal chain of ideals from $\emptyset$ to $S$. Then the gaps of $C$ are in one-to-one correspondence with the $\mathscr{J}$-classes of $S$.*

For example if $S = (\mathbb{Q}, \min)$, with $\mathbb{Q}$ the rationals, then the (non-empty) ideals of $S$ are of the form $(-\infty, r)$ with $r \in \mathbb{R}$ or $(-\infty, q]$ with $q \in \mathbb{Q}$. The (non-empty) ideals of $S$ form a chain isomorphic to the subset of $\mathbb{R} \times \{0, 1\}$, with lexicographical order, consisting of $\mathbb{R} \times \{0\} \cup \mathbb{Q} \times \{1\}$ via $(-\infty, r) \mapsto (r, 0)$

---

[1] Unpublished.

and $(-\infty, q] \mapsto (q, 1)$. The gaps are of the form $(-\infty, q) \nearrow (-\infty, q]$ and correspond with the one point $\mathscr{J}$-classes of the semilattice $\mathbb{Q}$.

Notice that we need an algebraic lattice and not just a complete lattice (or even a continuous lattice, see Definition 6.2.9) for Proposition 6.1.3 to hold. Indeed, if we consider $[0, 1]$ then there are no gaps as each interval contains a rational number.

### 6.1.2 Some lattice terminology

We consider here some definitions and terminology concerning lattices (including some already considered in this text); our terminology mostly follows that of McKenzie [203] but other names are sometimes used in the literature. In Chapter 7, we will study these concepts in detail for **PV** leading to several new results.

**Definition 6.1.5.** *Let $L$ be a lattice.*

1. *$c \in L$ is **compact** if $c \leq \bigvee X$ implies $c \leq \bigvee F$ for some finite subset $F$ of $X$.*
2. *$c \in L$ is **strictly compact** if $c = \bigvee X$ implies $c = \bigvee F$ for some finite subset $F$ of $X$.*
3. *$c \in L$ is **co-compact** if whenever $c \geq \bigwedge X$, then $c \geq \bigwedge F$ for some finite subset $F$ of $X$.*
4. *$c \in L$ is **strictly co-compact** if $c = \bigwedge X$ implies $c = \bigwedge F$ with $F$ a finite subset of $X$.*
5. *$\ell \in L$ is **strictly join irreducible** (sji) if whenever $\ell = \bigvee X$, there exists $x \in X$ such that $\ell = x$.*
6. *$\ell \in L$ is **join irreducible** (ji) if $\ell \leq \bigvee X$ implies there exists $x \in X$ such that $\ell \leq x$.*
7. *$\ell \in L$ is called **strictly finite join irreducible** (sfji) if $\ell = \bigvee F$ with $F$ finite implies $\ell = f$ for some $f \in F$.*
8. *$\ell \in L$ is called **finite join irreducible** (fji) if $\ell \leq \bigvee F$ with $F$ finite implies $\ell \leq f$ for some $f \in F$.*
9. *$m \in L$ is called **strictly meet irreducible** (smi) if $m = \bigwedge X$ implies that there exists $x \in X$ such that $m = x$.*
10. *$m \in L$ is called **meet irreducible** (mi) if $m \geq \bigwedge X$ implies that $m \geq x$ for some $x \in X$.*
11. *$m \in L$ is called **strictly finite meet irreducible** (sfmi) if whenever $m = \bigwedge F$ with $F$ finite, there exists $f \in F$ such that $m = f$.*
12. *$m \in L$ is **finite meet irreducible** (fmi) if $m \geq \bigwedge F$ with $F$ finite implies $m \geq f$.*

Finite meet irreducible elements are called by some authors *primes* in analogy with ring theory. For each of the above definitions, one can verify that the strict version is a consequence of the non-strict version and the finite version is implied by the infinite version.

**Exercise 6.1.6.** Let $L$ be an algebraic lattice and $\ell \in L$. Prove the following:

1. $\ell$ is sji if and only if $\ell$ is sfji and compact;
2. $\ell$ is ji if and only if $\ell$ is fji and compact.

One of the most popular classes of lattices is that of distributive lattices.

**Definition 6.1.7 (Distributive lattice).** *A lattice $L$ is distributive if*

$$a \wedge (b \vee c) = (a \wedge b) \vee (a \wedge c)$$

*for all $a, b, c \in L$.*

**Exercise 6.1.8.** Show that in a distributive lattice $a \vee (b \wedge c) = (a \vee b) \wedge (a \vee c)$.

If $X$ is a topological space, then the meet of the lattice $\mathcal{O}(X)$ of open subsets of $X$ distributes over infinite joins leading to the following definition.

**Definition 6.1.9 (Frame).** *A complete lattice $L$ is called a frame if it satisfies the infinite distributive law*

$$a \wedge \bigvee X = \bigvee_{x \in X} a \wedge x \tag{6.3}$$

*for any subset $X \subseteq L$.*

Frames also go under the name complete Heyting algebras. A *morphism of frames* is a map preserving arbitrary sups and finite infs. For instance, a continuous map $f \colon X \to Y$ of topological spaces gives rise to a frame morphism $f^{-1} \colon \mathcal{O}(Y) \to \mathcal{O}(X)$. The opposite of the category of frames is the category of *locales*, which is the main object of study in pointless topology [97, 147, 186]. However, **PV** and the other lattices we have been considering are not frames. They enjoy instead a weaker property.

**Definition 6.1.10 (Meet and join continuity).** *Let $L$ be a complete lattice. Then $L$ is said to be meet-continuous if, for each $\ell \in L$, the operator $\ell \wedge (-)$ is continuous, i.e., the infinite distributive law (6.3) holds for all directed sets $X$. One says $L$ is join-continuous if $L^{op}$ is meet-continuous.*

The terminology join-continuous is somewhat unfortunate, but seems to be well established. Because many of the lattices of interest to us are meet-continuous, we state some of the basic properties of such lattices.

**Proposition 6.1.11.** *Let $L$ be a meet-continuous complete lattice. Then $c \in L$ is compact if and only if $c$ is strictly compact.*

*Proof.* It is clear compactness implies strict compactness. For the converse, let $c$ be strictly compact and assume that $c \leq \bigvee X$. Let $D$ be the collection of finite joins of elements of $X$. Then $D$ is directed and $\bigvee X = \bigvee D$. Because $L$ is meet-continuous and $D$ is directed,

$$c = c \wedge \bigvee D = \bigvee_{d \in D} (c \wedge d).$$

Because $c$ is strictly compact, $c = \bigvee_{d \in F}(c \wedge d) \leq \bigvee F$ for some finite subset $F \subseteq D$. Because $D$ is directed and $F$ is finite, in fact $c \leq d$ for some $d \in F$. But $d$ is a finite join of elements of $X$, establishing that $c$ is compact.    □

**Proposition 6.1.12.** *Let $L$ be a meet-continuous lattice. Then $\ell \in L$ is ji if and only if it is both sji and fji.*

*Proof.* Necessity being clear, suppose that $\ell$ is both sji and fji. Suppose further that $\ell \leq \bigvee X$. Let $D$ be the collection of finite joins of elements of $X$. Then $\ell \leq \bigvee X = \bigvee D$. Also $D$ is directed. So by meet-continuity

$$\ell = \ell \wedge \bigvee D = \bigvee_{d \in D} (\ell \wedge d).$$

Thus, as $\ell$ is sji, we have $\ell = \ell \wedge d \leq d$ for some $d \in D$. But $d = \bigvee F$ for some finite subset $F$ of $X$. Because $\ell$ is fji, we have $\ell \leq x$ for some $x \in F \subseteq X$. Thus we have shown that $\ell$ is ji, as required.    □

Figure 6.1 provides a partial Hasse diagram of the concepts we have been considering. The next lemma will be used in Chapter 7 to give examples of fji pseudovarieties.

**Lemma 6.1.13.** *Let $L$ be a complete lattice and suppose that $D \subseteq L$ is a directed subset of fji elements. Then $\bigvee D$ is fji. A dual result holds true for fmi and downwards directed sets.*

*Proof.* Let $d = \bigvee D$ and suppose $d \leq m \vee n$. Suppose $d \not\leq m$; then $\ell \not\leq m$ for some $\ell \in D$. Because $\ell$ is fji, $\ell \leq n$. Let $\ell' \in D$ and let $k \in D$ be an upper bound for $\ell$ and $\ell'$. If $k \leq m$, then $\ell \leq m$, a contradiction. Because $k$ is fji, we conclude that $k \leq n$. Hence $\ell' \leq n$. Because $\ell'$ was arbitrary, we conclude $d \leq n$, as desired.    □

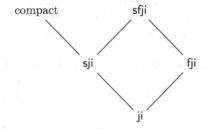

**Fig. 6.1.** Relationship between lattice theoretic concepts in an algebraic lattice

We shall reprise the study of these notions in Chapter 7.

*Question 6.1.14.* A lattice is said to be *semi-distributive* at an element $\ell$ if $\ell \wedge x = \ell \wedge y$ implies $\ell \wedge x = \ell \wedge (x \vee y)$. Because meets and joins are continuous in **PV**, if **PV** is semi-distributive at **V** and **W** $\leq$ **V**, then there is a largest pseudovariety $\widehat{\mathbf{W}}$ such that $\mathbf{V} \cap \widehat{\mathbf{W}} = \mathbf{W}$. For instance, **PV** is semi-distributive at **G** and if **H** is a pseudovariety of groups, then $\overline{\mathbf{H}}$ is the maximum pseudovariety with $\mathbf{G} \cap \overline{\mathbf{H}} = \mathbf{H}$. Reilly and Zhang proved that **PV** is semi-distributive at the pseudovariety of bands **B** [260]. It follows from a result of G. Higman [217, 54.24] that **PV** is not semi-distributive at certain pseudovarieties of groups. At which pseudovarieties **V** is **PV** semi-distributive?

**Exercise 6.1.15.** Let $L$ be a complete lattice. The *spectrum* Spec $L$ of $L$ is the set of fmi elements of $L$ other than $\mathsf{T}$. Define the *hull-kernel* topology on Spec $L$ by taking as the closed sets all sets of the form $V(\ell) = \{m \in \text{Spec } L \mid m \geq \ell\}$ with $\ell \in L$.

1. Show that the hull-kernel topology is indeed a topology.
2. For $\ell \in L$, set $D(\ell) = \text{Spec } L \setminus V(\ell)$. Show that $D \colon L \to \mathcal{O}(L)$ is a surjective map preserving all sups and finite infs. A frame is said to have *enough points* if this map is also injective.

## 6.2 Continuous Lattices

There is an important generalization of an algebraic lattice that will play a role in our work, namely that of a continuous lattice [97]. In this section, we follow closely [97], but our presentation is more condensed. The results are not new, but we have new applications in the context of **PV** and **Cnt(PV)**.

**Definition 6.2.1 (Way below).** *Let $L$ be a poset and $k, \ell \in L$. Then $k$ is said to be way below $\ell$, denoted $k \ll \ell$, if whenever $\ell \leq \bigvee D$ with $D$ a directed set, there exists $d \in D$ such that $k \leq d$.*

By taking $D = \{\ell\}$, we see that $k \ll \ell$ implies $k \leq \ell$.

**Exercise 6.2.2.** Show that in a join semilattice $k \ll \ell$ if and only if whenever $\ell \leq \bigvee X$, there is a finite subset $F \subseteq X$ such that $k \leq \bigvee F$.

**Exercise 6.2.3.** Suppose $x \leq k \ll \ell \leq z$. Prove $x \ll z$.

A good example of the notion of "way below" is provided by the following proposition.

**Proposition 6.2.4.** *Let $k \leq \ell$ and suppose that $k$ is compact. Then $k \ll \ell$.*

**Exercise 6.2.5.** Prove Proposition 6.2.4.

In fact, compact elements can be characterized as follows.

**Proposition 6.2.6.** *Let $L$ be a join semilattice. Then $k \in L$ is compact if and only if $k \ll k$. That is, $k$ is compact if and only if whenever $k \leq \bigvee D$ with $D$ directed, $k \leq d$ some $d \in D$.*

*Proof.* The necessity of $k \ll k$ follows from Proposition 6.2.4. The sufficiency is proved as follows. Suppose $k \leq \bigvee X$. Let

$$ Y = \{ \bigvee F \mid F \subseteq X, \; |F| < \infty \}. $$

Clearly, $\bigvee X = \bigvee Y$, so $k \leq \bigvee Y$. But $Y$ is directed, so there exists $y \in Y$ such that $k \leq y$, as $k \ll k$. But $y = \bigvee F$ for some finite subset of $X$. Thus $k$ is compact.                                                    □

*Remark 6.2.7.* In posets admitting directed joins, called *domains*, Proposition 6.2.6 is taken as the definition of being compact. See [97, 147] for more details.

**Exercise 6.2.8.** Let $L$ be a lattice and $m \in L$. Show that the set

$$ m^{\Downarrow} = \{ \ell \in L \mid \ell \ll m \} $$

is closed under joins of pairs. Conclude that $m^{\Downarrow}$ is directed. Show that if $\ell \in m^{\Downarrow}$ and $y \leq \ell$, then $y \in m^{\Downarrow}$.

Now we reach the important notion of a continuous lattice, due to D. Scott [97].

**Definition 6.2.9 (Continuous lattice).** *Let $L$ be a complete lattice. Then $L$ is called a continuous lattice if each element $\ell$ of $L$ is a directed supremum of elements way below it. That is, $\ell = \bigvee D$ where $D$ is directed and $d \in D$, implies $d \ll \ell$.*

It is immediate from Proposition 6.2.4 that each algebraic lattice is continuous. But the class of continuous lattices is much bigger (it contains for instance the interval $[0, 1]$) and has better closure properties [97]. Observe, by Exercise 6.2.8, $L$ is a continuous lattice if and only if each element is a join of elements way below it.

The key idea behind continuous lattices is that they have a reasonable theory of computation via "finite approximations." For algebraic lattices, finite approximation comes from the fact that each element is a directed supremum of compact objects; in the general case, the notion of "way below" plays the role of compactness.

For example, the unit interval $[0, 1]$ with its usual order is a continuous lattice but not an algebraic lattice. The only compact element is 0. For if $a > 0$, then $a$ is the sup of all elements strictly below it, but $a$ is not below any finite join. But $y \ll x$ if and only if $y < x$ or $y = x = 0$.

Suppose $X$ is a locally compact topological space and consider the complete lattice of open subsets of $X$ with the determined meet. Then an open set $U$ is way below an open set $V$ if and only if there exists a compact set $C$ (in the topological sense) such that $U \subseteq C \subseteq V$ [97, Prop. I-1.4].

**Exercise 6.2.10.** Prove that in an algebraic lattice, one has $\ell \ll m$ if and only if there exists a compact element $k$ such that $\ell \leq k \leq m$.

Let us show that every continuous lattice, in particular every algebraic lattice, is meet-continuous.

**Proposition 6.2.11.** *A continuous lattice is meet-continuous. In particular, every algebraic lattice is meet-continuous.*

*Proof.* Let $L$ be a continuous lattice. Suppose that $\ell \in L$ and $D$ is a directed set. Clearly,

$$\ell \wedge \bigvee D \geq \bigvee_{d \in D} (\ell \wedge d).$$

To show the converse, set $m = \ell \wedge \bigvee D$ and let $c \ll m$. Then $m \leq \bigvee D$ and so by the definition of $c \ll m$, we must have $d_0 \in D$ such that $c \leq d_0$. But then

$$c \leq m \wedge d_0 \leq \ell \wedge d_0 \leq \bigvee_{d \in D} (\ell \wedge d). \tag{6.4}$$

Because $m$ is the supremum of the elements way below it, we may conclude

$$\ell \wedge \bigvee D = m \leq \bigvee_{d \in D} (\ell \wedge d)$$

completing the proof.                                                     $\square$

### 6.2.1 Philosophical discussion on continuous lattices

The following subsection is an informal discussion of continuous lattices and the spectral theory of lattices, none of which shall be used explicitly later in the text.

With a ring $R$, a topological space called the *spectrum* can be associated in several different technical ways. For a commutative ring, one could take the spectrum as the set of prime ideals with the Zariski topology.

Specifically for a commutative ring $R$, let the spectrum be the set of prime ideals, viewed as a subset of the algebraic lattice of all ideals of the ring. Then the (Jacobson/Zariski) spectrum is exactly the family of fmi elements of the distributive algebraic lattice of all radical ideals of $R$, where a radical ideal is an intersection of prime ideals.

We now turn to the spectrum of operators on a Hilbert space. Let $C$ be a $C^*$-algebra (see [31]). Consider the *closed* two-sided ideals of $C$. This complete

436	6 Algebraic Lattices, Continuous Lattices and Closure Operators

lattice is not algebraic in general, but it is a continuous lattice (and moreover a frame [97]). Also $IJ = I \cap J$ for closed ideals.

Now the abstract spectrum of a continuous lattice $L$ is "really" the maximal image satisfying the infinite distributive law (meaning distributivity plus meet continuity, which equals Brouwerian lattice equals Heyting algebra equals frame equals locale equals "pointless" topological space), and this goes back to M. Stone in the 1930s and is determined by the fmi elements [97, 147, 341, 342]. Then identifying two elements of $L$ if and only if they have the same set of fmi above them gives the maximal image satisfying the distributive law (as basically $L$ is inf-generated by its fmi elements if and only if it satisfies this law). See [97, Chapter V].

In the case of *separable* $C^*$-algebras, primitive ideals (kernels of irreducible representations) are exactly the fmi (primes) of the distributive continuous lattice of all closed ideals and are inf-generating. Thus the abstract spectral theory of the distributive continuous lattice of all closed ideals of the $C^*$-algebra is the traditional primitive ideal spectrum. That this spectrum is a locally compact (not necessarily $T_2$) $T_0$-space follows from the fact that this is true for all distributive continuous lattices.

However, see [31, Chapter 4, p. 82] for a discussion of the inadequacy of this approach in the non-commutative setting (the space is not $T_2$, etc.) and that one should go to the Mackey program for $C^*$-algebras of making the spectrum the set of equivalence classes of the irreducible representations with "natural" Borel structure. So measures, Borel structures and descriptive set theory and such are needed.

The general idea is that the abstract spectral theory of a complete lattice will put one "in the ballpark," but then specifics for the situation must be introduced and utilized. This will be true for us also. See Chapter 8. This all ties up with the motivation for considering quantales [303].

Continuous lattices have enough sfmi elements to inf-generate [97, Cor. I-3.10]. For an algebraic lattice the smi elements inf-generate (Theorem 7.1.6), but this is not true in $[0, 1]$.

## 6.3 Closure and Kernel Operators, Ideals and Morphisms

In this section, we are concerned with certain types of morphisms between complete lattices. We begin with the definition of closure and kernel operators.

**Definition 6.3.1.** *Let $L$ be poset.*

1. *A map $p: L \rightarrow L$ is called a **projection** if $p$ is order preserving and idempotent.*
2. *A projection $c$ is called a **closure operator** if it is increasing, that is, $\ell \leq c(\ell)$, all $\ell \in L$.*

3. A projection $k$ is called a **kernel operator** if it is decreasing, that is, $k(\ell) \leq \ell$, for all $\ell \in L$.

The canonical examples are the following. If $X$ is a topological space, then the map $c \colon 2^X \to 2^X$ sending a subset $Y$ to its topological closure is a closure operator. The map $k \colon 2^X \to 2^X$ sending a subset $Y$ to its interior is a kernel operator. Other examples were mentioned in Chapter 2.

A main source of closure and kernel operators is the theory of Galois connections (cf. Section 1.1.2). Recall that if $P$ and $Q$ are partially ordered sets, then a *Galois connection*

$$ P \xleftarrow{\ g\ }\xrightarrow[\ d\ ]{} Q \tag{6.5} $$

consists of a pair of order preserving maps $g \colon P \to Q$ and $d \colon Q \to P$ so that

$$ d(t) \leq s \iff t \leq g(s). \tag{6.6} $$

Equivalently, a Galois connection is an adjunction of categories between $P$ and $Q$, viewed as categories in the natural way [185]; here $g$ is the right adjoint and $d$ is the left adjoint. As such, $g$ is **inf** and $d$ is **sup** and they uniquely determine each other (see [97, Prop. O-3.2]). The following summarizes [97, Thm. O-3.6, Proposition O-3.7].

**Proposition 6.3.2.** *Let* (6.5) *be a Galois connection. Then:*

1. *$dg$ is a kernel operator and $gd$ is a closure operator;*
2. *$gdg = g$ and $dgd = d$;*
3. *$g$ is injective if and only if $d$ is surjective, if and only if $dg = 1$; moreover, this occurs precisely when $g(s) = \max d^{-1}(s)$;*
4. *$d$ is injective if and only if $g$ is surjective, if and only if $gd = 1$; moreover, this occurs precisely when $d(t) = \min g^{-1}(t)$.*

*Proof.* We begin with 1. Because $g(s) \leq g(s)$, we have by (6.6) that $dg(s) \leq s$. Similarly, $d(t) \leq d(t)$ implies $t \leq gd(t)$. Thus to establish 1 we just need to show that $d$ and $g$ are idempotent. We do this by showing 2 holds. By what we have just seen $g(s) \leq gd(g(s))$ as $gd$ is increasing. But $g(dg(s)) \leq g(s)$ as $dg$ is decreasing. So $g(s) = gdg(s)$. Similarly $dgd = d$.

To verify 3, clearly $dg = 1$ implies $d$ is surjective and $g$ is injective. Suppose that $g$ is injective. Then $gdg = g$ implies $dg = 1$. If $d$ is surjective, then $dgd = d$ implies $dg = 1$. If $dg = 1$, then $dg(s) = s$ so $g(s) \in d^{-1}(s)$. If $r \in d^{-1}(s)$, then $d(r) = s$ implies $r \leq g(s)$ by (6.6). Thus $g(s) = \max d^{-1}(s)$. Conversely, if $g(s) = \max d^{-1}(s)$, then $dg(s) = s$ and so $dg = 1$.

Statement 4 is dual to statement 3 and we leave it to the reader.     □

**Exercise 6.3.3.** Let $d \colon P \twoheadrightarrow Q$ be an onto map of partially ordered sets such that $\max d^{-1}(q)$ exists for all $q \in Q$. Show that if one defines $g \colon Q \twoheadrightarrow P$ by $g(q) = \max d^{-1}(q)$, then $d$ and $g$ form a Galois connection with $d$ the left adjoint and $g$ the right adjoint. Also state the dual result.

The following special case of Freyd's Adjoint Functor Theorem [185] will be used without comment throughout.

**Theorem 6.3.4 (Adjoint Functor Theorem).** *Let $L$ and $N$ be complete lattices and suppose that $g\colon L \to N$ is* **inf** *(respectively, $d\colon N \to L$ is* **sup***). Then there is a left adjoint $d\colon N \to L$ (respectively, right adjoint $g\colon L \to N$).*

*Proof.* Define $d\colon N \to L$ by

$$d(n) = \bigwedge \{\ell \in L \mid n \leq g(\ell)\}. \tag{6.7}$$

Suppose $n \leq g(\ell)$. Then (6.7) yields $d(n) \leq \ell$. Conversely, if $d(n) \leq \ell$, then

$$g(\ell) \geq g(d(n)) = g\left( \bigwedge \{\ell' \in L \mid g(\ell') \geq n\} \right) = \bigwedge_{\{\ell' \in L \mid g(\ell') \geq n\}} g(\ell') \geq n$$

because $g$ is **inf**, establishing (6.6). The dual result is proved similarly, define

$$g(\ell) = \bigvee \{n \in N \mid d(n) \leq \ell\}. \tag{6.8}$$

We leave the details to the reader.                                             □

Let $L$ be a complete lattice and $M$ a subset. It is rare that $M$ will be closed under both infs and sups of $L$. But notice that if $M$ is closed under arbitrary sups, then $M$ is a complete lattice (in itself as a poset) but is *not* necessarily closed under infs of $L$. For example, suppose that $X$ is a topological space, and $L$ is the complete lattice of *all* subsets of $X$ under inclusion, so a complete lattice under $\cup$ and $\cap$. Let $M$ be the set of *open* subsets of $S$, so $M$ is sup-closed, *not* inf-closed, and $(M, \cup)$ yields the determined meet operator

$$\bigwedge_{\mathrm{det}} Y = \mathrm{Interior}\left( \bigcap Y \right).$$

Dual remarks hold for meet-closed subsets.

First let us define the notion of a morphism in the continuous lattice sense, which will then lead us to a definition of a subalgebra.

**Definition 6.3.5 (CL-morphism).** *Let $f\colon S \to T$ be a morphism of posets. Then $f$ is a morphism in the continuous lattice sense, or more succinctly a CL-morphism, if it is* **inf** *and continuous (see Definition 1.1.4).*

The Adjoint Functor Theorem shows that a CL-morphism of complete lattices has a left adjoint. Following [97], we call a subset $M$ of a complete lattice $L$ a *subalgebra* if the inclusion map is a CL-morphism, or equivalently, $M$ is closed under all infs and under directed sups, both calculated in $L$. One then has that $M$ is a complete lattice with the determined join. The assumption of being closed under directed sups means that directed sups in $M$ calculated in $L$ or $M$ coincide. However, it is not the case that $x, y \in M$ implies that $x \vee_L y = x \vee_M y$. Of course, $x \vee_L y \leq x \vee_M y$, but the former need not belong to $M$.

### 6.3.1 Homomorphism and substructure theorems

We now wish to state and prove homomorphism and substructure theorems for the various types of morphisms we have been considering. First we must consider closure operators. The following proposition is standard from lattice theory; see [97, Chapter O.3]. However, it plays a very fundamental role in constructing lower bounds for group complexity of semigroups and automata; see [128].

**Proposition 6.3.6.** *Let $L$ be a complete lattice and $c: L \to L$ be a closure operator. Let $\iota: c(L) \to L$ be the inclusion and $c^0: L \to c(L)$ the corestriction. There results a Galois connection*

$$L \xleftarrow{\iota} c(L).$$
$$\xrightarrow{c^0}$$

*Hence $c(L)$ is closed under all infs, and so is a complete lattice with the same inf as $L$ but with the determined sup. Moreover, $c^0: L \to c(L)$ is **sup**.*

*Conversely, if $M \subseteq L$ is an inf-closed subset of $L$, then there is a unique closure operator $c: L \to L$ such that $c(L) = M$. A dual result holds for kernel operators.*

*Proof.* Suppose that $c: L \to L$ is a closure operator. Let $c^0: L \to c(L)$ be the corestriction and let $\iota: c(L) \to L$ be the inclusion. We claim that $c^0$ and $\iota$ form a Galois connection with $c^0$ the left adjoint and $\iota$ the right adjoint. Indeed, for $s \in c(L)$ and $t \in L$, we have $\iota(s) \geq t$ implies

$$s = c(s) \geq c(t) = c^0(t).$$

Conversely, if $s \geq c^0(t)$, then

$$\iota(s) = s \geq c^0(t) = c(t) \geq t.$$

It follows that $\iota$ preserves infs (that is, $c(L)$ is closed under all infs) and $c^0$ preserves all sups. Thus $c(L)$ is a complete lattice with meets induced from $L$ and its determined join.

Conversely, let $M$ be an inf-closed subset of $L$. So $M$ is a complete lattice with the inf from $L$ and its determined join. Let $\iota: M \to L$ be the inclusion. Then by the Adjoint Functor Theorem, $\iota$ has a left adjoint $d: L \to M$. If we set $c = \iota d$, then $c$ is a closure operator by Proposition 6.3.2. Because $\iota$ is injective, $d$ is surjective by Proposition 6.3.2 and so $c(L) = M$. The uniqueness of $c$ follows from the fact that $c = \iota c^0$ and that $c^0$ must be the left adjoint of $\iota$ and hence $c^0 = d$ showing that $c = \iota d$.    $\square$

We remark that the closure operator $c$ defined in the above proof via the Adjoint Functor Theorem (cf. (6.7)) has the following explicit description, which is classical:

$$c(\ell) = \bigwedge\{m \in M \mid \ell \leq m\}. \tag{6.9}$$

The next result is a homomorphism theorem for the category **Sup** and provides a converse to Proposition 6.3.6.

**Theorem 6.3.7 (Homomorphism theorem for Sup).** *Let* $d\colon L_1 \twoheadrightarrow L_2$ *be a surjective* **sup** *map between complete lattices. Then there exists a closure operator* $r\colon L_1 \to L_1$ *such that* $d|_{r(L_1)}$ *is an isomorphism and* $dr = d$. *The corestriction* $r^0\colon L_1 \twoheadrightarrow r(L_1)$ *is* **sup** *(for the determined join of* $r(L_1)$ *).* *Moreover, if the right adjoint* $g\colon L_2 \hookrightarrow L_1$ *is continuous, then* $r$ *is continuous.*

*Proof.* Because $L_1$ and $L_2$ are complete, the adjoint functor theorem says that $d$ has right adjoint $g\colon L_2 \to L_1$. Then $r = gd$ is a closure operator by Proposition 6.3.2 and $dr = dgd = d$. If, in addition, $g$ is continuous then so is $r$, as $d$ is **sup**. Moreover, because $d$ is surjective, we have by Proposition 6.3.2 that $dg = 1$. We claim $h = d|_{r(L_1)}$ is an isomorphism with inverse $g$. Because $g = gdg = rg$, the image of $g$ is contained in $r(L_1)$. We already know that $hg = dg = 1$. Conversely, $gh = gd|_{r(L_1)} = r|_{r(L_1)} = 1$. The fact that $r^0$ is **sup** is a consequence of Proposition 6.3.6.    □

The substance of the previous two results is that the quotients of a complete lattice $L$ under **sup** morphisms correspond precisely to the images of $L$ under closure operators.

Next we turn to substructures in the category **Sup**. The following theorem is just the dual of Proposition 6.3.6.

**Theorem 6.3.8 (Substructure theorem for Sup).** *Let* $L$ *be a complete lattice and* $S$ *be a subset of* $L$. *Then* $S$ *is a sup-closed subset of* $L$ *if and only if it is the image of a kernel operator* $k$ *on* $L$.

So **sup** injections into a complete lattice $L$ are in correspondence with kernel operators on $L$. Now we turn to the analogous results for CL-morphisms. The homomorphism theorem in this context is a bit more complicated than in the setting of **sup** maps because the combination of being **inf** and continuous is not exactly dual to **sup**; see [97, Thm. I-2.15].

**Definition 6.3.9 (Congruence).** *An equivalence relation* $R$ *on a complete lattice* $L$ *is called a congruence if its graph* $\#R \subseteq L \times L$ *is a subalgebra of* $L \times L$.

**Theorem 6.3.10 (Homomorphism theorem for CL-morphisms).** *Let* $L$ *be a complete lattice and let* $R$ *be an equivalence relation on* $L$. *Then the following conditions are equivalent:*

1. *$R$ is a congruence;*
2. *There exists a continuous kernel operator* $k$ *on* $L$ *whose associated equivalence relation on* $L$ *is* $R$;

3. *The set $L/R$ admits a complete lattice structure making $L \twoheadrightarrow L/R$ an onto CL-morphism.*

*Proof.* To see that 1 implies 2, let $R(x)$ be the equivalence class of $x$ for $x \in L$. Define a map $k \colon L \to L$ by $k(x) = \bigwedge R(x)$. Because the graph of $R$ is inf-closed, we have $k(x) \in R(x)$. Hence $k(x)$ is the smallest element of $R(x)$ and so $k(x) \leq x$ and $k = k^2$. Because $R$ is closed under finite meets, it is a congruence on the $\wedge$-semilattice $L$. Hence if $x \leq y$, then $x \wedge y = x$ and so $R(x) \wedge R(y) \subseteq R(x \wedge y)$. Thus

$$k(x) = k(x \wedge y) \leq k(x) \wedge k(y) \leq k(y).$$

It follows that $k$ is order preserving. Thus $k$ is a kernel operator.

We must now verify that $k$ is continuous. Suppose $D$ is a directed subset of $L$. Let $d_0 = \bigvee D$ and $d_1 = \bigvee k(D)$. Because $k$ is a kernel operator, clearly $d_1 \leq k(d_0)$. For the reverse inequality, first observe that for all $d \in D$ we have $d \mathrel{R} k(d)$. Also the set $\{(d, k(d)) \mid d \in D\}$ is directed in the graph of $R$. Hence, as the graph of $R$ is a subalgebra of $L \times L$,

$$(d_0, d_1) = \bigvee_{d \in D} \{(d, k(d))\} \in \operatorname{graph}(R).$$

Thus $k(d_0) \leq d_1$, as desired.

Next we check $R$ is the equivalence relation associated to $k$. First of all, if $k(x) \mathrel{R} k(y)$, then $k(x) = k(y)$ because for any element $\ell \in L$, $k(\ell)$ is the smallest element of its equivalence class. As $x \mathrel{R} k(x)$ and $y \mathrel{R} k(y)$, we have

$$x \mathrel{R} y \iff k(x) \mathrel{R} k(y) \iff k(x) = k(y).$$

This completes the proof that 1 implies 2.

For 2 implies 3, the corestriction $k^0 \colon L \to k(L)$ factors as the quotient $g \colon L \twoheadrightarrow L/R$ followed by a bijection. We can transport via the bijection the order structure of $k(L)$ to $L/R$, so it suffices to show that $k(L)$ is a complete lattice and that $k^0$ is a CL-morphism. By the dual of Proposition 6.3.6, $k(L)$ is sup-closed — and hence a complete lattice with the determined meet — and $k^0$ is **inf**. Because $k$ is continuous and sups in $k(L)$ coincide with sups in $L$, we have $k^0 \colon L \to k(L)$ is continuous.

For 3 implies 1, suppose $g \colon L \twoheadrightarrow L/R$ is the canonical surjection. Then $g \times g \colon L \times L \to L/R \times L/R$ is a CL-morphism. But $\#R = (g \times g)^{-1}(\operatorname{diag}(L/R))$. Hence $\#R$ is a subalgebra of $L \times L$ by Exercise 6.3.11 below. □

**Exercise 6.3.11.** Let $g \colon L \to N$ be a CL-morphism of complete lattices. Let $K$ be a subalgebra of $N$. Show that $g^{-1}(K)$ is a subalgebra of $L$.

It will take some preparatory work before we can give the substructure theorem for CL-morphisms. The following proposition [97, Lemma O-3.11] almost decomposes arbitrary projections as a union of a closure and a kernel operator.

**Proposition 6.3.12.** *Let $L$ be a complete lattice and $p\colon L \to L$ be a projection. Define the following two subsets of $L$:*

$$L_c = \{x \in L \mid x \leq p(x)\} \tag{6.10}$$
$$L_k = \{x \in L \mid x \geq p(x)\}. \tag{6.11}$$

*(Note that $L_c \cup L_k$ need not equal $L$, see 2 below.) Then:*

1. *$L_c$ is closed under arbitrary sups and $L_k$ is closed under arbitrary infs (calculated in $L$);*
2. *If $p_c\colon L_c \to L_c$ and $p_k\colon L_k \to L_k$ are the restrictions (which are well defined), then $p_c$ is a closure operator, $p_k$ is a kernel operator and*

$$p_c(L_c) = p_k(L_k) = p(L) = L_k \cap L_c; \tag{6.12}$$

3. *If $p$ is **sup** (respectively, continuous), then $L_k$ and $p(L)$ are closed under sups (respectively, directed sups). Dually, if $p$ preserves (downwards directed) infs, then $L_c$ and $p(L)$ are closed under (downwards directed) infs.*

*Proof.* For 1, if $\mathsf{B}$ is the bottom of $L$, then clearly $\mathsf{B} \leq p(\mathsf{B})$ so $L_c$ contains the empty join. Suppose $\ell_i \in L_c$, $i \in I$. Then we have the inequalities

$$\bigvee_{i \in I} \ell_i \leq \bigvee_{i \in I} p(\ell_i) \leq p\left(\bigvee_{i \in I} \ell_i\right)$$

where the first inequality follows because $\ell_i \leq p(\ell_i)$ and the second because $p$ is order preserving. This shows that $\bigvee_{i \in I} \ell_i \in L_c$. The result for $L_k$ is dual.

Statement 2 we leave as an exercise. For 3, we just handle the case $p$ is continuous. Let $D \subseteq p(L)$ be directed. Then

$$p\left(\bigvee D\right) = \bigvee p(D) = \bigvee D$$

the first equality holding because $p$ is continuous, the second because $p = p^2$. Thus $p(L)$ is closed under directed sups. Suppose now that $D \subseteq L_k$ is directed. Then one easily checks

$$\bigvee_{\ell \in D} \ell \geq \bigvee_{\ell \in D} p(\ell) = p\left(\bigvee_{\ell \in D} \ell\right)$$

the first inequality holding by the definition of $L_k$ and the second equality holding because $p$ is continuous. So $\bigvee D \in L_k$, as was required.   $\square$

**Exercise 6.3.13.** Prove statement 2 and the dual version of statement 3 of Proposition 6.3.12.

**Theorem 6.3.14 (Substructure theorem for CL-morphisms).** *Let $L$ be a complete lattice and $M$ a meet-closed subset. Then $M$ is a subalgebra if and only if the associated closure operator $c\colon L \to L$ with $c(L) = M$ from (6.9) is continuous.*

*Proof.* The third item of Proposition 6.3.12 shows that if $c$ is continuous, then $M$ is a subalgebra. Conversely, suppose $M$ is a subalgebra. Then the inclusion $\iota\colon M \to L$ is continuous. But the corestriction $c^0\colon L \to M$ is left adjoint to $\iota$ by Proposition 6.3.6 and hence is **sup**. Thus $c = \iota c^0$ is continuous.    $\square$

A continuous closure operator on $2^X$ is ofttimes referred to as an *algebraic closure operator* [203]. The following theorem and its corollary have already been used to good effect in Chapter 2; see also [97, Prop. I-4.13]. Recall that the set of compact elements of an algebraic lattice $L$ is denoted $K(L)$.

**Theorem 6.3.15 (Continuous closure operator theorem).** *Let $L$ be an algebraic lattice and suppose that $c\colon L \to L$ is a continuous closure operator. Then $c(L)$ is an algebraic lattice where the meet is induced from $L$ and the join is determined. Moreover, $K(c(L)) = c(K(L))$.*

*Proof.* By Theorem 6.3.14, $c(L)$ is a subalgebra. We shall denote the determined join of $c(L)$ by $\bigvee_{c(L)}$. Let us begin by showing $c(K(L)) \subseteq K(c(L))$ in order to get our hands on some compact elements of $c(L)$. So suppose $k \in K(L)$. By Proposition 6.2.6, it suffices to show that if $c(k) \leq \bigvee_{c(L)} D$ with $D \subseteq c(L)$ directed, then $c(k) \leq d$ some $d \in D$. Theorem 6.3.14 implies that $c(L)$ is closed under directed sups, so $\bigvee_{c(L)} D = \bigvee D$. Thus, because $c$ is increasing,

$$k \leq c(k) \leq \bigvee D.$$

Compactness of $k$ yields $d \in D$ such that $k \leq d$. Then

$$c(k) \leq c(d) = d$$

as required. Thus we have shown that $c(K(L)) \subseteq K(c(L))$.

We can now show that $c(L)$ is algebraic. Indeed, let $\ell \in c(L)$. Then there is a directed set $D \subseteq K(L)$ such that $\ell = \bigvee D$. Hence

$$\ell = c(\ell) = c\left(\bigvee D\right) = \bigvee c(D)$$

because $c$ is continuous. But $c(D) \subseteq K(c(L))$ by what we just proved. Also, $c(D)$ is clearly directed as $c$ is order preserving. Thus $L$ is algebraic.

Finally, we prove $K(c(L)) \subseteq c(K(L))$. Suppose $k \in K(c(L))$. We just saw that $k = \bigvee X$ where $X$ is a directed set of elements from $c(K(L))$. As $k$ is compact, $k \leq x$ some $x \in X$. But clearly $x \leq k$. So $k = x \in c(K(L))$, completing the proof of the theorem.    $\square$

The following corollary was used to show that **GMC(PV)** and **Cnt(PV)** are algebraic lattices in Chapter 2.

**Corollary 6.3.16.** *Suppose that we have an algebraic lattice $L_1$ and a surjective* **sup** *morphism $d\colon L_1 \twoheadrightarrow L_2$ such that the right adjoint $g\colon L_2 \hookrightarrow L_1$ of $d$ is continuous. Then $L_2$ is an algebraic lattice and $d(K(L_1)) = K(L_2)$.*

*Proof.* By Theorem 6.3.7 there is a continuous closure operator $r\colon L_1 \to L_1$ such that $d|_{r(L_1)}$ is an isomorphism and $dr = d$. Theorem 6.3.15 then tells us that $r(L_1)$ is an algebraic lattice with compact elements $r(K(L_1))$. Applying the isomorphism $d|_{r(L_1)}$ we deduce that $L_2$ is an algebraic lattice with set of compact elements $K(L_2) = d|_{r(L_1)} r(K(L_1)) = d(K(L_1))$.    □

Let us interpret Theorem 6.3.15 for the case of a subalgebra $L$ of $2^X$. Let $c\colon 2^X \to 2^X$ be the associated continuous closure operator with image $L$. An element $\ell \in L$ is said to be *finitely generated* if there is a finite subset $Y \subseteq X$ such that $c(Y) = \ell$. Because the finite subsets are precisely the compact subsets of $2^X$, Theorem 6.3.15 says that the compact elements in $L$ are the finitely generated ones.

For example, the compact elements of **PV** are the pseudovarieties generated by a finite number of semigroups $\{S_1, \dots, S_n\}$. Of course this pseudovariety can also be generated by the single semigroup $S_1 \times \cdots \times S_n$. Membership is always decidable for a compact pseudovariety (provided one is given a finite collection of generators) by a result of Birkhoff [85, 217]. In fact, we saw earlier (Exercise 1.3.12) how to explicitly realize the free $X$-generated semigroup in $(S)$ as a subsemigroup of $S^{S^X}$. Hence one can test whether a finite semigroup $T$ belongs to $(S)$ by checking whether $T$ divides $S^{S^T}$. This is a doubly exponential algorithm.

Similarly, the compact elements of **PVRM** (or **PVRM**$^+$) are the finitely generated pseudovarieties of relational morphisms. Again these are the same as the pseudovarieties generated by a single relational morphism $f\colon S \to T$. It was shown in Theorem 3.5.16 that such are decidable by a similar means to the Birkhoff result.

The following exercise shows that algebraic lattices are closed under products and subalgebras. They are not however closed under images via CL-morphisms. We shall see later that the class of continuous lattices is exactly the closure of the class of algebraic lattices under CL-morphic images.

**Exercise 6.3.17.** Show that algebraic lattices are closed under formation of products and subalgebras.

The following lemma is useful for the next exercise.

**Lemma 6.3.18.** *Suppose $L$ is a continuous lattice, $c\colon L \to L$ is a continuous closure operator and $\ell \ll \ell'$ in $L$. Then $c(\ell) \ll c(\ell')$ in $c(L)$.*

*Proof.* By Theorem 6.3.14 $c(L)$ is a subalgebra of $L$ and so closed under directed sups. Suppose $D \subseteq c(L)$ is directed and $c(\ell') \le \bigvee_{c(L)} D$. Then

$$\ell' \le c(\ell') \le \bigvee_{c(L)} D = \bigvee D$$

and so there exists $d \in D$ such that $\ell \leq d$. But then $c(\ell) \leq c(d) = d$. Thus $c(\ell) \ll c(\ell')$ in $c(L)$, as required.    □

**Exercise 6.3.19.** This exercise is to show that continuous lattices are closed under subalgebras and products.

1. Show that the class of continuous lattices are closed under products;
2. Show, by imitating the proof of Theorem 6.3.15, that if $L$ is a continuous lattice and $M$ is a subalgebra, then $M$ is a continuous lattice.

## 6.3.2 Some further facts about algebraic and continuous lattices

In this subsection, we collect some facts from [97] that will help us to further understand the relationship between algebraic and continuous lattices.

A useful characterization of continuous lattices is via their ideal structure. Let $L$ be a poset. Then a subset $I \subseteq L$ is called an *ideal* (or a *lattice ideal* if we are trying to emphasize that we are speaking of ideals in the sense of lattice theory) if it is directed and if $a \in I$ and $b \leq a$ implies $b \in I$. This notion is essentially the dual of a filter. If $L$ is a join semilattice, then the condition of being directed in the definition of an ideal can be replaced by the conditions $I \neq \emptyset$ and $a, b \in I$ implies $a \vee b \in I$. The set $\mathsf{Id}(L)$ of all ideals of a join semilattice $L$ is an algebraic lattice with intersection as the inf and with the determined sup. The compact elements are the principal ideals. A *principal ideal* is a set of the form

$$x^{\downarrow} = \{y \mid y \leq x\}$$

with $x \in L$. Notice that the join of a directed set of ideals is the set-theoretic union.

**Exercise 6.3.20.** Let $L$ be a lattice with top and bottom. Show that a proper ideal $\mathfrak{p}$ of $L$ is fmi if and only if there is a map $\varphi \colon L \to \{0,1\}$ preserving all finite infs and sups (including empty ones) with $\mathfrak{p} = 0\varphi^{-1}$.

Let us now assume $L$ is a complete lattice. The map $d \colon \mathsf{Id}(L) \to L$ given by $I \mapsto \bigvee I$ is the left adjoint to the map $g \colon L \to \mathsf{Id}(L)$ given by $x \mapsto x^{\downarrow}$. In particular $d$ is **sup** and $g$ is **inf**.

**Exercise 6.3.21.** Let $L$ be a complete lattice. Verify that

$$\mathsf{Id}(L) \xleftarrow{\quad g \quad} L \qquad \xrightarrow{\quad d \quad}$$

given by $d(I) = \bigvee I$ and $g(x) = x^{\downarrow}$ is a Galois connection.

Recall from Exercise 6.2.8 the notation

$$\ell^{\Downarrow} = \{c \in L \mid c \ll \ell\} \tag{6.13}$$

and that $\ell^{\Downarrow}$ is directed. With this notation, $L$ is a continuous lattice if and only if, for all $\ell \in L$, one has $\ell = \bigvee \ell^{\Downarrow}$.

**Lemma 6.3.22.** *Let $L$ be a complete lattice. Then $\ell^{\Downarrow}$ is an ideal of $L$; moreover, if $I$ is an ideal with $\ell \leq \bigvee I$, then $\ell^{\Downarrow} \subseteq I$.*

*Proof.* The fact that $\ell^{\Downarrow}$ is an ideal is immediate from Exercise 6.2.8. Let $I$ be an ideal with $\ell \leq \bigvee I$. We first observe that $I$ is directed by the definition of an ideal. So if $c \ll \ell$, then $\ell \leq \bigvee I$ implies $c \leq i$ some $i \in I$. Therefore $c \in I$ by the definition of an ideal. Thus $\ell^{\Downarrow} \subseteq I$.

The following can be found in [97, I-1.10].

**Proposition 6.3.23.** *Let $L$ be a complete lattice and let $d \colon \mathsf{Id}(L) \to L$ be given by $d(I) = \bigvee I$. Then the following are equivalent:*

1. *$L$ is a continuous lattice;*
2. *For all $\ell \in L$, $\ell^{\Downarrow}$ is the smallest ideal $I$ with $\ell \leq \bigvee I$;*
3. *For each $\ell \in L$, there is a smallest ideal $I$ with $\ell \leq \bigvee I$;*
4. *The **sup** map $d$ has a left adjoint;*
5. *The map $d$ is both **inf** and **sup**.*

*Proof.* Clearly 4 and 5 are equivalent. Because $d$ is onto, 4 and 5 are equivalent to 3. Clearly 2 implies 3. To see that 3 implies 2, let $I$ be the smallest ideal with $\ell \leq \bigvee I$. Then $\ell^{\Downarrow} \subseteq I$ by Lemma 6.3.22. For the reverse inclusion, suppose that $D$ is a directed set with $\ell \leq \bigvee D$. Consider the ideal $J$ generated by $D$; it is easily verified to consist of all elements below an element of $D$. Then $\ell \leq \bigvee D = \bigvee J$, so $I \subseteq J$. Hence every element of $I$ is below some element of $D$. It follows that each element of $I$ is way below $\ell$ and so $I \subseteq \ell^{\Downarrow}$.

Clearly 2 implies 1 because we then have $\ell \leq \bigvee \ell^{\Downarrow} \leq \ell$ and so $\ell = \bigvee \ell^{\Downarrow}$. For 1 implies 2, we have, by the definition of a continuous lattice, that $\ell = \bigvee \ell^{\Downarrow}$. An application of Lemma 6.3.22 shows that $\ell^{\Downarrow}$ has the desired minimality property.  □

For $L = [0, 1]$, one can show that $\mathsf{Id}(L)$ is isomorphic to the algebraic lattice $\{0, 1\} \times (0, 1] \cup \mathsf{B}$ with lexicographic order and that the **sup** map $d$ is the projection to $[0, 1]$ (where $\mathsf{B}$ goes to 0).

**Exercise 6.3.24.** Verify the above assertion.

We have already remarked that $[0, 1]$ is a continuous lattice, but not an algebraic lattice. Note however $\{0, 1\} \times (0, 1] \cup \mathsf{B}$ is an algebraic lattice with lexicographic order and the map projecting to $[0, 1]$ (with $\mathsf{B}$ mapping to 0) is a surjective CL-morphism.

Let $X$ be a set. Then $2^X$ denotes the lattice of all functions from $X$ to $\{0,1\}$ with the pointwise operations induced from the two-element Boolean algebra $\{0,1\}$. We denote by $[0,1]^X$ the Hilbert cube over $X$ consisting of all functions $X$ to $[0,1]$, again viewed as a lattice with the pointwise operations. The following combines [97, Thms. I-4.16, IV-3.3].

**Proposition 6.3.25.** *Let $L$ be a complete lattice. Then:*

1. *$L$ is algebraic if and only if $L$ is a subalgebra of $2^X$ for some set $X$. Moreover, one can take $X$ to be the set of compact elements of $L$.*
2. *$L$ is a continuous lattice if and only if $L$ is a subalgebra of $[0,1]^X$ for some set $X$.*
3. *$L$ is algebraic if and only if $L$ has enough CL-morphisms into $\{0,1\}$ to separate points.*
4. *$L$ is a continuous lattice if and only if $L$ has enough CL-morphisms into $[0,1]$ to separate points.*

*Proof.* We prove only 1. Because $2^X$ is algebraic, so are its subalgebras by Exercise 6.3.17. Conversely, if $L$ is algebraic with set $K$ of compact elements, then we embed $L$ in $2^K$ as follows. Send $\ell \in L$ to the set

$$I_\ell = \{k \in K \mid k \le \ell\} = \ell^\downarrow \cap K.$$

Certainly $I_\ell \le I_{\ell'}$ if and only if $\ell \le \ell'$ as $L$ is algebraic. It is clear that if $X \subseteq L$, then $\bigcap_{\ell \in X} I_\ell = I_{\bigwedge X}$. Also, if $D \subseteq L$ is a directed set, we claim

$$\bigcup_{d \in D} I_d = I_{\bigvee D}. \tag{6.14}$$

The left-hand side of (6.14) is clearly contained in the right-hand side. For the converse, let $k \in I_{\bigvee D}$. Then $k \le \bigvee D$. Because $k$ is compact, $k \le d$ for some $d \in D$. Hence $k \in I_d$ with $d \in D$. This establishes (6.14).    □

A slight variation of the above proposition is [97, Thm. I-2.11, Cor. I-4.18].

**Proposition 6.3.26.** *The following statements hold:*

1. *The class of algebraic lattices is the smallest class of complete lattices containing the two-element Boolean algebra $\{0,1\}$ closed under all products and subalgebras (but not under CL-morphic images);*
2. *The class of continuous lattices is the smallest class of complete lattices containing $\{0,1\}$ and closed under arbitrary products, subalgebras and CL-morphic images.*

*Proof.* The first item is immediate from Proposition 6.3.25. For 2, closure under products and subalgebras is Exercise 6.3.19.

Let us begin by showing that the image of a continuous lattice under a CL-morphism is a continuous lattice. Suppose that $g \colon L \twoheadrightarrow N$ is an onto CL-morphism with $L$ a continuous lattice. Then $g$ has a left adjoint $d \colon N \to L$, which is **sup**. Moreover, $gd = 1$. Let $n \in N$. We claim first that

$$n = \bigvee_{c \ll d(n)} g(c). \tag{6.15}$$

Indeed, the set of elements $c \ll d(n)$ is a directed set (as it is closed under finite sups by Exercise 6.2.8). Thus, because $g$ and $L$ are continuous, we see

$$n = gd(n) = g\left(\bigvee_{c \ll d(n)} c\right) = \bigvee_{c \ll d(n)} g(c).$$

Next we claim that if $c \ll d(n)$, then $g(c) \ll n$. Indeed, suppose $n \leq \bigvee X$ with $X$ a directed subset of $N$. Then

$$d(n) \leq d\left(\bigvee X\right) = \bigvee d(X)$$

where the last equality follows because $d$ is **sup**. But $d(X)$ is directed, so $c \ll d(n)$ implies $c \leq d(x)$ for some $x \in X$. Hence

$$g(c) \leq gd(x) = x$$

establishing $g(c) \ll n$. This, combined with (6.15), establishes that $N$ is a continuous lattice.

To finish the proposition, it suffices to observe that every continuous lattice is the CL-morphic image of an algebraic lattice. In fact, if $L$ is a continuous lattice, then $L$ is a CL-morphic image of the algebraic lattice $\mathsf{Id}(L)$ by Proposition 6.3.23. □

In order to obtain some insight into the idempotents of $\mathbf{Cnt}(\mathbf{PV})$, we are led to consider some important results from [97]. Let $\alpha\colon L \to L$ be a function. Then the *fixed-point set* of $\alpha$ is

$$\mathrm{Fix}(\alpha) = \{\ell \in L \mid \alpha(\ell) = \ell\}.$$

The following important theorem is due to Tarski [97, Thm. O-2.3].

**Theorem 6.3.27 (Tarski's Fixed-Point Theorem).** *Let $L$ be a complete lattice and let $\alpha\colon L \to L$ be order preserving. Then $\mathrm{Fix}(\alpha)$ is non-empty and is a complete lattice under the restricted ordering inherited from $L$.*

*Proof.* Let $M = \{\ell \in L \mid \ell \leq \alpha(\ell)\}$. Clearly, the bottom of $L$ belongs to $M$. If $X \subseteq M$, then, for all $x \in X$, we have $x \leq \alpha(x) \leq \alpha(\bigvee X)$. Thus $\bigvee X \leq \alpha(\bigvee X)$ and hence $M$ is closed under all sups, and in particular is a complete lattice. If $m \in M$, then $m \leq \alpha(m)$ implies $\alpha(m) \leq \alpha(\alpha(m))$ and so $\alpha$ leaves $M$ invariant. Let $F = \{m \in M \mid \alpha(m) \leq m\}$. A dual argument to the one above shows that $F$ contains the top of $M$ and is closed under the determined inf of $M$. Thus $F$ is a complete lattice (and in particular non-empty). But clearly $F = \mathrm{Fix}(\alpha)$. □

Of course, this theorem would be trivial if $\alpha$ was assumed **sup** or **inf**. The surprising fact is that order preserving is enough. However, the reader should be warned that neither the meet nor join of $\mathrm{Fix}(\alpha)$ need be the restricted meet or join from $L$. That is, we are not claiming that $\mathrm{Fix}(\alpha)$ is closed under either meets or joins.

Let us consider an example from semigroup theory. Let $\alpha$ be the order preserving, non-continuous map from the (2.12). That is,

$$\alpha(\mathbf{V}) = \begin{cases} \mathbf{DS} \vee (B_2) & \text{if } \mathbf{V} = \mathbf{DS} \\ \mathbf{V} & \text{otherwise.} \end{cases}$$

Clearly, $\mathrm{Fix}(\alpha) = \mathbf{PV} \setminus \mathbf{DS}$. According to the Tarski fixed point theorem, $\mathbf{PV} \setminus \mathbf{DS}$ is a complete lattice. In fact, it is easy to see that $\mathbf{PV} \setminus \mathbf{DS}$ is closed under arbitrary meets from $\mathbf{PV}$. The supremum of a subset $X \subseteq \mathbf{PV} \setminus \mathbf{DS}$ is as in $\mathbf{PV}$, unless the result is $\mathbf{DS}$ in which case the join in $\mathbf{PV} \setminus \mathbf{DS}$ is $\mathbf{DS} \vee (B_2)$. We remark that $\mathbf{PV} \setminus \mathbf{DS}$ is not meet-continuous and so not a continuous or algebraic lattice. Indeed, let $\mathbf{DS} = \bigvee \mathbf{V}_i$ with $\mathbf{V}_1 \leq \mathbf{V}_2 \leq \cdots$ a chain of compact pseudovarieties. Then

$$\left( \bigvee_{\det} \mathbf{V}_i \right) \cap (B_2) = (B_2)$$

whereas

$$\bigvee_{\det} (\mathbf{V}_i \cap (B_2)) = \mathbf{DS} \cap (B_2) < (B_2).$$

The key point here is that $\mathbf{DS}$ is smi (see Definition 6.1.5). One can generalize this to any smi pseudovariety in the obvious manner.

If $\alpha$ is order preserving and idempotent, then the fixed point set is the image and Theorem 6.3.27 applies. However under stronger hypothesis, a stronger conclusion is true. The following theorem [97, Cor I-2.3], due to Dana Scott, is *very important* for this chapter and a major reason for writing [97]. Compare this with Theorem 6.3.15.

**Theorem 6.3.28 (D. Scott).** *If $L$ is a continuous lattice and $\alpha \colon L \to L$ is a continuous projection, then $\alpha(L)$ is a continuous lattice relative to the induced order.*

*Proof.* Recall from Proposition 6.3.12 items 1 and 3 that the set

$$L_k = \{x \in L \mid \alpha(x) \leq x\}$$

is closed under infs and directed sups. Hence it is a subalgebra and thus a continuous lattice by Exercise 6.3.19; in fact, it is the image under the associated continuous closure operator on $L$ by Theorem 6.3.14. Let $\alpha_k \colon L_k \to L_k$ be the restriction of $\alpha$. Then $\alpha_k$ is a kernel operator with $\alpha_k(L) = \alpha(L)$. But $\alpha_k$ is continuous because $\alpha$ is continuous and $L_k$ is closed under directed sups and arbitrary infs. Thus we can apply the next lemma (note that up to this point "continuous" could be replaced by "algebraic" but not in the next lemma).                                                                       □

The next lemma is from [97].

**Lemma 6.3.29.** *If $\alpha \colon L \to L$ is a continuous kernel operator on a continuous lattice $L$, then $\alpha(L)$ is a continuous lattice in the induced order.*

*Proof.* By the dual result for kernel operators of Proposition 6.3.6, the core-striction $\alpha^0 \colon L \to \alpha(L)$ is **inf**. As $\alpha(L)$ is closed under directed sups by item 3 of Proposition 6.3.12, the continuity of $\alpha$ implies that $\alpha^0$ is a CL-morphism. Proposition 6.3.26 item 2 then shows that $\alpha(L)$ is a continuous lattice.    □

Theorem 6.3.28 is not true in general if "continuous" is replaced by "algebraic." In fact, we shall see momentarily that every continuous lattice is the image of an algebraic lattice under a continuous projection. We say that a complete lattice $R$ is a *retract* of a complete lattice $L$ if there is a continuous projection $p \colon L \to L$ with $p(L) = R$. The following result is also well-known (cf. [97, Cor I-4.18]).

**Theorem 6.3.30.** *A complete lattice $L$ is continuous if and only if it is a retract of an algebraic lattice.*

*Proof.* By Theorem 6.3.28, a retract of an algebraic lattice is a continuous lattice. For the converse, we know from Proposition 6.3.26 that if $L$ is a continuous lattice, then there is an algebraic lattice $N$ and an onto CL-morphism $g \colon N \twoheadrightarrow L$. Let $d \colon L \hookrightarrow N$ be the left adjoint. Then $dg$ is a kernel operator on $N$, by Proposition 6.3.2, and $dg(N) \cong L$ (cf. Theorem 6.3.10).    □

There is not an exact dual of Theorem 6.3.28. When we dualize we must demand preservation of all infs, not just downwards directed ones [97].

**Proposition 6.3.31.** *If $L$ is a continuous lattice and $\alpha \colon L \to L$ is an* **inf** *projection, then $\alpha(L)$ is closed under infs and is a continuous lattice.*

*Proof.* Clearly, $\alpha(L)$ is closed under infs. Define $L_c$ as per Proposition 6.3.12. Then by the self-same proposition, $L_c$ is closed under infs and sups so is a continuous lattice by Exercise 6.3.19. The map $\alpha_c \colon L_c \to L_c$ (cf. Proposition 6.3.12) is an **inf** closure operator. But, by Proposition 6.3.6, the corestriction $\alpha_c^0 \colon L_c \to \alpha_c(L_c)$ is **sup**. Thus $\alpha_c^0$ is continuous, in fact a CL-morphism, and so $\alpha(L) = \alpha_c(L_c)$ is a continuous lattice by item 2 of Theorem 6.3.26.    □

### 6.3.3 Continuous lattices and q-theory

We now want to show that any continuous lattice with countable weight is isomorphic to the image of a projection in $\mathbf{Cnt}(\mathbf{PV})$. The following definition is from [97, Chapter III]. The reader should recall the notation $\ell^{\Downarrow}$ from (6.13).

**Definition 6.3.32 (Basis).** *Let $L$ be a continuous lattice. A subset $B$ of $L$ is called a* basis *if, for all $\ell \in L$:*

1. $\ell^{\Downarrow} \cap B$ *is directed;*
2. $\ell = \bigvee(\ell^{\Downarrow} \cap B)$.

The definition of a continuous lattice says that $L$ is a basis for $L$. The *weight* $w(L)$ of a continuous lattice $L$ is the minimum cardinality of a basis.

The following is part of [97, Prop. III-4.2]; it uses the interpolation property for continuous lattices [97, Thm I-1.9].

**Proposition 6.3.33 (Basis Characterization).** *Let $B$ be a subset of a continuous lattice $L$. Then $B$ is a basis for $L$ if and only if every element of $L$ is a supremum of some directed subset of $B$.*

**Exercise 6.3.34.** Show that if $L$ is an algebraic lattice, then every basis for $L$ contains the set $K(L)$ of compact elements of $L$.

It follows from Exercise 6.3.34 that if $L$ is an algebraic lattice, then $w(L)$ is the cardinality of $K(L)$.

The above proposition shows that if $B$ is a basis for $L$, so is the join-subsemilattice generated by $B$. Hence if $L$ is infinite, the weight of $L$ is the same as the minimal cardinality of a join-subsemilattice that generates $L$ under sups.

We now prove [97, Prop. III-4.3]. The reader should compare with Proposition 6.3.23, which is the special case where $B = L$.

**Proposition 6.3.35.** *Let $L$ be a continuous lattice and $B$ be a basis for $L$ so that $B$ is a join semilattice. Then $g \colon \mathsf{Id}(B) \to L$ given by $g(I) = \bigvee I$ is both an onto* **sup** *and* **inf** *map. Moreover, $\mathsf{Id}(B)$ is an algebraic lattice with*

$$K(\mathsf{Id}(B)) = \{b^{\downarrow} \mid b \in B\} \cong B.$$

*Proof.* Because $B$ is a basis, the map $g$ is onto. To see that $g$ is **sup**, we construct a right adjoint $h \colon L \to \mathsf{Id}(B)$ given by $h(\ell) = \ell^{\downarrow} \cap B$. Indeed, $g(I) \leq \ell$ if and only if $\ell$ is an upper bound for the elements of $I$, if and only if $I \subseteq h(\ell)$. Thus $g$ is the left adjoint of $h$ and hence **sup**.

To show that $g$ is **inf**, we construct a left adjoint $d \colon L \to \mathsf{Id}(B)$. It is defined by $d(\ell) = \ell^{\Downarrow} \cap B$. First of all, the definition of a basis and Exercise 6.2.3 shows that $d(\ell) \in \mathsf{Id}(B)$. Let $\ell \in L$ and $I \in \mathsf{Id}(B)$. Suppose that $d(\ell) \subseteq I$. Then, by definition of a basis,

$$\ell = \bigvee d(\ell) \leq \bigvee I = g(I).$$

Conversely, suppose $\ell \leq g(I) = \bigvee I$. If $b \ll \ell$ and $b \in B$, then because $I$ is directed, there exists $y \in I$ such that $b \leq y$. Thus $b \in I$, by the definition of an ideal. It follows that $d(\ell) \subseteq I$. We have thus proved that

$$d(\ell) \leq I \iff \ell \leq g(I)$$

establishing that $g$ is the right adjoint of $d$ and hence $g$ is **inf**.

The assertion that $\mathsf{Id}(B)$ is an algebraic lattice with compact elements the principal ideals is straightforward and left to the reader.    $\square$

The following is [97, Cor III-4.7] and refines Proposition 6.3.26 and Theorem 6.3.30.

**Corollary 6.3.36.** *Let $L$ be a continuous lattice of infinite weight. Then there is an algebraic lattice $N$ of the same weight and an onto map $g\colon N \to L$ that is both* **sup** *and* **inf***. Also there is a* **sup** *kernel operator $k\colon N \to N$ with $k(N) \cong L$.*

*Proof.* Let $B$ be a basis for $L$ of minimum cardinality. Because $B$ is infinite, if we close $B$ under finite sups, then it still is a basis of minimum cardinality, so we may assume without loss of generality that $B$ is a join-subsemilattice. Then we have that there is an onto map $g\colon \mathsf{Id}(B) \twoheadrightarrow L$ that is both **inf** and **sup**. Also $K(\mathsf{Id}(B)) \cong B$ and so $\mathsf{Id}(B)$ has the same weight as $L$. This establishes the first statement. If $d\colon L \twoheadrightarrow \mathsf{Id}(B)$ is the left adjoint, then $dg$ is a kernel operator by Proposition 6.3.2 and is **sup**, being the composition of two **sup** maps. It is easy to see that $\operatorname{Im} dg \cong L$ via $g$, cf. Theorem 6.3.10.    $\square$

Our next proposition is a more precise version of [97, Cor I-4.18]. It implies in particular that every continuous lattice of countable weight is an image of the power set of a countable set under a continuous projection.

**Proposition 6.3.37.** *Let $L$ be a continuous lattice with infinite basis $B$ of cardinality $w(L)$. Then there is a continuous projection $p\colon 2^B \to 2^B$ with $\operatorname{Im} p \cong L$.*

*Proof.* By Corollary 6.3.36, there is an algebraic lattice $N$ of the same weight as $L$ and a continuous kernel operator $r\colon N \to N$ with $r(N) \cong L$. In particular, $K(N)$ is in bijection with $B$. Thus, by Proposition 6.3.25, we can view $N$ as a subalgebra of $2^B$. Let $c\colon 2^B \to 2^B$ be the continuous closure operator with image $N$ (cf. Theorem 6.3.14) and $i\colon N \to 2^B$ the continuous inclusion ($N$ is a subalgebra). Let $c^0\colon 2^B \to N$ be the corestriction; it is also continuous. Then $p = irc^0\colon 2^B \to 2^B$ is a continuous projection with image isomorphic to $L$.    $\square$

Our next goal is to characterize those lattices occurring as images of **PV** under a continuous projection; they turn out to be precisely the continuous lattices of countable weight. This is one of our main motivations for considering continuous lattices in this text.

**Theorem 6.3.38.**

1. *Let $\alpha \in \mathbf{Cnt}(\mathbf{PV})$ be a projection. Then $\alpha(\mathbf{PV})$ is a continuous lattice of countable weight.*
2. *Conversely, if $L$ is any continuous lattice of countable weight, then there exists a projection $\alpha \in \mathbf{Cnt}(\mathbf{PV})$ such that $\alpha(\mathbf{PV}) \cong L$.*

*Proof.* The lattice **PV** is algebraic and so Theorem 6.3.28 tells us that $\alpha(\mathbf{PV})$ is a continuous lattice. Because **PV** has countable weight, $\alpha(\mathbf{PV})$ has countable weight.

For 2, if $L$ is finite, then $L$ is algebraic and every element is compact. Hence $L$ embeds as a subalgebra of $2^L$, which in turn embeds in $2^B$ for $B$ a countably infinite set. If $L$ is infinite, then it has a countable basis $B$ and by Proposition 6.3.37, there is a continuous projection $p\colon 2^B \to 2^B$ such that Im $p \cong L$. Thus, in either case we can find a countably infinite set $B$ and a continuous projection $p\colon 2^B \to 2^B$ with Im $p \cong L$. The rest of the proof amounts to finding a copy of $2^B$ inside of **PV** to simulate $p$.

More precisely, let $B$ be the set of prime numbers and $p\colon 2^B \to 2^B$ a continuous projection. Let **VS** be the pseudovariety generated by all cyclic groups $\mathbb{Z}_q$ with $q$ ranging over all primes (**VS** stands for vector spaces). Then there is an isomorphism $\tau\colon [\mathbf{1}, \mathbf{VS}] \to 2^B$ given by sending a pseudovariety **V** to the set $X = \{q \in B \mid \mathbb{Z}_q \in \mathbf{V}\}$ (cf. [323]). Let $j\colon \mathbf{PV} \twoheadrightarrow \mathbf{VS}$ be the continuous projection given by $j(\mathbf{V}) = \mathbf{V} \cap \mathbf{VS}$. Consider now the projection $\pi\colon \mathbf{PV} \to \mathbf{PV}$ given by $\pi = \tau^{-1}p\tau j$. Clearly Im $\pi \cong$ Im $p \cong L$, finishing the proof. $\qquad\square$

**Corollary 6.3.39.** *There exists a projection $\alpha \in \mathbf{Cnt}(\mathbf{PV})$ with image isomorphic to the lattice $[0,1]$.*

*Proof.* The interval $[0,1]$ has a countable basis: the rational numbers. $\qquad\square$

**Proposition 6.3.40.** *The set of continuous lattices of countable weight is closed under countable direct products.*

*Proof.* Let $\{L_\alpha\}_{\alpha \in A}$ be a countable collection of countable weight continuous lattices and let $B_\alpha$ be a basis for $L_\alpha$, $\alpha \in A$. Let $B$ be the set of all $A$-tuples $(b_\alpha)$ such that $b_\alpha$ is the bottom in all but finitely many coordinates $A_0 \subseteq A$ and $b_\alpha \in B_\alpha$ for $\alpha \in A_0$. We claim that $B$ is a countable basis for $\prod_{\alpha \in A} L_\alpha$. Clearly $B$ is countable, so by Proposition 6.3.33 it suffices to show that every element of $L = \prod_{\alpha \in A} L_\alpha$ is a directed supremum of elements of $B$. As the set of finite subsets of $A$ is directed, one easily verifies that if $\ell = (\ell_\alpha) \in L$, then $\ell$ is the directed sup of the elements of $L$ that agree with $\ell$ in finitely many coordinates and are the bottom in all other coordinates. Hence, we may assume that $\ell$ is equal to the bottom in all but finitely many coordinates $A_0 \subseteq A$. Now it is clear $\ell$ is the directed sup of all elements from $B$ that are the bottom in all coordinates outside of $A_0$ and are below $\ell_\alpha$ for $\alpha \in A_0$. $\qquad\square$

**Corollary 6.3.41.** *There exists a projection $\alpha \in \mathbf{Cnt}(\mathbf{PV})$ satisfying the property $\alpha(\mathbf{PV}) \cong [0,1]^{\mathbb{N}}$.*

The next corollary seems amusing.

**Corollary 6.3.42.** *There is a projection $\alpha \in \mathbf{Cnt}(\mathbf{PV})$ with image isomorphic to $\mathbf{GMC}(\mathbf{PV})$ and similarly for $\mathbf{GMC}(\mathbf{PV})^+$.*

*Proof.* Because $\mathfrak{q}\colon \mathbf{PVRM} \to \mathbf{GMC(PV)}$ is an onto **sup** morphism with continuous right adjoint, Corollary 6.3.16 shows that $\mathbf{GMC(PV)}$ is an algebraic lattice of countable weight. The result then follows from Theorem 6.3.38. The case of $\mathbf{GMC(PV)}^+$ is handled similarly. $\qquad\square$

We have a theorem similar to Theorem 6.3.38 for the positive case, namely:

**Theorem 6.3.43.**

1. *Let $\alpha \in \mathbf{Cnt(PV)}^+$ be a projection. Then $\alpha(\mathbf{PV})$ is an algebraic lattice with $K(\alpha(\mathbf{PV})) = \alpha(K(\mathbf{PV}))$ countable.*
2. *Conversely, if $L$ is any algebraic lattice with its set $K(L)$ of compact elements countable, then there exists a projection $\alpha \in \mathbf{Cnt(PV)}^+$ such that $\alpha(\mathbf{PV}) \cong L \cup \mathbf{T}$, where $\mathbf{T}$ is an adjoined top.*

*Proof.* For 1, observe that because $\alpha \geq 1$, we have that $\alpha$ is in fact a continuous closure operator. Theorem 6.3.15 then guarantees that $\alpha(\mathbf{PV})$ is algebraic with compact elements $\alpha(K(\mathbf{PV}))$.

Turning to 2, let $B$ be the set of compact elements of $L$. By Proposition 6.3.25, we can view $L$ as a subalgebra of $2^B$. Hence, by Theorem 6.3.14, there is a continuous closure operator $c\colon 2^B \to 2^B$ with image isomorphic $L$. Let $\tau\colon [\{1\}, \mathbf{VS}] \to 2^B$ be the isomorphism from the proof of Theorem 6.3.38. Define a projection $\pi \in \mathbf{Cnt(PV)}^+$ by

$$\pi = \begin{cases} \tau^{-1}c\tau(\mathbf{V}) & \text{if } \mathbf{V} \subseteq \mathbf{VS} \\ \mathbf{Fsgs} & \text{else.} \end{cases}$$

The image is then isomorphic to $L$ with a new top adjoined. $\qquad\square$

An interesting question is which lattices can be of the form $\alpha(\mathbf{PV})$ for a projection from $\mathbf{GMC(PV)}$ or $\mathbf{GMC(PV)}^+$.

## 6.4 Topologies on Algebraic Lattices

The goal of this section is to motivate the usage of the term *continuous* for maps that preserve directed sups. To do this, we introduce a topology on any algebraic lattice such that the continuous order preserving maps between algebraic lattices with respect to this topology are precisely the continuous maps in the sense of preserving directed sups.

First we define the dual notion to a principal ideal. If $L$ is a lattice and $\ell \in L$, we set $\ell^{\uparrow} = \{x \in L \mid x \geq \ell\}$; it is called the *principal filter* generated by $\ell$. Set $\ell^{\nearrow} = \{x \in L \mid x \not\geq \ell\}$. The following topology is a special case of a topology studied in [97, Chapter III] called the *patch topology*.

**Definition 6.4.1 (Zero-dimensional topology).** *Let $L$ be an algebraic lattice with set of compact elements $K(L)$. Define a topology, called the zero-dimensional topology, by taking as a subbasis all sets of the form $k^{\uparrow}$ and $c^{\nearrow}$ with $k, c \in K(L)$.*

It is easy to check that a basic neighborhood is of the form

$$k_1^\uparrow \cap \cdots \cap k_n^\uparrow \cap c_1^\gamma \cap \cdots \cap c_m^\gamma.$$

In particular, we have the following fact.

**Fact 6.4.2.** *The zero-dimensional topology on an algebraic lattice $L$ is countably based if $K(L)$ is countable.*

To describe better the topology, we record here the following variation of Lemma 6.3.35.

**Lemma 6.4.3.** *Let $L$ be an algebraic lattice and $K(L)$ the set of compact elements. Then $\mathsf{Id}(K(L))$ and $L$ are isomorphic lattices via the maps $d\colon \mathsf{Id}(K(L)) \to L$ and $g\colon L \to \mathsf{Id}(K(L))$ given by $d(I) = \bigvee I$ and $g(\ell) = \ell^\downarrow \cap K(L)$.*

*Proof.* Clearly, $dg = 1$ as $L$ is algebraic. So it suffices to show that if $I$ is an ideal of $K(L)$ and $\ell = d(I) = \bigvee I$, then $g(\ell) = I$. By definition, $I \subseteq g(\ell)$. Conversely, if $k \in g(\ell)$, then $k \le \ell = \bigvee I$ and so by compactness of $k$, and because $I$ is directed, $k \le k'$ some $k' \in I$. Thus $k \in I$. So $I = g(\ell)$.    $\square$

Our next theorem justifies the name zero-dimensional topology.

**Theorem 6.4.4.** *Let $L$ be an algebraic lattice with set of compact elements $K(L)$. Then $L$ with the zero-dimensional topology is a profinite meet semilattice with identity, isomorphic to a closed meet subsemilattice of $\{0,1\}^{K(L)}$.*

*Proof.* We view $L$ as a subalgebra of $2^{K(L)}$ by identifying it with $\mathsf{Id}(K(L))$. We identify sets with their characteristic functions. Then $L$ is a meet-subsemilattice (with identity) of $\{0,1\}^{K(L)}$. We first show that the patch topology on $L$ coincides with the subspace topology. A subbasic neighborhood in $\{0,1\}^{K(L)}$ is of one of the following two forms. Fix $k \in K(L)$ and define:

$$\mathcal{N}_{1,k} = \{f\colon K(L) \to \{0,1\} \mid kf = 1\}, \quad \mathcal{N}_{0,k} = \{f\colon K(L) \to \{0,1\} \mid kf = 0\}.$$

It is then clear that $\mathcal{N}_{1,k} \cap \mathsf{Id}(K(L)) = k^\uparrow$ and $\mathcal{N}_{0,k} \cap \mathsf{Id}(K(L)) = k^\gamma$. Thus the zero-dimensional topology on $L$ is the induced topology.

It remains to show that $\mathsf{Id}(K(L))$ is a closed subset. Suppose $J \subseteq K(L)$ is not an ideal. Then one of the following happens: either there exist $c, k \in K(L)$ with $c \le k$, $k \in J$ and $c \notin J$; or there exist $c, k \in J$ such that $c \vee k \notin J$. In the first case, $\mathcal{N}_{1,k} \cap \mathcal{N}_{0,c}$ is a neighborhood of $J$ not containing any ideal. In the second case, $\mathcal{N}_{1,c} \cap \mathcal{N}_{1,k} \cap \mathcal{N}_{0,c\vee k}$ is a neighborhood of $J$ containing no ideals. Thus $\mathsf{Id}((K(L))$ is a closed subset of $\{0,1\}^{K(L)}$.    $\square$

*Remark 6.4.5.* In fact, the converse of Theorem 6.4.4 holds. Namely, if $L$ is a profinite meet semilattice with identity, then in fact $L$ is an algebraic lattice equipped with its zero-dimensional topology. The compact elements of $L$ are precisely those elements $c \in L$ for which the principal filter $c^\uparrow$ is clopen, and the fact that $L$ is algebraic corresponds with there being enough homomorphisms to finite semilattices to separate points. The tricky bit here is to prove that $L$ is a complete lattice by showing that downwards directed subsets of $L$ converge (viewed as a net) to their inf. In fact, one can prove that the category of profinite semilattices with identity is equivalent to the category of algebraic lattices with CL-morphisms. See [97, 134, 147] for details.

We now show that $K(L)$ is dense in $L$. In particular if $K(L)$ is countable, then $L$ is separable in the zero-dimensional topology.

**Proposition 6.4.6.** *If $\ell \in L$ and $\ell = \bigvee D$ with $D$ a directed set of elements, then $D$ (viewed as a net via the inclusion map) converges to $\ell$ in the zero-dimensional topology. As a consequence, $K(L)$ is dense in this topology.*

*Proof.* Let $V$ be a subbasic neighborhood of $\ell$. If $V = k^\uparrow$ with $k$ compact, then because $k \le \ell = \bigvee D$, we have that $k \le d_0$ some $d_0 \in D$. It follows, for $d \ge d_0$, we have $d \in V$. Suppose now that $V = k^{\nearrow}$. Then $\ell \not\ge k$ and hence, for all $d \in D$, we have $d \not\ge k$. Thus $D \subseteq V$. We conclude that $D$ converges to $\ell$.

The final statement is clear as every element of $L$ is a directed supremum of compact elements. □

One can define an analogue of the above topology for continuous lattices in general, but it will not be zero-dimensional unless the lattice is algebraic [97, Chapter III]. Although the zero-dimensional topology is extremely useful, it unfortunately has too many convergent nets for our purposes. This is due to the dual of Proposition 6.4.6, which is the content of the following exercise.

**Exercise 6.4.7.** Suppose that $D$ is a downwards directed subset of $L$. Viewing $D$ as a net in the natural way (that is indexed by the opposite order of $D$), show that $D$ converges to $\bigwedge D$.

Theorem 7.2.3 below shows that **PV** is not join-continuous meaning the join does not preserve downwards directed infs. Hence the join is not continuous in the zero-dimensional topology. This leads us to define a variant. One possible choice would be to use the Scott topology [97], but this topology is not Hausdorff. We instead introduce a Hausdorff topology by slightly altering the zero-dimensional topology to remove its inherent duality. This topology is a special case of the strong topology, going back to Grothendieck and Dieudonné; see [97, Exer. V-5.31] and the note [97, p. 429]. It can also be adapted to continuous lattices in general, but we stick to the case of algebraic lattices here.

First we remind the reader of the Alexandrov topology. Recall that a *downset* in a poset $P$ is a subset $X$ such that $y \le x \in X$ implies $y \in X$. For instance, a directed downset is what we have been calling an ideal.

**Definition 6.4.8 (Alexandrov topology).** *If $P$ is a partially ordered set, then the Alexandrov topology is the topology on $P$ whose open sets are the downsets.*

*Remark 6.4.9.* Some authors call this the dual Alexandrov topology and reserve the name Alexandrov topology for the topology in which the upsets are open.

Notice that each $p \in P$ has a unique smallest neighborhood in the Alexandrov topology, namely $p^{\downarrow}$. The strong topology is the join of the zero-dimensional topology and the Alexandrov topology. More precisely:

**Definition 6.4.10 (Strong topology).** *Let $L$ be an algebraic lattice. Define the strong topology to be the topology with subbasis all sets of the form $\ell^{\downarrow}$ with $\ell \in L$ and $c^{\uparrow}$ with $c \in L$ compact.*

**Lemma 6.4.11.** *Let $X \subseteq L$ be a downset. Then $X$ is open in the strong topology.*

*Proof.* One easily checks $X = \bigcup \{\ell^{\downarrow} \mid \ell \in X\}$ and hence is open.    □

Observe that the strong topology has *more* open sets than the zero-dimensional topology. Indeed, if $c \in L$ is compact, then $c^{\nearrow}$ is a downset and so open by Lemma 6.4.11. Hence the strong topology contains the subbasis for the zero-dimensional topology, and hence the whole topology. We may now conclude the strong topology is the smallest topology containing the Alexandrov topology and the zero-dimensional topology.

**Exercise 6.4.12.** Show that the sets of the form $\ell^{\downarrow} \cap V$ with $\ell \in L$ and $V$ open in the zero-dimensional topology form a basis for the strong topology.

We summarize some of the main properties of the strong topology in the next theorem. Recall first that a subset $Y$ of a topological space is called *discrete* if each point of $Y$ is both open and closed.

**Theorem 6.4.13.** *Let $L$ be an algebraic lattice, endowed with the strong topology. Then the following hold:*

1. *The strong topology on $L$ is Hausdorff;*
2. *A net $\{x_{\alpha}\}_{\alpha \in A}$ converges to $x$ in the strong topology if and only if $x_{\alpha} \to x$ in both the zero-dimensional and Alexandrov topologies, if and only if $x_{\alpha} \to x$ in the zero-dimensional topology and, for all $\alpha$ sufficiently large, $x_{\alpha} \leq x$;*
3. *A directed subset of $L$, viewed as a net via inclusion, converges to its supremum;*
4. *$K(L)$ is a discrete dense subset of $L$;*
5. *The strong topology on $L$ is separable if and only if $K(L)$ is countable;*

6. *If $K(L)$ is countable, then $L$ has a countable basis at each point for the strong topology (i.e., the strong topology is first countable);*
7. *A function $f: L_1 \rightarrow L_2$ between algebraic lattices preserves directed sups (i.e., is continuous in our usual usage) if and only if it is **order preserving** and continuous in the strong topology (on both lattices);*
8. *The lattice operations on $L$ are continuous in the strong topology;*
9. *The strong topology on $\mathbf{PV}$ is not compact.*

*Proof.* Because the zero-dimensional topology is Hausdorff and the strong topology has more open sets, it too must be Hausdorff, proving 1.

Because the strong topology is the join of the zero-dimensional topology and the Alexandrov topology, the first statement of 2 is clear. The second statement follows from the first and the observation that the unique smallest neighborhood of $\ell$ in the Alexandrov topology is $\ell^{\downarrow}$.

We immediately deduce 3 from 2 and Proposition 6.4.6 as a directed set is always below its supremum.

For 4, observe that for $k$ compact, $\{k\} = k^{\downarrow} \cap k^{\uparrow}$ is open. Because the strong topology on $L$ is Hausdorff, points are closed. Thus $K(L)$ is a discrete subset. The density of $K(L)$ is an immediate consequence of 3 as each element of $L$ is a direct supremum of compact elements.

For 5, recall that a topological space is said to be separable if it has a countable dense subset. Because $K(L)$ is discrete, any dense subset must contain $K(L)$. As $K(L)$ is dense by 4, we conclude $L$ is separable in the strong topology if and only if $K(L)$ is countable.

From the fact that $\ell^{\downarrow}$ is the unique smallest neighborhood of $\ell$ in the Alexandrov topology, it is immediate that a basis for the neighborhoods of $\ell \in L$ in the strong topology is given by all subsets of the form $\ell^{\downarrow} \cap V$ with $V$ open in the zero-dimensional topology. Because the zero-dimensional topology has a countable basis when $K(L)$ is countable, this establishes 6.

For 7, suppose first that $f: L_1 \rightarrow L_2$ is order preserving and continuous in the strong topology and let $\ell = \bigvee D$ with $D$ directed. Note that by 3, $D \rightarrow \ell$. On the other hand, because $f$ is order preserving, $f(D)$ is directed and so $f(D) \rightarrow \bigvee f(D)$ by another application of 3. Continuity of $f$ then implies $f(\ell) = \bigvee f(D)$

Suppose conversely $f: L_1 \rightarrow L_2$ preserves directed sups. Then $f$ is order preserving. Assume $\{x_\alpha\}_{\alpha \in A}$ is a net in $L_1$ converging to $x$. We must show $f(x_\alpha) \rightarrow f(x)$. First consider a subbasic neighborhood of $f(x)$ of the form $\ell^{\downarrow}$; so $f(x) \leq \ell$. Because $x_\alpha \leq x$ for all $\alpha$ sufficiently large enough (by 2) and $f$ is order preserving, $f(x_\alpha) \leq f(x) \leq \ell$, i.e., $f(x_\alpha) \in \ell^{\downarrow}$, for $\alpha$ large enough.

Next we consider a subbasic neighborhood of $f(x)$ of the form $c^{\uparrow}$ with $c$ compact. Write $x = \bigvee D$ with $D \subseteq K(L)$ directed. Then because $f$ preserves directed sups $c \leq f(x) = \bigvee f(D)$. Hence $c \leq f(d_0)$ for some $d_0 \in D$, as $c$ is compact and $f(D)$ is directed (as $f$ is order preserving). Now because $x \in d_0^{\uparrow}$, $d_0$ is compact and $x_\alpha \rightarrow x$, we have $x_\alpha \geq d_0$ for all $\alpha$ sufficiently large. Thus $f(x_\alpha) \geq f(d_0) \geq c$ for $\alpha$ sufficiently large, that is, $f(x_\alpha) \in c^{\uparrow}$ for all sufficiently large $\alpha$ (again using $f$ is order preserving). We conclude that $f(x_\alpha) \to f(x)$ by 2, as required.

We may immediately deduce 8 from 7 because algebraic lattices are meet-continuous by Proposition 6.2.11 and of course the join preserves directed sups.

For 9, let $\{p_1, p_2, \ldots, \}$ be an enumeration of the primes. Let $\mathbf{V}_i$ be the pseudovariety of abelian groups with order prime to $p_1, p_2, \ldots, p_i$. Then $\mathbf{V}_1 \geq \mathbf{V}_2 \geq \cdots$ and $\bigcap_{i=1}^{\infty} \mathbf{V}_i = \mathbf{1}$. So, by Exercise 6.4.7, $\mathbf{V}_i \to \mathbf{1}$ in the zero-dimensional topology. It follows from 2 that if $\{\mathbf{V}_i\}$ had a convergent subsequence in the strong topology, then that subsequence must converge to $\mathbf{1}$. But $\mathbf{1}$ is a compact pseudovariety and hence an isolated point in the strong topology by 4, so no such convergent subsequence can exist. If follows that $\mathbf{PV}$ is not compact in the strong topology. □

*Remark 6.4.14.* The hypothesis that $f$ is order preserving is necessary in 7. Indeed if $k \in L$ is compact, then $\delta_k \colon L \to \{0, 1\}$ given by

$$\delta_k(\ell) = \begin{cases} 1 & \ell = k \\ 0 & \ell \neq k \end{cases}$$

is continuous by 4. Clearly, $\delta_k$ is not order preserving unless $k = \mathsf{T}$.

The main consequence of Theorem 6.4.13 is that the maps between algebraic lattices that we have been calling continuous are precisely the continuous order preserving maps with respect to the strong topology. In particular, $\mathbf{PVRM}$, $\mathbf{CC}$, $\mathbf{GMC(PV)}$ and $\mathbf{Cnt(PV)}$ are all topological lattices and $\mathsf{q}$ is a continuous function in all the contexts where it is defined.

**Exercise 6.4.15.** A partially ordered set $P$ is said to be Noetherian if it has no infinite descending chains and no infinite anti-chains.

1. Show that if $X \subseteq \mathbf{PV}$ is Noetherian in the induced ordering, then every sequence in $X$ has a non-decreasing subsequence.
2. Deduce that if $X \subseteq \mathbf{PV}$ is closed and Noetherian, then it is sequentially compact, meaning every sequence has a convergent subsequence.
3. Show that a sequentially compact subspace of $\mathbf{PV}$ is closed and has no infinite descending chains.

*Question 6.4.16.* Is a subspace of $\mathbf{PV}$ sequentially compact in the strong topology if and only if it is closed and Noetherian?

# Notes

This chapter is essentially a crash course on algebraic and continuous lattices [97]; see also [147, Chapter VII]. The term *algebraic lattice* was apparently coined by Birkhoff. Continuous lattices were introduced by Scott [315];

the treatise [97] is the bible for the subject and also contains a wealth of references and historical information. The correspondence between algebraic lattices and profinite semilattices, as well as their duality with join semilattices, is the subject of [134]. All the essential ingredients of this duality can be found in [97, 147]. Continuous lattices correspond with a larger class of compact semilattices, known as Lawson semilattices [97, 147]. The strong, patch, Alexandrov and Scott topologies can all be found in [97]. Many authors refer to the topology in which all upsets are open as the Alexandrov topology and the topology in which all downsets are open as the dual Alexandrov topology. This choice is usually made so that the specialization order coincides with the original partial order.

In [97], the more general notions of algebraic and continuous domains are considered. An algebraic domain is a poset admitting directed joins such that each element is a directed join of compact elements and the compact elements below any given element form an ideal. Here one must use the formulation of compactness in terms of directed joins as the definition. A profinite poset is an algebraic domain, so for instance if $S$ is a profinite semigroup, then $S/\mathcal{J}$ is an algebraic domain. However, the converse is false because any set with the equality ordering is an algebraic domain. It has recently been observed by Steinberg's Ph.D. student, W. Hajji, that a compact inverse semigroup has enough finite dimensional irreducible representations to separate points if and only if it is an algebraic domain with respect to its natural partial order.

The realizability of any countable weight continuous lattice as the fixed-point lattice of an idempotent continuous operator on **PV** is original. As far as we know, the characterization of continuous maps in terms of the strong topology is also new. The Scott topology [97, 147] seems to be the more usual choice for studying maps preserving directed joins.

The spectral theory of lattices will play a bigger role in the next two chapters. Johnstone's book [147] is a readable introduction to spectral spaces, locales and pointless topology, as is [97, Chapter V]. See also [186] for connections with topos theory.

# 7

## The Abstract Spectral Theory of PV

The spectrum of a commutative ring is its space of prime ideals; the spectrum of a $C^*$-algebra is its maximal ideal space. Lattices also have a spectrum, going back to Stone [97, 147, 341, 342]. Recall that if $L$ is a lattice with top and bottom (not necessarily complete), then the set $\mathsf{Id}(L)$ of ideals of $L$ is a complete algebraic lattice, which is in fact distributive. A *prime ideal* of $L$ is a proper fmi element of $\mathsf{Id}(L)$. These are precisely the kernels of morphisms $L \to \{0, 1\}$ preserving finite (including empty) infs and sups [97, 147]. The *spectrum* of $L$ is the space $\operatorname{Spec} L$ of prime ideals with the *hull-kernel topology*, whose closed sets are of the form $V(I) = \{\mathfrak{p} \in \operatorname{Spec} L \mid \mathfrak{p} \supseteq I\}$ where $I \in \mathsf{Id}(L)$. If $L$ is a Boolean algebra, then the topology on $\operatorname{Spec} L$ is induced by the product topology on $\{0, 1\}^L$ (identifying a prime ideal with the associated homomorphism), and moreover it is a closed subspace and hence a profinite space. The lattice of clopen subsets of $\operatorname{Spec} L$ is isomorphic to $L$ itself. Conversely, if $X$ is a profinite space, then the set $K(X)$ of clopen subsets of $X$ is a Boolean algebra and $\operatorname{Spec} K(X) \cong X$. This is the classical Stone duality for Boolean algebras [61, 117, 147, 341, 342]. An important example for us, due to Almeida [7, Thm 3.6.1], is that if $\mathbf{V}$ is a pseudovariety of semigroups and $\mathbf{V}(A^+)$ is the Boolean algebra of $\mathbf{V}$-recognizable subsets of $A^+$, then $\operatorname{Spec} \mathbf{V}(A^+) \cong \widehat{F_{\mathbf{V}}}(A)$. More generally, Stone established a duality between distributive lattices and coherent spaces via $\operatorname{Spec} L$ [97, 147]. (A space is *coherent* if it is sober and its collection of compact open subsets is closed under finite meets and forms a basis for the topology.)

Because a prime ideal of a lattice is precisely the same thing as a proper fmi element of the complete lattice $\mathsf{Id}(L)$, it is natural to define the spectrum $\operatorname{Spec} L$ of a complete lattice $L$ to be the space of all fmi elements of $L$ except the top. We endow $\operatorname{Spec} L$ with the *hull-kernel topology* whose closed sets are of the form $V(\ell) = \{m \in \operatorname{Spec} L \mid m \geq \ell\}$ with $\ell \in L$ [97, 147]. This space can again be viewed as a space of homomorphisms. Let us view $\{0, 1\}$ as a complete lattice with the usual ordering. Recall that $\mathsf{T}$ denotes the top of a lattice. Notice that if $\ell \in L \setminus \mathsf{T}$ is fmi, then we can define an onto map

J. Rhodes, B. Steinberg, *The q-theory of Finite Semigroups*,
Springer Monographs in Mathematics, DOI 10.1007/978-0-387-09781-7_7,
© Springer Science+Business Media, LLC 2009

$\check{\ell} \colon L \to \{0,1\}$ preserving all sups and finite infs, i.e., a frame morphism, by

$$\check{\ell}(x) = \begin{cases} 0 & x \leq \ell \\ 1 & \text{else.} \end{cases}$$

Conversely, if $\varphi \colon L \to \{0,1\}$ is an onto map preserving all sups and finite infs, then $\ell = \bigvee \varphi^{-1}(0)$ is fmi, $\ell \neq \top$ and $\check{\ell} = \varphi$. If we endow $\{0,1\}$ with the *Sierpínski topology* in which the closed sets are $\{\emptyset, \{0\}, \{0,1\}\}$, then the hull-kernel topology is the induced topology from the product topology.

So by the abstract spectral theory of **PV**, we mean the study of the spectra of **PV** and **PV**$^{op}$. The current chapter is dedicated to furthering our understanding of mi, fmi, sfmi, ji, fji, and sfji pseudovarieties. We also begin to investigate these notions for some of our other favorite algebraic lattices like **PVRM** and **Cnt(PV)**.

Most of our efforts are focused on constructing non-trivial new examples of fji pseudovarieties (so the points of the spectrum of **PV**$^{op}$). Also we explore the "duality" between mi pseudovarieties and ji pseudovarieties. This chapter is very much a dialogue between abstract notion and concrete example as we introduce each concept for algebraic lattices in general and then see what it means for **PV**.

*Question 7.0.1.* Study the topological spaces Spec **PV** and Spec **PV**$^{op}$.

## 7.1 Birkhoff's Subdirect Representation Theorem

We first consider the abstract setting of algebraic lattices, beginning with an algebraic lattice version of Birkhoff's Subdirect Representation Theorem (see [203, Thm. 2.19]). Let us isolate a key lemma, which is the main idea (due to Birkhoff).

**Lemma 7.1.1.** *Let $L$ be a complete lattice, $\ell_0 \in L$ and $c \nleq \ell_0$ with $c$ compact. Then the set*

$$\mathsf{Exclude}(c; \ell_0^{\uparrow}) = \{\ell \mid \ell \geq \ell_0, \ c \nleq \ell\} = \ell_0^{\uparrow} \cap c^{\gamma} \qquad (7.1)$$

*has maximal elements. Moreover, any maximal element of $\mathsf{Exclude}(c; \ell_0^{\uparrow})$ is strictly meet irreducible.*

*Proof.* Note that $\ell_0$ belongs to the set (7.1) so it is not empty. First suppose $m$ is a maximal element of (7.1). Suppose further that $m = \bigwedge X$. Let $x \in X$ such that $x > m$. Then, by maximality of $m$, we must have $c \leq x$. Hence $m < m \vee c \leq x$. Thus not all $x \in X$ can be strictly greater than $m$. This proves that $m$ is strictly meet irreducible.

To see that (7.1) has maximal elements, we apply Zorn's Lemma. Suppose we have a chain $C$ of elements from (7.1). Then, because $c$ is compact, if

$c \leq \bigvee C$, then $c$ is below some element of $C$. But this cannot happen by the definition of $\mathsf{Exclude}(c; \ell_0^\uparrow)$. Hence $\bigvee C \in \mathsf{Exclude}(c; \ell_0^\uparrow)$. Thus by Zorn's Lemma, (7.1) has maximal elements.                                            $\square$

In the case of **PV** (respectively, **PVRM**), each compact element is generated by a single semigroup (relational morphism). So if $S$ is a finite semigroup and **V** is a pseudovariety, we denote the set of maximal elements of $\mathsf{Exclude}((S); \mathbf{V}^\uparrow)$ by $\mathsf{MaxExclude}(S; \mathbf{V}^\uparrow)$. In the case that $\mathbf{V} = \mathbf{1}$ is the trivial pseudovariety, we simply write $\mathsf{Exclude}(S)$ and $\mathsf{MaxExclude}(S)$. It is also convenient to define

$$\mathsf{Excl}(S) = \{T \in \mathbf{Fsgp} \mid (T) \in \mathsf{Exclude}(S)\} = \{T \in \mathbf{Fsgp} \mid S \nprec T^n, \; \forall n \geq 1\}.$$

The following special case is of extreme importance. Let $S$ be a finite semigroup such that the pseudovariety $(S)$ generated by $S$ is fji (equivalently, is ji by Exercise 6.1.6). In this case, we shall abuse notation and say that $S$ is fji; this is the same as saying that if $S \in \mathbf{V} \vee \mathbf{W}$, then $S \in \mathbf{V}$ or $S \in \mathbf{W}$. We claim that, for such $S$, $\mathsf{MaxExclude}(S)$ contains a unique element.

**Theorem 7.1.2.** *Let $S$ be a finite semigroup. The following are equivalent:*

*1. $S$ is fji;*
*2. $\mathsf{Excl}(S)$ is a pseudovariety of semigroups;*
*3. $|\mathsf{MaxExclude}(S)| = 1$.*

*Moreover, if the above equivalent conditions hold, then $\mathsf{Excl}(S)$ is mi. Conversely, every proper mi pseudovariety is of the form $\mathsf{Excl}(S)$ for some fji semigroup $S$. In particular, there are only countably many mi pseudovarieties and they all have decidable membership problem.*

*Proof.* Suppose first that $S$ is fji. We show that $\mathsf{Excl}(S)$ is a pseudovariety. By definition, $\mathsf{Excl}(S)$ is closed under taking divisors. Thus to show that it is a pseudovariety, we just need to show that it is closed under formation of direct products. So suppose that $T_1, T_2 \in \mathsf{Excl}(S)$ but $S \in (T_1 \times T_2)$. Then $S \in (T_1) \vee (T_2)$ and hence, because $S$ is fji, we must have $S \in (T_i)$, some $i$. But this contradicts that $T_i \in \mathsf{Excl}(S)$. So $T_1 \times T_2 \in \mathsf{Excl}(S)$ and thus $\mathsf{Excl}(S)$ is a pseudovariety.

Next assume $\mathsf{Excl}(S)$ is a pseudovariety and **V** is pseudovariety not containing $S$. To show $\mathbf{V} \leq \mathsf{Excl}(S)$, it suffices to show that every compact subpseudovariety of **V** is contained in $\mathsf{Excl}(S)$; that is, we may assume that **V** is compact, say $\mathbf{V} = (T)$. But then, as $S \notin \mathbf{V}$, by definition $T \in \mathsf{Excl}(S)$ and so $\mathbf{V} \leq \mathsf{Excl}(S)$, as desired. Thus $\mathsf{MaxExclude}(S) = \{\mathsf{Excl}(S)\}$.

Finally, suppose that $|\mathsf{MaxExclude}(S)| = 1$; say its unique element is **W**. Let $\mathbf{V}_1, \mathbf{V}_2$ be pseudovarieties that do not contain $S$. Then because

$$\emptyset \neq \mathsf{MaxExclude}(S; \mathbf{V}_i^\uparrow) \subseteq \mathsf{MaxExclude}(S)$$

(using Lemma 7.1.1), we see that $\mathbf{V}_1, \mathbf{V}_2 \leq \mathbf{W}$ and hence $\mathbf{V}_1 \vee \mathbf{V}_2 \leq \mathbf{W}$. It follows that $S \notin \mathbf{V}_1 \vee \mathbf{V}_2$ and so $S$ is fji. This establishes the equivalence of 1–3.

To see that $\mathsf{Excl}(S)$ is mi, suppose that $\mathsf{Excl}(S) \geq \bigcap_\alpha \mathbf{V}_\alpha$. Then $S \notin \bigcap_\alpha \mathbf{V}_\alpha$ and hence $S \notin \mathbf{V}_\alpha$ for some $\alpha$. But for this $\alpha$, we have $\mathbf{V}_\alpha \leq \mathsf{Excl}(S)$. Conversely, suppose that $\mathbf{V}$ is a proper mi pseudovariety. Let $\mathscr{C}$ be the set of all pseudovarieties not contained $\mathbf{V}$. The set $\mathscr{C}$ is non-empty. We claim there is a smallest pseudovariety $\mathbf{W}$ in $\mathscr{C}$. Indeed, the fact that $\mathbf{V}$ is mi is equivalent to saying that $\mathscr{C}$ is closed under meets. Because $\mathbf{W}$ is not contained in $\mathbf{V}$, there is a semigroup $S \in \mathbf{W}$ that is not in $\mathbf{V}$. Then $(S) \not\leq \mathbf{V}$ and $(S) \leq \mathbf{W}$. Hence, by minimality of $\mathbf{W}$, we must have that $\mathbf{W} = (S)$, that is, $\mathbf{W}$ is generated by $S$. It now follows that a pseudovariety $\mathbf{U}$ is not contained in $\mathbf{V}$ if and only if $S \in \mathbf{U}$. Thus $\{\mathbf{V}\} = \mathsf{MaxExclude}(S)$ and so, by the first part of the theorem, $S$ is fji.

The last sentence is clear because there are only countably many pseudovarieties of the form $\mathsf{Excl}(S)$ and each is decidable, as for a given $T$, it is decidable whether $S \in (T)$.                                        □

This theorem establishes a duality between ji and mi elements in an algebraic lattice. We shall consider examples of fji semigroups later in this chapter, many of which are new. Let us also say that a semigroup $S$ is sfji if $(S)$ is sfji.

To give an example where $\mathsf{MaxExclude}(S)$ has multiple elements consider $\mathbb{Z}_6 \cong \mathbb{Z}_2 \times \mathbb{Z}_3$. It is not fji, whence $\mathsf{MaxExclude}(\mathbb{Z}_6)$ has more than one element, namely the pseudovarieties of semigroups whose subgroups have odd order and whose subgroups have order prime to 3.

We leave as an exercise the proof of the following abstract version of Theorem 7.1.2. The reader is referred to Section 6.3.3 for the definition of the weight of a lattice.

**Theorem 7.1.3.** *Let $L$ be an algebraic lattice with bottom $\mathsf{B}$ and let $c$ be a compact element of $L$. Writing $\mathsf{MaxExclude}(c)$ for $\mathsf{MaxExclude}(c; \mathsf{B}^\uparrow)$, the following are equivalent:*

*1. $c$ is fji;*
*2. $|\mathsf{MaxExclude}(c)| = 1$.*

*Moreover, if the above equivalent conditions hold, then the unique element of $\mathsf{MaxExclude}(c)$, denoted $\mathsf{Excl}(c)$, is mi. Conversely, every mi element of $L$ other than the top $\mathsf{T}$ is of the form $\mathsf{Excl}(c)$ for some fji compact element $c \in L$. In particular, the number of (proper) mi elements of $L$ is bounded by the weight of $L$.*

**Exercise 7.1.4.** Prove Theorem 7.1.3.

**Exercise 7.1.5.** Show that if $S$ is fji, then $\mathsf{Excl}(S) \vee (S)$ is the unique cover of $\mathsf{Excl}(S)$.

Now we return to the proof of the Subdirect Representation Theorem.

**Theorem 7.1.6 (Birkhoff's Subdirect Representation Theorem).** *In an algebraic lattice $L$, every element $\ell \in L$ is the meet of some set of strictly meet irreducible elements of $L$.*

*Proof.* Let $\ell \in L$ and set $X = \{x \in L \mid \ell \leq x \text{ and } x \text{ is smi}\}$; note that $X$ is non-empty because it contains the top of $L$. We show $\ell = \bigwedge X$. Setting $\ell_0 = \bigwedge X$, we clearly have $\ell \leq \ell_0$. As $L$ is an algebraic lattice, to prove that $\ell_0 \leq \ell$, it suffices to show that, for each compact element $c \leq \ell_0$, also $c \leq \ell$. We proceed by contradiction. Suppose $c \leq \ell_0$ but $c \not\leq \ell$. Consider $Y = \mathsf{Exclude}(c; \ell^{\uparrow})$. Lemma 7.1.1 assures us that $Y$ has maximal elements and that they are strictly meet irreducible. If $m$ is such a maximal element, then $m \in X$ and so $\ell_0 \leq m$. But then $c \leq \ell_0 \leq m$, contradicting $m \in Y$. This establishes the theorem. $\qquad\square$

*Remark 7.1.7.* The reason for the name of Theorem 7.1.6 is that if $A$ is a universal algebra (perhaps infinite) and $\mathsf{Cong}(A)$ is the congruence lattice of $A$, then $\mathsf{Cong}(A)$ is an algebraic lattice. It is easy to see that a congruence $R$ is smi if and only if $A/R$ is subdirectly indecomposable. Because the trivial congruence is a meet of smi congruences by Theorem 7.1.6, it follows that $A$ is a subdirect product of subdirectly indecomposable universal algebras.

An important property of smi pseudovarieties is that they can be defined by a single pseudoidentity.

**Proposition 7.1.8.** *Suppose that $\mathbf{V}$ is an smi pseudovariety. Then $\mathbf{V}$ can be defined by a single pseudoidentity.*

*Proof.* Let $\mathscr{E}$ be a set of pseudoidentities defining $\mathbf{V}$. For each $e \in \mathscr{E}$, let $\mathbf{V}_e = [\![e]\!]$. Then $\mathbf{V} = \bigcap \mathbf{V}_e$. As $\mathbf{V}$ is smi, $\mathbf{V} = \mathbf{V}_e$ for some $e \in \mathscr{E}$. $\qquad\square$

The converse of Proposition 7.1.8 does not hold. Indeed, the pseudovariety of aperiodic semigroups $\mathbf{A}$ can be defined by the single pseudoidentity $x^{\omega+1} = x^{\omega}$, but $\mathbf{A}$ is not even sfmi because $\mathbf{A} = \overline{(\mathbb{Z}_2)} \cap \overline{(\mathbb{Z}_3)}$ and neither of these pseudovarieties is equal to $\mathbf{A}$. Also $\mathbf{1} = [\![x = y]\!]$ and $\mathbf{1} = \mathbf{Com} \cap \mathbf{RZ}$.

A strengthening of Proposition 7.1.8 can be obtained when $\mathbf{V} = \mathsf{Excl}(S)$ for an fji semigroup $S$. Such pseudovarieties are precisely the proper mi pseudovarieties by Theorem 7.1.2. The second author learned of this strengthened result from J. Almeida while working on [17].

**Proposition 7.1.9.** *Let $S$ be an $n$-generated fji semigroup. Then $\mathsf{Excl}(S)$ can be defined by a single pseudoidentity in at most $n$ variables.*

*Proof.* By Theorem 7.1.2, $\mathsf{Excl}(S)$ is an mi pseudovariety. Proposition 7.1.8 implies that $\mathsf{Excl}(S)$ is defined by a single pseudoidentity $u = v$ in variables $A$. If $|A| \leq n$, we are done. In any event, because $S \not\models u = v$, we can find a substitution $\sigma \colon A \to S$ that does not satisfy $u\sigma = v\sigma$. Let $X$ be a generating set for $S$ with $n$ elements. Then we can find, for each $a \in A$, a word $w_a$ in $X^{+}$

so that $w_a$ maps in $S$ to $a\sigma$. Substituting $a$ by $w_a$ in $u$ and $v$ for each $a \in A$ results in a new pseudoidentity $u' = v'$ in variables $X$ that is not satisfied by $S$. Clearly, $u' = v'$ is a consequence of $u = v$. So we have

$$\mathsf{Excl}(S) = [\![u = v]\!] \le [\![u' = v']\!] \le \mathsf{Excl}(S).$$

Thus $\mathsf{Excl}(S) = [\![u' = v']\!]$, completing the proof.                                □

Let us mention a converse to the previous result.

**Proposition 7.1.10.** *Let* **V** *be a proper mi pseudovariety defined by a pseu-doidentity in $n$ variables. Then* **V** $= \mathsf{Excl}(S)$ *for an fji semigroup generated by at most $n$ elements.*

*Proof.* By Theorem 7.1.2, we know **V** $= \mathsf{Excl}(T)$ for an fji semigroup $T$. Suppose that **V** is defined by $u = v$ in $n$ variables. Then $T$ fails to satisfy $u = v$, and hence some at most $n$-generated subsemigroup $S$ of $T$ fails to satisfy this pseudoidentity. We claim $(S) = (T)$ and hence $S$ is fji and $\mathsf{Excl}(S) = \mathsf{Excl}(T)$. Clearly $S \in (T)$. For the converse, suppose $T \notin (S)$. Then $S \in \mathsf{Excl}(T)$ and so $S$ satisfies $u = v$. This contradiction shows $(S) = (T)$.                                □

**Corollary 7.1.11.** *Let* **V** *be a proper mi pseudovariety. Then the minimal number of variables in a pseudoidentity defining* **V** *is the same as the minimal number of generators of a semigroup $S$ so that* **V** $= \mathsf{Excl}(S)$.

*Remark 7.1.12.* Notice that if $S$ is an fji semigroup and **V** $= \mathsf{Excl}(S)$, then a pseudoidentity $e$ defines **V** if and only if **V** satisfies $e$ and $S$ does not satisfy $e$. Indeed, in this case **V** $\le [\![e]\!] \le \mathsf{Excl}(S) =$ **V**. We shall use this technique later in several examples.

Before turning to examples, we give a characterization of smi elements of an algebraic lattice analogous to Theorem 7.1.3.

**Proposition 7.1.13.** *Let $L$ be an algebraic lattice. Then the following are equivalent for $\top \ne \ell \in L$ (where $\top$ is the top of $L$):*

1. *$\ell$ is smi;*
2. *$\ell$ is the unique element of $\mathsf{MaxExclude}(c; \ell^\uparrow)$ for some compact $c$ with $c \not\le \ell$;*
3. *$\ell$ has a unique cover $\ell \nearrow \bar{\ell}$.*

*Moreover, if the above conditions hold, then the unique cover $\bar{\ell}$ can be described as $\bigwedge\{\ell_1 \in L \mid \ell_1 > \ell\}$ or as $\ell \vee c$ where $c$ is any compact element with $c \le \bar{\ell}$ and $c \not\le \ell$.*

*Proof.* We show that 1 implies 3 implies 2 implies 1. Suppose then that $\ell$ is smi. We claim $\bar{\ell} = \bigwedge\{\ell_1 \in L \mid \ell_1 > \ell\}$ is the unique cover of $\ell$. Clearly $\bar{\ell} \ge \ell$. By the definition of $\bar{\ell}$ and the fact that $\ell$ is smi, we cannot have $\bar{\ell} = \ell$. So $\bar{\ell} > \ell$. Also by construction of $\bar{\ell}$, every element $\ell_1 > \ell$ is above $\bar{\ell}$. So $\bar{\ell}$ is the unique cover of $L$.

**Table 7.1.** Atoms of **PV**

| Name | Description | Generator | Pseudoidentities |
|------|-------------|-----------|------------------|
| **Ab**($p$) | Abelian groups of exponent $p$ ($p$ prime) | $\mathbb{Z}_p$ | $[\![x^p = 1, xy = yx]\!]$ |
| **N**$_2$ | Null semigroups (nilpotent of index 2) | $N_2$ | $[\![xy = 0]\!]$ |
| **LZ** | Left zero semigroups | $2^\ell$ | $[\![xy = x]\!]$ |
| **RZ** | Right zero semigroups | $\overline{2}$ | $[\![xy = y]\!]$ |
| **Sl** | Semilattices | $U_1$ | $[\![x^2 = x, xy = yx]\!]$ |

Suppose now that $\ell$ has a unique cover $\bar{\ell}$. Because $L$ is algebraic, there is a compact element $c$ with $c \leq \bar{\ell}$ but $c \not\leq \ell$. We must then have $\bar{\ell} = \ell \vee c$. Consider now $E = \mathsf{Exclude}(c; \ell^\uparrow)$. If $\ell_1 \in E$, then $c \not\leq \ell_1$ and $\ell_1 \geq \ell$. But if $\ell_1 > \ell$, then $\ell_1 \geq \bar{\ell} \geq c$. Thus we must have $\ell_1 = \ell$, from which we conclude $E = \{\ell\} = \mathsf{MaxExclude}(c; \ell^\uparrow)$.

The implication 2 implies 1 is immediate from Lemma 7.1.1.            $\square$

### 7.1.1 Atoms

By an *atom* of a lattice, we mean a cover of the bottom. The atoms of **PV** are of course compact and are well-known. They are described in Table 7.1. In this table we use the following notation: $N_2 = \{a, 0\}$ where all products are 0 and if $A$ is a set, $A^\ell$ is the left zero semigroup with underlying set $A$ and $\overline{A}$ is the right zero semigroup with underlying set $A$.

**Exercise 7.1.14.** Prove that Table 7.1 is a complete list of the atoms of **PV**.

Abusing language, we often identify an atom with the generator in Table 7.1. Each of these atoms turns out to be fji. To see this, it suffices by Theorem 7.1.2 to show that the exclusions of the atoms are pseudovarieties. We also want to specify a pseudoidentity defining each of these exclusions. To achieve this, we shall need a specific element of a free procyclic semigroup.

Let $\pi$ be a set of primes and denote by $\pi'$ the complementary set of primes. Recall that if $G$ is a (pro)finite abelian group, then $G$ is an internal direct product $PQ$ where $P$ is a (pro-)$\pi$-group, called the $\pi$-*primary component* of $G$, and $Q$ is a (pro-)$\pi'$-group, called the $\pi'$-*primary component* of $G$. If $g \in G$, then one can uniquely write $g = g_1 g_2$ where $g_1 \in P$ and $g_2 \in Q$. We call $g_1$ the $\pi$-*component* of $g$ and $g_2$ the $\pi'$-*component*. For further information, the reader is referred to [298, Section 4.3].

The minimal ideal of the free procyclic semigroup $\widehat{x^+}$ is a free procyclic group generated by $x^{\omega+1}$. We want to get a hold of the $\pi'$-component of $x^{\omega+1}$.

**Definition 7.1.15.** *Let* $\pi = \{p_1, p_2, \dots\}$ *be a non-empty set of primes. Define*

$$x^{\pi^\omega} = \lim_{n \to \infty} x^{(p_1 \cdots p_n)^{n!}}. \tag{7.2}$$

*We will prove this limit exists and is independent of the ordering of* $\pi$ *in the next proposition. In particular, if* $p$ *is a prime, $x^{p^\omega} = \lim x^{p^{n!}}$. If $\pi'$ is* the complementary set of primes to $\pi$, then $x^{(\pi')^\omega}$ makes sense via (7.2); in particular, $x^{(p')^\omega}$ is well defined.

The following proposition is from [17].

**Proposition 7.1.16.** Let $\pi$ be a non-empty set of primes. Then $x^{\pi^\omega}$ is the $\pi'$-component of $x^{\omega+1}$ and so in particular generates the $\pi'$-component of $\overline{\langle x^{\omega+1}\rangle}$. Similarly, $x^{(\pi')^\omega}$ is the $\pi$-component of $x^{\omega+1}$.

*Proof.* Let $p_1, p_2, \dots$ be an enumeration of $\pi$ and let $S = \langle s \rangle$ be a finite cyclic semigroup with minimal ideal the cyclic group $K = \langle s^{\omega+1} \rangle$. Assume $s^{\omega+1} = s_1 s_2$ where $s_1$ is the $\pi$-component and $s_2$ is the $\pi'$-component of $s^{\omega+1}$. Set $i_n = (p_1 \cdots p_n)^{n!}$. We need to show that, for $n$ sufficiently large, $s^{i_n} = s_2$. Suppose that $S$ has order $\ell$. For $n \geq \ell$, clearly $i_n \geq \ell$ and so $s^{i_n}$ is in $K$. Next we compute

$$(s^{\omega+1})^{i_n} = (s^{i_n})^{\omega+1} = (s^{i_n})^\omega s^{i_n} = s^\omega s^{i_n} = s^{i_n}$$

where the last equality follows because $s^{i_n}$ is in the minimal ideal of $S$, which is a group with identity $s^\omega$. Thus without loss of generality, we may assume that $s = s^{\omega+1}$ generates a cyclic group of order $\ell$.

Suppose $s_1$ has order $j$ and $s_2$ has order $k$; so $j$ is divisible only by primes in $\pi$ and $k$ by primes in $\pi'$ and also $\ell = jk$. Let $r$ be largest index so that $p_r \mid j$. Choose $N = \max\{j, r, \varphi(k)\}$ where $\varphi$ is the Euler totient function (so $\varphi(n)$ is the number of integers in $[1, n]$ that are relatively prime to $n$). We claim that, for $n \geq N$, the equality $s^{i_n} = s_2$ holds. Because $n \geq \max\{j, r, \varphi(k)\}$ the following hold:

$$j \mid (p_1 \cdots p_n)^{n!} = i_n \quad \text{and} \quad \varphi(k) \mid n!$$

Indeed, if $p$ is a prime dividing $j$, then certainly $p$ is among the list $p_1, \dots, p_n$ as $n \geq r$; if $p^u$ is the largest power of $p$ dividing $j$, then evidently $u \leq j!$ and so $j \mid (p_1 \cdots p_n)^{n!}$ as claimed; obviously $n \geq \varphi(k)$ implies $\varphi(k) \mid n!$. Because $p_1 \cdots p_n$ is prime to $k$, Euler's Theorem (or the fact that the group of units of $\mathbb{Z}_k$ has order $\varphi(k)$) yields

$$i_n = (p_1 \cdots p_n)^{n!} \equiv 1 \bmod k.$$

Therefore, $s^{i_n} = s_1^{i_n} s_2^{i_n} = s_2$. This completes the proof. □

Notice that the limit does not depend on the enumeration of $\pi$ because the end result is the $\pi'$-component of $x^{\omega+1}$.

*Remark 7.1.17.* The way to remember the notation is that $\omega$ is the limit of $n!$ in the profinite completion of $(\mathbb{N}, +)$ so $\pi^\omega$ is intuitively the limit of "raising each element of $\pi$ to the $n!$-power." This will kill off the $p$-component, for each prime $p \in \pi$, leaving only the $\pi'$-component.

Table 7.2. Exclusion pseudovarieties of the atoms of **PV**

| Atom | Excl(Atom) | Pseudoidentity |
|------|-----------|----------------|
| $\mathbb{Z}_p$ | $\overline{\mathbf{G}}_{p'}$ | $[\![ x^{(p')^\omega} = x^\omega ]\!]$ |
| $N_2$ | **CR** | $[\![ x^{\omega+1} = x ]\!]$ |
| $\mathbf{2}^\ell$ | **EL** | $[\![ x^\omega (y^\omega x^\omega)^\omega = (y^\omega x^\omega)^\omega ]\!]$ |
| $\overline{\mathbf{2}}$ | **ER** | $[\![ (x^\omega y^\omega)^\omega x^\omega = (x^\omega y^\omega)^\omega ]\!]$ |
| $U_1$ | $\mathbb{LG}$ | $[\![ (x^\omega y x^\omega)^\omega = x^\omega ]\!]$ |

**Corollary 7.1.18.** *Let $\pi$ be a non-empty set of primes. Let $\mathbf{G}_\pi$ be the pseudovariety of $\pi$-groups. Then $\overline{\mathbf{G}}_\pi = [\![ x^{\pi^\omega} = x^\omega ]\!]$. In particular, the equalities $\overline{\mathbf{G}}_p = [\![ x^{p^\omega} = x^\omega ]\!]$ and $\overline{\mathbf{G}}_{p'} = [\![ x^{(p')^\omega} = x^\omega ]\!]$ hold.*

*Proof.* By Proposition 7.1.16, an element $s$ of a finite semigroup satisfies $s^{\pi^\omega} = s^\omega$ if and only if the $\pi'$-component of $s^{\omega+1}$ is trivial, i.e., if and only if $\langle s \rangle \in \overline{\mathbf{G}}_\pi$. As a semigroup belongs to $\overline{\mathbf{G}}_\pi$ if and only if its cyclic subsemigroups do, this completes the proof.  □

The exclusion pseudovarieties of the atoms and their pseudoidentities are recorded in Table 7.2. The exclusion of $\mathbb{Z}_p$ is handled by Corollary 7.1.18. Let us deal with the exclusion of $\overline{\mathbf{2}}$. The remaining verifications will be left to the reader.

**Proposition 7.1.19.** *The pseudovariety* Excl($\overline{\mathbf{2}}$) = **ER** *is defined by the pseudoidentity* $[\![ (x^\omega y^\omega)^\omega x^\omega = (x^\omega y^\omega)^\omega ]\!]$.

*Proof.* Theorem 4.8.3 tells us Excl($\overline{\mathbf{2}}$) = **ER**. Because $(x^\omega y^\omega)^\omega x^\omega$ and $(x^\omega y^\omega)^\omega$ always map to $\mathscr{R}$-equivalent idempotents of the idempotent-generated subsemigroup of any semigroup, evidently **ER** satisfies this pseudoidentity. On the other hand $\overline{\mathbf{2}}$ fails it by taking $x = \overline{0}$ and $y = \overline{1}$. This completes the proof in light of Remark 7.1.12.  □

**Exercise 7.1.20.** Verify that the rest of Table 7.2 is correct.

The observation in the exercise below is a useful criterion to exclude a pseudovariety from being fmi.

**Exercise 7.1.21.** Suppose that $L$ is a lattice and that $\ell \in L$ is fmi. Suppose that $\ell_1, \ell_2 \in L$ are such that $\ell_1 \wedge \ell_2 = $ B. Then show that $\ell \geq \ell_i$ some $i$. Deduce that any fmi pseudovariety must contain either **G** or **A** and hence no fmi pseudovariety can be compact or even locally finite. In fact an fmi pseudovariety cannot satisfy a non-trivial semigroup identity.

**Exercise 7.1.22.** Show that any fmi pseudovariety must contain a pseudovariety of $p$-groups for some prime $p$ and hence cannot be aperiodic.

**Exercise 7.1.23.** Show that **G** is not sfmi.

Let us prove that $\mathbf{Cnt}(\mathbf{PV})$ has no atoms.

**Proposition 7.1.24.** *The lattice* $\mathbf{Cnt}(\mathbf{PV})$ *has no atoms.*

*Proof.* Recall that $\delta(S,T)$ is the operator that sends $\mathbf{V}$ to $(S)$ if $T \in \mathbf{V}$ and otherwise sends $\mathbf{V}$ to $\mathbf{1}$. Proposition 2.2.2 shows that every continuous operator on $\mathbf{PV}$ is a join of elements of the form $\delta(S,T)$. Hence any atom of $\mathbf{Cnt}(\mathbf{PV})$ is of the form $\delta(S,T)$ with $S$ non-trivial. So assume $\delta(S,T)$ is an atom. Let $T'$ be any semigroup not in $(T)$. Then $\delta(S, T \times T') < \delta(S,T)$. This contradiction completes the proof. □

The authors have a proof that **CC** does not have atoms, which will appear in a forthcoming paper. See the chapter notes for further information on the next question.

*Question 7.1.25.* What are the atoms of $\mathbf{Cnt}(\mathbf{PV})^+$, $\mathbf{GMC}(\mathbf{PV})^+$, $\mathbf{PVRM}^+$ and $\mathbf{CC}^+$?

### 7.1.2 Elementary examples of smi decompositions

Table 7.3 provides some examples of decompositions of pseudovarieties as intersections of smi pseudovarieties. In the table, $S_{ab\neq 0}$ denotes the four element semigroup with the presentation $\langle a, b \mid a^2 = b^2 = ba = 0 \rangle$.

Notice that all of the pseudovarieties on the left-hand side of an equation in Table 7.3. are not sfmi (and hence not fmi). We already saw this was the case for **A**. Replacing $\bigcap_{p \text{ prime}} \mathsf{Excl}(\mathbb{Z}_p)$ by **A** and $\bigcap_{q\neq p \text{ prime}} \mathsf{Excl}(\mathbb{Z}_q)$ by $\overline{\mathbf{G}}_p$, we obtain the remaining pseudovarieties as finite intersections.

**Table 7.3.** Some smi decompositions

$$\mathbf{A} = \bigcap_{p \text{ prime}} \mathsf{Excl}(\mathbb{Z}_p)$$

$$\mathbf{Sl} = \mathsf{Excl}(N_2) \cap \mathsf{Excl}(2^\ell) \cap \mathsf{Excl}(\overline{2}) \cap \left( \bigcap_{p \text{ prime}} \mathsf{Excl}(\mathbb{Z}_p) \right)$$

$$\mathbf{LZ} = \mathsf{Excl}(N_2) \cap \mathsf{Excl}(U_1) \cap \mathsf{Excl}(\overline{2}) \cap \left( \bigcap_{p \text{ prime}} \mathsf{Excl}(\mathbb{Z}_p) \right)$$

$$\mathbf{G}_p = \mathsf{Excl}(N_2) \cap \mathsf{Excl}(2^\ell) \cap \mathsf{Excl}(\overline{2}) \cap \mathsf{Excl}(U_1) \cap \bigcap_{q\neq p \text{ prime}} \mathsf{Excl}(\mathbb{Z}_q)$$

$$\mathbf{N} = \mathsf{Excl}(U_1) \cap \mathsf{Excl}(2^\ell) \cap \mathsf{Excl}(\overline{2}) \cap \left( \bigcap_{p \text{ prime}} \mathsf{Excl}(\mathbb{Z}_p) \right)$$

$$\mathbf{N}_2 = \mathsf{Excl}(U_1) \cap \mathsf{Excl}(2^\ell) \cap \mathsf{Excl}(\overline{2}) \cap \left( \bigcap_{p \text{ prime}} \mathsf{Excl}(\mathbb{Z}_p) \right)$$
$$\cap \left( \bigcap \mathsf{MaxExclude}(S_{ab\neq 0}; \mathbf{N}_2^\uparrow) \right)$$

We have yet to give an example of an smi pseudovariety that is not mi. Let us rectify this situation. The authors have recently shown, inspired by a suggestion of M. V. Sapir, that if $w_1, w_2 \in \{a, b\}^+$ have equal length and use both letters, then $[\![w_1 = w_2]\!]$ is smi; in particular, **Com** is smi. Because mi pseudovarieties do not satisfy identities (as $\mathbf{G} \cap \mathbf{A} = \mathbf{1}$ implies any mi pseudovariety contains $\mathbf{G}$ or $\mathbf{A}$), none of these pseudovarieties are mi. These results will appear in a forthcoming article. Let us deal with **Com** here.

**Theorem 7.1.26.** *The pseudovariety* **Com** *of commutative semigroups is smi but not mi. Its unique cover is* $\mathbf{Com} \vee [\![xyz = 0]\!]$.

*Proof.* By Proposition 7.1.13, it suffices to show that $\mathbf{Com} \vee [\![xyz = 0]\!]$ is the unique cover of **Com**. So let $\mathbf{Com} < \mathbf{V}$. Then there is a semigroup $S \in \mathbf{V}$ and elements $s, t \in S$ such that $st \neq ts$. We may assume without loss of generality that $s, t$ in fact generate $S$. Let $F$ be a free commutative 3-nilpotent semigroup on $\{a, b\}$; so $F = \{a, b, a^2, b^2, ab = ba, 0\}$ and all products of length 3 vanish. Then consider the subsemigroup $T$ of $F \times S$ generated by $(a, s)$ and $(b, t)$. The set $I = (\{0\} \times S) \cap T$ is an ideal of $T$ and we claim $T/I \in \mathbf{V}$ is a free 3-nilpotent semigroup on two generators. Indeed, any product of length 3 is 0 in $T/I$. By considering the $F$-coordinate, we see that $(a, s), (b, t), (a, s)^2, (b, t)^2, 0$ are distinct and none of them are $(a, s)(b, t)$ or $(b, t)(a, s)$. On the other hand, the $S$-coordinate distinguishes these latter two elements. The result now follows because $[\![xyz = 0]\!]$ is clearly generated by the free object on two generators as any non-zero element of a free 3-nilpotent semigroup is represented by a word with at most two letters.                                                □

It is important to find other examples of smi pseudovarieties. Do there exist compact ones?

*Question 7.1.27.* Give more examples of smi, fmi, sfmi pseudovarieties that are not mi. In particular, can a compact pseudovariety be smi? This is related to the existence of atoms in $\mathbf{Cnt}(\mathbf{PV})^+$.

In general, we shall be more interested in joins than in meets for both **PV** and **PVRM** because meets in this context are easy to determine. However, there are some interesting examples. A pseudovariety is called *monoidal* if it is generated by monoids. The results of [364] show that if **W** is a pseudovariety of semigroups then

$$\ell 1_D \mathsf{q}(\mathbf{W}) \leq \bigwedge \{\mathbf{V} * \mathbf{W} \mid \mathbf{V} \neq \mathbf{1} \text{ is a monoidal pseudovariety}\}.$$

## 7.2 Locally Dually Algebraic Lattices

If $(L, \leq, \wedge, \vee)$ is a (complete) lattice, then its *dual lattice* $L^{op}$ is the (complete) lattice $(L, \geq, \vee, \wedge)$ obtained by reversing the order. The goal of this section is to show that the dual of Birkhoff's Subdirect Representation Theorem holds for **PV**. Because **PV**$^{op}$ is not algebraic, there is work to be done. This will necessitate discussing infinite semigroups and Birkhoff varieties so we temporarily drop our implicit assumption of finiteness.

The concept of algebraic lattice is not self-dual and so we are led to the following definition.

**Definition 7.2.1 (Dually algebraic lattice).** *We shall call an algebraic lattice $L$ dually algebraic if $L^{op}$ is also an algebraic lattice.*

The key example of a dually algebraic lattice for us will be the interval $[\mathbf{1}, \mathbf{V}]$ for a locally finite pseudovariety $\mathbf{V}$. To prove this, first we must discuss Birkhoff varieties of semigroups. Recall that a (Birkhoff) *variety of semigroups* is a class of semigroups closed under taking arbitrary products, subsemigroups and quotients. The set $\mathfrak{Var}$ of all semigroup varieties is a lattice, being the set of closed subsets for the closure operator $\mathbb{HSP}$ on **Sgp**, but it is not an algebraic lattice. Indeed, it is not even meet-continuous and hence cannot be a continuous, let alone algebraic, lattice by Proposition 6.2.11.

**Proposition 7.2.2.** *The lattice $\mathfrak{Var}$ of semigroup varieties is not meet-continuous and hence neither algebraic nor continuous.*

*Proof.* Let $\mathfrak{Com}$ be the variety of commutative semigroups and, for $n \in \mathbb{N}$, let $\mathfrak{Ab}(n)$ be the variety of abelian groups of exponent dividing $n$. Then the set $D = \{\mathfrak{Ab}(n) \mid n \text{ is odd}\}$ is directed and $\bigvee D = \mathfrak{Com}$ as $\mathbb{N} \ll \mathbb{Z}_3 \times \mathbb{Z}_5 \times \cdots$. In particular, $\mathfrak{Ab}(2) \cap \bigvee D = \mathfrak{Ab}(2)$, but $\bigvee_{n \text{ odd}}(\mathfrak{Ab}(2) \cap \mathfrak{Ab}(n)) = \mathbf{1}$ where $\mathbf{1}$ is the trivial variety.    □

It turns out however that $\mathfrak{Var}^{op}$ *is* an algebraic lattice. To see this, we must discuss the varietal equational theory. References are [46, 61]. Let $X$ be a countable set. Then an *identity* or *equation* over $X$ is a formal equality $u = v$ between words $u, v \in X^+$ or equivalently an element of $X^+ \times X^+$ (we abuse the distinction between such). As usual, a semigroup $S$ *satisfies* $u = v$, written $S \models u = v$, if whenever $\sigma \colon X \to S$ is a substitution, then $u\sigma = v\sigma$ where $\sigma$ is also used for the unique extension of $\sigma$ to $X^+$. If $E$ is a set of identities, then $S \models E$ if $S$ satisfies every identity in $E$. One can associate to $E$ the variety $[E] = \{S \in \mathbf{Sgp} \mid S \models E\}$. Conversely, given a variety $\mathfrak{V}$, we can associate to it the set

$$\mathsf{Eq}(\mathfrak{V}) = \{u = v \in X^+ \times X^+ \mid \forall S \in \mathfrak{V},\ S \models u = v\}.$$

It is a well-known theorem of Birkhoff [61] that

$$\mathfrak{V} \subseteq \mathfrak{W} \iff \mathsf{Eq}(\mathfrak{V}) \supseteq \mathsf{Eq}(\mathfrak{W}). \tag{7.3}$$

Thus a variety is determined by the equations it satisfies. The reversal of inequalities in (7.3) shows us that the dual lattice to the lattice $\mathfrak{Var}$ of varieties should be the set of equationally closed subsets of $X^+ \times X^+$.

More precisely, if $E$ is a set of equations, then $u = v \in X^+ \times X^+$ is a *consequence* of $E$ if, for all semigroups $S$ such that $S \models E$, also $S \models u = v$. If this is the case, we write $E \models u = v$. The completeness theorem of equational logic [7, Thm. 1.4.6] says that $E \models u = v$ if and only if there is a derivation of $u = v$ from $E$. The compactness theorem of equational logic says that if $E$ is an infinite set and $E \models u = v$, then there is a finite subset $E_0$ of $E$ such that $E_0 \models u = v$ (just take a derivation of $u = v$ from $E$ and let $E_0$ be the equations used in the derivation). We can define a closure operator $C$ on $2^{X^+ \times X^+}$ by the formula

$$C(E) = \{u = v \in X^+ \times X^+ \mid E \models u = v\}.$$

The closed subsets under $C$ are called *equationally closed*; the lattice of such subsets is denoted $\mathfrak{Eq}$. We claim that $C$ is continuous. To see this it suffices to check that the equationally closed subsets are closed under directed unions. But this is clear from the compactness of equational logic: if $u = v$ is a consequence of $\bigcup D$ with $D$ a directed set of equationally closed subsets, then by compactness $u = v$ is a consequence of some finite set of equations from $\bigcup D$ and hence is a consequence of some element $d \in D$. Thus $u = v \in d$ and so $u = v \in \bigcup D$. We conclude that $C$ is continuous and hence $\mathfrak{Eq}$ is an algebraic lattice by Theorem 6.3.15.

It is clear from the definition that if $E$ is a set of equations, then $C(E)$ is the largest set of equations defining the variety $[E]$. Also, it is immediate from the definitions that $\mathsf{Eq}(\mathfrak{V})$, for a variety $\mathfrak{V}$, is equationally closed. It now follows easily from (7.3) that sets of the form $\mathsf{Eq}(\mathfrak{V})$ are precisely the equationally closed sets and that $\mathfrak{Eq} \cong \mathfrak{Var}^{op}$.

An advantage of working with the dual of $\mathfrak{Var}$ is that equations are relatively easy to work with. Also, for varieties the meet operation is simple (intersection) whereas the join operation is complicated. But for $\mathfrak{Eq}$ the situation reverses, namely:

$$\mathsf{Eq}(\mathfrak{V} \vee \mathfrak{W}) = \mathsf{Eq}(\mathfrak{V}) \cap \mathsf{Eq}(\mathfrak{W})$$
$$\mathsf{Eq}(\mathfrak{V} \wedge \mathfrak{W}) = C(\mathsf{Eq}(\mathfrak{V}) \cup \mathsf{Eq}(\mathfrak{W})).$$

We now verify that $\mathbf{PV}$ is not join-continuous and hence $\mathbf{PV}^{op}$ cannot even be a continuous lattice, let alone an algebraic lattice.

**Theorem 7.2.3.** *The lattice $\mathbf{PV}$ is not join-continuous and hence its dual is not an algebraic lattice. In fact, $\mathbf{PV}^{op}$ is not a continuous lattice.*

*Proof.* By the dual to Proposition 6.2.11, if $\mathbf{PV}^{op}$ were a continuous lattice, then $\mathbf{PV}$ would be join-continuous. We prove that it is not. Let $p_1, p_2, \ldots$ be an enumeration of the primes. Let $\mathbf{G}_{nil,i}$ be the pseudovariety of nilpotent groups with order prime to $\{p_1, \ldots, p_i\}$. Then

$$\mathbf{G}_{nil,1} > \mathbf{G}_{nil,2} > \cdots > \bigwedge_{i \in \mathbb{N}} \mathbf{G}_{nil,i} = \mathbf{1}.$$

Let **ACom** be the pseudovariety of aperiodic commutative semigroups. Then it is well-known that $\mathbf{ACom} \vee \mathbf{G}_p \geq \mathbf{N}$ for any prime $p$ [2]. Indeed, we show that if $A$ is any finite set, then the free nilpotent semigroup on $A$ of index $n$, denoted $F_{\mathbf{N}_n}(A)$, belongs to $\mathbf{ACom} \vee \mathbf{G}_p$. A nilpotent semigroup $N$ is said to have index $n$ if $N^n = 0$, but $N^{n-1} \neq 0$. The pseudovariety $\mathbf{N}_n$ consists of all nilpotent semigroups of index at most $n$. Because the free group on $A$ is residually a $p$-group [184, 217], there is an $A$-generated group $G \in \mathbf{G}_p$ such that any two words in $A$ of length $n-1$ are sent to distinct elements of $G$. Let $C_{n,1}$ be the cyclic semigroup $\langle x \mid x^n = x^{n+1} \rangle$ and consider the subsemigroup $S$ of $C_{n,1} \times G$ consisting of all pairs $(x^{|w|}, [w]_G)$ with $w \in A^+$. It is easily verified

$$S/((\{x^n\} \times G) \cap S) \cong F_{\mathbf{N}_n}(A)$$

showing (as $n$ is arbitrary) that $\mathbf{N} \leq \mathbf{ACom} \vee \mathbf{G}_p$. Because $\mathbf{G}_{nil,i}$ contains $\mathbf{G}_{p_{i+1}}$, we conclude that

$$\mathbf{N} \leq \bigwedge_{i \in \mathbb{N}} (\mathbf{ACom} \vee \mathbf{G}_{nil,i}).$$

On the other hand **N** contains non-commutative semigroups, so

$$\mathbf{N} \nleq \mathbf{ACom} = \mathbf{ACom} \vee \bigwedge_{i \in \mathbb{N}} \mathbf{G}_{nil,i}.$$

This proves that **PV** is not join-continuous, as desired.    □

We remark that similarly, the semidirect product operator $\mathbf{V} * (-)$ does not preserve downwards directed infs.

**Proposition 7.2.4.** *The operator* $\mathbf{Sl} * (-)$ *does not preserve downwards directed infs.*

*Proof.* Let $\mathbf{G}_{nil,i}$ be as above. It is well-known that, for any non-trivial pseudovariety of groups $\mathbf{H}$, $B_2^I \in \mathbf{Sl} * \mathbf{H}$. Hence $B_2^I \in \bigwedge_{i \in \mathbb{N}}(\mathbf{Sl} * \mathbf{G}_{nil,i})$. On the other hand,

$$B_2^I \notin \mathbf{Sl} * \mathbf{1} = \mathbf{Sl} * \bigwedge \mathbf{G}_{nil,i}$$

because $\mathbf{Sl} * \mathbf{1} \subseteq \mathbb{L}\mathbf{Sl}$ (see Corollary 4.1.32) and $B_2^I$ is not a semilattice.    □

**Exercise 7.2.5.** Prove that $B_2^I \in \mathbf{Sl} * \mathbf{H}$ for any non-trivial pseudovariety of groups $\mathbf{H}$.

Now we try and discuss why the dual of **PV** is not an algebraic lattice and how to "fix" the problem. Let $X$ be a countable set and $\widehat{X^+}$ be the free profinite semigroup on $X$. A *pseudoidentity* is, as usual, a formal equality $u = v$ of elements of $\widehat{X^+}$ or equivalently an element of $\widehat{X^+} \times \widehat{X^+}$. One can define, analogously to above, the definition of satisfaction and consequences of pseudoidentities. A pseudoidentity $u = v$ is a consequence of a set of pseudoidentities $E$ if $S \models E$ implies $S \models u = v$ for every finite semigroup $S$. Also we can associate to each set of pseudoidentities $\mathcal{E}$ the pseudovariety $[\![\mathcal{E}]\!]$ of semigroups satisfying $\mathcal{E}$. Conversely, to each pseudovariety $\mathbf{V}$ we can associate the set $\mathsf{Eq}(\mathbf{V})$ of pseudoidentities satisfied by all elements of $\mathbf{V}$. Reiterman's Theorem then yields, in analogy to (7.3):

$$\mathbf{V} \subseteq \mathbf{W} \iff \mathsf{Eq}(\mathbf{V}) \supseteq \mathsf{Eq}(\mathbf{W}). \tag{7.4}$$

Again one can define a closure operator $\overline{C}$ on $2^{\widehat{X^+} \times \widehat{X^+}}$ by closing under consequences and one obtains that $\mathbf{PV}$ is dual to the lattice $\mathbf{Eq}$ of equationally closed sets of pseudoidentities. The difference here is that there is *no* compactness theorem for pseudoidentities. It is not true that every consequence of an infinite set of pseudoidentities is a consequence of a finite subset. Indeed, $x^\omega = x^{\omega+1}$ is a consequence of the set of pseudoidentities $\mathcal{E} = \{x^{(p')^\omega} = x^\omega \mid p \text{ prime}\}$, but it is not a consequence of any finite subset of $\mathcal{E}$[1]. There are no derivations in the context of pseudoidentities. Because of the topology on $\widehat{X^+} \times \widehat{X^+}$, there is a limiting process involved in obtaining consequences. (The closest there is to compactness in this context is that every pseudovariety can be defined by pseudoidentities where each side involves only elements of finite support (content in the terminology of [7]).) Thus directed unions of equationally closed subsets need not be equationally closed and this is why we do not have join-continuity for $\mathbf{PV}$.

Locally speaking, however $\mathbf{PV}$ is a dually algebraic lattice. To see what we mean by this, we remark there is a natural map from $\mathfrak{Var}$ to $\mathbf{PV}$. Given a variety $\mathfrak{V}$, define $\mathfrak{V}^f$ to be the *finite trace* of $\mathfrak{V}$, that is, the finite members of $\mathfrak{V}$. Clearly, $\mathfrak{V}^f$ is a pseudovariety and the map $\mathfrak{V} \mapsto \mathfrak{V}^f$ is **inf**. The left adjoint is given by $\mathbf{V} \mapsto \mathbb{HSP}(\mathbf{V})$ (where $\mathbb{HSP}(\mathbf{V})$ is the variety generated by $\mathbf{V}$). In terms of the duals, the map $\mathfrak{V} \mapsto \mathfrak{V}^f$ corresponds to the inclusion of $\mathfrak{Eq}$ into $\mathbf{Eq}$. More informally, if $\mathfrak{V} = [E]$, then $\mathfrak{V}^f = [\![E]\!]$ where we view the set of identities $E$ as a set of pseudoidentities. The map $\mathbf{V} \mapsto \mathbb{HSP}(\mathbf{V})$ corresponds to the map that takes an equationally closed set of pseudoidentities to its subset of identities (that is, the intersection with $2^{X^+ \times X^+}$). The map $\mathfrak{V} \mapsto \mathfrak{V}^f$ is *not* continuous because $\mathbf{Com} = \left( \bigvee_{n \in \mathbb{N}} \mathfrak{Ab}(n) \right)^f$ whereas $\mathbf{Ab} = \bigvee_{n \in \mathbb{N}} \mathfrak{Ab}(n)^f$.

Not every pseudovariety is the finite trace of a variety. For instance, $\mathbf{G}$, $\mathbf{A}$ and $\mathbf{N}$ all satisfy no non-trivial identities and so any one of them generates the variety of all semigroups. Thus the map $\mathbf{V} \mapsto \mathbb{HSP}(\mathbf{V})$ is not injective and so the finite trace map is not surjective. The image of the finite trace map is precisely the class of so-called *equational pseudovarieties*. In summary we have a Galois connection

$$\mathbf{V} \underset{\mathbb{HSP}}{\overset{(-)^f}{\rightleftarrows}} \mathfrak{Var}$$

---

[1] This example was pointed out to us by K. Auinger.

in which both maps are neither injective nor surjective.

It is not even the case that the map $\mathfrak{V} \mapsto \mathfrak{V}^f$ preserves finite joins. The problem is that a finite semigroup may divide the direct product of infinite semigroups from the two factors. For instance, in [2] a finite set of identities $E$ is constructed so that the pseudovariety $\mathbf{V} = \mathbf{Com} \vee [\![E]\!]$ does not have decidable membership problem. Suppose taking finite traces preserved joins. Since $\mathbf{Com} = \mathfrak{Com}^f$, where $\mathfrak{Com}$ denotes the variety of commutative semigroups, and $[\![E]\!] = [E]^f$, we would then have $\mathbf{V} = [\![\mathscr{E}]\!]$ where $\mathscr{E} = \mathsf{Eq}(\mathfrak{Com}) \cap \mathsf{Eq}([E])$. Clearly, $\mathscr{E}$ is recursively enumerable because the varietal equational theory of a finite set of equations is recursively enumerable by the completeness of equational logic (as opposed to the case of the pseudovarietal equational theory of $[\![E]\!]$, which is proved not to be recursively enumerable in [2]). Hence the set of finite semigroups not belonging to $[\![\mathscr{E}]\!]$ is recursively enumerable (if a finite semigroup does not belong to $[\![\mathscr{E}]\!]$, we will eventually discover this by finding an identity of $\mathscr{E}$ that it fails cf. [19, 20]). On the other hand, because $[\![E]\!]$ is decidable as a pseudovariety of semigroups, as is $\mathbf{Com}$, the pseudovariety $\mathbf{Com} \vee [\![E]\!]$ is clearly recursively enumerable. This shows that if $\mathbf{Com} \vee [\![E]\!] = [\![\mathscr{E}]\!]$, then it has decidable membership, a contradiction to the results of [2]. We conclude that $\mathbf{Com} \vee [\![E]\!] \neq [\![\mathscr{E}]\!]$ and hence the operation of taking finite traces does not commute with the join operation.

There is a setting, though, in which things work. A variety $\mathfrak{V}$ is called *locally finite* if every finitely generated semigroup in $\mathfrak{V}$ is finite. In this case, if $A$ is a finite set, the relatively free semigroup $F_{\mathfrak{V}}(A)$ generated by $A$ in $\mathfrak{V}$ is finite. Notice that the finite trace of a locally finite variety is locally finite in the sense defined earlier for pseudovarieties. Conversely, if $\mathbf{V}$ is a locally finite pseudovariety, then $\mathbb{HSP}(\mathbf{V})$ is a locally finite variety and $\widehat{F_{\mathbf{V}}}(A) = F_{\mathbb{HSP}(\mathbf{V})}(A)$ for finite sets $A$. Hence $\mathbb{HSP}(\mathbf{V})^f = \mathbf{V}$.

**Exercise 7.2.6.** Show that a pseudovariety $\mathbf{V}$ is locally finite if and only if $\mathbb{HSP}(\mathbf{V})$ is locally finite and that in this case $\mathbb{HSP}(\mathbf{V})^f = \mathbf{V}$.

If $\mathfrak{V}$ and $\mathfrak{W}$ are varieties and $A$ is a set, then $F_{\mathfrak{V} \vee \mathfrak{W}}(A)$ is the subdirect product of $F_{\mathfrak{V}}(A) \times F_{\mathfrak{W}}(A)$ generated by the diagonal embedding of $A$. Hence, the collection of locally finite varieties is closed under finite joins. It is, in fact, an ideal in $\mathfrak{Var}$ because any subvariety of a locally finite variety is obviously locally finite. Similarly, the collection of locally finite pseudovarieties is an ideal in the lattice **PV**. The discussion above shows that the finite trace maps the ideal of locally finite varieties isomorphically to the ideal of locally finite pseudovarieties. If $S$ is a finite semigroup, $(S)$ is locally finite. Thus every compact element of **PV** belongs to the ideal of locally finite pseudovarieties.

**Exercise 7.2.7.** Show that the finite trace restricts to an order isomorphism between the ideals of locally finite varieties and locally finite pseudovarieties.

Let $\mathfrak{V}$ be a locally finite variety and $\mathbf{V} = \mathfrak{V}^f$ be the finite trace. Then $[\mathbf{1}, \mathbf{V}] \cong [\mathbf{1}, \mathfrak{V}]$ where $\mathbf{1}$ denotes the trivial variety. Exercise 6.1.2 shows that

$[\mathbf{1}, \mathbf{V}]$ is algebraic, being an interval in the algebraic lattice $\mathbf{PV}$. On the other hand, Birkhoff's Theorem, applied to $\mathfrak{V}$, shows that $[\mathbf{1}, \mathfrak{V}]^{op}$ is an algebraic lattice isomorphic to the interval $[\mathsf{Eq}(\mathfrak{V}), 2^{X^+ \times X^+}]$ of $\mathfrak{Eq}$ (or equivalently, one can do the equational theory relative to $\mathfrak{V}$ by using identities over $F_{\mathfrak{V}}(X) \times F_{\mathfrak{V}}(X)$). This discussion establishes that if $\mathbf{V}$ is a locally finite pseudovariety, then the interval $[\mathbf{1}, \mathbf{V}]$ is a dually algebraic lattice. In particular, for any compact element $c \in \mathbf{PV}$, the interval $[\mathbf{1}, c]$ is a dually algebraic lattice. This leads to the following notion.

**Definition 7.2.8 (Locally dually algebraic lattice).** *An algebraic lattice $L$ is called locally dually algebraic if, for each compact element $c \in L$, the interval $[\mathsf{B}, c]$ is a dually algebraic lattice.*

By Exercise 6.1.2, the interval $[\mathsf{B}, c]$ is algebraic in any algebraic lattice, so the dual statement is what need not hold. Of course a dually algebraic lattice is automatically a locally dually algebraic lattice.

**Proposition 7.2.9.** *The lattices $\mathbf{PV}$ and $\mathbf{PVRM}$ are locally dually algebraic lattices.*

*Proof.* The argument has already been given for $\mathbf{PV}$. The argument for $\mathbf{PVRM}$ is similar, but one uses the equational theory from Chapter 3 and the analogous theory for varieties of relational morphisms (defined in the obvious way). The details are left to the reader. □

Being a locally dually algebraic lattice is strong enough to obtain the dual version of Birkhoff's Subdirect Representation Theorem.

**Theorem 7.2.10 (Subdirect Representation Theorem for Joins).** *Let $L$ be a locally dually algebraic lattice. Then each element of $L$ is a join of a set of strictly join irreducible elements.*

*Proof.* Let $\ell \in L$. Then $\ell = \bigvee \{c \in K(L) \mid c \leq \ell\}$. Hence it suffices to consider the case that $\ell = c$ with $c$ compact. Then, because $[\mathsf{B}, c]$ is dually algebraic, Birkhoff's Subdirect Representation Theorem (Theorem 7.1.6) applied to $[\mathsf{B}, c]^{op}$ shows $c$ can be written as a join of a set of sji elements of $[\mathsf{B}, c]$. But it is easy to check that any sji element in $[\mathsf{B}, c]$ is also sji in $L$. This establishes the result. □

**Exercise 7.2.11.** Verify that if $L$ is a complete lattice and $\ell \in L$, then an element $\ell' \leq \ell$ is sji in $[\mathsf{B}, \ell]$ if and only if it is sji in $L$.

If $L$ is a locally dually algebraic lattice, $c \in K(L)$ and $k$ is a co-compact element of $[\mathsf{B}, c]$ with $k \neq c$, then we define

$$\mathsf{Exclude}(k; c^{\downarrow}) = \{\ell \in L \mid k \not\geq \ell, \ \ell \leq c\}. \tag{7.5}$$

This is the dual of (7.1). For a locally dually algebraic lattice, $\mathsf{Exclude}(k; c^{\downarrow})$ contains minimal elements, and any such element is sji, by the dual of

Lemma 7.1.1. We use $\mathsf{MinExclude}(k; c^{\downarrow})$ for the set of minimal elements. In the case of **PV** we have that $c$ is generated by a finite semigroup $S$ and $k$ can be taken as the pseudovariety corresponding to an equationally closed class in $[\mathsf{Eq}(\mathbb{HSP}(S)), 2^{X^+ \times X^+}]$ generated by a finite set of equations $E$. Thus we use the notation $\mathsf{Exclude}(E; S)$ and $\mathsf{MinExclude}(E; S)$ in this context. Notice that $\mathsf{MinExclude}(E; S) = \bigcup_{e \in E} \mathsf{MinExclude}(e; S)$. With this notation, we have the following dual result to Proposition 7.1.13.

**Proposition 7.2.12.** *Let $L$ be locally dually algebraic lattice. Then the following are equivalent for $\ell \in L$:*

1. *$\ell$ is sji;*
2. *$\ell$ is compact and for some co-compact element $k$ of $[\mathsf{B}, \ell]$, we have that $\ell$ is the unique element of $\mathsf{MinExclude}(k; \ell^{\downarrow})$;*
3. *$\ell$ is compact and there is a unique element $m \leq \ell$ such that $m \nearrow \ell$.*

*Proof.* Exercise 6.1.6 shows that $\ell$ must be compact in order to have a chance of being sji. So assume $\ell$ is compact. It is then easy to see that 1–3 hold in $L$ if and only if they hold in $[\mathsf{B}, \ell]$. But $[\mathsf{B}, \ell]$ is a dually algebraic lattice and so the equivalence follows from the dual of Proposition 7.1.13.     □

We recall that compact elements always cover some element (Proposition 6.1.3). The above proposition has the following reformulation for **PV**.

**Proposition 7.2.13.** *Let **V** be a pseudovariety of semigroups. Then the following are equivalent:*

1. ***V** is sji;*
2. ***V** $= (S)$ for a semigroup $S$ and there exists an equation $e$ such that **V** is the unique element of $\mathsf{MinExclude}(e; S)$;*
3. ***V** is finitely generated and has a unique maximal proper pseudovariety.*

*If **W** is the unique maximal proper pseudovariety, then the equation $e$ can be taken to be any equation satisfied by **W** but not by **V**. In this case, **W** $=$ **V** $\cap \llbracket e \rrbracket$.*

We now present another description of the unique maximal proper subpseudovariety.

**Proposition 7.2.14.** *Let **V** be an sji pseudovariety. Then its unique maximal proper subpseudovariety is the set*

$$\{T \in \mathbf{Fsgp} \mid (T) < \mathbf{V}\}. \tag{7.6}$$

*Proof.* Let **W** be the unique maximal proper subpseudovariety of **V**. If $T$ belongs to (7.6), then clearly $T \in \mathbf{W}$. If $T$ does not belong to (7.6), then either $T \notin \mathbf{V}$ and hence $T \notin \mathbf{W}$, or $(T) = \mathbf{V}$, in which case again $T \notin \mathbf{W}$.     □

For an fji semigroup, the unique maximal proper subpseudovariety admits a particularly nice description.

**Proposition 7.2.15.** *Suppose that $S$ is an fji semigroup. Then the unique maximal proper subpseudovariety of $(S)$ is $(S) \cap \mathsf{Excl}(S)$. In particular, it is decidable.*

*Proof.* Every proper pseudovariety of $(S)$ is contained in $\mathsf{Excl}(S)$, from which the first statement of the proposition is obvious. A semigroup $T \in (S) \cap \mathsf{Excl}(S)$ if and only if $T \in (S)$, but $S \notin (T)$. This is decidable by a well-known result Birkhoff. $\qquad\square$

Compactness turns out to be necessary for an sfji pseudovariety to have a maximal proper subpseudovariety.

**Proposition 7.2.16.** *Let $L$ be an algebraic lattice and suppose $\ell \in L$ is not compact. If $\ell$ covers some element of $L$, then $\ell$ is not sfji.*

*Proof.* Suppose that $\ell$ is sfji and covers an element $\bar{\ell}$ of $L$. We show that $\ell$ is compact. Because $\bar{\ell} < \ell$ there is a compact element $c$ with $c \leq \ell$ and $c \not\leq \bar{\ell}$. Hence $\ell = \bar{\ell} \vee c$. Because $\ell$ is sfji, either $\ell = \bar{\ell}$ or $\ell = c$. Because the first case is impossible, we conclude that $\ell = c$ and so $\ell$ is compact. $\qquad\square$

Showing that a non-compact pseudovariety is sfji is then a good way to show that it has no maximal proper subpseudovariety. This principle has been exploited by Margolis, Sapir and Weil [195]. We state it as a corollary to Proposition 7.2.16, although it is really a reformulation in a special case.

**Corollary 7.2.17.** *Let $\mathbf{V}$ be an sfji pseudovariety that is not finitely generated. Then $\mathbf{V}$ has no maximal proper subpseudovariety.*

We shall later need the following variant of Lemma 6.1.13 to show that certain pseudovarieties are fji.

**Lemma 7.2.18.** *Let $\mathbf{V}$ be a pseudovariety such that, for each $S \in \mathbf{V}$, there exists $S' \in \mathbf{V}$ such that:*

*1. $S \in (S')$*
*2. $S' \in \mathbf{V}_1 \vee \mathbf{V}_2$ implies $S \in \mathbf{V}_1$ or $S \in \mathbf{V}_2$.*

*Then $\mathbf{V}$ is fji.*

*Proof.* Let $S_0, S_1, S_2, \ldots$ be an enumeration of the elements of $\mathbf{V}$ with $S_0$ trivial. Set $T_0 = S_0$ and define, for $i > 0$, $T_i = (T_{i-1} \times S_i)'$. By 1, we have

$$\forall i, \ T_i \in (T_{i+1}) \text{ and } S_i \in (T_i). \tag{7.7}$$

Suppose $\mathbf{V} \leq \mathbf{V}_1 \vee \mathbf{V}_2$ and $\mathbf{V} \not\leq \mathbf{V}_1$. Then $S_i \notin \mathbf{V}_1$ for some $i$. Let $j \geq i$; then $S_i \in (T_j)$ by (7.7), so $T_j \notin \mathbf{V}_1$. Because $T_j \in \mathbf{V}_1 \vee \mathbf{V}_2$, we deduce from 2 that $T_{j-1} \times S_j \in \mathbf{V}_2$, $j \geq i$. Hence $T_k \in \mathbf{V}_2$ for $k \geq i - 1$. We conclude, using (7.7), that $\mathbf{V} \leq \mathbf{V}_2$, as desired. $\qquad\square$

Notice that in 2, $S'$ appears on the left of the implication and $S$ appears on the right.

## 7.3 A Brief Survey of Join Irreducibility

This section gives a brief survey of ji, fji, sji and sfji pseudovarieties. First we consider some elementary examples of fji semigroups. Some of these examples are new. This is followed by a survey of some results for non-compact pseudovarieties. In a subsequent section, we present a highly non-trivial class of fji semigroups called Kovács-Newman semigroups [287].

### 7.3.1 First examples of fji semigroups

The reader should recall that in an algebraic lattice an element is ji if and only if it is fji and sji, if and only if it is compact and fji; see Exercise 6.1.6. Throughout this section, we shall use extensively Remark 7.1.12 without explicit reference.

We already saw that the atoms $\mathbb{Z}_p$, $N_2$, $\overline{\mathbf{2}}$, $\mathbf{2}^\ell$ and $U_1$ are fji, and determined their exclusion pseudovarieties in Table 7.2. Recall that a semigroup $S$ is $\rtimes$-*prime* if $S \prec T_1 \rtimes T_2$ implies $S \prec T_i$, some $i$. Of course a $\rtimes$-prime semigroup is fji. According to Theorem 4.1.45 the $\rtimes$-prime semigroups are the simple groups and the divisors of $U_2$.

*Example 7.3.1 ($U_2$).* In Section 4.8, we saw that excluding $U_2$ yields the pseudovariety $\mathbb{LER}$. This latter pseudovariety is mi and must be definable by a single pseudoidentity in three variables according to the general theory. In fact

$$\mathbb{LER} = [\![ ((z^\omega x z^\omega)^\omega (z^\omega y z^\omega)^\omega)^\omega (z^\omega x z^\omega)^\omega = ((z^\omega x z^\omega)^\omega (z^\omega y z^\omega)^\omega)^\omega ]\!].$$

This pseudoidentity is clearly satisfied by $\mathbb{LER}$. Conversely, it is failed by $U_2$, as is seen by taking $z = I$ and $x, y$ to be the two right zeroes.

The unique maximal proper subpseudovariety of $(U_2)$ is the pseudovariety $(U_2) \cap \mathsf{Excl}(U_2) = (U_2) \cap \mathbb{LER}$. Now $U_2$ is an $\mathscr{L}$-trivial band, so any element of this intersection is a locally $\mathscr{R}$-trivial, $\mathscr{L}$-trivial band and hence locally a semilattice. So any such semigroup satisfies the identities $x^2 = x$, $yxy = xy$ and $zxzyz = zyzxz$. Consequently, $(U_2) \cap \mathbb{LER} \leq [\![ x^2 = x, xyz = yxz ]\!]$. On the other hand, if $A$ is a finite alphabet, then the free object $F$ on $A$ in $[\![ x^2 = x, xyz = yxz ]\!]$ has the property that any two words over $A^+$ with the same content (i.e., letters) and the same last letter are equal in $F$. Hence $F \ll U_1^A \times \overline{A}$ and so $F \in \mathbf{Sl} \vee \mathbf{RZ}$. It follows

$$(U_2) \cap \mathbb{LER} \leq [\![ x^2 = x, xyz = yxz ]\!] \leq \mathbf{Sl} \vee \mathbf{RZ} \leq (U_2) \cap \mathbb{LER}$$

and so $\mathbf{Sl} \vee \mathbf{RZ}$ is the unique maximal proper subpseudovariety of $U_2$. The Hasse diagram for the interval $[\mathbf{1}, (U_2)]$ can be found in Figure 7.1.

In fact, $(U_2)$ is precisely the pseudovariety $\mathbf{RRB}$ of $\mathscr{L}$-trivial bands (also called *right regular bands*), as the following well-known proposition shows.

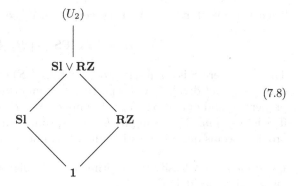

Fig. 7.1. Hasse diagram for the interval $[\mathbf{1}, (U_2)]$

**Proposition 7.3.2.** *The equality* $(U_2) = \mathbf{RRB}$ *holds.*

*Proof.* Let $F$ be the free right regular band on the finite alphabet $A$. It is easy to see using the identities $x^2 = x$ and $yxy = xy$ that $F$ consists of all non-empty linear words over $A$ (linear here means no repeated letters). Multiplication in $F$ is given by $u \cdot v = (uv)\rho$ where $\rho \colon A^+ \to A^+$ is the function that takes a word $w \in A^+$ to the linear word obtained by removing all repetitions as one scans $w$ from *right to left*. Details can be found, for instance, in [57, 58]. So to show $(U_2) = \mathbf{RRB}$, it suffices to show that if $u, v \in A^+$ are distinct linear words, then there is a homomorphism $\varphi \colon A^+ \to U_2$ separating $u$ and $v$, because then it will follow $F \in (U_2)$. Suppose $u = u_1ax$ and $v = v_1bx$ where $a \neq b \in A$ and $u_1, v_1, x \in A^*$. Define $\varphi$ by $a\varphi = \bar{0}$, $b\varphi = \bar{1}$ and $c\varphi = I$ for $c \in A \setminus \{a, b\}$. Because $u$ and $v$ have no repeated letters, all letters of $x$ are sent to $I$, and so $u\varphi = \bar{0}$, $v\varphi = \bar{1}$. This completes the proof. $\square$

**Exercise 7.3.3.** Show that $L \subseteq A^*$ is $\mathbf{RRB}$-recognizable if and only if it is a Boolean combination of languages of the form $A^*aB^*$ where $B \subseteq A \setminus \{a\}$.

Our next example hearkens back to Chapter 2.

*Example 7.3.4 ($B_2$).* We saw back in Lemma 2.2.3 that the Brandt semigroup $B_2$ is fji. In fact, we saw $\mathsf{Excl}(B_2) = \mathbf{DS}$ and that moreover, if $S \notin \mathbf{DS}$, then $B_2 \prec S^2$. Hence $\mathbf{DS}$ is mi and can be defined by a pseudoidentity in two variables. We claim $\mathbf{DS} = [\![((xy)^\omega(yx)^\omega(xy)^\omega)^\omega = (xy)^\omega]\!]$. Indeed, because $(xy)^\omega, (yx)^\omega$ are $\mathscr{J}$-equivalent idempotents, clearly $\mathbf{DS}$ satisfies this pseudoidentity. On the other hand, $B_2$ fails this pseudoidentity by taking $x = E_{12}$ and $y = E_{21}$. The unique maximal proper subpseudovariety of $(B_2)$ is $(B_2) \cap \mathbf{DS} = (B_2) \cap \mathbf{J}$ (as $B_2$ is aperiodic with commuting idempotents and an aperiodic element of $\mathbf{DS}$ with commuting idempotents is $\mathscr{J}$-trivial).

Define the semigroup $V$ by

$$V = \mathscr{M}^0\left(\{1\}, 2, 2, \begin{pmatrix} 1 & 1 \\ 0 & 1 \end{pmatrix}\right). \tag{7.9}$$

Then $V \notin \mathbf{DS}$, but $B_2 \not\prec V$. However, $B_2 \prec V^2$ so we have

$$\mathsf{Excl}(B_2) \neq \{S \in \mathbf{FSgp} \mid B_2 \not\prec S\}.$$

The case where $S$ is an fji semigroup with $\mathsf{Excl}(S)$ consisting precisely of those semigroups not divided by $S$ is particularly important. A semigroup $S$ has this property if and only if $S \prec T_1 \times T_2$ implies $S \prec T_i$, some $i$, that is, if and only if $S$ is $\times$-*prime*. For example, $U_2$ is $\times$-prime, but $B_2$ is not. The following problem seems quite interesting and also appears as [7, Problem 26].

*Question 7.3.5.* Classify the $\times$-prime finite semigroups. Is the property of being $\times$-prime decidable?

In turns out the semigroup $V$ is $\times$-prime, and hence fji, as is well-known.

*Example 7.3.6 (V).* It is well-known the semigroup $V$ (7.9) is $\times$-prime and

$$\mathbf{EDS} = \{S \in \mathbf{FSgp} \mid V \not\prec S\}.$$

Let us prove this. Clearly $V \notin \mathbf{EDS}$. We are left with showing if $S \notin \mathbf{EDS}$, then $V \prec S$. By Lemma 4.13.1, there must be a regular $\mathscr{J}$-class $J$ of $S$ with $J^0 \notin \mathbf{EDS}$ (in particular, $J$ is not the minimal ideal). Thus we may without loss of generality assume that $S = J^0$ is 0-simple. We claim that $\mathsf{AGGM}_J(S) \notin \mathbf{EDS}$. Suppose to the contrary that it belongs to $\mathbf{EDS}$. Because $S/\mathsf{AGGM} = \mathsf{AGGM}_J(S)$, we have $\Gamma_J \colon S \to \mathsf{AGGM}_J(S)$ is an $\mathbb{LG}$-morphism. Restricting $\Gamma_J$ to $\langle E(S) \rangle$ yields

$$\langle E(S) \rangle \in \mathbb{LG} \text{ⓜ} \mathbf{DS} = \mathbb{LG} \text{ⓜ} (\mathbb{LG} \text{ⓜ} \mathbf{Sl}) = \mathbb{LG} \text{ⓜ} \mathbf{Sl} = \mathbf{DS},$$

a contradiction. Thus we may assume without loss of generality that $S$ is a 0-simple, aperiodic generalized group mapping semigroup (that is, a congruence-free 0-simple semigroup). If $\overline{2}$ were not a subsemigroup of $S$, then we would have $S \in \mathbf{ER} \leq \mathbf{EDS}$, which is not the case. So $S$ has two distinct $\mathscr{R}$-equivalent idempotents. By Proposition 4.7.14, $S \cong \mathscr{M}^0(\{1\}, A, B, C)$ where $C$ has no identical rows or columns. The existence of distinct $\mathscr{R}$-equivalent idempotents means there is column $a$ that has two ones, say in rows $b, b'$. Because no two rows are identical, there must be a column $a'$ where $b$ and $b'$ differ, yielding a submatrix

$$\begin{pmatrix} 1 & 1 \\ 0 & 1 \end{pmatrix}$$

(up to reordering the rows and columns). Thus $V \prec S$, completing the proof.

We conclude $\mathbf{EDS}$ is mi and can be defined by a pseudoidentity in two variables as $V$ can be generated by $(2,1,1)$ and $(1,1,2)$. The following two-variable basis for $\mathbf{EDS}$ is due to Lee [180], although our proof is simpler:

$$\mathbf{EDS} = [\![ ((x^\omega y)^\omega (yx^\omega)^\omega)^\omega = (x^\omega yx^\omega)^\omega ]\!]. \tag{7.10}$$

To see that $V$ fails the pseudoidentity in (7.10), just take $x = (2, 1, 1)$ and $y = (1, 1, 2)$. Then the left-hand side is 0 whereas the right-hand side is $x$. It remains to verify that **EDS** satisfies this pseudoidentity. Let us set $e = x^\omega$. Then observe that $(ey)^\omega \mathscr{R} (eye)^\omega$ and $(ye)^\omega \mathscr{L} (eye)^\omega$, as follows easily from the fact $(ey)^\omega e = (eye)^\omega = e(ye)^\omega$. Stability then yields that a semigroup in **EDS** satisfies $(ey)^\omega (ye)^\omega \mathscr{H} (eye)^\omega$ and hence $((x^\omega y)^\omega (yx^\omega)^\omega)^\omega = (x^\omega yx^\omega)^\omega$, as required.

The pseudovariety generated by $V$ is quite interesting as well. The following result is well-known (see, for instance, [116]).

**Theorem 7.3.7.** *The equality* $(V) = \mathbf{Sl} * \mathbf{RZ}$ *holds. In particular, every 0-simple aperiodic semigroup belongs to* $(V)$ *and hence* $\mathbf{Sl} * \mathbf{RZ}$ *is the pseudovariety generated by 0-simple aperiodic semigroups.*

*Proof.* Corollary 4.12.21 shows that every 0-simple aperiodic semigroup (including $V$ of course) belongs to $\mathbf{Sl} * \mathbf{RZ}$. Because $(V)$ is compact, and hence defined by identities, to show the inclusion $\mathbf{Sl} * \mathbf{RZ} \leq (V)$ it suffices to show that if $u = v$ is an identity satisfied by $V$, then $\mathbf{Sl} * \mathbf{RZ} \models u = v$.

Suppose first $u, v \in A^+$ are such that:

(a) $u$ and $v$ begin with the same letter;
(b) $u$ and $v$ end with the same letter;
(c) $u$ and $v$ have the same factors of length 2.

Then we show that $\mathbf{Sl} * \mathbf{RZ} \models u = v$. So let us consider a homomorphism $\varphi \colon A^+ \to E \rtimes R$ where $R \in \mathbf{RZ}$ and $E \in \mathbf{Sl}$. We write $E$ additively. Suppose, for $a \in A$, that $a\varphi = (a\varphi_1, a\varphi_2)$. If $w = a_1 \cdots a_n$ with the $a_i \in A$, then

$$
\begin{aligned}
w\varphi &= (a_1\varphi_1, a_1\varphi_2)(a_2\varphi_1, a_2\varphi_2) \cdots (a_n\varphi_1, a_n\varphi_2) \\
&= (a_1\varphi_1 + a_1\varphi_2 a_2\varphi_1 + \cdots + (a_1\varphi_2 \cdots a_{n-1}\varphi_2)a_n\varphi_1, a_1\varphi_2 \cdots a_n\varphi_2) \\
&= (a_1\varphi_1 + a_1\varphi_2 a_2\varphi_1 + a_2\varphi_2 a_3\varphi_1 + \cdots + a_{n-1}\varphi_2 a_n\varphi_1, a_n\varphi_2).
\end{aligned}
$$

Because addition in $E$ is idempotent and commutative, it follows that $w\varphi$ is determined by its first letter $a_1$, its last letter $a_n$ and its set of factors

$$\{a_1 a_2, a_2 a_3, \ldots, a_{n-1} a_n\}$$

of length 2. Hence if $u, v \in A^+$ satisfy (a)–(c), then $u\varphi = v\varphi$.

To complete the proof, it suffices to show that if $u, v \in A^+$ do not satisfy (a)–(c), then $V \not\models u = v$. It will then follow that every element of $\mathbf{Sl} * \mathbf{RZ}$ satisfies all the identities satisfied by $V$, and so belongs to $(V)$. If $u$ ends in $a$ and $v$ ends in $b$ with $a \neq b$, then the map $\varphi \colon A^+ \to V$ that sends $a$ to $(2, 1, 2)$ and all other letters to $(2, 1, 1)$ satisfies $u\varphi \neq v\varphi$. Similarly, if $u$ begins with $a$ and $v$ begins with $b \neq a$, then the map $\varphi \colon A^+ \to V$ given by $a\varphi = (1, 1, 1)$ and $x\varphi = (2, 1, 1)$ for $x \neq a$ has $u\varphi \neq v\varphi$. Suppose now that $u$ and $v$ have different factors of length 2. Without loss of generality, we may assume $u$ has a factor of length 2 that is not a factor of $v$. If $u$ has a factor $aa$ and $v$ does not, define $\varphi\colon A^+ \to V$ by $a\varphi = (1,1,2)$ and $b\varphi = (2,1,1)$ if $b \neq a$. Then $u\varphi = 0$ and $v\varphi \neq 0$. So again $u\varphi \neq v\varphi$. Finally, if $u$ has a factor $ab$ with $a \neq b$ and $v$ does not have a factor $ab$, define $\varphi\colon A^+ \to V$ by $a\varphi = (2,1,2)$, $b\varphi = (1,1,1)$ and $c\varphi = (2,1,1)$ for $c \in A \setminus \{a,b\}$. Then $u\varphi = 0$ and $v\varphi \neq 0$. This concludes the proof that $V$ only satisfies identities such that (a)–(c) hold and hence $\mathbf{Sl} * \mathbf{RZ} \leq (V)$, as required. $\qquad\qquad\square$

Theorem 7.3.7 admits the following corollary [116].

**Corollary 7.3.8.** *Let* $\mathbf{H}$ *be a pseudovariety of groups. Then the pseudovariety generated by* 0-*simple semigroups with maximal subgroup in* $\mathbf{H}$ *is* $(\mathbf{H}\vee\mathbf{Sl})*\mathbf{RZ}$.

*Proof.* Theorem 7.3.7 handles the case $\mathbf{H}$ is trivial, so assume $\mathbf{H} \neq 1$. Every 0-simple semigroup with maximal subgroup in $\mathbf{H}$ belongs to $(\mathbf{H} \vee \mathbf{Sl}) * \mathbf{RZ}$ by Corollary 4.12.21. Conversely, by Exercise 1.2.40, the operator $(-) * \mathbf{RZ}$ preserves non-empty sups so

$$(\mathbf{H} \vee \mathbf{Sl}) * \mathbf{RZ} = (\mathbf{H} * \mathbf{RZ}) \vee (\mathbf{Sl} * \mathbf{RZ}).$$

Exercise 4.12.22 shows that $\mathbf{H} * \mathbf{RZ}$ consists of the simple semigroups with maximal subgroup in $\mathbf{H}$, whereas Theorem 7.3.7 shows that $\mathbf{Sl} * \mathbf{RZ}$ is generated by 0-simple aperiodic semigroups. This completes the proof. $\qquad\square$

Notice that $(\mathbf{H} \vee \mathbf{Sl}) * \mathbf{RZ}$ is decidable if $\mathbf{H}$ is decidable because $\mathbf{RZ}$ is locally finite and $\mathbf{H} \vee \mathbf{Sl}$ is local [151].

We now present a new family of examples of fji semigroups (as far as we know). The techniques involved here are more sophisticated than in the previous examples. The theory of generalized group mapping semigroups plays a key role here, a recurring theme in this chapter.

*Example 7.3.9 (Brandt plus a cyclic group).* Let $p$ be a prime number. Consider the semigroup of $p \times p$ matrices

$$BZ_p = \{C^k, E_{ij}, 0 \mid 1 \leq i,j,k \leq p\}$$

where $C$ is the cyclic permutation matrix

$$C = \begin{pmatrix} 0 & 1 & 0 & \cdots & 0 \\ 0 & 0 & 1 & \cdots & 0 \\ \vdots & & & \ddots & \vdots \\ 1 & 0 & 0 & \cdots & 0 \end{pmatrix} \qquad\qquad (7.11)$$

and the $E_{ij}$ are the standard matrix units. So $BZ_p = \mathbb{Z}_p \cup B_p$ where $B_p$ is the $p \times p$ Brandt semigroup. One easily verifies $CE_{ij} = E_{i-1\,j}$ and $E_{ij}C = E_{i\,j+1}$ where indices are taken modulo $p$.

Notice that $BZ_2$ is the symmetric inverse monoid $I_2$ on two symbols. Our goal is to prove that $\mathbb{L}\mathbf{G} \,\textcircled{m}\, \overline{\mathbf{G}}_{p'} = \mathsf{Excl}(BZ_p)$, where $\mathbf{G}_{p'}$ is the pseudovariety of groups with order prime to $p$. In particular, $BZ_p$ is fji and $\mathbb{L}\mathbf{G} \,\textcircled{m}\, \overline{\mathbf{G}}_{p'}$ is mi. As this result is non-trivial, we state it as a theorem.

**Theorem 7.3.10.** *Let $p$ be prime. Then $BZ_p$ is an fji semigroup and*

$$\mathsf{Excl}(BZ_p) = \mathbb{L}\mathbf{G} \, \textcircled{m} \, \overline{\mathbf{G}}_{p'}$$

*holds.*

*Proof.* We need to show that a semigroup $S \in \mathbb{L}\mathbf{G} \, \textcircled{m} \, \overline{\mathbf{G}}_{p'}$ if and only if $BZ_p \notin (S)$. Clearly, $BZ_p$ is generalized group mapping with aperiodic 0-minimal ideal $B_p$. Because $BZ_p \notin \overline{\mathbf{G}}_{p'}$, Theorem 4.6.50 implies $BZ_p \notin \mathbb{L}\mathbf{G}\textcircled{m}\overline{\mathbf{G}}_{p'}$. Therefore, $BZ_p \in (S)$ implies $S \notin \mathbb{L}\mathbf{G} \, \textcircled{m} \, \overline{\mathbf{G}}_{p'}$.

Conversely, suppose $S \notin \mathbb{L}\mathbf{G}\textcircled{m}\overline{\mathbf{G}}_{p'}$. Then by Theorem 4.6.50, we must have $S\mathscr{I}' \notin \overline{\mathbf{G}}_{p'}$. Because $S\mathscr{I}'$ is a subdirect product of generalized group mapping semigroups with aperiodic 0-minimal ideals, it suffices by Theorem 4.6.50 to prove the following:

*Claim.* If $S$ is a generalized group mapping semigroup with aperiodic 0-minimal ideal and $S \notin \overline{\mathbf{G}}_{p'}$, then $BZ_p \in (S)$.

To prove the claim, suppose $S$ is a generalized group mapping semigroup, which is not group mapping, and $S \notin \overline{\mathbf{G}}_{p'}$. We show $BZ_p$ divides a power of $S$. Because $S$ contains a non-$p'$-group, by Cauchy's Theorem $S$ must have a group element $g$ of order $p$. Let $e = g^p$. By Proposition 4.6.56, we may replace $S$ by $eSe$ and so assume without loss of generality that $S$ is a monoid with identity $e$ (and so $g$ belongs to the group of units of $S$). Because $S$ is non-trivial, it must have a 0 (a rectangular band does not act faithfully on both the right and left of itself). Let $I$ be the 0-minimal ideal of $S$ and $J = I \setminus \{0\}$. Because $S$ acts faithfully on the right of $I$, there is an element $x \in J$ with $xg \neq x$. As $xg \mathscr{R} x$ and $J^0$ is aperiodic, this implies, in fact, $L_x \neq L_{xg}$. By Proposition 4.7.14, there is then an $\mathscr{R}$-class $R$ of $J$ such that exactly one of $R \cap L_x$ and $R \cap L_{xg}$ contains an idempotent. Replacing $g$ by $g^{-1}$ if necessary, we may assume $R \cap L_x$ contains an idempotent $f$. Because $fg \in R_f \cap L_{xg} = R \cap L_{xg}$, it cannot be an idempotent. From $f \neq fg$ and $p$ being prime, we deduce the elements $fg^i$ with $0 \leq i \leq p - 1$ are pairwise distinct $\mathscr{R}$-equivalent elements of $J$. Let $n$ be the number of idempotents among this collection of elements; so $1 \leq n < p$ as $f$ is idempotent and $fg$ is not idempotent. Suppose $E = \{fg^{i_1}, \ldots, fg^{i_n}\}$ is the set of idempotents from $\{fg^i \mid 0 \leq i \leq p - 1\}$, where $0 = i_1 < i_2 < \cdots < i_n \leq p - 1$. We shall prove $BZ_p \prec S^n$.

Set $\mathbf{g} = (g, \ldots, g)$ and $\mathbf{f} = (fg^{i_1}, \ldots, fg^{i_n})$. Then $\mathbf{g}$ is a group element of order $p$ and $\mathbf{f}$ is an idempotent. Let $T = \langle \mathbf{g}, \mathbf{f} \rangle$ and let $K$ be the ideal of all elements of $T$ that are not $\geq_{\mathscr{J}}$-above $\mathbf{f}$. Our aim is to prove $T/K \cong BZ_p$. The key step in executing this task is to show that, for $1 \leq i \leq p - 1$, the element $\mathbf{f}\mathbf{g}^i$ is not an idempotent. It will then follow $(\mathbf{f}\mathbf{g}^i)^2 \in K$, for $1 \leq i \leq p - 1$, because if $fg^k$ is not an idempotent, it squares to 0 (as in an aperiodic 0-simple semigroup the square of any non-idempotent is 0). Indeed, it is immediate that $\mathbf{f}\mathbf{g}^i$ is an idempotent if and only if $Eg^i = E$. The set of such $g^i$ is a subgroup

$H$ of $\langle g \rangle \cong \mathbb{Z}_p$. Moreover, because $fg$ is not idempotent, $g \notin H$ and so $H$ is the trivial subgroup. This establishes that $\mathbf{fg}^i$ is not an idempotent, $1 \leq i \leq p-1$, and hence $(\mathbf{fg}^i)^2$ belongs to $K$.

As a consequence, observe that if $1 \leq i \leq p-1$, then $\mathbf{fg}^i\mathbf{f} \in K$ because $\mathbf{fg}^i\mathbf{f} \not{\mathcal{J}} (\mathbf{fg}^i)^2 \in K$. It follows that in $T/K$, for $1 \leq i, j, k, \ell \leq p$,

$$(\mathbf{g}^{-i}\mathbf{fg}^j)(\mathbf{g}^{-k}\mathbf{fg}^\ell) = \begin{cases} \mathbf{g}^{-i}\mathbf{fg}^\ell & j = k \\ 0 & \text{else.} \end{cases}$$

It is now straightforward to verify the assignment $C^j \mapsto \mathbf{g}^j$, $E_{ij} \mapsto \mathbf{g}^{-i}\mathbf{fg}^j$, $0 \mapsto 0$ gives an isomorphism of $BZ_p$ with $T/K$. This completes the proof $\mathsf{Excl}(BZ_p) = \mathbb{L}\mathbf{G} \textcircled{m} \overline{\mathbf{G}}_{p'}$.                                    □

Notice that the proof shows $I_2 = BZ_2$ is $\times$-prime because the orbit $\{f, fg\}$ contains exactly one idempotent. On the other hand, $BZ_p$ is not $\times$-prime for $p > 2$. Indeed, consider the semigroup $S = \mathbb{Z}_p \cup \mathscr{M}^0(\{1\}, p, p, J_p - I_p)$ where $J_p$ is the $p \times p$ matrix of all ones, $I_p$ is the $p \times p$ identity matrix and the actions of the generator $\mathbb{Z}_p$ by left and right translations are both given by the matrix $C$ from (7.11); the linked equations boil down to $C(J_p - I_p) = J_p - C = (J_p - I_p)C$. Then $S$ is a generalized group mapping semigroup with aperiodic 0-minimal ideal and it has a $p$-subgroup, so $S \notin \mathbb{L}\mathbf{G} \textcircled{m} \overline{\mathbf{G}}_{p'}$. Also $S$ is subdirectly indecomposable, and its unique minimal congruence is the Rees congruence associated to its minimal ideal. The above proof shows $BZ_p \prec S^{p-1}$ (and one can show that you cannot get away with a smaller exponent). Clearly if $p > 2$, then $BZ_p \not\prec S$ because $S$ has no proper subsemigroup containing both a cyclic group of order $p$ and at least three idempotents, and $S$ does not map onto $BZ_p$ by the above observation on its unique minimal congruence.

**Exercise 7.3.11.** Show that $BZ_p \not\prec S^n$ for $n < p - 1$, where $S$ is the semigroup above.

Because $\mathbb{L}\mathbf{G} \textcircled{m} \overline{\mathbf{G}}_{p'}$ is mi it can be defined by a single pseudoidentity in two variables (as $BZ_p$ is two-generated). A general method was given in [259] to obtain pseudoidentities for $\mathbb{L}\mathbf{G} \textcircled{m} \mathbf{V}$, but it seems that one would obtain two pseudoidentities by this method. The reader is referred to Definition 7.1.15 and Corollary 7.1.18 for the meaning of $x^{(p')^\omega}$. We claim

$$\mathbb{L}\mathbf{G} \textcircled{m} \overline{\mathbf{G}}_{p'} = \left[\!\left[\left((x^\omega y)^\omega x^{(p')^\omega}(x^\omega y)^\omega\right)^\omega = (x^\omega y)^\omega\right]\!\right].$$

To see that $\mathbb{L}\mathbf{G} \textcircled{m} \overline{\mathbf{G}}_{p'}$ satisfies this pseudoidentity, let $S \in \mathbb{L}\mathbf{G} \textcircled{m} \overline{\mathbf{G}}_{p'}$ and let $J$ be the $\mathscr{J}$-class of $(x^\omega y)^\omega$. Because $\mathsf{AGGM}_J(S) \in \overline{\mathbf{G}}_{p'}$ (by Theorem 4.6.50) and $x^{(p')^\omega}$ generates a cyclic group of order a $p$-power with identity $x^\omega$, we must have $\Gamma_J(x^{(p')^\omega}) = \Gamma_J(x^\omega)$, where $\Gamma_J \colon S \to \mathsf{AGGM}_J(S)$ is the canonical projection.

Now $(x^\omega y)^\omega x^\omega (x^\omega y)^\omega = (x^\omega y)^\omega (x^\omega y)^\omega = (x^\omega y)^\omega \in J$ and so by the very definition of $\Gamma_J$, we must have

$$(x^\omega y)^\omega x^{(p')^\omega} (x^\omega y)^\omega \in J.$$

Thus $\left((x^\omega y)^\omega x^{(p')^\omega} (x^\omega y)^\omega\right)^\omega = (x^\omega y)^\omega$ by stability, as required.

Next we show this pseudoidentity is not satisfied by $BZ_p$. Let $x = C$ and $y = E_{11}$. Then $x^{(p')^\omega} = x$ and $x^\omega = 1$. So $x^\omega y = E_{11}$ and hence the left-hand side is $(E_{11} C E_{11})^\omega$, whereas the right-hand side is $E_{11}$. But $E_{11} C E_{11} = E_{12} E_{11} = 0$. This completes the proof.

*Remark 7.3.12.* The intuition behind the above example is that if $S$ is generalized group mapping with aperiodic 0-minimal ideal $J^0$ and $g \in S$ is a group element of order $p$, then there is a row of the structure matrix $C$ with a 1 in position $b$ but a 0 in position $bg$ (recall $g$ acts on the right of the $\mathscr{L}$-classes). However, other $bg^i$ may also be 1. By taking a large enough tensor power $P$ of $C$ (corresponding with taking powers of $S$), we can guarantee there is a row and a $g$-orbit on the columns so that this row contains a single 1 in that orbit. The linked equations then force a copy of the identity matrix to appear in $P$, with $g$ acting on it in the appropriate way.

In light of the above example, it is quite natural to ask the following questions.

*Question 7.3.13.* Is $I_n$, the symmetric inverse monoid, fji for all $n \geq 2$? A positive answer would imply **ESl** is fji, in light of Ash's Theorem [32] (our Theorem 4.17.31) and Lemma 6.1.13. How about sfji? This may be easier.

*Question 7.3.14.* If $(\mathbf{n}, G)$ is a permutation group, let $S_{(\mathbf{n},G)}$ be the action semigroup of the partial transformation inverse semigroup $(\mathbf{n}, G \cup B_n)$ where we identify $B_n$ with the rank one partial bijections of $\mathbf{n}$. Theorem 4.7.21 implies $S_{(\mathbf{n},G)}$ is subdirectly indecomposable. Of course, $BZ_p = S_{(\mathbf{p}, \mathbb{Z}_p)}$. The question is: when is $S_{(\mathbf{n},G)}$ fji? One needs at least that the permutation group is subdirectly indecomposable as a permutation group.

It also seems tempting to conjecture that the full transformation monoid $T_n$ is fji.

*Question 7.3.15.* Is the full transformation monoid $T_n$ an fji semigroup for all $n \geq 2$? If so this would give a new proof that **FSgp** is fji by Lemma 6.1.13 (a proof that **FSgp** is fji appears later in this chapter). How about sfji? This may be easier.

In our next collection of examples, the exclusion pseudovarieties do not seem to be well-known. In fact, we can only describe them by giving the defining pseudoidentities. They all make use of an idempotent in the minimal ideal of a free profinite semigroup. Methods to construct such idempotents as limits of computable sequences can be found in [22, 23, 260].

*Example 7.3.16.* Consider the transformation semigroup $\overline{(\mathbf{3}, \{e, f\})}$, where

$$e = \begin{pmatrix} 0 & 1 & 2 \\ 0 & 1 & 0 \end{pmatrix} \text{ and } f = \begin{pmatrix} 0 & 1 & 2 \\ 0 & 1 & 1 \end{pmatrix}.$$

Denote by $W$ the action semigroup. Then $W = \{e, f, \overline{0}, \overline{1}, \overline{2}\}$, $e$ and $f$ are $\mathscr{L}$-equivalent idempotents and the minimal ideal of $W$ consists of the constant maps. Let $\epsilon$ be an idempotent in the minimal ideal of $\widehat{\{x, y, z\}}^+$. Let

$$\mathbf{W} = \left[\!\left[ \left((\epsilon z)^\omega (xy^\omega)^\omega\right)^\omega = \left((\epsilon z)^\omega y^\omega (xy^\omega)^\omega\right)^\omega \right]\!\right].$$

We claim that $W$ is fji, in fact $\times$-prime, and $\mathsf{Excl}(W) = \mathbf{W}$.

First we observe that $W$ fails this pseudoidentity: map $x$ to $e$, $y$ to $f$ and $z$ to $\overline{2}$. Then the left-hand side evaluates to $\overline{2}e = \overline{0}$, whereas the right-hand side evaluates to $\overline{2}f = \overline{1}$. Conversely, suppose that $S$ fails this pseudoidentity. We show that $W \prec S$. This will in particular imply $W$ is $\times$-prime. Consider a substitution of $x, y, z$ failed by $S$. Abusing notation, we identify the variables with their images. Replacing $S$ by the subsemigroup generated by $x, y, z$, we may assume without loss of generality that $x, y, z$ generate $S$. Then $(\epsilon z)^\omega$ maps to an idempotent $q$ of the minimal ideal $J$ of $S$. Consider $E = (xy^\omega)^\omega$ and $F = y^\omega (xy^\omega)^\omega$. These are $\mathscr{L}$-equivalent idempotents of $S$. Now in $J$, we have that $(qE)^\omega \neq (qF)^\omega$. As $qE \,\mathscr{R}\, qF$, this means $L_{qE} \neq L_{qF}$. We claim $L_q \neq L_{qE}$. Indeed, $q \,\mathscr{L}\, qE$ implies $qF \,\mathscr{L}\, qEF = qE$, a contradiction. Similarly, $L_q \neq L_{qF}$. Set $L_0 = L_{qE}$, $L_1 = L_{qF}$ and $L_2 = L_q$. It follows that in $(J/\mathscr{L}, \mathsf{RLM}_J(S))$, the subset $\{L_0, L_1, L_2\}$ is invariant under the action of $e' = E\mu_J^R$, $f' = F\mu_J^R$, while $qE\mu_J^R = \overline{L}_0$, $qF\mu_J^R = \overline{L}_1$ and $q\mu_J^R = \overline{L}_2$ where $\overline{L}_i$ is the constant map sending all $\mathscr{L}$-classes to $L_i$. Moreover, $L_i e' = L_{ie}$ and $L_i f' = L_{if}$, for $i = 0, 1, 2$. Thus $W \prec \mathsf{RLM}_J(S) \prec S$.

The semigroup $W$ is aperiodic. A natural question is the following:

*Question 7.3.17.* Is there an increasing sequence of compact fji aperiodic pseudovarieties $(S_1) \leq (S_2) \leq \cdots$ with $\bigcup (S_i) = \mathbf{A}$? If so, then $\mathbf{A}$ would be fji by Lemma 6.1.13.

The following family of semigroups is well-known [7].

*Example 7.3.18 ($K_p$).* Given an integer $n > 1$, consider the semigroup

$$K_n = \mathscr{M}\left(\mathbb{Z}_n, 2, 2, \begin{pmatrix} 1 & 1 \\ 1 & g \end{pmatrix}\right)$$

where we view $\mathbb{Z}_n$ as a multiplicative cyclic group $\langle g \rangle$. We show that if $p$ is prime, then $K_p$ is $\times$-prime and hence fji. A similar argument allows one to replace $\mathbb{Z}_p$ by $\mathbb{Z}_{p^n}$ for $n > 1$.

Let $e$ be an idempotent in the minimal ideal of $\widehat{\{x, y\}}^+$ (in particular, $e$ will have both $x$ and $y$ in its support). Recalling Definition 7.1.15, set

$$\mathbf{V} = \left[\!\left[ \left((xex)^\omega (yey)^\omega\right)^{(p')^\omega} = \left((xex)^\omega (yey)^\omega\right)^\omega \right]\!\right].$$

We claim $\mathsf{Excl}(K_p) = \mathbf{V}$. More precisely, we show that if $S \notin \mathbf{V}$, then $K_p \prec S$. This implies $K_p$ is $\times$-prime.

First note that $K_p$ fails this pseudoidentity. Set $x = (1,1,2)$ and $y = (2,1,1)$. Then the left-hand side is $(1,g,1)$ whereas the right-hand side is $(1,1,1)$. Conversely, suppose $S$ is a semigroup failing this pseudoidentity. Just like in the previous example, we may assume $S$ is generated by elements $x$ and $y$ failing the pseudoidentity via the canonical projection. Abusing notation, we write $e$ for the image of $e$ in $S$. Set $e_1 = (xex)^\omega$ and $e_2 = (yey)^\omega$. Notice $e_1 e_2$ belongs to the minimal ideal of $S$ and we are assuming $(e_1 e_2)^{(p')^\omega} \neq (e_1 e_2)^\omega$; in particular, $e_1 e_2$ is not idempotent, in fact $p$ divides the order of $e_1 e_2$. Let $T$ be the subsemigroup generated by $e_1, e_2$. First note that $R_{e_1} \neq R_{e_2}$ and $L_{e_1} \neq L_{e_2}$, as in either of these cases, we would have $e_1 e_2$ is idempotent. Now $T$ is completely simple. If we identify the maximal subgroup $G$ of $T$ with the $\mathscr{H}$-class of $e_1 e_2$ and we take $\ell_1 = (e_1 e_2)^\omega = r_1, \ell_2 = e_1, r_2 = e_2$ as the representatives for the Rees coordinates (as per the proof of Theorem A.4.15), then we obtain an isomorphism

$$T \cong \mathscr{M}\left(G, 2, 2, \begin{pmatrix} 1 & 1 \\ 1 & e_1 e_2 \end{pmatrix}\right).$$

The isomorphism takes $h \in G$ to $(1,h,1)$, $e_1$ to $(1,1,2)$ and $e_2$ to $(2,1,1)$. Consider the subsemigroup

$$T' = \mathscr{M}\left(\langle e_1 e_2 \rangle, 2, 2, \begin{pmatrix} 1 & 1 \\ 1 & e_1 e_2 \end{pmatrix}\right).$$

Because $p$ divides the order of $e_1 e_2$, it follows $K_p$ is a quotient of $T'$. This completes the proof.

**Exercise 7.3.19.** Show that if $p$ is prime, then $K_{p^n}$ is $\times$-prime, for $n \geq 1$ and if $e$ is an idempotent in the minimal ideal of $\widehat{\{x,y\}^+}$, then

$$\mathsf{Excl}(K_{p^n}) = \left[\!\left[ \left(\left((xex)^\omega (yey)^\omega\right)^{(p')^\omega}\right)^{p^{n-1}} = \left((xex)^\omega (yey)^\omega\right)^\omega \right]\!\right]$$

holds.

Our last example of an fji semigroup in this section considers augmented groups.

*Example 7.3.20 (Augmentation of $\mathbb{Z}_p$).* Recall $\mathbb{Z}_p{}^\sharp$ is the action monoid of $(\mathbb{Z}_p, \mathbb{Z}_p)$. We claim that $\mathbb{Z}_p{}^\sharp$ is $\times$-prime and

$$\mathsf{Excl}(\mathbb{Z}_p{}^\sharp) = \left[\!\left[ \left((x^\omega ex^\omega)^\omega x^{(p')^\omega}\right)^\omega = (x^\omega ex^\omega)^\omega \right]\!\right]$$

where $e$ is an idempotent in the minimal ideal of $\widehat{\{x,y\}^+}$ and $x^{(p')^\omega}$ is as in Definition 7.1.15. First note that $\mathbb{Z}_p{}^\sharp$ fails this pseudoidentity by mapping $x$ to the generator of $\mathbb{Z}_p$ and $y$ to $\overline{1}$. Indeed, $e$ will then be some constant map, say $\overline{k}$, and so the left-hand side will be $\overline{k+1}$ (where $k+1$ is understood modulo $p$), whereas the right-hand side will be $\overline{k}$.

Conversely, if $S$ fails this pseudoidentity, we show $\mathbb{Z}_p{}^\sharp \prec S$. As in the previous examples, we may assume $x, y$ generate $S$ and the canonical surjection fails the pseudoidentity in question. By localizing at $x^\omega$ and setting $g = x^{(p')^\omega}$, we may assume that $g$ belongs to the group of units, and $(eg)^\omega \neq e$ for some idempotent $e$ of the minimal ideal of $S$. In particular, $g$ is a non-trivial element of order a $p$-power. Because $eg \,\mathcal{R}\, e$, we have $L_{eg} \neq L_e$. Let $T$ be the right letter mapping semigroup of $\langle e, g \rangle$ with respect to its minimal ideal. Then $T = \langle g' \rangle \cup J$ where the minimal ideal $J$ is a right zero semigroup with at least two right zeroes and $g'$ generates a cyclic group of units of $p$-power order, acting faithfully and transitively on the right of $J$. Let $H$ be the stabilizer of the image of $e$. Then $H$ is a proper subgroup of $\langle g' \rangle$ and so is contained in a subgroup $K$ of index $p$. Now it is easy to see that defining $a \equiv b$ if $aK = bK$ is a congruence on $T$ and the quotient is isomorphic to $\mathbb{Z}_p{}^\sharp$. This completes the proof.

**Exercise 7.3.21.** Show that $\mathbb{Z}_{p^n}{}^\sharp$ is $\times$-prime and

$$\mathsf{Excl}(\mathbb{Z}_{p^n}{}^\sharp) = \left[\!\left[ \left( (x^\omega e x^\omega)^\omega (x^{(p')^\omega})^{p^{n-1}} \right)^\omega = (x^\omega e x^\omega)^\omega \right]\!\right]$$

where $e$ is an idempotent in the minimal ideal of $\widehat{\{x,y\}^+}$.

We shall make some further headway on Question 7.3.5 and the study fji pseudovarieties in Section 7.4. Let us now present an example of a sji pseudovariety that is not ji.

**Proposition 7.3.22.** *The pseudovariety* $[\![xyz = 0, x^2 = 0]\!]$ *is sji but not ji.*

*Proof.* Let $\mathbf{V} = [\![xyz = 0, x^2 = 0]\!]$. To show that $\mathbf{V}$ is sji it suffices, by Proposition 7.2.13, to show that $\mathbf{V}$ is finitely generated and has a unique maximal subpseudovariety. For $n > 0$, let $F_\mathbf{V}(A)$ be the free semigroup generated by $A$ in $\mathbf{V}$. It is the Rees quotient of $A^+$ by the ideal consisting of all words that are either of length greater than two or are the square of a generator. We claim that if $|A| > 2$, then $F_\mathbf{V}(A)$ is a subdirect product of copies of $S = F_\mathbf{V}(\{a, b\})$. It will follow that $\mathbf{V}$ is generated by the finite semigroup $S$. Indeed, suppose $u, v \in A^+$ are words representing distinct elements of $F_\mathbf{V}(A)$. Without loss of generality, we may assume $v \neq 0$. In particular, $v$ has at most two letters (necessarily distinct) and so we may assume that $0 \neq v$ belongs to $S \subseteq F_\mathbf{V}(A)$. Now we define a map from $F_\mathbf{V}(A)$ to $S$ by sending each letter not in $v$ to $0$ and sending each letter in $v$ to itself. If $u$ is zero or contains a letter not in $v$, then $u$ is sent to zero and so separated from $v$. Otherwise $u$

is sent to itself (as is $v$) and so this map still separates $u$ and $v$. The desired subdirect product decomposition is now plain.

We claim that the unique maximal proper pseudovariety of $\mathbf{V}$ is $\mathbf{V} \cap \mathbf{Com}$. Indeed, suppose $N \in \mathbf{V}$ and that $x, y \in N$ are such that $xy \neq yx$. In particular, $x \neq y$ and neither of $x$ or $y$ are zero. Also neither $x$ nor $y$ can be $yx$ or $xy$ as these latter elements commute with everything. We show that $S$ embeds in $N^2$. Consider the map $\varphi \colon S \to N \times N$ induced by sending $a$ to $(x, y)$ and $b$ to $(y, x)$. We must show that $\varphi$ is injective. Because $S = \{a, b, ab, ba, 0\}$, it suffices to show the subsemigroup of $N \times N$ generated by $(x, y)$ and $(y, x)$ has five elements. Clearly, it consists of $\{(x, y), (y, x), (xy, yx), (yx, xy), (0, 0)\}$ so it suffices to see that these elements are distinct. But this is immediate from the fact that $xy \neq yx$ and the observations above.

We have so far established that $\mathbf{V}$ is sji. It is not fji, because as was already seen in the proof of Theorem 7.2.3, $\mathbf{V} \leq \mathbf{G}_p \vee \mathbf{ACom}$ for any prime $p$. But neither $\mathbf{G}_p$ nor $\mathbf{ACom}$ contains $S$.                                                         □

*Question 7.3.23.* Given a finite set of computable pseudoidentities (say identities if you like), is it decidable whether $[\![E]\!]$ is sfji? A similar question can be asked for fji or for any of our other lattice theoretic adjectives.

Let us conclude this section with some further illustrations of the concepts we have been studying in the chapter with respect to compact pseudovarieties.

*Example 7.3.24.* Consider the symmetric group $S_3$ on three symbols. The unique pseudovariety contained in $(S_3)$ and not satisfying $xy = yx$ is $(S_3)$. So $(S_3)$ is sji by Proposition 7.2.13. The unique maximal subpseudovariety is $(\mathbb{Z}_6)$.

*Example 7.3.25.* Let $T_3$ be the full transformation semigroup on three letters. Then there are at least two elements of $\mathsf{MinExclude}(xy = yx; T_3)$, namely $(S_3)$ and $\mathbf{RZ}$.

**Exercise 7.3.26.** Calculate all the elements of $\mathsf{MinExclude}(xy = yx; T_3)$.

### 7.3.2 Non-compact fji and sfji pseudovarieties

In this section, we consider further examples of fji and sfji pseudovarieties, in particular, non-compact examples. First, we mention that whereas Theorem 7.2.10 asserts that each pseudovariety is a join of sji pseudovarieties, it is not the case that this join must be a finite join. For instance, let $\mathbf{Ab}$ be the pseudovariety of abelian groups. The structure theorem for finite abelian groups immediately implies that an sji pseudovariety of abelian groups must consist of abelian $p$-groups for a prime $p$. Hence no finite join of sji pseudovarieties can yield all abelian groups. A similar argument shows that the pseudovariety $\mathbf{G}_{nil}$ of finite nilpotent groups cannot be written as a finite join of sji pseudovarieties.

The following result is [7, Exercise 9.34]. The proof uses semigroup identities, so let us spell out our terminology. By a semigroup identity, we mean a formal equality between elements of a free semigroup. It is non-trivial if both sides are not the same word.

**Theorem 7.3.27.** *The pseudovariety* **N** *is sfji.*

*Proof.* First observe that any proper subpseudovariety of **N** satisfies a non-trivial semigroup identity. Indeed, $\widehat{F_{\mathbf{N}}}(A) = A^+ \cup 0$ with $0$ as the one-point compactification [7,25]. If $u = 0$ is a pseudoidentity over **N** with $u \in A^+$ and $a$ is a letter, then $u = 0$ has as a consequence $a^{|u|} = 0$, which in turn implies the non-trivial semigroup identity $a^{|u|} = a^{|u|+1}$. Thus every non-trivial pseudoidentity on **N** implies a non-trivial semigroup identity. Suppose for the moment that we can prove that any two non-trivial semigroup identities have a common non-trivial consequence and let $\mathbf{V}_1, \mathbf{V}_2 < \mathbf{N}$ be proper subpseudovarieties. Then, for each of $i = 1, 2$, there is a non-trivial semigroup identity $e_i$ so that $\mathbf{V}_i \models e_i$. If $e_1, e_2 \models e$ with $e$ a non-trivial semigroup identity, then $\mathbf{V}_1 \vee \mathbf{V}_2 \models e$ and hence $\mathbf{V}_1 \vee \mathbf{V}_2 \neq \mathbf{N}$. The theorem then follows from the next lemma. □

**Lemma 7.3.28 (Dean-Evans [78]).** *Any two non-trivial semigroup identities have a common non-trivial consequence.*

*Proof.* We shall use several times the easily verified fact that if $W \subseteq A^+$ is a set of words of equal length, then $W$ freely generates a free subsemigroup $W^+$ of $A^+$.

Let $u = v$ be a non-trivial identity over the alphabet $A$. Let $A'$ be a disjoint copy of $A$ and let $u' = v'$ be the corresponding identity in $A'$. Then consider the identity $uv' = vu'$ over $A \cup A'$. This is clearly a non-trivial consequence of $u = v$ in which both sides have the same length. So without loss of generality, assume $|u| = |v|$. Choose a set of words $W$ in $\{a, b\}^+$, all of equal length, in bijection with $A$. Denote by $\psi$ the extension of this bijection to an injective homomorphism $\psi \colon A^+ \hookrightarrow \{a, b\}^+$. Then $u\psi = v\psi$ is a non-trivial consequence of $u = v$ such that both sides have the same length. We conclude that every non-trivial identity has a non-trivial consequence in two variables in which both sides have the same length.

Let $u_1 = u_2$ and $v_1 = v_2$ be non-trivial identities over $\{a, b\}^+$ with both sides having equal length. Then $v_1, v_2$ freely generate a free subsemigroup of $\{a, b\}^+$. Let $\varphi \colon \{a, b\}^+ \hookrightarrow \{a, b\}^+$ be the injective endomorphism given by $a \mapsto v_1$, $b \mapsto v_2$. Then $u_1 \neq u_2$ implies that $u_1\varphi \neq u_2\varphi$ and so the identity $e$ given by $u_1\varphi = u_2\varphi$ is non-trivial. Clearly, $e$ is a consequence of $u_1 = u_2$. On the other hand, because $|u_1| = |u_2|$, it follows that $e$ is a consequence of $v_1 = v_2$ (as assuming this latter identity and replacing $v_2$ by $v_1$ in both sides of $u_1\varphi = u_2\varphi$ results in the trivial identity $v_1^{|u_1|} = v_1^{|u_2|}$). □

**Exercise 7.3.29.** Show that any subset $W \subseteq A^+$ of words of equal length freely generates a free subsemigroup.

**Corollary 7.3.30.** *The pseudovariety* **N** *is sfji, but neither fji nor sji.*

*Proof.* Theorem 7.3.27 shows that **N** is sfji. But we have already seen that $\mathbf{N} \leq \mathbf{ACom} \vee \mathbf{G}_p$, for any prime $p$, in the proof of Theorem 7.2.3, although **N** is contained in neither of the factors. So **N** is not fji. Also, **N** is not compact, or even locally finite, so it cannot be sji.  □

*Remark 7.3.31.* The proof of Theorem 7.3.27 shows that the Birkhoff variety of all semigroups is sji, which is the main result of [78].

On the other hand, Almeida proved:

**Theorem 7.3.32 (Almeida [5, 7]).** *The pseudovariety* **J** *is not sfji.*

*Question 7.3.33.* Give a proof of Theorem 7.3.32 that does not use pseudoidentities.

The following is one of the main results of [195]. We shall generalize this result shortly (although there are two cases that we do not obtain).

**Theorem 7.3.34 (Margolis, Sapir and Weil [195]).** *Let* **H** *be a pseudovariety of groups closed under extension (i.e.,* $\mathbf{H} \circ \mathbf{H} = \mathbf{H}$*). Then* $\overline{\mathbf{H}}$ *is sfji. In particular, the pseudovarieties of all finite semigroups and of all aperiodic semigroups are sfji.*

We now wish to provide some examples of fji pseudovarieties of groups. In order to do this we need some preliminary results relating join irreducibility in **PV** to join irreducibility in the lattice $[\mathbf{1}, \mathbf{G}]$.

**Lemma 7.3.35.** *Let* $L$ *be a complete lattice and let* $k \colon L \to L$ *be a* **sup** *kernel operator. Let* $\ell \in k(L)$. *Let* $\mathscr{P}$ *be any of the properties: ji, sji, fji, sfji. Then* $\ell$ *is* $\mathscr{P}$ *in* $L$ *if and only if it is* $\mathscr{P}$ *in* $k(L)$.

*Proof.* Because is $k(L)$ is closed under sups, clearly if $\ell$ is $\mathscr{P}$ in $L$, then it is $\mathscr{P}$ in $k(L)$. We handle the case of ji; the other cases are similar. Suppose $\ell$ is ji in $k(L)$ and $\ell \leq \bigvee_{i \in I} \ell_i$ with $\ell_i \in L$. Then

$$\ell = k(\ell) \leq k\left(\bigvee_{i \in I} \ell_i\right) = \bigvee_{i \in I} k(\ell_i).$$

So $\ell \leq k(\ell_i) \leq \ell_i$, some $i \in I$.  □

**Proposition 7.3.36.** *The map* $k \colon \mathbf{PV} \to \mathbf{PV}$ *given by* $\mathbf{V} \mapsto \mathbf{V} \cap \mathbf{G}$ *is a* **sup** *kernel operator with image* $[\mathbf{1}, \mathbf{G}]$. *Hence if* $\mathscr{P}$ *is any of the properties ji, sji, fji or sfji, then* $\mathbf{H} \in [\mathbf{1}, \mathbf{G}]$ *is* $\mathscr{P}$ *in* $[\mathbf{1}, \mathbf{G}]$ *if and only if it is* $\mathscr{P}$ *in* **PV**.

*Proof.* It is evident that $k$ is a kernel operator. To show that it is **sup** it suffices to demonstrate that the map $\mathbf{H} \mapsto \overline{\mathbf{H}}$ mapping $[\mathbf{1}, \mathbf{G}]$ to **PV** is right adjoint to the corestriction of $k$ to $[\mathbf{1}, \mathbf{G}]$. But clearly

$$\mathbf{V} \subseteq \overline{\mathbf{H}} \iff \mathbf{V} \cap \mathbf{G} \subseteq \mathbf{H}$$

so the two maps indeed form a Galois connection. The second statement follows from the first and Lemma 7.3.35.  □

The next theorem follows from a result of Auinger and the second author [37, 40] in conjunction with Proposition 7.3.36. Recall that if **V** is a pseudovariety, then $\mathbf{P}(\mathbf{V})$ is the pseudovariety of semigroups generated by power semigroups of semigroups in **V**.

**Theorem 7.3.37 (Auinger, Steinberg [40]).** *Let* **H** *be a pseudovariety of groups such that* $\mathbf{P}(\mathbf{H}) = \mathbf{J} \circ \mathbf{H}$. *Then* **H** *is fji.*

To give examples of when this theorem applies, we need the notion of a locally extensible pseudovariety of groups [37, 40].

**Definition 7.3.38 (Locally extensible).** *A pseudovariety of groups* **H** *is called locally extensible if, for each* $G \in \mathbf{H}$, *there is a non-trivial group* $C$ *such that the unitary wreath product* $C \wr G$ *belongs to* **H**.

Clearly, any non-trivial extension closed pseudovariety is locally extensible: in particular **G**, $\mathbf{G}_p$ and $\mathbf{G}_{sol}$ are all locally extensible. However many other pseudovarieties are locally extensible. For instance, the pseudovariety of groups with square-free exponent is locally extensible [37, 40]. So are pseudovarieties of the form $\mathbf{H} \circ \mathbf{K}$ where **H** is extension closed (or even locally extensible). If $\pi = \prod_{p \text{ prime}} p^{n_p}$, with $0 \leq n_p \leq \infty$, is a supernatural number [298, p. 33], [323] and $\mathbf{G}(\pi)$ is the pseudovariety of all finite groups with exponent dividing $\pi$, then $\mathbf{G}(\pi)$ is locally extensible if and only if $\pi$ is not a natural number [37, 40].

It is shown in [40] that any locally extensible pseudovariety of groups **H** satisfies $\mathbf{P}(\mathbf{H}) = \mathbf{J} \circ \mathbf{H}$ and hence, by the above theorem, is fji. We give here a direct proof for the locally extensible case, due to the second author, although the idea is directly taken from [40].

**Theorem 7.3.39 (Auinger, Steinberg [40]).** *Let* **H** *be a locally extensible pseudovariety of groups. Then* **H** *is fji. In particular,* **G**, $\mathbf{G}_p$ *and* $\mathbf{G}_{sol}$ *are fji.*

*Proof.* Let **H** be locally extensible. By Proposition 7.3.36, we need to show that if $\mathbf{V}_1$ and $\mathbf{V}_2$ are pseudovarieties of groups and $\mathbf{H} \leq \mathbf{V}_1 \vee \mathbf{V}_2$, then $\mathbf{H} \leq \mathbf{V}_i$, some $i = 1, 2$. Suppose that in fact $\mathbf{H} \nleq \mathbf{V}_i$, $i = 1, 2$. We show that $\mathbf{H} \nleq \mathbf{V}_1 \vee \mathbf{V}_2$.

Indeed, by Reiterman's Theorem [261] in the context of pseudovarieties of groups, we can find elements of a free profinite group $\pi_1, \pi_2$ such that

$\mathbf{V}_i \models \pi_i = 1$ and $\mathbf{H} \not\models \pi_i = 1$, $i = 1, 2$. Without loss of generality we may assume that $\pi_1, \pi_2$ have disjoint alphabets $A_1, A_2$, respectively. Let $a$ be a letter not belonging to $X = A_1 \cup A_2$ and let $Z = X \cup \{a\}$. Then $\mathbf{V}_2 \models a^{-1}\pi_2 a = 1$. It is straightforward to verify that $\mathbf{V}_1 \vee \mathbf{V}_2 \models \pi_1 a^{-1}\pi_2 a = a^{-1}\pi_2 a \pi_1$. Indeed, both sides are trivially true in both $\mathbf{V}_1$ and $\mathbf{V}_2$ and hence in $\mathbf{V}_1 \vee \mathbf{V}_2$. We show that

$$\mathbf{H} \not\models \pi_1 a^{-1}\pi_2 a = a^{-1}\pi_2 a \pi_1$$

or equivalently that

$$\mathbf{H} \not\models a = \pi_2^{-1} a \pi_1^{-1} a^{-1} \pi_2 a \pi_1. \tag{7.12}$$

This will establish that $\mathbf{H} \not\leq \mathbf{V}_1 \vee \mathbf{V}_2$ by Reiterman's Theorem.

Let $G \in \mathbf{H}$ be a relatively free $Z$-generated group such that $G \not\models \pi_i = 1$, $i = 1, 2$. Such exists by our assumptions on the $\pi_i$. By [37, Prop. 7.6], local extensibility implies that there is a prime $p$ such that the group $G^{\mathbf{Ab}(p)}$ (see (4.46)) belongs to $\mathbf{H}$. We show $H = G^{\mathbf{Ab}(p)}$ does not satisfy the pseudoidentity in (7.12).

Suppose that $H$ did satisfy this pseudoidentity. Choose words $w_i \in FG(A_i)$ such that $[w_i]_H = [\pi_i]_H$, $i = 1, 2$. Then we are assuming that

$$[a]_H = [w_2^{-1} a w_1^{-1} a^{-1} w_2 a w_1]_H. \tag{7.13}$$

Set $w = w_2^{-1} a w_1^{-1} a^{-1} w_2 a w_1$ in $FG(Z)$. As $[a]_H = (1 \xrightarrow{a} [a]_G, [a]_G)$ and $[w]_H = [a]_H$, we have $[w]_G = [a]_G$ and $w$ labels a path from $1$ to $[a]_G$ in the Cayley graph $\Gamma(G, A)$ of $G$ traversing the geometric edge $e = 1 \xrightarrow{a} [a]_G$ one time modulo $p$ (where forward traversals are counted positively and backwards traversals are counted negatively). Because $a$ does not appear in $w_1, w_2$, the only candidates for $e$ are

$$[w_2^{-1}]_G \xrightarrow{a} [w_2^{-1} a]_G \tag{7.14}$$

$$[w_2^{-1} a w_1^{-1} a^{-1}]_G \xrightarrow{a} [w_2^{-1} a w_1^{-1}] \tag{7.15}$$

$$[w_2^{-1} a w_1^{-1} a^{-1} w_2]_G \xrightarrow{a} [w_2^{-1} a w_1^{-1} a^{-1} w_2 a]_G. \tag{7.16}$$

In the case of (7.14), we obtain $1 = [w_2^{-1}]_G$ contradicting that $G \not\models \pi_2 = 1$. In the case of (7.15), we obtain $1 = [w_2^{-1} a w_1^{-1} a^{-1}]_G$. Because $G$ is relatively free on $Z$, we can substitute $1$ for the variables $\{a\} \cup A_1$ obtaining $1 = [w_2^{-1}]_G$, again contradicting our choice of $G$. Finally, if (7.16) is $e$, then $1 = [w_2^{-1} a w_1^{-1} a^{-1} w_2]$ and so $1 = [w_1^{-1}]_G$, contradicting that $G \not\models \pi_1 = 1$. Thus all cases lead to a contradiction showing that (7.13) cannot hold and so (7.12) does hold, as required. $\square$

*Question 7.3.40.* Find a proof of Theorem 7.3.39 that does not use pseudoidentities.

The above result can be used to show that the semidirect product operator does not preserve joins in the right-hand variable, but first a lemma.

**Lemma 7.3.41.** *Let* $\mathbf{H}_1, \mathbf{H}_2$ *be pseudovarieties of groups. Then the monoids in* $\mathbf{H}_1 * \mathbf{H}_2$ *are precisely the groups belonging to* $\mathbf{H}_1 \circ \mathbf{H}_2$.

*Proof.* Of course $\mathbf{H}_1 \circ \mathbf{H}_2 \leq \mathbf{H}_1 * \mathbf{H}_2$. Conversely, let $M$ be a monoid in $\mathbf{H}_1 * \mathbf{H}_2$. Then, without loss of generality, we may assume $M \leq H_1 \rtimes H_2$ where $H_1 \in \mathbf{H}_1$ and $H_2 \in \mathbf{H}_2$ (but $H_2$ need not act on $H_1$ by automorphisms). Let $\pi \colon H_1 \rtimes H_2 \to H_2$ be the projection and let $N = 1\pi^{-1} \cap M$. Then $N$ is a submonoid of $M$ containing all the idempotents of $M$. It follows $N$ embeds in $H_1$ (cf. Lemma 4.1.31) and so $M$ has a unique idempotent, and hence is a group. Thus $\pi|_M \colon M \to H_2$ is a group homomorphism with kernel $N \in \mathbf{H}_1$ and so $M \in \mathbf{H}_1 \circ \mathbf{H}_2$.    $\square$

**Proposition 7.3.42.** *Let* $\mathbf{G}_2, \mathbf{G}_3, \mathbf{G}_5$ *be the pseudovarieties of respectively* $2, 3, 5$-*groups. Then*

$$(\mathbf{G}_2 * \mathbf{G}_3) \vee (\mathbf{G}_2 * \mathbf{G}_5) < \mathbf{G}_2 * (\mathbf{G}_3 \vee \mathbf{G}_5). \tag{7.17}$$

*Proof.* That the left-hand side is contained in the right-hand side is clear. It is not hard to see using Lemma 7.3.41 that the monoids in the left-hand side are the elements of $(\mathbf{G}_2 \circ \mathbf{G}_3) \vee (\mathbf{G}_2 \circ \mathbf{G}_5)$, whereas the monoids in the right-hand side are the elements of $\mathbf{G}_2 \circ (\mathbf{G}_3 \vee \mathbf{G}_5)$. So if equality held in (7.17), then

$$(\mathbf{G}_2 \circ \mathbf{G}_3) \vee (\mathbf{G}_2 \circ \mathbf{G}_5) = \mathbf{G}_2 \circ (\mathbf{G}_3 \vee \mathbf{G}_5). \tag{7.18}$$

But the right-hand side of (7.18) is clearly a locally extensible pseudovariety of groups and so fji by Theorem 7.3.39. Therefore, it cannot be expressed as a join of two proper subpseudovarieties.    $\square$

We end this section with an answer to a question on joins raised by Almeida [7, Section 7.1, pg. 205]. Recall from (1.26) that there is a Galois connection

$$\mathbf{PV} \quad \overset{\mathbb{L}\mathbf{V} \longleftarrow \mathbf{V}}{\underset{\mathbf{V} \longmapsto \mathbf{V} \cap \mathbf{FMon}}{\xrightarrow{\hspace{3cm}}}} \quad \mathbf{MPV}$$

where **MPV** is the algebraic lattice of pseudovarieties of monoids. In particular, $\mathbb{L}$ preserves all infs. Almeida asked whether it preserves non-empty sups. He indicates that it is unlikely to be the case, but points out there is a difficulty in proving this because, for local pseudovarieties $\mathbf{V}$, one has $\mathbb{L}\mathbf{V} = \mathbf{V} * \mathbf{D}$ by the Delay Theorem [364] and $(-) * \mathbf{D}$ is $\sup_{\mathbf{B}}$ (that is, preserves non-empty sups). In fact, this difficulty is the key to finding a counterexample. Suppose that $\mathbf{V}$ and $\mathbf{W}$ are local pseudovarieties of monoids. Then there results the equalities

$$\mathbb{L}\mathbf{V} \vee \mathbb{L}\mathbf{W} = (\mathbf{V} * \mathbf{D}) \vee (\mathbf{W} * \mathbf{D}) = (\mathbf{V} \vee \mathbf{W}) * \mathbf{D}. \tag{7.19}$$

Now the Delay Theorem [364] implies that a non-trivial pseudovariety $\mathbf{U}$ of monoids is local if and only if $\mathbf{U} * \mathbf{D} = \mathbb{L}\mathbf{U}$. Thus $(\mathbf{V} \vee \mathbf{W}) * \mathbf{D} = \mathbb{L}(\mathbf{V} \vee \mathbf{W})$

if and only if $\mathbf{V} \vee \mathbf{W}$ is local. So in light of (7.19), to find a counterexample to $\mathbb{L}$ being $\mathbf{sup}$, we just need two local pseudovarieties of monoids whose join is not local. The pseudovarieties $\mathbf{A}$ and $\mathbf{G}$ are well-known to be local [364]. However, the second author showed [333, Thm. 2.2] that $\mathbf{g}(\mathbf{A} \vee \mathbf{G})$ is not even finite vertex rank (that is, it has no basis of pseudoidentities using graphs with a bounded number of vertices). Therefore, $\mathbb{L}\mathbf{A} \vee \mathbb{L}\mathbf{G} \neq \mathbb{L}(\mathbf{A} \vee \mathbf{G})$. We have thus proved:

**Proposition 7.3.43.** *The operator $\mathbb{L}$ does not preserve joins of pairs.*

# 7.4 Kovács-Newman Semigroups

In this section, we study a special class of $\times$-prime (and hence fji) semigroups that enjoy a strengthened form of subdirect indecomposability. A preliminary version of this section appeared in [287]. In this section all semigroups are assumed to be finite.

## 7.4.1 Kovács-Newman groups

We begin by refining slightly a result of Kovács and Newman [164–166, 217] showing that certain pseudovarieties of groups are ji. By Proposition 7.3.36, we may restrict our attention to the group setting.

**Definition 7.4.1 (Kovács-Newman group).** *We call a non-trivial group $G$ a Kovács-Newman group (or KN group for short) if it has the following property: whenever there is a diagram*

$$G \xleftarrow{\varphi} H \ll G_1 \times G_2 \tag{7.20}$$

*$\varphi$ factors through one of the projections.*

Because it is clear that $G \in \mathbf{H}_1 \vee \mathbf{H}_2$ if and only if there is a diagram as in (7.20) with $G_1 \in \mathbf{H}_1$ and $G_2 \in \mathbf{H}_2$, it follows that if $G$ is a KN group, then $G$ is $\times$-prime and so $(G)$ is fji and hence ji. Observe that if $G$ is a KN group and $G \in (H)$, then $G$ divides $H$. Indeed, $G$ must divide a product of copies of $H$ and so, being a KN group, it divides $H$. In particular, two non-isomorphic KN groups *cannot* generate the same pseudovariety.

We remark that there are ji pseudovarieties of groups that are not generated by KN groups. In fact, if $p$ is a prime, then $\mathbb{Z}_p$ is not a KN group, but $(\mathbb{Z}_p)$ is ji. To see that $\mathbf{Ab}(p) = (\mathbb{Z}_p)$ is ji, suppose $\mathbb{Z}_p$ divides $G_1 \times G_2$ with $G_i \in \mathbf{H}_i$, $i = 1, 2$. Then $p$ divides $|G_1| \cdot |G_2|$ and hence $|G_i|$ for some $i$. Thus $\mathbb{Z}_p$ is a subgroup of $G_i$ and so $\mathbb{Z}_p \in \mathbf{H}_i$; we conclude $\mathbb{Z}_p$ is $\times$-prime and hence $\mathbf{Ab}(p)$ is ji. The following proposition shows that KN groups do not have abelian normal subgroups (and hence no element of $\mathbf{Ab}(p)$ can be a

KN group). In [287], we were only able to prove that KN groups have a trivial center. The stronger result is due to L. G. Kovács.[2]

**Proposition 7.4.2 (Kovács).** *Let $G$ be a KN group. Then $G$ cannot have a non-trivial abelian normal subgroup. In particular, no solvable group is a KN group.*

*Proof.* Let $G$ be a group with a non-trivial abelian normal subgroup $A$ and let $\pi\colon G \to G/A$ be the natural projection. Let $H \subseteq G \times G$ be the congruence $\ker \pi$; that is $H = \{(g_1, g_2) \in G \times G \mid g_1\pi = g_2\pi\}$. Since $A$ is abelian, the conjugation action of $G$ on $A$ induces an action of $G/A$ on $A$ and so we can form the semidirect product $A \rtimes G/A$ with respect to this action. Consider the following two homomorphisms: $\psi_1\colon H \twoheadrightarrow G$ given by $(g_1, g_2)\psi_1 = g_2$ and $\psi_2\colon H \twoheadrightarrow A \rtimes G/A$ given by $(g_1, g_2)\psi_2 = (g_1 g_2^{-1}, g_2 A)$. It is clear that $\psi_1$ is an onto homomorphism. Turning to $\psi_2$, we compute

$$
\begin{aligned}
((g_1, g_2)(h_1, h_2))\psi_2 &= (g_1 h_1, g_2 h_2)\psi_2 \\
&= (g_1 h_1 h_2^{-1} g_2^{-1}, g_2 h_2 A) \\
&= (g_1 g_2^{-1} g_2 (h_1 h_2^{-1}) g_2^{-1}, g_2 h_2 A) \\
&= (g_1 g_2^{-1}, g_2 A)(h_1 h_2^{-1}, h_2 A) \\
&= (g_1, g_2)\psi_2 (h_1, h_2)\psi_2.
\end{aligned}
$$

One easily verifies $\ker \psi_1 = A \times \{1\}$ and $\ker \psi_2 = \{(a, a) \mid a \in A\}$. As these two normal subgroups of $H$ have trivial intersection, $H \ll G \times (A \rtimes G/A)$ with $\psi_1$ and $\psi_2$ as the respective projections.

Consider now the homomorphism $\varphi\colon H \twoheadrightarrow G$ given by $(g_1, g_2)\varphi = g_1$. Then $\ker \varphi = 1 \times A$, which is contained in neither $\ker \psi_1$ nor $\ker \psi_2$. We conclude $\varphi$ factors through neither $\psi_1$ nor $\psi_2$. This completes the proof that $G$ is not a KN group.    $\square$

Kovács and Newman essentially showed that there are a large number of KN groups. Clearly, any KN group must be monolithic: just take $\varphi = 1_G$ in (7.20).

**Exercise 7.4.3.** Verify that a KN group is monolithic.

Let us establish some notation. If $\psi\colon H \to G$ is a homomorphism and $N \lhd G$ is a normal subgroup, then $H$ acts on $N$ by first applying $\psi$ and then acting by conjugation. Let $C_H(N)$ denote the centralizer of $N$ under this action; it is a normal subgroup of $H$. Notice that $\ker \psi \leq C_H(N)$, in fact, $C_H(N) = C_G(N)\psi^{-1}$. Centralizers of minimal normal subgroups shall play an important role in this section thanks to Theorem 7.4.5 below.

**Lemma 7.4.4.** *Suppose $G$ is a monolithic group with non-abelian monolith $M$. Then $C_G(M)$ is trivial.*

---

[2] Personal communication.

*Proof.* If $C_G(M)$ is non-trivial, then as it is a normal subgroup, we must have $M \leq C_G(M)$. But this implies that $M$ is abelian. $\qquad\square$

The following theorem on lifting minimal normal subgroups and their centralizers was inspired by the results of Kovács and Newman [164–166,217], but first appeared in this explicit form in [287]. Our original proof was adapted from [217, Chapter 5, Section 3]. The elegant proof presented here is due to L. G. Kovács.[3]

**Theorem 7.4.5.** *Let $G$ be a monolithic group with monolith $M$. Suppose that one has a diagram as in (7.20). Let $\tau\colon G \to G/C_G(M)$ be the natural quotient map. Then $\varphi\tau$ factors through one of the direct product projections.*

*Proof.* Consider a diagram as per (7.20). Let $N_1$ be the kernel of the projection from $H$ onto $G_1$ and $N_2$ the kernel of the projection from $H$ onto $G_2$. Let $N = \ker\varphi$. Then $N_1 \cap N_2 = \{1\}$ and hence if $n_1 \in N_1$ and $n_2 \in N_2$, then $[n_1, n_2] \in N_1 \cap N_2 = \{1\}$ showing that $n_1$ and $n_2$ commute. If $N$ contains $N_i$ for some $i$, then $\varphi$ — and hence $\varphi\tau$ — factors through one of the projections. So assume that $N$ contains neither $N_1$ nor $N_2$. Then $N_1\varphi$ and $N_2\varphi$ are non-trivial normal subgroups of $G$ and therefore both contain $M$. Because $N_1\varphi$ and $N_2\varphi$ commute element-wise, it follows that $N_1\varphi, N_2\varphi \leq C_G(M)$. We conclude that $N_1, N_2 \leq C_H(M) = \ker\varphi\tau$, completing the proof. $\qquad\square$

As a consequence we obtain the following characterization of Kovács-Newman groups, one direction of which appeared in [287].

**Theorem 7.4.6.** *Let $G$ be a monolithic group. Then $G$ is a Kovács-Newman group if and only if its monolith is non-abelian.*

*Proof.* Proposition 7.4.2 shows that a KN group has a non-abelian monolith. The converse is immediate from Theorem 7.4.5 and Lemma 7.4.4. $\qquad\square$

**Corollary 7.4.7 (Kovács-Newman [164–166, 217]).** *Each distinct monolithic group with non-abelian monolith generates a **distinct** join irreducible pseudovariety.*

Another useful fact about minimal normal subgroups that we shall take advantage of later is the following.

**Lemma 7.4.8.** *Let $G$ be a group and $M \lhd G$ be a minimal normal subgroup. Then there exists a normal subgroup $N \lhd G$ such that $N \cap M = 1$ and $G/N$ is monolithic with monolith $MN/N \cong M$. Moreover, $C_G(M) = C_G(MN/N)$.*

*Proof.* Let $N$ be a maximal normal subgroup such that $N \cap M = 1$. Let $N < K \lhd G$; then $K \cap M \neq 1$. Because $M$ is minimal, we conclude $M \leq K$. It follows that $MN/N$ is the unique minimal normal subgroup of $G/N$. Because $M \cap N = 1$, we have $M \cong M/(M \cap N) \cong MN/N$.

---

[3] Personal communication.

Clearly, $C_G(M) \le C_G(MN/N)$. For the converse, suppose $m \in M$ and $g \in C_G(MN/N)$. Then $g^{-1}mg = mn$ with $n \in N$. So

$$n = m^{-1}(g^{-1}mg) \in M \cap N = 1.$$

We conclude $g \in C_G(M)$.                                                  $\square$

## Applications

We use the above results to provide some fji group pseudovarieties, including a weaker version of Theorem 7.3.39 (but with a more elementary proof).

Let $S_n$ be the symmetric group on $n$ letters and $A_n \lhd S_n$ the alternating group. Then $S_n$ is monolithic with monolith the simple non-abelian group $A_n$, for $n \ge 5$, and hence is a KN group by Theorem 7.4.6. Thus $\mathbf{G} = \bigvee(S_n)$ is fji by Lemma 6.1.13, as $\langle S_5 \rangle \le \langle S_6 \rangle \le \cdots$.

**Exercise 7.4.9.** Verify that, for $n \ge 5$, the symmetric group $S_n$ is monolithic.

The authors are indebted to J. Dixon for pointing out the following example. Set $G_i = \mathbb{PSL}(2, 2^i)$; the $G_i$ are simple non-abelian groups and $G_i \le G_j$ whenever $i \mid j$, so Lemma 6.1.13 shows $\mathbf{H} = \bigvee(G_i)$ is fji. Because, for any prime $q > 2$, the $q$-subgroups of $G_i$ are abelian, $\mathbf{H}$ is a proper fji pseudovariety of groups.

We now give a structural proof of a weaker version of Theorem 7.3.39.

**Corollary 7.4.10.** *Suppose $\mathbf{H}$ is a pseudovariety of groups such that, for each $G \in \mathbf{H}$, there is a simple non-abelian group $H$ with the unitary wreath product $H \wr G \in \mathbf{H}$. Then $\mathbf{H}$ is finite join irreducible.*

*Proof.* We use Lemma 7.2.18. Let $G \in \mathbf{H}$ and let $H$ be a simple non-abelian group such that $W = H \wr G \in \mathbf{H}$. Set $M = H^G \lhd W$; we claim that $M$ is a minimal normal subgroup. Indeed, suppose $1 \ne f \in M$; we show that the normal closure $N$ of $f$ is $M$. Conjugating by an element of $G$, we may assume $1f \ne 1$. Because $H$ has trivial center, there exists $h \in H$ such that $h^{-1}(1f)h \ne 1f$. Define $m \in M$ by $1m = h$, $h'm = 1$ for $h' \in H \setminus \{1\}$. Set $k = f(m^{-1}fm)^{-1}$; then $1k \ne 1$, $h'k = 1$ all $h' \in H \setminus \{1\}$ and also $k \in N$. Because $H$ is simple, it now follows that $K = \{m \in M \mid hm = 1, \forall h \in H \setminus \{1\}\} \le N$. But $\langle g^{-1}Kg \mid g \in G \rangle = M$.

By Lemma 7.4.8, there is a normal subgroup $N \lhd W$ such that: $N \cap M = 1$, $G_0 = W/N$ is monolithic with monolith $MN/N \cong M$ and $C_W(MN/N) = C_W(M)$. Because $G$ acts faithfully on $M$, $G \cap C_W(M) = 1$. Thus

$$G \cap N \le G \cap C_W(MN/N) = G \cap C_W(M) = 1$$

and so $G \le G_0$. Clearly, 1 of Lemma 7.2.18 is satisfied. Because $M$ is non-abelian, $G_0$ is a KN group and so 2 is also satisfied.                    $\square$

### 7.4.2 Kovács-Newman semigroups

Before defining Kovács-Newman semigroups, we recall some basic facts about finite semigroups. Each semigroup $S$ has a minimal ideal $K(S)$, sometimes called its *kernel*, which consists of a single $\mathscr{J}$-class and hence is a (completely) simple semigroup. Moreover, if $\varphi\colon S \twoheadrightarrow T$ is an onto homomorphism, then $K(S)\varphi = K(T)$.

**Exercise 7.4.11.** Suppose that $\varphi\colon S \twoheadrightarrow T$ is an onto homomorphism of finite semigroups. Prove $K(T) = K(S)\varphi$.

Notice that $S$ acts on both the left and right of $K(S)$. If $S$ is generalized group mapping with $K(S)$ as the distinguished $\mathscr{J}$-class, then we shall say $S$ is *generalized group mapping over its kernel*. Similar terminology will be used for the other types of mapping semigroups considered in Chapter 4.

Recall that a completely simple semigroup $S$ is always isomorphic to a regular Rees matrix semigroup $\mathscr{M}(G, A, B, C)$. It follows easily from [171, Chapter 8, Fact 2.22] or Section 5.5 that a non-trivial simple semigroup $S$ is group mapping if and only if $G$ is non-trivial and no two rows of $C$ are proportional on the left and no two columns of $C$ are proportional on the right.

**Exercise 7.4.12.** Prove that a non-trivial simple semigroup $S$ is group mapping if and only if $G$ is non-trivial and no two rows of $C$ are proportional on the left and no two columns of $C$ are proportional on the right.

In particular, any non-trivial group is group mapping. We now define the notion of a KN semigroup.

**Definition 7.4.13 (Kovács-Newman semigroup).** *A non-trivial semigroup $S$ is a Kovács-Newman semigroup (KN semigroup) if whenever there is a diagram*

$$S \overset{\varphi}{\leftarrow} T \ll T_1 \times T_2 \tag{7.21}$$

*$\varphi$ factors through one of the projections.*

As in the case of groups, it is clear that if $S$ is a KN semigroup, then $S$ is $\times$-prime and $(S)$ is ji; moreover, non-isomorphic KN semigroups generate *distinct* pseudovarieties. Also a KN semigroup must be subdirectly indecomposable.

It is not clear *a priori* that a KN group is a KN semigroup. This will be a consequence of our main result stating: a finite semigroup is a KN semigroup if and only if it is group mapping over its kernel and the maximal subgroup of its kernel is a KN group (that is a monolithic group with non-abelian monolith). In particular, each KN group is a KN semigroup.

To prove this, we shall need a special case of [171, Chapter 8, Prop. 3.28], which highlights the importance of group mapping semigroups by saying that homomorphisms to group mapping semigroups "have kernels."

**Proposition 7.4.14.** *Suppose that $\varphi\colon T \twoheadrightarrow S$ is a surjective homomorphism and that $S$ is group mapping over $K(S)$. Let $H$ be a maximal subgroup of $K(T)$ and suppose $\psi\colon T \twoheadrightarrow T'$ is a surjective homomorphism such that $\ker\psi|_H \leq \ker\varphi|_H$. Then $\varphi$ factors through $\psi$.*

*Proof.* Suppose that $t_1, t_2 \in T$ and $t_1\psi = t_2\psi$; we need to show that $t_1\varphi = t_2\varphi$. Because $S$ is group mapping, to do this it suffices, by Lemma 4.6.23, to show that, for all $k_1, k_2 \in K(S)$,

$$k_1(t_1\varphi)k_2 = k_1(t_2\varphi)k_2.$$

Because $K(S) = K(T)\varphi$, this is equivalent to showing, for all $j_1, j_2 \in K(T)$,

$$(j_1 t_1 j_2)\varphi = (j_1 t_2 j_2)\varphi.$$

Set $t_i' = j_1 t_i j_2$, $i = 1, 2$. First observe that $t_1' \,\mathscr{H}\, t_2'$ (by stability) and so they belong to the same maximal subgroup $G$ of $K(T)$. By, say Green's Lemma or Rees's Theorem, there exist $x, y, x', y' \in K(T)$ such that $u \mapsto xuy$ is a bijection from $G$ to $H$ with inverse given by $v \mapsto x'vy'$. Let $K = \ker\varphi|_H$ and $N = \ker\psi|_H$; so $N \leq K$ by hypothesis. Because $t_1\psi = t_2\psi$, it follows that $t_1'\psi = t_2'\psi$. Thus $(xt_1'y)\psi = (xt_2'y)\psi$ and so $xt_1'yN = xt_2'yN$. Because $N \leq K$, we have $xt_1'yK = xt_2'yK$, that is, $(xt_1'y)\varphi = (xt_2'y)\varphi$. Thus

$$t_1'\varphi = (x'xt_1'yy')\varphi = (x'xt_2'yy')\varphi = t_2'\varphi$$

completing the proof.    □

The following theorem, along with Theorems 7.4.6 and 7.4.22, can be viewed as one of the principal results of this section.

**Theorem 7.4.15.** *Let $S$ be a semigroup that is group mapping over $K(S)$ and such that $K(S)$ has a maximal subgroup $G$ that is a Kovács-Newman group (i.e., is monolithic with non-abelian monolith). Then $S$ is a Kovács-Newman semigroup. In particular, every KN group is a KN semigroup.*

*Proof.* Suppose we have a diagram as in (7.21). Let $\pi_i\colon T \to T_i$ be the projections, $i = 1, 2$. Because $K(S) = K(T)\varphi$, Lemma 4.6.10 implies that there is a maximal subgroup $H \leq K(T)$ such that $H\varphi = G$. Let $K = \ker\varphi|_H$; then $G = H/K$. Set $N_i = \ker\pi_i|_H$. Because $H \ll H\pi_1 \times H\pi_2$ and $G$ is a KN group, $N_i \leq K$ for some $i$. Proposition 7.4.14 then implies $\varphi$ factors through $\pi_i$.    □

We now aim to prove the converse. This answers a question raised in [287]. The proof proceeds via a series of reductions.

**Lemma 7.4.16.** *Let $S$ be a Kovács-Newman semigroup. Then $S$ cannot have a zero.*

*Proof.* Suppose $S$ has a 0. Let $T = S\Delta \cup (S \times \{0\}) \cup (\{0\} \times S) \ll S \times S$. Define a homomorphism $\varphi\colon T \twoheadrightarrow S$ by $(s, s)\varphi = s$, $(s, 0)\varphi = 0$, $(0, s)\varphi = 0$. It is evident that $\varphi$ does not factor through either projection.    □

By Corollary 4.7.7, a subdirectly indecomposable semigroup without zero falls into one of three mutually exclusive classes: it can either be group mapping, right letter mapping or left letter mapping over its kernel.

We begin with the group mapping case. Let $S$ be a non-trivial group mapping semigroup over its kernel $K(S)$ and suppose that the maximal subgroup $G$ of $K(S)$ is not a KN group. The semigroup $S$ acts on the right of the set $B$ of left ideals of $K(S)$ and the associated faithful transformation semigroup is $(B, \mathrm{RLM}_{K(S)}(S))$. One easily checks that elements of $K(S)$ act by constant maps on $B$ (by stability). Take a Rees matrix representation $\mathscr{M}(G, A, B, C)$ for the simple semigroup $K(S)$. Suppose $a_0 \in A$ is the $\mathscr{R}$-class of our choice of maximal subgroup and $b_0 \in B$ is its $\mathscr{L}$-class. We may assume all the entries of the column of $C$ associated to $a_0$ are the identity of $G$. Then $S$ embeds in the wreath product of transformation semigroups $G \wr (B, \mathrm{RLM}_{K(S)}(S))$. The embedding can be chosen to take $(a, g, b) \in K(S)$ to $(f, \bar{b})$ where $b'f = (b'Ca)g$ and $\bar{b}$ is the constant map taking $B$ to $b$. In particular, the embedding restricted to $G$, viewed as living in the $\mathscr{H}$-class $a_0 \cap b_0$, sends $g \in G$ to $(\bar{g}, \bar{b}_0)$ where $\bar{g} \colon B \to G$ is the constant map sending each element of $B$ to $g$. See Proposition 4.6.42 for details. The following result is a KN semigroup analogue of Theorem 4.7.10.

**Lemma 7.4.17.** *Let $S$ be a group mapping semigroup over its kernel. Then $S$ is a Kovács-Newman semigroup if and only if the maximal subgroup $G$ of $K(S)$ is a Kovács-Newman group.*

*Proof.* The sufficiency follows from Theorem 7.4.15. For necessity, suppose that $G$ is not a KN group. We continue to use the notation in the previous paragraph. Because $G$ is not a KN group, there exists a diagram (7.20) such that $\varphi$ does not factor through either projection $\pi_i \colon H \to G_i$.

Define maps $\tilde{\pi}_i \colon H \wr (B, \mathrm{RLM}_{K(S)}(S)) \to G_i \wr (B, \mathrm{RLM}_{K(S)}(S))$, $i = 1, 2$, by $(f, s)\tilde{\pi}_i = (f\pi_i, s)$. It is immediate $\tilde{\pi}_1, \tilde{\pi}_2$ induce a subdirect embedding

$$H \wr (B, \mathrm{RLM}_{K(S)}(S)) \ll (G_1 \wr (B, \mathrm{RLM}_{K(S)}(S))) \times (G_2 \wr (B, \mathrm{RLM}_{K(S)}(S))).$$

Also there is a map $\tilde{\varphi} \colon H \wr (B, \mathrm{RLM}_{K(S)}(S)) \to G \wr (B, \mathrm{RLM}_{K(S)}(S))$ given by $(f, s)\tilde{\varphi} = (f\varphi, s)$. Let $T = S\tilde{\varphi}^{-1}$ and let $T_i = T\tilde{\pi}_i$, $i = 1, 2$. Then we have a subdirect embedding $T \ll T_1 \times T_2$ and a map (abusing notation) $\tilde{\varphi} \colon T \twoheadrightarrow S$. We show that $\tilde{\varphi}$ does not factor through either projection. Let $\tilde{H} = \{(\bar{h}, \bar{b}_0) \mid h \in H\}$ (where $\bar{h}$ is the constant map to $h$). Then $\tilde{H} \cong H$ and $\tilde{H} \leq T$ (being in the inverse image of $G$). Moreover, $\tilde{\varphi}$ takes $\tilde{H}$ onto $G$ via $(\bar{h}, \bar{b}_0) \mapsto (\overline{h\varphi}, \bar{b}_0)$. Analogously, we can define $\tilde{G}_1 \cong G_1$ and $\tilde{G}_2 \cong G_2$ by $\tilde{G}_i = \{(\bar{g}, \bar{b}_0) \mid g \in G_i\} \leq T_i$ and the $\tilde{\pi}_i$ map $\tilde{H}$ to $\tilde{G}_i$ via $(\bar{h}, \bar{b}_0) \mapsto (\overline{h\pi_i}, \bar{b}_0)$, $i = 1, 2$. As $\varphi$ factors through neither $\pi_1$ nor $\pi_2$, we may now conclude that $\tilde{\varphi} \colon T \twoheadrightarrow S$ does not factor through either projection.     $\square$

We now turn to the right/left letter mapping case.

**Lemma 7.4.18.** *Let $S$ be a right letter mapping or left letter mapping semigroup over its kernel. Then $S$ is not Kovács-Newman.*

*Proof.* We just handle the case where $S$ is right letter mapping; the left letter mapping case is dual. In this case, $K(S)$ is a right zero semigroup with at least two elements and $S$ acts faithfully on the right of it. On the other hand $S$ must act trivially on the left of $K(S)$ by stability. Let $K(S) = \{c_1, \ldots, c_\ell\}$. If $\ell = 1$, then $S$ must be trivial and hence not a KN semigroup; so assume $\ell > 1$. Let us suppose for the moment that we can find a non-negative integer $n$ and three elements $x, y, z \in K(S)^n$ satisfying the following properties:

(S1) There is no position $1 \leq i \leq n$ where all three elements are $c_1$;
(S2) Given two elements $u \neq v$ from $\{x, y, z\}$ and a pair of not necessarily distinct elements $c_i, c_j$ from $K(S)$, there is a position $k$ where $u$ has $c_i$ and $v$ has $c_j$.

We shall choose $n$ and give a construction of such elements at the end of the proof.

Let $\Delta \colon S \to S^n$ be the diagonal map. Consider the subsemigroup $T$ of $S^n$ consisting of $S\Delta$ and the elements of the form $xs$, $ys$, $zs$ with $s$ in $S^\bullet$ (where we view $S$ as acting diagonally on the right of $K(S)^n$).

*Claim.* Any two distinct elements of the set $\{xS^\bullet, yS^\bullet, zS^\bullet\}$ have intersection contained in $S\Delta$.

*Proof.* Let $u \neq v$ be from $\{x, y, z\}$. First, suppose that $us = vs$ for some $s \in S$. If $s \in K(S)$, then $us = vs$ belongs to $S\Delta$, as $K(S)$ acts on $K(S)^n$ by constant maps, and we are done. If $s \notin K(S)$, then because $S$ is right letter mapping, $s$ does not act as a constant on $K(S)$. So $c_i s \neq c_j s$ some $i, j$. By (S2) there is a position $k$ where $u$ has $c_i$ and $v$ has $c_j$. Then clearly $xs$ and $ys$ differ in position $k$. This is a contradiction.

Now suppose that $us = vt$ with $s \neq t$. Then because $(K(S), S^\bullet)$ is a faithful transformation monoid, there exists $c_j \in K(S)$ such that $c_j s \neq c_j t$. By (S2) there is then a coordinate $k$ where both $u$ and $v$ have $c_j$. Clearly, $us$ and $vt$ differ in coordinate $k$, again leading to a contradiction. This establishes the claim. □

Fix coordinates $i, j, k$ such that $x$ has $c_1$ in coordinate $i$, $y$ has $c_1$ in coordinate $j$ and $z$ has $c_1$ in coordinate $k$; such exist by (S2). Now define a map $\varphi \colon T \twoheadrightarrow S$ by $s\Delta \mapsto s$, for $s \in S$, and by mapping elements in $xS^\bullet$ to their $i^{th}$-coordinate, mapping elements of $yS^\bullet$ to their $j^{th}$-coordinate and elements of $zS^\bullet$ to their $k^{th}$-coordinate.

*Claim.* The map $\varphi$ is a well-defined onto homomorphism.

*Proof.* Each pair from the collection $\{S\Delta, xS^\bullet, yS^\bullet, zS^\bullet\}$ has intersection contained in $S\Delta$ by the previous claim. As the various projections to the coordinates all agree on $S\Delta$, we conclude that $\varphi$ is well defined. It is clearly onto.

It remains to prove that $\varphi$ is a homomorphism. Clearly, $\varphi|_{S_\Delta}$ is a homomorphism. Because $K(S)$ consists of constant maps, if $u, v \in \{x, y, z\}$ and $s, t \in S^{\bullet}$, then $(us)(vt) = vt$ and so $((us)(vt))\varphi = c_1 t$. On the other hand, $(us)\varphi(vt)\varphi = (c_1 s)(c_1 t) = c_1 t$. If $t \in S$, $s \in S^{\bullet}$ and $u \in \{x, y, z\}$, then $(t\Delta)(us) = us$ and so $((t\Delta)(us))\varphi = c_1 s$. Also $(t\Delta)\varphi(us)\varphi = tc_1 s = c_1 s$, again as $K(S)$ consists of constant maps.

Finally we must verify that if $u \in \{x, y, z\}$, $s \in S^{\bullet}$ and $t \in S$, then $(us)\varphi(t\Delta)\varphi = ((us)(t\Delta))\varphi$. Indeed, $(us)\varphi(t\Delta)\varphi = c_1 st$. On the other hand $(us)(t\Delta) = u(st) \in uS^{\bullet}$ so its image under $\varphi$ is $c_1 st$.   □

If $S$ were a $KN$ semigroup, then an easy induction argument shows that $\varphi$ must factor through one of the projections from $T$ to $S$. But this is not the case. Indeed, by definition we have

$$x\varphi = c_1, \ \ y\varphi = c_1, \ \ z\varphi = c_1.$$

But by (S1) there is no coordinate where all three of $x, y, z$ are $c_1$. This shows that $S$ cannot be a KN semigroup, as desired.

It now remains to choose $n$ and construct $x, y, z$. Set $n = \ell^2$ (recall: $|K(S)| = \ell > 1$) and define $x$, $y$ and $z$ as follows:

$$x = (c_1, c_1, \ldots, c_1, c_2, c_2, \ldots, c_2, \ldots, c_\ell, c_\ell \ldots, c_\ell)$$
$$y = (c_1, c_2, \ldots, c_\ell, c_1, c_2, \ldots, c_\ell, \ldots, c_1, c_2, \ldots, c_\ell)$$
$$z = (c_\ell, c_{\ell-1}, \ldots, c_1, c_1, c_\ell, \ldots, c_2, \ldots, c_{\ell-1}, c_{\ell-2}, \ldots, c_\ell).$$

For example if $\ell = 3$, then

$$x = (c_1, c_1, c_1, c_2, c_2, c_2, c_3, c_3, c_3)$$
$$y = (c_1, c_2, c_3, c_1, c_2, c_3, c_1, c_2, c_3)$$
$$z = (c_3, c_2, c_1, c_1, c_3, c_2, c_2, c_1, c_3).$$

Let us establish some notation in order to give precise formulas. If $i$ is an integer, then $i \pmod{\ell}$ will denote the unique element of the interval $[1, \ell]$ that is congruent to $i$ modulo $\ell$. Then, for $1 \le i \le \ell^2$, one easily verifies that the following formulas apply:

$$x_i = c_{\lceil \frac{i}{\ell} \rceil} \tag{7.22}$$

$$y_i = c_{i \pmod{\ell}} \tag{7.23}$$

$$z_i = c_{\lceil \frac{i}{\ell} \rceil - i \pmod{\ell}}. \tag{7.24}$$

We remark that the sum of the indices on the right-hand sides of (7.23) and (7.24) is equal to index of the right-hand side of (7.22) modulo $\ell$. To verify (S1) observe that $x$ only has $c_1$ in the first $\ell$ coordinates. But among the first $\ell$ coordinates, $y$ has $c_1$ only in the first coordinate, whereas $z$ has $c_1$ only in the $\ell^{th}$ coordinate. This proves that (S1) holds.

We now turn to (S2). Let $c_i, c_j \in K(S)$. Then $x$ has $c_i$ in position $(i-1)\ell+j$ whereas $y$ has $c_j$ in this position. Let $k = j + i \pmod{\ell}$. Then $y$ has $c_i$ in position $\ell(k-1)+i$ whereas $z$ has $c_j$ in this position. For the final verification observe that if $x$ has $c_r$ in position $k$ and $y$ has $c_s$ in position $k$, then $z$ has $c_m$ in position $k$ where $m = r - s \pmod{\ell}$. So now let $s = i - j \pmod{\ell}$. We know that there is a position $k$ such that $x$ has $c_i$ in this position whereas $y$ has $c_s$ in this position by a previous case. Hence $z$ has $c_j$ in position $k$, because $j = i - s \pmod{\ell}$. This completes the proof that (S2) is satisfied and hence the proof of the lemma.                                                     □

Putting together Theorems 7.4.6 and 7.4.15 and Lemmas 7.4.16–7.4.18, we obtain the following result characterizing Kovács-Newman semigroups.

**Theorem 7.4.19.** *A finite semigroup is a Kovács-Newman semigroup if and only if it is group mapping over its kernel and the maximal subgroup of the kernel is monolithic with non-abelian monolith.*

### 7.4.3 Applications: Join irreducibility of $\overline{\mathbf{H}}$

Our first application is the following theorem, which can be viewed as a warmup exercise. It generalizes a result of Almeida [6] where strict finite join irreducibility is obtained.

**Theorem 7.4.20.** *The pseudovariety* **CS** *of (completely) simple semigroups is finite join irreducible.*

*Proof.* A subdirectly indecomposable simple semigroup must be a right zero semigroup, a left zero semigroup or a group mapping simple semigroup by Corollary 4.7.7. A non-trivial group $G$ can be embedded into the group mapping simple semigroup $S = \mathscr{M}(G, 2, 2, C)$ with structure matrix

$$C = \begin{pmatrix} 1 & 1 \\ 1 & g \end{pmatrix}$$

where $1 \neq g \in G$. Moreover, the two element left and right zero semigroups divide $S$, so we may conclude that **CS** is generated by the collection of all group mapping simple semigroups that are not groups.

Let $\mathscr{M}(G, A, B, C)$ be a group mapping simple semigroup, which is not a group, and suppose $G \leq S_n$ with $n \geq 5$. Then

$$\mathscr{M}(G, A, B, C) \leq \mathscr{M}(S_n, A, B, C)$$

(where we now view $C$ as a matrix over $S_n$) and the latter semigroup is group mapping. It follows that **CS** is generated by group mapping simple semigroups with structure group $S_n$, $n \geq 5$. Each such semigroup generates a ji pseudovariety by Theorem 7.4.19, as $S_n$ is monolithic with non-abelian monolith.

We now show that given two such semigroups

$$T_1 = \mathscr{M}(S_n, A, B, C), \quad T_2 = \mathscr{M}(S_j, A', B,' C')$$

there is another such a semigroup containing them both. First observe, that we may assume $n = j$ by replacing the smaller index by the larger one.

Next we construct a matrix $P$ over $S_n$ as follows. Without loss of generality, we may assume that $C$ has at least as many rows as $C'$. Then add to $C'$ as many rows of 1's as needed in order to obtain a matrix $C''$ with the same number of rows as $C$. Let $P = \begin{pmatrix} C & C'' \end{pmatrix}$. No two rows of $P$ are proportional on the left, as no two rows of $C$ are proportional on the left; however $P$ may have some columns proportional on the right. So we identify the proportional columns of $P$ to obtain a new matrix $P'$. The resulting Rees matrix semigroup $T$ with structure matrix $P'$ is group mapping over $S_n$ and so a KN semigroup. Because multiplying a column on the right by a scalar and changing the order of the columns does not change a Rees matrix semigroup, $T$ contains a copy of $T_1$ and $T_2$.

We may conclude that **CS** is a directed supremum of ji pseudovarieties and so an application of Lemma 6.1.13 establishes the theorem.  □

Recall that a semigroup $S$ is said to be an *ideal extension* of $B$ by $T$ if $B$ is an ideal of $S$ and $S/B = T$. For a semigroup $S$, we use $S^0$ to denote $S$ with an adjoined zero even if $S$ already had one.

**Lemma 7.4.21.** *Let $S$ be a semigroup, $G$ a non-trivial group and $1 \neq g \in G$. Then there is a semigroup $S(G,g)$ such that:*

1. *$S(G,g)$ is an ideal extension of $K(S(G,g))$ by $S^0$;*
2. *$G$ is the maximal subgroup of $K(S(G,g))$;*
3. *$S(G,g)$ is group mapping over $K(S(G,g))$.*

*Moreover, if $\varphi \colon G \to H$ is a surjective homomorphism and $g\varphi \neq 1$, then there is a natural surjective morphism $\widetilde{\varphi} \colon S(G,g) \twoheadrightarrow S(H, g\varphi)$ such that $\widetilde{\varphi}$ is injective on $S$ and $\widetilde{\varphi}|_G = \varphi$ (identifying $G$ and $H$ with appropriate maximal subgroups).*

*Proof.* We first construct a semigroup $S_0(G,g)$ meeting the requirements 1 and 2. Construct $K(G,g) = \mathscr{M}(G, A, B, P)$ as follows. Let $A$ be a set in bijection with $S^I \times S^I$ via $(s,t) \mapsto a_{s,t}$ and $B$ be a set in bijection with $S^I$ via $s \mapsto b_s$. Define a matrix $P \colon B \times A \to G$ by

$$b_s P a_{s',t} = \begin{cases} g & ss' = t \\ 1 & ss' \neq t. \end{cases}$$

We now form $S_0(G,g) = S \cup K(G,g)$ where multiplication of elements of $K(G,g)$ by elements of $S$ is defined as follows:

$$(a_{s_1,t}, h, b_{s_2})s = (a_{s_1,t}, h, b_{s_2s}), \text{with } s_1, s_2, t \in S^I, \ s \in S, \ h \in G$$
$$s(a_{s_1,t}, h, b_{s_2}) = (a_{ss_1,t}, h, b_{s_2}), \text{with } s_1, s_2, t \in S^I, \ s \in S, \ h \in G.$$

It is an exercise in the linked equations (5.6) to verify $S_0(G, g)$ is a semigroup with $K(S_0(G, g)) = K(G, g)$. Indeed, checking the linked equations in our setting amounts to verifying that $b_{s_2s}Pa_{s_1,t} = b_{s_2}Pa_{ss_1,t}$. But both sides are $g$ if $s_2ss_1 = t$ and 1 otherwise. Set $S(G, g) = \mathsf{GGM}_{K(G,g)}(S_0(G, g))$. Then $S(G, g)$ satisfies 2 and 3 by Proposition 4.6.31. We need to show that 1 is also satisfied. To prove this, it suffices to show that $\gamma_{K(G,g)}\colon S_0(G, g) \to S(G, g)$ does not identify elements of $S$. Suppose $s, s' \in S$ are distinct. Then

$$(a_{I,I}, 1, b_I)s(a_{I,s}, 1, b_I) = (a_{I,I}, 1, b_s)(a_{I,s}, 1, b_I) = (a_{I,I}, g, b_I)$$
$$(a_{I,I}, 1, b_I)s'(a_{I,s}, 1, b_I) = (a_{I,I}, 1, b_{s'})(a_{I,s}, 1, b_I) = (a_{I,I}, 1, b_I)$$

and so $s\gamma_{K(G,g)} \neq s'\gamma_{K(G,g)}$, as desired.

Suppose now $\varphi\colon G \twoheadrightarrow H$ is a surjective homomorphism with $g\varphi \neq 1$. The map $\Phi\colon S_0(G, g) \twoheadrightarrow S_0(H, g\varphi)$ defined by $s\Phi = s$ for $s \in S$ and $(a, g', b)\Phi = (a, g'\varphi, b)$ for $(a, g', b) \in K(G, g)$ is clearly a surjective homomorphism and $\Phi|_G = \varphi$. Because $K(S_0(G, g))$ is the minimal $\mathscr{J}$-class mapping onto $K(S_0(H, g\varphi))$, Proposition 4.6.31 immediately yields an onto homomorphism $\widetilde{\varphi}\colon S(G, g) \twoheadrightarrow S(H, g\varphi)$ such that

$$
\begin{array}{ccc}
S_0(G, g) & \overset{\Phi}{\twoheadrightarrow} & S_0(H, g\varphi) \\
\downarrow & & \downarrow \\
S(G, g) & \underset{\widetilde{\varphi}}{\twoheadrightarrow} & S(H, g\varphi)
\end{array}
\tag{7.25}
$$

commutes. From the commutativity of (7.25) and item 1 for $S(H, g\varphi)$, it immediately follows that $\widetilde{\varphi}$ is injective on $S$ and $\widetilde{\varphi}|_G = \varphi$.    □

Let **H** be a pseudovariety of groups. If **V** is a pseudovariety, set $\mathbf{V}(\mathbf{H}) = \mathbf{V} \cap \overline{\mathbf{H}}$. Our main application is the following theorem and its corollaries.

**Theorem 7.4.22.** *Suppose that* **H** *is a pseudovariety of groups containing a non-nilpotent group. Let* **V** *be a pseudovariety of semigroups containing a non-trivial semilattice and closed under ideal extensions of elements of* $\mathbf{CS}(\mathbf{H})$ *by elements of* **V**. *Then* **V** *is finite join irreducible.*

*Proof.* First note that **V** is closed under the operation of adjoining a (new) zero as it contains a non-trivial semilattice and $S \cup \{0\} = S \times U_1/(S \times \{0\})$.

We claim that **H** contains a monolithic group $G$ with non-central monolith $M$. Indeed, let $G \in \mathbf{H}$ be a non-nilpotent group of minimal order. Clearly, $G$ must be subdirectly indecomposable; let $M$ be its monolith. By choice of $G$, $G/M$ is nilpotent. If $M$ were central, then $G$ would be a central extension with nilpotent quotient and hence nilpotent, contradicting the choice of $G$.

Let $S \in \mathbf{V}$ and let $g \notin C_G(M)$. Set $S' = S(G, g)$ as per Lemma 7.4.21. By hypothesis, $S' \in \mathbf{V}$. Clearly, 1 of Lemma 7.2.18 holds; we show that 2 holds.

Suppose $S' \in \mathbf{V}_1 \vee \mathbf{V}_2$. Then there is a diagram $S' \overset{\varphi}{\leftarrow} T \ll T_1 \times T_2$ with $T_i \in \mathbf{V}_i$, $i = 1, 2$. As in the proof of Theorem 7.4.15, $K(T)\varphi = K(S')$ and there is a maximal subgroup $H$ of $K(T)$ with $H\varphi = G$. Also $H \ll H\pi_1 \times H\pi_2$, where $\pi_i \colon T \twoheadrightarrow T_i$ is the projection, $i = 1, 2$. Set $N_i = \ker(\pi_i|_H$, $i = 1, 2$. By Theorem 7.4.5, $N_i \leq C_H(M)$, some $i$, say $i = 1$.

Let $\rho \colon G \twoheadrightarrow G/C_G(M)$ be the projection and consider the induced map $\widetilde{\rho} \colon S(G, g) \twoheadrightarrow S(G/C_G(M), g\rho)$ as per Lemma 7.4.21; note that $g\rho \neq 1$. Then $N_1 \leq C_H(M) = \ker(\varphi\widetilde{\rho})|_H$, so, by Proposition 7.4.14, the quotient $\varphi\widetilde{\rho} \colon T \twoheadrightarrow S(G/C_G(M), g\rho)$ factors through $\pi_1$. Because $S \leq S(G/C_G(M), g\rho)$, it follows $S$ divides $T_1$ and so $S \in \mathbf{V}_1$. This completes the proof that 2 of Lemma 7.2.18 holds, establishing the theorem. $\qquad\square$

**Corollary 7.4.23.** *The pseudovarieties* **FSgp**, **CR** *and* **DS** *are all finite join irreducible. More generally, suppose* **H** *is a pseudovariety of groups containing a non-nilpotent group. Then* $\overline{\mathbf{H}}$, **CR(H)** *and* **DS(H)** *are all finite join irreducible. Hence none of these pseudovarieties has a maximal proper subpseudovariety.*

Because the only nilpotent pseudovarieties of groups closed under semidirect product are the trivial pseudovariety and pseudovarieties of the form $\mathbf{G}_p$ ($p$ prime), we have recovered all of the join results of Margolis, Sapir and Weil [195] (our Theorem 7.3.34) with the exceptions of $\mathbf{A} = \overline{\mathbf{1}}$ (aperiodic semigroups) and $\overline{\mathbf{G}}_p$. In fact, our results (when they apply) are stronger because we establish the property fji rather than sfji.

*Question 7.4.24.* Is it true that $\overline{\mathbf{H}}$ is sfji for every pseudovariety of groups **H**? Consider the same question for fji.

By Theorem 7.1.2, the mi pseudovarieties are obtained by excluding the fji semigroups and so we now have many new examples of such pseudovarieties using KN semigroups.

*Question 7.4.25.* Let $S$ be a KN semigroup with KN maximal subgroup $G$ in its minimal ideal. Describe $\mathsf{Excl}(S)$ in terms of $\mathsf{Excl}(G)$. In particular, if $\mathsf{Excl}(G) = [\![u = v]\!]$, then give a pseudoidentity defining $\mathsf{Excl}(S)$.

## 7.5 Irreducibility for the Semidirect Product

In this section, we consider analogues of irreducibility for the semidirect product. Let us begin with a trivial observation. Notice that a pseudovariety **V** is ji if and only if whenever $\mathbf{V} \leq \mathbf{V}_1 \vee \mathbf{V}_2 \vee \cdots$ one has $\mathbf{V} \leq \mathbf{V}_i$, some $i$. Indeed, this property implies fji and compactness; compactness follows because $\mathbf{V} = (S_1) \vee (S_2) \vee \cdots$ where $\mathbf{V} = \{S_1, S_2, \ldots\}$. Because ji pseudovarieties are

precisely the compact fji pseudovarieties, this establishes our claim. Similar remarks, of course, apply to sji. This leads to the following definitions (compare
with Definition 6.1.5).

**Definition 7.5.1.** *Let* **V** *be a pseudovariety. Then* **V** *is*

1. ***strictly semidirect irreducible (s\*i)*** *if* $\mathbf{V} = \bigcup_{n \geq 1} \mathbf{V}_n * \cdots * \mathbf{V}_2 * \mathbf{V}_1$,
   *for some sequence* $\mathbf{V}_1, \mathbf{V}_2, \ldots$, *implies* $\mathbf{V} = \mathbf{V}_i$ *for some i;*
2. ***semidirect irreducible (\*i)*** *if* $\mathbf{V} \leq \bigcup_{n \geq 1} \mathbf{V}_n * \cdots * \mathbf{V}_2 * \mathbf{V}_1$, *for some*
   *sequence* $\mathbf{V}_1, \mathbf{V}_2, \ldots$, *implies* $\mathbf{V} \leq \mathbf{V}_i$ *for some i;*
3. ***strictly finite semidirect irreducible (sf\*i)*** *if* $\mathbf{V} = \mathbf{V}_1 * \mathbf{V}_2$ *implies*
   $\mathbf{V} = \mathbf{V}_i$ *for some* $i = 1, 2;$
4. ***finite semidirect irreducible (f\*i)*** *if* $\mathbf{V} \leq \mathbf{V}_1 * \mathbf{V}_2$ *implies* $\mathbf{V} \leq \mathbf{V}_i$ *for*
   *some* $i = 1, 2$.

The next proposition collects some first properties concerning these notions.

**Proposition 7.5.2.**

1. *A \*i pseudovariety is ji and an f\*i pseudovariety is fji.*
2. *A pseudovariety* **V** *is \*i if and only if it is compact and f\*i.*
3. *If* **V** *is sf\*i (respectively s\*i) and* $\mathbf{V} * \mathbf{V} = \mathbf{V}$, *then* **V** *is sfji (respectively
   sji).*

*Proof.* We prove 1 and 2 simultaneously. Clearly, \*i implies f\*i, whereas con-
versely compactness and f\*i implies \*i. If **V** is f\*i and $\mathbf{V} \leq \mathbf{V}_1 \vee \mathbf{V}_2$, then
$\mathbf{V} \leq \mathbf{V}_1 * \mathbf{V}_2$ and so $\mathbf{V} \leq \mathbf{V}_i$ some $i = 1, 2$. Thus **V** is fji. Because ji is the conjunction of fji and compact, to finish the proof of 1 and 2, it suffices to show a
\*i pseudovariety is compact. Let **V** be \*i and suppose $\mathbf{V} = \{S_1, S_2, \ldots\}$. Then
$\mathbf{V} \leq \bigcup (S_n) * \cdots * (S_1)$ and so $\mathbf{V} = (S_i)$ for some $i$.

For 3, we just handle the case of sf\*i, the other case being similar. If
$\mathbf{V} = \mathbf{V}_1 \vee \mathbf{V}_2$, then $\mathbf{V} = \mathbf{V}_1 * \mathbf{V}_2$, as $\mathbf{V} = \mathbf{V} * \mathbf{V}$. Thus $\mathbf{V} = \mathbf{V}_i$ some $i = 1, 2$,
as required.                                                                    □

The assumption on closure under semidirect product is truly needed in 3.
For instance, $\mathbf{Sl} \vee \mathbf{RZ}$ is patently not sfji. But one easily checks that it is
sf\*i, as it does not contain the semidirect product of any two of its non-trivial
subpseudovarieties. See the Hasse diagram in (7.8).

**Exercise 7.5.3.** Verify $\mathbf{Sl} \vee \mathbf{RZ}$ is sf\*i.

We remark that fji does not imply f\*i. For instance, we saw **CS** is fji in
Theorem 7.4.20. But it is well-known $\mathbf{CS} = \mathbf{G} * \mathbf{RZ}$ (see Exercise 4.12.22)
so **CS** is not even sf\*i. Margolis, Sapir and Weil [195] proved the following
irreducibility result.

**Theorem 7.5.4 (Margolis-Sapir-Weil).** *Let* **H** *be a pseudovariety of groups
closed under extension. Then* $\overline{\mathbf{H}}$ *is strictly finite semidirect irreducible. In particular,* **FSgp** *and* **A** *are strictly finite semidirect irreducible.*

The Prime Decomposition Theorem leads to a complete classification of the *i pseudovarieties.

**Theorem 7.5.5.** *The non-trivial semidirect irreducible pseudovarieties are* $(Q)$, *with $Q$ a finite simple group,* $(U_2) = \mathbf{RRB}$, $(U_1) = \mathbf{Sl}$ *and* $(\overline{2}) = \mathbf{RZ}$.

*Proof.* Because *i pseudovarieties are precisely the compact f*i pseudovarieties, we analyze only compact pseudovarieties. Clearly, any ⋊-prime semigroup generates a *i pseudovariety, so by Theorem 4.1.45, each of the above-named pseudovarieties is *i. Conversely, suppose $(S)$ is f*i with $S$ non-trivial. By the Prime Decomposition Theorem, $S$ divides an iterated semidirect product of its finite simple group divisors and copies of $U_2$. Thus $(S) = (Q)$ with $Q$ a simple group, or $S \in (U_2)$. If $(S) = (U_2)$, there is nothing more to prove. On the other hand, because $\mathbf{Sl} \vee \mathbf{RZ}$ is the unique maximal proper subpseudovariety of $(U_2)$ by Example 7.3.1 and $(S)$ is fji by Proposition 7.5.2, we conclude $(S) = \mathbf{Sl}$ or $(S) = \mathbf{RZ}$.    □

This theorem shows that whereas ji and ⋊-prime are quite different concepts, ⋊-prime and *i are essentially the same thing (up to generating the same pseudovariety).

Lemma 6.1.13 admits the following analogue in this context, whose proof we omit.

**Lemma 7.5.6.** *The join of a directed set of f*i pseudovarieties is f*i.*

Because there are not many f*i semigroups, we do not get as much out of this result as we did in the join setting. Nonetheless, it is enough to show that the pseudovariety of finite groups is f*i.

**Theorem 7.5.7.** *The pseudovariety $\mathbf{G}$ of groups is finite semidirect irreducible.*

*Proof.* There is a well-known embedding of $S_n$ into $A_{2n}$ by the "doubling trick." It follows that $\mathbf{G} = \bigvee(A_n)$. Because $A_n$ is simple for $n \geq 5$, we conclude that $\mathbf{G}$ is finite semidirect irreducible from the previous lemma.    □

**Exercise 7.5.8.** Prove $S_n \leq A_{2n}$.

*Question 7.5.9.* Describe the semigroup structure of $(\mathbf{PV}, *)$. It is known that the semigroup of Birkhoff varieties of groups with the semidirect product as multiplication is a free monoid [217]. But this is not the case for $\mathbf{PV}$, which has many idempotents. Some results can be found in [369].

*Question 7.5.10.* Determine the s*i, sf*i and f*i pseudovarieties.

## 7.6 Irreducibility for Pseudovarieties of Relational Morphisms

Because $\mathbf{PVRM}^+$ is a monoid with respect to composition, it is natural to consider the analogous definitions in this context.

**Definition 7.6.1.** *Let* $\mathsf{V} \in \mathbf{PVRM}^+$. *Then* $\mathsf{V}$ *is*

1. *strictly irreducible if* $\mathsf{V} = \bigcup_{n \geq 1} \mathsf{V}_n \cdots \mathsf{V}_2 \mathsf{V}_1$, *where* $\mathsf{V}_1, \mathsf{V}_2, \ldots$ *is a sequence of positive pseudovarieties, implies* $\mathsf{V} = \mathsf{V}_i$ *for some* $i$;
2. *irreducible if* $\mathsf{V} \leq \bigcup_{n \geq 1} \mathsf{V}_n \cdots \mathsf{V}_2 \mathsf{V}_1$, *where* $\mathsf{V}_1, \mathsf{V}_2, \ldots$ *is a sequence of positive pseudovarieties, implies* $\mathsf{V} \leq \mathsf{V}_i$ *for some* $i$;
3. *strictly finite irreducible if* $\mathsf{V} = \mathsf{V}_1 \mathsf{V}_2$, *implies* $\mathsf{V} = \mathsf{V}_i$ *for some* $i = 1, 2$;
4. *finite irreducible if* $\mathsf{V} \leq \mathsf{V}_1 \mathsf{V}_2$, *implies* $\mathsf{V} \leq \mathsf{V}_i$ *for some* $i = 1, 2$.

**Proposition 7.6.2.** *A positive pseudovariety* $\mathsf{V}$ *is irreducible if and only if it is compact and finite irreducible.*

*Proof.* It is immediate that a compact and finite irreducible pseudovariety is irreducible. Conversely, an irreducible pseudovariety $\mathsf{V}$ is clearly finite irreducible. Let $\varphi_1, \varphi_2, \ldots$ be an enumeration of $\mathsf{V}$. Then

$$\mathsf{V} \leq \bigcup (\varphi_n)(\varphi_{n-1}) \cdots (\varphi_1)$$

(as we are working with positive pseudovarieties of relational morphisms). Thus $\mathsf{V} = (\varphi_i)$ some $i$. $\qquad\square$

In light of Proposition 7.6.2, a relational morphism $\varphi \colon S \to T$ is said to be *irreducible* if $(\varphi)$ is irreducible.

*Question 7.6.3.* Classify the irreducible relational morphisms. The Prime Decomposition Theorem for relational morphisms implies that an irreducible relational morphism $\varphi \colon S \to T$ must either have a non-trivial congruence-free monoid $M$ in the Mal'cev kernel and belong to $(M)_K$, or belong to $\ell 1_K$. Are the irreducible relational morphisms precisely the relations of this sort?

Notice that an MPS $\varphi$ need not be irreducible. For instance, the collapsing morphism $c_{B_2} \colon B_2 \to 1$ is an MPS. The MPS Decomposition Theorem (or the arguments in Section 2.6.3) imply $\varphi \in \mathbf{Sl}_D(\ell 1_D)^{op}$ although $\varphi$ belongs to neither $\mathbf{Sl}_D$ nor $(\ell 1_D)^{op}$ as it is routine to verify that the local semigroup $K_{c_{B_2}}(1, 1)$ is isomorphic to $B_2$. This example shows why relational morphisms are needed in the MPS Decomposition Theorem, as opposed to just homomorphisms.

## 7.7 The Abstract Spectral Theory of Cnt(PV)

It turns out that the spectra of $\mathbf{Cnt(PV)}$ and $\mathbf{Cnt(PV)}^{op}$, that is, the sets of finite meet and join irreducible elements of $\mathbf{Cnt(PV)}$, can be understood in the most part from knowledge of the corresponding spectral theory for $\mathbf{PV}$ and $\mathbf{PV}^{op}$ (and a little bit of topology).

In Proposition 2.2.2, the compact elements of $\mathbf{Cnt(PV)}$ were characterized as the finite joins of the operators $\delta(S,T)$ where $S$ and $T$ are finite semigroups. Let us extend the definition of $\delta$ to pseudovarieties.

**Proposition 7.7.1.** *Let $T$ be a finite semigroup. Then there is a* **sup** *order embedding $\delta(-,T)\colon \mathbf{PV} \to \mathbf{Cnt(PV)}$ defined by*

$$\delta(\mathbf{V},T)(\mathbf{W}) = \begin{cases} \mathbf{V} & T \in \mathbf{W} \\ \mathbf{1} & else. \end{cases}$$

*Proof.* Continuity of $\delta(\mathbf{V},T)$ is clear because if $\mathbf{W} = \bigvee D$ with $D$ directed, then $T \in \mathbf{W}$ if and only if $T$ belongs to some element of $D$. Now

$$\delta\left(\bigvee \mathbf{V}_i, T\right)(\mathbf{W}) = \begin{cases} \bigvee \mathbf{V}_i & T \in \mathbf{W} \\ \mathbf{1} & else \end{cases}$$

$$= \bigvee \delta(\mathbf{V}_i, T)(\mathbf{W})$$

establishing that $\delta(-,T)$ is **sup**. Because **sup** maps are order preserving, we just need to show $\delta(\mathbf{V}_1,T) \leq \delta(\mathbf{V}_2,T)$ implies $\mathbf{V}_1 \leq \mathbf{V}_2$ to complete the proof. But this follows immediately by evaluating both sides at **FSgp**.  □

Proposition 2.2.2 says that each compact element of $\mathbf{Cnt(PV)}$ is a finite join of operators of the form $\delta(\mathbf{V},T)$ with $\mathbf{V}$ compact and $T$ a finite semigroup. From Proposition 7.7.1, we obtain a characterization of sji elements of $\mathbf{Cnt(PV)}$.

**Proposition 7.7.2.** *An operator $\alpha \in \mathbf{Cnt(PV)}$ is sji if and only if there is an sji pseudovariety $\mathbf{V}$ and a finite semigroup $T$ such that $\alpha = \delta(\mathbf{V},T)$.*

*Proof.* Suppose first $\mathbf{V}$ is sji and $\delta(\mathbf{V},T) = \bigvee \alpha_i$. Then $\mathbf{V} = \delta(\mathbf{V},T)((T)) = \bigvee \alpha_i((T))$ and so $\mathbf{V} = \alpha_i((T))$ for some $i$, as $\mathbf{V}$ is sji. Because $\alpha_i \leq \delta(\mathbf{V},T)$, we need only establish the reverse inequality. If $\mathbf{W}$ is a pseudovariety with $T \notin \mathbf{W}$, then $\delta(\mathbf{V},T)(\mathbf{W}) = \mathbf{1} \leq \alpha_i(\mathbf{W})$. If $T \in \mathbf{W}$, then

$$\delta(\mathbf{V},T)(\mathbf{W}) = \mathbf{V} = \alpha_i((T)) \leq \alpha_i(\mathbf{W}).$$

Thus $\delta(\mathbf{V},T) \leq \alpha_i$, as required. We conclude $\delta(\mathbf{V},T)$ is sji.

For the converse, suppose $\alpha \in \mathbf{Cnt(PV)}$ is sji. Proposition 2.2.2 shows $\alpha = \bigvee \delta(\mathbf{V}_i,T_i)$ where the $\mathbf{V}_i$ are compact pseudovarieties. Thus $\alpha = \delta(\mathbf{V},T)$ for some compact pseudovariety $\mathbf{V}$. We claim $\mathbf{V}$ is sji. Indeed, suppose $\mathbf{V} = \bigvee \mathbf{V}_i$. Then $\delta(\mathbf{V},T) = \bigvee \delta(\mathbf{V}_i,T)$ by Proposition 7.7.1. Hence $\delta(\mathbf{V},T) = \delta(\mathbf{V}_i,T)$, some $i$. But then $\mathbf{V} = \mathbf{V}_i$ by another application of Proposition 7.7.1. This completes the proof.  □

As a corollary, we obtain that $\mathbf{Cnt}(\mathbf{PV})$ is sup-generated by its sji elements.

**Corollary 7.7.3.** *Every element of $\mathbf{Cnt}(\mathbf{PV})$ can be expressed as a supremum of sji elements.*

*Proof.* We saw in Proposition 2.2.2 that elements of the form $\delta(\mathbf{V}, T)$ with $\mathbf{V}$ compact generate $\mathbf{Cnt}(\mathbf{PV})$ under join. So it suffices to show that $\delta(\mathbf{V}, T)$ is a join of sji elements. Because $\mathbf{PV}$ is a locally dually algebraic lattice, Theorem 7.2.10 expresses $\mathbf{V} = \bigvee \mathbf{V}_i$ with the $\mathbf{V}_i$ sji pseudovarieties. Then $\delta(\mathbf{V}, T) = \bigvee \delta(\mathbf{V}_i, T)$ by Proposition 7.7.1 and each $\delta(\mathbf{V}_i, T)$ is sji by Proposition 7.7.2. $\qquad\qquad\square$

An entirely analogous argument serves to describe the ji elements of $\mathbf{Cnt}(\mathbf{PV})$. The proof is left to the reader.

**Proposition 7.7.4.** *The ji elements of $\mathbf{Cnt}(\mathbf{PV})$ are precisely the operators of the form $\delta(\mathbf{V}, T)$ with $\mathbf{V}$ a ji pseudovariety and $T$ a finite semigroup.*

**Exercise 7.7.5.** Prove Proposition 7.7.4.

From Theorem 7.1.3, we know that in an algebraic lattice there is a duality between ji elements and mi elements, namely each mi element is the exclusion of a ji element. This allows us to determine the mi elements of the algebraic lattice $\mathbf{Cnt}(\mathbf{PV})$ in terms of the mi elements of $\mathbf{PV}$.

Let $\mathbf{V}$ be a pseudovariety and $T$ a finite semigroup. We consider a dual construction to $\delta$. Define $\delta^*(\mathbf{V}, T) \in \mathbf{Cnt}(\mathbf{PV})$ by

$$\delta^*(\mathbf{V}, T)(\mathbf{W}) = \begin{cases} \mathbf{V} & \mathbf{W} \le (T) \\ \mathbf{FSgp} & \text{else.} \end{cases}$$

It is easy to verify $\delta^*(\mathbf{V}, T)$ is indeed continuous.

**Exercise 7.7.6.** Verify $\delta^*(\mathbf{V}, T)$ is continuous.

**Proposition 7.7.7.** *The mi elements of $\mathbf{Cnt}(\mathbf{PV})$ are the operators of the form $\delta^*(\mathbf{V}, T)$ where $\mathbf{V}$ is an mi pseudovariety and $T$ is a finite semigroup.*

*Proof.* The mi elements of $\mathbf{Cnt}(\mathbf{PV})$ are of the form $\mathsf{Excl}(\alpha)$ where $\alpha$ is ji by Theorem 7.1.3. Now $\alpha = \delta((S), T)$ with $S$ an fji semigroup by Proposition 7.7.4. Let $\mathbf{V} = \mathsf{Excl}(S)$; this is an mi pseudovariety. We claim $\mathsf{Excl}(\alpha) = \delta^*(\mathbf{V}, T)$. First note $\delta((S), T) \nleq \delta^*(\mathbf{V}, T)$ because

$$\delta((S), T)((T)) = (S) \nleq \mathsf{Excl}(S) = \mathbf{V} = \delta^*(\mathbf{V}, T)((T)).$$

Suppose now that $\delta((S), T) \nleq \gamma$. Then there exists a pseudovariety $\mathbf{V}_0$ such that $T \in \mathbf{V}_0$ and $(S) \nleq \gamma(\mathbf{V}_0)$, that is, $\gamma(\mathbf{V}_0) \le \mathsf{Excl}(S)$. Now if $\mathbf{W} \le (T)$, then $\gamma(\mathbf{W}) \le \gamma((T)) \le \gamma(\mathbf{V}_0) \le \mathsf{Excl}(S) = \mathbf{V}$. We conclude $\gamma \le \delta^*(\mathbf{V}, T)$, establishing the proposition. $\qquad\qquad\square$

Next we turn to sfji and fji operators. Here the situation is significantly more complicated. Nonetheless, we will take our cue from the fact that so far ji, sji and mi pseudovarieties take on at most two values! First let us dispense with the case of constant maps.

**Proposition 7.7.8.** *Let $\mathbf{V}$ be a pseudovariety and $q_{\mathbf{V}}$ the associated constant map on $\mathbf{PV}$. Then $q_{\mathbf{V}}$ is sfji (respectively fji) if and only if $\mathbf{V}$ is sfji (respectively fji).*

*Proof.* We just handle the sfji case. Suppose first $q_{\mathbf{V}}$ is sfji and $\mathbf{V} = \mathbf{V}_1 \vee \mathbf{V}_2$. Then $q_{\mathbf{V}} = q_{\mathbf{V}_1} \vee q_{\mathbf{V}_2}$ and so $q_{\mathbf{V}} = q_{\mathbf{V}_i}$, some $i = 1, 2$. Plainly we must then have $\mathbf{V} = \mathbf{V}_i$. Conversely, suppose $\mathbf{V}$ is sfji and $q_{\mathbf{V}} = \alpha \vee \beta$. Then $\mathbf{V} = \alpha(\mathbf{1}) \vee \beta(\mathbf{1})$ so we may assume $\mathbf{V} = \alpha(\mathbf{1})$. Hence if $\mathbf{W}$ is any pseudovariety, we have $q_{\mathbf{V}}(\mathbf{W}) = \mathbf{V} = \alpha(\mathbf{1}) \le \alpha(\mathbf{W})$ and hence $q_{\mathbf{V}} \le \alpha$. Because $\alpha \le q_{\mathbf{V}}$, this completes the proof. $\square$

Let $\mathbf{V}$ be a pseudovariety and let $\alpha \in \mathbf{Cnt(PV)}$. Define a continuous operator on $\mathbf{PV}$ by

$$
\alpha_{\mathbf{V}\prime}(\mathbf{W}) = \begin{cases} 1 & \mathbf{W} \le \mathbf{V} \\ \alpha(\mathbf{W}) & \text{else.} \end{cases}
$$

**Lemma 7.7.9.** *The operator $\alpha_{\mathbf{V}\prime}$ is continuous.*

*Proof.* Clearly, $\alpha_{\mathbf{V}\prime}$ is order preserving. Suppose $\mathbf{W} = \bigvee_{d \in D} \mathbf{W}_d$ with the $\mathbf{W}_d$ directed. Then, trivially, $\bigvee_{d \in D} \alpha_{\mathbf{V}\prime}(\mathbf{W}_d) \le \alpha_{\mathbf{V}\prime}(\mathbf{W})$. For the converse, if $\mathbf{W} \le \mathbf{V}$, then $\alpha_{\mathbf{V}\prime}(\mathbf{W}) = 1 = \bigvee_{d \in D} \alpha_{\mathbf{V}\prime}(\mathbf{W}_d)$ as each $\mathbf{W}_d \le \mathbf{V}$. If $\mathbf{W} \not\le \mathbf{V}$, then $\mathbf{W}_{d_0} \not\le \mathbf{V}$ some $d_0 \in D$, and hence $\mathbf{W}_d \not\le \mathbf{V}$ for $d \ge d_0$. Therefore, we compute

$$
\alpha_{\mathbf{V}\prime}(\mathbf{W}) = \alpha(\mathbf{W}) = \bigvee_{d \in D} \alpha(\mathbf{W}_d) = \bigvee_{d \ge d_0} \alpha(\mathbf{W}_d)
$$
$$
= \bigvee_{d \ge d_0} \alpha_{\mathbf{V}\prime}(\mathbf{W}_d) \le \bigvee_{d \in D} \alpha_{\mathbf{V}\prime}(\mathbf{W}_d)
$$

completing the proof. $\square$

Next we present a join decomposition of an element of $\mathbf{Cnt(PV)}$. We use here $\circ$ for composition.

**Lemma 7.7.10.** *Let $\alpha \in \mathbf{Cnt(PV)}$. Then $\alpha = \alpha \circ (\mathbf{V} \cap (-)) \vee \alpha_{\mathbf{V}\prime}$.*

*Proof.* Assume first $\mathbf{W} \not\le \mathbf{V}$. Then $\alpha(\mathbf{V} \cap \mathbf{W}) \le \alpha(\mathbf{W})$ and $\alpha_{\mathbf{V}\prime}(\mathbf{W}) = \alpha(\mathbf{W})$, so the right-hand side evaluated at $\mathbf{W}$ yields $\alpha(\mathbf{W})$. On the other hand, if $\mathbf{W} \le \mathbf{V}$, then the right-hand side evaluates to $\alpha(\mathbf{W}) \vee \mathbf{1} = \alpha(\mathbf{W})$. This establishes the lemma. $\square$

Our next goal is to show that an sfji operator takes on at most one non-trivial value. First we establish a lemma on operators taking on at least two non-trivial values.

**Lemma 7.7.11.** *Let $\alpha \in \mathbf{Cnt}(\mathbf{PV})$ take on at least two non-trivial values. Then there exist $\mathbf{V}_1 < \mathbf{V}_2$ with $1 < \alpha(\mathbf{V}_1) < \alpha(\mathbf{V}_2)$.*

*Proof.* Suppose $1 \neq \alpha(\mathbf{W}_1) \neq \alpha(\mathbf{W}_2) \neq 1$. Without loss of generality, assume $\alpha(\mathbf{W}_2) \not\leq \alpha(\mathbf{W}_1)$; in particular, notice $\mathbf{W}_2 \not\leq \mathbf{W}_1$. Then

$$1 < \alpha(\mathbf{W}_1) < \alpha(\mathbf{W}_1) \vee \alpha(\mathbf{W}_2) \leq \alpha(\mathbf{W}_1 \vee \mathbf{W}_2)$$

so taking $\mathbf{V}_1 = \mathbf{W}_1$, $\mathbf{V}_2 = \mathbf{W}_1 \vee \mathbf{W}_2$ does the job.     □

As a corollary of the above lemmas, we can show that any sfji operator takes on at most one non-trivial value.

**Corollary 7.7.12.** *Let $\alpha \in \mathbf{Cnt}(\mathbf{PV})$ be sfji. Then $\alpha$ takes on at most one non-trivial value.*

*Proof.* Suppose $\alpha \in \mathbf{Cnt}(\mathbf{PV})$ takes on at least two non-trivial values. By Lemma 7.7.11, we can find $\mathbf{V}_1 < \mathbf{V}_2$ so that $1 < \alpha(\mathbf{V}_1) < \alpha(\mathbf{V}_2)$. An application of Lemma 7.7.10 expresses $\alpha = \alpha \circ (\mathbf{V}_1 \cap (-)) \vee \alpha_{\mathbf{V}_1^\lambda}$. On the one hand, $\alpha(\mathbf{V}_1 \cap \mathbf{V}_2) = \alpha(\mathbf{V}_1) < \alpha(\mathbf{V}_2)$ shows $\alpha \circ (\mathbf{V}_1 \cap (-)) < \alpha$; on the other hand, $\alpha_{\mathbf{V}_1^\lambda}(\mathbf{V}_1) = 1 < \alpha(\mathbf{V}_1)$, so we also have $\alpha_{\mathbf{V}_1^\lambda} < \alpha$. This shows that $\alpha$ is not sfji.     □

Let us now characterize continuous maps $\alpha$ that take on exactly one non-trivial value, say $\mathbf{V}$. We remind the reader that we view **PV** as a topological space with the strong topology, see Definition 6.4.10. An order preserving operator on **PV** is continuous in this topology if and only if it is continuous in the sense of preserving directed joins by Theorem 6.4.13. The reader is reminded that every upset is closed in the strong topology.

**Proposition 7.7.13.** *Let $\alpha \in \mathbf{Cnt}(\mathbf{PV})$ have image $\{\mathbf{1}, \mathbf{V}\}$ with $\mathbf{V} \neq \mathbf{1}$. Let $U = \alpha^{-1}(\mathbf{V})$. Then $U$ is a proper, non-empty clopen upset in the strong topology. Conversely, if $\mathbf{V} \neq \mathbf{1}$ and $U$ is a non-empty clopen upset, then $\check{U}_\mathbf{V} \colon \mathbf{PV} \to \mathbf{PV}$ defined by*

$$\check{U}_\mathbf{V}(\mathbf{W}) = \begin{cases} \mathbf{V} & \mathbf{W} \in U \\ \mathbf{1} & else \end{cases}$$

*is continuous with image $\{\mathbf{1}, \mathbf{V}\}$.*

*Proof.* Because $\alpha$ is order preserving, if $\mathbf{U} \in U$ and $\mathbf{U} \leq \mathbf{W}$, then we have $\mathbf{V} = \alpha(\mathbf{U}) \leq \alpha(\mathbf{W}) \leq \mathbf{V}$, and so $\mathbf{W} \in U$. Thus $U$ is an upset and hence automatically closed. Now **PV** is Hausdorff and $\alpha$ is continuous, whence $\alpha^{-1}(\mathbf{1})$

is closed. But $U$ is the complement of $\alpha^{-1}(\mathbf{1})$, so $U$ is clopen. Because the image of $\alpha$ is $\{\mathbf{1}, \mathbf{V}\}$, $U$ is both proper and non-empty.

For the converse, $\check{U}_{\mathbf{V}}$ is clearly order preserving as $U$ is an upset. To prove $\check{U}_{\mathbf{V}}$ is continuous, we observe that because **PV** is Hausdorff, $\{\mathbf{1}, \mathbf{V}\}$ is discrete. Thus it suffices to verify $\check{U}_{\mathbf{V}}\colon \mathbf{PV} \to \{\mathbf{1}, \mathbf{V}\}$ is continuous, where the codomain has the discrete topology. But this follows immediately from $U$ being clopen. The assumption that $U$ is proper and non-empty says exactly that the image of $\check{U}_{\mathbf{V}}$ is $\{\mathbf{1}, \mathbf{V}\}$.    □

Recall that a closed subspace of a topological space $X$ is said to be *irreducible* if it cannot be expressed as a union of two proper closed subsets. Also recall that the Alexandrov topology on a poset takes the downsets as the open sets, and hence the upsets as closed sets.

**Theorem 7.7.14.** *An operator $\alpha \in \mathbf{Cnt(PV)}$ is sfji if and only if*

1. $\alpha = q_{\mathbf{V}}$ *where* **V** *is an sfji pseudovariety; or*
2. $\alpha = \check{U}_{\mathbf{V}}$ *where:*
    - (a) **V** *is a non-trivial sfji pseudovariety;*
    - (b) $U \subseteq \mathbf{PV}$ *is a proper, non-empty clopen upset of* **PV** *such that the subspace $U \cap K(\mathbf{PV})$ is irreducible in the poset $K(\mathbf{PV})$ of compact elements of* **PV** *equipped with the Alexandrov topology.*

*Proof.* The case where $\alpha$ is a constant map was already handled in Proposition 7.7.8. By Corollary 7.7.12 and Proposition 7.7.13, in order for a non-constant operator to be sfji, it must be of the form $\check{U}_{\mathbf{V}}$ for some non-trivial pseudovariety **V**, where $U$ is a clopen proper, non-empty upset of **PV**. So it suffices to show that (a) and (b) are the necessary and sufficient conditions on $U$ and **V** to obtain an sfji operator. We remark that $U \cap K(\mathbf{PV})$ is an upset, and so closed in the Alexandrov topology on $K(\mathbf{PV})$.

Suppose first that $\alpha = \check{U}_{\mathbf{V}}$ is sfji. If $\mathbf{V} = \mathbf{V}_1 \vee \mathbf{V}_2$, then one readily verifies $\alpha = \check{U}_{\mathbf{V}_1} \vee \check{U}_{\mathbf{V}_2}$ and hence $\alpha = \check{U}_{\mathbf{V}_i}$, some $i = 1, 2$. But then

$$\mathbf{V} = \alpha(\mathbf{FSgp}) = \check{U}_{\mathbf{V}_i}(\mathbf{FSgp}) = \mathbf{V}_i$$

and thus $\mathbf{V} = \mathbf{V}_i$. We conclude **V** is sfji. Suppose next that $U \cap K(\mathbf{PV}) = U_1 \cup U_2$ where $U_1$ and $U_2$ are closed in the Alexandrov topology on $K(\mathbf{PV})$. Then $U_1$ and $U_2$ are upsets. Define $\alpha_i\colon K(\mathbf{PV}) \to \mathbf{PV}$ by

$$\alpha_i(\mathbf{W}) = \begin{cases} \mathbf{V} & \mathbf{W} \in U_i \\ \mathbf{1} & \text{else} \end{cases}$$

for $i = 1, 2$ (where **W** is compact). Then as $U_i$ is an upset, a straightforward check shows $\alpha_i$ is order preserving and hence extends uniquely to a continuous operator on **PV**, also denoted $\alpha_i$, by Proposition 2.2.2. We claim $\alpha = \alpha_1 \vee \alpha_2$. To verify this, it suffices to check that both sides agree on compact pseudovarieties. So suppose **W** is compact. If $\mathbf{W} \in U \cap K(\mathbf{V}) = U_1 \cup U_2$, then $\alpha(\mathbf{W}) = \mathbf{V}$

and also $(\alpha_1 \vee \alpha_2)(\mathbf{W}) = \mathbf{V}$. If $\mathbf{W} \notin U$, then $\alpha(\mathbf{W}) = \mathbf{1} = \alpha_1(\mathbf{W}) = \alpha_2(\mathbf{W})$ and so, in fact, $\alpha = \alpha_1 \vee \alpha_2$ on all compact pseudovarieties. Thus $\alpha = \alpha_i$, some $i = 1, 2$. But from this, one immediately obtains $U \cap K(\mathbf{V}) = U_i$ by the definition of $\alpha_i$, establishing that $U \cap K(\mathbf{V})$ is irreducible.

Next suppose that $\mathbf{V}$ is sfji and $U \cap K(\mathbf{PV})$ is irreducible. We show that $\alpha = \check{U}_{\mathbf{V}}$ is sfji. Suppose $\alpha = \alpha_1 \vee \alpha_2$. Because $\mathbf{V}$ is sfji, we have $\mathbf{V} = \alpha(\mathbf{W})$ if and only if $\alpha_i(\mathbf{W}) = \mathbf{V}$ for some $i = 1, 2$. In other words,

$$U = \alpha^{-1}(\mathbf{V}) = \alpha_1^{-1}(\mathbf{V}) \cup \alpha_2^{-1}(\mathbf{V}). \tag{7.26}$$

Setting $U_i = \alpha_i^{-1}(\mathbf{V}) \cap K(\mathbf{PV})$, for $i = 1, 2$, we have each $U_i$ is an upset in $K(\mathbf{PV})$ and hence closed in the Alexandrov topology. Because $U \cap K(\mathbf{PV}) = U_1 \cup U_2$ by (7.26), we conclude $U \cap K(\mathbf{PV}) = U_i$, some $i$. We claim $\alpha = \alpha_i$. Because $\alpha_i \leq \alpha$, to prove this it suffices to show that $\alpha(\mathbf{W}) \leq \alpha_i(\mathbf{W})$ for any compact pseudovariety $\mathbf{W}$. So let $\mathbf{W}$ be compact. If $\mathbf{W} \notin U$, then $\alpha(\mathbf{W}) = \mathbf{1}$ and there is nothing to prove. If $\mathbf{W} \in U$, then $\mathbf{W} \in U \cap K(\mathbf{PV}) = U_i$ and so $\alpha(\mathbf{W}) = \mathbf{V} = \alpha_i(\mathbf{W})$. This completes the proof.  $\square$

There is of course an analogous result for fji operators, which we formulate but whose proof we omit.

**Theorem 7.7.15.** *An operator $\alpha \in \mathbf{Cnt}(\mathbf{PV})$ is fji if and only if*

1. $\alpha = q_{\mathbf{V}}$ *where $\mathbf{V}$ is an fji pseudovariety; or*
2. $\alpha = \check{U}_{\mathbf{V}}$ *where:*
   *(a) $\mathbf{V}$ is a non-trivial fji pseudovariety;*
   *(b) $U \subseteq \mathbf{PV}$ is a proper, non-empty clopen upset of $\mathbf{PV}$ such that whenever $U \cap K(\mathbf{PV})$ is contained in the union $U_1 \cup U_2$ of closed subspaces of $K(\mathbf{PV})$ in the Alexandrov topology, $U \cap K(\mathbf{PV}) \subseteq U_i$, some $i = 1, 2$.*

**Exercise 7.7.16.** Adapt the proof of Theorem 7.7.14 to prove Theorem 7.7.15.

*Question 7.7.17.* Study the abstract spectral theory of $\mathbf{GMC}(\mathbf{PV})$. What are the mi, smi, sfmi, fmi, ji, fji, sji, sfmi elements of $\mathbf{GMC}(\mathbf{PV})$?

## Notes

The spectral theory of lattices is the subject of [97, Chapter 5] and [147]. The standard results on algebraic lattices were taken from [203]. The notion of a locally dually algebraic lattice, and the basic properties of such, is novel to the best of our knowledge.

Shortly before publication, we obtained some new results about atoms that will appear in a forthcoming paper. We know that $\mathbf{CC}$ has no atoms. The atoms of $\mathbf{Cnt}(\mathbf{PV})^+$ are in bijection with compact smi pseudovarieties, and we do not know if the latter exist! The atoms of $\mathbf{PVRM}$ are $(1_{U_1}), (1_{\{a,b\}^\ell}), (1_{\bar{2}})$;

the atoms of $\mathbf{GMC(PV)}$ are the images of these under $\mathsf{q}$. We do not know anything about atoms of $\mathbf{CC^+}$ and $\mathbf{PVRM^+}$.

The known results on fji semigroups and pseudovarieties are attributed to their authors to the best of our knowledge. The fact that the exclusion of $V$ is $\mathbf{EDS}$ is a piece of semigroup folklore. The remaining results on this subject are new as far as we know. It is our opinion that the study of fji semigroups will be particularly fruitful and that they have not received enough attention in the past.

The notion of a Kovács-Newman group is inspired by the series of papers of Kovács and Newman [164–166, 217]. Their semigroup analogues were first studied by the authors in [287]; the final classification is new. The authors used the trick from the proof of Lemma 7.4.17 to show that a closed subgroup of a free profinite semigroup is a projective profinite group [290].

The study of pseudovarieties of semigroups that are irreducible with respect to the semidirect product is implicit in [169]; see also [195] where irreducibility with respect to the Mal'cev product is considered, as well. The study of irreducibility of morphisms is implicit in [293]. The results of Section 7.7 are new, but inspired by ideas from [97].

## Part IV

Quantales, Idempotent Semirings, Matrix
Algebras and the Triangular Product

# 8

# Quantales

Ordered semigroups have come to play an increasingly important role in Finite Semigroup Theory. For instance, Pin [228, 229] has shown that pseudovarieties of ordered monoids arise naturally in Formal Language Theory via the syntactic ordered semigroup. Recently, and more related to this work, Polák [242–244] has considered pseudovarieties of semilattice-ordered semigroups. A finite semilattice-ordered semigroup is a finite quantale in the sense of [303] or an idempotent semiring [100, 160]. These have an important relationship with Conway's universal automata [72] and language equations [242–244], as well as strong ties with tropical geometry [302]. More on idempotent semirings will appear in Chapter 9.

Both the semigroups **PVRM** and **Cnt(PV)** are ordered monoids. Moreover, they are also complete lattices, and there is a certain amount of compatibility between the multiplication and the lattice operations. This leads us to the notion of a quantale [303]. Classically, a quantale is a complete lattice with a semigroup multiplication such that left and right translations are **sup**. Morphisms between quantales are **sup** semigroup homomorphisms. However, **PVRM** and **Cnt(PV)** satisfy only a weaker form of the quantale axioms, and so we allow in this work a laxer notion of a quantale, which should perhaps more aptly be called a continuous quantale. Each quantale morphism is determined by a quantic nucleus, which is a closure operator that is also a dual prehomomorphism. The key example for us of a quantale morphism is $\mathsf{q}\colon \mathbf{PVRM} \to \mathbf{GMC(PV)}$. The associated quantic nucleus takes $V$ to $\max(V\mathsf{q})$.

Quantales (quantum locales) are part of the subject of pointless topology. Stone's Duality Theorem [341, 342] between Boolean algebras and compact totally disconnected Hausdorff spaces can be viewed as the first theorem in this subject. Studying profinite spaces is like doing algebraic geometry in the category of Boolean algebras or Boolean rings (the spectrum functor goes from coordinate ring to spaces reversing arrows). Hence profinite semigroups should correspond to bialgebras and profinite groups to Hopf algebras. The comultiplication turns out to be related to quantic nuclei. In particular, we put

J. Rhodes, B. Steinberg, *The q-theory of Finite Semigroups*,
Springer Monographs in Mathematics, DOI 10.1007/978-0-387-09781-7_8,
© Springer Science+Business Media, LLC 2009

a bialgebra structure on the Boolean ring of all recognizable subsets of $A^*$ and reformulate Eilenberg's Variety Theorem [85] in the language of bialgebras.

## 8.1 Ordered Semigroups and Quantales

Let us first recall the notion of an ordered semigroup.

**Definition 8.1.1.** *An ordered semigroup is a semigroup $S$ equipped with a partial order $\leq$ such that left and right translation are order preserving.*

Recent work of Pin and Weil [228–230, 236] [231, 237, 238] has revealed the importance of ordered semigroups in Finite Semigroup Theory and Formal Language Theory, in particular to understanding classes of recognizable sets that are not closed under complementation. See also [40, 325]. Here we are interested in quantales. Our definition of a quantale is slightly more general than the classical definition [303].

**Definition 8.1.2 (Quantale).** *A quantale is an ordered semigroup $(S, \leq)$, which is a complete lattice with respect to $\leq$ and such that, for all $s \in S$, the functions $x \mapsto s \cdot x$ and $x \mapsto x \cdot s$ are continuous; that is, left and right translations are continuous.*

Sometimes we will refer to such a quantale as a **cnt · cnt** quantale. More generally, from Chapter 1, we have the adjectives **cnt** (an abbreviation for continuous), **sup** and **sup$_B$**. If **x** and **y** are any two of these adjectives, then by an **x · y** quantale, we mean an ordered semigroup in which each left translation is a **y** map and each right translation is an **x** map. In other words the multiplication is **x** in the left variable and **y** in the right variable. So, for instance, a **cnt · sup** quantale is partially ordered semigroup such that, for each $s \in S$, the map $x \mapsto s \cdot x$ is **sup** and the map $x \mapsto x \cdot s$ is continuous.

With these conventions, the definition of quantale used in [303] is what we call a **sup · sup** quantale. Our notion of a quantale is to the classical notion as meet-continuous lattices are to frames. The semilattice-ordered semigroups (or idempotent semirings), studied by Polák [242–244] in the context of Formal Language Theory, are precisely the finite **sup · sup** quantales. Applications of quantales to graphs and networks can be found in [62]. Quantales are also related to the regular algebra of Conway [72]. Other applications of quantales in Computer Science can be found in [263]. Quantales also arise in non-commutative geometry [172, 173, 264, 303]. The following proposition gives several examples of quantales.

**Proposition 8.1.3 (Examples of Quantales).**

1. $(\mathbf{CC}, \subseteq, \odot)$ and $(\mathbf{CC}^+, \subseteq, \odot)$ are quantales.
2. $(\mathbf{PVRM}, \subseteq, \odot)$ and $(\mathbf{PVRM}^+, \subseteq, \odot)$ are quantales.
3. $(\mathbf{Cnt}(\mathbf{PV}), \leq, \cdot)$ and $(\mathbf{GMC}(\mathbf{PV}), \leq, \cdot)$ are **sup · cnt** quantales.

4. $(\mathbf{Cnt}(\mathbf{PV})^+, \leq, \cdot)$ and $(\mathbf{GMC}(\mathbf{PV})^+, \leq, \cdot)$ are $\sup_{\mathrm{B}} \cdot \mathrm{cnt}$ quantales.
5. $(2^{\mathrm{RM}}, \subseteq, \cdot)$ is a $\sup \cdot \sup$ quantale.
6. For a semigroup $S$, the triple $(P(S), \subseteq, \cdot)$ is a $\sup \cdot \sup$-quantale.
7. Let $L$ be a complete lattice and $\mathbf{Sup}(L)$ be the monoid of all $\mathbf{sup}$ maps from $L$ to $L$. Then $(\mathbf{Sup}(L), \leq_{\mathrm{pw}}, \cdot)$ is a $\sup \cdot \sup$ quantale.
8. Let $A$ be a $C^*$-algebra and $R(A)$ the set of closed right ideals of $A$. Then $(R(A), \subseteq, \cdot)$ (with $I \cdot J = \overline{IJ}$) is a $\sup \cdot \sup$ quantale.
9. $(\mathbf{PV}, \subseteq, *)$ is a $\sup_{\mathrm{B}} \cdot \mathrm{cnt}$ quantale.
10. If $L$ is a meet-continuous lattice, then $(L, \leq, \wedge)$ is a quantale. It is $\sup \cdot \sup$ if and only if $L$ is a frame.

*Proof.* We start with 1 and 2. Let $\mathsf{R} \in \mathbf{CC}$ and let $\{\mathsf{K}_i\}$ be a directed set from $\mathbf{CC}$. Because $\bigvee(\mathsf{R} \odot \mathsf{K}_i) \leq \mathsf{R} \odot \bigvee \mathsf{K}_i$, we just need to consider the other direction. As directed sups in $\mathbf{CC}$ are just unions, if $f$ belongs to the right-hand side, then $f \subseteq_s hg$ with $h \in \mathsf{R}$ and $g \in \mathsf{K}_i$ some $i$. Hence $f \in \bigvee(\mathsf{R} \odot \mathsf{K}_i)$. The argument for right translation is similar. The same argument also applies to $\mathbf{CC}^+$, $\mathbf{PVRM}$ and $\mathbf{PVRM}^+$.

We now handle 3 and 4. Suppose $\alpha \in \mathbf{Cnt}(\mathbf{PV})$ and $\{\beta_i\}$ is a directed set from $\mathbf{Cnt}(\mathbf{PV})$. Let $\mathbf{V}$ be a pseudovariety. Then because the ordering is pointwise, the $\beta_i(\mathbf{V})$ form a directed set. Hence, because $\alpha$ is continuous,

$$\alpha\left(\bigvee \beta_i(\mathbf{V})\right) = \bigvee \alpha\beta_i(\mathbf{V}).$$

Because the join in $\mathbf{Cnt}(\mathbf{PV})$ is pointwise, this shows that

$$\alpha\left(\bigvee \beta_i\right) = \bigvee(\alpha\beta_i).$$

Suppose now that $\alpha \in \mathbf{Cnt}(\mathbf{PV})$ and $\{\beta_i\}$ is an arbitrary collection of elements from $\mathbf{Cnt}(\mathbf{PV})$. Because sups are taken pointwise in $\mathbf{Cnt}(\mathbf{PV})$, for $\mathbf{V} \in \mathbf{PV}$, we have

$$\left(\bigvee \beta_i\right)\alpha(\mathbf{V}) = \bigvee(\beta_i(\alpha(\mathbf{V}))).$$

Hence $(\bigvee \beta_i)\alpha = \bigvee(\beta_i\alpha)$, as required. The arguments for $\mathbf{Cnt}(\mathbf{PV})^+$, $\mathbf{GMC}(\mathbf{PV})$ and $\mathbf{GMC}(\mathbf{PV})^+$ are similar, only observe that in the positive situation, multiplication does not commute with the empty sup.

We leave 5 and 6 as exercises. The proof of 7 is similar to 3. One can find 8 in [303]. For 9, we have already seen in Chapter 2 that if $\mathbf{V} \in \mathbf{PV}$, then the operator $\mathbf{V} * (-)$ is continuous. It is also clear that if $\mathbf{W} \in \mathbf{PV}$, then the operator $(-) * \mathbf{W}$ is continuous. Thus it suffices to show that if $\mathbf{V}_1, \mathbf{V}_2$ are pseudovarieties, then

$$(\mathbf{V}_1 \vee \mathbf{V}_2) * \mathbf{W} = (\mathbf{V}_1 * \mathbf{W}) \vee (\mathbf{V}_2 * \mathbf{W}). \tag{8.1}$$

Because the right-hand side of (8.1) is contained in the left-hand side, we just need the other direction. But if $S_1, S_2$ and $T$ are semigroups, then $(S_1 \times S_2) \wr T$ embeds in $(S_1 \wr T) \times (S_2 \wr T)$ by Exercise 1.2.13. This implies the equality in (8.1). Finally, 10 is clear. $\qquad\square$

**Exercise 8.1.4.** Prove 5 and 6 from Proposition 8.1.3.

*Remark 8.1.5.* If $L$ is a compact meet semilattice with identity, then $L$ is automatically a meet-continuous lattice and hence a quantale [97, 147].

Let us provide some methods for creating quantales.

**Definition 8.1.6.** *Let $Q$ be a partially ordered semigroup. Define $Q^{+\infty}$ to be $Q \cup \{+\infty\}$ where $+\infty$ is a new top for the order and a new zero for the multiplicative structure. Dually define $Q^0$ to be $Q \cup \{0\}$ where $0$ is a new bottom for the order and a zero for the multiplication.*

**Proposition 8.1.7.** *If $Q$ is an $\mathbf{x} \cdot \mathbf{y}$-quantale, where $\mathbf{x}, \mathbf{y} \in \{\mathbf{Cnt}, \mathbf{sup_B}\}$, then $Q^{+\infty}$ and $Q^0$ are also $\mathbf{x} \cdot \mathbf{y}$-quantales. Moreover, if $Q$ is any partially ordered semigroup, then in $Q^0$ left and right translations preserve empty sups.*

*Proof.* We handle the case $Q^{+\infty}$. Clearly, $Q^{+\infty}$ is a semigroup and a complete lattice. The key point is that if $\emptyset \neq X \subseteq Q^{+\infty}$, then

$$\bigvee X = +\infty \iff +\infty \in X.$$

So if $\bigvee X = +\infty$, then for all $s \in Q^{+\infty}$

$$\left(\bigvee X\right) \cdot s = +\infty = \bigvee(X \cdot s)$$

and dually. Also if $s = +\infty$, then for all $X \subseteq Q^{+\infty}$,

$$\left(\bigvee X\right) \cdot s = +\infty = \bigvee(X \cdot s)$$

and dually. So the only cases remaining are when $X \subseteq Q$ and $s \in Q$, which are handled by the fact that $Q$ is an $\mathbf{x} \cdot \mathbf{y}$-quantale.

The argument for $Q^0$ is left to the reader. The last statement is clear. $\square$

**Exercise 8.1.8.** Complete the proof of Proposition 8.1.7.

**Corollary 8.1.9.** *If $Q$ is a $\mathbf{sup} \cdot \mathbf{sup}$-quantale, then so is $(Q^{+\infty})^0$.*

**Exercise 8.1.10.** Recall that a map of lattices is called $\mathbf{inf}_T$ if it preserves all non-empty infima (it need not preserve the top). Let $Q_1, Q_2$ be $\mathbf{x} \cdot \mathbf{y}$-quantales and suppose that $f \colon Q_1 \to Q_2$ is $\mathbf{inf}_T$. Then the extended map $\bar{f} \colon Q_1^{+\infty} \to Q_2^{+\infty}$ given by $f$ on $Q_1$ and by $f(+\infty) = +\infty$ is $\mathbf{inf}$.

We now associate to each group a quantale. This construction will be used in the next chapter.

**Definition 8.1.11 ($G^\natural$).** *Let $G$ be a group and set $G^\natural = (G^{+\infty})^0$.*

Here we view $G$ being partially ordered by equality.

**Proposition 8.1.12.** *If $G$ is a group, then $G^{\natural}$ is a* **sup** $\cdot$ **sup** *quantale.*

*Proof.* Clearly, $G^{\natural}$ is a complete lattice. The multiplication preserves empty sups because 0 is both the bottom and a multiplicative zero. Let $\emptyset \neq X \subseteq G^{\natural}$ and $s \in G^{\natural}$. We show $s \cdot \bigvee X = \bigvee s \cdot X$. The dual equality is proved similarly.

If $X = \{0\}$, this is clear; else $s \cdot \bigvee X = s \cdot \bigvee (X \setminus \{0\})$ and $\bigvee s \cdot X = \bigvee s \cdot (X \setminus \{0\})$. So without loss of generality, we may assume $0 \notin X$. Also we may assume $s \neq 0$, as again the desired result is trivial otherwise. If $X$ is a singleton set, there is nothing to prove. So assume $X$ contains at least two non-zero elements. Then $\bigvee X = +\infty$. Now if $s \in G$, then $s \cdot X$ also has at least two non-zero elements and so $\bigvee(s \cdot X) = +\infty = s \cdot \bigvee X$, as required. If $s = +\infty$, then $s \cdot X = \{+\infty\}$ and so again $\bigvee(s \cdot X) = +\infty = s \cdot \bigvee X$. This completes the proof. $\qquad\square$

## 8.2 Green's Preorders vs. the Quantale Ordering

It is natural to compare Green's preorders with the order of the quantale. We commence with the following well-known fact concerning ordered monoids [190, 229, 348], which already has been exploited several times in this text.

**Proposition 8.2.1.** *Let $M$ be an ordered monoid satisfying $m \leq 1$ for each $m \in M$. Then, for all $m, n \in M$,*

$$m \leq_{\mathscr{J}} n \implies m \leq n.$$

*Conversely, if $e \in M$ is an idempotent, then*

$$e \leq m \in M \implies e \leq_{\mathscr{H}} m.$$

*In particular, $M$ is $\mathscr{J}$-trivial. A dual result holds when 1 is the least element.*

*Proof.* Because $m \leq_{\mathscr{J}} n$, there exist $u, v \in M$ such that $m = unv$. But then $m = unv \leq 1n1 = n$, as desired.

Suppose $e \leq m$ with $e$ idempotent. Then

$$e = e^2 \leq em \leq e1 = e.$$

So $e = em$ and hence $e \leq_{\mathscr{L}} m$. A dual argument establishes $e \leq_{\mathscr{R}} m$. $\qquad\square$

There is also the following observation relating the natural order on idempotents and the semigroup order [40]. Recall that the set of idempotents of a semigroup is ordered by $e \leq f$ if $ef = e = fe$, i.e., $e \leq_{\mathscr{H}} f$.

**Proposition 8.2.2.** *Let $(M, \preceq)$ be an ordered monoid. Then the natural order on idempotents coincides with the order $\preceq$ if and only if $e \preceq 1$ for each idempotent $e$ of $M$. Dually, the natural order on the idempotents is the reverse of the order $\preceq$ if and only if $1 \preceq e$ for all idempotents $e$ of $M$.*

*Proof.* Suppose first that the natural order coincides with $\preceq$ for idempotents. Then as $e \leq 1$ for any idempotent $e$, it follows $e \preceq 1$. Conversely, if $e \preceq 1$ for each idempotent $e$ and if $e, f$ are idempotents with $e \leq f$, then we have that $e = ef \preceq 1f = f$. The second assertion is dual.    $\square$

It follows $\mathbf{Cnt}(\mathbf{PV})^+$ is $\mathscr{J}$-trivial and the $\mathscr{J}$-order refines the reverse of the pointwise ordering; moreover, on idempotents the $\mathscr{J}$-order is the reverse of the pointwise ordering. More generally, if $\alpha$ is idempotent, then $\beta \leq \alpha$ if and only if $\alpha \leq_{\mathscr{H}} \beta$. Similar remarks apply to $\mathbf{GMC}(\mathbf{PV})^+$.

Recall that:

$$\mathbf{Cnt}(\mathbf{PV})^- = \{\alpha \in \mathbf{Cnt}(\mathbf{PV}) \mid \alpha \leq 1\}$$
$$\mathbf{GMC}(\mathbf{PV})^- = \{\alpha \in \mathbf{GMC}(\mathbf{PV}) \mid \alpha \leq 1\}.$$

For these quantales, the $\mathscr{J}$-order refines the pointwise ordering and these orders coincide for idempotents. More generally, if $\alpha$ is idempotent, then $\alpha \leq \beta$ if and only if $\alpha \leq_{\mathscr{H}} \beta$.

*Question 8.2.3.* What does the $\mathscr{J}$-order look like in $\mathbf{Cnt}(\mathbf{PV})^+$? Give an example of when $\alpha \leq \beta$ but $\beta \not\leq_{\mathscr{J}} \alpha$ (or prove there is no such example). Consider the analogous questions for $\mathbf{GMC}(\mathbf{PV})^+$, $\mathbf{Cnt}(\mathbf{PV})^-$ and $\mathbf{GMC}(\mathbf{PV})^-$.

For an example of a quantale in which the $\mathscr{J}$-order is very far from the quantale ordering, consider $Q = \mathbb{Z}^\natural$. Then $Q$ is a commutative $\mathbf{sup} \cdot \mathbf{sup}$-quantale. The $\mathscr{J}$-order and $\leq$ agree except for on pairs of the form $(+\infty, n)$ and $(n, +\infty)$, $n \in \mathbb{Z}$, where they are reversed. Namely, we have $+\infty \leq_{\mathscr{J}} n$ but $n \leq +\infty$.

**Exercise 8.2.4.** Let $L$ be any complete lattice with at least two elements and define a multiplication on $L$ by endowing $L \backslash \{\mathsf{B}\}$ with the right zero semigroup structure and making $\mathsf{B}$ an adjoined 0. Prove that $L$ is a $\mathbf{sup} \cdot \mathbf{sup}$-quantale. Notice that all of $L \setminus \{\mathsf{B}\}$ lies in one $\mathscr{R}$-class, so the relationship between the $\mathscr{R}$-order and $\leq$ is almost non-existent.

One of the most important examples of a quantale is the power set $P(S)$ of a semigroup $S$, ordered by inclusion and equipped with the usual setwise product. For instance, if $G$ is a finite group, then the idempotents of $P(G)$ are the subgroups and so the identity $\{1\}$ is the smallest idempotent. The dual of Proposition 8.2.2 then shows that the natural order on the idempotents of $P(G)$ is the reverse inclusion order. If $M$ is a monoid, then Proposition 8.2.1 shows that the submonoid $P_1(M)$ of all subsets of $M$ containing 1 is $\mathscr{J}$-trivial [190]. It is shown in Henckell [121] that if $S$ is a finite semigroup and if $G = \{X_1, \ldots, X_k\} \subseteq P(S)$ is a non-trivial group, then each $X_i \in G$ satisfies $X_i \subseteq \bigcup G$, but $X_i >_{\mathscr{H}} \bigcup G$.

We collect here some standard notions from the theory of quantales and relate them to Green's preorders. See [303] for more on these notions.

**Definition 8.2.5 (Right, left, two-sided quantale).** *A quantale* $Q$ *is termed right-sided if, for all* $q \in Q$, $q \cdot \mathsf{T} \leq q$. *Left-sided is defined dually. A quantale is said to be two-sided if it is both left- and right-sided.*

**Proposition 8.2.6.** *Let* $Q$ *be a quantale that happens to be a monoid. Then the following are equivalent:*

1. $1 = \mathsf{T}$;
2. $Q$ *is right-sided;*
3. $Q$ *is left-sided;*
4. $Q$ *is two-sided.*

*Proof.* Clearly 4 implies 3 and 2. Suppose $Q$ is right-sided. Then we have $1 = 1 \cdot 1 \leq 1 \cdot \mathsf{T} \leq 1$. Thus $1 = 1 \cdot \mathsf{T} = \mathsf{T}$. The implication 3 implies 1 is similar. The implication 1 implies 4 is trivial. $\square$

Our next proposition relates Green's preorders to the order in a right-sided (left-sided, two-sided) quantale, generalizing Proposition 8.2.1 (for the case of quantales).

**Proposition 8.2.7.** *Let* $Q$ *be a right-sided quantale. Then* $a \leq_{\mathscr{R}} b$ *implies* $a \leq b$ *and so* $Q$ *is* $\mathscr{R}$*-trivial. A dual result holds for left-sided quantales. If* $Q$ *is two-sided, then* $a \leq_{\mathscr{J}} b$ *implies* $a \leq b$ *and so* $Q$ *is* $\mathscr{J}$*-trivial.*

*Proof.* Suppose $Q$ is right-sided and $a \leq_{\mathscr{R}} b$. Then either $a = b$ or $a = bx$ some $x \in Q$. Then $a = bx \leq b\mathsf{T} \leq b$. The remaining verifications are similar. $\square$

Observe that $\mathbf{Cnt}(\mathbf{PV})^-$ is two-sided. We now turn to the dual notions.

**Definition 8.2.8 (Right, left, two-sided positive).** *A quantale* $Q$ *is said to be right positive if* $a \cdot \mathsf{B} \geq a$. *Left positive quantales are defined dually. The conjunction of being right and left positive is called being two-sided positive.*

Notice that $\mathbf{Cnt}(\mathbf{PV})^+$ is two-sided positive. We formulate the dual propositions to Propositions 8.2.6 and 8.2.7. The proofs are left to the reader.

**Proposition 8.2.9.** *Let* $Q$ *be a quantale that happens to be a monoid. Then the following are equivalent:*

1. $1 = \mathsf{B}$;
2. $Q$ *is right positive;*
3. $Q$ *is left positive;*
4. $Q$ *is two-sided positive.*

We remark that in a $\mathbf{sup} \cdot \mathbf{sup}$ quantale, the bottom is automatically a multiplicative zero and so in non-trivial situations, such a quantale can never be right, left or two-sided positive.

**Proposition 8.2.10.** *Let* $Q$ *be a right positive quantale. Then* $a \leq_{\mathscr{R}} b$ *implies* $b \leq a$ *and so* $Q$ *is* $\mathscr{R}$*-trivial. A dual result holds for left positive quantales. If* $Q$ *is two-sided, then* $a \leq_{\mathscr{J}} b$ *implies* $b \leq a$ *and so* $Q$ *is* $\mathscr{J}$*-trivial.*

## 8.3 Homomorphism and Substructure Theorems for Quantales

We now wish to consider homomorphism theorems for quantales. To discuss homomorphism and substructure theorems for quantales, we must first define the category in which we are working. In [303], **sup·sup** quantales are considered and so the morphisms are **sup** maps that are also semigroup morphisms. In this book, we allow more general objects, **cnt·cnt** quantales. However, we still only allow **sup** maps that are also semigroup homomorphisms as arrows. Because quantales are semigroups, we return to the convention of writing morphisms on the right of the variable.

In the following, if $Q$ is an **x·y** quantale, then all other quantales constructed from it are assumed to be of the same type.

**Definition 8.3.1 (Quantic nucleus).** *Let $Q$ be a quantale. Then a map $j: Q \to Q$ is termed a quantic nucleus (respectively, quantic co-nucleus) if it is a closure operator (respectively, kernel operator) such that*

$$aj \cdot bj \le (a \cdot b)j. \tag{8.2}$$

A map satisfying (8.2) is called a dual prehomomorphism in [176].

**Lemma 8.3.2.** *Let $j: Q \to Q$ be a quantic nucleus. Then*

$$(a \cdot b)j = (a \cdot bj)j = (aj \cdot b)j = (aj \cdot bj)j.$$

*Proof.* Because $j$ is a closure operator,

$$(a \cdot b)j \le (a \cdot bj)j \le (aj \cdot bj)j$$

and similarly

$$(a \cdot b)j \le (aj \cdot b)j \le (aj \cdot bj)j.$$

But, by (8.2),

$$(aj \cdot bj)j \le (a \cdot b)jj = (a \cdot b)j.$$

This establishes the lemma.    □

With quantic nuclei and co-nuclei in hand, we are now ready for the quotient and substructure theorems for quantales [303].

**Theorem 8.3.3 (Homomorphism theorem for quantales).** *If $j$ is a quantic nucleus on a quantale $Q$, then $Qj$ is a quantale with multiplication:*

$$a \star b = (a \cdot b)j.$$

*Moreover, the corestriction $j^0: Q \twoheadrightarrow Qj$ is a quantale morphism. Conversely, given a surjective morphism $d: Q \twoheadrightarrow Q'$ of quantales, there is a quantic nucleus $j: Q \to Q$ such that $d$ induces an isomorphism of quantales from $Qj$ to $Q'$.*

*Proof.* Proposition 6.3.6 implies $j^0$ is **sup**. Let us show that the multiplication on $Qj$ is associative. Let $a, b, c \in Qj$. We use Lemma 8.3.2 repeatedly, as well as the fact that $j$ fixes $Qj$. Then:

$$(a \star b) \star c = (a \cdot b)j \star c = ((a \cdot b)j \cdot c)j$$
$$= ((a \cdot b) \cdot c)j = (a \cdot (b \cdot c))j$$
$$= (a \cdot (b \cdot c)j)j = a \star (b \cdot c)j$$
$$= a \star (b \star c).$$

Because $j$ is order preserving, as is multiplication in $Q$, it is obvious $Qj$ is an ordered semigroup. Let us prove if $Q$ is **cnt** $\cdot$ **cnt**, then so is $Qj$. Similar proofs apply to **x** $\cdot$ **y**-quantales. Let $D$ be a directed subset of $Qj$. Recall that $\bigvee_{Qj} D = (\bigvee D) j$, where the latter join is taken in $Q$. If $a \in Qj$, then

$$a \star \bigvee_{Qj} D = \left( a \cdot \bigvee_{Qj} D \right) j = \left( a \cdot \left( \bigvee D \right) j \right) j$$

$$= \left( a \cdot \bigvee D \right) j = \left( \bigvee_{d \in D} a \cdot d \right) j$$

$$\leq \left( \bigvee_{d \in D} (a \cdot d)j \right) j = \left( \bigvee_{d \in D} a \star d \right) j$$

$$= \bigvee_{Qj} a \star d$$

where the last join runs over $d \in D$. But clearly, $a \star d \leq a \star \bigvee_{Qj} D$ for all $d \in D$, establishing the reverse inequality and proving that $Qj$ is a quantale.

For the converse, let $g: Q' \to Q$ be the right adjoint of $d$. The proof of Theorem 6.3.7 shows that $j = dg$ is a closure operator on $Q$ and $Qj \cong Q'$ via $d$ as complete lattices. We just need to show that $j$ is a quantic nucleus and that $d|_{Qj}$ is a semigroup homomorphism (where multiplication on $Qj$ is as above). Indeed, let $a, b \in Qj$. Then applying $d = dgd = jd$ yields

$$(aj \cdot bj)d = ajd \cdot bjd = ad \cdot bd = (a \cdot b)d.$$

Using that $g$ is the right adjoint of $d$, we deduce from the previous equation $aj \cdot bj \leq (a \cdot b)dg = (a \cdot b)j$. Therefore, $j$ is a quantic nucleus. To verify $d$ is a homomorphism, we compute

$$(a \star b)d = (a \cdot b)jd = (a \cdot b)dgd = (a \cdot b)d = ad \cdot bd,$$

as required.                                                                    □

*Remark 8.3.4.* If $d: Q \twoheadrightarrow Q'$ is an onto morphism of quantales, then the quantic nucleus is given by $qj = \max qdd^{-1}$.

We also state the substructure theorem. The proof will be left to the reader (or see [303]). If $Q$ is a quantale, we say that $S \subseteq Q$ is a *subquantale* if $S$ is a subsemigroup of $Q$ and the inclusion map is a morphism of quantales; equivalently $S$ must be closed under sups and multiplication.

**Theorem 8.3.5 (Substructure theorem for quantales).** *Let $Q$ be a quantale. Then $S \subseteq Q$ is a subquantale if and only if there is a quantic co-nucleus $j \colon Q \to Q$ with image $S$.*

**Exercise 8.3.6.** Prove Theorem 8.3.5.

*Remark 8.3.7 (Frames).* We briefly discuss the notion of a frame. If $X$ is a topological space with collection of opens sets $\mathcal{O}(X)$, then $\mathcal{O}(X)$ is a complete lattice with $\subseteq$ as the order and union as the supremum. The determined meet is $\bigwedge U_a = \mathrm{Int}\left(\bigcap U_a\right)$. Finite meets are just intersections. It is easy to verify that if we take intersection as the multiplication, then $\mathcal{O}(X)$ is a **sup $\cdot$ sup** quantale. In general, if $L$ is a complete lattice that is a **sup·sup** quantale with respect to the meet as multiplication, then $L$ is called a *frame* [97, 186, 303]. These form the basis for the field of so-called "pointless" topology. In fact, the dual category to frames is the category of locales. The origin of the name quantale is quantum locale [303].

Recall that a topological space is called *sober* if each irreducible closed subset admits a generic point. The following is [97, Thm. V-5.6], where compactness does not include the Hausdorff axiom.

**Proposition 8.3.8.** *If $X$ is a sober topological space, then the complete lattice $\mathcal{O}(X)$ is a continuous lattice if and only if $X$ is locally compact.*

### 8.3.1 Examples in the context of semigroup theory

By Theorem 2.3.9, $\mathsf{q} \colon \mathbf{CC} \twoheadrightarrow \mathbf{Cnt}(\mathbf{PV})$ is an onto morphism of quantales. We have in fact the Galois connection:

$$\mathbf{CC} \underset{\mathsf{q}}{\overset{M}{\longleftrightarrow}} \mathbf{Cnt}(\mathbf{PV}).$$

The associated quantic nucleus $j = \mathsf{q}M$ (recall that with our conventions for quantales we apply $\mathsf{q}$ first) is continuous and has image isomorphic with $\mathbf{Cnt}(\mathbf{PV})$. Its members are the maximal continuously closed classes from each of the equivalence classes of the congruence associated to $\mathsf{q}$. Theorem 2.3.7 shows that $Rj$ is an equational Birkhoff continuously closed class. Also, (2.15) implies that, for $\alpha \in \mathbf{Cnt}(\mathbf{PV})$, the continuously closed class $\alpha M$ has decidable membership if and only if $\alpha(\mathbf{V})$ is decidable for each *compact* pseudovariety $\mathbf{V}$.

*Question 8.3.9.* Suppose R ∈ **CC** has decidable membership. Must Rq send compact pseudovarieties to decidable pseudovarieties? A positive answer would show that the quantic nucleus $j$ above preserves decidability. Notice that if R ∈ **BCC** has decidable membership, then $Rq(\mathbf{V})$ is decidable for each compact pseudovariety **V**. Indeed, if $\mathbf{V} = (T)$, then $S \in Rq(\mathbf{V})$ if and only if the canonical relational morphism $\rho\colon S \to F_{(T)}(S \times T)$ belongs to R.

We now turn to the more interesting case for us. Recall that **PVRM** is a subsemigroup of **CC**. It is not however a subquantale as the inclusion is only continuous, not **sup** (note: **PVRM** is a subalgebra of **CC** — the inclusion is **inf**). On the other hand we do have an onto quantale morphism:

$$q\colon \mathbf{PVRM} \twoheadrightarrow \mathbf{GMC(PV)}. \tag{8.3}$$

More precisely we have the Galois connection:

$$\mathbf{PVRM} \underset{q}{\overset{\max}{\rightleftarrows}} \mathbf{GMC(PV)}.$$

The quantic nucleus $j = q \max$ is continuous. Because $q$ is **inf**, as is max, it follows that $j$ is **inf** and hence a CL-morphism.

We now wish to show that in this setting the quantic nucleus does preserve decidability.

**Proposition 8.3.10.**

1. *If* R ∈ **PVRM** *has decidable membership, then* $Rq(\mathbf{V})$ *is decidable for each compact pseudovariety* **V**.
2. *Let* $\alpha \in \mathbf{GMC(PV)}$. *Then* $\max(\alpha)$ *has decidable membership if and only if* $\alpha(\mathbf{V})$ *is decidable for each compact pseudovariety* **V**.

*Proof.* Item 1 follows immediately from Proposition 2.3.26 and the fact, due to Birkhoff, that the free objects in compact pseudovarieties are computable.

For 2, if $\max(\alpha)$ is decidable, then as $\alpha = \max(\alpha)q$, it follows by 1 that $\alpha(\mathbf{V})$ is decidable for each compact pseudovariety **V**. Conversely, if $\alpha(\mathbf{V})$ is decidable for each compact pseudovariety **V**, then $\max(\alpha)$ is decidable by Proposition 2.3.32.                                                                               □

As a corollary of Proposition 8.3.10, we obtain that the quantic nucleus associated to (8.3) preserves decidability.

**Corollary 8.3.11.** *The quantic nucleus associated to* (8.3) *preserves decidability.*

An important example is the following. Let $Q$ be a **sup** · **sup**-quantale. Then there is a natural surjective morphism of quantales $\bigvee\colon P(Q) \twoheadrightarrow Q$. The right adjoint $g\colon Q \to P(Q)$ is given by $qg = q^{\downarrow}$. The associated quantic nucleus $j\colon P(Q) \to P(Q)$ takes $X \subseteq Q$ to $(\bigvee X)^{\downarrow}$.

*Example 8.3.12 (Henckell's expansion).* Let us consider another example, related to Henckell's expansion [54, 124, 130] and the Schützenberger product [85, 230]. Let $M$ be a monoid. We define a product $\amalg$ on $P(M \times M)$ as follows. First, for $(a, b), (c, d) \in M \times M$, define

$$(a, b) \amalg (c, d) = \{(1, abcd), (a, bcd), (ab, cd), (abc, d), (abcd, 1)\}. \qquad (8.4)$$

One can then extend (8.4) to arbitrary subsets $X, Y \subseteq M \times M$ by

$$X \amalg Y = \bigcup \{x \amalg y \mid x \in X, y \in Y\}.$$

**Proposition 8.3.13.** *Let $M$ be a monoid. Then $(P(M \times M), \subseteq, \amalg)$ is a sup · sup quantale.*

*Proof.* We begin with associativity. Suppose $X, Y, Z \in P(M \times M)$. Then one can easily check that

$$(X \amalg Y) \amalg Z = \bigcup \{[(a, b) \amalg (c, d)] \amalg (e, f) \mid (a, b) \in X, (c, d) \in Y, (e, f) \in Z\}$$
$$X \amalg (Y \amalg Z) = \bigcup \{(a, b) \amalg [(c, d) \amalg (e, f)] \mid (a, b) \in X, (c, d) \in Y, (e, f) \in Z\}.$$

Thus it suffices to verify that $[(a, b) \amalg (c, d)] \amalg (e, f) = (a, b) \amalg [(c, d) \amalg (e, f)]$. But both sides are clearly $\{(1, abcdef), (a, bcdef), (ab, cdef), \dots, (abcdef, 1)\}$.

It is immediate from the definition that if $X_1 \subseteq Y_1$ and $X_2 \subseteq Y_2$, then $X_1 \amalg X_2 \subseteq Y_1 \amalg Y_2$, so $P(M \times M)$ is an ordered semigroup. Also, one can deduce directly from the definition that

$$\left( \bigcup_{\alpha \in A} X_\alpha \right) \amalg Y = \{x \amalg y \mid x \in \bigcup_{\alpha \in A} X_\alpha, y \in Y\}$$
$$= \bigcup_{\alpha \in A} \{x \amalg y \mid x \in X_\alpha, y \in Y\}$$
$$= \bigcup_{\alpha \in A} (X_\alpha \amalg Y),$$

which together with its dual shows that $(P(M \times M), \subseteq, \amalg)$ is a **sup · sup**-quantale. $\square$

Let $M$ be a monoid and $\mu \colon M \times M \to M$ the multiplication map. Then there is an induced onto map $\mu \colon P(M \times M) \twoheadrightarrow P(M)$, abusing notation, given by $X \mapsto X\mu$.

**Proposition 8.3.14.** *The map $\mu$ is a quantale morphism.*

*Proof.* If $f \colon X \to Y$ is any map, the direct image $f \colon P(X) \to P(Y)$ is always sup. Because $P(M \times M)$ is generated by the singletons as a quantale, it suffices to show that $\mu$ is a homomorphism when restricted to products of singletons. So let $(a, b), (c, d) \in M \times M$. Then $(a, b)\mu(c, d)\mu = abcd$. On the other hand,

$$[(a, b) \amalg (c, d)]\mu = \{(1, abcd), (a, bcd), (ab, cd), (abc, d), (abcd, 1)\}\mu = abcd$$

completing the proof. $\square$

So there is a Galois connection

$$P(M \times M) \underset{\mu}{\overset{g}{\leftrightarrows}} P(M) \tag{8.5}$$

where $Xg = \{(a,b) \mid ab \in X\}$. The quantic nucleus associated to $\mu$ is the map $j \colon P(M \times M) \to P(M \times M)$ given by

$$Xj = \{(a,b) \mid \exists (c,d) \in X \text{ with } ab = cd\}.$$

If $M$ is a **sup·sup** quantale, we have the composed map $P(M \times M) \twoheadrightarrow M$ given by $X \mapsto \bigvee X\mu$. The left adjoint is given by

$$mg = \{(a,b) \in M \times M \mid ab \leq m\}.$$

The associated quantic nucleus $j \colon P(M \times M) \to P(M \times M)$ is given by $Xj = \{(a,b) \in M \times M \mid \exists (c,d) \in X \text{ with } ab \leq cd\}$.

**Exercise 8.3.15.** Let $e \neq \top$ be an idempotent of $\mathbf{Cnt}(\mathbf{PV})^+$ and define a map $j \colon \mathbf{Cnt}(\mathbf{PV})^+ \to \mathbf{Cnt}(\mathbf{PV})^+$ by setting

$$\alpha j = \begin{cases} \alpha & \alpha \leq e \\ \top & \alpha \nleq e. \end{cases}$$

Show that $j$ is a quantic nucleus. Show the associated congruence is the Rees congruence associated to the ideal $I = \{\gamma \in \mathbf{Cnt}(\mathbf{PV})^+ \mid \gamma \nleq e\}$.

## 8.4 The Bialgebra of Regular Languages

Example 8.3.12 leads us to the bialgebra of a profinite monoid. Recall that a *Boolean algebra* [61, 97, 117, 147] is a distributive lattice $L$ with top 1 and bottom 0 such that each element $a$ has a complement $\neg a$ satisfying

$$a \wedge \neg a = 0 \quad \text{and} \quad a \vee \neg a = 1.$$

One can view $L$ as a semiring by taking $\vee$ as addition and $\wedge$ as multiplication (semirings are the subject of our next chapter). A *morphism of Boolean algebras* is a map preserving finite sups and infs (including empty ones). Denote by **BoolA** the category of Boolean algebras.

One can associate to a Boolean algebra $L$ a Boolean ring $L'$ by defining addition to be the symmetric difference $a + b = (a \wedge \neg b) \vee (\neg a \wedge b)$ and multiplication to be the meet $ab = a \wedge b$. Recall that a *Boolean ring* is a ring in which each element is idempotent. Such a ring is automatically commutative of characteristic 2. Conversely, if $R$ is a Boolean ring, one can define a Boolean algebra $R'$ by setting $a \wedge b = ab$, $a \vee b = a + b + ab$ and $\neg a = 1 - a$. Clearly,

$R'' \cong R$ and $L'' \cong L$. Boolean algebra homomorphisms correspond perfectly with ring homomorphisms and hence the categories **BoolA** and **BoolR** of Boolean algebras and Boolean rings are equivalent. For this reason, we shall use the same notation $L$ for both the Boolean algebra $L$ and the Boolean ring $L'$. In this chapter, we use $+$ for symmetric difference and $\vee$ for the join.

Let **PSet** be the category of profinite spaces. Then *Stone duality* [117,147, 341,342] asserts that **PSet**$^{op}$ is equivalent to **BoolA** (and hence to **BoolR**). The correspondence sends $L$ to $\mathrm{Spec}\, L$ discussed in the introduction of Chapter 7. In fact, if we view $L$ as a Boolean ring, then $\mathrm{Spec}\, L$ is the usual Zariski spectrum of prime ideals of the commutative ring $L$ with the Zariski topology. If $f\colon L_1 \to L_2$ is a homomorphism, then $f^*\colon \mathrm{Spec}\, L_2 \to \mathrm{Spec}\, L_1$ sends a prime ideal $P$ of $L_2$ to $Pf^{-1}$. Conversely, if $X$ is a profinite space, let $K(X)$ denote the Boolean algebra of clopen subsets of $X$. If $f\colon X \to Y$ is a continuous map, then $f^*\colon K(Y) \to K(X)$ defined by $Af^* = Af^{-1}$ is a morphism of Boolean algebras. Details can be found in [61, Chapter IV] and [97,117,147].

**Exercise 8.4.1.** Let $f\colon X \to Y$ be a continuous map of profinite spaces. Show that $f$ is injective (respectively surjective) if and only if $f^*\colon K(Y) \to K(X)$ is surjective (respectively injective).

**Exercise 8.4.2.** Let $X = \varprojlim_D X_\alpha$ with the $X_\alpha$ profinite spaces. Show that $K(X) = \varinjlim_D K(X_\alpha)$ (the direct limit).

**Exercise 8.4.3.** Let $X = \varprojlim_D X_\alpha$ with the $X_\alpha$ an inverse quotient system of profinite spaces. Use Stone duality and the previous exercise to prove if $f\colon X \to Y$ is a continuous map with $Y$ finite, then $f$ factors through the projection $\pi_\alpha\colon X \to X_\alpha$ for some $\alpha \in D$. Hint: The maps of the direct limit $K(X) = \varinjlim_D K(X_\alpha)$ are injective and $K(Y)$ is finite.

**Exercise 8.4.4.** Show that every finite Boolean algebra is projective. Conclude that every finite discrete space is injective in the category of profinite spaces.

**Exercise 8.4.5.** Use the previous exercise to prove that the conclusion of Exercise 8.4.3 remains true even if the inverse system is not an inverse quotient system.

Let us fix the two-element field $\mathbb{F}_2$ as our base field for the rest of this section. Notice that the tensor product (over $\mathbb{F}_2$) is the coproduct in the category **BoolR** of Boolean rings and so $\mathrm{Spec}(L_1 \otimes L_2) \cong \mathrm{Spec}\, L_1 \times \mathrm{Spec}\, L_2$ and $K(X \times Y) = K(X) \otimes K(Y)$.

**Exercise 8.4.6.** Verify $\mathrm{Spec}(L_1 \otimes L_2) \cong \mathrm{Spec}\, L_1 \times \mathrm{Spec}\, L_2$ and $K(X \times Y) = K(X) \otimes K(Y)$.

The following notion of a bialgebra goes back to Hopf [351].

**Definition 8.4.7 (Bialgebra).** *Let $L$ be a unital commutative algebra over $\mathbb{F}_2$. Let $\eta\colon \mathbb{F}_2 \to L$ be the inclusion map (called the unit) and $\mu\colon L \otimes L \to L$ be the multiplication. Then a bialgebra structure on $L$ is a unital algebra homomorphism $\Delta\colon L \to L \otimes L$, called a comultiplication, which is coassociative. Coassociativity means the diagram*

$$
\begin{array}{ccc}
L & \xrightarrow{\ \ \Delta\ \ } & L \otimes L \\
{\scriptstyle \Delta}\downarrow & & \downarrow{\scriptstyle 1_L \otimes \Delta} \\
L \otimes L & \xrightarrow[\ \Delta \otimes 1_L\ ]{} & L \otimes L \otimes L
\end{array}
$$

*commutes. The bialgebra is said to be counital if there is a unital algebra homomorphism $\varepsilon\colon L \to \mathbb{F}_2$, called the counit, such that the diagram*

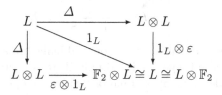

*commutes. By a Boolean bialgebra, we mean a bialgebra whose underlying algebra is a Boolean ring.*

*Remark 8.4.8.* Usually the definition of a bialgebra requires $\eta\varepsilon = 1$, but this comes for free when the ground field is $\mathbb{F}_2$.

Homomorphisms of (counital) bialgebras are defined in the natural way. Namely, if $L_1$, $L_2$ are bialgebras, then a unital algebra homomorphism $\varphi\colon L_1 \to L_2$ is a *bialgebra morphism* if $\Delta(\varphi \otimes \varphi) = \varphi\Delta$. A morphism $\varphi$ of counital bialgebras is required to respect the counit, that is, $\varepsilon = \varphi\varepsilon$.

Our goal is to show that a profinite semigroup (monoid) is the same thing as a Boolean (counital) bialgebra. Afterwards, we shall describe the comultiplication explicitly. This will in turn lead to a counital bialgebra structure on the set of recognizable subsets of $A^*$, which is quite related to the Henckell expansion.

**Exercise 8.4.9.** Let $L$ be a Boolean ring with multiplication $\mu\colon L \otimes L \to L$. Show that $\mu^*$ is the diagonal map $\Delta\colon \operatorname{Spec} L \to \operatorname{Spec} L \times \operatorname{Spec} L$. Hint: Observe $(X \times Y)\Delta^{-1} = X \cap Y$.

**Theorem 8.4.10.** *The category* **PSgp** *of profinite semigroups is equivalent to the opposite of the category of Boolean bialgebras. The category* **PMon** *of profinite monoids is equivalent to the opposite of the category of counital Boolean algebras.*

*Proof.* By Theorem 3.1.51, a profinite semigroup (monoid) is just a semigroup (monoid) object in the category **PSet** of profinite spaces. It follows that under Stone duality, profinite semigroups (monoids) correspond to (counital) Boolean bialgebras. It is easy to check that semigroup (monoid) homomorphisms correspond to (counital) bialgebra morphisms under dualization.     □

**Exercise 8.4.11.** A counital bialgebra $L$ is called a *Hopf algebra* if there is a linear map $S\colon L \to L$, called an *antipode*, such that the diagram

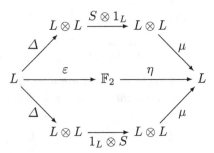

commutes. Show that the Boolean counital bialgebra associated to a profinite monoid $M$ is a Hopf algebra if and only if $M$ is a profinite group, in which case the antipode is the Stone dual of the inverse operation on the group.

**Exercise 8.4.12.** Let $S = \varprojlim_D S_\alpha$ with the $S_\alpha$ profinite semigroups. Show that $K(S) = \varinjlim_D K(S_\alpha)$ where the direct limit is taken in the category of bialgebras.

**Exercise 8.4.13.** Let $S = \varprojlim_D S_\alpha$ with $\{S_\alpha\}$ an inverse quotient system of profinite semigroups. Use Stone duality and the previous exercise to prove if $f\colon S \to T$ is a continuous homomorphism with $T$ a finite semigroup, then $f$ factors via a homomorphism through the projection $\pi_\alpha\colon S \to S_\alpha$ for some $\alpha \in D$.

Let us now study the bialgebra of a profinite semigroup. First note that if $S$ is a monoid, then the counit $\varepsilon\colon K(S) \to \mathbb{F}_2$ is given by

$$X\varepsilon = \begin{cases} 1 & 1 \in X \\ 0 & \text{else} \end{cases}$$

as it is the dual of the unit map $1 \to S$ sending 1 to 1. To understand the comultiplication, we first consider the case of a finite semigroup $S$. Let $\mu\colon S \times S \to S$ be the multiplication. Then $K(S) = P(S)$ and $K(S \times S) = P(S \times S)$. The comultiplication is given by

$$X\Delta = X\mu^{-1} = \{(s, s') \mid ss' \in X\}.$$

Notice that this is exactly the map $g$ from (8.5)! Now suppose $S$ is a profinite semigroup. Let $X \in K(S)$ and let $\equiv_X$ be the syntactic congruence associated to $X$, so $s \equiv_X s'$ if, for all $u, v \in S^I$,

$$usv \in X \iff us'v \in X.$$

The proof of Theorem 3.1.51 shows that $\equiv_X$ is a clopen congruence and so the syntactic morphism $\pi_X \colon S \to S_X = S/\!\equiv_X$ is continuous with $S_X$ finite. The commutative diagram

$$
\begin{array}{ccc}
S \times S & \xrightarrow{\ \mu\ } & S \\
\downarrow{\scriptstyle \pi_X \times \pi_X} & & \downarrow{\scriptstyle \pi_X} \\
S_X \times S_X & \xrightarrow[\ \mu\ ]{} & S_X
\end{array}
$$

asserting that $\pi_X$ is a homomorphism dualizes to

$$
\begin{array}{ccc}
K(S) & \xrightarrow{\ \Delta\ } & K(S \times S) \\
\uparrow{\scriptstyle \pi_X^*} & & \uparrow{\scriptstyle (\pi_X \times \pi_X)^*} \\
P(S_X) & \xrightarrow[\ \Delta\ ]{} & P(S_X \times S_X).
\end{array}
\tag{8.6}
$$

Because $X = X\pi_X\pi_X^{-1} = X\pi_X\pi_X^*$, commutativity of (8.6), together with our computation for the finite case, shows that

$$X\Delta = \sum \{a\pi_X^{-1} \times b\pi_X^{-1} \mid a,b \in S_X,\, ab \in X\pi_X\}. \tag{8.7}$$

Recall that a subset of the free monoid $A^*$ is said to be *recognizable* or *regular* if it is saturated by a finite index congruence. The recognizable subsets of $A^*$ form a Boolean algebra $\mathrm{Rec}(A^*)$. According to [7, Thm 3.6.1], $\mathrm{Rec}(A^*) \cong K(\widehat{A^*})$. The isomorphism takes $L \in \mathrm{Rec}(A^*)$ to $\overline{L}$ and $K \in K(\widehat{A^*})$ to $K \cap A^*$. In particular, if $\varphi \colon \widehat{A^*} \to M$ is a continuous homomorphism with $M$ finite, then $\overline{L\varphi} = L\varphi$. Consequently, if we view $\mathrm{Rec}(A^*)$ as an $\mathbb{F}_2$-algebra with symmetric difference as addition and intersection as multiplication, then by (8.7) and the above discussion it has the structure of a counital bialgebra described as follows.

**Definition 8.4.14 (Bialgebra of regular languages).** *The Boolean ring* $\mathrm{Rec}(A^*)$ *of recognizable subsets of* $S^*$ *is a counital bialgebra where the counit* $\varepsilon \colon \mathrm{Rec}(A^*) \to \mathbb{F}_2$ *is given by*

$$
L\varepsilon = \begin{cases} 1 & 1 \in L \\ 0 & \text{else} \end{cases}
$$

*and the comultiplication* $\Delta \colon \mathrm{Rec}(A^*) \to \mathrm{Rec}(A^*) \otimes \mathrm{Rec}(A^*)$ *is given by*

$$L\Delta = \sum \{a\pi_L^{-1} \otimes b\pi_L^{-1} \mid ab \in L\pi_L\} \tag{8.8}$$

*where* $\pi_L \colon A^* \to M_L$ *is the syntactic morphism.*

Of course, (8.8) can be expressed in terms of left and right quotients of $L$.

**Exercise 8.4.15.** Express (8.8) in terms of left and right quotients of $L$.

For example, if $w \in A^*$, then $\{w\}$ is recognizable and

$$\{w\}\Delta = \sum_{uv=w} \{u\} \otimes \{v\},$$

which again is closely related to the Henckell expansion from Example 8.3.12.

Let **FreeMon** be the full subcategory of **Mon** consisting of finitely generated free monoids, and let **BiAl** be the category of counital bialgebras. Then Rec is a functor from **FreeMon** to **BiAl**$^{op}$ where if $f\colon A^* \to B^*$ is a homomorphism, then $f^*\colon \mathrm{Rec}(B^*) \to \mathrm{Rec}(A^*)$ is given by $Lf^* = Lf^{-1}$. Indeed, $f^*$ is the Stone dual of the continuous extension $f\colon \widehat{A^*} \to \widehat{B^*}$ of $f$, which is a monoid homomorphism.

Of course, if **V** is a pseudovariety of monoids, then by identifying the collection **V**$(A^*)$ of **V**-recognizable subsets of $A^*$ with $K(\widehat{F_V}(A))$ as per [7, Thm 3.6.1], we obtain a counital bialgebra structure on **V**$(A^*)$ analogous to the one in Definition 8.4.14; in fact **V**$(A^*)$ is a subbialgebra of $\mathrm{Rec}(A^*)$ (corresponding with $\widehat{F_V}(A)$ being a quotient of $\widehat{A^*}$). Moreover, if **V** is a pseudovariety of groups, then **V**$(A^*)$ is in fact a Hopf algebra by Exercise 8.4.11.

Again **V**$(-)$ is a functor from **FreeMon** to **BiAl**$^{op}$, in fact a subfunctor of Rec. So the Eilenberg Correspondence [85] functors us from pseudovarieties into bialgebras! In fact, it is easy to verify that a variety of languages is precisely the same thing as a subfunctor of Rec and so there is an order isomorphism between the lattice **PV** and the lattice of subfunctors of Rec.

**Exercise 8.4.16.** Prove that if $F\colon$ **FreeMon** $\to$ **BiAl**$^{op}$ is a subfunctor of Rec, then the family $\{F(A^*)\}$ is a variety of languages in the sense of [85, Chapter VII]. Conversely, show that if $\mathscr{V} = \{\mathscr{V}(A^*)\}$ is a variety of languages, then the assignment $A^* \mapsto \mathscr{V}(A^*)$ is the object part of a unique subfunctor of Rec.

*Question 8.4.17.* Study the counital bialgebras $\mathrm{Rec}(A^*)$, $\mathbf{A}(A^*)$ and $\mathbf{J}(A^*)$. Also study the Hopf algebra of group languages $\mathbf{G}(A^*)$.

## 8.5 Matrix Quantales and an Embedding Theorem

In this section, we show that if $L$ is an algebraic lattice with set of compact elements $K$, then there is a natural CL-morphism and semigroup embedding

$$\mathbf{Cnt}(L) \hookrightarrow M_K(\mathbb{B})$$

where $\mathbb{B}$ is the two-element Boolean lattice $\{0,1\}$ with $0 \leq 1$ (viewed as a quantale using the meet as multiplication). We show conversely that $M_{\mathbb{N}}(\mathbb{B})$ embeds as an ordered subsemigroup of $\mathbf{Cnt}(\mathbf{PV})$.

## 8.5.1 Matrix quantales

Let $N$ be an index set and $Q$ a **sup·sup** quantale. Then $M_N(Q)$ is the set of all functions $A: N \times N \to Q$. We think of $M_N(Q)$ as all $N \times N$ matrices over $Q$. Clearly, $M_N(Q)$ can be made into a complete lattice under the pointwise ordering (and so meets and joins are also pointwise). Matrix multiplication is defined in the usual way: if $A, B: N \times N \to Q$, then

$$(AB)_{ij} = \bigvee_{k \in N} A_{ik} B_{kj}. \tag{8.9}$$

The proof of the following proposition is similar to the situation for usual matrix multiplication and we leave it to the reader as an exercise.

**Proposition 8.5.1.** *Let $Q$ be a **sup·sup** quantale and $N$ an index set. Then $M_N(Q)$ is a **sup·sup** quantale.*

**Exercise 8.5.2.** Prove Proposition 8.5.1.

Just as a matrix algebra in ring theory is the endomorphism ring of a free module, there is an analogous result in this context: $M_N(\mathbb{B}) \cong \mathbf{Sup}(2^N)$ via the map $F: M_n(\mathbb{B}) \to \mathbf{Sup}(2^N)$ given by

$$(AF)(X) = \bigcup_{j \in X} \{i \mid A_{ij} = 1\}.$$

**Proposition 8.5.3.** *The map $F: M_N(\mathbb{B}) \to \mathbf{Sup}(2^N)$ defined above is an isomorphism of **sup·sup**-quantales.*

*Proof.* First we check that $F$ is well defined. Suppose $A \in M_N(\mathbb{B})$. We verify that $AF$ is **sup**. Suppose $X = \bigcup Z$ where $Z \subseteq 2^N$. Then

$$(AF)(X) = \bigcup_{j \in X} \{i \mid A_{ij} = 1\}$$

$$= \bigcup_{Y \in Z} \bigcup_{j \in Y} \{i \mid A_{ij} = 1\}$$

$$= \bigcup_{Y \in Z} (AF)(Y).$$

Thus $AF \in \mathbf{Sup}(2^N)$.

The map $F$ is order preserving because if $A \leq B$, then

$$(AF)(X) = \bigcup_{j \in X} \{i \mid A_{ij} = 1\} \subseteq \bigcup_{j \in X} \{i \mid B_{ij} = 1\} = (BF)(X).$$

To see that it is an order embedding, suppose $AF \leq BF$. Let $i \in N$ and suppose $A_{ij} = 1$. Then we have $i \in AF(\{j\}) \subseteq BF(\{j\})$. Thus $B_{ij} = 1$. We conclude $A \leq B$ and so $F$ is an order embedding.

To see that $F$ is onto, let $\alpha \colon 2^N \to 2^N$ be **sup**. Define $A \in M_N(\mathbb{B})$ by

$$A_{ij} = \begin{cases} 1 & i \in \alpha(j) \\ 0 & \text{else.} \end{cases}$$

Then we have

$$(AF)(X) = \bigcup_{j \in X} \{i \mid A_{ij} = 1\} = \bigcup_{j \in X} \alpha(j) = \alpha(X)$$

where the last equality follows because $\alpha$ is **sup**.

To see that $F$ is **sup**, let $\{A_r\}_{r \in J}$ be a collection of $N \times N$ matrices. Then

$$\left[ (\bigvee_{r \in J} A_r) F \right](X) = \bigcup_{j \in X} \{i \mid (\bigvee_{r \in J} A_r)_{ij} = 1\}$$

$$= \bigcup_{j \in X} \bigcup_{r \in J} \{i \mid (A_r)_{ij} = 1\}$$

$$= \bigcup_{r \in J} \bigcup_{j \in X} \{i \mid (A_r)_{ij} = 1\}$$

$$= \bigcup_{r \in J} (A_r F)(X)$$

establishing $(\bigvee A_r)F = \bigvee(A_r F)$.

Finally, we show $F$ is a monoid homomorphism. Let $I$ be the identity matrix. Then $(IF)(X) = \bigcup_{j \in X} \{i \mid I_{ij} = 1\} = X$. Let $A, B \in M_N(\mathbb{B})$. To verify that $F$ is a semigroup homomorphism, we compute:

$$[(AB)F](X) = \bigcup_{j \in X} \{i \mid (AB)_{ij} = 1\}$$

$$= \bigcup_{j \in X} \{i \mid \exists k \in N : A_{ik} = 1, B_{kj} = 1\}$$

$$= \bigcup_{k \in (BF)(X)} \{i \mid A_{ik} = 1\}$$

$$= (AF)[(BF)(X)],$$

as required.                                                                $\square$

### 8.5.2  An embedding theorem

We want to show that if $L$ is an algebraic lattice with set $K$ of compact elements, then $\mathbf{Cnt}(L)$ embeds in $M_K(\mathbb{B})$. By Proposition 8.5.3, it suffices to embed $\mathbf{Cnt}(L)$ in $\mathbf{Sup}(2^K)$. Recall that $L \cong \mathsf{Id}(K)$ and $\mathsf{Id}(K)$ is a subalgebra of $2^K$. The inclusion $\iota \colon L \to 2^K$ takes $\ell$ to $\ell^{\downarrow} \cap K$ and is a CL-morphism.

Because a sup-map on $2^K$ is determined by what it does to $K$, viewed as singletons, given $\alpha \in \mathbf{Cnt}(L)$, we can define $\tilde{\alpha} \in \mathbf{Sup}(2^K)$ by $\tilde{\alpha}(k) = \iota\alpha(k)$ for $k \in K$ and then extending linearly. Also observe that the join in $\mathbf{Sup}(2^K)$ is pointwise. The meet is not pointwise, but it is pointwise on singletons.

**Exercise 8.5.4.** Verify that the join in $\mathbf{Sup}(2^K)$ is pointwise and the meet is pointwise on singletons.

**Theorem 8.5.5.** *Let $L$ be an algebraic lattice with set of compact elements $K$ and let $M\colon \mathbf{Cnt}(L) \to \mathbf{Sup}(2^K)$ be given by $\alpha \mapsto \tilde{\alpha}$ (defined above). Then $M$ is a semigroup homomorphism, a CL-morphism and an order embedding. Hence $\mathbf{Cnt}(L)$ is isomorphic to an ordered subsemigroup of $M_K(\mathbb{B})$.*

*Proof.* To see that $M$ is an order embedding, we observe

$$
\begin{aligned}
\alpha \le \beta &\iff \alpha(k) \le \beta(k), \forall k \in K \\
&\iff \iota\alpha(k) \le \iota\beta(k), \forall k \in K \\
&\iff \tilde{\alpha}(k) \le \tilde{\beta}(k), \forall k \in K \\
&\iff \tilde{\alpha} \le \tilde{\beta}.
\end{aligned}
$$

Next we verify that $M$ is continuous. Suppose that $\{\alpha_d\}_{d\in D}$ is a directed subset of $\mathbf{Cnt}(L)$. Then using the continuity of $\iota$ we have, for $k \in K$,

$$
\iota\left[\left(\bigvee_D \alpha_d\right)(k)\right] = \iota\left(\bigvee_D \alpha_d(k)\right) = \bigvee_D (\iota\alpha_d(k)) = \left(\bigvee_D \tilde{\alpha_d}\right)(k)
$$

and so $(\bigvee_D \alpha_d)M = \bigvee_D(\alpha_d M)$, as required.

Let us turn to preservation of meets. We use that the meet in $\mathbf{Cnt}(L)$ is pointwise on compact elements and that the meet in $\mathbf{Sup}(2^K)$ is pointwise on singletons. Let $\{\alpha_i\}_{i\in I}$ be a collection of continuous operators on $L$. Then, for $k \in K$, we compute using that $\iota$ preserves infs

$$
\iota\left[\left(\bigwedge_I \alpha_i\right)(k)\right] = \iota\left(\bigwedge_I \alpha_i(k)\right) = \bigwedge_I (\iota\alpha_i(k)) = \left(\bigwedge_I \tilde{\alpha_r}\right)(k)
$$

yielding $(\bigwedge_I \alpha_i)M = \bigwedge_I(\alpha_i M)$.

Finally, we verify that $M$ is a homomorphism. Let $k \in K$ and suppose $\alpha, \beta \in \mathbf{Cnt}(L)$. Then we have

$$
\tilde{\alpha}\tilde{\beta}(k) = \tilde{\alpha}(\iota\beta(k)) = \tilde{\alpha}(\beta(k)^{\downarrow} \cap K) = \bigcup_{x\in\beta(k)^{\downarrow}\cap K} \tilde{\alpha}(x) = \bigcup_{x\in\beta(k)^{\downarrow}\cap K} \iota\alpha(x)
$$

because $\tilde{\alpha}$ is **sup**. Because $\beta(k)^{\downarrow} \cap K$ is directed and $\iota$ is continuous, the right-hand side of the above equation equals

$$\iota\left(\bigvee_{x\in\beta(k)\downarrow\cap K}\alpha(x)\right),$$

which is in turn $\iota(\alpha(\beta(k))) = \widetilde{\alpha\beta}(k)$ because $\alpha$ is continuous. This completes the proof. $\square$

We remark that $M$ does not send the identity operator to the identity map, so it is not a monoid homomorphism. Also, it does not preserve empty sups because the constant map to the bottom of $L$ is not sent to the constant map to the empty set.

If we consider the case of $\mathbf{Cnt}(\mathbf{PV})$, we obtain an embedding into $M_{\mathbb{N}}(\mathbb{B})$ by identifying the set of compact pseudovarieties with the natural numbers. We state this as a corollary.

**Corollary 8.5.6.** *There is an order embedding of* $\mathbf{Cnt}(\mathbf{PV})$ *into* $M_{\mathbb{N}}(\mathbb{B})$ *that is also a semigroup and CL-morphism.*

We now go in the reverse direction and establish a semigroup embedding of $M_{\mathbb{N}}(\mathbb{B})$ into $\mathbf{Cnt}(\mathbf{PV})$. Denote by $\mathbf{VS}$ the pseudovariety of elementary abelian groups; that is, the pseudovariety generated by all cyclic groups of prime order. Let $p_1, p_2, \ldots$ be an enumeration of the primes. Then $[\mathbf{1}, \mathbf{VS}] \cong 2^{\mathbb{N}}$. More precisely, the map sending $X \subseteq \mathbb{N}$ to $\bigvee_{i\in X} \mathbf{Ab}(p_i)$ is a lattice isomorphism, where we recall that $\mathbf{Ab}(p)$ denotes the pseudovariety of abelian groups of exponent $p$.

**Lemma 8.5.7.** *There is an order embedding* $\Psi\colon \mathbf{Cnt}([\mathbf{1}, \mathbf{VS}]) \to \mathbf{Cnt}(\mathbf{PV})$ *given by* $(\alpha\Psi)(\mathbf{V}) = \alpha(\mathbf{V} \cap \mathbf{VS})$. *Moreover,* $\Psi$ *is a quantale morphism, but not a monoid morphism.*

*Proof.* Because $\alpha\Psi$ restricts to $\alpha$ on $[\mathbf{1}, \mathbf{VS}]$, the map is clearly an order embedding. The identity map on $[\mathbf{1}, \mathbf{VS}]$ is not sent to the identity on $\mathbf{PV}$. To see that $\Psi$ is a homomorphism, observe that

$$[(\alpha\beta)\Psi](\mathbf{V}) = \alpha\beta(\mathbf{V} \cap \mathbf{VS}) = \alpha(\beta\Psi(\mathbf{V}) \cap \mathbf{VS}) = (\alpha\Psi\beta\Psi)(\mathbf{V}).$$

To see that $\Psi$ is **sup**, we compute

$$\left[\left(\bigvee \alpha_i\right)\Psi\right](\mathbf{V}) = \left(\bigvee \alpha_i\right)(\mathbf{V} \cap \mathbf{VS}) = \bigvee \alpha_i(\mathbf{V} \cap \mathbf{VS}) = \bigvee(\alpha_i\Psi)(\mathbf{V})$$

showing that $(\bigvee \alpha_i)\Psi = \bigvee \alpha_i\Psi$. $\square$

Because $\mathbf{Sup}(2^{\mathbb{N}}) \subseteq \mathbf{Cnt}(2^{\mathbb{N}})$, composing the map $F$ from Proposition 8.5.3 (for $N = \mathbb{N}$) with $\Psi$ from Lemma 8.5.7 (after identifying $[\mathbf{1}, \mathbf{VS}]$ with $2^{\mathbb{N}}$), we obtain:

**Theorem 8.5.8.** *There is a quantale morphism, order embedding of* $M_{\mathbb{N}}(\mathbb{B})$ *into* $\mathbf{Cnt}(\mathbf{PV})$.

We obtain as a corollary of Corollary 8.5.6 and Theorem 8.5.8:

**Corollary 8.5.9.** *The isomorphism classes of (ordered) subsemigroups of* $\mathbf{Cnt}(\mathbf{PV})$ *and* $M_{\mathbb{N}}(\mathbb{B})$ *coincide.*

# Notes

Although we have slightly generalized the definition of a quantale, the basic results — the homomorphism and substructure theorems — remain the same as in the classical setting [303]. We have tried to include some examples to indicate why this concept should be interesting to semigroup theorists. The next chapter should be even more convincing. The basic guide to quantales is [303]. It turns out that certain involutive quantales are closely connected to inverse semigroups and étale groupoids [264]; this should be explored further. In [304], quantaloids are studied. These are the category analogues of quantales, so a unital quantale is a one-object quantaloid. Any sort of derived or kernel category theorem for quantales (and the results of the next chapter suggest that there should be such) will be sure to make use of quantaloids.

The results on bialgebras of profinite semigroups and the bialgebra structure on $\mathrm{Rec}(A^*)$ and $\mathbf{V}(A^*)$ seems to be new as far as we know. Of course, they echo the fact that the category of affine group schemes and finitely generated (commutative) Hopf algebras are opposites. Also, one should compare with the relationship between pro-algebraic groups and (commutative) Hopf algebras [316]; the fact that a Hopf algebra is the directed union of its finitely generated Hopf subalgebras implies that the dual object to a Hopf algebra is a pro-algebraic group; of course, a similar relationship holds between bialgebras and pro-algebraic monoids. The reformulation of Eilenberg's correspondence as being between pseudovarieties of semigroups and subfunctors of Rec is novel to this chapter. The bialgebra approach to profinite semigroups seems to offer a promising avenue for future research. Recently, Gehrke, Grigorieff and Pin [96] have independently viewed $\widehat{A^*}$ as a dual structure to $\mathrm{Rec}(A^*)$ via residuation theory. They then use Priestley duality [97,147,247] to develop an equational theory for lattices of recognizable languages. Their approach does not, however, encompass profinite semigroups in general, as ours does.

# 9

# The Triangular Product and Decomposition Results for Semirings

This chapter introduces a decomposition and complexity theory for finite **sup · sup** quantales, i.e., finite quantales in the classical sense [303], which is the only type we shall consider in this chapter. Plotkin's triangular product [239, 240, 376] is adopted as the analogue of the wreath product for this theory. The appropriate notion of an irreducible quantale is defined. The main results, Theorems 9.4.1 and 9.4.10, establish that any finite quantale admits a triangular product decomposition into irreducibles, which are in fact matrix algebras over power semigroups of its Schützenberger groups. This allows us to define a notion of complexity for finite quantales. Polák has already considered pseudovarieties of finite quantales with applications to automata and formal languages [241–244]. With the introduction of a product, it should be possible to develop a deeper theory.

A finite **sup · sup** quantale is the same thing as a finite idempotent semiring. Because much of our theory works in the more general context of semirings, we shall phrase our results in this language. Some basic references on semirings are [100, 120, 160]. Applications of semirings to automata and formal languages can be found in [48, 72, 84, 305].

The chapter ends with some applications to the complexity of power semigroups, improving on results of the first author and Fox [91]. In particular, our new machinery is used to compute the exact complexity of the power semigroup of an inverse semigroup and to show that the complexity of the power semigroup $P(T_n)$ of the full transformation semigroup grows at the same rate as the central binomial coefficient $\binom{n}{\lfloor \frac{n}{2} \rfloor}$.

## 9.1 Semirings

A *semiring* is a 4-tuple $(S, +, \cdot, 0)$ such that $(S, +, 0)$ is a commutative monoid, $(S, \cdot)$ is a semigroup, and the following axioms hold:

1. If $x, y, z \in S$, then $x(y + z) = xy + xz$;

J. Rhodes, B. Steinberg, *The q-theory of Finite Semigroups*,
Springer Monographs in Mathematics, DOI 10.1007/978-0-387-09781-7_9,
© Springer Science+Business Media, LLC 2009

2. If $x, y, z \in S$, then $(y + z)x = yx + zx$;
3. For all $x \in S$, $x0 = 0 = 0x$.

We normally just call the semiring $S$ rather than writing the 4-tuple. If $(S, \cdot)$ has an identity, then $S$ is called a *semiring with unit* or a *unital semiring*. The most important example of a semiring for us is any **sup · sup** quantale. Here we take the join as the addition.

Homomorphisms of semirings are defined in the natural way. A mapping $\varphi \colon R \to S$ between semirings is a *semiring homomorphism* if:

- $0\varphi = 0$;
- $(r + r')\varphi = r\varphi + r'\varphi$;
- $(rr')\varphi = r\varphi r'\varphi$,

for all $r, r' \in R$. A *unital semiring homomorphism* is a semiring homomorphism preserving the multiplicative identity.

If $S$ is a semiring, a *right module* $M$ over $S$ is a commutative monoid $(M, +)$ with a right action $M \times S \to M$ satisfying:

- For all $m_1, m_2 \in M$, $s \in S$, $(m_1 + m_2)s = m_1 s + m_2 s$;
- For all $m \in M$, $s_1, s_2 \in S$, $m(s_1 + s_2) = ms_1 + ms_2$;
- For all $m \in M$, $s_1, s_2 \in S$, $m(s_1 s_2) = (ms_1)s_2$;
- For all $s \in S$, $m \in M$, $0s = 0 = m0$.

If $S$ has a unit, then the action is called *unitary* if $m1 = m$ for all $m \in M$. Left modules are defined dually. If $R$ and $S$ are semirings and $M$ is a left $R$-module and a right $S$-module, then it is called an *R-S-bimodule* if $(rm)s = r(ms)$ for all $r \in R$, $m \in M$, $s \in S$. We use the notation $(M, R)$ to indicate that $M$ is a right $R$-module. Similar notations are used for left modules and bimodules.

Homomorphisms of (right) $R$-modules are defined in the obvious way and form a category. We use $\mathsf{Hom}_R(M, N)$ to denote the set of $R$-module morphisms from $M$ to $N$. It is a commutative monoid with pointwise operations. Direct sums of $R$-modules are defined in the obvious way: so $M \oplus N = M \times N$ with pointwise operations.

One can also define the tensor product of a right $R$-module $M$ with a left $R$-module $N$. A map $f \colon M \times N \to A$, with $A$ a commutative monoid, is called *R-bilinear* if

$$(m + m')fn = mfn + m'fn$$
$$mf(n + n') = mfn + mfn'$$
$$0fn = 0 = mf0$$
$$mrfn = mfrn$$

for all $r \in R$, $m, m' \in M$ and $n, n' \in N$ (where we write a function of two variables in between the two variables). The universal bilinear map for $M$ and $N$ is the natural map $M \times N \to M \otimes_R N$ given by $(m, n) \mapsto m \otimes n$, where $M \otimes_R N$ is the *tensor product* of $M$ and $N$. If the ground semiring $R$

is understood, one simply writes $M \otimes N$. This universal property serves to define the tensor product, but one can also define it as the free commutative monoid on $M \times N$ modulo the smallest congruence $\sim$ such that:

$$((m + m'), n) \sim (m, n) + (m', n)$$
$$(m, (n + n')) \sim (m, n) + (m, n')$$
$$(0, n) \sim 0 \sim (m, 0)$$
$$(mr, n) \sim (m, rn) \quad \text{[slide rule]}.$$

The congruence class of $(m, n)$ is denoted $m \otimes n$. Such an element is called an *elementary tensor*. We remark that if $M$ additionally has the structure of an $S$-$R$-bimodule (respectively, $N$ has the structure of an $R$-$S$-bimodule), then $M \otimes_R N$ has the structure of a left $S$-module (respectively, right $S$-module) induced by $s(m \otimes n) = sm \otimes n$ (respectively, $(m \otimes n)s = m \otimes ns$). See [157] for more on homological algebra in the context of semirings.

For the rest of this chapter, $k$ will always denote a commutative unital semiring. A semiring $S$ is called a $k$-*algebra* if $S$ is a unitary $k$-$k$-bimodule such that, for all $c \in k$, $s \in S$, one has $cs = sc$. Every semiring, for instance, is an $\mathbb{N}$-algebra in a natural way so talking about $k$-algebras does not restrict us. Notice that if $R$ and $S$ are $k$-algebras, then $R \otimes_k S$ has the natural structure of a $k$-algebra where the multiplication is given on elementary tensors by $(r \otimes s)(r' \otimes s') = rr' \otimes ss'$.

The most important type of semiring for us is an idempotent semiring.

**Definition 9.1.1 (Idempotent semiring).** *An idempotent semiring is a semiring for which the operation of addition is idempotent.*

If $\mathbb{B}$ denotes the two-element Boolean semiring $\{0, 1\}$ with addition given by $1 + 1 = 1$, then every idempotent semiring is a $\mathbb{B}$-algebra in a natural way. Conversely, any $\mathbb{B}$-algebra must have an idempotent addition since $s + s = (1 + 1)s = 1s = s$.

A finite $\sup \cdot \sup$ quantale is the same thing as a finite idempotent semiring. For this chapter, we will use the name *idempotent semiring* for these objects. In the context of idempotent semirings, we shall use $0$ for the bottom and $+$ for the addition. However, we shall occasionally revert to lattice notation when we are trying to emphasize order theoretic aspects.

If $R$ is a semiring, we use $RX$ to denote the *free $R$-module* on $X$. It consists of all finite formal $R$-linear combinations of elements of $X$. Formally, it consists of all finitely supported functions $f \colon X \to R$. We identify $f$ with the formal sum $\sum_{x \in X} xf \cdot x$. For instance if $X$ is a finite set, then the free $\mathbb{B}$-module on $X$ is just the power set $P(X)$ with union as the addition. In general, $\mathbb{B}X$ consists of the finite subsets of $X$. The functor $X \mapsto RX$ is the left adjoint to the forgetful functor from $R$-modules to sets.

**Exercise 9.1.2.** Show that $k[X \times Y] \cong kX \otimes_k kY$ via the map $(x, y) \mapsto x \otimes y$ (viewing $kY$ as a left $k$-module in the natural way).

If $S$ is an arbitrary semiring, then the set $M_n(S)$ of $n \times n$ matrices over $S$ is a semiring with the usual operations. If $S$ is a $k$-algebra, then so is $M_n(S)$. In particular, $M_n(k)$ is a $k$-algebra.

If $R$ is a $k$-algebra, then $R$ with an adjoined unit, denoted $R^I$, is $R \oplus k$ with multiplication given by $(r, c)(r', c') = (rr' + rc' + cr', cc')$; the identity is $(0, 1)$. Any $k$-homomorphism of $R$ into a unital $k$-algebra extends uniquely to $R^I$ in a way that sends the identity of $R^I$ to the identity. That is, $R \mapsto R^I$ is left adjoint to the forgetful functor from unital $k$-algebras to $k$-algebras.

When speaking about a $k$-algebra $R$, it is natural to mean by a right $R$-module a $k$-$R$-bimodule. Similar considerations apply to left $R$-modules. What we mean by an $R$-$S$-bimodule $M$ is then clear. In this context, homomorphisms of right $R$-modules are also expected to be $k$-linear. For example, if $S$ is an idempotent semiring, then an $S$-module is a join semilattice with minimum equipped with a right action of $S$ by maps preserving finite sups.

**Definition 9.1.3 (Semigroup algebra).** *If $S$ is a semigroup, the semigroup algebra of $S$ over $k$ is*

$$kS = \{f \colon S \to k \mid sf = 0 \text{ for all but finitely many values of } s \in S\}.$$

*Typically one writes elements of $kS$ in the form $f = \sum_{s \in S} c_s s$ where $c_s = sf$. Multiplication is then defined via the familiar convolution product:*

$$\sum_{s \in S} c_s s \cdot \sum_{t \in S} d_t t = \sum_{u \in S} \left( \sum_{st = u} c_s d_t \right) u.$$

For example, if $S$ is any finite semigroup, then the semigroup algebra $\mathbb{B}S$ is nothing more than the power set $P(S)$ with union as addition and setwise multiplication as the product. Hence we shall mostly retain the traditional notation $P(S)$ for this semiring. Notice that $k(S^I) \cong (kS)^I$ and so we may unambiguously write $kS^I$. The functor $S \mapsto kS$ is the left adjoint of the forgetful functor from $k$-algebras to semigroups. Hence any semigroup homomorphism from $S$ into a $k$-algebra extends uniquely to $kS$. The algebra $kX^+$ is called the *free $k$-algebra* on $X$. It plays a pivotal role in Schützenberger's theory of rational and algebraic power series [48, 65, 305], which is probably the most significant realm of applications of non-commutative semirings to date. The theory of commutative idempotent semirings is a part of what is called tropical geometry and is currently a hot research topic [302].

**Exercise 9.1.4.** Let $S$ and $T$ be monoids. Verify $k(S \times T) \cong kS \otimes_k kT$ via the map induced by $(s, t) \mapsto s \otimes t$.

If $S$ is a semigroup, then a $k$-module $M$ is called an $S$-*module* if there is a right action of $S$ on $M$ by $k$-module endomorphisms. It is straightforward to verify such an action extends to give a right $kS$-module structure to $M$

and conversely any $kS$-module $M$ can be viewed as an $S$-module by restriction. Left $S$-modules and $S$-$T$-bimodules ($S$ and $T$ semigroups) are defined similarly.

We shall need a module theoretic analogue of a standard result about faithful transformation semigroups [85] (see Exercise 4.1.16).

**Lemma 9.1.5.** *Suppose that $M$ and $N$ are $k$-modules such that $(M, S)$ and $(N, T)$ are faithful right modules, where $S$ and $T$ are semigroups ($k$-algebras). Moreover, assume there is a $k$-isomorphism $\rho \colon M \to N$ and a function $\psi \colon S \to T$ such that $ms\rho = m\rho s\psi$, all $m \in M$, $s \in S$. Then $\psi$ is an injective homomorphism of semigroups ($k$-algebras).*

*Proof.* First we show that $s\psi$ is the unique element $\widetilde{s} \in T$ such that $ms\rho = m\rho\widetilde{s}$ for all $m \in M$. Indeed, if $\widetilde{s}$ is such element, then

$$n\widetilde{s} = n\rho^{-1}\rho\widetilde{s} = (n\rho^{-1}s)\rho = n\rho^{-1}\rho s\psi = ns\psi$$

for all $n \in N$ and so $\widetilde{s} = s\psi$ by faithfulness. Then we have, for all $m \in M$, that $m\rho s\psi s'\psi = ms\rho s'\psi = mss'\rho$, and so $s\psi s'\psi = (ss')\psi$ by uniqueness. In the case that $S$ and $T$ are $k$-algebras, then

$$m\rho(s\psi + s'\psi) = m\rho s\psi + m\rho s'\psi = ms\rho + ms'\rho = (m(s + s'))\rho$$

establishing $(s + s')\psi = s\psi + s'\psi$. Similarly, if $c \in k$ and $s \in S$, then

$$(m(cs))\rho = (cms)\rho = c(ms)\rho = cm\rho(s\psi) = (m\rho)c(s\psi)$$

for all $m \in M$. Thus $(cs)\psi = c(s\psi)$, as required. $\qquad\square$

In the setting of Lemma 9.1.5, we say that $(M, S)$ *embeds* in $(N, T)$ and write $(M, S) \le (N, T)$. The following is a semiring analogue of both a standard fact from ring theory and of Proposition 8.5.3.

**Proposition 9.1.6.** *Let $X$ be a finite set of cardinality $n$ and $R$ a unital semiring. Then $\mathsf{Hom}_R(RX, RX) \cong M_n(R)$.*

**Exercise 9.1.7.** Adapt the usual proof from ring theory to prove Proposition 9.1.6.

**Corollary 9.1.8.** *Let $S$ be a finite semigroup of order $n$. Then $kS$ embeds in $M_{n+1}(k)$.*

*Proof.* Because $kS$ acts faithfully on the right of the free $k$-module $kS^I$ by $k$-endomorphisms, Proposition 9.1.6 then gives the desired result. $\qquad\square$

We now prove an analogue of the characterization of the structure of ideals in matrix algebras over rings. In fact, the proof is the same once the language of ideals is stripped away. More precisely, we prove that if $S$ is a unital semiring, then $M_n(S)$ and $S$ have isomorphic congruence lattices.

**Theorem 9.1.9.** *Let $S$ be a unital semiring. Then every quotient morphism from $M_n(S)$ is of the form $\overline{\varphi}\colon M_n(S) \twoheadrightarrow M_n(T)$ where $\overline{\varphi}$ is induced by a quotient morphism $\varphi\colon S \twoheadrightarrow T$. In particular, $S$ and $M_n(S)$ have isomorphic congruence lattices.*

*Proof.* Let $\equiv$ be a congruence on $M_n(S)$. Let $E_{ij} \in M_n(S)$ denote the standard matrix unit and $I$ the identity matrix. We identify $S$ with $S \cdot I$. Let $\sim$ be the congruence on $S$ induced by $\equiv$. We claim that $\equiv$ coincides with the congruence associated to the natural projection $\rho\colon M_n(S) \twoheadrightarrow M_n(S/\sim)$. Indeed, let $A = (a_{ij})$ and $B = (b_{ij})$. Suppose first that $A \equiv B$. Then we have

$$a_{ij}I = \sum_{k=1}^{n} E_{ki}AE_{jk} \equiv \sum_{k=1}^{n} E_{ki}BE_{jk} = b_{ij}I$$

and so $a_{ij} \sim b_{ij}$ all $i, j$. We conclude that $A\rho = B\rho$. Conversely, if $A\rho = B\rho$, then $a_{ij} \sim b_{ij}$ all $i, j$, that is, $a_{ij}I \equiv b_{ij}I$ for all $i, j$. But then

$$A = \sum_{i,j} E_{i1}(a_{ij}I)E_{1j} \equiv \sum_{i,j} E_{i1}(b_{ij}I)E_{1j} = B$$

showing that $A \equiv B$ and completing the proof.    $\square$

Because being congruence-free or subdirectly indecomposable depends only on the congruence-lattice, we obtain:

**Corollary 9.1.10.** *Suppose that $R$ is a congruence-free unital semiring. Then $M_n(R)$ is also congruence-free. If $R$ is subdirectly indecomposable, then so is $M_n(R)$.*

We shall require one further result about matrix semirings, which essentially says that if $R$ is a $k$-algebra, then $M_n(R) \cong R \otimes_k M_n(k)$.

**Proposition 9.1.11.** *Let $R$ be a $k$-algebra. Let $B_n(R)$ be the subsemigroup of $M_n(R)$ consisting of all matrices of the form $rE_{ij}$ with $1 \le i, j \le n$ and $r \in R$. Suppose $A$ is a $k$-algebra and $\varphi\colon B_n(R) \to A$ is a semigroup homomorphism such that:*

*1. $[(r_1 + r_2)E_{ij}]\varphi = (r_1E_{ij})\varphi + (r_2E_{ij})\varphi$, for $r_1, r_2 \in R$;*
*2. $(crE_{11})\varphi = c(rE_{11})\varphi$, for $c \in k$, $r \in R$.*

*Then $\varphi$ extends uniquely to a $k$-algebra homomorphism $\overline{\varphi}\colon M_n(R) \to A$ by the formula*

$$(a_{ij})\overline{\varphi} = \sum_{i,j}(a_{ij}E_{ij})\varphi.$$

*As a consequence, $R \otimes_k M_n(k) \cong M_n(R)$ via the map $r \otimes E_{ij} \mapsto rE_{ij}$.*

*Proof.* We leave the final statement to the reader, as we will not use it in the sequel. Because $M_n(R)$ is spanned as a $k$-algebra by $B_n(R)$, uniqueness is clear. First we show $\overline{\varphi}$ is a $k$-linear map. Indeed, for $c_1, c_2 \in k$,

$$[c_1(a_{ij}) + c_2(b_{ij})]\overline{\varphi} = \sum_{i,j}[(c_1 a_{ij} + c_2 b_{ij})E_{ij}]\varphi$$

$$= c_1 \sum_{i,j}(a_{ij}E_{ij})\varphi + c_2 \sum_{i,j}(b_{ij}E_{ij})\varphi$$

$$= c_1[(a_{ij})\overline{\varphi}] + c_2[(b_{ij})\overline{\varphi}]$$

as required.

Next, to verify $\overline{\varphi}$ is a multiplicative homomorphism, we compute

$$(a_{ij})\overline{\varphi}(b_{ij})\overline{\varphi} = \sum_{i,j}(a_{ij}E_{ij})\varphi \sum_{i,j}(b_{ij}E_{ij})\varphi$$

$$= \sum_{i,k,\ell,j}(a_{ik}E_{ik})\varphi(b_{\ell j}E_{\ell j})\varphi$$

$$= \sum_{i,k,\ell,j}(a_{ik}b_{\ell j}E_{ik}E_{\ell j})\varphi$$

$$= \sum_{i,j}\sum_{k=1}^{n}(a_{ik}b_{kj}E_{ij})\varphi$$

$$= \sum_{i,j}\left[\left(\sum_{k=1}^{n}a_{ik}b_{kj}\right)E_{ij}\right]\varphi$$

$$= [(a_{ij})(b_{ij})]\overline{\varphi}.$$

This completes the proof. $\qquad\square$

**Exercise 9.1.12.** Verify that $r \otimes E_{ij} \mapsto rE_{ij}$ induces an isomorphism of $R \otimes_k M_n(k)$ with $M_n(R)$.

## 9.1.1 Ideals and quotient modules

Even though ideals do not play the same role in the theory of semirings as they do in ring theory, they are quite important nonetheless. In particular, they lie at the heart of our decomposition results.

**Definition 9.1.13 (Ideal).** *An ideal in a $k$-algebra $S$ is a $k$-submodule $I$ of $S$ such that $SI \cup IS \subseteq I$. We sometimes write semiring ideal to distinguish the notion from the other types of ideals considered in this text.*

Equivalently, an ideal is an $S$-$S$-sub-bimodule of $S$. Of course, 0 belongs to any ideal. Left and right ideals are defined analogously. Notice that the image of an ideal under an onto semiring homomorphism is again an ideal.

If $I$ is an ideal of a semiring $S$, then we can define a congruence on $S$ by defining $s \equiv t \bmod I$ if there exist $x, y \in I$ such that $s + x = t + y$. In this case, we say $s$ is congruent to $t$ modulo $I$.

**Proposition 9.1.14.** *Let $S$ be a $k$-algebra and $I$ an ideal. Then congruence modulo $I$ is a $k$-algebra congruence.*

*Proof.* Clearly, congruence modulo $I$ is reflexive and symmetric. Assume that $s \equiv t \bmod I$ and $t \equiv u \bmod I$; so there exist $x, y, z, w \in I$ such that $s + x = t + y$ and $t + z = u + w$. Then

$$s + x + z = t + y + z = t + z + y = u + w + y$$

and hence $s \equiv u \bmod I$ as $x + z, w + y \in I$. This establishes that congruence modulo $I$ is an equivalence relation. Assume that $s \equiv t \bmod I$, $u \in S$ and $c \in k$; so $s + x = t + y$ for some $x, y \in I$. Then $u + s + x = u + t + y$ and hence $u + s \equiv u + t \bmod I$. Also $us + ux = ut + uy$, and because $ux, uy \in I$, we have $us \equiv ut \bmod I$. The proofs that $su \equiv tu \bmod I$ and $cu \equiv ct \bmod I$ are similar. This completes the proof.                                           $\square$

We write $S/I$ for the quotient of $S$ by the congruence in Proposition 9.1.14 (this should not be confused with the Rees quotient). Following ring theoretic tradition, we shall write $s + I$ for the congruence class of $s$ modulo $I$; however, the reader should be cautioned that this will in general be bigger than the subset $s + I$ of $P(S)$. Clearly, congruence modulo $I$ is the smallest congruence identifying $I$ with 0 and so $S/I$ has an obvious universal property.

**Proposition 9.1.15.** *Let $\varphi \colon S \to T$ be a homomorphism of $k$-algebras and let $I$ be an ideal of $S$ with $I \subseteq 0\varphi^{-1}$. Then $\varphi$ factors uniquely through the projection $S \twoheadrightarrow S/I$.*

**Exercise 9.1.16.** Prove Proposition 9.1.15.

Unlike in ring theory, not every congruence on a semiring is congruence modulo an ideal. For instance, if $S$ is a non-trivial semigroup, the natural map $\varphi \colon P(S) \to \mathbb{B}$ sending all non-empty sets to 1 and the empty set to 0 is a homomorphism satisfying $0\varphi^{-1} = \{\emptyset\}$. Because $\varphi$ is not injective, it follows easily that $\ker \varphi$ is not the congruence associated to an ideal quotient.

We shall need in the sequel the notion of a contracted semigroup algebra.

**Definition 9.1.17 (Contracted semigroup algebra).** *Let $S$ be a semigroup with a zero. Then the contracted semigroup algebra $k_0 S$ is the quotient $kS/k0$ where $k0$ is the ideal of all multiples of 0.*

If $S$ is a finite semigroup with zero, the contracted semigroup algebra $\mathbb{B}_0(S)$ can be identified with the collection of subsets of $S$ containing 0, which we shall denote $P_0(S)$. The functor $S \mapsto k_0 S$ is the left adjoint of the forgetful functor from $k$-algebras to semigroups with zero. Hence any semigroup homomorphism from $S$ into a $k$-algebra mapping 0 to 0 extends uniquely to $k_0 S$.

**Exercise 9.1.18.** Show the map $P(S) \to P_0(S)$ given by $X \mapsto X \cup \{0\}$ induces an isomorphism $\mathbb{B}_0(S) \cong P_0(S)$.

**Exercise 9.1.19.** Let $G$ be a group. Show that $k_0 \mathcal{M}^0(G, n, n, I_n) \cong M_n(kG)$.

Let us relate Rees quotients with ideal quotients of semirings.

**Proposition 9.1.20.** *Let $S$ be a semigroup and $I$ an ideal of $S$. Then $kI$ is a semiring ideal of $kS$ and $kS/kI \cong k_0[S/I]$.*

*Proof.* Clearly, $kI$ is an ideal of $kS$. The quotient $S \twoheadrightarrow S/I$ followed by the inclusion $S/I \hookrightarrow k_0[S/I]$ induces an onto homomorphism $\varphi \colon kS \to k_0[S/I]$. Evidently $kI \subseteq 0\varphi^{-1}$ and so, by Proposition 9.1.15, it suffices to show that $f\varphi = g\varphi$ implies $f + kI = g + kI$ for $f, g \in kS$. Let us write

$$f = \sum_{s \in S \setminus I} c_s s + \sum_{x \in I} c_x x,$$

$$g = \sum_{s \in S \setminus I} d_s s + \sum_{x \in I} d_x x.$$

Then $\sum_{s \in S \setminus I} c_s s = f\varphi = g\varphi = \sum_{s \in S \setminus I} d_s s$. As $\sum_{x \in I} c_x x, \sum_{x \in I} d_x x \in kI$, we conclude $f + kI = g + kI$. This completes the proof. $\square$

Alternatively, one can observe that $S \to kS/kI$ and $S \to k_0[S/I]$ are both universal maps of $S$ into a $k$-algebra sending $I$ to 0.

### Ideals in idempotent semirings

Recall that a subset of a lattice is called an ideal, or *lattice ideal*, if it is closed under finite sups and is a downset. If $\varphi \colon S \to T$ is a homomorphism of idempotent semirings, then $0\varphi^{-1}$ is plainly an ideal in both the semiring sense and the lattice sense. But not every ideal in the semiring sense is a downset (as we shall see momentarily). However, the following proposition provides the relationship between these concepts.

**Proposition 9.1.21.** *Let $S$ be an idempotent semiring and let $I$ be a semiring ideal of $S$. Set*
$$I^{\downarrow} = \{s \in S \mid \exists x \in I \text{ such that } s \leq x\}.$$
*Then $I^{\downarrow}$ is both a lattice and a semiring ideal and congruence modulo $I$ coincides with congruence modulo $I^{\downarrow}$.*

*Proof.* If $s \leq x$, $s' \leq x'$ with $x, x' \in I$, then $s + s' \leq x + x' \in I$ and so $s + s' \in I^{\downarrow}$. Because $I^{\downarrow}$ is evidently a downset, we conclude it is a lattice ideal. Also if $s \in I^{\downarrow}$ with $s \leq x \in I$, then $s's \leq s'x \in I$ and $ss' \leq xs' \in I$, so $s's, ss' \in I^{\downarrow}$. Thus $I^{\downarrow}$ is a semiring ideal. Clearly, $I \subseteq I^{\downarrow}$, so it remains to show that $s + I^{\downarrow} = s' + I^{\downarrow}$ implies $s + I = s' + I$. Let $y, y' \in I^{\downarrow}$ so that $s + y = s' + y'$. Then $y \leq x$, $y' \leq x'$ some $x, x' \in I$. But then

$$s + x + x' = s + y + x + x' = s' + y' + x + x' = s' + x + x'$$

and $x + x' \in I$. Therefore $s + I = s' + I$, completing the proof.     □

Let us give an example to show that the image under a surjective morphism of a semiring ideal, which is a downset (and hence a lattice ideal), need not be a lattice ideal. Consider the subsemirings of $M_2(\mathbb{B})$:

$$R = \left\{ \widetilde{0} = \begin{pmatrix} 1 & 0 \\ 1 & 1 \end{pmatrix}, \overline{0} = \begin{pmatrix} 0 & 0 \\ 1 & 1 \end{pmatrix}, \overline{1} = \begin{pmatrix} 0 & 0 \\ 0 & 1 \end{pmatrix}, I = \begin{pmatrix} 1 & 0 \\ 0 & 1 \end{pmatrix}, 0 = \begin{pmatrix} 0 & 0 \\ 0 & 0 \end{pmatrix} \right\}$$

$$S = \left\{ \overline{0} = \begin{pmatrix} 1 & 1 \\ 1 & 1 \end{pmatrix}, I = \begin{pmatrix} 1 & 0 \\ 1 & 1 \end{pmatrix}, \overline{1} = \begin{pmatrix} 1 & 0 \\ 1 & 0 \end{pmatrix}, 0 = \begin{pmatrix} 0 & 0 \\ 0 & 0 \end{pmatrix} \right\}.$$

It is straightforward to verify that $S \cong \overline{2}^I \cup \{0\}$ (the flip-flop with adjoined zero) with the order $\overline{0} > I > \overline{1} > 0$. On the other hand, $R$ is an inflation of $\overline{2}^I \cup \{0\}$ where $\overline{0}$ is inflated to $\{\overline{0}, \widetilde{0}\}$ with $\widetilde{0}\overline{0} = \overline{0} = \overline{0}\widetilde{0}$. The lattice ordering is given by the Hasse diagram in Figure 9.1. The set $J = \{\overline{0}, \overline{1}, 0\}$ is a semiring ideal and a lattice ideal in $R$. Consider the surjective map $\varphi \colon R \twoheadrightarrow S$ that identifies $\widetilde{0}$ and $\overline{0}$. It is easy to verify that $\varphi$ is a semiring homomorphism. The image of $J$ is the semiring ideal $\{\overline{0}, \overline{1}, 0\}$ of $S$, which is not a lattice ideal because $I < \overline{0}$. Notice that $S = U_2 \cup 0$ is a quotient of an idempotent semiring of upper triangular matrices. This implies that $S$ is not $\triangle$-irreducible in the sense to be defined later in this chapter.

Our next proposition describes the quantic nucleus associated to the quotient modulo an ideal in a finite idempotent semiring.

**Proposition 9.1.22.** *Let $S$ be a finite idempotent semiring and let $I$ be a semiring ideal. Let $x = \bigvee I$. Then the quantic nucleus $j \colon S \to S$ associated to the quotient $S \to S/I$ is given by $sj = s + x$. In particular, $s + I = s' + I$ if and only if $s + x = s' + x$.*

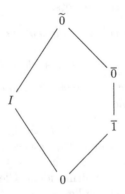

**Fig. 9.1.** Hasse diagram for $R$

*Proof.* By finiteness, $x \in I$ and so $s + I = s + x + I$. Also, if $s + I = s' + I$, then there exist $y, y' \in I$ with $s + y = s' + y'$. By choice of $x$, we know $x \geq y$. But then $s + x = s + y + x = s' + y' + x$ and so $s + x \geq s'$. Thus $s + x$ is the largest element in the congruence class of $s$ and so $j$ indeed gives the associated quantic nucleus and $s + I = s' + I$ if and only if $s + x = s' + x$. $\square$

## Quotient modules

A similar notion of quotient can be defined for modules as well and will be used frequently in the sequel. Let $S$ be a semigroup or semiring. Suppose that a $k$-module $M$ is an $S$-module. Let $N$ be an $S$-submodule. Then we can define a congruence on $M$ by $m \equiv m' \bmod N$ if $m + n = m' + n'$ for some $n, n' \in N$. A verification similar to Proposition 9.1.14 shows that this is an $S$-module congruence. The quotient is denoted $M/N$, and the congruence class of $m$ is denoted $m + N$. A universal property similar to the one in Proposition 9.1.15 can be established for such quotient modules.

**Exercise 9.1.23.** State and prove analogues of Propositions 9.1.14 and 9.1.15 for modules.

If $L, M, N$ are $S$-modules, we shall say that

$$0 \longrightarrow L \xrightarrow{f} M \xrightarrow{g} N \longrightarrow 0$$

is a *short exact sequence* if $L \xrightarrow{f} M$ is injective and $M \xrightarrow{g} N$ is surjective with $\ker g$ being congruence modulo $Lf$.

A crucial example we shall need later is the following.

*Example 9.1.24 (Direct sums).* Let $M$ and $N$ be $S$-modules. Then

$$0 \longrightarrow M \xrightarrow{\iota_M} M \oplus N \xrightarrow{\pi_N} N \longrightarrow 0 \tag{9.1}$$

is a short exact sequence where $\iota_M$ is the inclusion $m \mapsto (m, 0)$ and $\pi_N$ is the projection to $N$. Indeed, let us verify that the (9.1) is exact. We need to show that $(m, n)\pi_N = (m', n')\pi_N$ if and only if $(m, n) + M\iota_M = (m', n') + M\iota_M$. Suppose first $(m, n)\pi_N = (m', n')\pi_N$, so $n = n'$. Then $(m, 0), (m', 0) \in M\iota_M$ and

$$(m, n) + (m', 0) = (m + m', n) = (m' + m, n) = (m', n) + (m, 0)$$

and so $(m, n) \equiv (m', n') \bmod M\iota_M$. Conversely, if $(m, n) \equiv (m', n') \bmod M\iota_M$, then there exist $m_1, m_2 \in M$ with $(m, n) + (m_1, 0) = (m', n') + (m_2, 0)$. But this implies $n = n'$. This establishes the exactness of (9.1).

## 9.2 The Triangular Product

In this section, we introduce the triangular product [239,240,376]. We propose that the triangular product should play the role of the wreath product in the decomposition theory of semirings. Before describing this product, we consider triangular constructions in general.

Suppose $R$ and $S$ are $k$-algebras. If $M$ is an $R$-$S$-bimodule, then it is straightforward to verify that

$$\nabla(R, M, S) = \begin{pmatrix} S & 0 \\ M & R \end{pmatrix}$$

is a $k$-algebra with usual matrix addition and multiplication and usual scalar multiplication.

**Exercise 9.2.1.** Verify that $\nabla(R, M, S)$ is a $k$-algebra and that $\nabla(R, -, S)$ is a functor from the category of $R$-$S$-bimodules to the category of $k$-algebras.

There is a more general construction that encompasses the previous one on the multiplicative level. Let $M$ be a semigroup admitting a left action by a semigroup $S$ and a right action by a semigroup $T$ that commute with each other. We write $M$ additively, although we do not assume that the binary operation is commutative. Then their *triple product* [85] is

$$\nabla(S, M, T) = \begin{pmatrix} T & 0 \\ M & S \end{pmatrix}$$

with multiplication given by

$$\begin{pmatrix} t & 0 \\ m & s \end{pmatrix} \begin{pmatrix} t' & 0 \\ m' & s' \end{pmatrix} = \begin{pmatrix} tt' & 0 \\ mt' + sm' & ss' \end{pmatrix}.$$

One can easily check that the triple product is a semigroup. If $S$, $M$ and $T$ are monoids and the actions are unitary and by monoid endomorphisms, then $\nabla(S, M, T)$ is a monoid with the identity matrix as the identity.

**Exercise 9.2.2.** Verify that the triple product is a semigroup.

**Exercise 9.2.3.** Verify that the double semidirect product $S \bowtie T$ is the sub-semigroup of $\nabla(T, S, T)$ consisting of all elements of the form $\begin{pmatrix} t & 0 \\ s & t \end{pmatrix}$.

We shall need later in our decomposition theory for idempotent semirings a standard lemma concerning the Mal'cev kernel of the projection from a triple product (see Definition 5.3.1).

**Lemma 9.2.4.** *Let $\nabla(S, M, T)$ be a triple product and let*

$$\pi \colon \nabla(S, M, T) \to T \times S$$

*be the projection to the diagonal. Let $\mathbf{V}$ be a pseudovariety of semigroups containing $M$. Then $\pi \in (\mathbb{L}\mathbf{V}, \mathbf{1})$.*

*Proof.* Let $N$ be a submonoid of $\nabla(S, M, T)$ with identity $\begin{pmatrix} e & 0 \\ m_0 & f \end{pmatrix}$ mapping to the idempotent $(e, f)$. To prove the lemma, it suffices to construct an embedding $\theta \colon N \to M$. Define $\begin{pmatrix} e & 0 \\ m & f \end{pmatrix} \theta = fme$. Observe first that

$$\begin{pmatrix} e & 0 \\ m & f \end{pmatrix} = \begin{pmatrix} e & 0 \\ m_0 & f \end{pmatrix} \begin{pmatrix} e & 0 \\ m & f \end{pmatrix} \begin{pmatrix} e & 0 \\ m_0 & f \end{pmatrix} = \begin{pmatrix} e & 0 \\ m_0 e + fme + fm_0 & f \end{pmatrix}$$

and so $\theta$ is injective. To see that $\theta$ is a homomorphism, we observe that

$$\begin{pmatrix} e & 0 \\ m & f \end{pmatrix} \begin{pmatrix} e & 0 \\ m' & f \end{pmatrix} = \begin{pmatrix} e & 0 \\ me + fm' & f \end{pmatrix}$$

has image $f(me + fm')e = fme + fm'e$ under $\theta$. It follows that $\theta$ is a homomorphism.                                                        $\square$

Similarly to what happens in Corollary 2.4.16, one can use $\theta$ to show that $\pi$ belongs to the pseudovariety of relational morphisms $(\llbracket xyz = xwz \rrbracket \textcircled{m} \mathbf{V}, \mathbf{1})$ if $M \in \mathbf{V}$.

**Exercise 9.2.5.** Let $\nabla(S, M, T)$ be a triple product and let

$$\pi \colon \nabla(S, M, T) \to T \times S$$

be the projection to the diagonal. Let $\mathbf{V}$ be a pseudovariety of semigroups containing $M$. Prove $\pi \in (\llbracket xyz = xwz \rrbracket \textcircled{m} \mathbf{V}, \mathbf{1})$.

A special case of the triple product in the semiring context, which is of great importance, is the triangular product of modules. This notion, in the case of rings, was introduced by Plotkin [239,240,376] and basically *axiomatizes* the lower triangular form one obtains from an invariant subspace for a representation; see Section 9.3 for details. First observe that if $(M, R)$ and $(N, S)$ are right modules, then $\mathsf{Hom}_k(N, M)$ is an $S$-$R$-bimodule where $n(fr) = (nf)r$ and $n^s f = (ns)f$.

**Definition 9.2.6 (Triangular product).** *Suppose that $R$ and $S$ are $k$-algebras and that $(M, R)$ and $(N, S)$ are right modules. Then their triangular product is the $k$-algebra*

$$\triangle((M, R), (N, S)) = \nabla(S, \mathsf{Hom}_k(N, M), R) = \begin{pmatrix} R & 0 \\ \mathsf{Hom}_k(N, M) & S \end{pmatrix}.$$

*Moreover, $(M \oplus N, \triangle((M, R), (N, S)))$ is a right module with action given by*

$$(m, n) \begin{pmatrix} r & 0 \\ f & s \end{pmatrix} = (mr + nf, ns). \tag{9.2}$$

*In addition, if the actions of $R$ and $S$ are faithful, then so is the action (9.2).*

**Exercise 9.2.7.** Verify (9.2) is an action and is faithful if each of the original actions was faithful.

One can define analogously the triangular product in the case $R$ and $S$ are semigroups, but in this case one only obtains a semigroup acting on $M \oplus N$. The triangular product of modules is associative, and in fact one can define an $n$-fold triangular product. Let us proceed to do so.

**Definition 9.2.8 ($n$-fold triangular product).** *If $(M_1, R_1), \ldots, (M_n, R_n)$ are right modules (where the $R_i$ are $k$-algebras), their triangular product is*

$$\triangle((M_1, R_1), \ldots (M_n, R_n)) = \begin{pmatrix} R_1 & 0 & \cdots & 0 \\ \mathrm{Hom}_k(M_2, M_1) & R_2 & \cdots & 0 \\ \vdots & & \vdots & \ddots & \vdots \\ \mathrm{Hom}_k(M_n, M_1) & \mathrm{Hom}_k(M_n, M_2) & \cdots & R_n \end{pmatrix}.$$

*The triangular product is a $k$-algebra with usual matrix operations where if $f \in \mathrm{Hom}_k(M_i, M_\ell)$ and $g \in \mathrm{Hom}_k(M_\ell, M_j)$, then the composition $fg$ belongs to $\mathrm{Hom}_k(M_i, M_j)$. It acts on the right of $\bigoplus_{i=1}^n M_i$ where elements of this latter module are viewed as row vectors.*

There is an analogous definition for $k$-modules over semigroups. Let us remark that if $(M, R)$ and $(N, S)$ are right modules, then $(M \oplus N, R \times S)$ embeds in $\triangle((M, R), (N, S))$ as the diagonal. The associativity of the triangular product of group representations is proved in [376]. We leave the rather tedious proof in our context to the reader.

**Proposition 9.2.9.** *The triangular product of modules is associative. Any way of bracketing to form triangular products of $(M_1, R_1), \ldots, (M_n, R_n)$ results in the $n$-fold triangular product $\triangle((M_1, R_1), \ldots (M_n, R_n))$.*

**Exercise 9.2.10.** Prove Proposition 9.2.9.

**Exercise 9.2.11.** Let $(M_1, R_1), \ldots, (M_n, R_n)$ and $(M_1', R_1'), \ldots, (M_n', R_n')$ be right modules. Show $\triangle((M_1, R_1), \ldots, (M_n, R_n)) \times \triangle((M_1', R_1'), \ldots, (M_n', R_n'))$ embeds in $\triangle((M_1 \oplus M_1', R_1 \times R_1'), \ldots, (M_n \oplus M_n', R_n \times R_n'))$.

## Duality

Let $R$ be $k$-algebra. If $M$ is a right $R$-module, then $M^* = \mathrm{Hom}_k(M, k)$ is a left $R$-module called the *dual module* to $M$. Hence $M^*$ is a right $R^{op}$-module via the action $m(gr) = (mr)g$ (so $gr = {}^r g$ in our left action notation). Elements of $M^*$ are called *functionals* on $M$. If $f \colon M \to N$ is a morphism of $k$-modules, then there is an *adjoint* morphism $f^* \colon N^* \to M^*$ given by $(g)f^* = fg$. Let us say that $k$ is *separative* if, for every $k$-module $M$, there are enough functionals on $M$ to separate points. That is, given distinct elements $m, m' \in M$, there is a functional $f\colon M \to k$ such that $mf \neq m'f$. This is equivalent to saying the natural map $M \to M^{**}$ given by $m \mapsto e_m$ is injective, where $fe_m = mf$. Here $e_m$ stands for evaluation at $m$. For instance, fields are separative. One of the key properties of separative semirings is contained in the following proposition.

**Proposition 9.2.12.** *Let $k$ be separative. Then the $k$-linear map $f \mapsto f^*$ from $\mathsf{Hom}_k(M,N) \to \mathsf{Hom}_k(N^*,M^*)$ is injective.*

*Proof.* First we claim if $m \in M$, then $e_m f^{**} = e_{mf}$. Indeed,

$$g(e_m f^{**}) = g(f^* e_m) = (gf^*)e_m = (fg)e_m = mfg = g e_{mf}.$$

Suppose now $f^* = g^*$. Then $f^{**} = g^{**}$ and so $e_{mf} = e_{mg}$ all $m \in M$. As the map $n \mapsto e_n$ is injective, we conclude $mf = mg$ for all $m$, that is, $f = g$. □

Next we show that the semiring of primary interest to us, the Boolean semiring, is separative.

**Proposition 9.2.13.** *The Boolean semiring $\mathbb{B}$ is separative.*

*Proof.* Let $L$ be a $\mathbb{B}$-module, so $L$ is a $\vee$-semilattice with minimum. For each $\ell \in L$, define a functional $\check{\ell}\colon L \to \mathbb{B}$ by

$$(\ell')\check{\ell} = \begin{cases} 0 & \ell' \leq \ell \\ 1 & \text{else.} \end{cases}$$

Plainly $\check{\ell}$ is a $\mathbb{B}$-linear map. Suppose now that $\ell \neq \ell' \in L$. Without loss of generality, we may assume that $\ell' \not\leq \ell$. Then $(\ell)\check{\ell} = 0$, whereas $(\ell')\check{\ell} = 1$. Thus there are enough functionals to separate points. □

As a consequence, we obtain the following result. The reader should note the reversal of the order of the factors when one takes duals. It corresponds with the fact in representation theory that when taking duals, submodules become quotient modules and quotient modules become submodules.

**Proposition 9.2.14.** *Let $k$ be separative and let $R$ and $S$ be $k$-algebras. Consider right modules $(M,R)$, $(N,S)$. Then $\triangle((M,R),(N,S))^{op}$ embeds in $\triangle((N^*,S^{op}),(M^*,R^{op}))$.*

*Proof.* Define $\Psi\colon \triangle((M,R),(N,S))^{op} \to \triangle((N^*,S^{op}),(M^*,R^{op}))$ by sending

$$\begin{pmatrix} r & 0 \\ f & s \end{pmatrix} \longmapsto \begin{pmatrix} s & 0 \\ f^* & r \end{pmatrix}.$$

Injectivity of $\Psi$ is immediate from Proposition 9.2.12. We need to show that $\psi$ is a homomorphism. Let us use $\diamond$ for the product in $S^{op}$ and $R^{op}$.

First we compute:

$$\begin{pmatrix} r & 0 \\ f & s \end{pmatrix} \Psi \begin{pmatrix} r_0 & 0 \\ f_0 & s_0 \end{pmatrix} \Psi = \begin{pmatrix} s & 0 \\ f^* & r \end{pmatrix} \begin{pmatrix} s_0 & 0 \\ f_0^* & r_0 \end{pmatrix}$$

$$= \begin{pmatrix} s \diamond s_0 & 0 \\ f^* s_0 + {}^r f_0^* & r \diamond r_0 \end{pmatrix}$$

$$= \begin{pmatrix} s_0 s & 0 \\ f^* s_0 + {}^r f_0^* & r_0 r \end{pmatrix}$$

and

$$\left[ \begin{pmatrix} r_0 & 0 \\ f_0 & s_0 \end{pmatrix} \begin{pmatrix} r & 0 \\ f & s \end{pmatrix} \right] \Psi = \left[ \begin{pmatrix} r_0 r & 0 \\ f_0 r + {}^{s_0} f & s_0 s \end{pmatrix} \right] \Psi$$

$$= \begin{pmatrix} s_0 s & 0 \\ (f_0 r + {}^{s_0} f)^* & r_0 r \end{pmatrix}.$$

So it suffices to prove $f^* s_0 + {}^r f_0^* = (f_0 r + {}^{s_0} f)^*$. Because

$$(f_0 r + {}^{s_0} f)^* = (f_0 r)^* + ({}^{s_0} f)^*$$

we just need to show $(f_0 r)^* = {}^r f_0^*$ and $({}^{s_0} f)^* = f^* s_0$. So let $g \colon M \to k$ be a functional and let $n \in N$. Then

$$n[g(f_0 r)^*] = (n(f_0 r))g = ((n f_0) r)g = (n f_0)(gr) = n[(gr) f_0^*] = n[g^r f_0^*]$$

and

$$n[g(f^* s_0)] = n[(g f^*) s_0] = (n s_0)(g f^*) = (n s_0) f g = n^{s_0} f g = n[g({}^{s_0} f)^*].$$

This completes the proof.                                                          □

### 9.2.1 A wreath product embedding

We now wish to relate triangular products with wreath products. This serves in part to motivate why we think of the triangular product as the appropriate substitute for the wreath product in the context of semirings. The next theorem, expressing a triple product as a combination of semidirect and reverse semidirect products, and its corollary generalize of a result of the second author and Kambites [154].

**Theorem 9.2.15.** *Suppose that the triple product $\nabla(S, M, T)$ is defined. Then we can form the reverse semidirect product $T \ltimes M$. Moreover, $S$ acts on the left of $T \ltimes M$ via $s(t, m) = (t, sm)$. Dually, we can form $M \rtimes S$ and $T$ acts on the right of $M \rtimes S$ via $(m, s)t = (mt, s)$. One then has isomorphisms*

$$\nabla(S, M, T) \cong (T \ltimes M) \rtimes S \cong T \ltimes (M \rtimes S).$$

*Proof.* We just handle the first isomorphism, the second being dual. First we verify $S$ acts on the left of $T \ltimes M$. Indeed, if $s \in S$, $m, m' \in M$, $t, t' \in T$, then

$$
\begin{aligned}
s((t,m))(t',m')) &= s(tt', mt' + m') \\
&= (tt', s(mt' + m)) \\
&= (tt', (sm)t' + sm') \\
&= (t, sm)(t', sm') \\
&= (s(t,m))(s(t',m')).
\end{aligned}
$$

Define $\varphi \colon \nabla(S, M, T) \to (T \ltimes M) \rtimes S$ by

$$
\begin{pmatrix} t & 0 \\ m & s \end{pmatrix} \varphi = ((t,m), s).
$$

Clearly, $\varphi$ is a bijection, so it suffices to check that $\varphi$ is a homomorphism. Now on the one hand,

$$
\left[ \begin{pmatrix} t & 0 \\ m & s \end{pmatrix} \cdot \begin{pmatrix} t' & 0 \\ m' & s' \end{pmatrix} \right] \varphi = \begin{pmatrix} tt' & 0 \\ mt' + sm' & ss' \end{pmatrix} \varphi = ((tt', mt' + sm'), ss').
$$

On the other hand,

$$
\begin{pmatrix} t & 0 \\ m & s \end{pmatrix} \varphi \begin{pmatrix} t' & 0 \\ m' & s' \end{pmatrix} \varphi = ((t,m), s)((t', m'), s') = ((t,m) \cdot s(t', m'), ss')
$$

$$
= ((t,m) \cdot (t', sm'), ss')
$$

$$
= ((tt', mt' + sm'), ss').
$$

This completes the proof. $\qquad\qquad\qquad\qquad\qquad\qquad\qquad\qquad\qquad\quad$ $\square$

As a corollary, we obtain a wreath product embedding.

**Corollary 9.2.16.** *Let $(L, S)$ and $(M, T)$ be right modules. Then*

$$
\big( L \oplus M, \triangle((L,S),(M,T)) \big) \leq (L, S \ltimes L) \wr (M, T)
$$

*where $S \ltimes L$ acts on $L$ by $\ell_0(s, \ell) = \ell_0 s + \ell$.*

*Proof.* There is an action of $T$ on $S \ltimes \operatorname{Hom}_k(M, L)$ by $t(s, f) = (s, {}^t\!f)$ and

$$
\triangle((L,S),(M,T)) \cong (S \ltimes \operatorname{Hom}_k(M, L)) \rtimes T
$$

according to Theorem 9.2.15. Now $S \ltimes \operatorname{Hom}_k(M, L)$ embeds in $(S \ltimes L)^M$ by sending $(s, f)$ to $g_{(s,f)}$ where $mg_{(s,f)} = (s, mf)$. Moreover, the left actions of $T$ on $S \ltimes \operatorname{Hom}_k(M, L)$ and $(S \ltimes L)^M$ commute with this embedding because

$$
m^t g_{(s,f)} = mtg_{(s,f)} = (s, mtf) = (s, m^t\!f) = mg_{(s,{}^t\!f)} = mg_{t(s,f)}.
$$

Hence there results an embedding

$$\Delta((L,S),(M,T))) \le (S \ltimes L)^M \rtimes T = (S \ltimes L) \wr (M,T)$$

via the map

$$\begin{pmatrix} s & 0 \\ f & t \end{pmatrix} \longmapsto (g_{(s,f)}, t).$$

It remains to check that the actions agree.

First let us verify $S \ltimes L$ acts on $L$ as indicated. For $\ell_0, \ell \in L$ and $s, s' \in S$,

$$(\ell_0(s,\ell))(s',\ell') = (\ell_0 s + \ell)(s',\ell') = \ell_0 ss' + \ell s' + \ell'$$

whereas

$$\ell_0((s,\ell)(s',\ell')) = \ell_0(ss', \ell s' + \ell') = \ell_0 ss' + \ell s' + \ell'$$

so the action is indeed well defined.

Now we compare the actions:

$$(\ell, m)\begin{pmatrix} s & 0 \\ f & t \end{pmatrix} = (\ell s + mf, mt)$$

whereas

$$(\ell, m)(g_{(s,f)}, t) = (\ell \cdot mg_{(s,f)}, mt) = (\ell \cdot (s, mf), mt) = (\ell s + mf, mt).$$

This completes the proof.                                                    □

### 9.2.2 The Schützenberger product

Let us show that the Schützenberger product of two finite semigroups is an example of a triangular product. Let $S, T$ be semigroups and $k$ a fixed semiring. Then the *Schützenberger product* (see, for instance, [230]) is defined as follows. One can make $k(S^I \times T^I)$ into an $S$-$T$-bimodule by defining

$$s_0 \cdot \sum_{s \in S^I, t \in T^I} c_{(s,t)}(s,t) = \sum_{s \in S^I, t \in T^I} c_{(s,t)}(s_0 s, t)$$

$$\left( \sum_{s \in S^I, t \in T^I} c_{(s,t)}(s,t) \right) \cdot t_0 = \sum_{s \in S^I, t \in T^I} c_{(s,t)}(s, tt_0)$$

for $s_0 \in S$, $t_0 \in T$. Then the Schützenberger product is

$$\Diamond_k(S,T) = \nabla(S, k(S^I \times T^I), T) = \begin{pmatrix} T & 0 \\ k(S^I \times T^I) & S \end{pmatrix}.$$

We remark that it is often useful to think of $k(S^I \times T^I)$ as $kS^I \otimes_k kT^I$. In fact, if one defines the tensor product of $k$-algebras according to the usual universal property, then it is easy to verify that $k(S^I \times T^I) \cong kS^I \otimes_k kT^I$. More generally, if $A$ and $B$ are $k$-algebras, then one can make $A^I \otimes_k B^I$ into an $A$-$B$ bimodule in the obvious way. Hence, one should consider $\Diamond_k(A, B) = \nabla(A, A^I \otimes_k B^I, B)$ as the Schützenberger product of $A$ and $B$.

To describe the Schützenberger product of finite semigroups as a triangular product, we begin by describing the appropriate modules. First observe that $kS$ can be viewed on the level of $k$-modules as $k^S$. From this point-of-view, the natural right action of $S$ on $k^S$ is given by $f^s(s') = f(ss')$. If one translates this back to $kS$ viewed as formal sums, one obtains the following notion.

**Definition 9.2.17 (Adjoint translation).** *Let $S$ be a semigroup and $k$ a semiring. Define, for $s \in S$, a linear operator $s^{-1}$ on $kS^I$ by*

$$s^{-1}s_0 = \sum_{s' \in S^I, ss' = s_0} s'.$$

*Then $(st)^{-1} = t^{-1}s^{-1}$ and there is a resulting right action of $kS$ on $kS^I$, called the adjoint translation action, denoted $(kS^I, S)^\vee$. The linear extension to $kS$ is denoted $(kS^I, kS)^\vee$.*

We remark that the adjoint translation action is just the dual (or adjoint) of the left regular representation of $kS$ on $kS^I$.

**Proposition 9.2.18.** *Let $S$ and $T$ be finite semigroups. Then*

$$\Diamond_k(S, T) \cong \Delta((kT^I, kT), (kS^I, kS)^\vee).$$

*Proof.* First observe that $\mathrm{Hom}_k(kS^I, kT^I) \cong (kT^I)^{S^I}$ because $kS^I$ is free on $S^I$. We define a bijection $\alpha : k(S^I \times T^I) \to (kT^I)^{S^I}$ by

$$\sum_{s \in S^I, t \in T^I} c_{(s,t)}(s, t) \longmapsto \left[ s \mapsto \sum_{t \in T^I} c_{(s,t)} t \right].$$

It is easy to check that $\alpha$ is a bijection; in fact, viewing $kT^I$ as the set $k^{T^I}$ and $k(S^I \times T^I)$ as the set $k^{S^I \times T^I}$, then we are just writing down the usual bijection between $(k^{T^I})^{S^I}$ and $k^{S^I \times T^I}$. Clearly, $\alpha$ is a $k$-module isomorphism. If we can show that $\alpha$ is an isomorphism of $kS$-$kT$-bimodules, then by the functoriality of the triple product, we shall obtain the desired isomorphism.

First observe that if $t_0 \in T$, then

$$\left( \sum_{s \in S^I, t \in T^I} c_{(s,t)}(s, t) \right) \cdot t_0 = \sum_{s \in S^I, t \in T^I} c_{(s,t)}(s, tt_0).$$

Now $\displaystyle\sum_{s \in S^I, t \in T^I} c_{(s,t)}(s, tt_0)\alpha$ sends $s$ to $\displaystyle\sum_{t \in T^I} c_{(s,t)} tt_0$, which is exactly

$$\sum_{s \in S^I, t \in T^I} c_{(s,t)}(s, t)\alpha \cdot t_0.$$

This shows that the right action of $T$ commutes with $\alpha$. Next suppose $s_0 \in S$. Then

$$s_0 \cdot \left( \sum_{s \in S^I, t \in T^I} c_{(s,t)}(s,t) \right) = \sum_{s \in S^I, t \in T^I} c_{(s,t)}(s_0 s, t),$$

which maps under $\alpha$ to the function whose value at $s \in S$ is

$$\sum_{\substack{s' \in S^I, t \in T^I \\ s_0 s' = s}} c_{(s',t)} t. \tag{9.3}$$

But $s_0^{-1} s = \displaystyle\sum_{s' \in S^I, s_0 s' = s} s'$, so (9.3) is just the value of

$$\left( \sum_{s \in S^I, t \in T^I} c_{(s,t)}(s,t) \right) \alpha$$

on $s_0^{-1} s$; that is the value of $^{s_0}\!\left( \sum_{s \in S^I, t \in T^I} c_{(s,t)}(s,t) \right) \alpha$ on $s$. This completes the proof that $\alpha$ is a bimodule isomorphism.    □

Recall that if $S$ is a semigroup, then the Henckell-Schützenberger expansion of $S$ is the subsemigroup of $\Diamond_k(S, S)$ generated by all matrices of the form

$$\begin{pmatrix} s & 0 \\ (I, s) + (s, I) & s \end{pmatrix}$$

with $s \in S$ [130]. It is therefore natural to define the Henckell-Schützenberger expansion of a $k$-algebra $R$ to be the subalgebra of $\Diamond_k(R, R)$ generated by all matrices of the form

$$\begin{pmatrix} r & 0 \\ I \otimes r + r \otimes I & r \end{pmatrix}$$

with $r \in R$.

## 9.3 The Triangular Decomposition Theorem

This section provides a decomposition of the semigroup algebra of a finite semigroup into a triangular product of matrix algebras over the group algebras of its Schützenberger groups. This is the semiring analogue of the representation theoretic results of Munn and Ponizovskiĭ [18, 68, 95, 210, 211, 245, 297]. Some of the results of this section apply equally well to infinite semirings, so for the moment we drop our restrictions on finiteness.

Let us begin with some intuition. Let $S$ be a semigroup acting by endomorphisms on the right of a vector space $V$ and let $W$ be an $S$-submodule. Choose a vector space complement $W'$ for $V$; so $V = W \oplus W'$ as vector spaces, but $W'$ need not be an $S$-submodule. By choosing a basis for $V$ adapted to this direct sum decomposition, we can put the associated matrix representation of $S$ into block lower triangular form:

$$s \mapsto \begin{pmatrix} s\rho_W & 0 \\ s\rho' & s\rho_{V/W} \end{pmatrix}$$

where $\rho_W$ is the restriction of the original representation to the invariant subspace $W$, $\rho_{V/W}$ is the induced quotient representation and $s\rho'$ is a certain matrix, which can actually be viewed as a linear map from $V/W$ to $W$. This essentially puts $S$ into the triangular product $\triangle((W,S),(V/W,S))$, where of course we can now replace $S$ by the quotient via the kernel of the actions if we so desire. This idea is essentially the old Jordan-Hölder principle.

To make this work for arbitrary semirings, we have to axiomatize the decomposition $V = W \oplus W'$. The key ingredients are the projection morphism $\pi_1 \colon V \to W$ (which is an $S$-module morphism) and a vector space morphism $\pi_2 \colon V/W \to V$ splitting the quotient map (which need not be an $S$-module morphism). If $\varphi \colon V \to V/W$ is the projection, then the essential properties are: $\pi_2\varphi = 1_{V/W}$; $\pi_2\pi_1 = 0$; and $\pi_1 + \varphi\pi_2 = 1_V$.

Let us fix a base commutative semiring $k$ with unit. Define a short exact sequence of $k$-modules

$$0 \longrightarrow L \longrightarrow M \xrightarrow{\varphi} N \longrightarrow 0 \tag{9.4}$$

(where we view $L$ as a submodule of $M$) to be *split* if there are $k$-module morphisms $\pi_1 \colon M \to L$ and $\pi_2 \colon N \to M$ (called *splitting maps*) such that:

(S1) $\pi_2\varphi = 1_N$;
(S2) $\pi_2\pi_1 = 0$;
(S3) $\pi_1 + \varphi\pi_2 = 1_M$.

Sometimes we say that (9.4) is split over $k$ to emphasize the base semiring.

**Proposition 9.3.1.** $\pi_1|_L = 1_L$.

*Proof.* By (S3), if $\ell \in L$, then $\ell = \ell\pi_1 + \ell\varphi\pi_2 = \ell\pi_1$ because $\ell\varphi = 0$. $\qquad \square$

In ring theory, to get the sequence to split, it suffices to have $\pi_1$ satisfying Proposition 9.3.1 or $\pi_2$ satisfying (S1), as the other map can then be defined using subtraction. Let us show that our definition of a split sequence correctly axiomatizes a direct sum splitting.

**Proposition 9.3.2.** *A short exact sequence* $0 \longrightarrow L \longrightarrow M \xrightarrow{\varphi} N \longrightarrow 0$ *is split if and only if there is a $k$-isomorphism $\rho \colon M \to L \oplus N$ such that the diagram*

$$0 \longrightarrow L \longrightarrow M \xrightarrow{\ \varphi\ } N \longrightarrow 0$$

$$0 \longrightarrow L \longrightarrow L \oplus N \xrightarrow{\ \pi_N\ } N \longrightarrow 0$$

with vertical maps: identity on $L$, $\rho$ on the middle, identity on $N$.    (9.5)

*commutes, where $\pi_N$ is the projection.*

*Proof.* Suppose first the sequence is split. Let $\pi_1 \colon M \to L$ and $\pi_2 \colon N \to M$ be the splitting maps. Define $\rho \colon M \to L \oplus N$ by $m\rho = (m\pi_1, m\varphi)$. It is evident that $\rho$ is a $k$-morphism. We claim $(\ell, n) \mapsto \ell + n\pi_2$ is an inverse to $\rho$. Indeed, $m\pi_1 + m\varphi\pi_2 = m$ by (S3). Conversely, recalling $\pi_1|_L = 1_L$ (Proposition 9.3.1), we compute

$$(\ell + n\pi_2)\pi_1 = \ell\pi_1 + n\pi_2\pi_1 = \ell \qquad \text{by (S2)}$$
$$(\ell + n\pi_2)\varphi = \ell\varphi + n\pi_2\varphi = n \qquad \text{by (S1)}$$

thereby establishing that $\rho$ is invertible. From the definition, $\rho\pi_N = \varphi$. On the other hand, if $\ell \in L$, then $\ell\rho = (\ell\pi_1, \ell\varphi) = (\ell, 0)$ by an application of Proposition 9.3.1. This proves (9.5) commutes.

Conversely, suppose that $\rho$ exists. Let $\pi_1 = \rho\pi_L$ and $\pi_2 = \iota_N\rho^{-1}$ where $\iota_N \colon N \to L \oplus N$ is the inclusion $n \mapsto (0, n)$ and $\pi_L \colon L \oplus N \to L$ is the projection. We verify (S1)–(S3). First $\pi_2\varphi = \iota_N\rho^{-1}\varphi = \iota_N\rho^{-1}\rho\pi_N = 1_N$ as (9.5) commutes. This settles (S1). For (S2), $\pi_2\pi_1 = \iota_N\rho^{-1}\rho\pi_L = \iota_N\pi_L = 0$. Finally, suppose $m\rho = (\ell, n)$. Then $m\pi_1 = \ell$ and $m\varphi = n$. So

$$m\rho = (\ell, 0) + (0, n) = \ell\rho + m\varphi\iota_N.$$

Therefore, $m = m\rho\rho^{-1} = \ell + m\varphi\iota_N\rho^{-1} = m\pi_1 + m\varphi\pi_2$, verifying (S3).    □

Let us give an important example of a split exact sequence, generalizing the vector space case. Again $k$ will always stand for a unital commutative semiring.

**Proposition 9.3.3.** *Let $X$ be a set and $Y \subseteq X$. Then the short exact sequence*

$$0 \longrightarrow kY \longrightarrow kX \xrightarrow{\ \varphi\ } kX/kY \longrightarrow 0$$

*is split.*

*Proof.* Because $X = Y \uplus (X \setminus Y)$ (disjoint union) and the functor $X \mapsto kX$ is left adjoint to the forgetful functor, we must have $kX \cong kY \oplus k[X \setminus Y]$. Under this isomorphism, the quotient $kX \to kX/kY$ turns into the projection $kY \oplus k[X \setminus Y] \to k[X \setminus Y]$ (cf. Example 9.1.24). The result now follows from Proposition 9.3.2.    □

**Exercise 9.3.4.** Prove directly that $k[X \setminus Y]$ is isomorphic to $kX/kY$ via the map $\sum_{x \in X \setminus Y} c_x x \longmapsto \sum_{x \in X \setminus Y} c_x x + kY$.

*Remark 9.3.5.* So far, we have been considering only $k$-modules, and it is important that we only ask for sequences to split as sequence of $k$-modules, even if there is some $k$-algebra $S$ lurking in the background. We will never require splitting of $S$-modules.

We are now in a position to prove a semiring analogue of a result of Plotkin in the case of algebras over a field [239, 376] by using the notion of exact sequences split over $k$. This theorem is our main source of triangular product decompositions, so far.

**Theorem 9.3.6 (Triangular Decomposition Theorem).** *Let $k$ be a commutative semiring with unit. Let $S$ be a semigroup ($k$-algebra) and let*

$$0 \longrightarrow L \longrightarrow M \overset{\varphi}{\longrightarrow} N \longrightarrow 0$$

*be a short exact sequence of $S$-modules that splits over $k$. Then there is a $k$-module isomorphism $\rho\colon M \to L \oplus N$ and a function $\psi\colon S \to \Delta((L,S),(N,S))$ such that $ms\rho = m\rho s\psi$ for all $m \in M$, $s \in S$. In particular, if $M$ is a faithful $S$-module and $(L, S_L)$, $(N, S_N)$ are the faithful quotient modules, then $S$ is a subsemigroup ($k$-subalgebra) of $\Delta((L, S_L),(N, S_N))$.*

*Proof.* Let $\pi_1, \pi_2$ be the splitting maps. The proof of Proposition 9.3.2 shows that $\rho\colon M \to L \oplus N$ given by $m\rho = (m\pi_1, m\varphi)$ is a $k$-module isomorphism. Next we define

$$\psi\colon S \to \Delta((L,S),(N,S))$$

by setting, for $s \in S$,

$$s\psi = \begin{pmatrix} s & 0 \\ f_s & s \end{pmatrix}$$

where $f_s \in \mathsf{Hom}_k(N, L)$ is given by $nf_s = (n\pi_2 s)\pi_1$. Suppose that $m \in M$. Then $ms\rho = ((ms)\pi_1, (ms)\varphi)$. On the other hand, we compute

$$m\rho s\psi = (m\pi_1, m\varphi) \begin{pmatrix} s & 0 \\ f_s & s \end{pmatrix}$$

$$= (m\pi_1 s + m\varphi f_s, m\varphi s)$$

$$= (m\pi_1 s + (m\varphi\pi_2 s)\pi_1, m\varphi s).$$

Therefore, it suffices to show that $(ms)\pi_1 = m\pi_1 s + (m\varphi\pi_2 s)\pi_1$. Well, by (S3) we have $m = m\pi_1 + m\varphi\pi_2$ and so $ms = m\pi_1 s + m\varphi\pi_2 s$. Hence, we obtain

$$(ms)\pi_1 = (m\pi_1 s + m\varphi\pi_2 s)\pi_1 = m\pi_1 s + (m\varphi\pi_2 s)\pi_1$$

where the last equality uses that $L$ is an $S$-submodule and $\pi_1|_L = 1_L$ (Proposition 9.3.1). This proves the first statement of the theorem. The second follows immediately from Lemma 9.1.5. $\qquad\square$

Our next theorem is an important consequence of the triangular decomposition theorem, which underlies the results of Munn and Ponizovskiĭ [68, 210, 211, 245, 297] on representations of semigroups over fields and the results of Fox and Rhodes [91] on the complexity of power semigroups. See also Plotkin's work on decomposing linear automata [240].

**Theorem 9.3.7 (Ideal Decomposition Theorem).** *Let $k$ be a commutative semiring with unit and $S$ a semigroup with an ideal $J$. Let $(kJ, S_{kJ})$, respectively, $(kJ, (kS)_{kJ})$ be the associated faithful modules. Then*

$$(kS^I, kS) \leq \triangle((kJ, (kS)_{kJ}), (k_0[S^I/J], k_0[S/J])) \text{ and} \qquad (9.6)$$

$$(kS^I, S) \leq \triangle((kJ, S_{kJ}), (k_0[S^I/J], S/J)). \qquad (9.7)$$

*Proof.* The short exact sequence of $kS$-modules

$$0 \longrightarrow kJ \longrightarrow kS^I \longrightarrow kS^I/kJ \longrightarrow 0$$

splits over $k$ by Proposition 9.3.3. The result is then immediate from the Triangular Decomposition Theorem (Theorem 9.3.6), the isomorphism $kS^I/kJ \cong k_0[S^I/J]$ (Proposition 9.1.20) and the observation that $(k_0[S^I/J], k_0[S/J])$ and $(k_0[S^I/J], S/J)$ are the respective faithful quotients of $(k_0[S^I/J], kS)$ and $(k_0[S^I/J], S)$. $\qquad\square$

Recall [68, Section 2.6] that a finite semigroup $S$ admits an ideal series

$$S = I_0 \supsetneq I_1 \supsetneq \cdots \supsetneq I_n$$

where $I_n$ is the minimal ideal and such that the factors $I_j/I_{j+1}$ are precisely the principal factors $J^0$ where $J$ runs over the $\mathscr{J}$-classes of $S$ (where $I_n/I_{n+1}$ is interpreted as the principal factor associated to the minimal ideal). Such a series is called a *principal series*. The following elementary exercise is needed to perform the inductive step in our next corollary to Theorem 9.3.7.

**Exercise 9.3.8.** Suppose $(N, T)$ embeds in $(N', T')$ (as per Lemma 9.1.5). Show that $(M \oplus N, \triangle((M, S), (N, T))) \leq (M \oplus N', \triangle((M, S), (N', T')))$.

We now state the desired corollary. The reader is referred to Definition 4.6.28 for the meaning of $\mathsf{RM}_J(S)$.

**Corollary 9.3.9.** *Let $S$ be a finite semigroup and $S^I = I_0 \supsetneq I_1 \supsetneq \cdots \supsetneq I_n$ be a principal series for $S^I$. Let $J_m$ be the $\mathscr{J}$-class corresponding to $I_m/I_{m+1}$ (where $J_n = I_n$). Identifying $kI_m/kI_{m+1}$ with $kJ_m$, there exist embeddings*

$$(kS^I, kS) \leq \triangle((kJ_n, A_n), (kJ_{n-1}, A_{n-1}), \dots, (kJ_0, A_0))$$

$$(kS^I, S) \leq \triangle((kJ_n, \mathsf{RM}_{J_n}(S)), (kJ_{n-1}, \mathsf{RM}_{J_{n-1}}(S)), \dots, (kJ_0, \mathsf{RM}_{J_0}(S)))$$

*where $A_m$ denotes the $k$-span of $\mathsf{RM}_{J_m}(S)$ in $\mathsf{End}_k(kJ_m)$, for $m \geq 0$.*

It remains only to describe the module $(kJ_m, A_m)$. This requires a lemma.

**Lemma 9.3.10.** *Let $G$ be a group and $X$ a free left $G$-set. Let $T$ be a transversal for the orbits of $G$ on $X$. The $kX$ is a free left $kG$-module on $T$, that is, $kX \cong kG[T]$.*

*Proof.* The fancy proof is $kX \cong k[G \times T] \cong kG \otimes_k kT \cong kG[T]$, where all the isomorphisms are left $kG$-isomorphisms. We proceed with a direct proof. Clearly, $kX$ is a left $kG$-module by first extending linearly the left $G$-action on $X$ to $kX$ and then extending linearly to $kG$. Let $T$ be a transversal for the orbits of $G$ on $X$. For $x \in X$, let $g_x$ be the unique element of $G$ with $g_x^{-1}x \in T$. Then

$$\sum_{x \in X} c_x x = \sum_{t \in T} \left[ \sum_{x \in Gt} (c_x g_x) \right] t$$

and so $T$ spans $kX$ as a $kG$-module. To show linear independence, suppose

$$\sum_{t \in T} \left[ \sum_{g \in G} c_{g,t} g \right] t = \sum_{t \in T} \left[ \sum_{g \in G} d_{g,t} g \right] t \tag{9.8}$$

with the $c_{g,t}, d_{g,t} \in k$. Let $t \in T$, $g \in G$ and set $x = gt$. Because the action of $G$ on $X$ is free, $x$ uniquely determines and is determined by $g$ and $t$. Thus comparing the coefficients of $x$ on both sides of (9.8) yields $c_{g,t} = d_{g,t}$. We conclude that $T$ is a basis for the $kG$-module $kX$.  □

As a consequence, we can determine the $A_m$. We shall need to make use of the Schützenberger group (see [68, Section 2.4] or Section A.3.1 for the definition).

**Lemma 9.3.11.** *Let $J$ be a $\mathscr{J}$-class of a finite semigroup $S$ with Schützenberger group $G$. Suppose $J$ has $r$ $\mathscr{R}$-classes and $b$ $\mathscr{L}$-classes. Let $A$ be the $k$-span of $\mathsf{RM}_J(S)$ in $\mathsf{End}_k(kJ)$. Then there is a $k$-isomorphism $f \colon kJ \to (kG^b)^r$ and an embedding $\psi \colon A \to M_b(kG)$ so that $(kJ, A) \cong ((kG^b)^r, A\psi\Delta)$, where $\Delta$ denotes, as usual, the diagonal mapping.*

*Proof.* As $J$ is the disjoint union of its $\mathscr{R}$-classes, $kJ$ is the direct sum of the $kR$ as where $R$ ranges over the $\mathscr{R}$-classes of $J$. Moreover, by Green's Lemma all these right modules $(kR, \mathsf{RM}_J(S))$ are isomorphic. Consequently, $(kJ, \mathsf{RM}_J(S)) \cong (kR^r, \mathsf{RM}_J(S)\Delta)$. So it suffices to study $(kR, \mathsf{RM}_J(S))$.

Now $G$ acts freely on the left of $R$ and hence $kR$ is a free left $kG$-module on a transversal $T$ for the action of $G$ on $R$ by Lemma 9.3.10. But the orbit space $G\backslash T$ is in bijection with the set of $\mathscr{L}$-classes of $S$ in $R$ (see Theorem A.3.11); hence $|T| = b$. Moreover, the left $G$-action on $R$ commutes with the right action of $S$ by Theorem A.3.11. Therefore, the action of $\mathsf{RM}_J(S)$ on $kR$ is by $kG$-endomorphisms. Using the basis $T$, we obtain an isomorphism $(kR, \mathsf{End}_{kG}(kR)) \cong (kG^b, M_b(kG))$. This isomorphism gives in turn a representation $\rho\colon \mathsf{RM}_J(S) \to M_b(kG)$ — in fact, $\rho$ is none other than the classical Schützenberger representation of $S$ on $J$ by row monomial matrices over $G$ [68, Section 3.5]. Extending $\rho$ linearly yields the desired conclusion. □

**Exercise 9.3.12.** Verify that the representation $\rho\colon \mathsf{RM}_J(S) \to M_b(kG)$ constructed in the above proof is the classical Schützenberger representation by row monomial matrices over $G$ associated to the $\mathscr{J}$-class $J$ [68, Section 3.5].

Lemma 9.3.11 allows us to state a version of Corollary 9.3.9 in coordinates.

**Corollary 9.3.13.** *Let $S$ be a finite semigroup and let*

$$S^I = I_0 \supsetneq I_1 \supsetneq \cdots \supsetneq I_n$$

*be a principal series for $S^I$. Let $J_m$ be the $\mathscr{J}$-class corresponding to $I_m/I_{m+1}$ (where $J_n = I_n$). Let $a_m, b_m$ be the number of $\mathscr{R}$-, $\mathscr{L}$-classes in $J_m$, respectively and let $G_m$ be the Schützenberger group of $J_m$. Then*

$$(kS^I, kS) \le \triangle\left(([kG_n^{b_n}]^{a_n}, M_{b_n}(kG_n)), \ldots, ([kG_0^{b_0}]^{a_0}, M_{b_0}(kG_0))\right)$$

$$(kS^I, S) \le \triangle\left(([kG_n^{b_n}]^{a_n}, \mathsf{RM}_{J_n}(S)), \ldots, ([kG_0^{b_0}]^{a_0}, \mathsf{RM}_{J_0}(S))\right)$$

*where the action of $M_{b_i}(kG_i)$, respectively $\mathsf{RM}_{J_i}(S)$, on $[kG_i^{b_i}]^{a_i}$ is diagonal.*

This result can be viewed as a "Prime Decomposition Theorem" for $k$-algebras because it breaks them into triangular products of "simpler" $k$-algebras, namely matrix algebras over group algebras of maximal subgroups. In the next section, we make this precise for the case of idempotent semirings.

In the case that $k$ is the Boolean semiring $\mathbb{B}$, Corollary 9.3.13, in conjunction with the wreath product decomposition Corollary 9.2.16, yields the decomposition result of [91]. Stronger results will be obtained in Section 9.5.

*Remark 9.3.14.* Notice that it follows from Corollary 9.3.13 that any finite $\mathscr{R}$-trivial semigroup can be represented by lower triangular matrices for any commutative semiring $k$ with unit as the maximal subgroups are trivial and each $\mathscr{J}$-class has one $\mathscr{L}$-class. If $k$ is separative, then it follows that any finite $\mathscr{L}$-trivial semigroup can also be represented by lower triangular matrices over $k$ by Proposition 9.2.14. The semigroups that divide a semigroup of lower triangular Boolean matrices are precisely the semigroups of dot-depth two [233]. It is a long-standing open problem to compute membership in the pseudovariety of dot-depth two semigroups. It is hoped that the techniques of this chapter will help make progress in this direction.

*Question 9.3.15.* Is it decidable whether a finite semigroup divides a semigroup of lower triangular Boolean matrices? In other words, is dot-depth two decidable?

## 9.4 The Prime Decomposition Theorem for Idempotent Semirings

In this section, we formulate the Prime Decomposition Theorem for idempotent semirings and prove that it is a decomposition into irreducibles. We once again impose our assumption of finiteness throughout the section.

An idempotent semiring $S$ is said to *divide* an idempotent semiring $R$ if there is a subsemiring $U$ of $R$ mapping onto $S$ via a semiring homomorphism. In this case we write, as usual, $S \prec R$. One can then define a *pseudovariety of idempotent semirings* to be a class of finite idempotent semirings closed under taking finite direct products and divisors. Such things have been introduced and studied by Polák [241–244], where a connection with formal language theory is made.

### 9.4.1 The Prime Decomposition Theorem

We would like to propose the triangular product as the substitute for the wreath product in the theory of idempotent semirings. In what follows, all modules and triangular products over idempotent semirings are taken in the categories of $\mathbb{B}$-modules and $\mathbb{B}$-algebras. So if $S$ is an idempotent semiring, then a right $S$-module is a join semilattice with minimum admitting a right action of $S$ by **sup** maps. With this in mind, let us restate Corollary 9.3.13 as a Prime Decomposition Theorem for idempotent semirings (or quantales).

**Theorem 9.4.1 (Prime Decomposition Theorem: Quantales).** *Let $S$ be a finite idempotent semiring and let*

$$S^I = I_0 \supsetneq I_1 \supsetneq \cdots \supsetneq I_n$$

*be a principal series for the semigroup $S^I$. Let $J_m$ be the $\mathscr{J}$-class corresponding to $I_m/I_{m+1}$ (where $J_n = I_n$) with Schützenberger group $G_m$. Suppose $a_m$ and $b_m$ are the number of $\mathscr{R}$- and $\mathscr{L}$-classes of $J_m$, respectively. Then*

$$S \prec \triangle \left( ([P(G_n)^{b_n}]^{a_n}, M_{b_n}(P(G_n))), \ldots, ([P(G_0)^{b_0}]^{a_0}, M_{b_0}(P(G_0)))\right) \quad (9.9)$$

*where the action of $M_{b_i}(P(G_i))$ on $[P(G_i)^{b_i}]^{a_i}$ is diagonal.*

*Proof.* We know that $P(S)$ embeds in the right-hand side of (9.9) by Corollary 9.3.13. The identity map $S \to S$ extends to a surjective semiring morphism $P(S) \twoheadrightarrow S$ by the universal property, completing the proof. □

We shall confirm shortly that, for a group $G$, the matrix algebra $M_n(P(G))$ is irreducible with respect to the triangular product and so the above decomposition is indeed a prime decomposition.

If $\mathbf{V}$ and $\mathbf{W}$ are pseudovarieties of idempotent semirings, then we define their *triangular product* $\mathbf{V}\triangle\mathbf{W}$ to consist of all finite semirings dividing a triangular product $\triangle((M, R), (N, S))$ where $M$ and $N$ are finite modules and $R \in \mathbf{V}, S \in \mathbf{W}$. Recall that $\triangle((M, R), (N, S))$ refers to the semiring and not the module. Because the underlying base semiring in this section is always $\mathbb{B}$, we omit it from the notation for hom sets.

**Proposition 9.4.2.** *If $\mathbf{V}$ and $\mathbf{W}$ are pseudovarieties of idempotent semirings, then $\mathbf{V} \triangle \mathbf{W}$ is a pseudovariety.*

*Proof.* It is clearly closed under formation of divisors, so we just need to check finite products. The triangular product of two copies of the trivial module over the trivial semiring is the trivial semiring, so closure under empty products holds. Suppose now that $R_i \prec \triangle((M_i, S_i), (N_i, T_i))$, for $i = 1, 2$. We show

$$R_1 \times R_2 \prec \triangle((M_1 \oplus M_2, S_1 \times S_2), (N_1 \oplus N_2, T_1 \times T_2)). \qquad (9.10)$$

To do this, it suffices to embed $\triangle((M_1, S_1), (N_1, T_1)) \times \triangle((M_2, S_2), (N_2, T_2))$ into the right-hand side of (9.10). Given a pair of matrices

$$\begin{pmatrix} s_1 & 0 \\ f_1 & t_1 \end{pmatrix}, \begin{pmatrix} s_2 & 0 \\ f_2 & t_2 \end{pmatrix}$$

with $s_i \in S_i$, $t_i \in T_i$ and $f_i \in \mathsf{Hom}(N_i, M_i)$, $i = 1, 2$, we send it to the matrix

$$\begin{pmatrix} (s_1, s_2) & 0 \\ f & (t_1, t_2) \end{pmatrix}$$

where $f \in \mathsf{Hom}(N_1 \oplus N_2, M_1 \oplus M_2)$ is given by $(n_1, n_2)f = (n_1 f_1, n_2 f_2)$. It is a straightforward exercise, which we leave to the reader, to verify that this assignment is an embedding of idempotent semirings. $\qquad \square$

*Remark 9.4.3 (On associativity of the triangular product).* The associativity of the semidirect product of pseudovarieties is at first sight a direct consequence of the associativity of the wreath product of transformation semigroups [85]. One would therefore expect the same thing to occur for the triangular product of pseudovarieties. There is, however, a major obstacle. A key property of the wreath product of transformation semigroups, used in proving associativity, is that $(X, S) \prec (X', S')$ and $(Y, T) \prec (Y', T')$ implies

$$(X, S) \wr (Y, T) \prec (X', S') \wr (Y', T'). \qquad (9.11)$$

We observe that the analogous result does not seem to hold for the triangular product. Suppose $M'$ is a $\mathbb{B}$-submodule of $M$ and $L$ is a $\mathbb{B}$-module. Then there is a natural map $\mathsf{Hom}(M, L) \to \mathsf{Hom}(M', L)$ induced by restriction, but it need not be onto. In fact, $L$ is injective [185] in the category of $\mathbb{B}$-modules if and only if the functor $\mathsf{Hom}(-, L)$ is exact. On the other hand, if $M$ is a $\mathbb{B}$-module and $L'$ is a $\mathbb{B}$-quotient module of $L$, then there is a natural map $\mathsf{Hom}(M, L) \to \mathsf{Hom}(M, L')$, but again the map need not be onto. The functor $\mathsf{Hom}(M, -)$ being exact is equivalent to $M$ being projective [185] in the category of $\mathbb{B}$-modules. This is a barrier to proving an analogue of (9.11). Notice that all objects in the category **Set** are projective and injective, so there are no problems for transformation semigroups. For instance, if $X'$ is a subset of $X$ then the natural map $Y^X \to Y^{X'}$ given by restriction is onto, as any map $X' \to Y$ can be extended to $X$.

We believe that the triangular product of pseudovarieties of idempotent semirings is not associative. If it were, then the closure of $(\mathbb{B})$ under the triangular product should be contained inside the dot-depth 2 aperiodic semigroups. But iterating 2-fold Schützenberger products of copies of $\mathbb{B}$ should give aperiodic idempotent semirings of arbitrary dot-depth and we saw already that the 2-fold Schützenberger product is a triangular product.

*Question 9.4.4.* In [241–244], Polák sets up a correspondence between pseudovarieties of idempotent semirings and certain classes of languages. What is the effect of the triangular product on languages? It should be similar to the effect of the Schützenberger product in the theory of varieties of languages.

**Proposition 9.4.5.** *Let* **V** *and* **W** *be pseudovarieties of idempotent semirings. Then* $(\mathbf{V} \triangle \mathbf{W})^{op} = \mathbf{W}^{op} \triangle \mathbf{V}^{op}$.

*Proof.* Because $\mathbb{B}$ is separative, this is immediate from Proposition 9.2.14.   □

*Question 9.4.6.* What is the smallest pseudovariety of idempotent semirings closed under triangular product and containing $\mathbb{B}$? One would guess it should consist of all idempotent semirings whose underlying semigroup is aperiodic.

### 9.4.2 Irreducibility for the triangular product

Our eventual goal is to define a notion of complexity for finite idempotent semirings, following the model of group complexity. We begin with a study of the notion of irreducibility with respect to the triangular product.

**Definition 9.4.7 ($\triangle$-irreducible).** *An idempotent semiring $R$ is called $\triangle$-irreducible if whenever $R \prec \triangle((M, S), (N, T))$, with $S$ and $T$ idempotent semirings, then $R \prec S$ or $R \prec T$.*

An immediate consequence of Proposition 9.2.14 is the self-duality of the notion of $\triangle$-irreducibility.

**Proposition 9.4.8.** *An idempotent semiring $S$ is $\triangle$-irreducible if and only if its opposite semiring $S^{op}$ is $\triangle$-irreducible.*

An important open question is to classify all $\triangle$-irreducible idempotent semirings.

*Question 9.4.9.* Classify all $\triangle$-irreducible idempotent semirings.

In the current section, we shall make some significant progress toward this endeavor. In particular, we shall prove the following theorem (see Definition 8.1.11 for the meaning of $G^\natural$).

**Theorem 9.4.10 (Triangular Irreducibility Theorem).** *For any $n \geq 1$, the following idempotent semirings are irreducible: $M_n(\mathbb{B})$, $M_n(P(G))$ for a non-trivial group $G$ and $M_n(G^\natural)$ for a non-trivial monolithic group $G$.*

We shall prove the irreducibility results of the theorem one at a time. Given the theorem statement, and the usual intuition that a semiring should be "Morita equivalent" to its matrix amplifications, it seems natural to ask the following question and to formulate the subsequent definition.

*Question 9.4.11 (Margolis).* Is it true that if $S$ is $\triangle$-irreducible, then so is $M_n(S)$?

**Definition 9.4.12 (Basic $\triangle$-irreducible semiring).** *A $\triangle$-irreducible semiring is termed basic if it is not isomorphic to a matrix semiring $M_n(R)$ with $n \geq 2$ and $R$ a $\triangle$-irreducible idempotent semiring.*

*Question 9.4.13.* Classify the basic $\triangle$-irreducible semirings.

The idempotent semirings $\mathbb{B}$, $P(G)$ and $G^\natural$ (the latter for $G$ monolithic, non-trivial) are basic $\triangle$-irreducibles. To see that they are basic, observe that $\mathbb{B}$, $P(G)$ and $G^\natural$ are block groups, and so cannot be isomorphic to $M_n(R)$ with $n \geq 2$ for any unital semiring (as $\overline{2}$ embeds in $M_n(R)$ for $n \geq 2$ and $R$ unital). On the other hand, $M_n(R)$ is not unital if $R$ is not unital. Because $\mathbb{B}$, $P(G)$ and $G^\natural$ are unital, we conclude that they are indeed basic $\triangle$-irreducibles. We should mention that $\mathbb{B}$ is $P(\{1\})$, but the arguments for it turn out differently than for non-trivial groups so we treat it separately.

In order to establish that various idempotent semirings are $\triangle$-irreducible, we need some tools. Let us begin with a well-known proposition.

**Proposition 9.4.14.** *Let $G$ be a torsion group (not necessarily finite). Then $G$ does not admit any non-trivial partial order compatible with multiplication.*

*Proof.* Suppose $g_1 \leq g_2$ and set $g = g_1 g_2^{-1}$. Then $g \leq 1$ and so $g^n \leq 1$ for all $n$. Thus the submonoid generated by $g$ is $\mathscr{J}$-trivial by Proposition 8.2.1. But it is also a group since $G$ is torsion. Therefore $g = 1$ and so $g_1 = g_2$. Thus equality is the only compatible partial ordering on $G$.    $\square$

The following observation is also useful.

**Lemma 9.4.15.** *Let $S$ be a finite ordered semigroup and suppose $s \in S$ satisfies $s^2 \leq s$. Then:*

1. $s \geq s^2 \geq s^3 \geq \cdots \geq s^\omega = s^{\omega+1}$;
2. *The interval $[s^\omega, s]$ is a nilpotent semigroup.*

*Proof.* By induction $s^{n+1} \leq s^n$ for all $n$ and so if $s^N = s^\omega$, then $s^\omega = s^N \leq s$. Clearly, if $x_1, \ldots, x_n \in [s^\omega, s]$, then $s^\omega = (s^\omega)^n \leq x_1 \cdots x_n \leq s^n \leq s$. Thus $[s^\omega, s]$ is a subsemigroup and $[s^\omega, s]^N = \{s^\omega\}$, whence $[s^\omega, s]$ is nilpotent with zero $s^\omega$. In particular, $s^{\omega+1} = s^\omega$. This proves 1 and 2.  □

The next proposition is a useful technical tool for proving irreducibility of matrix semirings.

**Proposition 9.4.16.** *Let* $\varphi \colon Q \twoheadrightarrow Q'$ *be an onto morphism of finite idempotent semirings. Then* $0\varphi^{-1}$ *enjoys the following properties:*

1. *It is a semiring ideal;*
2. *It is a downset closed under all sups (i.e., a lattice ideal);*
3. *It has a unique maximum element* $x$;
4. $x^\omega = x^{\omega+1}$ *and* $x \geq x^2 \geq x^3 \geq \cdots \geq x^\omega$
5. *For all* $q \in Q$, $qx, xq \leq x$ *and* $qx^\omega, x^\omega q \leq x^\omega$;
6. *The interval* $[x^\omega, x]$ *is a nilpotent semigroup.*

*Proof.* Statements 1 and 2 are trivial, whereas 3 is an immediate consequence of 2. We may deduce 4 and 6 from Lemma 9.4.15 because $x \geq x^2$ by definition of $x$. For 5, we have by 1 that $xq, qx \in 0\varphi^{-1}$ and hence, by 3, are less than $x$. The second inequality follows from the first via multiplication by $x^\omega$ as $x^\omega = x^{\omega+1}$.  □

Once again we impose our standing assumption of finiteness. A crucial ingredient for establishing irreducibility of matrix semirings is the following lemma on "lifting" Brandt semigroups.

**Lemma 9.4.17.** *Let* $\varphi \colon R \twoheadrightarrow M_n(Q)$ *be an onto morphism of idempotent semirings with* $Q$ *unital and* $n \geq 2$. *Let* $E_{ij}$, $1 \leq i, j \leq n$, *be the standard matrix units of* $M_n(Q)$. *Then there exist elements* $e_{ij} \in R$, $1 \leq i, j \leq n$, *with* $e_{ij}\varphi = E_{ij}$ *and* $e_{ij}e_{jk} = e_{ik}$, *all* $i, j, k$. *Moreover, one can take each* $e_{ij}$ *to be a* $\leq_{\mathscr{J}}$*-minimal preimage of* $E_{ij}$.

*Proof.* Let $J$ be the $\mathscr{J}$-class of the $E_{ij}$ and let $J'$ be a $\leq_{\mathscr{J}}$-minimal $\mathscr{J}$-class of $R$ with $J'\varphi \subseteq J$. Then $J'$ is regular and $J'\varphi = J$ by Lemma 4.6.10. Choose for each $i$ an idempotent $e_{ii} \in J' \cap E_{ii}\varphi^{-1}$; such exist because if $\eta_{ii} \in J' \cap E_{ii}\varphi^{-1}$, then so is $\eta_{ii}^\omega$ by minimality. Next choose a preimage $e_{1i}$ of $E_{1i}$ in $R_{e_{11}} \cap L_{e_{ii}}$ (such exists by Lemma 4.6.10) and let $e_{i1}$ be an inverse of $e_{1i}$ in $R_{e_{ii}} \cap L_{e_{11}}$. As $e_{i1}\varphi$ is an inverse of $E_{1i}$ in $R_{E_{ii}} \cap L_{E_{11}}$, necessarily $e_{i1}\varphi = E_{i1}$. Notice that $e_{i1}e_{1i} = e_{ii}$ so there is no ambiguity in defining $e_{ij} = e_{i1}e_{1j}$. Then $e_{ij}\varphi = E_{ij}$ all $i, j$ and

$$e_{ij}e_{jk} = e_{i1}e_{1j}e_{j1}e_{1k} = e_{i1}e_{11}e_{1k} = e_{i1}e_{1k} = e_{ik}$$

as required.  □

We now establish that $M_n(\mathbb{B})$ is $\triangle$-irreducible. This is a strong analogue of the simplicity of matrix algebras over a field. The proof is a model for many of the proofs to come. Recall from Proposition 9.1.11 that if $R$ is an idempotent semiring, then $B_n(R)$ is the subsemigroup of $M_n(R)$ consisting of all matrices of the form $rE_{ij}$ with $r \in R$.

**Theorem 9.4.18.** *The idempotent semiring $M_n(\mathbb{B})$ is $\triangle$-irreducible for all $n \geq 1$.*

*Proof.* Suppose first $n = 1$. Clearly, $\mathbb{B}$ fails to divide an idempotent semiring $R$ if and only if the multiplicative semigroup of $R$ is nilpotent. One easily verifies that the triangular product of nilpotent semirings is nilpotent. Thus $\mathbb{B}$ is $\triangle$-irreducible.

For the case $n \geq 2$, let $\varphi \colon R \twoheadrightarrow M_n(\mathbb{B})$ be a surjective semiring homomorphism with $R$ a subsemiring of a triangular product $\triangle((M, S), (N, T))$. Choose $e_{ij}$ as per Lemma 9.4.17. Let $I = 0\varphi^{-1}$ and set $x = \bigvee I$, as per Proposition 9.4.16; note $I$ is a semiring ideal. Set $\epsilon_{ij} = e_{ij} + x^\omega$. First observe

$$\epsilon_{ij}\varphi = e_{ij}\varphi + x^\omega\varphi = E_{ij} + 0 = E_{ij}.$$

By item 5 of Proposition 9.4.16, we have

$$\epsilon_{ij}\epsilon_{jk} = (e_{ij} + x^\omega)(e_{jk} + x^\omega) = e_{ik} + e_{ij}x^\omega + x^\omega e_{jk} + x^\omega = \epsilon_{ik} \qquad (9.12)$$

as $e_{ij}x^\omega, x^\omega e_{jk} \leq x^\omega$, whereas if $j \neq k$, then $\epsilon_{ij}\epsilon_{k\ell} \in I$.

Let

$$\epsilon_{ij} = \begin{pmatrix} s_{ij} & 0 \\ f_{ij} & t_{ij} \end{pmatrix} \quad \text{and} \quad x = \begin{pmatrix} s_x & 0 \\ f_x & t_x \end{pmatrix};$$

consequently,

$$x^\omega = \begin{pmatrix} s_x^\omega & 0 \\ f_{x^\omega} & t_x^\omega \end{pmatrix}$$

some $f_{x^\omega} \in \mathsf{Hom}(N, M)$. Let us suppose that $M_n(\mathbb{B})$ divides neither $S$ nor $T$; we shall derive a contradiction.

Denote by $\pi_S$, $\pi_T$ the respective projections of $\triangle((M, S), (N, T))$ to $S$ and $T$. Without loss of generality, we may assume they are onto. Hence $I\pi_S$ and $I\pi_T$ are semiring ideals. Define a homomorphism $\psi \colon B_n(\mathbb{B}) \to S/I\pi_S$ by $(E_{ij})\psi = s_{ij} + I\pi_S$ and $0\psi = 0 + I\pi_S$. It is immediate from (9.12) (and the line following it) that $\psi$ satisfies the conditions of Proposition 9.1.11 and so can be extended to a semiring homomorphism $\overline{\psi} \colon M_n(\mathbb{B}) \to S/I\pi_S$. Because $\mathbb{B}$ is congruence-free, Theorem 9.1.10 implies either $\overline{\psi}$ is injective, or $\overline{\psi}$ is the zero homomorphism. Because we are assuming $M_n(\mathbb{B}) \nmid S$, we must be in the latter case. Then $s_{ij} + I\pi_S = E_{ij}\overline{\psi} = 0 + I\pi_S$ for all $i, j$. Because $\bigvee(I\pi_S) = (\bigvee I)\pi_S = x\pi_S = s_x$, Proposition 9.1.22 implies $s_{ij} + s_x = s_x$, that is, $s_{ij} \leq s_x$. On the other hand, $s_x^\omega \leq s_{ij}$ by definition of $\epsilon_{ij}$. So $s_{ij} \in [s_x^\omega, s_x]$, all $1 \leq i, j \leq n$. But Lemma 9.4.15 shows that $[s_x^\omega, s_x]$ is a nilpotent semigroup, as $x^2 \leq x$ implies $s_x^2 \leq s_x$. Because the $s_{ij}$ are regular elements of this semigroup, we conclude $s_{ij} = s_x^\omega$, all $1 \leq i, j \leq n$.

A similar argument shows that $t_{ij} = t_x^\omega$ for all $i, j$, yielding

$$\epsilon_{ij} = \begin{pmatrix} s_x^\omega & 0 \\ f_{ij} & t_x^\omega \end{pmatrix}.$$

Now $(s_x^\omega, t_x^\omega)$ is an idempotent. Let $\pi \colon \triangle((M, S), (N, T)) \to S \times T$ be the projection to the diagonal. Then every non-zero element of the subsemiring $Q$ generated by the $\epsilon_{ij}$ belongs to $(s_x^\omega, t_x^\omega)\pi^{-1}$, which is locally a semilattice by Lemma 9.2.4. Therefore, the multiplicative semigroup of $Q$ is aperiodic. But $Q$ maps onto $M_n(\mathbb{B})$, whose group of units is the symmetric group $S_n$, providing our sought after contradiction and establishing the $\triangle$-irreducibility of $M_n(\mathbb{B})$.    $\square$

**Corollary 9.4.19.** *If $S$ is an idempotent semiring of order $n$, then $S$ divides a $\triangle$-irreducible idempotent semiring, namely $M_{n+1}(\mathbb{B})$.*

*Proof.* We know that there is an onto morphism of idempotent semirings $\eta \colon P(S) \twoheadrightarrow S$ and that $P(S)$ embeds in $M_{n+1}(\mathbb{B})$ by Corollary 9.1.8.    $\square$

Next, we wish to prove that if $G$ is a finite, non-trivial monolithic group, then $G^\natural$ is $\triangle$-irreducible (see Definition 8.1.11). To do this we first establish a strong lifting property of groups.

**Lemma 9.4.20 (Groups lift).** *Let $\varphi \colon Q' \twoheadrightarrow Q$ be a quotient morphism of finite idempotent semirings. Let $G$ be a (multiplicative) subgroup of $Q$. Then there is a subgroup $G'$ of $Q'$ such that $\varphi|_{G'} \colon G' \to G$ is an isomorphism. In particular, $Q'$ contains an isomorphic copy of $G$.*

*Proof.* By Proposition 4.1.44, there is a subgroup $H$ of $Q'$ with $H\varphi = G$. Let $N = \ker \varphi|_H$ and set $G' = \{\sum_{n \in N} nh \mid h \in H\}$. Define $\psi \colon H \to G'$ by $h\psi = \sum_{n \in N} nh$. We first show $\psi$ is an onto homomorphism with $\psi\varphi = \varphi$. Trivially $\psi$ is onto. Observe, using idempotence of addition, that

$$h_1\psi h_2\psi = \sum_{n \in N} nh_1 \sum_{n \in N} nh_2 = \sum_{n, n' \in N} nh_1 n' h_2$$

$$= \sum_{n, n' \in N} n(h_1 n' h_1^{-1}) h_1 h_2 = \sum_{n \in N} nh_1 h_2 = (h_1 h_2)\psi.$$

Finally, $h\psi\varphi = \left(\sum_{n \in N} nh\right)\varphi = \sum_{n \in N} h\varphi = h\varphi$.

From $\psi\varphi = \varphi$, we have $\ker\psi \leq \ker\varphi|_H = N$. On the other hand, for $n' \in N$, $n'\psi = \sum_{n \in N} nn' = \sum_{n \in N} n = 1\psi$, so $N = \ker\psi$. It then follows that $\varphi|_{H\psi}$ is an isomorphism. But $H\psi = G'$. This completes the proof.    $\square$

As a consequence, if $G$ is a group, then $P(G)$ is a projective idempotent semiring.

**Corollary 9.4.21.** *Let $G$ be a finite group. Then $P(G)$ is a projective finite idempotent semiring. That is, if $Q$ is a finite idempotent semiring and $\varphi\colon Q \twoheadrightarrow P(G)$ is a surjective homomorphism, then there is a homomorphism $\psi\colon P(G) \to Q$ splitting $\varphi$.*

*Proof.* By Lemma 9.4.20, there is a homomorphism $\alpha\colon G \to Q$ such that $\alpha\varphi = 1_G$. By the universal property of $P(G)$, this extends to a homomorphism $\psi\colon P(G) \to Q$ splitting $\varphi$.                                  $\square$

**Corollary 9.4.22.** *Let $G$ be a non-trivial finite group and let $\psi\colon P(G) \twoheadrightarrow G^\natural$ be the natural projection. Then $\ker\psi$ is the largest congruence on $P(G)$ whose associated quotient morphism is injective on $G$. As a consequence, if $Q$ is a finite idempotent semiring, then $G^\natural \prec Q$ if and only if $Q$ contains a subgroup isomorphic to $G$.*

*Proof.* Clearly, $\psi$ is injective on $G$. First observe that if $\equiv$ is a proper congruence on $P(G)$, then no element $X \neq \emptyset$ of $P(G)$ can satisfy $X \equiv \emptyset$. Indeed, if $g \in X$, then $g = g \cup \emptyset \equiv g \cup X = X \equiv \emptyset$. Then $1 = g^{-1}g \equiv g^{-1}\emptyset = \emptyset$ and hence all of $P(G)$ is equivalent to $\emptyset$. Suppose $\equiv$ is a congruence whose associated quotient morphism is injective on $G$. Then $\emptyset$ is in a $\equiv$-class of its own. So to show $\equiv \subseteq \ker\psi$, it suffices to show that the $\equiv$-class of each element of $G$ is a singleton.

Suppose $g \equiv X$. Then by our assumptions and the above $|X| \geq 2$. Choose $g' \in X$ with $g' \neq g$. Then $g' \cup g \equiv g' \cup X = X \equiv g$ and so in $P(G)/\equiv$ we have $[g] \geq [g']$, where $[Y]$ denotes the equivalence class of a subset $Y$. It follows from Proposition 9.4.14 that $[g] = [g']$. So, by injectivity on $G$, we have $g = g'$, a contradiction. Thus each $\equiv$-class of an element of $G$ is a singleton, as required.

Next assume $G^\natural \prec Q$. Then there is a subsemiring $Q' \leq Q$ mapping onto $G^\natural$. An application of Lemma 9.4.20 shows that $Q'$, and hence $Q$, contains an isomorphic copy of $G$.

Conversely, suppose $G$ is a subgroup of $Q$. By the universal property, there is a homomorphism $\varphi\colon P(G) \to Q$ extending the inclusion. As $\varphi$ is injective on $G$, the projection $\psi\colon P(G) \twoheadrightarrow G^\natural$ factors through $\varphi\colon P(G) \to P(G)\varphi$ and hence $G^\natural \prec Q$.                                  $\square$

Our next result shows that if $G$ is non-trivial and monolithic, then $G^\natural$ is $\triangle$-irreducible. The following sharper result will be used later.

**Lemma 9.4.23.** *Let $G$ be a monolithic subgroup of $\triangle((M, S), (N, T))$. Then either the projection to $S$ or the projection to $T$ is injective on $G$.*

*Proof.* Set $Q = \triangle((M, S), (N, T))$ and let $\pi_S\colon Q \to S$, $\pi_T\colon Q \to T$ be the projections. By Lemma 9.2.4, the projection $\pi\colon Q \to S \times T$ to the diagonal belongs to $(\mathbb{LSl}, \mathbf{1})$ and hence $\pi|_G$ is injective. But $G\pi \ll G\pi_S \times G\pi_T$ and thus, as $G$ is monolithic, either $\pi_S$ or $\pi_T$ is injective on $G$.                                  $\square$

**Corollary 9.4.24.** *Let $G$ be a finite group. Then the idempotent semiring $G^\natural$ is $\triangle$-irreducible if and only if $G$ is non-trivial monolithic.*

*Proof.* Suppose first $G$ is non-trivial monolithic and $G^\natural \prec \triangle((M,S),(N,T)) = Q$. Then by Corollary 9.4.22, $G$ is isomorphic to a subgroup $G'$ of $Q$. By Lemma 9.4.23, $G'$ embeds in either $S$ or $T$. Another application of Corollary 9.4.22 then yields $G^\natural \prec S$ or $G^\natural \prec T$.

Next we observe

$$\{1\}^\natural \cong \left\{ \begin{pmatrix} 1 & 0 \\ 1 & 1 \end{pmatrix}, \begin{pmatrix} 1 & 0 \\ 0 & 1 \end{pmatrix}, \begin{pmatrix} 0 & 0 \\ 0 & 0 \end{pmatrix} \right\}$$

and hence is not $\triangle$-irreducible.

Finally, if $G \neq \{1\}$ is not monolithic, then $G \ll G_1 \times G_2$ where $G_1$, $G_2$ are proper quotients of $G$. Then $G$ embeds in $P(G_1) \times P(G_2)$ and hence $G^\natural \prec P(G_1) \times P(G_2)$ by Corollary 9.4.22. But $G^\natural \nprec P(G_i)$, $i = 1, 2$. Indeed, if such a division existed, then $G$ must embed in $P(G_i)$ by Corollary 9.4.22. But it is well-known that the maximal subgroups of $P(G_i)$ are of the form $N_{G_i}(H)/H$ where $H \leq G_i$ and hence are divisors of $G_i$. Because $G_i$ is a proper divisor of $G$, this shows $G$ cannot embed in $P(G_i)$, completing the proof. $\square$

**Exercise 9.4.25.** Let $G$ be a finite group. Verify that the maximal subgroups of $P(G)$ are of the form $N_G(H)/H$ where $H \leq G$ and $N_G(H)$ is the normalizer of $H$ in $G$.

Let $S$ be an idempotent semiring. Then Corollary 9.3.13 gives a triangular product decomposition of $S$ into matrix algebras over power semigroups of Schützenberger groups of $S$. Our next goal is to show that power semigroups of groups are $\triangle$-irreducible; afterwards we turn to matrices over power sets of groups. The first step is to show that power sets of groups are subdirectly indecomposable. Actually, what we need is the description of the unique minimal congruence.

**Proposition 9.4.26.** *Let $G$ be a finite group of order $n \geq 2$. Then $P(G)$ is subdirectly indecomposable. The unique minimal congruence identifies precisely the subsets of size at least $n - 1$.*

*Proof.* Let $n = |G|$. Let $\equiv$ be the equivalence relation on $P(G)$ that identifies all subsets of size at least $n - 1$ (and leaves all other elements intact). It is straightforward to verify $\equiv$ is a congruence. We claim it is the unique minimal congruence. Indeed, let $\sim$ be a non-trivial congruence on $P(G)$ and let $A \neq B \subseteq G$ with $A \sim B$; say $A \nsubseteq B$. Then $A + B \sim B + B = B$ and $A + B \neq B$. Thus, without loss of generality, we may assume $A \supsetneq B$. Then $G = A + G \setminus A \sim B + G \setminus A \neq G$. Thus we may assume without loss of generality that $A = G$ and $B \subsetneq G$. Let $X$ contain all the elements of $G \setminus B$ except one. Then $G = G + X \sim B + X$ and the right-hand side is a set with $n - 1$ elements. Because the singletons act transitively on the collection of $(n - 1)$-element subsets of $G$, whereas $G$ is fixed by each singleton, we conclude $\equiv \subseteq \sim$, as was required. $\square$

**Lemma 9.4.27.** *Let $G$ be a non-trivial finite group and suppose there is an embedding $P(G) \leq \triangle((M,S),(N,T))$. Then one of the projections from $P(G)$ to $S$ or to $T$ is injective.*

*Proof.* Let $|G| = n$ and suppose $P(G) \leq \triangle((M,S),(N,T))$. Denote by $\pi_S \colon P(G) \to S$ and $\pi_T \colon P(G) \to T$ the projections. We may assume without loss of generality that they are surjective and so $S$ and $T$ are semirings with unit. Suppose neither projection is injective; we shall derive a contradiction. Proposition 9.4.26 tells us that if $X \subseteq G$ with $|X| = n - 1$, then $X\pi_S = G\pi_S$ and $X\pi_T = G\pi_T$.

We write the embedding of $P(G)$ into $\triangle((M,S),(N,T))$ as

$$X \longmapsto \begin{pmatrix} s_X & 0 \\ f_X & t_X \end{pmatrix} \tag{9.13}$$

and observe that if $|X| = n - 1$, then $s_X = s_G$, $t_X = t_G$. Notice $s_1$ and $t_1$ are the identities of $S$ and $T$, respectively. We want to show that we can replace $M$ by $Ms_1$ and $N$ by $Nt_1$ to make the actions unitary (and hence make the actions of $S$ and $T$ on $\mathsf{Hom}(N,M)$ unitary). Indeed, define $\alpha \colon \mathsf{Hom}(N,M) \to \mathsf{Hom}(Nt_1, Ms_1)$ by $f\alpha = {}^{t_1}fs_1$. This is clearly a linear map. It is also a bimodule morphism. Indeed, ${}^t({}^{t_1}fs_1) = {}^{t_1}({}^tf)s_1$ and $({}^{t_1}fs_1)s = {}^{t_1}(fs)s_1$ as $s_1$ and $t_1$ are identities. Functoriality of the triple product $\nabla$ in the middle variable provides a morphism $\psi \colon P(G) \to \triangle((Ms_1,S),(Nt_1,T))$, which we claim is injective. To see this, note that

$$X\psi = \begin{pmatrix} s_X & 0 \\ {}^{t_1}f_X s_1 & t_X \end{pmatrix}.$$

But because $P(G)$ has identity 1, there results an equality

$$\begin{pmatrix} s_X & 0 \\ f_X & t_X \end{pmatrix} = \begin{pmatrix} s_1 & 0 \\ f_1 & t_1 \end{pmatrix}\begin{pmatrix} s_X & 0 \\ f_X & t_X \end{pmatrix}\begin{pmatrix} s_1 & 0 \\ f_1 & t_1 \end{pmatrix} = \begin{pmatrix} s_X & 0 \\ f_1 s_X + {}^{t_1}f_X s_1 + {}^{t_X}f_1 & t_X \end{pmatrix}.$$

Thus $X$ is determined by $s_X$, $t_X$ and ${}^{t_1}f_X s_1$ and hence $\psi$ is injective.

Resetting notation for $M$ and $N$, we now have $P(G) \leq \triangle((M,S),(N,T))$ where: $S$ and $T$ have identities $s_1$ and $t_1$, respectively; $(M,S)$, $(N,T)$ are unitary; and if $X \subseteq G$ with $|X| = n - 1$, then $s_X = s_G$, $t_X = t_G$. We shall proceed to draw a contradiction.

*Claim.* If $X \in P(G)$, then $f_X = {}^{t_X}f_1$.

*Proof.* Because $t_X = \sum_{x \in X} t_x$ and $f_X = \sum_{x \in X} f_x$, it suffices to prove the claim for $X = \{g\}$ with $g \in G$. From $1 = g^n$ and the actions being unitary

$$\begin{pmatrix} s_1 & 0 \\ f_1 & t_1 \end{pmatrix} = \begin{pmatrix} s_g & 0 \\ f_g & t_g \end{pmatrix}^n = \begin{pmatrix} s_1 & 0 \\ \sum_{j=0}^{n-1} t_g^j f_g s_g^{n-1-j} & t_1 \end{pmatrix}.$$

By considering $j = n-1$, we obtain $f_1 \geq {}^{t_g^{n-1}} f_g$ (as the action of $S$ is unitary). Because the action of $T$ on $\mathrm{Hom}(N, M)$ is order preserving and unitary, we obtain ${}^{t_g} f_1 \geq {}^{t_g^n} f_g = {}^{t_1} f_g = f_g$.

For the reverse inequality, observe that

$$\begin{pmatrix} s_g & 0 \\ f_g & t_g \end{pmatrix} = \begin{pmatrix} s_g & 0 \\ f_g & t_g \end{pmatrix} \begin{pmatrix} s_1 & 0 \\ f_1 & t_1 \end{pmatrix} = \begin{pmatrix} s_g & 0 \\ f_g s_1 + {}^{t_g} f_1 & t_g \end{pmatrix}$$

implies $f_g \geq {}^{t_g} f_1$. This establishes the claim. □

If $X \subseteq G$ with $|X| = n - 1$, then $s_X = s_G$, $t_X = t_G$ and the claim implies $f_X = {}^{t_X} f_1 = {}^{t_G} f_1 = f_G$. This contradicts (9.13) being an embedding. □

**Theorem 9.4.28.** *Let $G$ be a finite group. Then $P(G)$ is $\triangle$-irreducible.*

*Proof.* If $G = \{1\}$, then $P(G) = \mathbb{B}$ and we are done by Theorem 9.4.18. Suppose $G$ is non-trivial and $P(G)$ divides $\triangle((M, S), (N, T))$. By Corollary 9.4.21, we may assume $P(G) \leq \triangle((M, S), (N, T))$. Lemma 9.4.27 then yields $P(G) \prec S$ or $P(G) \prec T$. □

We next turn to matrices over $P(G)$ and $G^\natural$. Although the two proofs have many ingredients in common, we prove the results separately in order to make the proof more transparent.

**Theorem 9.4.29.** *Let $G$ be a finite group. Then, for all $n \geq 1$, $M_n(P(G))$ is $\triangle$-irreducible.*

*Proof.* The cases $G$ is trivial and $n = 1$ have already been handled, so we assume $|G|, n \geq 2$. Let $Q$ be a subsemiring of $\triangle((M, S), (N, T))$ such that there is a surjective homomorphism $\varphi \colon Q \twoheadrightarrow M_n(P(G))$. As usual we may assume the projections $\pi_S \colon Q \to S$ and $\pi_T \colon Q \to T$ are surjective. Let $I = 0\varphi^{-1}$ and set $x = \bigvee I$. Choose $e_{ij}$ as per Lemma 9.4.17 so that the $e_{11}$ is a $\leq_{\mathscr{J}}$-minimal preimage of $E_{11}$ under $\varphi$. Set $R = e_{11} Q e_{11}$. Then $R\varphi = E_{11} M_n(P(G)) E_{11} = P(G) E_{11}$. By $\leq_{\mathscr{J}}$-minimality of $e_{11}$, it follows that the group of units of $R$ maps onto the group of units $G E_{11} \cong G$ of $P(G) E_{11} \cong P(G)$ (cf. Lemma 4.6.10). Lemma 9.4.20 implies that there is a splitting $\alpha \colon G E_{11} \to R$, which extends to a semiring morphism $P(G) E_{11} \to R$, also denoted $\alpha$, by the universal property of $P(G)$.

We now wish to modify $\alpha$ to obtain a map $\beta$ with image at least as big as $x^\omega$ in the order. Set $X E_{11} \beta = X E_{11} \alpha + x^\omega$, for $\emptyset \neq X \in P(G)$. Of course, we take $\emptyset \beta = 0$ in order to make it a $\mathbb{B}$-linear map. Notice for $X, Y \neq \emptyset$

$$
\begin{aligned}
(X E_{11})\beta (Y E_{11})\beta &= (X E_{11}\alpha + x^\omega)(Y E_{11}\alpha + x^\omega) \\
&= (X E_{11})\alpha(Y E_{11})\alpha + (X E_{11})\alpha x^\omega + x^\omega(Y E_{11})\alpha + x^\omega \\
&= (XY E_{11})\alpha + x^\omega = (XY E_{11})\beta
\end{aligned}
$$

where the penultimate equality uses Proposition 9.4.16. It follows $\beta$ is a semi-ring homomorphism. Also, for $X \neq \emptyset$, $(XE_{11})\beta\varphi = (XE_{11}\alpha + x^\omega)\varphi = XE_{11}$ since $\alpha$ is a splitting and $x^\omega\varphi = 0$. Thus $\beta$ is a splitting of $\varphi$.

Next we wish to prove

$$(e_{11} + x^\omega)(XE_{11}\beta)(e_{11} + x^\omega) = XE_{11}\beta. \tag{9.14}$$

This is clear if $X = \emptyset$. Otherwise, recalling $XE_{11}\beta = XE_{11}\alpha + x^\omega$, we compute

$$\begin{aligned}
(e_{11} + x^\omega)(XE_{11}\alpha + x^\omega)(e_{11} + x^\omega) &= \big(e_{11}(XE_{11})\alpha + e_{11}x^\omega + x^\omega(XE_{11})\alpha \\
&\quad + x^\omega\big)(e_{11} + x^\omega) \\
&= (XE_{11}\alpha + x^\omega)(e_{11} + x^\omega) \\
&= (XE_{11})\alpha e_{11} + (XE_{11})\alpha x^\omega \\
&\quad + x^\omega e_{11} + x^\omega \\
&= XE_{11}\alpha + x^\omega
\end{aligned}$$

where we have used several times that $\alpha$ has image in $R = e_{11}Qe_{11}$ and Proposition 9.4.16.

Let us define, for $1 \leq i,j \leq n$ and $X \in P(G)$,

$$\epsilon_{ij}(X) = (e_{i1} + x^\omega)(XE_{11}\beta)(e_{1j} + x^\omega).$$

Then $\epsilon_{ij}(X)\varphi = E_{i1}(XE_{11})E_{1j} = XE_{ij}$. Clearly, $\epsilon_{ij}(X \cup Y) = \epsilon_{ij}(X) + \epsilon_{ij}(Y)$. Notice $\epsilon_{11}(X) = (XE_{11})\beta$ by (9.14), while $\epsilon_{ij}(\emptyset) = 0$. Next we want to verify

$$\epsilon_{ij}(X)\epsilon_{k\ell}(Y) = \begin{cases} \epsilon_{i\ell}(XY) & j = k \\ \text{an element of } I & j \neq k. \end{cases} \tag{9.15}$$

The second case is clear. For the first case, we begin by observing that the computation in (9.12) yields $(e_{1j} + x^\omega)(e_{j1} + x^\omega) = e_{11} + x^\omega$. Thus

$$\begin{aligned}
\epsilon_{ij}(X)\epsilon_{j\ell}(Y) &= (e_{i1} + x^\omega)(XE_{11}\beta)(e_{1j} + x^\omega)(e_{j1} + x^\omega)(YE_{11}\beta)(e_{1\ell} + x^\omega) \\
&= (e_{i1} + x^\omega)(XE_{11}\beta)(e_{11} + x^\omega)(YE_{11}\beta)(e_{1\ell} + x^\omega) \\
&= (e_{i1} + x^\omega)(XYE_{11}\beta)(e_{1\ell} + x^\omega) \\
&= \epsilon_{i\ell}(XY)
\end{aligned}$$

where we used (9.14) in the penultimate equality.

Let

$$\epsilon_{ij}(X) = \begin{pmatrix} s_{ij}(X) & 0 \\ f_{ij}(X) & t_{ij}(X) \end{pmatrix} \quad \text{and} \quad x = \begin{pmatrix} s_x & 0 \\ f_x & t_x \end{pmatrix},$$

whence

$$x^\omega = \begin{pmatrix} s_x^\omega & 0 \\ f_{x^\omega} & t_x^\omega \end{pmatrix}$$

some $f_{x^\omega} \in \mathrm{Hom}(N, M)$. In particular, we have $(XE_{11})\beta\pi_S = s_{11}(X)$ and $(XE_{11})\beta\pi_T = t_{11}(X)$ by the line preceding (9.15).

The main trick now is to localize $x$ at the idempotent $E_{11}\beta$. Set $z = (E_{11}\beta)x(E_{11}\beta)$. Notice $x^\omega = (x^\omega)^3 \leq (E_{11}\beta)x(E_{11}\beta) = z \leq x$ (where the last inequality uses Proposition 9.4.16). Also note that

$$z^2 = (E_{11}\beta)x(E_{11}\beta)(E_{11}\beta)x(E_{11}\beta) \leq (E_{11}\beta)x(E_{11}\beta) = z \qquad (9.16)$$

where the inequality follows from Proposition 9.4.16.

*Claim.* For all $X \in P(G)$, the equality $(XE_{11})\beta z = z = z(XE_{11})\beta$ holds.

*Proof.* It suffices to consider the case $X = g \in G$ by the distributive law. Now

$$
\begin{aligned}
(gE_{11}\beta)z &= (gE_{11}\beta)(E_{11}\beta)x(E_{11}\beta) \\
&= (E_{11}\beta)(gE_{11}\beta)x(E_{11}\beta) \leq (E_{11}\beta)x(E_{11}\beta) = z
\end{aligned}
\qquad (9.17)
$$

as $(gE_{11}\beta)x \leq x$ by Proposition 9.4.16. Applying (9.17) to $g^{-1}E_{11}$ yields

$$z = E_{11}\beta z = (gE_{11})\beta(g^{-1}E_{11})\beta z \leq (gE_{11}\beta)z.$$

This completes the proof $(XE_{11})\beta z = z$ for $X \in P(G)$. The other equality is proved dually. $\qquad\square$

Now define a semiring homomorphism $\gamma\colon P(G)E_{11} \to Q$ by $\emptyset\gamma = 0$ and

$$XE_{11}\gamma = XE_{11}\beta + z = \epsilon_{11}(X) + z$$

for $X \neq \emptyset$, where the last equality follows from the line preceding (9.15). To see that $\gamma$ is a multiplicative homomorphism, observe that, for $X \neq \emptyset \neq Y$,

$$
\begin{aligned}
(XE_{11}\beta + z)(YE_{11}\beta + z) &= (XE_{11})\beta(YE_{11})\beta + XE_{11}\beta z + z(YE_{11})\beta + z^2 \\
&= (XYE_{11})\beta + z
\end{aligned}
$$

where the last equality uses the claim and (9.16). It is clear $\gamma$ is an additive morphism. Moreover, $\gamma$ is a splitting of $\varphi$ because if $\emptyset \neq X$, then $XE_{11}\gamma\varphi = XE_{11}\beta\varphi + z\varphi = XE_{11}$ as $\beta$ is a splitting and $z \in I$. In particular, $(P(G)E_{11})\gamma \cong P(G)$ and so Lemma 9.4.27 implies that either $\pi_S$ or $\pi_T$ is injective on $(P(G)E_{11})\gamma$. We assume $\pi_S$ is injective; the other case is handled identically.

The rest of the proof follows along the lines of the proof of Theorem 9.4.18, but with a twist. Recall $I\pi_S$ is a semiring ideal of $S$ and $B_n(P(G))$ is the $P(G)$-span of the $n \times n$ matrix units in $M_n(P(G))$. Define a map $\psi\colon B_n(P(G)) \to S/I\pi_S$ by $(XE_{ij})\psi = s_{ij}(X) + I\pi_S$. It is immediate from (9.15) (and the remarks preceding it) that $\psi$ satisfies the conditions of Proposition 9.1.11 and hence extends to a semiring homomorphism $\overline{\psi}\colon M_n(P(G)) \to S/I\pi_S$. It follows from Theorem 9.1.9 that there is a congruence $\equiv$ on $P(G)$ so that $\ker\overline{\psi}$ is the congruence associated to the induced map $M_n(P(G)) \to M_n(P(G)/\equiv)$. We show that $\equiv$ is the trivial congruence, from whence it follows $M_n(P(G)) \prec S$. First observe that the proof of Corollary 9.4.22 shows that if $X \equiv \emptyset$ for some $X \neq \emptyset$, then all elements of $P(G)$ are congruent. Hence, as $|G| \geq 2$, it suffices to show that $X \equiv Y$ implies $X = Y$ for all $X, Y \neq \emptyset$. Suppose $X \equiv Y$ with $X, Y \neq \emptyset$. Then we must have

$$s_{11}(X) + I\pi_S = (XE_{11})\overline{\psi} = (YE_{11})\overline{\psi} = s_{11}(Y) + I\pi_S.$$

Because $\bigvee(I\pi_S) = (\bigvee I)\pi_S = s_x$, Proposition 9.1.22 provides the equality $s_{11}(X) + s_x = s_{11}(Y) + s_x$. But recall $s_{11}(Z) = ZE_{11}\beta\pi_S$ all $Z \in P(G)$. So we have $XE_{11}\beta\pi_S + s_x = YE_{11}\beta\pi_S + s_x$. Multiplying both sides of this equality on the left and right by $E_{11}\beta\pi_S$ and recalling that $z = (E_{11}\beta)x(E_{11}\beta)$ yields the middle equality of

$$(XE_{11})\gamma\pi_S = (XE_{11}\beta + z)\pi_S = (YE_{11}\beta + z)\pi_S = (YE_{11})\gamma\pi_S.$$

Because $\gamma\pi_S$ was assumed injective, we obtain $X = Y$, as was desired. This completes the proof that $M_n(P(G))$ is $\triangle$-irreducible.    □

Now we turn to the case of $G^\natural$. The proof is very similar to the previous one and so we omit some of the details.

**Theorem 9.4.30.** *Let $G$ be a non-trivial finite monolithic group. Then, for all $n \geq 1$, $M_n(G^\natural)$ is $\triangle$-irreducible.*

*Proof.* The case $n = 1$ has already been dealt with, so we assume $n \geq 2$. Let $Q$ be a subsemiring of $\triangle((M, S), (N, T))$ such that there is a surjective homomorphism $\varphi \colon Q \twoheadrightarrow M_n(G^\natural)$. Once again we may assume the projections $\pi_S \colon Q \to S$ and $\pi_T \colon Q \to T$ are surjective. Let $I = 0\varphi^{-1}$ and set $x = \bigvee I$.

Now we proceed exactly as in the previous proof, but this time $\alpha$, $\beta$ and $\gamma$ are only defined on $GE_{11}$ and we can only define $\epsilon_{ij}(g)$ for $g \in G$. We define $z$ as in the previous proof and retain the notation $s_{ij}(g)$. Lemma 9.4.23 implies one of the projections is injective on $GE_{11}\gamma$. We may assume that $\pi_S$ is injective on $GE_{11}\gamma$; the other case is dealt with similarly.

Using the analogue of (9.15) in this context, we can define a homomorphism $\psi \colon B_n(P(G)) \to S/I\pi_S$ by $XE_{ij} \mapsto \sum_{g \in X} s_{ij}(g) + I\pi_S$ satisfying the conditions of Proposition 9.1.11. Thus $\psi$ extends to a semiring homomorphism $\overline{\psi} \colon M_n(P(G)) \to S/I\pi_S$. It follows that the congruence associated to $\overline{\psi}$ is induced by a congruence $\equiv$ on $P(G)$ as per Theorem 9.1.9. To show that $M_n(G^\natural) \prec S/I\pi_S$, it suffices (by Corollary 9.4.22) to show that $\equiv$ is injective on $G$. Suppose $g \equiv g'$. Then we have

$$s_{11}(g) + I\pi_S = (gE_{11})\overline{\psi} = (g'E_{11})\overline{\psi} = s_{11}(g') + I\pi_S.$$

Equivalently, $gE_{11}\beta\pi_S + I\pi_S = g'E_{11}\beta\pi_S + I\pi_S$. Because $s_x = \bigvee I\pi_S$, Proposition 9.1.22 yields $gE_{11}\beta\pi_S + s_x = g'E_{11}\beta\pi_S + s_x$. As in the previous proof we multiply each side of this equation on both the left and right by $E_{11}\beta\pi_S$ to obtain $(gE_{11}\beta + z)\pi_S = (g'E_{11}\beta + z)\pi_S$ or, equivalently, $gE_{11}\gamma\pi_S = g'E_{11}\gamma\pi_S$. Because $\gamma\pi_S$ was assumed injective, we obtain $g = g'$, as required. This completes the proof.    □

We have now established Theorem 9.4.10. In particular, the Prime Decomposition Theorem for Quantales (Theorem 9.4.1) is truly a decomposition into primes.

Let us mention that if $S$ is $\triangle$-irreducible, then the set $\mathsf{Excl}(S)$ of idempotent semirings $Q$ such that $S \not\prec Q$ is a pseudovariety of idempotent semirings *closed* under triangular product. It is natural to try and determine what are the exclusion classes corresponding to the known $\triangle$-irreducible semigroups. The following exercises and questions address this.

**Exercise 9.4.31.** Prove that $\mathsf{Excl}(\mathbb{B})$ consists of all idempotent semirings whose underlying multiplicative semigroup is nilpotent.

**Exercise 9.4.32.** If $G$ is a non-trivial monolithic group, prove that $\mathsf{Excl}(G^{\natural})$ consists of all idempotent semirings whose underlying multiplicative semigroup does not contain a copy of $G$.

**Exercise 9.4.33.** If $G$ is a non-trivial group, then $\mathsf{Excl}(P(G))$ consists of all idempotent semirings that do not have $P(G)$ as a subsemiring.

*Question 9.4.34.* Compute the exclusion class of $M_n(Q)$ where $Q$ is one of $\mathbb{B}$, $G^{\natural}$ with $G$ non-trivial monolithic or $P(G)$ with $G$ a non-trivial group.

## 9.5 Complexity of Idempotent Semirings

We are now ready to define the complexity of an idempotent semiring. Unlike the case of $\rtimes$-prime semigroups, $\triangle$-irreducible semirings can be highly non-trivial. In particular, every idempotent semiring divides a $\triangle$-irreducible semiring of the form $M_n(\mathbb{B})$. Thus the issue of what complexity to give an irreducible is more delicate in this theory.

The definition of complexity for idempotent semirings is given in terms of a hierarchical complexity function. The notion of a hierarchical complexity function for idempotent semirings can be formalized in exactly the same way as for semigroups, cf. Section 4.3.

**Definition 9.5.1 (Complexity function).** *Let $S$ be a finite idempotent semiring. Define the complexity $c_q(S)$ to be the minimum of the quantity $c(R_1) + \cdots + c(R_n)$ over all divisions of the form*

$$S \prec \triangle((M_1, R_1), \ldots (M_n, R_n)) \tag{9.18}$$

*where the $R_i$ are $\triangle$-irreducible.*

We must show that $c_q$ is a hierarchical complexity function taking on only finite values.

**Theorem 9.5.2.** *The complexity $c_q$ is a hierarchical complexity function satisfying the following two properties:*

1. $c_q(S) < \infty$ for every idempotent semiring $S$;
2. $c_q(S) = c(S)$ if $S$ is $\triangle$-irreducible.

*Proof.* First we show that $c_q(-)$ is a hierarchical complexity function satisfying 1 and 2. Corollary 9.4.19 shows that $c_q(S) < \infty$, establishing 1. The trivial idempotent semiring $\{0\}$ divides $\mathbb{B}$, which is $\triangle$-irreducible, and so $c_q(\{0\}) = 0$. Clearly, $S \prec T$ implies $c_q(S) \leq c_q(T)$. From (9.10), we obtain $c_q(S \times T) = \max\{c_q(S), c_q(T)\}$. Property 2 is a consequence of the definition of $c_q(S)$ and of $\triangle$-irreducibility. Indeed, if $S$ is $\triangle$-irreducible, then by definition $c_q(S) \leq c(S)$. On the other hand, given a decomposition (9.18), $S$ must divide $R_i$ some $i$ and so $c(S) \leq c(R_i) \leq c(R_1) + \cdots + c(R_n)$. We conclude $c_q(S) = c(S)$, establishing 2.    $\square$

Alternatively, one could use the two-sided complexity function $C$ in place of $c$ in Definition 9.5.1 to obtain a self-dual complexity function $C_q$ for idempotent semirings in the sense that $C_q(S^{op}) = C_q(S)$. Because $\triangle$-irreducibility is a self-dual notion, this seems a natural choice.

*Question 9.5.3.* Given an oracle deciding group complexity, is $c_q$ computable? We ask the same question for $C_q$.

To obtain an estimate on the complexity of an idempotent semiring from the Prime Decomposition Theorem, we need to compute the group complexity of $M_n(P(G))$. This was done for the trivial group by the first author [273] and in general by the first author and Fox [91].

**Theorem 9.5.4 (Fox-Rhodes).** *Let $G$ be a group. Then*

$$c(M_n(P(G))) = n - 1 + c(G).$$

*Proof.* The monoid $RM_n(G)$ of $n \times n$ row monomial matrices over $G$ is clearly a submonoid of $M_n(P(G))$. Because $RM_n(G) \cong G \wr (\mathbf{n}, T_n)$, Theorem 4.12.32 yields $c(RM_n(G)) = n - 1 + c(G)$ and hence $c(M_n(P(G))) \geq n - 1 + c(G)$. To establish the reverse inequality, we use essentially the same argument as [362, Example 6.2]. Clearly, $M_1(P(G)) = P(G)$. If $G$ is trivial, then $P(G) = \mathbb{B}$ is aperiodic and so has complexity 0. If $G$ is non-trivial, an application of Proposition 4.1.7 yields $P(G) \prec (P_1(G) \cup \{\emptyset\}) \rtimes G$ where $P_1(G)$ is the subsemigroup of elements of $P(G)$ containing 1. Because $(P_1(G), \subseteq)$ is an ordered monoid with the identity as the smallest element, it is $\mathscr{J}$-trivial by Proposition 8.2.1. Hence $P(G) \in \mathbf{J} * \mathbf{G}$ and so $c(P(G)) = 1$.

Assume by induction that $c(M_{n-1}(P(G))) = n - 2 + c(G)$ for $n \geq 2$. It will be convenient in this proof to label entries by elements of $\mathbf{n}$. For $K \subseteq \mathbf{n}$, let $I_K$ be the subidentity matrix with ones in the diagonal entries corresponding to elements of $K$. In particular, $I_{\mathbf{n}}$ is the identity matrix. Let $I$ be the two-sided ideal of $M_n(P(G))$ generated by $I_{\mathbf{n-1}}$. By the Ideal Theorem,

$$c(M_n(P(G))) \leq c(I) + c(M_n(P(G))/I).$$

It is easy to see that $I_{n-1}M_n(P(G))I_{n-1} \cong M_{n-1}(P(G))$ and so an application of Proposition 4.12.20 yields

$$c(I) = c(M_{n-1}(P(G))) = n - 2 + c(G).$$

Thus it suffices to show that $c(M_n(P(G))/I) \leq 1$. We do this by verifying that $\overline{2}$ is not a subsemigroup of $M_n(P(G))/I$ and applying Corollary 4.8.5. The key point to verify is the following claim.

*Claim.* Let $A \in M_n(P(G)) \setminus I$ be an idempotent. Then $A \geq I_{\mathbf{n}}$.

*Proof.* First suppose that $A_{kk} = 0$ for some $k \in \mathbf{n}$. Let $B$ be the matrix:

$$B_{ij} = \begin{cases} 1 & i = j \neq k \\ A_{ij} & i = k \\ 0 & \text{else.} \end{cases}$$

Direct computation (using the idempotence of $A$) shows that $A = BI_{\mathbf{n}\setminus\{k\}}A$. Indeed, one readily verifies

$$(BI_{\mathbf{n}\setminus k}A)_{ij} = \sum_{\ell \neq k} B_{i\ell}A_{\ell j}. \tag{9.19}$$

In particular, if $i \neq k$, the right-hand side of (9.19) becomes $A_{ij}$ as $B_{i\ell} \neq 0$ only when $i = \ell$, in which case it is 1. On the other hand, if $i = k$, the right-hand side becomes

$$\sum_{\ell \neq k} A_{i\ell}A_{\ell j} = \sum_{\ell \in \mathbf{n}} A_{i\ell}A_{\ell j} = A_{ij}$$

where the first equality uses that $A_{ik} = A_{kk} = 0$ and the last equality uses that $A^2 = A$.

Because $I_{\mathbf{n}\setminus\{k\}} \not\leq I_{n-1}$ (conjugate by the permutation matrix corresponding to the transposition $(0k)$), we see that $A \in I$, a contradiction. Consequently, $A_{kk} \neq 0$ for all $k \in \mathbf{n}$. From $A^2 = A$, we obtain $A_{kk} \geq A_{kk}^2$. Therefore, $A_{kk} \geq A_{kk}^\omega \neq 0$. Because $A_{kk}^\omega$ is a non-empty subsemigroup of the finite group $G$, it must be a subgroup and hence contain the identity. Thus $A_{kk} \geq 1$, for all $k \in \mathbf{n}$, that is, $A \geq I_{\mathbf{n}}$. □

The result follows easily now from the claim. If $E, F$ form a right zero subsemigroup of $M_n(P(G))/I$ (and hence are non-zero), then using the claim $F = EF \geq EI_{\mathbf{n}} = E$ and similarly $E \geq F$. Thus $M_n(P(G))/I$ excludes $\overline{2}$ as a subsemigroup, as required. This completes the proof. □

To bound the complexity of an idempotent semiring, we need the following observation. The inverse operation on a group $G$ extends to an involution on $P(G)$ by $A \mapsto A^{-1}$ where $A^{-1} = \{a^{-1} \mid a \in A\}$. Clearly, $(AB)^{-1} = B^{-1}A^{-1}$ so $P(G) \cong P(G)^{op}$. Consequently,

$$M_n(P(G))^{op} \cong M_n(P(G)^{op}) \cong M_n(P(G))$$

where the first isomorphism uses the transpose.

**Theorem 9.5.5.** *Let $S$ be an idempotent semiring. Then $c_q(S)$ is bounded by both the number of non-zero $\mathscr{L}$-classes and the number of non-zero $\mathscr{R}$-classes of the multiplicative semigroup of $S$.*

*Proof.* First we show that the number of non-zero $\mathscr{L}$-classes of $S$ is an upper bound. By Theorem 9.4.1 and Theorem 9.4.10 we have

$$c_q(S) \leq \sum_{i=1}^{n} c(M_{b_i}(P(G_i)))$$

where $S^I$ has $\mathscr{J}$-classes $J_1, \ldots, J_n$, $b_i$ is the number of $\mathscr{L}$-classes in $J_i$ and $G_i$ is the Schützenberger group of $J_i$. In particular, for the $\mathscr{J}$-classes of $I$ and $0$, we have $c(M_1(\mathbb{B})) = 0$. In all other cases, we have

$$c(M_{b_i}(P(G_i))) = b_i - 1 + c(G_i) \leq b_i$$

by Theorem 9.5.4. This gives the desired bound. The bound in terms of $\mathscr{R}$-classes comes from applying the above analysis to $S^{op}$ and applying the duality of Proposition 9.2.14 and the remark immediately preceding the theorem.  $\square$

Of course, the program we have been carrying out for idempotent semirings can be carried out for $k$-algebras over a general (finite) commutative unital semiring $k$. Plotkin did this to some extent for the case where $k$ is a field [240]. This is a good research project.

*Question 9.5.6.* Study $\triangle$-irreducibility over more general semirings.

## 9.5.1 Applications to the group complexity of power semigroups

This section applies our results on the triangular product to the complexity of power semigroups. In particular, we compute the complexity of the power semigroup of an inverse semigroup and obtain an asymptotically tight bound on the complexity of the power semigroup of the full transformation monoid, the latter result improving on a result of [91]. Once again, all semigroups are assumed to be finite.

A primitive form of Theorem 9.3.7, using wreath products (essentially Corollary 9.2.16), was obtained in [91], leading to some complexity bounds for $P(S)$. Using the Prime Decomposition Theorem for idempotent semirings, we obtain a much tighter result.

**Proposition 9.5.7.** *Let $S$ and $T$ be semigroups and let $(M, S)$ and $(N, T)$ be $\mathbb{B}$-modules. Then*

$$c\big(\triangle((M, S), (N, T))\big) = \max\{c(S), c(T)\}.$$

*Proof.* The projection $\triangle((M,S),(N,T)) \to S \times T$ belongs to $(\mathbb{LSl},\mathbf{1})$ by Lemma 9.2.4, and hence is aperiodic. By the Fundamental Lemma of Complexity, we conclude

$$c\left(\triangle((M,S),(N,T))\right) = c(S \times T) = \max\{c(S), c(T)\}$$

as required.                                                                        $\square$

The associativity of the triangular product and induction then yield:

**Corollary 9.5.8.** *Let $S_1, \dots, S_n$ be semigroups and $(M_1, S_1), \dots, (M_n, S_n)$ be $\mathbb{B}$-modules. Then*

$$c\left(\triangle((M_1, S_1), \dots, (M_n, S_n))\right) = \max\{c(S_1), \dots, c(S_n)\}.$$

Now applying Corollary 9.3.13 to $P(S)$ and using Theorem 9.5.4 and Corollary 9.5.8, as well as applying these same results to $P(S^{op}) = P(S)^{op}$, along with the duality in Proposition 9.2.14, we obtain the following upper bound to the complexity of a power semigroup.

**Theorem 9.5.9.** *Suppose $S$ is a finite semigroup. Let $\ell_g, r_g$ be the maximum number of $\mathscr{L}$-classes, respectively $\mathscr{R}$-classes, in a $\mathscr{J}$-class of $S$ with nontrivial Schützenberger group and let $\ell_a, r_a$ be the maximum number of $\mathscr{L}$-classes, respectively $\mathscr{R}$-classes, in a $\mathscr{J}$-class of $S$ with trivial Schützenberger group. Let $\ell = \max\{\ell_g, \ell_a - 1\}$ and $r = \max\{r_g, r_a - 1\}$. Then the upper bound $c(S) \leq \min\{\ell, r\}$ holds.*

Let us highlight the following consequence of this theorem.

**Corollary 9.5.10.** *The complexity of $P(S)$ is bounded by both the maximum number of $\mathscr{L}$-classes and the maximum number of $\mathscr{R}$-classes in a non-zero $\mathscr{J}$-class of $S$.*

Our first application is to compute the complexity of the power semigroup of an inverse semigroup. This is a new result.

**Theorem 9.5.11.** *Let $S$ be an inverse semigroup. Let $J$ be a $\mathscr{J}$-class of $S$ with maximal subgroup $G_J$ and let $n_J$ be the number of idempotents in $J$. Setting $\alpha(J) = n_J - 1 + c(G_J)$, one has the equality*

$$c(P(S)) = \max\{\alpha(J) \mid J \in S/\mathscr{J}\}. \tag{9.20}$$

*In particular, complexity is computable for power semigroups of inverse semigroups.*

*Proof.* Because $n_J$ is the number of $\mathscr{L}$-classes in the $\mathscr{J}$-class $J$, Theorem 9.5.9 implies $c(P(S)) \leq \max\{\alpha(J) \mid J \in S/\mathscr{J}\}$. For the reverse inequality, let $J$ be a $\mathscr{J}$-class of $S$ and set $I = SJS$ and $V = I \setminus J$. Then

$$P(S) \succ P(I)/P(V) \cong P_0(I/V) \cong M_{n_J}(P(G_J))$$

as $I/V \cong \mathscr{M}^0(G_J, n_J, n_J, I_{n_J})$ (cf. Exercise 9.1.19). By Theorem 9.5.4

$$\alpha(J) = c(M_{n_J}(P(G_J))) \leq c(P(S))$$

thereby completing the proof of (9.20).                                      □

As a corollary we obtain a formula for the complexity of $P(I_n)$.

**Corollary 9.5.12.** *Let $I_n$ be the symmetric inverse monoid of degree $n$. Then* $c(P(I_1)) = 0$, $c(P(I_2)) = 1$ *and*

$$c(P(I_n)) = \binom{n}{\lfloor \frac{n}{2} \rfloor}$$

*for $n \geq 3$.*

*Proof.* The cases $n = 1, 2$ can be computed directly from Theorem 9.5.11. For $n \geq 3$, we have $\lceil \frac{n}{2} \rceil \geq 2$. Now if $J_k$ is the rank $k$ $\mathscr{J}$-class of $I_n$, then $\alpha(J_k) = \binom{n}{k}$ for $2 \leq k \leq n$ and $\alpha(J_k) = \binom{n}{k} - 1$ for $k = 0, 1$. Because the central binomial coefficient $\binom{n}{\lfloor \frac{n}{2} \rfloor} = \binom{n}{\lceil \frac{n}{2} \rceil}$ is the largest one, the result follows from Theorem 9.5.11.                                      □

The first author and Fox proved in [91]

$$\binom{n-1}{\lfloor \frac{n-1}{2} \rfloor} \leq c(P(T_n)) \leq 2^n - n - 1.$$

Using our new techniques, we are able to bring the upper bound in line with the lower bound.

**Theorem 9.5.13.** *Let $T_n$ be the full transformation semigroup of degree $n$ and let $n \geq 2$. Then*

$$\binom{n-1}{\lfloor \frac{n-1}{2} \rfloor} \leq c(P(T_n)) \leq \binom{n}{\lfloor \frac{n}{2} \rfloor}. \tag{9.21}$$

*Proof.* The number of $\mathscr{L}$-classes in the rank $k$ $\mathscr{J}$-class of $T_n$ is $\binom{n}{k}$ and so $c(P(T_n)) \leq \binom{n}{\lfloor \frac{n}{2} \rfloor}$ by Corollary 9.5.10. For the lower bound, observe that $I_{n-1} \leq T_n$ by viewing $n-1$ as a sink or zero: more precisely, if $\sigma \in I_{n-1}$, then the corresponding full transformation agrees with $\sigma$ on its domain and sends $n-1$ and the complement of the domain of $\sigma$ to $n-1$. Thus $P(I_{n-1}) \leq P(T_n)$ and Corollary 9.5.12 yields the lower bound except for when $n = 2, 3$. Clearly $c(P(T_2)) \geq 1$. When $n = 3$, observe that the rank 1 $\mathscr{J}$-class of $I_2$ is embedded into the rank 2 $\mathscr{J}$-class of $T_3$, which has non-trivial maximal subgroup $S_2$. Thus the rank 2 $\mathscr{J}$-class of $T_3$ has a $2 \times 2$ identity submatrix in its structure matrix and so we can find $M_2(P(S_2)) = P_0(\mathscr{M}^0(S_2, 2, 2, I_2))$ as a divisor of $P(T_3)$. This gives the lower bound of 2 for $n = 3$.                                      □

We would like to thank M. Putcha for suggesting the explicit use of $I_{n-1}$ in obtaining the lower bound. The approach in [91] only uses it implicitly.

It is straightforward to verify the lower bound and the upper bound in (9.21) differ asymptotically by a factor of 2, so the bounds in Theorem 9.5.13 are asymptotically tight. Stirling's formula yields

$$\binom{n}{\lfloor \frac{n}{2} \rfloor} \sim \frac{2^n}{\sqrt{\pi \left(\frac{n-1}{2}\right)}}$$

and so $c(P(T_n))$ is $\Theta\left(\dfrac{2^n}{\sqrt{n-1}}\right)$.

The following consequence of the Ideal Decomposition Theorem improves on a result of [91].

**Proposition 9.5.14.** *Let $S$ be a semigroup and $I$ an ideal of $S$. Then*

$$c(P(S)) = \max\{c(P(S)_{P(I)}), c(P_0(S/I))\}$$

*where $P(S)_{P(I)}$ is the quotient of $P(S)$ by the kernel of its action on the right of $P(I)$.*

The following questions were raised in [91].

*Question 9.5.15.* Let $S = \mathcal{M}^0(G, A, B, C)$ be a 0-simple semigroup.

1. Is $c(P(S)) = \iota(C) - 1 + c(G)$ where $\iota(C)$ is the largest size of an identity submatrix of $C$ (up to reordering rows and columns and renormalizing)? The right-hand side is a lower bound by the argument in Theorem 9.5.11. In [91], it is shown that an upper bound is $\tau(C) - 1 + c(G)$ where $\tau(C)$ is the largest size of a triangular submatrix of $C$.
2. Recall that the action of $S$ on the right of $G \times B$ extends to an action on the free left $P(G)$-module generated by $B$ and hence yields a representation $\rho\colon S \to M_B(P(G))$ (which concretely speaking is the classical Schützenberger representation by row monomial matrices). This in turn yields a representation $\widetilde{\rho}\colon P(S) \to M_B(P(G))$. What is the relationship between $c(P(S))$ and $c(P(S)\widetilde{\rho})$?

*Question 9.5.16 (Fox-Rhodes).* What is the largest size of an identity submatrix of the structure matrix for the rank $k$ $\mathcal{J}$-class of $T_n$?

*Question 9.5.17.* What is the exact complexity of $P(T_n)$?

*Question 9.5.18.* If $S$ is a semigroup and $I$ is an ideal, what is the relationship between $c(P(S))$ and $c(P(I))$, $c(P_0(S/I))$? Proposition 9.5.14 gives some indication, but $c(P(S)_{P(I)})$ and $P(I)$ do not seem so clearly related. In [91] it is shown that if $S$ is a monoid and $I = S \setminus G$ where $G$ is the group of units, then $c(P(S)) \leq c(P(I)) + c(P_0(S/I))$.

The next question draws its inspiration from the new results obtained via the triangular product.

*Question 9.5.19.* Is $c(P(S)) = \max\{c(P_0(J^0)) \mid J \in S/\mathscr{J}\}$?

Our final question is admittedly a bit vague: it is intended to spur the development of more of ring theory in the context of idempotent semirings. The reader should consult [157].

*Question 9.5.20.* Develop homological algebra for idempotent semirings. Find a homological definition of $c_q$ and $C_q$. Relate this to dot-depth.

# Notes

The preliminary material on semirings presented at the beginning of the chapter consists of obvious generalizations of ring theory and can be found in any book on semirings [100, 120, 160]. Semiring theory has enjoyed some popularity among computer scientists, mostly due to the work of Schützenberger on formal power series in non-commuting variables [48, 65, 84, 305]. The work of Polák [241–244] also deserves mention. Recently, tropical geometry [302] has brought idempotent semirings to the attention of the general mathematical community.

The deeper material in this chapter begins with the triangular product of Plotkin [239, 240, 376], an ingenious axiomatization of the Jordan-Hölder composition series from ring theory. The results from Section 9.3 onwards are new, excepting those results specifically attributed to other authors. Many of the decomposition results for the case of a power semigroup were inspired by the unpublished work of Fox and Rhodes [91], but these authors used wreath products thereby resulting in a much looser decomposition from the point of view of complexity. The introduction of the triangular product is also essential for creating a complexity theory of idempotent semirings as there does not seem to be a wreath product of semirings.

The exact computation of the complexity of the power semigroup of an inverse semigroup is novel to this chapter and justifies all the work on semirings for those who might only be interested in semigroups. The asymptotically tight estimate for the complexity of the power semigroup of the full transformation semigroup is also new; the lower bound is from [91].

A natural research project is to develop some sort of derived category or kernel category theory for the triangular product. Quantaloids [304] should somehow be involved.

This chapter is encouraging in that it would seem to indicate that much more of non-trivial ring theory goes over to semirings than one would expect. Possibly homology and cohomology should be the next step given the construction of the triangular product. Also perhaps an analogue of quiver theory [94] should be developed in this context.

# A

# The Green-Rees Local Structure Theory

The goal of this appendix is to give an admittedly terse review of the Green-Rees structure theory of stable semigroups (or what might be referred to as the local theory, in comparison with the semilocal theory of Section 4.6). More complete references for this material are [68, 139, 171].

## A.1 Ideal Structure and Green's Relations

If $S$ is a semigroup, then $S^I = S \cup \{I\}$, where $I$ is a newly adjoined identity. If $X, Y$ are subsets of $S$, then $XY = \{xy \mid x \in X, y \in Y\}$.

**Definition A.1.1 (Ideals).** *Let $S$ be a semigroup. Then:*

1. *$\emptyset \neq R \subseteq S$ is a right ideal if $RS \subseteq R$;*
2. *$\emptyset \neq L \subseteq S$ is a left ideal if $SL \subseteq L$;*
3. *$\emptyset \neq I \subseteq S$ is an ideal if it is both a left ideal and a right ideal.*

If $s \in S$, then $sS^I$ is the *principal right ideal* generated by $s$, $S^I s$ is the *principal left ideal* generated by $s$ and $S^I s S^I$ is the *principal ideal* generated by $s$. If $S$ is a monoid, then $S^I s = Ss$, $sS^I = sS$ and $S^I s S^I = SsS$.

A semigroup is called *left simple*, *right simple*, or *simple* if it has no proper, respectively, left ideal, right ideal, or ideal. The next proposition is straightforward; the proof is left to the reader.

**Proposition A.1.2.** *Let $I, J$ be ideals of $S$. Then*

$$IJ = \{ij \mid i \in I, \ j \in J\}$$

*is an ideal and $\emptyset \neq IJ \subseteq I \cap J$. Consequently, the set of ideals of $S$ is closed under finite intersection.*

**Corollary A.1.3.** *A finite semigroup $S$ has a unique minimal ideal.*

*Proof.* Proposition A.1.2 implies that the intersection of all ideals of $S$ is again an ideal; clearly it is the unique minimal ideal.    □

Note that if a semigroup has a zero, then its minimal ideal is $\{0\}$. In this context, the notion of minimal ideal is not so useful, and so we introduce the notion of a 0-minimal ideal.

**Definition A.1.4 (0-minimal ideal).** *A minimal non-zero ideal of a semigroup is called a 0-minimal ideal. We also, by convention, consider the minimal ideal of the trivial semigroup to be a 0-minimal ideal.*

With this definition, the minimal ideal of a semigroup without zero is considered to be a 0-minimal ideal. This convention is somewhat non-standard, but is convenient for stating results uniformly for semigroups with 0 and without 0. Note that 0-minimal ideals do not have to be unique. Next we introduce Green's relations [68, 108, 171]. They are an essential ingredient in semigroup theory.

**Definition A.1.5 (Green's relations).** *Let $S$ be a semigroup and $s, t \in S$. Green's equivalence relations $\mathscr{R}$, $\mathscr{L}$, $\mathscr{H}$ and $\mathscr{D}$ are defined as follows:*

- $s \mathrel{\mathscr{R}} t \iff sS^I = tS^I$;
- $s \mathrel{\mathscr{L}} t \iff S^I s = S^I t$;
- $s \mathrel{\mathscr{J}} t \iff S^I s S^I = S^I t S^I$;
- $\mathscr{H} = \mathscr{R} \cap \mathscr{L}$;
- $\mathscr{D} = \mathscr{R} \vee \mathscr{L}$.

*The $\mathscr{R}$-class of an element $s \in S$ is typically denoted by $R_s$. Similar meanings can be ascribed to $L_s$, $J_s$, $H_s$ and $D_s$.*

**Exercise A.1.6.** Show that $\mathscr{L} \circ \mathscr{R} = \mathscr{R} \circ \mathscr{L}$ and deduce from this

$$\mathscr{L} \circ \mathscr{R} = \mathscr{D} = \mathscr{R} \circ \mathscr{L}.$$

All of Green's relations coincide for a commutative semigroup. Note that

$$\mathscr{H} \subseteq \mathscr{L}, \mathscr{R} \subseteq \mathscr{D} \subseteq \mathscr{J}.$$

We shall see $\mathscr{D} = \mathscr{J}$ for stable (in particular, finite) semigroups. Green's relations are more naturally derived from his preorders.

**Definition A.1.7 (Green's preorders).** *Green's preorders are defined by:*

- $s \leq_{\mathscr{R}} t \iff sS^I \subseteq tS^I$;
- $s \leq_{\mathscr{L}} t \iff S^I s \subseteq S^I t$;
- $s \leq_{\mathscr{J}} t \iff S^I s S^I \subseteq S^I t S^I$;
- $s \leq_{\mathscr{H}} t \iff s \leq_{\mathscr{L}} t$ and $s \leq_{\mathscr{R}} t$.

Green's relations are the equivalence relations associated to his preorders. Therefore, for instance, $\leq_{\mathscr{R}}$ induces a partial order on the set of $\mathscr{R}$-classes of $S$. An element $e$ of a semigroup is *idempotent* if $e^2 = e$. Often, if $e, f$ are idempotents, then $e \leq_{\mathscr{H}} f$ is abbreviated to $e \leq f$. Observe that $e \leq f$ if and only if $ef = e = fe$ and that this is a partial order on the set of idempotents.

**Exercise A.1.8.** Show that if $e$ is an idempotent, then $s \leq_{\mathscr{L}} e$ if and only if $se = s$, and that $s \leq_{\mathscr{R}} e$ if and only if $es = s$. Conclude that if $e, f$ are idempotents, then $e \leq f$ if and only if $ef = e = fe$.

**Proposition A.1.9.** *Green's relation $\mathscr{R}$ is a left congruence and $\mathscr{L}$ is a right congruence.*

*Proof.* Suppose $s \mathscr{R} t$ and $x \in S$. Then $sS^I = tS^I$ and so $xsS^I = xtS^I$. Therefore $\mathscr{R}$ is a left congruence. The proof for $\mathscr{L}$ is dual. □

Similarly one can show that the preorder $\leq_{\mathscr{R}}$ is stable under left multiplication and the preorder $\leq_{\mathscr{L}}$ is stable under right multiplication.

**Exercise A.1.10.** Verify this last assertion.

The following notion, due to von Neumann in the context of ring theory [375], plays a fundamental role in semigroup theory.

**Definition A.1.11 (Regular element).** *An element $s$ of a semigroup $S$ is (von Neumann) regular if there exists $t \in S$ such that $sts = s$, i.e., $s \in sSs$. A semigroup is said to be regular if each of its elements is regular.*

It is immediate that idempotents are regular. In a group, all elements are regular.

**Exercise A.1.12.** Verify that if $s$ is regular then $S^I s = Ss$, $sS^I = sS$ and $S^I s S^I = SsS$.

**Definition A.1.13 (Inverse).** *Two elements $s, s'$ of a semigroup are said to be inverse to each other if $ss's = s$ and $s'ss' = s'$.*

Having an inverse is equivalent to being regular as the next proposition shows.

**Proposition A.1.14.** *Let $s \in S$ be a regular element. Then $s$ has an inverse.*

*Proof.* Suppose $s = sts$. Setting $s' = tst$ yields

$$ss's = ststs = sts = s$$
$$s'ss' = tststst = tstst = tst = s'.$$

Thus $s$ and $s'$ are inverses. □

Inverse elements allow for a description of $\mathscr{D}$-equivalence of idempotents analogous to von Neumann-Murray equivalence of projections in operator algebras [31].

**Proposition A.1.15.** *Let $S$ be a semigroup. Two idempotents $e, f \in S$ are $\mathscr{D}$-equivalent if and only if there exist inverse elements $x, x' \in S$ with $e = xx'$ and $f = x'x$. More precisely, if $e, f \in S$ are idempotents and $x \in S$, then $e \mathrel{\mathscr{R}} x \mathrel{\mathscr{L}} f$ if and only if there exists an inverse $x'$ of $x$ such that $xx' = e$ and $x'x = f$, in which case $e \mathrel{\mathscr{L}} x' \mathrel{\mathscr{R}} f$.*

*Proof.* We begin with the second statement, as the first is an immediate consequence of it. If $x, x'$ are inverses, then it is straightforward to verify $x \mathrel{\mathscr{R}} xx' \mathrel{\mathscr{L}} x'$ and $x \mathrel{\mathscr{L}} x'x \mathrel{\mathscr{R}} x'$. This handles the if direction. For the only if direction, suppose $e \mathrel{\mathscr{R}} x \mathrel{\mathscr{L}} f$. Then we can find $u, v \in S^I$ with $xu = e$ and $vx = f$. Set $x' = fue$. Then

$$xx'x = x(fue)x = xux = ex = x$$
$$x'xx' = (fue)x(fue) = fuxue = fue = x'$$

and so $x, x'$ are inverses. Moreover, $xx' = xfue = xue = e$ and $x'x = fuex = fux = vxux = vex = vx = f$ and so $e \mathrel{\mathscr{L}} x' \mathrel{\mathscr{R}} f$, completing the proof.    □

A semigroup is called an *inverse semigroup* if each element has a unique inverse. A *block group* is a semigroup in which each element has at most one inverse, or equivalently a semigroup in which each regular element admits a unique inverse.

In general, Green's relations on a subsemigroup do not coincide with Green's relations on the ambient semigroup. For instance, if $B_2$ is the semigroup of $2 \times 2$ matrix units and the zero matrix, then $E_{11} \mathrel{\mathscr{J}} E_{22}$ in $B_2$, but in the subsemigroup $\{E_{11}, E_{22}, 0\}$ they are not $\mathscr{J}$-equivalent. However, the following is true.

**Proposition A.1.16.** *Let $S$ be a semigroup and $T \leq S$ a subsemigroup. Suppose $t_1, t_2 \in T$ are regular. Then $t_1 \mathrel{\mathscr{K}} t_2$ in $T$ if and only if $t_1 \mathrel{\mathscr{K}} t_2$ in $S$ where $\mathscr{K}$ is any of $\mathscr{R}, \mathscr{L}$ or $\mathscr{H}$.*

*Proof.* We just handle $\mathscr{R}$ as the case of $\mathscr{L}$ is dual, and the result for $\mathscr{H}$ follows from those for $\mathscr{L}$ and $\mathscr{R}$. Clearly, $t_1 \mathrel{\mathscr{R}} t_2$ in $T$ implies $t_1 \mathrel{\mathscr{R}} t_2$ in $S$. For the converse, suppose $t_1 \mathrel{\mathscr{R}} t_2$ in $S$ and choose by regularity elements $x_1, x_2 \in T$ such that $t_i x_i t_i = t_i$, $i = 1, 2$. Then $e_i = t_i x_i$ is an idempotent of $T$ with $e_i \mathrel{\mathscr{R}} t_i$ in $T$ for $i = 1, 2$. It thus suffices to show $e_1 \mathrel{\mathscr{R}} e_2$ in $T$. But $e_1 \mathrel{\mathscr{R}} e_2$ in $S$ implies $e_1 e_2 = e_2$ and $e_2 e_1 = e_1$. Thus $e_1 \mathrel{\mathscr{R}} e_2$ in $T$, as required.    □

## A.2 Stable Semigroups

For the rest of this appendix, we shall be interested in stable semigroups. This class of semigroups include finite semigroups, compact semigroups (Proposition 3.1.10) and commutative semigroups. Algebraic semigroups are also stable [250, 262]. Although at first sight the definition may seem bizarre, it is somehow the crucial property that makes finite semigroup theory "work."

**Definition A.2.1 (Stability).** *A semigroup $S$ is called stable if both*

$$s \, \mathcal{J} \, sx \iff s \, \mathcal{R} \, sx \text{ and also } s \, \mathcal{J} \, xs \iff s \, \mathcal{L} \, xs.$$

The archetypical example of an unstable semigroup is the bicyclic monoid. It is the monoid of transformations of $\mathbb{N}$ generated by $\sigma, \tau \colon \mathbb{N} \to \mathbb{N}$ given by $\sigma(x) = x + 1$ and

$$\tau(x) = \begin{cases} x - 1 & x > 0 \\ 0 & x = 0. \end{cases}$$

Equivalently, it is the semigroup generated by a unilateral shift on a separable Hilbert space and its adjoint. The next exercise gives the reader a chance to become familiar with the notion of stability.

**Exercise A.2.2.** Suppose $S$ is a stable semigroup.

1. Show that $xsx \, \mathcal{J} \, x$ if and only if $xsx \, \mathcal{H} \, x$.
2. Show that if $e \in S$ is an idempotent, then $eSe \cap J_e = H_e$.
3. Show that if $S$ is a monoid with identity 1, then $J_1 = H_1$, which in turn is the group of units of $S$. Deduce that the non-units of $S$ form an ideal.
4. Show that a congruence-free stable monoid is either a simple group or is isomorphic to $(\{0, 1\}, \cdot)$.
5. Show that if $e, f$ are idempotents of $S$ with $e \, \mathcal{J} \, f$ and $e \le f$, then $e = f$.
6. Show that the bicyclic monoid is not stable.
7. Show that the bicyclic monoid does not embed in a stable semigroup.
8. Show that a regular semigroup is stable if and only if it does not contain an isomorphic copy of the bicyclic monoid.

**Exercise A.2.3.** Let $S$ be a non-empty finite semigroup. Prove that $S$ contains an idempotent. Hint: Let $T$ be a minimal (non-empty) subsemigroup of $S$. Show that $tT = T = Tt$ for all $t \in T$. Deduce that if $tx = t$, then $x = x^2$.

The following result is a special case of Proposition 3.1.10.

**Theorem A.2.4.** *Finite semigroups are stable.*

*Proof.* Clearly, $s \, \mathcal{R} \, st$ implies $s \, \mathcal{J} \, st$. Suppose $s \, \mathcal{J} \, st$. Evidently $st \le_{\mathcal{R}} s$, so we are left with establishing the reverse inequality. Because $s \, \mathcal{J} \, st$, we can find $x, y \in S^I$ such that $s = xsty$. One then has $s = x^n s(ty)^n$ for all $n > 0$. Let $k$ be a positive integer such that $(ty)^k$ is idempotent. Then

$$s(ty)^k = x^k s(ty)^k (ty)^k = x^k s(ty)^k = s$$

so $st(y(ty)^{k-1}) = s(ty)^k = s$. Thus $s \le_{\mathcal{R}} st$, as was required. The argument for $\mathcal{L}$ is dual. $\qquad \square$

**Fig. A.1.** $\mathscr{L}$-classes and $\mathscr{R}$-classes of a $\mathscr{J}$-class intersect in stable semigroups

Stability implies Green's relations $\mathscr{J}$ and $\mathscr{D}$ coincide.

**Corollary A.2.5.** *Let $S$ be a stable semigroup. Then $\mathscr{J} = \mathscr{D}$. More precisely, the following are equivalent for $s$, $t \in S$:*

1. $s \mathscr{J} t$;
2. *there exists $r \in S$ such that $s \mathscr{L} r \mathscr{R} t$;*
3. *there exists $z \in S$ such that $s \mathscr{R} z \mathscr{L} t$;*
4. $s \mathscr{D} t$.

*Proof.* Suppose first $s \mathscr{J} t$. Then there exist $u, v \in S^I$ such that $usv = t$. Hence $us \mathscr{J} s$, $sv \mathscr{J} s$ and so $us \mathscr{L} s$, $sv \mathscr{R} s$ by stability. Because $\mathscr{R}$ is a left congruence, $s \mathscr{L} us \mathscr{R} usv = t$ and dually, because $\mathscr{L}$ is a right congruence, $s \mathscr{R} sv \mathscr{L} usv = t$ thereby establishing that 1 implies 2 and 3. Clearly 2 and 3 imply 4, and 4 implies 1. This completes the proof.  □

What does all this mean about the structure of a stable semigroup? In any semigroup, each $\mathscr{J}$-class $J$ is a disjoint union of $\mathscr{R}$-classes and also a disjoint union of $\mathscr{L}$-classes. If $L$ is an $\mathscr{L}$-class and $R$ is an $\mathscr{R}$-class of $J$, then either $L \cap R = \emptyset$ or $L \cap R$ is an $\mathscr{H}$-class. In a stable semigroup, this intersection is never empty. Indeed, suppose $x \in R$, $y \in L$. Then $x \mathscr{J} y$ implies there exists $r$ such that $x \mathscr{R} r \mathscr{L} y$. Thus $r \in L \cap R$, see Figure A.1.

The picture of a $\mathscr{J}$-class $J$ of a stable semigroup is something like an eggbox where the rows represent the $\mathscr{R}$-classes of $J$, the columns represent the $\mathscr{L}$-classes and the boxes represent the $\mathscr{H}$-classes, see Figure A.2.

As an example, consider the full transformation monoid $T_n$ of degree $n$. We view $T_n$ as acting on the right of $\{0, \ldots, n-1\}$. If $f \in T_n$, define ker $f$ to be the equivalence relation given by $(x, y) \in$ ker $f$ if $xf = yf$. Define the *rank* of $f$ by $\mathrm{rk}(f) = |\mathrm{Im}\, f|$.

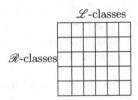

**Fig. A.2.** Eggbox picture of a $\mathscr{J}$-class of a stable semigroup

**Exercise A.2.6.** Let $f, g \in T_n$. Show:

1. $f \mathscr{L} g$ if and only if $\operatorname{Im} f = \operatorname{Im} g$;
2. $f \mathscr{R} g$ if and only if $\ker f = \ker g$;
3. $f \mathscr{J} g$ if and only if $\operatorname{rk}(f) = \operatorname{rk}(g)$.

## A.3 Green's Lemma and Maximal Subgroups

The following is known as Green's Lemma [108] and is used throughout this book, often without comment.

**Lemma A.3.1 (Green's Lemma).** *Let $S$ be a semigroup.*

1. *Let $s \mathscr{L} t$ and $u, v \in S^I$ be such that $us = t$, $vt = s$. Then $\varphi \colon R_s \to R_t$ defined by $x\varphi = ux$ and $\psi \colon R_t \to R_s$ defined by $y\psi = vy$ are inverse bijections. Moreover, if $x, x' \in R_s$ with $x \mathscr{H} x'$, then $x\varphi \mathscr{H} x'\varphi$.*
2. *Let $s \mathscr{R} t$ and $u, v \in S^I$ be such that $su = t$, $tv = s$. Then $\varphi \colon L_s \to L_t$ defined by $x\varphi = xu$ and $\psi \colon L_t \to L_s$ defined by $y\psi = yv$ are inverse bijections. Moreover, if $x, x' \in L_s$ with $x \mathscr{H} x'$, then $x\varphi \mathscr{H} x'\varphi$.*
3. *If $s \mathscr{L} r \mathscr{R} t$ and $u, v, w, z \in S^I$ are such that $us = r$, $vr = s$, $rw = t$ and $tz = r$, then $\varphi \colon H_s \to H_t$ given by $x\varphi = uxw$ and $\psi \colon H_t \to H_s$ given by $y\psi = vyz$ are inverse bijections.*

*Proof.* We prove only 1, as 2 is dual and 3 follows directly from 1 and 2. Let $x \in R_s$. Because $\mathscr{R}$ is a left congruence, $x \mathscr{R} s$ implies $ux \mathscr{R} us = t$. So $x\varphi = ux \in R_t$, establishing that $\varphi$ is well defined. Similarly, $\psi$ is well defined. Let us show that $\varphi\psi = 1_{R_s}$. If $x \in R_s$ then $x = sz$, for some $z \in S^I$. So

$$x\varphi\psi = ux\psi = vux = vusz = vtz = sz = x.$$

A similar verification shows $\psi\varphi = 1_{R_t}$.

Suppose now that $x, x' \in R_s$ and $x \mathscr{H} x'$. Then there exist $z, z' \in S^I$ such that $zx = x'$, and $z'x' = x$. So $(uzv)ux = uz(x\varphi\psi) = uzx = ux'$ and $(uz'v)ux' = uz'x' = ux$. Thus $x\varphi = ux \mathscr{L} ux' = x'\varphi$. But $x\varphi, x'\varphi \in R_t$, so in fact $x\varphi \mathscr{H} x'\varphi$, as required. $\square$

In a stable semigroup, Green's Lemma basically says left multiplication bijectively maps rows to rows and right multiplication bijectively maps columns to columns in the eggbox picture (provided one does not leave the $\mathscr{J}$-class). See Figure A.3.

As a consequence of Green's Lemma, each row (respectively column) has the same number of boxes and all boxes are the same size.

**Corollary A.3.2.** *All $\mathscr{H}$-classes of a $\mathscr{J}$-class $J$ of a finite semigroup have the same cardinality. Similarly, all $\mathscr{L}$-classes of $J$ have the same size and all $\mathscr{R}$-classes of $J$ have the same size. All $\mathscr{L}$-classes of $J$ contain the same number of $\mathscr{H}$-classes and all $\mathscr{R}$-classes of $J$ contain the same number of $\mathscr{H}$-classes.*

| | | $vt, sw$ | | | $s, vtz$ | |
|---|---|---|---|---|---|---|
| | | | | | | |
| | | $t, usw$ | | | $r$ | |
| | | | | | | |

**Fig. A.3.** Item 3 of Green's Lemma

The next lemma is classical.

**Lemma A.3.3.** *A non-empty semigroup is a group if and only if it is both left and right simple.*

*Proof.* Clearly, a group is both left and right simple. For the converse, suppose $S$ is left and right simple. Let $t \in S$. Choose $e, f \in S$ such that $te = t$ and $ft = t$. Then if $s \in S$, $s = zt$ for some $z \in S$. So $se = zte = zt = s$. Similarly $fs = s$. Thus $e$ is a right identity and $f$ is a left identity. We conclude $f = fe = e$ and so $S$ is a monoid with identity $e$.

If $s \in S$, then there exist $z, z' \in S$ such that $zs = e$ and $sz' = e$. Thus $z$ is a left inverse for $s$ and $z'$ is right inverse for $s$. Therefore, $z = zsz' = z'$ and so $s$ is invertible. Hence $S$ is a group.    $\square$

Notice if $G$ is a subgroup (with identity $e$) of a semigroup $S$, then $G \subseteq H_e$. We are now ready to prove a result of Green, which in particular identifies the maximal subgroups of a semigroup as the $\mathscr{H}$-classes of idempotents [68, 108, 171].

**Theorem A.3.4 (Green).** *Let $S$ be a semigroup and $s \in S$. Then the following are equivalent:*

1. $H_s$ *is a group;*
2. $H_s$ *contains an idempotent;*
3. $s^2 \in H_s$;
4. *there exist $x, y \in H_s$ such that $xy \in H_s$.*

*If $S$ is stable, these are in addition equivalent to:*

5. $s^2 \mathscr{J} s$;
6. *there exist $x, y \in H_s$ such that $xy \in J_s$;*
7. *there exist $x \in L_s$, $y \in R_s$ such that $xy \in J_s$;*
8. *for all $x \in L_s$, $y \in R_s$, one has $xy \in J_s$.*

*Proof.* Clearly, 1 implies 2 as the identity of a group is an idempotent. To deduce 3 from 2, let $s \mathscr{H} e$ with $e$ an idempotent. Then $se = s$ by Exercise A.1.8. Let $t \in S^I$ with $st = e$. Then $s = se = sst = s^2t$. Thus $s^2 \mathscr{R} s$. Similarly, $s^2 \mathscr{L} s$ and so $s^2 \in H_s$, as required.

For 3 implies 4, take $x = y = s$. To establish 4 implies 1 we use Lemma A.3.3. Suppose $x, y \in H_s$ are such that $xy \in H_s$. We first prove

$H_s$ is a subsemigroup. If $a, b \in H_s$, then we can find $u, w, z \in S^I$ such that $a = ux$, $xyw = x$ and $bz = y$. Then $abzw = ayw = uxyw = ux = a$, so $ab \; \mathscr{R} \; a$. Similarly, $ab \; \mathscr{L} \; b$. So $ab \in R_a \cap L_b = R_s \cap L_s = H_s$. Thus $H_s$ is a subsemigroup.

To prove $H_s$ is left simple, let $a \in H_s$. Because $H_s$ is a subsemigroup, $a^2 \in H_s$. As $aa = a^2$, Green's Lemma implies $\varphi \colon H_a \to H_{a^2}$ given by $x\varphi = ax$ is a bijection. As $H_a = H_s = H_{a^2}$, we obtain $aH_s = H_s\varphi = H_s$. A dual argument shows that $H_s$ is right simple. Therefore, $H_s$ is a group.

The implications 3 implies 5 and 4 implies 6, which in turn implies 7, are trivial. For 5 implies 3, observe that $s^2 \; \mathscr{J} \; s$ yields $s^2 \; \mathscr{L} \; s$ and $s^2 \; \mathscr{R} \; s$ by stability. For 7 implies 4, suppose $x \in L_s$, $y \in R_s$ and $xy \in J_s$. By stability, $xy \; \mathscr{R} \; x$ so $x = xyt$ for some $t \in S^I$. Also $xy \; \mathscr{L} \; y$ by stability, and hence $yt \; \mathscr{L} \; xyt = x$ as $\mathscr{L}$ is a right congruence. By stability $yt \; \mathscr{R} \; y$ so $yt \in R_y \cap L_x = H_s$. Because $s \; \mathscr{L} \; x$, we can write $s = ux$ with $u \in S^I$. Then $s(yt) = uxyt = ux = s$. Therefore, $s(yt) \in H_s$ yielding 4.

Clearly 8 implies 7. We prove 3 implies 8. Indeed, if $x \in L_s$, $y \in R_s$, then there exist $u, v \in S^I$ with $s = ux$, $s = yv$. Then $uxyv = s^2 \in H_s$ and so $s \leq_{\mathscr{J}} xy \leq_{\mathscr{J}} x \leq_{\mathscr{J}} s$. Therefore, $xy \in J_s$, as required. $\qquad\square$

**Definition A.3.5 (Maximal subgroup).** *If $e \in S$ is an idempotent, then $H_e$ is called the maximal subgroup of $S$ at $e$. It is the largest subgroup of $S$ with identity $e$.*

**Exercise A.3.6.** Show that if $e$ is an idempotent of a semigroup $S$, then $H_e$ is the group of units of the monoid $eSe$.

The next theorem is one of the principal results in the structure theory of stable semigroups [68, 171].

**Theorem A.3.7.** *Let $J$ be a $\mathscr{J}$-class of a stable semigroup $S$. Then the following are equivalent:*

1. *$J$ contains a subgroup;*
2. *$J$ contains an idempotent;*
3. *$J$ contains a regular element;*
4. *All elements of $J$ are regular;*
5. *The $\mathscr{R}$-class of each element of $J$ contains an idempotent;*
6. *The $\mathscr{L}$-class of each element of $J$ contains an idempotent;*
7. *$J^2 \cap J \neq \emptyset$.*

*Proof.* It is clear that 1 implies 2, which in turn implies 3 because idempotents are regular. To see that 3 implies 4, we first show that if $s \in J$ is regular, then $L_s$ consists entirely of regular elements. Choose $t \in S$ such that $s = sts$. Let $y \in L_s$. There exist $u, v \in S^I$ such that $uy = s$, $vs = y$. Hence $ytuy = yts = vsts = vs = y$. As $tu \in S$, we conclude $y$ is regular. A similar argument establishes each element of $R_s$ is regular.

Now suppose $y \ \mathscr{J} \ s$ with $s$ regular. Then there exists $r \in S$ such that $s \ \mathscr{L} \ r \ \mathscr{R} \ y$. From $s \ \mathscr{L} \ r$, we conclude $r$ is regular. Hence $r \ \mathscr{R} \ y$ implies $y$ is regular.

For 4 implies 5, let $s \in J$ be regular and choose $t \in S$ with $sts = s$. Then $stst = st$ and so $st$ is an idempotent. Clearly $st \ \mathscr{R} \ s$, as $(st)s = s$. A dual argument shows that 4 implies 6.

The implications 5 implies 7 and 6 implies 7 are trivial as any idempotent $e \in J$ belongs to $J^2 \cap J$. For 7 implies 1, let $x, y \in J$ with $xy \in J$. Theorem A.3.4 shows that the $\mathscr{H}$-class $L_x \cap R_y$ is a group.    □

**Definition A.3.8 (Regular $\mathscr{J}$-class).** *A $\mathscr{J}$-class satisfying the equivalent conditions of Theorem A.3.7 is called a regular $\mathscr{J}$-class. A non-regular $\mathscr{J}$-class is called a null $\mathscr{J}$-class.*

For example, every element of $T_n$ is regular because there is an idempotent of every possible rank. The next theorem shows that all maximal subgroups in a regular $\mathscr{J}$-class of a stable semigroup are isomorphic.

**Theorem A.3.9.** *Let $S$ be a semigroup and let $e, f \in S$ be $\mathscr{D}$-equivalent idempotents. Then the maximal subgroups $H_e$ and $H_f$ are isomorphic. More precisely, if $e \ \mathscr{R} \ x \ \mathscr{L} \ f$ and $x'$ is an inverse to $x$ with $xx' = e$, $x'x = f$ as per Proposition A.1.15, then $\varphi \colon H_e \to H_f$ given by $s\varphi = x'sx$ is an isomorphism of groups with inverse $\psi \colon H_f \to H_e$ given by $t\psi = xtx'$.*

*Proof.* Green's Lemma implies that $\varphi$ and $\psi$ are well-defined inverse bijections. It therefore suffices to verify that $\varphi$ is a homomorphism. Indeed, $s\varphi s'\varphi = x'sxx's'x = x'ses'x = x'ss'x = (ss')\varphi$ because $s \ \mathscr{H} \ e$.    □

**Corollary A.3.10.** *The maximal subgroups at $\mathscr{J}$-equivalent idempotents of a stable semigroup are isomorphic.*

### A.3.1 The Schützenberger group

For a null $\mathscr{J}$-class, the role of the maximal subgroup of a regular $\mathscr{J}$-class is played by a "phantom" group, called the *Schützenberger group*. This group is the subject of [68, Section 2.4] and [171, Chapter 7, Prop. 2.8], but we quickly review some facts that do not appear in these references in the precise form that we need them.

For a subset $X$ of a semigroup $S$, set $\mathrm{Stab}(X) = \{s \in S^I \mid Xs \subseteq X\}$. Let $H$ be an $\mathscr{H}$-class of a semigroup $S$. Then $\mathrm{Stab}(H)$ acts on the right of $H$. Denote by $\Gamma_R(H)$ the quotient of $\mathrm{Stab}(H)$ by the kernel of the action. Recall that a right permutation group $(X, G)$ is called *regular* if $xg = x$ implies $g = 1$, for $x \in X$ and $g \in G$. In this case one also says $G$ acts *freely* on $X$.

**Theorem A.3.11.** *Let $H$ be an $\mathscr{H}$-class of $S$ contained in the $\mathscr{L}$-class $L$. Then the following hold:*

1. *If $H'$ is an $\mathscr{H}$-class of $L$, then $\mathrm{Stab}(H) = \mathrm{Stab}(H')$;*
2. *$\mathrm{Stab}(H) \leq \mathrm{Stab}(L)$ and hence $\mathrm{Stab}(H)$ acts on the right of $L$;*
3. *The quotient of $\mathrm{Stab}(H)$ by the kernel of its action on $L$ is $\Gamma_R(H)$; consequently, if $H$ and $H'$ are $\mathscr{L}$-equivalent $\mathscr{H}$-classes, then $\Gamma_R(H) \cong \Gamma_R(H')$;*
4. *$(L, \Gamma_R(H))$ is a regular permutation group and the orbits are precisely the $\mathscr{H}$-classes of $L$ and in particular $\Gamma_R(H)$ acts transitively on $H$;*
5. *If $S$ is a stable semigroup, then the action of $\Gamma_R(H)$ is by endomorphisms of the left action of $S$ on $L$ by partial transformations defined by*

$$ s \cdot x = \begin{cases} sx & sx \in L \\ undefined & else \end{cases} $$

*for $s \in S$ and $x \in L$.*

*Proof.* Suppose $s \in \mathrm{Stab}(H)$ and let $h_0 \in H$ be fixed. Then $h_0 s \in H$ and so $h_0 st = h_0$ for some $t \in S^I$. According to Green's Lemma, right translation by $s$ and by $t$ induce mutually inverse permutations of $L$ preserving $\mathscr{H}$. In particular, $t \in \mathrm{Stab}(H)$ and $\mathrm{Stab}(H)$ is contained in the stabilizer of every other $\mathscr{H}$-class of $L$. This implies immediately 1 and 2. The quotient of $\mathrm{Stab}(H)$ by its action on $L$ yields a permutation group as we saw above that $s \in \mathrm{Stab}(H)$ acts as a permutation with inverse $t \in \mathrm{Stab}(H)$. Let us verify that this permutation group is regular. Suppose $x \in L$ and $xs = x$ with $s \in \mathrm{Stab}(H)$. If $y \in L$, then $y = ux$ some $u \in S^I$ and hence $ys = uxs = ux = y$. This establishes regularity. Moreover, if $s, s' \in \mathrm{Stab}(H)$, then $s$ and $s'$ induce the same permutation of $L$ if and only if $h_0 s = h_0 s'$ and so two elements act the same on $L$ if and only if they act the same on $H$. Thus the quotient of $\mathrm{Stab}(H)$ by the kernel of its action on $L$ is precisely $\Gamma_R(H)$.

As we already saw that the action of each element of $\mathrm{Stab}(H)$ preserves $\mathscr{H}$, each orbit is clearly contained in an $\mathscr{H}$-classes of $L$. Suppose $\ell, \ell' \in L$ are $\mathscr{H}$-equivalent. Then we can find $s, s' \in S^I$ so that $\ell s = \ell'$ and $\ell' s' = \ell$. Green's Lemma implies that $s, s' \in \mathrm{Stab}(H)$ and so the orbits are precisely the $\mathscr{H}$-classes.

For the final statement, we ask the reader to verify that we have indeed defined an action. Notice if $g \in \mathrm{Stab}(H)$ and $x \in L$, then $x \mathscr{L} xg$ and so $sx \mathscr{L} sxg$ for any $s \in S$. Thus $s \cdot x$ is defined if and only if $s \cdot (xg)$ is defined in which case $(s \cdot x)g = s \cdot (xg)$ by associativity. This completes the proof.    $\square$

The group $\Gamma_R(H)$ is called the *right Schützenberger group* of $H$. Because $(H, \Gamma_R(H))$ is a transitive regular permutation group, it follows easily that $H$ and $\Gamma_R(H)$ have the same cardinality. One can define dually the *left Schützenberger group* $\Gamma_L(H)$ and the dual of Theorem A.3.11 holds. Now we wish to show that $\Gamma_R(H) \cong \Gamma_L(H)$. Because the left action of $\Gamma_L(H)$ on $H$ clearly commutes with the right action of $\Gamma_R(H)$, this is an immediate consequence of the following lemma, whose proof we leave as an exercise.

**Lemma A.3.12.** *Let $G'$ and $G$ be groups with transitive regular permutation actions on the left and right of a set $X$, respectively, which commute. Let $x_0 \in X$ be a fixed element. Then $G'$ and $G$ are anti-isomorphic via the map $\gamma$ sending $g \in G$ to the unique element $g\gamma \in G'$ such that $g\gamma x_0 = x_0 g$.*

**Exercise A.3.13.** Prove Lemma A.3.12.

**Corollary A.3.14.** *Let $H$ be an $\mathcal{H}$-class of a $\mathcal{D}$-class $D$ of a semigroup $S$. Then $\Gamma_R(H) \cong \Gamma_L(H)$ and if $H'$ is any $\mathcal{H}$-class of $D$, then $\Gamma_R(H) \cong \Gamma_R(H')$.*

*Proof.* The first assertion is immediate from Lemma A.3.12 because (inner) left and right translations commute with each other. The second assertion follows from the first, together with Theorem A.3.11(3) and its dual.    □

So if $J$ is a $\mathcal{J}$-class of a stable semigroup, then there is a well-defined (up to isomorphism) Schützenberger group of $J$. It remains to show that if $J$ is a regular $\mathcal{J}$-class, then the Schützenberger group is the maximal subgroup.

**Proposition A.3.15.** *Let $H$ be a maximal subgroup of $S$. Then $\Gamma_R(H) \cong H$.*

*Proof.* Let $e$ be the identity of $H$. Clearly, $H \leq \mathrm{Stab}(H)$ and if $h \in H$, then in $(H, \Gamma_R(H))$ we have $eh = h$. It follows that $H$ can be identified with its image in $\Gamma_R(H)$. If $s \in \Gamma_R(H)$ and $es = h$ with $h \in H$, then $es = h = eh$ and so $s = h$ by regularity. Thus $H \cong \Gamma_R(H)$, as required.    □

## A.4 Rees's Theorem

Rees's Theorem [258], characterizing stable 0-simple semigroups, can be considered the first major theorem in semigroup theory. See also [350] for the special case of a finite simple semigroup.

**Definition A.4.1 (0-simple).** *A semigroup $S$ with zero is called 0-simple if it has no non-zero proper ideals and also $S^2 \neq 0$. A semigroup $S$ for which $S^2 = 0$ is called null.*

The following proposition is straightforward so we omit the proof.

**Proposition A.4.2.** *Let $s \in S$. Then $S^I s S^I \setminus J_s$ is an ideal of $S^I s S^I$ unless it is empty.*

*Remark A.4.3.* If $K$ is the minimal ideal of a semigroup $S$ and $s \in K$, then $S^I s S^I \subseteq K$ and hence $S^I s S^I = K$, for all $s \in K$. In particular $K$ is a $\mathcal{J}$-class and $S^I s S^I \setminus K = \emptyset$. In fact, the only time $S^I s S^I \setminus J_s = \emptyset$ is when $J_s$ is the minimal ideal of $S$.

The class of 0-simple semigroups admits the following alternative characterization.

**Proposition A.4.4.** *A semigroup $S$ is $0$-simple if and only if $SaS = S$, for all $0 \neq a \in S$.*

*Proof.* If $S$ is $0$-simple, then $S^2 \neq \{0\}$. Let $a \in S \setminus \{0\}$. Then $S^I a S^I$ is an ideal containing $a$, so $S^I a S^I = S$ by $0$-simplicity. Also $S^2$ is a non-zero ideal, so $S^2 = S$. Therefore, $S = S^2 = S^3$ and hence

$$0 \neq SSS = (S^I a S^I)(S^I a S^I)(S^I a S^I) \subseteq SaS.$$

Thus $SaS$ is a non-zero ideal of $S$, whence $SaS = S$.

Next assume $SaS = S$, for all non-zero elements $a \in S$. In particular, if $a \neq 0$, then $S = SaS \subseteq S^2$ and so $S^2 \neq 0$. Also if $I \subseteq S$ is a non-zero ideal and $0 \neq a \in I$, then $S = SaS \subseteq SIS \subseteq I$ and thus $I = S$. We conclude $S$ is $0$-simple. $\qquad \square$

As a consequence, $S \setminus \{0\}$ is a $\mathscr{J}$-class in a $0$-simple semigroup $S$. Part of the importance of $0$-simple semigroups stems from the following fact.

**Proposition A.4.5.** *The minimal ideal of a semigroup is simple. A regular $0$-minimal ideal of a semigroup with zero is $0$-simple.*

*Proof.* Let $I$ be the minimal ideal of a semigroup $S$. Suppose $J$ is an ideal of $I$. Then $IJI$ is an ideal of $S$, so by minimality $I \subseteq IJI \subseteq J$. We conclude $I = J$. Next assume $I$ is a regular $0$-minimal ideal of $S$. Then $I^2 \neq 0$ by regularity. Let $0 \neq J \subseteq I$ be an ideal. Any ideal in a regular semigroup is regular and so $0 \neq J = J^3 \subseteq IJI$. Because $IJI$ is an ideal of $S$, by $0$-minimality $I \subseteq IJI \subseteq J$. This completes the proof. $\qquad \square$

The semigroups appearing in the next definition are classically known as the principal factors [68], although we shall not make extensive use of this terminology.

**Definition A.4.6 (Principal factor).** *If $J = J_s$ is a $\mathscr{J}$-class, the principal factor associated to $J$ is the semigroup*

$$J^0 = \begin{cases} S^I s S^I / (S^I s S^I \setminus J_s) & \text{if } J_s \text{ is not the minimal ideal} \\ J_s \cup \{0\} & \text{else.} \end{cases}$$

The principal factor $J^0$ can alternatively be described as the semigroup with underlying set $J \cup \{0\}$ and with multiplication given by

$$x \cdot y = \begin{cases} xy & \text{if } xy \in J \\ 0 & \text{else.} \end{cases}$$

The next proposition explains the terminology null $\mathscr{J}$-class and shows that principal factors associated to regular $\mathscr{J}$-classes are $0$-simple.

**Proposition A.4.7.** *Let $S$ be a stable semigroup and let $J$ be a $\mathscr{J}$-class of $S$. Then $J$ is a regular $\mathscr{J}$-class if and only if $J^0$ is 0-simple. If $J$ is a null $\mathscr{J}$-class, then $J^0$ is null.*

*Proof.* Because $(J^0)^2 \neq 0$ is equivalent to $J^2 \cap J \neq \emptyset$, Theorem A.3.7 shows that $J^0$ is null if and only if $J$ is a null $\mathscr{J}$-class. In particular, if $J^0$ is 0-simple, then $J$ is regular.

Now suppose $J$ is regular. To show $J^0$ is 0-simple, it suffices by Proposition A.4.4 to show that, for all $x \in J$, $J^0 x J^0 = J^0$. Let $t \in J$. Then we can find $u, v \in S^I$ with $uxv = t$. Because $J$ is regular, $x$ is regular, so there exists $s \in S$ such that $x = xsx$. Hence $t = uxv = uxsxv = (uxs)x(sxv)$ and $uxs$, $sxv \in S$. Now $t \leq_{\mathscr{J}} uxs, sxv \leq_{\mathscr{J}} x$. So $uxs$, $sxv \in J$. Thus $t \in J^0 x J^0$, as required. $\qquad\square$

The previous proposition shows that "locally" (meaning in a $\mathscr{J}$-class) a semigroup is either null or 0-simple.

**Corollary A.4.8.** *A stable 0-simple semigroup is regular.*

*Proof.* Indeed, $S = J^0$ where $J = S \setminus \{0\}$ is a $\mathscr{J}$-class of $S$, so the previous proposition applies to show that $S$ is regular. $\qquad\square$

Our next goal is to prove Rees's Theorem [68, 171, 258], which describes stable 0-simple semigroups up to isomorphism. The key notion is that of a Rees matrix semigroup.

**Definition A.4.9 (Rees matrix).** *A Rees matrix is a map $C: B \times A \to G^0$ where $A$ and $B$ are non-empty sets, $G$ is a group and $G^0 = G \cup \{0\}$ with 0 an adjoined zero. We adopt the symmetric notation $bCa$ for the value of $C$ on $(b, a)$.*

Denote by $E_{ab}$, for $a \in A$ and $b \in B$, the $A \times B$ elementary matrix unit with 1 in position $(a, b)$ and 0 elsewhere, where 1 is the identity of $G$.

**Definition A.4.10 (Rees Matrix Semigroup).** *Let $G$ be a group and let $C: B \times A \to G^0$ be a Rees matrix. The Rees matrix semigroup with sandwich matrix $C$ is the set*

$$\mathscr{M}^0(G, A, B, C) = \{gE_{ab} \mid g \in G^0, \ a \in A, \ b \in B\}$$

*with multiplication, for $X, Y \in \mathscr{M}^0(G, A, B, C)$, given by $X \diamond Y = XCY$.*

The underlying set of $\mathscr{M}^0(G, A, B, C)$ consists of all $A \times B$ matrices over $G^0$ with at most one non-zero entry. Associativity follows from the associativity of matrix multiplication. Notice that

$$g_1 E_{a_1 b_1} \diamond g_2 E_{a_2 b_2} = (g_1(b_1 C a_2)g_2)E_{a_1 b_2}$$

and so $\mathscr{M}^0(G, A, B, C)$ is indeed closed under the multiplication.

Oftentimes it is convenient to identify $\mathscr{M}^0(G, A, B, C)$ with $(A \times G \times B) \cup \{0\}$ via the correspondence sending $gE_{ab}$, with $g \in G$, to $(a, g, b)$ and 0 to 0. Under this bijection, the multiplication rule becomes

$$(a_1, g_1, b_1)(a_2, g_2, b_2) = \begin{cases} (a_1, g_1(b_1 C a_2)g_2, b_2) & b_1 C a_2 \neq 0 \\ 0 & \text{else.} \end{cases} \tag{A.1}$$

Sometimes the matrix $C$ is called the *structure matrix* or the *sandwich matrix* of $\mathscr{M}^0(G, A, B, C)$. If $C \colon B \times A \to G$, then $\mathscr{M}^0(G, A, B, C) \setminus \{0\}$ is a subsemigroup, denoted by $\mathscr{M}(G, A, B, C)$.

**Definition A.4.11 (Regular Rees matrix).** *A Rees matrix is called regular if it has no zero rows or columns.*

**Proposition A.4.12.** *Let $C \colon B \times A \to G^0$ be a Rees matrix. Then the following are equivalent for $S = \mathscr{M}^0(G, A, B, C)$:*

1. *$S$ is regular;*
2. *$C$ is regular;*
3. *$S$ is 0-simple.*

*Moreover, if 1–3 hold, then:*

(a) *$S/\mathscr{R} = \{a \times G \times B \mid a \in A\}$;*
(b) *$S/\mathscr{L} = \{A \times G \times b \mid b \in B\}$;*
(c) *$S/\mathscr{H} = \{a \times G \times b \mid a \in A, \ b \in B\}$;*
(d) *The $\mathscr{H}$-class $H = a \times G \times b$ is a group if and only if $bCa \neq 0$; if $H$ is a group, its identity is $(a, (bCa)^{-1}, b)$ and it is isomorphic to $G$ via*

$$g \longmapsto (a, (bCa)^{-1}g, b).$$

*Proof.* Assume $S$ is regular and suppose $a \in A$, $b \in B$. Let $s = (a, 1, b)$ and choose $t = (a', g', b')$ such that $sts = s$. Then

$$(a, 1, b)(a', g', b')(a, 1, b) = s \neq 0$$

and so $bCa' \neq 0 \neq b'Ca$. We conclude $C$ is regular.

Next assume $C$ is regular. By Proposition A.4.4, to prove $S$ is 0-simple, we need to show that $S(a, g, b)S = S$ for all $(a, g, b) \in S \setminus 0$. So let $(a', g', b') \in S \setminus 0$ and choose $a'' \in A$, $b'' \in B$ such that $bCa'' \neq 0 \neq b''Ca$. Then

$$(a', g'(b''Ca)^{-1}, b'')(a, g, b)(a'', (bCa'')^{-1}g^{-1}, b') = (a', g', b')$$

and hence $S(a, g, b)S = S$. Thus $S$ is 0-simple.

Suppose $S$ is 0-simple. We show that $C$ is regular. Let $a \in A$ and $b \in B$. Because $S$ is 0-simple, $S(a, 1, b)S = S$ and so

$$(a, 1, b) = (a', g', b')(a, 1, b)(a'', g'', b'')$$

for appropriate choices of $a'$, $a''$, $b'$, $b''$, $g'$ and $g''$. But then $b'Ca \neq 0 \neq bCa''$, and so $C$ is regular.

Suppose now $C$ is regular; we show $S$ is regular. Let $(a, g, b) \in S$ and choose $a' \in A$, $b' \in B$ with $bCa' \neq 0 \neq b'Ca$. Then

$$(a, g, b)(a', (bCa')^{-1}g^{-1}(b'Ca)^{-1}, b')(a, g, b) = (a, g, b)$$

and so $\mathcal{M}^0(G, A, B, C)$ is regular, as required.

We now turn to (a)–(d). If $s = (a, g, b)$, then it follows directly from (A.1) that $R_s \subseteq a \times G \times B$. For the converse, suppose $(a, g', b') \in a \times G \times B$. Choose $a''$ such that $bCa'' \neq 0$. Then

$$(a, g, b)(a'', (bCa'')^{-1}g^{-1}g', b') = (a, g', b').$$

Hence $R_s = a \times G \times B$. The descriptions of $S/\mathcal{L}$ and $S/\mathcal{H}$ are handled similarly.

Next we turn to the case of when $H = a \times G \times b$ is a group. By Theorem A.3.4, $H$ is a group if and only if it contains an idempotent. But

$$(a, g, b)(a, g, b) = \begin{cases} (a, g(bCa)g, b) & bCa \neq 0 \\ 0 & \text{else.} \end{cases}$$

So $H$ contains an idempotent if and only if $bCa \neq 0$, in which case the idempotent is $(a, (bCa)^{-1}, b)$. Clearly, $g \mapsto (a, (bCa)^{-1}g, b)$ is a bijection from $G$ to $H$. It is a homomorphism as the computation

$$(a, (bCa)^{-1}g_1, b)(a, (bCa)^{-1}g_2, b) = (a, (bCa)^{-1}g_1g_2, b)$$

shows. □

**Corollary A.4.13.** *Let $S$ be a regular Rees matrix semigroup. Then $S$ is stable.*

**Exercise A.4.14.** Prove Corollary A.4.13.

We now prove Rees's Theorem [68, 171, 258].

**Theorem A.4.15 (Rees).** *Let $S$ be a stable semigroup. Then $S$ is 0-simple if and only if there exist a group $G$, sets $A$, $B$ and a regular Rees matrix $C \colon B \times A \to G^0$ such that $S \cong \mathcal{M}^0(G, A, B, C)$. Similarly, a stable semigroup $S$ is simple if and only if $S \cong \mathcal{M}(G, A, B, C)$ where $C \colon B \times A \to G$.*

*Proof.* We just handle the 0-simple case, as the simple case is a consequence. Corollary A.4.13 provides one direction. For the other, Corollary A.4.8 tells us $S$ is regular. Let $J$ be the non-zero $\mathcal{J}$-class of $S$ and let $e \in J$ be an idempotent (guaranteed by Theorem A.3.7). Let $G = H_e$, $A = S/\mathcal{R}$ and $B = S/\mathcal{L}$. Theorem A.3.4 says that $G$ is a group.

| e, g |  |  |  | $\ell_b$ |  |  |
|------|--|--|--|---------|--|--|
|      |  |  |  |         |  |  |
| $r_a$ |  |  |  | $r_a g \ell_b$ |  |  |
|      |  |  |  |         |  |  |

**Fig. A.4.** Rees coordinates

Choose for each $a \in A$ and $b \in B$, $r_a \in a \cap L_e$ and $\ell_b \in b \cap R_e$. Notice $e\ell_b = \ell_b$ and $r_a e = r_a$. Green's Lemma tells us $x \mapsto r_a x \ell_b$ is a bijection from $G = H_e$ to $a \cap b$. Hence each non-zero element of $S$ can be written uniquely in the form $r_a x \ell_b$ with $x \in G$. See Figure A.4.

To motivate the definition of the sandwich matrix $C$, notice for $g, g' \in G$

$$(r_a g \ell_b)(r_{a'} g' \ell_{b'}) = r_a (g \ell_b r_{a'} g') \ell_{b'}. \tag{A.2}$$

Now if $\ell_b r_{a'} \neq 0$, then $\ell_b r_{a'} \in J$. Stability then yields $\ell_b r_{a'} \mathcal{R} \ell_b \mathcal{R} e$ and $\ell_b r_{a'} \mathcal{L} r_{a'} \mathcal{L} e$, i.e., $\ell_b r_{a'} \in H_e = G$. In any event $\ell_b r_{a'} \in G^0$.

From (A.2) and the ensuing discussion, it is clear we should define $C : B \times A \to G^0$ by $bCa = \ell_b r_a \in G^0$. We then define an isomorphism $\psi : \mathcal{M}^0(G, A, B, C) \to S$ by $0 \mapsto 0$ and $(a, g, b) \mapsto r_a g \ell_b$. The discussion above shows that $\psi$ is a bijection. To see that it is a homomorphism, we compute

$$\big((a, g, b)(a', g', b')\big)\psi = r_a (g \ell_b r_{a'} g') \ell_{b'}.$$

A comparison with (A.2) yields $\psi$ is a homomorphism. Proposition A.4.12 now implies $C$ is a regular Rees matrix, completing the proof of the theorem.    $\square$

Let us show how Rees's Theorem determines the structure of stable left simple semigroups.

**Lemma A.4.16.** *Let $S$ be a stable simple semigroup whose idempotents form a subsemigroup. Then $S \cong \mathcal{M}(G, A, B, C)$ where $C$ is a $B \times A$ matrix all of whose entries are the identity of $G$.*

*Proof.* Each $\mathcal{H}$-class of a stable simple semigroup must be a group, so we can choose $r_a$ and $\ell_b$ to be idempotents in the proof of Rees's Theorem. Then $bCa = \ell_b r_a$ is an idempotent of $H_e$, and hence is $e$. This proves the lemma.    $\square$

**Corollary A.4.17.** *Let $S$ be a stable semigroup. Then $S$ is left simple if and only if $S \cong L \times G$ with $L$ a left zero semigroup and $G$ a group. Dually, $S$ is right simple if and only if $S \cong G \times R$ with $G$ a group and $R$ a right zero semigroup.*

*Proof.* Trivially, if $L$ is a left zero semigroup and $G$ is a group, then $L \times G$ is left simple. For the converse, we first show the idempotents of $S$ form a subsemigroup. Indeed, if $e, f \in S$ are idempotents, then $e \mathcal{L} f$ and so

$ef = e$, $fe = f$ by Exercise A.1.8. So Rees's Theorem, Lemma A.4.16 and Proposition A.4.12 show that $S \cong \mathcal{M}(G, A, \{b\}, C)$ where $C$ is a matrix all of whose entries are the identity of $G$. Let $L$ have underlying set $A$ with the left zero multiplication. Then $(a, g, b) \mapsto (a, g)$ gives an isomorphism $S \to L \times G$. The right simple case is dual.    $\square$

**Definition A.4.18 (Rees coordinatization).** *If $S$ is a stable semigroup and $J$ is a regular $\mathcal{J}$-class, then a Rees coordinatization of $J$ is an isomorphism $\psi\colon J^0 \to \mathcal{M}^0(G, A, B, C)$.*

Optimal Rees coordinatizations are studied in Section 4.13. This brings to an end our brief introduction to the Green-Rees local structure theory for stable semigroups.

*Remark A.4.19.* A simple compact semigroup is always isomorphic to a Rees matrix semigroup $\mathcal{M}(G, A, B, C)$ where $G$ is a compact group, $A, B$ are compact sets and $C\colon B \times A \to G$ is continuous [135]. The situation for 0-simple semigroups is more complicated as 0 might not be an isolated point.

# B

## Tables on Preservation of Sups and Infs

The reader is referred to Section 1.1.2 for nomenclature on types of maps between posets. In the following table, $\mathbf{V}$ is a pseudovariety of semigroups.

**Table B.1.** Preservation properties of products on **PV**

| Operator | Order Properties | Type of Operator |
|---|---|---|
| $\mathbf{V} \cap (-)$ | continuous, $\mathbf{inf}_T$ | $\mathbf{GMC(PV)}^-$ |
| $\mathbf{V} \vee (-)$ | $\mathbf{sup}_B$ | $\mathbf{GMC(PV)}^+$, $\mathbf{GMC(PV)}_1^\rho$ |
| $\mathbf{V} * (-)$ | continuous | $\mathbf{GMC(PV)}^+$ |
| $\mathbf{V} ** (-)$ | continuous | $\mathbf{GMC(PV)}^+$ |
| $\mathbf{V} \text{ⓜ} (-)$ | continuous | $\mathbf{GMC(PV)}^+$ |
| $(-) \cap \mathbf{V}$ | continuous, $\mathbf{inf}_T$ | $\mathbf{GMC(PV)}^-$ |
| $(-) \vee \mathbf{V}$ | $\mathbf{sup}_B$ | $\mathbf{GMC(PV)}^+$, $\mathbf{GMC(PV)}_1^\rho$ |
| $(-) * \mathbf{V}$ | $\mathbf{sup}_B$ | $\mathbf{GMC(PV)}_{L1}^\rho$ |
| $(-) ** \mathbf{V}$ | $\mathbf{sup}_B$ | $\mathbf{GMC(PV)}_{L1}^\rho$ |
| $(-) \text{ⓜ} \mathbf{V}$ | continuous, $\mathbf{inf}$ | $\mathbf{GMC(PV)}_{L1}^\rho$ |

In the next table, $\mathsf{V}$ denotes a pseudovariety of relational morphisms.

**Table B.2.** Preservation properties of products on **PVRM**

| Operator | Order Properties |
|---|---|
| $\mathsf{V} \cap (-)$ | continuous, $\mathbf{inf}_T$ |
| $\mathsf{V} \vee (-)$ | $\mathbf{sup}_B$ |
| $\mathsf{V} \odot (-)$ | continuous |
| $(-) \cap \mathsf{V}$ | continuous, $\mathbf{inf}_T$ |
| $(-) \vee \mathsf{V}$ | $\mathbf{sup}_B$ |
| $(-) \odot \mathsf{V}$ | continuous, $\wedge$-map |

In the following table, V represents a continuously closed class.

**Table B.3.** Preservation properties of products on **CC**

| OPERATOR | ORDER PROPERTIES |
|---|---|
| $V \cap (-)$ | continuous, $\inf_T$ |
| $V \vee (-)$ | $\sup_B$ |
| $V \odot (-)$ | continuous |
| $(-) \cap V$ | continuous, $\inf_T$ |
| $(-) \vee V$ | $\sup_B$ |
| $(-) \odot V$ | continuous |

In our final table, $\alpha$ denotes a continuous operator. Because all infs and sups of **GMC(PV)** coincide with infs and sups taken in **Cnt(PV)**, the corresponding table for **GMC(PV)** is identical. We use here ∘ for composition of operators.

**Table B.4.** Preservation properties of products on **Cnt(PV)**

| OPERATOR | ORDER PROPERTIES |
|---|---|
| $\alpha \wedge (-)$ | continuous, $\inf_T$ |
| $\alpha \vee (-)$ | $\sup_B$ |
| $\alpha \circ (-)$ | continuous |
| $(-) \wedge \alpha$ | continuous, $\inf_T$ |
| $(-) \vee \alpha$ | $\sup_B$ |
| $(-) \circ \alpha$ | $\sup$, $\wedge$-map |

**Exercise B.0.1.** Prove all assertions in the tables and find counterexamples to properties that are not claimed to be preserved.

# List of Problems

Some, but not all, of these problems appear in the text as questions.

**Problem 1.** What is $\mathbf{V} * \mathbf{1}$? It is easy to see that if $S$ is a semigroup in $\mathbf{V} * \mathbf{1}$, then the quotient of $S$ by the kernel of its action on the right of itself belongs to $\mathbf{V}$. The converse holds if $\mathbf{gV}$ is definable by pseudoidentities over strongly connected graphs. Is the converse always valid?

**Problem 2.** What is $\mathbf{V} ** \mathbf{1}$? Let $S$ be a semigroup and define a congruence on $S$ by $s \equiv s'$ if, for all $s_L, s_R \in S$, one has $s_L s s_R = s_L s' s_R$. It is easy to see that if $S \in \mathbf{V} ** \mathbf{1}$, then $S/\equiv \in \mathbf{V}$. The converse holds if $\mathbf{gV}$ is definable by pseudoidentities over strongly connected graphs. Does it hold in general?

**Problem 3.** Give an element of $\mathbf{PVRM}^+$ analogous to $\mathbf{V}_D$ or $\mathbf{V}_K$ yielding $\Diamond(-)$ under $\mathsf{q}$ and find a basis of pseudoidentities for it.

**Problem 4 (Tilson Question).** Is $\mathbf{V}_D = \min(\mathbf{V} * (-))$?

**Problem 5.** Is it true that a pseudovariety of semigroups $\mathbf{V}$ is decidable if and only if $\mathbf{gV}$ is decidable?

**Problem 6.** Find a natural axiom to add to the definition of a pseudovariety of semigroupoids to ensure definability by path pseudoidentities.

**Problem 7 (Tilson Question Version 2).** Is $\mathbf{V}_K = \min(\mathbf{V} ** (-))$?

**Problem 8.** Is it true that if $\mathbf{H}, \mathbf{K}$ are decidable pseudovarieties of groups, then $\mathbf{H} \vee \mathbf{K}$ is decidable?

**Problem 9.** Find a natural lattice in $\mathbf{CC}^+$ mapping to $\mathbf{GMC}^\varrho_{\mathbb{L}1}(\mathbf{PV})$ under $\mathsf{q}$.

**Problem 10.** Compute $\min(\alpha)$ for any operator $\alpha$ that is not a constant operator, or of the form $\mathbf{V} \cap (-)$.

**Problem 11.** Suppose K and L are equational continuously closed classes. Is K ⊙ L equational?

**Problem 12.** Suppose K and L are Birkhoff continuously closed classes. Is K ⊙ L Birkhoff?

**Problem 13.** Give an example showing that meets are not pointwise in the lattice **GMC(PV)**$^+$.

**Problem 14.** Find a projective basis for $\mathbf{V}_D$.

**Problem 15 (Almeida-Weil Basis Question).** Find an example of pseudovarieties **V** and **W** so that pseudoidentities in Theorem 3.7.15 (i.e., [27, Thms. 5.2 and 5.3]) do *not* define $\mathbf{V} * \mathbf{W}$. For such an example, give an explicit example of a member of the basis from Theorem 3.8.18 that is not a consequence of the pseudoidentities in Theorem 3.7.15.

**Problem 16.** Is it decidable whether a finite semigroup is projective?

**Problem 17 (Complexity Problem).** Is the group complexity function $c$ computable? More precisely, is there a Turing machine that given a finite semigroup $S$ by its multiplication table as input can output $c(S)$?

**Problem 18.** Determine the pointwise meet of all local upper bounds to complexity. Find more local upper bounds.

**Problem 19.** A finite semigroup $S$ is critical if each of its proper divisors has strictly smaller complexity. What are the possible posets for the $\mathscr{J}$-order of a critical semigroup? By [115], any finite poset with minimum is the $\mathscr{J}$-order of a finite semigroup. We ask the same question for critical semigroups with respect to two-sided complexity $C$ or dot-depth.

**Problem 20 (Henckell-Rhodes).** Is it true that the problem of deciding whether a semigroup has complexity one reduces to the case of a semigroup with at most 3 non-zero $\mathscr{J}$-classes? More generally, is it true that the problem of deciding whether a semigroup has complexity $n$ reduces to the case of a semigroup with at most $n + 2$ non-zero $\mathscr{J}$-classes?

**Problem 21 (Schützenberger's Question).** Does the pseudovariety $\mathbf{A} \vee \mathbf{G}$ have decidable membership?

**Problem 22.** Given a finite semigroup $S$, is the semidirect product closure of $(S)$ decidable? How about the closure under the two-sided semidirect product?

**Problem 23.** Study the complexity function $\kappa$ associated to the operators $\alpha = \mathbf{A} \,\textcircled{m}\, (-)$ and $\beta = \mathbf{G}^{\mathbf{N}} \,\textcircled{m}\, (-)$.

**Problem 24.** Is the complexity function for $\mathbf{A}$ associated to the operators $\alpha = \mathbf{Sl} ** (-)$ and $\beta = \mathbb{L}\mathbf{1} \,\textcircled{m}\, (-)$ computable? How exactly does it compare to dot-depth?

**Problem 25.** Find a proof of the classical Prime Decomposition Theorem using the Derived Category Theorem and some factorization theorem for relational morphisms.

**Problem 26.** Is the operator $\alpha = \mathbf{A} ** (-)$ idempotent? We suspect that it is not. This would be equivalent to $\mathbf{A} \textcircled{m} (-) = \mathbf{A} ** (-)$. It is known $\alpha^2 = \alpha^3$. If $PT_n$ is the semigroup of partial transformations of an $n$ element set and $I$ is the aperiodic ideal of maps of rank at most 1, then we suspect $PT_n \notin \mathbf{A} ** (PT_n/I)$ although the projection $PT_n \to PT_n/I$ is an aperiodic morphism.

**Problem 27.** Find an easier proof than the one in [362] that $c(\widehat{S}^{\mathscr{L}}) = c(S)$.

**Problem 28.** Suppose that $S$ is a semigroup such that the longest chain of non-zero $\mathscr{J}$-classes of $S$ has length at most 2. Must $C(S) \leq 1$?

**Problem 29.** Is the two-sided complexity function $C$ computable? Start with a semigroup with at most 3 non-zero $\mathscr{J}$-classes.

**Problem 30.** Is there some sort of Presentation Lemma to describe the existence of a cross section with respect to $\mathbb{L}\mathbf{G} \textcircled{m} \mathbf{V}$?

**Problem 31.** Are there arbitrarily large numbers $n$ such that there is a semigroup $S$ with $c(S) = n = C(S)$? Probably there are.

**Problem 32.** What is the two-sided complexity of the semigroup of binary relations on an $n$ element set?

**Problem 33.** A lattice is said to be *semi-distributive* at an element $\ell$ if $\ell \wedge x = \ell \wedge y$ implies $\ell \wedge x = \ell \wedge (x \vee y)$. Because meets and joins are continuous in $\mathbf{PV}$, if $\mathbf{PV}$ is semi-distributive at $\mathbf{V}$ and $\mathbf{W} \leq \mathbf{V}$, then there is a largest pseudovariety $\widehat{\mathbf{W}}$ such that $\mathbf{V} \cap \widehat{\mathbf{W}} = \mathbf{W}$. For instance, $\mathbf{PV}$ is semi-distributive at $\mathbf{G}$ and if $\mathbf{H}$ is a pseudovariety of groups, then $\overline{\mathbf{H}}$ is the maximum pseudovariety with $\mathbf{G} \cap \overline{\mathbf{H}} = \mathbf{H}$. Reilly and Zhang proved that $\mathbf{PV}$ is semi-distributive at the pseudovariety of bands $\mathbf{B}$ [260]. It follows from a result of G. Higman [217, 54.24] that $\mathbf{PV}$ is not semi-distributive at certain pseudovarieties of groups. At which pseudovarieties $\mathbf{V}$ is $\mathbf{PV}$ semi-distributive?

**Problem 34.** Is a subspace of $\mathbf{PV}$ sequentially compact in the strong topology if and only if it is closed and Noetherian?

**Problem 35.** What are the atoms of $\mathbf{Cnt(PV)}^+$, $\mathbf{GMC(PV)}^+$, $\mathbf{PVRM}^+$ and $\mathbf{CC}^+$?

**Problem 36.** Give more examples of smi, fmi, sfmi pseudovarieties that are not mi. In particular, can a compact pseudovariety be smi? This is related to the existence of atoms in $\mathbf{Cnt(PV)}^+$.

**Problem 37.** Classify the ×-prime finite semigroups. Is the property of being ×-prime decidable?

**Problem 38.** Is $I_n$, the symmetric inverse monoid, fji for all $n \geq 2$? A positive answer would imply **ESl** is fji, in light of Ash's Theorem [32] and Lemma 6.1.13. How about sfji? This may be easier.

**Problem 39.** If $(\mathbf{n}, G)$ is a permutation group, let $S_{(\mathbf{n},G)}$ be the action semigroup of the partial transformation inverse semigroup $(\mathbf{n}, G \cup B_n)$ where we identify $B_n$ with the rank one partial bijections of $\mathbf{n}$. Theorem 4.7.21 implies $S_{(\mathbf{n},G)}$ is subdirectly indecomposable. Of course, $BZ_p = S_{(\mathbf{p},\mathbb{Z}_p)}$. The question is: when is $S_{(\mathbf{n},G)}$ fji? One needs at least that the permutation group is subdirectly indecomposable as a permutation group.

**Problem 40.** Is the full transformation monoid $T_n$ an fji semigroup for all $n \geq 2$? If so this would give a new proof that **FSgp** is fji by Lemma 6.1.13. How about sfji? This may be easier.

**Problem 41.** Is there an increasing sequence of compact fji aperiodic pseudovarieties $(S_1) \leq (S_2) \leq \cdots$ with $\bigcup(S_i) = \mathbf{A}$? If so, then $\mathbf{A}$ would be fji by Lemma 6.1.13.

**Problem 42.** Given a finite set of computable pseudoidentities (say identities if you like), is it decidable whether $[\![E]\!]$ is sfji? A similar question can be asked for fji or for any of our other lattice theoretic adjectives.

**Problem 43.** Are the complexity pseudovarieties $\mathbf{C}_n$ fji or sfji? We ask the same question for $\mathbf{C}_n \cap \mathbf{DS}$ and $\mathbf{C}_n \cap \mathbf{CR}$.

**Problem 44.** Study the topological spaces Spec $\mathbf{PV}$ and Spec $\mathbf{PV}^{op}$.

**Problem 45.** Give a proof of Theorem 7.3.32 that does not use pseudoidentities.

**Problem 46.** Find a proof of Theorem 7.3.39 that does not use pseudoidentities.

**Problem 47.** Is it true that $\overline{\mathbf{H}}$ is sfji for every pseudovariety of groups $\mathbf{H}$? Consider the same question for fji.

**Problem 48.** Let $S$ be a KN semigroup with KN maximal subgroup $G$ in its minimal ideal. Describe $\mathsf{Excl}(S)$ in terms of $\mathsf{Excl}(G)$. In particular, if $\mathsf{Excl}(G) = [\![u = v]\!]$, then give a pseudoidentity defining $\mathsf{Excl}(S)$.

**Problem 49.** Describe the semigroup structure of $(\mathbf{PV}, *)$. It is known that the semigroup of Birkhoff varieties of groups with the semidirect product as multiplication is a free monoid [217]. But this is not the case for $\mathbf{PV}$, which has many idempotents. Some results can be found in [369].

**Problem 50.** Determine the s*i, sf*i and f*i pseudovarieties.

**Problem 51.** Classify the irreducible relational morphisms. The results of Chapter 5 imply that an irreducible relational morphism $\varphi \colon S \to T$ must either have a non-trivial congruence-free monoid $M$ in the Mal'cev kernel and belong to $(M)_K$, or belong to $\ell 1_K$. Are the irreducible relational morphisms precisely the relations of this sort?

**Problem 52.** Study the abstract spectral theory of $\mathbf{GMC(PV)}$. What are the mi, smi, sfmi, fmi, ji, fji, sji, sfmi elements of $\mathbf{GMC(PV)}$?

**Problem 53.** What does the $\mathscr{J}$-order look like in $\mathbf{Cnt(PV)}^+$? Give an example of when $\alpha \leq \beta$ but $\beta \not\leq_{\mathscr{J}} \alpha$ (or prove there is no such example). Consider the analogous questions for $\mathbf{GMC(PV)}^+$, $\mathbf{Cnt(PV)}^-$ and $\mathbf{GMC(PV)}^-$.

**Problem 54.** Suppose $\mathsf{R} \in \mathbf{CC}$ has decidable membership. Must $\mathsf{Rq}$ send compact pseudovarieties to decidable pseudovarieties? Notice that if $\mathsf{R} \in \mathbf{BCC}$ has decidable membership, then $\mathsf{Rq}(\mathbf{V})$ is decidable for each compact pseudovariety $\mathbf{V}$. Indeed, if $\mathbf{V} = (T)$, then $S \in \mathsf{Rq}(\mathbf{V})$ if and only if the canonical relational morphism $\rho \colon S \to F_{(T)}(S \times T)$ belongs to $\mathsf{R}$.

**Problem 55.** Study the counital bialgebras $\mathrm{Rec}(A^*)$, $\mathbf{A}(A^*)$ and $\mathbf{J}(A^*)$. Also study the Hopf algebra of group languages $\mathbf{G}(A^*)$.

**Problem 56.** Is it decidable whether a finite semigroup divides a semigroup of lower triangular Boolean matrices? In other words, is dot-depth two decidable?

**Problem 57.** In [241–244], Polák sets up a correspondence between pseudovarieties of idempotent semirings and certain classes of languages. What is the effect of the triangular product on languages? It should be similar to the effect of the Schützenberger product in the theory of varieties of languages.

**Problem 58.** What is the smallest pseudovariety of idempotent semirings closed under triangular product and containing $\mathbb{B}$? One would guess it should consist of all idempotent semirings whose underlying semigroup is aperiodic.

**Problem 59.** Classify all $\triangle$-irreducible idempotent semirings.

**Problem 60 (Margolis).** Is it true that if $S$ is $\triangle$-irreducible, then so is $M_n(S)$?

**Problem 61.** Classify the basic $\triangle$-irreducible semirings.

**Problem 62.** Compute the exclusion class of $M_n(Q)$ where $Q$ is one of $\mathbb{B}$, $G^\natural$ with $G$ non-trivial monolithic or $P(G)$ with $G$ a non-trivial group.

**Problem 63.** Given an oracle deciding group complexity, is $c_q$ computable? We ask the same question for $C_q$.

**Problem 64.** Study $\triangle$-irreducibility over more general semirings.

**Problem 65 (Fox-Rhodes).** Let $S = \mathcal{M}^0(G, A, B, C)$ be a 0-simple semigroup.

1. Is $c(P(S)) = \iota(C) - 1 + c(G)$ where $\iota(C)$ is the largest size of an identity submatrix of $C$ (up to reordering rows and columns and renormalizing)? The right-hand side is a lower bound by the argument in Theorem 9.5.11. In [91], it is shown that an upper bound is $\tau(C) - 1 + c(G)$ where $\tau(C)$ is the largest size of a triangular submatrix of $C$.
2. Recall that the action of $S$ on the right of $G \times B$ extends to an action on the free left $P(G)$-module generated by $B$ and hence yields a representation $\rho\colon S \to M_B(P(G))$ (which concretely speaking is the classical Schützenberger representation by row monomial matrices). This in turn yields a representation $\widetilde{\rho}\colon P(S) \to M_B(P(G))$. What is the relationship between $c(P(S))$ and $c(P(S)\widetilde{\rho})$?

**Problem 66 (Fox-Rhodes).** What is the largest size of an identity submatrix of the structure matrix for the rank $k$ $\mathscr{J}$-class of $T_n$?

**Problem 67.** If $S$ is a semigroup and $I$ is an ideal, what is the relationship between $c(P(S))$ and $c(P(I))$, $c(P_0(S/I))$? Proposition 9.5.14 gives some indication, but $c(P(S)_{P(I)})$ and $P(I)$ do not seem so clearly related. In [91], it is shown that if $S$ is a monoid and $I = S \setminus G$ where $G$ is the group of units, then $c(P(S)) \leq c(P(I)) + c(P_0(S/I))$.

**Problem 68.** Is $c(P(S)) = \max\{c(P_0(J^0)) \mid J \in S/\mathscr{J}\}$?

**Problem 69.** Develop homological algebra for idempotent semirings. Find a homological definition of $c_q$ and $C_q$. Relate this to dot-depth.

**Problem 70.** Develop a derived/kernel category theory for the triangular product.

**Problem 71.** Let $S$ be a finite semigroup. Is

$$c(S) = \max\{c(T) \mid T \in (S),\ T \text{ is fji}\}?$$

Consider the analogous problem for other hierarchical complexity functions such as two-sided complexity or dot-depth. This question is essentially asking if $c$ passes to $\mathrm{Spec}\,\mathbf{PV}^{op}$.

**Problem 72.** Transfer the results of this book to Formal Language Theory via the Eilenberg Correspondence [85]. Is there a quantized version of the Eilenberg Correspondence? We have a notion of an $A$-generated relational morphism. Perhaps each pseudovariety of relational morphisms is generated by canonical relational morphisms of $A$-generated syntactic semigroups where $A$ is any finite alphabet? What objects are recognized by relational morphisms? Is there a connection with Straubing's $\mathscr{C}$-pseudovarieties [347]?

**Problem 73.** Let $S$ be a finite semigroup and let $P = [\mathbf{1}, (S)] \cap K(\mathbf{PV})$ be the join semilattice of compact subpseudovarieties of $(S)$. This is a countable poset. What countable linear orders can occur as chains in $P$? Both $(\mathbb{N}, \leq)$ and $(\mathbb{N}, \geq)$ can occur [309].

**Problem 74.** Relate two-sided complexity to composition of bimachines [84, 284].

# References

1. M. Aguiar and S. Mahajan. *Coxeter groups and Hopf algebras*, volume 23 of *Fields Institute Monographs*. American Mathematical Society, Providence, RI, 2006. With a foreword by Nantel Bergeron.
2. D. Albert, R. Baldinger, and J. Rhodes. Undecidability of the identity problem for finite semigroups. *The Journal of Symbolic Logic*, 57(1):179–192, 1992.
3. D. Allen, Jr. and J. Rhodes. Synthesis of classical and modern theory of finite semigroups. *Advances in Mathematics*, 11(2):238–266, 1973.
4. J. Almeida. Implicit operations on finite $\mathscr{J}$-trivial semigroups and a conjecture of I. Simon. *Journal of Pure and Applied Algebra*, 69(3):205–218, 1991.
5. J. Almeida. On direct product decompositions of finite $\mathscr{J}$-trivial semigroups. *International Journal of Algebra and Computation*, 1(3):329–337, 1991.
6. J. Almeida. On finite simple semigroups. *Proceedings of the Edinburgh Mathematical Society. Series II*, 34(2):205–215, 1991.
7. J. Almeida. *Finite semigroups and universal algebra*, volume 3 of *Series in Algebra*. World Scientific Publishing Co. Inc., River Edge, NJ, 1994. Translated from the 1992 Portuguese original and revised by the author.
8. J. Almeida. A syntactical proof of locality of **DA**. *International Journal of Algebra and Computation*, 6(2):165–177, 1996.
9. J. Almeida. Hyperdecidable pseudovarieties and the calculation of semidirect products. *International Journal of Algebra and Computation*, 9(3-4):241–261, 1999. Dedicated to the memory of Marcel-Paul Schützenberger.
10. J. Almeida. Dynamics of implicit operations and tameness of pseudovarieties of groups. *Transactions of the American Mathematical Society*, 354(1):387–411 (electronic), 2002.
11. J. Almeida. Profinite groups associated with weakly primitive substitutions. *Fundamental' naya i Prikladnaya Matematika*, 11(3):13–48, 2005. Translation in *J. Math. Sci. (N. Y.)* 144(2):3881–3903, 2007.
12. J. Almeida, A. Azevedo, and L. Teixeira. On finitely based pseudovarieties of the form $\mathbf{V} * \mathbf{D}$ and $\mathbf{V} * \mathbf{D}_n$. *Journal of Pure and Applied Algebra*, 146(1):1–15, 2000.
13. J. Almeida, A. Azevedo, and M. Zeitoun. Pseudovariety joins involving $\mathscr{J}$-trivial semigroups. *International Journal of Algebra and Computation*, 9(1):99–112, 1999.

14. J. Almeida, J. C. Costa, and M. Zeitoun. Tameness of pseudovariety joins involving R. *Monatshefte für Mathematik*, 146(2):89–111, 2005.

15. J. Almeida and M. Delgado. Tameness of the pseudovariety of abelian groups. *International Journal of Algebra and Computation*, 15(2):327–338, 2005.

16. J. Almeida and A. Escada. On the equation $\mathbf{V} * \mathbf{G} = \mathcal{E}\,\mathbf{V}$. *Journal of Pure and Applied Algebra*, 166(1-2):1–28, 2002.

17. J. Almeida, S. W. Margolis, B. Steinberg, and M. V. Volkov. Characterization of group radicals with an application to Mal'cev products. Work in progress, 2006.

18. J. Almeida, S. W. Margolis, B. Steinberg, and M. V. Volkov. Representation theory of finite semigroups, semigroup radicals and formal language theory. *Transactions of the American Mathematical Society*, to appear.

19. J. Almeida and B. Steinberg. On the decidability of iterated semidirect products with applications to complexity. *Proceedings of the London Mathematical Society. Third Series*, 80(1):50–74, 2000.

20. J. Almeida and B. Steinberg. Syntactic and global semigroup theory: a synthesis approach. In *Algorithmic problems in groups and semigroups (Lincoln, NE, 1998)*, Trends Math., pages 1–23. Birkhäuser Boston, Boston, MA, 2000.

21. J. Almeida and B. Steinberg. Rational codes and free profinite monoids. Preprint, 2008.

22. J. Almeida and M. V. Volkov. Profinite identities for finite semigroups whose subgroups belong to a given pseudovariety. *Journal of Algebra and its Applications*, 2(2):137–163, 2003.

23. J. Almeida and M. V. Volkov. Subword complexity of profinite words and subgroups of free profinite semigroups. *International Journal of Algebra and Computation*, 16(2):221–258, 2006.

24. J. Almeida and P. Weil. Free profinite semigroups over semidirect products. *Izvestiya Vysshikh Uchebnykh Zavedeniĭ. Matematika*, 39(1):3–31, 1995.

25. J. Almeida and P. Weil. Relatively free profinite monoids: an introduction and examples. In *Semigroups, formal languages and groups (York, 1993)*, volume 466 of *NATO Adv. Sci. Inst. Ser. C Math. Phys. Sci.*, pages 73–117. Kluwer Acad. Publ., Dordrecht, 1995.

26. J. Almeida and P. Weil. Free profinite $\mathcal{R}$-trivial monoids. *International Journal of Algebra and Computation*, 7(5):625–671, 1997.

27. J. Almeida and P. Weil. Profinite categories and semidirect products. *Journal of Pure and Applied Algebra*, 123(1-3):1–50, 1998.

28. J. Almeida and M. Zeitoun. The pseudovariety **J** is hyperdecidable. *RAIRO Informatique Théorique et Applications. Theoretical Informatics and Applications*, 31(5):457–482, 1997.

29. J. Almeida and M. Zeitoun. Tameness of some locally trivial pseudovarieties. *Communications in Algebra*, 31(1):61–77, 2003.

30. J. Almeida and M. Zeitoun. The equational theory of $\omega$-terms for finite $\mathcal{R}$-trivial semigroups. In *Semigroups and languages*, pages 1–22. World Sci. Publ., River Edge, NJ, 2004.

31. W. Arveson. *An invitation to $C^*$-algebras*. Springer-Verlag, New York, 1976. Graduate Texts in Mathematics, No. 39.

32. C. J. Ash. Finite semigroups with commuting idempotents. *Australian Mathematical Society. Journal. Series A. Pure Mathematics and Statistics*, 43(1):81–90, 1987.

33. C. J. Ash. Inevitable graphs: a proof of the type II conjecture and some related decision procedures. *International Journal of Algebra and Computation*, 1(1):127–146, 1991.

34. K. Auinger. A new proof of the Rhodes type II conjecture. *International Journal of Algebra and Computation*, 14(5-6):551–568, 2004. International Conference on Semigroups and Groups in honor of the 65th birthday of Prof. John Rhodes.

35. K. Auinger, T. E. Hall, N. R. Reilly, and S. Zhang. Congruences on the lattice of pseudovarieties of finite semigroups. *International Journal of Algebra and Computation*, 7(4):433–455, 1997.

36. K. Auinger and B. Steinberg. On the extension problem for partial permutations. *Proceedings of the American Mathematical Society*, 131(9):2693–2703 (electronic), 2003.

37. K. Auinger and B. Steinberg. The geometry of profinite graphs with applications to free groups and finite monoids. *Transactions of the American Mathematical Society*, 356(2):805–851 (electronic), 2004.

38. K. Auinger and B. Steinberg. Constructing divisions into power groups. *Theoretical Computer Science*, 341(1-3):1–21, 2005.

39. K. Auinger and B. Steinberg. A constructive version of the Ribes-Zalesskiĭ product theorem. *Mathematische Zeitschrift*, 250(2):287–297, 2005.

40. K. Auinger and B. Steinberg. On power groups and embedding theorems for relatively free profinite monoids. *Mathematical Proceedings of the Cambridge Philosophical Society*, 138(2):211–232, 2005.

41. K. Auinger and B. Steinberg. Varieties of finite supersolvable groups with the M. Hall property. *Mathematische Annalen*, 335(4):853–877, 2006.

42. K. Auinger and P. G. Trotter. Pseudovarieties, regular semigroups and semidirect products. *Journal of the London Mathematical Society. Second Series*, 58(2):284–296, 1998.

43. B. Austin, K. Henckell, C. Nehaniv, and J. Rhodes. Subsemigroups and complexity via the presentation lemma. *Journal of Pure and Applied Algebra*, 101(3):245–289, 1995.

44. B. Banaschewski. The Birkhoff theorem for varieties of finite algebras. *Algebra Universalis*, 17(3):360–368, 1983.

45. T. Bandman, G.-M. Greuel, F. Grunewald, B. Kunyavskiĭ, G. Pfister, and E. Plotkin. Identities for finite solvable groups and equations in finite simple groups. *Compositio Mathematica*, 142(3):734–764, 2006.

46. G. M. Bergman. *An invitation to general algebra and universal constructions*. Henry Helson, Berkeley, CA, 1998.

47. G. M. Bergman. Every finite semigroup is embeddable in a finite relatively free semigroup. *Journal of Pure and Applied Algebra*, 186(1):1–19, 2004.

48. J. Berstel and C. Reutenauer. *Rational series and their languages*, volume 12 of *EATCS Monographs on Theoretical Computer Science*. Springer-Verlag, Berlin, 1988.

49. P. Bidigare, P. Hanlon, and D. Rockmore. A combinatorial description of the spectrum for the Tsetlin library and its generalization to hyperplane arrangements. *Duke Mathematical Journal*, 99(1):135–174, 1999.

50. J.-C. Birget. Arbitrary vs. regular semigroups. *Journal of Pure and Applied Algebra*, 34(1):57–115, 1984.

51. J.-C. Birget. Iteration of expansions—unambiguous semigroups. *Journal of Pure and Applied Algebra*, 34(1):1–55, 1984.

52. J.-C. Birget. The synthesis theorem for finite regular semigroups, and its generalization. *Journal of Pure and Applied Algebra*, 55(1-2):1–79, 1988.

53. J.-C. Birget, S. Margolis, and J. Rhodes. Semigroups whose idempotents form a subsemigroup. *Bulletin of the Australian Mathematical Society*, 41(2):161–184, 1990.

54. J.-C. Birget and J. Rhodes. Almost finite expansions of arbitrary semigroups. *Journal of Pure and Applied Algebra*, 32(3):239–287, 1984.

55. M. J. J. Branco. The kernel category and variants of the concatenation product. *International Journal of Algebra and Computation*, 7(4):487–509, 1997.

56. J. N. Bray, J. S. Wilson, and R. A. Wilson. A characterization of finite soluble groups by laws in two variables. *The Bulletin of the London Mathematical Society*, 37(2):179–186, 2005.

57. K. S. Brown. Semigroups, rings, and Markov chains. *Journal of Theoretical Probability*, 13(3):871–938, 2000.

58. K. S. Brown. Semigroup and ring theoretical methods in probability. In *Representations of finite dimensional algebras and related topics in Lie theory and geometry*, volume 40 of *Fields Inst. Commun.*, pages 3–26. Amer. Math. Soc., Providence, RI, 2004.

59. T. C. Brown. An interesting combinatorial method in the theory of locally finite semigroups. *Pacific Journal of Mathematics*, 36:285–289, 1971.

60. J. A. Brzozowski and I. Simon. Characterizations of locally testable events. *Discrete Mathematics*, 4:243–271, 1973.

61. S. Burris and H. P. Sankappanavar. *A course in universal algebra*, volume 78 of *Graduate Texts in Mathematics*. Springer-Verlag, New York, 1981.

62. B. Carré. *Graphs and networks*. The Clarendon Press Oxford University Press, New York, 1979. Oxford Applied Mathematics and Computing Science Series.

63. J. H. Carruth, J. A. Hildebrant, and R. J. Koch. *The theory of topological semigroups*, volume 75 of *Monographs and Textbooks in Pure and Applied Mathematics*. Marcel Dekker Inc., New York, 1983.

64. J. H. Carruth, J. A. Hildebrant, and R. J. Koch. *The theory of topological semigroups. Vol. 2*, volume 100 of *Monographs and Textbooks in Pure and Applied Mathematics*. Marcel Dekker Inc., New York, 1986.

65. N. Chomsky and M. P. Schützenberger. The algebraic theory of context-free languages. In *Computer programming and formal systems*, pages 118–161. North-Holland, Amsterdam, 1963.

66. A. H. Clifford. Matrix representations of completely simple semigroups. *American Journal of Mathematics*, 64:327–342, 1942.

67. A. H. Clifford. Basic representations of completely simple semigroups. *American Journal of Mathematics*, 82:430–434, 1960.

68. A. H. Clifford and G. B. Preston. *The algebraic theory of semigroups. Vol. I*. Mathematical Surveys, No. 7. American Mathematical Society, Providence, RI, 1961.

69. A. H. Clifford and G. B. Preston. *The algebraic theory of semigroups. Vol. II*. Mathematical Surveys, No. 7. American Mathematical Society, Providence, RI, 1967.

70. E. Cline, B. Parshall, and L. Scott. Finite-dimensional algebras and highest weight categories. *Journal für die Reine und Angewandte Mathematik*, 391:85–99, 1988.

71. R. S. Cohen and J. A. Brzozowski. Dot-depth of star-free events. *Journal of Computer and System Sciences*, 5:1–16, 1971.

72. J. H. Conway. *Regular algebra and finite machines.* Chapman & Hall, London, 1971.
73. J. C. Costa. Free profinite $\mathscr{R}$-trivial, locally idempotent and locally commutative semigroups. *Semigroup Forum,* 58(3):423–444, 1999.
74. J. C. Costa. Free profinite semigroups over some classes of semigroups locally in $\mathscr{D}\mathbf{G}$. *International Journal of Algebra and Computation,* 10(4):491–537, 2000.
75. J. C. Costa. Free profinite locally idempotent and locally commutative semigroups. *Journal of Pure and Applied Algebra,* 163(1):19–47, 2001.
76. T. Coulbois. Free product, profinite topology and finitely generated subgroups. *International Journal of Algebra and Computation,* 11(2):171–184, 2001.
77. D. F. Cowan, N. R. Reilly, P. G. Trotter, and M. V. Volkov. The finite basis problem for quasivarieties and pseudovarieties generated by regular semigroups. I. Quasivarieties generated by regular semigroups. *Journal of Algebra,* 267(2):635–653, 2003.
78. R. A. Dean and T. Evans. A remark on varietes of lattices and semigroups. *Proceedings of the American Mathematical Society,* 21:394–396, 1969.
79. M. Delgado. Abelian pointlikes of a monoid. *Semigroup Forum,* 56(3):339–361, 1998.
80. M. Delgado, V. H. Fernandes, S. Margolis, and B. Steinberg. On semigroups whose idempotent-generated subsemigroup is aperiodic. *International Journal of Algebra and Computation,* 14(5-6):655–665, 2004. International Conference on Semigroups and Groups in honor of the 65th birthday of Prof. John Rhodes.
81. M. Delgado, S. Margolis, and B. Steinberg. Combinatorial group theory, inverse monoids, automata, and global semigroup theory. *International Journal of Algebra and Computation,* 12(1-2):179–211, 2002. International Conference on Geometric and Combinatorial Methods in Group Theory and Semigroup Theory (Lincoln, NE, 2000).
82. M. Delgado, A. Masuda, and B. Steinberg. Solving systems of equations modulo pseudovarieties of abelian groups and hyperdecidability. In J. André, V. H. Fernandes, M. J. J. Branco, G. Gomes, J. Fountain, and J. C. Meakin, editors, *Semigroups and formal languages,* pages 57–65. World Sci. Publ., Hackensack, NJ, 2007.
83. P. M. Edwards. Eventually regular semigroups. *Bulletin of the Australian Mathematical Society,* 28(1):23–38, 1983.
84. S. Eilenberg. *Automata, languages, and machines. Vol. A.* Academic Press, New York, 1974. Pure and Applied Mathematics, Vol. 58.
85. S. Eilenberg. *Automata, languages, and machines. Vol. B.* Academic Press, New York, 1976. With two chapters ("Depth decomposition theorem" and "Complexity of semigroups and morphisms") by Bret Tilson, Pure and Applied Mathematics, Vol. 59.
86. S. Eilenberg and M. P. Schützenberger. On pseudovarieties. *Advances in Mathematics,* 19(3):413–418, 1976.
87. G. Z. Elston. Semigroup expansions using the derived category, kernel, and Malcev products. *Journal of Pure and Applied Algebra,* 136(3):231–265, 1999.
88. G. Z. Elston and C. L. Nehaniv. Holonomy embedding of arbitrary stable semigroups. *International Journal of Algebra and Computation,* 12(6):791–810, 2002.
89. J. A. Erdos. On products of idempotent matrices. *Glasgow Mathematical Journal,* 8:118–122, 1967.

90. D. G. Fitz-Gerald. On inverses of products of idempotents in regular semi-groups. *Australian Mathematical Society. Journal. Series A. Pure Mathematics and Statistics*, 13:335–337, 1972.

91. C. Fox and J. Rhodes. The complexity of the power set of a semigroup. Technical Report PAM-217, Center for Pure and Applied Mathematics, Math. Dept., Univ. of California at Berkeley, Berkeley, California, 1984.

92. P. Gabriel. Unzerlegbare Darstellungen. I. *Manuscripta Mathematica*, 6:71–103; correction, ibid. 6 (1972), 309, 1972.

93. P. Gabriel. Indecomposable representations. II. In *Symposia Mathematica, Vol. XI (Convegno di Algebra Commutativa, INDAM, Rome, 1971)*, pages 81–104. Academic Press, London, 1973.

94. P. Gabriel and A. V. Roiter. *Representations of finite-dimensional algebras.* Springer-Verlag, Berlin, 1997. Translated from the Russian, with a chapter by B. Keller, Reprint of the 1992 English translation.

95. O. Ganyushkin, V. Mazorchuk, and B. Steinberg. On the irreducible representations of a finite semigroup. Preprint, 2008.

96. M. Gehrke, S. Grigorieff, and J.-E. Pin. Duality and equational theory of regular languages. In *Automata, languages and programming*, volume 5216 of *Lecture Notes in Comput. Sci.*, pages 246–257. Springer, Berlin, 2008.

97. G. Gierz, K. H. Hofmann, K. Keimel, J. D. Lawson, M. Mislove, and D. S. Scott. *Continuous lattices and domains*, volume 93 of *Encyclopedia of Mathematics and its Applications*. Cambridge University Press, Cambridge, 2003.

98. R. Gitik, S. W. Margolis, and B. Steinberg. On the Kurosh theorem and separability properties. *Journal of Pure and Applied Algebra*, 179(1-2):87–97, 2003.

99. R. Gitik and E. Rips. On separability properties of groups. *International Journal of Algebra and Computation*, 5(6):703–717, 1995.

100. K. Głazek. *A guide to the literature on semirings and their applications in mathematics and information sciences.* Kluwer Academic Publishers, Dordrecht, 2002. With complete bibliography.

101. D. Gorenstein, R. Lyons, and R. Solomon. *The classification of the finite simple groups*, volume 40 of *Mathematical Surveys and Monographs*. American Mathematical Society, Providence, RI, 1994.

102. D. Gorenstein, R. Lyons, and R. Solomon. *The classification of the finite simple groups. Number 2. Part I. Chapter G*, volume 40 of *Mathematical Surveys and Monographs*. American Mathematical Society, Providence, RI, 1996. General group theory.

103. D. Gorenstein, R. Lyons, and R. Solomon. *The classification of the finite simple groups. Number 3. Part I. Chapter A*, volume 40 of *Mathematical Surveys and Monographs*. American Mathematical Society, Providence, RI, 1998. Almost simple $K$-groups.

104. D. Gorenstein, R. Lyons, and R. Solomon. *The classification of the finite simple groups. Number 4. Part II. Chapters 1–4*, volume 40 of *Mathematical Surveys and Monographs*. American Mathematical Society, Providence, RI, 1999. Uniqueness theorems, With errata: *The classification of the finite simple groups. Number 3. Part I. Chapter A* [Amer. Math. Soc., Providence, RI, 1998; MR1490581 (98j:20011)].

105. D. Gorenstein, R. Lyons, and R. Solomon. *The classification of the finite simple groups. Number 5. Part III. Chapters 1–6*, volume 40 of *Mathematical Surveys*

*and Monographs*. American Mathematical Society, Providence, RI, 2002. The generic case, stages 1–3a.

106. D. Gorenstein, R. Lyons, and R. Solomon. *The classification of the finite simple groups. Number 6. Part IV*, volume 40 of *Mathematical Surveys and Monographs*. American Mathematical Society, Providence, RI, 2005. The special odd case.

107. R. L. Graham. On finite 0-simple semigroups and graph theory. *Mathematical Systems Theory. An International Journal on Mathematical Computing Theory*, 2:325–339, 1968.

108. J. A. Green. On the structure of semigroups. *Annals of Mathematics. Second Series*, 54:163–172, 1951.

109. R. I. Grigorchuk, V. V. Nekrashevich, and V. I. Sushchanskiĭ. Automata, dynamical systems, and groups. *Proceedings of the Steklov Institute of Mathematics*, 231(4):128–203, 2000.

110. P.-A. Grillet. *Semigroups*, volume 193 of *Monographs and Textbooks in Pure and Applied Mathematics*. Marcel Dekker Inc., New York, 1995. An introduction to the structure theory.

111. D. Groves and M. Vaughan-Lee. Finite groups of bounded exponent. *The Bulletin of the London Mathematical Society*, 35(1):37–40, 2003.

112. M. Hall, Jr. A topology for free groups and related groups. *Annals of Mathematics. Second Series*, 52:127–139, 1950.

113. M. Hall, Jr. *The theory of groups*. The Macmillan Co., New York, 1959.

114. T. E. Hall. On regular semigroups. *Journal of Algebra*, 24:1–24, 1973.

115. T. E. Hall. The partially ordered set of all *J*-classes of a finite semigroup. *Semigroup Forum*, 6(3):263–264, 1973.

116. T. E. Hall, S. I. Kublanovskii, S. Margolis, M. V. Sapir, and P. G. Trotter. Algorithmic problems for finite groups and finite 0-simple semigroups. *Journal of Pure and Applied Algebra*, 119(1):75–96, 1997.

117. P. R. Halmos. *Lectures on Boolean algebras*. Van Nostrand Mathematical Studies, No. 1. D. Van Nostrand Co., Inc., Princeton, NJ, 1963.

118. J. Hartmanis. Loop-free structure of sequential machines. *Information and Computation*, 5:25–43, 1962.

119. J. Hartmanis. Further results on the structure of sequential machines. *Journal of the Association for Computing Machinery*, 10:78–88, 1963.

120. U. Hebisch and H. J. Weinert. *Semirings: algebraic theory and applications in computer science*, volume 5 of *Series in Algebra*. World Scientific Publishing Co. Inc., River Edge, NJ, 1998. Translated from the 1993 German original.

121. K. Henckell. Pointlike sets: the finest aperiodic cover of a finite semigroup. *Journal of Pure and Applied Algebra*, 55(1-2):85–126, 1988.

122. K. Henckell. Blockgroups = powergroups: a consequence of Ash's proof of the Rhodes type II conjecture. In *Monash Conference on Semigroup Theory (Melbourne, 1990)*, pages 117–134. World Sci. Publ., River Edge, NJ, 1991.

123. K. Henckell. Idempotent pointlike sets. *International Journal of Algebra and Computation*, 14(5-6):703–717, 2004. International Conference on Semigroups and Groups in honor of the 65th birthday of Prof. John Rhodes.

124. K. Henckell. Stable pairs. *International Journal of Algebra and Computation*, to appear.

125. K. Henckell, S. Lazarus, and J. Rhodes. Prime decomposition theorem for arbitrary semigroups: general holonomy decomposition and synthesis theorem. *Journal of Pure and Applied Algebra*, 55(1-2):127–172, 1988.

630     References

126. K. Henckell, S. W. Margolis, J.-E. Pin, and J. Rhodes. Ash's type II theorem, profinite topology and Mal'cev products. I. *International Journal of Algebra and Computation*, 1(4):411–436, 1991.

127. K. Henckell and J. Rhodes. The theorem of Knast, the $PG = BG$ and type-II conjectures. In *Monoids and semigroups with applications (Berkeley, CA, 1989)*, pages 453–463. World Sci. Publ., River Edge, NJ, 1991.

128. K. Henckell, J. Rhodes, and B. Steinberg. Complexity is decidable: the lower bound. Work in progress.

129. K. Henckell, J. Rhodes, and B. Steinberg. Aperiodic pointlikes and beyond. *International Journal of Algebra and Computation*, to appear.

130. K. Henckell, J. Rhodes, and B. Steinberg. A profinite approach to stable pairs. *International Journal of Algebra and Computation*, to appear.

131. B. Herwig and D. Lascar. Extending partial automorphisms and the profinite topology on free groups. *Transactions of the American Mathematical Society*, 352(5):1985–2021, 2000.

132. P. J. Higgins. Categories and groupoids. *Reprints in Theory and Applications of Categories*, (7):1–178 (electronic), 2005. Reprint of the 1971 original [*Notes on categories and groupoids*, Van Nostrand Reinhold, London] with a new preface by the author.

133. P. M. Higgins. *Techniques of semigroup theory*. Oxford Science Publications. The Clarendon Press, Oxford University Press, New York, 1992. With a foreword by G. B. Preston.

134. K. H. Hofmann, M. Mislove, and A. Stralka. *The Pontryagin duality of compact 0-dimensional semilattices and its applications*. Springer-Verlag, Berlin, 1974. Lecture Notes in Mathematics, Vol. 396.

135. K. H. Hofmann and P. S. Mostert. *Elements of compact semigroups*. Charles E. Merr ll Books, Inc., Columbus, Ohio, 1966.

136. C. H. Houghton. Completely 0-simple semigroups and their associated graphs and groups. *Semigroup Forum*, 14(1):41–67, 1977.

137. J. M. Howie. The subsemigroup generated by the idempotents of a full transformation semigroup. *Journal of the London Mathematical Society. Second Series*, 41:707–716, 1966.

138. J. M. Howie. Idempotents in completely 0-simple semigroups. *Glasgow Mathematical Journal*, 19(2):109–113, 1978.

139. J. M. Howie. *Fundamentals of semigroup theory*, volume 12 of *London Mathematical Society Monographs. New Series*. The Clarendon Press, Oxford University Press, New York, 1995. Oxford Science Publications.

140. R. P. Hunter. Certain finitely generated compact zero-dimensional semigroups. *Australian Mathematical Society. Journal. Series A. Pure Mathematics and Statistics*, 44(2):265–270, 1988.

141. B. Huppert. *Endliche Gruppen. I.* Die Grundlehren der Mathematischen Wissenschaften, Band 134. Springer-Verlag, Berlin, 1967.

142. M. Jackson. Finite semigroups whose varieties have uncountably many subvarieties. *Journal of Algebra*, 228(2):512–535, 2000.

143. M. Jackson. Small inherently nonfinitely based finite semigroups. *Semigroup Forum*, 64(2):297–324, 2002.

144. M. Jackson. Finite semigroups with infinite irredundant identity bases. *International Journal of Algebra and Computation*, 15(3):405–422, 2005.

145. M. Jackson and R. McKenzie. Interpreting graph colorability in finite semi-groups. *International Journal of Algebra and Computation*, 16(1):119–140, 2006.

146. M. Jackson and O. Sapir. Finitely based, finite sets of words. *International Journal of Algebra and Computation*, 10(6):683–708, 2000.

147. P. T. Johnstone. *Stone spaces*, volume 3 of *Cambridge Studies in Advanced Mathematics*. Cambridge University Press, Cambridge, 1986. Reprint of the 1982 edition.

148. P. R. Jones. Monoid varieties defined by $x^{n+1} = x$ are local. *Semigroup Forum*, 47(3):318–326, 1993.

149. P. R. Jones. Profinite categories, implicit operations and pseudovarieties of categories. *Journal of Pure and Applied Algebra*, 109(1):61–95, 1996.

150. P. R. Jones and S. Pustejovsky. A kernel for relational morphisms of categories. In *Semigroups with applications (Oberwolfach, 1991)*, pages 152–161. World Sci. Publ., River Edge, NJ, 1992.

151. P. R. Jones and M. B. Szendrei. Local varieties of completely regular monoids. *Journal of Algebra*, 150(1):1–27, 1992.

152. P. R. Jones and P. G. Trotter. Locality of **DS** and associated varieties. *Journal of Pure and Applied Algebra*, 104(3):275–301, 1995.

153. M. Kambites. On the Krohn-Rhodes complexity of semigroups of upper triangular matrices. *International Journal of Algebra and Computation*, 17(1):187–201, 2007.

154. M. Kambites and B. Steinberg. Wreath product decompositions for triangular matrix semigroups. In J. André, V. H. Fernandes, M. J. J. Branco, G. Gomes, J. Fountain, and J. C. Meakin, editors, *Semigroups and formal languages*, pages 129–144. World Sci. Publ., Hackensack, NJ, 2007.

155. I. Kapovich and A. Myasnikov. Stallings foldings and subgroups of free groups. *Journal of Algebra*, 248(2):608–668, 2002.

156. J. Karnofsky and J. Rhodes. Decidability of complexity one-half for finite semigroups. *Semigroup Forum*, 24(1):55–66, 1982.

157. Y. Katsov. Toward homological characterization of semirings: Serre's conjecture and Bass's perfectness in a semiring context. *Algebra Universalis*, 52(2-3):197–214, 2004.

158. A. S. Kechris. *Classical descriptive set theory*, volume 156 of *Graduate Texts in Mathematics*. Springer-Verlag, New York, 1995.

159. O. G. Kharlampovich and M. V. Sapir. Algorithmic problems in varieties. *International Journal of Algebra and Computation*, 5(4-5):379–602, 1995.

160. M. Kilp, U. Knauer, and A. V. Mikhalev. *Monoids, acts and categories*, volume 29 of *de Gruyter Expositions in Mathematics*. Walter de Gruyter & Co., Berlin, 2000. With applications to wreath products and graphs, A handbook for students and researchers.

161. S. C. Kleene. Representation of events in nerve nets and finite automata. In *Automata studies*, Annals of mathematics studies, no. 34, pages 3–41. Princeton University Press, Princeton, NJ, 1956.

162. D. J. Kleitman, B. R. Rothschild, and J. H. Spencer. The number of semigroups of order $n$. *Proceedings of the American Mathematical Society*, 55(1):227–232, 1976.

163. R. Knast. Some theorems on graph congruences. *RAIRO Informatique Théorique*, 17(4):331–342, 1983.

164. L. G. Kovács and M. F. Newman. Cross varieties of groups. *Proceedings of the Royal Society Series A*, 292:530–536, 1966.

165. L. G. Kovács and M. F. Newman. Minimal verbal subgroups. *Proceedings of the Cambridge Philosophical Society*, 62:347–350, 1966.

166. L. G. Kovács and M. F. Newman. On critical groups. *Australian Mathematical Society. Journal. Series A. Pure Mathematics and Statistics*, 6:237–250, 1966.

167. M. Krasner and L. Kaloujnine. Produit complet des groupes de permutations et problème d'extension de groupes. I. *Acta Universitatis Szegediensis. Acta Scientiarum Mathematicarum*, 13:208–230, 1950.

168. K. Krohn, R. Mateosian, and J. Rhodes. Methods of the algebraic theory of machines. I. Decomposition theorem for generalized machines; properties preserved under series and parallel compositions of machines. *Journal of Computer and System Sciences*, 1:55–85, 1967.

169. K. Krohn and J. Rhodes. Algebraic theory of machines. I. Prime decomposition theorem for finite semigroups and machines. *Transactions of the American Mathematical Society*, 116:450–464, 1965.

170. K. Krohn and J. Rhodes. Complexity of finite semigroups. *Annals of Mathematics. Second Series*, 88:128–160, 1968.

171. K. Krohn, J. Rhodes, and B. Tilson. *Algebraic theory of machines, languages, and semigroups.* Edited by Michael A. Arbib. With a major contribution by Kenneth Krohn and John L. Rhodes. Academic Press, New York, 1968. Chapters 1, 5–9.

172. D. Kruml, J. W. Pelletier, P. Resende, and J. Rosický. On quantales and spectra of $C^*$-algebras. *Applied Categorical Structures. A Journal Devoted to Applications of Categorical Methods in Algebra, Analysis, Order, Topology and Computer Science*, 11(6):543–560, 2003.

173. D. Kruml and P. Resende. On quantales that classify $C^*$-algebras. *Cahiers de Topologie et Géométrie Différentielle Catégoriques*, 45(4):287–296, 2004.

174. G. Lallement. *Semigroups and combinatorial applications.* John Wiley & Sons, New York-Chichester-Brisbane, 1979. Pure and Applied Mathematics, A Wiley-Interscience Publication.

175. G. Lallement. Augmentations and wreath products of monoids. *Semigroup Forum*, 21(1):89–90, 1980.

176. M. V. Lawson. *Inverse semigroups.* World Scientific Publishing Co. Inc., River Edge, NJ, 1998. The theory of partial symmetries.

177. M. V. Lawson. *Finite automata.* Chapman & Hall/CRC, Boca Raton, FL, 2004.

178. M. V. Lawson, S. W. Margolis, and B. Steinberg. Expansions of inverse semigroups. *Journal of the Australian Mathematical Society*, 80(2):205–228, 2006.

179. B. Le Saëc, J.-E. Pin, and P. Weil. Semigroups with idempotent stabilizers and applications to automata theory. *International Journal of Algebra and Computation*, 1(3):291–314, 1991.

180. E. W. H. Lee. On a simpler basis for the pseudovariety **EDS**. *Semigroup Forum*, 75(2):477–479, 2007.

181. J. Leech. $\mathcal{H}$-coextensions of monoids. *Memoirs of the American Mathematical Society*, 1(issue 2, 157):1–66, 1975.

182. J. Leech. The structure of a band of groups. *Memoirs of the American Mathematical Society*, 1(issue 2, 157):67–95, 1975.

183. M. Loganathan. Cohomology of inverse semigroups. *Journal of Algebra*, 70(2):375–393, 1981.

184. R. C. Lyndon and P. E. Schupp. *Combinatorial group theory*. Classics in Mathematics. Springer-Verlag, Berlin, 2001. Reprint of the 1977 edition.

185. S. Mac Lane. *Categories for the working mathematician*, volume 5 of *Graduate Texts in Mathematics*. Springer-Verlag, New York, second edition, 1998.

186. S. Mac Lane and I. Moerdijk. *Sheaves in geometry and logic*. Universitext. Springer-Verlag, New York, 1994. A first introduction to topos theory, Corrected reprint of the 1992 edition.

187. S. W. Margolis. $k$-transformation semigroups and a conjecture of Tilson. *Journal of Pure and Applied Algebra*, 17(3):313–322, 1980.

188. S. W. Margolis and J. C. Meakin. $E$-unitary inverse monoids and the Cayley graph of a group presentation. *Journal of Pure and Applied Algebra*, 58(1):45–76, 1989.

189. S. W. Margolis and J. C. Meakin. Free inverse monoids and graph immersions. *International Journal of Algebra and Computation*, 3(1):79–99, 1993.

190. S. W. Margolis and J.-E. Pin. Power monoids and finite $\mathscr{J}$-trivial monoids. *Semigroup Forum*, 29(1-2):99–108, 1984.

191. S. W. Margolis and J.-E. Pin. Varieties of finite monoids and topology for the free monoid. In *Proceedings of the 1984 Marquette conference on semigroups (Milwaukee, Wis., 1984)*, pages 113–129, Milwaukee, WI, 1985. Marquette University.

192. S. W. Margolis and J.-E. Pin. Expansions, free inverse semigroups, and Schützenberger product. *Journal of Algebra*, 110(2):298–305, 1987.

193. S. W. Margolis and J.-E. Pin. Inverse semigroups and extensions of groups by semilattices. *Journal of Algebra*, 110(2):277–297, 1987.

194. S. W. Margolis and J.-E. Pin. Inverse semigroups and varieties of finite semigroups. *Journal of Algebra*, 110(2):306–323, 1987.

195. S. W. Margolis, M. Sapir, and P. Weil. Irreducibility of certain pseudovarieties. *Communications in Algebra*, 26(3):779–792, 1998.

196. S. W. Margolis, M. Sapir, and P. Weil. Closed subgroups in pro-**V** topologies and the extension problem for inverse automata. *International Journal of Algebra and Computation*, 11(4):405–445, 2001.

197. S. W. Margolis and B. Tilson. An upper bound for the complexity of transformation semigroups. *Journal of Algebra*, 73(2):518–537, 1981.

198. D. B. McAlister. Representations of semigroups by linear transformations. I, II. *Semigroup Forum*, 2(4):283–320, 1971.

199. D. B. McAlister. Groups, semilattices and inverse semigroups. I, II. *Transactions of the American Mathematical Society*, 192:227–244; ibid. 196 (1974), 351–370, 1974.

200. D. B. McAlister. Regular semigroups, fundamental semigroups and groups. *Australian Mathematical Society. Journal. Series A*, 29(4):475–503, 1980.

201. D. B. McAlister. Semigroups generated by a group and an idempotent. *Communications in Algebra*, 26(2):515–547, 1998.

202. J. McCammond and J. Rhodes. Geometric semigroup theory. *International Journal of Algebra and Computation*, to appear.

203. R. N. McKenzie, G. F. McNulty, and W. F. Taylor. *Algebras, lattices, varieties. Vol. I*. The Wadsworth & Brooks/Cole Mathematics Series. Wadsworth & Brooks/Cole Advanced Books & Software, Monterey, CA, 1987.

204. R. McNaughton. Algebraic decision procedures for local testability. *Mathematical Systems Theory. An International Journal on Mathematical Computing Theory*, 8(1):60–76, 1974.

205. R. McNaughton and S. Papert. *Counter-free automata.* The M.I.T. Press, Cambridge, Mass.–London, 1971. With an appendix by William Henneman, M.I.T. Research Monograph, No. 65.

206. R. McNaughton and H. Yamada. Regular expressions and state graphs for automata. *IEEE Transactions on Electronic Computers,* 9:39–47, 1960.

207. A. R. Meyer. A note on star-free events. *Journal of the Association for Computing Machinery,* 16:220–225, 1969.

208. A. Minasyan. Separable subsets of GFERF negatively curved groups. *Journal of Algebra,* 304(2):1090–1100, 2006.

209. E. F. Moore. Gedanken-experiments on sequential machines. In *Automata studies,* Annals of mathematics studies, no. 34, pages 129–153. Princeton University Press, Princeton, NJ, 1956.

210. W. D. Munn. On semigroup algebras. *Mathematical Proceedings of the Cambridge Philosophical Society,* 51:1–15, 1955.

211. W. D. Munn. Matrix representations of semigroups. *Mathematical Proceedings of the Cambridge Philosophical Society,* 53:5–12, 1957.

212. K. S. S. Nambooripad. Structure of regular semigroups. I. *Memoirs of the American Mathematical Society,* 22(224):vii+119, 1979.

213. K. S. S. Nambooripad. The natural partial order on a regular semigroup. *Proceedings of the Edinburgh Mathematical Society. Series II,* 23(3):249–260, 1980.

214. C. L. Nehaniv. Cascade decomposition of arbitrary semigroups. In *Semigroups, formal languages and groups (York, 1993),* volume 466 of *NATO Adv. Sci. Inst. Ser. C Math. Phys. Sci.,* pages 391–425. Kluwer Acad. Publ., Dordrecht, 1995.

215. C. L. Nehaniv. Monoids and groups acting on trees: characterizations, gluing, and applications of the depth preserving actions. *International Journal of Algebra and Computation,* 5(2):137–172, 1995.

216. V. Nekrashevych. *Self-similar groups,* volume 117 of *Mathematical Surveys and Monographs.* American Mathematical Society, Providence, RI, 2005.

217. H. Neumann. *Varieties of groups.* Springer-Verlag New York, 1967.

218. G. A. Niblo. Separability properties of free groups and surface groups. *Journal of Pure and Applied Algebra,* 78(1):77–84, 1992.

219. W. R. Nico. Wreath products and extensions. *Houston Journal of Mathematics,* 9(1):71–99, 1983.

220. N. Nikolov and D. Segal. On finitely generated profinite groups. I. Strong completeness and uniform bounds. *Annals of Mathematics. Second Series,* 165(1):171–238, 2007.

221. N. Nikolov and D. Segal. On finitely generated profinite groups. II. Products in quasisimple groups. *Annals of Mathematics. Second Series,* 165(1):239–273, 2007.

222. K. Numakura. Theorems on compact totally disconnected semigroups and lattices. *Proceedings of the American Mathematical Society,* 8:623–626, 1957.

223. L. O'Carroll. Inverse semigroups as extensions of semilattices. *Glasgow Mathematical Journal,* 16(1):12–21, 1975.

224. J.-E. Pin. *Varieties of formal languages.* Foundations of Computer Science. Plenum Publishing Corp., New York, 1986. With a preface by M.-P. Schützenberger, Translated from the French by A. Howie.

225. J.-E. Pin. A topological approach to a conjecture of Rhodes. *Bulletin of the Australian Mathematical Society,* 38(3):421–431, 1988.

226. J.-E. Pin. Topologies for the free monoid. *Journal of Algebra*, 137(2):297–337, 1991.

227. J.-E. Pin. **BG = PG**: a success story. In *Semigroups, formal languages and groups (York, 1993)*, volume 466 of *NATO Adv. Sci. Inst. Ser. C Math. Phys. Sci.*, pages 33–47. Kluwer Acad. Publ., Dordrecht, 1995.

228. J.-E. Pin. Eilenberg's theorem for positive varieties of languages. *Izvestiya Vysshikh Uchebnykh Zavedeniĭ. Matematika*, 39(1):80–90, 1995.

229. J.-E. Pin. Syntactic semigroups. In *Handbook of formal languages, Vol. 1*, pages 679–746. Springer, Berlin, 1997.

230. J.-E. Pin. Algebraic tools for the concatenation product. *Theoretical Computer Science*, 292(1):317–342, 2003. Selected papers in honor of Jean Berstel.

231. J.-E. Pin, A. Pinguet, and P. Weil. Ordered categories and ordered semigroups. *Communications in Algebra*, 30(12):5651–5675, 2002.

232. J.-E. Pin and C. Reutenauer. A conjecture on the Hall topology for the free group. *The Bulletin of the London Mathematical Society*, 23(4):356–362, 1991.

233. J.-E. Pin and H. Straubing. Monoids of upper triangular matrices. In *Semigroups (Szeged, 1981)*, volume 39 of *Colloq. Math. Soc. János Bolyai*, pages 259–272. North-Holland, Amsterdam, 1985.

234. J.-E. Pin, H. Straubing, and D. Thérien. Locally trivial categories and unambiguous concatenation. *Journal of Pure and Applied Algebra*, 52(3):297–311, 1988.

235. J.-E. Pin and P. Weil. Profinite semigroups, Mal'cev products, and identities. *Journal of Algebra*, 182(3):604–626, 1996.

236. J.-E. Pin and P. Weil. Polynomial closure and unambiguous product. *Theory of Computing Systems*, 30(4):383–422, 1997.

237. J.-E. Pin and P. Weil. Semidirect products of ordered semigroups. *Communications in Algebra*, 30(1):149–169, 2002.

238. J.-E. Pin and P. Weil. The wreath product principle for ordered semigroups. *Communications in Algebra*, 30(12):5677–5713, 2002.

239. B. I. Plotkin. Triangular products of pairs. *Latvijas Valsts Univ. Zinātn. Raksti*, 151:140–170, 1971. Certain Questions of Group Theory (Proc. Algebra Sem. No. 2, Riga, 1969/1970).

240. B. I. Plotkin, L. J. Greenglaz, and A. A. Gvaramija. *Algebraic structures in automata and databases theory*. World Scientific Publishing Co. Inc., River Edge, NJ, 1992.

241. L. Polák. Syntactic semiring of a language (extended abstract). In *Mathematical foundations of computer science, 2001 (Mariánské Lázně)*, volume 2136 of *Lecture Notes in Comput. Sci.*, pages 611–620. Springer, Berlin, 2001.

242. L. Polák. Syntactic semiring and language equations. In *Implementation and application of automata*, volume 2608 of *Lecture Notes in Comput. Sci.*, pages 182–193. Springer, Berlin, 2003.

243. L. Polák. Syntactic semiring and universal automaton. In *Developments in language theory*, volume 2710 of *Lecture Notes in Comput. Sci.*, pages 411–422. Springer, Berlin, 2003.

244. L. Polák. A classification of rational languages by semilattice-ordered monoids. *Universitatis Masarykianae Brunensis. Facultas Scientiarum Naturalium. Archivum Mathematicum*, 40(4):395–406, 2004.

245. I. S. Ponizovskiĭ. On matrix representations of associative systems. *Matematicheskiĭ Sbornik. Novaya Seriya*, 38(80):241–260, 1956.

246. R. Pöschel, M. V. Sapir, N. W. Sauer, M. G. Stone, and M. V. Volkov. Identities in full transformation semigroups. *Algebra Universalis*, 31(4):580–588, 1994.

247. H. A. Priestley. Representation of distributive lattices by means of ordered stone spaces. *The Bulletin of the London Mathematical Society*, 2:186–190, 1970.

248. M. S. Putcha. Semilattice decompositions of semigroups. *Semigroup Forum*, 6(1):12–34, 1973.

249. M. S. Putcha. Algebraic monoids whose nonunits are products of idempotents. *Proceedings of the American Mathematical Society*, 103(1):38–40, 1988.

250. M. S. Putcha. *Linear algebraic monoids*, volume 133 of *London Mathematical Society Lecture Note Series*. Cambridge University Press, Cambridge, 1988.

251. M. S. Putcha. Complex representations of finite monoids. *Proceedings of the London Mathematical Society. Third Series*, 73(3):623–641, 1996.

252. M. S. Putcha. Monoid Hecke algebras. *Transactions of the American Mathematical Society*, 349(9):3517–3534, 1997.

253. M. S. Putcha. Complex representations of finite monoids. II. Highest weight categories and quivers. *Journal of Algebra*, 205(1):53–76, 1998.

254. M. S. Putcha. Hecke algebras and semisimplicity of monoid algebras. *Journal of Algebra*, 218(2):488–508, 1999.

255. M. S. Putcha. Semigroups and weights for group representations. *Proceedings of the American Mathematical Society*, 128(10):2835–2842, 2000.

256. M. S. Putcha. Products of idempotents in algebraic monoids. *Journal of the Australian Mathematical Society*, 80(2):193–203, 2006.

257. D. Quillen. Higher algebraic $K$-theory. I. In *Algebraic $K$-theory, I: Higher $K$-theories (Proc. Conf., Battelle Memorial Inst., Seattle, Wash., 1972)*, pages 85–147. Lecture Notes in Math., Vol. 341. Springer, Berlin, 1973.

258. D. Rees. On semi-groups. *Mathematical Proceedings of the Cambridge Philosophical Society*, 36:387–400, 1940.

259. N. R. Reilly and S. Zhang. Operators on the lattice of pseudovarieties of finite semigroups. *Semigroup Forum*, 57(2):208–239, 1998.

260. N. R. Reilly and S. Zhang. Decomposition of the lattice of pseudovarieties of finite semigroups induced by bands. *Algebra Universalis*, 44(3-4):217–239, 2000.

261. J. Reiterman. The Birkhoff theorem for finite algebras. *Algebra Universalis*, 14(1):1–10, 1982.

262. L. E. Renner. *Linear algebraic monoids*, volume 134 of *Encyclopaedia of Mathematical Sciences*. Springer-Verlag, Berlin, 2005. Invariant Theory and Algebraic Transformation Groups, V.

263. P. Resende. Quantales, finite observations and strong bisimulation. *Theoretical Computer Science*, 254(1-2):95–149, 2001.

264. P. Resende. Étale groupoids and their quantales. *Advances in Mathematics*, 208(1):147–209, 2007.

265. C. Reutenauer. **N**-rationality of zeta functions. *Advances in Applied Mathematics*, 18(1):1–17, 1997.

266. J. Rhodes. Some results on finite semigroups. *Journal of Algebra*, 4:471–504, 1966.

267. J. Rhodes. A homomorphism theorem for finite semigroups. *Mathematical Systems Theory. An International Journal on Mathematical Computing Theory*, 1:289–304, 1967.

268. J. Rhodes. The fundamental lemma of complexity for arbitrary finite semigroups. *Bulletin of the American Mathematical Society*, 74:1104–1109, 1968.

269. J. Rhodes. Algebraic theory of finite semigroups. Structure numbers and structure theorems for finite semigroups. In K. Folley, editor, *Semigroups (Proc. Sympos., Wayne State Univ., Detroit, Mich., 1968)*, pages 125–162. Academic Press, New York, 1969.

270. J. Rhodes. Characters and complexity of finite semigroups. *Journal of Combinatorial Theory*, 6:67–85, 1969.

271. J. Rhodes. Proof of the fundamental lemma of complexity (weak version) for arbitrary finite semigroups. *Journal of Combinatorial Theory. Series A*, 10:22–73, 1971.

272. J. Rhodes. Axioms for complexity for all finite semigroups. *Advances in Mathematics*, 11(2):210–214, 1973.

273. J. Rhodes. Finite binary relations have no more complexity than finite functions. *Semigroup Forum*, 7(1-4):92–103, 1974. Collection of articles dedicated to Alfred Hoblitzelle Clifford on the occasion of his 65th birthday and to Alexander Doniphan Wallace on the occasion of his 68th birthday.

274. J. Rhodes. Proof of the fundamental lemma of complexity (strong version) for arbitrary finite semigroups. *Journal of Combinatorial Theory. Series A*, 16:209–214, 1974.

275. J. Rhodes. Kernel systems—a global study of homomorphisms on finite semigroups. *Journal of Algebra*, 49(1):1–45, 1977.

276. J. Rhodes. Infinite iteration of matrix semigroups. I. Structure theorem for torsion semigroups. *Journal of Algebra*, 98(2):422–451, 1986.

277. J. Rhodes. Infinite iteration of matrix semigroups. II. Structure theorem for arbitrary semigroups up to aperiodic morphism. *Journal of Algebra*, 100(1):25–137, 1986. With an appendix by Jerrold R. Goodwin.

278. J. Rhodes. Survey of global semigroup theory. In *Lattices, semigroups, and universal algebra (Lisbon, 1988)*, pages 243–269. Plenum, New York, 1990.

279. J. Rhodes. Monoids acting on trees: elliptic and wreath products and the holonomy theorem for arbitrary monoids with applications to infinite groups. *International Journal of Algebra and Computation*, 1(2):253–279, 1991.

280. J. Rhodes. Flows on automata. Preprint, 1995.

281. J. Rhodes. Undecidability, automata, and pseudovarieties of finite semigroups. *International Journal of Algebra and Computation*, 9(3-4):455–473, 1999. Dedicated to the memory of Marcel-Paul Schützenberger.

282. J. Rhodes. c is decidable. Preprint, 2000.

283. J. Rhodes. *Applications of automata theory and algebra via the mathematical theory of complexity to biology, physics, psychology, philosophy, and games.* World Scientific Press, in press. With a foreword by Morris W. Hirsch.

284. J. Rhodes and P. Silva. Turing machines and bimachines. *Theoretical Computer Science*, 400(1):182–224, 2008.

285. J. Rhodes and B. Steinberg. Pointlike sets, hyperdecidability and the identity problem for finite semigroups. *International Journal of Algebra and Computation*, 9(3-4):475–481, 1999. Dedicated to the memory of Marcel-Paul Schützenberger.

286. J. Rhodes and B. Steinberg. Profinite semigroups, varieties, expansions and the structure of relatively free profinite semigroups. *International Journal of Algebra and Computation*, 11(6):627–672, 2001.

287. J. Rhodes and B. Steinberg. Join irreducible pseudovarieties, group mapping, and Kovács-Newman semigroups. In *LATIN 2004: Theoretical informatics*, volume 2976 of *Lecture Notes in Comput. Sci.*, pages 279–291. Springer, Berlin, 2004.

288. J. Rhodes and B. Steinberg. Krohn-Rhodes complexity pseudovarieties are not finitely based. *Theoretical Informatics and Applications. Informatique Théorique et Applications*, 39(1):279–296, 2005.

289. J. Rhodes and B. Steinberg. Complexity pseudovarieties are not local; type II subsemigroups can fall arbitrarily in complexity. *International Journal of Algebra and Computation*, 16(4):739–748, 2006.

290. J. Rhodes and B. Steinberg. Closed subgroups of free profinite monoids are projective profinite groups. *The Bulletin of the London Mathematical Society*, 40(3):375–383, 2008.

291. J. Rhodes and B. Tilson. A reduction theorem for complexity of finite semigroups. *Semigroup Forum*, 10(2):96–114, 1975.

292. J. Rhodes and B. Tilson. Local complexity of finite semigroups. In *Algebra, topology, and category theory (collection of papers in honor of Samuel Eilenberg)*, pages 149–168. Academic Press, New York, 1976.

293. J. Rhodes and B. Tilson. The kernel of monoid morphisms. *Journal of Pure and Applied Algebra*, 62(3):227–268, 1989.

294. J. Rhodes and B. R. Tilson. Lower bounds for complexity of finite semigroups. *Journal of Pure and Applied Algebra*, 1(1):79–95, 1971.

295. J. Rhodes and B. R. Tilson. Improved lower bounds for the complexity of finite semigroups. *Journal of Pure and Applied Algebra*, 2:13–71, 1972.

296. J. Rhodes and P. Weil. Decomposition techniques for finite semigroups, using categories. I, II. *Journal of Pure and Applied Algebra*, 62(3):269–284, 285–312, 1989.

297. J. Rhodes and Y. Zalcstein. Elementary representation and character theory of finite semigroups and its application. In *Monoids and semigroups with applications (Berkeley, CA, 1989)*, pages 334–367. World Sci. Publ., River Edge, NJ, 1991.

298. L. Ribes and P. Zalesskii. *Profinite groups*, volume 40 of *Ergebnisse der Mathematik und ihrer Grenzgebiete. 3. Folge. A Series of Modern Surveys in Mathematics [Results in Mathematics and Related Areas. 3rd Series. A Series of Modern Surveys in Mathematics]*. Springer-Verlag, Berlin, 2000.

299. L. Ribes and P. Zalesskii. Profinite topologies in free products of groups. *International Journal of Algebra and Computation*, 14(5-6):751–772, 2004. International Conference on Semigroups and Groups in honor of the 65th birthday of Prof. John Rhodes.

300. L. Ribes and P. A. Zalesskii. On the profinite topology on a free group. *The Bulletin of the London Mathematical Society*, 25(1):37–43, 1993.

301. L. Ribes and P. A. Zalesskii. The pro-$p$ topology of a free group and algorithmic problems in semigroups. *International Journal of Algebra and Computation*, 4(3):359–374, 1994.

302. J. Richter-Gebert, B. Sturmfels, and T. Theobald. First steps in tropical geometry. In *Idempotent mathematics and mathematical physics*, volume 377 of *Contemp. Math.*, pages 289–317. Amer. Math. Soc., Providence, RI, 2005.

303. K. I. Rosenthal. *Quantales and their applications*, volume 234 of *Pitman Research Notes in Mathematics Series*. Longman Scientific & Technical, Harlow, 1990.

304. K. I. Rosenthal. *The theory of quantaloids*, volume 348 of *Pitman Research Notes in Mathematics Series*. Longman, Harlow, 1996.

305. A. Salomaa and M. Soittola. *Automata-theoretic aspects of formal power series*. Springer-Verlag, New York, 1978. Texts and Monographs in Computer Science.

306. M. V. Sapir. Inherently non-finitely based finite semigroups. *Matematicheskiĭ Sbornik. Novaya Seriya*, 133(175)(2):154–166, 270, 1987.

307. M. V. Sapir. Problems of Burnside type and the finite basis property in varieties of semigroups. *Izvestiya Akademii Nauk SSSR. Seriya Matematicheskaya*, 51(2):319–340, 447, 1987.

308. M. V. Sapir. Sur la propriété de base finie pour les pseudovariétés de semigroupes finis. *Comptes Rendus des Séances de l'Académie des Sciences. Série I. Mathématique*, 306(20):795–797, 1988.

309. M. V. Sapir. On Cross semigroup varieties and related questions. *Semigroup Forum*, 42(3):345–364, 1991.

310. S. Satoh, K. Yama, and M. Tokizawa. Semigroups of order 8. *Semigroup Forum*, 49(1):7–29, 1994.

311. M. P. Schützenberger. 𝒟 représentation des demi-groupes. *Comptes Rendus de l'Académie des Sciences. Série I. Mathématique*, 244:1994–1996, 1957.

312. M.-P. Schützenberger. Sur la représentation monomiale des demi-groupes. *Comptes Rendus de l'Académie des Sciences. Série I. Mathématique*, 246:865–867, 1958.

313. M. P. Schützenberger. On finite monoids having only trivial subgroups. *Information and Control*, 8:190–194, 1965.

314. M. P. Schützenberger. Sur le produit de concaténation non ambigu. *Semigroup Forum*, 13(1):47–75, 1976/77.

315. D. Scott. Continuous lattices. In *Toposes, algebraic geometry and logic (Conf., Dalhousie Univ., Halifax, N. S., 1971)*, pages 97–136. Lecture Notes in Math., Vol. 274. Springer, Berlin, 1972.

316. J.-P. Serre. Groupes proalgébriques. *Institut des Hautes Études Scientifiques. Publications Mathématiques*, (7):67, 1960.

317. J.-P. Serre. *Trees*. Springer Monographs in Mathematics. Springer-Verlag, Berlin, 2003. Translated from the French original by John Stillwell, Corrected 2nd printing of the 1980 English translation.

318. I. Simon. Piecewise testable events. In *Automata theory and formal languages (Second GI Conf., Kaiserslautern, 1975)*, pages 214–222. Lecture Notes in Comput. Sci., Vol. 33. Springer, Berlin, 1975.

319. I. Simon. Locally finite semigroups and limited subsets of a free monoid. Unpublished manuscript, 1978.

320. I. Simon. A short proof of the factorization forest theorem. In *Tree automata and languages (Le Touquet, 1990)*, volume 10 of *Stud. Comput. Sci. Artificial Intelligence*, pages 433–438. North-Holland, Amsterdam, 1992.

321. J. R. Stallings. Topology of finite graphs. *Inventiones Mathematicae*, 71(3):551–565, 1983.

322. B. Steinberg. On pointlike sets and joins of pseudovarieties. *International Journal of Algebra and Computation*, 8(2):203–234, 1998. With an addendum by the author.

323. B. Steinberg. Monoid kernels and profinite topologies on the free abelian group. *Bulletin of the Australian Mathematical Society*, 60(3):391–402, 1999.

324. B. Steinberg. Semidirect products of categories and applications. *Journal of Pure and Applied Algebra*, 142(2):153–182, 1999.

325. B. Steinberg. Polynomial closure and topology. *International Journal of Algebra and Computation*, 10(5):603–624, 2000.
326. B. Steinberg. PG = BG: redux. In *Semigroups (Braga, 1999)*, pages 181–190. World Sci. Publ., River Edge, NJ, 2000.
327. B. Steinberg. A delay theorem for pointlikes. *Semigroup Forum*, 63(3):281–304, 2001.
328. B. Steinberg. Finite state automata: a geometric approach. *Transactions of the American Mathematical Society*, 353(9):3409–3464 (electronic), 2001.
329. B. Steinberg. Inevitable graphs and profinite topologies: some solutions to algorithmic problems in monoid and automata theory, stemming from group theory. *International Journal of Algebra and Computation*, 11(1):25–71, 2001.
330. B. Steinberg. On algorithmic problems for joins of pseudovarieties. *Semigroup Forum*, 62(1):1–40, 2001.
331. B. Steinberg. Inverse automata and profinite topologies on a free group. *Journal of Pure and Applied Algebra*, 167(2-3):341–359, 2002.
332. B. Steinberg. A modern approach to some results of Stiffler. In *Semigroups and languages*, pages 240–249. World Sci. Publ., River Edge, NJ, 2004.
333. B. Steinberg. On an assertion of J. Rhodes and the finite basis and finite vertex rank problems for pseudovarieties. *Journal of Pure and Applied Algebra*, 186(1):91–107, 2004.
334. B. Steinberg. On aperiodic relational morphisms. *Semigroup Forum*, 70(1):1–43, 2005.
335. B. Steinberg. Möbius functions and semigroup representation theory. *Journal of Combinatorial Theory. Series A*, 113(5):866–881, 2006.
336. B. Steinberg. Möbius functions and semigroup representation theory. II. Character formulas and multiplities. *Advances in Mathematics*, 217(4):1521–1557, 2008.
337. B. Steinberg. Maximal subgroups of the minimal ideal of a free profinite monoid are free. *Israel Journal of Mathematics*, to appear.
338. B. Steinberg. A structural approach to the locality of pseudovarieties of the form **LH** ⓜ **V**. *International Journal of Algebra and Computation*, to appear.
339. B. Steinberg and B. Tilson. Categories as algebra. II. *International Journal of Algebra and Computation*, 13(6):627–703, 2003.
340. P. Stiffler, Jr. Extension of the fundamental theorem of finite semigroups. *Advances in Mathematics*, 11(2):159–209, 1973.
341. M. H. Stone. The theory of representations for Boolean algebras. *Transactions of the American Mathematical Society*, 40(1):37–111, 1936.
342. M. H. Stone. Applications of the theory of Boolean rings to general topology. *Transactions of the American Mathematical Society*, 41(3):375–481, 1937.
343. H. Straubing. Aperiodic homomorphisms and the concatenation product of recognizable sets. *Journal of Pure and Applied Algebra*, 15(3):319–327, 1979.
344. H. Straubing. Families of recognizable sets corresponding to certain varieties of finite monoids. *Journal of Pure and Applied Algebra*, 15(3):305–318, 1979.
345. H. Straubing. Finite semigroup varieties of the form **V** * **D**. *Journal of Pure and Applied Algebra*, 36(1):53–94, 1985.
346. H. Straubing. *Finite automata, formal logic, and circuit complexity*. Progress in Theoretical Computer Science. Birkhäuser Boston, Boston, MA, 1994.
347. H. Straubing. On logical descriptions of regular languages. In *LATIN 2002: Theoretical informatics (Cancun)*, volume 2286 of *Lecture Notes in Comput. Sci.*, pages 528–538. Springer, Berlin, 2002.

348. H. Straubing and D. Thérien. Partially ordered finite monoids and a theorem of I. Simon. *Journal of Algebra*, 119(2):393–399, 1988.
349. H. Straubing and D. Thérien. Weakly iterated block products of finite monoids. In *LATIN 2002: Theoretical informatics (Cancun)*, volume 2286 of *Lecture Notes in Comput. Sci.*, pages 91–104. Springer, Berlin, 2002.
350. A. Suschkewitsch. Über die endlichen Gruppen ohne das Gesetz der eindeutigen Umkehrbarkeit. *Mathematische Annalen*, 99(1):30–50, 1928.
351. M. E. Sweedler. *Hopf algebras*. Mathematics Lecture Note Series. W. A. Benjamin, Inc., New York, 1969.
352. S. Talwar. Morita equivalence for semigroups. *Australian Mathematical Society. Journal. Series A. Pure Mathematics and Statistics*, 59(1):81–111, 1995.
353. S. Talwar. Strong Morita equivalence and a generalisation of the Rees theorem. *Journal of Algebra*, 181(2):371–394, 1996.
354. S. Talwar. Strong Morita equivalence and the synthesis theorem. *International Journal of Algebra and Computation*, 6(2):123–141, 1996.
355. M. L. Teixeira. On semidirectly closed pseudovarieties of aperiodic semigroups. *Journal of Pure and Applied Algebra*, 160(2-3):229–248, 2001.
356. D. Thérien. On the equation $x^t = x^{t+q}$ in categories. *Semigroup Forum*, 37(3):265–271, 1988.
357. D. Thérien. Two-sided wreath product of categories. *Journal of Pure and Applied Algebra*, 74(3):307–315, 1991.
358. D. Thérien and A. Weiss. Graph congruences and wreath products. *Journal of Pure and Applied Algebra*, 36(2):205–215, 1985.
359. B. Tilson. Decomposition and complexity of finite semigroups. *Semigroup Forum*, 3(3):189–250, 1971/72.
360. B. Tilson. Complexity of two-$\mathcal{J}$ class semigroups. *Advances in Mathematics*, 11(2):215–237, 1973.
361. B. Tilson. On the complexity of finite semigroups. *Journal of Pure and Applied Algebra*, 5:187–208, 1974.
362. B. Tilson. *Complexity of semigroups and morphisms*, chapter XII, pages 313–384. In Eilenberg [85], 1976.
363. B. Tilson. *Depth decomposition theorem*, chapter XI, pages 287–312. In Eilenberg [85], 1976.
364. B. Tilson. Categories as algebra: an essential ingredient in the theory of monoids. *Journal of Pure and Applied Algebra*, 48(1-2):83–198, 1987.
365. B. Tilson. Type II redux. In *Semigroups and their applications (Chico, Calif., 1986)*, pages 201–205. Reidel, Dordrecht, 1987.
366. B. Tilson. Presentation lemma... the short form. Unpublished manuscript, 1995.
367. B. Tilson. Modules. Unpublished manuscript, 2005.
368. B. R. Tilson. Appendix to "Algebraic theory of finite semigroups." On the $p$-length of $p$-solvable semigroups: Preliminary results. In K. Folley, editor, *Semigroups (Proc. Sympos., Wayne State Univ., Detroit, Mich., 1968)*, pages 163–208. Academic Press, New York, 1969.
369. A. V. Tishchenko. The ordered monoid of semigroup varieties with respect to a wreath product. *Fundamental′naya i Prikladnaya Matematika*, 5(1):283–305, 1999.
370. P. G. Trotter and M. V. Volkov. The finite basis problem in the pseudovariety joins of aperiodic semigroups with groups. *Semigroup Forum*, 52(1):83–91,

1996. Dedicated to the memory of Alfred Hoblitzelle Clifford (New Orleans, LA, 1994).

371. M. Vaughan-Lee. *The restricted Burnside problem*, volume 8 of *London Mathematical Society Monographs. New Series*. The Clarendon Press, Oxford University Press, New York, second edition, 1993.

372. M. V. Volkov. The finite basis property of varieties of semigroups. *Akademiya Nauk SSSR. Matematicheskie Zametki*, 45(3):12–23, 127, 1989.

373. M. V. Volkov. On a class of semigroup pseudovarieties without finite pseudoidentity basis. *International Journal of Algebra and Computation*, 5(2):127–135, 1995.

374. M. V. Volkov. The finite basis problem for finite semigroups. *Scientiae Mathematicae Japonicae*, 53(1):171–199, 2001.

375. J. von Neumann. *Continuous geometry*. Foreword by Israel Halperin. Princeton Mathematical Series, No. 25. Princeton University Press, Princeton, NJ, 1960.

376. S. M. Vovsi. *Triangular products of group representations and their applications*, volume 17 of *Progress in Mathematics*. Birkhäuser Boston, MA, 1981.

377. J. H. M. Wedderburn. Homomorphism of groups. *Annals of Mathematics. Second Series*, 42:486–487, 1941.

378. P. Weil. Products of languages with counter. *Theoretical Computer Science*, 76(2-3):251–260, 1990.

379. P. Weil. Closure of varieties of languages under products with counter. *Journal of Computer and System Sciences*, 45(3):316–339, 1992.

380. P. Weil. Profinite methods in semigroup theory. *International Journal of Algebra and Computation*, 12(1-2):137–178, 2002. International Conference on Geometric and Combinatorial Methods in Group Theory and Semigroup Theory (Lincoln, NE, 2000).

381. H. Wielandt and B. Huppert. Arithmetical and normal structure of finite groups. In *Proc. Sympos. Pure Math., Vol. VI*, pages 17–38. American Mathematical Society, Providence, RI, 1962.

382. S. You. The product separability of the generalized free product of cyclic groups. *Journal of the London Mathematical Society. Second Series*, 56(1):91–103, 1997.

383. Y. Zalcstein. Locally testable languages. *Journal of Computer and System Sciences*, 6:151–167, 1972.

384. Y. Zalcstein. Locally testable semigroups. *Semigroup Forum*, 5:216–227, 1972/73.

385. Y. Zalcstein. Group-complexity and reversals of finite semigroups. *Mathematical Systems Theory. An International Journal on Mathematical Computing Theory*, 8(3):235–242, 1974/75.

386. P. Zeiger. Yet another proof of the cascade decomposition theorem for finite automata. *Mathematical Systems Theory. An International Journal on Mathematical Computing Theory*, 1(3):225–228, 1967.

387. P. Zeiger. Yet another proof of the cascade decomposition theorem for finite automata: Correction. *Mathematical Systems Theory. An International Journal on Mathematical Computing Theory*, 2(4):381, 1968.

388. E. I. Zel'manov. Solution of the restricted Burnside problem for groups of odd exponent. *Izvestiya Akademii Nauk SSSR. Seriya Matematicheskaya*, 54(1):42–59, 221, 1990.

389. E. I. Zel'manov. Solution of the restricted Burnside problem for 2-groups. *Matematicheskiĭ Sbornik*, 182(4):568–592, 1991.

# Table of Pseudovarieties

| NOTATION | NAME | PSEUDOIDENTITIES |
|---|---|---|
| **1** | Trivial pseudovariety | $[\![x = y]\!]$ |
| **A** | Aperiodic semigroups | $[\![x^{\omega+1} = x^{\omega}]\!]$ |
| **Ab** | Abelian groups | $[\![x^{\omega} = 1, xy = yx]\!]$ |
| **Ab**$(n)$ | Abelian groups of exponent $n$ | $[\![x^n = 1, xy = yx]\!]$ |
| **ACom** | Aperiodic commutative | $[\![x^{\omega+1} = x^{\omega}, xy = yx]\!]$ |
| **B** | Bands | $[\![x^2 = x]\!]$ |
| **C**$_n$ | Level $n$ complexity | |
| **Com** | Commutative | $[\![xy = yx]\!]$ |
| **CR** | Completely regular | $[\![x^{\omega+1} = x]\!]$ |
| **CS** | Completely simple | $[\![(xy)^{\omega}x = x]\!]$ |
| **D** | Delay | $[\![xy^{\omega} = y^{\omega}]\!]$ |
| **DS** | Regular $\mathscr{J}$-classes are ssgps. | $[\![((xy)^{\omega}(yx)^{\omega}(xy)^{\omega})^{\omega} = (xy)^{\omega}]\!]$ |
| **ER** | Idem.-gen. ssgp. $\mathscr{R}$-trivial | $[\![(x^{\omega}y^{\omega})^{\omega}x^{\omega} = (x^{\omega}y^{\omega})^{\omega}]\!]$ |
| **G** | Groups | $[\![x^{\omega} = 1]\!]$ |
| **G**$_{nil}$ | Nilpotent groups | See [10] |
| **G**$_p$ | $p$-groups | $[\![x^{p^{\omega}} = 1]\!]$ |
| **G**$_{\pi}$ | $\pi$-groups | $[\![x^{\pi^{\omega}} = 1]\!]$ |
| **G**$_{\pi'}$ | $\pi'$-groups | $[\![x^{(\pi')^{\omega}} = 1]\!]$ |
| **G**$_{sol}$ | Solvable groups | See [45, 56] |
| **G**$^{\mathbf{N}}$ | Nilpotent extensions of groups | $[\![x^{\omega} = y^{\omega}]\!]$ |
| **J** | $\mathscr{J}$-trivial | $[\![(xy)^{\omega} = (yx)^{\omega}, x^{\omega+1} = x^{\omega}]\!]$ |
| **K** | Reverse delay | $[\![x^{\omega}y = x^{\omega}]\!]$ |
| **K**$_n$ | Level $n$ two-sided complexity | |
| **L** | $\mathscr{L}$-trivial | $[\![y(xy)^{\omega} = (xy)^{\omega}]\!]$ |
| **L1** | Locally trivial | $[\![x^{\omega}yx^{\omega} = x^{\omega}]\!]$ |
| **LG** | Local groups | $[\![(x^{\omega}yx^{\omega})^{\omega} = x^{\omega}]\!]$ |
| **LRB** | Left regular (=$\mathscr{R}$-trivial) band | $[\![x^2 = x, xyx = xy]\!]$ |
| **LS** | Left simple | $[\![xy^{\omega} = x]\!]$ |

| | | |
|---|---|---|
| **LS**$^{\mathrm{N}}$ | Nilpotent exts. of left simple | $[\![ x^\omega y^\omega = x^\omega ]\!]$ |
| **LSl** | Local semilattices | $[\![ (x^\omega y x^\omega)^2 = x^\omega y x^\omega,$ |
| | | $\quad x^\omega y x^\omega z x^\omega = x^\omega z x^\omega y x^\omega ]\!]$ |
| **LZ** | Left zero | $[\![ xy = x ]\!]$ |
| **N** | Nilpotent semigroups | $[\![ x^\omega = 0 ]\!]$ |
| **R** | $\mathscr{R}$-trivial | $[\![ (xy)^\omega x = (xy)^\omega ]\!]$ |
| **RB** | Rectangular bands | $[\![ x^2 = x, xyx = x ]\!]$ |
| **RRB** | Right regular ($=\mathscr{L}$-trivial) band | $[\![ x^2 = x, yxy = xy ]\!]$ |
| **RS** | Right simple | $[\![ x^\omega y = y ]\!]$ |
| **RS**$^{\mathrm{N}}$ | Nilpotent exts. of right simple | $[\![ x^\omega y^\omega = y^\omega ]\!]$ |
| **RZ** | Right zero | $[\![ xy = y ]\!]$ |
| **Sl** | Semilattices | $[\![ x^2 = x, xy = yx ]\!]$ |

# Table of Operators and Products

| NAME | NOTATION | REFERENCE |
|---|---|---|
| Semidirect product | $*$ | Example 2.4.9 |
| Two-sided semidirect product | $**$ | Example 2.4.14 |
| Mal'cev product | $\textcircled{m}$ | Example 2.4.4 |
| Generalized Mal'cev product | $(\mathbf{V}, \mathbf{W})\ \textcircled{m}\ (-)$ | Example 2.4.5 |
| Join | $\vee$ | Example 2.4.7 |
| Intersection | $\cap$ or $\wedge$ | Example 2.4.17 |
| Constants | $q_{\mathbf{V}}$ | Example 2.4.2 |
| Identity | $1_{\mathbf{PV}}$ | Example 2.4.1 |
| Old wreath product | $\circ$ | Example 2.4.24 |
| Local operator | $\mathbb{L}$ | Example 2.4.27 |
| Idempotents | $\mathbf{E}$ | Example 2.4.28 |
| Regular $\mathscr{D}$-classes | $\mathbf{D}$ | Example 2.4.26 |
| Power operator | $\mathbf{P}$ | Example 2.4.25 |
| Schützenberger product | $\Diamond$ | Example 2.4.18 |
| Regular elements | $\mathbf{R}$ | Example 2.4.29 |
| Semidirect closure | $()^{\omega}$ | Example 2.4.31 |

# Table of Implicit Operations

| Symbol | Description | Defining Sequence |
|---|---|---|
| $x^\omega$ | Idempotent in $\overline{\langle x \rangle}$ | $\lim x^{n!}$ |
| $x^{\omega+1}$ | Generator of the kernel of $\overline{\langle x \rangle}$ | $\lim x^{n!+1}$ |
| $x^{\omega-1}$ | Inverse of $x^{\omega+1}$ | $\lim x^{n!-1}$ |
| $x^{p^\omega}$ | $p'$-component of $x^{\omega+1}$ | $\lim x^{p^{n!}}$ |
| $x^{\pi^\omega}$ | $\pi'$-component of $x^{\omega+1}$ | $\lim x^{(p_1\cdots p_n)^{n!}}$, $p_i \in \pi$ |
| $x^{(\pi')^\omega}$ | $\pi$-component of $x^{\omega+1}$ | $\lim x^{(p_1\cdots p_n)^{n!}}$, $p_i \in \pi'$ |

# Index of Notation

# Author Index

# Index